Lecture Notes in Electrical Engineering

Volume 99

Xiaofeng Wan (Ed.)

Electrical Power Systems and Computers

Selected Papers from the 2011 International Conference on Electric and Electronics (EEIC 2011) in Nanchang, China on June 20–22, 2011, Volume 3

Springer

Xiaofeng Wan
Nanchang University
Xuefu Avenue 999
New Honggutan Zone
Jiangxi
Nanchang, 330047
China
E-mail: xiaofengwan417@gmail.com

ISBN 978-3-642-21746-3 e-ISBN 978-3-642-21747-0

DOI 10.1007/978-3-642-21747-0

Lecture Notes in Electrical Engineering ISSN 1876-1100

Library of Congress Control Number: 2011929495

Typeset & Cover Design: Scientific Publishing Services Pvt. Ltd., Chennai, India.

Printed on acid-free paper

9 8 7 6 5 4 3 2 1

springer.com

EEIC 2011 Preface

The present book includes extended and revised versions of a set of selected papers from the International Conference on Electric and Electronics (EEIC 2011), held on June 20-22 , 2011, which is jointly organized by Nanchang University, Springer, and IEEE IAS Nanchang Chapter.

The goal of EEIC 2011 is to bring together the researchers from academia and industry as well as practitioners to share ideas, problems and solutions relating to the multifaceted aspects of Electric and Electronics.

Being crucial for the development of Electric and Electronics, our conference encompasses a large number of research topics and applications: from Circuits and Systems to Computers and Information Technology; from Communication Systems to Signal Processing and other related topics are included in the scope of this conference. In order to ensure high-quality of our international conference, we have high-quality reviewing course, our reviewing experts are from home and abroad and low-quality papers have been refused. All accepted papers will be published by Lecture Notes in Electrical Engineering (Springer).

EEIC 2011 is sponsored by Nanchang University, China. Nanchang University is a comprehensive university which characterized by "Penetration of Arts, Science, Engineering and Medicine subjects, Combination of studying, research and production". It is one of the national "211" Project key universities that jointly constructed by the People's Government of Jiangxi Province and the Ministry of Education. It is also an important base of talents cultivation□ scientific researching and transferring of the researching accomplishment into practical use for both Jiangxi Province and the country.

Welcome to Nanchang, China. Nanchang is a beautiful city with the Gan River, the mother river of local people, traversing through the whole city. Water is her soul or in other words water carries all her beauty. Lakes and rivers in or around Nanchang bring a special kind of charm to the city. Nanchang is honored as 'a green pearl in the southern part of China' thanks to its clear water, fresh air and great inner city virescence. Long and splendid history endows Nanchang with many cultural relics, among which the Tengwang Pavilion is the most famous. It is no exaggeration to say that Tengwang Pavilion is the pride of all the locals in Nanchang. Many men of letters left their handwritings here which tremendously enhance its classical charm.

Noting can be done without the help of the program chairs, organization staff, and the members of the program committees. Thank you.

EEIC 2011 will be the most comprehensive Conference focused on the various aspects of advances in Electric and Electronics. Our Conference provides a chance for academic and industry professionals to discuss recent progress in the area of Electric and Electronics. We are confident that the conference program will give you detailed insight into the new trends, and we are looking forward to meeting you at this world-class event in Nanchang.

EEIC 2011 Organization

Honor Chairs

Prof. Chin-Chen Chang	Feng Chia University, Taiwan
Prof. Jun Wang	Chinese University of Hong Kong, HongKong

Scholarship Committee Chairs

Chin-Chen Chang	Feng Chia University, Taiwan
Jun Wang	Chinese University of Hong Kong, HongKong

Scholarship Committee Co-chairs

Zhi Hu	IEEE IAS Nanchang Chapter, China
Min Zhu	IEEE IAS Nanchang Chapter, China

Organizing Co-chairs

Jian Lee	Hubei Normal University, China
Wensong Hu	Nanchang University, China

Program Committee Chairs

Honghua Tan	Wuhan Institute of Technology, China

Publication Chairs

Wensong Hu	Nanchang University, China
Zhu Min	Nanchang University, China
Xiaofeng Wan	Nanchang University, China
Ming Ma	NUS ACM Chapter, Singapore

Contents

A Study on Equivalent Circuit of Short Wavelength Coplanar Waveguide Employing Periodic Ground Structure on Silicon RFIC

Jeong-Gab Ju*, Young-Bae Park, Bo-Ra Jung, Jang-Hyeon Jeong, Eui-Hoon Jang, and Young Yun

Department of Radio Communication and Engineering, Korea Maritime University, #1 Dongsam-dong, Youngdo-Ku, Busan 606-791, Korea
yunyoung@hhu.ac.kr

Abstract. In this paper, equivalent circuit of short wave length coplanar waveguide employing Periodic Ground Structure on Silicon (PGSS) were investigated using theoretical analysis. Equivalent circuits for the PGSS cell were extracted, and all lumped circuit parameters were expressed by closed form equation. A fairly good agreement between calculated and measured results were observed from 0 to 25 GHz. Above results indicate that the proposed equivalent circuit can be efficiently used up to K band.

Keywords: Coplanar waveguide, PGSS, Silicon, RFIC, LED.

1 Introduction

With the evolution of silicon CMOS device process technology, demands for fully-integrated CMOS RFIC, including all matching components, have increased in the wireless communication systems market [1]-[7]. However, bulky passive components such as conventional impedance transformers and dividers have been fabricated outside of RFIC owing to their large sizes, because the conventional RF transmission line shows long wavelength [2], [6]. To solve this problem, several papers dealing with short wavelength of periodic structure have been published on silicon and GaAs substrate [8]-[12]. However, an extensive investigation of equivalent circuit of periodic structure on silicon substrate has not been performed yet.

In this paper, using theoretical and experimental analysis, basic characteristics of the coplanar waveguide employing PGSS was investigated for application to a development of miniaturized on-chip passive components. In addition, for simplification of circuit design process, equivalent circuits for the PGSS cell were extracted, and all lumped circuit parameters were expressed by closed-form equation. For application to miniaturized on-chip component, Wilkinson power divider and

* Jeong-Gab Ju received the B.S. degree in Radio Communication and Engineering from Korea Maritime University, 2010, respectively, and is currently working toward the M.S. degree at Korea Maritime University.

X. Wan (Ed.): Electrical Power Systems and Computers, LNEE 99, pp. 1–7.
springerlink.com

impedance transformer were developed using coplanar waveguide employing PGSS. Using the PGSS, they were highly miniaturized compared with conventional ones.

2 Structure of Coplanar Waveguide Employing PGSS

Fig.1 shows a top view and corresponds to a cross-sectional view of the coplanar waveguide employing PGSS. As shown in Fig.1, PGSS exists at the interface between the SiO_2 film and the silicon substrate, and it was electrically connected to top-side ground planes (GND planes) through the contacts. Therefore, PGSS was grounded through GND planes. As is well known, a conventional coplanar waveguide without PGSS has only a periodical capacitance C_a per unit length, while the coplanar waveguide employing PGSS has an additional capacitance, C_b as well as C_a, owing to PGSS. As shown in this figure, C_b is capacitance between the line and PGSS. Therefore, we can see that the coplanar waveguide with PGSS exhibits much lower characteristic impedance(Z_0) and shorter guided-wavelength(λ_g) than conventional one, because Z_0 and λg are inversely proportional to the periodical capacitance, in other words, $Z_0=(L/C)^{0.5}$ and $\lambda_g=1/[f\bullet(LC)^{0.5}]$[13].

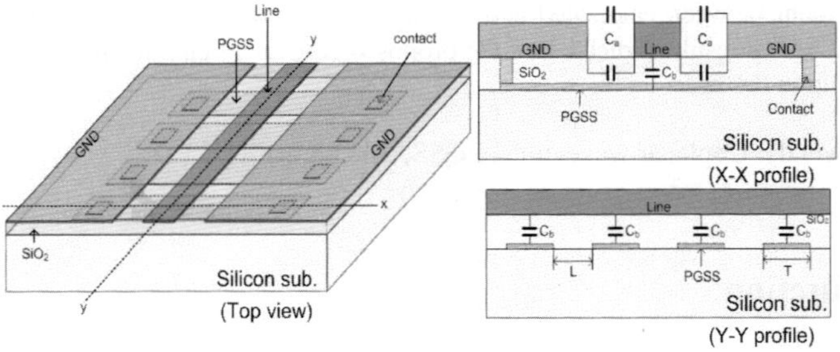

Fig. 1. Structure of coplanar waveguide employing PGSS

3 Equivalent Circuit Analysis of Coplanar Waveguide Employing PGSS

Fig. 2 shows the equivalent circuit of unit cell (small square box), which corresponds to the equivalent circuit of the Nth unit section of the periodic structure surrounded by square box. C_b corresponds to the capacitance between top line and PGSS, which is shown in Fig. 2, and it is proportional to the cross area W•T of line of line and PGSS (As shown in Fig. 2, W and T are the width of top lines and the periodic strips of PGSS, respectively). Rg and Lg are resistance and inductance originating from the loss and current flow of the periodic strip of PGSS with width T, respectively. Cf corresponds to the capacitance between PGSS and silicon substrate. RS is the loss resistance of silicon substrate. C_a is coupling capacitance between line and top ground metal. RL and L_{ind} are resistance and inductance originating from original coplanar waveguide's top line.

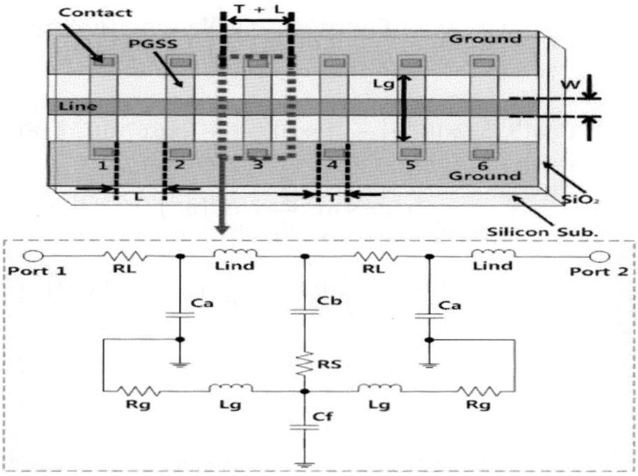

Fig. 2. Equivalent circuit for a unit cell of coplanar waveguide employing PGSS

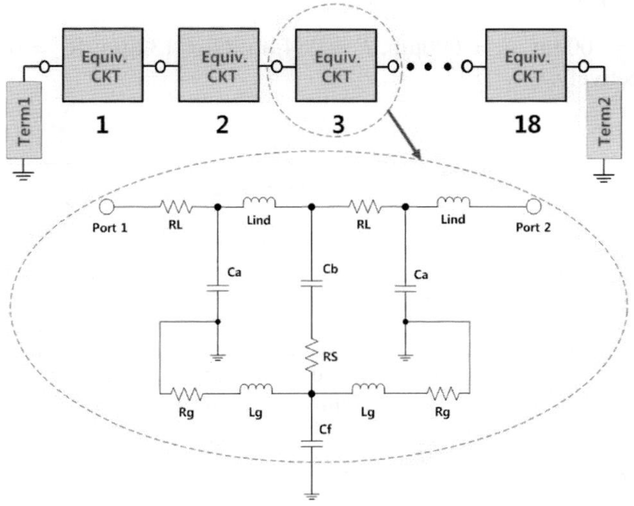

Fig. 3. Equivalent circuit for coplanar waveguide employing PGSS

Equivalent circuit for single cell of coupled CPW employing PGSS is shown in Fig. 2, and the whole equivalent circuit is shown in Fig. 3. As shown in Fig. 3, a number of the equivalent circuits of unit section are connected to each other. Closed form equation for all lumped elements of equivalent circuit shown in Fig. 3 can be given by,

$$C_a = \left[0.0532 - \left(\frac{T}{G}\right)\times 0.7613 + \left(\frac{T}{G}\right)^2 \times 3.0314\right](pF) \quad (1)$$

$$C_b = \left[3.3332\times10^{-4} + \left(\frac{T}{d_i}\right)\times 5\times10^{-6} - \left(\frac{T}{d_i}\right)^2 \times 6.67\times10^{-9}\right](pF) \quad (2)$$

$$C_f = 9.61\times\frac{T}{d_s}\times10^{-1}(pF) \quad (3)$$

$$L_{ind} = \left[0.0473 - \left(\frac{T}{W}\right)\times 0.142 + \left(\frac{T}{W}\right)^2 \times 0.1293\right](nH) \quad (4)$$

$$L_g = 1.77\times\frac{l_s}{T}\times10^{-3}(nH) \quad (5)$$

$$R_g = 906\times\frac{T}{l_s}(\Omega) \quad (6)$$

Where, d_i and d_s are thickness of SiO_2 film and thickness of semiconducting substrate. Also l_s and G are length of PGSS strip and length of between line and ground.

3.1 Measured and Calculated Insertion Loss S_{21} for Coplanar Waveguide Employing PGSS

Note that d_i = 100nm, d_s = 600μm, l_s = 284μm, G = 132μm, R_L = 0.0683 Ω , R_S = 100 Ω , respectively.

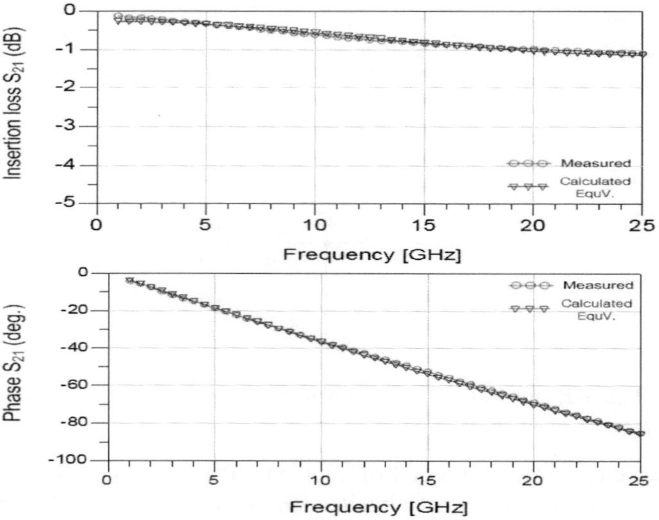

Fig. 4. Measured and calculated S_{21} for coplanar waveguide employing PGSS(T=5μm)

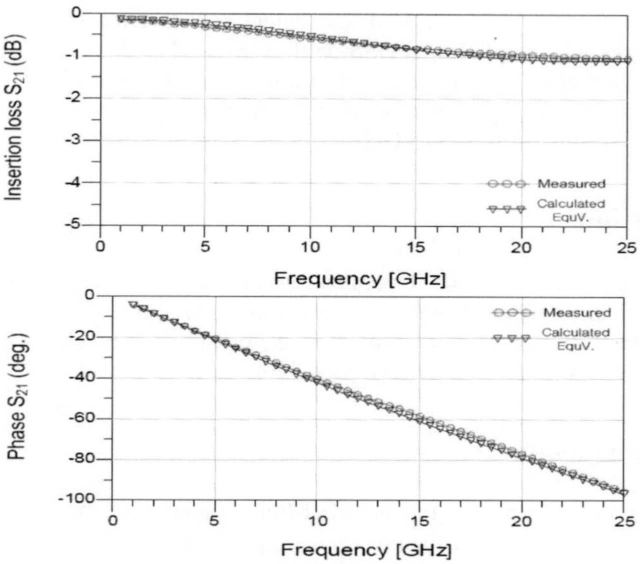

Fig. 5. Measured and calculated S_{21} for coplanar waveguide employing PGSS($T=10\mu m$)

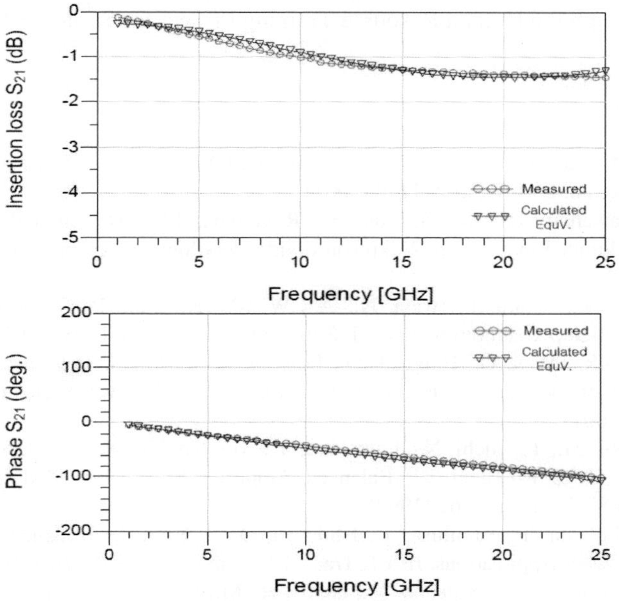

Fig. 6. Measured and calculated S_{21} for coplanar waveguide employing PGSS($T=20\mu m$)

Fig. 5 and 6 show the measured and calculated data for insertion loss S_{21} for coplanar waveguide employing PGSS with various T values. For the calculation result, equivalent circuit of Fig. 3 and closed form equations of (1)-(6) were used. As shown in these figures, we can observe a fairly good agreement between calculated and measured results from 0 to 25 GHz.

4 Conclusion

In this work, equivalent circuits of coplanar waveguide employing PGSS were studied using theoretical analysis. Concretely, equivalent circuits for the PGSS cell were extracted, and all lumped circuit parameters were expressed by closed-form equation. The calculated results showed a fairly good agreement with measured ones.

Acknowledgement

This research was supported by the MKE(the Ministry of Knowledge Economy), Korea, under the ITRC(Information Technology Research Center) support program supervised by the NIPA(National IT Industry Promotion Agency)(NIPA-2011-C1090-1121-0015). This work was financially supported by the Ministry of Knowledge Economy (MKE) and the Korea Industrial Technology Foundation (KOTEF) through the Human Resource Training Project for Strategic Technology.

References

1. Zargari, M., Su, D.: Challenges in designing CMOS wireless systems on a chip. IEICE Trans. Electron. E90-C, 1142–1148 (2007)
2. Chongcheawchamnam, M., Siripon, N., Rovertson, I.D.: Design and performance of improved lumped-distributed Wilkinson divider topology. Electron. Lett. 37(8), 501–503 (2001)
3. Matsunaga, K., Miura, I., Iwata, N.: A CW 4-W Ka-Band Power Amplifier Utilizing MMIC Multichip Technology. IEEE J. Solid State Circuits 35, 1293–1297 (2000)
4. Webster, D.R., Ataei, G., Haigh, D.G.: Low-Distortion MMIC Power Amplifier Using a New Form of Derivative Superposition. IEEE Trans. Microwave Theory and Tech. 49, 328–332 (2001)
5. Itoh, Y., Nii, M., Takeuchi, N., Tsukahara, Y., Kurebayashi, H.: MMIC/Super-MIC/MIC-Combined C- to Ku-Band 2W Balanced Amplifier Multi-Chip Module. IEICE Trans. Electron. E80-C(6), 757–762 (1997)
6. Yun, Y., Fukuda, T., Kunihisa, T., Ishikawa, O.: A High Performance Down converter MMIC for DBS Applications. IEICE Trans. Electron. E84-C(11), 1679–1688 (2001)
7. Yun, Y., Nishijima, M., Katsuno, M., Ishida, H., Minagawa, K., Nobusada, T., Tanaka, T.: A Fully-Integrated Broadband Amplifier MMIC Employing a Novel Chip Size Package. IEEE Trans. Microwave Theory Tech. 50, 2930–2937 (2002)
8. Yun, Y., et al.: Miniaturized on-chip branch-line coupler employing periodically arrayed grounded-strip structure for application to silicon RFIC. Microwave Journal 52(12), 82–90 (2009)

9. Yun, Y.: Miniaturized, low impedance ratrace fabricated by microstrip line employing PPGM on MMIC. IEE Electronics Letters 40(9), 540–541 (2004)
10. Yun, Y.: A Novel Microstrip Line Structure Employing a Periodically Perforated Ground Metal and Its Application to Highly Miniaturized and Low Impedance Passive Components Fabricated on GaAs MMIC. IEEE Transactions On Microwave Theory and Technique 53(6), 1951–1959 (2005)
11. Yun, Y., Lee, K.S., Kim, C.R., Kim, K.M., Jung, J.W.: Basic RF Characteristics of the Microstrip Line Employing Periodically Perforated Ground Metal and Its Application to Highly Miniaturized On-Chip Passive Components on GaAs MMIC. IEEE Transactions On Microwave Theory and Technique 54(10), 3805–3817 (2006)
12. Yun, Y., Jung, J.W., Kim, K.M., Kim, H.C., Chang, W.J., Ji, H.G., Ahn, H.K.: Experimental Study on Isolation Characteristics between Adjacent Microstrip Lines Employing Periodically Perforated Ground Metal for Application to Highly Integrated GaAs MMICs. IEEE Microwave and Wireless Components Letters 17(10), 703–705 (2007)
13. Pozar. D.M.: Microwave engineering. Addison-Wesley, 3nd edn., ch. 8 (2005)

Fault Diagnosis of Rolling Bearing Based on Wavelet Packet Transformation

Zhitong Jiang, Chengfei Zhu, Guanqing Chang, and Hongxing Chang

Research Center of Integrated Information
Institute of Automation, Chinese Academy of Sciences
No.95 Zhongguancun East Road, Beijing, P.R. China
{zhitong.jiang,chengfei.zhu,guanqing.chang,
hongxing.chang}@ia.ac.cn

Abstract. The faults of rolling bearings frequently occur in rotary machinery, therefore the rolling bearings fault diagnosis is a very important research project. In this paper, a method of pattern recognition for fault diagnosis of rolling bearing is proposed, which is based on wavelet packet transformation combined with Statistics. Firstly, the wavelet packet analysis is utilized to divide the dynamic signal of rolling bearings, and the features information of rolling bearing's dynamic signal is picked up, secondly, the extracted features are classified into several categories, and databases are built for each category. Finally, the new picked-up signals are compared with the standard signals in database, and then whether the rolling bearings have defects is diagnosed.

Keywords: Rolling bearings; Fault diagnosis; Vibration signal; Wavelet packet.

1 Introduction

Rolling bearings are the most applied components in vast majority of rotating machines, and they have the function of load and load transfer. According to statistics, many of the railway rotating machinery faults are caused by bearing failure [1]. Once the bearing has fault, it may cause heavy economic loss. So the study of the rolling bearings condition monitoring and fault diagnosis is essential. There are many methods for fault diagnosis of rolling bearing, in which vibration signal detection analysis is a very broadly application, such as spectrum analysis, resonance demodulation method, and so on [2-5].

Wavelet analysis came forth in the 1980s. Since then, people began to use wavelet analysis to deal with vibration signals and achieved some favorable effects [6-8]. Yet wavelet packet analysis is the promotion of multi-resolution analysis. It is a much more sophisticated than the wavelet analysis method. Wavelet packet analysis carries on the frequency band to multi-level divisions, and analyzes the high-frequency unit which wavelet analysis doesn't do. It can improve the time-frequency resolution, so it is widely used in signal processing [9].

In this paper, wavelet packet analysis and statistical methods are combined to diagnose rolling bearings fault. First, we use the wavelet packet analysis to carry on

X. Wan (Ed.): Electrical Power Systems and Computers, LNEE 99, pp. 9–14.
springerlink.com

the division of the rolling bearing's dynamic signal, then by statistics knowledge, we sum up the proportion of the energy of various frequency band of eligible rolling bearings account for total energy, make the corresponding form, regard it as standard data. Finally, we compare the testing bearing's energy spectrum with the standard data, and determine whether the bearing has fault.

2 Wavelet Packet Transformation

We use the accelerometer to get the bearing vibration acceleration signal. Since bearing failure makes the energy of vibration signals decreases in some frequency bands, and increases in others, the bearing fault could be analyzed according to the energy of the frequency components which could be acquired from bearing vibration signal.

As the wavelet packet analysis has the feature of multi-resolution, it is used to decompose bearing vibration signals in this paper, and then, we can get the energy of different frequency bands of the bearing vibration signal. The course of wavelet packet decomposition is shown in Fig.1.

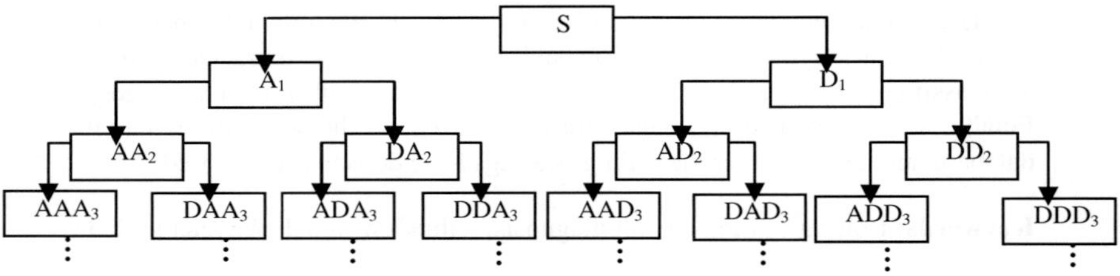

Fig. 1. Decomposition tree of wavelet packet

In Fig.1, A denotes low frequency, B denotes high frequency, and the subscript number is the level of wavelet packet decomposition.

The specific algorithm of wavelet packet which is used to pick-up the characteristics of rolling bearings fault is as follows:

1. Using wavelet packet to decompose vibration signal of bearing into 6 layers, the original signal is expressed by S, (*i,j*) denotes the *j*-th node of the *i*-th layer, the decomposition coefficient is expressed as X, and (1,0) is represented by X_{10}, the rest can be deduced by analogy.
2. We may use the function S=Wprcoef(t,N) to reconstruct wavelet decomposition coefficients [10]. N denotes the reconstructed nodes. Then the total signal can be expressed as:

$$S=S(6,0)+S(6,1)+\ldots+S(6,62)+S(6,63) \tag{1}$$

   Assuming the lowest frequency component in the vibration signal is 0, the highest one is 1, then the frequency area of picked-up signal S(6,j) (j=0,1…63) can be stated as: S_{60}:0～1/64; S_{61}:1/64～2/64; S_{62}:2/64～3/64…S_{63}:63/64～1.
3. Calculate the total energy of each frequency band signal:

$$E_{6j}=\int\left|s_{3j}(t)\right|^2dt=\sum_{k=1}^{n}\left|x_{jk}\right|^2 \tag{2}$$

x_{jk} (j=0,1…,63,k=1,…,n) denotes values of discrete points of reconstructed signalS_{6j}.

4. Construct eigenvector. If the system has failure, it would greatly affect the signal energy of various frequency bands. Therefore, the energy is taken as the elements to construct an eigenvector T:

$$T=\left[E_{(6,0)}, E_{(6,1)}, E_{(6,2)} \dots, E_{(6,62)}, E_{(6,63)}\right] \tag{3}$$

While the energy is big, $E_{(6,j)}(j = 0,1,2 \dots 63)$ is usually a big numerical value, it will bring some discommodious place in data processing. So the normalization processing is used here to make an improvement to the eigenvector T.

$$E=\sum_{j=0}^{63}\left|E_{(6,j)}\right| \tag{4}$$

$$T' = \left[E_{(6,0)}/E, E_{(6,1)}/E, E_{(6,2)}/E \dots, E_{(6,62)}/E, E_{(6,63)}/E\right] \tag{5}$$

T' is the normalized eigenvector in equation(5).

5. Using the method of system identification, the eigenvalues of eigenvectors are confirmed whether they are in normal or failure mode, and the math model is built with the knowledge of statistics.
6. Establish the mapping relationship between energy changes and the physical components.

3 Experiments and Result Analysis

The eligible and faulty rolling bearings are analyzed in this paper, based on the theory above, according to a large number of experiments, the frequency - energy diagrams can be calculated, as follows:

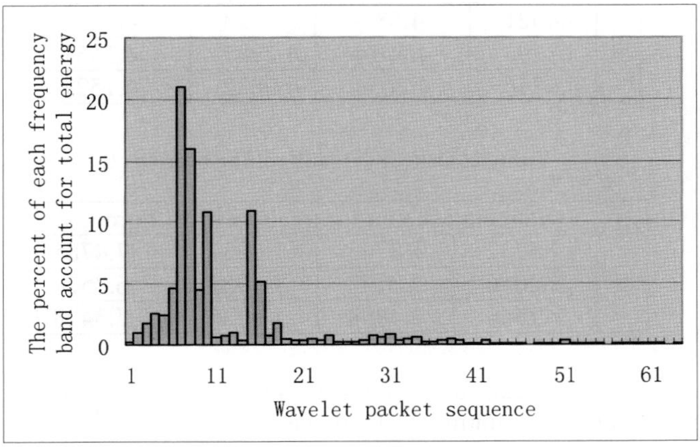

Fig. 2. Energy-spectrum graph of the eligible rolling bearing

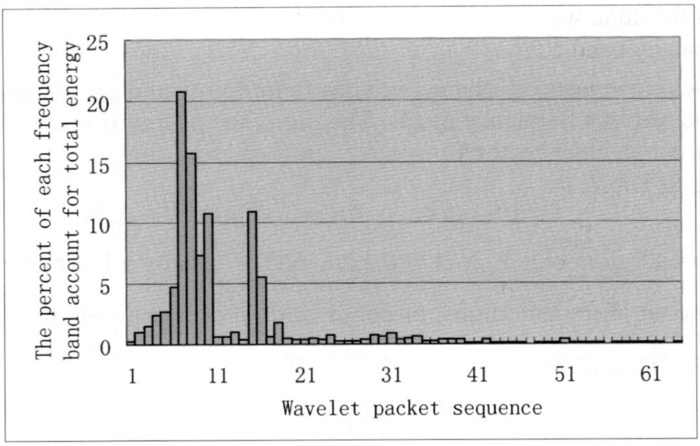

Fig. 3. Energy-spectrum graph of the faulty rolling bearing

Fig.2 and Fig.3 reflect energy distribution in each frequency band of the eligible and faulty bearings. In order to distinguish the two types of bearings, the energy of them are compared in each corresponding frequency band, and the frequency band in which the energy has changed biggest to analyze. There are two main reasons for selecting the frequency band mentioned above. Firstly, it may contain more fault information. Secondly, there are more differences in these frequency bands, we can distinguish the two types of bearings more distinctly.

From Fig.2 and Fig.3, we can see that the biggest change is in the 9-th frequency band. In the experiments, the percent that the energy of the 9-th frequency band accounts for total energy is shown as follows:

Table 1. Data of the eligible rolling bearing

n	Percent account for total energy(12 groups)			
9	4.4521	4.5874	4.7723	4.4327
	4.3893	4.5014	4.2985	4.3857
	4.6346	4.2654	4.3632	4.7523

Table 2. Data of the fault rolling bearing

n	Percent account for total energy(12 groups)			
9	7.5732	7.8743	7.6237	7.4785
	8.2016	7.0143	7.4052	6.8528
	7.7546	7.4894	7.5236	7.3428

According to Tab.1, we can calculate the arithmetic mean of the energy percent of the 9-th frequency band accounting for total energy about eligible bearings:

$$\overline{x_9} = \frac{\sum_{i=1}^{12} x_i}{n} = \frac{4.4521 + 4.5874 + \cdots + 4.7523}{12} = 4.4862 \tag{6}$$

By Bessel formula, we can calculate the standard deviation of the energy percent of the 9-th frequency band about eligible bearings:

$$\sigma_9 = \sqrt{\frac{\sum_{i=1}^{12}(x_i-\overline{x_9})^2}{n-1}} = \sqrt{\frac{(4.4521-4.4862)^2+\cdots+(4.5874-4.4862)^2}{12-1}} = 0.1673 \quad (7)$$

According to Tab.2, we can get the two values about fault bearings as the same:

$$\overline{x_9}' = \frac{\sum_{i=1}^{12} x_i}{n} = \frac{7.5732+7.8743+\cdots+7.3428}{12} = 7.5112 \quad (8)$$

$$\sigma_9' = \sqrt{\frac{\sum_{i=1}^{12}(x_i-\overline{x_9})^2}{n-1}} = \sqrt{\frac{(7.5732-7.5112)^2+\cdots+(7.3428-7.5112)^2}{12-1}} = 0.3575 \quad (9)$$

For the same bearing, the result of each measurement may be different. The change can be accepted if it is in a certain range, and this range is called the tolerance range.

$$\left|\frac{x-\overline{x}}{\sigma}\right| \le k \quad (10)$$

When k=2(that is 4σ), the tolerance range of the energy percent of the 9-th frequency band accounting for total energy about eligible bearings is 4.1516%~4.8208%, and the tolerance range about the faulty bearings is 6.7962%~8.2262%.

4 Conclusion

The rolling bearings fault diagnosis is a very important research project, in this paper, the method of wavelet packet transform and statistical methods is presented to diagnose rolling bearings faults. When the picked-up information changes, the energy of vibration signal changes as well, this can be indicated accurately by the eigenvectors which is composed of wavelet packet frequency decomposition signals. The eigenvectors is put into classifier to recognize and classify. This method is a new approach to the intelligent diagnosis technology of rolling bearings.

References

1. Zhong, B., Hunag, R.: Introduction To Machine Fault Diagnosis. Machinery industry press (2006)
2. Vas, P.: Parameter Estimation, Condition Monitoring and Diagnosis of Electrical Machines. Clarendron Press (2006)
3. Nandi, S., Toliyat, H.A.: Conditon Monitoring and Fault Diagnosis of Electrical Machines-A Review. IEEE Industry Applications Conference 1, 197–204 (1999)
4. Willsky, A.: A Survey of Design Method for Failure Dectection in Dynamic Systems. Automatica 12, 601–611 (1976)
5. Pang, P., Ding, G.: Wavelet-based Diagnostic Model for Rotating Machinery Subject to Vibration Monitoring. In: Chinese Control Conference, pp. 303–306 (2008)
6. Shen, S., Ying, H., Liu, J.: Passing vibration diagnosing using wavelet transform. Vibration and Shock 18(2), 1–5 (1999)

7. He, X., Shen, Y., Zhang, X.: An application of continuous wavelet transform to fault diagnosis of rolling element bearing. Mechanical science and Technology 20(7), 571–574 (2001)
8. Wang, L., Wang, C., Cai, Z.: Early fault diagnosis of the rolling bearing using wavelet transformation. Chinese Journal of Applied Mechanics 16(2), 95–100 (1999)
9. Liu, T., Xiangli, Z., Jun, Z.: The Introduction of Applied Wavelet Analysis. National defence industrial press (2006)
10. Newland, D.E.: Wavelet Analysis of Vibration. Journal of Vibration and Acoustics 16, 409–416 (1994)

Study on UV Detection of High-Voltage Discharge Based on the Optical Fiber Sensor

Yuanzhe Xu[1] and Hongzhen Chen[2]

[1] Qiongzhou University Sanya, China
[2] Northeast Dianli University Jilin, China
xuyuanzhe@263.net, chhzh118@126.com

Abstract. The main reason for the deterioration of electrical equipment insulation is the high- voltage discharge. So, real-time monitoring of electrical equipment discharge is the key to ensure the safe operation of equipment. Using the UV signal in solar blind area as characteristic parameter, this paper designs the Optical Fiber Sensor and analyze its spectral response characteristics. The lens coupling system is designed to improve the detection of light coupling efficiency. The efficiency is improved by the simulation and optimization of ZEMAX. Simulation detecting tests on needle plate discharge is carried out in the laboratory, and it presents that the detection system can accurately detect the UV radiation of the discharge and the strength of the discharge to linear response. The detection system has high sensitivity, and strong anti-interference. It can be used for online real-time monitoring of high voltage discharge.

Keywords: high voltage discharge, Optical Fiber Sensor, UV detection, lens coupling.

1 Introduction

The main reason for the deterioration of electrical equipment insulation is the high-voltage discharge. According to the statistics, in the accidents of the entire power system, the insulation accident next to the thunderbolt accident had accounted for the second place. The insulation accident, which involves a wide range and causes huge economic losses and power cut for a long time, is a major threat to the safe supply of electric power, as in [1-3]. Therefore, making timely and accurate judgments to the insulation condition of equipment is the key to ensure the safe operation of electrical equipment.

The primary assessment mean of the insulation condition is to detect the discharge strength of electrical device. The main methods of high-voltage discharge detection are the pulse current method, UV imaging method, infrared imaging method, radio frequency detection method and ultrasound method. The pulse current method is commonly used to detect discharge, which is produced by detecting the pulse current of discharge to judge discharge condition. But the detecting device, which can't be completely isolated, can be influenced by the external disturbances in practice, as in [4-8]. In recent years, some scholars use the UV signal as characteristic quantity to detect discharge and have achieved certain results, as in [9-11]. But the research on

X. Wan (Ed.): Electrical Power Systems and Computers, LNEE 99, pp. 15–22.
springerlink.com

improving UV light intensity of the discharge is little, and pulses in the UV pulse method depend on the incident intensity of UV light. This paper designs the Optical Fiber Sensor, and uses UV Optical Fiber for signal transmission. The lens coupling system is designed to improve the efficiency of UV detection.

2 The UV Detection Principle of High-Voltage Discharge

According to the corona discharge mechanism of electrical equipment, the process of high-voltage discharge will radiate light waves, sound waves, ozone and UV radiation etc. Most of the wavelength of UV light are in the range of 280nm-400nm by analyzing the spectral of corona discharge[12-13]. There is a small part of the wavelength in 240nm-280nm, the range of which is called solar blind area, while the ozone in the atmosphere absorb the UV wavelength less than 280nm in sunlight. Therefore, using the particular UV sensor can detect the UV signal at solar blind area, and remove the interference of UV light in sunlight. The UV light produced by the high-voltage equipment is detected to determine the discharge strength, early forecast the insulation down, tear, dirty development of the equipment, and ensure the safe operation of electrical equipment.

The UV sensor acquires the UV light signals and converts them into current signal. Then, the amplification of the current signals are compared with a predetermined number to obtain pulses. Analyzing the intensive degree of pulses is to determine the strength of the discharge, which is called UV-pulse method. The essence of UV detection is to detect the discharge energy of parts of insulation damage. It can be drawn the formula as follows by calculating the formula *Boltzman* in Plasma physics and the formula Debye shielding, as in [13].

$$P = \int P_{hz}\, dV = \int 1.57 \times 10^{-40} n_e^2 z \sqrt{T_e}\, dV \quad . \tag{1}$$

Where:

$$n_e = n_\infty \exp(e\phi / T_e) \quad . \tag{2}$$

$$\phi = \phi_0 \exp(-|x| / \lambda_D) \quad . \tag{3}$$

$$\lambda_D = \sqrt{\left(\frac{\varepsilon_0 T_e}{e^2 n_\infty}\right)} \quad . \tag{4}$$

P is the bremsstrahlung power during the air discharge of the surface of insulator, n_e is the electron density, n_∞ is the electron density far from the insulator, ϕ is the space potential, ϕ_0 is the potential of the insulator, λ_D is the Debye length, x is the distance between the point of the space charge and insulator, V is the profile of all the UV radiation points.

It is reflected the energy of the partial discharge by detecting UV pulses. When the insulation of high-voltage equipment is aging or damage, the corona discharge increase, thus affecting the discharge pulses at the position of insulation aging or damaged. As the pulse signals are changing, it can be detected for judging the insulation condition of equipment.

3 The Overall Design of Detection System to High-Voltage Discharge

According to the principle of UV pulse method, the detection system to high-voltage discharge consists of the Optical Fiber Sensor, the signal processing circuit, the datas acquisition. etc, of which, the Optical Fiber Sensor includes the ray filter, lens, the UV Optical Fiber and photomultiplier tube(PMT). The UV signals are produced by the high-voltage discharge of the power equipment, and then received by the Optical Fiber Sensor. The UV signals are converted into current signals. After amplification and filtering, the current signals are carried on A/D conversion, and the required datas are collected by data acquisition card and sent to PC. As the signal transmission channel, the UV Optical Fiber has the advantages of a low loss, light weight, free from electromagnetic interference, good insulation and so on. Therefore, the detection device has the characteristics of more fast and intuitive, remote and non-contact measurement, strong anti-interference, and high sensitivity. The diagram of the detection system of high-voltage discharge is shown in Figure 1.

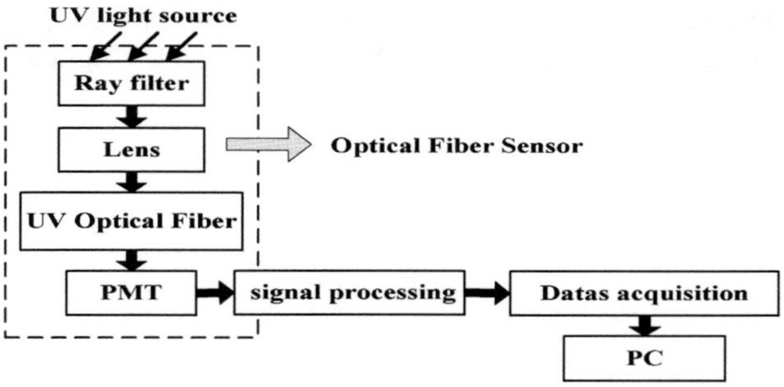

Fig. 1. The principle diagram of the detection system of high-voltage discharge

4 The Design and Characteristic Analysis of the Transmission System of the Optical Fiber

4.1 The Selection of the Ray Filter and Sensor

If the UV light in the sun enter the detector, it will cause interference and impact on the test result. So we need to add a ray filter matched the wavelength of the detector to eliminate the interference from other frequencies of light. When a ray filter is selected, it's taken into account the spectral range of the required UV light and the UV light of the interference resource. According to the experimental request, the homemade HB285 about narrowband UV filter is used in this experiment. Its light transmissivity is

10%～20% within the entire scope of luminous flux. The center wavelength is 254nm, right in the solar blind region, and the semi-Wave width is 20nm.

PMT, which based on the photoelectric effect, is photoelectric detector. The main characteristics of it are high sensitivity, good stability, fast response and low noise, especially for detecting weak light. In this experiment, the HAMAMATSU R7154 PMT is used, and its response wavelength are 160~320 nm, the maximum response wavelength of 254nm. But it does not fully meet the requirements of detecting spectrum, because the response wavelength has the part of more than 280nm. Therefore, the front should add the ray filter to change its response range for working at solar blind area.

4.2 The Characteristic Analysis of the Optical Fiber Sensor

The characteristic of the Optical Fiber Sensor is mainly decided by the characteristic parameters of the ray filter and PMT themselves. The light response coefficient of the sensor is defined $\gamma = S_\lambda \tau_\lambda$. S_λ is the spectral sensitivity of PMT. τ_λ is the light transmissivity of the ray filter. According to the measured light transmissivity of the ray filter and the spectral response parameter of PMT, the values are acquired to 5nm interval point by point, and then the corresponding parameters multiplied for obtaining the actual response curve at the entirely spectral, as shown in Figure 2. As the response curve shown, the range of 242nm~270nm is a narrow rectangular window. For this band of incident light, the output of sensor is approximately linear. The sensor are not reflected for the other wavelengths. Therefore, the design of the Optical Fiber Sensor can filter out UV light in sunlight, and accurately detect the UV light at solar blind area.

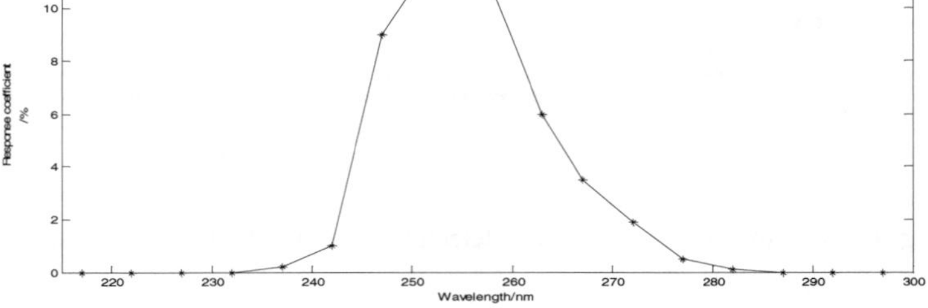

Fig. 2. The response curve of the Optical Fiber Sensor

5 The Design and Optimization of the Lens Coupling System

5.1 The Design of Coupled System

In order to improve the light intensity of the incident UV light, the optical coupling efficiency need to be improved. Generally, the ways of the optical fiber coupling have the direct coupling, single-lens coupling and lens coupling, etc. The efficiency of direct coupling and single-lens coupling is very low. The design uses the coupling system of combining a inverted telescope by Galileo and single-lens coupling, as shown in Figure 3.

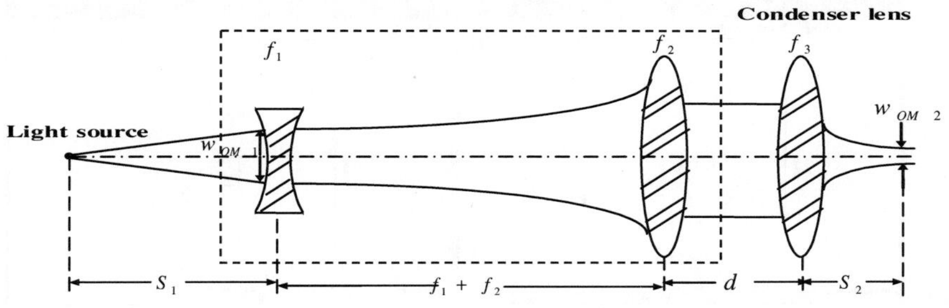

Fig. 3. The coupling system diagram of the spatial light and the optical fiber

5.2 The ZEMAX Simulation and Optimization of the Coupled System

The initial datas of the coupling lens are input to the optical design software ZEMAX. The incident wavelength select the UV light of 240nm-280nm. By adjusting parameter structure, the coupled system is optimized to obtain the two-dimensional imaging map, as shown in Figure 4. From figure 4, we can see that the aberration have been significantly improved and the focus is more effective, after optimized.

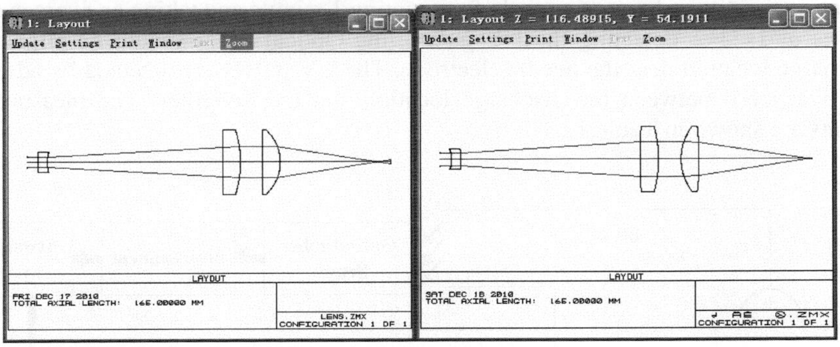

Fig. 4. The two-dimensional images of the before and after optimization

The plotdiagram of the before and after optimization to the imaging is simulated by ZEMAX, as shown in Figure 5. It can be seen from the figure, the number of the light spot at the center is small and the diffuse plaque is large before the system is optimizied. After optimizied, most of the light spot assembly reach the central point. The diffuse plaque become small. The light intensity of imaging at the center has been greatly improved. For the optimized coupled system, the coupling efficiency can be calculated by ZEMAX. The core diameter of the used UV fiber is 56 μm. calculated By ZEMAX The coupling efficiency is $T = 91.5\%$.

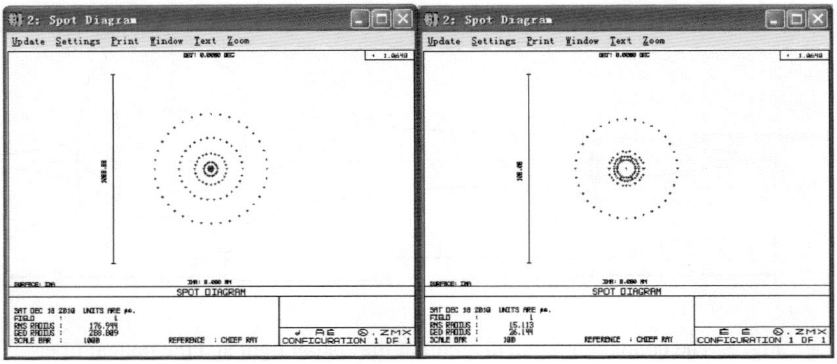

Fig. 5. The plotdiagram of the before and after optimization to the imaging

6 The Test and Analysis Needle Plate Discharge

The schematic diagram of the experiment is shown in Figure 6. In the experiment, the distance between the needle and the plate electrode is 5cm. The needle electrode is semicircle of 1mm diameter. The plate electrode is circular steel plate of 25cm diameter and the surface and the edges of it is smooth. When discharge pulses are counted, the counting time is set 1s, continuously for counting 1 minute and taking an average to get pulses of 1s. For detecting the UV pulses, the Optical Fiber should be aligned the discharge location near the needle electrode. The UV pulses are detected by adjusting the distance d between the discharge location and the ray filter. The measurement results are shown in Table 1.

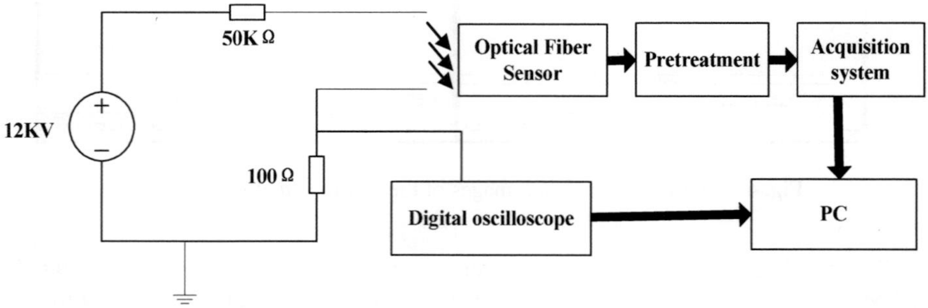

Fig. 6. The schematic diagram of discharge test to the needle plate

The test datas in table 1 are simulated to get the fitting curve. According to the curve, it can be known that the pulses of UV discharge and detecting distance are approximately the inversely-proportional relationship. With increasing the detecting distance, the pulses reduce and the energy of UV radiation decrease significantly. When the distance is more than 2m, the discharge energy drop obviously. Therefore, the detection will be better, when the distance is below 1m.

Table 1. Test datas

Distance(m) / Pulses	1	2	3	4	5	6	7	8
The first measurement	571	144	58	37	25	15	11	8
The second measurement	579	149	64	41	27	19	15	10
The third measurement	569	142	55	35	23	14	10	6
The average measurement	573	145	59	38	25	16	12	8

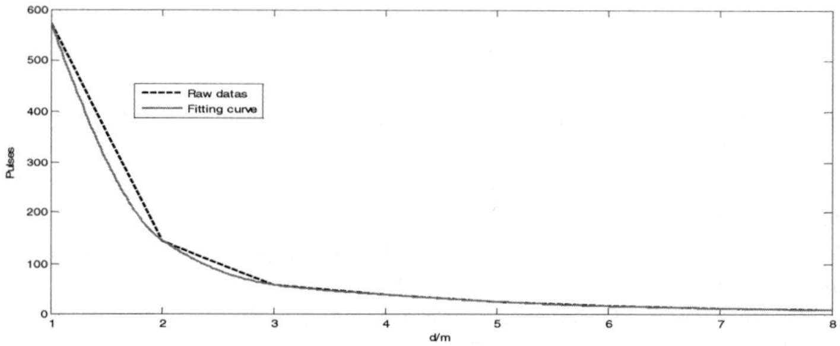

Fig. 7. The relationship between the discharge pulses and discharge distance

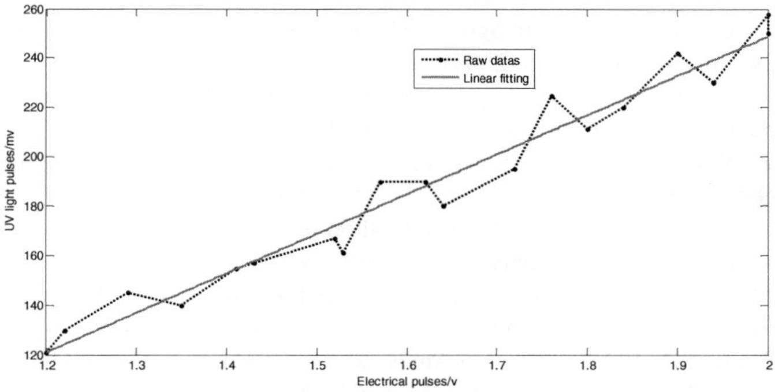

Fig. 8. The amplitude relationship of the UV light pulses and electrical pulses

We use Tektronix DPO3052 Digital Oscilloscope to measure the voltage to the 100Ω resistor with no sense. The voltage can indirectly reflect the pulse current of the discharge. The signal of electrical pulses and light pulses are simultaneously acquired, as shown in Figure 8. According to the figure, the amplitude of the light pulses and electric pulses has a good corresponding linear relationship. Detecting the light pulses can well reflect the variation of the current pulses. So, the detection of light pulses can have a good reflection to the discharge strength.

7 Conclusion

The paper designs the Optical Fiber Sensor to detect the UV light in solar blind area. The UV Optical Fiber as the signal transmission channel and the UV pulse method are used to detect the high-voltage discharge. In the laboratory, simulation detecting tests on needle plate discharge is carried out, and it presents that the detection system can accurately detect the UV radiation of the discharge and the strength of the discharge to linear response. It has high sensitivity, and strong anti-interference. It can be used for online real-time monitoring of high voltage discharge.

References

1. Yan, Z., Zhu, D.: Technology of high-voltage insulation. China Electric Power Press, Beijing (2007)
2. Wu, G.: Theory and practice of state testing to electrical equipment. Tsinghua University Press, Beijing (2005)
3. Zhang, J.: Technology and application of UV detection to high-voltage discharge. China Electric Power Press, Beijing (2009)
4. Guo, J., Wu, G., Zhang, X.: Current situation and development of partial discharge detection technology. J. Electrotechnical Society 2(20), 29–34 (2005)
5. Xu, Y., Yu, M., Cao, X.: Comparison with light pulse method and the electrometric method of partial discharge. J. High Voltage Engineering 27(4), 3–5 (2001)
6. Si, W., Li, j., Yuan, P.: Research situation and development of optical detection method to partial discharge. J. high-voltage electrical appliances 44(3), 261–264 (2008)
7. Chen, X., Jiang, Y.: Comparative study on two ultra-wideband detection technique of partial discharge. J. High Voltage Engineering 27(6), 6–8 (2001)
8. Dai, L.: Appication of UV imaging technology on high-voltage discharge detection. J. Power Systems 27(20), 97–98 (2003)
9. Li, Q., Chen, T., Wang, J.: Application of UV Pulse Method on the UHV Discharge Detection. High Voltage Engineering 32(12), 26–29 (2006)
10. Zhang, Z., Wang, K., Tang, J.: Detection Method for Corona Discharge of Transformers on Ultraviolet Pulse. J. Power Systems 2(34), 84–88 (2010)
11. Yin, Z., Fang, D., Xuechen, L.: Optical method in micro-DBD discharge characteristics. J. Spectroscopy and Spectral Analysis. 23(3), 607–608 (2003)
12. Zhao, W., Zhang, X., Jiang, J.: Spectral Analysis of corona discharge to tip plate. J.Spectroscopy and Spectral Analysis 23(5), 9,552,957 (2003)
13. Wang, W., Liu, D., Wu, Y.: Spectral Analysis on reactor distribution of high energy electron density along Air corona discharge. J. Molecular Science 15(3), 1,252,128 (1999)

Automatic Kidney Segmentation Using Gaussian Mixture Model on MRI Sequences

Evgin Goceri

Department of Computer Engineering, Faculty of Engineering,
Pamukkale University, Denizli, 20020,Turkey
evgingoceri@yahoo.com

Abstract. Robust kidney segmentation from MR images is a very difficult task due to the especially gray level similarity of adjacent organs, partial volume effects and injection of contrast media. In addition to different image characteristics with different MR scanners, the variations of the kidney shapes, gray levels and positions make the identification and segmentation task even harder. In this paper, we propose an automatic kidney segmentation approach using Gaussian mixture model (GMM) that adapts all parameters according to each MR image dataset to handle all these challenging problems. The efficiency in terms of the segmentation performance is achieved by the estimation of the GMM parameters using the Expectation Maximization (EM) method. The segmentation approach is compared to k-means method. The results show that the model based probabilistic segmentation technique gives better performance for both low contrast images and atypical kidney shapes where several algorithms fail on abdominal MR images.

Keywords: Kidney segmentation, MRI, Gaussian mixture model.

1 Introduction

Kidney segmentation is a challenging task in MR datasets because of the partial volume problem, intensity inhomogenity, leakage of contrast agent to adjacent organs, high signal to noise rate, more artifacts and a low gradient response. In literature, kidney segmentation methods using CT images can be classified into four main groups. The first group is knowledge based methods. Kobashi and Shapiro [1] have proposed a knowledge based identification using the CT image properties and anatomical information. The second group is region growing based methods. One of them has been explained in [2]. The authors have obtained the left and right kidney regions coarsely as ROIs, applied an adaptive region growing using initial seed points and a threshold value and then they have done region modification for accurate segmentation. Pohle and Tönnies [3], [4], [5] have proposed a region growing method that automatically optimizes the homogeneity criterion using the characteristics of the area to be segmented. Yan and Wang [6] have developed another region growing method by estimating kidney position, using labeling algorithm and obtaining seeds with mathematical morphology to extract the kidney regions. The third group is shape based approaches that are generally with active contours. A non-rigid registration based active shape model has been proposed for kidney segmentation in [7]. Another shape based level set method has been developed by applying the Chen-Vese level set model and the

X. Wan (Ed.): Electrical Power Systems and Computers, LNEE 99, pp. 23–29.
springerlink.com

connected component analysis to get separated components and then the size feature has been used to find the kidneys [8]. Finally, the fourth group of methods for kidney segmentation is deformable model based approaches. Tsagaan and Shimizu have proposed a B-spline based deformable model by incorporating mean and variation of the organ into the objective function in order to fit the model [9], [10].

When we search for kidney segmentation methods using MR images, we see that a shape aided kidney extraction method by matching a co-focus elliptical model to a binary mask using optimization in the parametric space has been proposed in MR urography by Yang Tang [11]. Another method is again a shape based and has been performed using level sets [15]. The authors have used Dynamic Contrast Enhanced Magnetic Resonance Imaging (DCE-MRI) and obtained a mean shape model for kidneys using extracted kidney shapes with level set functions as the first step. The second step in the proposed approach is to obtain an intensity model using several initial curve functions on the test images and to perform the registration of the intensity model and the shape model. Although this approach seems successful, the accuracy of the result depends on the estimated model shapes by using the training datasets.

Gray level values of the liver that is the adjacent organ of the right kidney are very similar. This similarity and noises reduces the performance of the thresholding methods. Gradient based approaches are vulnerable for weak edges. The accuracy of the region growing based methods depends on the seed point location and similarity metrics. Different modality settings and contrast media injection cause not only the kidneys but also all other organs to have different gray level values for different image sequences or even in different slices of the same patient data set. In addition to this difficulty, the anatomical structures of the kidneys in different image slices are different and their shapes can vary for each patient. Therefore, the shape based kidney segmentation may not always be sufficient. Although the method in [7] seems to be effective, the training of active shape models is required to model the expected shape so the obtained results depend on the training datasets. Therefore, a robust and efficient method that is GMM [12] based segmentation using the EM [13] algorithm on MR image sequences for healthy kidney segmentation is proposed in this paper. Mathematical binary morphologic operations are used for boundary refinement. Segmentation of each slice is performed iteratively for each patient rather than using a common parameter from all patient data sets because parameter values are different for each patient. The robustness and efficiency of the method is due to the capability of dealing with the contrast variations and different shaped kidney regions. These capabilities are provided by the iterative processing of each slice, the better segmentation performance of the GMM fitting approach than traditional thresholding and k-means based methods. There is not any method in literature that uses this approach to get handle all challenging difficulties mentioned above to our knowledge.

The remaining of this paper is organized as follows. Section 2 presents the properties of the patient data sets. Section 3 describes the proposed method for kidney segmentation on MR image sequences. Section 4 gives the experimental results and discussions. Finally, conclusions are identified in Section 5.

2 Dataset Descriptions

In the present study, upper abdominal MRI datasets have been used which are obtained from 3 different patients using a 1.5 Tesla MRI device (Gyroscan Intera,

Philips, ACS-NT, Best, The Netherlands) located in Dokuz Eylül University Radiology Department. The examined 16 bit DICOM images are fat suppressed T2-weighted (TR/TE, 1600/70 ms; flip angle, 90°; slice thickness, 8 mm) SPIR images in the axial plane with a resolution of 256x256.

3 Proposed Method

In our approach, the kidney image segmentation is performed by fitting a mixture model that is a composition form of several Gaussian distributions to intensity histograms. Model fitting is known as the procedure for estimating the unknown parameters. The first step to apply this approach to our MRI datasets is to select an initial kidney image. The initial kidney image is selected in which the kidneys are identified easily and they are seemed clearer so that the separation of them from other adjacent organs is easier (Fig. 1.a). The initial kidney image is generally the middle slice of the MRI sequence. The kidney segmentation process continues from the initial kidney image to the beginning of the dataset and from the initial kidney image to the end of the dataset by detecting kidney regions on each slice. Because, in addition to the visualization difficulties due to patients movements and breathing during the scanning, the intensity similarities and unclear boundaries between the adjacent organs and kidneys make the kidney detection and segmentation more difficult (Fig. 1.b). Therefore, the information, which is obtained from the initial kidney slice, is used to extract kidney from the next slice and this iterative process is applied for all slices.

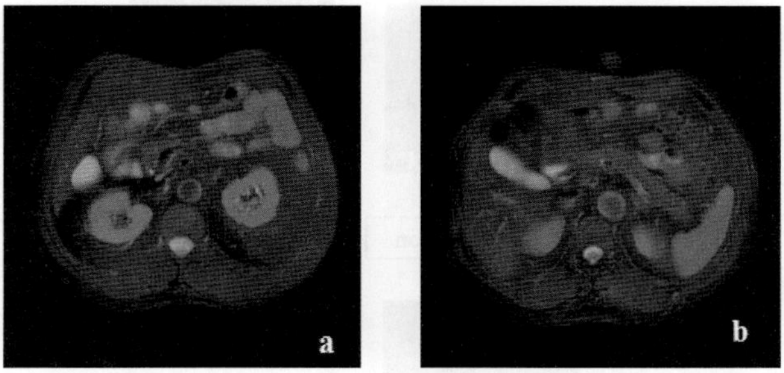

Fig. 1. Example initial kidney slice, where the border is very clear (a), and another slice where the border between the liver and the right kidney is not clear (b)

The initial kidney image is segmented automatically into five clusters using the probabilistic model based unsupervised clustering method. Each component of the mixture model corresponds to a different organ. Thresholds are determined as intensity values between the peaks of the intensity distributions of the organs as intensity values at which the probability distributions of the organs are equal. The intensity ranges between successive thresholds determine the intensity ranges of the corresponding organs. The pixels are clustered according to the intensity range in which they fall. The process in the GMM fitting approach is to estimate the GMM parameters, which are the prior probabilities, mean vectors and covariance matrices, so that the estimated Gaussian distribution follows the histogram as closely as possible.

It is seemed that the kidney regions are always clustered together with the spine and the bright parts of other organs into the brightest cluster when the GMM is fitted to image intensity histograms (Fig.2.b). Therefore, we use prior anatomic knowledge to find kidney locations. Kidneys are inside the ribs and located in the bottom-left and bottom-right side of abdominal images. After the detection of the image boundaries, the horizontal and vertical axis lengths are calculated and then the spine is used as a landmark to find kidney positions. To identify the right and left kidney, seed regions, which include some part of the kidneys, are selected at the right and left hand side of the spine after the spine position is detected. Then, binary morphological image reconstruction is applied by using the selected seed regions as the marker (Fig.2.a) and the initial kidney image as the mask (Fig.2.b). The output of this process gives us the segmented kidneys from the image (Fig.2.c).

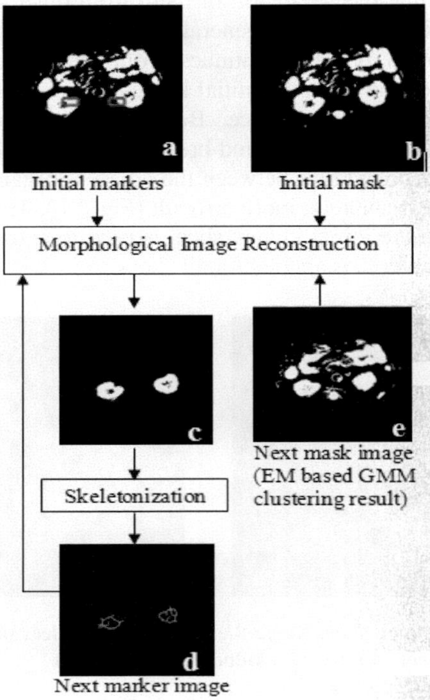

Fig. 2. Initial marker images (seed regions) in the red frames for image reconstruction (a); Initial mask image obtained by EM based GMM clustering of the original slice in Fig. 1.a (b); Segmented kidneys (c); Skeletons of the segmented kidneys used as markers for the next slice (d); Next slice as the next mask image (e)

Kidneys are segmented on each slice after the skeletonization step of the segmented kidney image. The skeletonization process is a thinning technique to reduce all other objects by removing boundaries in an image, but does not break objects apart, obtains lines without changing the essential structure of the image [16]. The skeleton of this segmented kidney is used as the new marker to segment kidneys from both the next and previous slice. The skeleton image of the previously segmented kidney image is used as the marker for each preceding slice. The necessary mask images for each slice are the brightest regions of their own, which are obtained automatically using the model based segmentation. Similarly, the skeleton image of the next segmented kidney image is used as the marker for each next slice. The image reconstruction process continues iteratively for other slices to detect and segment kidneys. The iteration ends automatically if the slice to be processed has not any kidney. Because each slice is clustered into five clusters, therefore if a slice without kidney comes then it can not be clustered into five clusters and the iteration ends.

4 Results and Discussions

Example results are shown with the original images (Fig. 3.a), when the proposed approach is applied using k-means method (Fig. 3.b) and using the EM based GMM method (Fig. 3.c).

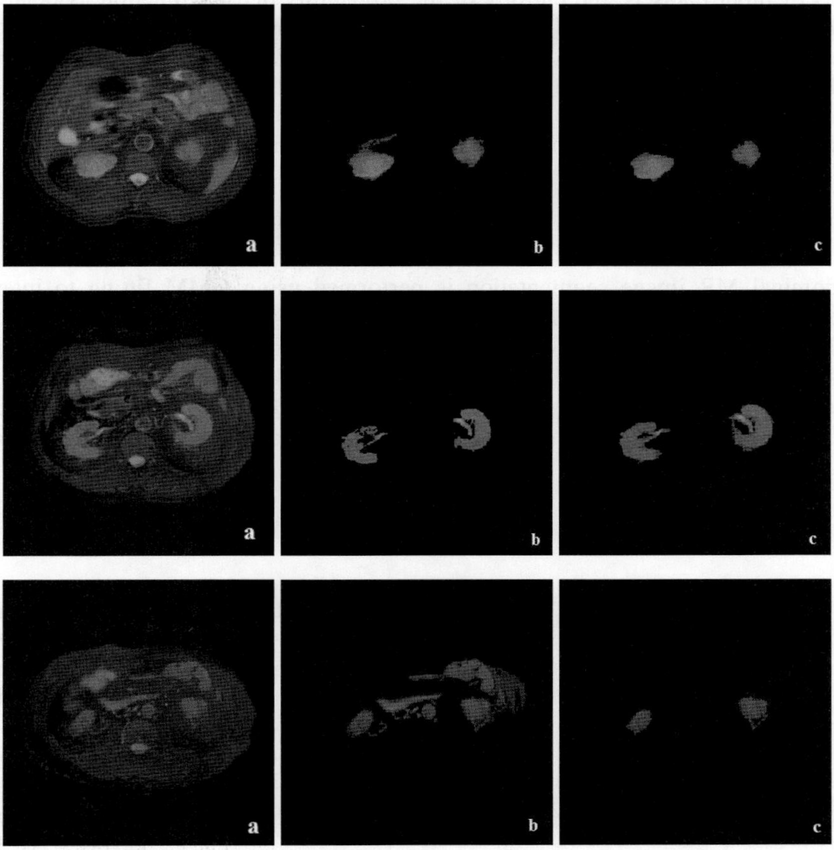

Fig. 3. Original slice (a); Result by using k-means (b); Result by using EM based GMM clustering (c)

It can be seen from the above results that the proposed segmentation method has the ability to classify to outline the anatomical structures on the used datasets. The results of k-means method have some misclassification areas because of the inhomogeneous regions of the acquisitions and the partial volume effects. K-means makes hard decisions since each data point is assigned to a single cluster. The GMM makes soft decisions since each data point can yield a posterior probability, which indicates that each data point has some probability of belonging to each cluster.

In the original k-means algorithm, the distance is calculated using each data element and center for each iteration. Therefore, the required computational time of this algorithm depends on the number of data elements, number of iterations and number of clusters so this method is computationally expensive.

Although the EM algorithm yields the ML solution, the drawback of the EM algorithm is its sensitivity to the selection of the initial parameters. In addition, the resulting mixture depends on the number of selected components. We have used a fixed kernel number for our abdominal SPIR images. In order to generalize the application and find the optimum number of components of the mixture for other kinds of input datasets, the entropy of the probability density function associated to each kernel can be estimated to check the Gaussianity of the underlying probability density function.

The GMM parameters are recomputed iteratively starting with initial values until convergence. The accuracy of the proposed segmentation algorithm depends on how much close the probabilistic model to the intensity distributions.

5 Conclusions

Abdominal MR image segmentation is performed using GMM fitting to intensity histograms with the EM algorithm in order to segment kidney regions. An advantage of the Gaussian model based clustering is to provide a rigorous approach to assess the cluster numbers and the role of each variable in the clustering process. Experiments on different abdominal MR images showed that this GMM approach is less affected by noise, more appropriate than k-means clustering and gives better classification results.

References

1. Kobashi, L., Shapiro, M.: Knowledge-based Organ Identification from CT Images. Pattern Recognition 28, 475–491 (1995)
2. Lin, D.T., Lei, C.C., Hsiung, S.Y.: An Efficient Method for Kidney Segmentation on Abdominal CT Images. In: 8th Australian and New Zealand Intelligent Information Systems Conference, Sydney, Australia, pp. 75–82 (2003)
3. Pohle, R., Tönnies, K.D.: A New Approach for Model-Based Adaptive Region Growing in Medical Image Analysis. In: Skarbek, W. (ed.) CAIP 2001. LNCS, vol. 2124, pp. 238–246. Springer, Heidelberg (2001)
4. Pohle, R., Toennies, K.D.: Segmentation of Medical Images Using Adaptive Region Growing. In: Proceedings of the Medical Imaging Conference of SPIE, vol. 4322, pp. 1337–1346 (2001)

5. Pohle, R., Toennies, K.D.: Self-learning Model-based Segmentation of Medical Images. Image Processing & Communication 7, 97–113 (2001)
6. Yan, G., Wang, B.: An Automatic Kidney Segmentation from Abdominal CT Images. In: International Conference on Intelligent Computing and Intelligent Systems (ICIS), pp. 280–284. IEEE Press, Xiamen (2010)
7. Spiegel, M., Hahn, D.A., Daum, V., Wasza, J., Hornegger, J.: Segmentation of Kidneys Using a New Active Shape Model Generation Technique Based on Non-rigid Image Registration. Computerized Medical Imaging and Graphics 33, 29–39 (2009)
8. Huang, C.L., Kuo, L.Y., Huang, Y.J., Lin, Y.H.: Shape-based Level Set Method for Kidney Segmentation on CT Image. In: 22nd Conference on Computer Vision, Nantou, Taiwan (2009)
9. Tsagaan, B., Shimizu, A., Kobatake, H., Kunihisa, M., Hanzawa, Y.: Segmentation of Kidney by Using a Deformable Model. In: International Conference on Image Processing (ICIP), pp. 1059–1062. IEEE Press, Thessaloniki (2001)
10. Tsagaan, B., Shimizu, A., Kobatake, H., Miyakawa, K.: An automated segmentation method of kidney using statistical information. In: Dohi, T., Kikinis, R. (eds.) MICCAI 2002. LNCS, vol. 2488, pp. 556–563. Springer, Heidelberg (2002)
11. Tang, Y., Jackson, H., Lee, S., Nelson, M., Moats, R.A.: Shape-aided Kidney Extraction in MR Urography. In: 31st Annual International Conference on Engineering in Medicine and Biology Society (EMBS), pp. 5781–5784. IEEE Press, Minneapolis (2009)
12. Shental, N., Bar-Hillel, A., Hertz, T., Weinshall, D.: Computing Gaussian Mixture Models with EM Using Equivalence Constraints. In: Advances in Neural Information Processing System, vol. 15, pp. 465–473 (2003)
13. Dempster, A., Laird, N., Rubin, D.: Maximum Likelihood from Incomplete Data via the EM Algorithm. Journal of the Royal Statistical Society 39, 1–38 (1977)
14. Duda, R.O., Hart, P.E., Stork, D.G.: Pattern Classification, 2nd edn. John Wiley and Sons Inc., New York (2000)
15. Abdelmunim, H., Farag, A.A., Miller, W., AbdelGhar, M.: A Kidney Segmentation Approach from DCE-MRI Using Level Sets. In: Conference on Computer Vision and Pattern Recognition Workshops (CVPRW), pp. 1–6. IEEE Press, Anchorage (2008)
16. Blum, H.A.: Transformation for Extracting New Descriptors of Shapes. In: Wathen-Dunn, W. (ed.) Models for the Perception of Speech and Visual Form, pp. 362–380. MIT Press, Cambridge (1967)

A Power Controller with Load Current Sensing for the Lundell Automotive Alternator

Paiboon Nakmahachalasint and Werachai Pattanapiboon

Department of Electrical and Computer Engineering, Faculty of Engineering, Thammasat University, Klongluang, Pathumthani, 12120 Thailand
npaiboon@engr.tu.ac.th

Abstract. A power controller for the Lundell automotive alternator is proposed with load current sensing. The power controller determines exactly how much power the alternator will need to generate, based on the actual load current measurement and the programmable battery charging profile. To validate the effectiveness of the proposed power controller, experiments are performed under the conditions that the battery charging current can be controlled to zero and to a uniform charge level.

Keywords: Lundell automotive alternator, power controller, current sensing.

1 Introduction

Higher bus voltage requirements and more power demands in hybrid/electric vehicles have recently motivated the development of new automotive alternator designs. For example, a 42-V, 3.4-kW Lundell alternator with switched-mode rectifier (SMR) was developed in [1] and a 200V, 4-kW interior permanent-magnet alternator with SMR was investigated in [2]. The SMR improves the power capability and efficiency of the alternator by matching the alternator output voltage to the actual output voltage so that the maximum possible power is always obtained from the alternator [3].

Most commercial automotive alternators in the present markets, however, are conventional 14-V Lundell-type altenators. In addition, the control capability of the built-in bridge rectifier and voltage regulator included with the Lundell alternator as shown in Fig. 1 are far from achieving the maximum possible power. This is because the bridge rectifier has nearly fixed the alternator output voltage to the actual output voltage [4].

The Lundell alternator is a wound-field three-phase synchronous generator with claw-pole rotor design [5]. A simple equivalent circuit model for the Lundell alternator is shown in Fig. 2. In a wound-field generator, the output voltage or current can be controlled by varying the field current (i_{field}) that in turn varies the line-to-neutral voltage back emf magnitude V_s which can be written as

$$V_s = k n_{alt} i_{field} \ . \tag{1}$$

where k is the machine constant and n_{alt} is the alternator speed.

X. Wan (Ed.): Electrical Power Systems and Computers, LNEE 99, pp. 31–37.
springerlink.com

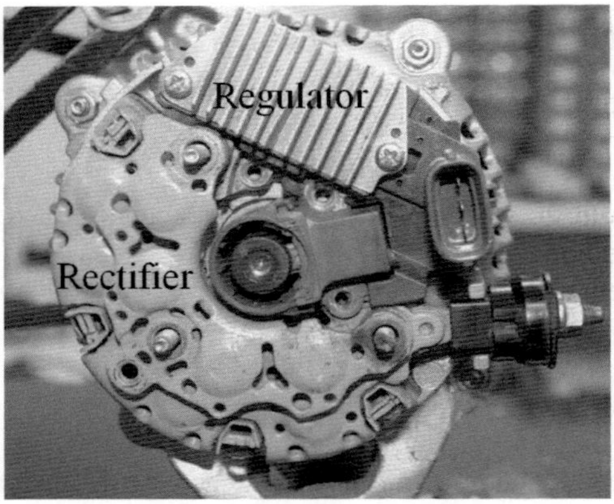

Fig. 1. Rectifier and regulator included with the Lundell alternator.

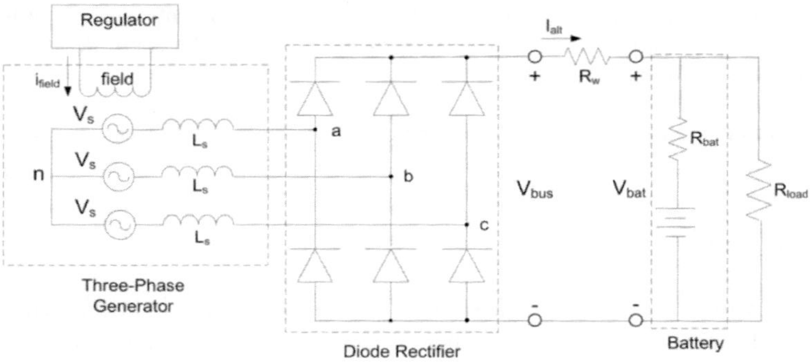

Fig. 2. A simple equivalent circuit model for the Lundell alternator.

Over a wide range of speeds, the bus voltage (V_{bus}) must be regulated to stay within a specified variation, normally in the range of 11–16V [6]. The voltage regulation is achieved by applying a pulse-width modulated (PWM) voltage across the field winding to weaken the average value of i_{field}, thus restraining V_{bus} from exceeding its upper limit at high speeds.

The bridge rectifier has advantages over the SMR in terms of simplicity and low cost. However, because of the lack of power control capability of the voltage regulator, the conventional Lundell alternator cannot avoid generating excessive or insufficient supply to the load. In this paper, a power controller is proposed for the Lundell alternator with current detection, thus enabling the power control capability to be accomplished.

This paper is organized as follows. The proposed power control system for the Lundell alternator is designed and implemented in Section 2. In Section 3, experimental results are presented to verify the effectiveness of the proposed power control scheme. Finally, some conclusions are given about the overall design.

2 Design and Implementation

The control system of the proposed power controller as shown Fig. 3 is composed of two PWM controllers: the voltage controller and the current controller. The controller outputs provide gate drive signals for two power MOSFETs connected in series with each other and the field winding of the alternator. A battery for energy storage and a dc electrical load are connected in parallel with the dc bus. The free-wheeling diode allows the field current a circulating path when each transistor is turned off.

In the diagram of Fig. 3, the voltage controller compares the bus voltage reference (V_{bus}^{*}) with its actual value (V_{bus}). The actual V_{bus} is measured using a simple voltage divider circuit where as the reference V_{bus}^{*} is the upper limit of the 14-V bus voltage. Note that the function of this voltage controller is similar to that of the voltage regulator found in a conventional automotive charging system.

To achieve power control capability, the current controller compares the alternator output current command (I_{alt}^{*}) with its measured counterpart (I_{alt}). The measured I_{alt} is actually the analog output voltage signal of the first Hall-effect current sensor attached to the dc bus. Meanwhile, the desired level of output current I_{alt}^{*} produced by the alternator is found using the following equation.

$$I_{alt}^{*} = I_{load} + I_{charge} \ . \tag{2}$$

where I_{load} is the actual load current measured by the second Hall-effect current sensor and I_{charge} is the preprogrammed battery charging current.

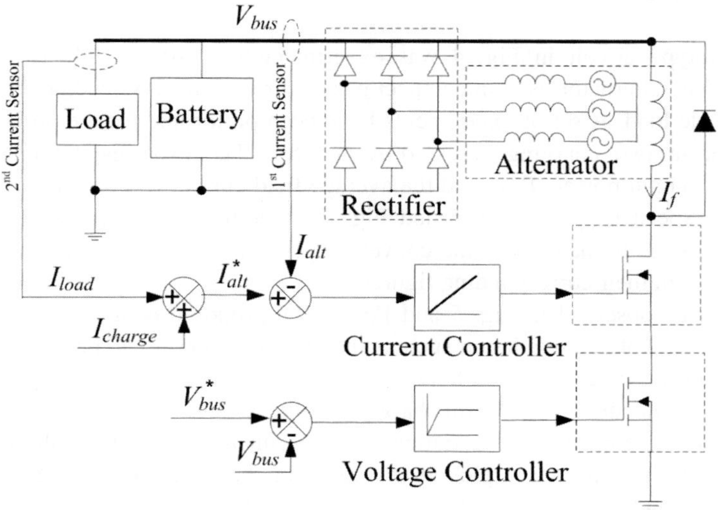

Fig. 3. Control system diagram of the proposed power controller.

In the schematic of Fig. 4, the aforementioned PWM controllers, MOSFETs, and Hall-effect current sensors are implemented with SG3524N [7], IRF540N [8], and ACS754SCB-200 [9], respectively.

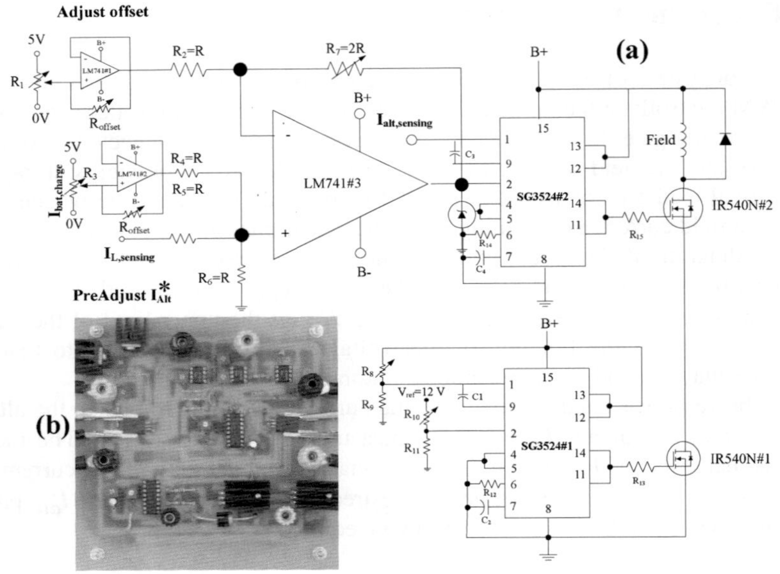

Fig. 4. Schematic (a) and prototype photograph (b) of the proposed power controller.

3 Experimental Verifications

The prototype circuit in Fig. 4 was experimentally verified with a 14-V 500-W Lundell-type alternator used in compact passenger cars, a 12-V lead-acid battery, and a 150-W dc load. A set of experiments have been preformed for the proposed power controller in two charging cases: 0 A and 5A. The experimental results of the alternator output power (P_{alt}) and the average field current (I_f) are plotted versus the alternator speed (n_{alt}) in Fig. 5 and Fig. 6. Additional P_{alt} and I_f data obtained from the same alternator with the conventional voltage regulator are also plotted for comparison in their corresponding figures.

As can be observed in Fig. 5 and Fig. 6, the proposed control scheme offers an edge over the voltage regulation such as the produced power can be better utilized by the load in the case when the battery is fully charged and the battery lifetime can possibly be extended in a uniform charging profile.

In view of the bus voltage variation, the experiment results of the alternator power versus alternator speed are plotted in Fig. 7 in the case when the bus voltage is unregulated and regulated at 13.5 V, 14.5 V, and 15.5 V. It is evident from Fig. 7 that the output power can increase/decrease significantly, although the bus voltage varies within a small range.

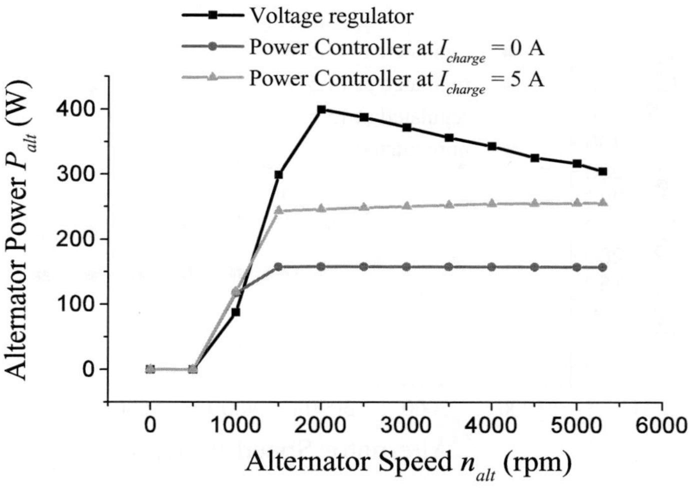

Fig. 5. Experimental results of the alternator power versus the alternator speed in the case when the voltage regulator is used and the power controller is used.

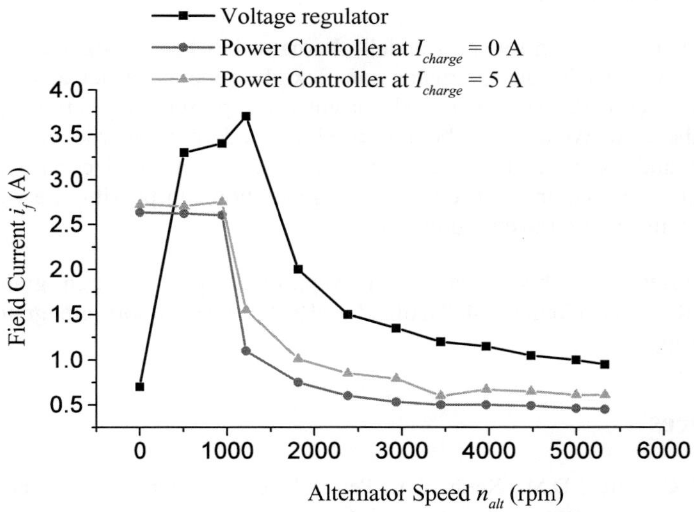

Fig. 6. Experimental results of the average field current versus the alternator speed in the case when the voltage regulator is used and the power controller is used.

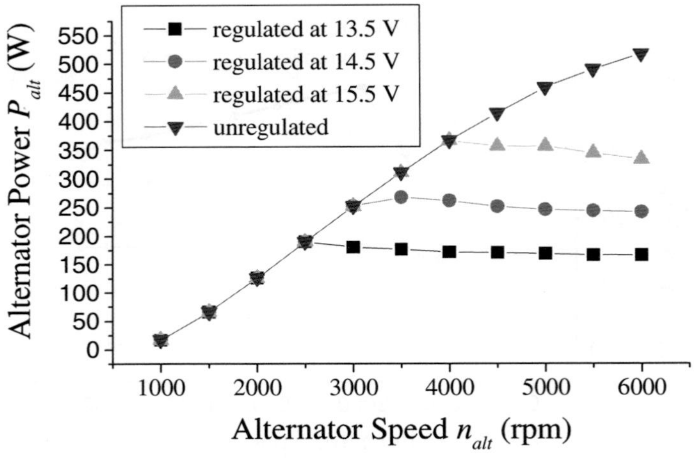

Fig. 7. Experimental results of the alternator power versus the alternator speed in the case when the bus voltage is unregulated and regulated at 13.5 V, 14.5 V, and 15.5 V.

4 Conclusion

A power control system for the Lundell alternator has been designed so that, over a wide range of speeds, the alternator can flexibly generate exact amount of output current to meet with the actual load current for a prescribed charging profile. To validate the effectiveness of the proposed method, a prototype circuit has been developed and experimentally verified. The use of proposed control scheme can ensure that excessive or insufficient power generation can possibly be well balanced by properly managing the charging profile.

Acknowledgments. This research was supported by a research grant from the National Research Council of Thailand (NRCT). The authors are grateful for the financial support.

References

1. Tang, S.C., Otten, D.M., Keim, T.A., Perreault, D.J.: Design and Evaluation of a 42-V Automotive Alternator with Integrated Switched-Mode Rectifier. IEEE Trans. Energy Convers 25, 983–992 (2010)
2. Liaw, C.Z., Whaley, D.M., Soong, W.L., Ertugrul, N.: Investigation of Inverterless Control of Interior Permanent-Magnet Alternators. IEEE Trans. Ind. Appl. 42, 536–544 (2006)
3. Perreault, D.J., Caliskan, V.: Automotive Power Generation and Control. IEEE Trans. Power Electron. 19, 618–630 (2004)
4. Ceuca, I.E., Tulbure, A., Ileana, I.: Experimental Research in Simulation of the Automotive Alternators and Solutions for Extend Voltage Level. In: 15th International Symposium for Design and Technology of Electronics Packages, pp. 215–220. IEEE Press, New York (2009)

5. Perreault, D.J., Afridi, K.K., Khan, I.A.: Automotive Applications of Power Electronics. In: Rashid, M.H. (ed.) Power Electronic Handbooks, pp. 791–828. Academic Press, New York (2001)
6. SAE Electronic Systems Committee: Recommended environment practices for electronic equipment design. Society of Automotive Engineers, Warrendale (1978)
7. STMicroelectronics, `http://www.st.com`
8. International Rectifier, `http://www.irf.com`
9. Allegro MicroSystems, `http://www.allegromicro.com`

An Iterative Sub-pixel Interpolation Centroid Algorithm

Zhang Yan and Peng Qingyu

Department of Computer Science, Jinan University, Guangzhou 510632, China;
Key Laboratory of Optoelectronic Information and Sensing Technologies of Guangdong Higher Educational Institutes, Jinan University,
Guangzhou 510632, China;
Sino-France Joint Laboratory for Astrometry, Dynamics and Space Science Joint Laboratory,
Jinan University, Guangzhou 510632, China
32324946@qq.com, pengqy@pub.guangzhou.gd.cn

Abstract. An improvement has been done in the sub-pixel interpolation algorithm proposed by Quine et al. by using iterative technique to obtain an accurate solution. Simulation results shows that the accuracy of the new iterative algorithm compared with the original one is significantly better.

Keywords: Star centroid, Sub-pixel interpolation, Iteration, Image analysis.

1 Introduction

Although the Hubble Space Telescope (HST) is not affected from the atmosphere, but its imaging system still has diffraction limitation and relatively short focal length, the FWHM of stars in the image is less than 2 pixels, thus the image is under-sampled. This under-sampled image will cause system error in positional measurement of stars [1]. Under-sampled images also appear in space navigation, ground-based short focal length telescope imaging and other actual situations, and therefore research of high precision centering algorithm for under-sampled image is of practical significance.

Anderson and King proposed ePSF method [1] to deal with the positional measurement of HST under-sampled images, but it was very complex. Quine et al. proposed a sub-pixel interpolation centroid algorithm [2]. Gray values of star were assumed to have a Gaussian distribution, and then a nonlinear equation could be derived by the brightest pixel and the next brightest neighboring pixel. The lower root of the equation was the center of the image. This method is simple, and it is a new way to solve the problem of high precision location in under-sampled image, but there still exits system error due to ignoring higher order terms.

This paper proposed a new iterative algorithm based on the method of Quine et al. Section 2 is sub-pixel interpolation centering theory and section 3 is the simulation experiments and results. The last section is the conclusion.

X. Wan (Ed.): Electrical Power Systems and Computers, LNEE 99, pp. 39–46.
springerlink.com

2 Theory of Sub-pixel Interpolation Algorithm

According to Quine et al. [1], it is assumed that a point source incident on the surface of CCD has a Gaussian intensity distribution, and the CCD pixels have an even intensity sensitivity across their active surface areas. Then the intensity reading of a particular pixel k in the CCD may be expressed as the integral over the active pixel area K :

$$I_k = \frac{I}{2\pi\sigma_x\sigma_y}\iint \exp\left\{-\frac{(x-x_0)^2}{2\sigma_x^2}-\frac{(y-y_0)^2}{2\sigma_y^2}\right\}dxdy\ , \tag{1}$$

where I is the total intensity of light, (x_0, y_0) is the point-source center offset, and (σ_x,σ_y) is the standard deviation of the 2D Gaussian (Gaussian width). Since the integration is over a rectangular area, it can be separated into a product:

$$I_k = I\frac{1}{\sqrt{2\pi}\sigma_x}\int_{x_1}^{x_2}\exp\left\{-\frac{(x-x_0)^2}{2\sigma_x^2}\right\}dx\frac{1}{\sqrt{2\pi}\sigma_y}\int_{y_1}^{y_2}\exp\left\{-\frac{(y-y_0)^2}{2\sigma_y^2}\right\}dy. \tag{2}$$

Write $g(x_0,\sigma_x,x_1,x_2)$ and $g(y_0,\sigma_y,y_1,y_2)$ as follows:

$$\begin{aligned} g(x_0,\sigma_x,x_1,x_2) &= \frac{1}{\sqrt{2}\sigma_x}\int_{x_1}^{x_2}\exp\left\{-\frac{(x-x_0)^2}{2\sigma_x^2}\right\}dx\ , \\ g(y_0,\sigma_y,y_1,y_2) &= \frac{1}{\sqrt{2}\sigma_y}\int_{y_1}^{y_2}\exp\left\{-\frac{(y-y_0)^2}{2\sigma_y^2}\right\}dy\ , \end{aligned} \tag{3}$$

then,

$$I_k = \frac{I}{\pi}g(x_0,\sigma_x,x_1,x_2)g(y_0,\sigma_y,y_1,y_2)\ , \tag{4}$$

Where g can be expressed as the difference between two error functions:

$$g(x_0,\sigma_x,x_1,x_2) = \frac{\sqrt{\pi}}{2}\left\{erf\left(\frac{1}{\sqrt{2}\sigma_x}(x_2-x_0)\right)-erf\left(\frac{1}{\sqrt{2}\sigma_x}(x_1-x_0)\right)\right\}, \tag{5}$$

where x_1 , x_2 are the pixel boundaries in x direction, x_0 is the centroid location, and σ_x is the standard deviation of the incident light. The error function is written as:

$$erf(x)=\frac{2}{\sqrt{\pi}}(x-\frac{x^3}{3}+\frac{x^5}{5\cdot 2!}-\frac{x^7}{7\cdot 3!}+\frac{x^9}{9\cdot 4!}-\cdots). \tag{6}$$

The error function in Eq.(6) is absolutely convergent for all x, but requiring $x<1$ to give good convergence within a few terms [3].

Star centroid can be separated by axis. Take x-axis, for example, the intensity reading of two neighboring pixels are written as I_1, I_2:

$$I_1=\frac{I}{\pi}g(x_0,\sigma_x,a,b)g(y_0,\sigma_y,e,f)=Mg(y_0,\sigma_y,a,b), \tag{7}$$

$$I_2=\frac{I}{\pi}g(x_0,\sigma_x,c,d)g(y_0,\sigma_y,e,f)=Mg(y_0,\sigma_y,c,d), \tag{8}$$

where a,b,c,d are the two pixel boundaries in x-axis, and e,f the boundaries in y-axis. Because the y-axis integrations of the two pixels are equal, the two equations can be combined to eliminate M:

$$g_1-\frac{I_1}{I_2}g_2=0, \tag{9}$$

where $g_1=g(x_0,\sigma_x,a,b)$ and $g_2=g(x_0,\sigma_x,c,d)$. Since g is expressed as the difference between two error functions, we can approximately write g as:

$$\begin{aligned} g_1=&\left\{\frac{1}{\sqrt{2}\sigma_x}(b-x_0)-\frac{1}{3}\left(\frac{1}{\sqrt{2}\sigma_x}(b-x_0)\right)^3+\frac{1}{5\cdot 2!}\left(\frac{1}{\sqrt{2}\sigma_x}(b-x_0)\right)^5-\cdots\right\} \\ &-\left\{\frac{1}{\sqrt{2}\sigma_x}(a-x_0)-\frac{1}{3}\left(\frac{1}{\sqrt{2}\sigma_x}(a-x_0)\right)^3+\frac{1}{5\cdot 2!}\left(\frac{1}{\sqrt{2}\sigma_x}(a-x_0)\right)^5-\cdots\right\} \end{aligned}. \tag{10}$$

We can collate Eq.(10) as:

$$\begin{aligned} g_1=&\sum_{n=1}^{\infty}\frac{(-1)^{n-1}(b^{2n-1}-a^{2n-1})}{(2n-1)(n-1)!(\sqrt{2}\sigma)^{2n-1}}+x_0\cdot\sum_{n=1}^{\infty}\frac{(-1)^{n-1}(b^{2n}-a^{2n})}{n!(\sqrt{2}\sigma)^{2n+1}} \\ &+x_0^{\,2}\cdot\sum_{n=1}^{\infty}\frac{(-1)^{n}(b^{2n-1}-a^{2n-1})}{(n-1)!(\sqrt{2}\sigma)^{2n+1}}+x_0^{\,3}\cdot\sum_{n=1}^{\infty}\frac{(-1)^{n}(2n+1)(b^{2n}-a^{2n})}{3\cdot n!(\sqrt{2}\sigma)^{2n+3}} \\ &+x_0^{\,4}\cdot\sum_{n=1}^{\infty}\frac{(-1)^{n-1}(2n+1)(b^{2n-1}-a^{2n-1})}{6\cdot(n-1)!(\sqrt{2}\sigma)^{2n+3}}+\cdots\cdots \end{aligned}. \tag{11}$$

Each series of Eq.(11) is absolutely convergent. Similarly, we can also get $g(x_0, \sigma_x, c, d)$. Taking the results into Eq.(9) and truncating the expansion to the forth order in x_0, then Eq.(9) may be rewritten as:

$$Ax_0^4 + Bx_0^3 + Cx_0^2 + Dx_0 + E = 0 , \tag{12}$$

with coefficients,

$$A = A_1 - A_2 \frac{I_1}{I_2}, B = B_1 - B_2 \frac{I_1}{I_2}, C = C_1 - C_2 \frac{I_1}{I_2}, D = D_1 - D_2 \frac{I_1}{I_2}, E = E_1 - E_2 \frac{I_1}{I_2}, \tag{13}$$

where A_1, B_1, C_1, D_1, E_1 are the expansion coefficient for the first pixel, and A_2, B_2, C_2, D_2, E_2 are the equivalent coefficients for the second pixel.

The brightest pixel in the image is chosen as I_2, and I_1 is the next brightest neighboring pixel. Then if the third order and the forth order in Eq.(13) are ignored, solving the roots of the quadratic by formula, the lower quadratic root gives the centroid offset. This is the method of Quine et al.

We improved the method above. Take the lower quadratic root into the higher terms $Ax^4 + Bx^3$ and subtract it, the first iteration solution is obtained. Then take the solution into Eq.(12) and repeat, this is the iterative algorithm. In the implementation, the iteration will be terminated when the absolute values of deviations between two solutions is small then 0.0001 pixel.

3 Simulation Experiments and Results

3.1 Comparison of Quadratic Roots and Iterative Algorithm

It is assumed that the point source incident on the surface of CCD has a Gaussian distribution, and the fill-ratio is 100%. By referring to Eq.(1) and Eq.(5), simulated data was generated using the error function, then we used sub-pixel interpolation of Quine et al. to find the center of a star. At the same time, the iterative algorithm was also implemented. In this paper, we took x direction for example.

Since the fill-ratio of CCD is 100%, then the right edge b of the first pixel and the left edge c of the second pixel are equal. The roots of quadratic by formula (called quadratic roots algorithm) and iteration (called iterative algorithm) were used to find the center respectively. The results of the two algorithms are showed in Fig.1. Obviously, if the star center is close to the pixel center (such as *Centroid Offset* < 0.1), the results of the two algorithms are equal, but when the deviation of the two centers are becomes larger (such as *Centroid Offset* > 0.25), the result of iterative algorithm is significantly better.

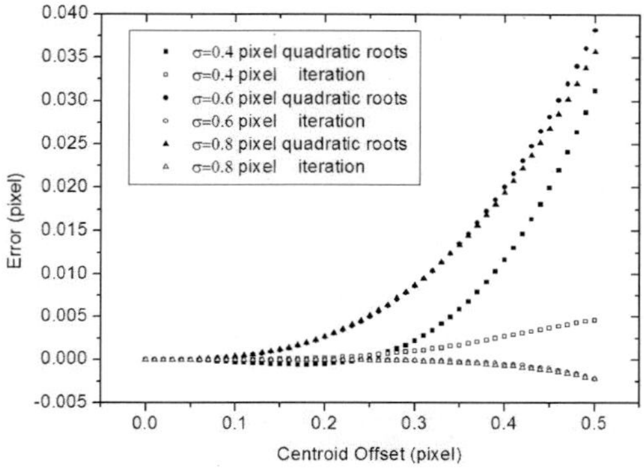

Fig. 1. Results of the two algorithms in noise-free image when $I = 1000$ and $\sigma (= \sigma_x)$ has different values.

3.2 Relationship between Light Intensity and Centering Accuracy

Generally, the brighter the star is, the greater the signal to noise ratio (SNR) is. And then the centering accuracy should be higher. In order to study the relationship between the light intensity of the star and the sub-pixel interpolation centroid accuracy in actual situation, we added Poisson noise on noise-free data. Give σ a fixed value, and I takes different values. The results of centering by iterative algorithm are showed in Fig.2 when $\sigma = 0.5$ and I =100, 1000, 10000, 100000 are adopted respectively. Obviously, if the total light intensity is larger, the dispersion of centering is smaller, and there is no obvious systematic error. Fig.3 is the root mean square error corresponding to the centering error when I has different values.

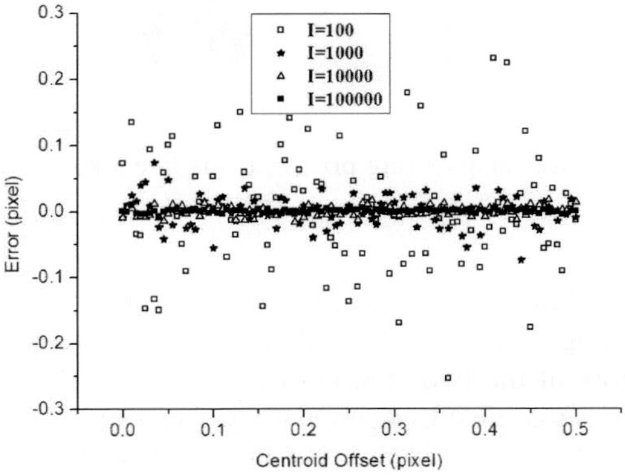

Fig. 2. Dispersion of finding center by iterative algorithm when $\sigma = 0.5$ and I =100, 1000, 10000, 100000 are adopted.

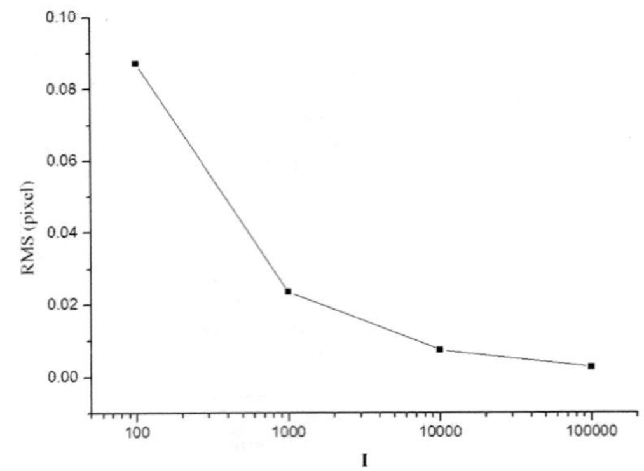

Fig. 3. The root mean square error corresponding to the centering error when I has different values.

3.3 Results of wo Pixels and Two Rows of Pixels

Similarly to the approach of Quine et al, in order to improve SNR, we accumulate all intensity of pixels of a row where the brightest pixel is and accumulate all intensity of pixels of the row next to it. Replacing the two pixels with the two cumulative sums, and then find the center using the two algorithms. Fig.4 shows the two integration regions.

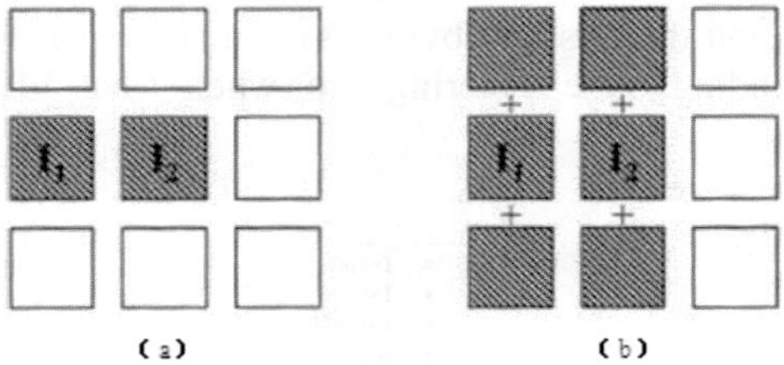

Fig. 4. (a) is integration over neighboring pixels , (b) is integration over rows of neighboring pixels.

The iterative algorithm was used to find the star center. Fig.5 is the result of the two approaches dealing with the data without noise, where $I = 1000$, $\sigma = 0.5$. We can see that the errors of the two approaches are very small and exactly equal. But when the Poisson noise is added, accumulating pixels in a row is much better than using single pixel (Fig.6). Fig.7 is the root mean square error corresponding to the centering error when $\sigma = 0.5$ and I =100, 1000, 10000, 100000 are adopted respectively.

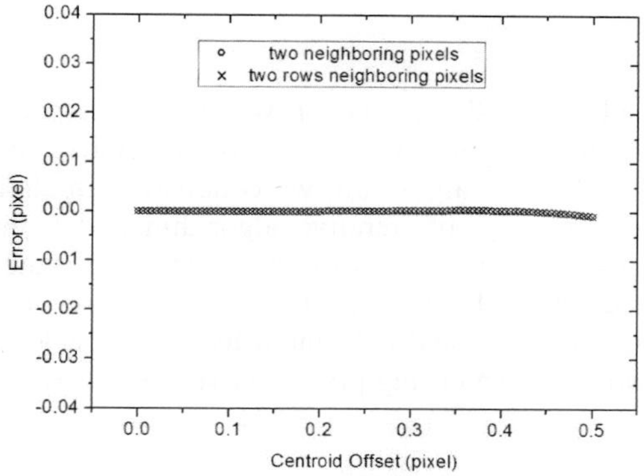

Fig. 5. Results from the two approaches when dealing with the data without noise, where $I = 1000$ and $\sigma = 0.5$.

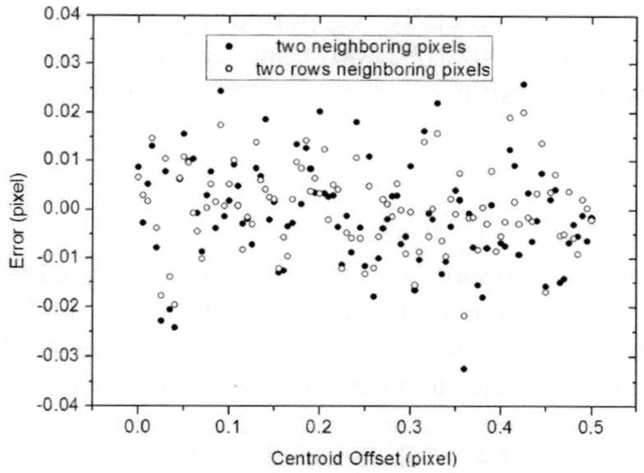

Fig. 6. Results from the two approaches in the simulation when Poisson noise is added.

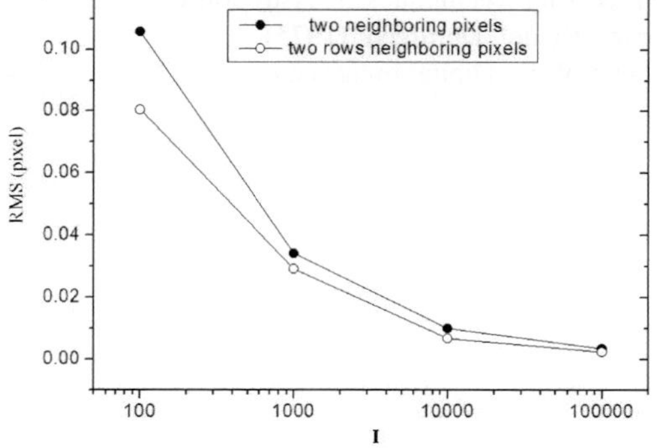

Fig. 7. The root mean square error corresponding to the centering error when $\sigma = 0.5$ and I =100, 1000, 10000, 100000 are adopted.

4 Conclusions

This paper described the basic theory of sub-pixel interpolation technique, and improves Quine's solution algorithm. We proposed iterative technique. In order to study the accuracy of the new algorithm, we conducted simulation experiments and compared the centering errors of iterative algorithm and Quine's algorithm. The results showed that the new iterative algorithm was more accurate than Quine's.

In addition, for the actual situation, in this paper we also discussed the impact of the Gaussian standard deviation and light intensity on the centering accuracy, and the accuracy of using two pixels and using pixels in two rows were compared.

Acknowledgements

We are most grateful to ZHANG Qingfeng, MENG Xiaohua, Li Yan and Li Zhan in Department of Computer Science of Jinan University for giving suggestions to this paper. This project was financially supported by the National Natural Science Foundation of China (Grant No 10973007) and partly by the Fundamental Research Funds for the Central Universities.

References

1. Anderson, J., King, I.R.: Toward High Precision Astrometry with WFPC2.I.Deriving an Accurate Point-Spread Function. In: PASP, 2000, vol. 112, pp. 1360–1392 (2000)
2. Quine, B. M., Tarasyuk, V., Mebrahtu, H., et al.: Determining star-image location: A new sub-pixel interpolation technique to process image centroids. Comput. Phys. Commun. 177, 700–706 (2007)
3. Abramowitz, M., Stegun, I.A.: Handbook of Mathematical Functions, (P297). Dover Publications, New York (1972)
4. Stone, R.C.: A comparison of digital centering algorithms. AJ 97, 1227–1237 (1989)
5. van Altena, W.F., Auer, L.H.: Digital image centering I. In: de Jager, C., Nieuwenhuijzen, H. (eds.) Image Processing Techniques in Astronomy, vol. 54, p. 411. Astrophysics and Space Science Library, Reidel, Dordrecht (1975)
6. Auer, L.H., van Altena, W.F.: Digital image centering II. AJ 83, 531–536 (1978)

Electronic Sealed-Bid Auctions with Incoercibility

Chongzhi Gao[1,2], Zheng-an Yao[3], Dongqing Xie[1], and Baodian Wei[4]

[1] School of Computer Science, Guangzhou University , Guangzhou 510006, China
[2] Key Laboratory of Network Security and Cryptology, Fujian Normal University, Fuzhou 350007, China
[3] College of Mathematic and Computational Science, Sun Yat-Sen University, GuangZhou 510275, China
[4] Department of Electronics and Communication Engineering, Sun Yat-sen University, Guangzhou 510006, China

Abstract. Electronic sealed-bid auctions are a method to establish the price of goods through the internet while keeping the bids secret during the bidding phase. In this paper, our concern is incoercibility of an auction protocol. This paper gives the first security model of incoercible sealed-bid auctions. By using undeniable signatures and deniable encryptions as building blocks, an efficient sealed-bid auction protocol with incoercibility is also proposed. Furthermore, our construction is proven to be secure in the random oracle model.

Keywords: Electronic Sealed-bid Auctions, Incoercibility, Undeniable Signatures, Deniable Encryptions.

1 Introduction

Electronic sealed-bid auctions are a method to establish the price of goods through the internet while keeping the bids secret during the bidding phase. To-date, there have been many research works on the sealed-bid auctions [7, 9, 11, 14, 6, 10, 13, 8, 2]. Most of these works have focused on the secrecy, verifiability and undeniability of the auction protocols. In this paper, our concern is incoercibility. That is, we want the auction protocols remain secure while a coercer can force a bidder show his random coins in the bidding phase. This paper gives the first security model of incoercible sealed-bid auctions and also proposes an efficient construction.

It is worth noting that the incoercibility in this paper is different from the receipt-freeness in [1, 5]. Receipt-freeness is a property to prevent bid rigging. The auction protocol with this property ensures that anyone, even if the bidder himself, must not be able to prove any information about the bidding price to any party. Receipt-freeness can be viewed as a kind of incoercibility because a coercer can not obtain a proof of a bidder's price. [1, 5] use interaction between the auctioneer and bidders in the bidding phase to achieve the receipt-freeness.

X. Wan (Ed.): Electrical Power Systems and Computers, LNEE 99, pp. 47–54.
springerlink.com

However, to achieve the incoercibility introduced in this paper, we need no interaction in the bidding phase.

1.1 Organization

The rest of this paper is organized as follows. In Section 2, we give some preliminaries. Section 3 gives the security model of electronic sealed-bid auctions with incoercibility, and presents an efficient construction whose security is proven in the random oracle model. Section 4 concludes the paper.

2 Preliminaries

Two main building blocks are used in our incoercible sealed-bid auction protocol. They are undeniable signatures and deniable encryptions.

Undeniable Signature [4]. An undeniable signature scheme $\mathcal{US} = (\mathsf{G}_S, \mathsf{Sig}, \pi)$ is a tuple of algorithms, where

- G_S is the key generation algorithm that takes a security parameter as input and outputs a public/private key pair (pk, sk). We write the execution of this algorithm as $(pk, sk) \leftarrow \mathsf{G}_S$.
- Sig is the signing algorithm that takes a private key sk and a message m as inputs, and outputs a signature σ. We write the execution of this algorithm as $\sigma \leftarrow \mathsf{Sig}_{sk}(m)$.
- π is a interactive verification protocol that takes a signature σ, a message m, and a public key pk as inputs, and outputs 0 or 1 for reject or accept respectively. We write the execution of this algorithm as $b \leftarrow \pi(\sigma, m, pk)$.

Properties: An undeniable signature scheme $\mathcal{US}$ should satisfy the following properties. *Unforgeability:* The scheme should be existential unforgeable under adaptive chosen message attacks. *Invisibility:* A cheating verifier, given only a signer's public key, a message, and an undeniable signature, cannot decide whether the signature is valid for the message or not. *Soundness:* A cheating signer cannot prove a valid signature invalid (nonrepudiation), or an invalid signature valid (false claim of origin).

Deniable Encryption [3]. A δ-sender-deniable encryption scheme $\mathcal{DE} = (\mathsf{G}_E, \mathsf{Enc}, \mathsf{Dec}, \phi)$ is a tuple of algorithms, where

- G_E is the key generation algorithm that takes a security parameter as input and outputs a public/private key pair (pk, sk). We write the execution of this algorithm as $(pk, sk) \leftarrow \mathsf{G}_E$.
- Enc is the encryption algorithm that takes a public key pk, a plaintext m and a random value r as inputs, and outputs a ciphertext c. We write the execution of this algorithm as $c \leftarrow \mathsf{Enc}_{pk}(m, r)$.
- Dec is the decryption algorithm that takes a private key sk and a ciphertext c as inputs, and outputs a plaintext m. We write the execution of this algorithm as $m \leftarrow \mathsf{Dec}_{sk}(c)$.

- ϕ is a faking algorithm, which takes a message m, a random value r and another message m' ($\neq m$) as inputs, outputs a string r'. We write the execution of this algorithm as $r' \leftarrow \phi(m, r, m')$. Additionally, pk or sk may be required as input to ϕ if they are needed.

Properties: A deniable encryption scheme $\mathcal{DE}$ should satisfy the following properties. *Correctness:* The receiver should be able to decrypt the correct value (except, perhaps, with negligible probability of error). *Security:* the protocol should be semantically secure against chosen ciphertext attacks. *Deniability:* For any $m_0, m_1 \in M$ where M is the plaintext space, let $(pk, sk) \leftarrow \mathsf{G}_E$, and r, r_0 are randomly chosen from the random input space. Let $c = \mathsf{Enc}_{pk}(m_0, r)$, $r_1 = \phi(pk, m_0, r, m_1)$. The random variables

$$(m_1, r_1, c) \text{ and } (m_1, r_0, \mathsf{Enc}_{pk}(m_1, r_0))$$

are δ-close for any PPT(an abbreviation for "probabilistic polynomial-time") distinguisher $\mathcal{A}$. More formally, We define an adversary $\mathcal{A}$'s advantage breaking the sender-deniability of $\mathcal{DE}$ as

$$\mathrm{Adv}_{\mathcal{DE}}^{\mathcal{A},sd} = \Pr\left[\begin{array}{l} b = b' : (pk, sk) \leftarrow \mathcal{K}; b \leftarrow \{0,1\}; r, r_0 \leftarrow R; r_1 \leftarrow \phi(pk, m_0, r, m_1); \\ ch \leftarrow (m_1, r_b, \mathsf{Enc}_{pk}(m_1, r_b)); b' \leftarrow \mathcal{A}(pk, m_0, ch) \end{array}\right] - \frac{1}{2}.$$

The deniable encryption scheme $\mathcal{DE}$ is δ-sender-deniable if for any PPT adversary $\mathcal{A}$, $\mathrm{Adv}_{\mathcal{DE}}^{\mathcal{A},sd}$ is less than δ.

Remark 1. Until now, the most efficient deniable encryption scheme expands 1 bit plaintext to a ciphertext message [3]. Exploring efficient and non-interactive deniable encryption schemes with a short message expansion is still an open problem.

3 Sealed-Bid Auctions with Incoercibility

3.1 Security Model

A sealed-bid auction protocol Σ, which has three phases, is executed among some bidders and an auctioneer. We assume there are t bidders $B_1, B_2, ..., B_t$.

- The first is *initializing phase*, where the public parameters and keys of all parties are generated. We write an execution of this phase as $(PK, sk^{(i)}, sk_T) \leftarrow \mathsf{INI}$ where PK includes all the public keys, $sk^{(i)}$ is party B_i's private key and sk_T is the auctioneer's private key.
- The second is *bidding phase*, where all bidders set their (sealed) bidding price and use some given steps to commit the price. We write an execution of this phase as $CMT^{(i)} \leftarrow \mathsf{BID}(PK, sk^{(i)}, tp^{(i)}, r^{(i)})$ where $tp^{(i)}$ is the party B_i's true bidding price, $r^{(i)}$ is B_i's random input and $CMT^{(i)}$ is the commitment of his bidding price.
- The third phase is *opening phase*. In this phase, the highest price is revealed while other prices are still kept secret. We write an execution of this phase as $tp \leftarrow \mathsf{OPEN}$ where tp is the highest bidding price.

A sealed-bid auction protocol must satisfy the following properties.

- **Secrecy:** All bidding prices except winning price must be kept secret even from the auctioneer. Formally, We define an adversary $\mathcal{A}$'s advantage breaking the secrecy of the auction protocol Σ as

$$\mathrm{Adv}_{\Sigma}^{\mathcal{A},se} = \Pr\left[\begin{array}{l} p' < tp \text{ and } p' \in \mathcal{P} : (PK, sk^{(i)}, sk_T) \leftarrow \mathsf{INI}; \\ CMT^{(i)} \leftarrow \mathsf{BID}(PK, sk^{(i)}, tp^{(i)}, r^{(i)}); tp \leftarrow \mathsf{OPEN}; p' \leftarrow \mathcal{A}(st) \end{array}\right]$$

  where $\mathcal{P} = \{tp^{(1)}, ..., tp^{(t)}\}$ and st is all the transcripts obtained by $\mathcal{A}$ while $\mathcal{A}$ watches the execution of the protocols $\mathsf{INI}, \mathsf{BID}$ and OPEN. A sealed-bid auction protocol Σ has the property of secrecy if for any PPT adversary $\mathcal{A}$, $\mathrm{Adv}_{\Sigma}^{\mathcal{A},se}$ is negligible.
- **Verifiability:** Anyone must be able to verify the correctness of the auction.
- **Undeniability:** No bidder is able to deny his bidding price except for a negligible probability.

Furthermore, a δ-incoercible sealed-bid auction protocol with 2^l-ambiguousness must satisfy:

δ-Incoercibility(with 2^l-ambiguousness, $l \in \mathbb{N}$). Before opening, for any bidder B_i, a coercer can not discriminate the true bidding price from other $2^l - 1$ "fake" prices, which are chosen by the bidder, except that the coercer is the auctioneer. A *coercer* is a party who can coerce the bidder to show his random coins used in the bidding step. Formally, for any i, the probability

$$\Pr\left[\begin{array}{l} p'^{(i)} = tp^{(i)} : (PK, sk^{(i)}, sk_T) \leftarrow \mathsf{INI}; CMT^{(i)} \leftarrow \mathsf{BID}(PK, sk^{(i)}, tp^{(i)}, r^{(i)}); \\ (\tilde{tp}^{(i)}, \tilde{r}^{(i)}) \leftarrow \mathsf{FORGE}(PK, sk^{(i)}, tp^{(i)}, r^{(i)}); p'^{(i)} \leftarrow \mathcal{A}(st, \tilde{tp}^{(i)}, \tilde{r}^{(i)}) \end{array}\right] - \frac{1}{2^l}$$

should be less than δ where st is all the transcripts obtained by $\mathcal{A}$ while $\mathcal{A}$ watches the execution of the protocols $\mathsf{INI}, \mathsf{BID}$ and $(\tilde{tp}^{(i)}, \tilde{r}^{(i)})$ is consistent with the transcript st.

3.2 Construction: A Sealed-Bid Auction Protocol with 2-Ambiguousness

Let A be an auctioneer and $B_1, B_2, \ldots, B_t$ be bidders. Let $\mathcal{US} = (\mathsf{G}_S, \mathsf{Sig}, \pi)$ be an undeniable signature scheme whose message space is $\{0,1\}^v$, and $\mathcal{DE} = (\mathsf{G}_E, \mathsf{Enc}, \mathsf{Dec}, \phi)$ be a deniable encryption scheme whose plaintext space is $\{0,1\}$. Let $h : \{0,1\}^* \rightarrow \{0,1\}^v$ be a collision free hash function.

Initialization. Let $(pk_s^{(i)}, sk_s^{(i)}) \leftarrow \mathsf{G}_S$ $(i = 1, \ldots, t)$, ant let $(pk_e, sk_e) \leftarrow \mathsf{G}_E$. $(pk_s^{(1)}, \ldots, pk_s^{(t)}, pk_e)$ is the public key. $sk_s^{(i)}$ is B_i's private signature key, and sk_e is A's private decryption key.

Bidding. Assume x_i is B_i's bidding price for a good whose index is w.

(1) B_i chooses a fake price x_i'.

(2) B_i randomly chooses r_i, r_i' as auxiliary random input, computes $e = \mathsf{Enc}_{pk_e}(1, r_i)$, $\sigma = \mathsf{Sig}_{sk_s^{(i)}}(h(x_i \| w \| e))$, and $e' = \mathsf{Enc}_{pk_e}(0, r_i')$, $\sigma' = \mathsf{Sig}_{sk_s^{(i)}}(h(x_i' \| w \| e'))$. B_i then publishes the two value $\sigma \| e, \sigma' \| e'$ in random order. Without loss of generality, we assume B_i publishes $\sigma_0 \| e_0, \sigma_1 \| e_1$ where $\{\sigma_0, \sigma_1\} = \{\sigma, \sigma'\}$ and $\{e_0, e_1\} = \{e, e'\}$.

Remark 2. If B_i doesn't want to commit to a fake price, he just chooses σ' as a random value.

Oppening. Auctioneer A and $B_1, B_2, \ldots, B_t$ iterate following steps for prices $j = n, n-1, \ldots, 1$.

(1) A executes $b_0^{(j)} \leftarrow \pi(\sigma_0, h(j \| w \| e_0), pk_s^{(i)})$, $b_1^{(j)} \leftarrow \pi(\sigma_1, h(j \| w \| e_1), pk_s^{(i)})$ together with B_i.
(2) If $b_k^{(j)} = 1,\ k \in \{0, 1\}$, A executes $d_k \leftarrow \mathsf{Dec}(e_k)$. If $d_k = 1$, A then announces that the winning bidder is bidder B_i and winning price is j.
(3) In step 2, if $b_k^{(j)} = 0,\ \forall k \in \{0, 1\}$ or no obtained d_k equals to 1, A is convinced that no bidder bids at price j. Then auctioneer decreases j by 1 and repeats the above steps.

3.3 Security

- **Secrecy:** By the invisibility of the undeniable signature scheme $\mathcal{US}$, a verifier including the auctioneer can not know any other bidding price except the winning price.

Formally, the property of secrecy is described in the following theorem.

Theorem 1. *The incoercible auction protocol described above has the property of secrecy in the random oracle model [12], provided that the utilized undeniable signature scheme is invisible.*

Proof. To prove the theorem, we will prove the following: "If there exists a PPT algorithm $\mathcal{A}$ which breaks our protocol's secrecy with advantage ϵ, then we can construct an algorithm $\mathcal{B}$ which breaks the invisibility of an undeniable signature scheme $\mathcal{US}$ with advantage ϵ/t ."

Suppose $\mathcal{B}$ is given pk_s, a public key generated by the key generation algorithm of an undeniable signature scheme $\mathcal{US}$.

Given (m, σ) as a challenge, $\mathcal{B}$ works as follows to determine whether σ is m's signature:

Setup: $\mathcal{B}$ randomly chooses a $i \in \{1, .., t\}$ and sets up an auction protocol as in the initializing phase of Section 3.2, except that the i-th auctioneer's public key is given by pk_s.
Challenge:
1) $\mathcal{B}$ simulates $B_1, .., B_t$'s execution for a good w as in the bidding phase, except that the simulation for B_i is as follows. $\mathcal{B}$ chooses $tp^{(i)} < max\{tp^{(1)},$

.., $tp^{(i-1)}, tp^{(i+1)}, .., tp^{(t)}\}$, computes e, e' as in the bidding phase, and sets σ' as a random value. $\mathcal{B}$ then publishes the two value $\sigma \| e, \sigma' \| e'$ in random order on behalf of B_i. Without loss of generality, we also assume $(\sigma_0 \| e_0, \sigma_1 \| e_1)$ is published. Note that σ is obtained by $\mathcal{B}$ as his challenge and $(\sigma_0 \| e_0, \sigma_1 \| e_1)$ is distributed exactly as in a real auction protocol.
2) $\mathcal{B}$ simulates the opening phase. Note that since $tp^{(i)} < max\{tp^{(1)}, .., tp^{(i-1)}, tp^{(i+1)}, .., tp^{(t)}\}$, $\mathcal{B}$ can perfectly complete the simulation even if he does not know sk_s.
The simulation of hash oracle: When $\mathcal{A}$ submits his hash oracle queries to $h(\cdot)$, $\mathcal{B}$ responds the queries as usual except responding $h(tp^{(i)} \| w \| e)$ as m.
Guess: $\mathcal{A}$ outputs its price guess p'. If $p' = tp^{(i)}$, then $\mathcal{B}$ outputs 1 indicating that σ is a signature of m, otherwise $\mathcal{B}$ outputs 0.

Now we analyze the probability that $\mathcal{B}$ breaks the invisibility of the undeniable signature scheme $\mathcal{US}$. Because $\mathcal{B}$ selects i at random and simulates the execution of the auction protocol perfectly, the probability of $p' = tp^{(i)}$ equals to $1/t$. Thus the probability that $\mathcal{B}$ successfully outputs his guess is ϵ/t, which is also a non-negligible probability. ■

- **Verifiability:** Everyone can verify the signature (with the help of bidder), but only the auctioneer can decrypt the encryption tags e_0, e_1. To achieve Verifiability, after the step of opening, the private decryption key of the auctioneer should be published. And once the private key of the auctioneer is published, the auctioneer should immediately update his private key.
- **Undeniability:** By the nonreputation property of the undeniable signature, and the correctness of the deniable encryption, the bidder can only deny his bidding price except for a negligible probability.
- **Incoercibility:** The above scheme is 2-ambiguous. When a coercer forces the bidder to show his random coins in the bidding step, the bidder can use the faking algorithm of the deniable encryption to show the plaintext as 0 or 1 as his will. Thus the coercer doesn't know which price is the true bidding price.

Formally, we have the following theorem.

Theorem 2. *The incoercible auction protocol described above is δ-incoercibility with 2-ambiguousness, provided that the utilized deniable encryption scheme is δ-sender-deniable.*

Proof. To prove the theorem, we will prove the following: "If there exists a PPT algorithm $\mathcal{A}$ which breaks our protocol's incoercibility with advantage δ, then we can construct an algorithm $\mathcal{B}$ which breaks the sender-deniability of a deniable encryption scheme $\mathcal{DE}$ with advantage δ ."

Suppose $\mathcal{B}$ is given pk_e, a public key generated by the key generation algorithm of a deniable encryption scheme $\mathcal{DE}$.

Given $(m_0, m_1, \tilde{r}, c)$ as a challenge where $m_0, m_1 \in \{0, 1\}$ and $c = \mathsf{Enc}_{pk_e}(m_1, \tilde{r})$, $\mathcal{B}$ works as follows to determine whether c is m_0 or m_1's ciphertext.

Setup: $\mathcal{B}$ sets up an auction protocol as in the initializing phase of Section 3.2, except that the auctioneer's public key is given by pk_e.
Challenge:
1) $\mathcal{B}$ simulates $B_1, .., B_t$'s execution for a good w as in the bidding phase, except that the simulation for B_i is as follows. $\mathcal{B}$ chooses $tp^{(i)}$ and $\tilde{tp}^{(i)}$ for true price and fake price respectively. If $m_1 = 1$, then let $e = c$ and compute $e' = \mathsf{Enc}_{pk_e}(0, r_i')$ where r_i' is randomly chosen, otherwise let $e' = c$ and compute $e = \mathsf{Enc}_{pk_e}(1, r_i)$ where r_i is also randomly chosen. Then $\mathcal{B}$ simulates the remaining bidding phase as in the real auction protocol.
2) When $\mathcal{A}$ coerces B_i to reveal his bidding price and random local input used in the bidding phase, $\mathcal{B}$ just outputs $(tp^{(i)}, 1, r_i, e)$ and $(\tilde{tp}^{(i)}, 0, r_i', e')$ to show that $tp^{(i)}$ is the true bidding price.
Guess: $\mathcal{A}$ outputs its price guess p'. If $p' = tp^{(i)}$, then $\mathcal{B}$ outputs 1 indicating that c is a ciphertext of m_1, otherwise $\mathcal{B}$ outputs 0.

It is easy to analyze that $\mathcal{B}$'s success probability is the same with $\mathcal{A}$'s success probability. ■

Extension to Sealed-bid Auction Protocols with 2^l-ambiguousness. Applying a deniable encryption scheme whose plaintext space is $\{0,1\}^l$ to the above basic auction protocol, we get a sealed-bid auction protocol with 2^l- ambiguousness.

4 Conclusion

We introduce a new type of sealed-bid auctions with incoercibility, where a coercer may force the bidders to show their random coins used in the bidding phase. An efficient protocol satisfying the incoercibility is also proposed. Our construction is proven to be secure in the random oracle model.

Acknowledgement

This paper is supported by Natural Science Foundation of China(60903165, 60803135, 10871222, 10971234, 11026227), Natural Science Foundation of Guangdong Province of China(9151064007000004), Open Funds of Key Lab of Fujian Province University Network Security and Cryptology(09A008), Fundamental Research Funds for the Central Universities(10lgpy31), and The National High Technology Research and Development Program of China(2009AA01Z420).

References

1. Abe, Suzuki: Receipt-free sealed-bid auction. In: ISW: International Workshop on Information Security, LNCS. Springer, Heidelberg (2002)
2. Bansal, N., Chen, N., Cherniavsky, N., Rurda, A., Schieber, B., Sviridenko, M.: Dynamic pricing for impatient bidders. ACM Transactions on Algorithms 6(2), 726–735 (2010)

3. Canetti, R., Dwork, C., Naor, M., Ostrovsky, R.: Deniable encryption. In: Kaliski Jr., B.S. (ed.) CRYPTO 1997. LNCS, vol. 1294, pp. 90–104. Springer, Heidelberg (1997)
4. Chaum, D., van Antwerpen, H.: Undeniable signatures. In: Brassard, G. (ed.) CRYPTO 1989. LNCS, vol. 435, pp. 212–216. Springer, Heidelberg (1990)
5. Chen, Lee, Kim: Receipt-free electronic auction schemes using homomorphic encryption. In: Lim, J.-I., Lee, D.-H. (eds.) ICISC 2003. LNCS, vol. 2971, Springer, Heidelberg (2004)
6. Chida, Kobayashi, Morita: Efficient sealed-bid auctions for massive numbers of bidders with lump comparison. In: ISW: International Workshop on Information Security. LNCS. Springer, Heidelberg (2001)
7. Franklin, M.K., Reiter, M.K.: The design and implementation of a secure auction service. In: Proceedings of the IEEE Symposium on Research in Security and Privacy, Oakland, CA, May 1995, pp. 2–14. IEEE Computer Society, Technical Committee on Security and Privacy, IEEE Computer Society Press (1995)
8. Ha, J., Zhou, J., Moon, S.-J.: An improved double auction protocol against false bids. In: Katsikas, S.K., López, J., Pernul, G. (eds.) TrustBus 2005. LNCS, vol. 3592, pp. 274–287. Springer, Heidelberg (2005)
9. Harkavy, M., Tygar, J.D., Kikuchi, H.: Electronic auctions with private bids. In: Proceedings of the 3rd USENIX Workshop on Electronic Commerce, pp. 61–74 (1998)
10. Juels, Szydlo: A two-server, sealed-bid auction protocol. In: Blaze, M. (ed.) FC 2002. LNCS, vol. 2357, Springer, Heidelberg (2003)
11. Kikuchi, H., Harkavy, M., Tygar, J.D.: Multi-round anonymous auction protocols. In: Proceedings of the First IEEE Workshop on Dependable and Real-Time E-Commerce Systems, pp. 62–69. Springer, Heidelberg (1998)
12. Kurosawa, K., Schmidt-Samoa, K.: New online/Offline signature schemes without random oracles. In: Yung, M., Dodis, Y., Kiayias, A., Malkin, T. (eds.) PKC 2006. LNCS, vol. 3958, pp. 330–346. Springer, Heidelberg (2006)
13. Peng, Boyd, Dawson: A multiplicative homomorphic sealed-bid auction based on goldwasser-micali encryption. In: Won, D.H., Kim, S. (eds.) ICISC 2005. LNCS, vol. 3935, Springer, Heidelberg (2006)
14. Suzuki, K., Kobayashi, K., Morita, H.: Efficient sealed-bid auction using hash chain. In: Won, D. (ed.) ICISC 2000. LNCS, vol. 2015, p. 183. Springer, Heidelberg (2001)

Methods of Superior Design for the Full Scale Output of Piezoresistive Pressure Sensors

Ruirui Han, Zhaohua Zhang, Tianling Ren, Huiwang Lin, and Bo Pang

Institute of Microelectronics, Tsinghua University,
Beijing, 100084, P.R.C
Tsinghua National Laboratory for Information Science and Technology,
Beijing 100084, P.R.C
hrr09@mails.tsinghua.edu.cn, Rentl@tsinghua.edu.cn

Abstract. Sensitivity is one of the most important parameters for piezoresistive pressure sensors. It is usually through superior design of the full scale output of pressure sensors to achieve high sensitivity of the devices and meet the requirement for certain application. Two kinds of methods of evaluating the full scale output of pressure sensors are discussed .Both of them are based on finite element analysis (FEA) and integration of stress difference with respect to certain path, which are realized by ANSYS. In addition, results of these two methods are coincident with each other. The full scale output of the pressure sensor by simulation is 42.996mv while the best result from experiment is 43.112mv. For all the experiment results, relative errors are limited to 2.5%. Therefore the experiment results show good agreement with the simulation results.

Keywords: piezoresistive pressure sensors, full scale output, finite element analysis (FEA), integration of stress difference.

1 Introduction

Piezoresistance effect and principles of thin-film mechanics are the major theories for design and fabrication of piezoresistive pressure sensors .However, it is impossible to calculate accurate stress values of all points on the surface of the film and then to deduce the resistance variation of the piezoresistors. Thanks to the methods of finite element analysis and the appearance of FEA tools. Such methods for obtaining the resistance variation of piezoresistors on the film of the pressure sensor have been provided by papers published previously[1][2], mostly based on the following formula:

$$\frac{\Delta R}{R} = \frac{\pi_l \sum_1^n \sigma_{li} v_i + \pi_t \sum_1^n \sigma_{ti} v_i}{\sum_i^n v_i} \quad (1)$$

where ΔR is the deviation of the resistance, R is the zero-stress resistance, π_l and π_t are the longitudinal and transverse piezoresistance coefficient, v_i is the volume of the ith element, and σ_{li} and σ_{ti} are the longitudinal and transverse stress of the ith

X. Wan (Ed.): Electrical Power Systems and Computers, LNEE 99, pp. 55–61.
springerlink.com

element respectively. Although this method has been already applied into practice during the design process of pressure sensors, it is somewhat tedious and therefore less efficient to accomplish such summation.

In this paper two simple and reliable methods based on FEA and integral arithmetic are presented. With these methods, it is possible to achieve convenient and effective design of the full scale output of pressure sensors, making full use of the FEA software ANSYS.

2 Design Theory

2.1 The Output of Wheatstone Bridge Comprised of Piezoresistors

Wheatstone bridge is applied to the network of pressure sensors in order to get high sensitivity and minor zero output, which is shown in Fig.1.Obviously, the output voltage of the network can be calculated with Eq. (2) when no pressure is applied on the square diaphragm. In the case when pressure is loaded, the output voltage can be expressed by Eq. (3).

$$V_0 = \frac{R_2R_4 - R_1R_3}{(R_1 + R_2) \times (R_3 + R_4)} V_B \tag{2}$$

$$V_{out} = \frac{(R_2 + \Delta R_2)(R_4 + \Delta R_4) - (R_1 + \Delta R_1)(R_3 + \Delta R_3)}{(R_1 + \Delta R_1 + R_2 + \Delta R_2) \times (R_3 + \Delta R_3 + R_4 + \Delta R_4)} V_B \tag{3}$$

Assume that $R_1 = R_2 = R_3 = R_4 = R$, $\Delta R_2 = -\Delta R_1 = \Delta R_4 = -\Delta R_3 = \Delta R$, Eq. (4) is obtained:

$$V_{out} = \frac{\Delta R}{R} V_B \tag{4}$$

According to piezoresistance effect [3], and if the piezoresistor is located in [011] direction on [100] facet of p type silicon, then $\Delta R/R = 0.5\pi_{44}(\sigma_l - \sigma_t)$ and V_{out} can also be expressed by Eq.(5) :

$$V_{out} = \frac{\Delta R}{R} V_B = 0.5\pi_{44}(\sigma_l - \sigma_t) V_B \tag{5}$$

However, stress values vary at different points of the diaphragm, so it is necessary to find out the equivalent value for the stress difference across the whole diaphragm. Therefore, Eq. (5) is modified as Eq.(6):

$$V_{out} = \frac{\Delta R}{R} V_B = 0.5\pi_{44}\overline{(\sigma_l - \sigma_t)} V_B \tag{6}$$

In conclusion, in order to evaluate the full scale output of the pressure sensor, it is critical to calculate either the resistance deviation of the piezoresistor or the equivalent value for the stress difference of the whole diaphragm when full scale pressure is applied on.

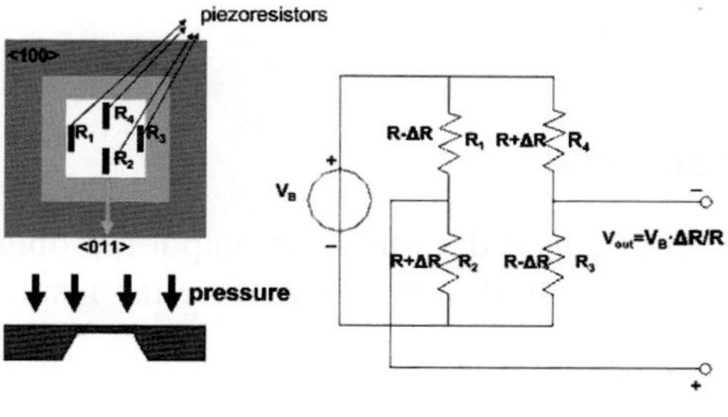

Fig. 1. Schematic view of the sensor and its network

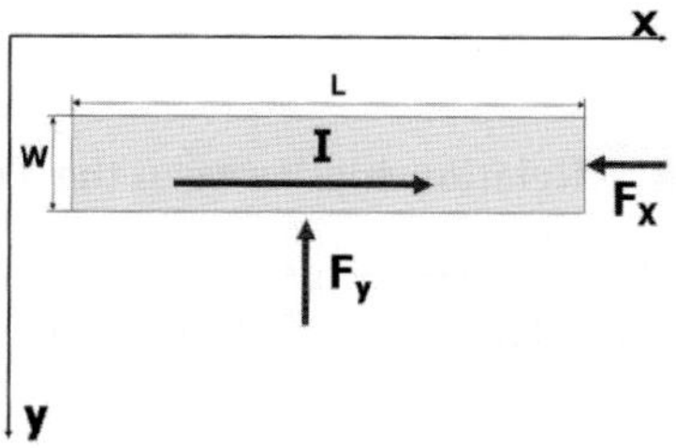

Fig. 2. The location of the piezoresistor

2.2 Method to Calculate $\frac{\Delta R}{R}$

As is shown in Fig.2, a piezoresistor is located along the X direction in the XY plane. The thickness in the Z direction is ignored for its extremely minor value. The resistance of the piezoresistor is R and the sheet resistance is $R_{\square}$. The longitudinal and transverse stress of the piezoresistor are σ_x and σ_y.Compared with the length of the piezoresistor , the width is also very small; therefore we can assume that σ_x and σ_y change in the X direction rather than Y.So Eq.(7),(8) are obtained:

$$\Delta R(X) = \frac{R_{\square}}{W}\left(\pi_x\sigma_x + \pi_y\sigma_y\right)dx \tag{7}$$

$$\Delta R = \frac{R_{\square}}{W}\int_0^L (\pi_x\sigma_x + \pi_y\sigma_y)\,dx = \frac{0.5\pi_{44}R_{\square}}{W}\int_0^L (\sigma_x - \sigma_y)\,dx \tag{8}$$

And R can be expressed as:

$$R = R_{\square}\frac{L}{W} \tag{9}$$

So Eq.(10) is obtained:

$$\frac{\Delta R}{R} = \frac{\pi_{44}}{2L}\int_0^L (\sigma_x - \sigma_y)\,dx \tag{10}$$

2.3 Method to Calculate $\overline{(\sigma_l - \sigma_t)}$

The equivalent value for the stress difference can be the average of the stress difference distributed on the surface of the piezoresistor[4], which is calculated by the following formula:

$$\overline{\sigma_l - \sigma_t} = \frac{\int_0^L (\sigma_x - \sigma_y)\, dx}{L} \tag{11}$$

2.4 Relationship between the Two Methods

Although these two kinds of methods to calculate the full scale output are obtained from different points of view, they are inherently the same; therefore the results are surely coincident with each other.

Table 1. Simulation results of Re1, Re2 and Re3

Re_1	Re_2	Re_3
0.23584E-03	0.14978E-02	0.17842E-02

3 Simulation

The simulation of the stress distribution on the surface of the diaphragm and the integration of stress difference with respect to certain path are both accomplished with the finite element analysis software ANSYS. The main steps of the simulation include modeling, meshing, applying loads and so on. Fig.3 shows the contour plot of the stress difference on the surface of the diaphragm from the simulation.

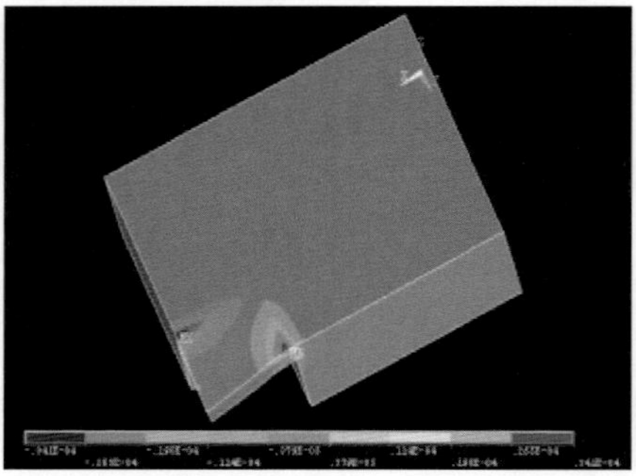

Fig. 3. Contour plot of the stress difference

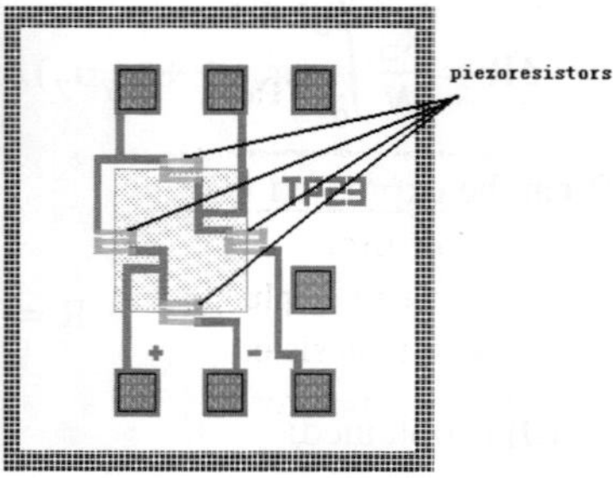

Fig. 4. Layout of a pressure sensor unit

As is shown in Fig. 4, the piezoresistors are usually designed into muti-strips in order to make full use of the high stress area of the diaphragm. Therefore it is necessary to accomplish integral operation along several paths. Based on the concrete data (including the size of the film, the location and the size of the piezoresistors, as well as the full scale pressure loaded) in practice, we obtain the results of integration along three paths respectively: Re_1, Re_2 and Re_3, of which Re_2 is shown by ANSYS plot results and list results in Fig.5. All these data are summarized in Table1.

Put these data into the formulas deduced previously with either of the two methods provided, π_{44} is determined by the surface ion concentration of the piezoresistors that can be simulated with related software and be controlled in the step of boron implanting during the fabrication process. Then we figure out the full scale output of the pressure sensor which is expressed as Eq.(12). The corresponding parameters are listed in Table2 and the voltage applied is 3V. During the process of simulation, one or several of these parameters can be adjusted in order to meet the requirement for the full scale output of the pressure sensor.

$$V_{out} = \frac{\Delta R}{R} V_B = 0.5\pi_{44}\overline{(\sigma_l - \sigma_t)} V_B = 42.996\text{mv} \tag{12}$$

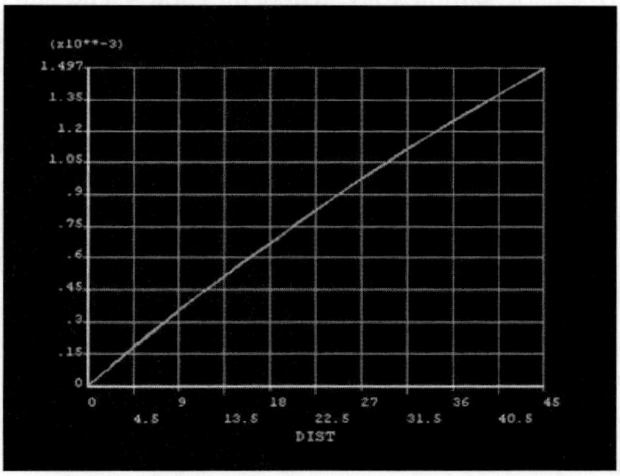

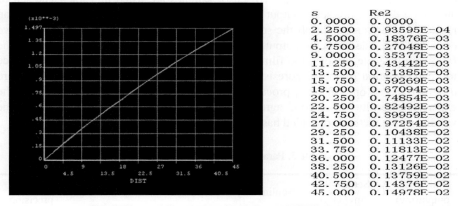

S	Re2
0.0000	0.0000
2.2500	0.93595E-04
4.5000	0.18376E-03
6.7500	0.27048E-03
9.0000	0.35377E-03
11.250	0.43442E-03
13.500	0.51385E-03
15.750	0.59269E-03
18.000	0.67094E-03
20.250	0.74854E-03
22.500	0.82492E-03
24.750	0.89959E-03
27.000	0.97254E-03
29.250	0.10438E-02
31.500	0.11133E-02
33.750	0.11813E-02
36.000	0.12477E-02
38.250	0.13126E-02
40.500	0.13759E-02
42.750	0.14376E-02
45.000	0.14978E-02

Fig. 5. Plot and list results of Re2 from ANSYS simulation

Table 2. Parameters needed for the simulation

parameter	value
Thickness of the film	20um
Width of the square diaphragm	300um
Length of the piezoresistor	90um
Width of the piezoresistor	10um
Full scale pressure loaded	1MPa
Dose of the boron implanting	4e14cm^{-3}
Energy of the boron implanting	80KeV

4 Fabrication

The fabrication process of the sensor is mainly based on the application of SOI wafers and ICP etching technics. The main steps of fabrication are: (1)Formation of the piezoresistors by p implanting,(2)Realization of a good ohmic contact between the piezoresistors and the aluminum layer by p^+ implanting,(3)Deposition and etching of the aluminium to form the electrodes,(4)Formation of the pressure reference cavity from backside by ICP etching with automatic stop at the buried SiO_2 layer,(5)Bonding of silicon and glass.

5 Results and Discussion

Each of the 10 sensor samples has been tested under the pressure from 0 to 1 MPa and then back from 1 to 0 MPa for three round voyages and the voltage applied is 3.00V. Output voltage of the sensors for several fixed pressure values are recorded and processed by MATLAB. Results for one of those sensors are shown in Fig.6. Parameters of this pressure sensor including full scale output, sensitivity, nonlinearity, hysteresis, repeatability and total precision are also calculated and listed in Table3, which helps to get a better understanding of the performance of the sensor. Statistics of the full scale output for all the 10 sensor samples are listed in Table4.The sensor numbered 4 has a full scale output of 43.112mv that is very close to the simulation result. Absolute errors for all the sensor samples are limited to1.075mv and relative errors to 2.5%.The full scale output of the sensor numbered 8 is a little higher for the following possible reasons: the film of this sensor is thinner than others; the surface ion concentration of the piezoresistors is lower; the size of the piezroresistors are different and so on. After all, process variation can't be avoided. In conclusion, the experiment results are in good agreement with the simulation results; therefore the validity of the methods provided has been verified.

Table 3. Parameters of the pressure sensor

Full scale output(mv)	Sensitivity (mv/KPa)	Nonlinearity	Hysteresis	Repeatability	Total precision
42.860	0.0429	0.1393%	0.1091%	0.2522%	0.308%

Table 4. Statistics of test results for all the pressure senor samples

code	Full scale output by experiment(mv)	Full scale output by simulation(mv)	Absolute error (mv)	Relative error
1#	43.196		0.200	0.465%
2#	42.264		0.732	1.702%
3#	42.860		0.136	0.316%
4#	43.112		0.116	0.270%
5#	42.744	42.996	0.252	0.586%
6#	42.163		0.833	1.937%
7#	42.581		0.415	0.965%
8#	44.071		1.075	2.500%
9#	42.531		0.465	1.081%
10#	42.545		0.451	1.049%

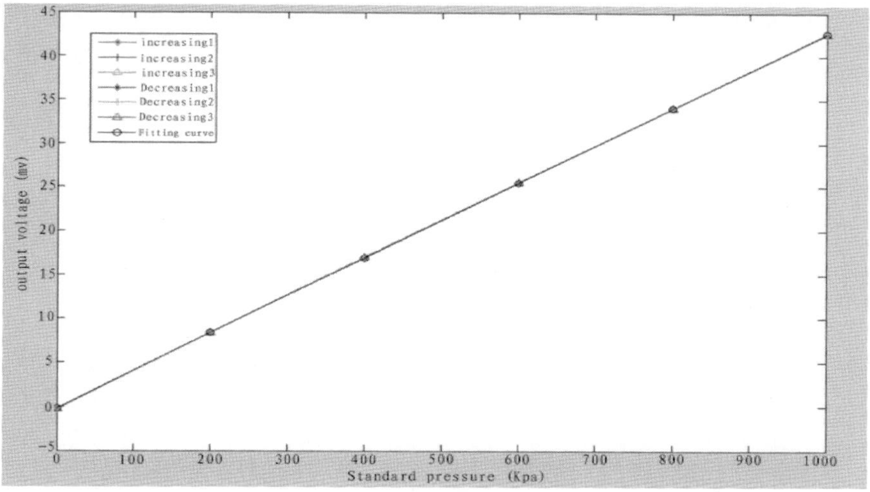

Fig. 6. Test results of the pressure sensor

6 Conclusion

Methods based on finite element analysis and mathematical integration are provided for evaluating the full scale output of the piezoresistive pressure sensor. With the methods, the simulation can be more accurate and easy to conduct and therefore be significantly helpful to realize the superior design of the pressure sensors.

Acknowledgement

The work is supported by National Natural Science Foundation of China, 863 project of China (2006AA04Z372) and National Key Project of Science and Technology of China(2009ZX02023-001-3).

References

1. Krondorfer, R., Kim, Y.K., Kim, J., et al.: Finite element simulation of package stress in transfer molded MEMS pressure sensors. Microelectron Reliab. 44, 1995 (2004)
2. Tao, C., Zhaohua, Z., Tianling, R., et al.: A novel dual-functional MEMS sensor integrating both pressure & temperature units. Journal of Semiconductor 31(7) (2010)
3. Smith, C.S.: Piezoresistance effect in germanium and silicon. Phys. Rev. 94, 42 (1954)
4. Yao, C., Xu, K., Ma, W.: The mechanical model and geometric factors of piezoresistive pressure sensors. Process Automation Instrumentation 10(8), 20–24 (1989)

A Design of Passively Omnibearing Directivity Acoustic Cylinder Array

Haiyan Wang*, Jun Bai, Xiaohong Shen, and Fu-zhou Yang

College of Marine Engineering, Northwestern Polytechnical University, Xi'an 710072, China
hywang@nwpu.edu.cn

Abstract. A acoustic cylinder array of passively omnibearing directivity is designed, including receiving element, transmission and control system, whose characters are: firstly there are $N \times M$ signal receiving elements in the array, and N linear arrays well-proportioned arrangement, meanwhile the first element of each array lain at the top of acoustic cylinder array constitute a M elements uniform circle array; Secondly, located in the same transversal surface array elements connect with the input ports of selecting switch electrocircuit, simultaneously the output ports of latter connected with receiver electrocircuit; At last, electrocircuit generating beam forming weights link to output terminus of receiver electrocircuit by multiplier, whose output data going to processor through data output bus. The design achieves the capability of acoustic target location omnidirectional and tridimensional. It overcomes the default of location accuracy low caused by array aperture restriction of acoustic detection equipment, furthermore improves the three-dimensiona accuracy on target detection.

Keywords: acoustic, cylinder array, control system.

1 Introduction

In the underwater passive target location field, ship noise is broadband and mainly distributes between 100Hz and 5 kHz [1][2]. So the wave length is too long, as a result that the number of array elements is limited in the small volume detection equipment, and degrade the location precise of array. Under modern technical condition, array selected is relatively simple, which mainly adopts method of layout circular ring shape array in the front of the device to locate target. However, the plane array or volume array on the top usually lead detection field angle too small only as to detect in a certain area, thus limit its capability. In order to overcome the modern technical defaults due to precise degrading and limitation detection area, we designed an underwater acoustic cylinder with the omni-bearing location performance, which includes receiving, transmission and control system. It improves the target detection precise and realizes the omni-bearing location of vessel.

* Haiyan Wang (1981.09 -), male, han, ShanDong province of China, professor, Ph.d. supervisor, committed to underwater communication and signal processing research.

X. Wan (Ed.): Electrical Power Systems and Computers, LNEE 99, pp. 63–71.
springerlink.com

2 Structure Design of Small Aperture Cylinder Array

This paper designs a kind omni-bearing location cylinder array, whose detailed part showed in figure 1.

The cylinder array includes the receiving elements, transmission and control system, whose characteristic is that the cylinder array includes $N \times M$ signal receiving elements, where N notes the number of linear arrays, and M notes the elements number of each linear array. The N linear arrays homogeneous distribute along the circle of underwater mine outside wall, and the location of first element belonged to each linear array is on the top of underwater mine.

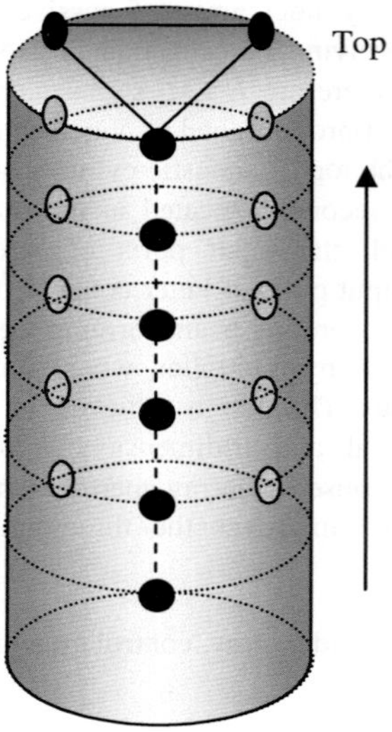

Fig. 1. Array structure

3 Transmission and Control System

In figure 2, the transmission and control system includes digital signal processor, the total M one by N selected switch circuit, N channels parallel selected switch circuit, M 6 channels parallel Analog to Digital Converters (ADC). The digital signal processor includes Beamforming weights generator, Target estimate direction and Channel controller. N elements belonged to N linear array and located at the identical cross section of mine body connect to the same input terminus one by N selected switch circuit. On the top of underwater mine, the N elements connect to one input terminus of N channels parallel selected switch circuit. The output terminus due to total M one by N selected switch circuit absolutely connects to M receiver circuit. Beamforming weight generator circuit connects to output terminus of receiver circuit through multiplier, and the output data of multiplier going to processor through data output bus. The output terminus of processor connect the beamforming weight generator circuit with total M one by N selected switch circuit and input terminus controlled of N channels parallel selected switch circuit. The output terminus of N channels parallel selected switch circuit connects to input terminus of frontal N receiver circuit. Here something about connection need explanation is: all switch input Numbers link to the element in the same Numbers linear array in sequence. The connection relation between total M one by N selected switch circuit as well as N channels parallel selected switch circuit and elements of receiving array satisfy the matrix bellow:

$$D_{N\times M} = \begin{bmatrix} d_{11} & d_{12} & \cdots & d_{1M} \\ d_{21} & d_{22} & \cdots & d_{2M} \\ \vdots & \vdots & \ddots & \vdots \\ d_{N1} & d_{N2} & \cdots & d_{NM} \end{bmatrix} \tag{1}$$

Where $d_{ij}(0 < i \le N, 0 < j \le M)$ notes the j th element of i th row, and d_{i1} notes the element due to the top of underwater mine. d_{ij} represents the element connect with the i th input port of the j th one by N selected switch. d_{i1} notes the element connect with the i th input port of N channels parallel selected switch circuit, simultaneously the i th output port connect with input terminus of i receiver circuit. Distance of each element below $\lambda/2$, λ notes the wave length corresponding to the smallest receiving signal frequency of whole array. Further more, $N \ge 3$ and is smaller than or equals to M, $M \ge 3$.

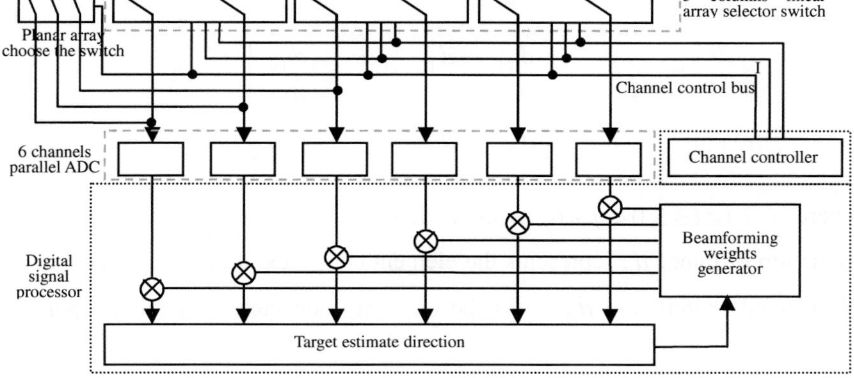

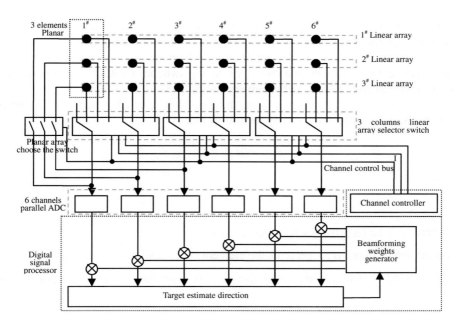

Fig. 2. System design of small aperture cylinder array

4 Design Examples of Small Aperture Cylinder Array System

4.1 Underwater Mine Volume Array Display

Passive signal receiving volume array due to underwater mine detecting target own total 18 signal receiving elements, including each 6 elements homogenous distributed linear array belonged to 3 rows. The 6 elements homogenous distributed linear array display downwards from the mine top along the lateral generatrix direction of mine body. 3 rows linear array homogenous distribute outside the mine, and each 6 homogenous linear elements due to mine top compose the plane array vertical to each row linear array. The cylinder of 6 homogenous distributed linear array belonged to each row divides to 3 parts, each row realize 120 degrees location about space area target pitch attitude. The display of array shows in figure 1.

4.2 Control System of Underwater Mine Volume Array

Underwater mine volume array control system consists of control circuit, plane array selected switch circuit, and linear array selected circuit and Beamforming weights generated circuit. The aperture realized in figure 1. Display array on the mine body due to figure 1, and the same number(1# showed) of each 6 elements homogenous distributed liner array due to 3 rows connect with one channel of dual 4-channel analog switch CD4052 to make up 1 by 3 circuit worked for outputting. 6 channels output signal through 3 pieces dual 4-channel analog switches CD4052 arrive at 6 channels signal receivers, and the output signal of receivers go through weighted processor to the signal sample system of processor. The weighted processor controlled by Beamforming weight generated equipment of processor, in order to weight each channel's signal.

Simultaneously, each 1# element of every 6 homogenous linear elements due to 3 rows connect with four double directions analog switches CD4066, and 3 output port of CD4066 isolate connect with receivers among 1~3.

Six 1 by 3 selected switch circuits, three parallel selected switch circuits and the elements of receiving array satisfy the follow matrix:

$$D_{3\times 6} = \begin{bmatrix} d_{11} & d_{12} & \cdots & d_{16} \\ d_{21} & d_{22} & \cdots & d_{26} \\ d_{31} & d_{32} & \cdots & d_{36} \end{bmatrix} \tag{2}$$

Where d_{ij} $(0<i\le 3, 0<j\le 6)$ notes j element of i row, and d_{i1} notes the element on the top of mine. d_{ij} represents the element connected with i input port of j 1 by N selected switch, and d_{i1} notes the element connected with i input port of total

N channels parallel selected switch circuits, meanwhile the i output terminus connected with input port of i receiver.

4.3 Location Working Model of Underwater Mine Volume Array

The model shows in figure 3 and 4. Control signal INH=1, and 3 elements of plane array connect with receiver. The processor receives plane signal to evaluate the azimuth angle when target signal arrives. As 3 elements of plane array justify the area of appearing target, the control signal changes to INH=0, then disconnect the plane array and receiver. After that control AB signal, and select the 6 elements of linear array due to target area: AB=01, selecting 1# linear array; AB=10, selecting 2# linear array; AB=11, selecting 3# linear array. The processor receive signal selected by linear array, and evaluate pitch attitude if target signal arrive. Beamforming weight generated circuit produces optimum Beamforming weights through processor control included in the control system circuit.

Control switch CD4066 includes 4 independent analog switches, and each analog swich owns input, output, control terminus, which the input terminus and output terminus can exchange. When control terminus set by high voltage, the switch connects, otherwise it disconnects. The resistance is about dozens of ohm as the switch connects. However, we consider incomplete circuit as large impedance when switch disconnects. The analog switch can transmit digital and analog signal, and the largest frequency of analog signal is 40MHz.Between each switch, the disturbance is small, whose typical value is -50dB.

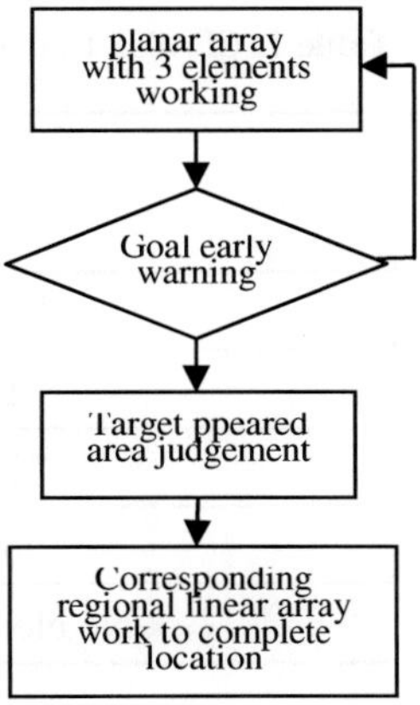

Fig. 3. Working mode

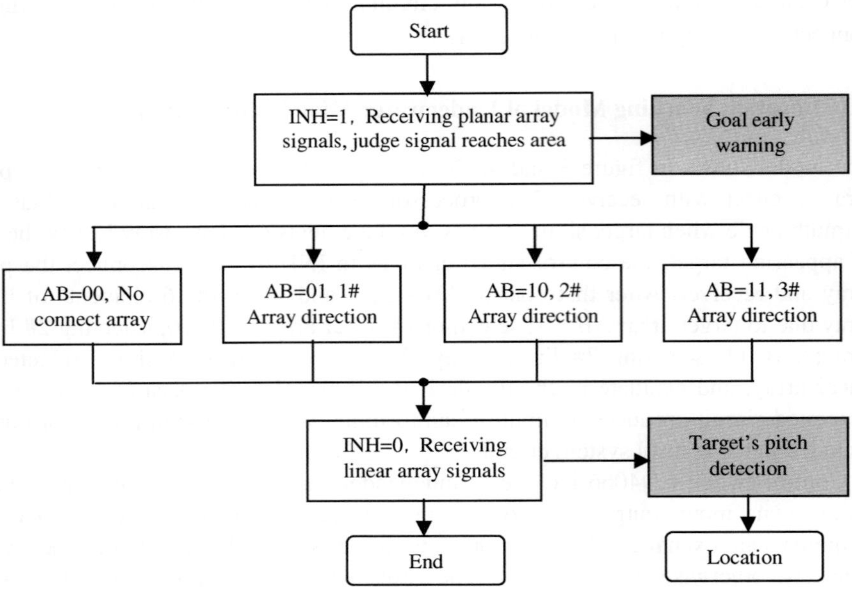

Fig. 4. Working process

CD4052 equals to dual four-channel switch. Input address encode AB determine the channel connected with CD4052. The values show in table 1.

Table 1. CD4052 truth table

Input states			Through channels
INH	B	A	
0	0	0	"0" X、 "0" Y
0	0	1	"1" X、 "1" Y
0	1	0	"2" X、 "2" Y
0	1	1	"3" X、 "3" Y
1	ф	ф	No element is connected

4.4 Measurement Performance of Small Aperture Cylinder Array

Figure 5, and figure 6 respectively show beam results due to 6 elements homogenous linear array and 3 elements homogenous linear array. The 3dB beam width of 6 elements is ±8°, respectively 26°of 3 elements.

Use 2kHz frequency as the basic point of element distance, and suppose the distance is $d = \lambda/2$, where $\lambda = c/f$ notes length of incident plane wave. Sound velocity c=1500 m/s . Introduce MUSIC method to evaluate incident angel of signal received by 6 elements linear array, and the result shows in figure 7. It indicates: beam angel identical to array normal direction through measurement and correction of 6 elements linear array distance, and width of 3dB beam angle controlled within ±8° by arithmetic compensation.

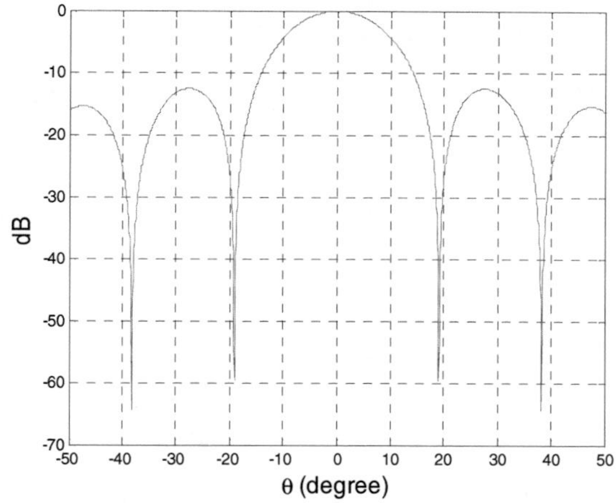

Fig. 5. Beamforming of 6 elements homogenous linear array at cylinder array lateral

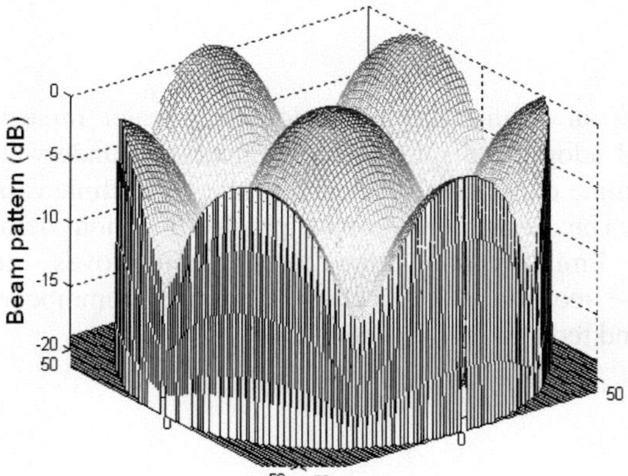

Fig. 4. Beamforming of 3 elements plane array on the cylinder array top

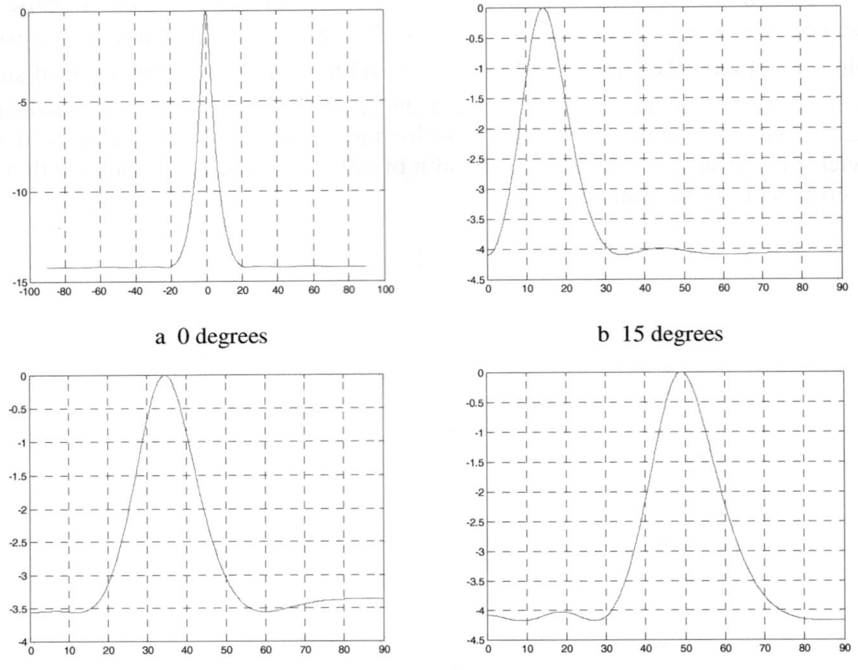

a 0 degrees

b 15 degrees

c 30 degrees

d 45 degrees

Fig. 7. Beam angle measurement of 6 elements homogenous linear array at cylinder array lateral

5 Conclusion

This paper proposes an omni-bearing location underwater mine volume array. For technology proposal adopting display array on lateral of underwater mine body, we realize underwater mine capability of omni-bearing, three-dimensional location to the underwater target. The design overcomes precise location default resulted from aperture displayed limitation on mine top, and improves the accurate target identification. So the acoustic cylinder array of passively omnibearing directivity has actual application and technical reference value.

References

1. Shi, G.-Z., Hu, J.-C.: Theoretical analysis of the ship noise demodulation line spectrum structure Source: Harbin Gongcheng Daxue Xuebao. Journal of Harbin Engineering University 28 (suppl. n), 138–142 (2007)
2. Foote, K.G.: Underwater acoustic technology: Review of some recent developments. In: OCEANS (2008)

3. He, Z., Ma, Y.: Directivity calculation with experimental verification for a conformal array of underwater acoustic transducers. Shengxue Xuebao/Acta Acustica 32(3), 270–274 (2007)
4. Solomon, I.S.D., Knight, A.J.: Array processing of underwater acoustic sensors using weighted Fourier integral method. In: IEEE Signal Processing Workshop on Statistical Signal and Array Processing, SSAP, pp. 707–711 (2000)
5. Zetterberg, V., Pettersson, M.I., Tegborg, L., Claesson, I.: Passive scattered array positioning method for underwater acoustic source. In: OCEANS (2006)

A Multitarget Passive Recognition and Location Method Fusing SVM and BSS

Jun Bai*, Haiyan Wang, Xiaohong Shen, and Zhao Chen

College of Marine Engineering, Northwestern Polytechnical University,
Xi'an 710072, China
baijunwait@163.com

Abstract. A multitarget passive recognition and location method which fuses SVM and blind signal processing technique is proposed in this paper. Its characters are: Sampling data via multitarget information receiving array at first; And then getting separated signal and matrix by blind signal separation (BSS) to these data; Completing classification of each separated signal by using decision tree support vector machine (SVM) multitarget recognition process to the separated signal; Obtaining direction information of each signal by blind deconvolution location algorithm based on array model to the separated matrix at the same time; Finally, realizing target recognition and location by synthesizing targets information of the classification and direction. This paper studies technique principle of this method, gives a detailed implement step and proves its validity by multitarget recognition and location experiment of measured ship-radiated noise.

Keywords: SVM, Blind signal, Multitarget, Recognition, location.

1 Introduction

SVM is a recently developed novel general knowledge discovery method which has a good performance in classification. SVM is constructed on structural risk minimization principle of statistics learning theory. Its main idea is to find a super plane in higher dimensional space for the separation of two classes in order to guarantee a minimum classification error rate for classification problem of two classes.

BSS is originated from a feedback neural network model and a learning algorithm based on Hebbian learning rule [1]. As the only hypothetical condition of BSS is the statistic independence of the source signals, which makes blind signal processing a widely used signal processing method [2][3]. Especially when it is difficult or even impossible to set up a transmission channel model between the source and the array, BSS becomes the only feasible signal processing method. It is of greater advantage to adopt blind signal processing technique for separation of multitarget from the same direction [4][5], it can utilize array to distinguish class or even direction of a target in underwater environment with multitarget at the same time. How to realize target

* BaiJun (1981.09 -), male, han, MinQin county, GanSu province of china. PhD, committed to underwater signal processing research.

X. Wan (Ed.): Electrical Power Systems and Computers, LNEE 99, pp. 73–81.
springerlink.com

recognition and location technique simultaneously is the key of technological application.

In the existing techniques, SVM is mainly pure recognition of certain or multiple unknown targets, when multitarget signal are randomly mixed, it can not complete recognition mission[6][7]. As for the separated signal from blind signal, due to its uncertainty of separation, the separated signal and the source signal are usually not one-to-one correspondences. But the separation does not correspond to recognition. So it is meaningless to classification of passive target signal.

In order to overcome the disadvantages of the existing technique, a multitarget passive recognition and location method which fuses SVM and blind signal processing technique is proposed in this paper. It combines SVM classification, recognition and blind separation, blind deconvolution location technique of multitarget, and realizes synchronized multitarget passive recognition and location.

2 Multitarget Information Separation, Location and Recognition System

This paper combines multitarget blind separation, blind deconvolution location technique of blind signal and SVM technique of targets classification, and realizes synchronizing targets recognition and location. The structure of this system is shown in figure 1: Sampling data via multitarget information receiving array at first; And then getting separated signal and matrix using these data through BSS; Completing classification of each separated signal using decision tree SVM multitarget recognition process to the separated signal; Obtaining direction information of each signal by blind deconvolution location algorithm based on array model to the separated matrix at the same time; Final, realizing target recognition and location by synthesizing targets information of the classification and direction.

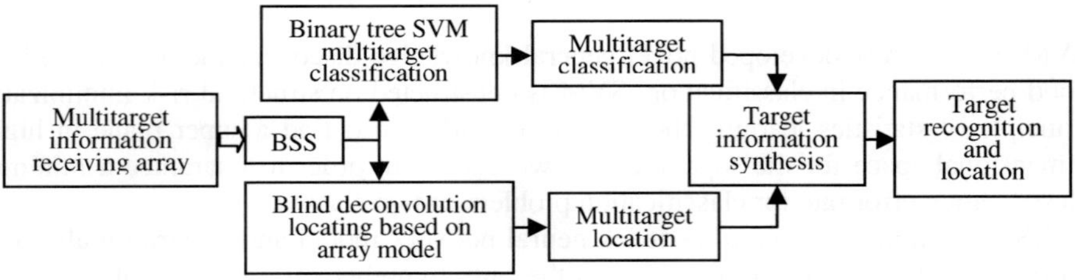

Fig. 1. Functional block diagram

2.1 Blind Separation Multitarget Locating Technique Based on Constant Beamwidth Array Model

The principle of an array receiving target radiated noise for locating is to estimate direction and distance of a target by distinguishing the time delay of a signal arriving at each element.

For an arbitrary array of M elements which receives d far field signals of frequency between $[f_l, f_h]$, the outputs of the k th element at time t is as follows:

$$x_k(t)=\sum_{i=1}^{d}\omega_k(f_i)g_k(\theta_i)s_i[t-\tau_k(\theta_i)]+n_k(t),\ k=1,2,\cdots,M \tag{1}$$

Where $\omega_k(f_i)$ is the constant beamwidth weighting coefficient of the array, and it is a function of the incident frequency $f_l \le f_i \le f_h$. Make the beamwidth of the array adaptively constant in the frequency range $[f_l, f_h]$ by weighting, so as to eliminate the error caused by different incident frequencies. $\tau_k(\theta_i)$ is the time delay between the reference element and the k th element, θ_i is the parameter of $\tau_k(\theta_i)$, and is the incident angle of each signal, and can be interpreted as target direction information of radiated signal. Transform equation (1),

$$\tilde{X}(t)=\begin{bmatrix}\sum_{i=1}^{d}\omega_1(f_i)g_1(\theta_i)\tilde{s}_i(t)e^{-j2\pi f_i\tau_1(\theta_i)}\\ \sum_{i=1}^{d}\omega_2(f_i)g_2(\theta_i)\tilde{s}_i(t)e^{-j2\pi f_i\tau_2(\theta_i)}\\ \cdots\cdots\\ \sum_{i=1}^{d}\omega_M(f_i)g_M(\theta_i)\tilde{s}_i(t)e^{-j2\pi f_i\tau_M(\theta_i)}\end{bmatrix}+\begin{bmatrix}\tilde{n}_1(t)\\ \tilde{n}_2(t)\\ \cdots\cdots\\ \tilde{n}_M(t)\end{bmatrix}=\sum_{i=1}^{d}a(\theta_i)\tilde{s}_i(t)+\tilde{N}(t) \tag{2}$$

Where,

$$a(\theta_i)=[\omega_1(f_i)g_1(\theta_i)e^{-j2\pi f_i\tau_1(\theta_i)},\omega_2(f_i)g_2(\theta_i)e^{-j2\pi f_i\tau_2(\theta_i)},\cdots,\omega_M(f_i)g_M(\theta_i)e^{-j2\pi f_i\tau_M(\theta_i)}]^T \tag{3}$$

It is obvious that time delay information of the signal, namely direction information is fully included in $a(\theta_i)$, and $a(\theta_i)$ is column vector of array manifold $A(\Theta)$. So direction information in blind separation array model is converted from parameter of the source signal to parameter of $A(\Theta)$, finally the equation is $\tilde{X}(t)=A(\Theta)\tilde{S}(t)$ After doing complex blind separation to $\tilde{X}(t)$, $\tilde{Y}(t)=W(\Theta)\tilde{X}(t)$. here, $\tilde{Y}(t)$ is complex restored signal and is analytic signal of the source signal with uncertainty of the arrange. It can be deduced that the product of plural order hybrid matrix $W(\Theta)$ and complex hybrid matrix $A(\Theta)$ is a permutation matrix. So it can be seen that the direction information is included in the inverse matrix of $W(\Theta)$ and also has uncertainty. Summarizing the above theory, a procedure of blind separation target direction estimation method based on array model is proposed:

① Implement constant beamwidth weighting for each channel of the receiving array according to the frequency band range of the received signal;

② Convert the received mixed signal to analytic signal and separate the analytic signal by utilizing instantaneous mixing complex value blind separation algorithm;

③ Solve analytic signal of the source signal and complex solution mixed matrix by applying iterative algorithm;

④ Obtain inverse matrix of the complex solution mixed matrix which includes direction information, and estimate direction of the signal according to different array manifold of the receiving array.

2.2 Technique of SVM Realizing Multi-mode Recognition

SVM decision tree can classify multiple classes for multitarget. It firstly classifies all classes into two subclasses, and then divides the subclasses into two secondary subclasses; circulate until obtaining an individual class, then it can get an upside down binary classification tree. This paper mainly completes classification recognition of three classes of underwater targets, so we adopt the algorithm namely SVM decision tree method to realize underwater multitarget. The detailed method is to set up classification decision tree with two-stage SVM. The first stage SVM realizes classification of class I and class II, III. The second stage realizes classification recognition of class II and class III. It realizes classification recognition of 3 classes of underwater targets though this SVM classification decision tree. Combine all the contents above we can obtain flow chart of the classification procedure, see Figure 2.

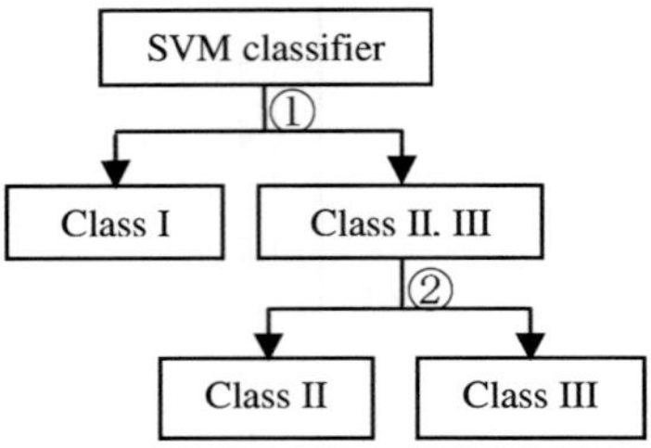

Fig. 2. Decision tree SVM classifier

Different inner product kernel functions in SVM will form different algorithm, there are three kinds of kernel functions that are most frequently investigated. One is polynomial kernel function $K(x,x_i)=[(x\cdot x_i)+1]^q$, q is the order, it has good global feature[8] and strong extrapolate ability, the lower order of the kernel function is, the stronger extrapolate ability it will have, but it does not mean that the higher the order is, the better the learning ability of the polynomial kernel function will be. As for the interpolation problem, when the order is high there may be runge phenomenon, i.e. comparatively large truncation error at either end of the interpolation region. Thus in the SVM, we must choose appropriate polynomial kernel function so as to get an SVM with good learning and extrapolate ability. The second one is radial basis function $K(x,x_i)=\exp\left\{-\frac{|x-x_i|^2}{\sigma^2}\right\}$, this kind of kernel function has strong local feature, its interpolate ability decreases as parameter σ increases. So, we also need to choose appropriate parameters to make the SVM have a good performance in practical applications. The third one take Sigmoid function as inner product, $K(x,x_i)=\tanh(v(x\cdot x_i)+c)$, at this time realization of SVM is multilayer perceptron with a hidden layer, the number of nodes of the hidden layer is adaptively determined by the algorithm, and algorithm has no local minimum point problem which bothers the neural network method.

The choice of all parameters in this paper is based on repeating experiment, and finally choose the one with the best experiment result. As for the choice of kernel function, ① choose Gaussian radial basis function, ② choose polynomial kernel function.

2.3 Multitarget Information Separation, Positioning and Recognition Method

This paper fuses SVM and blind signal processing technique and presents passive recognition and positioning method for multitarget, its procedure is as follows:

Step 1: Received signal of N elements array is $X(t) = A(\Theta)S(t) + N(t)$, where $A(\Theta) = [a(\theta_1), a(\theta_2), \cdots, a(\theta_d)]$ is the receiving array manifold, $a(\theta_i) = [g_1(\theta_i)e^{-j2\pi f_i \tau_1(\theta_i)}, g_2(\theta_i)e^{-j2\pi f_i \tau_2(\theta_i)}, \cdots, g_M(\theta_i)e^{-j2\pi f_i \tau_M(\theta_i)}]^T$ is the directivity function. $S(t) = [s_1(t), s_2(t), \cdots, s_d(t)]^T$ is d source signals, $N(t) = [n_1(t), n_2(t), \cdots n_M(t)]^T$ is the received noise; f_i is the frequency of source signal s_i; g_m is weight coefficient of the mth element, τ_m is time delay of s_i arriving at the mth element, θ_i is its arriving angle;

Step 2: Convert the received signal $X(t)$ to complex signal matrix $\tilde{X}(t)$ by Hilbert transformation;

Step 3: Use blind separation model $\widetilde{Y}(t) = W(\Theta)\widetilde{X}(t)$ to convert matrix transposition in the blind signal processing natural gradient algorithm to conjugate transpose $\mathbf{W}(k+1) = \mathbf{W}(k) + \mu(k)[\mathrm{I} + \mathbf{g}(\mathbf{y}(k))\mathbf{y}^{\mathrm{H}}(k)]\mathbf{W}(k)$. Do blind separation for the complex signal matrix $\tilde{X}(t)$ and obtain analytic signal $\widetilde{Y}(t)$ of source signal $S(t)$ and complex solution mixed plural matrix $\mathbf{W}(k+1)$; where the constructed iterative equation of learning factor is $\mu(k) = \lambda / a\tan(\gamma k)$, λ is the initial value of learning factor, γ is its final convergence value and k is the iteration number.

Step 4: Take the real part of the analytic signal $\widetilde{Y}(t)$ from blind separation to get real signal testing sample data $Y(t) = y_1(t), y_2(t), \cdots, y_d(t)$;

Step 5: Construct classifier for source signals sample data of all kinds of signals $s_1(t), s_2(t), \cdots, s_d(t)$ to the source signal $S(t) = [s_1(t), s_2(t), \cdots, s_d(t)]^T$ respectively by SVM classification method, and make identification for the test samples $y_1(t), y_2(t), \cdots, y_d(t)$ obtained respectively With each classifier, complete the data sample classification;

Step 6: Take the inverse of the complex solution mixing matrix $\mathbf{W}(k+1)$ which contains direction information obtained from step 4, we can get mixing matrix $\widehat{A}(\Theta)$;

Step 7: Attain $a(\widehat{\theta}_1), a(\widehat{\theta}_2), \cdots, a(\widehat{\theta}_d)$ corresponding to the mixing matrix $\widehat{A}(\Theta)$ by calculating according to array manifold $A(\Theta) = [a(\theta_1), a(\theta_2), \cdots, a(\theta_d)]$ of the receiving array. $\widehat{\theta}_1, \widehat{\theta}_2, \cdots, \widehat{\theta}_d$ are the arriving angles of the signal corresponding to $y_1(t), y_2(t), \cdots, y_d(t)$;

Step 8: Synchronize the results of step 5 and 6 we can estimate class and corresponding arriving angle of the sources signal $S(t) = [s_1(t), s_2(t), \cdots, s_d(t)]^T$.

Procedures of the SVM classification method described in step 5 are:

① Choose N^1 groups of data of the known source signal respectively, and obtain M-dimensional classification eigenvector set of each known source signal as training samples for target recognition. Utilize SVM to construct multitarget classifier;

② Acquire M-dimensional classification eigenvector $(Z_1, Z_2, \cdots Z_M)$ of each separated signal after blind separation, and use them as training samples for the target. Adopt decision tree classification method, and use the obtained multitarget classifier to do recognition for the testing samples in turn, finally complete ship classification.

3 Experimental Verification

3.1 Multitarget Information Separation and Positioning

Experiments are aimed real ship radiated noise and still choose three classes of targets in figure 6 which are ship, submarine and merchant ship from top to bottom respectively. Figure 7 shows analytic signal from the source signal by Hilbert transformation, lateral axis stands for its real part and vertical axis is its imaginary part. Figure 8 gives its power spectrum.

Other experimental conditions are: complex mixing matrix $\mathbf{A} = \begin{bmatrix} 1 & 1 & 1 \\ e^{-j\tau_2(\theta_1)} & e^{-j\tau_2(\theta_2)} & e^{-j\tau_2(\theta_3)} \\ e^{-j\tau_3(\theta_1)} & e^{-j\tau_2(\theta_2)} & e^{-j\tau_2(\theta_3)} \end{bmatrix}$; $d = 0.5\lambda_{\max}$; data sample length is N=2000, the sample frequency is $f_S = 2\text{KHz}$, $\lambda = 0.001$, $\gamma = 0.1$, $N = 4000$ is the iteration number, $W(1) = I$.

Time domain waveform of three targets (ship, submarine and merchant ship each has one) is shown in Figure 3a, it is 5000 sample data taking form real ship radiated noise samples after down sampling. Use a uniform circle array as receiver, choose *M*=3, direction of arrivals of the targets are set to be 0°, 5° and 10°. Target location result of complex blind separation algorithm is shown in Figure 6, the estimated direction of arrivals are -0.220°, 5.232° and 10.278° respectively, its error to the ture values are 0.220°, 0.232° and 0.278°. Their order uncertainty is caused by the order uncertainty of blind separation.

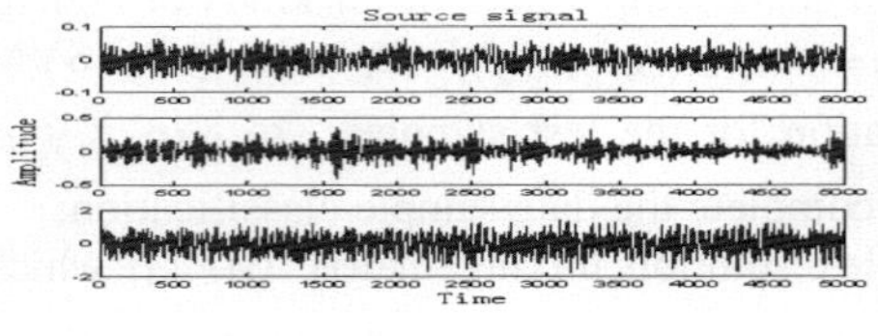

Fig. 3a. Source signals

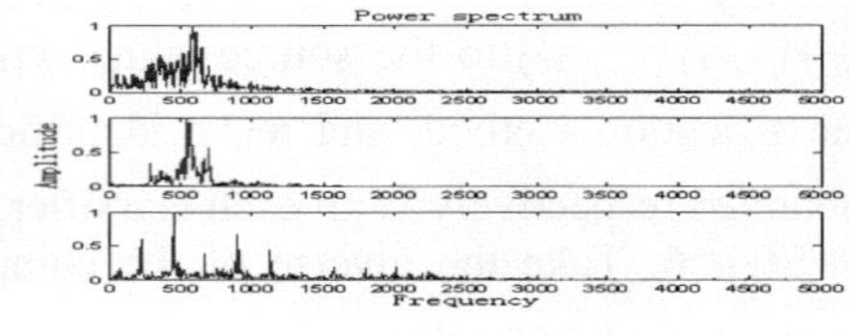

Fig. 3b. Source signal power spectrum

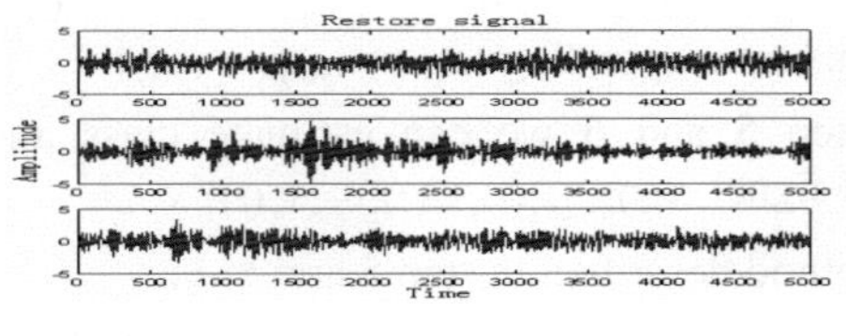

Fig. 4a. Restore signals

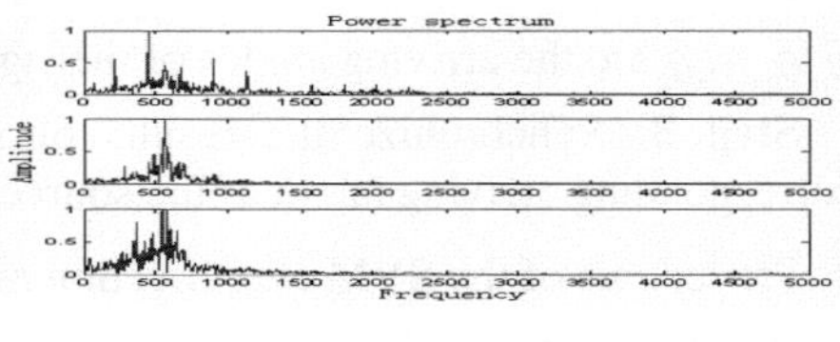

Fig. 4b. Restore signal power spectrum

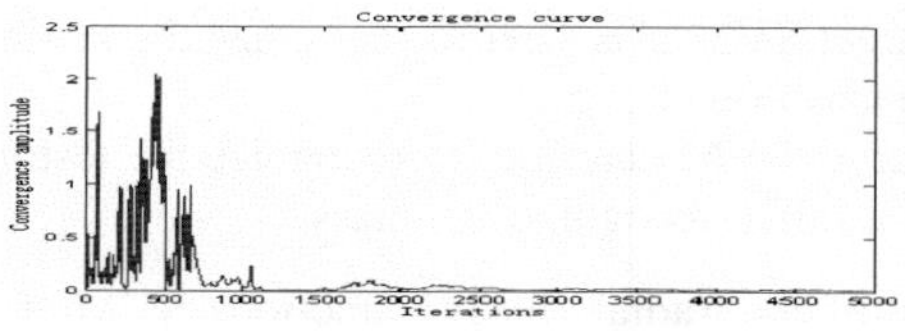

Fig. 5. CMNG convergence mark quantity curve

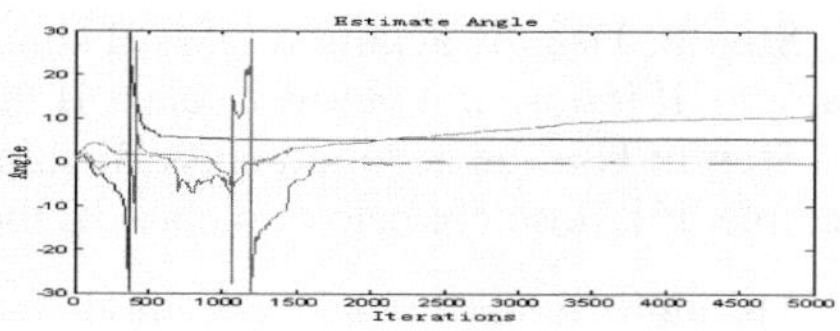

Fig. 6. Direction estimation results

It can be seen from figure 5-4 and 5-6 that the source signal and restored signal are one-to-one correspondences, they are $y_1 \to s_3$, $y_2 \to s_2$ and $y_3 \to s_1$ respectively. Further more, the similarity coefficients of the restored signal and source signal are $\rho = \begin{bmatrix} 0.0032 & 0.0078 & \underline{\underline{0.4930}} \\ -0.0263 & \underline{\underline{0.7713}} & -0.0128 \\ \underline{\underline{1.0000}} & 0.1324 & -0.1761 \end{bmatrix}$. By such power spectrum similarity coefficients, we can also konw that $y_1 \to s_3$, $y_2 \to s_2$ and $y_3 \to s_1$. It is obvious that the source signals are well recovered, and at the same time we get accurate target direction estimation.

3.2 Information Recognition of Multitarget Separation

For recognition of ship radiated noise, this paper uses the following method for extraction of the classification eigenvector:

① Calculate 1½-dimensional spectrum of the ship radiated noise bellow 2kHz, and divide it into M^1 uniformly-spaced subbands. Integrate each subband respectively to get its energy, and finally obtain 1½-dimensional spectrum subband distribution eigenvector $(Z_1^1, Z_2^1, \cdots, Z_{M^1}^1)$ of M^1 -dimensional ship radiated noise.

② Calculate 2½-dimensional spectrum of the ship radiated noise bellow 2kHz, and divide it into M^2 uniformly-spaced subbands. Integrate each subband respectively to get its energy, and finally obtain 2½-dimensional spectrum subband distribution eigenvector $(Z_1^1, Z_2^1, \cdots, Z_{M^1}^1)$ of M^2-dimensional ship radiated noise.

③ Implement level M^3 wavelet decomposition to the ship radiated noise bellow 2kHz, and calculate the spectrum energy of each subband respectively to obtain scale-energy eigenvector $(Z_1^3, Z_2^3, \cdots, Z_{M^3}^3)$ of the ship radiated noise;

④ Combine the subband energy distribution eigenvectors from step 1 to 3 to form a $(M^1+M^2+M^3)$ dimensional synthetic eigenvector of ship radiated noise: $(Z_1^1, Z_2^1, \cdots, Z_{M^1}^1, Z_1^2, Z_2^2, \cdots, Z_{M^2}^2, Z_1^3, Z_2^3, \cdots, Z_{M^3}^3)$;

⑤ Use correlation matrix of eigenvector $(Z_1^1, Z_2^1, \cdots, Z_{M^1}^1, Z_1^2, Z_2^2, \cdots, Z_{M^2}^2, Z_1^3, Z_2^3, \cdots, Z_{M^3}^3)$ as production matrix, carry out K-L transform for it to turn a $(M^1+M^2+M^3)$-dimensional vector to a M-dimensional classification eigenvector $(Z_1, Z_2, \cdots, Z_M)$.

The procedure of the aforementioned decision tree classification method is:

Step a: Use the acquired class I ship target SVM classifier to recognize the testing sample. If it dose not below to class I then carry on step b;

Step b: Use the acquired class II ship target SVM classifier to recognize the testing sample. If it dose not below to class II then carry on step c;

Step c: Use the acquired class III ship target SVM classifier to recognize the testing sample. If it dose not below to class III then it fail to recognize any class.

Through experiment, we choose Gaussian radial basis function SVM for classification recognition between class I and class II,III, and choose polynomial kernel function SVM for classification recognition between class II and class III, it can get the best result. The optimal parameters for the time being are: the order of polynomial kernel function q=3, parameter of Gaussian radial basis function σ=0.5. Table 1 gives the experiment result:

Table 1. Classification recognition result of SVM classification decision tree for 3 kind of targets (σ=0.5, q=3)

Sample classes	Training sample set			Testing sample set		
	Samples number	Correct recognition number	Recognition ratio (%)	Samples number	Correct recognition number	Recognition ratio (%)
Class I	15	15	100	50	41	82
Class II	15	15	100	50	41	82
Class III	15	15	100	50	41	88

4 Conclusion

This paper fuses the latest signal processing method of multitarget information separation, location and recognition effectively, and develops advantages of blind source separation, blind deconvolution location and SVM technique respectively, and proposes ship multitarget classification and location technique especially appropriate for the underwater environment. We study its principle and give a detailed implement procedure, and validate its effectiveness through real ship radiated noise. It is important to point out that blind signal processing technique has its inherent superiority which avoids complicated analysis of signal transmission problem come from the complex and unstable underwater channel, so it is an applicable underwater multitarget location technique and provides classification samples for multitarget recognition at the same time, and also has wide development prospect.

References

1. Mitianoudis, N., Davies, M.E.: Audio source separation of convolutive mixtures. Speech and Audio Processing 11(5), 489–497 (2003)
2. Eriksson, J., Koivunen, V.: Complex-valued ICA using second order statistics. In: Proceedings of the 2004 IEEE Signal Processing Society Workshop on Machine Learning for Signal Processing 2004, pp. 183–191 (2004)
3. Kopriva, I.: Blind signal deconvolution as an instantaneous blind separation of statistically dependent sources. In: Davies, M.E., James, C.J., Abdallah, S.A., Plumbley, M.D. (eds.) ICA 2007. LNCS, vol. 4666, pp. 504–511. Springer, Heidelberg (2007)
4. Zhang, K., Chan, L.-W.: Convolutive blind source separation by efficient blind deconvolution and minimal filter distortion. Neurocomputing 73(13-15), 2580–2588 (2010)

5. Douglas, S.C., Sawada, H., Makino, S.: Natural gradient multichannel blind deconvolution and speech separation using causal FIR filters. IEEE Transactions on Speech and Audio Processing 13(1), 92–104 (2005)
6. Xu, T., He, D.-K.: Theory of hypersphere multiclass SVM Kongzhi Lilun Yu Yinyong. Control Theory and Applications 26(11), 1293–1297 (2009)
7. Cherkassky, V., Ma, Y.: Practical selection of SVM parameters and noise estimation for SVM regression. Neural Networks 17(1), 113–126 (2004)
8. Gao, H., Liu, W.: An improved SVM classifier ICIC Express Letters 3(4), 1001–1005 (2009)

A Novel rhenium(I) tricarbonyl Complex with 4,5-Diazafluoren-Drived Ligand: Synthesis, Spectroscopic and Theoretical Studies

Jiexiu Wang, Ya-qian Wu, and Feng Zhao

Jiangxi Key laboratory of Organic Chemistry, Jiangxi Science & Technology Normal University,
Nanchang, Jiangxi, China
zhf19752003@yahoo.com.cn

Abstract. A novel diimine ligand 9-(N-butyl)-4,5-diazafluoren(BADF) and corresponding rhenium tricarbonyl complex Re[(BADF)(CO)3Cl] (BADF-Re) have been synthesized. Photophysical behaviors are investigated by UV-vis absorption and fluorescence spectrometry. It is found that the triplet metal-to-ligand charge-transfer dπ(Re)-π*(N-N) (3MLCT) emission of BADF-Re centered at around 563nm. Theoretical calculation revel that the energy gap between the HOMO and LUMO is 2.42 eV, wich is similar to that (2.43 eV) obtained from the optical absorption spectra.

Keywords: Synthesis, Rhenium(I) complex, 4,5-diazafluoren, Theoretical sudy.

1 Introduction

During the past few decades, the coordination chemistry of rhenium has been intensively studied, mainly due to their versatile photophysical and photochemical properties. This class of complexes, with low-lying metal-to-ligand charge-transfer (MLCT) excited states and fairly long lifetimes, have been widely used as photosensitizers for a variety of reactions, including photoisomerization. Among the various Re(I) complexes reported in the literature[1-4], many researchers have focused on a particularly important class of complexes with the type of *fac*-[$Re(CO)_3$(N-N)(X)], where N-N = diimine and X =halides. The spectroscopic and redox behaviour of the Re(I) complexes are ligand dependent and can be tuned by varying the identity of their chelate ligands. Most attention has focussed on common chelating ligands such as bipyridine (bpy) and phenanthroline (phen) in which differences in electronic and steric properties of the ligands may lead to changes in the properties of their metal complexes. It is interesting to find that the synthetic rhenium complexes exhibited good photophysical and photochemical properties due to efficient sensitization by the ligand. Herein we synthesis a novel

X. Wan (Ed.): Electrical Power Systems and Computers, LNEE 99, pp. 83–88.
springerlink.com

ligand, 9-(N-butyl)-4,5-diazafluoren, and also its Re complex. The synthetic detail, spectroscopic and results are reported in this paper.

2 Experimental

2.1 Materials

All solvents were reagent grade, except for photophysical and photochemical measurements. 1,10-phenanthroline(Phen), 4-toluenesulfonic acid(PTSA) and n-butylamine were purchased from Aladdin-reagent Co. Rhenium pentacarbonyl chloride ($Re(CO)_5Cl$) was purchased from Acros Chemical Co. 4,5-diazafluoren-9-one(Dafo) was prepared using a reported procedure[5].

2.1.1 Synthesis of 9-(N-butyl)-4,5-Diazafluoren (BADF)

A mixture of 4,5-diazafluoren-9-one(0.200g, 1mmol), n-butylamine(0.088g, 1.2mmol) and 4-toluenesulfonic acid(PTSA) (0.019g, 0.1mmol) were dissolved in 10mL toluene. The mixture was heated to reflux for 8 h. Most of the solvent was then removed in a water bath under reduced pressure, and the precipitate formed was filtered. The crude product was purified by recrystallization twice from toluene to give the desired product as a pale yellow solid. Yield: 0.166g (70%). ^1H NMR (δ, $CDCl_3$, 400MHz): 8.80-8.82 (t, 2H), 8.00(m, 2H), 7.36-7.38(t, 2H), 4.16-4.20(t, 2H), 1.91-1.95(m, 2H), 1.55-1.56(m, 2H), 1.01-1.05(t, 3H).

2.1.2 Synthesis of BADF-Re

9-(N-butyl)-4,5-diazafluoren (0.066g, 0.276mmol), $Re(CO)_5Cl$ (0.100 g, 0.276mmol) and 10 mL toluene were heated to reflux under N_2 for 9 h. After the mixture was cooled to RT, the solvent was removed in a water bath under reduced pressure. After removal of the solvent, the residue was triply recrystallized from toluene. Yield: 0.11g (73%). ^1H NMR (δ, $CDCl_3$, 400MHz): 8.86-8.88(t, 2H), 8.20-8.22(t, 2H), 7.65-7.68(m, 2H), 4.23-4.25(t, 2H), 1.92-1.96(m, 2H), 1.55(m, 2H), 1.01-1.05(t, 3H).

2.2 Physical Measurements

^1H NMR spectra were obtained on a 400 MHz Bruker system, using tetramethylsilane (TMS) as internal reference. $CDCl_3$ was used as the solvents. UV-vis absorption spectra were measured using a Perkin Elmer Lambda-900 spectrophotometer. Fluorescence spectra were determined with a Hitachi F-4500 fluorescence spectrophotometer.

3 Computational Details

All calculations were performed with the Gaussian 03 program package employing the density functional theory (DFT)[6] with B3LYP functional[7]. Re ion was described with the LANL2DZ basis set, whereas 3-21G* basis set was used for C, H, N, O, and Cl atoms.

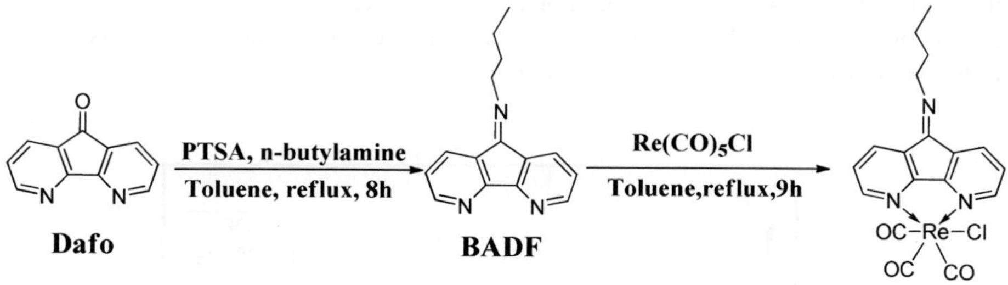

Scheme 1. Synthesis route of the Re complex.

4 Results and Discussion

4.1 Synthesis

The synthetic pathway of the Re-complex is as shown in **Scheme 1**. The diimine ligand 9-(N-butyl)-4,5-diazafluoren(**BADF**) was synthesized in two steps according to the modified literature procedure[8]. **BADF-Re** were prepared by the modified literature procedures[9]. The structures of the compounds were confirmed by ^{1}H NMR.

Table 1. Absorption and emission spectra parameters of BADF and BADF-Re in CH_2Cl_2 solution at RT.

Compound	medium	Absorption (λ, nm)	Emission λ_{max}(nm)
BADF	CH_2Cl_2	245, 304, 317	
BADF-Re	CH_2Cl_2	232, 323, 390	563

4.2 Physical Chemistry

The absorption spectra of compounds BADF and BADF-Re were measured, and the data were summarized in **Table 1**. As shown in **Figure 1**, the absorption spectra of compounds BADF and BADF-Re consist of several absorption bands in the 230-510 nm spectral region. By comparison to the absorption of the free BADF ligand, The intense absorption band of the rhenium complex, related to intraligand (π-π*) transitions, are observed in the UV spectral region (<350 nm). The relatively weak absorption bands in the range 350-510nm are tentatively attributed to an admixture of metal-to-ligand charge-transfer states, dπ(Re)-π*(N-N) (MLCT).

The emission spectra of BADF-Re in CH_2Cl_2 is also shown in **Figure 1**. Upon irradiation with 366nm light, the complex of BADF-Re emit strong emission bands at around 563 nm in dichloromethane solutions, which can be predominantly assigned to radiative transitions from ^{3}MLCT level [10].

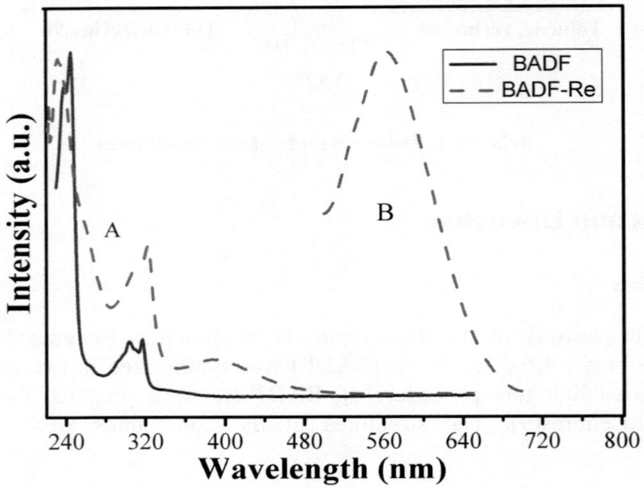

Fig. 1. UV-vis spectra(A) of BADF-Re and BADF and emission spectra(B) of BADF-Re in CH_2Cl_2 solution at RT.

4.3 Theoretical Calculation

The partial molecular orbital diagram of BADF-Re with several highest occupied and lowest unoccupied molecular orbital contours are shown in **Figure 2**. The HOMO of the Re(I) complex is mainly composed of the d Re orbital and the π-orbital localized on CO and Cl moieties. The LUMO of the Re(I) complex is mainly composed of the π* orbital localized on the diimine ligand. The value of the energy separation between the highest occupied molecular orbital (HOMO) and the lowest unoccupied molecular orbital (LUMO) equals to 2.42 eV, which is close to that (2.43 eV) obtained from the optical absorption spectra.

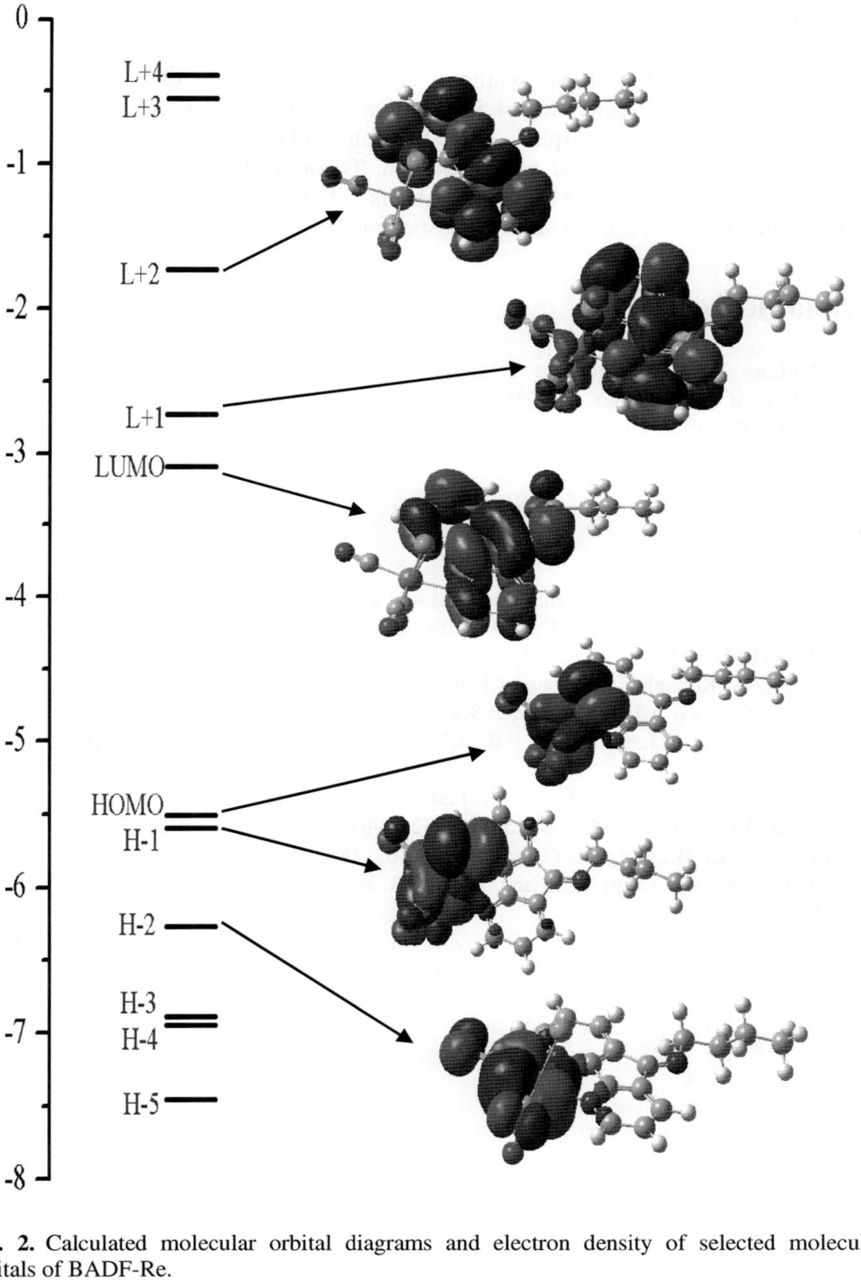

Fig. 2. Calculated molecular orbital diagrams and electron density of selected molecular orbitals of BADF-Re.

5 Conclusions

In summary, we reported the synthesis and characterization of a novel rhenium complex, BADF-Re, containing 4,5-diazafluoren derived ligand. It is found that BADF-Re present the triplet metal-to-ligand charge-transfer phosphorescent transitions (^{3}MLCT) center at around 563nm. Theoretical calculation revel that the energy gap between the HOMO and LUMO is 2.42 eV, wich is similar to that (2.43 eV) obtained from the optical absorption spectra.

Acknowledgements

The authors acknowledge the financial support from National Natural Science Foundation of China (No. 20903049). The computational studies reported here were done on the Guizhou University High Performance Computation Chemistry Laboratory (GHPCC).

References

1. Luong, J.C., Nadjo, L., Wrighton, M.S.J.: Am. Chem. Soc. 100, 5790-5795 (1978)
2. Schanze, K.S., MacQueen, D.B., Perkins, T.A., Cabana, L.A.: Coord. Chem. ReV. 122, 63-89 (1993)
3. Vogler, A., Kunkeley, H.: Coord. Chem. ReV. 200-202, 991-1008 (2000)
4. Sun, S.-S., Lees, A.J.J.: Am. Chem. Soc. 122, 8956-8967 (2000)
5. Henderson Jr., L.J., Fronczek, F.R., Cherry, W.R.: J. Am. Chem. Soc. 106, 5876-5879 (1984)
6. Runge, E., Gross, E.K.U.: Phys. Rev. Lett. 52, 997-1000 (1984)
7. Becke, A.D.: J. Chem. Phys. 98, 5648-5752 (1993)
8. Zhang, L., Li, B.: Inorganica Chimica Acta 362, 4857-4861 (2009)
9. Yam, V. W.-W., Yang, Y., Zhang, J., Chu, B. W.-K., Zhu, N.: Organometallics 20, 4911-4918 (2001)
10. Li, J., Si, Z.J., Liu, C.B., Li, C.N., Zhao, F.F., Duan, Y., Chen, P., Liu, S.Y., Li, B.: Semicond. Sci. Technol. 22, 553-556 (2007)

A Novel Data Embedding Method for High Payload Using Improved Pixel Segmentation Strategy

Wien Hong[1], Che-Lun Pan[2], Tung-Shou Chen[3], Wan-Yi Ji[1], and Yeon-Kang Wang[2]

[1] Department of Information Management, Yu Da University, Taiwan
[2] Department of Multimedia and Game Science, Yu Da University, Taiwan
[3] Deptartment of Computer Science and Information Engineering, National Taichung Institute of Technology, Taiwan
{wienhong,panpeter,ykwang}@ydu.edu.tw, tschen@ntit.edu.tw, wanyi.ji@google.com

Abstract. In this paper, we propose a novel data embedding method based on the pixel segmentation strategy with minimum distortion. The well-known exploring modification direction (EMD) embedding method achieves a high image quality; however, the corresponding embedding capacity is low. Some recent works extended the embedding capacity of EMD method by using various strategies. Inspired by Lee et al.'s work, the proposed method uses an indicator bit to improve the stego image quality under various payloads. The experimental results revealed that the proposed method not only greatly enhances the embedding capacity of EMD, but also maintains a very acceptable image quality.

Keywords: EMD, Data Embedding, Embedding Capacity.

1 Introduction

Data hiding imperceptibly embeds data into a cover media so that messages can be delivered secretly [1]. Digital images are often used as carriers to deliver messages. A cover image is an image that is used to carry data, and a stego image is the image that carried data. Embedding distortion is occurred when data are embedded into a cover image [2]. Generally, the distortion caused by data embedding should be as small as possible. A low-distorted stego image not only provides better image quality but also has smaller chance from being detected [3].

Based on the reversibility of the stego image, the data hiding method can be classified into reversible [4]-[7] and non-reversible [8]-[11]. Reversible data hiding methods offer the capability to recover the stego image to the original (cover) image. However, the embedding capacity is often smaller than those of non-reversible ones under the same payload. Therefore, when a large payload is crucial and the reversibility of the stego image is not required, the non-reversible methods can be applied to these applications.

LSB replacement is a method that embeds data by replacing the least significant bits of image pixels with message bits. Although this method is easy to implement

X. Wan (Ed.): Electrical Power Systems and Computers, LNEE 99, pp. 89–95.
springerlink.com

and has low CPU cost, it is widely known that the produced stego image has relatively low quality and is vulnerable to the detection of LSB-based steganalysis tools [12],[13]. In 2004, Chan and Cheng [8] proposed an optimal pixel adjustment process (OPAP) to reduce the embedding distortion caused by LSB replacement and had a significant improvement in image quality under the same payload. Mielikainen [9] in 2006 proposed a LSB Matching Revisited method in which a pixel pair carries two bits and only modifies one pixel a grayscale value at most. Inspired by Mielikainen, Zhang and Wang [10] proposed the exploring modification direction (EMD) method to further enhance the image quality by reducing the embedding distortion. EMD embeds a digit in base $2m+1$ into m pixels by adding or subtracting one grayscale unit of a pixel at most. Lin et al. [11] in 2010 further extent Zhang and Wang's work by calculating the best number of pixels as an embedding group for a given payload to minimize the image distortion, and had better image quality over EMD under the same payload. However, the maximum payload of EMD and Lin et al.'s works is only $\frac{1}{2}\log_2(2)$ bpp.

To further extend the payload of EMD, several works have been proposed recently. Lee et al. [14] in 2008 proposed a high payload data hiding method by using pixel segment strategy (PSS). Chao et al. [15] also proposed a diamond encoding method (DE) to extent the payload of EMD by using a diamond-shape neighborhood set as the guide to embed data. In DE, a pixel pair is used to carry a digit in base $2k^2+2k+1$, where k is the embedding parameter. When $k=1$, the embedding performance is equivalent to that of EMD.

In this paper, we modified Lee et al.'s method by redesigning PSS method. With the aid of an indicator bit, the embedding quality can be increased around 1.8 dB. The rest of this paper is organized as follows. In Section 2, Lee et al.'s method is briefly introduced. Section 3 describes the proposed method and gives a simple example to illustrate the proposed method. Section 4 presents the experimental results, and concluding remarks are made in the last section.

2 Relative Works

Lee et al. proposed a novel embedding method PSS that extends the payload of EMD efficiently by using pixel segmentation strategy. In EMD, a digit in base $2m+1$ can be carried by m pixels. PSS extends the payload of EMD by using a pixel pair as an embedding unit, and a digit in base $2n+1$ can be embedded into each embedding unit, where n is the total number of bits to be modified.

In PSS, pixels in the embedding unit are segmented into two area; namely vector of coordinate area (VCA) and vector of modification area (VMA). An embedding unit consists of two pixels. The bits that correspond to VCA in the first and second pixel are denoted by vca1 and vca2, respectively. Similarly, the bits that correspond to VMA in the first and second pixel are denoted by vma1 and vma2, respectively. VCA of an embedding unit is obtained by concatenating vca1 and vca2, and VMA is obtained by concatenating vma1 and vma2, as shown in Fig. 1. The VCA is served as a seed for generating random integers using a hash function, and VMA is used as a guide to conceal data.

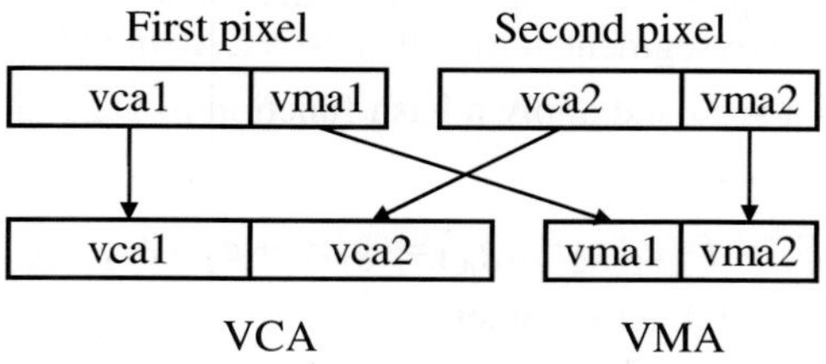

Fig. 1. Bits segmentation of the first and second pixel of an embedding unit.

To embed data, VCA are hashed to generate random integers. VMA is then replaced by a bit string that is calculated by an extraction function using these random integers as an input argument. The stego pixels are then obtained by converting VCA and VMA back to 8-bit pixel pair. To extract the embedded digit, the VCA and VMA of the stego pixels are obtained, and the random integers generated by using the VCA as the seed are calculated and modified according to VMA. The embedded digit is then extracted by applying the extraction function to the modified random integers.

3 The Proposed Method

This paper proposed a method PSS-IB to extend PSS method by using an indicator bit to improve the image quality. Although PSS method significantly increases the payload of EMD, it also distorted the image quality considerably. To eliminate this problem, the proposed PSS-IB method segments an indicator bit form VCA. The indicator bit is flipped if required to enhance the image quality. The segmentation of VCA and VMA of the proposed method is shown in Fig. 2.

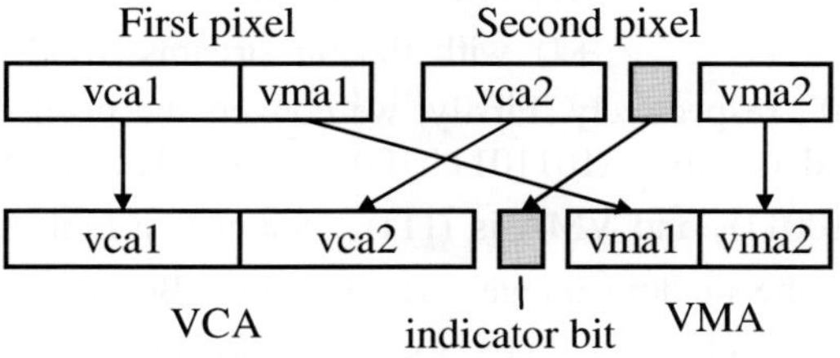

Fig. 2. VCA and VMA of the proposed method.

Let I be an eight-bit grayscale image, v_1 and v_2 be the number of bits that are used to represent vma1 and vma2, respectively. In this case, a digit in base $2(v_1+v_2)+1$ is embedded into each pixel pair. The detailed embedding procedure is listed as follows.

Step 1. Partition the cover image I into blocks of size 1×2, and associate each n-element logical outcome a bit string of length n, $n=v_1+v_2$.

Step 2. For each block, pixels are converted to their binary representation (totally 16 bits), and then are segmented the bits into VCA, an indicator bit, and VMA.

Step 3. Use VCA as a seed and apply a hash function to generate n random integers $g_1, g_2, \cdots, g_n$.

Step 4. Calculate $E_f = f(g_1, g_2, ..., g_n) = (g_1 \times 1 + g_2 \times 2 +, ..., g_n \times n) \bmod (2n+1)$, where E_f is the extraction function.

Step 5. If $E_f = S$, set the logical outcome to be $(g_1, g_2, ..., g_n)$. Otherwise, calculate $x = (S - E_f) \bmod (2n+1)$ and the logical outcome can be obtained by performing the following modification: If $x \le n$ the xth element of $(g_1, g_2, ..., g_n)$ is increased by one; otherwise, the $(2n+1-S)^{\text{th}}$ element of $(g_1, g_2, ..., g_n)$ is decreased by one.

Step 6. The VMA of the embedding unit is replaced by the associated bit stream of the calculated outcome.

Step 7. Flip the indicator bit to check whether the result is closer to the original pixel value. If yes, the indicator bit is flipped.

Step 8. Repeat Steps 2-7 until the entire secret data are embedded.

To extract the embedded digits, the stego pixel pair is segmented and converted to VCA and VMA. The integers $(g_1', g_2', ..., g_n')$ is obtained by using the hash function with the seed VCA. According to VMA, the modified logical outcome $(g_1'', g_2'', ..., g_n'')$ is determined. The embedded secret S can then be obtained by calculating $S = f(g_1'', g_2'', ..., g_n'') = (g_1'' \times 1 + g_2'' \times 2 +, ..., g_n'' \times n) \bmod (2n+1)$.

Here is a simple to illustrate the proposed method. Let the original pixel pair be $(183,155)$, $v_1 = 1$ $v_2 = 2$, and $S = 0_7$. In this case, $n = 3$ and there are 7 possible logical outcomes. Suppose we associate logical outcomes (g_1, g_2, g_3), $(g_1 - 1, g_2, g_3)$, $(g_1 + 1, g_2, g_3)$, $\cdots$, $(g_1, g_2, g_3 + 1)$ with the bit streams '011', '100', '010', '101', '001', '110', '000', respectively. Firstly, we convert the pixel pair into their binary representation, and obtained $(10110111{,}10011011)_2$. The VCA of this embedding unit is $(101101110011)_2$ and VMA is $(111)_2$. Suppose a hash function is applied to VCA and obtains the random integers $(57,197,33)$. Because $f(57,197,33) = 4 \ne S$, we have $x = (0-4) \bmod 7 = 3$. Since $x \le n$, the logical outcome is $(g_1, g_2, g_3 + 1)$ and the corresponding bit stream is '000'. Therefore, the VMA is replaced by '000'. The recovered binary representation of VCA and VMA is $(10110110{,}10011000)_2$ and the decimal value is (182,152). However, if the indicator bit is flipped, the decimal value is (182,156), which is closer to the original value $(183,155)$. Therefore, we replace the $(183,155)$ by (182,156), and the digit $S = 0_7$ is embedded.

To extract the embedded digit, The VCA $(101101110011)_2$ and VMA $(000)_2$ of the marked pixel pair (182,156) are obtained. The random integers (57,197,33) are calculated by using the hash function with the seed VCA. The associate logical outcome of VMA is $(g_1, g_2, g_3 + 1)$; therefore, the random integers are modified to $(57,197,34)$. Calculate $S = f_E(57,197,34)$, we obtain the embedded digit $S = 0_7$.

4 Experimental Results

In this section, several experiments were performed to demonstrate the embedding performance of the proposed method. Four standard test images obtained from USC-SIPI image database [16] were used in the experiments, as shown in Fig. 3. These test images are 8-bit, and the size of each image is 512×512.

(a) Lena (b) Baboon (c) Jet (d) Peppers

Fig. 3. Four test images.

The embedded data were generated by using a pseudo random number generator (PRNG). The peak signal to noise ratio (PSNR) was used to measure the image quality. PSNR is defied as

$$\text{PSNR} = 10 \log_{10} \frac{255^2}{\text{MSE}},$$

where MSE is the mean square error between the cover image and the stego image. A higher PSNR indicates that the stego image has better image quality.

Table 1 shows the comparison of the proposed method and Lee et al.'s method under the same number of VMA bits. The numbers listed in the left column and top row are the number of bits used to represent vma1 and vma2, respectively.

As can be seen in Table 1, the proposed method always has high PSNR under the same payload. The improvements are more significant when the payloads are higher. For example, at the payload 393,216 bits, the PSNRs of the proposed method and Lee et al.'s method are 48.05 dB and 46.38 dB, respectively. The improvement is 1.7 dB. However, when the payload is 655,360 bits, the PSNR improvement is 2.45 dB ($37.19 - 34.74$).

Table 1. Embedding performance compassion of PSS method and PSS-IB methods.

		1		2		3		4	
		PSS	PSS-IB	PSS	PSS-IB	PSS	PSS-IB	PSS	PSS-IB
1	PSNR	51.11	51.15	46.36	48.08	40.73	43.10	34.72	37.18
	Payload	262144 bits		393216 bits		524288 bits		655360 bits	
2	PSNR	46.38	48.05	44.14	46.26	46.38	48.05	44.14	46.26
	Payload	393216 bits		524288 bits		655360 bits		786432 bits	
3	PSNR	40.75	43.11	40.01	42.46	40.75	43.11	40.01	42.46
	Payload	524288 bits		655360 bits		786432 bits		917504 bits	
4	PSNR	34.74	37.19	34.56	37.00	34.74	37.19	34.56	37.00
	Payload	655360 bits		786432 bits		917504 bits		1048576 bits	

Figs. 4 (a) and (b) depict the payload versus PSNR of the Lena image of PSS and PSS-IB methods. In Fig. 4 (a), we set $v_1 = 1$ and $v_2 = 2$. In Fig. 4(b), $v_1 = 3$ and $v_2 = 4$.

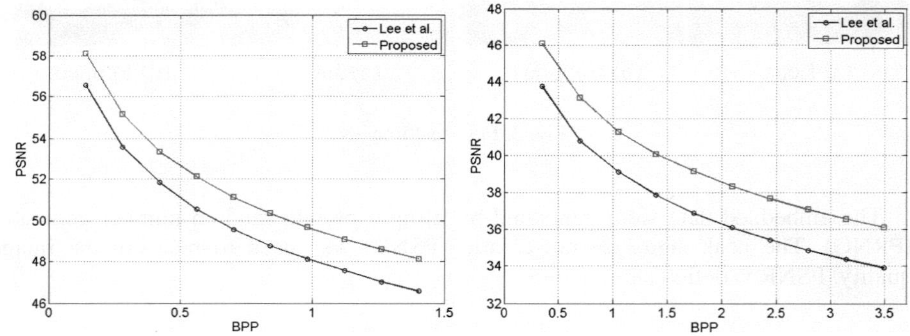

(a) $v_1 = 1, v_2 = 2$ (b) $v_1 = 3, v_2 = 4$

Fig. 4. Comparison of payload versus PSNR

Fig. 4 reveals that the proposed method always has image quality under various embedding rate. For example, when $v_1 = 1, v_2 = 2$ at 1 bpp, the image quality of the proposed PSS-IB method is around 48.08 dB, which is 2 dB higher than that of PSS method.

5 Conclusions

This paper proposed a modified version of Lee et al.'s method by inducing an indicator bit. The indicator bit is not part of VMA, and therefore does not join the hash process. Instead, the indicator is capable of providing a lower distortion to the cover image. Experimental results revealed that the proposed method effectively increases the image quality around 2 dB compared to Lee et al.'s method. Because our method produces a relatively high image quality, the proposed method can be used in applications where high quality stego image is demanded.

References

1. Provos, N., Honeyman, P.: Hide and Seek: An Introduction to Steganography. IEEE Security and Privacy 3, 32–44 (2003)
2. Cheddad, A., Condell, J., Curran, K., McKevitt, P.: Digital Image Steganography: Survey and Analysis of Current Methods. Signal Processing 90, 727–752 (2010)
3. Wang, H., Wang, S.: Cyber Warfare: Steganography vs. Steganalysis. Communications of the ACM 47(10), 76–82 (2004)
4. Ni, Z., Shi, Y.Q., Ansari, N., Su, W.: Reversible Data Hiding. IEEE Transactions on Circuits and Systems for Video Technology 16(3), 354–362 (2006)
5. Hong, W., Chen, T.S.: A Local Variance-Controlled Reversible Data Hiding Method Using Prediction and Histogram-Shifting. The Journal of Systems and Software 83(12), 2653–2663 (2010)
6. Hong, W., Chen, T.S.: Reversible Data Embedding for High Quality Images Using Interpolation and Reference Pixel Distribution Mechanism. Journal of Visual Communication and Image Representation 22(2), 131–140 (2011)
7. Thodi, D.M., Rodríguez, J.J.: Expansion Embedding Techniques for Reversible Watermarking. IEEE Transactions on Image Processing 16(3), 721–730 (2007)
8. Chan, C.K., Cheng, L.M.: Hiding Data in Images by Simple LSB Substitution. Pattern Recognition 37(3), 469–474 (2004)
9. Mielikainen, J.: LSB Matching Revisited. IEEE Signal Processing Letters 13(5), 285–287 (2006)
10. Zhang, X., Wang, S.: Efficient Steganographic Embedding by Exploiting Modification Direction. IEEE Communications Letters 10(11), 781–783 (2006)
11. Lin, K.Y., Hong, W., Chen, J., Chen, T.S., Chiang, W.C.: Data Hiding by Exploiting Modification Direction Technique Using Optimal Pixel Grouping. In: The 2010 International Conference on Education Technology and Computer (ICETC 2010), July 2010, vol. 3, pp. 121–123 (2010)
12. Fridrich, J., Goljan, M., Du, R.: Reliable Detection of LSB Steganography in Color and Grayscale Images. In: Proceedings of the International Workshop on Multimedia and Security, pp. 27–30 (2001)
13. Ker, A.D.: Steganalysis of LSB Matching in Grayscale Images. IEEE Signal Processing Letters 12(6), 441–444 (2005)
14. Lee, C.F., Chang, C.C., Wang, K.H.: An Improvement of EMD Embedding Method for Large Payloads by Pixel Segmentation Strategy. Image and Vision Computing 26(12), 1670–1676 (2008)
15. Chao, R.M., Wu, H.C., Lee, C.C., Chu, Y.P.: A Novel Image Data Hiding Scheme with Diamond Encoding. EURASIP Journal on Information Security, 2009, Article ID 658047 (2009)
16. USC-SIPI image database, `http://sipi.usc.edu/database`

Design of Wireless and Motor Control System for Amoeba-Like Robot

Rong-rong Qian[1,2,3], Min-zhou Luo[1,3], and Bing Li[1,2,3]

[1] University of Science and Technology of China,
Hefei 230027, China
[2] Institute of Intelligent Machines, Hefei Institutes of Physical Science,
Chinese Academy of Sciences,
Hefei 230031, China
[3] Institute of Advanced Manufacturing Technology,
Hefei Institutes of Physical Science,
Chinese Academy of Sciences, Changzhou 213164, China
vigour1022@yahoo.com.cn

Abstract. This paper presents the design of wireless and motor control system for Amoeba-like robot based on the radio frequency chip of nRF24L01. Atmega16 is used as a micro-controller which is in charge of configuring nRF24L01 to fulfill the wireless transmitting and receiving control information. The wireless communication module transmits the information which is collected by the robot into the upper machine. After that the upper machine offers relevant control instructions to the robot in order to control the motors of the robot. These instructions accomplish the robot's movement such as moving forwards and backward, turning the corner or entering the narrow hole and so on.

Keywords: Amoeba-like robot, wireless communication, motor control.

1 Introduction

More and more natural disasters bring the search-and-rescue robot into people's sights. The main task of the search-and-rescue robot can be described as locating the wounded by bringing the sensors into the relic or taking the place of rescue personnel or dogs to implement the rescue work. This requires the robot to acclimatize itself better and vary its mode with the change of the environment.

Observed through a microscope, the amoeba has a granular cytoplasm or protoplasm in the body of the cell and a non-granular, hyaline region at the tip of an advancing pseudopod[7]. It works by way of an elongated toroid which turns itself inside out in a single continuous motion, effectively generating the overall motion of the cytoplasmic streaming cytoplasmic tube [8]. In other words, the amoeba movement mechanism is contraction of the tail and expansion of the front-end and it is illustrated in figure 1. The design of amoeba-like robot is based on this mechanism.

X. Wan (Ed.): Electrical Power Systems and Computers, LNEE 99, pp. 97–103.
springerlink.com

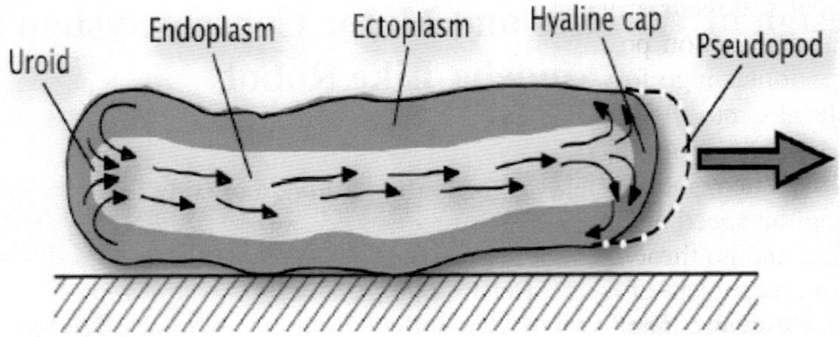

Fig. 1. The movement mechanism of amoeba[2]

2 The Introduction of Amoeba-Like Robot

In the macroscopic view, the implementation of cytoplasm turning inside out is difficult. Therefore, a kind of thin skin which is easy to bend is used and it is made into circle column shape. The liquid is injected into the cavity and this structure is called fluid filled toroid[9]. The appearance of robot is showed in figure 2. The robot is cylindrical when it is static and there are 16 driving rings fixed on the inner and outer skin of the robot. Each driving ring is made of anti-elastic ropes and used to compress the relative part of the skin. Every control loop has six nodes of framework and they are distributed respectively 30°, 90°, 150°, 210°, 270°, 330° with the motor shaft. These nodes are used for fixing the ropes and making the motor torque into the tangential force of the spherical body. Both ends of the open-loop non-elastic ropes fix at the front motor shaft and the rotation of the motors drives the non-elastic rope contracted. As figure 1 illustrated, the slender cycle skin can turn over continuously in the axial. During the process of turnover, the inner layer of skin turns into the outer or vice versa.

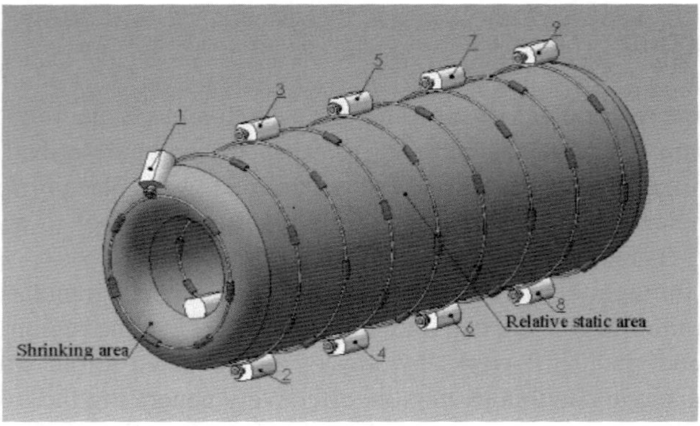

Fig. 2. The appearance of the robot

The robot's move is driven by the motors. The tail of skin is contracted by different degrees of contraction power from the constrictive anti-elastic ropes and this makes the flow endoplasm go forward along the direction of the least resistance. That means the tail shrinks into the inner surface which is low resistance. The convergent power from the non-elastic ropes force the inside and outside skins evert to make the robot go forward. This motion simulates the Amoeba movement mechanism effectively. Controlling the speed and direction of the motors can urge the robot to move forward, make a turn and go through the narrow gap and so on. The following briefly describes the control mechanism of moving forward.

The ropes on the inner and outer surface are in a relaxed state when the robot is motionless. The motor's whirling in positive direction of loop 1 which is near the edge of the tail pulls off its circle and spurs the loop 2 behind it. At this moment, loop 2 also shrinks but the power from it is less than loop 1. When loop 2 comes to the circular boundary it repeats loop 1's action. Meanwhile, the motor of the loop 1 continues rolling till it comes into the relatively static area. Then the MCU gives the reverse voltage to the motor to make it roll back so that the non-elastic rope loosens for the front-end expansion. The control loops located in the inner radius of the robot expand in the proper order to make the robot moving. The steady and high-efficient moving of the robot needs the effective controlling to the motors of the front and tail.

Based on the above analysis of the robot's function and structure, the amoeba-like robot's wireless communication control and motor control is presented in details as follows.

3 Control System

In order to detect the environment outside to make sure whether there are obstacles or vital signs in front of the robot while the robot is moving, there are several kinds of sensors such as pressure sensor, location sensor and ultrasonic integrated in the robot. In most cases, the amoeba-like robot works in the anti-structure environment, the traditional wire transmission is not applicable. For that reason the wireless communication control is adopted. The wireless link between the robot and the user interface has been implemented by utilizing 2.4 GHz nRF24L01 radio transceiver. The control part of the sensors and wireless communication is based on the Atmel's ATmega16 microcontroller running at 8 MHz. The robot moves according to the relevant command which is suitable for the outside conditions and sent by the upper machine. The command is to control the continuous current dynamo connected with the anti-elastic ropes and make the anti-elastic ropes straining. Some part of the robot's soft skin is compressed by the constrictive anti-elastic to impel the robot to move. The control system consists of several modules including data acquisition module, wireless communication module, terminal receiver module and control module etc. Figure 3 displays the block diagram of the system.

3.1 Design of the Wireless and Motor Control System

The wireless communication is composed of RF chip nRF24L01 and Atmega16, connected via SPI interface. The SPI bus system is a kind of synchronous serial peripheral interface using four lines such as serial clock line, master input or slave output, master output or slave input and active-low slave select line.

Atmega16 is an 8-bit COMS micro-controller based on the structure of the enhanced low-power AVR RISC. The 16-bit timer can generate high-precision, correct phase and frequency PWM waveform. Phase and frequency correct PWM mode is based on dual-slope operation and the symmetry of the waveform is very suitable for the motor control.

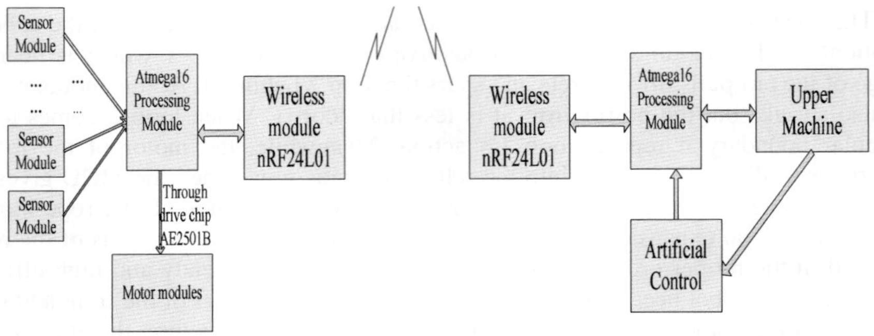

Fig. 3. Block diagram of the system

The single chip radio transceiver nRF24L01 works in 2.4~2.5GHz ISM band, and its power supply range is 1.9V to 3.6V, work temperature range is -40~+85℃, maximum data rate is 2Mbps. It has four work modes as follows power down mode, standby mode, ShockBurstTM mode and enhanced ShockBurstTM mode. Enhanced ShockBurstTM mode is a packet based data link layer. In a typical bidirectional link, one will let the terminating part acknowledge received packets from the originating part in order to make it possible to detect data loss. The features enable significant improvements of power efficiency for bi-directional and un-directional systems, without adding complexity on the host controller side. In the view of this easy and effective protocol, Enhanced ShockBurstTM mode is utilized in this paper. The connection of Atmega16 and nRF24L01 is showed in figure 4.

As described above about the robot moving forward, only the driving ring' s and shrinking area's motors are working at each time of the moving process and the efficiency is very low. Thus it is necessary to drive the relative static area's motor to achieve the efficient utilization. The motors' speed is uniform during the process above mentioned. The wireless communication is used to select the mode of the motors' running. It makes the robot running in the way of adapting to the external environments.

Micro DC reducer motor GM12-N20VA is used in this paper, the driver of the motor is AE2501B which is provided with forward and reverse functions and is ideal as a driver circuit for a motor. AE2501B has pins of motor forward output and motor reverse output, these two pins are controlled by the pins of motor forward input and motor reverse input. The waveform of the pins is showed in figure 5.

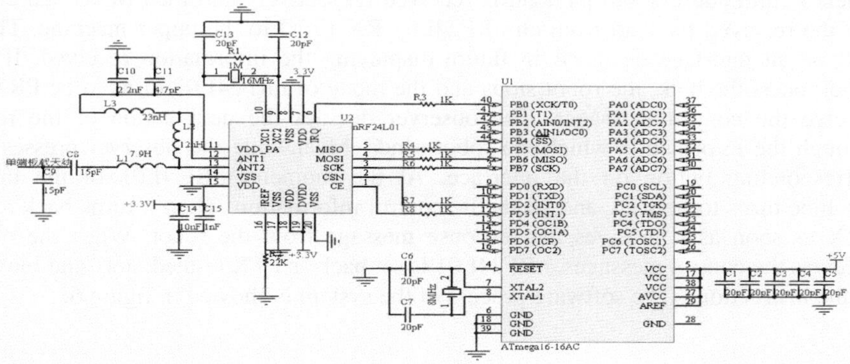

Fig. 4. Connection of Atmega16 and nRF24L01

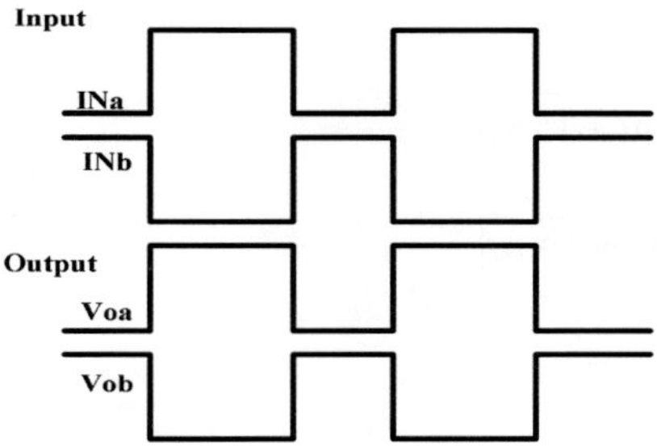

Fig. 5. Waveform of the AE2501B

Table 1. The correspondence of the received data and the way of robot moving

Data received	The way of robot moving
"0x00"	Stop run
"0x01"	Moving forward
"0x10"	Over obstacles

3.2 Design of Communication Software of the System

The single-chip Atmega16 of the robot is initialized and set the mode of nRF24L01 in the robot as PRX. Meanwhile the upper machine sets the nRF24L01 connected with it as PTX through one button on the user interface. The robot is static when it powers on. The sensors in the robot acquire the environmental information and send it to the upper machine. Then the upper machine sends a control commands according to the environment information to make the robot move. When the robot receives the control commands successfully, nRF24L01 of robot will be set as PTX by Atmega16 automatically and nRF24L01 of upper machine will be set as PRX. In the course of the robot moving forward, the sensors acquire the environmental information in real-time. In the process of information received, the interrupt IRQ informs Atmega16 when a valid address and payload is received respectively and then MCU can clock out the received payload from an nRF24L01 RX FIFO to the upper machine. There will be an interface designed in future displaying the information received. If the robot meets the balk, the robot stops and the mode of nRF24L01 turns to be PRX to receive the control messages. The observer decides the next action of the robot through the information which the robot sends. Meanwhile the observer presses the corresponding button on the interface. At the moment nRF24L01 of the upper machine turns to be PTX and sends the control information. Then it turns back to be PRX as soon as it receives the response message from the robot. When the robot receives the control messages, nRF24L01 is set back as PTX immediately and moving as the instructions. The software process of the system is showed in figure 6.

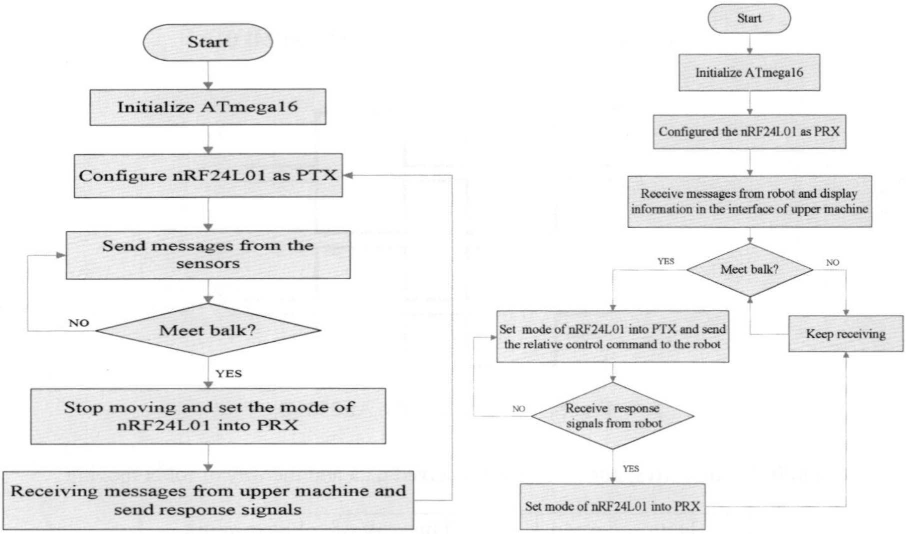

Fig. 6. Software process of the system

4 Experiments

Based on the analysis, the program is loaded to the single-chip and the joint debugging of the hardware and software is conducted. The wireless communication is accomplished and the motors are controlled to make the robot move. During the process of the wireless communication test, the button of the sender is pressed to make the LED monitor display '1' meanwhile the sender sends the data '01' to the receiver. When the receiver receives the data '01', the motor rotates in the positive direction. To the contrary, the motor rotates in the negative direction when receiving the data '10' and the LED monitor of the sender displays '2'. It is demonstrated that the wireless communication control and the control of the robot moving work well.

5 Conclusion and Future Work

The design of the wireless communication control system and mechanism of the robot movement are described in this paper. It approves that the robot of the flexible structure can move forward or turn around steadily through the experiment and simulation. Since the robot is a flexible structure, the ordinary rigid PCB circuits are no longer applicable. In the future work, the flexible circuit board will be planed to fabricate to accommodate the frame of the robot and the action of the robot will be further designed such as crossing the obstacles, drilling the holes and other activities.

Acknowledgment. This work is supported by the National Science Foundation of China under Grant (No. 60875069) and the development of multi-fingered search and rescue robot in collapsed building under Grant (No. 2007AA041502-3).

References

1. Hou, J., Gao, Y.: Greenhouse Wireless Sensor Network Monitoring System Design Based on Solar Energy. In: International Conference on Challenges in Environmental Science and Computer Engineering, vol. 274, pp. 475–479 (2010)
2. Chen, L.: A dynamic simulation of force in the movement of amoeba based on cosmos (2010)
3. Wang, B.: The Wireless Data Transmission System Design Based on nRF2401. In: International Conference on Challenges in Environmental Science and Computer Engineering, vol. 293, pp. 218–221 (2009)
4. Hiro, Yokoi, S., Mizuno, T., Akita, M., Kakazu, Y.: Amoeba Searching Behavior Model Using Vibrating Potential Field 7.26, 1297–1302 (1995)
5. Wang, X., Zhang, C., Yang, S.: Design of Wireless Video Communication System Used To Monitor And Control Plant Disease. In: International Conference on Challenges in Environmental Science and Computer Engineering, vol. 235, pp. 584–587 (2009)
6. Chan, H.-L., Chao, P.-K., Chen, Y.-C., Kao, W.-J.: Wireless Body Area Network for Physical-Activity Classification and Fall Detection 6.1, 157–160 (2008)
7. Cameron, I.: Amoeba proteus displays a walking form of locomotion. In: Cell Biology International, pp. 760–762 (2007)
8. Allen, R.D.: Biophysical Aspects of Pseudopium Formation and Retraction. In: The Biology of Amoeba, pp. 201–247 (1973)
9. Hou, J., Luo, M., Mei, T.: The Design And Control Of Amoeba-Like Robot. In: IEEE International Conference on Computer Application and System Modeling, V1-88–V1-91 (2010)

A Fast Frame Synchronous Methodology Used in the Chip Implementation for 10Gb/s FEC-Coded Ethernet Frame

Bin Zhang[1], Lijun Zhang[2], Wenshi Li[1], and Canyan Zhu[2]

[1] The School of Electronics & Information Engineering, Soochow University
Suzhou, China
[2] The School of Urban Rail Transportation, Soochow University
Suzhou, China
zhangbin09-25@163.com, {zhanglijun,lwshi,qiwuzhu}@suda.edu.cn

Abstract. This paper presents a chip implementation of fast frame synchronous scheme for forward error correction (FEC) layer of 10 Gb/s Ethernet frame. The fast frame synchronous methodology is achieved by changing the endian mode, the improved error trapper circuit can work in both syndrome generator mode and error trapper mode. One is the current syndrome generator; the other is the improved error trapper which works in syndrome generator mode. The two syndrome generators can work in parallel. When frame is synchronized, the error trapper can return to its normal working mode. A kind of network device that realizes the FEC functions is designed and a test network topology is set up to test and evaluate the FEC method. Experimental result shows that the frame synchronizing speed is twice that of the conventional method, while the hardware overhead is very small.

Keywords: Forward Error Correction (FEC), fast frame synchronous, endian mode, Ethernet.

1 Introduction

Ethernet is one of the most common digital networking technologies. Specified in the Institute of Electrical and Electronics Engineers (IEEE) standard 802.3, the technology has a large installed base of compatible network devices. Ethernet technology continues to evolve, with newer and faster variants, such as the Gigabit Ethernet, providing network speeds of 1 Gigabit per second [1-2]. Forward Error Correction (FEC) is one method for improving the bit error rate (BER) of a received signal with low signal to noise ratio [3-10]. FEC is a coding technique that uses additional, i.e. redundant symbols, as part of a transmission of a digital sequence through a physical channel. Because of the presence of redundancy, when errors corrupt the received signal, the receiver subsequently recognizes and corrects the errors without requesting data frame retransmission [11-12]. It is an error correction technique where in a receiving device

X. Wan (Ed.): Electrical Power Systems and Computers, LNEE 99, pp. 105–112.
springerlink.com

has the capability to detect and correct any block of symbols that contains fewer symbols than a predetermined number of error symbols [13]. Forward error correction layer of BASE-R physical sub-layer for 10 Gigabit Ethernet is proposed in IEEE 802.3-2008. The FEC sub-layer standard decreases the BER rate from 10^{-7} to 10^{-12} [14]. This standard is also suitable for 40 G/100 G Ethernet.

In order to compatible with different Ethernet application layers, the original head of FEC layer is compressed for FEC parity bits in this protocol, so that the frame length of this layer can be consistent with other application layers. The drawback is that it is very difficult to identify each frame boundary. In order to find out the frame boundaries, it will take a lot of time to synchronize with the transmit frames.

Synchronizing controller needs to get the correct frame start position, so that the frame can finish the synchronization. The drawback of conventional method is that it can only detect one frame boundary at one time during the procedure of finding the frame start position. Under the application environment of carrier Ethernet, it is necessary to improve the transport performance of the Ethernet physical channel by using a special method. Based on the frame structure defined in IEEE 802.3-2008, this paper presents a novel structure which can detect two frame boundaries at one time. Thus it greatly improves the FEC decoding speed.

In this paper, we propose a novel frame boundary detecting structure for forward error correction layer of 10 Gb/s Ethernet frame. By changing the endian mode, the improved error trapper circuit can work in both syndrome generator mode and error trapper mode. One is the current syndrome generator; the other is the improved error trapper which works in syndrome generator mode. The two syndrome generators can work in parallel by adding small hardware. When frame is synchronized, the error trapper can return to its normal working mode. So the frame boundary detecting speed can be accelerated by detecting two frame boundaries at one time, and the frame will be fast-synchronized.

The remainder of this paper is organized as follows. Section 2 addresses the proposed FEC decoding structure. Section 3 presents experimental result. Section 4 summarizes the conclusions from this paper.

2 Proposed Fast Frame Synchronous FEC Decoding Structure

In conventional scheme, due to pipelined structure in hardware, nearly half of frames will be given up, and the detecting time become longer as well as the frame synchronizing speed is very lower. This paper optimizes the FEC decoding structure and accelerates the frame synchronizing speed.

2.1 The Conventional FEC Decoding Circuit Structure

Fig.1 shows the conventional FEC decoding circuit structure. The common FEC decoder in chip implementation is composed of three parts: Syndrome Generator, Error Trapper module and Error Pattern Correction module.

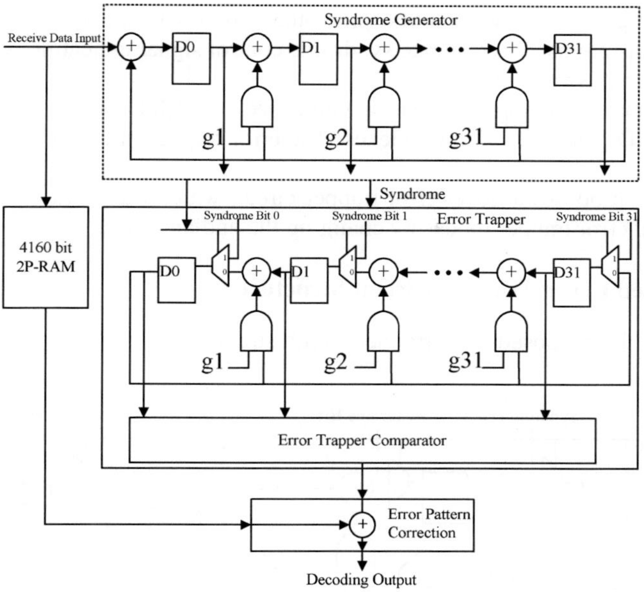

Fig. 1. FEC decoding circuit structure

Syndrome generator does mod operation for g(x), g(x) is the polynomial generated by FEC, and the remainder is the syndrome.g1-g31 are the polynomial coefficients. Since the FEC operation is in Galois field, all add/sub operation are related to exclusive-OR operation. So the circuit of syndrome generator is equivalent to a shifting subtract division circuit, and the remainder is syndrome. If syndrome is '0', there is no error of the input frame, and the parity checking is successful, then the data frame is synchronous.

If syndrome is not '0', then syndrome is sent to the error trapper. The error trapper realizes the ((.)/x) mod g(x) operation in Galois field. Each result will be compared with the FEC error pattern. If the comparison is correct, it means the error can be corrected. The error pattern and the error position will be recorded and sent to the error pattern correction block. The error pattern correction block reads the receiving data stored in cache and do exclusive-OR operation with the error pattern in error position. Thus decoding process is finished.

Only syndrome is working before the frame is in synchronous state. The error trapper and the error pattern correction block will only be started after the frame synchronization.

The syndrome generator includes 32 bit registers, 32 exclusive-OR gates and 31 AND gates. The inputs of each AND gate are MSB bit D31 and related coefficient gx. The output of each AND gate is then exclusive-ORed with the lower bit register, the result is sent to higher bit register. So the syndrome generator circuit is a 32 stage linear feedback shift register (LFSR). After all frame data input to the syndrome generator, all register outputs are then input to error trapper. If data which input to the syndrome generator are all one frame data, then all register outputs are invalid. At the same time, the syndrome generator also includes some other circuit which is not included in Fig. 1, and the circuits are used to generate syndrome control signal indicating whether current syndrome output is valid or not.

Conventional error trapper circuit is composed of 32 bit register, 31 exclusive-OR gates, 31 AND gates and 32 multiplexes. The error trapper in Fig. 1 is also a 32-stage LFSR.

This paper proposes a novel error trapper circuit which can work in both syndrome generator and error trapper modes by changing the endian mode.

2.2 Proposed Error Trapper Circuit Structure

Fig. 2 shows the proposed error trapper circuit structure.

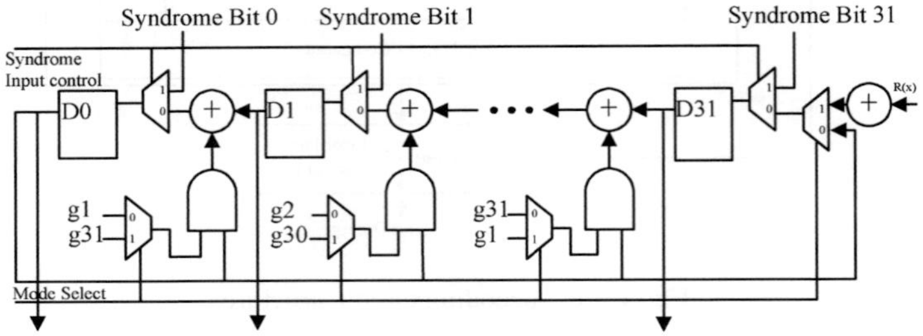

Fig. 2. Proposed error trapper circuit

In Fig. 2, the switch of little endian mode and big endian mode is controlled by endian controller. Mode select signal determines whether coefficient gx or g(32-x) is input to the AND gate.

When in normal mode, the mode select is set '0', the error trapper operates in little endian mode. The circuit function equals to the conventional error trapper. It receives the syndrome generated by syndrome generator in the order that the MSB is D31 and LSB is D0. The error pattern is in the same order.

When the mode select is set '1', and syndrome input control is set '0', the error trapper works in big endian mode. It realizes the function of syndrome generator. The error trapper's equivalent circuit is shown in Fig. 3 when it works in big endian mode.

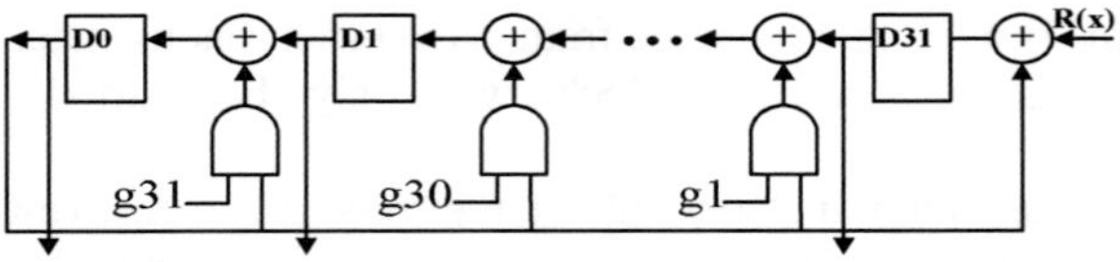

Fig. 3. Error trapper equivalent circuit in big endian mode

In Fig. 3, each parameter of AND gate node is inverted compared with that of normal mode. Codeword are input from the right side of LFSR. Syndromes are output in parallel from D0:D31 when FEC parity checking is finished. The MSB bit of output order is in D0 and LSB bit is in D31.

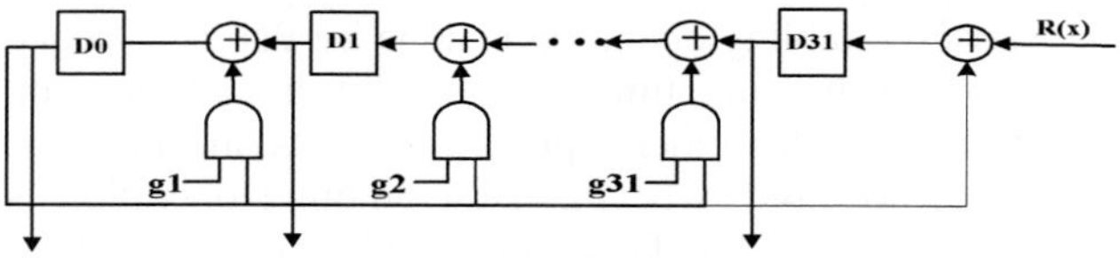

Fig. 4. Error trapper equivalent circuit in little endian mode

2.3 FEC Receiving Layer Structure with Fast Frame Synchronous Methodology

Fig. 5 shows the FEC receiving layer structure in different working mode with the modified error trapper integrated into the FEC decoder system. Fig. 5(a) is the structure when frame is not in synchronous state, and Fig. 5(b) is the structure when frame is in synchronous state.

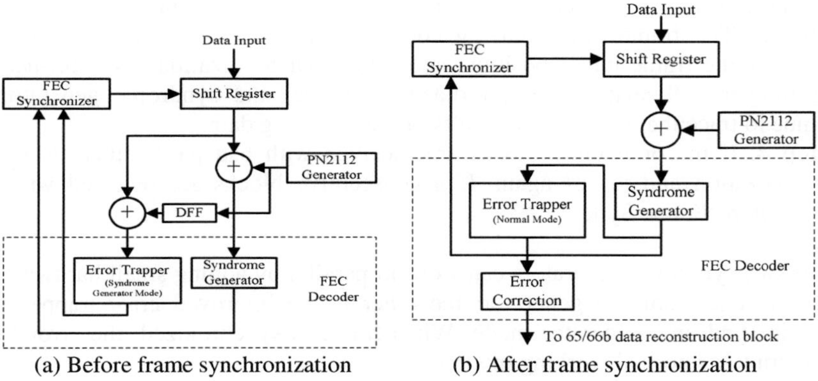

Fig. 5. The FEC fast frame synchronous structure

In Fig. 5, the FEC receiving structure includes a data stream synchronous system for Ethernet forward error correction. Fig. 5 can be used for both one bit input and multi-bit input. In Fig. 5(a), the modified error trapper works in syndrome generator mode. When one frame and one bit data (2113 bits) input to the FEC receiving layer during frame synchronizing operation, bit (0,2111) will be sent to syndrome generator and do syndrome parity check after it does exclusive-OR operation with PN-2112 sequence. Bit (1, 2112) will do exclusive-OR operation with one clock cycle delayed PN2112 sequence since it is one bit later than bit (0, 2111). Then it will be done syndrome check in error trapper which works in syndrome generator mode. After synchronization, as in Fig. 5(b), the error trapper restore to its normal working mode that is error pattern correction mode. It is the same as the current FEC decoding system.

The improved synchronous scheme is as following:

Step 1. Detect two consecutive supposed frame boundary positions. Suppose a frame boundary by shifting register, and define the supposed frame boundary to be the initial frame F1, descrambling operation is done by PN-2112 sequence generated by PN-2112 generator. At the same time, shift one bit of F1 frame boundary and the new position is defined to be the other supposed frame boundary. Suppose the frame which starts from this position to be F2. Descrambling operation is done by the delayed PN-2112 sequence which delays 1 bit. Syndrome generator does FEC check for descrambled F1, and error trapper which works in syndrome generator mode does FEC check for F2. If the two results fail, then shift the candidate frame to skip 2 bit position and try again.

Step 2. For a supposed frame boundary, the synchronizing controller will confirm the FEC check results of n consecutive frame according to the output results of syndrome generator. If the checking result of any one of the F1, F2 frame is correct, the synchronizing controller will trace the following n frames after the correct frame. If any one of the n frames has errors, the synchronizing controller will control the shift register to skip 2 bits, and then start the entire frame synchronizing process. If all FEC check results of the n consecutive frames are correct, then go to step 3.

Step 3. The frame synchronizing controller will take the correct frame initial position as the frame initial position, and frame synchronization is established. The error trapper will switch to error diagnosis mode, and syndrome generator will continue to trace the frame check results of the receiving data.

Step 4. If m consecutive blocks are received with bad parity, then drop frame synchronization and restart again. If n consecutive blocks are received with good parity, report 'Frame Sync'.

The two syndrome generators can work in parallel by adding small hardware. One is the current syndrome generator; the other is the improved error trapper which works in syndrome generator mode. When frame is synchronized, the error trapper can return to its normal working mode.

3 Experimental Results

The proposed structure is realized by Verilog HDL, and the chip has been fabricated in 65 nm CMOS process. In order to verify the FEC method, a kind of network device is designed to carry out the FEC method. An E1 over Gigabit Ethernet system is realized in the network. First, non FEC end-to-end performance is tested. Then, end-to-end performance using the FEC method is also tested by setting up the FEC configuration. Test result shows that the fast frame synchronous FEC method improves the frame boundary detecting speed greatly. Compared to conventional method, two syndrome generators can work in parallel mode with only small extra hardware overhead. It only needs to shift 1055 times in worst case, the decoding system will detect the correct frame boundary. Table 1 shows the end-to-end performance comparison result.

Table 1. End–to-end performance comparison result

Back ground flux input		622Mb/s	2.5Gb/s	10Gb/s
No FEC Method	Error code	Few	Many	Many
	AIS light	Blink	Blink	On
	Connection	OK	Fail	Fail
	Voice	Bad	-	-
Using FEC Method	Error code	Null	Null	Few
	AIS light	Off	Off	Off
	Connection	OK	OK	OK
	Voice	Good	Good	Good

In Table 1, "Connection" represents whether the telephone connection can be built or not. "AIS" is the alarm indication signal light in the device, and "Voice" represents the voice quality. From Table 1, it is clearly that end-to-end performance is greatly improved using the fast frame synchronous FEC method.

4 Conclusion

This paper presented a novel forward error correction decoding structure for 10 Gb/s Ethernet networks. With the fast frame synchronous methodology, the synchronous time is only half that of conventional method. Test result shows that the FEC decoding method improves the Ethernet performance obviously. In the meantime, the hardware overhead brought by the FEC method is very small.

References

1. Khermosh, L., Maislos, A., Haran, O.: Forward error correction coding in Ethernet networks. U.S. Patent, 7,343,540, March 11 (2008)
2. Pato, S., Monteiro, P., Silva, H.: Impact of Mode-partition Noise in the Performance of 10Gbit/s Ethernet Passive Optical Networks. In: 9th International Conference on Transparent Optical Networks ICTON 2007, Rome, July 1-5, pp. 67–70 (2007)
3. Lee, S.J., Goel, M., Zhu, Y., et al.: Forward error correction decoding for WiMAX and 3GPP LTE Modems. In: IEEE 42nd Asilomar Conference on Signals, Systems and Computers, October 26-29, pp. 1143–1147. Pacific Grove, CA (2008)
4. Zhong, Y., Yang, H., Prabhakar, A.: A Vlsi Implementation Of A Fec Decoding System For Dtmb(Gb20600-2006) Standard. In: 7th International Conference on ASICON 2007, Guilin, China, October 22-25, pp. 926–929 (2007)
5. Pato, S., Silva, H., Monteiro, P.: Forward error correction in 10 Gbits/s Ethernet passive optical networks. Journal of Optical Networking 8(1), 84–94 (2009)

6. Dai, J., Yu, S.H.: An adaptive forward error correction method for TDM over Ethernet. In: International Conference on Wireless Communications, Networking and Mobile Computing, Shanghai, China, September 21-25, pp. 1393–1397 (2007)
7. Akyildiz, I.F., Joe, I., Driver, H., Ho, Y.L.: A new adaptive FEC scheme for wireless ATM networks. In: Military Communications Conference, Boston, MA, October 18-21, pp. 277–281 (1998)
8. Schramm, P., Wachsmann, U.: Efficient system and method for forward error correction. U.S. Patent, 6,553,540, April 22 (2003)
9. Cole, R.M., Bishop, J.E.: System and method for forward error correction. U.S. Patent, 7,076,724, July 11 (2006)
10. Merritt, D.: Error Correction apparatus and method. U.S. Patent, 7,296,204, November 13 (2007)
11. Khermosh, L.: Method of ethernet frame forward correction initialization and auto-negotiation. U.S. Patent, 7,555,214, June 30 (2009)
12. Lin, S., Costello, D.: Error Control Coding: Fundamentals and Applications. Prentice Hall, Englewood Cliffs (1983)
13. Nuyen, H.C., Kramer, G., Hirth, R.E.: Method and apparatus for delineating data in an FEC-coded ethernet frame. U.S. Patent, 7,152,199, December 19 (2006)
14. IEEE 802.3 - 2008. IEEE standard for information technology - telecommunications and information exchange between systems-local and metropolitan area networks specific requirements, Part 3: Carrier sense multiple access with collision detection (CSMA/CD) access method and physical layer specifications (2008)

Simple Feedback with Priority List Radio Resource Scheduling Scheme for 3GPP LTE Networks

Modar Safir Shbat, Md. Rajibur Rahaman Khan, and Vyacheslav Tuzlukov

College of IT Engineering, Electronics Engineering Department
Kyungpook National University,
1370 Sankyuk-dong, Buk-gu, Daegu 702-701, South Korea
modboss80@gmail.com, rajibur_ckt@yahoo.com,
tuzlukov@ee.knu.ac.kr

Abstract. Self organizing network (SON) techniques, algorithms and eventually standards are critical steps in long term evolution (LTE) femtocell deployments. Third generation partnership project (3GPP) standardizes self-optimizing and self-organizing capabilities for LTE in order to provide the upcoming requirements of broadband mobility, development of SON, inter cell interference (ICI) elimination, fairness, and quality of service (QoS). The need of intelligent radio resource management (RRM) allows the network to have a full control over services and the physical radio blocks (PRBs). These features should be deeply studied under designing the signaling protocols and RRM algorithms. In this paper, important radio resource scheduling aspects are discussed. New feedback scheme for channel quality information (CQI) and a new PRBs scheduling method using a priority list created by the RRM algorithm are introduced satisfying the predefined requirements of the RRM in 3GPP LTE networks.

Keywords: Self organizing network (SON) in long term evolution (LTE), Radio resource management (RRM), Decreasing time algorithm (DTA), Simple feedback.

1 Introduction

Recent increase in mobile data usage and a demand of new applications have motivated the third Generation Partnership Project (3GPP) to work on the Long-Term Evolution (LTE) as the latest standard in the mobile network technology. Future network architectures must be efficiently flexible to support scalability, as well as reconfigurable network elements, in order to provide the best resource management solutions in hand under effective network employment with low cost. The ultimate target is to increase the valuable spectrum efficiency using more flexible and effective spectrum allocation and radio scheduling scheme to optimize the QoS, maximize system capacity, and satisfy the self-organizing network (SON) requirements.

Radio resources are scheduled every 1ms in 3GPP LTE network and different frequency bandwidths and/or aggregated bandwidths can be assigned to an individual user based on the channel condition and availability. Owing to rapidly and instantaneously

X. Wan (Ed.): Electrical Power Systems and Computers, LNEE 99, pp. 113–121.
springerlink.com

changing nature of radio channel quality, there must be a sufficiently fast scheduling algorithm to compensate the changing channel conditions. Before assigning the modulation technique and coding rate to user equipment (UE) by eNodeB (the base station (BS) in the LTE network) based on the transmission channel condition, there must be defined the physical radio resource blocks (PRBs). Thus, the task of scheduling and distribution of the PRBs in 3GPP LTE among users is a complicated process. This paper investigates the most important enabling technologies to support the radio resources scheduling and ICI cancelation such as frequency reuse and channel quality information (CQI), and presents a simple feedback concept form of the UE to the eNodeB, and also introduces a PRB scheduling method to be considered in the radio resource management (RRM) algorithm of LTE networks with the propose to achieve the balance or tradeoff between the simplicity, fairness, and ICI elimination. The rest of this paper is organized as follows: the system model and problem statement are discussed in Section 2. Section 3 introduces the new simple feedback and related physical radio blocks scheduling scheme. The introduced scheduling general analysis and problems are presented in Section 4. The conclusion remarks are discussed in Section 5.

2 The System Model and Problem Statement

None PRBs scheduling algorithms can solve all existing problems associated with the maximum number of users with available transmission services, also the limited and imperfect channel information used at the BS, QoS, and fairness problems. The most important requirements are the accurate channel state information (CSI) at the BS that is not easy to obtain according to limited feedback channel capacity. Additionally a fairness providing is the common wireless networks problem. CSI has a strong impact on the network performance metrics, in particular, the delay, which determines the QoS for many applications. We assume that K_{au} is the total number of active users in the cell, that says us about a feedback quality which is sufficiently good to perform the down link transmission. K_{fb} is the number of users that does feedback and we call such user the basic-feedback user. Some scheduling algorithm is needed to choose K_{ch} (the number of chosen users from the total number of users) to serve simultaneously. Evidently, the number of chosen users is between K_{au} and K_{fb}, i.e.

$$K_{au} \le K_{ch} < K_{fb} \tag{1}$$

Different kinds of parameters will affect the decision of the scheduling algorithms to choose the number of users. Some parameters can serve as the indicator of the feedback quality and its threshold. Choice of the number of users to serve based on the feedback quality requires an optimal transmission environment and decreases the possible number of users with good feedback quality to be served from the total number of users that really do and send feedback to the BS. We can always assume that K_{ch} is lesser K_{fb} in the current scheduling algorithms. The previous scenario works well against the fairness in the service and the maximum number of users that are under service. This means a default of the LTE system throughput and deterioration in spectral efficiency defined by the fixed number of feedback bits X, quality indicator ρ and its threshold δ.

The main idea to employ a frequency reuse is to assign the same frequency band in different cells that are usually far from each other to avoid high interference between neighboring cells. We can significantly improve the signal-to-interference-noise ratio (SINR) without using the same frequency band for neighboring cells [1]. Unfortunately, this improvement in SINR causes a reduction in the available spectrum per cell. The system capacity can be estimated using Shannon's formula [2]:

$$TP_k = \frac{BW}{K}\log_2(1+SINR_k), \tag{2}$$

where k is the reuse factor meaning that only $1/k$th part of the spectrum can be used by a single cell, BW is the LTE total bandwidth in Hz, $SINR_k$ is the $SINR$ with reuse k. $SINR$ is given by [3]:

$$SINR = \frac{P_r}{P_{intracell} + P_{intercell} + N_0}, \tag{3}$$

where P_r is the received power density from the user, $P_{intracell}$ is the interference that comes from users inside the cell, $P_{intercell}$ is the interference from neighboring cells, and N_0 is the noise power.

In order to have a beneficial frequency reuse, an appropriate tradeoff between the bandwidth and SINR is important to utilize the spectrum in efficient way by setting a frequency reuse factor to proper value and to maximize the cell/user throughput. The frequency reuse factor should be chosen according to intercell interference level that depends on the cell size. Powerful interference favors a high reuse factor and vice-versa. In this paper, a soft frequency reuse (SFR) is used. This technique consist of splitting the bandwidth into two parts, namely, the full reuse (FR) and partial reuse (PR) parts. The FR part uses the reuse factor equal to 1 and the PR part is allocated to the cell edge-users. This structure allows us to have two level allocation scheme (TLA), where the first level is the cell-level resource allocation (CRA) and the second level is the user-level resource allocation (URA). It means that the cell users are divided into two categories, namely, the cell centre user (CCU) and cell edge user (CEU). This classification can be done using the geometry factor G:

$$G = \frac{P_{serving}}{N + P_{non-serving}}, \tag{4}$$

where $P_{serving}$ is the total power generated by the connected BS, $P_{non-serving}$ is the total power received from all BSs served as the interference sources, and N is the portion of the power from BSs that can be modeled as AWGN.

SFR is the applying frequency reuse factor (FRF) of 1for CCUs and FRF of 3 to CEUs [4]. One third of the whole available bandwidth named the major segment can be used by CEUs where the packets should be sent with higher power. CCUs can access the entire physical radio resources with lower transmission power. To realize FRF of 3 for CEUs, the major segments among directly neighboring cells should be orthogonal (Fig.1). The power allocation for each type of users can be determined as:

$$P_{CCU} = \frac{S\,P}{(\alpha-1)T+S} = \frac{3P}{\alpha+2}, \tag{5}$$

$$P_{CEU} = \alpha\,P_{CCU}, \tag{6}$$

where S is the total number of subchannels in LTE system, T is the number of available subchannels for the CEUs, α is the power ratio between the subchannel used by CEU to the subchannel used by CCU, and P is the reference power signifying the uniform transmit power used by each subchannel in a classical reuse-1 system. We can see that when α equals 1, P_{CCU} is equal to P_{CEU}, and the SFR is a reuse-1 system. As $\alpha \rightarrow \infty$, P_{CCU} and P_{CEU} will converge to 0 and $3P$, respectively, and the SFR becomes a reuse-3 system.

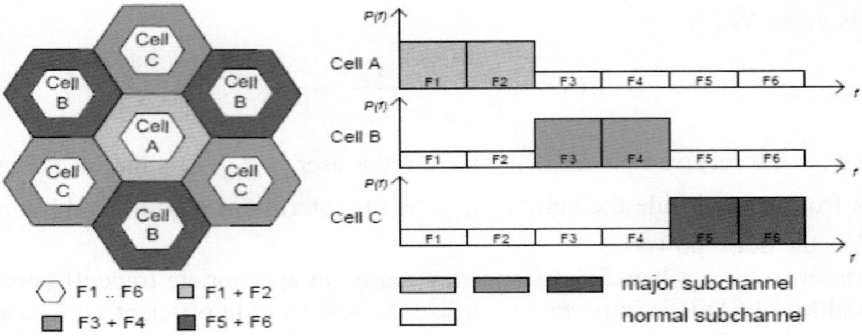

Fig. 1. Concept of the SFR scheme in LTE network.

The introduced SFR scheme (also called reuse 1/3) has low complexity and good performance for CEUs. Additionally, it has two main drawbacks, namely, the signaling overhead and overall loss of throughput. In the next section, we try to overcome these drawbacks.

3 The Limited Feedback and PRBs Scheduling Scheme

3.1 The Feedback Scheme

Since the channel quality information (CQI) has to be available at BS (eNodeB), the feedback information can be used for partitioning users. Another important topic here is the required number of feedback bits to cover and achieve optimal scenario for LTE system and, additionally, to reduce the signaling overhead problem. The number of feedback bits is the indicator of the feedback quality and is used by the BS (transmitter) to define the served users from the total number of users sending feedback to the BS. Based on the previous statement, we see that to apply any kind of scheduling scheme there is a need to evaluate the feedback quality and to decide the user should be served or not. In this case, the less the number of the feedback bits the less complexity and the best stability in the scheduling model.

In this case, each eNodeB receives only one bit from each user instead of full information about SINR (in the case of MIMO system, UE sends information about the SINR for the best beam of every antenna element and this feedback consists of $N_{real} + N_{integer}$ numbers). This bit indicates either SINR of the receive antenna is over a given value (threshold) or not. Now, the transmission by M beams of the BS transmitter is carried out using a single bit of feedback from each user which measures the SINR and compares it with a predetermined constant threshold δ. The threshold δ is considered as a network parameter known by the BS and all users. The only bit ("0" or "1") as a feedback from the user can inform the BS either SINR exceeds the threshold value or not. For pre-introduced LTE system, this threshold can be adjusted to indicate the CCUs at SINR $> \delta$ ("0"), otherwise – CEUs. Another scenario is to use two thresholds, δ_1 in the case of "0" feedback bit and CCUs, and δ_2 in the case of "1" feedback bit and CEUs. After receiving the previous simple feedback from all users, the BS will schedule a radio resource block or blocks for each user. The presented method is simple and ensures an effectiveness to decrease the signaling complexity of the network.

3.2 PRBs Scheduling Scheme

For each transmission time interval (TTI) and in each cell, the BS scheduler assigns the PRBs to the UE that have to be served. LTE network uses adaptive coding and modulation (ACM) per resource block. Thus, the scheduler determines the modulation type for the PRBs assignment process. The proposed scheme can be implemented as a scheduling algorithm creating a priority list. There are many possible strategies leading us to create the priority list. We consider only the decreasing-time algorithm (DTA). This algorithm handles the BS to put all users with the feedback when SINR $> \delta$ to be among CCUs, and with the feedback when SINR$< \delta$ to be among CEUs in the priority list.

DTA is based on a seemingly simple strategy, namely, at first, we should carry out the longer task transmission to a single user and save the shorter task in the last. DTA creates a priority of tasks in decreasing order of transmitting time, namely, the longest task is the first, and the shortest task is the last. Tasks with equal transmitting time can be listed in any order in the list. The priority list created by DTA is often called the decreasing-time list. There is a need to think about a level of fairness for the PRBs scheduling process. Different kind of fairness concepts are introduced, such as, a multiplication between each user throughput and an individual proportional fair factor [5] and the optimization method [6]. DTA strategy keeps all the users in the cell of the priority list and assigns PRBs. This procedure guarantees a certain minimum rate for each user. By this way, the overall system throughput is maximized while assuring that none of the users gets more than a certain maximum rate determined before. To calculate the expected throughput of any user, we can use the optimization method [6]. This model introduces the binary user/resource block assignment variable x that is "1" if the user m obtains resource block r and "0" otherwise. The expected throughput of the user m using the block r depends on the expected SINR. The expected SINR is derived from the latest SINR measurement. Thus, the expected throughput can be presented in the following form:

$$\hat{THR}_{m,r} = \Delta f \log_2(1 + \hat{SINR}_{m,r}), \tag{7}$$

where Δf is the resource block bandwidth. For the QoS criterion, we should take into account the guaranteed bit rate (GBR) as the only criterion under different services.

Based on user's GBR and CSI, the required number of PRBs for each user can be determined as [7]:

$$N_m = \frac{GBR_m}{M\ BW_{PRB}\ S_m} \quad (8)$$

$$S_m = \log_2(1 + \overline{SNIR_m}), \quad (9)$$

where $\overline{SNIR_m}$ is the average $SINR$ for user m over whole frequency band, S_m is the spectral efficiency of the user m, BW_{PRB} is the bandwidth of a PRB, M is the number of OFDM symbols in a PRB, and N_m *is* the required number of PRBs per TTI by the user m. The basic admission control criterion can be presented as the sum of PRBs per TTI required by new user requesting admission and the number of active users in the cell, and should be less than or equal to the total number of PRBs in the LTE system. This admission criterion can be presented in the following form:

$$\sum_{i=1}^{k} N_m + N_{new} \leq N_{total}. \quad (10)$$

4 The Scheduling Scheme General Analysis and Problems

4.1 DTA Problem

DTA ignores any information in the case if one or more users should be served earlier. For instance, if one or more users with long service time cannot begin the transmission process until the user with a very short service time is finished, then assigning the user with the short time will probably result in a shorter finishing time even though assigning the short time user violates the DTA. This problem can be solved by using the critical-path algorithm (CPA) that is based on a strategy similar to that of the DTA. This algorithm states: there is a need to serve the user with the largest critical serving time first and serve the user with shortest critical service time last. CPA creates the priority list by listing the users in decreasing order of critical service time. Users with equal critical service time can be listed in any order.

4.2 LTE Network with Enabled MIMO Technology

In the case of MIMO technology employment in LTE network, when the number of users equals to n with the fixed number of transmit antennas M and any number N receive antennas, and the beamforming of LTE system based on binary quantization of SINR, the system should achieve the typical scaling law of the sum rate (throughput) to use the proposed feedback scheme [8]. An important case should be considered in the scheme, namely, there is a small probability that at any time one beam (or PRB) is not requested from any user that means no user measures SINR on that beam. To solve this case we assume that the eNodeB assigns this unrequested beam with the purpose to establish a communication with a user picked up randomly from the total number of users in the cell. Let P_m be the probability of any beam m to be requested and be calculated by the following formula:

$$P_m = 1 - (F(\delta))^{\frac{nN}{M}}, \tag{11}$$

where δ is the network threshold, and *F(x)* is the cumulative distribution function of the SINR:

$$F(x) = 1 - \frac{e^{-x/\rho}}{(1+x)^{M+1}}, \tag{12}$$

ρ is the quality indicator. Now we can define the available throughput *R* in the following form:

$$R \geq MP_m \,\mathrm{E}\left[\log(1+S) \middle| S > \delta\right], \tag{13}$$

where *S* is the SINR identically distributed for all users and beams.

4.3 SFR Scheme

There are three major frequency reuse schemes that can be used in LTE networks to cancel ICI effects, namely, the hard frequency reuse (HFR) with the fixed frequency reuse factor (1 or 3 are popular); the partial frequency reuse (PFR); and the soft frequency reuse (SFR) that is a part of the discussed scheme in this paper. A simple LTE system level simulation shows us that the SFR has a good performance in terms of total throughput in the cell (Fig.2).

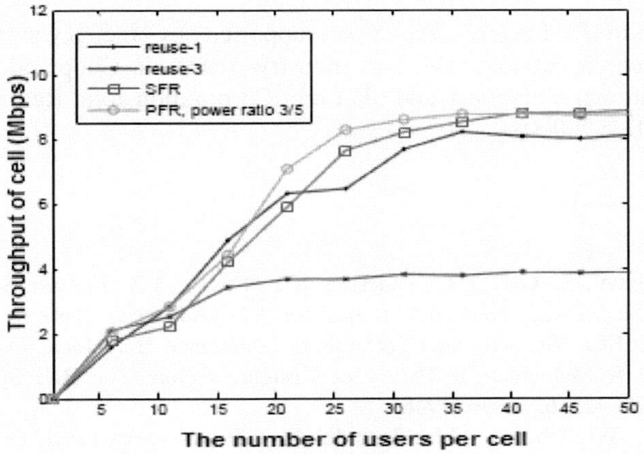

Fig. 2. Comparison between the frequency reuse schemes.

It is proposed in the 3GPP LTE networks that every 1 ms the radio resources in the cell should be scheduled. Thus, the way to speed up the scheduling process is very essential and important. SFR processing load is acceptable with good performance, especially for the cell edge users. Another frequency reuse schemes are introduced and some of them may have better performance than SFR, but there are some disadvantages in complexity and high processing load in the eNodeBs.

5 Conclusion

Based on engineering viewpoint, the best possible solution can be achieved only when elements of the radio network are properly configured and suitable radio resource management approaches/algorithms are applied for any PRBs scheduling scheme. In a multicell scenario, the intercell interference is the major limiting factor for the LTE system performance. This kind of interference becomes critical for the users that are near the cell-edge where the interference from neighboring cells is usually very high. SFR has a good performance in both the average cell throughput and cell edge user throughput. DTA and the simple feedback scheme work jointly to reduce the signaling overhead and speed up the scheduling process employing the simple feedback method instead of the geometry factor to distinguish the cell-center users (CCUs) and cell edge users (CEUs) between each other. Further improvement applying different PRBs assignment methods can be done in the form of semistatic versions of SFR. It means that the frequency resource configuration is adjusted on a time scale corresponding to definite interval, for example, some seconds or longer, that makes the resource partition adaptive to the traffic load variety. This procedure leads us to more complicated and more signaling and proceeding load in the system. For the proposed scheduling scheme, a deep analysis must be done to assure that we obtain an acceptable average scheduling delay, and are able to study the scheduling speed, the average throughput of the cell, and the achieved fairness level to evaluate the overall performance of the LTE system and find the weak points with the purpose to define appropriate scheduling scheme which can avoid these undesired problems.

Acknowledgments. This research was supported by the Kyungpook National University Research Grand, 2009, and Industry-Academic Cooperation Foundation, Kyungpook National University and SL Light Corporation Joint Research Grant (the Grant No. 201014590000).

References

1. Wang, Y., Kumar, S., Garcia, L., Pedersen, K.I., Kovács, I.Z., Frattasi, S., Marchetti, N., Mogensen, P.E.: Fixed Frequency Reuse for LTE-Advanced Systems in Local Area Scenarios. In: IEEE 69th Vehicular Technology Conference, Barcelona, Spain (2009)
2. Shannon, C.: A Mathematical Theory of Communication. The Bell System Technical Journal 27, 379–423, 623–656 (1948)
3. Krasniqi, B., Wrulich, M., Mecklenbräuker, C.F.: Network-Load Dependent Partial Frequency Reuse for LTE. In: 9th International Symposium on Communication and Information Technology (ISCIT 2009), pp. 672–676 (2009)
4. Xie, Z., Walke, B.: Enhanced Fractional Frequency Reuse to Increase Capacity of OFDMA Systems. In: NTMS 2009 the 3rd international conference on New technologies, mobility and security, NJ, USA (2009)
5. Wang, Q., Xu, J., Bu, Z.: Proportional-fair bit and power adaptation in multi-user OFDM systems. In: The IEEE International Symposium on Personal, Indoor and Mobile Radio Communications (PIMRC), Helsinki, Finland, pp. 1–4 (2006)

6. Bohge, M., Grossy, J., Wolisz, A.: Optimal Power Masking in Soft Frequency Reuse based OFDMA Networks. In: The European Wireless Conference, pp. 162–166 (2009)
7. Lu, Z., Tian, H., Sun, Q., Huang, B., Zheng, S.: An Admission Control Strategy for Soft Frequency Reuse Deployment of LTE Systems. In: The 7th IEEE Conference on Consumer communications and networking conference CCNC 2010 (2010)
8. Diaz, J., Osvaldo, S., Bar-Ness, Y.: How Many Bits of Feedback is Multiuser Diversity Worth in MIMO Downlink? In: IEEE Ninth International Symposium on Spread Spectrum Techniques and Applications, pp. 505–509 (2010)

Image Threshold Segmentation Technology Research Based on Adaptive Genetic Algorithm

Deying Gu[1] and Zhiliang Ren[2]

[1] Northeastern University at Qinhuangdao, Qinhuangdao, China
gdy0335@163.com
[2] Northeastern University, Shenyang 110819, China
z_l_ren@126.com

Abstract. On the basis of the OTSU methods study, the paper introduced adaptive genetic algorithm to optimize algorithm and achieve image segmentation. Experiment shows that the speed of the algorithm improves and the quality of segmentation is better. Lay the foundation for image recognition in the following.

Keywords: Image Segmentation, Adaptive Genetic Algorithm, OTSU Method.

1 Introduction

Image segmentation is pretreatment stage of pattern recognition. It is the key step from image processing to image analysis. Understanding in computer vision, including edge detection, feature extraction and target recognition, etc. all depends on image segmentation quality. The image processing and recognition effect is directly influenced by Image segmentation. Image segmentation is not only one of the most important problems in image processing, but also a typical problem in the study of computer vision. So far we have failed to find a general method to image segmentation, either an objective standard to judge the success of image segmentation [1].

Image segmentation is the process of dividing target into many regions of interest and extracting interested part of them. There are so many image segmentation methods, but it is hard to find a suitable method for each occasion. The segmentation methods commonly used include the threshold value method, edge detection and regional tracking method. The most commonly used method is the threshold value method. Threshold value method is to set a gray image threshold , the gray value which is less than the given threshold is set to 0, greater is set to 255. The transformation function expression of the threshold value method as follows:

$$f(x) = \begin{cases} 0, & x < \mathrm{t} \\ 255, & x > t \end{cases} \tag{1}$$

Where, t represents the specified threshold.

Using threshold value method in combination with other algorithm can get good segmentation results. Genetic algorithm is a parallel, efficient and global searching method. It is very suitable for large-scale searching space optimization because of its

X. Wan (Ed.): Electrical Power Systems and Computers, LNEE 99, pp. 123–130.
springerlink.com

inherent robustness, parallelism and adaptability. It has been widely used in many subjects and engineering fields. The application of computer vision is being valued. Using genetic algorithm in image segmentation can greatly shorten the time interval to search for the threshold, more effective especially when the searching space is larger.

This paper adopts the maximum infra-class variance method optimized by genetic algorithm to solve the threshold selection problem in image segmentation.

2 Algorithm

One of the most important problems in image segmentation is the selection of threshold value. Among the most commonly used threshold value selection methods such as the maximum infra-class variance method (OTSU method), the maximum histogram entropy method, the minimum error method, the gray histogram method, the occurrence matrix method, the correlation coefficient method and so on[2]. OTSU method is considered as one of the most outstanding method.

2.1 OTSU Method

The maximum infra-class variance method (OTSU method) put forward by OTSU in 1978 is a threshold value method attracted much more attention. It is derived on the theory of the judgment analysis or least square method. Based on gray histogram, this algorithm divide the image into target and background, then confirm the image segmentation threshold through the maximum infra-class variance.

According to the gray threshold value, pixel of image is divided into target and background parts. The pixel with gray value in the interval [0,t] belongs to the target part, and the pixel with gray value in the interval [t+1,L-1]belongs to the background part. Assume that represents the sum of grey value and Pi represents the probability that the gray value is i . Then:

$$p_i = \frac{g_i}{M \times N} \tag{2}$$

The probability of target and background part is:

$$\omega_0 = \sum_{i=0}^{t-1} p_i \tag{3}$$

$$\omega_1 = 1 - \omega_0 \tag{4}$$

The average gray of target and background is:

$$\mu_0 = \frac{\sum_{i=0}^{t-1} i p_i}{\omega_0} \tag{5}$$

$$\mu_1 = \frac{\sum_{i=t}^{L-1} ip_i}{\omega_1} \tag{6}$$

The average of the total pixels:

$$\mu = \omega_0\mu_0 + \omega_1\mu_1 = \sum_{i=0}^{L-1} ip_i \tag{7}$$

The optimal threshold is:

$$T = Arg \max_{0<t<L-1} \left[\omega_0(\mu_0 - \mu)^2 + \omega_1(\mu_1 - \mu)^2\right] = Arg \max_{0<t<L-1} \left[\omega_0\omega_1(\mu_0 - \mu_1)^2\right] \tag{8}$$

The expressions in brackets on the right side of the equation are actually infra-class variance. The target and background split out by the threshold t are two parts which constitute the whole image. As the variance is a measure of gray distribution uniformity, maximum variance between the target and background means maximum difference and minimum rate of mistake, this is the essence of OTSU method [3].

2.2 Genetic Algorithm

Genetic algorithm (GA) abstracted from the evolution of biology. Through the simulation of natural selection and genetic mechanism, it formed a kind of search algorithm with "generation + inspection" characteristics. GA is put forward earliest by professor Holland of Michigan University in 1962. Basic idea of GA is from Darwin's evolutionism and Mendel's genetic doctrine. Because of the global search ability, the more complex the problems needed to be solved is, the bigger superiority of GA is. In GA a problem resolution can be expressed as "chromosome". Particularly in standard genetic algorithm (SGA), it generally is expressed as binary code strings. The particular set of solution is called "population" and the variable of solution is called "gene". According to the survival of the fittest principle, it place the "population" in question "environment", then choose adaptive "chromosome" from the environment and produce new generation of "chromosome" group that is more adaptive through selection, crossover and mutation operation. Just like this, generation after generation, it finally comes out to the most adaptive individual of the environment and that is the optimal solution of the problem.

For SGA Goldberg provided its pseudo-code describe as follows: [4]

```
SGA Procedure
Begin
   T=0;
```

The expressions in brackets on the right side of the equation are actually infra-class variance. The target and background split out by the threshold t are two parts which constitute the whole image. As the variance is a measure of gray distribution uniformity, maximum variance between the target and background means maximum difference and minimum rate of mistake, this is the essence of OTSU method [3].

2.3 Genetic Algorithm

Genetic algorithm (GA) abstracted from the evolution of biology. Through the simulation of natural selection and genetic mechanism, it formed a kind of search algorithm with "generation + inspection" characteristics. GA is put forward earliest by professor Holland of Michigan University in 1962. Basic idea of GA is from Darwin's evolutionism and Mendel's genetic doctrine. Because of the global search ability, the more complex the problems needed to be solved is, the bigger superiority of GA is. In GA a problem resolution can be expressed as "chromosome". Particularly in standard genetic algorithm (SGA), it generally is expressed as binary code strings. The particular set of solution is called "population" and the variable of solution is called "gene". According to the survival of the fittest principle, it place the "population" in question "environment", then choose adaptive "chromosome" from the environment and produce new generation of "chromosome" group that is more adaptive through selection, crossover and mutation operation. Just like this, generation after generation, it finally comes out to the most adaptive individual of the environment and that is the optimal solution of the problem.

For SGA Goldberg provided its pseudo-code describe as follows: [4]

```
SGA Procedure
Begin
   T=0;
Initialize P (t);
   Evaluate P (t);
   While not finished do
   Begin
      T=t+1;
      Select P (t) form P (t-1);
      Reproduce pairs in P (t);
      Evaluate P (t);
   End
End
```

The image segmentation problem can be defined as an optimization problem. Take advantage of high efficiency of GA, it can search out the optimal image segmentation threshold which can obtain the top-quality segmentation or separate the target and background well.

2.4 Adaptive Genetic Algorithm Optimize OTSU Method

Optimizing or improving existing OTSU method with the global search capability of GA, we can fix the threshold. In order to put GA into image threshold segmentation discussed in this paper, several key problems are discussed as follows:

- Encode

  Adopt binary coding as the coding method. GA is not act directly on the question solution space, but on a coded representations space. Coding method will have a dramatic influence on the performance and efficiency of the algorithm. Due to there is a total of 256 gray level in gray image, we use 8 bits binary numbers to represent the associated chromosome string of threshold.

- Fitness Function

  Selected fitness function must be able to reflect the evolutionary degree that the individual may get or reach the optimum solution. It is the foundation of individual operation in GA and has a direct influence on the performance and efficiency of the algorithm. The paper adopts below formula as fitness function. It is equivalent to optimal threshold formula of OTSU method. The optimization of the OTSU algorithm is mainly represented here.

$$f = \max_{0<t<L-1} \left[\omega_0 \omega_1 (\mu_0 - \mu_1)^2\right] \tag{9}$$

- Genetic Parameters

  In SGA it is easy to appear premature that algorithm converge to a local optimal solution but cannot converge to the global optimal solution. To solve this problem, we can adopt different tactics to improve SGA algorithm. For instance, make improvement on coding method, on selection, crossover and mutation operators in iterative process etc. The paper adopts an improved adaptive genetic algorithm (AGA) to optimize OTSU method. AGA uses adaptive crossover and mutation probability to take the place of fixed value

Adopt the combination mechanism of ranking selection method and roulette selection method. Introduce the preliminary select mechanism which adds the optimal value of father generation into the competition of offspring generation. First put individuals in order.

According to their fitness from large to small, then make choice by the roulette method.

Define crossover probability p_c and mutation probability p_m as follows: [5]

$$p_c = \begin{cases} \dfrac{l_1(f_m - f_c)}{f_m - \overline{f}} & f \geq \overline{f} \\ l_2 & f < \overline{f} \end{cases} \tag{10}$$

$$p_m = \begin{cases} \dfrac{l_3(f_m - f_c)}{f_m - \overline{f}} & f \geq \overline{f} \\ l_4 & f < \overline{f} \end{cases} \tag{11}$$

In the formulas, f_m is the maximum fitness value; $\overline{f}$ is the maximum fitness value of each generation group; f_c is the larger fitness value of the two individuals to crossover; f is the fitness value of the individual to variation. The value of coefficient l_1 l_2 l_3 l_4 is in the interval $[0,1]$.

2.5 Flowchart

From the above, the flowchart of algorithm can be summed up as follows:

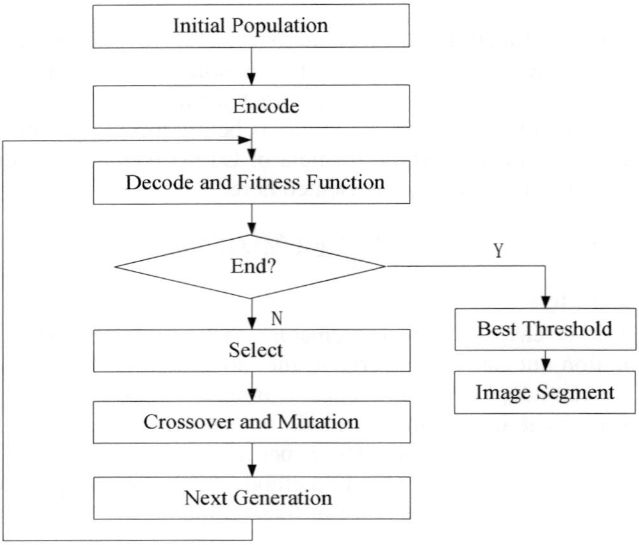

Fig. 1. The flowchart

3 Simulation

In order to validate algorithm, testing experiment is carried out on Intel(R) Core(TM) 2 2.00 GHz CPU. In the experiment, all programs are programmed with the C language and compiled by VC++6.0 software in windows 7 environment. In image segmentation experiments of 256×256 pixel grayscale Lena bitmap, OTSU method, OTSU method optimized by SGA (SGA OTSU method) and OTSU method optimized by AGA (AGA OTSU method) is used respectively.

The main parameters of GA are chromosome's length, string's length, the population size, the population of current generation, individual fitness, crossover probability, mutation probability, maximum generation and so on. In the experiments, initial population size is 20, the individuals generate randomly. Chromosome's length is 8. Total number of iterations is 50, the conditions for the termination is iterative end. The crossover probability and mutation probability is determined adaptively. Fig. 2 is the image segmentation result of bitmap Lena.

Fig. 2. The image segmentation

From left to right they are: original image, image segmentation by OTSU method, image segmentation by SGA OTSU method, image segmentation by AGA OTSU method. Table 1 shows the experiment parameters that can be used to examine algorithm quality:

Table 1. Segmentation Results Evaluation

	OTSU	SGA OTSU	AGA OTSU
threshold	101	106	116
time(s)	0.063	0.002	0.014
generation		13	24

Experiments show that:

- From Fig. 2, after segmentation by SGA OTSU method, contrast of the image is enhanced, the quality of segmentation is improved.
- From Table 1, OTSU method requires more time. The kernel of OTSU method is calculating infra-class variance, so calculating infra-class variance is necessary every time. The reason for AGA OTSU method requires more time than SGA OTSU method is that the premature of SGA OTSU method. It converges to a local optimal value.
- From Table 1, AGA OTSU method requires more iterations than SGA OTSU method. That is also because of the premature of SGA. Compare to SGA OTSU method, the convergence of AGA OTSU method is better, not easy to be premature.

4 Paper Submission

As a kind of optimized algorithm, GA is applied into image segmentation combine with OTSU method. It accelerates the convergence speed of algorithm, greatly shortens the threshold searching time and gets better segmentation quality. It improves segmentation algorithm efficiency and real-time. Because of the parallel processing ability and the global convergence characteristic, using GA in image segmentation provides a convenient for the follow-up image recognition.

References

1. Zheng, Y.J.: A Survey on Evaluation Methods for Image Segmentation. Pattern Recognition 29(8), 346–352 (1996)
2. Wu, Y.Q.: The Progress of Methods for Image Threshold Selection in Last Thirty Years (1962-1992). Journal of Data Acquisition & Procession 8(3), 193–201 (1993)

3. Huang, J.X., Liu, H., Huang, W.: A Threshold Selection Method of Image Segmentation Based on Genetic Algorithms. Journal of Nanjing Normal University (Engineering and Technology Edition) 7(1), 14–17 (2007)
4. Goldberg, D.: Genetic Algorithms in Search Optimization and Machine Learning. Addison-Wesley, Pearson, Reading, MA (1989)
5. Guo, Z., Chen, Y.Z.: Research of Threshold Methods for Image Segmentation. Journal of Communication University of China (Science and Technology) 115(2), 77–82 (2008)

Adaptive Sliding Mode Control of Networked Control Systems with Variable Time Delay

Yang Yin, Li Xia, Lizhong Song, and Mei Qian

College of Electric and Information Engineering, Naval University of Engineering, Wuhan, Hubei 430033, China
reeyan@163.com

Abstract. A new adaptive sliding mode control algorithm, based on the T-S fuzzy model, is proposed to deal with the problems of uncertainty and disturbances in networked control system (NCS) . There is variable time delay occurs between controller and actuator in the NCS. Firstly, Choosing the time delay as premise variables of the fuzzy system, the system with control time delay is transformed into the one without time delay based on the fuzzy fusion technology. And then, the global sliding surface is designed by the parallel distributed compensation technique. Adaptive reaching law and disturbances prediction methods are researched. Finally, with a certain variable time delay, the simulation results about DC motor illustrate that the proposed scheme is effective.

Keywords: networked control systems(NCS), Takagi-Sugeno model, Discrete sliding mode control, Adaptive prediction.

1 Introduction

The concept of Networked Control System (NCS) have been proposed and cause people's attention from the early 90s of last century. The characteristic of the system is that components (sensors, controllers, actuators) can exchange information through the network. Time delay, packet loss, packet transmission constraints and other issues of network related to control strategy would inevitably lead to changes. Many domestic and foreign scholars have done a lot of extensive and in-depth studies[1-4]. However, the literature considering the robustness when system has network variable time delay, uncertainty and and external interference simultaneity of the literature is not much.

Variable structure control is a system integrated control method, the most important feature is that provides a powerful deterministic control system design methods and simple structure of the controller to the uncertain object, and ensure the system has strong robustness and adaptability[5]. Variable structure control applied to the network control system is a new research ideas to improve system robustness. Literature [6] designed sliding mode predictive controller based on discrete system model and state estimation. This method use the buffer to make the uncertainty of variable time delay into determined time delay. This artificial expansion of the delay

X. Wan (Ed.): Electrical Power Systems and Computers, LNEE 99, pp. 131–138.
springerlink.com

reduced the system control performance. Literature [7] designed robust predictor based on sliding mode controller for the uncertainty of variable time delay system. Literature [8] designed Sliding mode controller based on the linear quadratic optimal methods for a class of multi-state delay and uncertainty of single-input linear control system. In the literature [7-8] controller design are based on the continuous system model, which does not match with the actual NCS, so study the discrete model directly more naturally. Literature [9] used discrete state estimation to construct delay compensation sliding hyperplane, but the proposed variable conditions $|s(k+1)|\leq|s(k)|$ can not reduce the chattering of sliding mode.

For time-varying delay exists between controller and actuator, this paper use discrete random variable delay as a prerequisite and the probability distribution of delay as the membership function to obtain no-delay global model based on T-S fuzzy methods. And in this model we design discrete adaptive variable sliding mode control strategy[10], while ensuring the system robustness and less chattering.

2 Model Established

A typical structure of networked control system is shown in Figure 1.

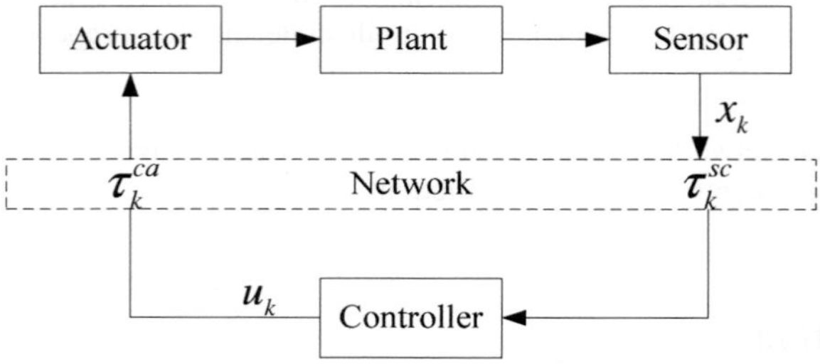

Fig. 1. Block diagram of a networked control system

Where τ_{sc} is the transmission delay of the sensor to the controller and τ_{ca} delay is the transmission delay of the controller to actuator. We assume that $\tau_{sc}=0$, such as sensor contains control unit the case. τ_{ca} is Variable. The generalized state equation of the controlled object is

$$x(k+1)=Ax(k)+Bu(k-\tau)+f(k) \tag{1}$$

Where $x\in R^n$, $u\in R^m$, $A\in R^{n\times n}$, $B\in R^{n\times m}$, $f\in R^n$. f represents the system uncertainties, including the system parameter uncertainty and external disturbances.

In this paper, Takagi-Sugeno fuzzy model is used to study NCS. NCS is essentially a variable time delay system with random characteristics. If $0\leq\tau\leq Mh$, the variable time delay system can be decomposed into a switching subsystem. The Fuzzy rules is

$$\begin{aligned} &IF \quad (m-1)h < \tau \le mh \\ &THEN \quad x(k+1) = Ax(k) + Bu(k-m) + f(k) \end{aligned} \tag{2}$$

We use discrete random variable delay as a prerequisite and the probability distribution of delay as the membership function to fuzzy fuse the subsystem, global model obtained.

$$x(k+1)=\sum_{m=1}^{M} \upsilon_m[Ax(k)+Bu(k-m)+f(k)] =Ax(k)+\sum_{m=1}^{M} \upsilon_m Bu(k-m)+f(k) \tag{3}$$

Where $\upsilon_m=P[(m-1)h<\tau\le mh]$, so

$$\sum_{m=1}^{M} \upsilon_m=1(0\le\upsilon_m\le1) \tag{4}$$

Introducing the transformation

$$z(k)=x(k)+\sum_{m=1}^{M} \upsilon_m \sum_{i=1}^{m} A^{i\text{-}m\text{-}1}Bu(k\text{-}i) \tag{5}$$

So

$$z(k+1)=x(k+1)+\sum_{m=1}^{M} \upsilon_m \sum_{i=1}^{m} A^{i-m-1}Bu(k+1-i) \tag{6}$$

Taking equation (3) into equation (6),

$$z(k+1)=Ax(k)+\sum_{m=1}^{M} \upsilon_m Bu(k-m)+\sum_{m=1}^{M} \upsilon_m \sum_{i=1}^{m} A^{i-m-1}Bu(k+1-i)+f(k)$$

$$=Ax(k)+\sum_{m=1}^{M} \upsilon_m\left[A^{-m}u(k)+\sum_{i=2}^{m+1} A^{i-m-1}Bu(k+1-i)\right]+f(k) =Ax(k)+\sum_{m=1}^{M} \upsilon_m\left[A^{-m}Hu(k)+\sum_{i=1}^{m} A^{i-m}Bu(k-i)\right]+f(k)$$

$$=A\left[x(k)+\sum_{m=1}^{M} \upsilon_m \sum_{i=1}^{m} A^{i-m-1}Bu(k-i)\right]+\sum_{m=1}^{M} \upsilon_m A^{-m}Bu(k)+f(k) =Az(k)+\sum_{m=1}^{M} \upsilon_m A^{-m}Bu(k)+f(k)$$

Noting $G=A, H=\sum_{m=1}^{M} \upsilon_m A^{-m}B$ the above equation can be expressed as

$$z(k+1)=Gz(k)+Hu(k)+f(k) \tag{7}$$

According to the principle of parallel distributed compensation, the f fuzzy rules for the subsystem m is

$$\begin{aligned} &IF \quad (m-1)h < \tau \le mh \\ &THEN \quad s_m(k) = C_m z(k) \end{aligned} \tag{8}$$

Fuzzy Fusing all the sliding surface to be globally

$$s(k)=\sum_{m=1}^{M} \upsilon_m C_m z(k)=Cz(k) \tag{9}$$

Where $C=\sum_{m=1}^{M} \upsilon_m C_m$.

3 Design of the Sliding Mode Controller

For system (7), using the following reaching law

$$s(k+1)=\begin{cases}\mu s(k)-\varepsilon T\,\mathrm{sgn}(s(k)), & |s(k)|>\Delta \\ -\mu s(k), & |s(k)|\le\Delta\end{cases} \tag{10}$$

where $\varepsilon>0, 0<\mu<1, \Delta=\dfrac{\varepsilon T}{1+\mu}$, Denoting

$$s(k+1)=\alpha s(k)+\beta\,\mathrm{sgn}(s(k)) \tag{11}$$

To simplify the problem, we assume that the system (7) satisfies the matching condition $f(k)=H\cdot d(k)$. Since H is related with subsystem m, therefore $d(k)$ will change with μ_m.

From equation (9) and (11) we can obtain be dynamic sliding mode function

$$s(k+1)=Cz(k+1)=CGz(k)+CH\left[u(k)+d(k)\right] \tag{12}$$

A global variable structure control law can be solved for the system (7) By equation (11) and (12).

$$u(k)=-d(k)-(CH)^{-1}[CGx(k)-\alpha s(k)+\beta\,\mathrm{sgn}(s(k))] \tag{13}$$

Lemma 1. *For the discrete reaching law (11), with any initial value* $s(0)\neq 0$, *when* $k\to\infty$, $|s(k)|\to 0$.

Proof. ① When $|s(k)|>\Delta$, equation (11) becomes

$$s(k+1)=\mu s(k)-\varepsilon T\,\mathrm{sgn}(s(k)) \tag{14}$$

So $[s(k+1)+s(k)][s(k+1)-s(k)]=(\mu^2-1)s(k)^2-2\mu\varepsilon T|s(k)|+\varepsilon^2T^2$

$$=(\mu^2-1)(|s(k)|-\frac{\varepsilon T}{1+\mu})(|s(k)|+\frac{\varepsilon T}{1-\mu})<0 \Rightarrow |s(k+1)|<|s(k)|$$

② When $|s(k)|\le\Delta$, at this time equation (11) becomes

$$s(k+1)=-\mu s(k) \tag{15}$$

Obviously satisfied $|s(k+1)|<|s(k)|$.

Comprehensive ① and ②, equation (11) always satisfy the convergence conditions. Proposition is proved.

Theorem 1. *In the case of the control law (13) being realized, for any non-zero initial value* $x(0)$ *, when* $k \to \infty$ *, the state trajectory of system (7) will reach and stabilize at a good sliding surface* $s(x)=0$.

Proof. In the sliding surface, the system state should meet $s(k+1)=s(k)=0$, so the equivalent control can be obtained

$$u_{eq} = -d(k) - (CH)^{-1} CGz(k) \tag{16}$$

Taking equation (6) to (7) can obtain the system slide equations.

$$z(k+1) = [I - H(CH)^{-1} C]Gz(k) \tag{17}$$

With the action of the control law (13) , motion equation of system (7) is

$$\begin{aligned} z(k+1) &= [I - H(CH)^{-1} C]Gz(k) + H(CH)^{-1}[\alpha s(k) - \beta \operatorname{sgn}(s(k))] \\ &= [I - H(CH)^{-1} C]Gz(k) + H(CH)^{-1} Q(k) \end{aligned} \tag{18}$$

Obviously, $Q(k)$ and the ideal reaching law (11) are equivalent. By the Lemma 1, when $k \to \infty$, $Q(k)$ tends to zero. Then the system equations (18) and sliding equation (17) are equivalent. Proposition is proved.

Theorem 1 shows that, control law (13) based on the reaching law (11) can make the system states reach the switching surface and converted to sliding movement. However the control law (13) can not be achieved with unknown items $d(k)$. So we design the disturbance predictor as follow.

$$\hat{d}(k) = \hat{d}(k-1) + \tilde{d}(k-1) + \lambda \tilde{d}(k-1) \tag{19}$$

where $\tilde{d}(k) = d(k) - \hat{d}(k)$ represent estimation error of uncertain parts. The actual control law

$$u(k) = -\hat{d}(k) - (CH)^{-1}[CGx(k) - \alpha s(k) + \beta \operatorname{sgn}(s(k))] \tag{20}$$

Theorem 2. *If The uncertain system (7) select the disturbance predictor (19) and control law (20), then the disturbance predictor meet*

$$\hat{d}(k) = \hat{d}(k-1) + (I + \lambda)(CH)^{-1}[s(k) - \alpha s(k-1) + \beta \operatorname{sgn}(s(k-1))] \tag{21}$$

Proof. Taking equation (20) into (7), closed-loop system dynamic equation can be obtained

$$z(k+1) = Gx(k) + H\tilde{d}(k) - H(CH)^{-1}[CGx(k) - \alpha s(k) + \beta \operatorname{sgn}(s(k))] \tag{22}$$

Both ends of The equation (22) left multiply matrix C and simplification

$$s(k+1)=\alpha s(k)-\beta \operatorname{sgn}(s(k))+CH\tilde{d}(k) \tag{23}$$

By equation (23) easy to get

$$\tilde{d}(k-1)=(CH)^{-1}[s(k)-\alpha s(k-1)+\beta \operatorname{sgn}(s(k-1))] \tag{24}$$

Taking it to equation (19), the equation (21) is obtained, the proposition is proved.

4 Simulation Research

A network DC servo motor control system, model such as (1), specific parameters are as follows.

$$A=\begin{bmatrix}1.1 & 0.1\\ 0 & 0.7\end{bmatrix},\quad B=\begin{bmatrix}0\\ 1\end{bmatrix},\ f(k)=\begin{bmatrix}0.1083\\ 0.7094\end{bmatrix}\omega(k)$$

$$\omega(k)=0.05+0.05\sin(2\pi k/50)\quad x(0)=[1,0.6]^T$$

Assuming $M=3$, $\upsilon_1=\upsilon_2=\upsilon_3=1/3$. Within ternary sample period fuzzy fusing system.

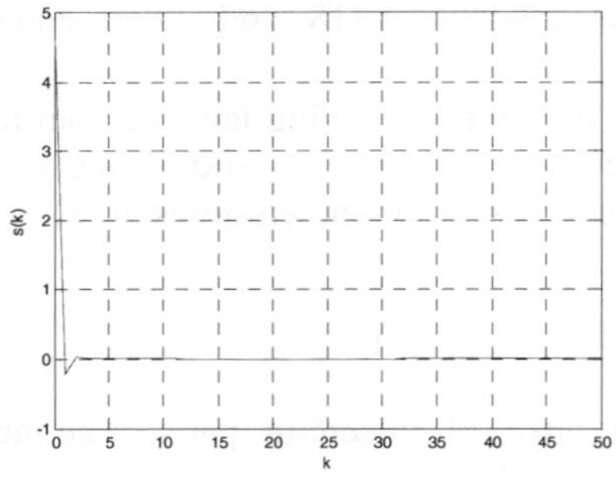

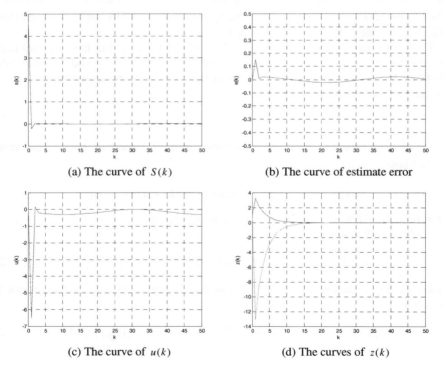

(a) The curve of $S(k)$

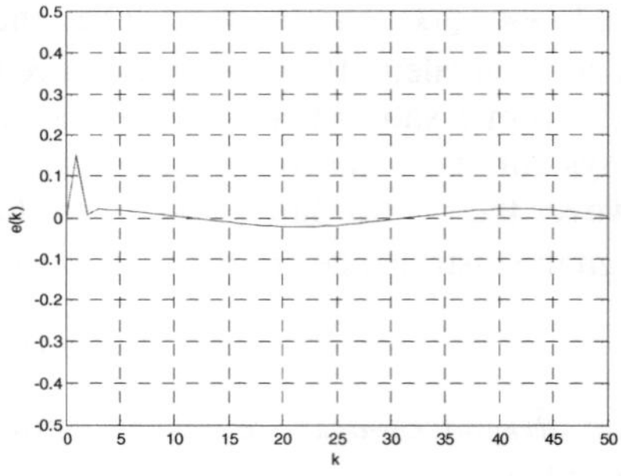

(b) The curve of estimate error

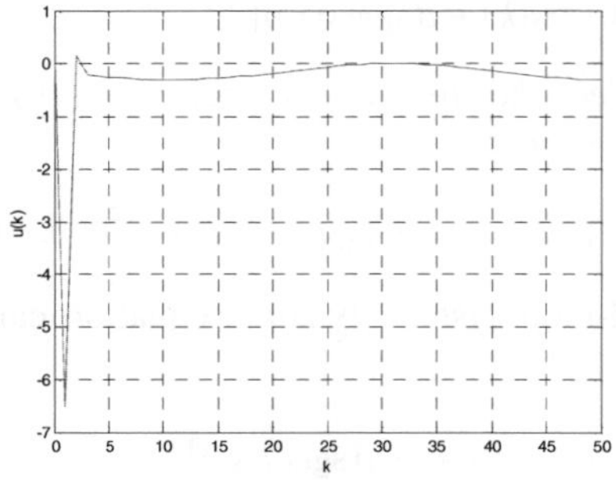

(c) The curve of $u(k)$

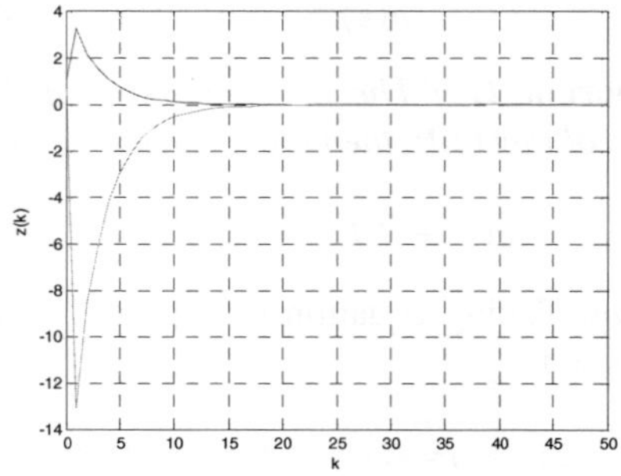

(d) The curves of $z(k)$

Fig. 2. Simulation results of a networked servo system

Then

$$z(k) = x(k) + \begin{bmatrix} -0.3248 \\ 2.1283 \end{bmatrix} u(k-1) + \begin{bmatrix} -0.1445 \\ 1.1565 \end{bmatrix} u(k-2) + \begin{bmatrix} -0.0433 \\ 0.4762 \end{bmatrix} u(k-3),$$

$$G=\begin{bmatrix} 1.1 & 0.1 \\ 0 & 0.7 \end{bmatrix}, H=\begin{bmatrix} -0.3248 \\ 2.1283 \end{bmatrix}, \; d(k)=3\omega(k)$$

Take $z(0)=[1,0.6]^T$, $C=[4,1]$, $\mu=0.1$, the simulation results shown in Figure. 2.

Horizontal axis in Figure 2. represent the number of steps. Vertical axis of Figure 2. followed by sliding surface, the disturbance estimation error, control volume and the size of the state. The figure shows that using this method system still has strong robustness when the time-varying delay in the network, the system parameter perturbation and external disturbance exists. And suppress the chattering of sliding mode control existence.

5 Conclusions

In this paper, adaptive sliding mode controller is design for a class of time-varying delay network system. The significant advantages of this method is that the global discrete random variables model based on T-S fuzzy model is lower conservative than deterministic control model. And adaptive sliding mode controller in the same time to ensure robust and effective reduce the chattering. Future research will focus on the optimal division of sub-systems for the actual network environment.

References

1. Zhang, W., Branicky, M.S., Phillips, S.M.: Stability of networked control systems. IEEE Control Systems Magazine 21(1), 84–99 (2001)
2. Walsh, G.C., Ye, H., Bushnell, L.: Stability analysis of networked control systems. IEEE Transactions on Control Systems Technology 10(3), 438–446 (2002)
3. Baillieual, J., Antsaklis, P.J.: Control and communication challenges in networked real-time system. Proc.IEEE 95(1), 9–28 (2007)
4. Li, H.-b., Sun, Z.-q., Sun, F.-c.: Networked control systems: an overview of state-of-the-art and the prospect in future research. Control Theory & Applications 27(2), 238–243 (2010)
5. Yao, Q.-h., Song, L.-z., Yan, s.-m.: Evolution and prospect of discrete variable structure control theory. Journal of Naval University of Engineering 16(6), 23–29 (2004)
6. Xiong, Y.-s., Yu, L., Xu, J.-m.: Design of sliding mode predicting controller for network control system. Electric Drive Automation 25(4), 39–40 (2003)
7. Roh, Y.-H., Oh, J.-H.: Robust stabilization of uncertain input-delay systems by sliding mode control with delay compensation. Automatica 35, 1861–1865 (1999)

8. Huang, M., Wang, X.-d.: Design of sliding mode control for networked system with uncertainties and multiple time delay. Journal of Shenyang University of Technology 26(5), 539–542 (2004)
9. Wang, Z., Guo, X.-j., Si, J.-f., Zhang, Q.: Discrete sliding mode control with time delay compensation for single input networked control systems. Systems Engineering and Electronics 28(8), 1237–1239 (2006)
10. Song, L.-z., Chen, S.-c., Yao, Q.h.: Discret evariable structure control algorithm based on sliding mode prediction. Control Theory &Applications 21(5), 826–829 (2004)

Design and Implement of a High-Frequency Bridge Type AC-AC Direct Converter for Contactless Power Transfer Systems

Xiaomei Chen, Wangqiang Niu, and Xiong Zhang

Academy of Science & Technology, Shanghai Maritime University,
Shanghai, 201306, China
media.28@163.com

Abstract. Focusing on the defects of traditional contactless power transfer systems with an AC-DC-AC power transition type, a high-frequency bridge type bidirectional switch AC-AC direct converter was presented to generate the high-frequency current in the primary circuit. Applying the principle of energy injection and free oscillation, the operating state of IGBTS were designed and the high frequency current in the primary circuit was obtained. The operation principle and control strategy of the converter was further analyzed in detail by the ac impedance analysis method. The converter was simulated by the software of MATLAB/SIMULINK, and the simulation results verify the correctness of analysis and the feasibility of the control strategy.

Keywords: Contactless Power Transfer, High frequency AC-AC Converter, Bridge Type Converter, SIMULINK simulation.

1 Introduction

The contactless power transmission(CPT)technology is a novel technology which is based on the principle of electromagnetic induction, and utilizes power electronics technology and control principle to realize electrical connection without wires[1-3]. To achieve high power transfer capability and efficiency, high frequency resonant circuits are often used in the CPT system [4]. The CPT system is composed of transmission devices, energy pick-up and control apparatus. A traditional structure of CPT systems is shown in Fig.1. In the traditional CPT system, the AC source powers electric equipments through rectification, high frequency inverter, loose coupling transformer and pick-up regulators. There are several shortcomings in the rectification stage, such as great power loss, high equipment cost and low reliability [5]. So, this paper proposed a new technology to resolve this problem.

This paper represents the transformation model of primary circuit in the contactless power transfer system. The existing AC-AC converter technique of high frequency mainly includes cycloconverter technique [6], matrix converter technology, low/high-frequency AC ring transform technique [7]. Because the output frequency of a cycloconverter is less than its input frequency, and the matrix converter technology

X. Wan (Ed.): Electrical Power Systems and Computers, LNEE 99, pp. 139–146.
springerlink.com

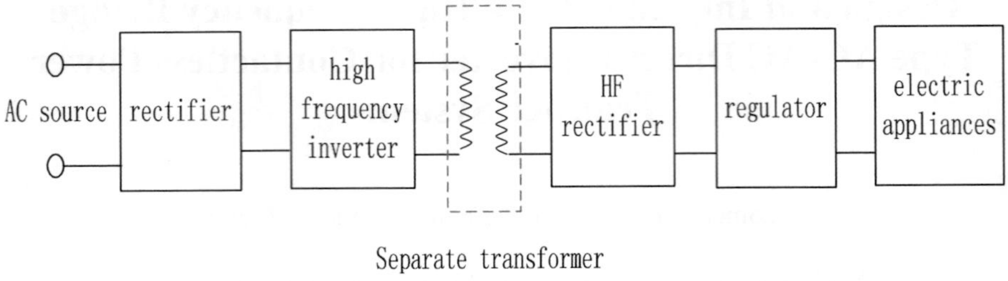

Fig. 1. The structure of traditional contactless power transfer system

has no electrical isolation, these two technologies are not suitable for non-contact power transmission systems. Paper [8] presents a model of BOOST type AC-AC direct converter, but it only realizes the dynamic adjustment and optimization of output voltage waveforms, and fails to achieve the function of frequency conversion. Paper [9] proposed an AC-AC direct transform structure that convert a low frequency source to a high frequency source, but its high-frequency AC output is partly stable. Aimed at an AC-DC-AC conversion problem, this paper presents a high frequency bridge type AC-AC direct transform technique which produces a high frequency current output in the primary circuit. The circuit structure of the converter and control strategy are analyzed, and further simulated by MATLAB platform.

2 Circuit Structure and Modeling Analysis

2.1 Circuit Structure

A circuit structure about high-frequency bridge type bidirectional switch AC - AC converter is shown in Fig. 2. It employs a AC power *V1*, RLC series resonance loop and bridge type switch (IGBT1-8) with anti-parallel body diodes to achieve dual directional power flow and energy injection. *V1* is the AC power with 50 Hz, the current of resonance loop is the output which is a high frequency alternating current in the CPT system.

2.2 Modeling Analysis

The referential positive direction of input voltage *V1* and load current are shown in Fig. 2. According to the internal mechanism of soft switch resonant circuit [1], this paper adopts ac impedance analysis method to make system model. Linear elements of resonance circuit contain R_p、C_p、L_p , supposing the source *V1* is a ideal sine power, R_p is equivalent resistance in the primacy circuit, which include reflected resistance in the secondary circuit . According to the circuit principle, the process is divided into two working mode: energy inject mode and free oscillation mode.

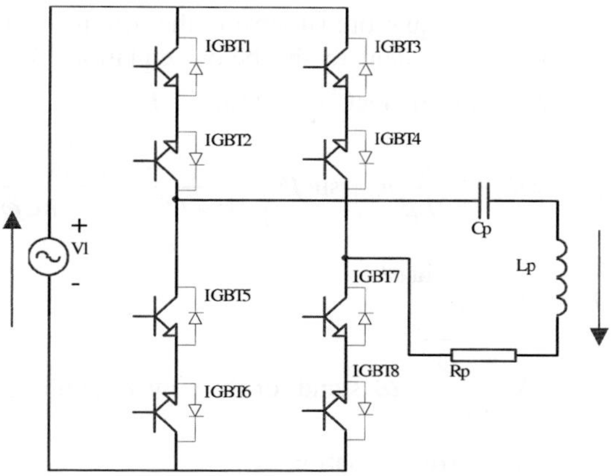

Fig. 2. Circuit of a high-frequency bridge type AC-AC direct converter

(1) energy inject mode

Energy injection process is divided into positive and reverse injection. In the initial condition, all switches are shut off, and the energy storage components in the resonant network are fully release. Energy begins forward direction inject when *V1*>0, switches 1,7 are both "on" or *V1*<0,and switches 4,6 are both "on". Energy begins reverse direction inject when *V1*>0, switches 3,5 are both "on" or *V1*<0, and switches 2,8 are both "on". The equivalent circuit of energy injection mode in the primary circuit is shown in Fig. 3:

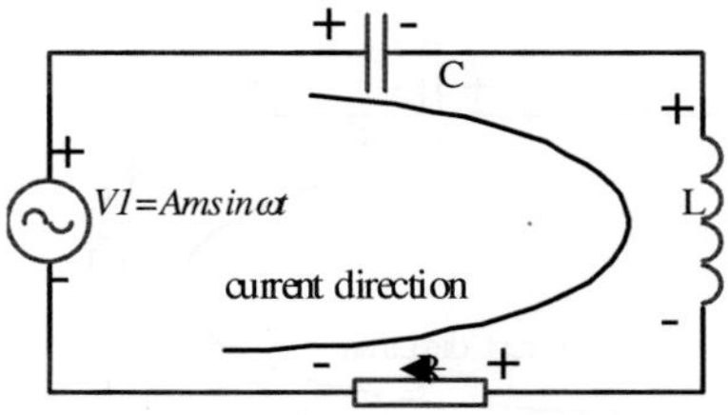

Fig. 3. The equivalent circuit of energy injection mode

According to the circuit principle, the mathematical equations of energy inject model are set up:

$$U_C + U_L + U_R = E \tag{1}$$

According to $\frac{1}{C}\int idt + L\frac{di}{dt} + Ri = E$, in which $E = A_m \sin \omega t$ and $i = C\frac{dU}{dt}$, We can get the equation (2) as below:

$$LC\frac{d^2U}{dt^2} + RC\frac{dU}{dt} + U = E \tag{2}$$

Solve the above differential equation: supposing the original state: $U(0)=u_o$, $U'(0)=0$, the equation (2) can be equivalent to the below equation (3) on the base of series circuit oscillating condition[10]: $(RC)^2-4LC<0$.

$$U(t)=e^{\alpha t}\left[u_0\cdot\cos(\beta t)+\left(\frac{R}{2L\beta}u_0\right)\cdot\sin\beta t\right]+\frac{-A_mRC\omega}{(1+LC\omega^2)^2-(RC\omega)^2}\cdot\cos\omega t$$
$$+\frac{A_m(1+LC\omega^2)}{(1+LC\omega^2)^2-(RC\omega)^2}\cdot\sin\omega t. \tag{3}$$

in which $\alpha=-\frac{R}{2L}, \beta=\frac{\sqrt{4\frac{L}{C}-R^2}}{2L}$, ω stand for the power angular frequency. And due to $i=C\frac{dU}{dt}$, we can get the current equation as:

$$i(t)=\alpha\cdot e^{\alpha t}\left[u_0\cdot\cos(\beta t)+\left(\frac{R}{2L\beta}\cdot u_0\right)\cdot\sin\beta t\right]+e^{\alpha t}\left[\begin{matrix}-u_0\cdot\beta\cdot\sin(\beta t)+\left(\frac{R}{2L\beta}\cdot u_0\right)\cdot\beta\cdot\\ \cos\beta t\end{matrix}\right]$$
$$+\frac{-A_mRC\omega^2}{(1+LC\omega^2)^2-(RC\omega)^2}\cdot\sin\omega t+\frac{A_m\omega(1+LC\omega^2)}{(1+LC\omega^2)^2-(RC\omega)^2}\cdot\cos\omega t \tag{4}$$

(2) free oscillation mode

RLC series resonance circuit will be in the free resonant modal after energy injection and the energy storage components be full of energy. The equivalent circuit of the free oscillation mode in the primary circuit is shown as Fig. 4:

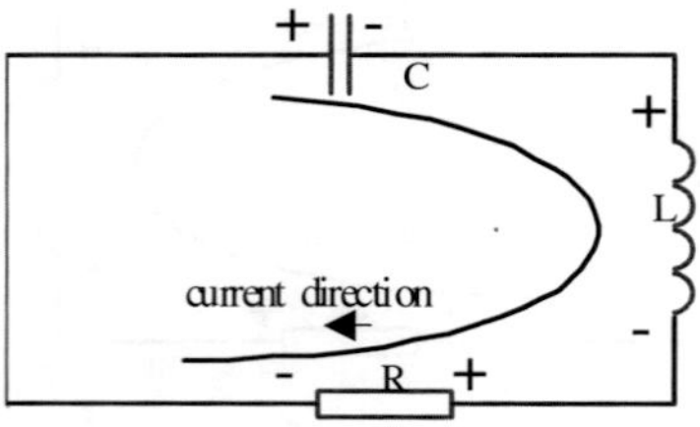

Fig. 4. The equivalent circuit of the free oscillation mode

According to the circuit principle, establish free oscillation condition mathematical model:

$$U_C+U_L+U_R=0. \tag{5}$$

That is: $\mathrm{LC}\frac{d^2U}{dt^2}+RC\frac{dU}{dt}+U=0$, we can get the equation (6) by $U=L\frac{di}{dt}$:

$$\mathrm{LC}\frac{d^2i}{dt^2}+RC\frac{di}{dt}+i=0. \tag{6}$$

The initial condition is $i(0)=i\big|_{t=0}=0, \frac{di}{dt}\big|_{t=0}=V\omega$, solve the homogeneous differential equation, then we can get:

$$i(t)=e^{\alpha t}\cdot\frac{V\omega}{\beta}\cdot\sin\beta t, \alpha=-\frac{R}{2L}, \beta=\frac{\sqrt{4\frac{L}{C}-R^2}}{2L}. \tag{7}$$

3 Control Strategy

According to the analysis of the above provisions, providing the coil output current in half a cycle contains energy injection modal and free oscillation modal, the process can be divided into four stages by the difference polarity of voltage source and coil current, and T means oscillation period.

When $V1>0$:

The first stage: in the time of nT～(n+1/4)T, switches 1 and 7 are "on", and the resonance circuit is injected in forward direction from the source $V1$.

The second stage: in the time of (n+1/4)T～(n+2/4)T, switch 6 and 7 are "on", resonance network current through the negative terminal of next bridge, until the resonance network current come back zero.

The third stage: in the time of (n+2/4)T～(n+3/4)T, switch 3 and 5 are "on", resonance network is injected in the reverse direction form the source $V1$.

The fourth stage: in the time of (n+3/4)T～(n+1)T, switch 5 and 8 are "on", resonance network current through the negative terminal of up bridge, until the resonance network feedback current come back zero.

The process is the same as the source $V1>0$ When $V1<0$. In conclusion, the on-off condition of each switch at each stage is shown as table 1:

According to the condition of each switch from the above table, switches can be controlled by drive circuit and controller circuit feedback the polarity of current and voltage.

Table 1. The operating state of IGBTs (× means switch off, ✓means switch on)

Source polarity	System modal	s1	s2	s3	s4	s5	s6	s7	s8
Vac>0	Energy forward inject	✓	×	×	×	×	×	✓	×
	Free oscillation	×	×	×	×	×	✓	✓	×
	Energy reverse inject	×	×	✓	×	✓	×	×	×
	Free oscillation	×	×	×	×	✓	×	×	✓
Vac<0	Energy forward inject	×	×	×	✓	×	✓	×	×
	Free oscillation	✓	×	×	✓	×	×	×	×
	Energy reverse inject	×	✓	×	×	×	×	×	✓
	Free oscillation	×	✓	✓	×	×	×	×	×

4 System Simulation Analysis

The frequency of the primary circuit alternating current is usually among 10-100kHz in the CPT system, this paper adopted a working frequency of 20kHz.

Table 2. Parameters of the CPT system

Parameters	Values
Vac	11V/50Hz
C_P	1.0μF
L_P	64μH
R_P	1.5Ω
f_P	20kHz

Based on the circuit diagram shown in Fig.2, computer simulations were carried out for the proposed high-frequency bridge type AC-AC direct converter with the circuit parameters in the table 2. In order to get power frequency ac output in the secondary circuit, input frequency resonant frequency and the resonant network have to meet the relationship: $\frac{f_P}{f_1} = 2N$, in which f_P means natural frequency of resonance network, f_1 is the source *V1* frequency.

According to the circuit diagram and the parameters, MATLAB/SIMULINK is used for this purpose to simulation. The results of simulation are showed below from Fig.5 to Fig.8, Fig.5 shows the driving wave of switch 2. The combined diagram of current wave and source wave in one input power cycle is shown in Fig.6, where the current wave shaped a sinusoidal envelope curve with the same frequency 50Hz as the source, thus it supplies the AC power to the secondary circuit, and the Fig.7 is the current local amplification figure, from which the oscillation forms sine curve with 20kHz and the amplitude changes with value *V1*.

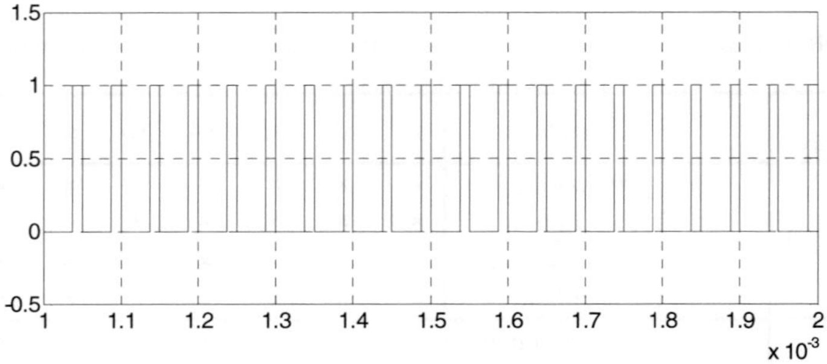

Fig. 5. The derive waveform of IGBT2

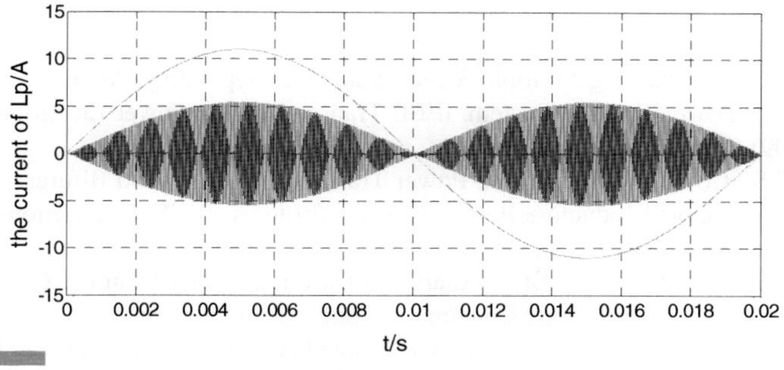

Fig. 6. Current waveform of Lp and voltage waveform of power source V

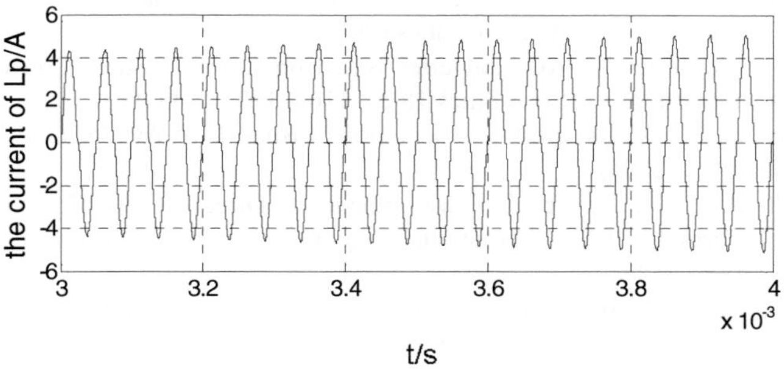

Fig. 7. Zoom of Lp's current waveform

5 Conclusions

In this paper, a high-Frequency bridge type AC-AC direct converter technique has been developed based on MATLAB/SIMULINK platform, and the new converter technique changes a low frequency of power source into a high frequency AC source, which has realized an AC-AC direct transition, what is more, this technology can be used in the CPT system. The high-frequency bridge type AC-AC direct converter has many unique features for removing the rectifier link, such as elimination of DC link capacitors, simplification of system structure, reduction the CPT system cost, and improving the system conversion efficiency. The theoretical analysis demonstrate the correctness of the proposed converter structure, and the simulation results indicate that the proposed converter technology works well and can serve as a good reference for other high frequency voltage/current generation.

Acknowledgments. This work is supported by The National High Technology Research and Development Program of China (2009AA043001), by Shanghai Leading Academic Discipline Project (S30602), and by Science & Technology Program of Shanghai Maritime University (20100082).

References

1. Tang, C.S.: Determining Multiple Steady-State ZCS Operating Points of a Switch-Mode Contactless Power Transfer System. IEEE Transactions On Power Electronics 24(2), 416–424 (2009)
2. Wang, C.S., Covic, A., Stielau, H.: Power Transfer Capability and Bifurcation Phenomena of Loosely Coupled Inductive Power Transfer Systems. IEEE Transactions on Industrial Electronics 51(1), 148–157 (2004)
3. Chao, L., Hu, A.P.: Steady State Analysis of a Capacitively Coupled Contactless Power Transfer System. IEEE Energy Conversion Congress and Exposition, 3233–3238 (2009)
4. Zhang, K., Pan, M.-c.: The Study Status Quo and Application Analysis on the Inductively Coupled Power Transfer Technology. Power Electronics 43(3), 76–78 (2009)
5. Sun, Y., Dong, J.: Forecast Control Strategy for Bidirectional Switch AC/ AC Converter. Journal of Chongqing Institute of Technology 23(3), 121–125 (2009)
6. Lei, L.: Research on AC/AC Converters with High Frequency AC Link. D. Nanjing University of Aeronautics and Astronautics (2004)
7. Mehrdad, K.: A Direct AC/AC Converter Based on Current-Source Converter Modules. IEEE Transaction on Power Electronics 18(5), 1168–1175 (2003)
8. Ren, Y., Zhang, Y.: Design and Implementation of Buck Type AC/AC converter. Research and Design, 21–25 (2004)
9. Li, H.L., Hu, A.P., Gao, J.F., et al.: Development of a Direct AC-AC Converter Based on a DSPACE Platform. In: IEEE International Conference on Power System Technology, vol. 10, pp. 1–6 (2006)
10. Wang, Z., Sun, Y.: A New Type AC/DC/AC Converter for Contactless Power Transfer System. Transactions Of China Electrotechnical Society 25(1), 84–89 (2010)

Impact of the Hybrid Multi-channel Multi-interface Wireless Mesh Network on ETX-Based Metrics Performance

Hassen Mogaibel*, Mohamed Othman,
Shamala Subramaniam, and Nor Asila Wati Abdul Hamid

Department of Communication Technology and Network, Universiti Putra Malaysia,
43400 UPM, Serdang, Selangor D.E., Malaysia
Tel.:+603-89466556; Fax:+603-89466576
Hassen.mogaibel@gmail.com,
{mothman,shamala,asila}@fsktm.upm.edu.my

Abstract. In this paper, we study the impact of multi-radio multi-channel Wireless Mesh Networks (WMNs) characteristics on the performance of Expected Transmission Count based metrics (ETX-based). Many characteristics of multi-radio multi-channel WMNs such as channel switching, an absence of a direct connection between neighbors, and difficulty of measuring the packet loss probability in the both forward and reverse direction may affect the performance of ETX-based metrics in hybrid multi-radio multi-channel WMNs compared to static multi-radio multi-channel WMNs. Through simulation experiments, we compared the performance of ETX-based metrics with Hop count metric. The experiment result shows the superior improvement of hope count metrics over ETX-based metrics in terms of packet delivery ratio, goodput, and end-to-end delay. From the results, we concluded that major causes of the poor performance of ETX-based metrics in a multi-radio multi-channel WMNs are the extra delay caused by interface switching between channels and the inaccurate value of the ETX-based metrics.

Keywords: Wireless mesh network, multi-radio, on-demand routing, multi-channel, multi-link.

1 Introduction

Recently, Wireless Mesh Networks (WMNs) technology has gained a lot of attention and became popular in wireless technology and industry field owing to their low cost, rapid development and offer broadband wireless access to the internet in places where wired infrastructure is not available or worthy to deploy [1]. WMNs are composed of mesh routers, which collect and relay the traffic generated by mesh client. Mesh routers are usually stationary and equipped with multiple radios. Mesh clients are typically mobile and relay on mesh routers to deliver

* Corresponding author.

X. Wan (Ed.): Electrical Power Systems and Computers, LNEE 99, pp. 147–160.
springerlink.com

data to the intended destination. One or more mesh router may have gateways functionally and provide connectivity to other networks such as internet access Fig. 1.

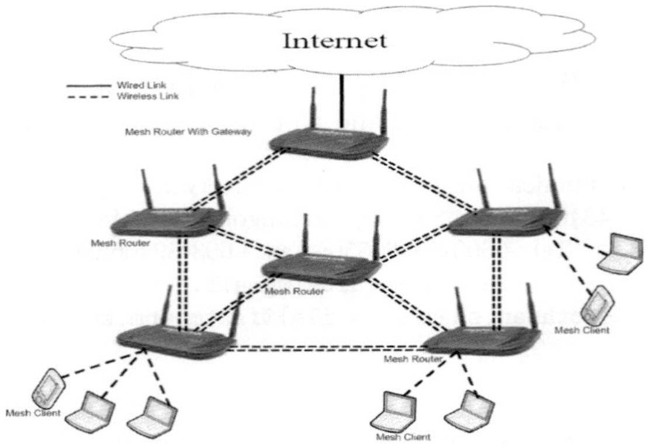

Fig. 1. Multi-radio wireless mesh networks.

Previous study was shown that the WMN backbone inherits some scalability problem in terms of throughput, delay and packet delivery ratio due to interferences [2-5]. WMN capacity is reduced by interference from concurrent transmissions. There are two types of interference that affect the throughput of WMN, intra-flow and inter-flow interferences. The intra-flow interference refers to the interference between intermediate nodes sharing same flow path, whereas, inter-flow interference refers to the interference between neighboring nodes competeing the same busy channel. These come from the half duplex of the radio and the broadcast nature of the wireless medium. Several approaches have been proposed to improve the WMNs capacity. One approach is that each mesh router uses a single radio interface that dynamically switches to a wireless channel with a different frequency band to communicate with different nodes. However, this approach increases the routing overhead due to a switching delay. A more practical approach uses multiple radio interfaces that are dedicated to non-overlapping channels. The IEEE 802.11 b/g and IEEE 802.11a standards define three and twelve non-overlapping channels (frequencies), which greatly increases WMN capacity due to avoidance of competition in transmitting data, avoidance of collisions in the same channel and elimination of interference [6, 7]. One of the most important design questions for multi-radio multi-channel networks is how to bind each radio interface to a channel in a way that maintains network connectivity. The following approaches have been proposed to solve the channel assignment problem in multi-radio multichannel WMNs:

1. Static channel assignment: In a static channel assignment approach, each interface is assigned to a channel for long time durations. Static assignment can be further classified into two types:

(a) Common channel approach: In this approach, the radio interfaces of all nodes in the network are assigned to common channels [8]. For example, if two interfaces are used at each node, then the two interfaces are assigned to the same two channels at every node. The advantage of this approach is that the connectivity of the network is the same as that of a single channel approach.

(b) Varying channel approach: In this approach, the radio interfaces in different nodes may be assigned to different channels [6, 9]. With this approach, it is possible that the length of the routes between nodes may increase, also, the network partitions may arise due to inability of different neighbors to communicate with each other unless they assign a common channel.

2. Dynamic channel assignment: Dynamic channel assignment approach allows any interface to be assigned any channel, and interfaces can frequently switch from one channel to another. Therefore, a network using such a strategy needs some kind of synchronization mechanisms to enable communication between nodes in the network. For example, such mechanisms may require all nodes to periodically visit a predetermined rendezvous channel to negotiate channels for the next phase of transmission. The benefit of dynamic assignment is the ability to switch an interface to any channel, thereby offering the potential to use utilizes the non-overlapping channel spectrum with few interface. However, the key challenges involve channel switching delays. The examples for this category are: Hyacinth [7], MCRP [10].
3. Hybrid Assignment In the hybrid approach, all the nodes are equipped with multi-radio interfaces in which the multiple radios are divided into two groups, fixed group and switchable group. In the fixed group, each radio interface is assigned a fixed channel for receiving packets, thereby, ensuring the network connectivity, while the switchable group can dynamically switch among the other data channels [11]. When a data transmission is required, the source node switches one radio interface of it's switchable group to the fixed channel of the destination node. Thus, the channel assignment for the fixed radios is the most important aspect of the hybrid approach. Hybrid assignment strategies are attractive as they allow simplified coordination algorithms supported by static assignment while retaining the flexibility of dynamic assignment.

The remainder of the paper is organized as follows: section two discusses the relevant ETX-based metric and the multi-channel multi-interface routing protocols. In section three we discussion the impact of the hybrid multi-channel multi-interface WMNs on ETX-based metrics. In section four, we provide the details of our simulation environment. Simulation results and their analysis are presented in section five with concluding remarks in selection six.

2 Related Works

Routing metrics are very important to network performance. Good routing metric should carry enough information about the link quality so that a node can choose the best path to reach the gateway. The recently proposed routing metrics for WMNs include Expected Transmission Count (ETX) [8], Expected Transmission Time (ETT) [12], Weighted Cumulative ETT (WCETT) [12], and Metric of Interference and Channel-switching (MIC) [13] were developed for static multi-radio multi-channel WMNs, where each mesh router's interface is statically assigned a channel for long time. Many researchers have been proposed multi-channel multi-interface routing protocol to improve the WMNs network performance. The Multi-Channel Routing protocol (MCR) [11], proposed a hybrid channel assignment strategy as well as it developed a new metric routing protocol based on ETX metric to include the cost of channel switching delay. Hyncian [7] is another dynamic channel assignment routing protocol proposed for multi-radio multi-channel WMNs, which divides the collision domain into a sub-collision domain. In this section, we describe the relevant work of ETX based metric and multi-radio multi-channel routing protocols.

2.1 LINK Quality Metrics

Several high quality metrics has been proposed for WMNs. Refer the reader to [14] for more details. Most of the ETX-based metrics proposed for WMNs based on ETX metric such as ETT, WCETT, MIC, and MCR. We call this metrics as ETX-based metrics.

ETX Metric. The ETX metric proposed to improve the hop count metric. The hop count metric does not distinguish between high and low quality links, instead it considers the path length for route selection. The ETX metric considers the path length as well as the effect of link loss ratio in selecting the route. The metric can be defined as the expected number of Medium Access Control (MAC) layer transmission that is needed for successfully delivering packet through a wireless link, see Eq. (1).

$$ETX = \frac{1}{d_f * d_r} \tag{1}$$

The parameter df is the forward delivery ratio of a link, i.e. the ratio of data frames successfully traversing the link in the forward direction. The parameter dr is the corresponding parameter for the reverse direction of the link. Both df and dr can be interpreted as the probability of successfully transmitting a data frame. In order for a data frame to be successfully transmitted and acknowledged, a successful transmission in both the forward and the reverse direction is required, with the corresponding probability of $df * dr$. The ETX value represents the expected number of transmissions, i.e. the inverse of the success probability of a single transmission. The ETX metric is generally measured using periodic link probe packets. The ratio of successfully received probes from a neighbor provides the reverse link delivery ratio dr. Similarly, the ratio of successfully received probes by that neighbor indicates the forward link delivery ratio df.

ETT Metric. In multi-radio backbone nodes, ETX performance is low, because it neither distinguishes between links of different rates such as WLAN a/b/g nor reduces the interferences between neighbors. To cope with these limitations, the EET proposed to capture the impact of link bandwidth heterogeneity the route selection. The ETT calculated as in Eq. (2).

$$ETT = ETX * \frac{PacketSize}{Bandwidth} \tag{2}$$

There are two approaches to calculate the link bandwidth, first approach [8], prefer to periodically estimate the bandwidth than uses fixed size, it used packet pair techniques to calculate bandwidth per link. The second approach to compute ETT is considered in [12]. The author estimates the loss probability by considering that IEEE 802.11 uses data and ACK frames. The idea is to periodically compute the loss rate of data and ACKnowledgement (ACK) frames to each neighbor. The former is estimated by broadcasting a number of packets of the same size as data frames, one packet for each data rate defined in IEEE 802.11. The latter is estimated by broadcasting small packets of the same size as ACK frames and sent at the basic rate that is used for ACKs. Note that broadcasting packets at higher data rates may require firmware modifications. However, it is possible to improve the network throughput by turned each interface to non-overlapping channel to reduce the interferences between channels. With the multi-channel, there are two issues:

- Inter-flow interference.
- Intra-flow interference.

The ETT does not capture these issues. Resulting in, the ETT may choose a path that only uses one channel instead of a path with more channel diversity.

WCETT Metric. To improve the ETT metric, the authors proposed WCETT that considers the intra-flow interference in the route selection. The WCETT is the first routing metric that explicitly takes the channel diversity into account. Consider channel diversity during selecting the path will reduce the links on the same channel along the path of a flow. Resulting in, reduce the contention time at MAC layer. WCETT can be defined as in Eq. (3).

$$WCETT(\rho) = (1-\beta) \sum_{link\iota\epsilon\rho}^{n} ETT_\iota + \beta * \max_{1<j<k}(x_j) \tag{3}$$

Where β is a tuneable parameter subject to [0-1]. The max x_j component counts the maximum number of times that the same channel appears along a path. It picks the intra-flow interference of a short path since it essentially gives low weights to paths that have more diverse channel assignments on their links and hence lower intra-flow. Although, WCETT solves the of intra-flow interferences problem by reducing the number of the links on the same channel, the metric could not consider the inter-flow interference which is related to the number of the nodes share the channel. All the above metrics developed based on the ETX metrics which means improves the ETX to work in different environments. For example, ETT metric has been developed to improve ETX to capture the data rate when the auto rate is enabled or the heterogeneity in the network card is existed. Furthermore, WCETT improves the ETX to work in static multi-interface multi-channel WMNs by maximizing the channel diversity along the path.

2.2 Multi-channel Multi-interface Routing Protocols

Several routing protocols have been proposed to join routing protocols with channel assignment. Deves [8] proposed first multi-radio multi-channel metric, which increases the diversity between channels to reduce the intra-flow interference. The protocol has proposed identical channel assignment i.e. assign channel one to interface one and channel two to interface two and so on. Such approach clearly preserves network connectively but does not reduce interference. The Hyacinth architecture [7] proposed distributed channel assignment based on spanning tree topology where the gateway is the root of the spanning tree. The protocol dedicates one interface channel for communicate with its parent node on the tree and other interfaces are configured as child for communication with its child nodes. A node can only switch one of it child interfaces while the parent Interface is associated with a unique child of its parent. One drawback of this protocol is that it considers only the common traffic where data is transmitted from source to gateway and vice versa. MCR [11] proposed to overcome Hyacinth's problem by considering local traffic as well as internet traffic. MCR classified the node interfaces into two kinds:

- Fixed Interface.
- Switchable Interface.

The protocol uses fixed channel as a common channel for communication between neighbors and the remaining considered as switchable interfaces. When a node wants to communicate with the neighbor, it looks into the table to find out the neighbor's fixed channel and switches one of the interfaces to that channel. To distribute fixed channels between neighbors, MCR uses a hello message to carry the fixed channel information. However, this protocol may not work well in a multi-flow transmission because of high switching interfaces and it does not utilize all non-overlapping channels as the static channel assignment uses. In [15], the authors proposed the local channel assignment (LCA) algorithm which adopts tree-based routing protocol for common traffic similar to Hyacinth. LCA algorithm solved Hyacinth conflict interface-channel assignment which is caused when a parent switches to the least load channel that may be in use by one of its children, the interface-channel assignment problem may cause recursive channel switching and delay. LCA solved the above problem by dividing the non-overlapping channel into groups and made each parent interface belong to one group which differ from its child interface group. Both protocols discussed earlier assign channels from node to node and every node in WMN assigns fixed channels which makes it different from our approach. The authors of [16] and

[17] proposed algorithms to minimize network interference. The first one uses a genetic approach to find the largest number that makes whole network connectivity, while minimizing network interference. The algorithm proposed in [17] is interference-aware, as it visits the links in decreasing order of the number of links falling in the interference range and selects the least used channel in that range. Assuming the knowledge of the set of connection requests to be routed, both an optimal algorithm based on solving a Linear Programming (LP) and a simple heuristic are proposed to route such requests given the link bandwidth availability determined by the computed channel assignment. The algorithm considers minimum-interference channel assignments that preserve k-connectivity. However, such approaches only focus on minimizing network interference which may decrease the network connectivity, in constructing our approach based on eliminating the interference for common traffic on WMNs while keeping the network connectivity. The Channel Assignment Ad hoc On-demand Distance Vector routing (CA-AODV) [18], has been proposed to assign channels within K hops in an ad hoc network, allowing for concurrent transmission on the neighboring links along the path and effectively reducing the intra-flow interference. However, such approach may not work well in WMNs where most of the traffic is directed toward the gateways and must pass through mesh routers.

3 Impact of the Hybrid Multi-channel Multi-interface WMNs on ETX-Based Metrics

Based on the discussion in section two, the stability of the mesh router is one of the most important characteristics that made measuring the quality of the link possible. This property made the link between two neighbors always available. However, this property is violated in two cases:

- When the node is a mobile node.
- When the node has an ability to switch the interface between channels.

For the first case, the researcher in [8] proved that the Hops metric performs better than the ETX in ad hoc scenario, where the node changes its location frequently. By investigating the similarities between ad hoc network and hybrid multi-channel multi-interface WMNs, we found that in both networks the link is not stable because of either the movement of the node or the switching of the interface between channels. The impact of WMNs multi-channel multi-radio characteristics on the ETX based metric performance can be summarized in the following:

1. Increasing the intra-flow interferences: The hybrid multi-channel multi-interface network divides the node interfaces into fixed interface and switchable interface, due to that, the communication between two nodes is not existed unless both nodes have same fixed channel or one of them switch its switchable interface to other fixed channel. The ETX probe message will be sent over the fixed interface without any extra delay caused by channel

switching. However, for the other channels, a copy of the broadcasting message will be put in each channel queue, waiting for their channel scheduling time. As result, the broadcasting message will not deliver on time and it may be lost due to time expiration. Based on that, the link connects two neighbors node through the fixed interface will provide a high quality link metric compared to the other links that use the switchable interface. In Fig. 2, the sold line connects the neighboring nodes using fixed interface (Fx_Interf), while the dash line connects the neighbors using switchable interface (Sw_Interf). In this figure, the path $a - b - d$ is select as the high quality path and thus increases the intra-flow interference.

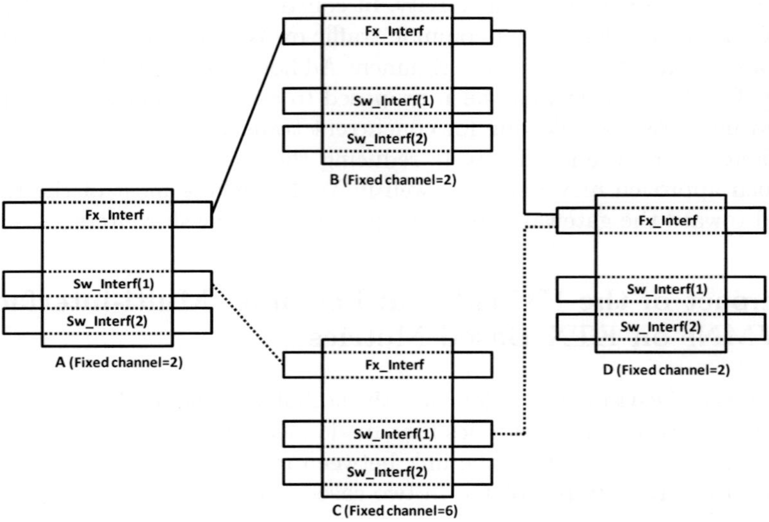

Fig. 2. Increasing the intra-flow interference.

2. Inaccurate value of ETX-based metrics: In the hybrid channel assignment protocol each node randomly select a channel for its fixed interface and this channel is called as fixed channel. The node broadcast probe packet on every channel in order to allowed the neighbors node to calculate the ETX metric. A copy of message are sent on fixed channel without any extra delay, while other copies are delayed by the channel switching delays, the delay caused by switching the interface from one channel to others. From above discussion, it is clear that the ETX value of switchable link, the link that connected the node with its neighbors using switchable interface, effectives by following parameters:
   - Number of the packet buffered on the channel queue.
   - Number of nodes using same channel.
   - And channel conflict that occurs when the two nodes switching to same channel at the same time.

All these parameters make the ETX value are inaccurate and the link connects two nodes using fixed channel always will has a good link quality as it will not affect by above parameters.

3. Difficult to calculate the reverse link direction: The ETX of a link from node X to a node Y on some channel, for example j, depends on the forward packet loss probability from X to Y on this channel, and the reverse packet loss probability from Y to X on the channel. In the single channel and static multi-channel WMNs, the calculation of EXT metric is possible, since the two nodes work on the same channel. However in hybrid multi-channel multi-interface WMNs both nodes may listen on different channels, which make it difficult to measure the reverse link on same channel that is used by the forward link. For example, the node Y measures the forward packet loss probability on channel two, while node X measures the reverse packet loss probability on different channel, channel three, see Fig 3.

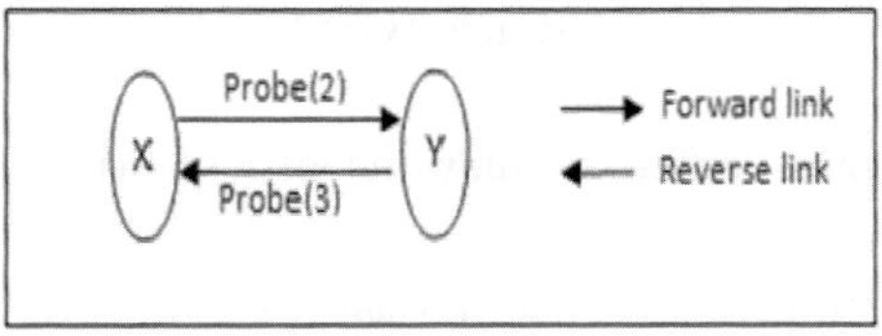

Fig. 3. Difficult to calculate the reverse link.

When using the previously described link layer protocol [11], a node X can measure the packet loss probability from its neighbors on its own fixed channel, channel two, as that is the only channel on which the node X is always listening and can correctly count the number of probe packets sent by its neighbors. During route discovery procedure, when a node Y receives a route request packet from a node X on Y's fixed channel j, the forward loss probability from X to Y on channel j is known (based on Y's earlier measurements on channel j), but the reverse loss probability from Y to X is not known.

4 Simulation Environment

The efficient of ETX-based metrics in comparison to hop count metric was evaluated using ns-2 [19]. A mesh network converge on area 1000 x 1000m was established using random distribution mesh router. Each mesh router was equipped with three wireless interfaces which dynamically turned to non-overlapping channel using hybrid channel allocation scheme [11]. For simplicity we disable the Address Resolution Protocol (ARP) messages. Concurrent UDP flows are established between randomly select source and gateway. The performances metrics are obtain by average the results from over thirty simulation runs for every experiment. The common parameters for all the simulation are listed on Table 1.

Table 1. Simulation parameters

Simulation time	150
Simulation area	1000×1000 meter^2
Propagation model	Two-ray ground reflection
Transmission range	250 meter
Traffic type	CBR(UDP)
Packet size	512 bytes
Packet rates	160 kbps
Number of nodes	100
Number of connection	50

Communication Model. IEEE 802.11 Distribution Coordination Function (DCF) [20] is used at Mac layer with channel switching delay 80 millisecond [11]. All packets are transmitted using the un-slotted Carrier sense multiple access protocol with collision avoidance (CSMA/CA).

Performance Metrics. The simulations provide the following four performance metrics:

– Packet Delivery Ratio: The ratio between the number of data packets successfully received by destination nodes and the total number of data packets sent by source nodes.
– Aggregate Goodput: The total number of application layer data bits successfully transmitted in the network per second.
– Packet Loss: The number of packets that were lost due to unavailable or incorrect routes, MAC layer collisions or through the saturation of interface queues.
– End-to-end delay of data packets: This is define as the delay between the time at which the data packet originated at source and the time it reaches the destination, and includes all possible delays caused by queuing for transmission at node, buffering the packet for detour, retransmission delays. This metric represents the quality of routing protocol.

5 Simulation Result and Discussion

To study the impact of hybrid multi-channel multi-interface WMNs characteristics on the performance of ETX-based metric, we conducted two different scenarios. In first scenario, we studied the behaviour of ETX-based metrics and HOPs on the low and high traffic load. And for the second scenario we varied the packet size.

5.1 Scenario 1: Varying Number of Flows

In this scenario, we varied the traffic load in the network by increasing the number of simultaneous connections between mesh routers and the gateway from 10

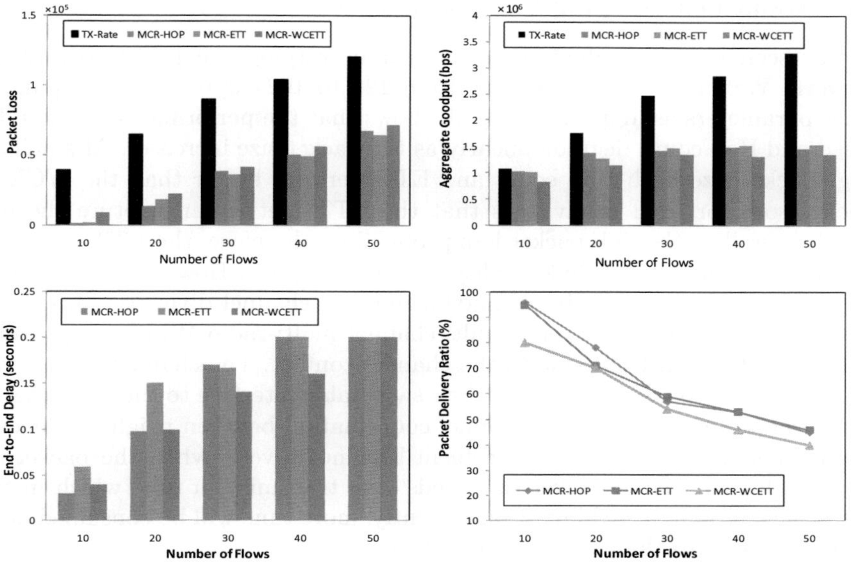

Fig. 4. Simulation results for scenario 1 (Varying the number of flows).

to 50 connections, with increment of 10 connections. The packet size for each connection was fixed to 512 bytes with 160kbps data rate. Fig. 4, shows that at low traffic load such as 10 or 20, the performance of Hop count metric is higher than the ETX-based metrics. This is because at the low traffic load, the aggregated throughput does not exceed the actual bandwidth. However, the ETX-based metrics select the path based in inaccurate ETX link value which may lead to increase the intra-flow interference. When the number of flows increased, both metrics Hop count and ETT perform almost the same. This is because the ETT metric can avoid routing the packet through congested area, while the Hop count does not have such an intelligence to avoid routing the packet through congested area. In contrast to the Hop count and ETT, the WCETT get poor performance even the number of the flow is low. This because WCETT selects the path based on maximizing the channel diversity along the path. As result, maximizing the channel diversity may lead to increase the packet collision due to that multiple nodes may switch their switchable interface to the current used channel at the same time. Other problem that degrades the performance of ETX-based metric is the hide terminal problem, which affect the goodput and increases the delay. This happen due to the collision caused by ETX probe packet sent by the hidden node especially when the traffic load exceeds the network bandwidth. Resulting in, packet retransmission and DCF back-off would decrease the network throughput and increase the end-to-end delay.

5.2 Scenario 2: Varying the Packet Size

In this scenario, we evaluated the impact of varying the packet size on the network. We varied the packet size from 128 to 1024 bytes while keeping the other parameters as in Table 1. Fig. 5, shows that the performance of the ETX-based and Hop count degrade linearly as the packet size increases. At small and large packet size both Hop count and ETT perform better than the WCETT. The reasons for such behavior is that the ETT metric can capture the path length as well as the link packet loss probability, therefore, the ETT selects the shortest path with good link quality as the best path. However, the WCETT selects the path with an aim of maximizing the channel diversity. Maximizing the channel diversity in hybrid multi-channel multi-radio WMNs may lead to increasing the packet loss due to the channel conflict. The channel conflict may happen when two nodes switching their switchable interface to the same channel at the same time. The lack of channel coordination between neighbors leads to channel conflict problem. This problem becomes severe, when the packet size is large; due to a large packet size needs long transmission time which in turn raises the collision probability. Consequently, more time will be consumed at the MAC layer due to the packet retransmission.

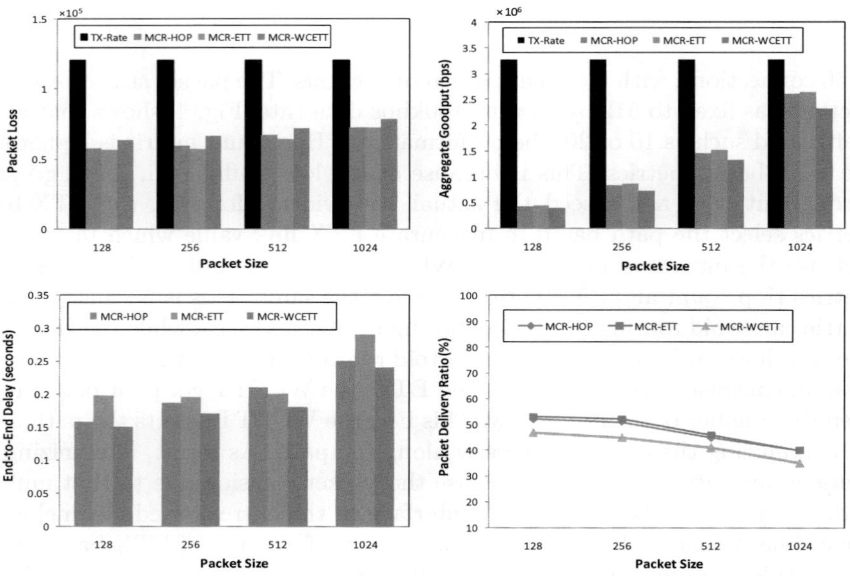

Fig. 5. Simulation results for scenario 2 (Varying the packet size).

6 Conclusions

In this paper we study the impact of hybrid multi-channel multi-interface WMNs characteristics on the performance of ETX-based metrics. Analysis and results show that the ETX-based metrics may not improve multi-channel multi-interface WMNs as it does in single channel and static multi-channel multi-interface WMNs. The major cause of such undesirable performance is the WMNs characteristics which make the ETX value inaccurate. As a results, new link metrics need to be developed take into account these characteristics as well as consider the link stability.

Acknowledgments. The research was partially supported by the Research University, RUGS No: 05-03-10-1039RU.

References

1. Akyildiz, I., Wang, X.: A survey on Wireless Mesh Networks. Communications Magazine 43(9), S23 - S30(2005)
2. Gupta, P., Kumar, P.: The Capacity of Wireless Networks. IEEE Transactions on Information Theory, 388–404 (March 2000)
3. Jiang, W., Liu, S., Zhu, Y., Zhang, Z.: Optimizing Routing Metrics for Large-scale Multi-radio Mesh Networks.: Wireless Communications. In: International Conference on Networking and Mobile Computing (WiCom), pp. 1550–1553 (2007)
4. Niculescu, D., Ganguly, S., Kim, K., Izmailov, R.: Performance of voip in a 802.11 Wireless Mesh Network. In: Proc. IEEE INFOCOM (2006)
5. Ramachandran, K.N., Belding, E.M., Almeroth, K.C., Buddhikot, M.M.: Interference-aware Channel Assignment in Multi-radio Wireless Mesh Networks. In: Proceedings of the 25th IEEE International Conference on Computer Communications INFOCOM 2006, pp. 1–12 (2006)
6. Raniwala, A., Gopalan, K., cker Chiueh, T.: Centralized Channel Assignment and Routing Algorithms for Multi-channel Wireless Mesh Networks. ACM Mobile Computing and Communications Review 8, 50–65 (2004)
7. Raniwala, A., cker Chiueh, T.: Architecture and Algorithms for an ieee 802.11-based Multi-channel Wireless Mesh Network. In: Proceedings of the 24th Annual Joint Conference of the IEEE Computer and Communications SocietiesINFOCOM 2005, vol. 3, pp. 2223–2234 (2005)
8. Draves, R., Padhye, J., Zill, B.: Routing in Multiradio, Multi-hop Wireless Mesh Networks. In: Proceedings of the 10th Annual International Conference on Mobile Computing and Networking MobiCom 2004, pp. 114–128 (2004)
9. Marina, M., Das, S.R., Subramanian, A.P.: A topology Control Approach for Utilizing Multiple Channels in Multi-radio Wireless Mesh Networks. Comput. Netw. 54, 241–256 (2010)
10. So, J., Vaidya, N.H.: A Routing Protocol for Utilizing Multiple Channels in Multi-Hop Wire-less Networks with a Single Transceiver. Technical report, Dept. of Computer Science and Coordinated Science Laboratory, University of Illinois at Urbana-Champaign (October 2004)

11. Kyasanur, P., Vaidya, N.: Routing and Link-layer Protocols for Multi-channel Multi-interface Ad hoc Wireless Networks. SIGMOBILE Mob. Comput. Commun. Rev. 10, 31–43 (2006)
12. De Couto, D.S.J., Aguayo, D., Bicket, J., Morris, R.: A High-Throughput Path Metric for Multi-Hop Wireless Routing. In: ACM Mobicom (2003)
13. Wang, Kravets, R.: Designing Routing Metrics for Mesh Networks. In: IEEE Wksp. Wireless Mesh Networks (September 2005)
14. Mogaibel, H., Othman, M.: Review of Routing Protocols and It's Metrics for Wireless Mesh Networks. In: Computer Science and Information Technology -Spring Conference IACSITSC 2009, pp. 62–70 (2009)
15. Kim, S.H., Suh, Y.J.: Local Channel Information Assisted Channel Assignment for Multi-channel Wireless Mesh Networks. In: IEEE Vehicular Technology Conference (VTC), pp. 2611–2615 (2008)
16. Chen, J., Jia, J., Wen, Y., zhao, D., Liu, J. : A genetic Approach to Channel Assignment for Multi-radio Multi-channel Wireless Mesh Networks. In: Proceedings of the First ACM/SIGEVO Summit on Genetic and Evolutionary Computation GEC 2009, pp. 39–46 (2009)
17. Tang, J., Xue, G., Zhang, W.: Interference-aware Topology Control and Qos Routing in Multi-channel Wireless Mesh Networks. In: Proceedings of the 6th ACM International Symposium on Mobile Ad hoc Networking and Computing MobiHoc 2005, pp. 68–77 (2005)
18. Gong, M.X., Midkiff, S.F., Mao, S.: On-demand Routing and Channel Assignment in Multi-channel Mobile Ad hoc Networks. Ad Hoc Netw. 7(1), 63–78 (2009)
19. NS. The Network Simulator (1989), http://www.isi.edu/nsnam/ns/
20. Wireless LAN Medium Access Control (MAC) and Physical Layer (PHY) Specifications. IEEE Standard 802.11 (June 1999)

Time-Spatial Recursive Denoising for LLL Images Based on Motion Detection

Xia Penghao[1], Zhang Junju[1,*], He Junfeng[2], Nie Weile[2], Xu Hui[1] and He Tingting[1]

[1]School of Electronic Engineering and Optic-electronic Technology, Nanjing University of Science and Technology, 200 Xiaolingwei, Nanjing 210094, China,
zj_w1231@163.com
[2] Xi'an Institute of Applied Optics, Xi'an 710065, China

Abstract. In LLL images, there is a lot of random flicker noise. When denoising in LLL image, the traditional time-domain recursive noise reduction will bring smear in moving images. An improved algorithm is needed. To improve the time-domain recursive filter, the block that contain moving target in images should not participate the recursive algorithm. Time-spatial recursive denoising for LLL images based on motion detection is proposed to improve the time-domain recursive noise reduction. The blocks that contain moving object and background are processed respectively. Finally, simulation results show the effectiveness of the proposed method.

Keywords: LLL, image processing, filter, time-spatial recursive, denoising.

1 Introduction

Compared with images in normal illumination, LLL (Low light level) image features for low signal-to-noise ratio, low contrast, low spatial resolution and blurred vision. Currently, many algorithms were used to reduce noise in LLL image, such as wavelet transform filtering, histogram equalization, median filtering and contrast stretching, etc. But for LLL images, because of the low signal-to-noise ratio and the randomness and diversity of noise, these methods did not achieve satisfied result.

In this paper, we proposed time-spatial recursive denoising for LLL images based on motion detection that combined temporal filter and spatial filter. In this algorithm, the recursive coefficient is determined by the movement of adjacent frames to reduce the smear effect. So we can combine the advantages of both and improve the image quality.

X. Wan (Ed.): Electrical Power Systems and Computers, LNEE 99, pp. 161–168.
springerlink.com

2 Time-Domain Recursive Filter

Time-domain filter includes multi-frame accumulation average filter and time domain recursive filter while spatial filter includes mean filtering, median filtering and so on. There is a lot of glimmer flickering random noise in LLL images, the noise has no relevance between frames with zero mean.

2.1 Time-Domain Recursive Filter

In 1971, the CBS proposed time-domain recursive filter that can effectively reduce dynamic image noise and improve the quality of image. The mathematical expression of the time-domain recursive filter is written as follow:

$$Y_n' = (1-k)Y_n + kY_{n-1}' \tag{1}$$

where Y_n is current frame and k is the filter coefficients, between 0 and 1. Y_n' is the output of the current frame image to be processed. Y_{n-1}' is the output of the last frame image. Here the current frame and the output of previous frame are averaged by respective weight. In fact, the output is a weighted average of the all previous frame.

Figure 1 show typical example. The left is the input image and the right is the image that was processed with k=0.875. Compared the two images, time-domain recursive filter can effectively filter out random flicker noise and improve image visibility. Figure 2 illustrates the frequency characteristics of the recursive filter of time domain.

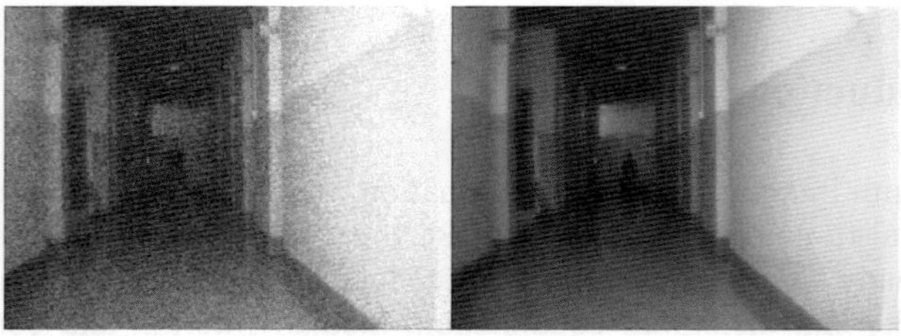

Fig. 1. LLL image and LLL image processed by time-domain recursive filtering.

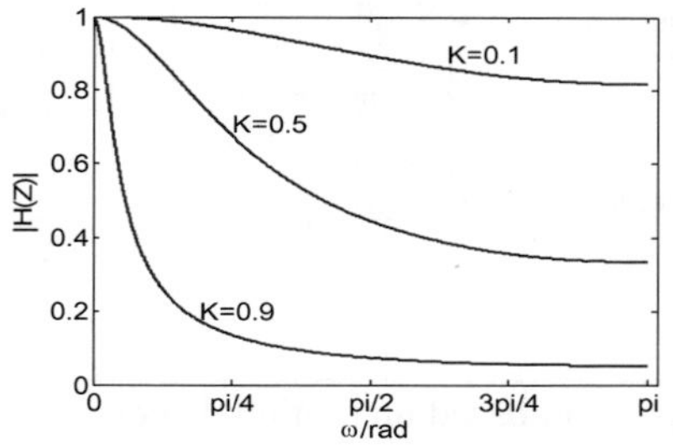

Fig. 2. Frequency characteristics of the recursive filter.

2.2 Selection Coefficients k in Recursive Filter

In general, we use a rate of motion between different frame images ε to describe the movement intensity. First we set a threshold T_h, when the difference of pixel gray value in corresponding position between different frames is greater than T_h, the pixel was thought of a dynamic pixel. Otherwise, the pixel was a static pixel. The number of dynamic pixels $|y_n - y_{n-1}| > T_h$ in current frame can be gotten from statistics. The number of these pixels is P_m. If the total number of pixels in an image is P, the frame rate of motion is defined as:

$$\varepsilon = \frac{P_m}{P} \; . \tag{2}$$

The larger ε is, the stronger the movement exists[2]. We can take the rate of motion ε between frames as the basis for selecting coefficients. There are many methods to select filter coefficient. Selecting filter coefficients should depend on the actual application conditions and image characteristics.

3 Time-Spatial Recursive Denoising Based on Motion Detection

Time-spatial recursive denoising based on motion detection is an innovative noise reduction algorithm. It improved current filter and combine the temporal filtering and spatial filtering brings together. The algorithm balances SNR improvement and motion blur. The algorithm mainly consists of the noise estimation, motion detection, time domain filtering and spatial filtering. The block diagram shown in Figure 3.

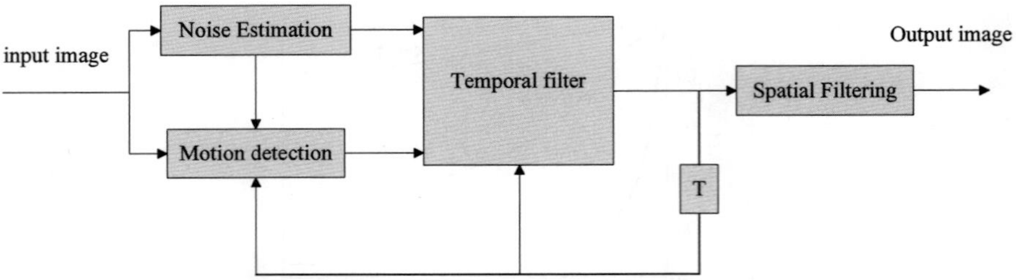

Fig. 3. A recursive algorithm based on motion detection in time-domain real-time

Among them, the noise estimation is used to measure the noise pollution of the image. Motion detection will divide the whole image into small regions and group these small regions into three types: matching with the previous frame, matching with the next frame and no match. Three sorts of regions are processed by different approaches. Most of the noise was effectively eliminated, avoiding motion blur at the same time. Finally, the image was processed by spatial filtering. Spatial filtering can suppress the residual noise, especially the noise that were not processed by time-domain filtering [3]. T is the time delay. The results of time-domain recursive were delayed by T .

3.1 Noise Estimation

According to the illumination environmental factors, the standard deviation of the noise was corrected with a empirical value p, $p > 1$. The noise level is calculated as follows.The noise standard deviation of a size $M \times N$ pixel of the image region is calculated as follows:

$$\mu = \frac{\sum_{m=1}^{M}\sum_{n=1}^{N} f(m,n)}{M \times N} \tag{3}$$

$$\sigma^2 = \frac{\sum_{m=1}^{M}\sum_{n=1}^{N} (f(m,n) - \mu)^2}{M \times N} \tag{4}$$

$$\delta = \sigma = \sqrt{\frac{\sum_{m=1}^{M}\sum_{n=1}^{N} (f(m,n) - \mu)^2}{M \times N}} . \tag{5}$$

where $f(m,n)$ is the gray value of pixel (m,n). μ is the mean of pixels in this region, σ^2 is the noise variance and δ is the standard deviation of the noise in this region.

The noise level is defined as follows:

$$G_n = p \times \delta . \tag{6}$$

3.2 Motion Detection

The algorithm takes the output of the previous image and the next frame image as reference frames. The two frames are most closely with the current frame. Therefore, it is possible to lower the blocking effect caused by inaccurate match. We select a threshold T_h as a standard of motion intensity and calculate the mean absolute difference MAD of a small corresponding region in the current and the next frames. MAD is calculated as follows :

$$MAD = \frac{\sum_{m=1}^{M}\sum_{n=1}^{N}\left|f_c(m,n) - f_r(m,n)\right|}{M \times N} . \tag{7}$$

$$T_h = G_n = p \times \delta . \tag{8}$$

where $f_c(m,n)$ is the pixel gray value in small region of the current frame, $f_r(m,n)$ is pixel gray value in a corresponding small region of the reference frame . $M \times N$ is the total number of pixels in a small region. T_h is the threshold of intensity of movement detection, which can be calculated with the noise level G_n . δ is the noise standard deviation. p is the correction factor, $p > 1$. Mean absolute difference MAD has two sources. One is caused by the target movement, the other is caused by noise. The essence of motion detection is to separate these two sources. The noise level G_n , defined in 3.2, is the basis to distinguish the two sources. If the difference is less than the noise level G_n , the difference is caused by noise. Such region does not contain a moving target. If the difference is greater than the noise level, the difference is caused by the movement and noise together. So the region contains the moving object.

The mean absolute difference of the current frame image and the output of previous frame is MAD_1 and the mean absolute difference of the current frame image and the next image is MAD_2 .In the first motion detection, MAD_1 was compared with threshold of motion intensity detection T_h . If $MAD_1 \le T_h$, the movement intensity in the region is little and this region will processed by time-domain filtering match with the last output frame(the first match) .

3.3 Filter Processing

Time-spatial recursive denoising based on motion detection contains filter in space-domain and filtering in time-domain. The specific processing method is shown as (9):

$$\begin{cases} y_n' = ky_{n-1}' + (1-k)y_n & \text{the first match} \\ y_n' = ky_{n+1}' + (1-k)y_n & \text{the second match}. \\ y_n' = y_n & \text{no match} \end{cases} \tag{9}$$

Where, y_n is the current frame. y_n' is the output of the current frame processed by time-domain filter. y_{n-1}' is the output of the previous frame. y_{n+1} is the next frame . k is the filter coefficient. The greater the value of k is, the stronger the effect of the filtering is. In the algorithm, for the first match, the region of current frame and output of the previous recursion took part in the recursion operation. For the second match, the output was gotten by means of weighted average from the region of current frame and the corresponding region of the next frame. For the case of no matching, the region was dealt with none treatment.

4 The Simulation Results

In order to verify time-spatial recursive denoising for LLL images based on motion detection, we capture a video in low-light environment, with a target running in view. In MALAB, the video was processed by the algorithm. First, the result was evaluated by subjective observing. Then result was evaluated by the indicator of goal clarity, a objective evaluation, to verify the superiority of this algorithm. In MATLAB simulation, the size of small region adopted this algorithm is 16×16, correction factor $p = 1.3$.

After the video was processed by recursive algorithm based on motion detection in time-domain, the 235th, 281th, 303th frame image were captured from the original video and the processed video.

Figure 4 shows original images on the left and processed images on the right. Noise has been reduced, particularly random flicker noise. The contrast was significantly improved , the scene is more clear, and the image quality is improved.

The 280th to 319th frame images from original video and processed video was gotten to calculate target definition and evaluate the effect of the algorithm. In this paper, gradient magnitude was used to evaluate the target clarity. Gradient magnitude can reflect the clarity of the image edge. The larger gradient magnitude is, the better clarity of the image has. We evaluated image target clarity of LLL television as follows: Target region was first selected to be evaluated, which region contains the edge information; Laplace operator of each pixel within the target region of the gradient was individually calculated, as type (10) shows. The target clarity D was obtained by accumulating the N-of the selected average gradient value, as shown in Equation (11).

$$g(x,y)=\begin{vmatrix}8f(x,y)-f(x-1,y-1)-f(x-1,y)-f(x-1,y+1)-f(x,y-1)-\\ f(x,y+1)-f(x+1,y-1)-f(x+1,y)-f(x+1,y+1)\end{vmatrix}\ . \tag{20}$$

$$D=\frac{1}{N}\sum_{i=1}^{N}g_i\ . \tag{11}$$

Where $f(x,y)$ is the gray value at the pixel (x,y) and $g(x,y)$ is the point of the gradient. Figure 4 shows evaluated results. As we can see that the clarity of the algorithm is better than the one of traditional time-domain recursive approach.

Fig. 4. Original images and processed images of low light level television.

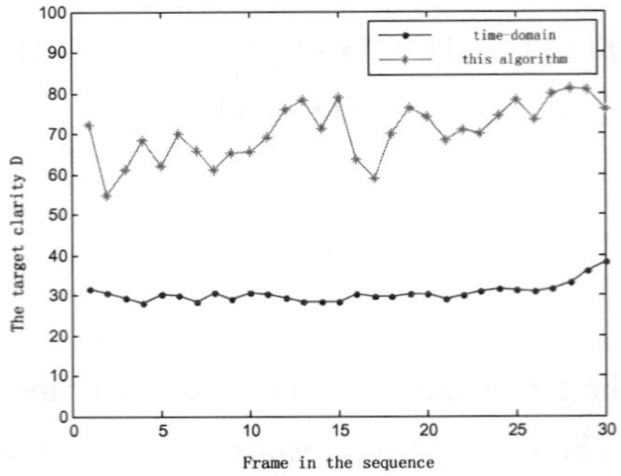

Fig. 5 Evaluation result of starget clarity in 50 ~ 79 image .

5 The References Section

This paper introduces time-spatial recursive denoising for LLL images based on motion detection. The algorithm combines the time-domain recursive noise reduction and spatial noise reduction. According to the results of movement intensity, blocks that contain the moving target adopt special approach in space domain effectively eliminating the large number of random flicker noise. What's more, the smear phenomenon caused by the traditional time-domain algorithm disappeared. The algorithm will be optimized and realized in FPGA to reach processing in real time.

References

1. Wang, Z., Bovik, A.C., Sheikh, H.R., Simoncelli, E.P.: Image Quality Assessment: From Error Visibility to Structural Similarity. IEEE Transactions on Image Processing 13(4), 600–612 (2004)
2. Zlokolica, V., Pizurica, A., Philips, W.: Recursive temporal denoising and motion estimation of video. In: International Conference on Image Processing(ICIP), pp. 1465–1468 (2004)
3. Bhagavathy, S., Llach, J.: IEEE International Conference on Adaptive Spatio Temporal Video Noise Filtering fo High Quality Applcation, vol. 1, pp. I-761-I-764 (2007)

Research of Social Insurance System Based on DW

Yuhua Feng[1], Ming Peng[2], and Longzhen Duan[1]

[1] Information Engineering School, Nanchang University
Nanchang 330031, P. R. China
[2] Broadcasting TV university of Jiangxi, 330046, P. R. China
Nanchang 330001, P. R. China
fengyuhua@ncu.edu.cn

Abstract. The paper analysis the tradition decision support system firstly, and then introduced the theory of data warehouse. In combination with the data and business's specialty of the original social insurance management system, the paper applies data warehouse to social insurance decision support system and gives the way of devise thinking about social insurance data warehouse. Finally, the further research direction is pointed out.

Keywords: data warehouse (DW), data source, and model.

1 Introduction

Decision support system (called DSS) is to assist decision-makers through the data, models and knowledge to human-computer interaction for semi-structured or non-structured decision-making computer applications. The basic structure of decision support system mainly consists of four parts, namely, the data part, the model part, reasoning part and the part of human-computer interaction; it can help to solve semi-structured and non-structured decision problems, and to man-machine dialogue, the main form of work as a system. The DSS goal is efficient, that is to find ways to make things run better as far as possible in order to enhance the capacity and effectiveness of decision-making; .The traditional method of establishing decision support system can be used extensively to build all kinds of DSS, including intelligence DSS, group DSS and general DSS, but the traditional method of DSS has some disadvantages. For example, separate components, numerous interfaces, and complicate system are difficult to design and realize, uneasy to expand, and so on. During exploring the DSS for the social insurance DSS in Jiangxi province, some similar problems are met. Although it is a general DSS, the social insurance DSS is a complicated and dynamic open system, which has many different data resource and complex method models. If we used the traditional method of establishing the DSS, the project is bound to have many above problems.

Now with the increased availability of data collected from the different sources and the implementation of enterprise-wise databases the amount of data that company is growing at a phenomenal rate. It becomes increasing important for the companies to be able to better manage their databases.

X. Wan (Ed.): Electrical Power Systems and Computers, LNEE 99, pp. 169–174.
springerlink.com

2 Data Warehouse Technology

2.1 The Theory of Data Warehouse

A data warehouse is a consolidated database, which contains a huge integrated amount of data organized around major subject areas of an organization that span over a period of time to serve a historic purpose. W.H. Inmon defines data warehouse as a collection of integrated, subject-oriented, time-variant and nonvolatile database designed to aid decision support functions. Data warehouse data come from numerous data sources. These data source are normally built to satisfy the day-to-day activities of an enterprise. Hence, the data contained in these operational data sources do not have time stamps and are updateable. A data warehouse is primarily built to integrate operational databases and other legacy systems over a long period of time for decision support and analytical data querying purpose.

It is apparent from the above definitions of a data warehouse that the data contained in a data warehouse are drawn from different sources. The different data sources might have been implemented on different computer hardware and software platforms. For our project, a branch of social insurance in Jiangxi province has a number of units.

2.2 Data Warehouse Construction

Data warehouse (DW) data originates from a variety of different sources. These could include: 1) The DW database needs to be designed and integrated in a way which will eliminate many of the inconsistencies which have evolved over the years in many of the legacy system Operational databases and local application data stores. 2) Meta data (technical and business information about the data) is an integral component of a robust Data Warehouse Infrastructure. Without this information, it will be extremely difficult for both administrators of the Data Warehouse and users of the data to know and understand the data means and its appropriate usage. Metadata is also vital for the administrators for change management and impact analysis.3) A metadata repository is required to maintain descriptive information of all available data in the information warehouse. The structure of the Metadata enables business users with easy retrieval and access to the required information in a manner which is easily understood in business terms.

The data quality of these data stores should be managed by a process of certification, by the owners of the data, to assure all interested users that the data has met the minimum threshold levels of acceptable quality. Important factors of quality, which need to be monitored, include timeliness and completeness of the data stored in the data warehouse. Performance indicators are required to enable monitoring.

Some important design characteristics of information warehouse data-stores which distinguish them from existing production operational data stores include: 1) *None Volatile:* Real time updates occur to selective data warehouse data stores. Most data stores are refreshed in batch, not less than every 24 hours. Time consistent context of data across different sources need to be maintained. 2) *Time Variant*: A 3 to 7 year time horizon for maintaining data is normal for the information warehouse. The 7 year retention is typically driven by regulatory requirements for the retention of data.

The data is periodic and maintained as a series of snapshots, taken as of some moment in time. The key structure of data tables must contain some element of time. 3) *Granular structure*: Data is maintained at various levels of granularity and summarization. Frequently access data can be rejoined and summarized to enable quick turnaround on queries and reports. Detailed and atomic level data will be maintained alongside summarized and pre-calculated data. New approaches to data storage are evolving such as “multi temperature” data storage to minimize costs associated with maintaining large and multi-year business data. The concept behind ‘multi-temperature’ data storage strategies is to optimize data access for more frequently used data and isolating infrequently accessed data. DW minimizes the need to maintain historical information within the operational application data stores. Operational data-bases in the production environment will only maintain historic information if it is absolutely required for processing in “transaction-based” production applications. Otherwise, all historical data beyond "current value" will be maintained in the DW data stores for access and use by business users for informational analysis and reporting purposes. Costs for storing history data will be optimized by using tables containing different levels of summarization. A successful approach in migrating towards an effectively architected enterprise warehouse environment is the one which requires much greater levels of involvement from business users than those typically required in the development of operational based applications in production. The best approach involves designing and building the warehouse data environment one increment at a time. This way, technical and business community staff can work closely together through a process of continuous iteration, to design and implement each component of the warehouse until the structure and content of the data, in each component, meets the satisfaction of the business. The starting point for the migration is the creation of a DW data model. Initially the model will include the definition and confirmation of subject areas (business and application specific) and high-level list of entities for the information warehouse data model. This level of the model will help to chunk out the planned warehouse data environment into components prioritized by business requirements, specific needs of business user groups, and the readiness of the users to move ahead with this initiative. The design of each enterprise warehouse component will involve a number of transformations and refine mend activities to the related areas of the DW.

3 Design for the Social Insurance Data Warehouse

We should design an efficacious data warehouse of social insurance firstly for a DSS of social insurance. We know data warehouse is data environments for analytical process, so we cannot use the traditional develop methods to design it.

3.1 Investigation of Data Sources

Data flows of data warehouse system begin with data sources. Present social insurance management systems are complex to integrate and commonly shared because of different named system, code standard and key code, which are huge obstacles to statistics and decision analysis.

Though our investigation, the data sources of current social insurance management system have some specially characteristics as the following:

- Many different data source

Data comes from numerous data sources. Because of different management system, the social insurance systems carry out respective. Every local has its independent online system environment, which lead to many problems, such as duplicate information, many different named rules and inconsistent data item. As far as our province is concerned, many relational database management systems involved Oracle Sybase SQL Server, FoxPro, etc. The relevant data is stored in institution of town, city or province.

- Data quantity

Social insurance has collected relative data for a long time, so now the data quantity has reached thousand mage while history data had reached to hundred GB; however, the valuable data resource has not been further exploited and utilized.

- Stored online for a long time

For example, the endowment insurance information is required to store for 50years, maybe for hundred years.

- Unstable data structure

Because of the instable insurance policy, the management mode and the traffic process have great mutability, which should reflect the data structure of management system.

3.2 Analysis of Decision Demands

Through investigation and research, some decision demands are following:

- Analysis data means process the data through data decomposition and data summarize, take the endowment insurance as example, average pension, the total of pension, the constitutor of pension, pension fee onetime.
- Prediction means that we can simply forecast analysis for future through researching the data's variant tendency based on current data. For example, the increase tendency of pension fee, the tendency of retirement life span, etc.

Suggestions for decision making means that we give some suggestions deepened on researching agedness and fund accumulation questions. For example, personal account interest rate, pension adjustment rate.

3.3 Design the Subject-Oriented Data Warehouse

The data structure of data warehouse is very different from the traditional database, since the data warehouse usually obtains valuable information from a great quantity data which be collected for over the many years. We organized the data with subject-oriented method.

Design the data warehouse always has two methods: star model and snow model, which are made of correlative face table and dimension tables. The star model is a relational database structure that fact tables in the middle and the dimension tables are around the fact table. Every dimension tables are related to fact table by key codes.

The snow model is a extend form of star model, dimension tables are divided into direct correlative main dimension table and the second dimension tables correlated to main dimension tables.

Now, current insurance data is involved institutions of town, city and province. If we want to analysis the fluctuant policyholder number or pension pay in some area from different aspect such as area, time, economic. So we design the pension, fund account are measure and the area, time, economic are dimension.

In this model, it included one fact table and three dimension tables. Which the fact table is the monthly account file through summarize and arrangement, and the time, area, economic type, trade became the four dimension tables. So we can build data cube and are very easy to cut data into slices and other operations from different dimensions.

With the business's development and decisions demand's variation, the structure of multi-dimension is variable, so it further meets the analytical requirement.

4 Conclusions and Future Work

In our country, the data warehouse is an ascendant field on data management technology and market, and has a bright prospect. When people focus on the main problems of current data warehouse, the new generation of data warehouse already be intentioned by researchers like "budding", such as object-oriented data warehouse, active data warehouse, dynamic query optimization, key mission data warehouse, etc. Anyhow, data warehouse is based on data management and utilization of comprehensive technology and solutions, and it will become the new round of database market growth, and also will become an important part of the next generation application system.

Data warehouse as a database management technology has developed in china for a short time, now there were some successful applications of data warehouse in some fields, but very lack in insurance. The paper applies data warehouse to social insurance decision support system and only tentatively gives the way of devise thinking about social insurance data warehouse, but many aspects of data warehouse deserve further probing and investigation.

References

1. Inmon, W.H.: Building the data warehouse, 2nd edn. John Wiley and sons Inc., New York (1996)
2. Pressmen, R.S.: Software Engineering: A practitioner's approach(5th). Machine press, meihong, china (translated); Hammer, M., Champy, J.: Reengineering Corpotation. HarperBusiness, NY (2001)
3. Gorla, N.: Features to consider in a data warehousing system. Communications of the ACM 46(11) (November 2003)

4. Microsoft Corporation, Data Warehouse Design Considerations, `http://www.microsoft.com/technet/prodtechnol/sql/2000/reskit/part5/c1761.mspx` (retrieved on January 2008)
5. Sen, A., Sinha, A.: A comparison of data Warehousing methodologies. Communications of the ACM 48(3) (March 2005)
6. Winter, R., Strauch, B.: Information requirements engineering for data warehouse systems. In: Proceedings of ACM symposium on applied computing. Nicosia, Cyprus (March 2004)
7. Zepeda, L., Celma, M., Zatarain, R.: A methodological framework for conceptual data warehouse design. In: Proceedings of the 43rd ACM southeast conference. Kennesaw, GA (March 2005)
8. Rifaie, M., Kianmehr, K.: Data Warehouse Architecture and Design. In: IEEE IRI 2008, Las Vegas, Nevada, July 13-15 (2008)

Adaptive Resource Discovery in Grid Computing Based on Reinforcement Learning

Mohammad Ali Jabraeil Jamali and Yalda Sani

Islamic Azad University, Shabestar Branch, Shabestar, Iran
m_jamali@itrc.ac.ir, sani.yalda@gmail.com

Abstract. Grid is a distributed computing environment. There are lots of resources in grid environment that are heterogeneous and geographically distributed. By receiving a resource request the resource discovery mechanism should return an appropriate resource if there exist one. Resource discovery is a challenging problem because of the heterogeneity and distribution of resources. In this paper, we propose and evaluate an adaptive resource discovery algorithm using reinforcement learning for grid computing that can be used for multi resource requests. The algorithm achieves the most suitable node that can satisfy the requested resource by using the past experience of agents. We compare our model with random walk resource discovery through simulation and the results show that the proposed algorithm provides higher success rate, less message passing and shorter response time. Also the algorithm leads to load balancing in whole grid.

Keywords: Grid Resource Discovery, Multi Resource Requests, Reinforcement Learning, Adaptive.

1 Introduction

As a new network computing platform, grid aims to construct an infrastructure that fully supports various resources sharing for different users. Grid can be considered as an environment integrating computing and resource, or a computing resource pool [1]. There are lots of resources in grid environment that are heterogeneous and geographically distributed. As one of the basic services of grid that building the connection between the resources users unknown and users, grid resource discovery is the process to find the suitable resource for users in grid [1]. Users are not interested in where resources actually are. Just by giving a description about resources they desire, the resource discovery mechanism will find a set of resources that match the user's description if there exists one [2]. In the environment, heterogeneous computational resources spread across geographically distributed areas worldwide. The resources such as storage space and CPU are dynamic, and nodes can enter or leave the system unpredictably [3].

Traditional resource discovery approaches relying on centralized or hierarchical policies cannot tolerate such an environment. In the resource discovery approaches relying on centralized policies, all nodes report their available resources to a central

X. Wan (Ed.): Electrical Power Systems and Computers, LNEE 99, pp. 175–181.
springerlink.com

grid node. When a node needs resources, it resorts to the central grid node for the information of resource providers who have its required resources. Since the central grid node needs to store all the information of available resources in the grid system, and needs to process the resource requests from all of the nodes in the system, it could easily become a bottleneck and is unable to efficiently process the resource requests, leading to low performance of the grid system. In the resource discovery approaches relying on hierarchical policies, all nodes are formed into a hierarchical structure with a number of levels. A node can ask for the information of available resources from the nodes in the above level [3]. The hierarchical approach, while being a popular, well-assessed technique for managing large repositories of quasi-static data in distributed systems (such as the Internet DNS), does not fit well the dynamic nature of resource availability data. This purpose can be better served by peer-to-peer approaches [4].

By considering the advantages of peer to peer architecture we use it in our work. A new resource discovery algorithm using reinforcement learning is introduced in this paper. In this algorithm by receiving more queries each node learn more information about its neighbors. And by using this information it can answer the query in a better way. This algorithm can be used in multi resource queries.

The paper is structured as follow. In section 2 we briefly review related work. In section 3 we introduce our proposed adaptive algorithm using reinforcement learning. Simulation and results are presented in section 4. Finally, section 5 is our conclusion.

2 Related Works

There are many resource discovery methods in grid environment. Some of these methods are using centralized or hierarchical architecture, but as mentioned in introduction these architectures have some disadvantages so the algorithms for peer to peer environment are preferred.

A resource discovery algorithm with probe feedback mechanism based on advanced reservation is introduced in [9]. The most important factor of the algorithm is that if the discovery failed it can rediscover the requested resource on its response message back way. So it provides more chances to find the resource.

In [10] a cashed based optimized random walk protocol for resource discovery in large scale dynamic grids is proposed. The protocol can save significant bandwidth and reduce the network bandwidth consumption.

Sanya Tangpongprasit et al. [8] propose a reservation algorithm to find an appropriate resource in a grid environment. In the forward path, the mechanism will check the local node. If the node has a resource that matches the request, it will be added into the request and reserved. The mechanism uses the experienced-based plus random rule to decide which node to forward the request. In the backward path, if there is more than one resource reserved, only the one added in the request can be chosen, and others should be released [2].

In [5] a simple cashed based mechanism is proposed for peer to peer resources. The algorithm is based on push-pull strategy and each peer maintains a local cache of messages and uses the information for routing and discovering a resource.

A scalable peer to peer based proximity-aware multi resource discovery scheme is presented in [3]. It collects the resource information of physically close nodes together, and maps resource requests from requesters to the resource information pool of its physically close nodes. In addition, it relies on a single DHT and achieves balanced resource discovery load distribution, enhancing the system scalability.

In [6] a large scale peer to peer grid system which employs an ant colony optimization algorithm to locate the required resources is proposed. This method avoids a large scale flat flooding and supports multi attribute range queries.

3 Adaptive Resource Discovery

3.1 Resource Classification

There are many types of resources with variety of attributes in grid environment. We can divide these resources in two types, Static resources and dynamic resources. Static resources such as operating system type, software and etc. Dynamic resources whose value is not stable and can change over the time, such as free memory size, cpu load, etc. queries for static resources are like "os = linux", so having the information of their existence is enough. But queries for dynamic resources are like "free memory size > 7 G byte" or "cpu speed > 2 GHz", and we should try to find a best fit resource to satisfy the request, so we divide the possible range of dynamic resources to some sub ranges and each of these sub ranges act as an individual resource.

For example if the possible range for memory size is 1Gbyte to 50 G byte, it can be divided in the following sub ranges: {< 1G}, {>1G, <5G}, {>5G, <10G}, {>10G, <15G}, {>15G, <20G}, ... , {>45, <50}

When a user send a request for memory-size of 7G bytes, the resource discovery algorithm tries to find resource {>5G, <10G} among provided resources in grid environment, if there isn't any resource {>5G, 10G<}, the algorithm starts to discover a resource that can satisfy the request such as {>10G, <15G}. On the other hand this mechanism prevents wasting grid resources by choosing the best fit resource for the requested resource.

3.2 Reinforcement Learning

Reinforcement learning is learning what to do, how to map situations to actions, so as to maximize a numerical reward signal. The learner is not told which actions to take, as in most forms of machine learning, but instead must discover which actions yield the most reward by trying them. Reinforcement learning uses a formal framework defining the interaction between agent and environment in terms of states, actions, and rewards [7]. In each step an agent take an action and for that action it receives a reward from environment. Depending on receiving rewards the agent learns to take actions with better rewards in next steps.

3.3 Resource Discovery Using Reinforcement Learning

In resource discovery, agents learn which node can satisfy the requested resource by using their past experience. N is the set of grid nodes and node $n \in N$ is the local node with $n' \in N$ as a neighbor. R is the set of provided resources in grid environment and $r \in R$ is a requested resource.

Each node n is an agent and has a Q-value table which stores the estimated value of neighbor nodes for requested resource r. Q-value indicates the efficiency of a node for requested resource r. By using this value, the agents can decide which neighbor can satisfy the requested resource and they send the request to that neighbor.

At first the table is empty. When a request for resource r is received by node n, for the first time there wouldn't be any data for resource r so the item r will be added to the table of node n and the request will be forwarded randomly to one of the neighbors of node n. By time passing and receiving more requests the table will grow and the value of nodes will be updated so the knowledge of agents about the environment will grow and they can choose the optimal actions.

Table 1. Q-value Table of node n

resources	Neighbors of node n			
	n'_1	n'_2	...	n'_k
r_1	$Q(r_1,n'_1)$	$Q(r_1,n'_2)$		$Q(r_1,n'_k)$
r_2	$Q(r_2,n'_1)$	$Q(r_2,n'_2)$		$Q(r_2,n'_k)$
r_3	$Q(r_3,n'_1)$	$Q(r_3,n'_2)$		$Q(r_3,n'_k)$
...				
r_m	$Q(r_m,n'_1)$	$Q(r_m,n'_2)$		$Q(r_m,n'_k)$

A resource request message will be received by local node n. The local node n checks whether there is a local matching resource for requested resource. If there is one, a matching response is sent to the user. If there isn't any matching resource, the local node n checks its Q-value table. Which neighbor n' who has the highest value for resource r will be chosen and the request message will be forwarded to that neighbor. For each action (message forwarding to a neighbor) there is a reward, so by getting this reward the value of that node increases or decreases. The reward is calculated using this formula:

$$Rew = D_{n'} + \beta\,(\,Load_{n,n'} + U_{r,n'}\,)$$

If the node n' has a resource to satisfy the request then n' is a destination node and $D_{n'}$ will be 100 so it increases the value of node n' for resource r. If n' is not a destination node the $D_{n'}$ will be equal to 0. β is a constant between -1 and 0.

Users are interested in getting their answers in a short time so a path with a low load is preferred. $Load_{n,n'}$ presents the average load of link between node n and n'. Choosing a neighbor with a low load link will have higher reward. Over the time this factor leads to load balancing in whole grid.

Another factor that is important in calculating the reward is the unavailability of the resource of a node or the node itself. $U_{r,n'}$ is the percentage of unavailability of resource r in node n' (consider that here, r is the local resource) that is equal to number of failed connections divided by total connections. Sometimes it is possible that node n' leaves the network for a while or the provider shots down the system or for a reason node n' can't satisfy the request for resources that it provided and etc. we call these situations failed connections.

In this point, after taking action and calculating the reward, the Q-value of node n' for resource r in value table of node n should be updated. The updating formula is:

$$Q_n(r,n') = Q_n(r,n') + \alpha\,[\,rew + \gamma\,\max_{n''} Q_{n'}(r,n'') - Q_n(r,n')]$$

In which α and γ are constants between 0 and 1. $Q_n(r,n')$ is the Q-value of node n' for resource r in the table of node n. rew is the calculated reward for choosing node n' as a next node. n'' is a neighbor node of node n' and $Q_{n'}(r,n'')$ is the Q-value of node n'' in Q-value table of node n'. If the node n' doesn't have any resource for user request, it forwards request to a neighbor. Each node who receives the resource request message will use the same algorithm until the matching resource is found or the TTL decreases to zero.

3.4 Multi Resource Requests

Sometimes users request more than one resource, such as "os-type = linux, ram-size = 2G". We call these requests multi resource requests. There are two cases for these requests. If all of the requested resources should be satisfied in the same node then the local node checks its Q-value table and chooses a neighbor whose Q-value for all of the requested resources is opposed to zero. In this case there will be more than one neighbor with these conditions. Then the request will be forwarded to all of those neighbors and they continue the same algorithm to find the all resources. But if the requested resources can be satisfied in different nodes, then the local node who receives the request, divides the multi resource request to single resource requests and forwards each of them to an individual neighbor node according to their Q-value. Because the aim of grid is resource sharing, we use the second case.

For example if the request "os-type = linux, ram-size = 2G" is received by node n, node n divides it in two individual requests, "os-type = linux" and "ram-size = 2G". Then for each of the requests, according to Q-value table of node n, a neighbor node with a highest Q-value will be chosen for forwarding the request, and the algorithm continues as described.

4 Simulation and Performance Evaluation

In this section we present the result of simulation of our algorithm. The environment consists of N=1000 nodes and R=200 resources. The graph generates randomly. We tried to simulate an environment that is close to real environment as possible. The nodes that share a large number of resources are fewer than nodes that share only one or two. Hence, the distribution of resources on nodes is decided by geometric distribution [8]. Links between nodes are all duplex. Every node generates its own requests during simulation period. The type of the requested resource is generating randomly for simulating the real word. The time of request generation in each node, is the poisson process. Requests can be single resource request or multi resource request. According to results when $\alpha = 0.1$, $\beta = 0.2$, $\gamma = 0.8$ and TTL= 20 the performance of our model achieves the best.

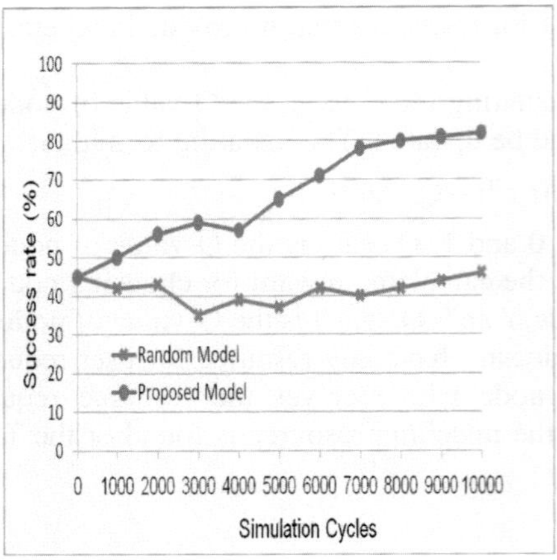

Fig. 1. Success rate

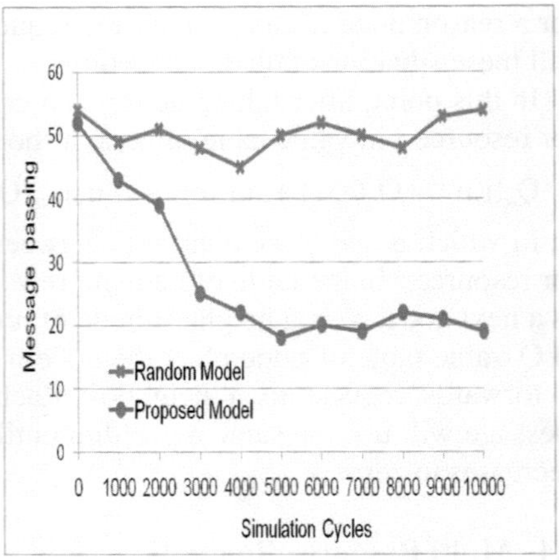

Fig. 2. Message passing

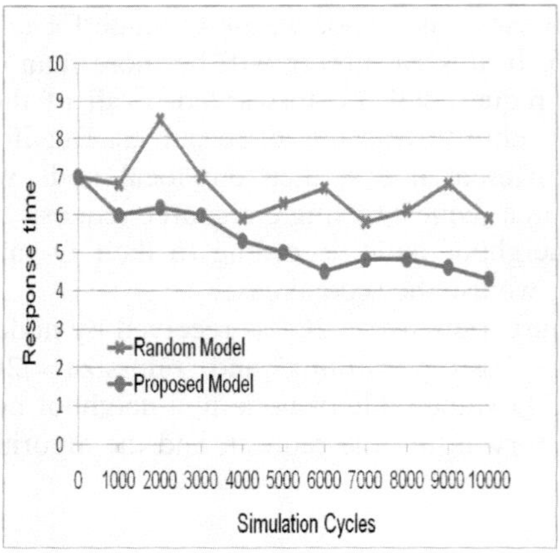

Fig. 3. Response time

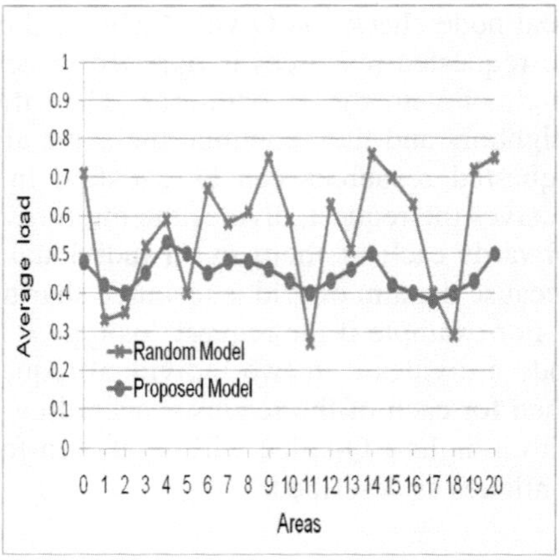

Fig. 4. Average load vs. area.

To evaluate the performance of our algorithm we compare it with the basic random walk resource discovery. We can see that the performance of proposed model at the beginning cycles of simulation is low because initially agents don't have any knowledge about the environment but by time passing and receiving more requests the agents collect more information about their neighbors so the algorithm works better and the success rate will improve greatly. As shown in figure1 at the beginning cycles, the success rates of both algorithms are close. But in continue there is a great difference. Figure2 shows that the proposed model decreases the number of forwarding messages for discovering a resource, so it has a great affection on reducing the grid traffic. From figure3 we can see that our model has a faster request response time which is one of the most important factors for users. The proposed algorithm leads to load balancing in whole grid. To show this easily, we divide the whole grid into 20 areas. Each area has a load which is the average load of links that exist in the area. Figure4 shows the load of areas after 10000 simulation cycles. The loads of areas are too close, that means the grid has a balanced load.

5 Conclusion

Rapid development of grids requires adaptive and efficient resource discovery approaches [3]. The heterogeneity and volatility of grid nodes and resources makes resource discovery a challenging problem. This paper presents an adaptive resource discovery algorithm using reinforcement learning. This algorithm also can be used for multi resource discovery. The preferred architecture is the peer to peer architecture. Each node is an agent. Agents can learn and adapt themselves with environment so for a requested resource they choose the most suitable resource among existing resources.

Results show that the success rate of the proposed algorithm is high and it reduces the discovery time because it uses the best and shortest path to discover a resource, also agents prevent using a path with a high traffic load for forwarding messages, so it achieves balanced load network.

References

1. Ma, S., Sun, X., Guo, Z.: A Resource Discovery Mechanism Integrating P2P and Grid. IEEE, New York(2010), doi: 978-1-4244-5540-9/10
2. Chang, R.-S., Hu, M.-S.: A resource discovery tree using bitmap for grids. Future Generation Computer Systems 26, 29–37 (2010)
3. Shen, H., Li, Z.: SPPS: a scalable p2p-based proximity-aware multi-resource discovery scheme for grids. IEEE, New York (2008), doi: 978-1-4244-2677-5/08
4. Messina, F., Pappalardo, G., Santoro, C.: HYGRA: A Decentralized Protocol for Resource Discovery and Job Allocation in Large Computational Grids. IEEE, New York(2010), doi: 978-1-4244-7755-5/10
5. Filali, I., Huet, F., Vergoni, C.: A Simple Cache Based Mechanism For Peer To Peer Resource Discovery In Grid Environments. In: Eighth IEEE International Symposium on Cluster Computing and the Grid (2008), doi: 10.1109/CCGRID.2008.110
6. Deng, Y., Wang, F., Ciura, A.: Ant colony optimization inspired resource discovery in P2P Grid systems. In: LLC 2008. Springer Science+Business Media (2008), doi: 10.1007/s11227-008-0214-0
7. Sutton, R.S., Barto, A.G.: Reinforcement Learning: An Introduction. A Bradford Book. MIT Press, Cambridge (1998)
8. Tangpongprasit, S., Katagiri, T., Kise, K., Honda, H., Yuba, T.: A time-to-live based reservation algorithm on fully decentralized resource discovery in grid computing. Parallel Computing 31(6), 529–543 (2005)
9. Cui, J., He, Y., Wu, L., Li, F.: A resource discovery algorithm with probe feedback mechanism based on advance reservation. In: Proceedings of the Fifth International Conference on Grid and Cooperative Computing, GCC 2006, October 2006, pp. 281–286 (2006)
10. Jeanvoine, E., Morin, C.: RW-OGS: an Optimized RandomWalk Protocol for Resource Discovery in Large Scale Dynamic Grids. In: 9th Grid Computing Conference. IEEE, New York (2008), doi: 978-1-4244-2579-2/08

Modeling and Control of Electro-Hydraulic-Controlled Stepping Cylinder for Mold Oscillation

Shenghao Zhou, Min Xiao, and Jinchun Song

Northeastern University, Shenyang, 110819

Abstract. Conventional electro-hydraulic servo control system is widely used for mold oscillation, but it is difficult for this system to construct a stable, low cost and high performance control system. This paper proposes an oscillation method driven by an electro-hydraulic-controlled stepping cylinder, and the stepping cylinder has the performances of less control links, easy maintenance, relatively low cost and high precision. The mathematic model of electro-hydraulic-controlled stepping cylinder is established, and through PWM control the sine curve and non-sinusoidal curve are simulated. The result of simulation demonstrates that the model is effective, and the proposed electro-hydraulic-controlled stepping cylinder system instead of conventional electro-hydraulic servo control system is feasible.

Keywords: Continuous casting crystallizer, stepping cylinder, non-sinusoidal curve, impulse frequency control.

1 Introduction

In recent years, the development and popularity of computer control techniques laid a foundation for the combination of electronic technology and the hydraulic technology, which greatly improved the functions and complex control ability of the hydraulic control system. In order to obtain a high speed casting performance and good surface quality, the vibration system of continuous casting crystallizer should be improved from the traditional mechanical vibration control to the electro-hydraulic servo control. The electro-hydraulic servo vibration control device can easily achieve a variety of waveforms vibration and monitor continuous casting process to display real-time vibration waveforms, and it also can modify vibration mode and parameters such as frequency, amplitude online [1], [2]. Mold oscillation electro-hydraulic servo vibration control system drives controlled object by the hydraulic mechanism mainly comprised by electro-hydraulic servo valve and servo cylinder, combining with modern computer measurement and control technology to realize controls. The whole system has such defects as relatively too much control links, high requirements for the stability of the control software and the appropriate hardware configuration, long cycle for technology upgrades. In addition, in order to ensure the security and reliability of the system, for hydraulic servo system, it requires a high oil cleanliness and reliable electromagnetic compatibility [3].

X. Wan (Ed.): Electrical Power Systems and Computers, LNEE 99, pp. 183–189.
springerlink.com

Along with the computer application in hydraulic servo control, digital servo control components and driving mechanism are extremely popular. Digital hydraulic servo drive technology can be divided into digital valve control technique and digital cylinder control technology. Stepping type digital valve is used stepping motor as electricity - machinery conversion components, the input signal is converted to output signal of the valve which is proportional with the number of steps, This type of valves have a high repeat accuracy, no hysteresis, no need to use D/A converter and linear amplifiers, etc., the disadvantage is slow response [4]. Digital cylinder is incremental digital control servo components, namely, converts the electrical signal used to control the stepping motor to mechanical displacement. Stepping motors can be controlled by a microcomputer or a programmable logic controller (PLC).Its working principle is that the controlling pulse sequence signals given by a microcomputer, drive stepping motor after amplified by the driving power, the microcomputer control the Stepping motor speed through the control of pulse, and thus to control the electro-hydraulic-controlled stepping cylinder's movement. The displacement of electro-hydraulic stepping cylinder is proportional to the total number of the control pulses, and the movement speed of electro-hydraulic stepping cylinder is proportional to the frequency of the control pulses. Digital hydraulic servo drive technology with adopting the combination of Stepping motor driving and screw driver of servo following with high reliability and high precision hydraulic, has the same high dynamic response with high accuracy under less control links, and it is relatively low cost compared to the traditional electro-hydraulic servo system, and avoids the problems in the traditional electro-hydraulic servo system such as complex maintenance, interference problems and zero shift [5-7].

2 Structure Design of Stepping Cylinder

Different from the standard servo cylinder components, the electro-hydraulic stepping cylinder is actually an assembly of the servo electro-hydraulic-controlled system. As shown in Fig.1. The left figure is the schematic drawing of a two-dimensional profile, showing the basic structure of the stepping cylinder; The right figure is the three-dimensional diagram of the stepping cylinder, which can dynamically display the working principle of the stepping cylinder [8] .It includes: cylinder, servo valve spool, precise ball screw nut, gear transmission, absolute encoders, five- phase stepping motor, etc. The connectors assorted with the stepping cylinder contain hydraulic oil source, stepping motor drive unit, encoders control unit, procedures controller, etc. Stepping motor rotating certain angles, opens the valve spool through the valve spool driven by the rotating shaft, which make the high-pressure oil flow into a side of the piston and push the piston till closing the opening in the valve opened earlier, complete a feedback servo operating cycle.

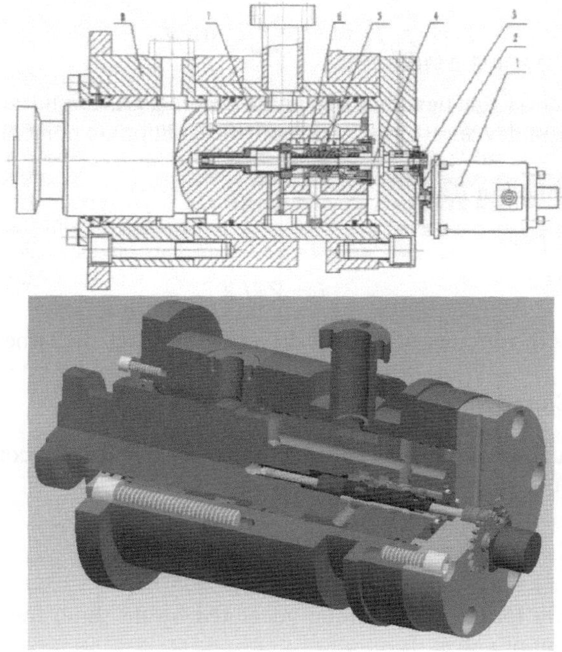

Fig. 1. Electro-hydraulic stepping hydraulic cylinder structure scheme 1- Stepping motor 2, 3-Gear 4- Rotating shaft 5- Spool 6- Valve cover 7- Piston 8- cylinder body

3 Mathematics Model Establishment of Stepping Cylinder

Electro-hydraulic-controlled cylinder contains five-phase stepping motor, gear transmission, servo valve, hydraulic cylinder and other components, and the following paper will make models for the above components.

3.1 Model of Five-Phase Stepping Motor

Adopt five-phase stepping motor whose driving means are below bridge chopper wave with constant frequency chopper wave and constant total flow, the equation of the voltage can be obtained:

$$V_k = i_k R + L_s \frac{di_k}{dt} + L_m \sum_{\substack{j=a \\ j \neq k}}^{e} \frac{di_j}{dt} + u_k(t) \qquad (k = a,b,c,d,e) \tag{1}$$

where V_k is the voltage of the k phase winding, R, L_s and L_m is the resistance, self-induction and mutual inductance of the phase windings respectively , $u_k(t)$ is the rotational voltage of the k phase winding.

The value of electromagnetic torque depends on the current of every phase winding, to assume that it is equal to the sum of the torque produced by every phase current in linear condition. Let the torque-angle characteristic is the sine waves, so

$$T = \sum_{k=a}^{e} T_k = -k_t \sum_{k=a}^{e} i_k \sin\theta_k \tag{2}$$

where T_k is the electromagnetic torque of k phase winding, k_t is torque coefficient, θ_k is the angular displacement deviation of the rotor from the equilibrium point of k phase torque.

3.2 Gear Transmission Ratio

Gear transmission ratio

$$i = Z_1 / Z_2 \tag{3}$$

where Z_1 and Z_2 is the number of teeth of driving wheel and idler wheel respectively.

3.3 Dynamics Equation of Slide Valve Spool

The dynamics equation of the slide valve core in electro-hydraulic-controlled stepping hydraulic cylinder is

$$T_L = J_L \frac{d^2\theta}{dt^2} + B_m \frac{d\theta}{dt} + T_f + T_x \tag{4}$$

where T_f is friction torque, T_x is torque generated by the axial force, B_m is viscous damping coefficient of valve.

3.4 Valve Controlling Cylinder Model

Stepping cylinder is a single out rod hydraulic cylinder, the control slide valve is symmetrical and four sides slide valve, form symmetric valve controlling asymmetrical cylinder. According to the valve port flow equation, hydraulic cylinder continuous equation and force balance equation can get valve controlling cylinder nonlinear state equation model as follows

$$\begin{cases} \dot{x}_1 = x_2 \\ \dot{x}_2 = \dfrac{1}{m_p}(A_1 x_3 - A_2 x_4 - F_f - F_L) \\ \dot{x}_3 = \begin{cases} \dfrac{1}{V_1 / K}(C_i (x_3 - x_4) - A_1 x_2 + R_0 \cdot x_v \sqrt{p_s - x_3}) & x_v \geq 0 \\ \dfrac{1}{V_1 / K}(C_i (x_4 - x_3) - A_1 x_2 + R_0 \cdot x_v \sqrt{x_3 - p_0}) & x_v < 0 \end{cases} \\ \dot{x}_4 = \begin{cases} \dfrac{1}{V_2 / K}(C_i (x_3 - x_4) - C_e x_4 + A_2 x_2 - R_0 \cdot x_v \sqrt{x_4 - p_0}) & x_v \geq 0 \\ \dfrac{1}{V_2 / K}(C_i (x_4 - x_3) - C_e x_4 + A_2 x_2 - R_0 \cdot x_v \sqrt{p_s - x_4}) & x_v < 0 \end{cases} \end{cases} \tag{5}$$

Let state variable $X=[x_1,x_2,x_3,x_4]^T=[x_p,\dot{x}_p,p_1,p_2]^T$, where x_p is the displacement of cylinder, p_1 and p_2 is the pressure of no rod cavity and rod cavity respectively, A_1 and A_2 is effective function area of no rod cavity and rod cavity respectively, V_1 and V_2 is instantaneous effective volume of no rod cavity and rod cavity respectively, C_i and C_e is the internal and external leakage coefficient of cylinder respectively, K is oil bulk modulus, m_p is load equivalent quality, F_f is friction of cylinder, F_L is suffered external load of cylinder, x_v is valve opening, R_0 is valve coefficient.

Integrated the models of stepping motor, gear shifting device, slide valve spool and valve control cylinders into together can get the mathematical model of nonlinear stepping cylinder.

4 Numerical Simulation Study

Fig.2 is the electro-hydraulic stepping hydraulic cylinder system structure. In the control of five phase stepping motor this paper use three-phase excitation PWM output way. According to the above the mathematical model of stepping cylinder, set up simulation model in MATLAB/SIMULINK environment, and combine the force that continuous casting crystallizer suffers to analysis model, then make sine wave vibration control and non-sinusoidal vibration control simulation for stepping cylinder.

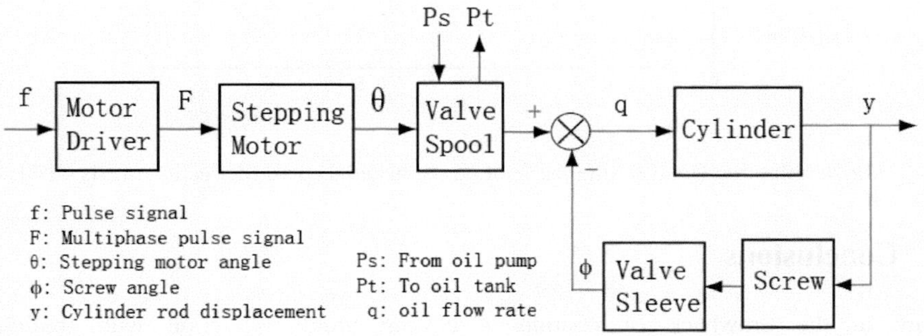

Fig. 2. Electro-hydraulic stepping hydraulic cylinder system structure

The simulation results shown in Fig.3 and Fig. 4, the system vibration is uniform and steady and the curve has high follow precision, there is no distortion, fluctuations or other undesirable phenomena. In the continuous casting mold oscillation curve, non-sinusoidal curve is gradually widely used due to two obvious advantages: one is that adopting the sine curve can reduce vibration frequency and improve the system reliability and service life, the other is that moving with a longer time can increase the casting speed [9]. The simulation results of the non-sinusoidal curve shows that the electro-hydraulic-controlled stepping hydraulic cylinder and traditional electro-hydraulic servo control system are comparable on the control accuracy and the effect.

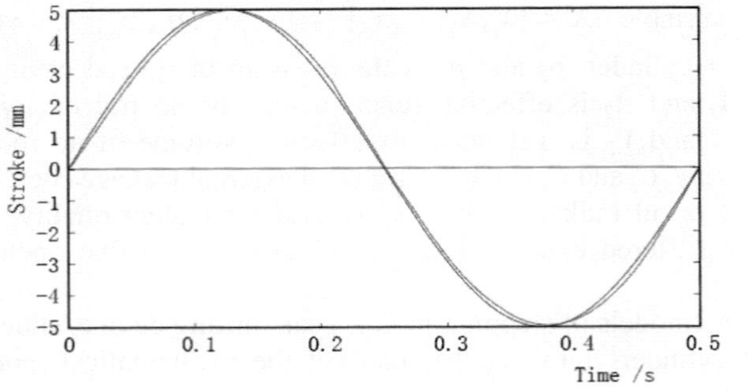

Fig. 3. Sine curve vibration control simulation curve

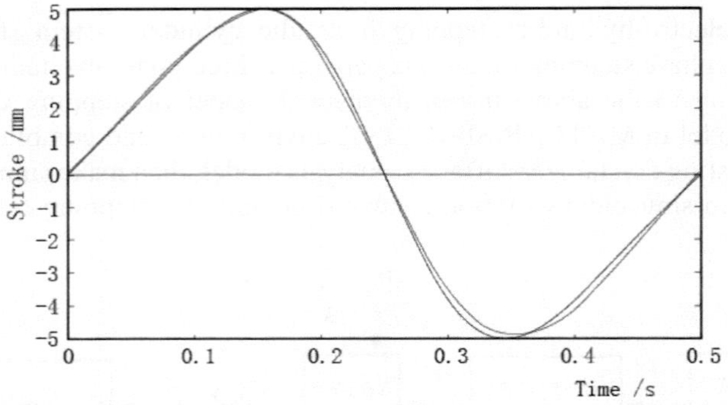

Fig. 4. Vibration Control Simulation of non-sinusoidal curve (deflection rate is 40%)

5 Conclusions

Aim at the drawback of continuous casting mold oscillation with traditional electro-hydraulic servo control system, this paper presents an electro-hydraulic-controlled stepping cylinder system. This system has these features such as less control links, easy maintenance, simple software control, relatively low cost and high accuracy and so on. Use mechanism method, This paper establishes the electro-hydraulic-controlled stepping hydraulic cylinder system model composed of elements such as five phase stepping motor, gear transmission device, servo valve spool, cylinder, etc. And sine curve vibration and the non-sinusoidal curve vibration of 40% deflection are simulated through the PWM control method. The simulation results show that the vibration of this system is symmetrical and steady and the curve has high follow precision with no distortion, fluctuations or other undesirable phenomena.The simulation results will be the theoretical basis for the corresponding experiments.

References

[1] Li, X., Zhang, D.: The Technique of the Mold Oscillation in Continuous Casting, pp. 84–88. Metallurgical Industry Press, Bejing (2000)

[2] Chen, D., Yang, W.: Development Of Electro-hydraulic Servo Oscillation System For Continuous Casting Mold, Steel, vol. 32 (suppl.), pp. 705-708 (1997)

[3] Song, J., Shu, D., Zhang, Z. (eds.): Hydraulic And Pneumatic Transmission. Science press (2006)

[4] Chen, B, Yi, M.-l.: Digital Technology Applied in Hydraulic System Hydraulics. Pneumatics & Seals No. 4 (2005)

[5] Li, W., Zhang, Y., Yu, L., Xiao, J.: Mold Non-sinusoidal Oscillating Mode Analye and Control of Continuous Casting Machine. In: ICEMI 2009, pp. 567–570 (2009)

[6] Bhattacharya, A.K., Debjani, S., et al.: Optimization of Continuous Casting Mould Oscillation Parameters in Steel Manufacturing Process Using GA. In: IEEE CEC 2007, pp. 3998–4004 (2007)

[7] Thomas, B.G.: Modelling of the Continuous Casting of Steel – Past, Present and Future. In: Electric Furnace Conf. Proc. vol. 59, pp. 3–30. ISS, Warrendale, PA (Phoenix, AZ) (2001); Also Metallurgical and Materials Transactions B, 33B(6), 795- 812 (December 2002)

[8] Institute of Hydraulic & Pneumatic Tech., `http://www.ihptneu.com`

[9] Xiaoming, W., et al.: Introduction on Development of Non-sinusoidal Oscillation Technology of Mould Controlled by Electro-hydraulic Servo System. In: Equipment technology, pp. 211–214 (2000)

Test Case Design Methods and Strategy for Software Component Test of Digital Safety System in Nuclear Power Plants

Zhicheng Zhang, Shuhui Zhang, Zhicai Ma, and Dongling Xu

Shanghai Nuclear Engineering Research & Design Institute
No. 29 Hongcao Road, Shanghai, China
zhangzhc@snerdi.com.cn

Abstract. Software component test is of great importance within the Independent Verification and Validation process of digital safety system used in nuclear power plants, and acts as the base of software engineer implementation. Now it has encountered the difficulty of test case generation, which needs to be more intensively investigated in nuclear energy industry. The paper describes three methods for test case generation according to the suggestion given by an IAEA's technical report, and proposes a strategy on how to use these methods in the software component test of digital safety system. This work will be also beneficial to the subsequent tests of digital safety system in nuclear power plants.

Keywords: Nuclear Power Plants, Digital Safety System, Software Engineering, Software Component Test, Test Case, Independent Verification and Validation.

1 Introduction

Nuclear power plays an important role in the energy layout in China as it is a kind of clean, efficient and safe new-style energy. The development of nuclear power has been speeded up during last five years, and all nuclear power plants (NPP) under construction in china employed digital safety systems, regardless of "second-generation and plus" or third-generation (e.g. AP1000, EPR).

Chinese government has pointed out, in China's 12th Five-Year Plan [1], that the nuclear power should be effectively developed on the basis of ensuring safety, which stresses that safety acts as the precondition of promoting nuclear power development in China. Recently, the Fukushima Nuclear Accident has drawn global attention to nuclear safety issues again, while digital safety systems play a vital role of ensuring the safety of nuclear power plants.

Safety system is "a system that is relied upon to remain functional during and following design basis events to ensure: (a) the integrity of the reactor coolant pressure boundary, (b) the capability to shut down the reactor and maintain it in a safe shutdown condition, or (c) the capability to prevent or mitigate the consequences of accidents that could result in potential off-site exposures" [2]. Traditional safety systems generally consist of analog components, and comply with redundancy and

X. Wan (Ed.): Electrical Power Systems and Computers, LNEE 99, pp. 191–200.
springerlink.com

diversity requirements, Single-failure criterion, and etc. to achieve defense-in-depth goal. On the other hand, different from traditional safety systems, the digital safety systems (i.e. computer-based safety systems) employ software much beside hardware. As software failures increase the probability of system common cause failure, they challenge the system's safety built on the redundancy and reliability of hardware. In order to achieve high reliability of the digital safety systems, the Independent Verification and Validation (IV&V) process [3] is performed throughout the development process of digital safety systems, which is the experience learned from software engineering by nuclear industry.

Software Component (SC) is the basic function module in digital safety systems, while Software Component Test (SCT) is an important activity in IV&V process. Test case design is the base and also the hardest work of SCT.

For the topic of designing effective test case, the related IEEE standards involved little and only gave the definition of the "test case" [4], while IEC standards recommended several test techniques (i.e. methods for test case design) for software module test in digital safety-related systems [5][6]. A technical report prepared by EPRI and Autoridad Regulatoria Nuclear provided suggestion on techniques used functional test during the V&V process for digital systems in NPP [7]. Although the above work could be referred when we designing test case for SCT of digital safety systems in NPP, further research is still needed as to find suitable solution for test techniques selection and combination, which is termed as design strategy in this paper.

In this paper, we first discussed three methods for test case design for SCT of digital safety systems in NPP according to the suggestion from a technical report by IAEA [8], and then proposed a design strategy complying with actual test requirements. The work will be also beneficial to the subsequent functional tests of digital safety system in NPP.

2 Background

2.1 Position of SCT

IV&V process, performed throughout the development process of system, can help to achieve high reliability and availability of the digital safety system. More details of IV&V process are described in IEEE Std 1012™-2004 [9].

SCT corresponds to the software design and implementation phases in the development process of the safety system, which is a very important activity of IV&V process, as shown in Fig. 1.

Testing of digital safety systems usually follows the bottom-up methodology, which means from lower level to higher level, from component test to system integration test. SCT locates the lowest level in the whole test tasks of digital safety system and acts as the base for the subsequent tests.

Fig. 1 illuminates the SCT's position in IV&V process and test tasks of the digital safety system.

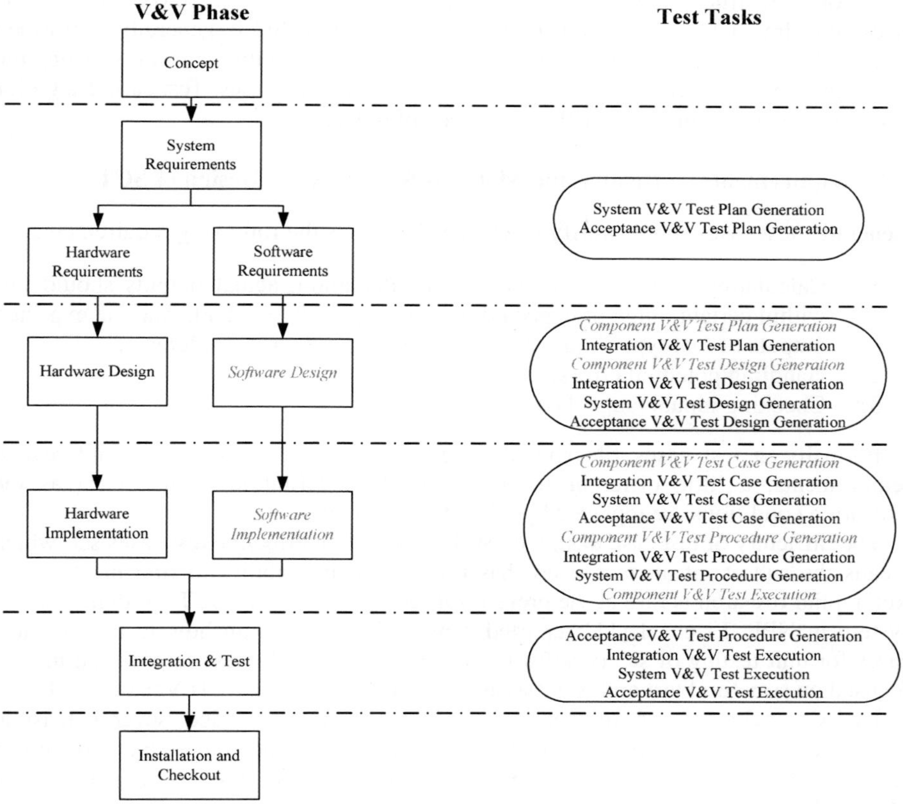

Fig. 1. SCT's Position in V&V Process and Test Tasks [9]

2.2 Definition of Test Case and Classification of SCT

SCT aims at verifying whether the SCs comply with the requirements in Software Requirements Specification (SRS). In this work "a set of test inputs, execution conditions, and expected results developed for a particular objective" [4] is called test case. Designing effective test case is always the most important work in the software testing.

According to its aim, SCT can be categorized as either functional test or structural test. The former verifies software components complying with specified functional requirements, and the test case is drawn from the requirements in SRS, which is based on using test inputs and checking test outputs. The test inputs are executed across the whole range while the test outputs are compared with expected results. Each test case is directly related with requirement. The design of test case for functional test is usually accomplished by test engineers, which is of the major concern in this paper. Hence the test case design of SCT in the paper means the test case design for function test of software components.

In contrast, structural test verifies the inter structure of SCs, therefore the test case is usually drawn from SC's structure. Structural test of SC is generally performed with the help of automated test-bed (e.g. LDRA Testbed®) with the result showing the coverage of SC's internal structure by executing all the test cases. Test case design for structural test is not included within the scope of this paper.

2.3 Requirements, Difficulty and Methods for Test Case Design of SCT

Generally, test cases should verify whether SC satisfies the following requirements:

1) Calculating correctly. For numerical calculation, actual outputs should lies within certain range of expected outputs (e.g. 100%±0.1%). And the expected outputs of real type should not be "0" as to avoid wrong judgment.
2) Handling potential errors;
3) Checking inputs range. [10]

Formally, a test cases file should specify inputs, expected outputs and related requirements, a little different from the standard definition of test case as the execution conditions are replaced by related requirements.

The difficulty of test case design of SCT is to find an effective test cases set, which means the subset, of all test cases, has the highest probability to find most errors. Referential research is rare at the present for test case design of SCT for digital safety system in NPP. Li et. al [11] reported a preliminary study on how to develop test cases for unit testing of safety software V &V. However, the test case design mainly focused on structural test and was assisted by automate testing tools VectorCAST.

IAEA's technical reports series No. 384 [8] recommended several feasible techniques for V&V of software related to NPP, and introduced three methods of test cases design for SCT. We use these methods as guidelines for test case design for SCT of digital safety systems, and they are:

(a) Equivalence partitioning;
(b) Boundary value analysis;
(c) Cause–effect graphing.

The detail description of these three methods could be found in the standard textbooks of software testing, e.g. ref.[12], so we describe little on the concept of each method, but introduce more about the consideration and applicability of using them in the actual SCT. Then the process of using these methods is demonstrated for test case design in a simple SCT.

3 Methods for Test Cases Design

3.1 Equivalence Partitioning

According to this method, we first divide the input domain into a certain number of equivalence classes, and then choose a few typical values from each class as the test inputs for test cases.

This method consists of two steps: (1) indentifying equivalence classes, and (2) generating test cases.

When an input is of Real, Short Integer or Long Integer type, the range of input could be a condition for dividing equivalence classes. Generally, only the maximum and minimum need to be considered for input of Short Integer or Long Integer type, while infinitesimal limit needs to be taken into account as well for Real type input.

If an input is of Boolean type or in the form of logical condition, we can indentify the equivalence classes by "True" and "False" of the input.

3.2 Boundary Value Analysis

As to verify that the SC behaves correctly across boundaries, we choose values directly on, above, and beneath the edges of equivalence classes, together with any likely error values. For example, the maximum of Short Integer type value is 32767, so 32766, 32767 and 32768 are chosen as the test inputs.

If we only need to check the input range, the consideration of output range could be omitted.

3.3 Cause–Effect Graphing

This method uses graph to analyze combinations of input circumstances, and then generates test cases in a systematic way. And it is applicable to check all the combination of input conditions.

Cause–effect graphing is a very useful method when SC has several Boolean inputs or the logical relation between inputs and outputs is a little complex. The truth table, usually included in SDD, could assist to generate decision table.

3.4 Strategy of Test Case Design

Each method mentioned above could supply a set of useful test cases, but not a quite complete set since each method has its own disadvantage. According to the suggestion by Glenford J. Myers [12] and the experience in SCT of digital safety system, we propose a reasonable strategy on how to use these methods in combination as follows,

1) Use cause–effect graphing method first if SC has inputs of Boolean type or its requirements contain condition combination of inputs, otherwise use equivalence partitioning method;
2) Use boundary value analysis method under any circumstance, especially when the inputs are of Real, Short Integer or Long Integer type;
3) Use equivalence partitioning method last to check the completeness of test cases set if cause–effect graphing method is employed first.

4 Application of the Strategy for Test Case Design

In this chapter, we demonstrate the application of the strategy introduced above, by designing test cases for SC UNITCONV.

4.1 Functional Requirements for UNITCONV

UNITCONV (i.e. Unit Conversion) shall convert the length from metric unit (cm) into imperial unit (inch) when needed. It simplifies the conversion of safety systems from NPP using metric unit to NPP using imperial unit. The input and output terminals are shown in Table.1.

Table 1. Input and Output Terminals of UNITCONV.

Terminal	Name	Type	Description
1	UNIT_IN	Input_Real	The length input for SC.
2	CONVERT	Input_Boolean	Input that determines whether converting the length from metric unit into imperial unit or not.
3	UNIT_OUT	Output_Real	The length output for SC.

Testable requirements in SRS are as follows,

[i] When CONVERT is "True", the SC UNITCONV shall convert the length from metric unit (cm) into imperial unit (inch), using the following equation (the equation has been simplified as it is just used as an example),

$$\text{UNIT\_OUT} = \text{UNIT\_IN} \times (2/5)\ . \qquad (1)$$

[ii] When CONVERT is "False", the output UNIT_OUT shall equal to the input UNIT_IN without any conversion

This SC has an implicit requirement, which we can identify as [iii], that the input of Real type needs range-checking, as data used in computer have a certain range. Assuming that the lowest and highest limits of Real type data are -1.0×10^{17} and 9.0×10^{18}, respectively, any input exceeding the range will be set to the nearby limit value. On the other hand, the infinitesimal limit of Real type data is 5.0×10^{-19}, and any input with an absolute value smaller than this will be set to 0. The test input here should not be 0 to avoid expected output being 0. The input range could be divided into five segments as shown in Fig.2.

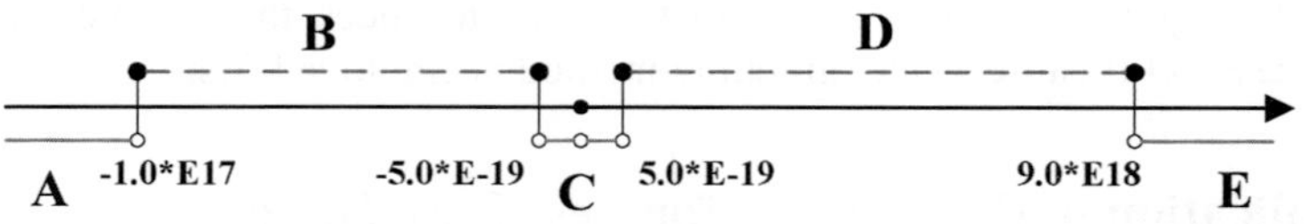

Fig. 2. Range Partitions of Real Type Input

4.2 Specific Steps of Test Case Design

Identifying Causes and Effects. As UNITCONV has a Boolean type input, we use cause–effect graphing method first. After analyzing the requirements, we get following Causes, Interim States and Effects (numbered from 1, 31 and 61 for easy identification respectively), which are shown in Table. 2.

Table 2. Causes, Interim States and Effects of UNITCONV.

Cause	Interim States	Effects
① CONVERT is True	㉛ UNIT_IN belongs to virtual range	⑹ output in imperial unit, and calculation right;
② UNIT_IN belongs to A	㉜ UNIT_IN exceeds virtual range	⑹ output in imperial unit, and calculation wrong;
③ UNIT_IN belongs to B		⑹ output in metric unit, and calculation right;
④ UNIT_IN belongs to C		⑹ output in metric unit, and calculation wrong;
⑤ UNIT_IN belongs to D		
⑥ UNIT_IN belongs to E		

Drawing Cause-effect Graph. We then convert Table 2 to a cause-effect graph according to the logical relationship in SRS. The graph is shown as follows, in which common logic symbols stands for causality, while the symbol O denotes that one and only one of causes (i.e. ②, ③, ④, ⑤ and ⑥) is "True".

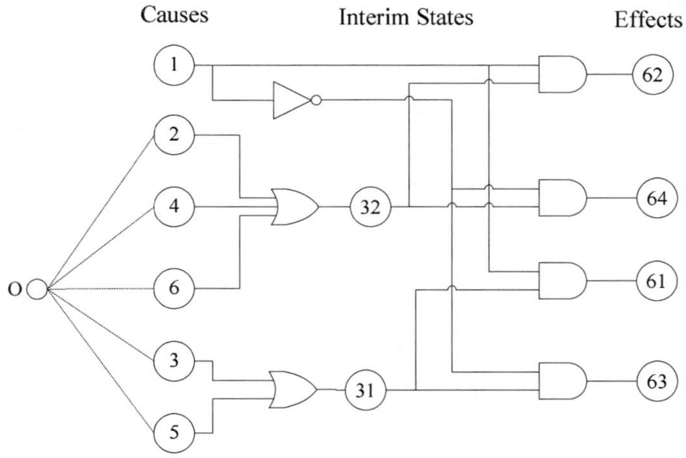

Fig. 3. Cause-Effect Graph for UNITCONV

Converting Graph into Decision Table. The cause-effect graph is then converted into a decision table by tracing the state conditions in the graph.

Table 3. Decision Table of UNITCONV.

	C1	C2	C3	C4	C5	C6	C7	C8	C9	C10
①	1	1	1	1	1	0	0	0	0	0
②	0	0	1	0	0	0	0	1	0	0
③	1	0	0	0	0	1	0	0	0	0
④	0	0	0	1	0	0	0	0	1	0
⑤	0	1	0	0	0	0	1	0	0	0
⑥	0	0	0	0	1	0	0	0	0	1
61	1	1	0	0	0	0	0	0	0	0
62	0	0	1	1	1	0	0	0	0	0
63	0	0	0	0	0	1	1	0	0	0
64	0	0	0	0	0	0	0	1	1	1

Generating Test Cases from Decision Table. Each column in the decision table can be converted into a corresponding test case. In the conversion process, boundary value analysis method should be used in combination, which means the test inputs of test cases should cover values directly on, above, and beneath the boundaries. Meanwhile, actual test requirements should be considered, for instance, test cases whose expected outputs are 0 should be modified or canceled. Table 4 shows the test inputs and expected outputs of test cases, in which Test case_1 to Test case_10 correspond to column 1 to column 10 of Table 3, and Test case_11 to Test case_14 are added to cover the boundaries. Test case_4 and Test case_9 will be canceled as the outputs are 0.

Table 4. Test Inputs and Expected Outputs of Test Cases.

	UNIT_IN	CONVERT	UNIT_OUT
Test case_1	-9.90E+16	1	-3.96E+16
Test case_2	1.30E-18	1	5.20E-19
Test case_3	-1.10E+17	1	-4.00E+16
Test case_4	*-4.90E-19*	*1*	*0*
Test case_5	9.10E+18	1	3.60E+17
Test case_6	-5.10E-19	0	-5.10E-19
Test case_7	8.90E+18	0	8.90E+18
Test case_8	-1.00E+18	0	-1.00E+17
Test case_9	*4.90E-19*	*0*	*0*
Test case_10	9.00E+19	0	9.00E+18
Test case_11	-1.00E+17	0	-1.00E+17
Test case_12	-5.00E-19	0	-5.00E-19
Test case_13	5.00E-19	0	5.00E-19
Test case_14	9.00E+18	0	9.00E+18

Supplementing Test Cases if Need. In the end, we use equivalence partitioning method to check the completeness of test cases. The equivalence classes are shown in the following table,

Table 5. Equivalence Classes of UNITCONV

Input & output condition	Valid equivalence classes	Invalid equivalence classes
Inputs are valid	UNIT_IN belongs to B; (1)	UNIT_IN belongs to A; (5)
	UNIT_IN belongs to D; (2)	UNIT_IN belongs to C; (6)
		UNIT_IN belongs to E; (7)
output in imperial unit	CONVERT=1; (3)	Not Available
output in metric unit	CONVERT=0; (4)	Not Available

All the valid equivalence classes in table 4 have been covered by test cases, and every invalid equivalence class has its own test case except the one (e.g. (6)) in which expected output could only be 0. Hence, no more test cases are needed.

Corresponding Test Cases to Requirements. We have generated the test inputs and expected outputs of test cases in the above work, and the remaining activity is to identify the relationship between requirements in SRS and test cases, which is recorded in another table.

Table 6. Relationship between Requiremens and Test Cases

Test Cases	Requirements
1, 2, 3, 5	[i]
6, 7, 11-14	[ii]
1-3,5-8, 10-14	[iii]

4.3 Conclusion

The test quality can not be guaranteed if we choose any number of random test cases, while exhaustive test will generally need infinite quantity of test cases. The test cases design strategy, which integrates equivalence partitioning method, boundary value analysis method and cause–effect graphing method, could be used to generate a set of effective test cases for SCT and get fairly good results.

5 Discussion and Prospect

Generally, SC should be designed to have simple functions without many requirements for the sake of reliability. Sometimes, however, for certain SC having complex functions or many requirements, orthogonal design method could be employed instead of cause-effect graphing method to generate test cases, as the latter method would lead to onerous work.

We can easily get a conclusion from the previous chapter that test case design for even a simple SC is not easy. However, digital safety system usually consists of dozens or even hundreds of SCs, of which some are of complex functions or with many requirements. It means a large amount of difficult work to design test cases for the whole SCs in the digital safety system, so to establish a practicable and effective strategy to guide test cases design is highly desirable.

The strategy, integrating equivalence partitioning method, boundary value analysis method and cause–effect graphing method, gives us a set of economic and effective methods to test cases for SCT of digital safety system in NPP, and helps us resolve the difficulties we encounter in SCT. Since some of the subsequent tests of digital safety system are also based on functional requirements, the strategy could be employed in these tests.

Acknowledgment. The authors wish to express their sincerely thanks to Mr. LU Shudong, who is a professional engineer in Shanghai Nuclear Engineering Research & Design Institute, as he gave valuable instruction on the writing of this paper.

References

1. The Central People's Government of the People's Republic of China, `http://www.gov.cn/2011lh/content_1825838_4.htm`
2. IEEE Std 603™-2009. IEEE Standard Criteria for Safety Systems for Nuclear Power Generating Stations (2009)
3. IEEE Std 7-4.3.2™-2003. IEEE Standard Criteria for Digital Computers in Safety Systems of Nuclear Power Generating Stations (2003)
4. IEEE Std 610.12-1990. IEEE Standard Glossary of Software Engineering Terminology
5. IEC 61508-3 -1997. Functional safety of electrical/electronic/programmable electronic safety-related systems. Part 3: Software requirements, pp. 26, 40, 44 (1997)
6. IEC 61508-6 -1997. Functional safety of electrical/electronic/programmable electronic safety-related systems. Part 6: Guidelines on the application of parts 2 and 3, p. 76 (1997)
7. Handbook for Verification and Validation of Digital Systems, EPRI, Palo Alto, CA, and Autoridad Regulatoria Nuclear, Buenos Aires, Argentina, TR-103291-CD, pp. 5-19 (1998)
8. IAEA TRS384. Verification and Validation of Software Related to Nuclear Power Plant Instrumentation and Control
9. IEEE Std 1012™-2004. IEEE Standard for Software Verification and Validation (2004)
10. M.J. Stofko: Software Program Manual for Common Q Systems, pp. 4–6
11. Li, D., Zhang, L.J., Feng, J.T.: Technique for Unit Testing of Safety Software Verification and Validation. Atomic Energy Science and Technology 42(6)
12. Myers, G.J.: The Art of Software Testing, 2nd edn.

Frequency Measurement and Tracking Algorithm for Integrated Grids

P. Meena[1], K. UmaRao[2], and Ravishankar Deekshit[3]

[1] Asst. Prof., Department of Electrical and Electronics,
BMS College of Engineering, Bangalore, India
meenabms@gmail.com
[2] Hod & Dean (Academic), Department of Electronics and Communication,
R.N.S.Institute of Technology, Bangalore, India
drumarao@yahoo.co.in
[3] Hod, Department of Electrical and Electronics,
BMS College of Engineering, Bangalore, India
ravi_yedatore@rediffmail.com

Abstract. The smart grid envisages providing quality power such that fault free operation of the digital devices that power the twenty first century economy is achieved. The future scenario for electricity consumers would be one of participation with demand control based on frequency linked real-time (RT) prices as well as real-time load control measures implemented at the consumer end for efficient use of electric power. This necessitates that the consumers have real time information with regard to nature of the voltage and fundamental frequency of power supply systems. Information about frequency under balanced operation of three phase supply can be obtained by using the space vector approach for representing the three phase quantities. The determination of the phase of resultant space vector at every instant is used to find out the frequency..Under conditions of unbalanced operation of three phase supply, frequency information cannot be extracted from the resultant space vector obtained under balanced conditions. In this case information about frequency can be obtained by extracting the positive or negative sequence components of the three phase unbalanced supply voltages and subsequently monitoring the phase of their resultant vector. This paper presents an effective method of measuring the fundamental frequency as well as tracking the frequency deviations under different situations during both balanced and unbalanced operation .of the power supply. The simulation results obtained clearly indicate the effectiveness of the method. The simplicity of the algorithm enables an easy implementation of the same in hardware using a pic microcontroller and is cost effective.

Keywords: Frequency measurement, frequency tracking, space vectors, positive sequence components, negative sequence components, balanced sag, unbalanced sag.

1 Introduction

The vision of countries for the next decade has been to provide clean and reliable power by working towards energy efficient systems .Smart grids are to replace

X. Wan (Ed.): Electrical Power Systems and Computers, LNEE 99, pp. 201–212.
springerlink.com

the existing electric grids. Increased quality of power with high levels of reliability is one of the outcomes envisaged through such endeavors. Specific approaches that the smart grid will bring include [1] PQ meters, System wide PQ monitoring, etc. A variety of methods for efficient monitoring of effective electric systems would be in place. New distributed generation (DG) devices such as fuel cells, micro turbines and micro grids that can provide clean power to sensitive loads would be in abundant use. Most of them are interconnected with the distribution network to supply power into the network as well as local loads.

One of the problems that occur in such set ups is when the power supply from the main utility is interrupted due to several reasons but the DG keeps supplying the power into the distribution networks. These kind of islanding conditions cause negative impacts on protection, operation and management of the distribution network. These conditions are easily detected by monitoring several parameters such as, voltage magnitude, phase displacement and frequency changes. In general, utilities maintain very close control of the power system frequency. Frequency variations that go beyond accepted limits for normal steady state operation of the power system are normally caused by faults on the bulk power transmission system, a large block of load being disconnected, or a large source of generation going off-line. The frequency variations can also arise due to harmonics injection.

In countries where demand exceeds supply, under-frequency poses a major problem for efficient grid operation. One way to dynamically control demand is to have frequency linked real-time prices. A real-Time balancing market, where the real-time price varies inversely with the system frequency, is described in [1],[2]. In such a market, producers and consumers can get the real-time price, simply by monitoring the frequency deviations themselves. The concept and implementation of Availability Based Tariff (ABT) is an excellent example in reducing demand based on frequency based pricing. This type of control by distribution not only results in significant benefit to companies and consumers but also helps the system operator in frequency regulation. By making the demand side participate in frequency regulation, such control can potentially facilitate an increase in renewable energy portfolio since renewable energy can be used under such situations. The results of simulation on a single air conditioner [2] shows that significant reduction in energy consumption can be achieved during severe frequency dips in a real-time market. Load control under such situations not only results in better frequency control but also lowers the real-time price of power.

The system frequency is a reflection of the quality of energy supply. Based on the fact that the system frequency at any time represents the result of the balance between generation and consumption, the size of the variations also reflect in real time how well the system is controlled. The phenomena of significant voltage and frequency deviations are typical for isolated power systems. In most cases these deviations are harmless. Long term voltage and frequency deviations can induce additional heat and energy loss in electrical machines, their overheating and result in decrease of their operational life. Frequency deviations can affect the operation of power electronic equipment that use controlled switching devices unless the control signals are derived from a signal that is phase locked with the applied voltage. Harmonics originating in customer equipment can cause power quality problems for other utility customers and the enforcement of guidelines on harmonic limits by utilities and state commissions would necessitate real time monitoring of frequency variations.

In [4] it was shown that by pricing of electric energy based on the value of the integral of the load active power measured by energy meters, the electric power utilities waste some revenues for the energy delivered to current harmonic generating customer and those who do not generate harmonics but are supplied with distorted and/or asymmetrical voltages are billed not only for the useful energy but also for the energy which may cause only harmful effects on their equipment. These two disadvantages of the present tariff could be eliminated if the energy account is based on the value of the integral of the active power of only the power system fundamental frequency component.

As per the review stated in [9], many solutions have been suggested for accurate tracking and measurement of supply frequency and most of the existing methods for frequency measurement are based on single-phase signals. Various numerical algorithms for power measurements are sensitive to frequency variations and need simultaneous frequency measurements. Various techniques have been developed to measure power system frequency. Amongst the various techniques available, the phasor based technique is much closer to the definition of instantaneous frequency. They demonstrate good dynamic responses which is an important feature for frequency measurement applications.. Most of the digital algorithms are based on the measurements of a single phase of the system. The utilization of Clarke's transformation in the estimation of power system frequency provides the classical single phase methods with more robustness, because the estimated frequency is computed using the information provided by the three-phase voltage. Clarke's α β transform is used to convert three-phase quantities to a complex quantity where the real part is the in-phase component; the imaginary part is the quadrature component. The methods using α β transformation, work well for balanced conditions .However, under unbalance conditions oscillations exist in the estimation of the resultant space vector. It is also stated that the Phase Locked Loop based on the synchronous reference frame (SRF-PLL) works well under most abnormal grid conditions; however, during unbalance, its performance becomes poorer if the input contains negative sequence. In this paper a simple method of frequency measurement and tracking is presented which gives excellent results under both balanced and unbalanced conditions of operation of power supply under dynamic conditions of short duration voltage disturbances such as sag and swell in supply voltage during the occurrence of three phase faults.

The paper is organized as follows. The proposed method is explained in section 2 followed by block diagram representation of the constituent blocks necessary for the measurement of the frequency under balanced conditions. This is followed by the block diagram for frequency measurement under unbalanced conditions of power supply which now includes an additional block in the beginning for the extraction of positive sequence components. The relevant mathematical transformations required for extracting the sequence components from an unbalanced three phase supply are then discussed. Simulations results are then presented for various cases of operation of power supply. This is then followed by the conclusion which highlights the merits of this method.

2 Frequency Estimation Method

In the proposed method, the frequency is measured by extraction of the phase angle of the resultant space vector of the three phase voltage signals. The technique of phase angle extraction at every instant used in this work is as per [6],applied for the determination of motor speed.

The three phase voltages namely,V_a $V_b$and V_c is transformed mathematically into three vectors, V_{sa}, V_{sb}, V_{sc} , along the three spatial axes displaced by 120°.The resultant of these spatial components is represented by a single equivalent rotating space vector [3] , rotating at an uniform angular velocity of ω radians per second, called the resultant space vector ,$\vec{V}_{res}$, which can be resolved into two phase components namely, α and β along a stationary reference frame.

These orthogonal components of the space vector, V_α and V_β,can be obtained from the knowledge of the original phasor values by means of a simple transformation given by,

$$\begin{bmatrix} V_\alpha \\ V_\beta \end{bmatrix} = \begin{bmatrix} 1 & -1/2 & -1/2 \\ 0 & \sqrt{3}/2 & -\sqrt{3}/2 \end{bmatrix} \begin{bmatrix} V_a \\ V_b \\ V_c \end{bmatrix} \qquad (1)$$

The resultant space vector is evaluated as,

$$|\vec{V}_{res}| = \sqrt{V_\alpha^{\,2} + V_\beta^{\,2}} \qquad (2)$$

The phase of the resultant space vector is to be determined in order to find out the value of frequency. The phase can be determined by passing the orthogonal components namely V_α andV_β, through low pass filters as shown below in Fig.1.

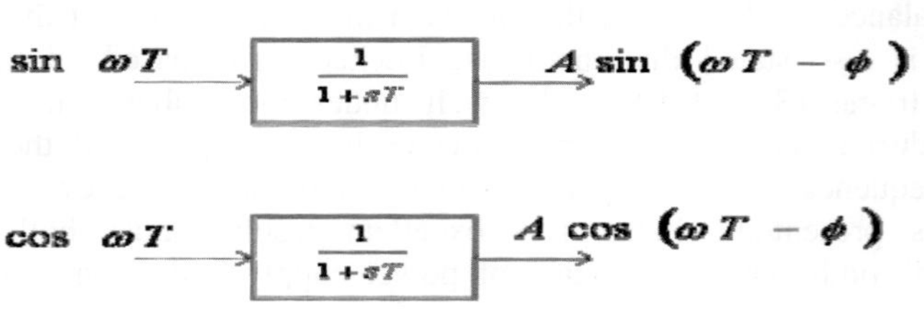

Fig. 1. Block Diagram indicating the nature of the output signals obtained when passed through low pass filters.

It is seen from Fig.1. that the output signals of both the low pass filters have the same amplitudes and phase shifts with respect to the input signals. Now using the identities, we get

$$\sin\omega T.A\sin(\omega T-\varphi)+\cos\omega T.A.\cos(\omega T-\varphi)=A\cos\varphi=\frac{1}{1+\omega^2T^2} \qquad (3)$$

$$\sin\omega T\,.A\cos(\omega T-\varphi)+\cos\omega T.A.\sin(\omega T-\varphi)=A\sin\varphi=\frac{-\omega T}{1+\omega^2T^2} \qquad (4)$$

Now, equation (4) ÷ (3) gives $-\omega T$
where T is the filter time constant. Thus the angular frequency of the signal ω and therefore the frequency of the signal in Hertz can be obtained.

2.1 Frequency Measurement and Tracking for a Balanced Three Phase Supply

Fig.2. indicates the block diagram of the method used in the detection of phase and hence the frequency variations of the voltages in a balanced three phase supply system during normal as well as fault conditions.

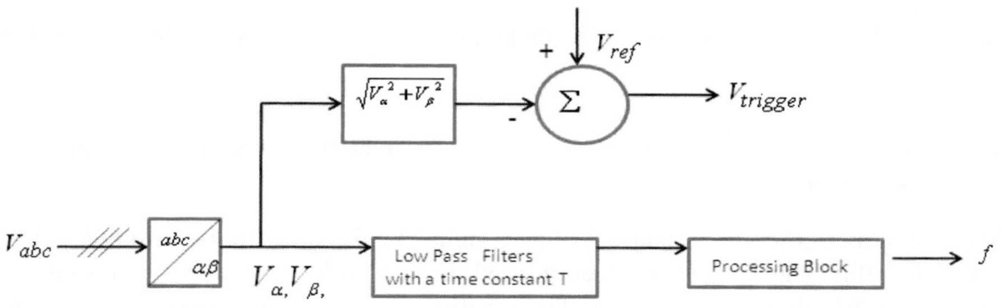

Fig. 2. Block diagram representation of the proposed method under three phase balanced conditions

The input signals in this case are three phase balanced supply voltages. $\mathbf{V_{abc}}$, These signals are then passed through a three phase to two phase transformation unit as per (1) in order to obtain their orthogonal components V_α and V_β with respect to a stationary reference frame. These orthogonal components are passed through the low pass filters. Applying equations (2),(3),(4) to the signals at the filter outputs, the frequency information of the input three phase signal is obtained.Fig.2 also shows the possible detection of occurrence of short duration voltage disturbances such as voltage sag or voltage swell. The voltage deviations are quantified and are reflected on the magnitude of the trigger signal $V_{trigger}$ obtained.[8]. The focus of this paper is on frequency detection.

2.2 Frequency Measurement and Tracking for an Unbalanced Three Phase Supply

Frequency measurement and tracking from the phase of the resultant space vector under unbalanced conditions of magnitude and phase of the three phase input voltage signals namelyV_a, $V_b$and V_c are not accurate and have a lot of oscillations. Under such conditions the extraction of positive and negative sequence components of the unbalanced three phase signal, is then used to determine the information about the phase and magnitude of the resultant of the balanced sequence components of the signal. The following block diagram indicates the method used in the detection of phase and hence frequency variations of the voltages in an unbalanced three phase supply during normal as well as fault conditions.

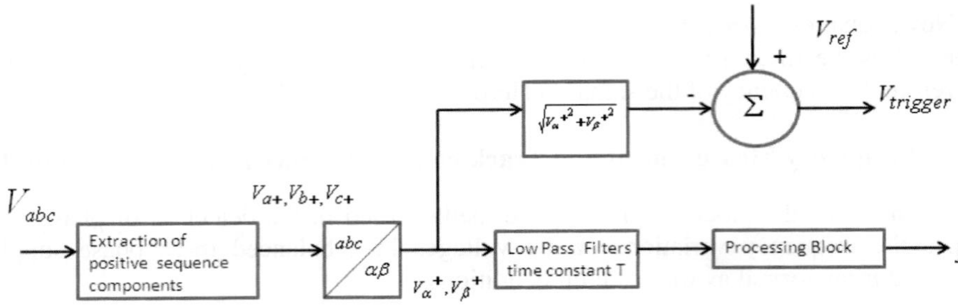

Fig. 3. Block diagram representation of the proposed method under three phase unbalanced conditions

2.3 Extraction of Sequence Components from a Three Phase Unbalanced Power Supply

Consider a periodic three phase voltage signal, V_a, V_b, V_c with or without neutral conductor, this signal can be decomposed in the following way: into sequence components which are well known. The positive sequence components are extracted as stated in [7].

$$\begin{bmatrix} V_a^+(t) \\ V_b^+(t) \\ V_c^+(t) \end{bmatrix} = \frac{-1}{3} \begin{bmatrix} -1 & 1\angle -60^\circ & 1\angle 60^\circ \\ 1\angle 60^\circ & -1 & 1\angle -60^\circ \\ 1\angle -60^\circ & 1\angle 60^\circ & -1 \end{bmatrix} \begin{bmatrix} V_a(t) \\ V_b(t) \\ V_c(t) \end{bmatrix} \quad (5)$$

The negative sequence components are extracted as follows,

$$\begin{bmatrix} V_a^-(t) \\ V_b^-(t) \\ V_c^-(t) \end{bmatrix} = \frac{-1}{3} \begin{bmatrix} -1 & 1\angle 60^\circ & 1\angle -60^\circ \\ 1\angle -60^\circ & -1 & 1\angle 60^\circ \\ 1\angle 60^\circ & 1\angle -60^\circ & -1 \end{bmatrix} \begin{bmatrix} V_a(t) \\ V_b(t) \\ V_c(t) \end{bmatrix} \quad (6)$$

These sequence components form a balanced set of three phasors. Therefore a two phase transformation similar to α, β is applied to the components .The frequency is detected from this decomposition after passing it through the low pass filters as per the block diagram in Fig.3.

3 Simulation Results Obtained

The whole system is simulated in Matlab/Simulink environment for different situations under both balanced and unbalanced conditions and the following are the observations made. First of all the method was tested for its performance by using a test signal $x = \cos(0.98\omega_0 t), f = 49\text{Hz}$.

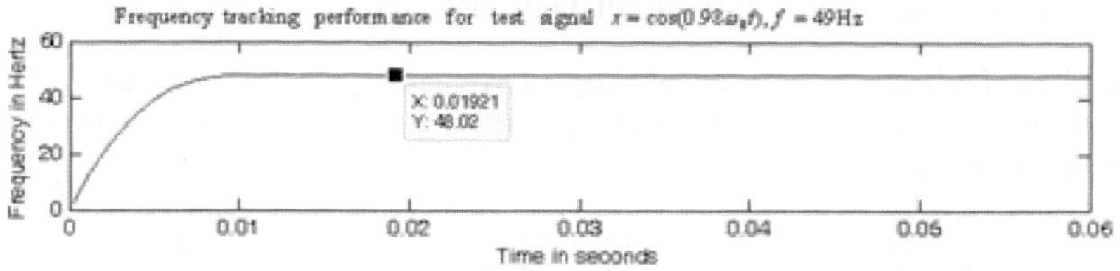

Fig. 4. Response of the method to frequency measurement of a signal $x = \cos(0.98\omega_0 t), f = 49Hz$.

Subsequently the following cases were considered and results obtained.

3.1 Case (a): Normal Conditions with Balanced Three Phase Voltages

The three phase balanced input voltage signal of 1p.u. in magnitude is subjected to the transformation as per (1) and the orthogonal axes α, β components on a stationary reference frame obtained. These components are then passed through the low pass filters as per Fig.2 and the frequency output measured.

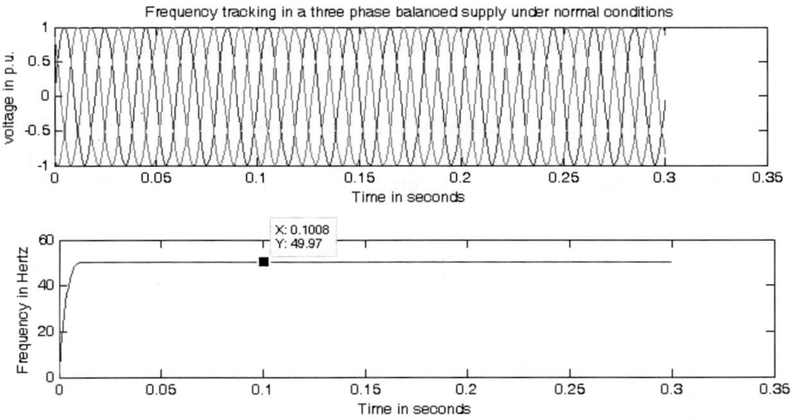

Fig. 5. Frequency measurement during three phase balanced normal operating conditions.

From Fig.5 we see that the method of frequency detection works for the above mentioned case. The time delay in effectively tracking the frequency variations by this method is found to be around 7msec and the final value settles at 49.97Hz (an error of 0.06%) for an input frequency of value 50Hz.

3.2 Case (b): Balanced Fault on All Three Phases

Three phase fault leads to an equal drop in voltage in all three phases [3].Balanced dips are due to three phase and three phase to ground faults.These situations are effectively tracked.

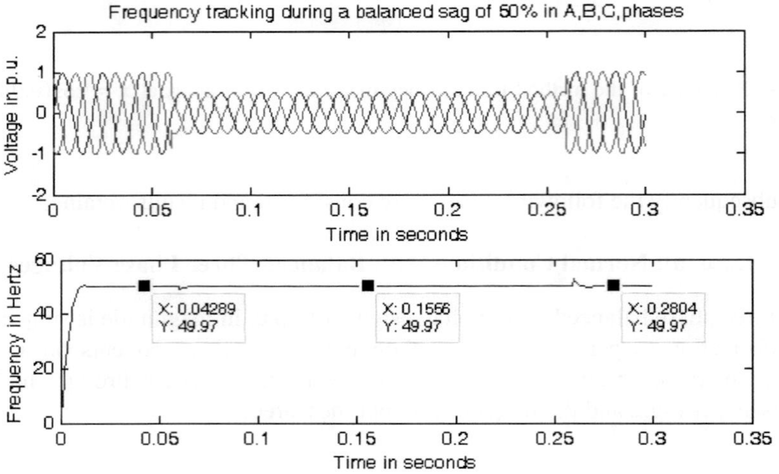

Fig. 6. Frequency measurement during a balanced three phase voltage sags of 50% on all the phases.

The results shown in Fig.6 indicate the effective tracking of phase variation both at the instants of occurrence and recovery of the sag. The frequency value during the sag is at the nominal value of 50Hz. The error in measurement of frequency by this method is found to be close to 0.06%

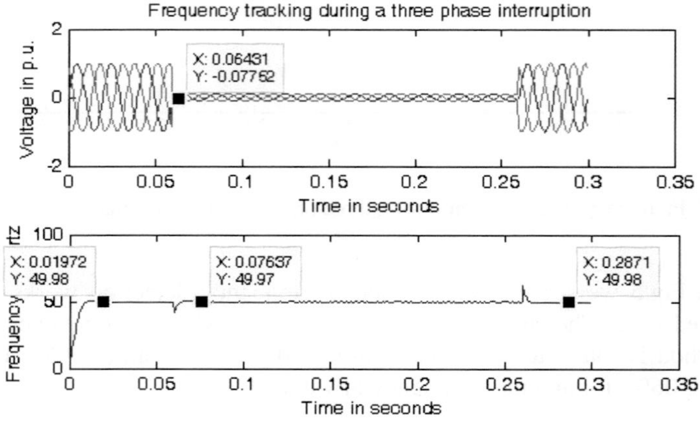

Fig. 7. Frequency measurement during a balanced three phase voltage sags of 50% on all the phases.

Fig.7 indicates the effectiveness of the method in measurement of frequency during an interruption when the voltage drops to a value of 10% of the rated voltage. This can occur due to power system faults, equipment failures, and control malfunctions.

3.3 Case (c): Unbalanced Condition

This situation is quite common at the point of common coupling at the distribution end. When the load distribution is not uniform across the three phases.Fig.8 shows the effectiveness of the method when negative sequence components are extracted. The results indicate accurate detection of frequency magnitude.

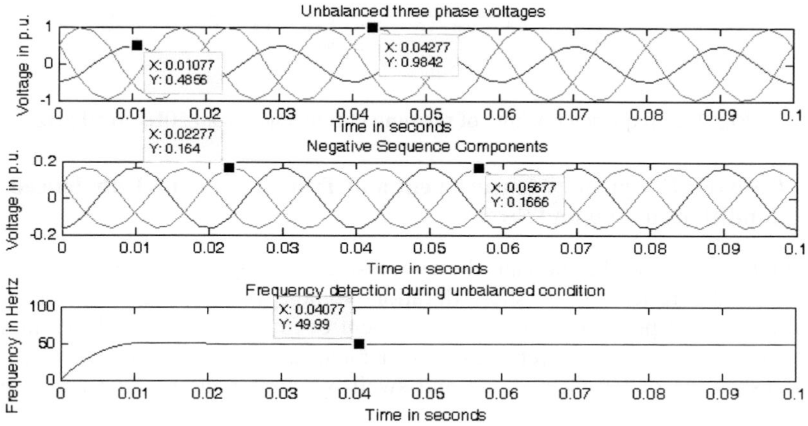

Fig. 8. Frequency measurement during unbalanced conditions in three phase supply

3.4 Case (d): Frequency Tracking during a Dynamic Change in Frequency

Fig.9 indicates the effective tracking of the step change in frequency and it is seen that the value stabilizes in 10msec.which is quite good. The response to tracking of ramping in frequency from 50Hz to 50.5Hz in0.8 seconds as indicated by the results in fig.10.The magnitude error is observed to be negligible.

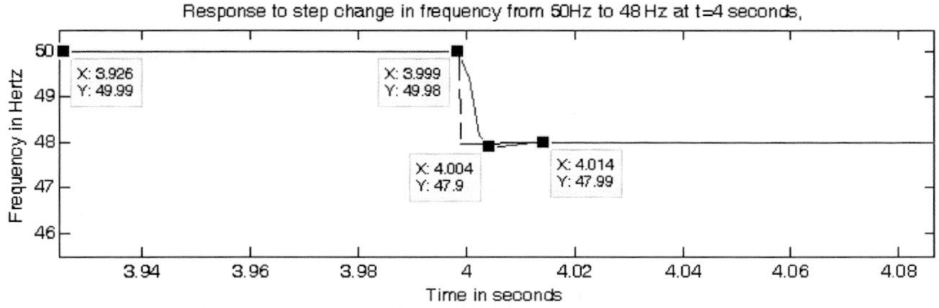

Fig. 9. Frequency tracking during a step change in frequency from 50Hz to 48Hz

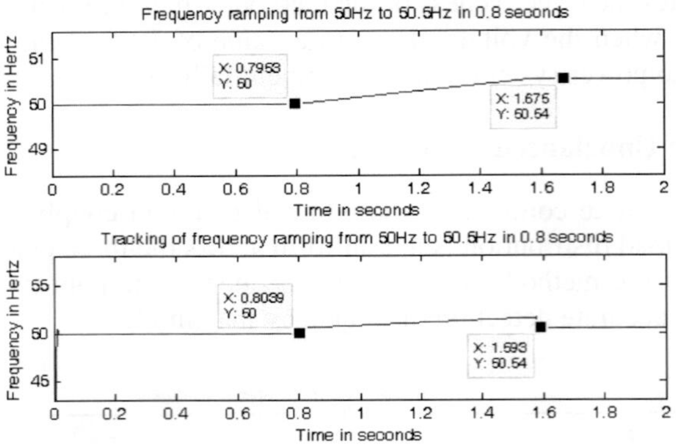

Fig. 10. Frequency tracking of ramping in frequency from 50Hz to 50.5Hz

3.5 Case (e): Frequency Measurement and Tracking during Unbalanced Conditions of Sag and Swell

The Figures 11 and 12 indicate the successive tracking of frequency. The figures also show the reflection of short duration voltage disturbance variations on the magnitude of the positive sequence component even in the case of different sags in phases A and B . The trigger signal generated as seen in Figure 12 also clearly distinguishes the occurrence of sag and swell by corresponding dip and rise in its magnitude.

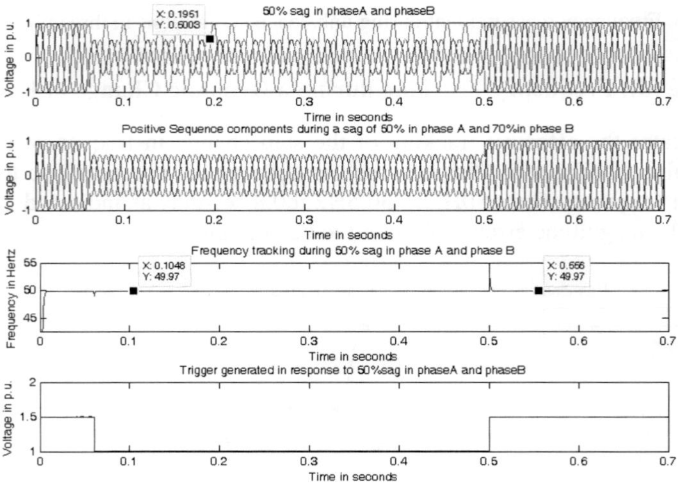

Fig. 11. Frequency tracking during unbalanced sag in phase A and phase B.

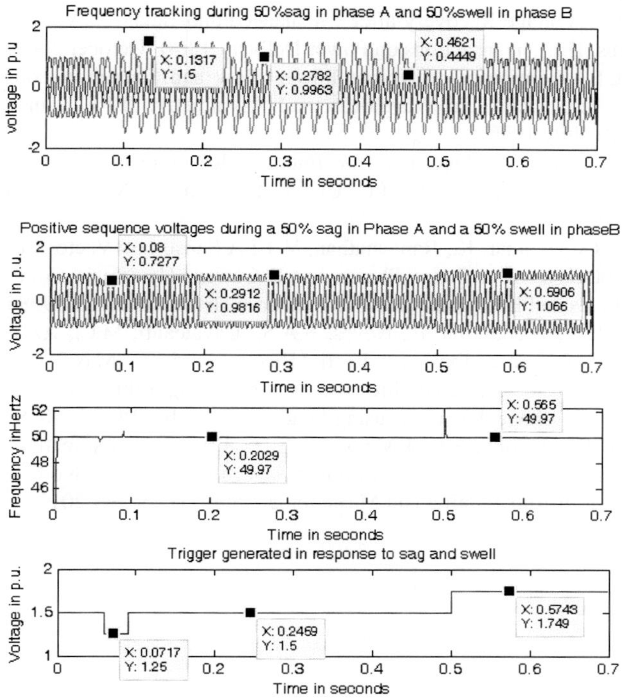

Fig. 12. Frequency tracking during unbalanced sag in phase A and phase B

4 Conclusion

The instantaneous frequency measurement and tracking method described in this paper is based on simultaneous sampling of the three phase voltage signals. The proposed method has exhibited excellent results in frequency detection during dynamic fault conditions. The proposed algorithm has been tested under different conditions of unbalance and found to be efficient for the successful measurement of the frequency. In addition, this approach is particularly suitable for frequency measurement and tracking when asymmetric sags and swells occur in the three phases. This algorithm is suitable for online applications. However, the limitations of this method would be in detecting multiple frequencies which would need a cascade of filters for the input signal before applying the mathematical transformation for obtaining the orthogonal components.

References

1. Chanana, S., Kumar, A.: Demand response by dynamic demand control using frequency linked real time rices. International Journal of Energy Sector Management 4(1), 44–58 (2010)
2. The modern Grid Strategy, `http://www.netl.doe.gov/moderngrid/`

3. Bollen, M.H.J.: Voltage Recovery after Unbalanced and Balanced Voltage Dips in Three-Phase Systems. IEEE Transactions on Power Delivery 18(4) (October 2003)
4. Czarnecki, L.S.: Comments on Active Power Flow and Energy Accounts in Electrical Systems with non sinusoidal Waveforms and Asymmetry. IEEE Transaction on Power Delivery 11(3) (July 1996)
5. Langella, R., Testa, A.: Algorithm for Energy Measurement at Positive Sequence of Fundamental Power Frequency. In: Under Unbalanced Non-Sinusoidal Conditions, PowerTech (2007)
6. Ebenezer, V., Gopakumar, K., Ranganathan, V.T.: A Sensorless Vector Control Scheme for Induction Motors using a Space Phasor based Current Hysteresis Controller. In: Power Electronic Drives and Energy Systems for Industrial Growth, PEDES 1998 (1998)
7. De Souza, H.E.P., Bradaschia, F., Neves, F.A.S., Cavalcanti, M.C., Azevedo, G.M.S., de Arruda, J.P.: IEEE Transactions on Industrial Electronics 56(5) (May 2009)
8. Meena, P., UmaRao, K., Deekshit, R.: A simple algorithm for Fast detection and quantification of Voltage deviations using Space Vectors. In: IPEC 2010, Singapore (2010)
9. Kušljevi´c, M.D., Tomi´c, J.J., Jovanovi´c, L.D.: Frequency Estimation of Three-Phase Power System Using Weighted-Least-Square Algorithm and Adaptive FIR Filtering. IEEE Transactions on Instrumentation and Measurement 59(2) (February 2010)

Cognitive Ultra-Wideband Key Technology Research

Li Zimu

Navy DaiBiaoShi in Beijing area military, Beijing, China, 100841
allen2004_y@yahoo.com.cn

Abstract. Cognitive radio technology is a kind of intelligent spectrum sharing technology, it can perceive spectrum environment selfly through certain method, look for unused spectrum (spectrum cavity), adjust the communications terminal parameter settings and realize the involvement of free spectrum resources, so as to improve spectrum efficiency, alleviate the purpose of spectrum resources nervous. Combing cognitive radio technology with ultra broadband wireless communications technologies, devise a new intelligent wireless system - cognitive ultra broadband wireless communication system (CUWB). In CUWB system, the selection of spectrum perception technology is vital, and restricts the whole system performance. Based on the noparametric estimation method, guarantee the accuracy of cases, realize the rapid estimate of interference temperature.

Keywords: cognitive radio, UWB, spectrum perception, noparametric estimation.

1 Introduction

With the new business, increasing frequency resource is becoming more and more nervous. Research shows that: in any given moment, all the spectrum people used accounts for only 2% ~6% of the available spectrum, in the meanwhile, spectrum resources nervous, a large number of spectrum are idle. The cognitive radio (CR) has the characteristics of perception, combes cognitive radio technology with ultra broadband wireless communications technologies, devises a new intelligent wireless system - cognitive ultra broadband wireless communication system (CUWB).

The CUWB system based on shared ideas, shared the bandwidth with existing traditional wireless technology narrow-band. In ultra-wideband frequency band, the existing communications system is called authorized users, for not affecting an authorized user's communication quality, cognitive users have lower priority than authorized user, only with authorized users not use authorized frequency or get an authorized user's permission, cognitive user can use authorized frequency band.

2 CUWB System Cognitive Cycle

CUWB system can perceive whether spectrum is used by authorized users in a very wide range. if spectrum is free, the cognitive users can spare according to rules. If this band temporary is occupied by authorized users again, cognitive users can select two

X. Wan (Ed.): Electrical Power Systems and Computers, LNEE 99, pp. 213–219.
springerlink.com

cognitive way to continue communication, first, a jump to backup frequency band, avoiding authorized, second, change power levels, avoid continuing to use the spectrum interference and launch main users. CR's main function is to estimate interference temperature and detection of frequency cavity, and with perception of spectrum information, according to dynamic spectrum allocation strategies, adaptively construct radiation masking, to constrain the firing pulse waveform and power. CUWB system has three functions: cognitive cycle inspection, analysis and adjustment.

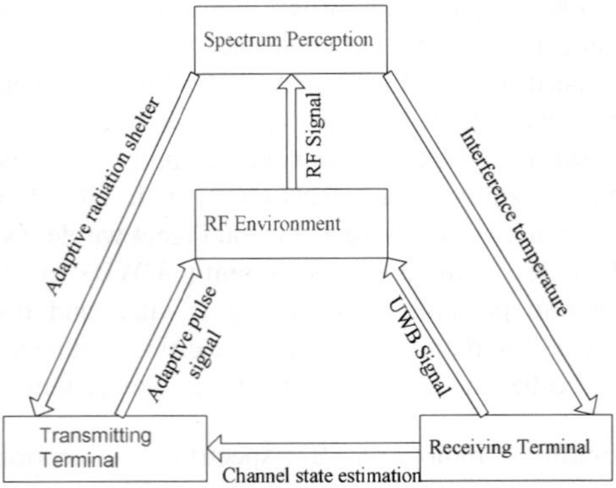

Fig. 1. CUWB system cognitive cycle

As shown in figure 1, cognitive circulation system process can be divided into the following several steps.

[1] spectrum perception module testing, rf signal interference temperature and detection spectrum cavity estimation.

[2] to mark spectrum cavity, identification according to user needs.

[3] transmitting terminal constructs dynamic radiation masking according to launch the perception.

[4]establish the dynamic transmission power spectrum allocation strategies, according to the receiver channel state information feedback control.

[5] adjust the transmission parameters, generate flexible adaptive pulse, for transmission.

[6] real-time detection of the spectrum environment, update, formulate fluffing spectrum empty list and alternative strategies.

[7] if authorized users appears, jump according to the strategy and repeat steps (3).

CUWB system can adjust dynamicly transmission parameters according to changes in the environment and business of UWB frequency, the blank of the dynamic within the optimal allocation of bandwidth.

3 Non-parameter Spectrum Perception Technology

In the cognitive cycle of cognitive radio systems, the first mission to the spectrum is testing, namely environmental information idle spectrum perception. The perception of spectrum environment is the premise and computed radiography (CR) was established in correct, only prerequisite perception and detection based on the spectrum environment free of the spectrum to occupy and communication. Spectrum perception includes two tasks: perception is to test whether a band authorized users within existing, judge whether the frequency signal in the idle state, to decide whether to occupy the band; second, periodic detection wireless spectrum environment changes, because authorized users within the band, CR users priority is below, when CR users occupy the frequency to communication, if authorized user appeared, CR users in the first time detect and in return channel with the fastest speed.

In CUWB system, the task is the spectrum continuously testing in UWB legal frequencies (3. IGHz ~10.6 GHz), to monitor of the spectrum environment, real-time perceive spectrum changes in the environment. Because the pulses UWB signal is a no carrier modulation, so the baseband signal based on carrier modulation signal detection method does not apply to UWB signal, and because the UWB signal bandwidth is wide, the traditional method of frequency domain here is difficult to achieve. Spectrum perception algorithm are mainly parameters spectrum estimation and non-parametric spectrum estimation two kinds. According to the characteristics of UWB signals, this paper do spectrum detection with MTM (multi window spectral estimation) nonparametric spectral, through the window function of spectrum testing brings energy decreases data leakage because of the introduction of truncated, using the singular value decomposition to remove noise interference and take temperature estimates.

In dealing with actual data, introduce window function of data truncation, to improve the frequency spectrum resolution and reduce spectrum leak, meanwhile would increase spectral estimation variance. Usually, spectral estimation algorithm will choose a single window function. Multi-window spectrum estimation algorithm uses the cluster data window to replace a single data window, composed of each data window of discrete Fourier transform (DFT)of the time series and forms spectrum estimation from the feature spectrum functions of the weighted average, therefore, it is a low variance, high resolution, simpler calculation spectral analysis method. In addition, in multi-window spectrum estimation algorithm, the data of window function cluster uses discrete oblate ellipsoid sequences (Slepian sequence) together with the best energy characteristic, each Slepian sequence applied to the sample data, and for broadband signal is concerned, multi-window spectrum estimation reached almost non-parameter spectrum estimator in Latin America - ROM bound (CRB), therefore, multi-window spectrum estimation is widely thought to be better than any non-parameter spectrum estimation method.

MTM has strong advantage estimates in low signal-to-noise signal, which is very suitable for signal analysis in nonlinear climate system under the the weak signal background of short sequences, high noise. it is a kind of spectrum estimation method in larger dynamic range signal and faster spectrum change, in climate research has been widely used.

Set $s(t)$ to actual data, W_t^k (k = 1, 2, 3... k) for k Slepian sequence, $Y_k(f)$ for the corresponding feature spectrum.

$$Y_k(f)=\sum_{t=1}^{N}W_t^{(k)}s(t)e^{-j2\pi f} \qquad k=1,2\cdots K \tag{1}$$

Type: N is for the receiver sample data length.

Spectral estimation Based on the characteristic spectrum $\hat{S}(f)$.

$$\hat{S}(f)=\frac{\sum_{k=1}^{K}\lambda_k(f)\left|Y_k(f)\right|^2}{\sum_{k=1}^{K}\lambda_k(f)} \tag{2}$$

Type: λ_k for the first k feature spectrum corresponding eigenvalues .

Figure 2 is $S_k(f)$ with classical periodic chart for more results, assuming there are two spectrum users are in communication, user 1 for relatively narrow-bandwidth signals, center frequency 8.8GHz, bandwidth 400MHz, user 2 for broadband signal, 4.3 center frequency, 2.2 GHz bandwidth, noise is white gaussian noise, signal-to-noise ratio is 5dB, the receiver for 5,000 sample datas.

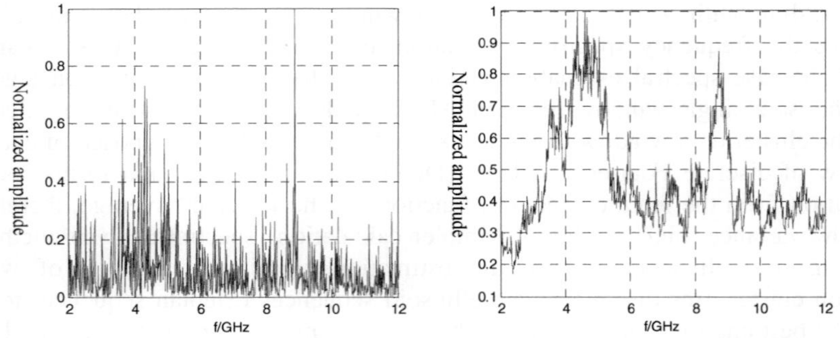

(a) Cycle map (b) Spectrum estimation

Fig. 2. Cycle with characteristic spectral spectrum

As can be seen from the graph, multiple orthogonal window superposition reduced lose information brought by adding windows, reduced the spectral estimation variance with not affect spectrum estimation deviation of cases, make spectrum estimation more smooth. MTM spectrum estimation process can be interpreted as maximum likelihood power spectral estimator. for the approximate, broadband signals are concerned, MTM spectrum estimation process is nearly optimal. In the power spectrum estimation, this method is widely considered superior to any other non-parameter spectrum estimation method.

4 Singular Value Decomposition

Singular value decomposition of feature spectrum operation can be used to denoise and estimate interference temperature. Because sensors and the distance and authorized users launch transmission environment is different, different received signals in amplitude, phase, signal-to-noise ratio, have a lot of differences. When the singular value decomposition to different sensors from the feature spectrum weighted. all feature spectrum are arranged in the following matrix $\mathrm{A}(f)$

$$\mathrm{A}(f)=\begin{bmatrix} w_1Y_1^{(1)}(f) & w_1Y_2^{(1)}(f) & \cdots & w_1Y_K^{(1)}(f) \\ w_2Y_1^{(2)}(f) & w_2Y_2^{(2)}(f) & \cdots & w_2Y_K^{(2)}(f) \\ \vdots & \vdots & & \vdots \\ w_MY_1^{(M)}(f) & w_MY_2^{(M)}(f) & \cdots & w_MY_K^{(M)}(f) \end{bmatrix} \tag{3}$$

Type: M is the number of the sensor inestigate area; w_i is the weight coefficient.

In matrix $\mathrm{A}(f)$, the row vector matrix Slepian says the same sensor different characteristics of sequence, column says the same spectral characteristics of different sensors Slepian sequence spectrum.

Use two orthogonal matrix U, V respectively on $\mathrm{A}(f)$, the results for transformation, transform is to get diagonal matrix Σ:

$$\mathrm{U}^{\mathrm{T}}\mathrm{AV}=\Sigma \tag{4}$$

$$\Sigma=\begin{bmatrix} \mathrm{S} & 0 \\ 0 & 0 \end{bmatrix} \quad \mathrm{S}=diag(\sigma_1,\sigma_2,\cdots,\sigma_r) \tag{5}$$

$$\mathrm{A}=\sum_{i=1}^{r}\sigma_i\,\mathrm{u}_i\mathrm{v}_i^{\mathrm{T}} \tag{6}$$

Type: σ_i is for matrix A singular value.

Matrix A can be considered as singular vectors outside doing the deposition, the weight namely is nozero singular value σ_i . all nonzero singular value in an arrangement, the biggest characteristic value $|\sigma_{\max}(f)|^2$ is selected or eigenvalue of maximum numbers as a linear combination for estimation of interference temperature.

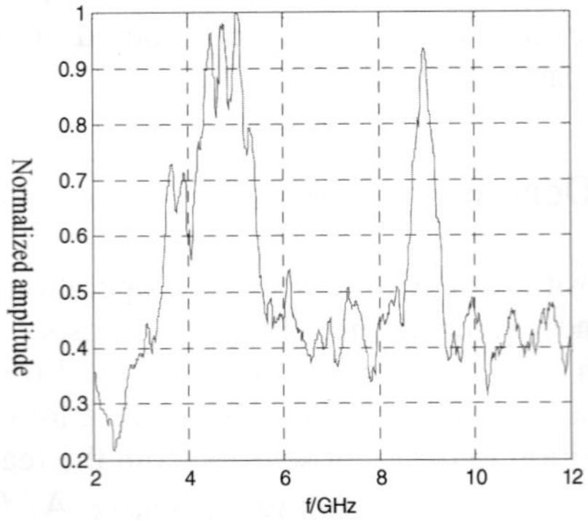

Fig. 3. Singular value transform

Figure 3 is the estimation of interference temperature after using singular value transform, assuming that different sensors hypothesis temperature of received signal SNR and decline is different. We can see, multi-window spectrum estimation has focused power characteristics, the existing signal has high peak, but due to feature spectrum peak smoothness, poor judgment threshold set against. After singular value decomposition operation later, interference temperature estimates have very good smoothness and stability, set corresponding interference temperature limit can judge the existing in the " frequency spectrum hole".

5 Conclusion

Wireless spectrum environment is changing, cognitive ultra-wideband systems must accurately and timely detect spectrum environment, seek spectrum hollow, realize dynamic spectrum access, no interference to authorized user communication, this is premise to guarantee the normal work of cognitive system . So it asked spectrum perception algorithm to accurate and efficient. Multi window spectrum estimation is closest to Latin America - ROM world's (CRB), its accuracy can satisfy the requirements of cognitive ultra-wideband systems, and singular value transform algorithm combining, reduced the system after of calculation and improve the detection efficiency.

References

1. Zhou, X.F., Yazdandoost, K.Y., Zhang, H.: Cognospectrum: Spectrum adaptation and evolution in cognitive ultra-wideband radio. In: IEEE International Conference on Ultra-Wideband (ICU), September 2005, pp. 713–718 (2005)
2. Cui, W.T., Hai, C., Li, X.Y.: Cognitive Ultra-Wideband Radio Based on Adpative Spectrum Allocation and Rake Receiver. In: 5th International Conference On Wireless Communications, Networking And Mobile Computing, vol. 1-8, pp. 1425–1428 (2009)

3. Wu, X.L., Sha, X.J., Li, C.: Pulse shaping for cognitive ultra-wideband communications. Wireless Communications & Mobile Computing 10, 772–786 (2011)
4. Zhao, C.L., Xu, F.M., Sun, X.B.: Capacity Analysis and Optimization of Cognitive Ultra Wideband Networks. China Communications 7(2), 109–116 (2010)
5. Zhang, H.G., Zhou, X.F., Yazdandoost, K.Y.: Multiple signal waveforms adaptation in cognitive ultra-wideband radio evolution. In: IEEE International Conference on Ultra-Wideband, Switzerland (2005)
6. Wang, S.B., Zhou, Z., Kwak, K.: An Emission Strategy of Cognitive Radio-Ultra Wideband Based on Interference Temperature model. In: 5th International Conference on Wireless Communications, Networking and Mobile Computing, Beijing, Peoples R China, vol. 1(8), pp. 1499–1502 (2009)

Realization of Sound Reproduction in a Specified Region Based on Transfer Function Interpolation

Jie Cao, Ming Wu, and Jun Yang

Key Laboratory of Noise and Vibration Research, Institute of Acoustics,
Chinese Academy of Sciences, Beijing, 100190, China
{mingwu,jyang}@mail.ioa.ac.cn

Abstract. In this paper, a multichannel sound reproduction system based on transfer function interpolation is developed. The system creates a directional sound field by weighting signals fed to the loudspeaker units. To calculate the weighting coefficients, the acoustic transfer functions between the input signal of each loudspeaker and the signal in the target zone should be measured in advance. Three types of interpolation techniques are investigated to relieve the workload of measurements for obtaining the acoustic transfer functions at the target points. Based on the Discrete Fourier Transform (DFT) interpolation, a sound reproduction system using loudspeaker array is realized and more than 20 dB is guaranteed for the side lobe suppression at the frequencies above 800 Hz.

Keywords: Multichannel sound reproduction, loudspeaker array, beam pattern, transfer function interpolation.

1 Introduction

Loudspeaker array can create arbitrary beam pattern by adjusting the weight of each loudspeaker, which draws much attention in sound field reproduction. Several techniques [1-8] can be used to calculate the weights of the loudspeaker array according to the sound field distribution/beam pattern needed. For example, an acoustic contrast control technique has been developed for arbitrary array type by maximizing the acoustic energy density ratio of the bright zone to the dark zone [6, 7]. Capon beamformer [9], Chebyshev beamformer [10], and many other algorithms [11, 12] derived from them usually can optimize some of the performance indicators such as main lobe response, side lobe level, array gain, robustness, and so on. However, the algorithms above are either applicable only to uniformed line arrays with isotropic elements, or complicated to design and need large computation. Moreover, for practical applications, the transfer functions between the signal of the loudspeakers and the signals of the target zone must to be measured advanced individually with much effort due to the inconsistent frequency responses between the loudspeaker units. In this paper, an acoustic transfer function interpolation technique is proposed to reduce the workload.

X. Wan (Ed.): Electrical Power Systems and Computers, LNEE 99, pp. 221–228.
springerlink.com

There are various interpolation methods, such as first-order linear, cubic spline, and DFT interpolation techniques [13, 14]. A multichannel sound reproduction system using loudspeaker array is developed in this paper, and three interpolation methods applied to interpolate the acoustic transfer functions are analyzed and compared. Furthermore, a sound reproduction system using acoustic contrast control based on the best performed fitting data is implemented, and experiments conducted in an anechoic chamber to evaluate the system performance using three interpolation methods. Finally, conclusions are summarized in Section 5.

2 Acoustic Contrast Control Technique

The acoustic contrast control technique can provide the maximum acoustic energy density ratio between a bright zone and a dark zone. The region of the sound propagation is the bright zone, and the rest area is the dark zone. The sound pressure vector can be expressed as

$$\mathbf{P} = \mathbf{w}^H \mathbf{H}\mathbf{x} . \tag{1}$$

where $\mathbf{x}$ is the input signals to the loudspeaker array, vector $\mathbf{w}$ contains the weight values corresponding to the loudspeaker units, and $\mathbf{H}$ is the acoustic transfer functions between loudspeaker inputs and sound pressure at a certain position. Hence, the acoustic energy for the bright zone can be expressed as

$$e_b = \frac{|x|^2}{N_b} \mathbf{w}^H \mathbf{H}_b{}^H \mathbf{H}_b \mathbf{w} \cdot \tag{2}$$

where $\mathbf{H}_b$ represents the transfer function matrix between the loudspeaker inputs and the field positions in the bright zone, N_b is the number of discrete points in the bright zone. Similarly, the acoustic energy for the whole zone can be expressed as

$$e_t = \frac{|x|^2}{N_t} \mathbf{w}^H \mathbf{H}_t{}^H \mathbf{H}_t \mathbf{w} \cdot \tag{3}$$

where $\mathbf{H}_t$ represents the transfer function matrix for the total zone, N_t is the number of discrete points in the total zone. A cost function for optimizing the acoustic energy density ratio of the bright zone to the total zone of interest is defined here as acoustic contrast β

$$\beta = \frac{N_t \mathbf{w}^H \mathbf{H}_b{}^H \mathbf{H}_b \mathbf{w}}{N_b \mathbf{w}^H \mathbf{H}_t{}^H \mathbf{H}_t \mathbf{w}} \cdot \tag{4}$$

The optimal solution that maximizes the acoustic contrast β is simply given as the eigenvector corresponding to the maximum eigenvalue of $(\mathbf{H}_t{}^H \mathbf{H}_t)^{-1} \mathbf{H}_b{}^H \mathbf{H}_b$.

Consider the sound reproduction using a loudspeaker array as shown in Fig. 1, We define the targeted region from -20° to 20° (white) as the bright zone, and the other area (gray) as the dark zone. In the simulation, the half-plane in front of the array is divided into 19 discrete areas from -90° to 90° with an interval of 10°. The acoustic transfer functions for the bright zone is denoted as $\mathbf{H}_b$, the acoustic transfer functions for the total zone as $\mathbf{H}_t$. Eq. (4) is used to obtain the eigenvector corresponding to the maximum eigenvalue of $(\mathbf{H}_t{}^H\mathbf{H}_t)^{-1}\mathbf{H}_b{}^H\mathbf{H}_b$, which is the weighting for the array.

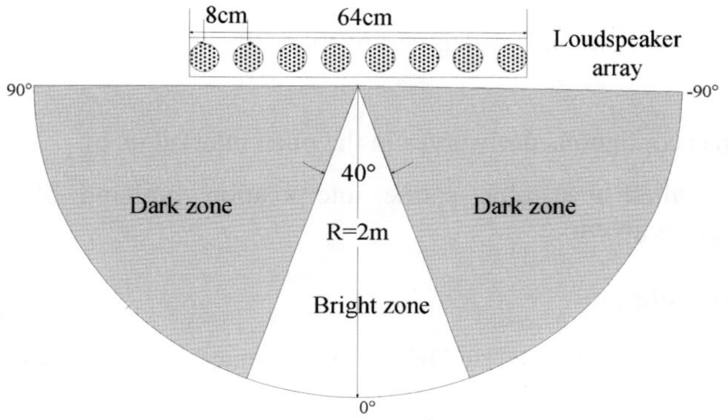

Fig. 1. Sound reproduction in a specified region (white for bright zone and gray for dark zone)

3 Interpolation Algorithms

Assume that $H_p(k_i), (i = 1, 2, \cdots M-1)$ are the acoustic transfer functions measured at different points k_i with an interval d, then the acoustic transfer function value at any point k can be obtained by interpolation technique. There kinds of interpolation methods are described as follows.

3.1 First-Order Linear Interpolation

If the function values at two adjacent points are known, the function values at points between these two points can be estimated by using the first-order linear interpolation method, which is expressed as

$$H(k) = H(mL+l) = (1-\frac{l}{L})H_p(mL) + \frac{l}{L}H_p((m+1)L), l \in [0, L-1], m \in [0, M-1]. \quad (5)$$

L is the interval between these two given points, which is defined as the interpolation factor.

3.2 Cubic Spline Interpolation

Cubic spline interpolation can improve the smoothness of the curve resulted from the linear interpolation method. A segment Δ for the points in the interval of [*a*, *b*] is given as

$$\Delta : a = k_0 < k_1 < \cdots < k_n = b . \tag{6}$$

If function $s(k)$ meets the following conditions:

(1) $s(k_j) = H_p(k_j), j = 0,1,2,\cdots n$;

(2) $s(k)$ is a polynomial of degree not more than three on each smaller interval $[k_{j-1}, k_j]$;

(3) $s(k)$ has continuous derivatives in the open interval (a, b).

Then $s(k)$ is called the cubic spline interpolation function of interval [*a*, *b*] corresponding to segment Δ.

3.3 DFT Interpolation

The interpolation method based on DFT is implemented in the following four steps:

(1) Determine the number of points *N* of the interpolated sequence based on the accuracy requirements;
(2) Assume $y_i = H_p(k_i), i = 1,2,\cdots M-1$, apply a DFT of dimension *M* to get a sequence $\{Y_i\}, i = 1,2,\cdots M-1$;
(3) Add zeros to $\{Y_i\}$ to get a new sequence $\{Z_i\}(i = 1,2,\cdots N-1)$, the values of $\{Z_i\}$ is given as

$$\{Z_i\} = \begin{cases} \{Y_0,\cdots,Y_{(M-1)/2-1}, 1/2Y_{(M-1)/2}, 0,\cdots,0, 1/2Y_{(M-1)/2}, Y_{(M-1)/2+1}\cdots,Y_{M-1}\}, & \text{if } M \text{ is odd} \\ \{Y_0,\cdots,Y_{M/2-1}, 0,\cdots,0, Y_{M/2},\cdots,Y_{M-1}\}, & \text{if } M \text{ is even} \end{cases} . \tag{7}$$

(4) Apply IDFT of dimension *N* to $\{Z_i\}(i = 1,2,\cdots N-1)$ to get a sequence $\{z_i\}(i = 1,2,\cdots N-1)$, the values corresponding to *k* are the approximation of function $y = H_p(k)$.

4 Experiments

To evaluate the performance of the three methods described in Section 3 when they are applied to interpolate the acoustic transfer functions of an actual system, an experimental device was setup in the anechoic chamber at Institute of Acoustics, Chinese Academy of Sciences. As shown in Fig.2.

Fig. 2. Experimental setup: a loudspeaker array with eight 8 cm equally spaced loudspeakers and a microphone 2 m away facing the middle of the array

4.1 Performance of the Three Acoustic Transfer Function Interpolation Methods

The acoustic transfer functions of each loudspeaker was measured every 20°. The acoustic transfer functions at other directions were interpolated every 10° by using first-order linear interpolation, cubic spline interpolation, and DFT interpolation. To evaluate the interpolation performance, acoustic transfer functions for the interpolated points are also measured. The interpolation results at 1k Hz are shown in Fig.3.

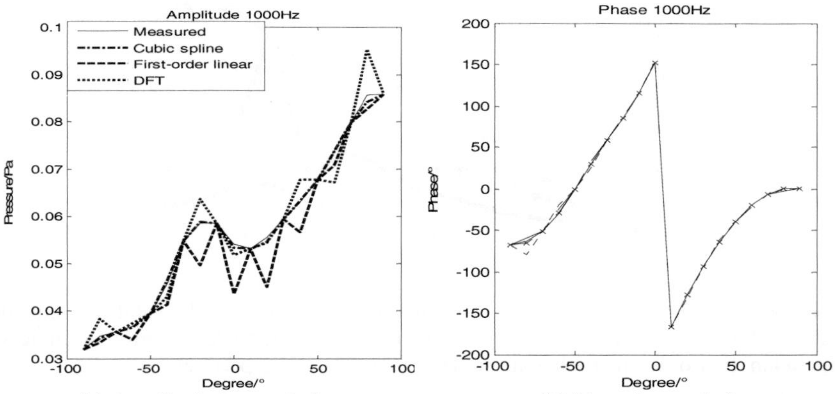

(a) Amplitude interpolation (b) Phase interpolation

Fig. 3. Three acoustic transfer function interpolation results compared to the measured results at 1k Hz.

From Fig.3, we can see that at 1k Hz the phase curve can be fitted very well while the amplitude curve has some differences when using the three interpolation methods. That's because the phase curve is a smooth slope, all the three methods can reach good results under this condition. In fact, the phases of the acoustic transfer functions on the whole frequency band of interest are all smooth slope, the differences of the three methods are quite small here. As for the amplitude interpolation, the results become much more complicated for the amplitude curve may not be very smooth. To analysis the three interpolation methods more accurately, the average interpolation error is defined as Eq. (8), and shown in Fig. 4.

The performance of each interpolation technique is evaluated by average interpolation error given as

$$E=\sqrt{((H_p(1)-H_o(1))^2+(H_p(2)-H_o(2))^2+\cdots+(H_p(N)-H_o(N))^2)/N}\ , \tag{8}$$

where H_p represents the fitted values, H_o represents the measured values, N is the number of fitted points.

Fig.4 shows that at the frequencies below 1800 Hz, the cubic spline interpolation outperforms the other two methods, while at higher frequencies, the DFT interpolation has the best performance. For the whole frequency range of interest, only the average interpolation error of the DFT interpolation is below 1 dB. The performances of these three interpolation methods varies from each other in that the first-order linear interpolation has poor interpolation accuracy while cubic spline interpolation can approach functions of second-order continuous derivatives with very small errors, and the interpolation accuracy of DFT interpolation method only relates to the accuracy of calculation.

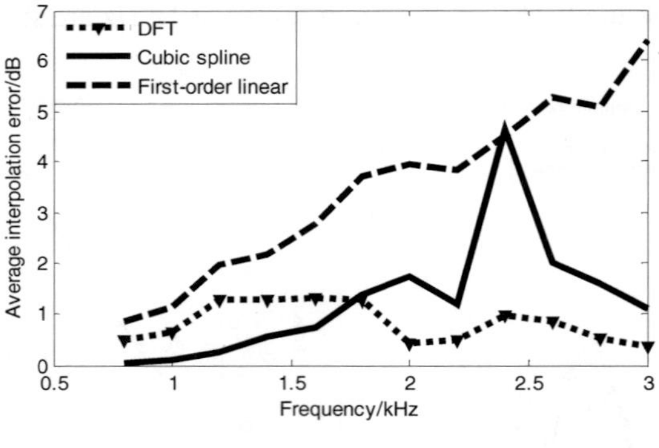

Fig. 4. Average interpolation errors of three interpolation methods. The solid line, dotted line, and dotted line with triangles in it represent the average interpolation error of cubic spline interpolation, first-order interpolation, and DFT interpolation respectively.

In terms of the average interpolation error, DFT interpolation is selected for fitting the transfer functions in the frequency range from 800 Hz to 3000 Hz with an interval of 50 Hz.

4.2 Pressure Distribution of Reconstructed Sound Field

In this subsection, we examine the results of sound field reproduction based on the interpolated acoustic transfer function matrix. The bright zone spreads from the middle of the loudspeaker array with an angle of 40° (see Fig. 1). The fitted transfer functions are represented by matrix $\mathbf{H}_p$ with a dimension of $M \times N$, where M

represents the number of loudspeakers and N represents the number of discrete frequencies. With the weight for each loudspeaker unit in the array calculated from Eq. (4), the sound pressures for different directions are calculated by using Eq. (1).

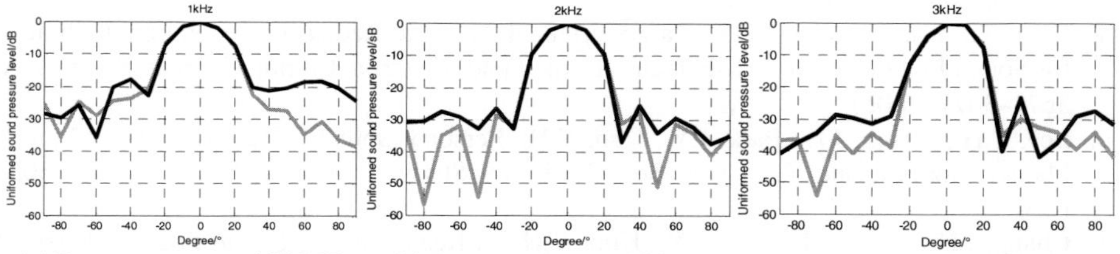

(a) Beam pattern at 1000 Hz (b) Beam pattern at 2000 Hz (c) Beam pattern at 3000 Hz

Fig. 5. Beam patterns obtained from experimental results and simulation results at (a) 1000 Hz, (b) 2000 Hz, and (c) 3000 Hz (Black for the experimental results, gray for the simulation results)

Fig.5 shows the measured pressure distribution and the simulated results at 1000, 2000, and 3000 Hz, respectively. We can see that the experimental results agree with the simulation result quite well. The sound energy is focused in the direction of 0° as expected. The side lobe is suppressed lower than -20 dB at 1000 Hz and lower than -30 dB above 2000 Hz. As the frequency increases, the pressure level of side lobe can be suppressed further. Therefore, we can believe that the sound field can be reconstructed with the fitted acoustic transfer functions by using interpolation technique.

5 Conclusion

In this paper, the acoustic transfer function interpolation is employed to calculate the weights of the loudspeaker array to create a directional sound field. Three interpolation techniques are analyzed and compared. Simulation results show that the DFT interpolation method has the best performance with average interpolation error below 1 dB. Based on the presented method, a directional audio delivery system is realized using a loudspeaker array with eight units. Experimental results show that the sound energy is focused at 0° direction with a beam width of 40°, more than 20 dB acoustic contrast is guaranteed at the frequencies above 800 Hz, which further proves the feasibility of using interpolation method to avoid large laborious measurements.

References

1. Choi, J.W., Kim, Y.H.: Generation of an Acoustically Bright Zone with an Illuminated Region Using Multiple Sources. J. Acoust. Soc. Am. 111, 1695–1700 (2002)
2. Wen, Y., Yang, J., Gan, W.S.: Strategies for an Acoustical-Hotspot Generation. IEICE Trans. Fund. Electron. Comm. Comput. Sci. E88-A, 1739–1746 (2005)

3. Wen, Y., Gan, W.S., Yang, J.: Nonlinear Least-square Solution to Array Pattern Synthesis Using Arbitrary Linear Array. Signal Processing 85, 1869–1874 (2005)
4. Wen, Y., Gan, W.S., Yang, J.: Application of Radiation Mode in Desired Sound Field Generation Using Loudspeaker Array. In: IEEE International Symposium on Circuits and Systems, pp. 3139–3142 (2005)
5. Wen, Y., Yang, J., Gan, W.S.: Target-oriented Acoustic Radiation Generation Technique for Sound Field Control. IEICE Trans. Trans. Fund. Electron. Comm. Comput. Sci. E89-A, 3671–3677 (2006)
6. Lee, C.H., Chang, J.H., Park, J.Y., Kim, Y.H.: Personal Sound System Design for Mobile Phone, Monitor, and Television Set; Feasibility Study. J. Acoust. Soc. Am. 122, 3053 (2007)
7. Chang, J.H., Lee, C.H., Park, J.Y., Kim, Y.H.: A Realization of Sound Focused Personal Audio System Using Acoustic Contrast Control. J. Acoust. Soc. Am. 125, 2091–2097 (2009)
8. Park, J.Y., Chang, J.H., Kim, Y.H.: Generation of Independent Bright Zones for a Two-Channel Private Audio System. J.A.E.S. 58, 382–393 (2010)
9. Capon, J.: High-resolution Frequency-wavenumber Spectrum Analysis. Proc. IEEE 57, 1408–1418 (1969)
10. Dolph, C.L., Riblet, H.J.: Discussing on: A Current Distribution for Narrow Broadside Arrays which Optimizes the Relationship Beamwidth and Side-lobe Level. In: Proc. IRE, vol. 35, pp. 489–492 (1947)
11. Yan, S., Ma, Y.: Design of FIR Beamformer with Frequency Invariant Patterns via Joint Optimizing the Spatial and Frequency Responses. In: Proc. ICASSP 2005, vol. 4, pp. 789–792 (2005)
12. Yan, S., Sun, H., Svensson, U.P., Ma, X., Hovem, J.M.: Optimal Modal Beamforming for Spherical Microphone Arrays. IEEE Trans. Audio, speech, and language processing 19, 361–371 (2011)
13. Hawkins, W.G.: FFT Interpolation for Arbitrary Factors: A Comparison to Cubic Spline Interpolation and Linear Interpolation. In: IEEE Nuclear Science Symposium and Medical Imaging Conference, vol. 3, pp. 1433–1437 (1994)
14. Bai, Y., Zhuang, H.: On the Comparison of Bilinear, Cubic Spline, and Fuzzy Interpolation Techniques for Robotic Position Measurements. IEEE Trans. Instruction and measurement 54, 2281–2288 (2005)

Deep-Sea Riser Fatigue Monitoring

Mengyang Zhu*, Haiyan Wang, Xiaohong Shen, Baojun Li, and Wanzheng Ning

Northwestern Polytechnical University, Youyixi Road 127#, Xi'an, China
weiyang4096@gmail.com

Abstract. This paper analyzes the fatigue monitoring parameter selection and long-distance wireless transmission in deepwater drilling riser fatigue monitoring technologies, and designs a deep-sea riser fatigue monitoring system based on MCU+DDS structure. This monitoring system can monitor riser Vortex induced vibration, riser stress and strain and current velocity of seawater in real time. This system can also process and compress collected data, and then transmit modulated data to surface via underwater acoustic channel. The prototype of monitoring system was tested, it is practicable and reliable.

Keywords: deep-sea riser, monitoring, acoustic communication.

1 Introduction

Riser is vital equipment for deepwater oil and gas extraction. It connects sub sea wellhead and oil platform and plays an important role in isolating see water, guiding drill, drilling fluid circulation, etc. Alternating stress caused by riser Vortex induced vibration (VIV) would accelerate the riser fatigue damage and make deepwater drilling riser fatigue failure prone which usually results in riser accident [1].

To ensure the safe operation of the riser and prevent riser fatigue and fracture, it is essential to monitor the fatigue parameters of deepwater riser [2]. The key parameters representing potential risks of riser fatigue are riser stress and strain, VIV and current velocity of seawater nearby.

How to achieve monitoring riser fatigue of with water depth of 3,000 meters is a very important research direction under the conditions of non-cable power supply.

2 Riser Fatigue Monitoring Model

2.1 Riser VIV Monitoring Model

Riser VIV monitoring mathematical model is as follows [3].

$$y(t) = \sum_{i=1}^{M} a_y^i \sin(2\pi f_i t + \phi_i) \tag{1}$$

* Deepwater Information Technology Laboratory, School of Marine Technology, Northwestern Polytechnical University (NPU).

X. Wan (Ed.): Electrical Power Systems and Computers, LNEE 99, pp. 229–236.
springerlink.com

In this formula, a_y^i is the riser VIV acceleration of all modes, f_i is the natural frequencies of each order riser VIV. Acceleration is the most important parameter in this formula. In engineering applications, riser VIV displacement monitoring can be calculated with this formula. In time domain, $a(t)$ --acceleration of riser VIV is related to $y(t)$ --riser VIV displacement.

$$y(t) = \int_0^t \int_0^t a(\tau) d\tau \, dt \tag{2}$$

Displacement measurement principle is shown in Figure 1.

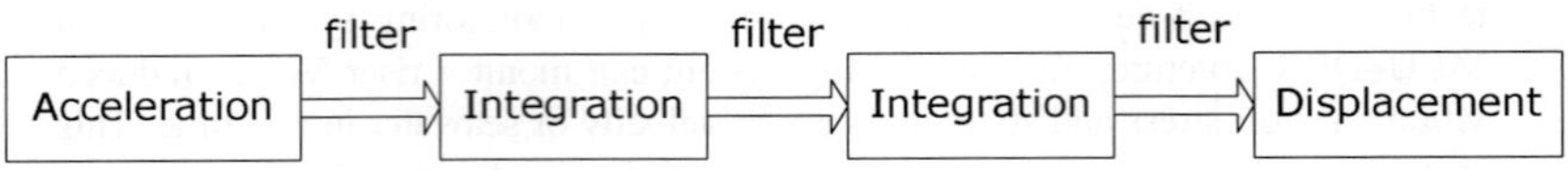

Fig. 1. Displacement measurement principle

2.2 Riser Stress and Strain Monitoring Model

FBG strain sensors are deployed outside the riser as Figure 2. The riser stress and strain monitoring model is as formula (3).

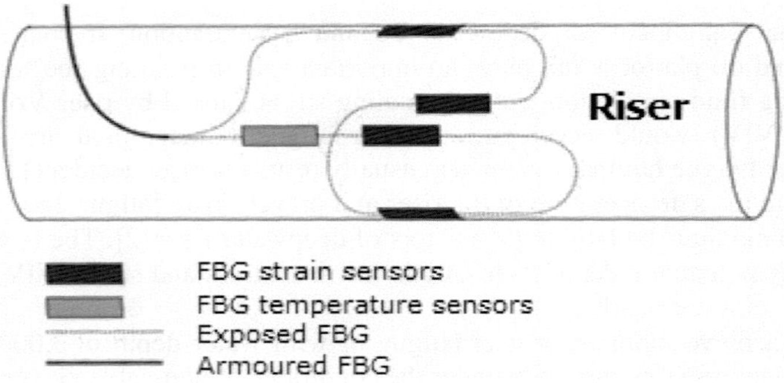

Fig. 2. FBG strain sensors deploying

$$\sigma_{\max} = \frac{E}{4K\lambda_B}\left[(\Delta\lambda_{B1} + \Delta\lambda_{B2} + \Delta\lambda_{B3} + \Delta\lambda_{B4}) + 2\sqrt{(\Delta\lambda_{B1} - \Delta\lambda_{B3})^2 + (\Delta\lambda_{B2} - \Delta\lambda_{B4})^2}\right] \tag{3}$$

When riser starts deforming, 4 FBG strain sensors will receive different strains respectively and these strains will cause center wave length shift. $\Delta\lambda_{B1}$, $\Delta\lambda_{B2}$, $\Delta\lambda_{B3}$ and $\Delta\lambda_{B4}$ are center wave length shift of each sensor; E is the modulus of riser; resonant wavelength λ_B is already known. So riser stress and strain monitoring are changed into center wave length shift monitoring.

2.3 Current Velocity of Seawater Monitoring Model

Acoustic Doppler velocimetry (ADV) is based upon the Doppler shift effect. Specific frequency sound pulses are transmitted by acoustic transducers and part of pulse energy is reflected or scattered by small particles in water. It causes scattering wave frequency changes. When the transducer is stationary, the Doppler shift of backscatter signal is as follows.

$$f_d = \frac{2 f_0 v}{c} \tag{4}$$

Where f_d is the difference between sending and receiving frequencies which is also the Doppler shift; f_0 is the emitted frequency, c is the speed of sound in seawater and v is the current velocity. Doppler frequency shift is proportional to the current velocity. If c and f_0 are known, current velocity can be derived from Doppler frequency shift.

3 Hardware System Design

Riser stress and strain, VIV and current velocity of seawater are most important parameters in riser fatigue assessment; they can be acquired by triaxial accelerometer, FBG strain sensor and ADV.

As a cable-free remote monitoring system, battery is used as the supporting power, and therefore the system has a strict power consumption requirement. Monitoring system is based on the "MCU+DSP" dual-CPU structure. As the main control unit of the monitoring system, MSP430F5438 collects and codes data from sensors; TMS320VC5509 analyzes and compresses those data. MCU is responsible to control system and exchange data, which makes DSP do its best work.

MCU+DDS structure underwater acoustic communication modulator is based on MSP430F5438 and AD9833, a DDS chip. This modulator can achieve BPSK modulation and signal transmission.

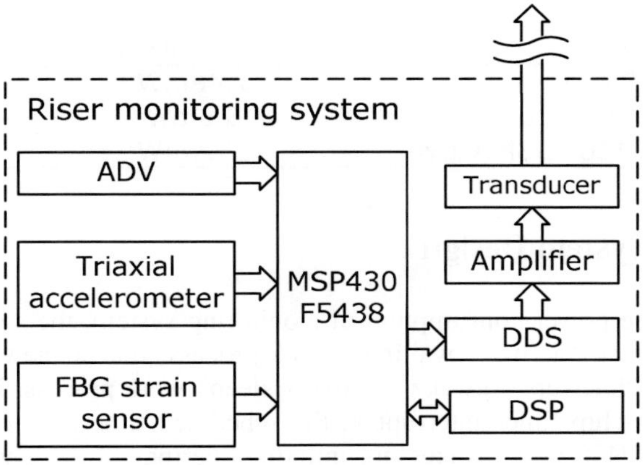

Fig. 3. System block diagram

To reduce system power consumption and extend the working life of the system, all the devices are low-power products. For example, active state power consumption of MSP430F5438 is only 140μA/MHz/3.0V [4] and power consumption of AD9833 is only 4.5mA/3V [5]. Other peripheral devices selected in this device are also chips that contain low-power mode or sleep mode, in order to achieve low power consumption of the system.

Main system schematic is shown in Figure 4.

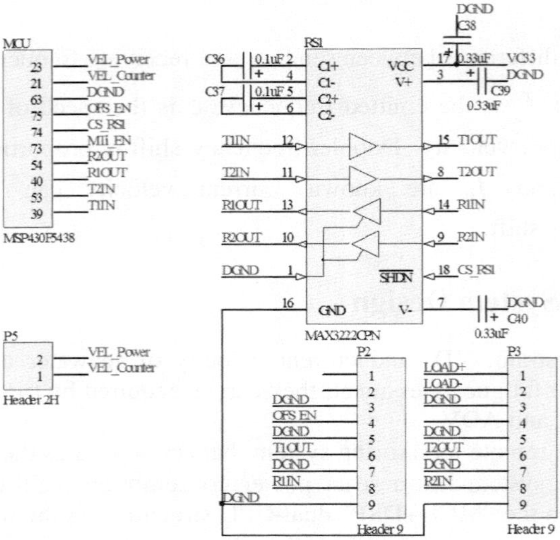

Fig. 4. Main system schematic

Actual system power consumption is shown in table 1.

Table 1. Actual system power consumption under different operating modes.

Mode	Avg. Power 0.5h(Active)/24h
Sleep	1.48mW
MCU Active	3.6mW
MCU+DSP Active	15.4mW

4 Software System Design

In order to reduce power consumption of monitoring system, the system processes is not the traditional "active - waiting" work process but a new one. The new work process is "interrupt - awake - active - sleep" work process. MSP430F5438 is waked up by on-chip timer interrupt at the scheduled time and it exits low-power mode. Then MSP430F5438 wakes up, initializes peripheral devices and sensors and starts collecting, storing, modulating and transmitting processed data.

After data transmitted, MSP430F5438 shuts down all peripheral devices and sensors, and then MSP430F5438 enters low-power mode to wait for next scheduled awake time.

In order to improve program modularity, system main function does not include any functional code; the sub-functions exist independently of each function in the form and main function calls module function if necessary.

System software block diagram is shown in Figure 5.

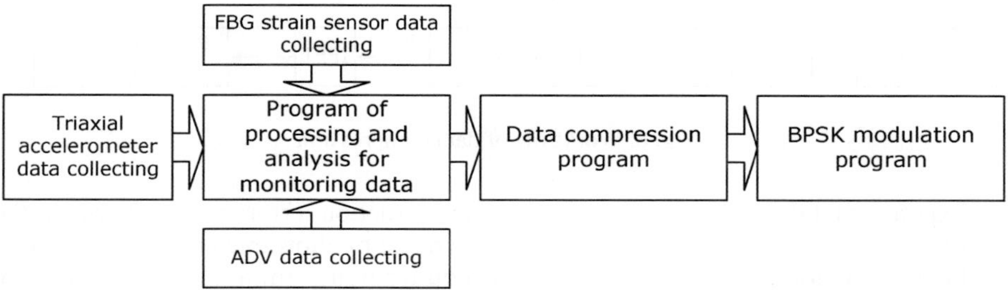

Fig. 5. System software block diagram

4.1 Program of Processing and Analysis for Monitoring Data

After collected sensor data delivered to DSP by MCU, data will be analyzed by DSP. Three different sensor data analysis DSP programs are based on formula (1), (2) and (3) respectively. The amounts of computation of these programs are very large, especially formula (2) requiring twice integrations, which makes DSP the prefect device to accomplish this kind of data processing.

4.2 Data Compression Program

There are a huge number of data collected by self-sensor. If all of them are sent up to water surface via underwater acoustic channel and collected by water surface receiver, the consumptions of system working hours and power would increase significantly.

To deal with this issue effectively, data compression tech is carried out and by using this, the amount of translated data decreases.

With the view to the riser monitoring system requirement for real-time, the Fourier transform of lossless compression is selected, which can use FFT to improve the speed of compression and reconstruction. It does FFT to the digital signal $x(n), 1 \le n \le N$, and gets frequency signal $F(k), 1 \le k \le N$. Then, it takes the non-zero values (or the larger absolute values) form $F(k)$, which is abbreviated as $F(m), 1 \le m \le M$, $M < N$. So, the data compression is realized.

In the receiver, the compressed data is received after the channel decoding. First, the signal $F(m)$ must be recovered to the frequency signal $\hat{F}(k), 1 \le k \le N$, which can use the interpolation method. Then, it does IFFT to $\hat{F}(k)$ and gets the reconstruction signal $\hat{x}(n)$. So the decompression is completed.

Workflow of data compression is shown in Figure 6.

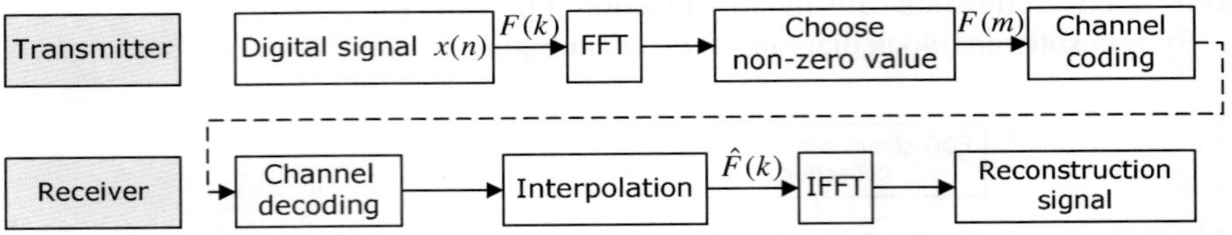

Fig. 6. Workflow of data compression

Experimental data shows that average reconstruction error is 0.017 when the compression ratio is 3:1, and this error is in the range of allowable error. Compression and reconstruction process only require fast Fourier transform and inverse transform and the computation speed of this algorithm is high, which makes this a real-time algorithm.

4.3 BPSK Modulation Program

Workflow of modulation: MSP430F5438 transmits a group of tuning words to AD9833 to wake up and initialize DDS; MSP430F5438 decides to enable phase register 0 or phase register 1 according to next binary code, and AD9833 outputs sine wave with initial phase 0 or Pi determined by phase register. Then hold this sine wave for a certain period. Loop the above until all the codes are modulated. Finally, shut down DDS.

This is a phase selection switch type BPSK modulator and the modulation principle is shown in Figure 7.

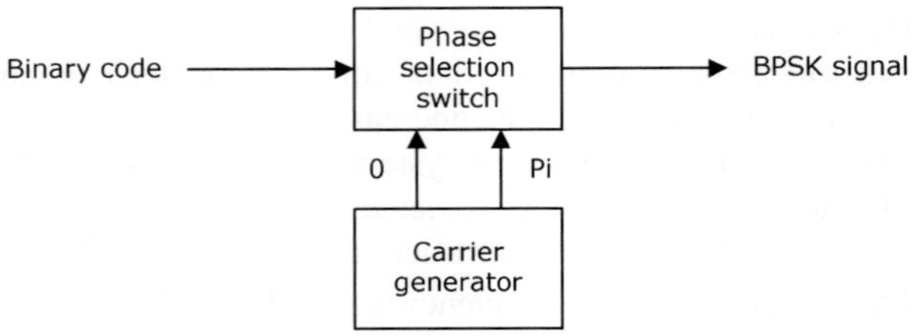

Fig. 7. Phase selection switch type BPSK modulator

The carrier frequency is 10 KHz and the data rate is up to 250/bps under the condition of 3000m communication distance.

5 Data Reception and Reconstruction

After processing and compressing data, date will be transmitted to surface via sound. Data will be received and acoustic signal will be amplified, filtered, A/D converted demodulated, reconstructed, displayed and stored by data reception and reconstruction system.

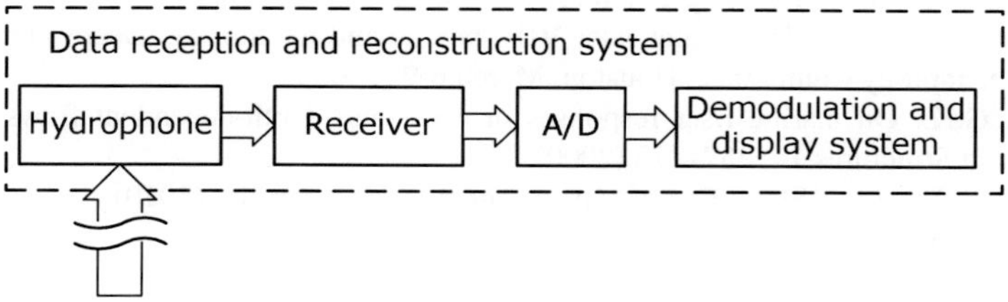

Fig. 8. Data reception and reconstruction system

To verify the data reception and reconstruction and to test if the system could receive and reconstruct the data correctly, the entire system was tested in a lake.

The extent of space between monitoring system and data reception & reconstruction system were 1800m and monitoring system transmitted 120 groups of monitoring data with the electrical power of 11W. Data reception and reconstruction system received all 120 group data, which meant that the receiving rate was 100% and the bit error rate was 0.05%.

Taking displacement monitoring as an example, root-mean-square displacement error was 3.08% and it met the design requirements.

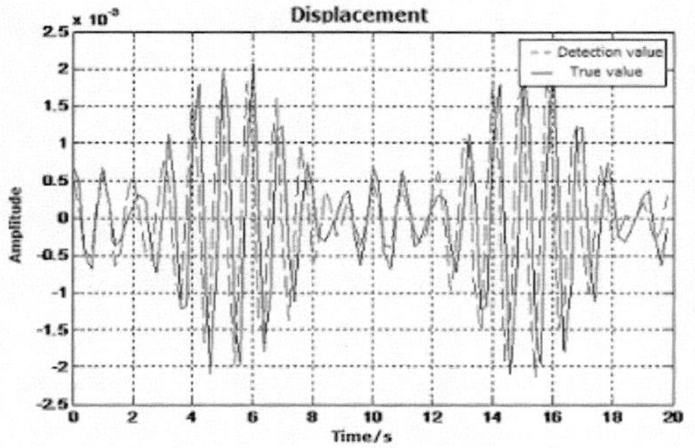

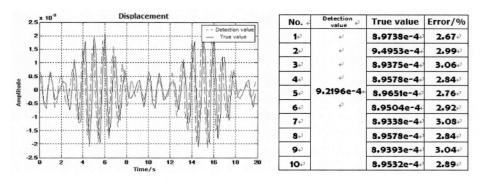

No.	Detection value	True value	Error/%
1		8.9738e-4	2.67
2		9.4953e-4	2.99
3		8.9375e-4	3.06
4		8.9578e-4	2.84
5	9.2196e-4	8.9651e-4	2.76
6		8.9504e-4	2.92
7		8.9338e-4	3.08
8		8.9578e-4	2.84
9		8.9393e-4	3.04
10		8.9532e-4	2.89

Fig. 9. Displacement monitoring result

6 Conclusion

In this paper, a deep-sea riser fatigue monitoring system is designed and a deep-sea riser fatigue monitoring system implementation process is introduced both from hardware design and system software design.

The prototype of deep-sea riser fatigue monitoring system was tested to prove the feasibility of the principle and the effective function of the prototype.

References

1. Li, B.: Research on Fatigue Monitoring Techniques of Deepwater Risers. Northwestern Polytechnical University Dissertation for Graduate Student (2010)
2. Lim, F., Howells, H.: Deepwater riser VIV, fatigue and monitoring. Deepwater Pipeline & Riser Technology Conference, Houston, March 6-9 (2000)
3. Furnes, G.K.: On marine riser responses in time- and depth-dependent flows. Journal of Fluids and Structures 14, 257–273 (2000)
4. TI,MSP430F543x, MSP430F541x Mixed Signal Microcontroller (2010)
5. Analog Devices, AD9833 datasheet (2010)

Forced Oscillations in Second Order SISO Systems

Jun Fu and A.P. Loh

Mechanical and Electrical Engineering
Wuhan Textile University
Wuhan Hubei 430073, P. R. China
fujun2008@sina.cn

Abstract. In the relay feedback system shown in Figure 1, when the magnitude of the forcing signal is larger enough, the phenomenon of forced oscillation occurs, that is, the system synchronizes itself automatically with the frequency of the external signal. Forced oscillation behaviors in First Order plus Time Delay system (FOPTD) has been discussed in many papers. But those behaviors in second order plus time delay system (SOPTD) is under discussed. In this paper, with analysis of the conditions of forced oscillation, the minimum magnitude of the external forcing signal for SOPTD system is obtained. Overdamped, underdamped, critically damped systems are discussed separately. Simulation results are given.

Keywords: Forced Oscillation, Second Order plus Dead Time system, relay feedback, auto-tuning.

1 Introduction

[1] presents the forced oscillation behaviors that are possible in relay feedback systems involving FOPDT plants. This was achieved by a steady state analysis of the plant's response to general switching signals. Then by using the super condition, all possible oscillation behaviours were analyzed. The minimum magnitude, R_{min} of the external forcing signal to achieve forced oscillations may also be numerically obtained.

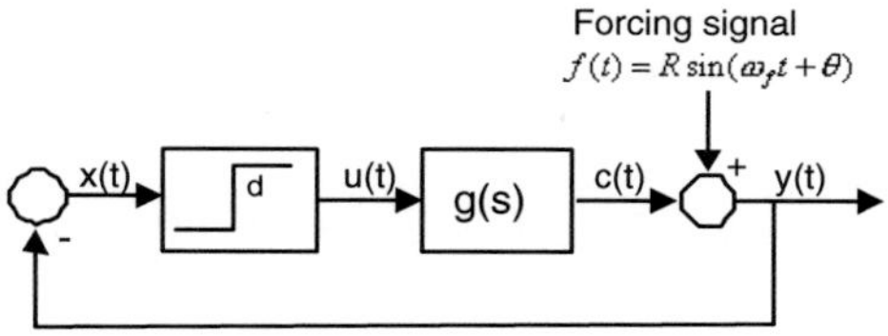

Fig. 1. Forced oscillation in SISO system.

X. Wan (Ed.): Electrical Power Systems and Computers, LNEE 99, pp. 237–244.
springerlink.com

In this paper, such results are extended to second order plus dead time (SOPDT) systems. The treatment is very similar. However, SOPDT, in contrast to FOPDT, systems are more complex to analyse because, firstly, their solutions are dependent on their first derivatives. Secondly, their solution expressions are dependent on the pole locations of the system corresponding to the different damping conditions that are possible. These damping conditions refer to the three cases where the damping ratios are $\zeta > 1$ for overdamped systems, $\zeta = 1$ for critically damped systems and $\zeta < 1$ for overdamped systems. The $C(t)$ expressions for each of these cases are rather different.

Recall from [2] that the necessary and sufficient conditions require that, for all i_0,

$$y(t_{i_0}) = y(t_{i_0+1}) = 0 \tag{1}$$

$$\dot{y}^{(p)}(t_{i_0}) = -\dot{y}^{(p)}(t_{i_0+1}) \qquad p = 1, 2, \ldots, (m-1) \tag{2}$$

$$y(t)(-1)^{i_0+1} < 0 \qquad t \in (t_{i_0}, t_{i_0} + T_f/2) \tag{3}$$

where $m = 2$ for SOPDT systems. (1) and (3) are equivalent to both

$$t_{i_0+1} = t_{i_0} + \frac{T_f}{2}, \qquad \forall i_0 \tag{4}$$

$$\dot{y}^{(p)}(t_{i_0}) = -\dot{y}^{(p)}(t_{i_0+1}) \qquad p = 1, 2, \ldots, (m-1). \tag{5}$$

In Section 2, the necessary and sufficient conditions for overdamped systems are given. In Sections 3 and 4, similar analyses are carried out for critically damped and underdamped systems, respectively. Finally, discussions and conclusions are given in Section 5. In all the analyses, it is assumed that $L < \frac{T_f}{2}$ and initial conditions are zero.

2 Overdamped Systems

In the configuration of Figure 1, $g(s)$ is assumed to have the following transfer function:

$$g(s) = \frac{\omega_n^2}{s^2 + 2\xi\omega_n s + \omega_n^2} e^{-Ls}, \ \xi > 1.$$

For this class of plants, with the assumption of zero initial conditions, the step response and its first derivative are:

$$h(t) = \begin{cases} 0 & t \in (0, L] \\ 1 + \frac{\omega_n}{2\sqrt{\xi^2-1}} \left[\frac{e^{-s_1(t-L)}}{s_1} - \frac{e^{-s_2(t-L)}}{s_2} \right] & t \geq L \end{cases} \tag{6}$$

$$\dot{h}(t) = \begin{cases} 0 & t \in (0, L] \\ \frac{\omega_n}{2\sqrt{\xi^2-1}} \left[e^{-s_2(t-L)} - e^{-s_1(t-L)} \right] & t \geq L \end{cases} \tag{7}$$

where $s_1 = \omega_n(\xi + \sqrt{\xi^2 - 1})$, $s_2 = \omega_n(\xi - \sqrt{\xi^2 - 1})$ are the pole magnitudes of $g(s)$.

As in Figure 1, recalling the general response, $C(t)$, to switching square waves, $u(t)$, we have

$$C(t) = \begin{cases} 0 & t \in [0, L) \\ -h(t) & t \in [L, t_1 + L) \\ -h(t) + 2h(t - t_1) & t \in [t_1 + L, t_2 + L] \\ -h(t) - 2\sum_{k=1}^{i}(-1)^k h(t - t_k) & t \in [t_i + L, t_{i+1} + L). \end{cases} \tag{8}$$

where $t_i,\ i = 1, 2, ...$ corresponds to the unknown switching times of $u(t)$.

At $t = t_{i_0}$, for some $i_0 >> 1$, i_0 odd, we have

$$C(t_{i_0}) = -C_0 - 2\sum_{k=1}^{i_0-1}(-1)^k h(t_{i_0} - t_k) \quad t_{i_0} \in [t_{i_0-1} + L, t_{i_0} + L)$$

$$= -1 - \frac{\omega_n}{\sqrt{\xi^2 - 1}}\sum_{k=1}^{i_0-1}(-1)^k \left[\frac{e^{-s_1(t_{i_0} - t_k - L)}}{s_1} - \frac{e^{-s_2(t_{i_0} - t_k - L)}}{s_2}\right]$$

$$\dot{C}(t_{i_0}) = \frac{\omega_n}{\sqrt{\xi^2 - 1}}\sum_{k=1}^{i_0-1}(-1)^k \left[e^{-s_1(t_{i_0} - t_k - L)} - e^{-s_2(t_{i_0} - t_k - L)}\right]$$

since $C_0 = 1$ for $g(s)$.

Suppose it is assumed that $t_{i_0+1} = t_{i_0} + \frac{T_f}{2}$, according to (8), we have

$$C(t_{i_0}) = -1 - \frac{\omega_n}{\sqrt{\xi^2 - 1}}\left(\frac{Z_1 e^{s_1 L}}{s_1} - \frac{Z_2 e^{s_2 L}}{s_2}\right) \tag{9}$$

$$C(t_{i_0+1}) = 1 - \frac{\omega_n}{\sqrt{\xi^2 - 1}}\left[\frac{e^{-s_1(T_f/2 - L)}}{s_1}(Z_1 - 1) - \frac{e^{-s_2(T_f/2 - L)}}{s_2}(Z_2 - 1)\right] \tag{10}$$

$$\dot{C}(t_{i_0}) = \frac{\omega_n}{\sqrt{\xi^2 - 1}}(Z_1 e^{s_1 L} - Z_2 e^{s_2 L}) \tag{11}$$

$$\dot{C}(t_{i_0+1}) = \frac{\omega_n}{\sqrt{\xi^2 - 1}}\left[e^{-s_1(T_f/2 - L)}(Z_1 - 1) - e^{-s_2(T_f/2 - L)}(Z_2 - L)\right] \tag{12}$$

where $Z_1,\ Z_2$ are defined as

$$Z_1 = \sum_{k=1}^{i_0-1}(-1)^k e^{-s_1(t_{i_0} - t_k)} \quad Z_2 = \sum_{k=1}^{i_0-1}(-1)^k e^{-s_2(t_{i_0} - t_k)} \tag{13}$$

The necessary switching condition, $y(t_{i_0}) = y(t_{i_0+1}) = 0$ implies that $C(t_{i_0+1}) = -C(t_{i_0})$, which further leads to the relationship

$$\frac{e^{s_1 L}}{s_1}[e^{-s_1 T_f/2}(Z_1 - 1) + Z_1] = \frac{e^{s_2 L}}{s_2}[e^{-s_2 T_f/2}(Z_2 - 1) + Z_2] \tag{14}$$

Since (14) contains two unknowns, Z_1 and Z_2, a second equation can be obtained using the first derivative of $C(t)$ as follows.

$$\dot{C}(t_{i_0+1}) = -\dot{C}(t_{i_0}) \tag{15}$$

which leads to

$$e^{s_1 L}[e^{-s_1 T_f/2}(Z_1 - 1) + Z_1] = e^{s_2 L}[e^{-s_2 T_f/2}(Z_2 - 1) + Z_2]. \tag{16}$$

Solving (14) and (16), we have

$$Z_1 = \frac{1}{1 + e^{s_1 T_f/2}} \quad Z_2 = \frac{1}{1 + e^{s_2 T_f/2}}. \tag{17}$$

Assuming that $t_{i_0+2} = t_{i_0+1} + \frac{T_u}{2}$ where T_u is unknown, the switching condition, (1), implies that

$$C(t_{i_0+2}) - R\sin[\omega_f(t_{i_0} + \frac{T_u}{2})] = 0$$

which leads to

$$1 + \frac{\omega_n}{\sqrt{\xi^2 - 1}}\left[\frac{e^{s_1 L}}{s_1}\frac{e^{s_1\frac{T_f - T_u}{2}}}{1 + e^{s_1 T_f/2}} - \frac{e^{s_2 L}}{s_2}\frac{e^{s_2\frac{T_f - T_u}{2}}}{1 + e^{s_2 T_f/2}}\right] + R\sin\omega_f\frac{T_u}{2}\cos\omega_f t_{i_0}$$

$$+ \cos\omega_f\frac{T_u}{2}\left\{1 + \frac{\omega_n}{\sqrt{\xi^2 - 1}}\left[\frac{e^{s_1 L}}{s_1}\frac{1}{1 + e^{s_1\frac{T_f}{2}}} - \frac{e^{s_2 L}}{s_2}\frac{1}{1 + e^{s_2\frac{T_f}{2}}}\right]\right\} = 0 \tag{18}$$

Since (18) has to be satisfied for any arbitrarily large R and t_{i_0}, we conclude that $T_u = T_f$. This result proves that as long as there exists two switching instants that satisfy (4), the subsequent switchings also satisfies (4). We, next, establish the steady state periodic waveform by proving that $C(t) = C(t + T_f)$ for all $t > t_{i_0}$. It follows from (8) that

$$C(t_{i_0} + \triangle t) = C(t_{i_0+2} + \triangle t)$$
$$= \begin{cases} -1 - \frac{\omega_n}{\sqrt{\xi^2-1}}\left[\frac{e^{-s_1(\triangle t - L)}}{s_1(1+e^{s_1 T_f/2})} - \frac{e^{-s_2(\triangle t - L)}}{s_2(1+e^{s_2 T_f/2})}\right] & \triangle t \in [0, L] \\ 1 + \frac{\omega_n}{\sqrt{\xi^2-1}}\left[\frac{e^{-s_1(\triangle t - \frac{T_f}{2} - L)}}{s_1(1+e^{s_1 T_f/2})} - \frac{e^{-s_2(\triangle t - \frac{T_f}{2} - L)}}{s_2(1+e^{s_2 T_f/2})}\right] & \triangle t \in [L, T_f/2] \end{cases} \tag{19}$$

since $t_{i_0+2} = t_{i_0} + T_f$.

The self-oscillating period can be obtained numerically by solving for T_f in the equation $C(t_{i_0}) = 0$, where $C(t_{i_0})$ is given by

$$C(t_{i_0}) = -1 - \frac{\omega_n}{\sqrt{\xi^2 - 1}}\left[\frac{e^{s_1 L}}{s_1(1 + e^{s_1 T_f/2})} - \frac{e^{s_2 L}}{s_2(1 + e^{s_2 T_f/2})}\right].$$

As in SOPDT systems, R_{min} can be obtained by finding the mininum R which satisfies the following inequality

$$C(t_{i_0} + \triangle t) + R\sin(\omega_f \triangle t + \phi) < 0, \quad \triangle t \in (0, \frac{T_f}{2}) \tag{20}$$

where $\phi = \sin^{-1}\frac{-C(t_{i_0})}{R}$.

Example 1: The plant and the external forcing signal, $f(t)$, are given by

$$g(s) = \frac{4}{s^2 + 6s + 4}e^{-s}, \ \xi = 1.5$$
$$f(t) = R\sin(\omega_f t).$$

At a frequency corresponding to $T_f = 10$ sec, from (20), we get: $|C(t_{i_0})| = 0.8921$, $R_{min} = 0.9421$. Firstly, $R = |C(t_{i_0})|$ is used. It can be seen from Figure 2 that forced oscillation did not occur. On the other hand, when the simulation is repeated using $R = R_{min} = 0.9421$, Figure 3 shows the resulting signals for $C(t)$ and $u(t)$. This verifies that the R_{min} derived from our equations is indeed accurate.

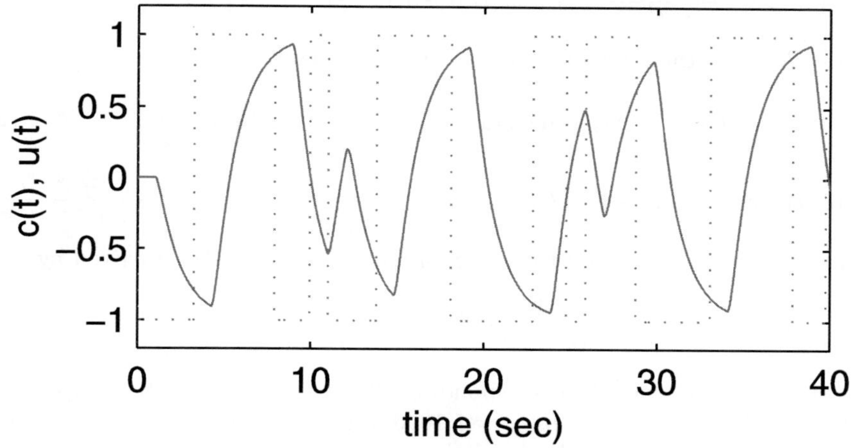

Fig. 2. Forced oscillations for plant in Example 2, $R = |C_{t_{i_0}}|$.

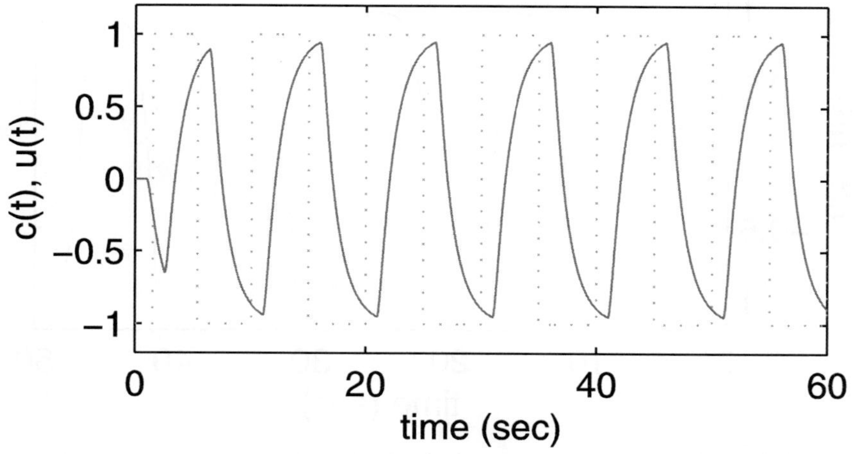

Fig. 3. Forced oscillations for plant in Example 2, $R = |R_{min}|.u(t)$(dotted line)

3 Critically Damped Systems

In the configuration of Figure 1, $g(s)$ is assumed to have the following transfer function:

$$g(s) = \frac{\omega_n^2}{(s+\omega_n)^2} e^{-Ls}.$$

Following the similar analysis in section 2, since $t_{io+2} = t_{io} + T_f$, we get:

$$C(t_{io} + \triangle t) = C(t_{io+2} + \triangle t)$$
$$= \begin{cases} -1 + 2\frac{e^{-\omega_n(\triangle t - L)}}{1+e^{\omega_n T_f/2}} \left[1 + \omega_n(\triangle t - L) + \frac{\omega_n \frac{T_f}{2} e^{\omega_n T_f/2}}{1+e^{\omega_n T_f/2}}\right] & \triangle t \in [0, L] \\ 1 - 2\frac{e^{-\omega_n(\triangle t - \frac{T_f}{2} - L)}}{1+e^{\omega_n T_f/2}} \left[1 + \omega_n(\triangle t - L) - \frac{\omega_n T_f/2}{1+e^{\omega_n t T_f/2}}\right] & \triangle t \in [L, T_f/2] \end{cases} \quad (21)$$

The R_{min} for the forced oscillation can be solved by

$$C(t_{i_0} + \triangle t) + R\sin(\omega_f \triangle t + \phi) < 0, \quad \triangle t \in (0, \frac{T_f}{2}) \quad (22)$$

where $\phi = \sin^{-1}\frac{-C(t_{i_0})}{R}$ and $C(t_{i_0}) = -1 + \frac{2e^{\omega_n L}}{1+e^{\omega_n T_f/2}}\left[1 - \omega_n L + \frac{\omega_n \frac{T_f}{2} e^{\omega_n T_f/2}}{(1+e^{\omega_n T_f/2})^2}\right]$.

Example 2: The plant and the external forcing signal, $f(t)$, are given by

$$g(s) = \frac{4}{(s+2)^2} e^{-s} \text{ and } f(t) = R\sin(\omega_f t).$$

Choose $T_f = 10$ second, the corresponding $C(t_{i_0})$ and R_{min} can be computed by (22), $|C(t_{i_0})| = 0.9940$, $R = R_{min} = 1.1640$. Figure 4 shows the resulting signals for $C(t)$ and $u(t)$. $u(t)$ is indeed switching at every 5 sec.

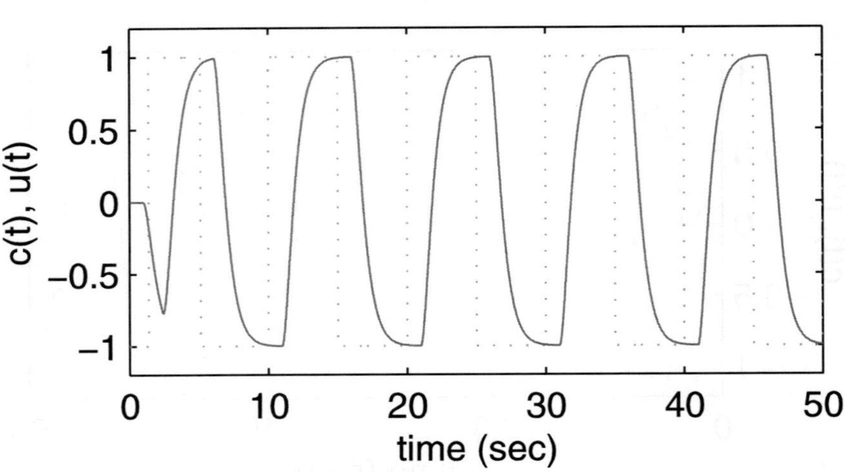

Fig. 4. Forced oscillations for plant in Example 3, $R = R_{min}$. $u(t)$(dotted line)

4 Underdamped Systems

In the configuration of Figure 1, $g(s)$ is assumed to have the following transfer function:

$$g(s) = \frac{\omega_n^2}{s^2 + 2\xi\omega_n s + \omega_n^2} e^{-Ls}$$

where L is the time delay, $0 < \xi < 1$.

Following the similar analysis in section 2, with further simplification, we obtain

$$C(t_{i_0} + \Delta t) = C(t_{i_0+2} + \Delta t)$$
$$= \begin{cases} -1 + Me^{-\xi\omega_n(\Delta t - L)} \sin[\omega_d(\Delta t - L) + \beta + \tan^{-1}\frac{Z_s}{Z_c}] & \Delta t \in [0, L] \\ 1 + Me^{-\xi\omega_n(\Delta t - \frac{T_f}{2} - L)} \sin[\omega_d(\Delta t - L) + \beta + \tan^{-1}\frac{Z_s}{Z_c - 1}] & \Delta t \in [L, \frac{T_f}{2}] \end{cases} \tag{23}$$

where $Z_s = \frac{e^{\xi\omega_n \frac{T_f}{2}} \sin \omega_d \frac{T_f}{2}}{e^{\xi\omega_n T_f} + 2e^{\xi\omega_n \frac{T_f}{2}} \cos \omega_d \frac{T_f}{2} + 1}$, $Z_c = \frac{e^{\xi\omega_n \frac{T_f}{2}} \cos \omega_d \frac{T_f}{2} + 1}{e^{\xi\omega_n T_f} + 2e^{\xi\omega_n \frac{T_f}{2}} \cos \omega_d \frac{T_f}{2} + 1}$, $\omega_d = \omega_n\sqrt{1-\xi^2}$, $\beta = \arctan\frac{\sqrt{1-\xi^2}}{\xi}$, $M = \frac{2}{\sqrt{1-\xi^2}}(e^{\xi\omega_n T_f} + 2e^{\xi\omega_n \frac{T_f}{2}} \cos \omega_d \frac{T_f}{2} + 1)^{-\frac{1}{2}}$.

Substituting (23) into (22), the R_{min} of the underdamped system can be solved by

$$C(t_{i_0} + \Delta t) + R\sin(\omega_f \Delta t + \phi) < 0, \quad \Delta t \in (0, \frac{T_f}{2}) \tag{24}$$

where $\phi = \sin^{-1}\frac{-C(t_{i_0})}{R}$ and $C(t_{i_0}) = -1 + Me^{\xi\omega_n L} \sin(\beta + \tan^{-1}\frac{Z_s}{Z_c} - \omega_d L)$.

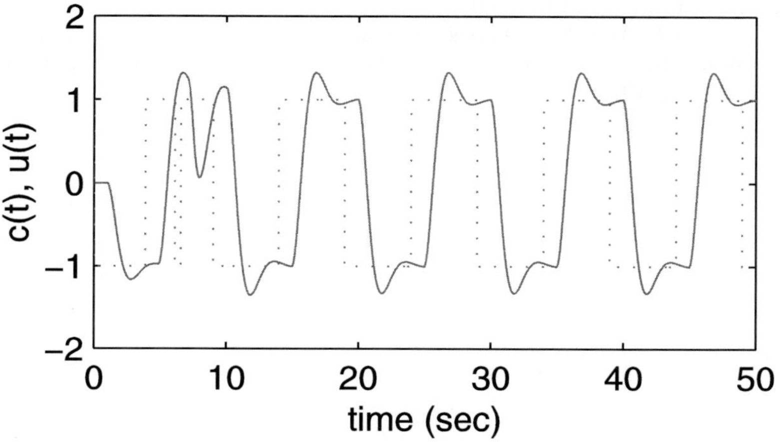

Fig. 5. Forced oscillations for plant in Example 4, $R = R_{min}$. $u(t)$(dotted line)

Example 3: The plant and the external forcing signal $f(t)$ are given by

$$g(s) = \frac{4}{s^2+2s+4}e^{-s},\ \xi = 0.5$$
$$f(t) = R\sin(\omega_f t)$$

Choose $T_f = 10$sec, based on (24), $|C(t_{i_0})| = 0.9578$, $R = R_{min} = 1.5778$. Figure 5 shows the resulting signals for $C(t)$ and $u(t)$. $u(t)$ is indeed switching at every 5 sec.

5 Discussion and Conclusion

In this paper, following the analysis of FOPDT systems proposed in [3], we have successfully extended some of the results to SOPDT systems. Three types of SOPDT systems were analysed separately and simulation results were given for all cases. The exact expressions of $C(t)$ when forced oscillation occurs are also given. Based on these results, the minimum R for forced oscillation can be solved numerically. Furthermore, the self-oscillation period of SOPDT systems can also be solved using $C(t_{i_0}) = 0$.

The results for SOPDT systems were only given for the case of external frequencies satisfying $L < T_f/2$. Results for higher frequencies were not presented because of the tedious nature of the solutions.

References

1. Lim, L.H., Loh, A.P., Fu, J.: Estimation of minimum conditions for forced oscillations in relay feedback systems. In: Proceedings of the 5th International Conference on Control and Automation, ICCA 2005, pp. 1262–1267 (2005)
2. Loh, A.P., Fu, J.: Forced oscillations in First Order Systems. In: Europe Control Conference, UK (September 2003)
3. Fu, J., Loh, A.P., Wei, Y.Y.: Forced Oscillations Conditions in Relay Feedback Control Systems. In: The Chinese Symposium on Information Science and Technology, CSIST 2010 (2010)

Research on Locomotive Evolution Based on Worm-Shaped Configuration of Self-reconfigurable Robot HitMSR II

Yanhe Zhu, Xiaolu Wang, Xindan Cui, Jingchun Yin, and Jie Zhao

State Key Laboratory of Robotics and System,
Harbin Institute of Technology, Harbin 150001, China
yhzhu@hit.edu.cn

Abstract. In this paper, we constructed a three-dimensional dynamic simulator for HitMSR II which is a module self-reconfigurable robot system composed of single-rotational-freedom modules. Downhill Simplex Optimization Algorithm were used to evolve the locomotion of the robot ensuring that it is possible to get access to the relatively more optimized locomotion gaits parameters after large enough numbers of iterative calculations under a certain evaluation function. What's more. We selected worm-shaped configuration to research on the validation of locomotive evolution both in simulation and experiment.

Keywords: self-reconfigurable robot, 3D control simulator, worm, locomotive evolution, Downhill Simplex Algorithm.

1 Introduction

Since DRRS (Dynamically Reconfigurable Robotic System) [1] was introduced in 1988 by Fukuda from Tokyo Institute of Technology, many successful self-reconfigurable robot systems have been developed because of the excellent performance and broad application prospect. In Mark Yin's PhD thesis[2], a self-reconfigurable robot was defined as a system composed of lots of modules, which can change it's configuration, move smoothly or complete corresponding mission by changing the connection state or position relative to each other without any outside assistance.

After years of research scholars from various countries on self-reconfigurable robot have discovered lots of methods for the motion planning and control. For example, harmonic oscillation control, gait table control, task-based control[3,4], digital hormonal control, central pattern generator(CPG) [7,8] etc.

Once the controller structure is determined, it is necessary to select optimized parameters to get a more effective motion[9]. There are some common methods to optimize robot's locomotion, for example, genetic algorithm, particle swarm optimization, simulated annealing algorithm and so on.

Daniel Marbach[10] uses Powell algorithm for the locomotive evolution of their developed self-reconfigurable robot, but the rules in the search process are too complex. Mtran's researchers [11] use genetic algorithms to evolve their robot's

X. Wan (Ed.): Electrical Power Systems and Computers, LNEE 99, pp. 245–252.
springerlink.com

locomotion, but genetic operator's process is too complicated, time consuming and occupy too much computer memory space.

In this paper, CPG (Central Pattern Generator) is used to generate actuation signals.Based on dynamic simulation environment in collaboration with optimization algorithm in MATLAB®, we explored the influence to worm-shaped configuration's locomotion by simulating CPG gaits in different parameters and use Downhill Simplex Algorithm [12,13] to realize the worm-shaped configuration's locomotive evolution.

2 HitMSR II System Overview

We use HitMSR II [14,15] as our experimental system, which was developed by State Key Laboratory of Robotics and System in Harbin Institute of Technology. It consists of computer, relay board and robot. The computer can plan motion gaits, simulate and transmit motion commands to relay board. The relay board can get the commands from PC and transmit them to robot via wireless communication. Robot can get the commands from relay board and execute them.

a) active module b) passive module

Fig. 1. The structure of HitMSR II

As shown in Fig.1, self-reconfigurable robot HitMSR II modules are divided into two different types, active module and passive module. The active module's surfaces have self-locking hook claws. Passive module can be connected stably by that claw. Each module has a rotational degree of freedom and four identical faces. Each module consists of two L-shaped thing directly connected by the motor output shaft, and the rotational scope is ± 90 °.

3 The 3D Dynamic Simulator for Self-reconfigurable Robot

In this paper, we constructed a joint simulation platform using virtual dynamic simulation environment software and numerical calculation software. Based on the simulation software MRDS® (Microsoft Robotics Developer Studio), we constructed a dynamic environment for self-reconfigurable robot using SPL® (Simulation

Programming Language), and call the MATLAB ® command window in the simulation process which can be used to get the data from the simulation environment to process and pass over dynamically calculated results in MATLAB to the simulation environment. The dataflow throughout the simulation system is shown in Fig. 2.

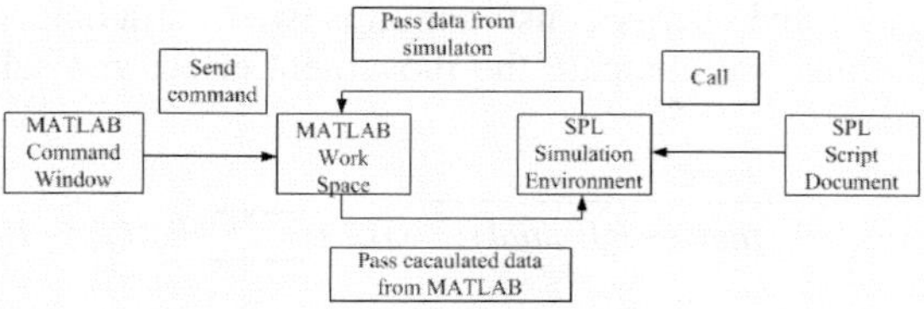

Fig. 2. Dataflow chart of joint simulation platform

4 Locomotive Evolution Based on Worm-Shaped Configuration Using Downhill Simplex Algorithm

4.1 Worm-Shaped Configuration

Self-reconfigurable robot can generate some typical configuration, for example, worm-shaped configuration. A worm-shaped configuration consists of four modules or more which are connected in a line and each module's rotational axis is parallel to others. Figure 3 shows the worm-shaped configuration robot in simulation environment. In this paper, we will explore the locomotive evolution based on worm-shaped configuration.

Fig. 3. Worm-shaped configuration in simulation environment

4.2 Locomotive Evolution

In this paper we evolve the locomotion ability of worm-shaped configuration by searching the best optimized parameters of the controller according to a defined evaluation function, and we use downhill simplex algorithm which needn't calculate numerical derivatives in the searching process.

In this optimization process, we set the distance in a certain period of time (Equation 1, equal to average speed) and energy consumption of the robot system (Equation 2) as the evaluation index.

$$distance = \sqrt{(P_0(final) - P_0(initial))^2} \tag{1}$$

$$E=\sum_{i}\sum_{k}(P_{2i}(kT_0)-\overline{P}_{2i})^2 \tag{2}$$

In the formula, $\overline{P}_{2i}=\frac{1}{N}\sum_{k=1}^{N}P_{2i}(kT_0)$, Pi (t) is the coordinates of the module i when time is t.T0 is the sampling period. N is the maximum number of samples.

Movement distance and energy consumption is two contradictory evaluation index. After many simulations, we compare the movement effect and select formula (3) as evaluation function.

$$f(\vec{P}_j)=-\sqrt{(P_0(final)-P_0(initial))_j^2}\times 0.7+\sum_{i}\sum_{k}(P_{2i}(kT_0)-\overline{P}_{2i})_j^2\times 0.3 \tag{3}$$

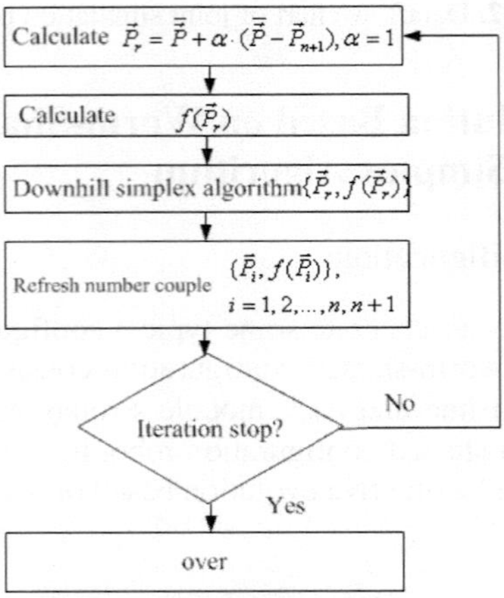

Fig. 4. Flowchart for the algorithm of locomotive evolution

First of all, we define the controller parameters as the solution space vector in downhill simplex algorithm, $\vec{P}_i=[E_i,\ \ \alpha_i,\ \ \tau_i,\ \ delay_i]^T$ .Then give it a initial value and generate drive signals of each module according to central pattern generator (CPG). After that, we apply the signals to the dynamic simulation environment and can get the initial locomotion status. Thus, based on the initial data we can evolve the locomotion ability using downhill simplex search algorithm. After several iterations, the evaluation function that can achieve the minimum solution in a set of controller parameters, that is, the optimization parameters of central pattern generator. The whole process flow chart is shown in figure 4.

4.3 Simulation Research on Locomotive Evolution of Worm-Shaped Configuration

According to the process stated above, we do simulation research on the locomotive evolution of worm-shaped configuration, the drive signals of each module are planned by the central pattern generator. Paper 8 gives the model of stable central pattern generator (CPG).

In this paper, based on 12 initial data, after 60 times iteration search, we obtain a better locomotion. Figure 5 shows the dive signals of CPG on first module, and the below one is the signal after optimization. The other module's drive signal is the same but for a certain time delay.

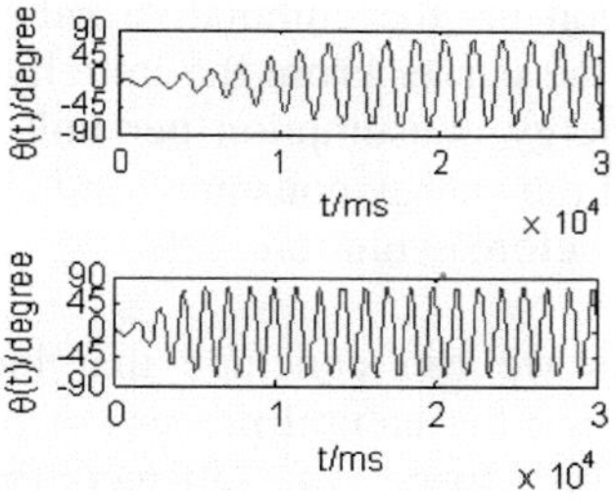

Fig. 5. CPG signals of first module before and after optimization for worm-shaped configuration

The trajectory of first module in the forward direction before and after optimization is shown in figure 6, the energy consumption of worm-shaped robot before and after optimization is shown in figure 7.

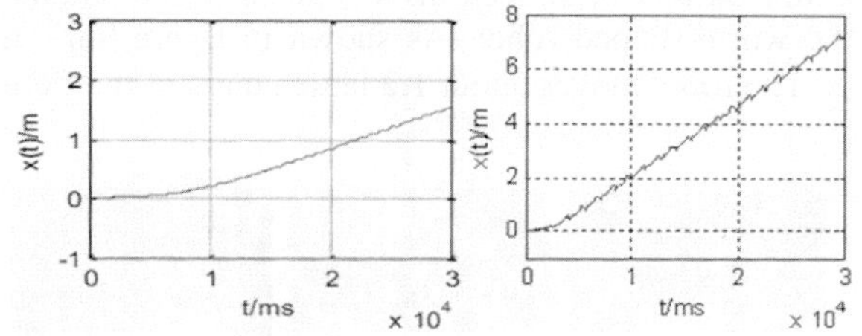

Fig. 6. The trajectory of first module in the forward direction before and after optimization

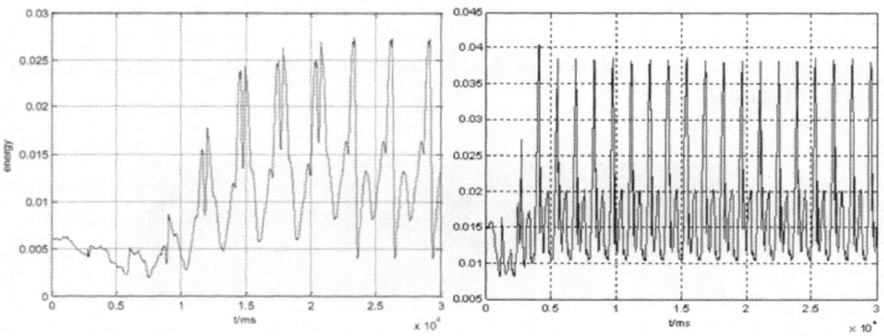

Fig. 7. The energy consumption of worm-shaped robot before and after optimization

We can see by comparing the locomotion before and after the evolution of the movement:

(1) In a period of 30s, the robot moves about 1.5m before optimization and up to 6.5m after optimization in the forward direction. The robot's movement speed increased as 4 times as before.

(2) the robot's total energy consumption increased after the locomotion evolution. The total energy consumption is the cumulative value of energy consumption of every moment recorded on the curve. From the initially approximately 125 to about 189. But if we consider energy consumption per unit distance, it is 29.1 per unit distance after evolution, and the energy consumption is reduced to 34.93% compared to 83.3 energy consumption before evolution.

From the above analysis we can conclude that locomotion evolution of self-reconfigurable robot simulating the natural process of biological evolution approach can effectively improve the speed of self-reconfigurable robot, and energy consumption per unit distance. As the evolutionary approach does not depend on the specific configuration of the robot, it can be easily extended to other configurations' locomotion evolution and applied online.

5 Verification Experiment on Locomotive Evolution

Apply the pre-optimized and optimized drive signals in the simulation environment respectively to the worm-shaped robot. As shown in figure 8,it's the pre-optimized locomotion status. The robot moves about 1.2 lattice units of the floor tile's length.

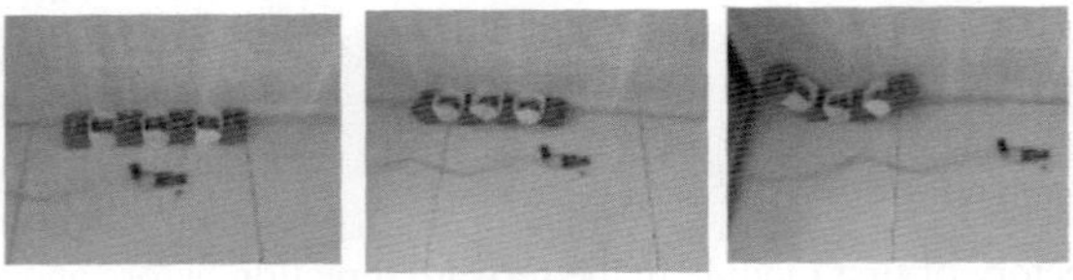

Fig. 8. Locomotion of worm-shaped configuration by pre-optimized CPG signals

The robot moves a distance of 4.2 lattice units using the optimized signals, as shown in figure 9.

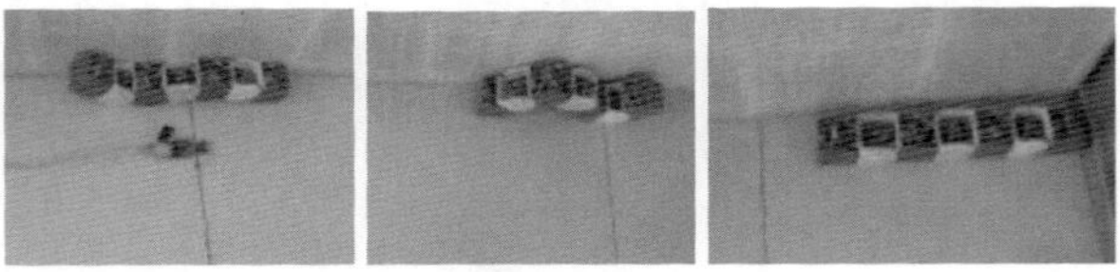

Fig. 9. Locomotion of worm-shaped configuration by optimized CPG signals

We can see a clear advantage in the movement distance of forward direction and the mass center of each module doesn't change a lot relative to the equilibrium position compared with the pre-optimized locomotion status. That means optimized locomotion does not consume too much more energy. This fact confirms the effectiveness of the optimization process.

6 Conclusion

In this paper, we use the downhill simplex algorithm to optimize the locomotion of worm-shaped configuration robot. Downhill simplex algorithm is a optimization algorithm which needn't calculate the derivative value, so it's especially for the evaluation function which set the displacement in the simulation environment as the variables. From the research on the locomotive evolution of worm-shaped robot, we verify the effectiveness of downhill simplex optimization algorithm.

Acknowledgments. The authors would like to thank the National Natural Science Foundation of China (60705027) and National 863 Program (2006AA04Z220) for their support of the project.

References

1. Fukuda, T., Nakagawa, S.: Dynamically reconfigurable robotic system. In: Proc. IEEE Int. Conf. Robotics and Automation, vol. 3, pp. 1581–1586 (1988)
2. Yim, M.: Locomotion with Unit-Modular Reconfigurable Robot. Ph.D. Thesis. Department of Mechanical Engineering, Stanford University, CA (1994)
3. Støy, K., Shen, W.-M., Will, P.M.: Using Role-Based Control to Produce Locomotion in Chain-Type Self-Reconfigurable Robots. IEEE/ASME Transactions On Mechatronics 7(4) (December 2002)
4. Støy, K., Shen, W.-M., Will, P.M.: A simple approach to the control of locomotion in self-reconfigurable robots. Robotics and Autonomous Systems 44, 191–199 (2003)
5. Shen, W.-M., Lu, Y., Will, P.: Hormone-Based Control for Self-Reconfigurable Robots. ACM, New York (2000), doi: 1-58113-230-1/00/6
6. Shen, W.-M., Salemi, B., Will, P.: Hormone-Inspired Adaptive Communicationand Distributed Control for CONRO Self-Reconfigurable Robots. IEEE Transactions On Robotics And Automation 18(5) (October 2002)
7. Ijspeert, A.J.: Central pattern generators for locomotion control in animals and robots:A review. Neural Networks 21, 642–653 (2008)
8. Kamimura, A., Kurokawa, H., Yoshida, E., Tomita, K., Murata, S., Kokaji, S.: Automatic Locomotion Pattern Generation for Modular Robots. In: Proceedings of the 2003 IEEE International Conference on Robotics &Automation, Taipei, Taiwan, September 14-19 (2003)
9. Dutta, A.: Reconfigurable Self-Replicating Robotics. Master of Science, School of Informatics University of Edinburgh (2007)
10. Marbach, D., Ijspeert, A.J.: Online Optimization of Modular Robot Locomotion. In: Proceedings of the IEEE International Conference on Mechatronics & Automation, Niagara Falls, Canada · (July 2005)

11. Kamimura, A., Kurokawa, H., Yoshida, E., Tomita, K., Murata, S., Kokaji, S.: Automatic Locomotion Pattern Generation for Modular Robots. In: Proceedings of the 2003 IEEE International Conference on Robotics &Automation, Taipei, Taiwan, September 14-19 (2003)
12. Hornby, G.S., Pollack, J.B.: Bodybrain coevolution using L-systems as a generative encoding. In: Genetic and Evolutionary Computation Conference, San Francisco, CA, pp. 868–875 (2001)
13. Yoshida, E., Murata, S., Kamimura, A., Tomita, K., Kurokawa, H., Kokaji, S.: Evolutionary Motion Synthesis for a Modular Robot using Genetic Algorithm. Journal of Robotics and Mechatronics 15(2), 227–237 (2003)
14. Yuhua, Z., Yanhe, Z., Jie, Z., Zongwei, R.: Self-reconfigurable moduar robot and it's motion plan. Journal of Jilin University(Engineering and Technology Edition) 4 (2007)
15. Yuhua, Z., Yanhe, Z., Jie, Z., Zongwei, R.: Metamorphic strategy based on dynamic meta-modules for a self-reconfigurable robot. High Technology Letters 3 (2008)

The Influence of the University Teachers' Psychological Contract on Knowledge Sharing*

Li Xiaojun**, Zhou Zongkui, and Zhang Yumei

Psychology College, Central China Norm University,
Wuhan, The People's Republic of China
lixiaojun7910@163.com

Abstract. 201 college teachers to study in Hubei Province, the use of the psycological contract, knowledge integration, cooperation, the three dimensional structure of satisfaction questionnaires as a research tool. Investigated the psychological contract and knowledge sharing relationship. The results show: the different dimensions of psychological contracts: relational contracts, balancing contracts and transactions the performance of knowledge-sharing contract on the different effects of relational contract in which knowledge sharing contract with the balance a significant role in promoting. The former is stronger than the latter, but the trading performance of knowledge-sharing contract for a significant negative impact. Results consistent with the initial assumption, for the psychological contract between the impact and knowledge sharing to provide theoretical foundation and empirical evidence.

Keywords: Psychological Contract, Knowledge-sharing, influence.

1 Introduction

Psychological contract was first pay attention to in social psychology area and then has been continuously extension by organization psychologists. Organizational psychologist Argyris (1960) used the concept "Work psychological contract" first, but psychological contract was clear defined by famous American management psychologist Schein. He thought psychological contract indicate to the whole expectation between members and mangers or others.

The content and forms of expression of psychological contract are very complex. In order to study it in a deep going way, researchers explored lots about its dimensions. Rousseau analyzed two typical dimensions from a survey on psychological contract of 129 business administration graduates. One dimension was called exchange contract, it refers to contract relationships based on economic exchange such as employees get high rewards, hortation or training opportunities for promotion through work overtime. The other one is relationship contract which based on social exchange. It reflects employees exchange work for a long time, to be trusty

* Supported by the Fundamental Research Funds by The Hubei Province Social Science Foundation Grant (Program No: [2010]022).

** Corresponding author.

X. Wan (Ed.): Electrical Power Systems and Computers, LNEE 99, pp. 253–258.
springerlink.com

and for long-term work guarantee[2]. In later study, Rousseau & Tijiorimala (1996) conducted a survey on American registered Nurse, the results showed psychological contract can be divided into three dimensions. Except Exchange and Relationship, psychological contract include Team Members dimension which refers to good interpersonal relation between employees and the organization. Some researcher call it Counterpoise dimension.

Nowadays, study on Psychological contract mainly use White Collar, MBA in big corporations, nurses, students and other special groups as research objects. As a group major task is knowledge impartation and creation, teachers receive receives increasing attention from psychological contract researchers. Some study found psychological contract of teachers has psychological, V two directivity and fuzziness and other characteristics. Reasonable use of psychological contract can help to improve teacher manage level[5]. Study on teachers found psychological contract can improve the manage efficiency of kindergartens effectively and arouse teachers' working enthusiasm. Psychological contract has compensation, dynamic and cohesive effect in kindergarten teachers management [6]. Study in High school found: Besides human resource contract, it is more important there exists psychological contract which plays a very important role between young teacher and school[7].

Knowledge sharing refers to the explicit and implicit knowledge of employees, teams and organizations share by various methods with others and transform organizations' knowledge wealth. One of the important aims of knowledge sharing is to improve both core competence that not only includes knowledge creative competence but also includes knowledge sharing and transformation competence and ability of acquisition knowledge and transition valuable knowledge.

Knowledge sharing is the major dynamic system of knowledge transmission. It can not only help finish teachers' mission effectively but also promote knowledge flow speed among teachers and improve knowledge utility. Knowledge sharing plays an important role either in students' knowledge acquisition or in teachers' teaching improving. And knowledge sharing also is an important job in teaching management.

The key to improve teachers' scientific research creativity is to set up an efficient knowledge sharing mechanism in teaching management. Management scientist Peter Drucker consider: The basic of knowledge society is knowledge but no longer capital or labour force now and in future[8]. Teachers are the major knowledge carrier in schools. Teachers who engaged in research scientific can bring value to society through their own originality, analyze, judgment, integration and design. And they are important resource and core competition of school.

Psychological contract and knowledge complement each other[9]. Psychological contract is the beginning of knowledge and help education organizations to retain teachers and encourage talent. And efficient knowledge sharing can improve Relation and Team dimension of psychological contract so that promote psychological contract. There are some study about teacher psychological contract now, but the number of explorations about relation of psychological contract and knowledge sharing still be few. And lacking of psychological contract study on high school teacher that have high knowledge and creativity. In addition, researchers think teachers have knowledge capital and exist psychological contract with school. But in nowadays, manage measures are mainly control and use teacher, still use unilateral leading encourage methods and seldom reflect such transformation. And how to improve high school teachers' knowledge sharing in knowledge society is one of reasons of developing this study.

2 Method

2.1 Participants

Through stratified cluster sampling, 230 teachers were selected from a high school in Hubei province. We grated 230 questionnaires, 216 were enrolled. 201 valid questionnaires, including 149 man and 52 women, were got.

2.2 Tools

The questionnaire mainly starts from follow levels: knowledge integration, job satisfaction and psychological contract. It includes 22 items. It includes knowledge integration, satisfaction of cooperation and psychological dimension. The questionnaire has good reliability and validity. Concordance coefficient α of total scale was 0.88, test-retest reliability r was 0.753. Adopted five grades scoring, the order score of "Strongly disagree"、"Disagree"、"No comment"、"Agree", "Strongly agree" answers was as follow: 1point, 2 points, 3 points, 4 points, 5 points.

We input, arranged and analyzed the statistics by SPSS13.0, we anglicized the statistics mainly with relevant and regression analysis.

3 Results

3.1 Descriptive Statistics

Descriptive statistics (e.g. Means, Standard Difference) of each variables show in Table 1.

Table 1. Descriptive statistics of three dimensions of psychological contract and knowledge sharing

	N	Mean	Std. Deviation
Relation contract	201	3.579	.743
Balance contract	201	4.139	.735
Exchange contract	201	3.397	.771
Knowledge sharing	201	4.323	.474

Table 2 show the relevant analysis of psychological contract and knowledge sharing. We can get the correlation coefficients of three dimensions of psychological contract and knowledge sharing from the table. The results manifested all variables exist positive correlation which are accord with the hypotheses. Among these correlations, the correlation of Relation Contract with Balance Contract and Exchange Contract are most effectively, were respectively 0.637 and 0.522. And the correlation of Exchange Contract with knowledge sharing was the lowest 0.043.

Table 2. Relevant analysis of three dimensions of psychological contract and knowledge sharing

	Relation contract	Balance contract	Exchange contract	Knowledge sharing
Relation contract	1			
Balance contract	.637***	1		
Exchange contract	.522***	.543**	1	
Knowledge sharing	.287***	.245**	.043*	1

Notes: * $p<0.05$ ** $p<0.01$ ***$p<0.001$.

We verified the relationship of psychological contract and knowledge sharing following. Coefficients of model 1 showed control variables had no effects on knowledge sharing. So this model was verified. Then added psychological contract on the basics of model 5, the model effected better and had greater explanation power.

Table 3. Stepwise regression analysis for knowledge sharing on psychological contract

		Model 1	Model 2
Control variables	gender	-.112	-.095
	age	.032	.017
	work age	-.029	.009
	educational background	-.977	-.130
Independent variables	Relation contract		321***
	Balance contract		.197*
	Exchange contract		-.255**
	R^2	.024	.157
	Adjusted R^2	.007	.123
	F	1.139	4.962***

3.2 Discussion

We found the three dimensions of psychological contract and knowledge existed significant correlation in this study. According to the results, we can draw a conclusion: The three dimensions of psychological contract affect knowledge sharing differently. Among these, there are obvious promotion of Relation contract and Balance contract in knowledge sharing and with former greater than in the latter. While Exchange contract showed negative effective effects on knowledge sharing.

Psychological contract between high school teachers and schools is a kind of teachers' tacit knowledge about school. Researchers introduced theories about psychological contract and sticky knowledge and classified sticky knowledge based on the viscosity of psychological contract and tacit knowledge. They analysed the characteristic of tacit knowledge, and put forward tacit knowledge transformation and sharing management countermeasures. In addition, they carried on aimed management according to different types of tacit knowledge. They thought many tacit knowledge namely implicit knowledge management were issues about psychological contract. Classification of tacit knowledge based on psychological contract revealed more microscopic activity mechanism. On this basic, integrating encouragement, cultural, trust, communication, study and empathy design etc factors to prop up a platform for Exchange, Relation and Counterpoise dimension of psychological contract. That was beneficial to the development and utilization of tacit knowledge especially sticky tacit knowledge, then to realize the transformation and sharing of tacit knowledge in cooperation. They thought employ relationship in nowadays was no longer traditional employees supplied faith, obey and credit to exchange working guarantee, promotions, training opportunity and organization support. Contract changed to employees accept long –term job, more responsibility, more skills, more press, more fuzzy roles' requirements and the organization supply higher reward, performance appraisal encouragement and a position increasingly.

Added to the model of knowledge sharing, concrete manifestation of relation dimension was it significantly effected on the realize and promotion of knowledge sharing and integration, obligatory, affective and instrumentality relation which were same with the relavant results[10]. Existing study showed that obligatory and affective relation positive effected on knowledge sharing will and integration ability, while instrumentality relation went against the effect.

From employees' knowledge sharing promotion degree, obligatory relation impacted knowledge sharing significantly while the effects of affective relation was weaker. It is probably because obligatory relation has more close relationship with roles' responsibility, organization system etc work factors.

Relative obligatory relation between each others can increase the our own and fellows' attention degree and responsibility. At the same time, it shows that knowledge sharing willing of high school employees in our country is more out of responsibility and obligatory, it is a response dominated by ethical criterion, institutional restriction and moral authority.

From function route, the function route of various dimensions on knowledge sharing willing and integration ability is different. For example, obligatory relation and instrumentality relation effect on three kinds of communication behaviors and then effects on employees' knowledge will. While affective relation doesn't influence employees' will by avoidance this path and partly through cooperation and competition communication path only.

4 Conclusion

Through introducing psychological contract into high school teachers management, studying and constructing coping mechanisms from psychological contract violation, job satisfaction and turnover intention three aspects, and then form a new management model. That will probably become a breakthrough in encourage and management of high school teachers.

Study on contents of psychological contract showed the type features of high school teachers' psychological contract generally was between develop and interpersonal type except obvious tendency of chiasma type.

References

1. Rousseau, D.M., Parks, J.M.: The contracts of individuals and organizations. In: Cummings, L.L., Staw, B.M. (eds.) Research in Organizational Behavior, vol. 15, pp. 1–43. JAI Press, Greenwich (1994)
2. Wang, S., Noe, R.A.: Knowledge sharing: A review and directions for future research. Human Resource Management Review 20, 115–131 (2010)
3. Rousseau, D.M.: Changing the deal while keeping the people. Academy of Management Executive 10, 50–61 (1996)
4. Zhang, J.m., Wang, X.: The research on psychological contract:present situation and prospect. Realistic approach 2, 94–97 (2010)
5. Li, J.:The research on psychological contract:in teacher managerment. Teaching and management 1, 33–35 (2011)
6. Shang, H.: The management of preschool teacher from a Perspective of Psychological Contract Introduction to Education 8, 64–65 (2010)
7. Tian, J.: Psychological contract in the professional development of young college application. The higher education of Jiangsu 1, 105–106 (2011)
8. Qian, Y.: The Research of Incentive Strategy for Knowledable Based on Phychological Contract. XIDIAN University (2009)
9. Thomas, D.: Changesin newcomers' psychological contracts during organizational socialization. Journal of Organizational Behavior (1998)
10. Shapiro, J.C., Kessler, L.: Consequences of The psychological for the employment relationship: A Large scale survey. Journal of management studies (2000)

Generator Parameters' Impact on Power System Stability and Their Engineering Testing Methods

Xiaoming Sun[1,3,4], Dabo Chen[2], Mengping Gao[3], Dichen Liu[4], and Tao Zhu[3]

[1] Postdoctoral Workstation of Yunnan Power Grid Corporation, Kunming, China
[2] Chongqing Water Resources and Electric Engineering College, Chongqing, China
[3] Yunnan Electric Power Dispatching Center, Kunming, China
[4] Electrical Engineering Department of Wuhan University, Wuhan, China

Abstract. The impact of the generator parameters (including the parameters of the generator model, excitation system and power system stabilizer) on power system's transient and dynamic stabilities is researched by simulation. From the tables and plots of the resultant data, some evident and useful rules are summarized. These rules can be directly applied to the generator parameters' engineering testing. Since complex theoretical analyses are circumvented, the testing procedure is simplified, and the efficiencies of the testing technicians are greatly promoted.

Keywords: Generator parameters, transient stability, dynamic stability, excitation system, power system stabilizer, automatic voltage regulator.

1 Introduction

Generator parameters, including both the parameters of the generator model and the parameters of the excitation system and the power system stabilizer (PSS), are closely linked with the power system's stability [1, 2]. The degree of the power system's stability is directly determined by the parameters' correctness and the fitness of their combinations. So checking and testing the generator parameters' correctness is one of the most important works before carrying on the stability computation of the power system. Once the generator parameters are mistaken, it is very likely to produce conclusions that are unfit to reality. But strictly testing the generator parameters' correctness is sometimes cumbersome and even impossible, because various field tests [3] of the generator should be carried out repeatedly, which is time-consuming and uneconomical. In addition, the number of the generators in a bulk power system is enormous; when the stability computation is carried out, which would take a great amount of time, a simple and feasible method is necessitated to determine the correctness of some specific generators' parameters quickly. Such a method has been wanted by the technicians at the electric power dispatching center for years, whose work is analyzing and arranging the operation modes of the power system.

This paper is based on the engineering experiences of the authors. From plenty of simulation experiments, the impacts of some important generator parameters on power system's transient and dynamic stabilities are illustrated, and some feasible and

X. Wan (Ed.): Electrical Power Systems and Computers, LNEE 99, pp. 259–276.
springerlink.com

rapid methods for testing the generator parameters are proposed accordingly. The rules that are summarized from the simulation experiments can be directly applied or referenced by the technicians dealing with power system stability analyses.

2 Generator Models and Their Parameters

The number of the generator parameters and their detailed definitions are correlated with specific models. For the versatility of the generator models, the following cases should be considered: the stator of the generator is with three phase windings; the rotator of the generator is with salient poles, excitation winding f, d-axis equivalent damping winding D, and 2 q-axis equivalent damping windings g and Q. Based on the Park transform, the per-unit equations of the generator under dq0 coordinates are as follows (because the magnetic field produced by 0-axis current i_0 in the stator windings is 0, it has no effect on the electric quantities of the rotator [4]; thus the equations related to 0-axis components are omitted in the equation):

$$\begin{cases} v_{\mathrm{d}} = -\psi_{\mathrm{q}} - R_{\mathrm{a}} i_{\mathrm{d}} \\ v_{\mathrm{q}} = \psi_{\mathrm{d}} - R_{\mathrm{a}} i_{\mathrm{q}} \\ v_{\mathrm{f}} = \mathrm{d}\psi_{\mathrm{f}}/\mathrm{d}t + R_{\mathrm{f}} i_{\mathrm{f}} \\ 0 = \mathrm{d}\psi_{\mathrm{D}}/\mathrm{d}t + R_{\mathrm{D}} i_{\mathrm{D}} \\ 0 = \mathrm{d}\psi_{\mathrm{g}}/\mathrm{d}t + R_{\mathrm{g}} i_{\mathrm{g}} \\ 0 = \mathrm{d}\psi_{\mathrm{Q}}/\mathrm{d}t + R_{\mathrm{Q}} i_{\mathrm{Q}} \end{cases}, \quad \begin{cases} \psi_{\mathrm{d}} = -X_{\mathrm{d}} i_{\mathrm{d}} + X_{\mathrm{ad}} i_{\mathrm{f}} + X_{\mathrm{ad}} i_{\mathrm{D}} \\ \psi_{\mathrm{f}} = -X_{\mathrm{ad}} i_{\mathrm{d}} + X_{\mathrm{f}} i_{\mathrm{f}} + X_{\mathrm{ad}} i_{\mathrm{D}} \\ \psi_{\mathrm{D}} = -X_{\mathrm{ad}} i_{\mathrm{d}} + X_{\mathrm{ad}} i_{\mathrm{f}} + X_{\mathrm{D}} i_{\mathrm{D}} \\ \psi_{\mathrm{q}} = -X_{\mathrm{q}} i_{\mathrm{q}} + X_{\mathrm{aq}} i_{\mathrm{g}} + X_{\mathrm{aq}} i_{\mathrm{Q}} \\ \psi_{\mathrm{g}} = -X_{\mathrm{aq}} i_{\mathrm{q}} + X_{\mathrm{g}} i_{\mathrm{g}} + X_{\mathrm{aq}} i_{\mathrm{Q}} \\ \psi_{\mathrm{Q}} = -X_{\mathrm{aq}} i_{\mathrm{q}} + X_{\mathrm{aq}} i_{\mathrm{g}} + X_{\mathrm{Q}} i_{\mathrm{Q}} \end{cases}, \tag{1}$$

where v_{d}, v_{q} and v_{f} are the voltages of d-axis, q-axis and excitation wingding f, respectively; i_{d}, i_{q}, i_{f}, i_{D}, i_{g} and i_{Q} are the currents of d-axis, q-axis, excitation wingding f, equivalent damping windings D, g and Q, respectively; R_{a}, R_{f}, R_{D}, R_{g} and R_{Q} are the resistors of one phase stator winding, excitation winding f, equivalent damping windings D, g and Q, respectively; ψ_{d}, ψ_{q}, ψ_{f}, ψ_{D}, ψ_{g} and ψ_{Q} are the total magnetic flux linkages of the fictitious d-axis and q-axis windings, excitation winding f, equivalent damping windings D, g and Q, respectively; X_{d}, X_{q}, X_{ad}, X_{aq}, X_{f}, X_{D}, X_{g} and X_{Q} are the d-axis and q-axis synchronous reactances, the armature reaction reactances of d-axis and q-axis windings, the reactances of excitation winding f, the reactances of equivalent damping windings D, g and Q, respectively. It should be noted that 2 assumptions are made in Eq. (1): 1) the electromagnetic transient processes are not considered, or the aperiodic component of the stator current is considered in another way, i.e. assume $\mathrm{d}\psi_{\mathrm{d}}/\mathrm{d}t \approx 0$ and $\mathrm{d}\psi_{\mathrm{q}}/\mathrm{d}t \approx 0$ [4]; 2) in equations related to v_{d} and v_{q}, assume the per-unit value of the electric angular velocity $\omega \approx 1$, which makes the equations linearized.

The total magnetic flux linkages ψ_{d}, ψ_{q}, ψ_{f}, ψ_{D}, ψ_{g} and ψ_{Q} in Eq. (1) are inconvenient to measure and use in practice, so some *practical variables* are introduced to indirectly represent these total magnetic flux linkages:

$$\begin{cases} E'_{\mathrm{d}} = -X_{\mathrm{aq}} \psi_{\mathrm{g}} / X_{\mathrm{g}} \\ E'_{\mathrm{q}} = X_{\mathrm{ad}} \psi_{\mathrm{f}} / X_{\mathrm{f}} \end{cases}, \quad \begin{cases} E''_{\mathrm{d}} = -X_{\mathrm{aq}} \left(X_{\sigma\mathrm{g}} \psi_{\mathrm{Q}} + X_{\sigma\mathrm{Q}} \psi_{\mathrm{g}} \right) / \left(X_{\mathrm{Q}} X_{\mathrm{g}} - X_{\mathrm{aq}}^2 \right) \\ E''_{\mathrm{q}} = X_{\mathrm{ad}} \left(X_{\sigma\mathrm{D}} \psi_{\mathrm{f}} + X_{\sigma\mathrm{f}} \psi_{\mathrm{D}} \right) / \left(X_{\mathrm{D}} X_{\mathrm{f}} - X_{\mathrm{ad}}^2 \right) \end{cases}, \tag{2}$$

where E'_d, E'_q and E''_d, E''_q are the d-axis, q-axis transient electromotive forces and the d-axis, q-axis subtransient electromotive forces, respectively; $X_{\sigma f}$, $X_{\sigma D}$, $X_{\sigma g}$ and $X_{\sigma Q}$ are the leakage reactances of excitation winding f, equivalent damping windings D, g and Q. All parameters' units in Eqs. (1) and (2) are in p.u. (per unit).

Again, some *practical parameters* are introduced to not only simplify the representation of the equations, but also make the equations' physical meanings clearer. More important, these parameters can be measured from experiments directly:

$$\begin{cases} T'_{d0} = X_f / R_f \\ T'_{q0} = X_g / R_g \end{cases}, \quad \begin{cases} T''_{d0} = \left(X_D - X_{ad}^2 / X_f \right) / R_D \\ T''_{q0} = \left(X_Q - X_{aq}^2 / X_g \right) / R_Q \end{cases}, \tag{3}$$

where T'_{d0}, T'_{q0} and T''_{d0}, T''_{q0} are the d-axis, q-axis open circuit transient time constants and the d-axis, q-axis open circuit subtransient time constants, respectively. All units are in s (second).

By virtue of Eqs. (2) and (3), the 6th-order practical model of the generator can be derived from Eq. (1) [4]:

$$\begin{cases} \begin{cases} v_d = E''_d + X''_q i_q - R_a i_d, \\ v_q = E''_q - X''_d i_d - R_a i_q, \end{cases} \\ T'_{d0} \dfrac{dE'_q}{dt} = \dfrac{X_{ad} v_f}{R_f} - \dfrac{X_d - X_{\sigma a}}{X'_d - X_{\sigma a}} E'_q + \dfrac{X_d - X'_d}{X'_d - X_{\sigma a}} E''_q - \dfrac{(X_d - X'_d)(X''_d - X_{\sigma a})}{X'_d - X_{\sigma a}} i_d, \\ T''_{d0} \dfrac{dE''_q}{dt} = \dfrac{X''_d - X_{\sigma a}}{X'_d - X_{\sigma a}} T''_{d0} \dfrac{dE'_q}{dt} - E''_q + E'_q - (X'_d - X''_d) i_d, \\ T'_{q0} \dfrac{dE'_d}{dt} = -\dfrac{X_q - X_{\sigma a}}{X'_q - X_{\sigma a}} E'_d + \dfrac{X_q - X'_q}{X'_q - X_{\sigma a}} E''_d + \dfrac{(X_q - X'_q)(X''_q - X_{\sigma a})}{X'_q - X_{\sigma a}} i_q, \\ T''_{q0} \dfrac{dE''_d}{dt} = \dfrac{X''_q - X_{\sigma a}}{X'_q - X_{\sigma a}} T''_{q0} \dfrac{dE'_d}{dt} - E''_d + E'_d - (X'_q - X''_q) i_q, \\ \dfrac{d\delta}{dt} = \omega - 1, \end{cases} \tag{4}$$

where $X_{\sigma a}$, X'_d, X'_q and X''_d, X''_q are the leakage reactance of the stator winding, the d-axis, q-axis transient reactances and the d-axis, q-axis subtransient reactances, respectively. The 6 equations are the voltage equations of the stator, the excitation winding f, the equivalent damping windings D, g and Q, and the motion equation of the stator, successively. From Eq. (4), each generator parameter's position in the formula and its corresponding function can be seen clearly.

3 Fast Testing of Generator Parameters in Different Models

By simplifying Eq. (4) to different extent (neglecting a certain number of windings or introducing some new assumptions), the 5th-order, 4th-order, 3rd-order and 2nd-order models of the generator can be obtained one by one. These models can be found in [4] and [5], so they are not listed out for brevity. From comparisons, it can be seen that in these models only one or two parameters' detailed definitions have certain discrepancies, and other parameters are just the same. But it is because of the discrepancies that some evident incorrectness of the parameters can be detected by simple and fast means. This is discussed respectively as follows.

1) The 6th-order model (the d-axis, q-axis windings, excitation winding f and equivalent damping windings D, g and Q are considered together; it is the detailed model for solid steam turbine or non-salient pole machine): $T'_{q0}>0$, $X_q \neq X'_q$.
2) The 5th-order model (the equivalent damping winding g is neglected from the 6th-order model; it is the detailed model for hydraulic turbine or salient pole machine): $T'_{q0}=0$, $X_q \neq X'_q$.
3) The 4th-order model (only the d-axis, q-axis windings, excitation winding f and equivalent damping winding g are considered; it is fit to describe the solid steam turbine): $T'_{q0}>0$, $X_q \neq X'_q$.
4) The 3rd-order model (only the d-axis, q-axis windings and excitation winding f are considered; it is fit to describe the salient pole machine when high computation accuracy is not required): $T'_{q0}=0$, $X_q = X'_q$.
5) The 2nd-order model (it is assumed that the excitation system is strong enough, and it can maintain the constancy of E'_d and/or E'_q): T'_{d0} = *a very big value*.
6) Both the salient and non-salient pole machines: $X_q \neq X'_d$.

The above are merely the qualitative testing methods, and the number of parameters that can be tested is extremely limited. Further, some quantitative testing criteria [4, 5] that are derived from engineering practice are listed in Table 1.

From Table 1, the following rules can be summarized: 1) $X_d \geq X_q \geq X'_q > X'_d > X''_q \geq X''_d > X_{\sigma a}$; 2) $T_J > T'_{d0} > T'_{q0} > T''_{d0} \geq T''_{q0}$.

Provided that the generator parameters are prominently deviating from the aforementioned 6 requirements, the above 2 rules and the reference ranges of Table 1, it is justified in doubting that the parameters are incorrect, and more careful testing means

Table 1. Quantitative testing criteria (reference range) for generator parameters.

Parameter name (unit)	Notation	Steam turbine	Hydraulic turbine
Synchronous reactances (p.u.)	X_d	1.0~2.3	0.6~1.5
	X_q	1.0~2.3	0.4~1.0
Transient reactances (p.u.)	X'_d	0.15~0.4	0.2~0.5
	X'_q	0.2~1.0	0.2~1.0
Subtransient reactances (p.u.)	X''_d	0.1~0.25	0.15~0.35
	X''_q	0.1~0.25	0.2~0.45
Open circuit transient time constants (s)	T'_{d0}	3.0~10.0	1.5~9.0
	T'_{q0}	0.5~2.0	0~2.0
Open circuit subtransient time constants (s)	T''_{d0}	0.02~0.05	0.01~0.05
	T''_{q0}	0.02~0.07	0.01~0.09
Stator leakage reactance (p.u.)	$X_{\sigma a}$	0.05~0.2	0.05~0.2
Stator resistor (p.u.)	R_a	0.001 5~0.005	0.001 5~0.005
Inertia time constant (s)	T_J	4.0~8.0	8.0~16.0

should be taken. However, it is a simple and fast method to test the generator parameters and is very suitable for the preliminary test of the newly obtained parameters.

4 Generator Parameters' Impacts on Power System's Stabilities

The testing criteria given in the previous section, however, are *necessary conditions* but not *sufficient conditions* to guarantee the power system's stability. Whether the power system can maintain stability or not, and the level of stability lie also on some important parameters and their combinations. This section illustrates this argument by plenty of simulation experiments. In order to enable the readers to reproduce the experimental results, the IEEE 9-node test system, a standard system, is selected as the simulation model. And PSD-BPA (ver. 4.2) [6] is the simulation software used.

4.1 IEEE 9-Node Test System Overview

The geographically interconnected diagram of the IEEE 9-node test system is shown in Fig. 1, which displays the power flow distribution under the normal (default) operation condition. The steady-state and transient parameters of the loads, buses, transmission lines and transformers can be found in [7], so this section only lists out the generator parameters (Table 2), and the units of the parameters are the same as those in Table 1. It should be noted that because $R_a \approx 0$, R_a is omitted from Table 2.

From the comparison of the generator parameters in Table 1 and Table 2, it can be inferred that GEN2 is a hydraulic turbine and GEN3 is a steam turbine. However, GEN2's X''_d and X''_q (indicated by shadings in Table 2) do not strictly comply with the reference range in Table 1. This means that Table 1 should only be referenced (to pay

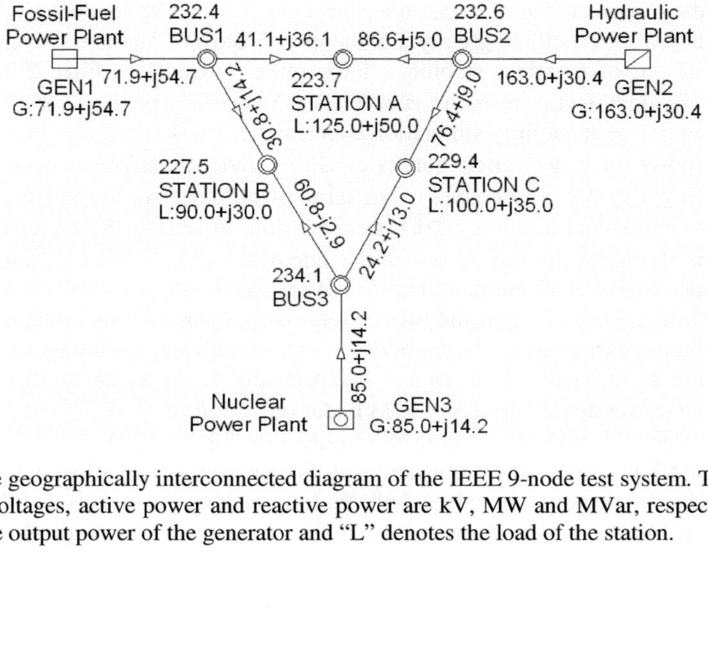

Fig. 1. The geographically interconnected diagram of the IEEE 9-node test system. The units of the node voltages, active power and reactive power are kV, MW and MVar, respectively. "G" denotes the output power of the generator and "L" denotes the load of the station.

Table 2. Generator parameters in IEEE 9-node test system.

	X_d	X_q	X'_d	X'_q	X''_d	X''_q
GEN1	0.1460	0.0969	0.0608	0.0969	0.0400	0.0600
GEN2	0.8958	0.8645	0.1189	0.1969	0.0890	0.0890
GEN3	1.3130	1.2580	0.1813	0.2500	0.1070	0.1070
	T'_{d0}	T'_{q0}	T''_{d0}	T''_{q0}	$X_{\sigma a}$	T_J
GEN1	8.9600	0	0.0400	0.0600	0.0336	47.280
GEN2	6.0000	0.5400	0.0330	0.0780	0.0521	12.800
GEN3	5.8900	0.6000	0.0330	0.0700	0.0742	6.0200

more attention to the parameters prominently deviating from Table 1) but not rigidly obeyed, because the combinations of the parameters are also crucial.

GEN1 is the balancing machine of the system, which is a $V\theta$-node ($\theta = 0°$) in simulation. Table 2 shows that most of the parameters of GEN1 (indicated by shadings) do not comply with the reference range in Table 1. But considering that so long as the output (active) power of the balancing machine is not a negative value or over the nominal value, the output power of the balancing machine can be arranged according to the actual demands freely, implying that the output power and parameters of the balancing machine have no great impacts on power system's stabilities. It is very easy to testify this argument by simulation experiments, so it can be said that the parameters of GEN1 in Table 2 are acceptable and are not unreasonable.

4.2 Transient Stability of Power System

From a good many simulation experiments, the authors find that the transient stability of the power system is very sensitive to the values of X''_d/X'_d and X''_q/X'_q. If one of these 2 values is greater than 0.95, under large disturbance, the power angle curve of the generator would oscillate greatly and damp very slowly, and even diverge (i.e. the generator is out of transient stability). Experimental results in Figs. 2 and 3 have illustrated this fact, so the correctness of X'_d, X''_d, X'_q and X''_q and their combinations can be tested by transient stability simulation experiments.

Fig. 2 shows the power angle curves of GEN3 under 2 different conditions – $X''_d/X'_d<0.95$, $X''_d/X'_d>0.95$ respectively. The large disturbance set in the experiment is a three-phase permanent fault on 220 kV transmission line BUS1–STATION B: at 0 s a three-phase short-circuit fault occurs on the side of STATION B; at 0.2 s the breakers on both sides trip to clear the fault, but do not reclose. Comparing the curves in Fig. 2 (a) and (b), it can be seen that the former returns to a smooth line quickly, while the latter oscillates ceaselessly and can not damp to a smooth line for a long time.

The large disturbance set in Fig. 3's experiment is the same as that in Fig. 2. Because the power angle curve of GEN3 under condition of $X''_q/X'_q<0.95$ is similar to Fig. 2(a), Fig. 3's experiment considers only the condition of $X''_q/X'_q>0.95$. But because GEN3 has been out of transient stability under this condition (i.e. out of step with GEN1), the power angle curve deviates too much and has exceeded the drawing ranges of BPA, the power angle curve can not be displayed properly. So the power

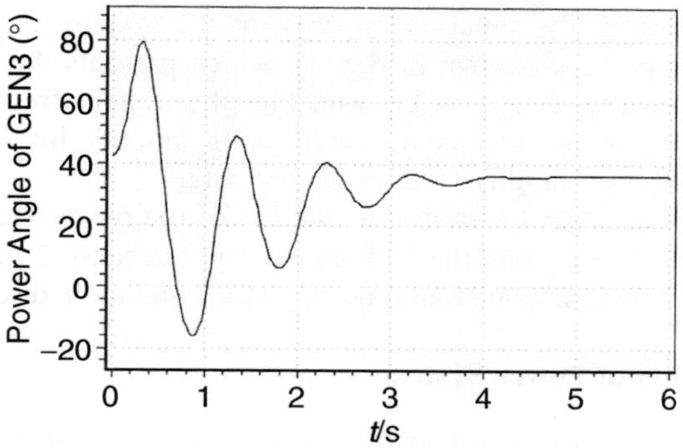

(a) $X''_d = 0.107\,0, X'_d = 0.181\,3, X''_d / X'_d \approx 0.590\,2 < 0.95.$

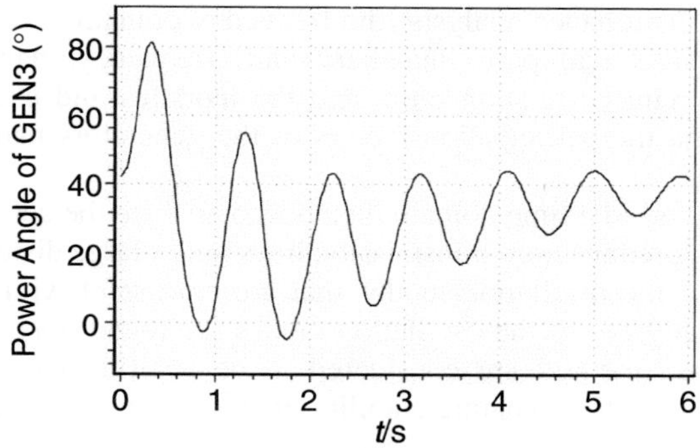

(b) $X''_d = 0.179\,0, X'_d = 0.181\,3, X''_d / X'_d \approx 0.987\,3 > 0.95.$

Fig. 2. The impact of X''_d/X'_d on the transient stability of the power system.

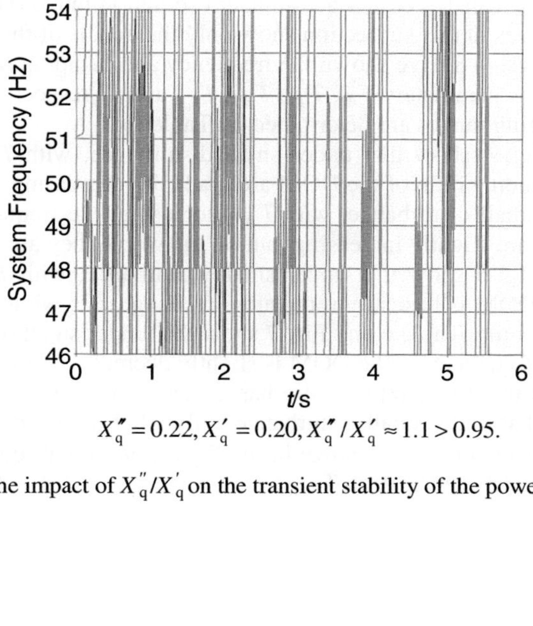

$X''_q = 0.22, X'_q = 0.20, X''_q / X'_q \approx 1.1 > 0.95.$

Fig. 3. The impact of X''_q/X'_q on the transient stability of the power system.

angle curve is replaced by the supervisory curve of the system frequency recorded during the simulation process (shown in Fig. 3), which presents the power system's status indirectly. From Fig. 3, it can be seen that the system frequency oscillates severely, and already has no chance to return to a smooth line, illustrating the generator is out of transient stability from a different angle.

In fact, it is very likely that the transient stability of the power system is not only sensitive to X''_d/X'_d and X''_q/X'_q. But the authors merely find these 2 values at present. As to other possible parameter combinations, they still await to be discovered later on.

4.3 Dynamic Stability of Power System

The commonly used procedures for analyzing the dynamic stability of the power system are as follows:

1) Use the small disturbance analysis (the frequency domain method) to calculate each oscillation mode's real part, imaginary part, frequency, damping ratio and electromechanical circuit correlation ratio, and the modulus and phase angle of the right eigenvector, and the participation factors of the generators participating in the oscillation.

2) Use Prony method (the time domain method) to analyze the active power curves of some important interconnection transmission lines under large disturbance.

3) Try to find out the oscillation modes that are consistent with those found by small disturbance analysis – if the oscillation modes are found, then the results from the frequency domain method are considered to be testified by the time domain method, and accordingly the dynamic stability analysis is considered to be rounded and effective.

According to the above procedures, at first, apply the small disturbance analysis to the IEEE 9-node test system. From the small disturbance analysis, 6 oscillation modes are obtained, but only 1 oscillation mode has an electromechanical circuit correlation ratio that is greater than 1, i.e. it is a dominant oscillation mode (DOM). For brevity, the tables and figures in this subsection show only the details of this oscillation mode. Tables 3–7 and Figs. 4–8 have shown the frequency and damping ratio changes of the DOM with GEN3's parameters T'_{d0}, T'_{q0}, T''_{d0}, T''_{q0} and T_J respectively (the generator parameters' changing ranges are determined by Table 1).

Table 3 and Fig. 4 show that under small disturbance, with T'_{d0} increasing, the frequency and damping ratio of the DOM are gradually decreasing. Table 4 and Fig. 5 show that under small disturbance, with T'_{q0} increasing, the frequency and damping ratio of the DOM are slightly increasing, but on the whole they are not sensitive to T'_{q0}. Table 5 and Fig. 6 show that under small disturbance, with T''_{d0} increasing, the frequency of the DOM is basically remaining unchangeable, while the damping ratio is gradually decreasing. Table 6 and Fig. 7 show that under small disturbance, with T''_{q0} increasing, the frequency of the DOM is slightly decreasing (it is not sensitive to T''_{q0} on the whole), while the damping ratio has a small tendency of increasing. Table 7 and Fig. 8 show that under small disturbance, only when T_J is smaller than a specific value (e.g. smaller than 6.5 s, as shown in the figure), are the frequency and damping ratio of the DOM fairly sensitive to T_J.

Table 3. The impact of T'_{d0} on DOM's characteristics under small disturbance.

T'_{d0} (s)	3.00	3.70	4.40	5.10	5.80	6.50
Frequency (Hz)	1.3364	1.3248	1.3159	1.3091	1.3038	1.2994
Damping ratio (%)	7.17	6.86	6.53	6.21	5.92	5.67
T'_{d0} (s)	7.20	7.90	8.60	9.30	10.0	
Frequency (Hz)	1.2959	1.2929	1.2904	1.2882	1.2862	
Damping ratio (%)	5.44	5.24	5.06	4.91	4.76	

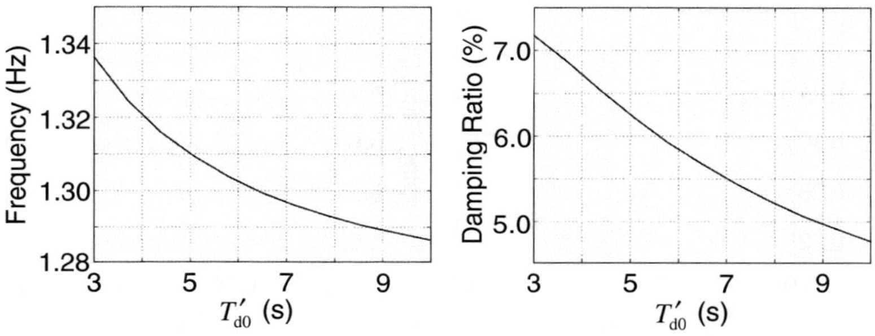

(a) The change of frequency. (b) The change of damping ratio.

Fig. 4. The curves of the impact of T'_{d0} on DOM's characteristics under small disturbance.

Table 4. The impact of T'_{q0} on DOM's characteristics under small disturbance.

T'_{q0} (s)	0.50	0.65	0.80	0.95	1.10	1.25
Frequency (Hz)	1.3028	1.3033	1.3039	1.3044	1.3049	1.3055
Damping ratio (%)	5.82	5.92	5.99	6.04	6.08	6.12
T'_{q0} (s)	1.40	1.55	1.70	1.85	2.00	
Frequency (Hz)	1.3057	1.3061	1.3063	1.3067	1.3069	
Damping ratio (%)	6.13	6.16	6.17	6.18	6.19	

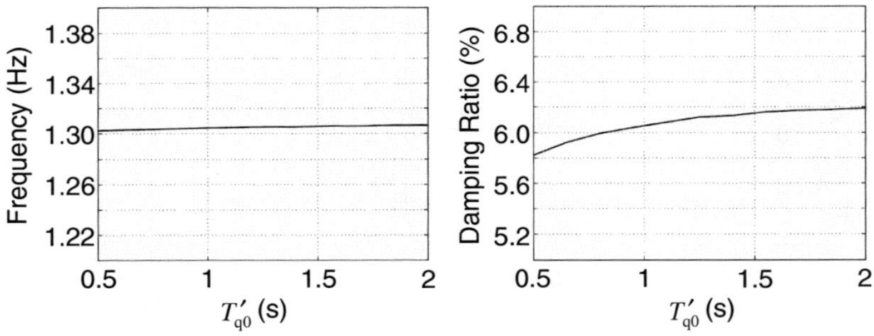

(a) The change of frequency. (b) The change of damping ratio.

Fig. 5. The curves of the impact of T'_{q0} on DOM's characteristics under small disturbance.

Table 5. The impact of T''_{d0} on DOM's characteristics under small disturbance.

T''_{d0} (s)	0.020	0.023	0.026	0.029	0.032	0.035
Frequency (Hz)	1.302 3	1.302 5	1.302 8	1.303 0	1.303 2	1.303 2
Damping ratio (%)	6.30	6.20	6.10	6.01	5.89	5.83
T''_{d0} (s)	0.038	0.041	0.044	0.047	0.050	
Frequency (Hz)	1.303 3	1.303 4	1.303 4	1.303 4	1.303 3	
Damping ratio (%)	5.74	5.65	5.57	5.49	5.41	

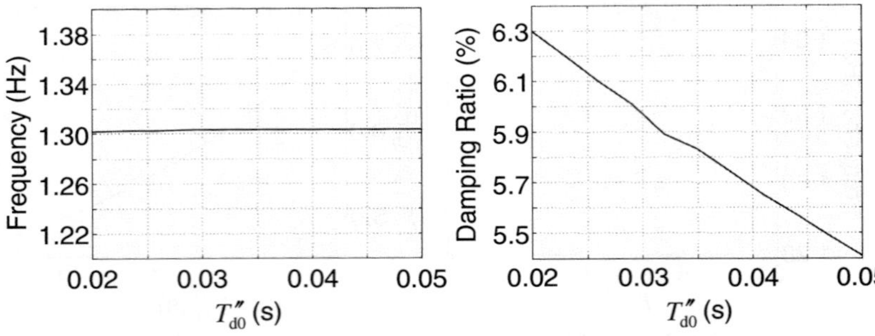

(a) The change of frequency. (b) The change of damping ratio.

Fig. 6. The curves of the impact of T''_{d0} on DOM's characteristics under small disturbance.

Table 6. The impact of T''_{q0} on DOM's characteristics under small disturbance.

T''_{q0} (s)	0.020	0.025	0.030	0.035	0.040	0.045
Frequency (Hz)	1.308 8	1.308 0	1.307 2	1.306 6	1.306 0	1.305 4
Damping ratio (%)	5.53	5.57	5.60	5.64	5.68	5.71
T''_{q0} (s)	0.050	0.055	0.060	0.065	0.070	
Frequency (Hz)	1.304 9	1.304 4	1.303 9	1.303 5	1.303 2	
Damping ratio (%)	5.75	5.79	5.82	5.85	5.89	

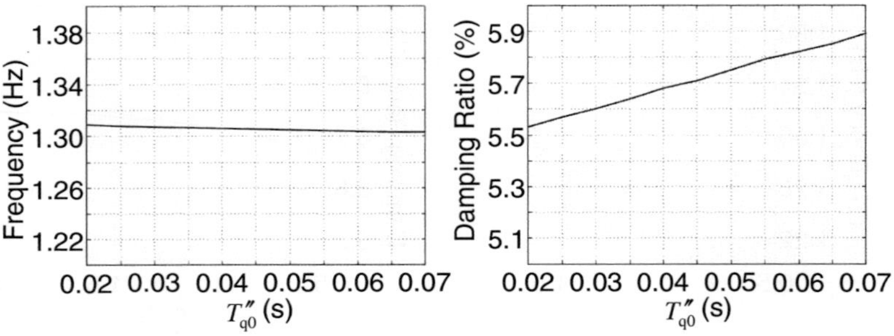

(a) The change of frequency. (b) The change of damping ratio.

Fig. 7. The curves of the impact of T''_{q0} on DOM's characteristics under small disturbance.

Table 7. The impact of T_J on DOM's characteristics under small disturbance.

T_J (s)	4.00	4.40	4.80	5.20	5.60	6.00
Frequency (Hz)	1.300 1	1.301 7	1.303 7	1.306 2	1.308 9	1.311 8
Damping ratio (%)	5.41	5.70	5.94	6.15	6.32	6.45
T_J (s)	6.40	6.80	7.20	7.60	8.00	
Frequency (Hz)	1.314 7	1.317 5	1.320 1	1.322 6	1.324 8	
Damping ratio (%)	6.53	6.59	6.61	6.61	6.60	

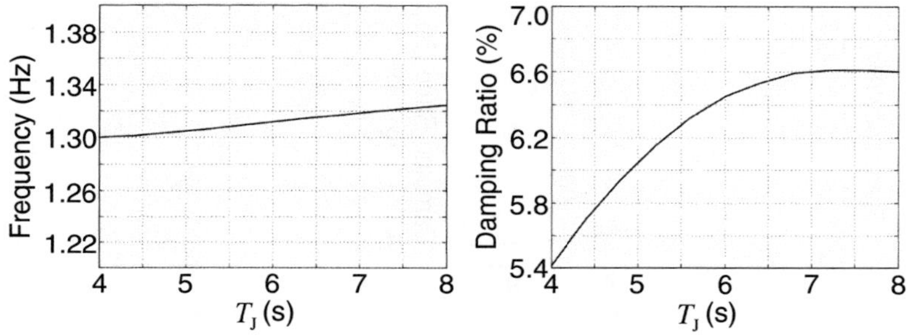

(a) The change of frequency. (b) The change of damping ratio.

Fig. 8. The curves of the impact of T_J on DOM's characteristics under small disturbance.

Next, use the time domain method (under large disturbance) to testify the experimental results obtained from small disturbance analysis. The large disturbance set in the experiment is the same three-phase permanent fault on 220 kV transmission line BUS1–STATION B as mentioned in Subsection 4.2. And the active power curve of the 220 kV transmission line BUS2–STATION A is analyzed by Prony method. Likewise, Tables 8–12 and Figs. 9–13 have shown the frequency and damping ratio changes of the DOM with GEN3's parameters T'_{d0}, T'_{q0}, T''_{d0}, T''_{q0} and T_J respectively under this condition.

Table 8 and Fig. 9 show that under large disturbance, with T'_{d0} increasing, the frequency of the DOM is gradually decreasing, while the damping ratio is firstly increasing and then (to about 6 s) decreasing. Table 9 and Fig. 10 show that under large disturbance, with T'_{q0} increasing, the frequency and damping ratio of the DOM are slightly oscillating across a straight line and their mean values are basically remaining unchangeable. Table 10 and Fig. 11 show that under large disturbance, with T''_{d0} increasing, the frequency of the DOM is not sensitive to T''_{d0} and is basically remaining unchangeable, while the damping ratio is firstly increasing and then (to about 0.035 s) decreasing. Table 11 and Fig. 12 show that under large disturbance, with T''_{q0} increasing, the frequency of the DOM is slightly decreasing, while the damping ratio has a tendency of increasing. Table 12 and Fig. 13 show that under large disturbance, only when T_J is smaller than a specific value (e.g. smaller than 6.5 s, as shown in the figure) is the frequency of the DOM fairly sensitive to T_J, while the damping ratio is firstly increasing and then (to about 4.7 s) decreasing.

Table 8. The impact of T'_{d0} on DOM's characteristics under large disturbance.

T'_{d0} (s)	3.00	3.70	4.40	5.10	5.80	6.50
Frequency (Hz)	1.366	1.341	1.309	1.241	1.245	1.152
Damping ratio (%)	14.100	16.299	18.870	20.834	28.129	17.284
T'_{d0} (s)	7.20	7.90	8.60	9.30	10.0	
Frequency (Hz)	1.142	1.137	1.133	1.130	1.127	
Damping ratio (%)	15.039	13.519	12.315	11.375	10.545	

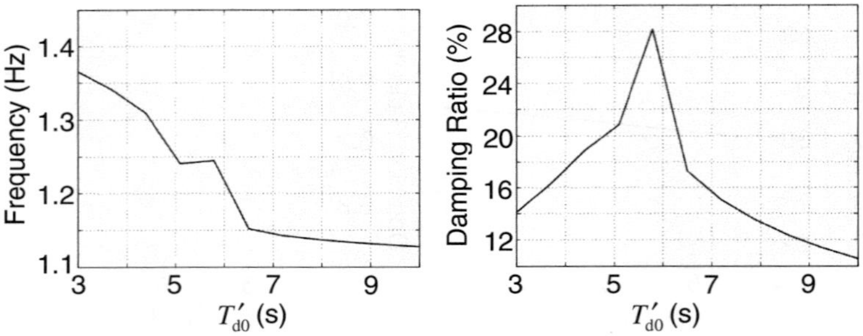

(a) The change of frequency. (b) The change of damping ratio.

Fig. 9. The curves of the impact of T'_{d0} on DOM's characteristics under large disturbance.

Table 9. The impact of T'_{q0} on DOM's characteristics under large disturbance.

T'_{q0} (s)	0.50	0.65	0.80	0.95	1.10	1.25
Frequency (Hz)	1.162	1.153	1.160	1.212	1.137	1.174
Damping ratio (%)	19.826	20.667	21.922	20.643	23.690	20.886
T'_{q0} (s)	1.40	1.55	1.70	1.85	2.00	
Frequency (Hz)	1.137	1.188	1.187	1.200	1.124	
Damping ratio (%)	23.974	21.453	21.711	20.818	24.803	

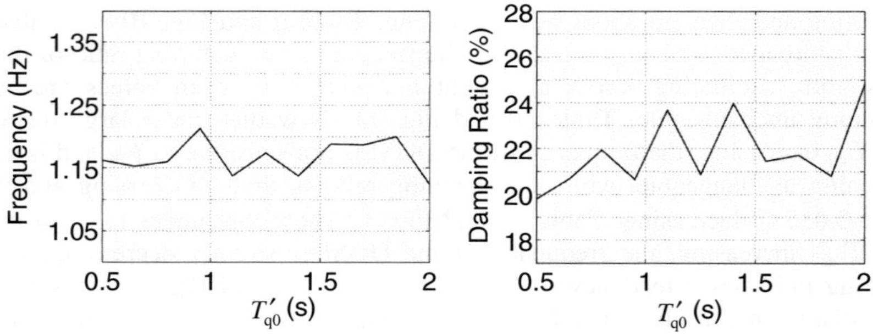

(a) The of change frequency. (b) The change of damping ratio.

Fig. 10. The curves of the impact of T'_{q0} on DOM's characteristics under large disturbance.

Table 10. The impact of T''_{d0} on DOM's characteristics under large disturbance.

T''_{d0} (s)	0.020	0.023	0.026	0.029	0.032	0.035
Frequency (Hz)	1.262	1.203	1.251	1.166	1.219	1.174
Damping ratio (%)	25.667	24.390	26.483	22.538	28.185	30.647
T''_{d0} (s)	0.038	0.041	0.044	0.047	0.050	
Frequency (Hz)	1.194	1.184	1.186	1.189	1.192	
Damping ratio (%)	31.080	18.083	17.250	16.728	16.095	

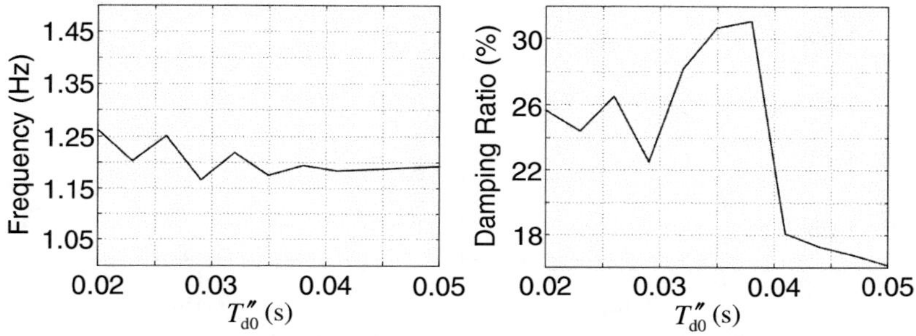

(a) The change of frequency. (b) The change of damping ratio.

Fig. 11. The curves of the impact of T''_{d0} on DOM's characteristics under large disturbance.

Table 11. The impact of T''_{q0} on DOM's characteristics under large disturbance.

T''_{q0} (s)	0.020	0.025	0.030	0.035	0.040	0.045
Frequency (Hz)	1.198	1.195	1.190	1.191	1.182	1.173
Damping ratio (%)	17.260	17.330	17.199	17.756	18.688	18.955
T''_{q0} (s)	0.050	0.055	0.060	0.065	0.070	
Frequency (Hz)	1.180	1.175	1.172	1.166	1.186	
Damping ratio (%)	17.896	19.287	20.799	21.297	29.204	

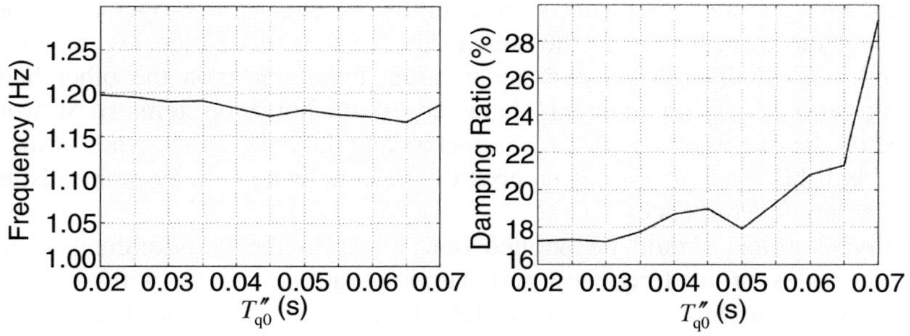

(a) The change of frequency. (b) The change of damping ratio.

Fig. 12. The curves of the impact of T''_{q0} on DOM's characteristics under large disturbance.

Table 12. The impact of T_J on DOM's characteristics under large disturbance.

T_J (s)	4.00	4.40	4.80	5.20	5.60	6.00
Frequency (Hz)	1.144	1.149	1.153	1.205	1.249	1.268
Damping ratio (%)	14.950	17.870	21.542	21.070	19.780	18.923
T_J (s)	6.40	6.80	7.20	7.60	8.00	
Frequency (Hz)	1.285	1.295	1.302	1.309	1.310	
Damping ratio (%)	17.844	17.029	16.399	15.782	15.445	

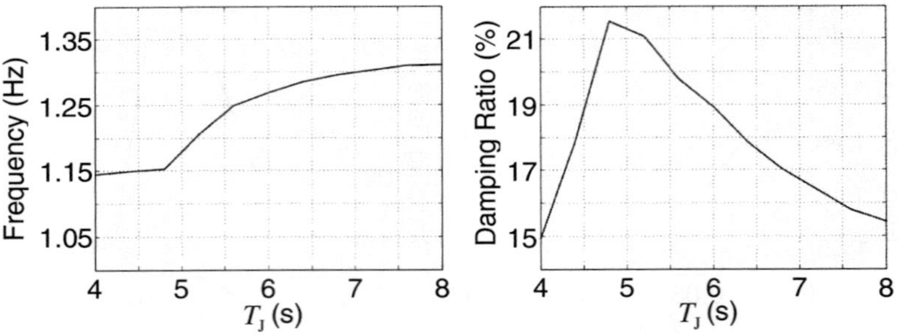

(a) The change of frequency. (b) The change of damping ratio.

Fig. 13. The curves of the impact of T_J on DOM's characteristics under large disturbance.

Now, comparing the experimental results obtained from the time domain method (under large disturbance) with those obtained from the frequency domain method (under small disturbance) one by one, the following facts can be seen: as to the frequency changes of the DOM with the generator parameters, the conclusions from both the time domain analysis and the frequency domain analysis are consistent with each other; while as to the damping ratio changes of the DOM with the generator parameters, except for T'_{q0} and T''_{q0}, the conclusions from the time domain analysis and the frequency domain analysis are all different. The reasons for this can be explained as follows: on one hand, the frequency of an oscillation mode is actually the characteristic frequency of the power system that is determined by the inherent structural characters of the power system and is irrelevant to the operation mode and the type of the disturbance, so the conclusions related to frequencies must conform both under small disturbance and under large disturbance; on the other hand, the damping ratio is not an inherent character of the power system, so it would be affected by the operation mode, the parameters and the disturbance types of the power system, and this leads to the differences of the conclusions under small disturbance and under large disturbance.

Finally, 3 points should be pointed out. 1) As to the non-dominant oscillation mode, the rules of its frequency and damping ratio changing with the generator parameters are similar to those of the DOM, so the conclusions obtained from the DOM are applicable to the non-dominant oscillation mode. 2) When the experimental results (rules) in this subsection are used to test the generator parameters, some parameters' impacts on the frequency and damping ratio of the oscillation mode are too small to be used to test the parameters. 3) When the damping ratio of the oscillation mode is used to test the generator parameters, the results obtained under small disturbance and under large disturbance should be analyzed separately.

5 Test of Excitation System Parameters and PSS Parameters

The adjustment of the excitation system and PSS plays a very important role in guaranteeing the stable operations of the generator and the power system, so the parameters of the excitation system and PSS are always considered together with the generator parameters, and are sometimes treated as a component part of the generator parameters. This section is dedicated to proposing 2 engineering methods for testing excitation system parameters and PSS parameters.

5.1 Excitation System Parameters Test

This subsection proposes a method for testing the excitation system parameters, which can be fell into 4 steps:

1) stop the operation of the tested generator's PSS;
2) separate the tested generator from the electric network;
3) adjust the reference voltage of the excitation system according to a specific function, e.g. step function or ramp function;
4) investigate whether the output voltage of the generator is able to track the reference voltage effectively – the rising edge is steep, the overshoot is small and the steady area is of no great oscillations.

A good way to accomplish step 4) is to compare the output voltage curve of the generator obtained from simulation experiment with another classic curve (obtained from the same simulation experiment of a generator which has correct excitation system parameters). If the differences of the 2 curves are fairly small, then the excitation system parameters can be primarily considered to be correct and effective; otherwise, the excitation system parameters would be unreasonable or ineffective, and further measures must be taken to find the mistakes or the parameters must be remeasured from field test.

Fig. 14 shows the simulation curves of the excitation system of GEN3 in IEEE 9-node test system. For comparison, a classic curve is superposed on the figure. From Fig. 14 (a) and (b), it can be seen that the differences between the experimental curves and the classic curves in both figures are fairly small, implying that under these 2 conditions the excitation system parameters are both correct and effective. The discrepancies between Fig. 14 (a) and (b) lie on the dynamic amplification coefficients of the automatic voltage regulators (AVR) [8] of the excitation system: the former's dynamic amplification coefficient is big, so the response of the output voltage of the generator is fast but the overshoot is relatively large; the latter's dynamic amplification coefficient is small, and the output voltage of the generator has no overshoot, but the response of the output voltage is very slow. Because the types of the curves, when the excitation system parameters are incorrect, are numerous but are easy to distinguish, so the curve samples are omitted for brevity.

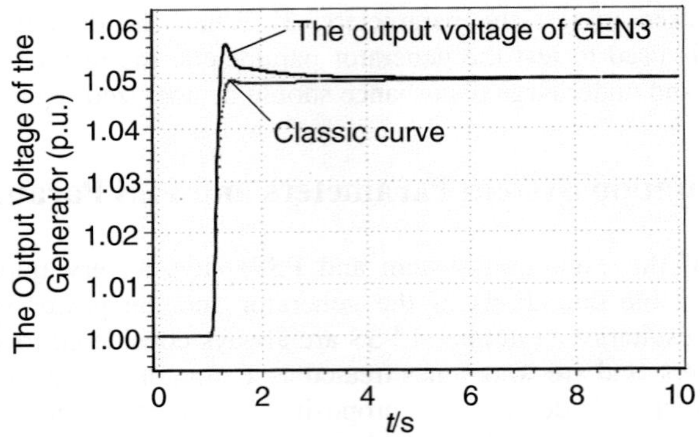

(a) The dynamic amplification coefficient of the AVR is big.

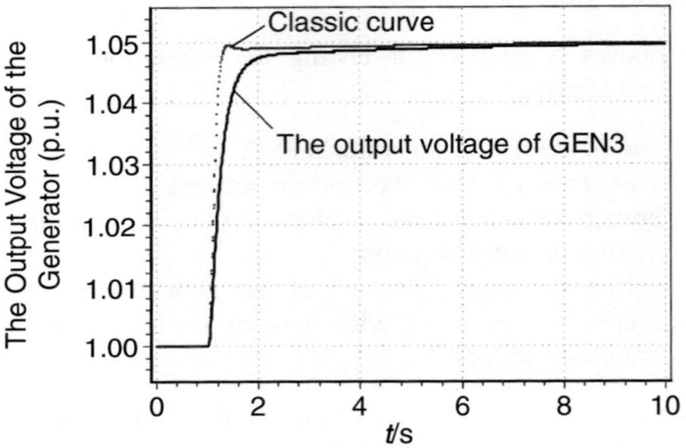

(b) The dynamic amplification coefficient of the AVR is small.

Fig. 14. The simulation testing of the excitation system.

5.2 PSS Parameters Test

In this subsection, the proposed method for testing the PSS parameters can also be fell into 4 steps:

1) set up the "double machines and double lines" simulation system as shown in Fig. 15(a) (all the parameters of GEN3 are the same as those in IEEE 9-node test system and the parameters of the other components are labeled in the figure);

2) set a three-phase permanent fault on the high-voltage bus of the transformer on one of the two transmission lines – at 0 s a three-phase short-circuit fault occurs, then at 0.1 s the breakers on both sides trip to clear the fault, but do not reclose;

3) switch on and off the PSS of GEN3 respectively;

4) compare the damping speed of the oscillation of GEN3's output power under these 2 conditions – if the damping speed of the oscillation of the output power is much faster when PSS is switched on than that when PSS is switched off, the PSS

parameters are considered to be correct and effective; otherwise, the PSS parameters would be incorrect or ineffective, and need to be remeasured by field test.

Fig. 15(b) shows the output power curves of GEN3 when PSS is switched on and switched off. From comparison, it can be inferred that the PSS parameters of GEN3 in IEEE 9-node test system are quite correct and effective, considering that when PSS is switched on the output power curve damps to a straight line much faster. Likewise, the curves when the PSS parameters are incorrect are omitted.

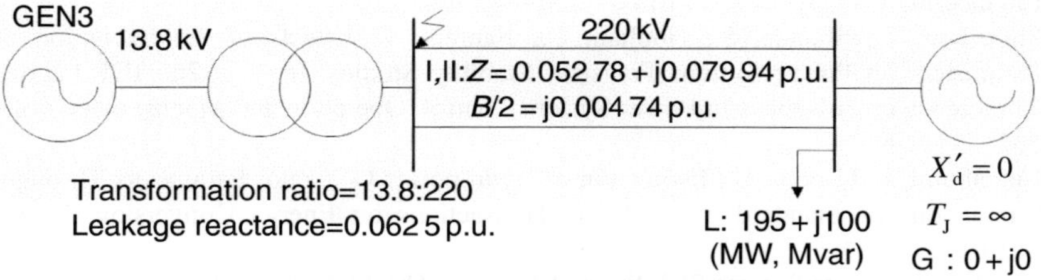

(a) Double machines and double lines simulation system.

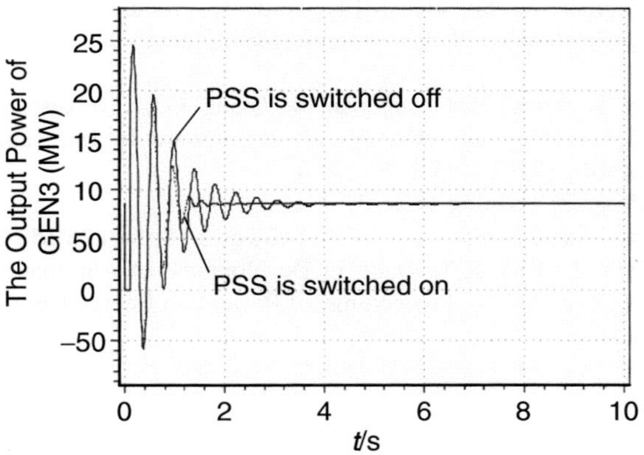

(b) The output power curve of the generator.

Fig. 15. The simulation testing of the PSS.

6 Conclusion

This paper summarizes the impact of the generator parameters (including the parameters of the generator model, excitation system and PSS) on power system's transient and dynamic stabilities from a good many simulation experiments. Since the experimenting process is closely related to the engineering practice and does not involve any complex theoretical analysis, the summarized rules and conclusions are very evident and practical, and can be directly referenced or applied by the technicians. Based on the rules and conclusions some feasible methods are proposed for testing the generator parameters quickly, which may greatly promote the working efficiencies of the technicians. Although the computational example used in this paper is very simple, the resultant conclusions are of generalization and can be easily testified by readers in their researching and working activities.

References

1. Aghamohammadi, M.R., Beik Khormizi, A., Rezaee, M.: Effect of Generator Parameters Inaccuracy on Transient Stability Performance. In: IEEE Asia-Pacific Power and Energy Engineering Conference (APPEEC), pp. 1–5 (2010)
2. Shouzhen, Z., Shande, S., Houlian, C., Jianmin, J.: Effects of the Excitation System Parameters on Power System Transient Stability Studies. In: The 2nd IET International Conference on Advances in Power System Control, Operation and Management, vol. 2, pp. 532–535 (1993)
3. Lidenholm, J., Lundin, U.: Estimation of Hydropower Generator Parameters Through Field Simulations of Standard Tests. IEEE Transactions on Energy Conversion 25, 931–939 (2010)
4. Kundur, P.: Power System Stability and Control. The McGraw-Hill Companies, Inc., New York (1994)
5. IEEE Power Engineering Society: IEEE Guide for Synchronous Generator Modeling Practices and Applications in Power System Stability Analyses. IEEE Std 1110-2002, pp. 1–72 (2003)
6. Wuzhi, Z., Xinli, S., Yong, T., Guangquan, B., Qiang, G.: New Research and Exploitation of Power System Small Signal Stability Analysis Software. In: International Conference on Electrical Engineering (ICEE), pp. 1–5 (2006)
7. PSD Software Program Training Manual, http://wenku.baidu.com/view/552a33e9856a561252d36f08.html
8. Hoong, C.S., Taib, T., Rao, K.S., Daut, I.: Development of Automatic Voltage Regulator for Synchronous Generator. In: Proceedings of Power and Energy Conference (PECon), pp. 180–184 (2004)

A New Strategy for Incremental Maintenance of Cover Quotient Cube

Peng Xiang-kai and Chen Fu-qiang

Guangdong Polytechnic Normal University,
Guangzhou, Guangdong, China
gspxk@gdin.edu.cn

Abstract. By establishing a new strategy for incremental maintenance of cover quotient cube, cover quotient cube is divided into several sub-sets, in determining whether to add or modify operation. With this approach, only some subsets of the cover quotient cube are accessed when there is an incremental maintenance operation. A new algorithm UpdateAddNew is provided with respect to the case when there is new record added to the base table. Experimental results show that the count of records needed to be accessed is only 85% of that with an approach based on full access.

Keywords: Quotient cube, Cover quotient cube, Incremental Maintenance, OLAP.

1 Overview

As we know data in the OLAP system is obtained by aggregation. When the OLTP system data is updated, the system needs the corresponding update operation. As for the data update operation, it consists of two steps, the first step is to update the dimension table and the base table, the second step is to update the data cube. This paper discusses a new approach for incremental maintenance of cover quotient cube [1]. The related research mainly includes: (1) Ki Yong Lee's refresh algorithm [2], this algorithm bases on delta cuboids, updates the number delta cuboid in the data cube, thus reducing the amount of computation and disk I/O times; (2) Cuiping Li, etc. analyze the possibly situations of the base table tuples to be updated, based on analysis of its characteristics, it impacts by type of coverage, covering categories. Implementation of the merger, split and update operations are given so as to improve the efficiency of incremental maintenance algorithm [3]. Cuiping Li and others also use Galois Lattice to expound the relationship between the set of tuples in the base table and that of in the cover quotient cube, on the basis of which puts forward such methods of incremental maintenance as distributive aggregate functions, algebraic aggregation function and holistic aggregate functions [4]; (3) Kim Yong Lee and other calculations improve the efficiency of incremental maintenance by calculating only a small part of the delta cuboids [5]; (4) Dong Jin, etc. reduce the cost of incremental maintenance through the use of extended multi-dimensional arrays, the use of its higher efficiency of random access [6].

X. Wan (Ed.): Electrical Power Systems and Computers, LNEE 99, pp. 277–281.
springerlink.com

The method for updating cover quotient cube usually includes two types: one is to regenerate the cover quotient cube, that is, the data changes in the base table to a certain extent, and regenerate the cover quotient cube. This approach has two shortcomings, firstly because the base table usually contains a large number of records, the cover quotient cube generated from the base table requires a lot of computation and disk I/O operations, so it is a larger workload, longer time consumption and requires more hardware resources to support the calculation process; Second, data changes in the base table take a long time to appear in the corresponding data cube. As the update operation will occupy part of the machine resources, thus reducing the efficiency of OLAP system. To avoid the side-effect, system administrators of the data warehouse often use the leisure time of the system, to re-obtain data from the OLTP system and update the OLAP system data. Another approach for updating the incremental maintenance that is on the basis of the original cube, by adding, modifying, updating one cell and get the new cover quotient cube, the time for updating is perhaps the leisure time, or working hours of the system. In this way, when the data within the base table of the OLAP system change, the maintenance work is relatively small. Although this method, to a certain extent, affects the ability of OLAP systems to provide services, the impact is less obvious. Thus it is often acceptable; Data changes in the OLTP system can reflected timely in the OLAP system by a series of operations, so the user can query the possibly new data from the data cube.

As can be seen from the above analysis, incremental maintenance method has low maintenance costs and enable more timely data cube to obtain update. Our study is aiming to present a new approach for incremental maintenance of cover quotient cube.

2 Research Ideas

As for incremental maintenance of cover quotient cube, the input is the original cover quotient cube and base table increment; the output is the new cover quotient cube. Here the content of the input and output both in the OLAP system is to be physically stored in the form of a disk file. Therefore, the incremental maintenance process for the cover quotient cube is to update the original disk file in order to make the disk file turn into a new disk file. In the existing method, this process is divided into three steps, the first step is to calculate the incremental base table T, that is $\triangle$T; the second step is to calculate incremental of the data cube $\triangle$DT based on $\triangle$T; the third step is to refresh the data cube. Because of the base tables and cube data files' large size, the time to maintain mainly by the time consumed in the disk I/O , so in the process of incremental maintenance, reducing disk I/O times will greatly improve the maintenance efficiency. The second step in the existing methods is searching the corresponding upper bound among the cover quotient cube, and comparing with the basic element increments of the base table to determine the need to add, delete, or update the existing upper bound, which resulting disk I/O times increase. Our idea is that the original cover cube is divided into some specific subset of the table when comparing the incremental tubles in the base table with the upper bound of the original cell, and only need to search it in some subsets, thereby reducing the disk I/O times, the second step improve the existing methods.

3 The Improved Method

The base table T (A1, Ai, An, M1, Mj, Mm), T's cube indicated by DT, T's cover quotient cube indicated by CQT, The theorem is as followed:

Theorem 1: As for base table T, suppose S1, S2 included in the CQT, the set of all the dimension attributes is Z, for Y∈Z, S1= {ub|ub∈CQT, ub[Y]for non *, ub [Z-Y] for non*}; S2= {ub|ub∈CQT, ub[Y] for non *, ub [Z-Y] for *}. If there is no ub1∈S1, ub1 [A] =a, so there is no ub2∈S2, thus ub2 [A] =a.

Proof: by contradiction. Suppose there exists ub2∈S2, thus ub2 [A] =a. As ub2 is an upper bound, and in the Z-Y for *, so there exists two different tuples t1and t2, thus t1[Z-Y]≠t2[Z-Y], at least in the Z-Y There is a dimension attribute A, thus t1[A]≠t2[A], so there are two cells c1 and c2 in DT, thus c1[Y+A]=t1[Y+A], c2[Y+A]=t2[Y+A], where UBc1 is the fine cell for the c1, it must meet the UBc1[A]≠*, thus UBc1[Z-Y]≠*, UBc1∈S1. Inconsistent with known conditions. Proved.

For example, suppose a base table T has three-dimensional properties A1, A2, A3, if there is an certain upper bound (3, *, *) in CQT, there are two different tuples A2 and A3 in T, at least one dimension attribute has a different value. Now let two tuples (3, 4, a) and (3, 5, b) exist in the T data cube, while the upper bound of (3,4, a) is 3 in A1 and 4 in A2, thus making the * at least in the upper bound of A2, A3.

We divided the above CQT into eight subsets as S111, S110, S101, S011, S100, S010, S001, S000, among of which S111 contains all the subsets of three-dimensional lattice that does not include *, S110 includes all maintenance property on the first two for not the *, and all the subset of the cell in the third dimension attributes for *,and so on. Theorem 1 shows that if the upper bound ub does not exist in the subset of S111, the value in the first dimensional attributes is a, thus it does not exist the upper bound in S110, S100, S101, the value in the first one-dimensional attributes is a, and so on. If there is no the upper bound ub in the subset S111, making the value in the first and second dimensional attributes is a, b, thus it does not exist the upper bound in S110. The value in first and second dimension attribute is a, b. Theorem 1 provides the inspiration for us with a new incremental maintenance algorithm.

When a tuple t is added to the base table, in order to update the cover quotient cube, we no longer to traverse the entire file to find the original table, what we should do is to search with more non-* subset of attributes. If we cannot find the corresponding upper bound, it shows that it does not exist the corresponding cell. For example, when a tuple (5,6,7) is added to the base table, we need to search and judge if there is exist (5,6,7), (5,6 ,*),(5, *, 7),(*, 6,7), (5 ,*,*),(*, 6 ,*),(*,*, 7) in cover quotient cube, and so on. If we can't search the ub in the S111, and making ub [A1] =5, therefore (5, *, *) isn't exist in whole cover quotient cube.

Accordingly, we focused on the base table and the proposed algorithm UpdateAddNew as follows:

Algorithm UpdateAddNew:
Input: cover quotient cube CQT, add the tuple (a, b, c)
Output: The new cover quotient cube

Steps:

Find out the upper bound in S111 (a, b, c), and to be updated, the new record was given when we did not search it.
In CQT, in the order of S111, S110, search (a, b, *);
if there is no qualified tuple in S111, stop the search. Find (a, b, *), at the same time, search(*, b, *);
// searching (a, b, *) and (*, b, *) in one time is to avoid the repeatedly traversing the same subset.
As for the combination of the subset (*, b, c), (a, *, c), find out the corresponding subset, and repeat step 2;
Return;

When deleting a single tuple (a, b, c) from the base table, you can also use a similar method. Firstly search (a, b, c) in S111, update or remove the upper bound, according to S111, S110 to find the corresponding tuple with (a, b), if the S111 will not be found eligible in the tuple, then S110 can not exist in the corresponding tuple. And so on. Update a base tuple t, you can decompose into deleting a tuple and then adding a tuple.

When adopting the star schema store data cubes, we often store all the dimension values in each dimension table. If a dimension value does not exist in the corresponding dimension table, then the corresponding upper bound of all possible lattice are not present in the original cover quotient cube, so this time, we can add the record directly, while the number of records to be traversed is 0.

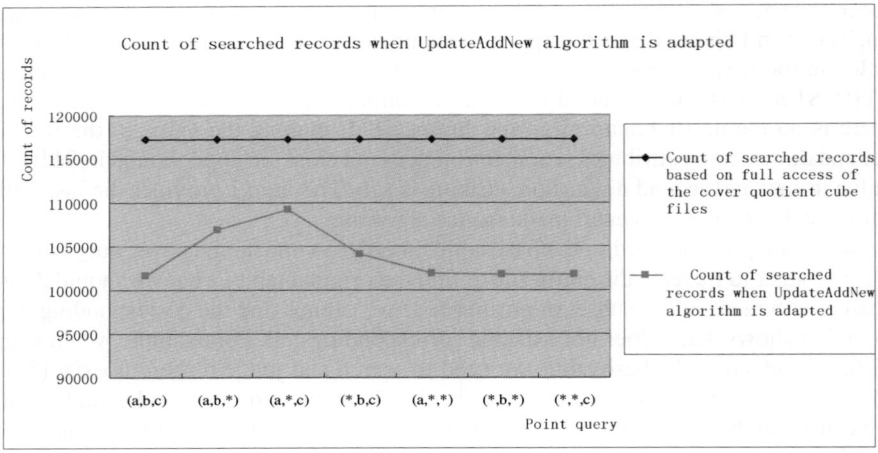

Fig. 1. Count of searched records when UpdateAddNew algorithm is adapted

4 Experimental Result

The experimental data set we use is the weather data set provided by Carbon Dioxide Information Analysis Center, U.S. Department of Energy [7], which contains weather data measured in a few decades within thousand test points. We extracted data subset containing one million records from the data set, in which the projection of the subset in the three-dimensional of the solar-altitude, longitude and present weather formed the properties of the base table, and then generates cover quotient cube of the base table. As for the new table, (a, b, c), (a, b,*), (a, *, c), (*, b, c), (a,*,*), (b,*,*), (*,*, c) to be added in the cover quotient cube, we can find out the corresponding number of records by searching, the resulting data shown in Figure 1. Experimental results show that the counts of records need to be accessed is only 85% of that with an approach based on full access.

5 Conclusion

In order to get higher efficiency for incremental maintenance of cover quotient cube, this paper puts forward an UpdateAddNew algorithm. The algorithm contains that the original cover quotient cube is divided into some specific subset of the table when comparing the increment of the basic element in the group and the upper bound of the original cell, only need to find it in some subset, thereby reducing the disk I/O times. Experimental results show that the algorithm by reducing the number of records to be traversed improve efficiency. Research work we should do in the next step is to focus on the table in the case of add, delete, modify and explore a new approach for incremental maintenance efficiency of cover quotient cube.

References

[1] Lakshmanan, L.V.S., Pei, J., Han, J.: Quotient Cube: How to Summarize the Semantics of a Data Cube. In: Proceedings of VLDB 2002, pp. 778–789 (2002)

[2] Lee, K.Y., Kim, M.-H.: Efficient Incremental Maintenance of Data Cubes. In: Proceedings of VLDB 2006, pp. 823–833 (2006)

[3] Li, C., Tung, K.-H., Wang, S.: Incremental Maintenance of Quotient Cube Based on Galois Lattice. Journal of Computer Science and Technology 19(3), 302–308 (2004)

[4] Li, C., Wang, S.: Efficient Incremental Maintenance for Distributive and Non-Distributive Aggregate Functions. Journal of Computer Science and Technology 21(1), 52–65 (2006)

[5] Lee, K.Y., Chung, Y.D., Kim, M.-H.: An efficient method for maintaining data cubes incrementally. Information Sciences: an International Journal 180(6), 928–948 (2010)

[6] Jin, D., Tsuji, T., Tsuchida, T., Higuchi, K.: An incremental maintenance scheme of data cubes. In: Haritsa, J.R., Kotagiri, R., Pudi, V. (eds.) DASFAA 2008. LNCS, vol. 4947, pp. 172–187. Springer, Heidelberg (2008)

[7] Hahn, C., et al.: Edited synoptic cloud reports from ships and land stations over the globe (DB/OL) (1982-2003), http://cdiac.ornl.gov/ftp/ndp026b

An Optimal Design for the Stackable Piezoelectric Power Generation Device and Its Rapid Energy-Saving Method

Ming-Han Hsu, Ching-Wu Wang*, and Tsun-Kai Hsu

Graduate Institute of Opto-Mechatronics
National Chung Cheng University
168 University Road, Minhsiung Township
Chiayi County 621, Taiwan
Tel.: 886-5-2722982; Fax: 886-5-2724036
melcww@ccu.edu.tw

Abstract. In this paper, the optimal stackable piezoelectric power generation device and its rapid energy-saving system was designed and fabricated. In order to increase the efficiency of conversion power from piezoelectric device, three two-layer stackable piezoelectric power generation devices stacked in different geometric structures were arranged and compared for obtaining a greater created output power. Evidence shows that the sample No.3 could generate the highest output instantaneous power (4.132mW) than both of samples No.2 (2.852μW) and No.1 (1.897μW), which acts as the optimal structural design. Finally, by utilizing optimal of two-layered stackable piezoelectric power generation module combined with the single-level buck converter rapid energy storage system fabricated in this work, the Ni-MH battery with capacitance of 100mAh could be fully-charged within two hours.

Keywords: Piezoelectric Power Generator, Stackable Piezoelectric Power Generation device, Rapid Storage, Energy-saving, and Energy Storage.

1 Introduction

Over the past few years, the development of renewable energy for different ambient energy sources has been an urgent issue. For solving the above problem, the action of creating renewable energy by wildly building the solar and wind power plants has been undoubtedly to become two major developments [1-2]. Nevertheless, the above two major efforts accordingly rely on a huge budget as well as a suitable working place to fully display their function efficiency. In recent years, the piezoelectric power generation devices having been explored by some researches were used to transform mechanical energy into electrical energy [3-9]. Unfortunately, in most of cases the output power created from piezoelectric power generation device is too small to be directly utilized. To solve such a problem, many researchers have proposed different methods such as using more efficient piezoelectric materials [10], and using different mechanical structures [11]. However, from the statement mentioned above, the

* Corresponding author.

X. Wan (Ed.): Electrical Power Systems and Computers, LNEE 99, pp. 283–290.
springerlink.com

complicate structure was limited the improvement efficiency. On the other hand, energy storage circuit has also been studied by some researchers [12]. But, many of the circuit were too high power consumption to increase the storage efficiency. Therefore, a novel of optimal stackable structure for piezoelectric power generation device to enhance its output power will be proposed. Besides, a rapid energy-saving system constituting with the optimal stackable piezoelectric power generation device will be also realized.

2 Experiments

The configuration of stackable piezoelectric power generation device combing with energy-saving system fabricated in this work was depicted in Fig.1. Among them the piezoelectric device to generate the output power by punching the machine, due to the output power is too small to use it. Hence, the output power was first through the bridge rectifier, and then stored by the temporary capacitor, finally adjusted quickly by buck converter to charge into the Ni-MH battery. In order to increase the created output power of stackable piezoelectric power generation device, three different designs of two-layered stackable piezoelectric power generation devices were consisted of piezoelectric device and cushion material as illustrated in Fig.2. The optimal two-layered stackable piezoelectric power generation device was figured out by comparing the output power of different two-layered stackable piezoelectric power generation devices. Furthermore, in order to improve the storage efficiency, three different designs of rapid energy-saving system implemented by using the IC of LTC3588-1 (Linear Technology Corp., USA) were displayed in Figs.3(a)~3(c). Finally, the optimal of two-layered stackable piezoelectric power generation device combing with rapid-saving system charge a Ni-MH battery with capacitance of 100mAh, and at the same time the charge time of three different designs of rapid energy-saving system was accurately analyzed and estimated.

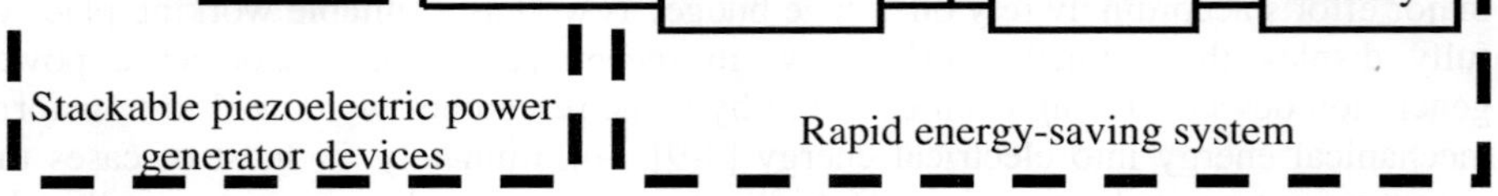

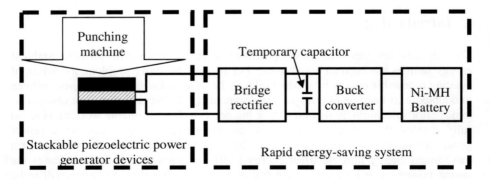

Fig. 1. The configuration of stackable piezoelectric power generation device combing with energy-saving system established in this work.

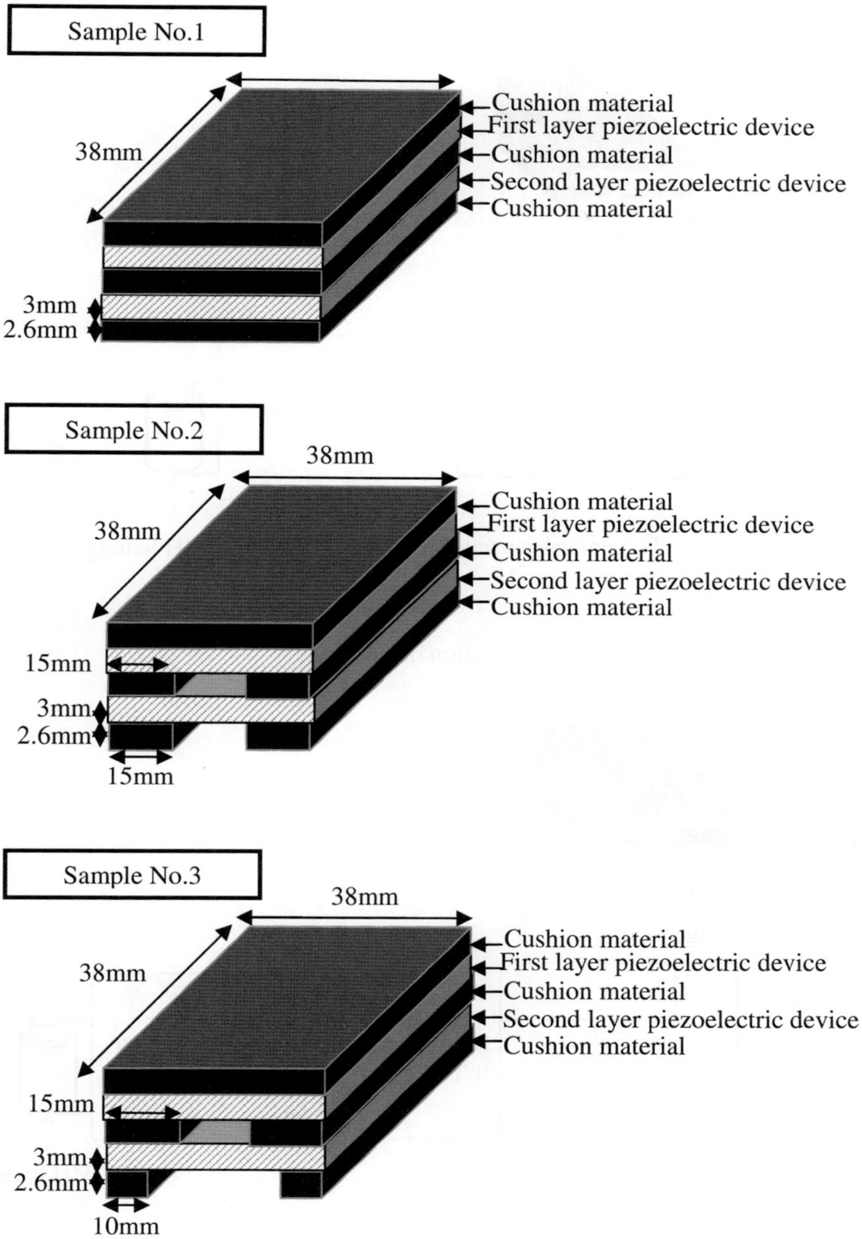

Fig. 2. Three kinds of two-layer stackable piezoelectric generation devices stacked in different geometric structures.

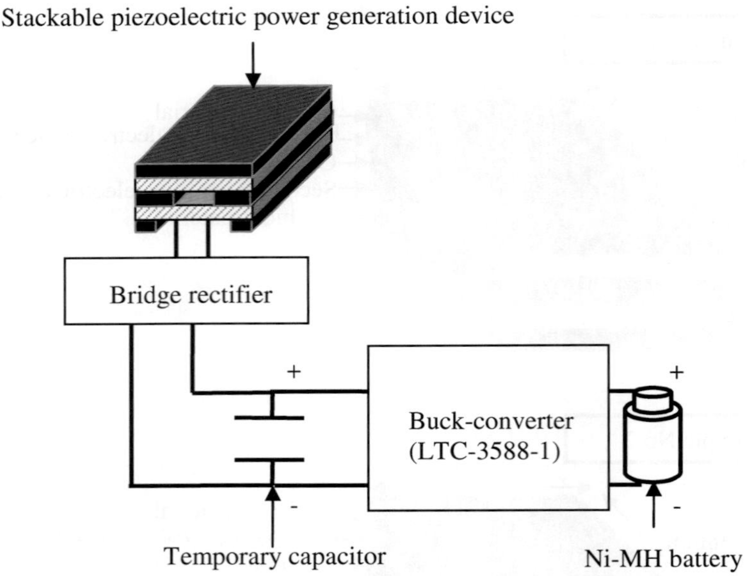

(a)

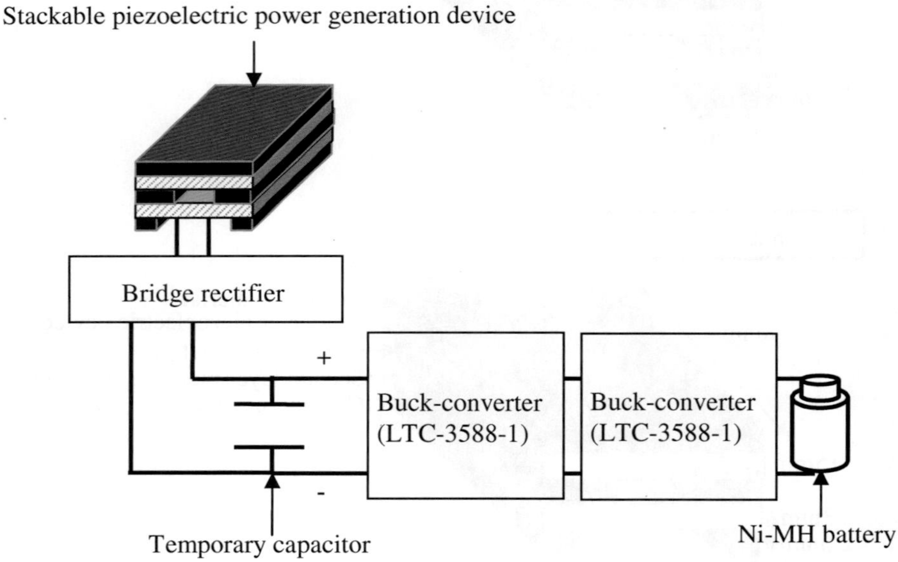

(b)

Fig. 3. The optimal stackable piezoelectric power generation device combing with rapid energy-saving system. (a): The rapid energy-saving system with single-level buck converter, (b): The rapid energy-saving system with series-level buck converter, and (c): The rapid energy-saving system with parallel-level buck converter.

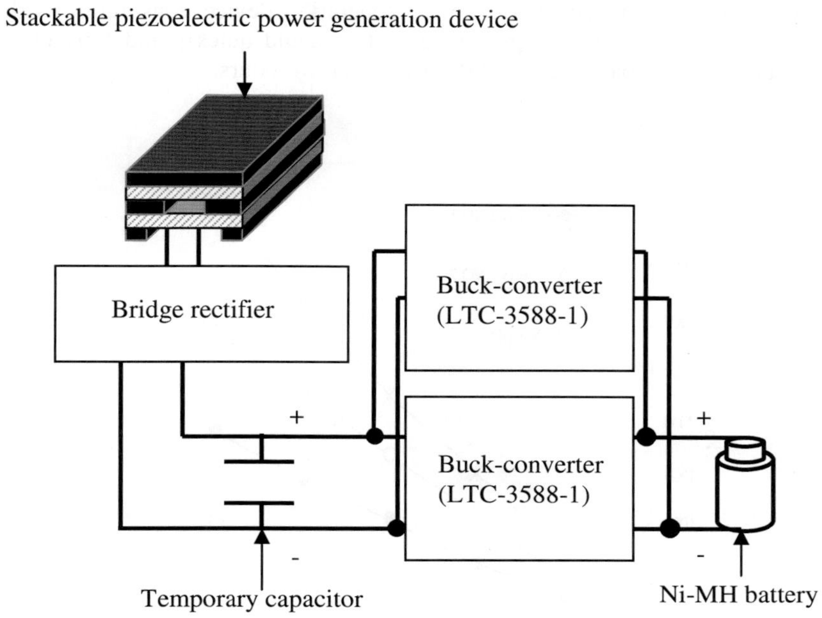

(c)

Fig. 3. *(continued)*

3 Results and Discussion

Fig. 4 shows the measurement of the output power for the two-layered stackable piezoelectric power generation devices with three different kinds of geometric structures that are under the same punching frequency but different pressure. The results indicated that the two-layered stackable piezoelectric power generation device of sample No.3 performs an optimal structure of two-layer stackable piezoelectric power generation device since sample No.3 could generate more output power (4132μW@500KgF) than that of sample No.1 (1897μW@500KgF) and sample No.2 (2852μW@500KgF). It is suggested that the piezoelectric device of sample No.3 could generate great deformation by varying the geometric structure of cushion material and thus create higher instantaneous output power. Next, charge time of three different designs of rapid energy-saving system were measured and shown in table 1. It is evident that the charge time of single-level buck converter energy-saving system was fastest than that in series-level buck converter and parallel-level buck converter. It is the fact that increasing the used number of IC (LTC3588-1) made the total power consumption rise, leading to the storage efficiency decay. Finally, the optimal of two-layered stackable piezoelectric power generation device constituting with the single-level buck converter energy-saving system could quickly and fully charge a Ni-MH battery with capacitance of 100mAh within two hours.

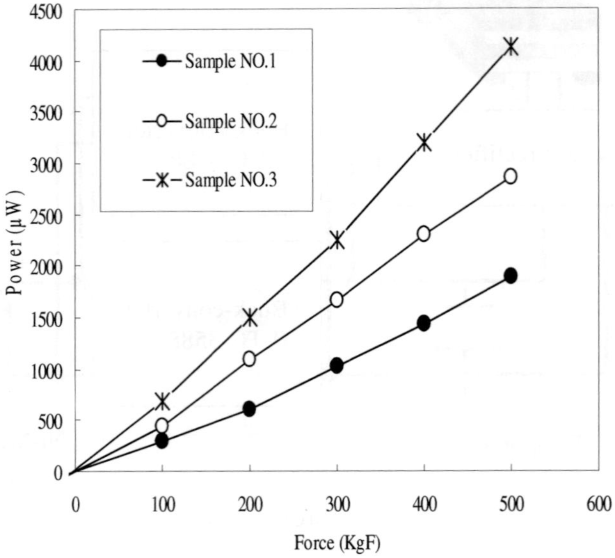

Fig. 4. The created output power by two-layer stackable piezoelectric power generation devices under the same punching frequency but different punching pressures.

Table 1. The charge time of three different energy-saving systems for Hi-MH battery.

	0.5hr	1 hr	1.5 hr	2 hr	2.5 hr
Single-level buck converter energy-saving system	26%	53%	78%	100%	-
Series-level buck converter energy-saving system	24%	38%	63%	81%	100%
Parallel-level buck converter energy-saving system	21%	42%	65%	79%	100%

4 Conclusions

In summary, it is demonstrated that the deformation of piezoelectric power generation device could be increased by varying the cushion material of geometric structure. The sample No.3 of two-layer stackable piezoelectric power generation device generated a great deformation could obviously achieve a higher instantaneous output power which acts the optimal structural. Furthermore, the rapid energy-saving system was constructed by using single-level buck converter energy-saving system. This is due to the fact that reducing the used number of IC made the total power consumption decrease, leading to the storage efficiency increasing. Finally, a Ni-MH battery with a capacitance of 100mAh could be fully-charged within two hours by utilizing optimal of two-layered stackable piezoelectric power generation device combined with the single-level buck converter energy-saving system fabricated in this work.

Acknowledgment

One of the authors (Ching-Wu Wang) gratefully acknowledges financial support from the National Science Council (NSC) in Taiwan under Contract No.: NSC99-2221-E-194-049.

References

[1] Hagedorn, K., Forgacs, C., Collins, S., Maldonado, S.: Design Considerations for Nanowire Heterojunctions in Solar Energy Conversion/Storage Applications. The Journal of Physical Chemistry C 114, 12010–12017 (2010)

[2] Leishman, J.G.: Challenges in modelling the unsteady aerodynamics of wind turbines. Wind Energy 5, 85–132 (2002)

[3] Sodano, H.A., Inman, D.J., Park, G.: Comparison of Piezoelectric Energy Harvesting Devices for Recharging Batteries. Journal of Intelligent Material Systems and Structures 16, 799–807 (2005)

[4] Hehn, T., Hagedorn, F., Manoli, Y.: Highly efficient energy extraction from piezoelectric generators. Procedia Chemistry 1, 1451–1454 (2009)

[5] Tabesh, A., Frechette, L.G.: A Low-Power Stand-Alone Adaptive Circuit for Harvesting Energy From a Piezoelectric Micropower Generator. IEEE Transactions on Industrial Electronics 57, 840–849 (2010)

[6] Guyomar, D., et al.: Synchronized switch harvesting applied to selfpowered smart systems: Piezoactive microgenerators for autonomous wireless transmitters. Sensors and Actuators 138, 151–160 (2007)

[7] Vullers, R.J.M., Schaijk, R.V., Doms, I., Hoof, C.V., Mertens, R.: Micropower energy harvesting. Solid-State Electronics 53, 684–693 (2009)

[8] Hu, Y., Xue, H., Hu, T., Hu, H.: Nonlinear Interface Between the Piezoelectric Harvesting Structure and the Modulating Circuit of an Energy Harvester with a Real Storage Battery. IEEE Transactions on Ultrasonics, Ferroelectrics, and Frequency Control 55, 148–160 (2008)

[9] Garbuio, L., Lallart, M., Guyomar, D., Richard, C., Audigier, D.: Mechanical energy harvester with ultralow threshold rectification based on SSHI nonlinear technique. IEEE Transactions on Industrial Electronics 56, 1048–1056 (2009)
[10] Anton, S.R., Sodano, H.A.: A review of power harvesting using piezoelectric materials (2003–2006). Smart Materials and Structures, 16, R1- 16, R1-R21 (2007)
[11] Wu, N., Wang, Q.: Repair of vibrating delaminated beam structures using piezoelectric patches. Smart Materials and Structures 19, 035027 (2010)
[12] D'hulst, R., Sterken, T., Puers, R., Deconinck, G., Driesen, J.: Power Processing Circuits for Piezoelectric Vibration-Based Energy Harvesters. IEEE Transactions on Industrial Electronics 57, 4170–4177 (2010)

The Improved Method for Unitary Space-Time Signal Sets Partitioning

Jin Wang

Faculty of Electrical Engineering, Tianjin University of Technology & Education, P.R. China
wangjinnn99@163.com

Abstract. It is necessary to design an optimal method of partitioning a unitary space-time (UST) signal set for trellis coded unitary space-time modulation (TC-USTM). The unique features of these UST signals have necessitated a different partitioning methodology from that of the conventional two dimensional constellations. In this letter, we suggest a well-ordered set partitioning through a novel subset-pairing strategy, for an arbitrary UST signal set. This way leads to a geometrically coherent partitioning, i.e., subsets of the same size (order) have identical intra-distance profiles. According to this partitioning, the resulting TC-USTM can get a minimum bit error probability.

1 Introduction

Unitary space-time modulation (USTM) [1], [2] has drawn increased attention for its potential in realizing high data rate transmission in a multiple-antenna wireless communication system, where the channel state information is unknown both at the transmitter and receiver. To further improve the spectrum efficiency of this non-coherent communication system, trellis-coded USTM (TC-USTM), a combination of trellis-coded modulation (TCM) [3] with USTM, has been reported in [5]-[11].

Error rate performance analysis in [5]-[11] has led to design criteria for TC-USTM operated in Rayleigh flat fading channels. However, how to systematically partition an *arbitrary* unitary space-time (UST) signal set has not been addressed. A systematic partitioning is important in that firstly we can see from the partitioning steps that an arbitrary UST signal set can satisfy the required rules. Secondly, a systematic partitioning approach renders the partitioning of a large-size signal set realizable.

In this letter, we formulate a novel systematic partitioning on UST signal sets by employing *subsetpairing*. That is, subsets in the same layer of the partitioning tree are paired to form a larger subset in the immediately higher layer, making use of a simple *integer-pairing* mechanism. Consequently, subsets of the same size have identical intra-distance profiles and the resulting TC-USTM can achieve a minimum bit error probability.

2 Set Partitioning Rules for TC-USTM

Let $\mathbf{\Phi}_L = \{\mathbf{\Phi}_l | l \in Z_L\}$ denote a UST signal set, where $Z_L = \{0, \cdots, L-1\}$. $\mathbf{\Phi}_l$ is a $T \times M\,(M \le T)$ unitary matrix, systematically formed by $\mathbf{\Phi}_l = \Theta^l \mathbf{\Phi}_0$, where $\mathbf{\Phi}_0$ is a unitary matrix and Θ is a $T \times T$ diagonal matrix with elements

X. Wan (Ed.): Electrical Power Systems and Computers, LNEE 99, pp. 291–298.
springerlink.com

$\Theta_{i,i} = e^{j2\pi u_i / L}$, $u_i \in Z_L, 1 \le i \le T$. The signal size is $L = 2^b$, where b=RT and R is the information rate in bits per symbol [1]. The distance metric between Φ_l and $\Phi_{l'}$, referred to as *distance* in this letter, is defined as $d_{l,l'} = \prod_{m=1}^{M} \left(1 - d_{l,l',m}^2\right)^{1/2M}$ [1], where $d_{l,l',m}, m = 1, M$ are singular values of the correlation matrix $\Phi_l^* \Phi_{l'}^{\ 2}$. Two properties can be easily proved for $d_{l,l'}$:

Property 1. $d_{l,l'}$ is determined by the signal *index difference*, defined as $\delta_{l,l'} = (l' - l) \bmod L$, *i.e.*, $d_{l,l+\delta} = d_{l'',l''+\delta}$ for $l \ne l'' \in Z_L$.

Property 2. $d_{l,l+\delta} = d_{l,l-\delta}$.

Definition. $Z_i^{(j)}$ and $Z_{i'}^{(j)}$ - $i \ne i'$ (correspondingly, $\varsigma_i^{(j)}$ and $\varsigma_{i'}^{(j)}$ are defined as *congruent subsets*, denoted as $Z_i^{(j)} \cong Z_{i'}^{(j)}$ (correspondingly, $\varsigma_i^{(j)} \cong \varsigma_{i'}^{(j)}$, if there exist $l \in Z_i^{(j)}$ and $l' \in Z_i^{(j)}$ such that $\Delta_i^{(j)}(l) \cong \Delta_{i'}^{(j)}(l')$.

We can infer that if $Z_i^{(j)} \cong Z_{i'}^{(j)}$, then *any* integer $a \in Z_i^{(j)}$ can be the reference index in $Z_i^{(j)}$, as correspondingly we can let $a \oplus_b \delta_{l,l'}$ as the reference in $Z_{i'}^{(j)}$. Therefore, $Z_{i'}^{(j)}$ is a "rotated" version of $Z_i^{(j)}$ by an integer $\delta_{l,l'}$. For example, in Fig. 2, $Z_1^{(3)} = \{1,5,9,13\}$ and accordingly, $\Delta_1^{(3)}(1) = \{4,8,12\}$, which is the same as that for $Z_0^{(3)} = \{0,4,8,12\}$. Therefore $Z_0^{(3)} \cong Z_1^{(3)}$ and $Z_0^{(3)} = Z_1^{(3)} \oplus_4 1$[3].

The average symbol error rate in the congruent subsets are identical. In fact, denoting $p(\Phi_l \to \Phi_{l'} | \Phi_l)$ the pairwise symbol error probability of mistaking Φ_l for $\Phi_{l'}$ when Φ_l is transmitted, we can evaluate the average symbol error rate $P_{symbol}(i)$ in subset $\varsigma_i^{(j)}$ by

$$P_{symbol}(i) = \sum_{l \in Z_i^{(j)}} \sum_{l' \in Z_i^{(j)}, l \ne l'} p(\Phi_l \to \Phi_{l'} | \Phi_l) \psi$$

$$= \frac{1}{2^{b-j+1}} \left\{ \sum_{\delta_{l,l'} \in \Delta_i^{(j)}(l)} p\left(\Phi_l \to \Phi_{l \oplus_b \delta_{l,l'}} \middle| \Phi_l\right) + \right.$$

$$\left. \sum_{\delta_{l,l''} \in \Delta_i^{(j)}(l)} \sum_{\delta_{l,l'} \in \Delta_i^{(j)}(l), \delta_{l,l'} \ne \delta_{l,l''}} p\left(\Phi_{l \oplus_b \delta_{l,l''}} \to \Phi_{l \oplus_b \delta_{l,l'}} \middle| \Phi_{l \oplus_b \delta_{l,l''}}\right) \right\} \quad (1)$$

Evidently $P_{symbol}(i)$ depends only on an arbitrary $l \in Z_i^{(j)}$ and the $\Delta_i^{(j)}(l)$ associated with it. As $p(\Phi_l \to \Phi_{l'} | \Phi_l)$ is only determined by the index difference $\delta_{l,l'}$ at high SNR [1], $P_{symbol}(i)$ are all equal for different *i*.

For TC-USTM with 2^{b-j+1} parallel paths, if signals in $\varsigma_i^{(j)}$ are assigned to these paths, the overall symbol error probability will be dominated by $P_{symbol}(i)$ and therefore be easily evaluated. Hence the following rule is expected to hold.

Rule 1. In layer-ψ/$1 \le j \le b$ ψ$Z_0^{(j)} \cong \cdots \cong Z_{2^{j-1}-1}^{(j)}$.

We let $A \oplus_b l$ denote the set formed by the modulo-2^b addition between each element in integer set – and an integer ψ

From [6], [9] the pairwise error event probability (PEP) P_{event} of deciding in favor of sequence $\{\hat{S}_0, \hat{S}_1, \cdots, \hat{S}_t, \cdots\}$ when $\{S_0, S_1, \cdots, S_t, \cdots\}, S_t, \hat{S}_t \in \Phi_L$ is transmitted, is upper bounded by

$$P_{event} \le \left(\frac{1}{2^l} \left(\frac{\rho T}{4M} \right)^{-MNl} \right) \cdot \left(\prod_{t \in \eta} d_t \right)^{-2MN} \tag{2}$$

where l is the *length* of the error event, *i.e.*, the number of *t*'s for which $S_t \ne \hat{S}_t$ψ and η is the set of these *t*'s. ρ is the signal to noise ratio (SNR) at each receiver antenna. d_t is the distance between S_t and $\hat{S}_t$.

Rule 2. For TC-USTM with 2^{b-j+1}, $2 \le j \le b$ parallel paths, $d_{\min}^{(j)}$ for the smallest subsets $\varsigma_i^{(j)}$ should be maximized and meanwhile $d_{\min}^{(j)} \ge d_{\min}^{(j-1)} \ge \cdots \ge d_{\min}^{(1)}$, and the values in this chain should decrease as slowly as possible.

3 A Systematic Set Partitioning for UST Signal Sets

In this section, we first introduce a systematic approach of partitioning Z_L, satisfying only Rule 1. We introduce Lemma 1(1.1 and 1.2) that defines two *basic* operations (Operation I and II) for partitioning a generic integer set $\mathcal{S}$ into congruent size-2 subsets. Then we introduce Lemma 2 (2.1 and 2.2) that proves that if these two operations are performed successively on $\mathcal{S}$, regardless of the sequence, the resulting subsets in the same layer would all be congruent. We summarize this finding in Theorem 1. After incorporating a search scheme satisfying Rule 2, we give a detailed proposition for an optimal set partitioning satisfying both Rule 1 and 2 at the end of this section.

Let us consider an integer group $S=\{0,2^p,2^p\cdot 2,\cdots,2^p\cdot(2^{B-p}-1)\}=2^p Z_{2^{B-p}}$ with group operation $\oplus_B$, where $0\le p\le B-1$. By choosing an appropriate pair of p and B, *any* integer group with size given as a power of 2 can be expressed as $\mathcal{S}$. Let $\Delta_S=\{2^p,2^p\cdot 2,\cdots,2^p\cdot(2^{B-p}-1)\}$ denote the integer difference set for $\mathcal{S}$, regardless of the reference integer $s\in S$. We introduce the following two lemmas that demonstrate a systematic approach to form the size-2 congruent subsets in $\mathcal{S}$, with an *arbitrary* integer $\Delta\in\Delta_S$ as the identical difference. Let $\delta=\dfrac{\Delta}{2^p}\in Z_{2^B-p}$ be the *normalized difference*, in the sense that δ is an element in the *continuous* integer group Z_{2^B-p}.

Lemma 1.1. For an odd integer δ, $\{2^{p+1}i,2^{p+1}i\oplus_B\Delta\}$, $i\in Z_{2^B-p-1}$ form the congruent size-2 subsets in $\mathcal{S}$.

Lemma 1.2. For an even integer δ, $\{2^{p+p'+1}i\oplus_B m,2^{p+p'+1}i\oplus_B\Delta\oplus_B m\}$, $i\in Z_{2^{B-p-p'-1}}$, $m\in 2^p Z_{2^{p'}}$ form the congruent size-2 subsets in $\mathcal{S}$, where $p',1\le p'\le B-p-1$, is chosen such that $\dfrac{\delta}{2^{p'}}$ is an odd integer.

Lemma 2.1. Given $S^{(k)}$, $k\ge 0$ and the resulting $R^{(k)}$ by Operation I, with a $\Delta\in\Delta_{S^{(k)}}$. Set $S^{(k+1)}\leftarrow R^{(k)}$. Then the congruent size-2^n generally $n\ge 1$ subsets in $S^{(k+1)}$ give rise to congruent size-2^{n+1} subsets in $S^{(k)}$.

Lemma 2.2. Given $S^{(k)}$, $k\ge 0$ and the resulting $R^{(k)}$ by Operation II, with a $\Delta\in\Delta_{S^{(k)}}$. Suppose K operations have been applied in subsequent subset-pairing and $R^{(k+K)}$ is the *redefined* reference set associated with this Operation II. Set $S^{(k+K+1)}\leftarrow R^{(k+K)}$. Then congruent size-$2^n$ generally $n\ge 1$ subsets $S^{(k+K+1)}$ in give rise to congruent size-subsets 2^{K+n+1} in $S^{(k)}$.

Theorem 1. Given S and $S^{(1)},S^{(2)},\cdots,S^{(k)}$, $k\ge 1$ which is a series of the generic forms on which Operation I or II has been applied. Then congruent size-2 subsets in $S^{(k)}$ lead to congruent subsets in *S*.

Proposition 1. The subset-pairing for an arbitrary for TC-USTM:

1) (*Initialization*) Set $k\leftarrow 0$, $S^{(k)}\leftarrow Z_L$, $\oplus\leftarrow\oplus_b$;

2) (*Subset-pairing*) Obtain $\Delta^* \in \Delta_{S^{(k)}}$ through (4) and determine the corresponding δ. Depending on δ, employ Operation I or Operation II based on $\oplus$ to form congruent size-2 subsets in $S^{(k)}$, resulting in $R^{(k)}$;

3) If $R^{(k)} = \{0\}$, go to step 4); otherwise set $k \leftarrow k+1$, $S^{(k)} \leftarrow R^{(k-1)}$ and go to step 2);

4) (*Redefinition*) If $|\varsigma(0)| = L$, go to step 5); otherwise redefine $R^{(k)}$ and determine Q for $\oplus_Q$ in the redefined $R^{(k)}$. Set $k \leftarrow k+1$, $S^{(k)} \leftarrow R^{(k-1)}$, $\oplus \leftarrow \oplus_Q$, then go to step 2);

5) (*Termination*) The subset-pairing procedure terminates.

We note that the above proposition is only appropriate for TC-USTM with two or without parallel paths. In the case of TC-USTM with 2^{b-j+1}, $j \le b-1$, *i.e.*, at least 4 parallel paths, Proposition 1 should be modified to satisfy Rule 2. Otherwise, we note that the search for an optimal Δ^* for layer-*j*, $j \le b-1$ is under the constraint that $d_{\min}^{(b)}$ for layer-*b* is maximized in priori. In other words, the searching space for the optimal Δ^* in layer-*j* is reduced and the true maximal $d_{\min}^{(j)}$ may be missed. To avoid this, we employ an exhaustive search for $d_{\min}^{(j)}$ by examining all the possible combinations of *congruent* size-$2^{(b-j+1)}$ subsets for layer-*j*. Then the reference set for the resulting size-$2^{(b-j+1)}$ subsets is assigned to $S^{(k)}$ in step 1) of Proposition 1.

4 Examples and Numerical Results

We now proceed to show how the partitioning tree for Φ_{16} $(T=4, M=2, R=1)$ is formed in Fig. 2 according to the Proposition. For brevity, we record every operation, and the data associated with it, in Table I. Note that the table was recorded in the sequence from the bottom row to the top row, in accordance with the subset-pairing sequence as indicated by the arrows in Fig. 2. In Table I, we start from layer-5, which comprises trivial size-1 subsets with $d_{\min}^{(5)}$ defined as ∞. This layer is not included in Fig. 2 for brevity. Once again, we want to point out that in layer-2, the 2 size-8 subsets would otherwise be $\{0,2,4,\cdots,14\}$ and $\{1,3,5,\cdots,15\}$ if following the partitioning method for the 16-ary PSK signal set.

The above partitioning results can be used for the TC-USTM of $(T=4, M=2, R=\frac{3}{4})$ with a $b/(b+1)$ encoder and we choose the trellis diagram as shown in Fig. 3 (a). For comparisons, we also consider two TC-USTM realizations with non-optimal set partitioning. In *Case* 1, signals in layer-5 are paired optimally $(\Delta^* = 8)$, however subsets in layer-4 are non-optimally paired by letting $\Delta = 2 \neq \Delta^* = 4$. The subsequent subset-pairing follows Proposition 1 and the trellis diagram is given in Fig. 3(b). In *Case* 2, we let $\Delta = 1 \neq \Delta^* = 8$ in layer-1 and the resulting trellis diagram is shown in Fig. 3(c).

In Fig. 4 we can see that TC-USTM with the optimal set partitioning has more than 7dB coding gain over Case 2 in high SNR. Compared with Case 1, it has only $\leq 1dB$ gain in moderate SNR and performs almost the same at high SNR. Here we see that $d_{\min}^{(4)}$ for size-2 subsets plays a more important role than $d_{\min}^{(j)}$, $j \leq 3$ when there are 2 parallel paths, which justifies that the two signals with the largest distance should be paired to form the size-2 subsets. The slight BER loss of Case 2 at moderate SNR and its high performance gain over Case 1 also justify the second part of Rule 2, *i.e.*, apart from $d_{\min}^{(4)}$, minimum distance $d_{\min}^{(3)}$ and $d_{\min}^{(2)}$ should also be maximized. In fact, in moderate SNR, longer-length error events instead of the shortest error events can be dominant. Therefore by the PEP upper bound in (2), Rule 2 can help maximize the *distance product* term in the second bracket.

The BER lower bound can be evaluated by the formulae given in [9]. At high SNR, bit errors associated with the shortest error event dominate the overall BER performance and we find that the simulation and the analytical results agree well. As TC-USTM with the optimal set partitioning and that in Case 1 have the same shortest error events, the lower bounds for both cases are identical.

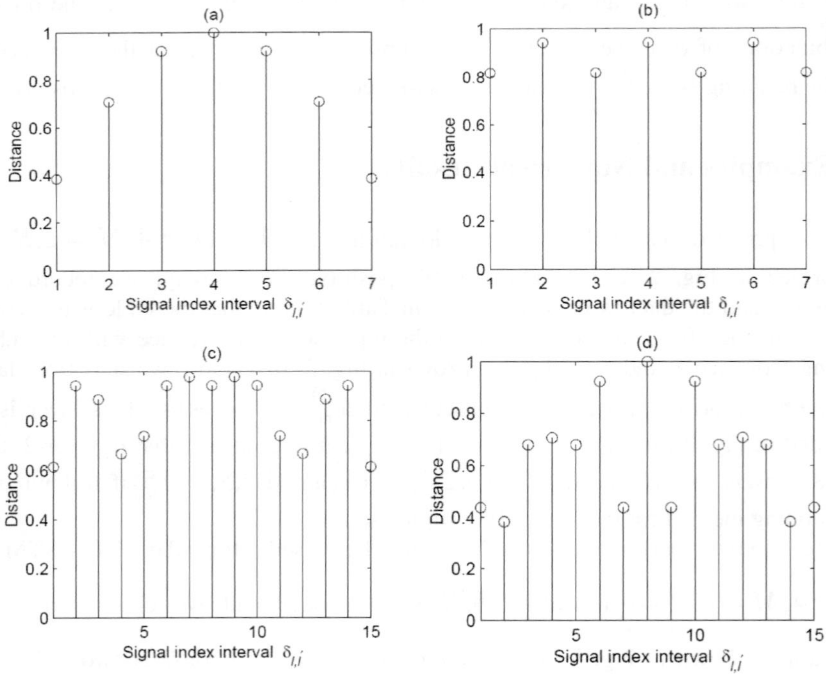

Fig. 1. Distance profiles P for four UST signal sets. (a) $\Phi_8(T=2, M=1, R=1.5)$ (b) $\Phi_8(T=3, M=1, R=1)$ (c) $\Phi_{16}(T=3, M=1, R=1.33)$ (d) $\Phi_{16}(T=4, M=2, R=1)$

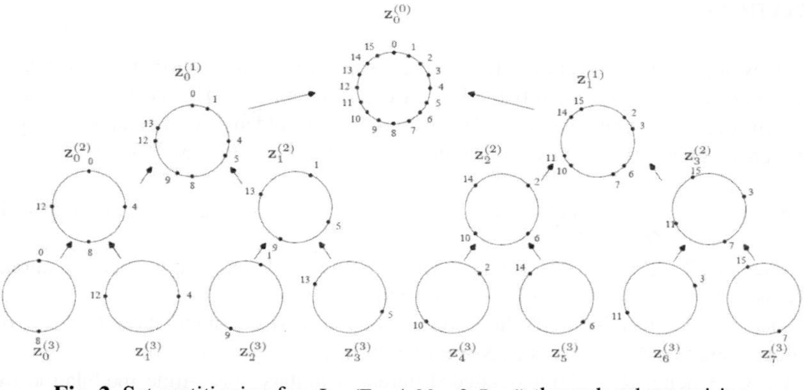

Fig. 2. Set partitioning for $\Phi_{16}(T=4, M=2, R=1)$ through subset-pairing

$\{0,\ 8\}\{4,12\}\{1,\ 9\}\{5,13\}$ $\{0,\ 8\}\{2,10\}\{1,\ 9\}\{3,11\}$
$\{2,10\}\{6,14\}\{3,11\}\{7,15\}$ $\{4,12\}\{6,14\}\{5,13\}\{7,15\}$
$\{5,13\}\{1,\ 9\}\{4,12\}\{0,\ 8\}$ $\{3,11\}\{1,\ 9\}\{2,10\}\{0,\ 8\}$
$\{7,15\}\{3,11\}\{6,14\}\{2,10\}$ $\{7,15\}\{5,13\}\{6,14\}\{4,12\}$
(a) (b)

$\{0,\ 1\}\ \{8,\ 9\}\ \{4,\ 5\}\ \{12,13\}$
$\{2,\ 3\}\ \{10,11\}\{6,\ 7\}\ \{14,15\}$
$\{12,13\}\{4,\ 5\}\ \{8,\ 9\}\ \{0,\ 1\}$
$\{14,15\}\{6,\ 7\}\ \{10,11\}\{2,\ 3\}$
(c)

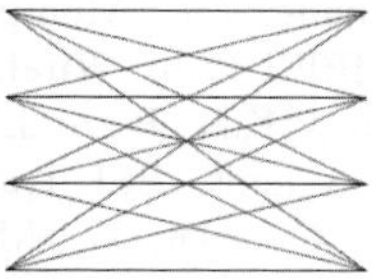

Fig. 3. 4-state trellis diagrams for TC-USTM employing $\Phi_{16}(T=4, M=2, R=1)$. Mapping is based on (a) optimal set partitioning (b) non-optimal set partitioning (Case 1). (c) non-optimal set partitioning (Case 2)

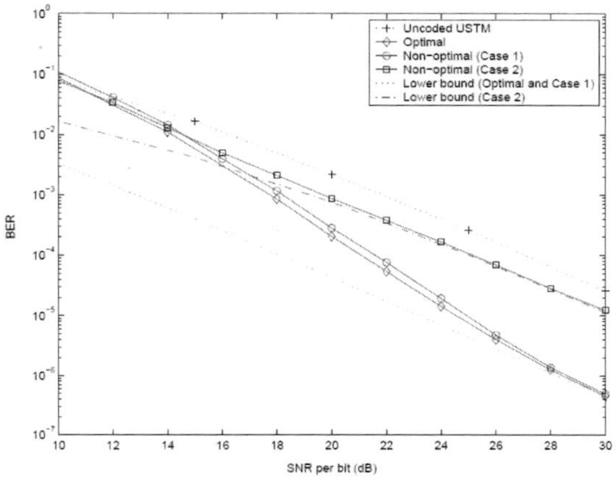

Fig. 4. BER comparison between TC-USTM $(T=4, M=2, R=0.75)$ with optimal set partitioning and non-optimal set partitioning

References

1. Hochwald, B.M., Marzetta, T.L.: Unitary space-time modulation for multiple-antenna communications in Rayleigh flat fading. IEEE Trans. Inform. Theory 46, 543–564 (2000)
2. Hochwald, B.M., Marzetta, T.L., Richardson, T.J., Sweldens, W., Urbanke, R.: Systematic design of unitary space-time constellations. IEEE Trans. Inform. Theory 46, 1962–1973 (2000)
3. Ungerboeck, G.: Channel coding with multilevel/phase signal. IEEE Trans. Inform. Theory 28, 55–66 (1982)
4. Forney, G.D.: Coset codes - Part I: Introduction and geometrical classification. IEEE Trans. Inform. Theory 34, 1123–1151 (1988)
5. Bahceci, I., Duman, T.M.: Trellis coded unitary space-time modulation. In: Proc. of IEEE GlobeCom, San Antonio,Texas, November 2001, pp. 1108–1112 (2001)
6. Tao, M., Cheng, R.S.: Trellis-coded differential unitary space-time modulation over flat fading channels. IEEE Trans. Commun. 51, 587–596 (2003)
7. Tao, M.: High rate trellis coded differential unitary space-time modulation via Super Unitarity. IEEE Trans. Wireless Commun. 5, 3350–3354 (2005)
8. Sun, Z., Tjhung, T.T.: Trellis-coded unitary space-time modulation. In: Proc. of 3G Wireless Conf., San Francisco, USA, May 2002, pp. 196–201 (2002)
9. Sun, Z., Tjhung, T.T.: On performance analysis and design criteria for trellis coded unitary space-time modulation. IEEE Commun. Letters 7, 156–158 (2003)
10. Sun, Z., Tjhung, T.T.: Multiple trellis coded unitary space-time modulation in Rayleigh flat fading. IEEE Trans. Wireless Commun. 3, 2335–2344 (2004)
11. Wu, Y., Lau, V.K.N., Pätzold, M.: Constellation deisgn for trellis coded unitary space-time modulation systems. IEEE Trans. Commun. 54, 1948–1959 (2006)
12. Biglieri, E., Divsalar, D., McLane, P.J., Simon, M.K.: Introduction to Trellis-Coded Modulation with Applications, Maxwell MacMillanInt. Editions, New York (1991)

The Application of Complex Torque Coefficient Method in Multi-generator System

Kun Xu, Chengyong Zhao, Jie Zhang, and Chunlin Guo

Key Laboratory of Power System Protection and Dynamic Security Monitoring and Control of Ministry of Education,
North China Electric Power University, Beijing Changping, 102206

Abstract. A multi-input and multi-output linear model of multi-machine power system was established based on perturbation analysis, and the complex torque coefficient of target generator was deduced in detail. From a mathematical perspective, the meaning of complex torque coefficient was explained. Meanwhile, the reasonability about non-target generators being equivalent to different models was discussed when investigating subsynchornous oscillation phenomenon in a multi-generator system. The results showed that the complex torque coefficient method was suitable for multi-generator system and the procedure of simplifying non-target generator as single rigid body or voltage sources with fixed frequency was inappropriate. Then a case study of IEEE second benchmark verified the conclusions above.

Keywords: complex torque coefficient, subsynchronous oscillation, multi-generator system.

1 Introduction

The complex torque coefficient method[1] is a Swiss scholar I. M. Canay proposed in 1982 in the research of Subsynchronous Oscillation(SSO) phenomenon. This method analyzes the shaft movement from electrical subsystem and mechanical subsystem, when investigating SSO phenomenon. By calculating the electrical complex torque coefficient and the mechanical complex torque coefficient, the method analysis the net damping of target generator in the vicinity of the natural frequencies to evaluate the risk of SSO of target generator. This method avoided the eigenvalue analysis, but as for theoretical basis of complex torque coefficient method and its applicability have not been approved yet.

It was considered that the complex torque coefficient method was only applicable to single generator infinite bus system, not to multi-generator system in literature [5]. However, in literature [6,10], the method was applied to multi-generator systems. In literature [5,7], the method was applied to multi-generator system by simplifying non-target generator as voltage source with fixed frequency. In this paper, the complex torque coefficient of target generator was deduced in detail, and the reasonability of substituting non-target generator with single rigid body or voltage source was analyzed. Finally, a case study based on IEEE Second Benchmark Model verified the results of analysis.

X. Wan (Ed.): Electrical Power Systems and Computers, LNEE 99, pp. 299–306.
springerlink.com

2 Basic Principles of the Complex Torque Coefficient Method

The Complex torque coefficient method is based on small disturbance analysis, when the rotor exhibit small oscillations of $\Delta\delta$, the mechanical torque and electromagnetic torque response can be expressed as:

$$\Delta T_m = K_m \Delta\delta + D_m \Delta\omega \quad (1)$$

$$\Delta T_e = K_e \Delta\delta + D_e \Delta\omega \quad (2)$$

The above formula contains two items, as is synchronous torque and damping torque respectively, Where K_m is the mechanical spring coefficient, D_m for the mechanical damping coefficient, K_e for an electrical spring coefficient, D_e for the electrical damping coefficient, And unit of $\Delta\delta$ is rad, units of other variables are p.u. K_m , D_m , K_e , D_e are variables in frequency domain. In the vicinity of natural frequency f_0 , if

$$D_m + D_e < 0 \quad (3)$$

suited, the system torsional vibration would be instability at the vicinity, and SSO can experience. Instead, the system torsional vibration is stable, no risk of SSO.

3 Deduction of Complex Torque Coefficient in Multi-generator System

When investigating SSO risk in a multi-generator system, the whole system can be considered as a liner, continuous time dynamic system, which can be described by state space model. After linearized the multi-generator system at its quiescent operating point, rotor load angle input vector is taken as $\Delta\delta = [\Delta\delta_1 \ \ ... \ \ \Delta\delta_n]^T$, generators' electromagnetic response for the output vector $\Delta T_E = [\Delta T_{e1} \ \ ... \ \ \Delta T_{en}]^T$. State vector X contains rotor load angles, winding currents, and other parameters. The state space based model of whole system is

$$\Delta\dot{X} = A\Delta X + B\Delta\delta \quad (4)$$

$$\Delta T_E = C\Delta X + D\Delta\delta \quad (5)$$

Where A, B, C, D, respectively, the system matrix, input matrix, output matrix and connecting matrix. In order to achieve complex torque expression, transform above formula into Laplace domain,

$$\Delta T_E = [C\frac{(sI - A)^*}{|sI - A|}B + D]\Delta\delta = G(s)\Delta\delta \quad (6)$$

Where I is the identity matrix, $|sI - A|$ the characteristic polynomial and $(sI - A)^*$ for the adjoint matrix. According to formula (6), the multi-generator system is equivalent to a Multi-Input Multi-Output(MIMO) linear system. When generator j is selected as study objects (namely target generator), How to establish the transfer function from input $\Delta\dot{\delta}_j$ to output $\Delta\dot{T}_{ej}$ with non-target generators equivalent to different models was discussed below.

1) Multi-mass shaft model of non-target generator with windings electromagnetic transient accounted.

Assuming that the multi-generator system contains n generators, and the *j*th generator oscillate with $\Delta\dot{\delta}_j = e^{j\xi\omega_0 t}$. To the *i*th non-target generator, there exists transfer function $(K_{mi} + j\xi D_{mi})^{-1}$ from output ΔT_{ei} to input $\Delta\delta_i$, when accounting its mechanical response characteristic (1), as is illustrated in fig 1. In this way, the ranks corresponding to non-target generator in formula (6) are eliminated. Thus figure1.becomes an single input and single output linear system, and it is easy to analyze the transfer function between electromagnetic torque $\Delta\dot{T}_{ej}$ and $\Delta\dot{\delta}_j$ of target generator. Which is written as:

$$\Delta\dot{T}_{ej} = K_{Ej}(\xi)\Delta\dot{\delta}_j \tag{7}$$

$K_{Ej}(\xi)$ denotes complex torque coefficient, and we can deduced complex torque coefficient of every generator similarly.

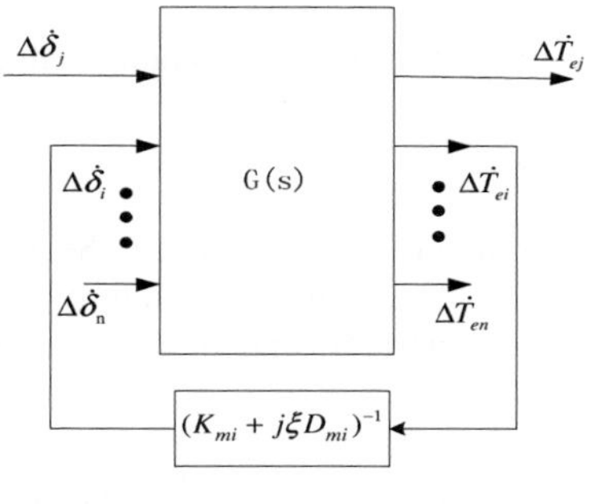

Fig. 1. Torque response diagram with shaft dynamic characteristic accounted

Because of the torsional modes of non-target generator shaft, $K_{Ej}(\xi)$ contains poles corresponding to mode frequencies. Meanwhile, the frequency response characteristic curve of $K_{Ej}(\xi)$ would also be affected by these poles. These poles play a key role in evaluate system SSO risk, which should not be overlooked. In addition, it is not difficult to analyze that the load angle deviation $\Delta\dot{\delta}_i$ of non-target generator is a linear representation of $\Delta\dot{\delta}_j$, as is

$$\Delta\dot{\delta}_i = f(\Delta\dot{\delta}_j) \tag{8}$$

2) Single rigid body model of non-target generator with windings electromagnetic transient accounted.

Under this circumstance, the analyzing process is similar to the above. Because non-target generator's shaft was simplified as a single rigid body, the transfer function $(K_{mi} + j\xi D_{mi})^{-1}$ no more reflects any torsional mode of shaft. So this equivalent model is not suitable for SSO investigation.

3) Voltage source model of non-target generator without accounting windings electromagnetic transient.

In literature [6,7], non-target generators were equivalent to voltage sources with fix frequencies, when applying complex torque coefficient method in multi-generator system. This procedure is equivalent to set the non-target generator input to be zero, i.e. $\Delta\dot{\delta}_i = 0$ and set the target generator load angle to oscillate by $\Delta\dot{\delta}_j = e^{j\xi\omega_0 t}$, as is illustrated in figure 2. Because of ignoring the mechanical characteristic and windings electromagnetic of non-target generator, this analyzing procedure can hardly yield an proper result.

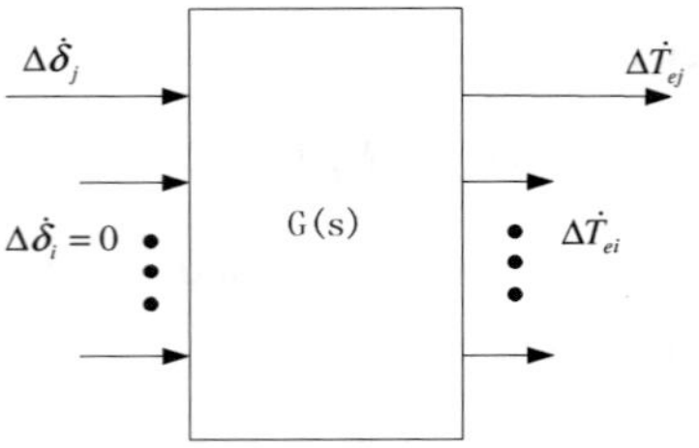

Fig. 2. Torque response diagram without accounting shaft dynamic characteristic

4 A Case Study

A case study was given based on case 2 of IEEE Second Benchmark Model[3] to verify the above discussion. The wiring structure is illustrated in Figure 3 respectively, the parameters of generators and shafts were detailed in literature [3].

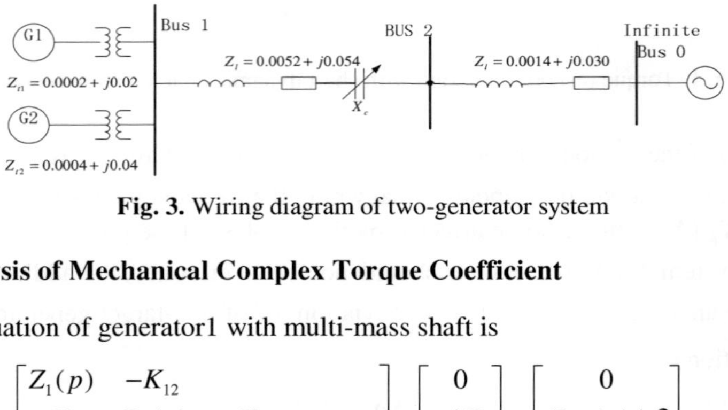

Fig. 3. Wiring diagram of two-generator system

4.1 Analysis of Mechanical Complex Torque Coefficient

Motion equation of generator1 with multi-mass shaft is

$$\begin{bmatrix} Z_1(p) & -K_{12} & & \\ -K_{12} & Z_2(p) & -K_{23} & \\ & -K_{23} & Z_3(p) & -K_{34} \\ & & -K_{34} & Z_4(p) \end{bmatrix} = \begin{bmatrix} 0 \\ -\Delta T_e \\ 0 \\ 0 \end{bmatrix} = \begin{bmatrix} 0 \\ K_{1M}(p)\Delta\delta \\ 0 \\ 0 \end{bmatrix} \quad (9)$$

Where $Z_i(p)=2H_ip^2+D_ip+K_{i-1,i}+K_{i,i+1}$, inertia constant H_i , self-damping constants D_i of mass i, and spring constants $K_{i,i+1}$ between mass i and mass i+1are in p.u. And p denotes Complex frequency domain operator.

We solve the equation (9) to get the mechanical complex torque coefficient:

$$K_{1M}(p)=-\frac{K_{12}^2}{Z_1(p)}+Z_2(p)-\frac{K_{23}^2}{Z_3(p)-\frac{K_{34}^2}{Z_4(p)}} \tag{10}$$

Similarly ,we can get the mechanical complex torque coefficient of generator 2:

$$K_{2M}(p)=-\frac{K_{12}^2}{Z_1(p)}+Z_2(p)-\frac{K_{23}^2}{Z_3(p)} \tag{11}$$

Setting $p=j\xi$, and making frequency scanning of equ (10)(11), it can get that unit1's natural torsional frequencies are 24.65Hz, 32.39Hz, 51.10Hz, and unit2's natural torsional frequencies are 24.65Hz, 44.99Hz. the two shaft shares the same torsional mode of 24.65Hz.

4.2 Analysis of Electrical Complex Torque Coefficient

Complex torque coefficient method is based on perturbation analysis, and for this paper, the mathematical model of system elements are in complex frequency domain. The mathematical model of generator in DQ reference takes damping windings and stator windings transient into account, detailed in literature [1], and transmission line model are in XY synchronous reference, which takes line inductance, capacitor transient characteristics into account and details are given in literature [9].

Based on elements' model, It's not hard to get input and output equation of two-generator system:

$$\begin{bmatrix}\Delta T_{e1}\\ \Delta T_{e2}\end{bmatrix}=\begin{bmatrix}ke11 & ke12\\ ke21 & ke22\end{bmatrix}\begin{bmatrix}\Delta\delta_1\\ \Delta\delta_2\end{bmatrix} \tag{12}$$

When generator 1 was selected as target generator oscillating by $\Delta\delta_1$, the torque response of generator 2 is

$$\Delta T_{e2}=-\Delta T_{m2}=-K_{M2}\Delta\delta_2 \tag{13}$$

By solving (12)(13), it gets that

$$\Delta T_{e1}=\left(ke11-\frac{ke12*ke21}{ke22+K_{M2}}\right)\Delta\delta_1=K_{e1}\Delta\delta_1 \tag{14}$$

Similarly K_{e2} is deduced. K_{e1}, and K_{e2} are the electrical complex torque coefficient of generators.

4.3 Simulation and Analysis

Built system model in PSCAD/EMTDC simulation platform as shown in figure 3. The compensation level of the transmission line between bus2 and bus 1 is set to 30%, generator 1 active power output P_1=60MW and generator 2 active power output P_2=70MW.

When small disturbance occurs at bus 1, the time domain simulation results are illustrated in figure 4. It can be observed that load angle and torsional torque of generator 1 is stable. Select generator 1 as the target-generator, and its complex torque characteristic was shown in figure 5. Curve De1, De2, De3 are corresponding to Non-target generator 2 using multi-mass model, using single rigid body model , and using voltage source model respectively. It is clear that only De1 reflects the common torsional mode of 24.65Hz. In terms of undamping, De1 shows a more serious undamping at the range from 25Hz to 50Hz, and De2 and De3 are relatively more conservative. But all of the three curves indicate a stable torsional oscillation, as is consistence with simulation results.

When the transmission line compensation level is set at 60%, and active power outputs unchanged with previous. the simulation results and damping characteristic of generator 1 are illustrated in figure 6 and 7 respectively. The meanings of De1, De2, De3 are the same with the above. According to simulation results, the generator 1 is torsional instable, but complex torque coefficient analysis results shows that only De1 reflect SSO risk at the vicinity of common torsional mode frequency(i.e. 24.65Hz). However, from De2 and De3 we yield an incorrect conclusion that target generator is risk free. The false result comes from ignoring the influence of non-target generator's mechanical characteristic.

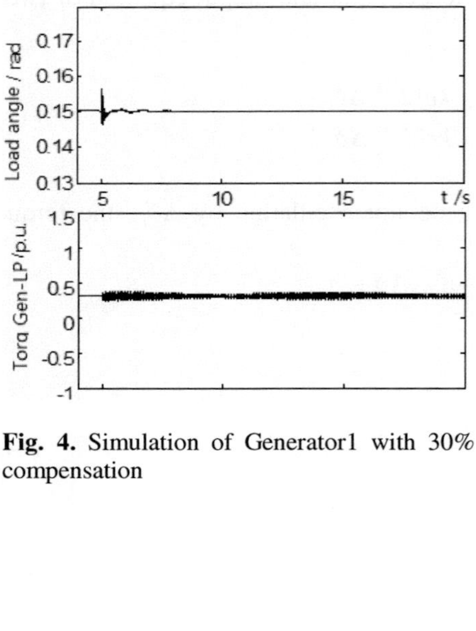

Fig. 4. Simulation of Generator1 with 30% compensation

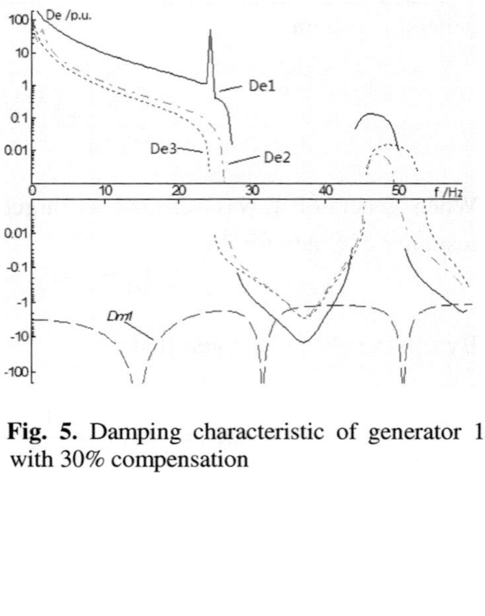

Fig. 5. Damping characteristic of generator 1 with 30% compensation

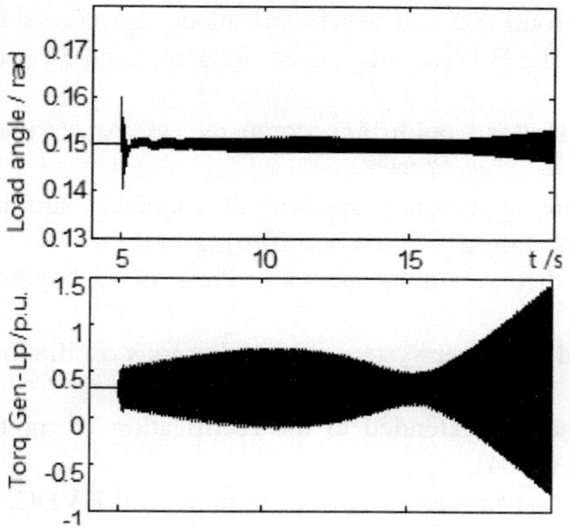

Fig. 6. Simulation of Generator1 with 60% compensation

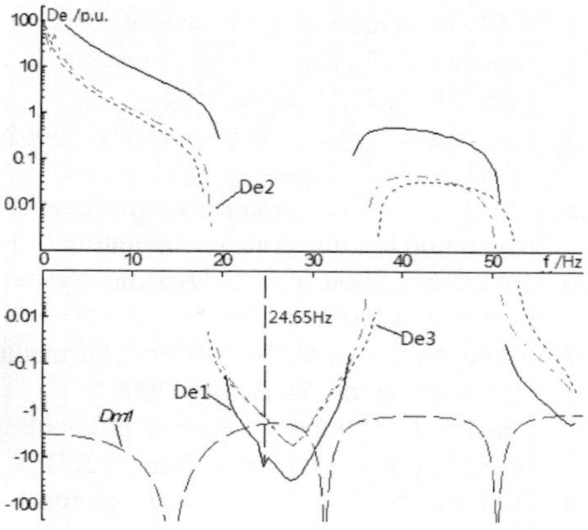

Fig. 7. Damping characteristic of generator 1 with 60% compensation

5 Conclusion

Complex torque coefficient is deduced specifically and the influences of different generator models are compared in this paper. Finally a case study on IEEE second benchmark model verifies these discussions. And the conclusions are:

1) A multi-generator system can be regarded as a multi-input and multi-output linear system based on perturbation analysis. When disturbance $\Delta\delta_i$ applied to target generator *j*, the load angle deviation $\Delta\dot{\delta}_i$ of non-target generator is a linear representation of $\Delta\dot{\delta}_j$, as is $\Delta\dot{\delta}_i = f(\Delta\dot{\delta}_j)$.
2) Any generator's electromagnetic torque increment can always be a linear representation of itself load angle increment, and complex torque coefficient method is suitable for SSO investigation in multi-generator system.
3) When evaluate the risk of SSO in a multi-generator system, the mechanical characteristic has unnegligible influence on the damping characteristic of target generator, especially the share common torsional modes. Thus accurate model of non-target generator should be retained.

References

1. Canay, I.M.: A novel approach to the torsional interaction and electrical damping of the synchronous machine, Part I and Part II. J. IEEE Trans. on Power Apparatus and System 101(10), 3630–3647 (1982)
2. IEEE subsynchronous resonance working group. First benchmark model for computer simulation of Subsynchronous resonance. J. IEEE Trans. on Power Apparatus and System 96(5),1565–1572 (1977)

3. IEEE subsynchronous resonance working group. Second benchmark model for computer simulation of subsynchronous resonance. J. IEEE Trans. on Power Apparatus and System PAS-104(5),1057–1066 (1985)
4. Pai, M.A.: Power system stability. North Holland publishing company, M. Amsterdam (1981)
5. Zheng, X.: The complex torque coefficient approach s applicability analysis and its realization by time domain simulation. J. Proceedings of the Csee 20(6) (2000)
6. Yixin, N., Shousui, C.: Dynamic system theory and its analysis. Press of Qstinghua, M.Beijing (2002)
7. Shijie, C., Yijia, C.: The theory and method of power system subsynchornous oscillation. Press of Science, M. Beijng (2009)
8. Fang, L.J.: Complex torque coefficients method extended to the rectification of multi-machine power systems. J. Relay 32(8), 5–8 (2004)
9. Junyong, W., Shijie, C.: A new approach to ssr problem in power system with HVDC. J. Automation of electric power systems 11(1), 20–25 (1996)
10. Yixin, N., Yanchun, W.: A Study of HVDC-Caused Subsynchronous Oscillations in Multimachine System. J. Proceedings of the Csee 13(2), 64–71 (1993)
11. Yixin, N., Yanchun, W.: The frequency scanning——complex torque ceofficient analysis of hvdc-caused shaft torsional oscillations in multimachine system. J Proceedings of Electric Power System and Automation 3(2), 44–55 (1991)

Design and Implementation of Cultural Relic Image Storage Based on SVG

Shen Hong and Ji Quan-zhi

Institute of Information Technology, Beijing Union University,
NO. 97 North Four Crossing West Road, Zhaoyang Distract, 100101
Beijing, China
shenhonghong@hotmail.com

Abstract. This paper proposes design and implementation method of SVG -based cultural relic image storage by each SVG entity modularized and structured. Among of which include: design on database and model of cultural relic image, SVG document framework, design and realization of combination and splitting algorithm, generation and processing methods of cultural relic image information. In the end, an overall description will be given through the relevant example.

Keywords: SVG, cultural relic image, database, storage, algorithm.

1 Overview on the Storage Design and Implementation Method of Cultural Relic Image

Database support is an important part in SVG technology applications. For the storage and display system of relic image, SVG relic image dynamically generated in the Web page is more conducive to human-computer interaction. However, the dynamic interface is not solved by a simply SVG document, but for more space to support the relic data. The problem of the speed cannot be resolved simply using the features or advantages of SVG documents while interacting with the users, and update and image processing cannot be demanded in time. Therefore, it normally takes a combination of a large number of SVG entities to complete. For the storage and display system of relic image, adopting the document form is certainly not an ideal solution for data management, but requires database support. How to save SVG document to the database which is the necessary precondition for the completion of system functions.

In the storage and display system of SVG-based relic image, using the relational database storage system which is not only save the graphic elements and attributes of relic image into the database, but so do the culture relic data that realize the saving and taking union between the data and image. [1, 2] In this way, the culture relic data is not only reused, but meets various queries. On one hand, through the use of relic data for dynamic database publishing can greatly reduced the document image data; on the other hand, by using directly generation of SVG documents, displaying the data can release the efficiency problem of graphics data publishing in the Web environment.

X. Wan (Ed.): Electrical Power Systems and Computers, LNEE 99, pp. 307–314.
springerlink.com

2 Design of Cultural Relic Image Database

2.1 Design on Graph-Element Library of Cultural Relic

When the relic image is being processed, due to the existence of vector graphics, bitmaps, text and other information, the use of SVG can be a good representation of this information. According to relic information and the features of SVG, storage model of relic image information based on the relational database will be established [3], mainly SVG graphics library and the SVG graph-element library.

(1) Storage of relic graph-element based on SVG
SVG graph-element storage is a part of cultural relic image storage, for the data of the relic images generate a a large number of basic SVG graphics file that consist of graph-elements, on the basis of which complex SVG graphics file constitute the new one by calling the basic SVG graphics. So the basic SVG graphics memory is very important.

Analyzing SVG graph-element need storage data, mainly the basic parameters of graph-element, such as graph-element identification, the coordinates, type, and line width, fill color and so on. However, the properties of each graph-element is different, if the uniform construction of the table is taken, the column properties of table will be designed so cumbersome that it is not conducive to search data. Therefore, we often select the table design by analyzing the parameters features of graph-element and some common properties these graph-element have, such as ID, graphic type, fill color, and a number of their own unique attributes, such as the origin coordinates of the circle radius. SVG graph-element will be stored in accordance with the common attributes, reducing the properties of a single table column, while the files are stored more clearly for the common properties of SVG graphics, it also can be stored in tables. The basic design of SVG graphics memory as follows:

To establish a graph-element form with common characteristics, the table attributes include: ID, type, fill color, line color and line width and so on. In the SVG document, which is realized by such mark command as <line>, <plotline>, <recto>, <polygon>, <circle>, <ellipse>, <path>, <text> and <g>etc. Summarize and classify as a relational table: MG (command, id, start, shape, color, event, meta-graph data, other), among of which, MG is the name of relational table that contains eight attributes: name of the command, logo id name, location coordinates, shape properties, color property, event attributes, meta-data and other properties.

(2) The complex SVG relic image storage
It is well known that the complicated picture is combined by changing the basic ones that consists of graph-element. According to the composition of SVG graphic entity and data representation, the storage ways include four relational tables. Among them, the main table CBG {id, start, shape, event, GDA, transform, other}, describes the basic information of SVG complex graphics entities, mainly include graphic identity id, position information, shape information, events, interactive information, graphics, reference data collection parameter address information GDA, graphic entity transformation parameters, and other information.

The role for graphic data collection citing the parameters address information GDA is to save the relational table of SVG complex graphical data and its related parameters.

as the main components of complex SVG graphics data is more complex, and it can not used a field to complete, hence according to graph-element reference type, saving graph-element and citing data will be completed by the three relational tables that is MG1, MG2 and MG3, using the GDA field and graph-element to cite data relational tables to associate.

The storage of SVG graphical entities is generally consisted of four relational tables. One master list and three attached lists, the master list will be cited parameters address information GDA fields through graphical data collection and associated with the attached list.

2.2 Design of Relic Graphic Library

In the SVG-based relic image storage and display system, you can sort out the SVG document, and split it into a SVG graphic entity, graph-elements, create SVG graphics library. Through the establishment of the standard SVG graphics library system, the expansion of SVG markers, a large number of SVG vector graphics file will be stored in the same library according to the decomposition of graph-element library and graphics library, for which it can reduce the size of SVG graphics, share the graphics library, and save time when a number of cultural relic vector graphic files transmitted on the network. So it is beneficial for a large number of SVG graphics file management [4] [[5] [6] [7].

According to SVG Standard [8] , we can see that a SVG file consists of the following components:

(1) XML header;
(2) SVG name space format;
(3) SVG frame format;
(4) SVG document name, description, retrieve information and global properties etc.
(5) SVG graph-element information;
(6) SVG entity graphic object.

Upon each SVG document, Part (1) is absolutely the same with Part (2), Part (3) is basically the same; for a dedicated graphics library, Part (3) may use the uniform format when designing the SVG graphics library , Part (5) only exist the SVG document, in a single SVG file the contents of this part belongs to the public information of the various graphic entity, but for the entire SVG relic image storage and display system, it is public graph-element library, and which is the most frequent information sharing in the system and also the core of the graphics library. Part (4) and Part (6) is not the same, but for a dedicated graphics library, Part (4) can be removed, Part 6 is the specific performance of each SVG document graphics, you can create SVG graphics library.

According to statistics [9], for a simple SVG document, (1), (2) and (3) accounted for about half the size of the entire document, a complex SVG document accounted for about 10%. For example, before the document of a graph remove Part (1), (2), (3), the file occupies 1198 bytes, when it remove them, it only takes a 890 bytes, which means SVG format occupies 298 bytes. For the vector graphics like SVG file, removing some of the SVG format, you can reduce to about 294 × n bytes. Therefore, stored Part (1),

(2), (3) in the public graph-element library, it is only read the data when each calls, which reducing the problem of reuse the public data, saving data storage space, greatly improving the efficiency of the system.

The design process of SVG graphics library is the process of decomposing SVG documents into the SVG framework document, SVG graph-element library and SVG graphics library, as shown in Figure 1, where SVG graph-element library is designed with SVG graph-element storage design, while the SVG graphics library with complex graphics storage design.

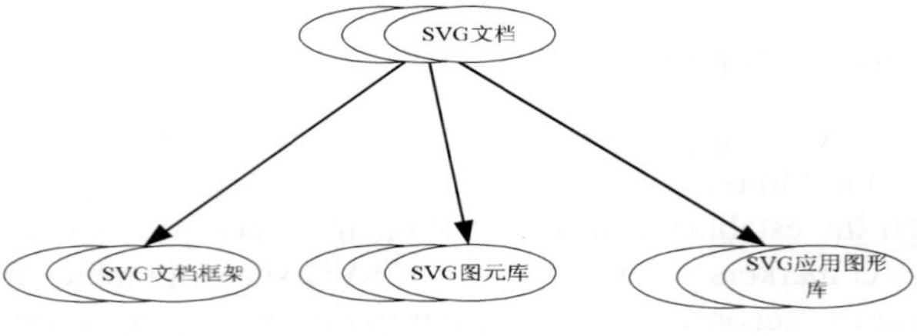

Fig. 1. SVG document decomposition diagram

2.3 Design and Realization of Combination and Splitting Algorithm on SVG-Based Cultural Relic Image

Input: SVG relic graphics

Output: SVG document framework, SVG graph-element library and SVG graphics library

Steps:

Step 1, use document structure method to deal with the SVG graphics library.

Step2, cleaning relic SVG graphics data, removing SVG graphics part (1) (2) (3), and preserve SVG graphics Part (4) information, create SVG file frameworks;

Step3, save each graph-element in the part (5) of SVG relic graphics into the SVG public graph-element library, when saving graph-elements, first check the public graph-element library, for example, the contents of Part (5) exists in the library, then it is no longer saved, while establishing the framework document, graph-element storage location information will be preserved in the SVG file frame;

Step 4, save each entity graphic in the part (5) of SVG relic graphics to the SVG graphics library, while preserving graphics entity, first inquiries on public graphics libraries, for example, if the contents of Part (6) exists in the library, then it is no longer saved, while establishing the framework document, storage location information of entity graphic will be preserved in the SVG file frame;

Step 5, output SVG document framework;

Step 6 Repeat Step 1 ~ 6 steps until all of the SVG relic graphics store in the library.

Split the SVG cultural relic graphics in accordance with the steps of split algorithm, and then save the contents of the corresponding to the system database and generate SVG document framework.

2.4 Design and Implementation of Recombination and Display Algorithm for SVG-Based Relic Image

When we display SVG document in the image storage and display system, it need to restore from the SVG document framework, SVG graph-element library and SVG graphics library. Reconstruction algorithm of SVG document is as follows:

Input: SVG document framework, SVG graph-element library and SVG graphics library

Output: SVG relic graphics

Specific steps are as follows:

Step 1, create SVG relic graphics, automatically add the first, second and third part information to SVG relic graphics;

Step 2, read SVG document framework, add the fourth part of the contents to SVG relic graphics

Step 3, read location information of each graph-element from SVG document framework, and then search for the graph-element library according to location information, combine the information of the graph-element library into graph-element and add to the SVG relic graphics.

Step 4, read location information of entity graphic from SVG document framework, and then search for the graphic library according to the graphic location information, combine the information of the graphic library into SVG entity graphic and add to the SVG relic graphics,

Step 5, output SVG relic graphics,

Step 6, If all the information on SVG relic graphics in SVG document framework is dealt with, then that is the end. Otherwise, repeat Step 1 ~ Step 5.

According to the steps of reconstruction algorithm, SVG relic graphics add the first, second and third part information, and then read out specific information from the corresponding graph-element library, combine graph-element information and compose SVG entity graphic and add to the SVG files.

Implementation process is as follows:

```
   objStreamWriter = File.CreateText(FILENAME)  // create SVG file
 sql = "select * from SVG file STA"   // read the 1、 2、 3 part information from the public
    information list
      dr = MG(sql)  // open database
      dr.Read()     // read database
      Do
        b = dr(1).ToString
        objStreamWriter.WriteLine(b)  // write the information in SVG file
      Loop While (dr.Read())
      conn.Close()
       For i = 1 To n
         Search_MG(Mid(GDA_MG(i), 1, 3), Right(GDA_MG(i), Len(GDA_MG(i)) - 4))  //
transfer graph-element information and find function, and combine them
```

```
Next i
    objStreamWriter.WriteLine("</g>")
    objStreamWriter.WriteLine("</svg>")
    objStreamWriter.Close()
```

2.5 Generation and Processing of Relic Image Information

Firstly, regarding the generation of SVG files, XmlWriter abstract method is being used in our system.

XmlWriter class is used to serialize XML documents that achieves the generation of XML documents by

Overriding each property and method. The main idea of this process is:

```
Define a temporary SVG, give the document path and file name
Write in various element,
objStreamWriter.WriteLine ("<? Xml version=""1.0""?>")
..................
conn.Open() // open the database
dr = sqlcom.ExecuteReader() // read the data
objStreamWriter.WriteLine(dr) // write the data in the temporary document
conn.Close() // close the database
```

Through the above steps, SVG relic graphics will be generated within the specified path. To display it, just embed SVG file in the HTML by <embed>, as shown in Figure 2.

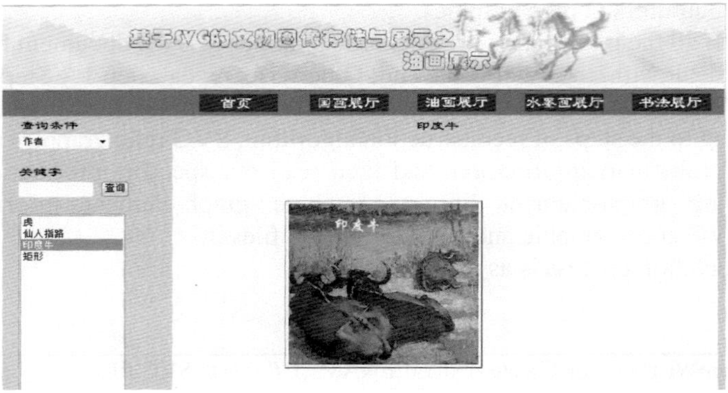

Fig. 2. SVG relic image display

2.6 Storage Examples

It is often introduced the background of relic works, title, author, and other related information in the cultural relic image storage and display system, such as the oil painting " Lion chart" by Chinese famous artist Xu Beihong, as shown in Figure 4-5, A lot of marks need to add in this painting so that the readers can have a comprehensive understanding of the works.

In figure 4-5, the Lions chart is divided into five parts. The shape of a lion, work background, age, author profiles, and works name were respectively introduced. In order to avoid boring to the readers, this five-part must be shown step by step through the mouse, the data is clearly involved in the Lion Chart which contains the lion image, the rectangular text display and other information. The way shown in figure 4-5 is clearly not possible to realize in the traditional web model, only can be fully realized in the SVG-based database. The specific storage is as follows:

(1) As a bitmap image need a large storage space, in order to improve the storage efficiency, we can decompose [10] as shown in Figure 4-6. The Lions chart is fully exhibited by using complex SVG entity graphics memory: the identity id, location, sub graph identification(multiple sub-diagrams; separated), other retrieve information, specific storage: x = "0" y = " 0 ", Lion 01; Lion 02; Lion 03; Lion 04; Lion 05;

(2) In the first part of the Lion chart, 01 is bitmap image, in general, the bitmap can be stored as a whole, saved as jpg format, all the jpg files classified according to the picture and stored in the directory, such as: C: \ Users \ butterfly \ Documents \ Visual Studio 2011 \ Websites \ WebSite5 \ pic \ shizi.jpg, In order to fast retrieval, establish a database table is needed for saving the key Information of the Lions 01, such as: bitmap id, shape attributes, graph-element reference address , events, and other information.

(3) Inputting the second, third and fourth part of the graph need to use the SVG entities to describe, and save SVG graph-elements, specifically including, graphics, id, location information, graphics rendering commands, text messages and so on. Among of which, Part 3 is similar with Part 4. After saved Part 3, Part 4 can be generated from Part 3 only by saving the transformation parameters.

(4) In the Part 5 of the "lion chart", name information can be stored in tables.

Figure 4-5 the lion shape is finally generated by the Web programming technologies (such as ASP.NET). The generated SVG code is organized as follows:

```
<? Xml version="1.0" encoding="UTF-8”?>
<!DOCTYPE svg PUBLIC "-//W3C//DTD SVG 20010904//EN"
"http://www.w3.org/TR/2001/REC-SVG-20010904/DTD/svg10.dtd" >
<svg width="100%" height="100%" id="id1" viewbox="0 0 300 300" onload
="initialize(evt)" id2="main" xmlns="http://www.w3.org/2000/svg"
 <g transform="translate(140,80)">
<script type="text/ecmascript">
… …   … …
 </script>
<image x="0" y="0" width="459" height="469"
xlink:href="C:\Users\butterfly\Documents\VisualStudio2008\WebSites\WebSite5\pic\shizi\b
g.jpg"/>
<image x="1” y="446.5” width="22" height="20" xlink: href=
"C:\Users\butterfly\Documents\VisualStudio2008\WebSites\WebSite5\pic\shizi\zhang.jpg"
……/>
<rect x="0" y="0"width="460" height="470"fill="none" stroke="blue" stroke-width="3"/>
… …   … …
… …   … …
<text x="40" y="80"style="fill: black;" font-size="28" font-family= "STXingkai">狮子
</text>
</g>
</svg>
```

References

1. Wen, J., Li, Y.: Design and Implementation Based on XML - SVG Space Database. Computer Engineering and Applicatio (18), 169–175 (2005)
2. Gan, Z., Li, Z., Peng, B.: Data Description Model Based on SVG Vector Graphics Editing System. Computer Engineering and Design 26(1), 270–273 (2005)
3. Yuan, J.: Research on Scalable Vector Graphics (SVG) Data Representation. Doctoral Dissertation, 2 (2008)
4. Wang, J., Chen, J., Qu, Z.: Design and Implementation Based on SVG Power Figure Yuan Library. Relay 36(8), 79–82 (2008)
5. Zhao, X., Zhang, Y., Zhu, Z.: The SVG Graphics Library Research and Design Based On RDBMS. Computer Engineering and Design 30(1), 225–227 (2009)
6. Yang, X., Cui, W.: The Research Based on Database Comprehensive and SVG Network Mapping Integrated Systems. Computer Application 29(1), 201–204 (2009)

Contextual Information Guided Image Categorization Algorithm

Hong Shen*, Tian_Gong Li, and Zhen-heng Zhang

Institute of Information Technology
Beijing Union University
Beijing, China
shenhonghong@126.com

Abstract. This paper proposes a method for scene categorization by integrating region contextual information into the popular Bag-of-Visual-Words approach. The Bag-of-Visual-Words approach describes an image as a bag of discrete visual words, where the frequency distributions of these words are used for image categorization. However, the traditional visual words suffer from the problem when faced these patches with similar appearances but distinct semantic concepts. This paper introduces an improved contextual CRF model to learn each visual word simultaneously depending on itself and the rest of the visual words in the same region. The experimental results on the three well-known datasets show that region contextual visual words indeed improves categorization performance compared to traditional visual words.

Keywords: Image categorization, Conditional Random Fields, Bag of visual words.

1 Introduction

This paper investigates scene categorization, which focuses on the task of assigning images to predefined categories. For example, an image may be categorized as a coast, office or street scene. Scene classification is an important problem for computer vision, and has received considerable attention in the recent past. A scene classification system can be broadly divided into two modules. The first defines the image representation, while the second delineates the classifier used for decision making. Early efforts at scene classification targeted binary problems, such as distinguishing indoor from outdoor scenes [2], etc. Subsequent researches mainly paid much attention to modeling a scene using the global statistical information, which was inspired by the literature on human perception. Oliva et al. [1] proposed a low dimensional global features to represent scenes. More recently, there have been some efforts to solve the problem in greater generality. One particular successful representation method was Bag-of-Visual-Words (BOVW), which relied on local

---

* Hong Shen, Vice Professor. Researcher of Information technology Institute Beijing Union University, Chief Director of Information Technology College Beijing University.

X. Wan (Ed.): Electrical Power Systems and Computers, LNEE 99, pp. 315–322.
springerlink.com

interest-points/regions descriptors and represented an image as the orderless collection of discrete visual words. As a simple but relative effective method, the clustering approach has shown its advantage in the visual words construction. However, the neglect of the relationship among patches in the image space makes it inevitably sacrifice certain discriminative capability. To address this problem, we combine the region contextual information to learn the visual words. The motive for this work inspired by the result of some psychophysical experiments that the constituent parts of a region do not exist in isolation, and the visual context (i.e., the spatial dependencies between parts) can be used to improve visual words definition.

To this end, this paper introduces the Conditional Random Fields to construct the Visual Words. Since this model simultaneously considers two factors, it can construct more semantic meaningful visual words representation. The model is adapted from Conditional Random Field [12] by incorporating the region information. This paper considers the contextual information among patches to reduce the limitation factors. Furthermore, in the aspect of context information, we utilize the region contextual relationship. The region contextual information has shown the more robust than absolute spatial layout with respect to partial occlusion, clutters, and changes in viewpoint and illumination. The frequency of these visual words in an image forms a histogram which is subsequently used in a scene categorization task via Support Vector Machine (SVM) classifier. The experimental results on the three well-known datasets show that the region contextual visual words indeed improve categorization performance compared to the traditional visual words.

The rest of the paper is organized as follows. The next Section introduces the Conditional Random Fields in the visual words construction process. We show the performance of our method on three datasets in Section 3. Finally, Section 4 concludes the paper.

2 The Proposed Method

Since the performance of Bag-of-Visual-Words method depends in a fundamental way on the visual words, the problem of effective design of these visual words has been gaining increasing attention in recent literature [5-8]. In general, thousands of visual words are used to achieve better performance. However, they may contain a large amount of information redundancy. Therefore, the researchers have attempted to find a more compact representation. Fei-fei et al. [16] and Bosch et al. [5] have respectively applied Latent Dirichlet Allocation (LDA) or probabilistic Latent Semantic Analysis (pLSA) model [6] to discover latent semantic concepts beyond the BOVW. Liu et al. [7] utilized Maximization of Mutual Information co-clustering approach to capture the latent semantic concepts. These approaches perform image categorization using the frequency distributions of the latent semantic concepts. Though the general approaches performed surprisingly well when visual feature of each patch was independently considered, their weakness also derived from being independent. To address the problem, this paper introduces the CRF to construct visual words. The aim of this model is to discover the compact representation about appearance features can meet the contextual relationship in the image space maximally. Two patches, indistinguishable from each other when analyzed independently, might be discriminated as belonging to the correct visual word with the help of context knowledge. Our work has shown that the region contextual relationship provides a complementary and effective source for visual words construction.

The inputs are natural scene images. We consider the two cues: appearance descriptor cue, in our case Scale Invariant Feature Transform (SIFT) features extracted from each regular segmentation patch, and a high-level guidance cue, region contextual information. The region contextual information is the spatial interaction between patches in the homogenous region which is obtained by JSEG algorithm segmenting image. The initial visual word of each patch is automatically defined by K-means algorithm as the input of CRF model. Subsequently, this region contextual information is used to build the novel potential function of this model. The output of this model is Contextual Visual Words. We expect to obtain a superior performance derived from the combination two properties, than the methods which only possess one of these two properties.

2.1 Conditional Random Field

The initialization data is $\{x^{(k)}, c^{(k)}\}$, where $x^{(k)} = \{x_1^k, x_2^k, ..., x_n^k\}$ are visual features of image patches, $c^{(k)} = \{c_1^k, c_2^k, ..., c_n^k\}$ are the corresponding visual words of the image patches obtained by K-means algorithm. k is the index of the image patch. In the graphic model, we take an image with six patches as an example. Two kinds of relationships are simultaneously considered in the model. The first relationship decides the association of a given image patch to a certain visual word ignoring its neighbors. This model defines the appearance potential function to represent this relationship in the low-level features space. The second relationship serves as a visual word dependent function. The nodes which connected with red lines in the green circle co-occur in the same region. The inter-dependence of nodes in the same region provides the region contextual information. This information is a high-level guidance used to modulate the associated cliques' definition in the region contextual potential function. In our model, we consider the graphic structure of nodes as a lattice with pair wise potentials.

Following the definition of the CRF model, the conditional probability for the visual word c_i given the patch x_k can be expressed as:

$$P(c_i \mid x_k) = \frac{1}{Z_x} \exp[-(\lambda \bullet \underbrace{A(c_i, x_k)}_{appearance\ potential} + \mu \bullet \underbrace{R(c_j, c_i, x_k)}_{region\ contextual\ potential})] \qquad (1)$$

where Z_x is a normalization factor, $A(c_i, x_k)$ is the appearance potential function of the patch x_k ; $R(c_j, c_i, x_k)$ is the region contextual potential function of patch

x_k; (c_j, c_i) indicates the indices of neighboring visual words c_j and c_i in the same region. $A(c_i, x_k)$ and $R(c_j, c_i, x_k)$ are pair-wise potential functions characterizing the observation-state associations and state-state interactions respectively. λ and μ are the learned weights. The appearance potential function is defined by the Euclidean distance between visual words c_i and the patch x_k, as follows:

$$A(c_i, x_k) = d^2(c_i, x_k) \tag{2}$$

The region contextual potential function models the pair-wise interaction between all neighboring visual words in the homogeneous region. This potential function is defined as follows:

$$R(c_j, c_i, x_k) = \sum_j d^2(c_j, c_i) \qquad c_j, c_i \in region \tag{3}$$

First, the JSEG algorithm [14] segments the image into regions with homogeneous chrominance component. Here, the accuracy in this step is not important, as the clear contours do not need for acquiring region contextual information. Then we exploit the spatial layout distribution of each visual word in this region to construct the regional information which is an important cue for the associated cliques' definition in CRF model. Finally, this function is defined by the sum of Euclidean distance between c_i and all visual words c_j which co-occur in the same region.

2.2 Region Contextual Visual Words

The traditional approaches independently learnt each visual word according to the visual feature, such that the patches in the same visual word may be in some sense "different" from one another. This may be appropriate for text but not for sensory data with large variety in appearance. To address this problem, we construct the visual words depends on the other visual words in the same region by CRF model. Take patch x_k as an example to illustrate the proposed approach, when most of rest patches in the same region are clustered into the visual word c_i, even if the low-level features of patch x_k exclude the possibility, our algorithm trades off the two effects and makes the x_k be labeled to c_i, and vice versa. Let (x_k, c_k) denote the pair of feature and visual words for the k -th image patches, where x_k is a Scale Invariant Feature Transform (SIFT) [13] feature vector of image patch and c_k is the corresponding traditional visual word automatically through K-means.

Thus $(x_1, c_1), (x_2, c_2), \ldots$ are the initializations in the proposed algorithm. In the second step, the region contextual information is obtained by the image region segmentation results. In particular, we unitize the neighboring visual words in the same region to describe the region contextual information of this visual word. The region contextual potential function is computed using Eq.3. In the third step, we calculate the integrated probabilities $P(c_i \mid x_k)$ using Eq.1. The update visual words are defined according to minimizing the probabilities of Eq.1. Detailed analysis of the definition of integrated probabilities, the distance in the feature space and the image space are both taken into account, the two parameters are used for keep balance between the two properties. The final visual words are obtained by iteratively updating step 3 until convergence.

2.3 Scene Classification

Bag-of-Visual Words models for image classification work by quantizing high-dimensional descriptors of local image patches into discrete visual words, representing images by frequency counts of the visual word indices contained in them, and then learning classifiers based on these frequency histograms. Having the image representation, we simply use a Support Vector Machine (SVM) classifier, which is the standard choice in the scene classification literature [3] [5]. SVM has been shown to be a powerful technique for discriminative learning [9-10]. It focuses on structural risk minimization by maximizing the decision margin. We apply SVM using the Radial Basis Function (RBF) as the kernel, which is $K(x_i, x_j) = \exp(-\gamma \|x_i - x_j\|^2)$ .We apply five-fold cross validation with a grid search by varying (S, γ) on the training set to find the best parameters to achieve the highest accuracy. Within the optimal parameters, it then assigns to the testing image the category label which is mostly represented.

3 Experiments

We experimentally compare the proposed method against the traditional Visual Words in the fifteen natural scene categories dataset. We start our experiments with an in-depth analysis of our methods on the set of fifteen natural scene categories. Each scene category is randomly divided into two separate image sets, i.e., 50 images for training and 50 images for testing, respectively. For classification, we use a SVM with a histogram intersection kernel. Specifically, we use libSVM, and use the built in one-versus-one approach for multi-class classification. We use 10-fold cross-validation on the train set to tune parameters of the SVM. The classification rate we report is the average of the per-class recognition rates. For image features, we compute all SIFT descriptors on 16×16 pixels of patches, computed over a dense grid sampled every 8 pixels. The dataset we consider is the Scene-15 dataset, which is compiled by several researchers [16]. The Scene-15 dataset consists of 4485 images spread over 15 categories. The fifteen scene categories contain 200 to 400 images each and range from natural scenes like mountains and forests to man-made environments like kitchens and offices.

We start the experiments with an in-depth analysis of the types of visual words and vocabulary size. For investigating how classification performance is affected by the number of visual words, we construct five differently sized vocabularies {50, 100, 300, 500, and 1000}. For vocabulary size of 50, it can be seen that RCVWs already outperform the VWs with a 300-vocabulary. The proposed method outperforms VWs for smaller vocabulary size, however for larger vocabularies VWs performs equally well. If the size of vocabulary is small, several different image patches will be clustered into the same visual word depended on their appearance features only. Nevertheless, there is much improvement in the classification performance by this paper, since it also considers the region contextual information provided by the neighbor patches. If the vocabulary size is relative large, each patch of the image will match to a single, unique visual word, which defies the purpose of the Bag-of-Visual Words method. Hence, the classification performances of the two methods are similar.

Fig. 1 shows the confusion table which is used to illustrate the performance of the approach. In the confusion table, the x-axis represents the results of the proposed approach for each scene category. The y-axis represents the ground truth categories of scenes. The orders of the scene categories are the same in both axes. Hence in the ideal case on should expect diagonal lines include all the testing data which show perfect discrimination power of the category model over all categories of scenes. The average classification accuracy of our approach, over all categories is 74%. A closer look at the confusion table reveals that the highest error occurs in the kitchen and store scene. Due to the fact that the JSEG algorithm over-segments the images of these two scenes, the number of region is much more than ones of the actual situation. This induces that the amount of patches in each region is relatively small for these scenes. So, the region contextual information can not improve the traditional visual words apparently. And the best performance is achieved in both the street and forest scene. It can be difficult to distinguish between street scene and highway scene, since the images from these two scenes are visually similar. It is needed to consider the high-level region contextual information as the importance cues to distinct them.

Table 1 compares the classification performance of the proposed method on 15-scene categories with existing results in the literature [7] and [11]. The classification performance of Lazebnik et al. [11] is 71.2%, which represents image also considers the spatial contextual into the basic "Bag-of-Visual Words" model. Comparison with it, the proposed method using the region contextual information performs the scene classification task better, achieving a rate of 74.1%. The region contextual information shows the robustness than the absolute spatial information with respect to partial occlusion, clutters, and changes in viewpoint and illumination. Compared to the MMI clustering approach [7], our performance also achieves improvement. Though they use the novel co-clustering method to obtain visual words, not any contextual information is considered into the visual words construction.

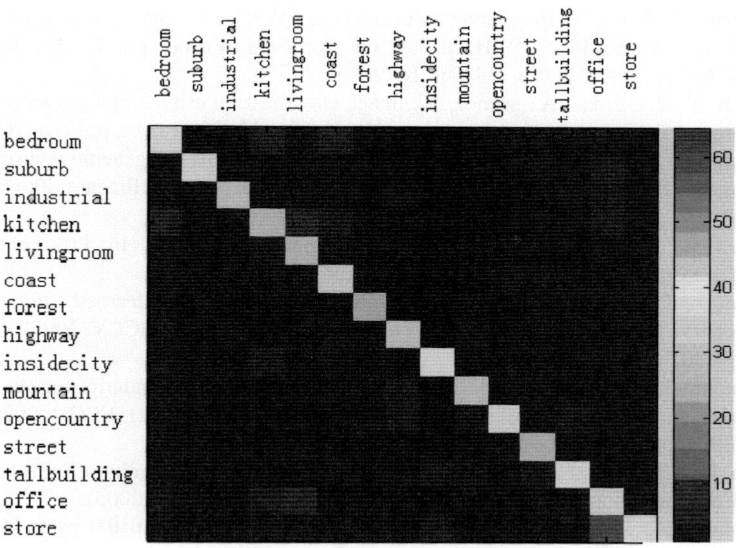

Fig. 1. The confusion table using 50 training from each category of 15-scene categories

Table 1. Comparison classification results for 15-categories with existing works

Scene-15	Our method	Liu et al. [7]	Lazebink et al. [11]
Accuracy	74.1%	72.1%	71.2%

4 Conclusions

This paper proposes a novel Contextual Visual Words to perform scene categorization task. Since the constituent patches of a region do not exist in the isolation, the region contextual information can well capture intrinsic property of patches. A study on the effectiveness of the proposed visual words with application to the scene classification performance was presented on the well-known datasets, and the experimental results indicated that the proposed method outperformed the traditional visual words.

References

1. Oliva, A., Torralba, A.: Modeling the shape of the scene: A holistic representation of the spatial envelope. International Journal of Computer Vision 42(3), 145–175 (2001)
2. Vogel, J., Schiele, B.: Semantic modeling of natural scenes for content-based image retrieval. International Journal of Computer Vision 72(2), 133–157 (2007)
3. Boutell, M., Luo, J., Brown, C.: Factor-graphs for region-based whole-scene classification. In: Proc. of IEEE Int. on Computer Vision and Pattern Recognition Workshop (CVPRW 2006), USA, p. 104 (2006)

4. Florent, M., Pedro, Q.: Integrating co-occurrence and spatial contexts on patch-based scene segmentation. In: Proc. of IEEE Int. on Computer Vision and Pattern Recognition Workshop (CVPRW 2006), USA, p. 14 (2006)
5. Bosch, A., Zisserman, A., Munoz, X.: Image classification using random forests and ferns. In: Proc.of IEEE Int. Conf. on Computer Vsion (ICCV 2007), Brazil, pp. 1–8 (2007)
6. Bosch, A., Zisserman, A.: Scene classification using a hybrid generative/discriminative approach. IEEE Trans. on Pattern Analysis and Machine Intelligence 30(4), 712–727 (2008)
7. Jingen, L., Mubarak, S.: Scene Modeling Using Co-Clustering. In: Proc. of IEEE Int. Conf. on Computer Vsion (ICCV 2007), Brazil, pp. 1–7 (2007)
8. Winn, J., Criminisi, A., Minka, T.: Object categorization by learned universal visual dictionary. In: Proc. of IEEE Int. Conf. on Computer Vsion (ICCV 2005), China, pp. 1800–1807 (2005)
9. Perronnin, F., Dance, C., Csurka, G., Bressan, M.: Adapted vocabularies for generic visual categorization. In: Leonardis, A., Bischof, H., Pinz, A. (eds.) ECCV 2006. LNCS, vol. 3954, pp. 464–475. Springer, Heidelberg (2006)
10. Jurie, F., Triggs, B.: Creating efficient codebooks for visual recognition. In: Proc. of IEEE Int. Conf. on Computer Vsion (ICCV 2005), China, pp. 604–610 (2005)
11. Lazebnik, S., Schmid, C., Ponce, J.: Beyond bags of features: Spatial pyramid matching for recognizing natural scene categories. In: Proc. of IEEE Int. Conf. on Computer Vision and Pattern Recognition (CVPR 2006), USA, pp. 2169–2178 (2006)
12. Lafferty, J., McCallum, A., Pereira, F.: Conditional random fields: probabilistic models for segmenting and labeling sequence data. In: Proc. of Int. Conf. on Machine Learning (ICML 2001), USA, pp. 282–289 (2001)
13. Lowe, D.: Distinctive image features from scale-invariant key points. International Journal of Computer Vision 60(2), 91–110 (2004)
14. Deng, Y., Manjunath, B.: Unsupervised segmentation of color-texture regions in images and video. IEEE Trans. on Pattern Analysis and Machine Intelligence 23(8), 800–810 (2001)
15. Jan, C., Gemert, V.: Kernelk codebooks for scene categorization. In: Forsyth, D., Torr, P., Zisserman, A. (eds.) ECCV 2008, Part III. LNCS, vol. 5304, pp. 696–709. Springer, Heidelberg (2008)
16. Fei-Fei, L., Perona, P.: A Bayesian hierarchical model for learning natural scene categories. In: Proc. of IEEE Int. Conf. on Computer Vision and Pattern Recognition (CVPR 2005), USA, pp. 524–531 (2005)

Design of an Improved Multiplier Unit for an Experimental RISC CPU

Ajay Joshi[1], Siew Lam[2], and Yee Chan[2]

[1] The University of the West Indies
Department of Electrical and Computer Engg,
St. Augustine, Trinidad and Tobago
Ajay.Joshi@eng.uwi.tt
[2] Multimedia University
Faculty of Engineering,
Cyberjaya, Malaysia

Abstract. An 8-bit RISC-CPU designed at gate level using completely custom based chip approach. CPU has an 8-bit integer unit and 16-bit floating point unit. The circuits are optimized by using more efficient algorithms. The algorithm discussed in this paper was applied for an efficient Multiplier design. An attempt has been made to improve conventional[6] algorithm. This paper discusses the design of an efficient Multiplier unit, with respect to its algorithm and VHDL implementation. The project was implemented using VHDL and simulated using Altera MaxPlus II simulation software which can map the design into Altera CPLD.

Keywords: Multiplier, CPU, simulation, algorithm, Floating point unit, VHDL.

1 Introduction

Paper focuses on the design and implementation of an improved and efficient Multiplier unit, which is a part of CPU with 8-bit integer unit. CPU has 4x16bit FPU registers, 16 bit data, address busses and 16-bit program counter. Data path is where most of the operations are done on by the processor's control unit. There are seven functional units, out of which 3 for FPU. This paper will discuss the design of a Multiplier unit. Logic of algorithm is discussed in detail and implementation block diagram along with the VHDL code and simulation results. The design and methodology is different than a generic one[3]. Main focus is on improved algorithm and relevant design. Amazing designs [4][5] helped us thinking different.

1.1 Rationale

An attempt has been made to improve the performance in terms of speed, number of states & space on chip. In our drive for improvement we made a detailed study of the original design. Original Algorithm is shown in the figure 1.

X. Wan (Ed.): Electrical Power Systems and Computers, LNEE 99, pp. 323–330.
springerlink.com

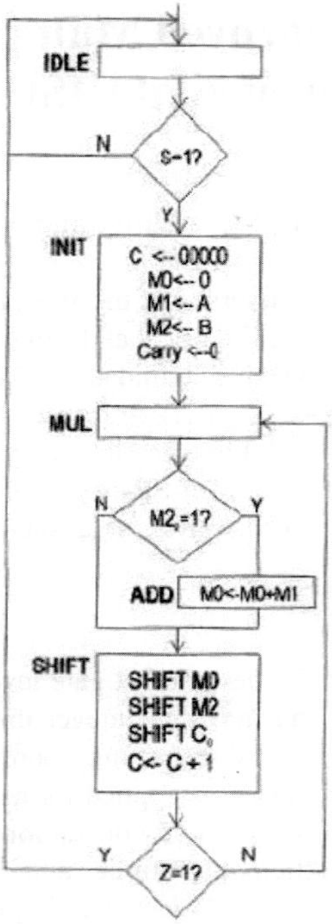

Fig. 1. Original Algorithm.

2 Design: Multiplier Unit

We have thought quiet a few different designs and finally decided to implement the one which we will discuss in detail. Improved algorithm tries to reduce the clock cycles needed to complete the multiplication process. As compared to original algorithm, few states are merged into one. New implementation is shown in the figure 2, showing new shift register implementation. This work builds upon our experience with a different block for same CPU [2].

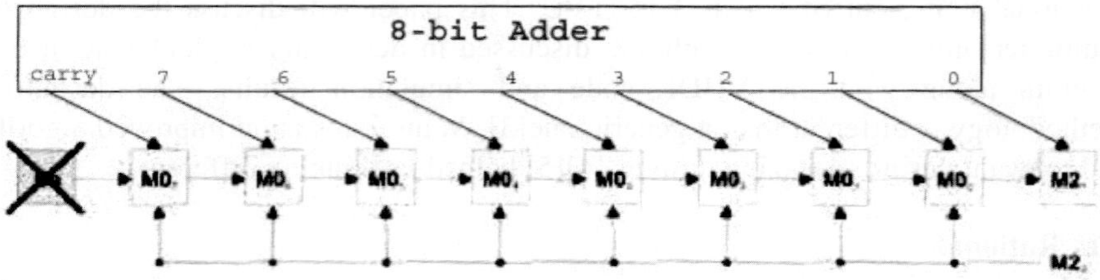

Fig. 2. New shift register implementation.

Here, the adder output is 'pre-shifted' before entering M0. Besides this, new approach does not need 'Ca' register to store the carry out result from adder, and then shift it into M0 during the next clock cycles. The new implementation is named as 'Algorithm 2'.

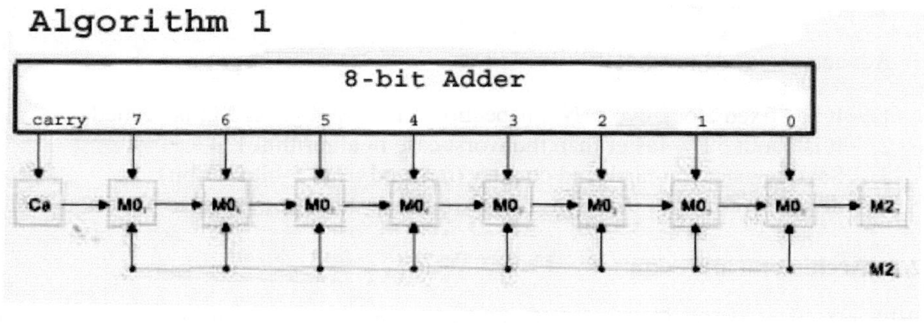

Fig. 3. Shift register implemented for algorithm 1

In algorithm 1 as seen in the figure 5., if $M2_0 = 0$, register M0 & M2 will acts as *right shift* registers. If, $M2_0 = 1$, the adder output will be loaded into M0 & Ca during the first clock. It will then right shift during second clock.

The IDLE and INIT state is similar to 1.

MUL0 sums indicate the start of the multiplication. It combines ADD and SHIFT states into 1.

MUL1 state is similar with MUL1 but it has an extra instruction C 6 C + 1. "Count" is placed in MUL1 instead of MUL0 to save 1-bit in the counter design. If the counter starts increment in MUL0, the final value after 9 clocks is 1000_1, whereas in MUL1, the final value is 111_2.

Example:

100 * 255 = 0110 0100 *11111111

From the Table 1., the final result is similar to 1, but it takes 9 fixed clocks.

Table 1. Multiplier operation with algorithm 2

Clock	M1	CARRY	M0	M2	Count	State
1	0110 0100	0	0000 0000	1111 1111	000	INIT
2	0110 0100	0	0011 0010	0111 1111	000	MUL0
3	0110 0100	0	0100 1011	0011 1111	001	MUL1
4	0110 0100	0	0101 0111	1001 1111	010	MUL1
5	0110 0100	0	0101 1101	1100 1111	011	MUL1
6	0110 0100	0	0110 0000	0110 0111	100	MUL1
7	0110 0100	0	0110 0010	0011 0011	101	MUL1
8	0110 0100	0	0110 0011	0001 1001	110	MUL1
9	0110 0100	0	0110 0011	1000 1100	111	MUL1

Algorithm 2 needs the following hardware:

1. One controller that is able to generate 4 state signals.
2. Three 8-bit registers for M0,M1 & M2 (Carry register isn't needed)
3. One 3-bit counter to count from 000 (0_{10}) to 111 (7_{10})
4. one 8-bit adder

Advantages of Algorithm 2:

1. It has fixed clock cycle. No inspection on '1' in register M2 is needed.
2. It is about 52 % faster than the worst case in algorithm 1.
3. Save register for carry and counter (reduced from 5-bit to 3-bit)
4. Simpler controller design as there are fewer states compared to algorithm 1.

2.1 Architectural Design

With reference to the figure 6, When MS=1, the multiplication process will begin. First, the INIT signal will go active low. This will reset the counter and register M0 to zero, and load data from source A and B from main register file into register M1and M2. When initialization is done, INIT will be deactivated (INIT = 1) during the next clock.

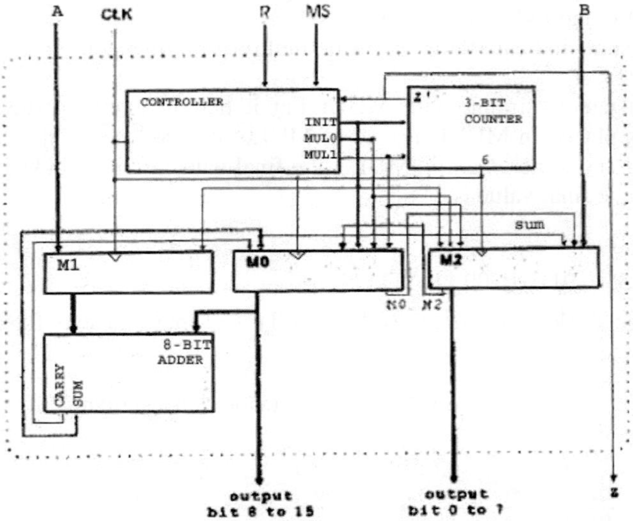

Fig. 4. Algorithm 2 architecture

After that, the multiplication process starts when MUL0 and MUL1 are activated. These signals enable M0 and M2 only because register M1 doesn't change its value throughout the multiplication process. The value of $M2_0$ is used to select the function of registers M0 and M2. If $M2_0 = 1$, M0 + M1 → MO. If $M2_0 = 0$, 'shift right' function is enabled.

The 3-bit counter starts to count down on every clock cycle when MUL1 is activated. Once the counter reaches '111', Z will change from 1 to 0 so that the controller changes its state to IDLE during the next clock cycle.

Clock signal only goes into the sequential circuit which has D flip flop in it, such as registers, controller and counter. Adder is a combinational circuit, so clock signal input is not required.

2.2 Controller Design

We have chosen the decoder approach in controller design as it is more compact than the one flip flop per state approach. The state table is shown as Table 2.

- The current state number is the values inside two D flip flops (DO and D1).
- The next state number is the values that will be loaded into the D flip flops during the next clock cycle.
- A 2 to 4 decoder takes in the D flip flops output and decodes the values into state signals, such as IDLE, INIT, MUD) and MUL1.
- Some logic circuit must present at the input port of D flip flops (DFF) in order for the decoder to generate the correct state signals.

Table 2. State table of algorithm 2

Current state		Inputs		Next state		Decoder output			
Name	#	S	Z	Name	#	IDLE	INIT	MUL0	MUL1
IDLE	00	0	x	IDLE	00	1	0	0	0
		1	x	INIT	01	1	0	0	0
INIT	01	x	x	MUL0	10	0	1	0	0
MUL0	10	x	x	MUL1	11	0	0	1	0
MUL1	11	x	0	MUL1	11	0	0	0	1
		x	1	IDLE	00	0	0	0	1

The 2 to 4- decoder has 4x NAND2 so that IDLE, INIT, MUL0 and MUL1 go active low when they are activated.

- Statistic Computation

With reference to our components statistics, the transistor count including the 2:4 decoder, the controller has:

NAND2: 5 units AND2 :2 unit
NAND3: 2 units DFF :2 units
NOT2 : 4 units
0.18 micron: 5(4) + 2(6) 4- 4(2) + 2(6) + 2(26) = 106 transistors.
0.50 micron: 5(4) + 2(6) + 4(2) + 2(8) + z(zs) = 112 transistors.
Power consumption:
0.18 micron: 108 transistors x 1.6 μW = 169.6 μW
0.50 micron: 112 transistors x 4.0 μW = 448.0 μW

2.3 VHDL Code and Simulation

Decoder is written in the dataflow style, to verify if the desired output is correct or not.

```
2:4 decoder VHDL
LIBRARY ieee;
USE ieee.std_1ogic_1164.all;
ENTITY dec24 IS
PORT ( i0,i1 : IN STD_LOGIC;
o            : OUT STD__LOGIC_.VECTOR (3 DOWNTO 0)
END ENTITY;

ARCHITECTURE d OF dec24 IS
SIGNAL n0, n1 : STD_LOGIc; -- buffer NOT in0, in1 resuTt
BEGIN
n0 <= NOT i0; n1 <= NOT i1;
0(3) <= n1 NAND ; -- IDLE state output for controller
0(2) <= n1 NAND ; -- INIT state output for controller
o(1) <= i1 NAND ; -- MUL0 state output for controller
o(0) <= i1 NAND ; -- MUL1 state output for controller
END d;
```

Multiply controller:

```
LIBRARY ieee;
USE ieee.std_1ogic_1164.all;
ENTITY mul_ctrl IS
      port ( r,s,z,c1k : IN STD_LOGIC;
      o : OUT STD_LOGIC_VECTOR (0 TO 3) );
END ENTITY;
ARCHITECTURE mul OF mul_ctrl IS
      SIGNAL q : STD_LOGIC_VECTOR (0 to 1); -- dff output buffer
      SIGNAL dec : STD_LOGIC_VECTOR (0 to 3); -- decoder output buffer
BEGIN
      PROCESS (r,c1k)
            BEGIN
                  IF r = '0' THEN -- during reset signal
                  q <= "00"; -- output of dff
            ELSE IF (clk'EVENT AND c1k = '1') THEN
                  CASE q IS
            WHEN "00" v> dec <= "0111"; -- output as IDLE
                  IF s = '1' THEN q <= "01"; -- change dff to 01
                  ELSE q <= "00"; -- dff remains unchanged
                  END IF;
            WHEN "0l" => dec <= "1011"; -- output as INIT
                  q <= "10": -- change dff to 10
            WHEN "10" => dec <= "1101"; -- output as MUL0
                  q <u "11"; -- change dff to 11
```

```
              WHEN "11" => dec <= "1110"; -- output as MUL1
                       IF z = '0' THEN q <= "00"; ~- change dff to 00
                       ELSE q <= "11"; -- dff remains unchanged
                       END IF
              WHEN OTHERS => NULL;
              END CASE;
            END IF;
          END IF;
        END PROCESS;
       o <= dec; -- transfer dec to output port
END mul;
```

Decoder Simulation

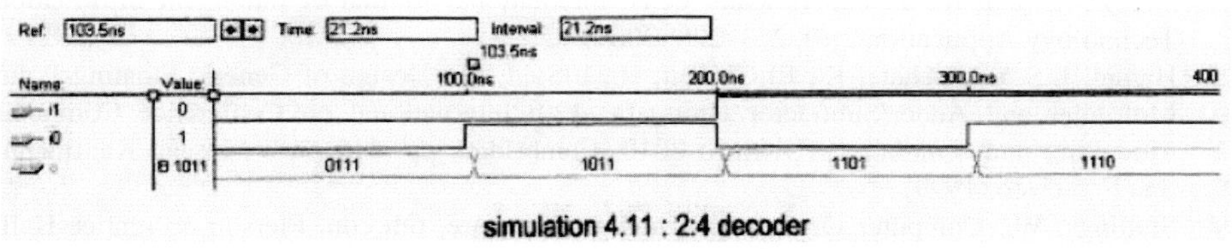

Fig. 5. Decoder simulation results

Multiplier Controller simulation

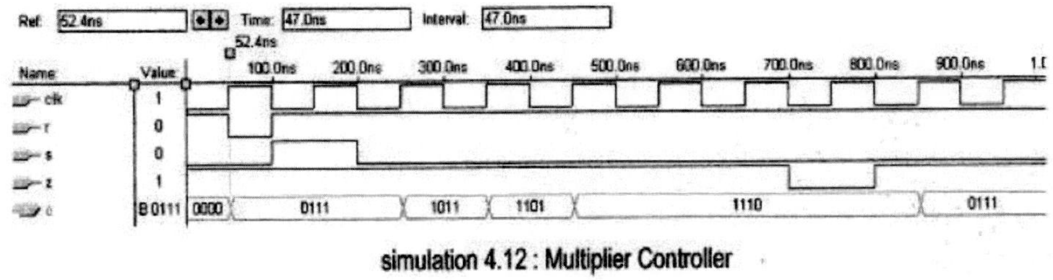

Fig. 6. Multiplier controller simulation results.

3 Conclusion

The decoder simulation result matched the desired input as seen from figure 5.

- When i1&i0 = ‘00’, output = “0111” which indicate IDLE’ state is active.
- When i1&i0 = ‘01’, output = “1011” which indicate INIT’ state is active.
- When i1&i0 = ‘10’, output = “1101” which indicate MUL0’ state is active.
- When i1&i0 = ‘11’, output = “1110” which indicate MUL1' state is active.

From the simulation result, the propagation delay is 3.5 ns only.

The Multiplier controller result matched the output figure 6, in the statue table.

- R is asserted when the CPU boots up to activate IDLE (0111).
- When S = 1, output goes 1011 on the next clock. This means IDLE is disabled and INIT is activated. It is followed by one MUL0 state (1101) and multiple MUL1 states (1110).
- When Z=0, multiplication ended and return to IDLE state during next clock.

References

1. Akkas, A., Schulte, M.J.: Dual-mode floating-point multiplier architectures with parallel operations. Journal of Systems Architecture 52, 549–562 (2006)
2. Joshi, A., Lam, S.L., Chan, Y.Y.: Algorithm & design of an efficient floating point ADD/SUB unit for an experimental CPU. International Journal of Intelligent Information Technology Application 2(6), 273–278 (2009)
3. Hamid, L.S.A., Shehata, K., El-Ghitani, H., ElSaid, M.: Design of Generic Floating Point Multiplier and Adder/Subtractor Units. In: 12th International on Conference Computer Modelling and Simulation (UKSim) 2010, Cambridge, pp. 615–618. Morgan Kaufmann, San Francisco (1999)
4. Stallings, W.: Computer Organization and Architecture, 6th edn. Pierson &Prentice-Hall, Englewood Cliffs (2003)
5. Even, G., Mueller, S.M., Seidel, P.M.: A dual precision IEEE floating-point multiplier. Integration, the VLSI Journal 29(2), 167–180 (2000), ISSN:0167-9260
6. Hida, Y., Li, X.S., Bailey, D.H.: Algorithms for Quad-Double Precision Floating Point Arithmetic. In: Proceedings of the 15th IEEE Symposium on Computer Arithmetic, p. 155 (2001)

A Fingerprint Image Enhancement Method Based on Contourlet Transform

Xiukun Yang, Yong Chai, and Zhigang Yang

College of Information and Communication Engineering, Harbin Engineering University, 150001 Harbin, China
yangxiukun@hrbeu.edu.cn, chaiyong@yahoo.cn, zgyang@hrbeu.edu.cn

Abstract. The quality of fingerprint image has great effect for its follow-up steps such as recognition. Therefore, an effective method is needed to enhance the collected fingerprint image. In this paper, a new fingerprint image enhancement approach combining contourlet transform and maximum modulus detection is proposed. Contourlet transform decomposes the input image into high-frequency part and low-frequency part. The maximum modulus detection is applied to the high-frequency image and the low-frequency image is used to compensate the high-frequency part. Experimental results demonstrate that the proposed method can effectively improve the quality of fingerprint image, increase the clarity of fingerprint ridges and valleys, and have better continuity.

Keywords: fingerprint image enhancement, maximum modulus, contourlet transform, low-frequency compensation.

1 Introduction

As the biological characteristics of human, fingerprints can be used for identification, and have been applied in many fields such as management, access control, finance, public security and network. The main requirement of fingerprint identification system is matching features of tested fingerprints and fingerprints stored in database. And extracting features from input fingerprint image reliably is the key process, so the quality of the input image is of much importance. In practice, the quality of resulting fingerprint image is usually not high due to a number of reasons such as changing of illumination environment, noise of collecting equipment, uncooperative collectors, and different skin conditions. Before the extraction of fingerprint features, it is necessary to process the fingerprint image properly, to enhance useful information, remove useless information and suit the requirements of follow-up steps. Therefore, a method to improve the quality of fingerprint image is highly needed.

At present, there is a variety of fingerprint enhancement research both in spatial domain and frequency domain. Hong L and A. Jain enhanced the fingerprint ridges by calculating the local direction and frequency using Gabor filters [1]. But it needs different filter templates of coefficients with different directions because of lacking multi-scale features in window size selection. Wavelet is a powerful information processing tool for multi-scale. Wei-Peng Zhang proposed an approach of texture filtering for fingerprint image enhancement based on wavelet transform [2]. However, two-dimensional

X. Wan (Ed.): Electrical Power Systems and Computers, LNEE 99, pp. 331–338.
springerlink.com

separable wavelet is constituted by one-dimensional wavelet, its basic function is isotropic, and only has limited directions, which make it impossible to detect the line and surface singularity of two-dimensional image but only the point singularity [3].

As a real representation of two-dimensional image, contourlet transform has attracted more and more attention of scholars in recent years and achieved very good effect in various areas of image processing. Contourlet transform not only has the characteristics of multi-resolution and time-frequency analysis which is the same as wavelet, but also has flexible characteristics of multi-directional and anisotropic [4-5]. Compared with wavelet transform, contourlet transform can set different directional filter banks and the number of directions for each of them, and then capture more directional information. Ibrahim described a method using nonsubsampled contourlet transform and directional Gaussian filtering [6], but it is complex to achieve. Guang-Quan Cheng proposed an approach reducing noises by coefficients modification after contourlet decomposing and joining ridges by orientation estimation [7]. Both of the methods above are focused on the processing of high-frequency part while ignoring the use of low-frequency information.

In this paper, to make full use of low-frequency information, a new fingerprint image enhancement algorithm based on the integration of contourlet transform and maximum modulus detection is proposed. Firstly, we decompose the fingerprint image by nonsubsampled contourlet transform, and then detect maximum modulus of the high-frequency subband. At last, low-frequency subband is used to compensate the high-frequency image.

2 Contourlet Transform

Contourlet transform is a multi-directional and multi-resolution representation of image, which can effectively express the important and complex geometry in visual information. As shown in Fig. 1, the supporting interval of contourlet base is like strip, which can make full use of the geometric regularity of original function, so the contourlet coefficients will be less than wavelet in describing a singular curve [8]. In fact, the strip structure of the base is an exemplification of directionality that means the base is anisotropic.

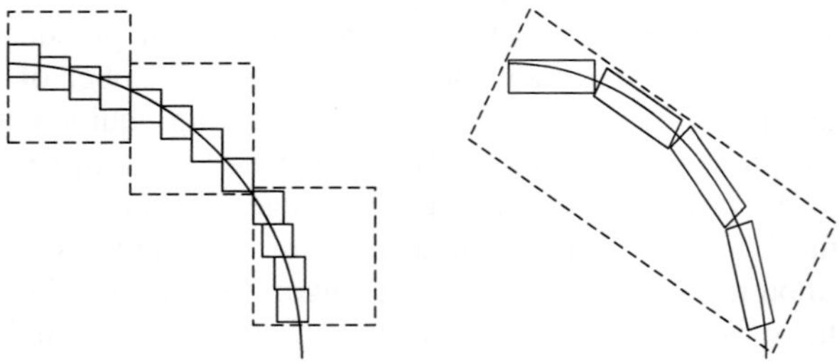

Fig. 1. Comparison of wavelet and contourlet transform

Contourlet transform consists of two main steps (as shown in Fig. 2): subband decomposition and directional filtering. Firstly, the input image is decomposed using Laplacian pyramid (LP) in order to capture singular point, and then the singular point in the same direction will be synthesized as one coefficient by directional filter banks (DFB). The LP and DFB of contourlet filter banks are inseparable, but they are performed separately [9], which make it more flexible for the directionality of contourlet transform.

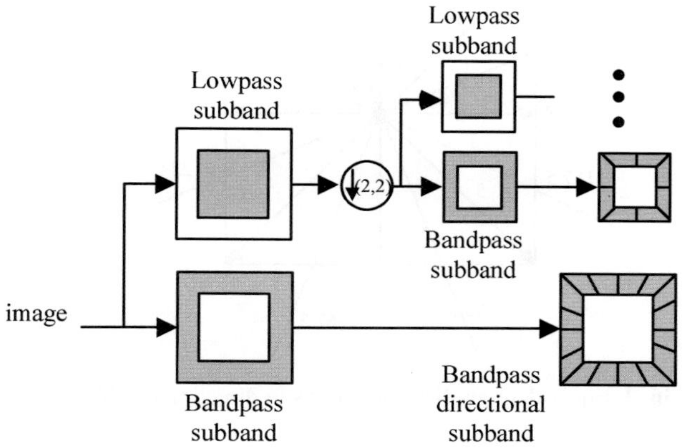

Fig. 2. Diagram of contourlet decomposition

In this paper, nonsubsampled contourlet transform is applied [10], which size of the output subband is the same as original image, so there is a one-to-one relationship between all of its output subbands and the original image pixels, that make it perform well in image denoising and enhancement [11].

3 Maximum Modulus Algorithm Based on Contourlet Transform

The traditional maximum modulus detection algorithm based on wavelet needs to determine the gradient direction of curves by calculating the phase angle of the modulus, while contourlet transform filter banks are inseparable, the direction information is enough with the coefficients of each directional subband, that the maximum modulus points in specific direction are contained in specific directional subbands. Therefore, the process of calculating maximum modulus based on contourlet transform is simplified as the following in [12].

Now set i scales of DFB decomposition and k directions in each scales as an example to illustrate how to determine whether a point $Coeffs_{j,k}^{i}(n)$ in a directional subband coefficients is the maximum modulus point or not. The parameter n is the coordinates of point in the subband of j scale and k direction. As shown in Fig. 3, there are eight directions of each point $Coeffs_{j,k}^{i}(n)$ in the directional subbands can be set as the equivalent gradient direction of edge in the direction of $\mathrm{Arg}Coeffs_{j,k}^{i}$ (argument of $Coeffs_{j,k}^{i}(n)$). Therefore, whether the modulus at one point is the local maximum modulus can be determined after comparing between $\mathrm{mod}\left[Coeffs_{j,k}^{i}(n)\right]$ (modulus of $Coeffs_{j,k}^{i}(n)$) and the modulus of two adjacent elements in the equivalent gradient direction $\mathrm{Arg}(\mathrm{grad}Coeffs_{j,k}^{i})$ (vertical direction of $\mathrm{Arg}Coeffs_{j,k}^{i}$).

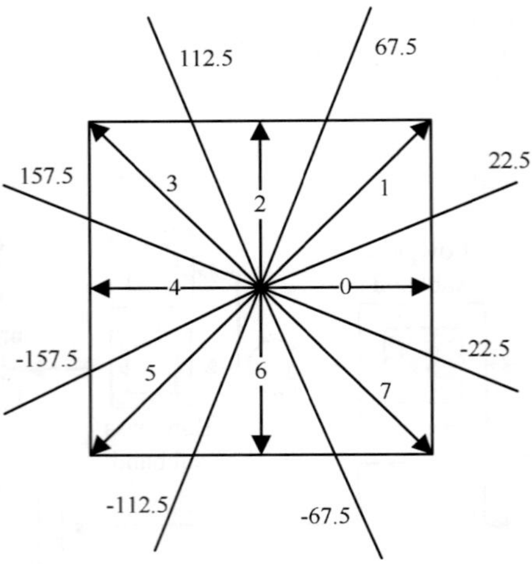

Fig. 3. Equivalent gradient directions in directional subband

From the above, the process of judgment can be realized by formula (1):

$$\operatorname{mod}[Coeffs^{i}_{j,k}(n_1,n_2)] = \begin{cases} 0 & \operatorname{mod}[Coeffs^{i}_{j,k}(n_1,n_2)] < \\ & \quad \operatorname{mod}[Coeffs^{i}_{j,k}(n_1 - r_1(n_1), n_2 - r_1(n_2))] \\ 0 & \operatorname{mod}[Coeffs^{i}_{j,k}(n_1,n_2)] < \\ & \quad \operatorname{mod}[Coeffs^{i}_{j,k}(n_1 - r_2(n_1), n_2 - r_2(n_2))] \\ \operatorname{mod}[Coeffs^{i}_{j,k}(n_1,n_2)] & \quad others \end{cases} \tag{1}$$

where the $r(n_1)$ $r(n_2)$ respectively denote the horizontal and vertical coordinate offset of the point which is compared with $Coeffs^{i}_{j,k}(n)$ in the direction of $\operatorname{Arg}(\operatorname{grad}Coeffs^{i}_{j,k})$.

4 Model Based on Contourlet Transform and Maximum Modulus Algorithm

In this paper, maximum modulus method is applied to detect the high-frequency subband of fingerprint image and the ridges of result are rich in detail but not continuous. Meanwhile, the dual threshold method is utilized to detect the low-frequency subband, though the processed image performs well in describing the overall contour of fingerprint, but there are less weak edges or detail. To take both advantages of high-frequency and low-frequency information, the low-frequency image is used to compensate the high-frequency part, the broken ridges after high-frequency processing are modified to link, finally a more comprehensive and accurate image of ridges is obtained.

The experimental procedures are as follows:

1. Contourlet decomposition. Compute the contourlet coefficient of the input image, set $Coeffs_0 = C_0$, while set $Coeffs_0 = 0$, keep $Coeffs_0$ unchanged.
2. High-frequency subband detection. Firstly, adaptive window hard threshold is applied to each directional subbands of each scales [13], then maximum modulus of the coefficients are detected, at last the inverse transformation of contourlet is done to the coefficient matrix to get high-frequency subband image.
3. Low-frequency subband detection. Calculate the gradient magnitude and direction of smooth low-frequency data C_0 using first-order partial derivatives finite difference, then non-maxima suppression is applied to the gradient magnitude, finally the ridges are detected and joined with dual-threshold algorithm and the binary low-frequency subband image is obtained.
4. Compensation. Compensate and link the ridges from high-frequency subband and low-frequency subband according to [14].

5 Experimental Results

The experimental environment is MatlabR2009b and the experimental images are grayscale fingerprint images from NIST-27 database. In this paper, two scales of nonsubsampled contourlet transform is applied and the directions in the scales are 8, 8 from coarser to finer. The '9-7' filters and 'pkva' filters are chosen for the LP stage and DFB stage respectively. The high threshold is set to 0.35 while the low threshold is 0.4 times higher than the high threshold in low-frequency subband detection.

The experimental results are shown in **Fig.** 4, where (a) and (c) are original fingerprint images with poor quality; (b) and (d) are enhanced fingerprint images by the proposed method.

The algorithms based on Gabor filtering method [1] and wavelet method [2] are compared with to further verify the effectiveness of this proposed method. The block size of Gabor filter is 16 × 16. The 'db4' is chosen for wavelet and two scales of decomposition are utilized. Comparison results are shown in Fig. 5. From Fig. 5(b), the fingerprint image enhanced by Gabor filter always has some extra noise in specific directions, which makes only a certain part of the fingerprint image can be effectively enhanced while superfluous noise is generated in other parts of it. From Fig. 5(c), the fingerprint information can be well detected and the contrast between ridges and valleys is increased by soft threshold denoising based on wavelet, but the ridges are not continuous, some useful information loses while enhancing. From Fig. 5(d), the quality of fingerprint image is obviously improved. It can be concluded from the comparison that the proposed algorithm compensates high-frequency subband image with low-frequency subband part, the noise of modified image has been effectively suppressed, the clarity of fingerprint ridges is improved, the ridges are more continuous and complete, and the image contrast is better than the others.

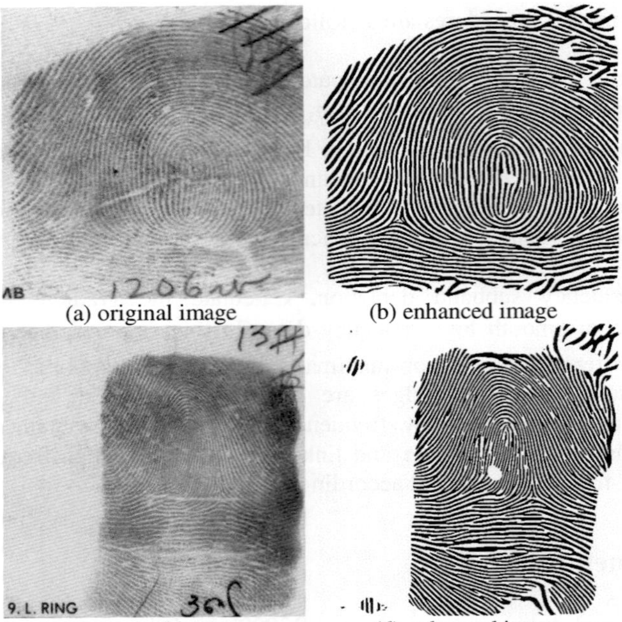

(a) original image (b) enhanced image

(c) original image (d) enhanced image

Fig. 4. Contourlet fingerprint image enhancement

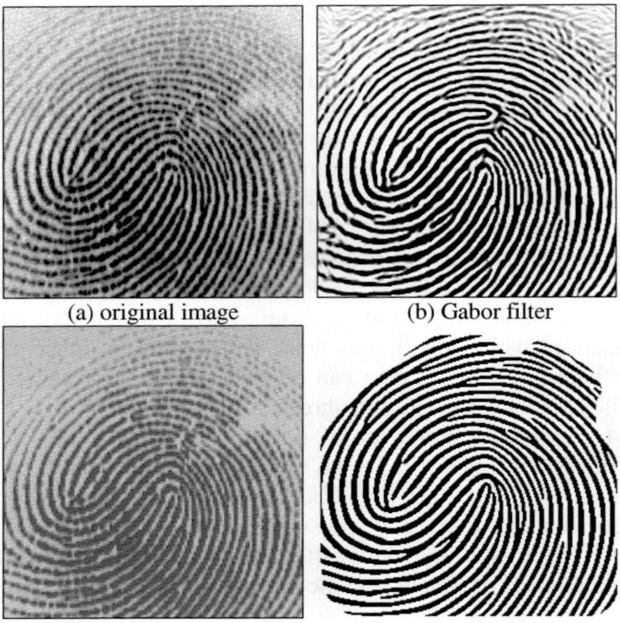

(a) original image (b) Gabor filter

(c) wavelet (d) proposed method

Fig. 5. Comparison of different fingerprint enhancement algorithms

6 Conclusion

Based on contourlet transform, this paper presents a fingerprint image enhancement algorithm. This algorithm not only enhances the edge characteristics of fingerprint image, but also suppresses noises by detecting the maximum modulus of high-frequency subband image. Meanwhile, the edge information of low-frequency image is effectively used through the compensation between high-frequency and low-frequency subband images. Experimental results demonstrate that the enhanced fingerprint image is rich in detail and has good continuity, the overall visual effect has been significantly improved, and the enhancement algorithm introduced in this paper is proved to be effective for fingerprint images with poor quality.

Acknowledgments

This work was supported in part by the National Science Foundation of Heilongjiang Province under Grant 42400621-1-09130 and the Fundamental Research Funds for the Central Universities under Grant HEUCFR1017.

References

1. Hong, L., Wan, Y.-f., Jain, A.: Fingerprint image enhancement: algorithm and performance evaluation. IEEE Transactions on Pattern Analysis and Machine Intelligence 20(8), 777–789 (1998)
2. Zhang, W.-p., Wang, Q.-r., Tang, Y.Y.: A wavelet-based method for fingerprint image enhancement. In: International Conference on Machine Learning and Cybernetics, vol. 4, pp. 1973–1977 (2002)
3. Starck, J.L., Candes, E.J., Donoho, D.L.: The curvelet transform for image denoising. IEEE Transactions on Image Processing 11(6), 670–684 (2002)
4. Do, M.N., Vetterli, M.: The contourlet transform: An efficient directional multiresolution image representation. IEEE Transactions on Image Processing 14(12), 2091–2106 (2005)
5. Do, M.N., Vetterli, M.: Framing pyramids. IEEE Transactions on Signal Processing 51(9), 2329–2342 (2003)
6. Ibrahim, M.T., Bashir, T., Guan, L.: Robust fingerprint image enhancement: an improvement to directional analysis of fingerprint image using directional Gaussian filter and non-subsampled contourlet transform. In: International Symposium on Multimedia, pp. 280–285 (2008)
7. Cheng, G.-Q., Cheng, L.-Z.: Adaptive fingerprint image enhancement with contourlet transform. Congress on Image and Signal, 261–264 (2008)
8. Do, M.N., Vetterli, M.: Contourlets: A directional multiresolution image representation. In: International Conference on Image Processing, vol. 1, pp. 357–360 (2002)
9. Po, D.D.-Y., Do, M.N.: Directional multiscale modeling of images using the contourlet transform. IEEE Transactions on Image Processing 15(6), 1610–1620 (2006)

10. Cunha, A.L., Zhou, J.-p., Do, M.N.: The nonsubsampled contourlet transform: theory, design, and applications. IEEE Transactions on Image Processing 15(10), 3089–3101 (2006)
11. Feng, P., Pan, Y.-j., Wei, B., Jin, W., Mi, D.-l.: Enhancing retinal image by the contourlet transform. Patten Recognition Letter 28, 516–522 (2007)
12. Zhang, Y.t., Meng, X.-f., Yin, Z.k., Wang, J.y.: Image Edge Detection Based on Contourlet Modulus Maxima. Journal of the CHINA Railway Society 30(5), 41–45 (2008)
13. Cheng, G.: Matlab image processing and application, 2nd edn. National Defense Industry Press, Beijing (2007)
14. Shang, Z.-g., Zhao, C.-h., Sun, Y., Liu, J.-m.: A new edge detection method based on nonsubsampled contourlet. Journal of Optoelectronics Laser 20(4), 525–529 (2009)

Compression Performance of Wide-Band Chirp Pulse in the Fractional Fourier Domain

Deng Bing[1], Xiao Mei-ping[2], and Wang Hong-xing[1]

[1] Naval Aeronautical and Astronautical University
Yantai, Shandong, 264001, China
navy_dbing@tom.com
[2] Hengyang Vocational Technical Secondary School
Hengyang, Hunan, 421008, China
147687230@qq.com

Abstract. Based on the fractional Fourier transform, a time-delay estimator of chirp signals was proposed by Tao in 2009. In this paper, compression performance of the time-delay estimator is discussed in the fractional Fourier domain, referring to linear frequency-modulated pulse compression in the time domain. We find: (a) the pulse-width compression ratio is the product of time duration and band width, denoted as D; (b) the peak amplitude ratio is about $D^{1/2}$ when the absolute value of chirp rate is large enough. The conclusions help to the further application of this time-delay estimator based on the FRFT.

Keywords: Chirp pulse, fractional Fourier transform, pulse compression, time delay.

1 Introduction

Pulse compression has been widely used in high-resolution radar signal processing. The representative one is linear frequency-modulated pulse compression, which obtains utilization not only in pulse compression radar but also in synthetic aperture radar. Along with the development of fractional Fourier transform theory, Tao et al proposed a time-delay estimator (TDE) of chirp signals in the Fractional Fourier Domain, and proved that this estimator has less computation cost and reliable accuracy theoretically equaling to the CRLB[1]. However, the pulse-width compression ratio and peak amplitude ratio have not been analyzed, which are the two important parameters to show the performance of linear frequency-modulated pulse compression. Since the time domain is the fractional Fourier domain[2], we discuss the pulse-width compression ratio and peak amplitude ratio of Tao's TDE in the fractional Fourier domain, referring to the linear frequency-modulated pulse compression.

In this paper, some characters of linear frequency-modulated pulse compression are depicted, as well as the conception of the fractional Fourier transform (FRFT). Then, the signal model is described, and corresponding derivation is given. After that, simulations are presented. Finally, conclusions are driven.

X. Wan (Ed.): Electrical Power Systems and Computers, LNEE 99, pp. 339–344.
springerlink.com

2 Preliminaries

Some characters of linear frequency-modulated pulse compression are as follows: (a) the pulse-width compression ratio is $D=TB$, where T is the pulse duration and B denotes the pulse bandwidth; (b) the peak amplitude ratio is $D^{1/2}$ [3].

The FRFT is the generalization of conventional Fourier. As a unified time-frequency transform, it has the linearity property, not suffering from cross-terms. What's more, the efficient fast digital algorithm of FRFT has been proposed, which has approximate computation complexity of classical FFT. Thus, the FRFT has already been used in signal analysis and reconstruction, signal detection and parameter estimation, filtering, neural network, pattern recognition, image processing, array signal processing, radar, sonar and communication[2, 4]. The FRFT is defined as[2]:

$$S_\alpha(u) = F_\alpha[s](u) = \begin{cases} \sqrt{\dfrac{(1-\mathrm{j}\cot\alpha)}{2\pi}}\, e^{\mathrm{j}\frac{u^2}{2}\cot\alpha} \displaystyle\int_{-\infty}^{+\infty} s(t) e^{\mathrm{j}\frac{t^2}{2}\cot\alpha - jut\csc\alpha}\, \mathrm{d}t & \alpha \neq n\pi \\ s(u) & \alpha = 2n\pi \\ s(-u) & \alpha = (2n\pm 1)\pi \end{cases} \tag{1}$$

where α is the FRFT order, F_α denotes the FRFT operator.

Let the received signal $\tilde{s}(t) = s(t-\tau)$, then

$$\tilde{S}_\alpha(u) = S_\alpha(u - \tau\cos\alpha) e^{\mathrm{j}\frac{\tau^2 \sin\alpha\cos\alpha}{2} - \mathrm{j}u\tau\sin\alpha} . \tag{2}$$

So the time delay τ can be obtained from the peak position difference of $\tilde{S}_\alpha(u)$ and $S_\alpha(u)$, the corresponding TDE is as follows[1]:

$$\hat{\tau} = \left(\arg\max_u \left|\tilde{S}_\alpha(u)\right| - \arg\max_u \left|S_\alpha(u)\right|\right) \cdot \sec\alpha . \tag{3}$$

3 Character of This TDE in the Fractional Fourier Domain

Let the transmitted chirp pulse is

$$s(t) = A\exp\left(\mathrm{j}2\pi f_0 t + \mathrm{j}\pi\mu t^2 + \mathrm{j}\varphi\right)\mathrm{rect}\left(\frac{t}{T}\right) \tag{4}$$

where $\mathrm{rect}\left(\frac{t}{T}\right)$ denotes the rectangle pulse with duration T. Without loss of generality, we assume f_0=0 and $\varphi = 0$, then

$$S_\alpha(u) = F_\alpha[s](u) = A\sqrt{\frac{(1-\mathrm{j}\cot\alpha)}{2\pi}}\mathrm{e}^{\mathrm{j}\frac{u^2}{2}\cot\alpha}\int_{-T/2}^{T/2}\mathrm{e}^{\mathrm{j}t^2\left(\frac{1}{2}\cot\alpha+\pi\mu\right)-jtu\csc\alpha}\mathrm{d}t \tag{5}$$

According to the FRFT theory[2], the optimal order is $\cot\alpha_m = -2\pi\mu$ to the chirp signal described as (4). Substitute (4) into (5)

$$\begin{aligned} S_{\alpha_m}(u) &= A\sqrt{\frac{(1-\mathrm{j}\cot\alpha_m)}{2\pi}}\mathrm{e}^{-\mathrm{j}\pi\mu u^2}\int_{-T/2}^{T/2}\mathrm{e}^{-jtu\csc\alpha_m}\mathrm{d}t \\ &= AT\frac{1}{\sqrt{2\pi}\cdot\sqrt{\sin\alpha_m}}\mathrm{e}^{\mathrm{j}\frac{\alpha_m}{2}+\mathrm{j}\frac{3\pi}{4}}\mathrm{e}^{-\mathrm{j}\pi\mu u^2}\frac{\sin\frac{Tu\csc\alpha_m}{2}}{\frac{Tu\csc\alpha_m}{2}} \end{aligned} \tag{6}$$

Let $\frac{T\csc\alpha_m}{2} = \gamma\pi$, we have

$$S_{\alpha_m}(u) = AT\frac{1}{\sqrt{2\pi}\cdot\sqrt{\sin\alpha_m}}\mathrm{e}^{\mathrm{j}\frac{\alpha_m}{2}+\mathrm{j}\frac{3\pi}{4}}\mathrm{e}^{-\mathrm{j}\pi\mu u^2}\,\mathrm{sinc}(\gamma u) \tag{7}$$

where the function $\mathrm{sinc}(u) = \frac{\sin\pi u}{\pi u}$. From (7), we find that $|S_{\alpha_m}(u)|$ reaches the maximum when u=0, so the TDE can be simplified as

$$\hat{\tau} = \arg\max_u \left|\tilde{S}_{\alpha_m}(u)\right|\cdot\sec\alpha_m\,. \tag{8}$$

3.1 Pulse-Width Compression Ratio

According to (2) and (8), we know that the pulse width is the main lobe width of $|S_{\alpha_m}(u)|$ after compressed in the fractional Fourier domain. Thus, from (7) we have

$$|S_{\alpha_m}(u)| = AT\frac{1}{\sqrt{2\pi}\cdot\sqrt{\sin\alpha_m}}\mathrm{sinc}(\gamma u)\,. \tag{9}$$

Since $\mathrm{sinc}\left(\frac{1}{2}\right) = \frac{2}{\pi}$, i.e., about -4dB, the -4dB main lobe width of $|S_{\alpha_m}(u)|$ equals $|1/\gamma|$, which is described as follows:

$$|1/\gamma| = \left|\frac{2\pi}{T\csc\alpha_m}\right| = \left|2\pi\frac{\sin\alpha_m}{T}\right|. \tag{10}$$

According to the TED defined by (8), after compressed, the pulse width should be

$$\left|2\pi\frac{\sin\alpha_m}{T}\cdot\sec\alpha_m\right| = \left|2\pi\frac{1}{\cot\alpha_m\cdot T}\right| = \left|2\pi\frac{1}{-2\pi\mu\cdot T}\right| = \frac{1}{B}. \tag{11}$$

where $B = |\mu T|$. This means that the pulse-width compression ratio of aforementioned TDE is $T/(1/B)=TB$, same as the linear frequency-modulated pulse compression.

3.2 Peak Amplitude Ratio

From (9), the peak amplitude ratio of $|S_{\alpha_m}(u)|$ to $s(t)$ is

$$\xi = \left|AT\frac{1}{\sqrt{2\pi}\cdot\sqrt{\sin\alpha_m}}\right| \Big/ A = T\frac{1}{\sqrt{2\pi}\cdot\sqrt{|\sin\alpha_m|}}. \tag{12}$$

When $\alpha_m \to 0$, $|\sin\alpha_m| \to 0$, $\sin\alpha_m \approx \alpha_m$, and $\cos\alpha_m \approx 1$. Since $\cot\alpha_m = -2\pi\mu$, we know that ξ is direct proportion to T and $|\mu|$. If $|\mu|$ is large enough,

$$\cot\alpha_m = -2\pi\mu = \frac{\cos\alpha_m}{\sin\alpha_m} \approx \frac{1}{\alpha_m}. \tag{13}$$

Substitute (13) into (12)

$$\xi \approx \frac{T}{\sqrt{2\pi}}\cdot\frac{1}{\sqrt{|\alpha_m|}} = \frac{T}{\sqrt{2\pi}}\cdot\sqrt{2\pi|\mu|} = \sqrt{T^2|\mu|} = \sqrt{D}. \tag{14}$$

From (14), we know that peak amplitude ratio is about $D^{1/2}$, also same as the linear frequency-modulated pulse compression, when $|\mu|$ is large enough.

4 Simulation

Let the sampling rate be 50MHz, T=10μs, B=10MHz. The discrete FRFT algorithm was proposed by Ozaktas[5]. From (11), we know that the pulse width of TDE after compressed should be $1/B$=0.1μs. That is to say, the distance resolution is about 15m. Four targets are set on 12000km, 12010km, 13000km and 13020km apart from the observer respectively. Obviously, target 1 and 2 can not be distinguished, whereas target 3 and 4 are resoluble. The simulation result of the TDE based on the FRFT is shown in Fig. 1, which matches with the theoretical derivation well.

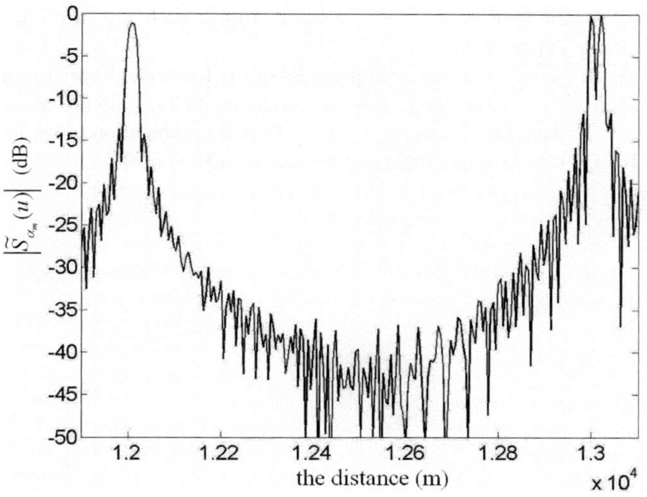

Fig. 1. The result of the TDE based on the FRFT.

5 Conclusions

Referring to linear frequency-modulated pulse compression in the time domain, we analyze the compression performance of the TDE proposed by Tao in the fractional Fourier domain. It is found that the pulse-width compression ratio is D (the product of time duration and band width), and the peak amplitude ratio is about $D^{1/2}$ when the absolute value of chirp rate is large enough. That is to say, the TDE has almost same performance as the linear frequency-modulated pulse compression. But the computing complexity of the former is less than the latter[1], which means the former has huge potential in high-resolution radar signal processing such as digital array synthetic aperture radar.

Acknowledgements

This work was supported in part by the National Natural Science Foundation of China under Grant No.60902054, China Postdoctoral Science Foundation Funded Project under Grant No.201003758 and No.20090460114, as well as the Tai Shan Scholars Program.

References

1. Tao, R., Li, X.-M., Li, Y.-L., et al.: Time Delay Estimation of Chirp Signals in the Fractional Fourier Domain. IEEE Trans. Signal Processing 57, 2852–2855 (2009)
2. Ozaktas, H.M., Kutay, M.A., Zalevsky, Z.: The fractional Fourier transform with Applications in Optics and Signal Processing. John Wiley & Sons, New York (2000)

3. Skolnik, M.I.: Introduction to radar systems (third edition), 3rd edn. McGraw-Hill Companies, New York (2001)
4. Tao, R., Deng, B., Wang, Y.: Research progress of the fractional Fourier transform in signal processing. Science in China (Ser.F, Information Science) 49, 1–25 (2006)
5. Ozaktas, H.M., Arikan, O., Kutay, M.A., et al.: Digital computation of the fractional Fourier transform. IEEE Trans. Signal Processing 44, 2141–2150 (1996)

A Modified Bit-Serial Montgomery Multiplier Algorithm in Fault Detection Method

M. Prabu[1] and R. Shanmugalakshmi[2]

[1] Research Scholar, Anna University Coimbatore, Tamil Nadu, India
prabu_pdas@yahoo.co.in
[2] Assistant Professor/CSE, Government College of Technology, Tamil Nadu, India
shanmuga_lakshmi@yahoo.co.in

Abstract. Elliptic Curve Cryptography is one of the major cryptographic algorithms which play an efficient role in cryptography and security fields. ECC makes a good conscientiousness for deployment of new level of architecture and design in those fields. In this article, a new modified architecture for the Montgomery algorithm is proposed. Montgomery multiplication is defined and derived from irreducible polynomial fields such as f(x). Here the fields can be estranged into two ways fixed and normal, a(x),b(x) are two fields elements in Galois Field used in prime number that is $GF(2^m)$. r(x) is a fixed element in $GF(2^m)$. In this article, first the bit serial Montgomery multiplier GF (2^m) is presented, then, a sequential based on circuit is added to avoid the power analysis based hackers with a consistent output. Complexities of the Montgomery multiplier in terms of gate operation and time delay of the circuit are investigated and found to be as good as or better than that of pervious bit serial architecture for the power analysis in the same field. We analyze result in graphical manner. Our modified bit-serial architecture proved the same level of output with the help of using logic gates. It produces same level of latency with different logic gates .The modified Elliptic curve based bit serial Montgomery architecture is computationally efficient and suitable for hardware implementations.

Keywords: Bit Serial Architecture, Montgomery Architecture, Polynomial Fields, Elliptic Curves.

1 Introduction

Concurrent error detection technique plays a massive role in Montgomery multipliers. It enumerates the test results gained, while the system is in working environment [9]. It consists of duplication of hardware, parity codes and time redundancy etc.; here we choose hardware duplication as an aspect of our proposal. The concurrent error technique is normally used to increase the reliability. It's a massive beneficial process for parity based on circuits, which require a high degree of reliability. Here architecture based on fault detection technique have also been proposed with a new dimensional. The proposed modified bit- serial architecture encompasses a massive part of polynomial

X. Wan (Ed.): Electrical Power Systems and Computers, LNEE 99, pp. 345–351.
springerlink.com

basis multiplier including different bit-serial and bit parallel architecture. Some extended bit- level Montgomery multiplications have been explained through the parity bit [1]. The Montgomery multiplication algorithm has received a lot of attention in finite field arithmetic see in example [2]. [5].and [12]. A new modular multiplication algorithm for integer has been proposed [10].

In this article, we propose a new modified bit-serial architecture for the Montgomery algorithm. It increases the complexity in hacker's performance through circuit and logic gates. Our goal is to maximize the circuit level and utilize the gates with same level of latency, which already explained in [3].

The summary of this article is as follows:

In section 2, firstly, a brief review on the Montgomery algorithm is stated. Design methodology and the Montgomery algorithms were presented in sections 3 and 4, respectively. Consecutively bit serial architectural design and the modified bit serial architectural design were explained in section 5 and 6. A comparison of our proposal with some others is made in section 7. A few concluding remarks are given in section 8.

2 Previous Work

In this section, we briefly analyze some basic concept, which are used throughout this article. We mainly concentrate on Montgomery algorithm and the bit serial Montgomery architectural design. Then, we consider the previous work of [2] error detection in Montgomery algorithm over GF (2^m).

2.1 Montgomery Algorithm

In arithmetic computation, Montgomery is an algorithm introduced in 1985 by peter Montgomery, this algorithm allows modular arithmetic to be performed efficiently when the modulus is large. A single application of the Montgomery algorithm is faster than "naïve" modular multiplication [15].

C= a x b mod n
C=a mod n
C=b mod n

The new modified bit serial architecture has all the capabilities of original architecture, but provides the better complexity for the hackers, who are willing to hack the system through power analysis. The reason for this complexity is in this new modified algorithm; all devices are interacted and implemented through logic gates. Each and every logic gates have a specific power value. Here we added logic gates such as NOT, AND, OR etc., in this modified bit-serial Montgomery algorithm. We include these logic gates and measure the power consumption value. It becomes tedious to trace out the performance of original architecture. A digit serial Montgomery multiplier algorithm is proposed in [11].which is based on the algorithm proposed in [5].

3 Design Methodology

To improve efficiency and complexity for the hackers, this algorithm is based on logic gates and circuit design level with the same latency and the latch power. Algorithm 1 shows the normal Montgomery algorithm and algorithm 2 refers the bit serial architectural design. Algorithm 3 illustrates the new modified Montgomery algorithm with architectural design. The procedure of architectural design is explained below.

It is similar to the bit–serial Montgomery algorithm but increases the complexity with power values through the latch.

Step 1: Take the analysis of bit-serial Montgomery algorithm with same latency (Algorithm 1).
Step 2: Add the new logic gate component to the architectural design
Step 3: Combine the design with (NOT) logic gate (Figure 1)
Step 4: Simplify the table. The final architecture is shown in (Figure 2) and the algorithm in (Algorithm3)

The design methodology helps to determine the latency and complexity through latch for modified bit serial Montgomery architecture. At the end the execution results are summarized in Table 1.

4 Montgomery Multiplier Algorithm

Montgomery multiplication is defined both in prime field GF(p) and in binary field GF(2n). GF(p) incorporates on number, GF(2n) incorporates on polynomial leading carry for arithmetic. Montgomery multiplication is defined as MonPro (a, b) =a.b. 2^{-n} (mod p)

Algorithm 1. Montgomery Multiplier Algorithm

Input A=(am-i............a1,a0)x
B=(bm-1..............b1,b0)x
P=(pm-1..........P1,P0)x
Q(x)=-p-1(x) modx1
Output: c(x)=A(x) B(x) x^{-n} mod P(x)

Elliptic curve based on Montgomery multiplier algorithm is an very efficient algorithm for security, there is no other blabbering against the elliptic curve based Montgomery algorithm. So many proofs were shown already in previous works. Algorithm 1, represents the basic polynomial based on Montgomery multiplier algorithm, which is purely based on polynomial aspects. Here, we incorporates four parameters such as A(x), B(x), C(x) and P(x). The A(x), B(x) and C(x) are the variables. The P(x) is the polynomial based mod property.

5 Bit- Serial Montgomery Algorithm

Algorithm 2: **Bit- Serial Montgomery Algorithm**
Inputs: A, B, F(x)
Output c= A.B.r^{-1} mod F(x)
Step 1: T:=0
Step 2: For i=0 to m-1
Step 3: T` := T+ bi A
Step 4: T`` := T`+t_0' F(x)
Step 5: T:= T``/x
Step 6: C:=T

Here, the Bit- serial Montgomery represents the name of the work with the help of logic gates such as 2 AND & 2 XOR gates. In algorithm 2, line 4 F(x) store the mode value for both the algorithm 2 and 3. (T'), and (-T) denotes the basic terminology of temporary variable, which is used to store the logic gates values.

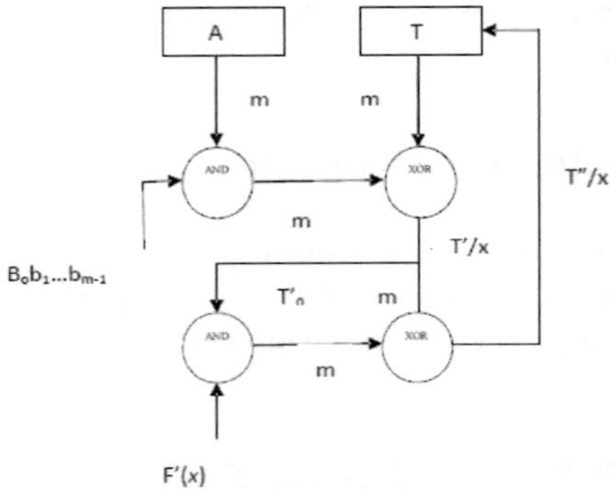

Fig. 1. Fault Detection Method Based Bit Serial Montgomery Algorithm

6 Modified Bit-Serial Montgomery Algorithm

Algorithm 2: Modified Bit-Serial Montgomery Algorithm

Inputs: A, B, F(*x*)
Output c= A.B.r^{-1} mod F(*x*)
Step 1: T:=0
Step 2: For i=0 to m-1
Step 3: T` := T+ bi A
Step 4: T` := T`(-T`)
Step 5: T``:= T`+$t_0$` F(*x*)
Step 6: T:= T``/x
Step 7: T``= T"(-T``)
Step 8: C:=T

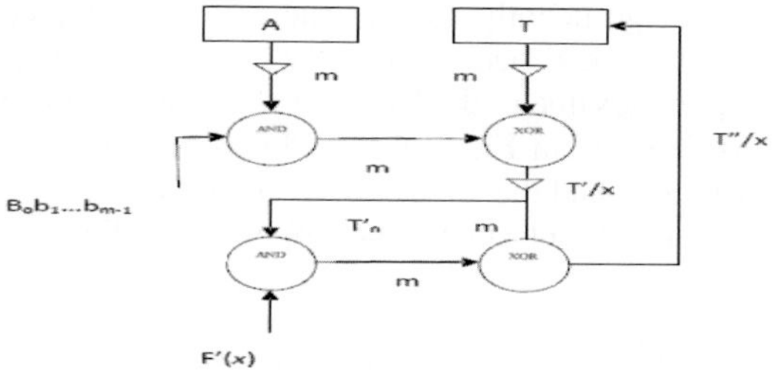

Fig. 2. Fault Detection Method Based Modified Bit Serial Montgomery Algorithm

The Modified algorithm fully loaded with latches and their logic gates. In algorithm 3, line 4, we can add the T (-T) as normal inverse process, which totally change the environment. After this change had been applied, another NOT gate is added to the same algorithm 3, line 7. It makes the changes in the previous environment.

7 Comparison and Discussion

In this discussion,

The two different architectures were compared, while executing the events through the combinational and sequential circuits through latches. Here we verified these architectures by using logical designs. The latch critical path delay and latency are compared in table [1].

To evaluate the improvement of the new modified architecture, we consider the time redundancy based on fault detection algorithm explained through the flowchart (based on Montgomery Algorithm).

$$H=h_{m-1} x^{m-1} +\ldots.+h_1x+h_0$$
$$x^m= f_{m-1}x^{m-1}+\ldots.f_1x+1$$
$$H.x^m \bmod F(x)$$
$$H.x^m= (h_{m-1} x^{m-1} +\ldots.+h_1x+h_{0)}$$
$$(f_{m-1}x^{m-1}+\ldots.f_1x+1) \bmod F(x)$$
$$x^{-1}=(x^{m-1}+f_{m-1}x^{m-2}+\ldots\ldots+f_1)$$
$$H.x^{-1}=(h_{m-1}x^{m-1}+\ldots..h_1x+h_0)$$
$$(x_{m-1}+f_{m-1}x^{m-2}+\ldots\ldots\ldots f_1) \bmod F(x)$$
$$= h_0.(x_{m-1}+f_{m-1}x^{m-2}+\ldots..f_1) \bmod f(x) + (h_{m-1}x^{m-2}+h_1x+h_1)$$
$$=h_0(x_{m-1})+(h_0f_{m-1}x^{m-2}+h_{m-1})x^{m-2}+h_0(f_1+h_1)\ldots\ldots\ldots\ldots\ldots\ldots\ldots\ldots\ldots.7$$

The time complexity and the area complexity of the finite filed multiplication are o(m2) and o(log 2 m).Note that [2] includes three of these multiplications, where two of them are in series in the critical path.

The area of complexity of (7) o(m) and its constant time is TA+TX. The multiplication of H(x) mod F(x) before comparison has the same area and complexities.

The modified architecture significantly reduces the area and time complexities of the error detection scheme [2].our modified architecture has the same capacity, while it has lower time and the complexities due to the Not gate performance. Now, we consider the bit-serial architecture as depicted in Fig 2, the modified bit serial architecture has the critical path of 2(TA+TX+TN) and the latency of m clock cycles.

Error detection in bit serial Montgomery multiplier over GF (2^m) are given below,

Table 1. Comparison between the Bit-serial Architecture, and Modified Bit-serial architecture

Architecture	#AND	#XOR	#NOT	#Latch	Critical path Delay	Latency
Bit- Serial Montgomery Multiplier	2m-1	2m-1	-	2m	2(TA+TX)	M
Modified Bit-serial Montgomery Multiplier	2m-1	2m-1	2m-1	2m	2(TA+TX+TN)	M

It can be seen that 2 XOR and 2 AND gates are used. And the latch and the latency value is equivalent to the latency even after adding one logical NOT gate.

8 Conclusion

In this article, we have considered error detection in Montgomery multiplication over binary extension fields. A new modified error detection technique is a bit-serial Montgomery over GF (2m) in elliptic curve field. We proved that our modified Montgomery architecture results in table [2] significantly produce the variation between the previously published algorithms. These modified architectures reduce the time and overheads for Montgomery multipliers. Especially it incorporates the logic gates in design phase. It increases the performance and makes it more complex and tedious for hackers' to break this architecture with consistent latency value. Besides, it also derives a polynomial based complex equation from the elliptic curve through modified bit-serial Montgomery architecture over GF (2m).Using a new Montgomery factors, we have proposed a modified bit-serial Montgomery multipliers, which is faster than the previously published Montgomery multipliers.

Acknowledgement

The authors thank anonyms reviewers for their fruitful comments and helpful comments to deliver this article.

References

1. Shantz, S.C.: From Euclid's GCD to Montgomery Multiplication to the Great Divide. Sun Microsystems Laboratories, 901 San Antonio Road, Palo Alto, California
2. Castryck, W., Galbraith, S., Farashahi, R.R.: Efficient arithmetic on elliptic curves using a mixed Edwards Montgomery representation. Springer Lecture Notes in Computer Science (LNCS). Springer, Heidelberg
3. Lee, Y.K., Verbauwhede, I.: A Compact Architecture for Montgomery Elliptic Curve Scalar Multiplication Processor. Springer Lecture Notes in Computer Science (LNCS). Springer, Heidelberg
4. Ibrahim, A., Gebali, F., El-Simary, H., Nassar, A.: High-performance, low-power architecture for scalable radix 2 montgomery modular multiplication algorithm. Canda Journal Electronics Computer Engineering 34(4) (Fall 2009)
5. Ahmadi, H.R., Afzali-Kusha, A.: Low-Power Low-Energy Prime-Field ECC Processor Based on Montgomery Modular Inverse Algorithm. In: 12th Euromicro Conference on Digital System Design / Architectures, Methods and Tools (2009), 978-0-7695-3782-5/09, doi: 10.1109/DSD.2009.140
6. Miyamoto, A., Homma, N., Aoki, T., Satoh, A.: SPA against an FPGA-Based RSA Implementation with a High-Radix Montgomery Multiplier, pp. 1847–1850, doi: 1-4244-0921-7/07
7. Mukaida, K., Takenaka, M., Torii, N., Masui, S.: Design of High-speed and Area-Efficient Montgomery Modular Multiplier for RSA Algorithm. In: Symposium On VLSl Circuits Digest of Technical Papers, pp. 320–323 (2004), doi: 0 7803-8287-010
8. Shin, J.-B., Kim, J., Lee-Kwang, H.: Optimisation of Montgomery modular multiplication algorithm for systolic arrays. Electronics Letters 34(79), 1830–1831 (1998)
9. Lai, J.-Y., Hung, T.-Y., Yang, K.-H., Huang, C.-T.: High-Performance Architecture for Elliptic Curve Cryptography over Binary Field, pp. 3033–3936, doi: 978-1-4244-5309-2/10
10. Homma, N., Miyamoto, A., Aoki, T., Satoh, A., Shamir, A.: Comparative Power Analysis of Modular Exponentiation Algorithms. IEEE Transaction on Computers 59(6), 795–807 (2010)
11. de Dormale, G.M., Bulens, P., Quisquater, J.-J.: An Improved Montgomery Modular Inversion Targeted for Efficient Implementation on FPGA. In: ICFPT 2004, pp. 441–444 (2004), doi: 0-7803-8652-3/04
12. McIvor, C., McLoone, M., McCanny, J.V.: Improved Montgomery modular inverse algorithm. Electronics Letters 40(18) (September 2, 2004)
13. Miyamoto, A., Homma, N., Aoki, T., Satoh, A.: Systematic Design of RSA Processors Based on High-Radix Montgomery Multipliers. IEEE Transactions on very large scale integration(VLSI) Systems 1, 1063–8210, doi: 10.1109/TVLSI.2010.2049037
14. Ibrahim, A.A., Elsimary, H.A., Nassar, A.M.: Design and Implementation of Scalable Low Power Radix-4 Montgomery Modular Multiplier. In: IEEE Conference,
15. Son, H.-K., Oh, S.-G.: Design and Implementation of Scalable Low-Power Montgomery Multiplier. In: Proceedings of the IEEE International Conference on Computer Design (ICCD 2004) (2004), doi: 1063-6404/04

An Intelligent Approach for Medium Term Hydropower Scheduling Using Ensemble Model

Thais Gama de Siqueira[1] and Ricardo Menezes Salgado[2]

[1] Institute of Science and Technology - University of Alfenas
Poços de Caldas – MG - Brazil
[2] Institute of Exact Sciences – Federal University of Alfenas
Alfenas – MG – Brazil
thaisgama@unifal-mg.edu.br, ricardomenezes@ieee.org

Abstract. The medium term hydropower scheduling (MTHS) problem involves an attempt to determine, for each time stage of the planning period, the amount of generation at each hydro plant which will maximize the expected future benefits throughout the planning period, while respecting plant operational constraints. Besides, it is important to emphasize that this decision-making has been done based mainly on inflow earliness knowledge. To perform the forecast of a determinate basin, it is possible to use some intelligent computational approaches. In this paper one considers the Dynamic Programming (DP) with the inflows given by their average values, thus turning the problem into a deterministic one which the solution can be obtained by deterministic DP (DDP). The performance of the DDP technique in the MTHS problem was assessed by simulation using the ensemble prediction models. Features and sensitivities of these models are discussed.

Keywords: Medium Term Hydropower Scheduling, Dynamic Programming, Inflow Forecast, Artificial Intelligence, Predictive Models, Ensembles.

1 Introduction

The medium term hydrothermal scheduling (MTHS) problem is quite complex due to some of their characteristics, specially the randomness of inflows 1, 3. For the case of single reservoir systems, stochastic dynamic programming (SDP) models have been widely suggested as a proper technique to solve MTHS due to its ability to handle stochastic inflows and nonlinear relations. Alternatives to stochastic models for MTHS can be developed through operational policies based on deterministic models. The advantage of such approaches is their ability to handle multiple reservoir systems without the need of any modeling manipulation.

Deterministic models assume that the future inflows are known along the planning period and determine the set of discharge and spillage decisions that will correspond to the optimal reservoir evolution for that pre-established inflow sequence 7. The optimization process is based on a previous knowledge of the future possibilities and its consequences, satisfying the Bellman optimality principle. Thus, the total optimal operation cost from stage t until the end of the planning period is obtained as the one that minimizes the sum of the present cost at stage t with the optimal future cost for

X. Wan (Ed.): Electrical Power Systems and Computers, LNEE 99, pp. 353–362.
springerlink.com

the remaining stages, which were previously determined. Therefore, the solution by Deterministic Dynamic Programming (DDP) is performed by the backward resolution of the following recursive equation at each stage.

To perform the forecast of a determinate basin, it is possible to use some computational approaches, such as: Neural Networks, Autoregressive Models, Genetic Programming, Neuro-Fuzzy Logic, and others, largely used to building models to solve time series forecasting problems [4] and [8]. In this paper, the inflow forecasting will be made using the ensemble technique. This approach was chosen because the ensemble was computed with the information from several forecasters combined into a single forecast. This strategy usually makes the prediction of the ensemble better than the individual forecasting thus, increasing the accuracy in obtaining the predictions.

Many papers have been done in investigating why and how an ensemble of learning machines works. The basic motivation is to find proper mechanisms for exploiting, instead of ignoring, the information constructed by each component throughout its learning process, in such a way as to produce a final model containing the best of each individual capability generated.

In this paper, the inflow predictions performed by ensemble were used in the simulation of the operational policies of DDP and its results are compared with predictions obtained by conventional models (based on individual predictors). In the results we can see that the costs were lower with results were obtained with the forecasts of the ensemble, thus these technique showed superior results if compared to model DDP fed inflows provided via traditional forecasting models.

This paper intends to be a contribution to the question of what kind of model would be more suitable for MTHS problems. The deterministic dynamic programming approach is implemented and compared on single hydro plant systems. Two different hydro plants located in different river basins in Brazil have been selected for the case studies performed. This paper is organized as follows: Section 2 describes the deterministic model used to solve the MTHS. In section 3 the forecasting methodology is described. Section 4 presents an application of the ensemble model in monthly inflow forecasting. Section 5 presents the case studies performed and section 6 states the conclusions.

2 Deterministic Model

For systems comprising a single hydro plant, the deterministic version of the MTHS can be formulated as the following nonlinear programming problem:

$$\min \sum_{t=1}^{T-1} \psi_t (D_t - Gh_t) \quad (1) \quad \text{Subject to:}$$

$$Gh_t = k.h_{lt}.q_t \quad \forall t \quad (2)$$

$$h_{lt} = \phi(x_t) - \theta(u_t) - pc \quad \forall t \quad (3)$$

$$x_t = x_{t-1} + (y_t - u_t)\gamma \quad \forall t \quad (4)$$

$$u_t = q_t + s_t \quad \forall t \quad (5)$$

$$\underline{x}_t \le x_t \le \overline{x}_t \quad \forall t \quad (6)$$

$$\underline{u}_t \le u_t \le \overline{u}_t \quad \forall t \quad (7)$$

$$\underline{q}_t \le q_t \le \overline{q}_t(h_{lt}) \quad \forall t \quad (8)$$

$$s_t \ge 0 \quad \forall t \quad (9)$$

where, t is the index of the time interval (month); T is the number of time intervals in the optimization period; ψ_t is the cost function associated with non-hydraulic complementary generation at interval t (in \$); Gh_t is the total hydro generation at interval t (in MW); D_t is the load demand at interval t (in average MW); x_t is the water storage at the end of interval t (in hm3); $\underline{x}_t$, $\overline{x}_t$ are bounds on minimum and maximum storage at interval t; u_t is the water release from the reservoir at interval t (in m3/s); $\underline{u}_t$, $\overline{u}_t$ are bounds on minimum and maximum water release from the reservoir at interval t; qt is the water discharge through the turbines at interval t (in m^3/s); $\underline{q}_t$, $\overline{q}_t$ are bounds on minimum and maximum water discharge at interval t; s_t is the water spillage from the reservoir at interval t (in hm3); k is a constant factor representing the product of water density, gravity acceleration and average turbine/generator efficiency (in $MW/(m^3/s)\,m$); ϕ_i is the forebay elevation function (in m); θ_i is the tailrace elevation function (in m); pc is the average penstock head loss of the hydro plant (in m); y_t is the incremental water inflow into the reservoir at interval t (in m3/s) and γ is a constant factor that converts flow from (m3/s) into (hm3/month);

The operational cost ψ_t represents the minimum cost of non-hydraulic complementary generation, such as, thermoelectric generation, imported from neighboring systems, or even load shortage. This cost is obtained by an economic dispatch and, as consequence, is a convex decreasing function of total hydro generation Gh_t for a given system load demand D_t.

Hydro generation at interval t, represented by (2), is a nonlinear function of the water storage in the reservoir, the water discharge through the turbines and the water spillage from the reservoir. Equality constraints in (4) represent the water balance in the reservoir, where terms such as evaporation and infiltration have not been considered for the sake of simplicity. Lower and upper bounds on variables, expressed by constraints (6)-(9) are imposed by the physical operational constraints of hydro plants, as well as the constraints associated with multiple uses of water.

2.1 Deterministic Dynamic Programming

Deterministic models assume that the future inflows are known along the planning period and determine the set of discharge and spillage decisions that will correspond to the optimal reservoir evolution for that pre-established inflow sequence.

Deterministic Dynamic Programming (DDP) 5-[6] can be used to solve problem (1)-(9) assuming for instance the future inflows given by their expected values $\overline{y}_t$. The optimization problem is divided into stages and at each stage the optimal control variable is chosen in order to minimize the cost function for each state of the system. The state variable is represented by the reservoir storage and the control variables by the water discharge and spillage.

The optimization process is based on a previous knowledge of the future possibilities and its consequences, satisfying the Bellman optimality principle 2. Thus, the total optimal operation cost from stage t until the end of the planning period is obtained as the one that minimizes the sum of the present cost at stage t with the optimal future cost for the remaining stages, which were previously determined. Therefore, the solution by DDP is performed by the backward resolution of the following recursive equation at each stage:

$$\alpha_{t-1}(x_{t-1}) = \min_{q_t} \{C_p + C_f\} \quad (10)$$

$$C_p = \psi_t(D_t - Gh_t) \quad (11)$$

$$C_f = \alpha_t(x_t) \quad (12)$$

$$x_t = x_{t-1} + (\overline{y}_t - q_t)\gamma \quad (13)$$

Note that the decision variable is the discharge q_t since the spillage s_t works as a slack variable that is different from zero only when is not possible to increase the discharge and store the excess of water. The solution process starts at the final stage T where the cost $\alpha_T(x_T)$ is supposed to be known. If not, the horizon can be extended in order to allow the adoption of a null terminal cost function with no influence on the solution over the planning period. The resolution goes back until the initial stage t=0 according to the recursive equation (10)-(13), where Cp is the present cost at stage t and C_f is the optimal future cost from the end of that stage until the end of the planning period. By this way, the decision rule in a DDP operation policy is given by decision tables witch provide optimal water discharge and the operational cost for each possible discrete state of the system, considering the inflows given by their average values.

3 Intelligent Forecasting Model

The forecasting methodology used in this paper is based on the technique of combining predictors, also known as ensemble. The goal is to use ensemble techniques for obtaining accurate forecasts and less variability in the historical inflow and with these predictions is possible to obtain reliable simulations for the hydrothermal planning.

3.1 Ensembles

The term "ensemble" is commonly used for the combination of a set of learning machines (hereafter referred to as components, models or predictors) that provides isolated solutions to the same task, generally obtained from different means (e.g., by employing different machine learning paradigms such as ANNs, decision trees, etc.) 13, 14 and 15. As pointed out by Sharkey [16], "combining a set of imperfect predictors can be thought of as a way of managing the recognized limitations of the individual predictors; each component is known to make errors, but they are combined in such a way as to minimize the effect of these errors".

In the construction process of an ensemble, there are three basic steps: generation of many different candidates to components, selection of components, and combination of their results. To implement this methodology, usually three data sets are needed: a data set for generating the candidates to components, another for selecting them, and yet another one to test the ensemble's performance. After generating several candidates to components, the selection of the best subset of those components is fulfilled. Generally, this selection is made aiming at decreasing the generalization error. The remaining step to construct an ensemble is to combine the components. In other words, the outputs of the selected components must be someway combined to generate a unique output for the ensemble. In what follows, additional details are provided concerning the design procedure adopted in this paper for ensembles. In order to solve the load forecasting problem, one adopts as candidates to compose an ensemble neural networks model, more specifically multilayer perceptrons neural networks and symbolic regression based on genetic programming.

Multilayer Perceptrons (MLPs): MLP architectures are known as the most frequently adopted neural network models for regression tasks. An MLP consists of n inputs nodes, h hidden layer nodes, and m output nodes connected in a feed-forward fashion via multiplicative weights that can be arranged in a weight matrix W. The MLP must be trained with historical data to find the appropriate values for the elements in matrix W, given the number of neurons in the hidden layer. In this paper, the learning algorithm employed is the well-known error back-propagation.

Genetic Programming and Symbolic Regression (GP): In conventional regression, one has to decide on the approximation function (can be an n-degree polynomial, non-polynomial, or a combination of both) and try to find the coefficients of this selected function. Constructing an approximation function can be a hard task. There is another form of regression called "symbolic regression" where, the aim is to find out a symbolic representation of a model, instead of only searching for coefficients of a predefined model. Genetic programming (GP) method introduced by Koza [17] can be used for the symbolic regression problem. GP searches for the model and coefficients of the model at the same time. In GP, individuals are represented as trees. Elements of the trees are functions and terminals. Terminals are the variables and the functions are operations applied to these variables forming the model together. Genetic programming is an iterative method and the first step in the genetic programming algorithm is the generation of an initial population either by using random compositions of the functions and terminals or by using a predefined strategy. In the next step, the termination condition is checked [18]. If the termination condition is reached, the process is ended and best result so far is reported.

Generation of the components: The main purpose of an ensemble is to provide performance improvement, which is obtained when a few requirements are satisfied by the candidates to components. However, the generation of components is still the most relevant and the most demanding phase of the whole design process. Here, the objective of the generation of components is to synthesize ANN's and GP's with good individual performance and outputs as dissimilar as possible.

Selection of Components: In this phase, the candidates already implemented are considered interesting according to their individual good performance and diversity. However, there is no theoretical foundation that can guarantee that those components will contribute positively; it is just an empirical conclusion. After generating the components, it is possible to select the best by assigning an individual performance criteria to each component, based on the correlation indices and error rates, choosing only the n best or until some stopping condition is satisfied.

In this work, one considers the same technique proposed in 9, where the components are ordered according to their mean square error (MSE). The 10 components that present a better performance when according to the performance of the best one are selected. This number was chosen based on experimental results with trial and test process.

The combined forecasting model: The idea of using a combined forecasting model is not recent. Since the pioneering works of Reid 10 and of Bates and Granger 11, several attempts have been conducted to indicate that a combination of forecasts often outperforms the forecasts obtained from a single source.

The main problem when combining forecasts can be described as follows. Suppose there are r forecasts such as $\hat{y}_1(t), \hat{y}_2(t), \ldots, \hat{y}_r(t)$. The point is how to combine these different forecasts into a single forecaster $\hat{y}_{ag}(t)$, which is intended to be a more accurate one. The general form of the model for such combination is defined as:

$$\hat{y}_{ag} = \sum_{i=1}^{r} w_i \hat{y}_i \tag{14}$$

where w_i denotes the assigned weight of $\hat{y}_i(t)$.

There are a variety of methods available to determine the weights used in the combined forecasts. First of all, the equal weights method, which uses an arithmetic average of the individual forecasts, is a relatively easy and robust method. However, since the components are diverse in terms of behavior and average performance, in practice a simple average may not be the best technique to use 12. In this work, one uses four strategies for the combination of individual forecasts. Table 1 presents the equations (A to D) that were used to combine the values of the predictions and find the forecast ensemble.

Table 1. Metrics for combining the results.

$AM = \dfrac{Y_1 + Y_2 + Y_3 + \ldots + Y_n}{n}$ (A)	$PM = \dfrac{\rho_1 Y_1 + \rho_2 Y_2 + \ldots + \rho_n Y_n}{\rho_1 + \rho_2 + \ldots + \rho_n}$ (B)
$GM = \sqrt[n]{Y_1.Y_2.Y_3.\ \ldots\ .Y_n}$ (C)	$QM = \sqrt{\dfrac{Y_1^2 + Y_2^2 + Y_3^2 + \ldots + Y_n^2}{n}}$ (D)

where *Yi* is the forecasting of the predictor *i*, *n* is the total number of predictors, *AM* is the arithmetic mean, *PM* is the pondered mean, *GM* is the geometric mean and *QM* is the quadratic mean of the individual forecasts.

4 Application of the Ensemble Model Inflow Forecasting

The proposed ensemble model was configured using several ANN's and GP's predictors, calibrated with different configurations. The ensemble was divided in three stages: generating and training the components, validating and selecting the components, and combining and predicting. The application involves short-term load forecasting. To determine the inflow of the next month an adjustment of the models is accomplished to deal with every month of the year on step ahead. Thus, for each time interval an individual forecasting model is adjusted. So, 12 forecasts will be accomplished for each component of the ensemble, aiming at obtaining the prediction of the whole inflow demand for each month of the year.

To create a forecast component from any one technique, one needs to select the parameters and information to adjust the component. One of the difficulties is that each forecasting technique has distinct characteristics and parameters that should be explored to find the combination of the best results. The choice of parameters is associated with the type of prediction model (ANN, GP, Autoregressive, and others) and the choice of inputs is related to the degree of correlation between inflows.

To create the forecast components uncorrelated, and increase the capacity of generalization of the ensemble, it was set random values for the parameters of each model. Tables 3 and 4 show the ranges of selection for each type of component, ANN and GP. These parameters were selected to create and configure the ensemble components.

Table 2. Set of parametrers ANN.

N. of input Patterns	[20 – 65]
N. of Lags	[5 – 12]
N. of Hidden Layers	[3 – 4]
N. of Neurons	[9 – 10]
Learning rate	[1.0 – 2.0]
Momentun rate	[0.4 - 0.6]
Activation Funcion	[tanh,logsig]

Table 3. Set of parametrers GP.

N. of input Patterns	[20 – 65]
N. of Lags	[5 – 12]
Math Operators	[+, -, *, /]
Type of Mutations	[M1, M2, M3, M4]

The selection of the best forecast components is based on the variation of the parameters previously defined. Five hundred prediction components are generated for each technique and the ten better are used in the ensemble forecasting. The inflow forecasted by the ensemble was estimated based on equations presented in Table 1. In this case, n represents the total number of components which is always assumed as 10. The number of components used to compose the ensemble was defined from empirical testing in a process of trial and error.

5 Results

The hydro plants were chosen for the case studies: Furnas in the Grande River, located in the southeastern region of Brazil; and Sobradinho in the São Francisco River, located in the northeastern region of Brazil. Some relevant information about these hydro plants is shown in Table 5. In order to get equilibrated hydrothermal systems, the thermal plant capacity, in MW, was considered equal to the installed capacity of the hydro plants, and the load demand was assumed constant and equal to half the installed capacity of the systems. The operational cost ψ_t was given by the following quadratic function:

$$\psi_t = 0.02(D_t - Gh_t)^2 \tag{15}$$

Table 4. Hydro plant operational data.

Hydro Plant	Installed Capacity (MW)	Storage Capacity (hm^3)	Discharge min/max (m^3/s)
Furnas	1312	17217	211 / 1624,3
Sobradinho	1050	28669	713 / 4226,2

The control policies were implemented and simulated in a monthly basis throughout the inflow historical sequence, which in this case begins in 1931. All the constraints presented in the formulation of the optimization problem have been considered in the simulation. The discretization adopted in the DP policy for state and control variables were 60 and 30, respectively. Tables 6 summarizes the simulation results for Furnas and Sobradinho hydro plants, respectively, in terms of average and standard deviation values of generation, and average operational cost. The results presents the simulation results obtained using the historical inflow, best forecasting individual component and the four ensembles performed during the decade of 50 (dry period).

Table 5. Simulation results for Furnas and Sobradinho hydro plant during the 50s.

		Generation (MW)		Cost ($)
		Average	Std. Dev.	Average
Historical		668.99	153.38	8.73e+003
Component 0		672.08	153.50	8.65e+03
Ensembles	*AM*	673.92	138.71	8.52e+03
	PM	673.81	147.24	8.57e+03
	QM	687.35	144.99	8.22e+03
	GM	659.87	135.17	8.86e+03

		Generation (MW)		Cost ($)
		Average	Std. Dev.	Average
Historical		580.46	199.46	5.19e+03
Component 0		564.07	199.76	5.51e+03
Ensembles	*AM*	589.24	189.96	4.96e+03
	PM	583.30	190.06	5.07e+03
	QM	598.24	190.52	4.80e+03
	GM	580.83	190.96	5.12e+03

As one could expect the better performance for Furnas and Sobradinho hydro plants are associated to the ensemble models, due to its sophisticated forecasting inflow modeling. The quadratic mean ensemble approach furnished the better results in terms of generation and operational cost. For both case studies the historical inflow records provide higher standard deviation associated to the generation, once the DP considers the inflow records average.

In the best situation for decade of 50 the minimum operational cost for Furnas and Sobradinho, respectively, were $ 8.22e+03 and $ 4.80e+03. The same conclusion behavior occurs if one compares the hydro power generation, but in this case, one considers the higher amount obtained from the simulation results.

Fig 1 show the water reservoir storage trajectory obtained with the simulation to ilustrate the performance of the best individual component, the best ensemble and the worst ensemble, for Furnas and Sobradinho hydro plants during the 70s. Overall, for both hydro plants considered the forecasting ensemble models are more cost efective than the individual ones.

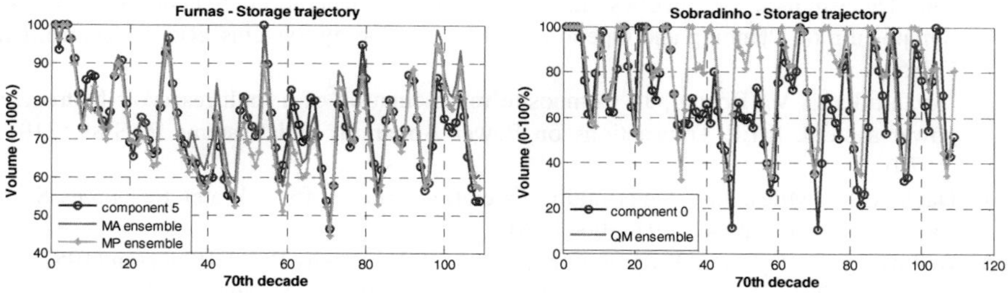

Fig. 1. Storage trajectory of Furnas and Sobradinho during the 70s.

6 Conclusions

This work has analyzed the performance of intelligent forecasting models for medium term hydrothermal scheduling. A deterministic dynamic programming model that determines the decision variables assuming the future inflows given by average values was considered in optimization.

In order to concentrate the analysis on the individual aspects of each hydro plant, the performance comparisons were performed for systems composed of a single hydro plant. The numerical results were obtained through simulation using historical inflow records and the inflows given by several inflow forecasting. Two hydro plants located in different river basins in Brazil were selected for the case studies and two different decades were considered to evaluate the performance of the ensemble models. The simulation results have shown that inflow forecasting approaches using ensemble models have provided higher performance, especially with respect to the operational cost and hydro generation. Summarizing, the simulation using some intelligent computational approaches presented improvement in the average hydro generation, since the goal of the MTHS is to minimize the total operational cost. Even considering critical periods the ensemble models had a coherent performance. Therefore, one can conclude that the ensemble models are more attractive than the individual forecasting ones.

Acknowledgment

This work was supported by the Research Foundation of the State of Minas Gerais (FAPEMIG).

References

1. Pereira, M.V.F.: Optimal Scheduling of Hydrothermal System – An Overview. In: IFAC Symposium on Planning and Operation of Electric Energy Systems, Rio de Janeiro, pp. 1–9 (1985)
2. Araripe Neto, T.A., Cotia, C.B., Pereira, M.V.F., Kelman, J.: Comparison of Stochastic and Deterministic Approches in Hydrothermal Generation Scheduling. In: IFAC Symposium on Planning and Operaion of Electric Energy Systems, Rio de Janeiro, Brazil (1985)
3. Arvanitidis, N.V., Rosing, J.: Composite representation of a multireservoir hydroelectric power system. IEEE Transactions on Power Apparatus and Systems PAS-89, 319–326 (1970)
4. Dagli, C.H., Miles, J.F.: Determining Operating Policies for a Water Resources System. Journal of Hydrology 47, 297–306 (1980)
5. Bellman, R.E.: Dynamic Programming. Princeton University Press, Princeton (1957)
6. Bertsekas, D.P.: Dynamic Programming: Deterministic and Stochastic Models. Academic Press, London (1987)
7. Stedinger, J.R., Sule, B.F., Loucks, D.P.: Stochastic Dynamic Programming Models for Reservoir Operation Optimization. Water Resources Research 20(11), 1499–1505 (1984)
8. Martinez, L., Soares, S.: Comparison between Closed-Loop and Partial Open-Loop Feedback Control Policies in Long Term Hydrothermal Scheduling. IEEE Transactions on Power Systems 17(2) (2002)
9. Perrone, M.P.: Improving regression estimates: Averaging methods for variance reduction with extensions to general convex measure optimization. PhD Thesis, Brown University (1993)
10. Reid, D.J.: Combining three estimates of gross domestic product. Economics 35, 431–444 (1968)
11. Bates, J.M., Granger, G.W.J.: The combination of forecasts. Operations Research Quaterly 20, 451–468 (1969)
12. Kang, B.H.: Unstable weights in the combination of forecasts. Management Science 32, 683–695 (1986)
13. Baxt, W.G.: Improving the accuracy of an artificial neural network using multiple differently trained networks. Neural Computation 4(5), 135–144 (1992)
14. Dietterich, T.G.: Ensemble methods in machine learning. In: Kittler, J., Roli, F. (eds.) MCS 2000. LNCS, vol. 1857, pp. 1–15. Springer, Heidelberg (2000)
15. Hansen, L., Salamon, P.: Neural network ensembles. IEEE Transactions on Pattern Analysis and Machine Intelligence 12(10), 993–1001 (1990)
16. Sharkey, A. (ed.): Combining artificial neural nets: Ensemble and modular multi-net systems. Springer, London (1999)
17. Koza, J.R.: Genetic Programming: On the Programming of Computers by Means of Natural Selection. MIT Press, Cambridge (1992)
18. Koza, J.R.: Survey of genetic algorithms and genetic programming. In: Proceedings of the Wescon 95 - Conference Record: Microelectronics, Communications Technology, Producing Quality Products, Mobile and Portable Power, Emerging Technologies, San Francisco, CA, November 7–9. IEEE, New York (1995)

Research on Grid Resource Scheduling Algorithm Based on Multi-attribute Constraints

Jing Chen[1] and Duanjun Chen[2]

[1] Department of Computer, College of Information Science and Engineering, Yanshan University, Qinhuangdao 066004, China
[2] Beijing Jiaotong University Haibin College, Huanghua 061100, China
xychenjing@ysu.edu.cn

Abstract. To improve the completed number of grid tasks, grid resource scheduling algorithm MCSA (Multi-attribute Constraints Scheduling Algorithm) was proposed. The accomplished process of grid resource scheduling algorithm was transformed multi-attribute constraints, according to the parametric resource information, task information and the thought of parameter sweep scheduling algorithm. Classifying different task weight value according to the priority of tasks and the given task weigh value by simulation experiment results. Scheduling algorithm MCSA were simulated by the GridSim toolkits according to different conditions, which show that algorithm MCSA is effective and stable in solving such kind of issues.

Keywords: Multi-attribute constraints, grid resource scheduling, task weight, GridSim.

1 Introduction

Grid task scheduling is important research direction in the field of grid computing[1]. It has been proved that the task scheduling is NP problem, according to researching the traditional task scheduling problem. Therefore, studying task scheduling algorithm, that is to find the approximate optimal solution usually. At present, the study of resource scheduling algorithm, different institutions broadly focus on three areas: the use of economic/market model scheduling algorithm[2-3], QoS-based scheduling algorithm[4-5], and based on the trust model scheduling algorithm[6-7]. QoS-based scheduling algorithm is to meet the demand the degree of resource consideration, and the present study focus on single or multi-QoS constraints, which can not satisfy the user's QoS demand without task attributes in the environment of grid system.

In this paper, the goal is to improve the efficiency of scheduling problem by integrating multi-QoS attribute constraints with task attributes, the proposed resource scheduling algorithm not only take into account the QoS requirements of the resource itself, but also fully integrated with the user's QoS constraints on the tasks.

X. Wan (Ed.): Electrical Power Systems and Computers, LNEE 99, pp. 363–368.
springerlink.com

2 The Description of Algorithm MCSA Based on Multi-QoS Attribute Constraints

In this paper, the set of resources *R* is defined as a local grid, and ignores the network topology of grid system, and assumes the resource nodes are all connected. The Reason for using this scheme is to research the key issues on the design and implementation of scheduling algorithm, to maximize avoid the impact because of the complexity of network topology.

Given the currently available resources in grid system is *m*, the grid resource collection $R = \{R_1, R_2, ..., R_m\}$, where resources $R_i = \{PEs, MIPS, Bandwidth\}$, the elements of the collection represent the number of processor, the performance of the processor performance, and the available network bandwidth. The resource description is abstracted the heterogeneous of performance (including the processor and network performance) and number, which represents the grid system can mask the characteristics of heterogeneous hardware.

From the user point of view, applying batch scheduling mode and parameters of scan task types as a scheduling algorithm in this paper. Defined a set of scheduled tasks: $Gridlets = \{j1, j2, ..., jm\}$, which represents a scheduled task *ji*, and $ji = \{Length, MIPS, Bandwidth, Deadline\}$. The meaning of parameter is the length of the grid task, users deadlines demand, the demand for bandwidth, and processing power of CPU MIPS Respectively.

To take full account of QoS parameters, in the resource management process, not only analyzes its qualitative, but also measure quantitatively the QoS parameters for the task scheduling algorithm. Some parameters for the resource can not be given directly to the quantitative value of their specific needs calculated by the corresponding value, therefore, quantization parameter is defined as follows:

1) Execution time = Computation time + transmission time.
2) Computation time = Computational tasks / CPU speed.
3) Ttransmission time = Data Size / network bandwidth.

3 Design Algorithm MCSA Based on Multi-attribute Constraints

3.1 The Design Idea of MCSA

MCSA scheduling algorithm is considering the Deadline of gridlets, CPU speed and task demands on network bandwidth constraints when send tasks. Given $Q1 = a \times D + b \times B + c \times C$, where *D*, *B* and *C* describe the task deadline, the network bandwidth limitations of task and CPU rate respectively.

The coefficient a, b and c indicate that the task requirements degree of three constraints. According to The demands of the user tasks, and supposing user QoS demands are as follows by simulation experiment : a> = 0.8 or b> = 0.8 or c> = 0.8 is the urgent task, a> = 0.6 or b> = 0.6 or c> = 0.6 is relative urgent task, a> = 0.3 or b> = 0.3 or c> = 0.3 is general requirement task, the remaining task is the required minimum. Scheduling algorithm takes into account the different tasks which demand different resources in the implementation of grid system. Therefore, dividing into four different set of tasks according to the task scheduling order, which is based on weight value $w = a + b + c$.

3.2 The Design and Implementation of MCSA

MCSA grid task scheduling algorithm divides task weight into four sets, according to the batch scheduling tasks of the difference degree demands. Setting its weight value from 0 to 1, a certain value based on deadline of gridlets, CPU or network bandwidth requirements. 0.8, 0.6 and 0.3, respectively, as the cutoff value, the task set is divided into four tasks. In each task, the use of deadline of gridlets for each task, the execution time of each task to do pre-judge, considering the task of deadline, bandwidth and CPU speed will be part of the three constraint information to filter out impossible task, which to increase the number of tasks performed, reduce user waiting time.

In MCSA algorithms, the matching goal of scheduling tasks and resources is that resources not only meet the bandwidth needs which make the task sent to the resource node implementation as soon as possible, but also allocate as far as possible high CPU resources to the urgent scheduling task be completed on schedule, in order to obtain a relatively satisfactory scheduling results. Its computational complexity is $O(m + n)$ by analyzing the MCSA scheduling algorithm.

4 Simulation Results and Analyse of Algorithm MCSA

4.1 Scheduling Environment

To make simulation results of scheduling algorithm in a simulated grid environment more general, three task scheduling algorithms are compared and realized, namely, the traditional Min-min scheduling algorithm Senior scheduling algorithm and MCSA scheduling algorithm.

In the implementation of the grid tasks simulation, applying the Gridsim random function packet, using a random number generator to produce four parameters, namely, the length of task, deadlines of gridlets, the task required network bandwidth and CPU rate, and resulting in a schedule grid task set. Task list was used in the experimental as shown in table 1.

Using Visual Modeler tool provided by the GridSim toolkit to generate a set of grid resources, including 8 of heterogeneous resources in the cluster nodes. Quantity and performance from the PE, and network bandwidth resources on the node, there are differences, thus resources and other characteristics of heterogeneous structure are well simulated in the environment of the grid. Consisted different tasks as a unit of a collection, and doing a number of groups scheduling simulation in this paper. Resource list was used in the experimental as shown in table 2.

Table 1. A batch of tasks

No.	Deadline	Bandwidth	Length	CPU
0	9.706675638311683	54.98297124326032	1174.297540178687	233
1	9.82004527440799	101.32586151541636	4791.087220126449	380
2	5.578956579513086	153.72984814664105	3248.908429470724	106
3	0.3148195783467722	182.17754377985864	4372.500952280463	328
4	3.997217728848441	24.305933734250228	3041.310064799452	249
5	3.47085154282879	90.66610517439469	4743.838802205148	376
6	9.305426483591496	141.4120303214384	1301.325262986264	279
7	0.7553418365431652	104.48128124132938	922.2435773346465	326
8	1.3861841546486864	90.8609128920165	2853.890094115909	427
9	5.97774768433705	100.39378563797914	104.2300328678891	338
10	7.423754857807951	30.11349544940296	4572.925396242926	415
11	1.5143863452802775	142.1295454501374	4152.901129104048	428
12	6.50744851245854	36.26355204858671	335.1236416689396	432
13	8.361166575349264	117.42944600873653	299.2715178796845	277

Table 2. List of simulated resources

Name	PEs	MIPSperPE	Bandwidth
R0	4	823	501
R1	5	617	605
R2	6	835	852
R3	7	917	951
R4	3	459	759
R5	2	552	656
R6	5	784	597
R7	7	1254	982

4.2 Analyzing the Scheduling Results

The number of task completion can be measured not only user satisfaction with the grid system, and is key performance indicators in the grid task scheduling system. Therefore, the number of task completion is tested by using the proposed algorithm MCSA. Experimental data from multiple data records, choosing the relevant data and averaging them, the experimental results obtained, Comparing the traditional Min-min scheduling algorithm[8], Senior algoritm[9] with MCSA algorithm which is shown in figure 4-1.

We choose different coefficient value to verify the effect of classifying task weight value. In this paper, the task execution time and task completed number were tested according to the given differnet value of a, b and c .The results shown in figure 2 and figure 3.

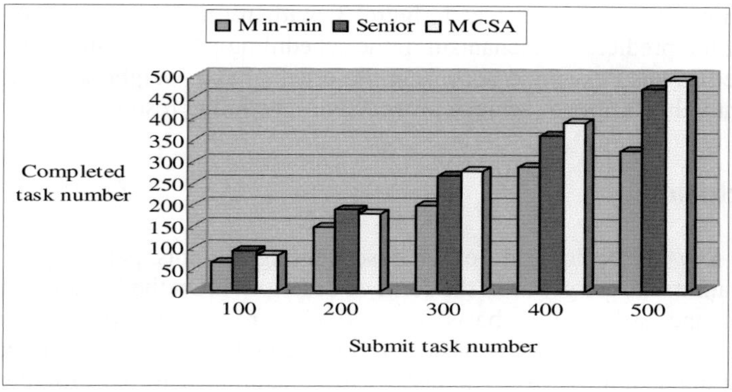

Fig. 1. Scheduling results of the different task sets

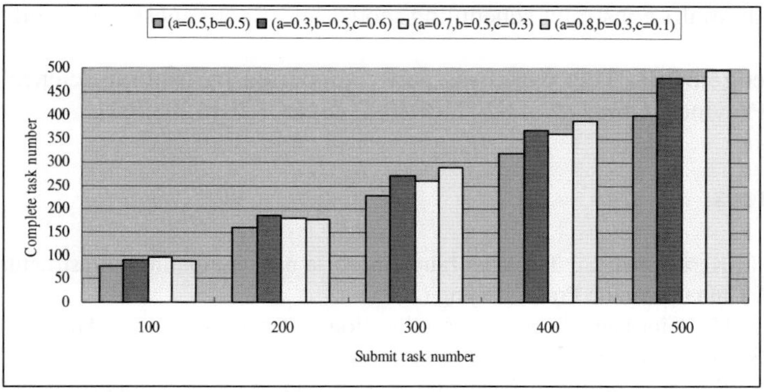

Fig. 2. Completed task number by calssifying different task weight value

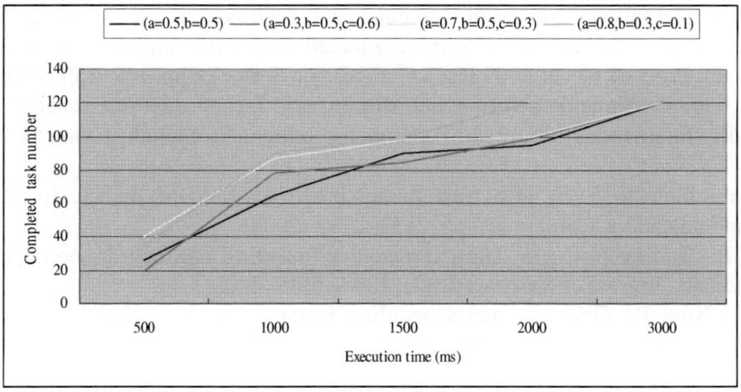

Fig. 3. Scheduling results of the different task classification in different deadline

It is knows form Figure 2 and figure 3, with refining the QoS constraint and combining the prediction mechanism in the scheduling process, submitting task to the task queue has been executed according the order of task weight, and classifying the task weight value is able to achieve prospective scheduling results.

5 Conclusion

QoS is a key factor in grid resource scheduling. In this paper, grid scheduling resources and tasks are given in the form of description.On the basis of multiple QoS constraints and task weight basis, respectively, from CPU speed, deadlines, and network bandwidth of three to consider proposed multi-attribute constrained resource scheduling algorithm MCSA. The tasks will be submitted to the different collection based on different weights, and integrating refined QoS parameters with classifying task weight value by experiment testing data, which the more tasks can be completed within the deadline. We use GridSim simulation tools to verify the MCSA scheduling algorithm in multi-constraint scheduling model for effectiveness and stability.

Acknowledgements. This work was partly supported by Natural ScienceFoundation of Hebei Province, China (No. F2011203092 & No. F2011203067).

References

1. Foster, I., Kesselman, C.: The grid: blueprint for a new computing infrastructure, 2nd edn., pp. 3–7. China Machine Press, Beijing (2005)
2. Caramia, M., Giordani, S.: Resource allocation in Grid computing: An economic model. WSEAS Transactions on Computer Research 1(3), 19–27 (2008)
3. Yang, L., Kai, H.: Consumer sharing policy constrained resource allocation method for grid system. In: The 9th International Symposium on Communications and Information Technology, vol. 9, pp. 721–725 (2009)
4. Chunlin, L., Layuan, L.: A multi-QoS guaranteed dynamic grid resource scheduling algorithm. International Journal of Computers and Applications 29(3), 245–252 (2007)
5. Engelbrecht, G., Benkner, S.: A service-oriented Grid environment with on-demand QoS support. In: Proceedings of 2009 IEEE Congress on Services, pp. 147–150 (2009)
6. Tao, F., Hu, Y.: An approach to manufacturing grid resource service scheduling based on trust-QoS. International Journal of Computer Integrated Manufacturing 22(2), 100–111 (2009)
7. Yiyu, Y., Junhua, T.: A grid trust model based on MADM theory. In: IEEE Global Telecommunications Conference, pp. 2133–2137 (2008)
8. Qing, D., Guoliang, C.: A Unified resource mapping strategy in computational grid environments. Journal of Software 13(7), 1303–1308 (2002)
9. Jing, C., Xun, P.: Grid resource scheduling strategy based on double QOS constraints. Computer Integrated Manufacturing Systems 14(8), 1571–1578 (2008)

Mining Equipment Line Detection and Remote Control System Base on PLC

Qiang Zhang[1,2,3], Qiushuang Song[2], and Li yong Tian[1]

[1] Mechanical engineering college of liaoning Technica University, Fuxin, 123000
[2] China national coal mining equipment co., LTD, Beijing, 100011
[3] State Key Laboratory of Structural Analysis for Industrial Equipment, Dalian University of Technology, Dalian 116023, P.R. China
Lgdjx042@126.com

Abstract. The improve of the relay control system according to the original mine compressors determine the programmable logic controller (PLC) as the core of the compressors control system. Detection and control the compressors through the collection of sensors, pressure, temperature, vibration, voltage and power factor, current signal processing and analysis. On-site monitoring use human-machine interface, PC use configuration software for remote monitoring, Design and finished mine compressor control system installation and commissioning.

Keywords: Mine compressors, Programmable logic controller(PLC), Human-machine interface, Configuration software.

1 Introduction

Along with the coal mine modernization development, the mine is getting higher and higher to the mine equipment request, to construct the essence secure mine has become the core of the coal mine production construction. Originally, the pine tree coal mine ground compressors stands uses relay control system to carry on the control to three air compressors, three air compressors are all the single screw rod type compressors, and all use the fashion of Y-△ and reduct voltage to start, each air compressor uses a start cabinet to start or stop the control respectively, and they all need to control artificially.

The control circuit which builds with the relay has the defect of reliability badly, not easy to maintain and not easy to monitor, it has not been able to adapt the current request. Now the urgent need is reliability high, easy to maintain, easy to operate, may to monitor, and the price is not high, such controller replaces the circuit which builds with the relay. Along with the electronic technology, the software technique, the control technology swift development, PLC develops swiftly and violently, the performance is very high, the price is more reasonable and it have a very big superiority to compared with the control circuit which builds with the relay.

X. Wan (Ed.): Electrical Power Systems and Computers, LNEE 99, pp. 369–374.
springerlink.com

2 Overall Design

The mine compressor control system take PLC as the control core, uses three collection and distribution type structural frame, PLC obtains the signal such as the pressure, the temperature, the vibration, the voltage, the electric current as well as the switch quantity condition from the compressor to through the digital quantity module and the simulation quantity module, and carries out the dependent program, carries on the logic synthesis judgment to the signal, then carries out the related instruction, like the starting, closing down, ultra limit warning and so on. The man-machine contact surface connected with the superior machine through communication way of Modbus which is supported by the PLC, thus achieve to each kind of data carries on the real time display, achieve the compressor control system's monitoring and the protection finally.

3 The System Hardware Designs

3.1 Sensor Choice

Specific to compressors request which must control to various sensors carries on the comparison, the temperature examination selects the wzpj-236 integration temperature transmitting instrument of the PT100 series, and we use the JK series pressure transmitting instrument to examine the pressure, then we choose the VB-Z9500-2-1 integration vibration transmitting instrument to carry on the examination to the electric motor about its vibration. The over-load protection, we select T series thermal relay to protect the compressors main engine as well as conditioner electric motor. The EDA9033A three-phase electrical parameter gathering module is selected to carry on the three-phase electrical parameter gathering.

3.2 PLC Lectotype

Regarding the switch quantity comes first, and taking partly simulation quantity control mine compressor supervisory system, should choose the simulation quantity input module which equiped with A/D transforms and has the simulation quantity output module which equiped with D/A transforms, matches with the corresponding sensor, the transmitting instrument and the drive equipment and choose the small PLC which have strong operable function. When carries on the small digital pneumatic analogue mixed system control, the Siemens S7-200 series PLC has the high cost performance, enforcement is also quite convenient. Therefore we use S7-200 series 224XP model PLC as the system CPU.

The mine compressor supervisory system output 25 spots altogether , because 224XP model PLC is 14 inputs, 10 outputs. It need to carry on the digital quantity module to carry on the expansion, through to various digital quantity expansion module comparison, uses two 8 inputs, 8 outputs EM223 modules to carry on the expansion.

The I/O analog quantity expansion module could be divided into three models which are EM231, EM232 and EM235SAN. The mine compressor supervisory system needs to carry on the examination to three compressors the simulation quantity to have the discharge temperature, the lubrication oil pressure, the exhaust pressure as well as the electrical machinery vibrates. Therefore altogether must examine 3 temperature signals, 6 pressure signals as well as 3 vibration signals, separately by the temperature transmitting instrument, the pressure transmitting instrument as well as the vibration transducer send the signal into PLC, chooses 4 inputs the EM231 simulation quantity expansion modules.

4 The Control System Designs

The mine compressor supervisory system may realize the following function to the pine tree mine's three compressors:

(1)To start the control;
(2)To stop the control. Including normal engine off, automatic protection urgent warning engine off and automatic protection delayed alarm engine off;
(3)To examinate the compressor each three-phase electrical parameter, including temperature, pressure, vibration, voltage, electric current and power factor examination;
(4) To carry on the scene monitoring through the human-machine interface;
(5) The superior machine carries on the long-distance monitoring through the soft panel.

4.1 PLC Control System

The PLC CPU input and output diagram is the three-phase asynchronous motor, which uses the star - triangle type voltage dropping starts, each compressor start movement needs to complete through two relay control, K1, K3, and K5, this three relays control three air compressor star start separately, then through K2, K4, K6 controls three compressor angle movement again, and simultaneously controls three compressor start/ stop indicating lamp, and K7 relay control alarm bell.

Expansion module power is supplied by the CPU, the output relay to supply power by the external connection DC24V power source. The 15 expansive switch quantities output distinction control export exhaust pressure excessively high indicating lamp, the exhaust hyperpyrexia indicating lamp, overload indicating lamp of the compressor main engine electric motor, overload alarm indicator lamp of the air-cooled motor and indicating lamp of the electric motor oversized libration.

The system has 12 simulation quantity input altogether, no simulation quantity output, so the simulation quantity expansion uses three 4 inputs, 0 outputs EM231 modules carry on the expansion, the simulation quantity input including three compressor export exhaust pressures, the discharge temperature, the lubrication oil pressure as well as the electrical machinery libration.

5 Control System Software Design

5.1 Program

PLC procedure is realized by STEP the 7-Micro/Win32 development software, it uses in Simens S7-200 series PLC, Siemens designed it for the S7-200 series PLC development specially; It is based on the Windows application software, its function is formidable, mainly uses for the user development control procedure, it may also take the real-time monitoring of user program executing state simultaneously.

Mine compressors supervisory system must be the signal which as well as the switch quantity condition examines according to the digital quantity module and the simulation quantity module, administer the dependent program, carry on the logic synthesis judgment to the signal, to decide that the normal operation or the engine warning off, its whole control procedure flow chart as shown in Figure1.

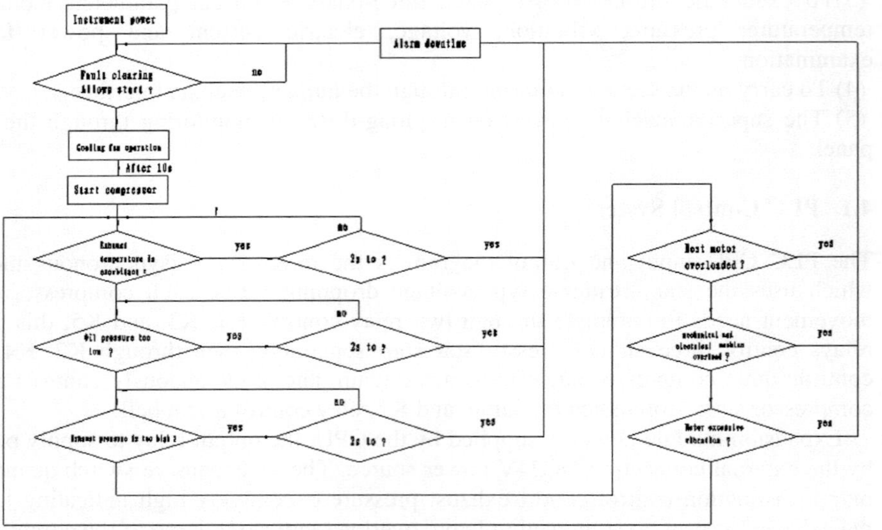

Fig. 1. Flow diagram of program

Pressure and temperature as well as the output signal of the electrical machinery libration are the 4~20mA electric current signals, so it needs to turn the electric current signal through A/D transformation the digital quantity signal. The pressure, the temperature and the electrical machinery libration A/D conversion formula is: 327246998

$$\text{actual value}=\frac{\text{observed value-6400}}{\text{32000-6400}}\times\text{measuring range} \tag{1}$$

Observed value--The simulation quantity signal which examined by the pressure, the temperature and the vibration transducer;

Actual value-- Actual pressure, temperature and undulating quantity when the compressor works;

Measuring range--Pressure, temperature, vibration transducer measuring range.

5.2 Human-Machine Interface Designs

Human-machine interface mainly demonstrate the data to have compressors' lubrication oil pressure (unit Kpa), the discharge temperature (unit ℃), the exhaust pressure (unit Kpa), the electrical machinery to vibrate (unit mm/s), the electrical machinery voltage (unit V), the electrical machinery electric current (unit A), whether the power factor and the cooling fan are normal work. After the touch screen establishment completes, it is connected the touch screen and PLC and the superior machine, in the touch screen has the corresponding connection PLC option, establishes the PLC communication parameter agreement which inside the communication parameter must connect then.

5.3 Superior Machine Configurations Design

Configuration software are the data acquisition and the process control special software, they are in the automatic observation system monitoring level first-level software platform and the development environment, can by the nimble diverse configuration way, but is not the programming provides the good user development contact surface and the simple and direct application method, its pre-establishment each kind of software module may achieve and completes the monitoring level each function easily, and can simultaneously support each kind of hardware factory the computer and the hardware equipment, controls the computer and the network system with the redundant reliable labor unifies, it may provide the software and hardware to the entire observation system with the complete connection, carries on the system integration.

The configuration contact surface plan may use in the development system toolbox each item of draw functional module and the palette completes, the map storage has provided the massive graphic object, through transfers in the map storage the graph part as well as the insertion point bitmap completes each picture plan. System overall examine picture is shown in Figure 5, before system initiation, the picture is at the static condition, when after the system starts the movement, the electric motor indicating lamp set to red, the conditioner forced draft fan ventilator starts to revolve, simultaneously the data demonstration which gathers from the scene each kind of detecting element in the picture relevant position, the manager may through the parameter establishment picture to three compressors, each parameter bound carry on the establishment, and may pull the menu through under carries on the inquiry to the single compressor real-time curve and the historical curve. When the data ultra restricts production lives the warning, carries on the analysis and the confirmation after the current warning, the contact warning and carries on the elimination to the corresponding breakdown.

The compressor start and stop may carry on the operation directly in the scene, may also carry on the long-distance operation through the superior machine configuration supervisory system, simultaneously the superior machine also carried on the overhaul and switch operation's setup while it is working.

6 Conclusions

Mine compressors control system based on PLC and configuration software has substituted for the original compress relay control system, it uses sensor gathering compressor each kind of parameter, and through PLC, Human-machine interface and the superior machine configuration software carried on the monitor and the control. PLC maintains conveniently, the movement is quick, the simplicity of operator, played the new role to compressors control and the aerodynamic force, raised the compressors control system automated level, the compressors pressed pressure, the temperature, the vibration, the voltage, the electric current, as well as the power factor was more precise, enhances the compressors security and the reliability while they are working. It has achieved the production requirements.

Acknowledgments. The study was partially financial supported by the program Ministry of Education Doctor Fundation of China(20060147001); Chinese coal machine equipment company scientific research foundation; Liaoning technology university outstanding youth scientific research fund; Liaoning technology university mechanical engineering institute outstanding young teacher scientific research plan; Liaoning technology university graduate student scientific research fund; Dalian Science and Technology University of structure state key laboratory open fund (G0818).

References

1. Taoqi-ke, Li, Z.: The State Monitor of Air Compressor Unite Base on Configuration Software and PLC. Equipment Manufacturing Technology (4), 57–59 (2008)
2. Wang, F., Chen, Y.-s., Zhang, S.-m.: Monitoring System of Coal Mine Air Compressor Based on PLC and Cofiguration King. Coal Mine Machinery 29(11), 176–177 (2008)

Signal Enhancement and Complex Signal Analysis of GPR Based on Hilbert-Huang Transform*

De-shan Feng[1,**], Cheng-shen Chen[1], and Kai Yu[1,2]

[1] School of Geosciences and Info-Physics, Central South University, Changsha, 410083, China
[2] Reconnaissance, Planning, Design & Research Institute, Ministry of Water Resources, Zhengzhou, 450003, China
Tel.: +86-13618474853
{fengdeshan,chenchengshen,CKYUKAI}@126.com

Abstract. The theory of Hilbert-Huang transform and complex signal analysis technology are described. The EMD decomposition method is applied to decompose GPR forward profile, the intrinsic mode function GPR figures of different frequency ranges from high to low are obtained. According to different exploration goals, the intrinsic mode function GPR figures are reconstructed, thereby achieving the purpose of enhancing GPR signals. Then, the measured GPR profiles of traffic channel of Heimi Peak Pumped Storage Power Station are selected. Firstly, EMD decomposition is carried out on the profile to remove a part of noises, then, Hilbert-Huang transform is utilized to calculate the complex signal of GPR profiles, and independent instantaneous parameters are drawn out. According to different geological conditions corresponding to GPR three 'instantaneous' information respectively, multi-parameters are comprehensively analyzed, the compression function of EMD decomposition on noise is combined to avoid interpretation bias caused by using single time interval profile analysis, the abnormal information can be better reflected, and the resolution precision of GPR data is improved.

Keywords: ground penetrating radar, empirical mode decomposition, complex signal analysis, Hilbert-Huang transform, intrinsic mode function.

1 Introduction

Hilbert - Huang transform is a new method for analyzing nonlinear and non-stationary signals, which was proposed by Huang N.E.[1~2] in 1998. It mainly includes empirical mode decomposition and Hilbert spectral analysis, which firstly decomposes signals into a number of intrinsic mode functions through utilizing EMD method, then acts Hilbert transform on every IMF, and obtains corresponding Hilbert instantaneous spectrum, and the multi-scale oscillation change characteristics of

---

* Foundation item: Projects(41074085,40804027) supported by the National Natural Science Foundation of China; Projects(09JJ3048) supported by the Hunan Natural Science Foundation.
** Corresponding author: Associate professor, PhD.

X. Wan (Ed.): Electrical Power Systems and Computers, LNEE 99, pp. 375–384.
springerlink.com

original signals are revealed through analyzing each component and its Hilbert spectrum. Hilbert-Huang transform is widely studied and applied in the fields such as system simulation[3], spectral data preprocessing[4], geophysics[5] and the like through being developed for more than ten years. Yu[6] and Xie[7] et al. applied Hilbert transform to convert ground penetrating radar real signal into complex signals, the instantaneous amplitude, instantaneous phase and instantaneous frequency waveform figures were extracted, and independent profiles of three parameters were formed, thereby improving the accuracy of radar interpretation. But they were based on Hilbert transform, the signals must be narrowband when Hilbert transform is used for calculating instantaneous parameters of signals. The GPR data often adopt broadband, and error physical interpretation can be caused through directly adopting Hilbert for calculating instantaneous parameters, such as negative frequency and the like. The paper firstly carries out EMD decomposition on GPR signal, thereby obtaining IMF components of the GPR signals, and then utilizes Hilbert transform to calculate GPR complex signal, three parameters of instantaneous amplitude, instantaneous phase and instantaneous frequency are also extracted, three independent profiles are formed, interpretation errors caused by using single time interval profile analysis can be avoided through multi-parameter waveform profile cross-referencing and comprehensive analysis, and the resolution precision of GPR data is improved while physical meaning is provided for instantaneous parameters.

2 Empirical Mode Decomposition Principle

Empirical mode decomposition is a key component of Hilbert-Huang transform, and it has three assumptions[1~2]:

a. Signals to be processed at least comprise one maximum value and one minimum value.

b. The time interval between the extreme points determines the characteristic time scale.

c. If the data sequence only contains a turning point, the extreme point can be determined through calculating one stage or multi-stage derivative, and the final result can be obtained through quadrature.

After the three assumptions are met, the empirical mode decomposition method believes that all signals can be made up of different intrinsic mode functions IMF, wherein any IMF can be linear or nonlinear. And the IMF must meet the following two conditions:

a. As for a column of data, the number of extreme points and zero-crossing point must be equal or at most difference of 1 point is allowed.

b. At any point, the average value of two envelope lines composed of local maximum points and minimum points should be zero. Every IMF can be considered as one mode function inherent in the signal.

Specific implementation steps of EMD can be described as the follows: setting signal sequence as $f(t)$, first of all, finding all maxima and minima of $f(t)$, obtaining the upper envelope line $u_1(t)$ and the lower envelope line $v_1(t)$ of $f(t)$ through cubic spline fitting, calculating the average values of the upper envelope line and the lower envelope line on each point, thereby obtaining one average value curve $m_1(t)$, and subtracting the obtained average value from original signal $f(t)$ to obtain one new data sequence $h_1(t)$:

$$m_1(t) = \frac{1}{2}(u_1(t) + v_1(t)) \ . \tag{1}$$

$$f(t) - m_1(t) = h_1(t) \ . \tag{2}$$

The new data sequence $h_1(t)$ is determined according to the IMF determinant conditions provided above. If $h_1(t)$ can not satisfy the determinant conditions, it can not be an IMF component sequence, thereby, the above treatment process should be repeated for n times, the obtained $h_k(t)$ can meet the determined condition of IMF, in the condition, $h_k(t)$ is (IMF) $c_1(t)$ of stage 1, and the component represents the highest frequency component in signal $f(t)$. Then the $c_1(t)$ is subtracted from the original signal $f(t)$, and the difference signal sequence $r_1(t)$ after high-frequency part is reduced can be obtained:

$$r_1(t) = f(t) - c_1(t) \ . \tag{3}$$

At this point, then the above steps can be repeated on $r_1(t)$ as raw data sequence to be processed, thereby obtaining the second IMF $c_2(t)$, the above operation should be repeated for n times, thereby obtaining n intrinsic mode function components. Of course, IMF can not be obtained endlessly, when $r_n(t)$ is turned into a constant or becomes a monotonic function, the operation can be stopped. Therefore, it can be obtained that the original signal $f(t)$ can be expressed as the sum of all IMF and surplus:

$$f(t) = \sum_{i}^{n} c_i(t) + r_n(t) \ . \tag{4}$$

In the formula, $r_n(t)$ is the resulting residual function (monotonic function), which represents the average trend of the signal. All IMF components reflect the characteristic scale of the signal, and the scale is from smallness to bigness in turn. Therefore, each IMF component correspondingly contains elements of different frequency bands from highness to lowness, which represents inherent mode characteristic of non-linear model, and it changes with the change of the signal itself. Obviously, EMD decomposition process is a "screening" process in fact, the decomposition is completely adaptive to the decomposed signal, which is different from wavelet transform which needs to pre-select wavelet bases.

3 Hilbert Transform Theory

Hilbert transform can effectively and truly obtain effective information contained in the signal, which is essentially an all-pass filter. After the input signal $f(t)$ passes through the filter $H(\omega)$, an output signal $\hat{f}(t)$ can be generated, if $H(\omega)$ has the following characteristics: (1) amplitude-frequency characteristics belong to all-pass type; (2) phase-frequency characteristic is -90 ° phase shift; both can be expressed as the follows after being considered uniformly:

$$H(\omega)=\begin{cases}+i,\omega<0\\-i,\omega>0\\0,\omega=0\end{cases}. \tag{5}$$

Then the filter output $\hat{x}(t)$ is called as Hilbert transform of $x(t)$. Obviously, $\hat{x}(t)$ and $x(t)$ are orthogonal. Filter $H(\omega)$ is called as the Hilbert filter. The Hilbert transform of $f(t)$ can be written as $\hat{f}(t)$ or $H\cdot f(t)$. $\hat{f}(t)$ is defined as the follows:

$$\hat{f}(t)=\frac{1}{\pi}\int_{-\infty}^{\infty}f(\tau)\left(\frac{1}{t-\tau}\right)d\tau=\frac{1}{\pi}\int_{-\infty}^{\infty}f(t-\tau)\frac{1}{\tau}d\tau=f(t)*\frac{1}{\pi t}. \tag{6}$$

4 Analysis Principle of Complex Signal

Complex signal analysis is also known as analytic signal analysis, namely, relevant information of recording trace is directly decomposed into a processing and interpretation technique of instantaneous amplitude, instantaneous phase and instantaneous frequency in the time domain. Before complex signal analysis is carried out, Hilbert transform must be carried out firstly. The input signal is set as $x(t)$ after EMD decomposition treatment, and the output signal after being filtered by the filter $H(\omega)$ is $\hat{x}(t)$. The $x(t)$ and its Hilbert transform are combined to form a complex signal[6~7], that is

$$u(t)=x(t)+i\hat{x}(t)=x(t)+ix(t)h(t)=x(t)\left[\delta(t)+i\frac{1}{\pi t}\right]. \tag{7}$$

In the formula: $u(t)$ is complex signal of $x(t)$, and also known as analytical signal. Since $x(t)$ can be decomposed into a trigonometric form, $x(t)=A(t)\cos\left[\omega_0 t+\varphi(t)\right]$, $\hat{x}(t)$ is set, it also can be expressed as $\hat{x}(t)=A(t)\sin\left[\omega_0 t+\varphi(t)\right]$ (wherein $\omega_0=2\pi f$), thereby the complex signal of $x(t)$ also can be expressed as the follows:

$$u(t)=x(t)+i\hat{x}(t)=A(t)\cos\left[\omega_0 t+\varphi(t)\right]+iA(t)\sin\left[\omega_0 t+\varphi(t)\right]$$
$$=A(t)e^{i\left[\omega_0 t+\varphi(t)\right]}=A(t)e^{i\theta(t)} \quad . \tag{8}$$

Obviously, $A(t)$ and $q(t)$ can be changed along with time. $A(t)$ is known as instantaneous amplitude of $u(t)$; $\theta(t)=\omega_0 t+\varphi(t)$ is known as instantaneous phase of $u(t)$, and the time change rate of the phase is as the follows:

$$S(t)=\frac{d\theta}{dt}=\omega_0+\frac{d\varphi(t)}{dt} \quad . \tag{9}$$

$S(t)$ is the instantaneous frequency of so-called $u(t)$. When the $\varphi(t)$ is unchanged or little changed, $\varphi'(t)$ can be regarded as zero or a constant C, that is, $S(t)=\omega_0+C$ is only related with the frequency.

The instantaneous amplitude, instantaneous phase and instantaneous frequency can be calculated as follows: firstly $\hat{x}(t)$ can be obtained through GPR recording trace $x(t)$ after EMD decomposition through Hilbert transform, and then the instantaneous amplitude can be calculated as the follows:

$$A(t)=\sqrt{x^2(t)+\hat{x}^2(t)} \quad . \tag{10}$$

It is function of time variable t, and is not related with phase $\theta(t)$. Instantaneous phase is as the follows:

$$\theta(t)=\arcsin\frac{\hat{x}(t)}{\sqrt{x^2(t)+\hat{x}^2(t)}} \quad . \tag{11}$$

or

$$\theta(t)=\operatorname{Im} Inf(t) \quad . \tag{12}$$

Instantaneous frequency $S(t)$ is the change rate of instantaneous phase function on time, namely, $\theta(t)$ is differentiated to obtain the follows:

$$S(t)=\frac{d\theta(t)}{dt}=\operatorname{Im}\left[\frac{1}{f(t)}\frac{df(t)}{dt}\right] \quad . \tag{13}$$

Three kinds of instantaneous information of instantaneous amplitude, instantaneous phase and instantaneous frequency of complex signal generally refer to a particular moment rather than average of one time section. The complex signal analysis of GPR signal recording trace $x(t)$ is used for analyzing and detecting parameters such as GPR signal energy, frequency, phase and the like.

5 Radar Data Processing Instances Based on H-H Transform

5.1 GPR Signal Reconstruction and Enhancement of EMD Decomposition

The earth is equivalent to a complex filtering system, after radar signals pass through the underground, not only the amplitude is reduced sharply, but also the waveform is different from the original waveform due to energy attenuation and noise interference. How to utilize weak signals in GPR profiles to recognize anomalous bodies is very important. In order to enhance the intensity of weak signals, time-varying gain is often used for correcting signal loss caused by wave-front expansion and dielectric absorption to achieve the improvement of the signal identifiability. It is feasible under ideal condition of high signal to noise ratio, however, when the signal to noise ratio of GPR profiles is lower, although the method can improve the signal intensity, the noise intensity also can be amplified simultaneously, many unnecessary reflection signals can appear in the profiles, the shallow signals are saturated, in essence, the interpretation accuracy can not be improved. EMD decomposition is utilized to separate high frequency section in GPR signals from low frequency section, signals with relative high and low frequencies can be overlapped according to actual needs, thereby enhancing high and low frequency signals.

In order to illustrate the effect of EMD which separates GPR deep and shallow signals and reconstructs for enhancement, the radar model is set as shown in Fig.1, three metal pipes are arranged on the upper portion of the concrete layer, and a rectangular empty area is arranged under the metal pipes. The top portion of the metal pipe is buried for 0.40m, the conductivity is 2.0×10^{8} S/m.The space among the mental pipes is 0.20m, the metal pipe radius is 0.20m, concrete dielectric constant is 6, the conductivity is 0.0001, rectangular empty area is 0.4m × 1.0m. 80 traces are set in simulation, 2048 sampling points are set on each trace, pulse waveform selects the Ricker wavelet, its frequency is 900 MHz, finite difference method is adopted to forward the model, and its forward synthesis profile is shown in Figure 2. It can be know through analysis of Fig.2 that the diffraction wave caused by three rounded steel meshes is very clear, the relatively deep rectangular abnormal body energy is weaker due to strong reflection shielding effect of the metal pipe and high-frequency radar rapid decay effect, and the rectangular zone is basically invisible under the condition of small gain in the forward gain profile. Fig.3 shows the intrinsic mode function Figure IMF1~ IMF4 obtained through decomposing original GPR forward profile through the use of EMD. Since EMD decomposition is a high-pass filtering process in fact, the high frequency components in the decomposition process of signals are slowly filtered, the low frequency signals separated subsequently rightly correspond to the anomaly features caused by abnormal body on the lower layer. Fig.3(a)~(b) refer to two front IMF components obtained after decomposing the original signal, steel bar abnormity can be prominently observed from the figure. Fig.3(c) shows that the IMF3 is obtained through decomposition, and weak abnormal signals are generated on the escaping zone position on the deep portion; Fig.3 (d) shows that the IMF4 is obtained through decomposition, and the escaping abnormity in the deep portion is prominently reflected.The effective parts of GPR signals are basically separated, thereby it belongs to deep and shallow part abnormal body figure and can not be reflected, the subsequent IMF components are not listed.

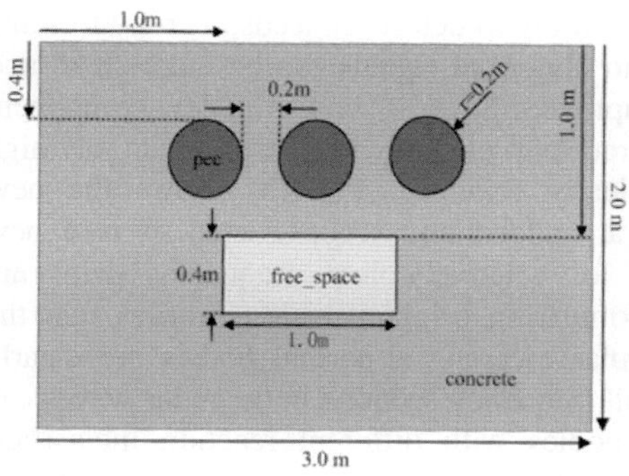

Fig. 1. Sketch map of radar model

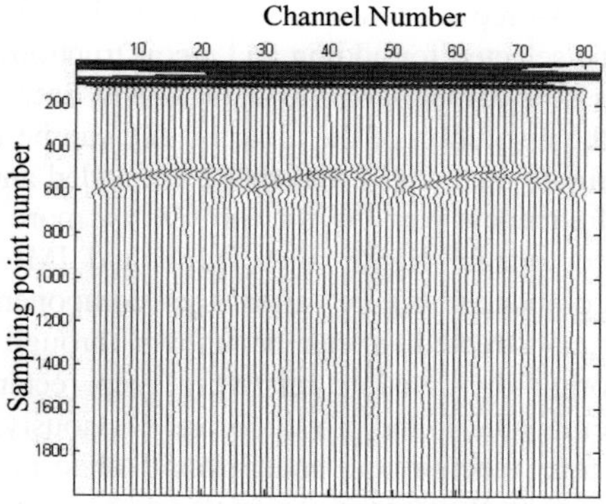

Fig. 2. Finite difference method radar forward profile

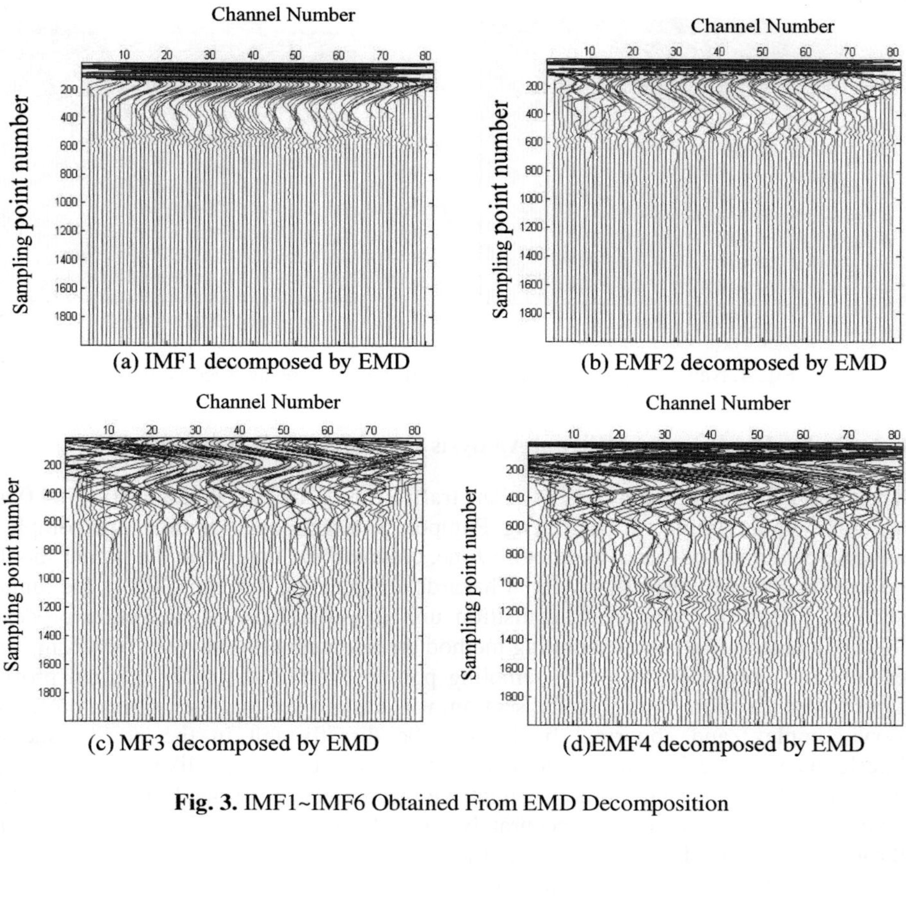

Fig. 3. IMF1~IMF6 Obtained From EMD Decomposition

Moreover, two or more relatively high or relatively low-frequency components can be selected for adding and reconstruction according to detection needs, thereby forming a new component, the shallow and deep part abnormal signals can be more clear, and the resolution of the radar profile can be improved. Fig.4(a) shows the new component figure after IMF1 and IMF2 are added and reconstructed, obviously, the relatively high frequency part in the profile is more clearly reflected. Fig.4(b) shows the new component figure after IMF3 and IMF4 are added and reconstructed, it is a new component figure after the components of relatively low frequency parts are reconstructed, it can be known through comparing original analog profiles that the originally-invisible relatively deep rectangular escaping abnormal bodies are clearly embodied in the profile figure. Obviously, all frequency sections in the radar waves can have reaction on underground abnormal bodies with different reaction intensifies, thereby the abnormal signals can be displayed more clearly in the profile figure after all components are reconstructed, thereby achieving the purpose enhancing the signals.

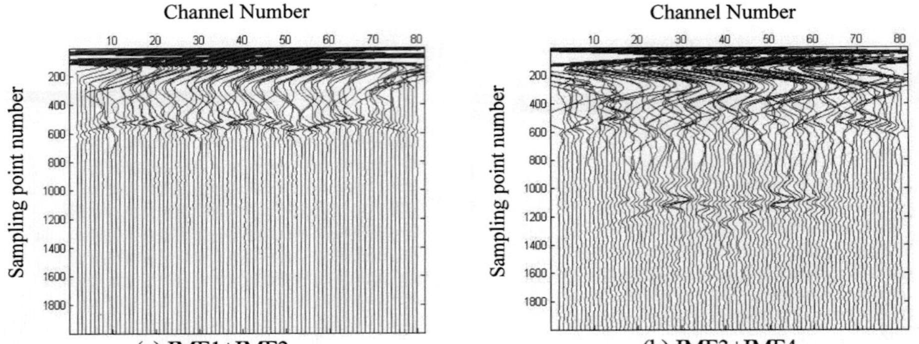

(a) IMF1+IMF2 (b) IMF3+IMF4

Fig. 4. New Component Figure after IMF Superimposing Reconstruction

5.2 Radar Data Complex Signal Analysis Based on EMD Decomposition

The rock geological conditions of plant traffic tunnel in J0+200.00m~J0+ 984.00m section without lining of Heimifeng Pumped Storage Power Station are complex, adverse geologic body such as fracture zone, crannies and the like in the tunnel should be explored clearly, thereby potential hazards are detected through GPR in the tunnel section. SIR-3000 radar data acquisition unit is adopted in the detection process, 900MHz antenna and point sampling method are adopted with dot pitch of 0.20m, and each scan is provided with 512 sampling points. Fig.5 is the radar original profile figure which is more typical in the section, it is easy to see that the original profile is very complex, and abnormal bodies can be directly felt in the profile due to interference of steel arch tunnel, barbed wire, cable and the like in the tunnel, however, the abnormal bodies cannot be accurately positioned. Obviously, the detection section can not be accurately judged and geologically interpreted only through the profile figure. Therefore, EMD decomposition should be carried out on the measured signals to remove high frequency noise.

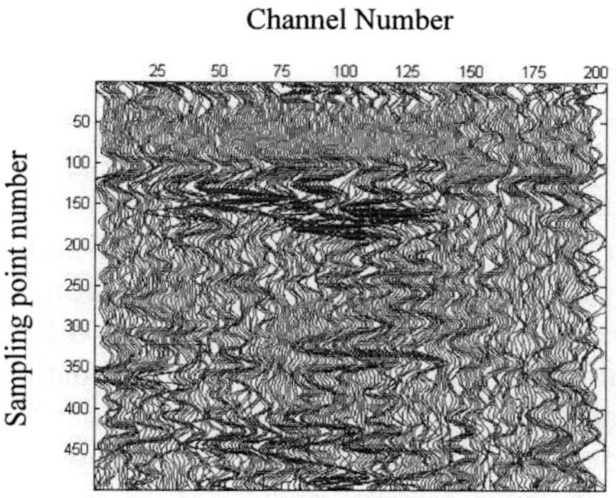

Fig. 5. Original profile of radar

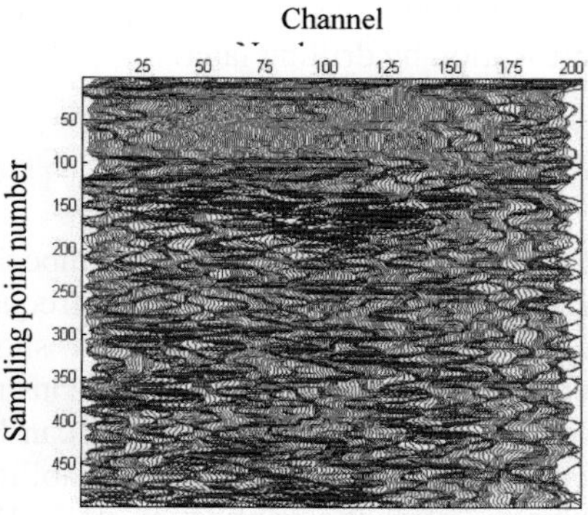

Fig. 6. Instantaneous amplitude profile

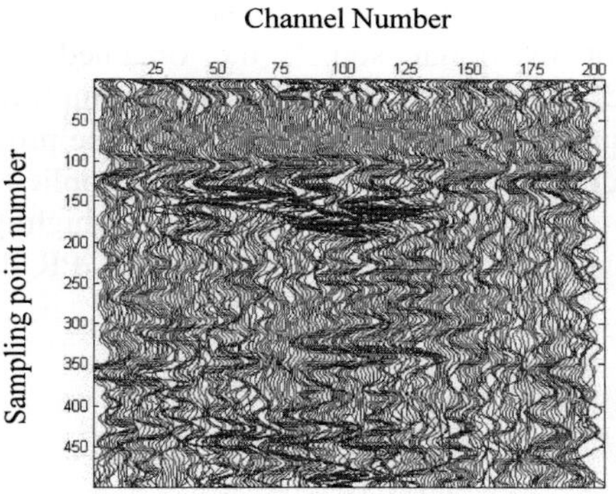

Fig. 7. Instantaneous phase profile

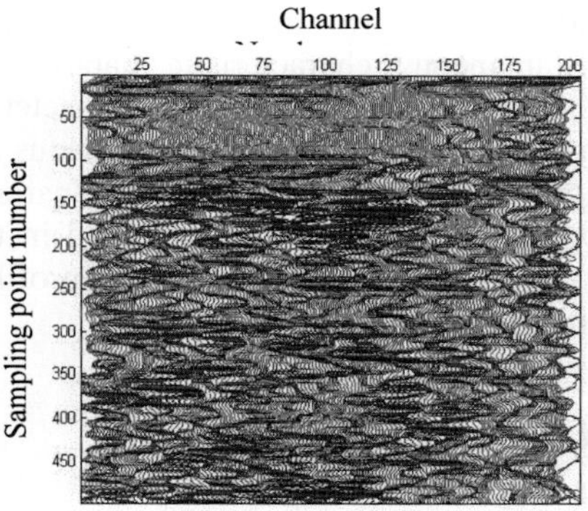

Fig. 8. Instantaneous frequency profile

Firstly, we removal of the first IMF by EMD, the abnormal information is clearer than that the original profile figure, but it still cannot reach the resolution for accurately guiding abnormal interpretation. Hilbert-Huang Transform is carried out on the signals after EMD decomposition, and the complex signal analysis is carried out to extract the instantaneous characteristic parameters on this basis. Fig.6 shows the reconstructed GPR instantaneous amplitude profile, Fig.7 shows the GPR instantaneous phase profile after reconstruction, and Fig.8 shows the GPR instantaneous frequency profile after reconstruction. It can be known through comparing three kinds of instantaneous profiles that the GPR profile after complex signal analysis and processing is more clear and intuitive with higher resolution, because they respectively correspond to the different physical parameters of subsurface media, so the appearances of profiles are different, the instantaneous phase spectrum effect is the best followed by instantaneous frequency spectrum in the three instantaneous parameter spectrums. We can clearly differentiate the position of fracture zones between trace No.50 and 125, and between sampling point No.150 and

200 in the three instantaneous profiles, and the results are completely meshed through verification by drilling later.

6 Conclusion

(1) The essence of EMD decomposition is summarized: which is equivalent to gradually separating out different frequency components in the GPR profile from high to low. The IMF radar profiles should be reconstructed according to different detection purposes, thereby the reconstruction IMF figures of abnormal information in the deep part or the shallow part can be more prominent, examples show that EMD decomposition has stronger adaptability corresponding to radar data with low signal to noise ratio, the application of EMD in noisy measured radar signals can better realize the removal of noise in noisy GPR data and achieve the purposes of enhancing GPR signal and improving the interpretation precision.

(2) EMD decomposition should be carried out on GPR data firstly, and then the Hilbert transform and complex signal analysis should be combined to obtain the instantaneous characteristic parameters of the radar signal, the obtained three instantaneous information characteristic figures of instantaneous amplitude, instantaneous phase and instantaneous frequency should be combined with the profile figure of original signals, multi-parameter analysis method should be applied to express the information contained in the radar profile, which can help to highlight anomalous features in the radar profile and to improve the precision of GPR data processing and interpretation.

References

1. Huang, N.E., Shen, Z., Long, S.R., Wu, M.C., Shih, H.H., Zheng, Q., Yen, N.C., Tung, C.C., Liu, H.H.: The empirical mode decomposition and the Hilbert spectrum for nonlinear and non-stationarity time series analysis. J. Proceedings of the Royal Society London. Ser A 454, 903–995 (1998)
2. Huang, N.E., Wu, M.C., Long, S.R., Shen, S.P., Qu, W.D., Gloersen, P., Fan, K.L.: A confidence limit for the empirical mode decomposition and the Hilbert spectral analysis. J. Proceedings of the Royal Society London. Ser A 31, 417–457 (2003)
3. Zheng, J.S., Wu, B.Y.: Research on EMD Method Based on Reproducing Kernels. J. Journal of System Simulation. 22(1), 188–190, 194 (2010)
4. Cai, J.H., Wang, X.C.: Near-Infrared spectrum pretreatment based on empirical mode decomposition. J. Acta Optica Sinica 30(1), 267–271 (2010)
5. Pi, H.M., Liu, C., Wang, D.: Using Hilbert-Huang transform to pick up instantaneous parameters of seismic signal. J. Oil Geophysical Prospecting 42(4), 418–424 (2007)
6. Xie, X.Y., Wan, M.H.: The application of complex signal in treatment of signal of GPR. J. geophysical exploration and chemical exploration 22(2), 108–112 (2000)
7. Yang, Q.F.: The instantaneous frequency analysis of GPR data using empirical mode decomposition. J. Coal Geology & Exploration 37(4), 64–67 (2009)

Granular Structure Model Based on Artificial Emotion

Jun Hu[1] and Chun Guan[2]

[1] School of Software, Nanchang University, Nanchang, Jiangxi, China
[2] School of Information Engineering, Nanchang University, Nanchang, Jiangxi, China
hujun@ncu.edu.cn

Abstract. Granular computing theory is an important tool for processing vague, incomplete, inaccurate information. Based on the granular computing theory, a new artificial emotion granular structure model is proposed in this paper. The model presents a new artificial emotion granular concept and defines some key factors. Finally, an experiment is given to calculate and verify the emotion granular structure model.

Keywords: Granular, Emotion, Model.

1 Introduction

Emotions are an essential part of human life; they influence how we think, adapt, learn, behave, and how humans communicate with others. It is clear that without the preferences reflected by positive and negative effects, our experiences would be a neutral gray [1]. The importance of emotion has been identified in human-like intelligence recently. Some neurological evidence proves that emotions do in fact play an important and active role in the human decision-making process [2].

The explanation for emotion in psychology is that emotions are evaluations for oneself or for relation status between agents and environment by human body. The emotional reaction of human beings is generated by stimulation of outside environment and affected by the mechanism of demand and requirement inside themselves. The interaction between the emotional process and the cognitive process may explain why humans excel at making decisions based on incomplete information. Emotions are seen occurring when the cognitive, physiological and motor/expressive components are usually more or less dissociated in serving separate functions as a consequence of a situation-event appraised as highly relevant for an individual [3].

While in the psychology field, emotions can be described in terms of desires and expectations. Some hormones sent by the brain sometimes inhibit pain. Inspired by these psychological models and the growing interest in AI, many models that simulate the human mind have been proposed. However, since the psychology of emotions is not yet complete at this time, it is not easy to find a computational model that describes the complete emotional concept. By the 1990s, the Japanese researchers were interested in a system that can communicate with humans. Emotions are regarded as one of the most important factors in communication [1].

Granular Computing(GrC) is a new way to simulate human thinking to help solve complicated problems, and involves all the theories, methodologies and techniques of

X. Wan (Ed.): Electrical Power Systems and Computers, LNEE 99, pp. 385–391.
springerlink.com

granularity, providing a powerful tool for the solution of complex problems, massive data mining, and fuzzy information processing [8]. In the past few years, there is a renewed and fast growing interest in GrC. Granular computing has begun to play important roles in bioinformatics, e-Business, security, machine learning, data mining, high-performance computing and wireless mobile computing in terms of efficiency, effectiveness, robustness and uncertainty [9].

Affective Computing is important to realize harmonious human-computer interactions and process intelligent information. To deal with the uncertainty problem in artificial emotion expression, a new artificial emotion granular structure model is proposed. The model presents a new artificial emotion granular concept and defines some key factors.

2 Basic Components of Granular Computing

The pager [8, 9, 15] gave an overview of Granular Computing. In modeling granular computing, the author focused on three basic components and their interactions as follows:

(1) Granules

Granules are regarded as the primitive notion of granular computing. A granule may be interpreted as one of the numerous small particles forming a larger unit. Collectively, they provide a representation of the unit with respect to a particular level of granularity. That is, a granule may be considered as a localized view or a specific aspect of a large unit. The size of a granule is considered as a basic property. Intuitively, the size may be interpreted as the degree of abstraction, concreteness, or detail. In the set-theoretic setting, the size of a granule can be the cardinality of the granule.

(2) Granulated views and levels

A level consists of entities called granules whose properties characterize and describe the subject matters of study, such as a real world problem, a theory, a design, a plan, a program, or an information processing system. Granules are formed with respect to a particular degree of granularity or detail. The granularity is reflected by the sizes of all granules involved. A granule in a higher level can be decomposed into many granules in a lower level, and conversely many granules in a lower level can be combined into one granule in a higher level.

(3) Hierarchies

Granules in different levels are linked by the order relations and operations on granules. The order relation on granules can be extended to granulated views (levels). A level is above another level if each granule in the former level is ordered before a granule in the latter level, and each granule in the latter level is ordered after a granule in the former level, under the order relation. The ordering of levels can be described by the notion of hierarchy. A hierarchy represents relationships between different granulated views, and explicitly shows the structure of granulation.

With the introduction of the three components, one can examine three types of structures for modeling their interactions. They are the internal structure of a granule, the collective structure of the all granules (i.e., the internal structure of a granulated view or level), and the overall structure of all levels. The three structures as a whole is referred to as the granular structure.

3 Artificial Emotion Granular Structure Model

3.1 Description of Abstract Granule

According to the main theory of GrC, formally, we can use a quadruple to describe a granule, namely, ((U, F, T), R). Let U be an universe of discourse; Let F be a set of attributes; T denotes the relation of granularity structure between all objects, which may be an order structure, a topological structure, as well as a general structure of graph or other structures; R denotes granulation rule, which may be equivalence relation, compatibility relation, indistinguishability relation, functionality relation, similarity relation, constraint relation, fuzzy relation, as well as compound relations consisted of a variety of relations.

3.2 Granular Structure Model

Definition 1. An Artificial Emotion Granular Structure Model is the following tuple:

GS=((P, A, S), R) ,

where

P is a set of all artificial emotion,
A is a set of all attributes of emotion,
S denotes the relation of structure between all emotions,
R denotes granulation rule.

On the basis of a hyperlinked structural graph of artificial emotion, we can form a granule of emotion as follows: (V, C, T, R) , where V only marks a node of emotion; C is a set of all attributes of the node, namely, $C=\{C_i|C_i$ is a attribute, $i=1,2,\cdots\}$, including happy, surprise, anger, fear, disgust, sadness, respiration, heart rate, body temperature, etc. T denotes the relation of structure of artificial emotion, which is given by T={Parent, Son}, then Parent is called a Granular set of father nodes, and Son is called a Granular set of child nodes; R denotes granulation rule.

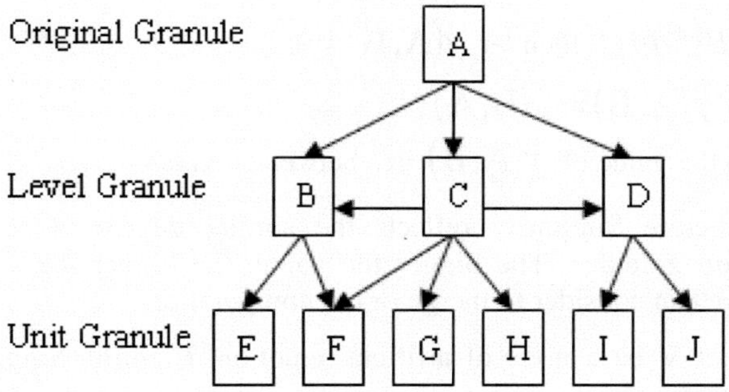

Fig. 1. Artificial Emotion Granular Structure

Original Granule is called the roughest granule; Unit Granule is called the thinnest granule; Level Granule is called the middle granule.

Definition 2. Let V be a node of artificial emotion. Granular of V is defined by

$$G(V) = \frac{f(Son)}{f(Parent)} = \left|\bigcup_{i=1}^{m} Son_i\right| \Big/ \left|\bigcup_{j=1}^{n} Parent_j\right|$$

where

f denotes the amount of information which are contained in a granule,

$|\bullet|$ denotes the cardinality of a set of the granule.

we can obtain the following properties:

(1) if V is a Original Granule, then let f(Son) be equal to f(Parent);
(2) if V is a Unit Granule, then let f(Son)=1;
(3) if V is a empty granule, then let G(V)=0;
(4) the value of G(V) lies between 0 and 1.

Emotion Granular is the measure of the amount of information which are contained in a emotion node. The bigger the value, the more abundant the amount of information, and the more important the node of emotion in all node structure.

Definition 3. For any A, B□GS, Emotion Structure Similarity of V is defined as follows:

$$\gamma(A,B) = \frac{1}{m \times n}\sum_{i=1}^{m}\sum_{j=1}^{n}\left|A(Son_i) \cap B(Son_j)\right| / \left|A(Son_i) \cup B(Son_j)\right|$$

where

$A(Son_i) \subseteq A$, $B(Son_j) \subseteq B$

we have the following properties:

(1) if A=B, then let $\gamma(A,B)=1$;

(2) if one does not exist intersection set between $A(Son_i)$ and $B(Son_j)$, then let $\gamma(A,B)=0$;

(3) $\gamma(A,B) = \gamma(B,A)$;

(4) the value of $\gamma(A,B)$ lies between 0 and 1.

Emotion Structure Similarity reflects the similar degree of structure between different emotion granules. The bigger the value, the higher the similar degree of structure, and we can consider to merge or cut computation.

Definition 4. Let V be a node of artificial emotion. Emotion Significance of V is defined as follows:

$$Sig(V) = \sum_{i=0}^{m} \alpha_i * Sig(son_v(i)),$$

where

α_i denotes the emotion significant weight of node i ;

if i=0, then let $Sig(V) = \sum_{l=1}^{n} \mathrm{k}_l^t \Big/ \sum_{h=1}^{m} \mathrm{K}_h^t$.

Emotion Significance indicates the significant degree of node V. The bigger the value, the more important the node, it should be mainly considered in artificial emotion computing.

4 Example

An artificial emotion Granular structure example is showed in Fig.2, A is a Original Granule, B, C, D, E, F are Level Granules, G, H, I, J are Unit Granules.

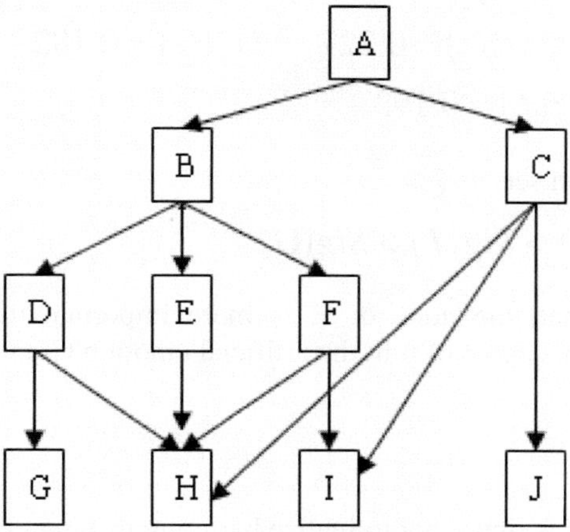

Fig. 2. Artificial Emotion Granular Structure Example

According to the Definition 1, the level of Granular Structure can be denoted as follows: A={ B={D={G ,H}, E={H}, F={H,I}}, C={H,I,J} } .

According to the Definition 2, with the knowledge of Set-theoretic Setting, Emotion Granular can be calculated as follows:

$$G(A) = 1,$$

$$G(B) = \frac{|D \cup E \cup F|}{|A|}$$

$$= \frac{|D| + |E| + |F| + |D \cap E \cap F| - |D \cap E| - |E \cap F| - |D \cap F|}{|B| + |C| - |B \cap C|}$$

$$= \frac{2+1+2+1-1-1-1}{3+3-2} = \frac{3}{4},$$

G(C)=3/4, G(D)=2/3, G(E)=1/3, G(F)=2/3, G(G)=1/2, G(H)=1/5, G(I)=1/3, G(J)=1/3.

Apparently, the values of Emotion Granular are degressive from top to bottom in the Granular Structure.

According to the Definition 3, Emotion Structure Similarity between node of D and F can be calculated as follows:

$$\gamma(D,F)=\frac{1}{2\times 2}\left(\frac{\{G\}\cap\{H\}}{\{G\}\cup\{H\}}+\frac{\{G\}\cap\{I\}}{\{G\}\cup\{I\}}+\frac{\{H\}\cap\{H\}}{\{H\}\cup\{H\}}+\frac{\{H\}\cap\{I\}}{\{H\}\cup\{I\}}\right)=\frac{1}{4}$$

Similarly, $\gamma(C,\mathrm{F})=1/3$. Obviously, the structure of C is more similar with F than D.

According to the Definition 4, the node of Emotion Significance can be given.

$Sig(A)=1$, $Sig(B)=0.363$, $Sig(C)=0.405$, $Sig(D)=0.227$, $Sig(E)=0.323$, $Sig(F)=0.227$, $Sig(G)=0.182$, $Sig(H)=0.311$, $Sig(I)=0.25$, $Sig(J)=0.193$.

From value, we can see

$$Sig(H)>Sig(I)>Sig(J)>Sig(G).$$

It is obviously that, the node of H is more important than the others, which dovetails with the key degree of it in the artificial emotion Granular structure.

5 Conclusion

Affective Computing is important to realize harmonious human-computer interactions and process intelligent information. Granular computing theory is an important tool for processing vague, incomplete, inaccurate information. Based on the granular computing theory, a new artificial emotion granular structure model is proposed in this paper. The model presents a new artificial emotion granular concept and defines some key factors. Finally, an experiment is given to calculate and verify the emotion granular structure model. In the future, we will research and extend the topic in following papers.

Acknowledgments. This work is supported by the Natural Science Foundation of Jiangxi, China (Grant No. 2008GZS0062).

References

1. Nahl, D.: Affective computing. Information Processing & Management 34(4), 510–512 (1998)
2. Jin, H., Gao, W.: The human facial combined expression recognition system. Chinese Journal of Computers 23(6), 202–208 (2000)

3. Song, Y., Jia, P.: A control architecture based on artificial emotion for anthropomorphic. Robot 26(6), 491–495 (2004)
4. Wang, Y., Yuan, B.: Pen-based gesture recognition in multi-modal human-computer interaction. Journal of Northern Jiaotong University (2), 10–13 (2001)
5. Picard, R.W.: Affective computing. MIT Press, London (1997)
6. Sloman, A., Croucher, M.: Why robots will have emotions. In: Proceedings IJCAI, Vancouver (1981)
7. Ward, R.D., Marsden, P.H.: Affective computing: prob1ems, reactions and intentions. Interacting with Computers 16(4), 707–713 (2004)
8. Zadeh, L.A.: Fuzzy sets and information granulation. Advances in fuzzy set theory and applications. North-Holland Publishing, Amsterdam (1979)
9. Hobbs, J.R.: Granularity, Proc. of the Ninth International Joint Conference on Artificial Intelligence, Los Angeles (1985)
10. Frijda, N.: Recognition of Emotion. Advances in Experimental Social Psychology, 167–223 (1969)
11. Pawlak, Z.: Rough sets. International Journal of Computer and Information Sciences 11, 341–356 (1982)
12. Liu, Q.: Rough Sets and Rough Reasoning, 3rd edn. Science Press, Beijing (2005)
13. Picard, R.W.: Affective computing. MIT Press, London (1997)
14. Chen, B., Sun, M., Zhou, M.: Granular Rough Theory: A representation semantics oriented theory of roughness. Applied Soft Computing 9(2), 786–805 (2009)
15. Han, J.C., Lin, T.Y.: Granular computing: Models and applications. International Journal of Intelligent Systems 25(2), 111–117 (2010)

Empirical Study on Company Scale Economies Based on the DEA Model

Chongming Liu and Yirong Jiang

North China Electric Power University, Beijing, China
liu-chongming@sohu.com, yirong1208@163.com

Abstract. DEA using data analysis has obvious advantages in company economies of scale. In this paper, a sample of listed electric power companies was used to establish the method of using DEA model to analyze, and point out this application model's data choose , indicator identify and assumptions of the condition. The present paper validates scale economy effects of the sample companies, including overall performance, technical performance and scale performance, which has great theoretical and practical significance to the study of company scale economies.

Keywords: DEA, Economies of scale, listed electric power companies.

1 Introduction

The definition of economies of scale is that reduction in cost of per unit resulting from increased production, realized through operational efficiencies. Economies of scale can be accomplished is that as production increases, the cost of producing each additional unit falls. It's investigated in the same level of technology under the premise, all factors of production increased to increasing yield, not only a factor of production changes on the yield of the law of diminishing marginal returns. Economies of scale is expressed increasing returns to scale, namely, the rate of increase greater than the income scale range after the scale of production expanded. Economies of scale also can be analyzed on two aspects, from the enterprises themselves and the scale of industry, the former is the internal economy, the latter is the external economy. Analysis of economies of scale has a certain value on studying of an industry, determining enterprise appropriate scale, making investment decisions and competing in the industry.

2 Document Summary

Duranton and Puga (2003) summarized the concentration of economic enhance corporate efficiency of the micro-mechanism. Enterprises gathered in the region can share an inseparable products or facilities, share increasing economies of scale from industry diversity and division of labor specialization and share the benefits of specialization and the total risk and so on [1]. Heru Margono, Subhash C. Sharma and Paul D. Melvin II (2010) make use of cost efficiency, economies of scale and technological progress to analyze Bank Indonesia's productivity from 1993 to 2000. By comparing the relationship between the efficiency and total cost, proposing the bank's best assets scale.

X. Wan (Ed.): Electrical Power Systems and Computers, LNEE 99, pp. 393–398.
springerlink.com

Technological advances result in cost reductions and economies of scale are more significant before the economic crisis [2].

Zhou Wen and Li Youai (1999) took from January 5, 1996 to the end of 1998 the Shanghai Stock exchange's 50 companies for the research sample, confirmed the hypothesis of scale [3]. Chen Junning, Ma Zhitian (2000) randomly selected 60 listed companies of varying scale from the Shanghai stock market, empirical research its 1997-1999 rate of return , and found the conclusion that Shanghai stock market existed scale [4]. Tong Wenjun (2004) took the economies of scale situation of China's banking from 1998-2003 for empirical analysis, found that with increasing the size of bank assets, asset expense ratio had a not significant decreasing trend. The studies suggest that economies of scale exist in the Chinese banking industry [5].

In summary, the measurement of the methods for economies of scale contain accounting analysis, survival of the fittest, entropy weight double point, the production function, translog cost function, unbounded analysis and data envelopment analysis . Now, empirical analysis of the literature, most use statistics and correlation analysis, few uses DEA data analysis. This paper intends to use DEA model, select the input and output of financial indicators, and research the listed companies on economies of scale. The reasons of selecting DEA method are: 1. No specific production function; 2. Can analysis scale with multi-input and multiple output freely; 3. Facilitate the identification of invalid unit size of the direction of improvement, 4. Has more strong practical operation and time of the adaptation.

3 Set Up DEA Model

DEA model is an effective method which compares the relative efficiency and effectiveness in multi-input, multiple output of the DMU in a number of the same type. This method does not require description explicit mathematical expression of the production system between the input and output relationship, but the application of linear programming theory is that separate the effective sample and non-valid samples. It is using a valid sample points constructed a piecewise linear hypersurface to form a production frontier.

Assuming decision making unit DMU $_j$ with n, the input of DMU is $x_j = (x_{1j}, x_{2j}, \ldots x_{mj})^T$,the output $y_j = (y_{1j}, y_{2j}, \ldots y_{sj})^T$, m is the number of the input indicators, s is the number of output indicators. $x_j >= 0$, $y_j >= 0$, j=1,2,…n, that is, its components non-negative and at least one is positive.

1 Based on the input of the evaluation of DMU overall efficiency with a non-Archimedean infinitesimal C²R model:

$$\begin{cases} \min(\theta - \varepsilon(\hat{e}^T s^- + e^T s^+)) \\ s.t. \sum_{j=1}^{n} \lambda_j T_j + s^- = \theta x_0 \\ \sum_{j=1}^{n} \lambda_j x_j - s^+ = y_0 \\ \lambda_j \geq 0; j = 1, 2 \cdots, n, s^- \geq 0, s^+ \geq 0 \\ \hat{e} = (1,1,\cdots,1)^T \in R^m, e = (1,1,\cdots,1)^T \in R^s \\ \varepsilon \text{is Archimedes infinitesimal} \end{cases} \qquad \text{(Formula 1)}$$

This model can evaluate technology and scale of the DMU's overall efficiency, called the total efficiency. Let the optimal solution for the formula 1: λ^*, s^{*-}, s^{*+}, σ^*, Are: ① If $\theta^* = 1$, then DMU_{j_0} is weak efficient; ② If $\theta^* = 1$, and $s^{*-} = 0$, $s^{*+} = 0$, then DMU_{j_0} is DEA efficient; ③ Order $\widehat{x_0} = \theta^* x_0 - s^{*-}$, $\widehat{y_0} = y_0 + s^{*+}$, then ($\widehat{x_0}, \widehat{y_0}$), as ($x_0$, y_0) in the efficient frontier projection surface , relative to the original n-DMU is effective; ④ If there is λ_j^*>(j=1,2,...m), so $\sum_{j=1}^{n} x_j^* = 1$ established, then the DMU is the economies of scale invariant, if $\sum_{j=1}^{n} \lambda_j^* < 1$, the DMU_{j_0} is increasing the economies of scale, and $\sum_{j=1}^{n} \lambda_j^* > 1$, The DMU_{j_0} is decreasing the economies of scale.

2 Based on the input of Evaluated pure technical efficiency of DMU with non-Archimedes infinitesimal C^2GS^2 model:

$$\begin{cases} \min(\sigma - \varepsilon(\hat{e}^T s^- + e^T s^+)) \\ s.t. \sum_j^n = {}_1 \lambda_j T_j + s^- = \sigma x_0 \\ \sum_j^n = {}_1 \lambda_j x_j - s^+ = y_0, \sum_{j=1}^n \lambda_j = 1 \\ \lambda_j \geq 0; j = 1, 2 \cdots, n, s^- \geq 0, s^+ \geq 0 \\ \hat{e} = (1, 1, \cdots, 1)^T \in R^m, e = (1, 1, \cdots, 1)^T \in R^s \\ \varepsilon \text{ is Archimedes infinitesimal} \end{cases} \qquad \text{(Formula 2)}$$

This model calculates the DMU'S technical efficiency, reflects DMU'S pure technical efficiency status, called pure technical efficiency. Let the optimal solution for the formula 2: λ^*, s^{*-}, s^{*+}, θ^*, are: ① If $\sigma^* = 1$, then DMU_{j_0} is weak efficient (pure technical); ② $\sigma^* = 1$, and $s^{*-} = 0$, $s^{*+} = 0$, then DMU_{j_0} is efficient (purely technical). So the scale of DMU efficiency is calculated as $s^* = \theta^* / \delta^*$.

4 Empirical Analysis of Economies of Scale in Power Generation Companies

4.1 Samples and Index Selection

The present paper selected 26 listed companies on power for samples as a decision-making unit before the year of 2000. These companies are with a positive value, have gone through years of a mature business, and have relatively large individual scales. The selected financial indicators firstly need to meet the evaluation aim, the input or output indicators are not obvious linear relations, and data-caliber has consistency, comparability and data availability. This paper selected the 26 listed companies on power for samples, input and output indicators are as follows:

Output indicators: main business income, net profit.

Input indicators: net fixed assets, first asset (including paid-in capital, capital surplus, surplus reserves and undistributed profits), main business costs.

4.2 The Calculation Results of DEA

Put the Original data into the C^2R model and the C^2GS^2 model, use the lingo software to program solve, obtain thermal power listed companies DEA overall efficiency, are technical efficiency and scale efficiency values in Table1.

Table 1. 26 Electric Companies DEA model results

ID	Company Name	θ^*	σ^*	s^*	$\sum \lambda_j$	Scale
1	Bao Xin New Energy Company	0.125	0.21	0.60	0.345	Ascending
2	ST Hui Tian Company	1.00	1.00	1.00	1.00	Unchanged
3	ST Neng Shan Company	1.00	1.00	1.00	1.00	Unchanged
4	Hua Yin Electric Power Company	1.00	1.00	1.00	1.00	Unchanged
5	Tong Bao Energy Company	0.281	0.38	0.75	4.329	Decreasing
6	Shen Zhen Energy	0.389	0.95	0.41	0.439	Ascending
7	Shen Nan Power A	0.516	0.61	0.84	0.938	Ascending
8	Zhang Ze Power	0.884	1.00	0.88	0.689	Ascending
9	Guo Dian Power	0.305	0.44	0.69	0.212	Ascending
10	Nei Meng Power	1.00	1.00	1.00	1.00	Unchanged
11	Fu Long Power	0.757	1.00	0.76	0.843	Ascending
12	Sui Heng Yun A	1.00	1.00	1.00	1.00	Unchanged
13	Yue Power A	1.00	1.00	1.00	1.00	Unchanged
14	Wan Neng Power	1.00	1.00	1.00	1.00	Unchanged
15	Gan Neng Share	1.00	1.00	1.00	1.00	Unchanged
16	Shen Neng Share	1.00	1.00	1.00	1.00	Unchanged
17	Le Shan Electric Power	0.385	1.00	0.39	0.589	Ascending
18	Ha Tou Shares	0.415	0.75	0.55	0.605	Ascending
19	Guo Tou Electric Power	0.456	1.00	0.46	0.245	Ascending
20	Jian Tou Energy	0.649	0.69	0.95	0.729	Ascending
21	Shao Neng Shares	1.00	1.00	1.00	1.00	Unchanged
22	Kai Di Electric Power	0.878	1.00	0.88	3.356	Decreasing
23	Guang Zhou Holding	1.00	1.00	1.00	1.00	Unchanged
24	San Xing Electric Power	1.00	1.00	1.00	1.00	Unchanged
25	San Xia Electric Power	1.00	1.00	1.00	1.00	Unchanged
26	Si Chuan Investment Energy	0.403	0.945	0.426	1.480	Decreasing

When $\theta^* \neq 1$, illustrating non-DEA efficient, from the table we can see a total of 13 companies. Select 13 listed companies on power in non-DEA efficient to do projection analysis, the results shown in Table 2.

Table 2. Non-DEA efficient projection results of listed companies on power (10 thousand Yuan)

ID	Company Name	Save the net fixed assets	asset savings	main business cost savings	main business income	net profit Increase
1	Bao Xin New Energy Company	43.434	610.931	296.139	81.609	276.514
5	Tong Bao Energy Company	436.170	131.660	253.231	0	0
6	Shen Zhen Energy	9.464	25.493	0.048	0	459.175
7	Shen Nan Power A	2.708	159.184	88.191	12.591	186.906
8	Zhang Ze Power	76.298	136.140	93.489	23.417	158.298
9	Guo Dian Power	88.835	276.567	2.794	0	275.914
11	Fu Long Power	19.302	67.401	94.486	63.139	153.098
17	Le Shan Electric Power	92.415	114.872	49.613	0	97.216
18	Ha Tou Shares	5.212	79.142	21.594	65.013	287.937
19	Guo Tou Electric Power	4.218	88.421	120.348	53.348	0
20	Jian Tou Energy	2.195	36.095	208.173	29.249	37.318
22	Kai Di Electric Power	52.859	52.329	183.295	88.298	238.472
26	Si Chuan Investment Energy	0.928	9.432	25.336	152.596	0

4.3 Analysis

DEA efficiency analysis. From Table 1, we can know in the 26 listed companies on thermal power which $\theta^{*}= 1$ the total of the company is 13, indicating that in the overall efficiency and pure technical efficiency frontier surface, accounting for 50% of the total sample, indicating that the overall efficiency of listed companies on power is medium. The number of less efficient and more efficient companies is basically the same. Except this 13 companies , where $\sigma^{*}= 1$ of 5 companies which are at the forefront of the surface in the pure technical efficiency, accounting for 19.23% of the total sample, indicating that the level of technology of listed companies on power is relatively low, therefore, the main factors affect the overall efficiency is the pure technical efficiency.

Economies of scale efficiency analysis. Traditional production theory presumes that production and operation in one phase from economies of scale invariant to diminishing economies of scale is efficiency. C^2R model gives the overall efficiency of 26 companies have RMS, also can use $\sum \lambda_i$ to measure the situation of listed companies' scale. Table 1 shows the scale of the company's situation: $\sum \lambda_i \langle 1$, there are 10, increasing economies of scale; $\sum \lambda_i = 1$, there are 13, economies of scale invariant; $\sum \lambda_i \rangle 1$, there are 3, decreasing economies of scale, accounting for 11.54% of samples, showing the state of diseconomies of scale. So that, the total economies of scale situation in the listed companies on power is better.

DMU'S projection in the production frontier. DEA methods provide basis to improve the efficiency of the company. The companies which are not the overall efficiency of the efficient frontier surface, through adjustment input indicators and output indicators finally achieve the overall efficiency of the relatively effective. Table 2 is calculated by the Non-DEA efficient of listed companies on power can save the amount of input and increase the amount of output that indicates achieve to economies of scale requirement, the sample companies have some room for improvement in both input and output. Taking the Bao Xin New Energy Company for example, it's fixed assets, first assets, mainly business cost should be separately reduced by 4.3434, 61.0931 and 29.6139 million Yuan, and it's main business revenue and net profit should increase 8.1609 and 27.6514 million Yuan, the results of operations can achieved the DEA efficient, so achieve the economies of scale.

5 Conclusion

Listed companies on power have lager scales and relatively stable performance. This paper uses DEA methods to empirical analyze listed companies on power, analyses the overall performance, technical performance and scale performance. The results show that the electric company are better on overall performance of the economies of scale, and propose methods on improving the direction of company in the economies of scale which include raising the utilization rate of fixed assets, controlling costs and expenses and improving operational efficiency and so on.

References

1. Duranton, G., Puga, D.: Micro-foundations of Urban Agglomeration Economies. NBER Working Paper, 67–69 (2003)
2. Margono, H., Sharma, S.C., Melvin II., P.D.: Cost Efficiency, Economies of scale, Technological progress and Productivity in Indonesian Banks. Journal of Asian Economics (1), 53–65 (2010)
3. Wen, Z., Youai, L.: The correlation of return rate, scale and market efficiency - on the Shanghai and Shenzhen Stock Market test. Modern Economic Science (1), 56–62 (1999)
4. Junning, C., Zhitian, M.: The stock market Empirical Research. Hua Zhong University of Science and Technology (Social Sciences) (10), 37–41 (2000)
5. Wenjun, T.: Bank development of the theory: Motivation and Performance. The world economy (22) (2004)
6. Jiatian, Y.: Horizontal Mergers economies of scale analysis. Coastal Enterprises and Science and Technology, 47–50 (2008)
7. Lin, G.: Financial scale fluctuations in the economy of the stabilizer is up: MS Thesis, Fu Dan University, Shanghai (2009)

Research on Electricity-Purchase Risk Optimization Model of Provincial Grid Companies in Provincial and Regional Markets Considering Price Fluctuations

Lijun Tan, Hua Zheng, and Xiaofei Li

School of Electrical and Electronic Engineering, North China Electric Power University, Beijing 102206, China

Abstract. Price fluctuation in the electricity market brings risk to electricity purchase. To reduce the risk effectively, this paper established a mean-variance risk measurement model for six in-and-out-of-province electricity markets based on the portfolio theory and the analysis of the distribution of provincial power gird market. On this basis, a risk optimization model for power purchasing was established and solved to get the reasonable allocation ratio of yearly electricity volume, monthly electricity volume and day-ahead electricity volume. Finally, through the calculation analysis using the model and simulation data from one provincial gird company, this paper proved the validity and practicability of the model.

Keywords: Electricity Purchasing Strategy, Price Fluctuations, Portfolio, Mean-variance Method, Risk Optimization.

1 Introduction

The marketization is the only way for power industry to lead to prosperity. In China, power grid companies supply power on the premise of safety and reliability, and then concentrate on profit-maximization. Due to a lot of factors which influence the profits of grid companies, such as price, balance between supply and demand, network security and so on, in electricity market environment, income risk problem which grid companies have to face becomes one of the hot issues. Along with the construction of our country UHV grid, the growth of the inter-provincial power transactions will accelerate, and the electricity quantity ratio of inter-province transactions in in-province transactions is increasing. How to optimize quantity ratio between the in-province and inter-province transactions in the electricity market environment to lower the risk of electricity purchases becomes a great concern of the grid companies.

Some research fruits of how to allocate the electricity quantity under multi-market environment already exist [1-5], Paper [6-7] classified the electricity purchasing market of the grid company into short-term and long-term markets, introduced some risk factors to research the optimal allocation of electricity, and analyzed the purchasing strategy when demand is determined and under the conditions of both independent and linear-correlated purchasing price in two markets. However, researches on electricity purchasing of provincial electricity companies are rare. Paper

X. Wan (Ed.): Electrical Power Systems and Computers, LNEE 99, pp. 399–405.
springerlink.com

[8] established a mathematical model of electricity purchasing cost for provincial electricity companies in 96 intraday trading periods, and the change in marginal electricity price brought by yearly, monthly and daily electricity allocation plan as well as deviation of load forecasting. However, the trading is classified by trading period instead of trading area.

In China, because of the differences of the economic development, natural resources and the level of electric power construction in different provinces, the generation cost also varies. Therefore, the region factor must be taken into account when provincial grid companies making purchase decisions. This paper, through the analysis of the power market and transaction types which provincial grid companies will face possibly in future electricity environment, based on portfolio theory, established a mean-variance risk measurement model for provincial grid companies in in-and-inter-province electricity purchases. Then, an optimization model is constructed and solved based on the former model to procure the reasonable allocation ratio of yearly electricity quantity, monthly electricity quantity and day-ahead electricity quantity of in-and-inter-province electricity purchases of grid companies. Finally, through the calculation analysis using the model and simulation data from one provincial gird company, this paper proves the validity of the model.

2 The Distribution of Provincial Electricity-Purchase Market

In the market environment, provincial grid companies can purchase the electricity in and out of the province, and these two markets can be further divided to 3 transaction markets which are yearly, monthly and day-ahead. As shown in figure 1 below.

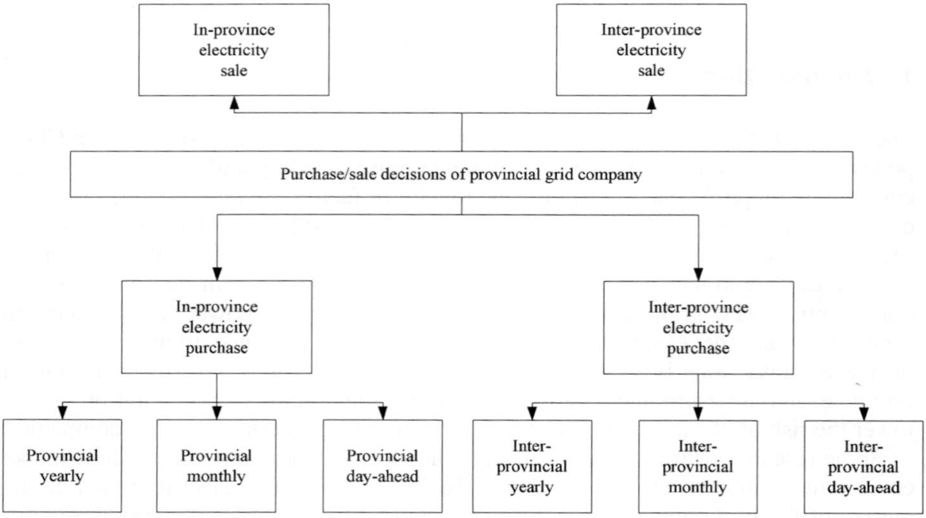

Fig. 1. Provincial and inter-provincial electricity markets of the provincial grid company

Generally speaking, price fluctuation risks faced by different cycle power transaction are different. By domestic and foreign power market practice experiences, price fluctuation of day-ahead is fiercer, price fluctuation of year is smooth, and price fluctuation of month is between two ahead. Thus different cycle power exchange transaction risks also vary, and this is suitable for the power industry production characteristics.

Provincial grid companies can meet the future expected basic demand through purchasing appropriate mid-and-long-term contract, which can reduce the power purchasing risk brought by short-term price fluctuations, and minimize the expenses and risks through optimal electricity transactions as well as maximize the profit of the electricity companies.

Assume that provincial grid companies can purchase electricity in provincial and inter-provincial market, and can also purchase specific quantity of electricity or contracts in advance according to supply and demand and price fluctuation in different markets, with the additional required electricity purchased from the more risky day-ahead markets. Provincial grid companies hope to find an optimized power purchase strategy to minimize the expense and risk. To achieve this goal, we can use the portfolio theory to get the optimized purchasing plan.

Provincial grid companies forecast the electricity demand first, and allocate the quantity needed to be purchased to each transaction market. Let the price of provincial yearly electricity trading market be P_1, and the expected value and variance be μ_1 and σ_1^2; let the price of provincial monthly electricity trading market be P_2, and the expected value and variance be μ_2 and σ_2^2; let the price of provincial day-ahead electricity trading market be P_3, and the expected value and variance be μ_3 and σ_3^2; let the price of out-of-province yearly electricity trading market be P_4, and the expected value and variance be μ_4 and σ_4^2; let the price of out-of-province monthly electricity trading market be P_5, and the expected value and variance be μ_5 and σ_5^2; let the price of out-of-province day-ahead electricity trading market be P_6 and the expected value and variance be μ_6 and σ_6^2; let the electricity sale price be P ,a constant value, because the fluctuation of sale price can be ignored. We also assume the purchasing ratio between these 6 markets is $x_1 : x_2 : x_3 : x_4 : x_5 : x_6$, and the line loss is γ, so we get the profit per unit electricity sold as following:

$$R(x) = P(1-\gamma) - (P_1x_1 + P_2x_2 + P_3x_3 + P_4x_4 + P_5x_5 + P_6x_6) \tag{1}$$

3 Provincial and Inter-provincial Electricity Purchasing Risk Optimization Model

According to the unit electricity profit model of the provincial grid companies, we can get the mean per unit electricity purchased $E[R(x,P)]$:

$$E[R(x)] = P(1-\gamma) - \mu_1 x_1 - \mu_2 x_2 - \mu_3 x_3 - \mu_4 x_4 - \mu_5 x_5 - \mu_6 x_6 \quad (2)$$

Then the variance of the i th market and the covariance between the i th and the j th market can be expressed as:

$$\sigma^2[R(x)] = \sigma^2[R(x,y)] = \mathbf{x}^{\mathrm{T}} \mathbf{\Sigma} \mathbf{x} \quad (3)$$

$\sum$ is the variance matrix

$$\Sigma = \begin{bmatrix} \sigma_{11} & \sigma_{12} & \cdots & \sigma_{16} \\ \sigma_{21} & \sigma_{22} & \cdots & \sigma_{26} \\ \cdots & \cdots & \cdots & \cdots \\ \sigma_{61} & \sigma_{62} & \cdots & \sigma_{66} \end{bmatrix} \quad (4)$$

According to the Markowitz portfolio theory [9], the consolidated utility function of expected return and risk can be used to evaluate the portfolio. The greater the expected return, the higher the value of the utility function. The greater the variance, the lower the value of the utility function. Therefore, equation (2), (3) can serve as the risk measurement model for the grid companies.

As we know, the higher the price level and the lower the fluctuation, the higher the utility. Thus we measure the electricity purchasing risk with the mean-variance method in risk decision, using function $U = k_1 E(R) - k_2 \delta^2$ to maximize the utility.

The K_1 and K_2 in the function are the risk aversion factors, which represent the level of risk aversion of the people making final decision. The ratio between K_1 and K_2 indicates the significance of the standard deviation relative to the mean. To solve the inter-provincial electricity purchasing allocation problem, we believe that the solution with the lowest risk is equivalent to the one with highest utility, or the solution with the maximum U which meet the following function is the one with the lowest risk:

$$\begin{cases} \max U = K_1 E[R(x)] - K_2 \delta^2 = P - \mu_1 x_1 - \mu_2 x_2 - \mu_3 x_3 \\ - \mu_4 x_4 - \mu_5 x_5 - \mu_6 x_6 - 0.5\sigma^2[R(x,y)] \\ s.t. \quad \sum_{i=1}^{6} x_i = 1, \quad x_i \geq 0, i = 1,2,3,4,5,6 \end{cases} \quad (5)$$

Here we give $K_1 = 1$, $K_2 = 0.5$, in accordance with the widely used utility function in finance:

$$U = E(R) - 0.5\delta^2 \quad (6)$$

The mean electricity prices of provincial yearly electricity trading market, provincial monthly electricity trading market, provincial day-ahead electricity trading market, out-of-province yearly electricity trading market, out-of-province monthly electricity trading market, out-of-province day-ahead electricity trading market are $\mu_1, \mu_2, \mu_3, \mu_4, \mu_5, \mu_6$.

Then the utility maximization model can be constructed:

$$\begin{cases} \max\ U = E[R(x)] - 0.5\delta^2 = P - \mu_1 x_1 - \mu_2 x_2 - \mu_3 x_3 \\ \quad - \mu_4 x_4 - \mu_5 x_5 - \mu_6 x_6 - 0.5\sum_{i=1}^{6}\sum_{j=1}^{6}\sigma_{ij}^2 \\ s.t.\ \sum_{i=1}^{6} x_i = 1, \quad x_i \geq 0 \quad i, j = 1,2,3,4,5,6 \end{cases} \tag{7}$$

This model is a convex quadratic programming optimization problem with equality constraint, so we can use Lagrange method to solve. And we can get the optimized portfolio by programming using Matlab.

4 Examples

Using the historical electricity purchasing data of some provincial grid company, we can do a purchasing strategy analysis using the above-mentioned model. Assume the electricity purchasing prices through 2001 to 2010 for this provincial grid company are shown as the following table:

Table 1. Electricity purchasing data of one provincial grid company Unit: Yuan/(MW•h)

year	provincial yearly	provincial monthly	provincial day-ahead	inter-provincial yearly	inter-provincial monthly	inter-provincial day-ahead
2001	315	317	320	317	318	322
2002	318	315	310	318	319	329
2003	314	318	332	317	318	335
2004	313	317	323	316	316	310
2005	315	318	313	317	319	308
2006	316	319	328	316	317	331
2007	316	316	327	318	317	309
2008	315	314	314	318	319	324
2009	317	318	312	319	317	301
2010	314	315	320	315	318	345

With the help of Matlab7.1, we can get the optimized portfolio of provincial and inter-provincial yearly, monthly, and day-ahead trading market. The optimized ratio is 68.43%、17.19%、2.84%、8.39%、3.15%、0.01% as shown in the following fig 2:

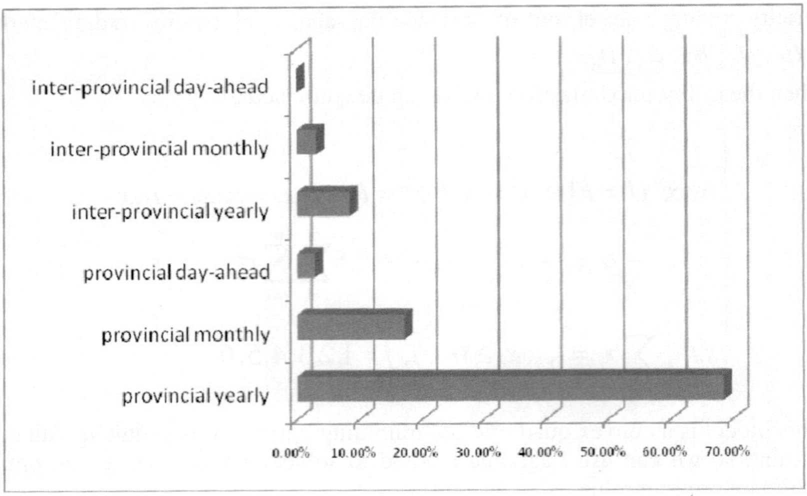

Fig. 2. Optimized portfolio of provincial and inter-provincial markets

The grid companies can purchase electricity in provincial and inter-provincial yearly, monthly, and monthly market according to the calculation results. At this time, the utility of the unit returns is the biggest. If the grid company has a greater risk appetite, the value of parameters K_2 will be greater. Assuming the value of parameters K_2 is 0.6 in this example, the optimized ratio is 59.85%、15.55%、03.09%、11.62%、9.83%、0.6%.

5 Conclusions

Markowitz portfolio theory provides the investment decision with the most fundamental and most complete framework. This paper used the Markowitz portfolio theory to analyze how the provincial grid companies can get a utility-maximized portfolio in provincial and inter-provincial yearly, daily and monthly markets, and constructed a mean-variance risk measurement model first, then based on that, a risk-optimized model. Through the example, a reasonable allocation ratio between the six contract markets is retrieved by solving the model in Matlab. The grid companies can adjust the parameters K_2 in the model according to their risk appetite. If the grid companies start to use the model to purchase electricity in provincial and inter-provincial markets, they can have lower risk, and reduce the losses these risks may bring.

References

1. Xu, W., Xia, Q., Kang, C.: Risk Assessment of Electricity Buyers and Sellers in Power Markets Based on Sequence Operation Theory. Automation of Electric Power Systems 32(24), 15–19 (2008)

2. Zhang, X.-s., Wang, S.-f., Guo, X.-l.: Research on evaluating risk of power market based on Monte Carlo simulation. Relay 35(18), 30–33 (2007)
3. Zhou, M., Nie, Y.-l., Li, G.-y., et al.: Long-term electricity purchasing scheme and risk assessment in power markets. Proceedings of the CSEE 26(3), 116–122 (2006)
4. Guo, J., Jiang, W., Tan, Z.-f.: Research on optimized power purchasing of power suppliers under risk condition. Power System Technology 28(11), 18–22 (2004)
5. Wang, R., Shang, J.-c., Feng, Y., et al.: Combined bidding strategy and model for power suppliers based on CVaR risk measurement techniques. Automation of Electric Power Systems 29(14), 5–9 (2005)
6. Yan, H., Yan, H.: Optimal energy purchases in deregulated California energy markets. Proceedings of IEEE power engineering society winter meeting, 1249–1254 (2000)
7. Ya'an, L., Guan, X.H.: Purchase allocation and demand bidding in electric power markets. IEEE Trans. on Power Systems 18(1), 106–112 (2003)
8. Chen, Q., Li, Y.: Study of electricity purchase strategies for provincial power companies. East China Electric Power 36(3), 13–16 (2008)
9. Bodie, Z., Kane, A., Marcus, A.J.: Investments, 5th edn. The McGraw-Hill Companies, New York (2002)

Automatic Speech Recognition in Real Environments: A Methodology for Evaluation of New Noises

David D. e Silva[1], Carlos A. Ynoguti[2], and Marcelo R. Stemmer[3]

[1] UDESC – Santa Catarina State University, Brazil
[2] INATEL – Institute National of Telecommunications, Brazil
[3] UFSC – Santa Catarina Federal University, Brazil
davidd.speech@gmail.com, ynoguti@inatel.br, marcelo@das.ufsc.br

Abstract. This paper presents a methodology to verify if the characteristics of the background noise are well represented in a given ASR system. The methodology is based on the PESQ vs. Recognition Rate relationship. It is shown that if a noise is not well represented in the training database, the inclusion of such noise can significantly improve the recognition rate. On the other hand, it is also shown that if the new noise is found to be adequately represented, the inclusion of such noise does not significantly improve the recognition rate. Thus, the methodology proposed here can be used to minimize the production cost of robust databases and improve recognition rate of speech signals in new noisy environments.

Keywords: PESQ, real environments, recognition rate, robust automatic speech recognition, training material.

1 Introduction

It's a well-known fact that mismatches between the training and recognition acoustic environments lead to degradation in the performance of automatic speech recognition systems. This issue becomes important especially in mobile applications, where the types and levels of background noise are unpredictable and may vary over time.

This paper focus on the following problem: given a noisy speech, how well is this noise represented in the training material, and therefore in the recognition system? This is an important issue because this knowledge allows a system to predict its performance given the noise conditions of the incoming audio signal.

On the other hand, it was shown in [1] that the relationship between the PESQ score [2] and the recognition rate can be modeled as a logistic function. Furthermore, for the Aurora-2 database [3], it was also shown in [1] that the noises could be grouped together with similar PESQ vs. Recognition Rate behaviors.

The experimental results shown that it's possible to predict if a new noise is well represented in the training database just by comparing the PESQ vs. Recognition Rate curves. Thus, being possible to identify the noise type and level, it would be possible to predict the recognition rate of an ASR system with advance. This knowledge is useful, for instance, in call center systems that would redirect the call to a human

X. Wan (Ed.): Electrical Power Systems and Computers, LNEE 99, pp. 407–414.
springerlink.com

attendee if the noise conditions make the predicted performance fall below a given threshold.

2 PESQ vs. Recognition Rate

In this section a brief explanation about the relation between the PESQ score and the recognition rate is provide in order to pave the discussions in this paper. In [1] it was shown that the relationship between the PESQ score and the recognition rate could be adequately modeled by the logistic function shown below:

$$f(x) = \left(\frac{1}{1+e^{b-ax}} - c \right) 100 . \tag{1}$$

In addition, it was also shown that it's possible to associate real world meanings for the a, b and c parameters. In short, we have: the parameter a defines the characteristic inclination of the curve, b defines a horizontal offset, and c defines a vertical offset. It was shown in [1] that each of these parameters can be associated to a performance measure of an RASR system: a can be viewed as "sensitivity of the system to PESQ variation", parameter b is related to the robustness of the system to noise, and parameter c is the average recognition rate in clean conditions.

Therefore, this model is not just a curve fitting for experimental data, but also a meaningful parametric model that can be used to analyze and predict the behavior of speech recognition systems in noisy conditions. Also, systems with similar curves can be grouped in the same "family", which means that their characteristics in terms of recognition rate are similar under similar noise conditions.

This work uses the following property: if the above statement is true, it can be possible to verify if the training material for a given ASR system has information about the acoustic environment in which the utterances are being collected. If the acoustic mismatch is higher than a certain threshold, then the system will probably have a poor performance. This issue is addressed in Section 5.

On the other hand, in [1], the values of a, b and c were estimated by varying their values and calculating the square error, an approach which is computationally inappropriate. Thus, an efficient solution for numeric computation of these parameters is proposed here, and is described in Section 4.

3 Experimental Setup

The Aurora-2 [3] was used as the speech database for this work. Although well known, it's worth a brief description: this database is composed of clean speech signals from the TI-Digits database [4], acoustically mixed with 8 noises: 4 noises for training (subway, babble, car and exhibition-hall), and 4 noises for tests (restaurant, street, airport and train-station).

3.1 Database

For this work four new noises were recorded and added to the Aurora-2 database: metal-cutting, tunnel-front, tunnel-inside and crowd-children. This was done in order to add more diversity to the analysis scenarios.

The metal cutting noise has very different characteristics compared to the noises in the Aurora-2 database. On the other hand, the remaining three have very similar characteristics when compared to the noises in the Aurora-2 database. These choices allow the study of the behavior of the ASR system in the presence of both similar and different noises.

The metal-cutting noise has been recorded while a dry multi-cutting saw was processing a ¾" thickness stainless steel block. The noise in front of the tunnel was obtained in a highway, around 15 m away from the tunnel's entrance and 2 m away from the track. The noise inside the tunnel was obtained at approximately the half of the tunnel's length. Finally, the crowd of children's noise was obtained at a school with a group of scholars between 5 and 12 years old at the classroom arrival time.

The microphone used was a flexible omni-directional with adjustable headset. All the noises were recorded with a sampling rate of 16 kHz and 16 bits quantization. The sampling rate was converted for 8 kHz, the same used in Aurora-2 database.

To generate the noisy utterances, the clean utterances from the TI-Digits database were mixed together with the new noises using the same SNRs used in the Aurora-2 database: -5 dB, 0 dB, 5 dB, 10 dB, 15 dB and 20 dB for the tests, and 5 dB, 10 dB, 15 dB and 20 dB for training. Before the mixing process, the noises have been passed by a G.712 filter [5], in the same way that in the Aurora-2 database.

3.2 Recognition Engine

This work was developed using the Front-End version 2.0 from WI007 [6], which consists of log-energy and twelve mel-cepstrum coefficients with their first and second derivatives. The HTK (Hidden Markov Model Toolkit) [7] was used for recognition (Back-End). The PESQ score were obtained using the software provided by the ITU [5].

4 Solution of the Logistic Function

One of the most widely used methods to solve the problem of curve fitting is the Maximum Likelihood Estimation (MLE). However, it does not perform well in cases where only a few points are available [8][9], exactly the case of the problem at hand (six points only).

In this section an algorithm to calculate the a, b and c parameters of the logistic function (1) from N experimental points is derived (N small).

The resulting algorithm is called the Initial Logistic Adjust Method (ILAM). The name ILAM comes from the necessity to use a logistic initial adjustment curve of subjective form, with only three points and the equations presented for the solution of (1).

The first step is to choose three points (x_i, y_i), $i = 1,2,3$, which adequately represent the "behavior" of the curve being parameterized. An adequate choice would be a first point in the beggining of the curve, another in the middle and the third near the end. Thus, the c parameter is calculated using (2):

$$\frac{(A_1 - A_2 K_1)}{K_2} + (A_1 - A_2)K_3 - A_3 = 0 \,. \tag{2}$$

where,

$$K_1 = \frac{x_1}{x_2} \,.\quad K_2 = 1 - \frac{x_1}{x_2} \,.\quad K_3 = \frac{x_3}{x_1 - x_2} \,.$$

$$A_n = \ln\left(\frac{100}{y_n + 100\,c} - 1\right) .\quad y_n = f(x_n) \text{ and } n=1,2,3 \,.$$

Equation (2) cannot be directly solved for c, so a numerical method must be used. In this case, Newton-Raphson method can be used, provided that the analysis interval guarantees a real solution. The following analysis shows that in the problem at hand, this condition is satisfied.

Analyzing the expression of A_n for real values:

$$y_n + 100\,c > 0 \Rightarrow c > -\frac{y_n}{100} \,. \tag{3}$$

$$\frac{100}{y_n + 100\,c} > 1 \Rightarrow c < 1 - \frac{y_n}{100} \,. \tag{4}$$

From (3) and (4) it can be concluded that:

$$-\frac{y_n}{100} < c < 1 - \frac{y_n}{100} \,. \tag{5}$$

As y_n varies in the range (0, 100), the allowed interval for c from equation (5) is (-1, 1). Remembering that c is the recognition rate of the ASR in clean conditions, and therefore is the [0,1) interval, it's clear that this condition is satisfied.

The parameter c is then calculated to some given tolerance ε, and the resulting value is applied to (6) and (7) to obtain the parameters a and b.

$$a = \frac{A_2 - A_1}{x_1 - x_2} \,. \tag{6}$$

$$b = \frac{A_1 - (x_1 / x_2) A_2}{1 - x_1 / x_2} \,. \tag{7}$$

With the values of a, b and c, it is possible to construct the fitting curve using (1) and calculate the square error between this curve and the experimental points using the equation (8):

$$lss = \sum_{i=1}^{n} \left(y_i - y_{ref}\right)^2 \,. \tag{8}$$

where:

lss – least squares sum;
n – number of points;
y_i – recognition rate from system under evaluation;
y_{ref} – recognition rate from reference curve.

It is important to note that this evaluation is made with all the N available experimental points, and not only for the initial 3 points.

If the resulting lss is less ou equal than a given threshold Δ, i.e. $lss \leq \Delta$, it's considered that a satisfactory logistic curve has been defined for the experimental points, and the algorithm is finished.

In the otherwise, the following steps were used to refine the values of a, b and c:

1. The experimental point (x_i, y_i) that has the bigger perpendicular distance to the adjusting curve, i.e., with biggest error, is computed. Note that this point is not necessarily one of the three original points selected for the previous step.
2. For this point, a gradual approximating from curve fitting is made, resulting in the lowest lss value possible with (x_i, y'_i). In this work, this approach was performed with a linear step obtained from the point of greatest error and the point closest to it. Thus, for each value of y', new parameters a, b, c and a new lss are calculated, and consequently, the "best" value for point (x_i, y'_i) is obtained.

This process is repeated until the condition $lss \leq \Delta$ is achieved. To prevent the algorithm from entering an infinite loop, the maximum number of iterations was limited to 20 (the convergence result was obtained with three to six iterations for Newton-Raphson method).

5 Results

The distances of the curves of the new noises (metal-cutting, tunnel-front, tunnel-inside, crowd-children) for each reference curve from Aurora-2 database (subway, babble, car and exhibition-hall) were calculated from the adjustment curves for all conditions [10]. The minimal distance (lss) between each curve of the new noises for each reference curve from Aurora-2 database, for all conditions, are shown in Table 1.

Table 1. Minimal values of lss between new noises and reference noises.

	New noises			
Reference noises	**Metal-cutting**	**Tunnel-front**	**Tunnel-inside**	**Crowd-children**
Subway	-	-	-	-
Babble	-	-	0.0157	-
Car	0.0662	-	-	0.0188
Exhibition-hall	-	0.0072	-	-

It was observed that the metal-cutting noise was the only one that did not reach the desired tolerance ($lss \leq \Delta$), which was defined as 0.05 ($\Delta = 0.05$) for all tests. The result of the least squares method [9] shows that the noise nearest to metal-cutting curve was the car noise curve.

Figure 1 shows the curves for the car noise and metal-cutting noise. The *lss* value calculated for these two curves was $lss = 0.0662$.

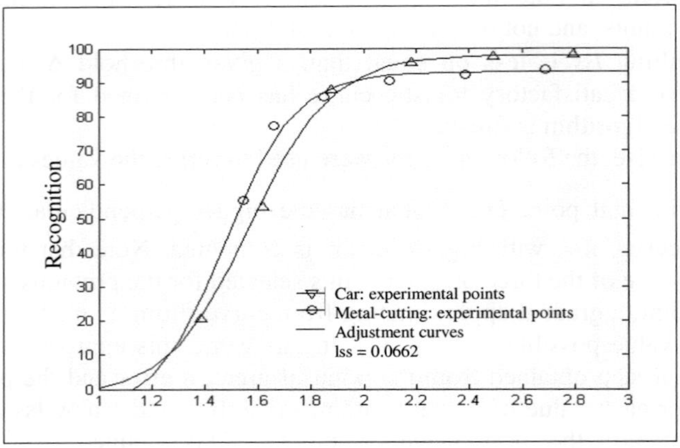

Fig. 1. Metal-cutting and Car curves for WI007 under multi-condition with ILAM.

Table 2 shows the *lss* of metal-cutting noise for all reference noises, considering WI007 under multi-condition, where the *lss* values have been smaller for this noise.

Table 2. Value of *lss* for metal-cutting.

Reference noises	lss
Subway	0.0823
Babble	0.1174
Car	0.0662
Exhibition-hall	0.1007

For the adopted criterion ($lss \leq 0.05$), it's evident that the characteristics of the metal-cutting noise are not well represented in the Aurora-2 database. When adding the characteristics of this type of noise to the training material of the database (without the interference of noise reduction filters), a 3.69% average better performance in recognition rate of the WI007 system was obtained.

Table 3 shows the recognition rate with and without the inclusion of metal-cutting noise in the training material, beyond details of the improvement for each SNR level, for WI007 system under multi-condition training [10].

Table 3. Recognition rate for metal-cutting.

Front-End WI007		Recognition rate (%) using metal-cutting noise		
Degradation		Original Aurora-1	Metal-cutting in training material	Improvement (%)
SNR Levels (dB)	**Clean**	98.68	98.68	-
	20	93.69	95.21	1.52
	15	92.18	94.34	2.16
	10	90.54	94.13	3.59
	5	85.90	91.16	5.26
	0	77.40	83.86	6.46
	-5	55.39	58.55	3.16
Average (20 to -5 dB)		82.52	86.21	3.69

The results from Table 3 are graphically shown in Figure 2.

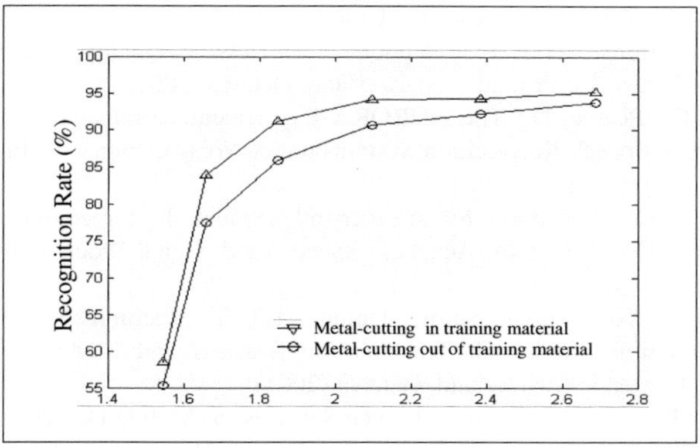

Fig. 2. WI007 system experimental performance with and without metal-cutting noise in training material.

6 Conclusions

The problem of acoustic mismatches between the training database and the environment in which an ASR system operates is addressed in this paper.

In this direction, this work presents two contributions: a) a method to calculate the parameters of the logistic function observed in equation (1) and b) a methodology to evaluate, through a mathematical criterion, the mismatch between the incoming noise and the noises used to train the recognition system.

The methodology was tested using the Aurora-2 database, augmented with four new noises: metal-cutting, tunnel-front, tunnel-inside and crowd-children. After calculating the mismatches between the noises in the Aurora-2 database and the new noises, the metal-cutting noise was identified as the only one whose acoustic characteristics are not adequately modeled by the ASR system ($lss > 0.05$). The inclusion of this noise in the training material led to an average 3.69% improvement in the recognition rate of the system, while the inclusion of other noises in the training material did not cause noticeable changes in the recognition rates.

Acknowledgements

The authors would like to thank UDESC for the financial support, UFSC/PPGEAS for the physical structure and INATEL for orientation and technical support.

References

1. Fraga, F.J., Ynoguti, C.A., Chiovato, A.G.: Further Investigations on the Relationship between Objective Measures of Speech Quality and Speech Recognition Rates in Noisy Environments. In: INTERSPEECH 2006, pp. 185–188 (2006)
2. International Telecommunication Union. ITU-T Recommendation P.862, ITU-T Recommendations, Series P: Telephone Transmission Quality, Telephone Installations, Local Line Networks, Printed in Switzerland, Geneva (2001)
3. Hirsch, H.G., Pearce, D.: The AURORA Experimental Framework for the Performance Evaluation of Speech Recognition Systems under Noisy Conditions. In: ISCA ITRW ASR (2000)
4. Leonard, R.G.: A database for speaker-independent digit recognition. In: Proc. IEEE International Conference on Acustics, Speech, and Signal Processing, vol. 3, pp. 4.211-4.214 (1984)
5. International Telecommunication Union. ITU-T Recommendation G.712, ITU-T Recommendations, Series G: Transmission Systems and Media, Digital Systems and Networks, Printed in Switzerland, Geneva (2001)
6. Pearce, D.: Enabling New Speech Driven Services for Mobile Devices: An overview of the ETSI Standards Activities for Distributed Speech Recognition Front-Ends. In: AVIOS 2000: The Speech Applications Conference (2000)
7. Young, S., Evermann, G., Kershaw, D., Moore, G., Odell, J., Ollason, D., Valtchev, V., Woodland, P.: HTK Book Version 3.1, Cambridge University Engineering Department (2001)
8. Cramer, J.S.: Econometric Applications of Maximum Likelihood Methods. Cambridge University Press, Cambridge (1986)
9. Minitab Software. Least Squares (LSXY) Estimates Versus Maximum Likelihood Estimates (MLE) - ID 767, Minitab Company Information, `http://www.minitab.com/en-US/support/answers/` (accessed on February 17, 2010)
10. Silva, D.D., Stemmer, M.R., Ynoguti, C.A.: Contributions to Robust Automatic Speech Recognition. Doctorate Study, Santa Catarina Federal University, DAS/PPGEAS (2010)

Calibration of Pressure Sensor Array Based on Information Fusion

Bing Guo[1] and Xiaohong Zeng[2]

[1] Chongqing Technology and Business Institute, Chongqing 400050, P.R. China
[2] Chongqing College of Electronic Engineering, Chongqing 401331, P.R. China
cq_guobing@tom.com, zengxiaohong111@126.com

Abstract. It is existence of cross-sensitivity factors such as the temperature, humidity and power fluctuations in pressure sensor array. To solve this problem, it is proposed an information fusion based on the pressure sensor array calibration algorithm. By this method can effectively eliminate the environmental temperature and voltage disturbances and other non-target parameters, and crosstalk between large-scale sensor array signal output of pressure sensor array characteristics. It is improved system stability and reliability. Experimental results show that the information fusion processing algorithms realize the precise calibration of the pressure sensor array, the effective sensor array measurement data to ensure the accuracy and reliability of the results.

Keywords: Pressure Sensor Array, Information Fusion, Calibration.

1 Introduction

Location of the sensor for robot tactile, pressure sensor data in the performance of the two-dimensional space of information, in addition to spatial data, the pressure distribution data, the use of tactile sensors can get access to information on a variety of physical objects in order to facilitate human-computer Information interaction. The research challenge is the uncertainty of how many effective sensor data processing [1, 2, 3, 4]. As a result of the sensor parameters from each other, to accept the role and effectiveness of the state is not the same [5, 6, 7], not only to consider the data obtained compensation, more importantly, how to deal with the integration of multi-sensor information [8, 9], so the calibration of sensor array data analysis and processing is to play a key role.

Multi-sensor array system in view of the diversity and complexity of data in the current study of robot tactile sensor array, the information fusion problem rarely involved, this combination of research and practice of information fusion is proposed based on the pressure sensor array Calibration method for the majority of quantitative calibration of the sensor array provides a good technical reference.

X. Wan (Ed.): Electrical Power Systems and Computers, LNEE 99, pp. 415–423.
springerlink.com

2 Pressure Sensor Array

Tactile pressure sensor array is a new type of tactile sensor array, providing reliable and accurate touch-sensitive information [10]. According to their spatial resolution, sensitivity, sensitive element, stability and other requirements of the different sensing principle can be appropriate and sensitive material, which is especially important for large area array sensors, sensitive material should have good flexibility. Smart clothes to cut production due to the array of clothing structure, flexible higher, so select a good flexible conductive rubber as sensitive materials [6]. Block through the production of clothing materials, components, and then connect with the combination method of sewing clothes, and ultimately cut into the robot tactile sensing can be dressed in costume.

The 8×8 matrix touch material components: the electrodes parallel to each other constitute the sensitive elements of the external leads, the (row) electrode and the lower (column) electrode perpendicular to the piezoresistive sensitive material in the middle, upper and lower electrodes is defined as the intersection of a sensitive tactile array unit, shown in Fig.1 and Fig.2. Tactile sensor array, the ranks of the electrode structure, its purpose is to reduce the sensor's external leads, to increase the stability and accuracy of the array [8].

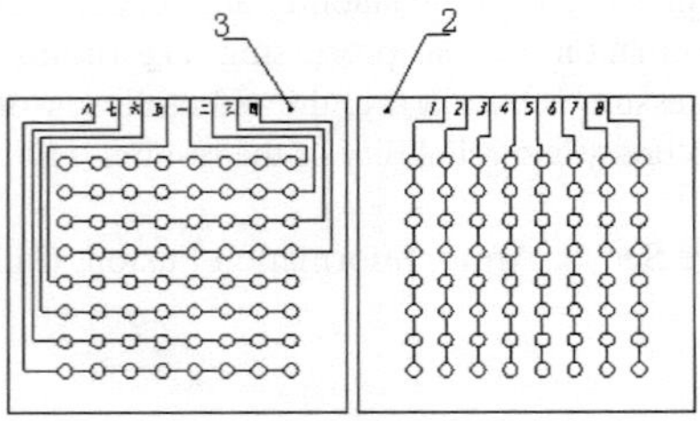

Fig. 1. Top and bottom electrodes tactile sensor array

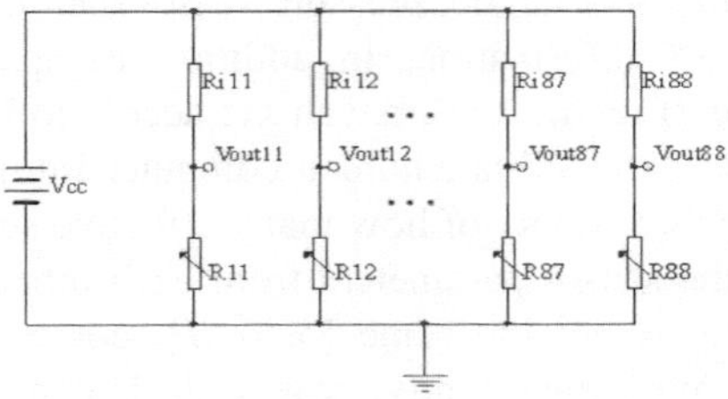

Fig. 2. Sensor array circuit

3 System Configuration and Architecture

The functional block diagram of basic system architecture is shown in Fig.3. The overall system is divided into five function units that are as follows.

1) Data Acquisition Unit, to acquire and store the raw tactile sensor information for later retrieval.

2) Data Pre-Processing Unit, determined by the current phase. Only appropriate sensor data is transferred to the arithmetic processing unit and converted raw sensor data into the needed data formats.

3) The Compensation Unit, to be known to give data that deviate from the actual values by a known relationship.

4) The Data Processing Unit, to fuse all appropriate sensor data into a coherent data unit.

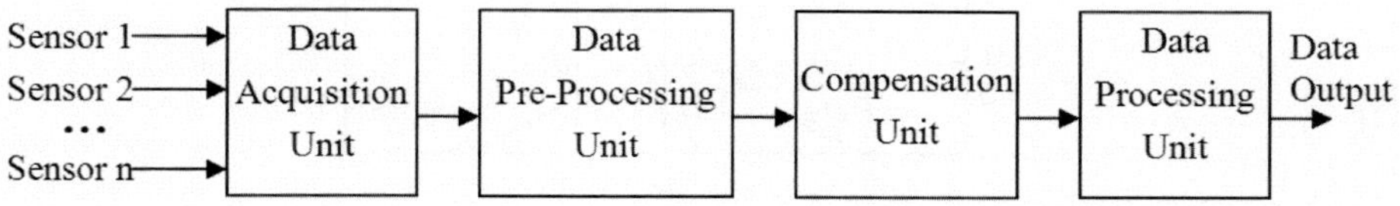

Fig. 3. Block diagram of system configuration

4 Fusion of Sensor Data

The idea of fusion is to merge two or more separate items into a single entity. The problems can be arised in this fusion process if any of the sensing data specified are uncertain or in conflict, implying that an error in a sensor has occurred [4]. Therefore the data fusion will entail detection, verification, and recovery of sensor errors.

4.1 Distance Matrix and Confidence Distance Measures

Measures of confidence describing sensors by probability distributing, represent a primary means of effective data fusion [11]. Based on calibrated sensor's characteristic curve, the error detection criterion is defined and the distance matrix is used as the criterion for detecting sensor errors [12].

For the tactile sensor array system, there consider two kinds of probability distributions Pi (x) and Pj(x) in Fig.4. P_{ij} is defined as the following

$$P_{ij} = P_i(x_j / x_i) \tag{1}$$

Similarly, Pji is defined as

$$P_{ji} = P_j(x_i / x_j) \tag{2}$$

Due to uncertainties derived from multiple sensors, it will need to find the interrelations among different sensors. If the values are close it may fuse them together, otherwise not to consider fusion.

Therefore it defines a new distance measure called as 'confidence distance measure', either d_{ij} or d_{ji} as a criterion, where

$$d_{ij} = 2\int_{x_i}^{x_j} p_i(x/x_i)dx = 2A \tag{3}$$

$$d_{ji} = 2\int_{x_j}^{x_i} p_j(x/x_j)dx = 2B \tag{4}$$

Here A or B is the area between sensor reading values x_i and x_j under $P_i(x)$ or $P_j(x)$. In general, $d_{ij} \neq d_{ji}$ (unless the standard deviation $\sigma_i = \sigma_j$) and $0 \leq d_{ij}, d_{ji} \leq 1$. As an extreme cases of d_{ij}, $x_i = x_j$ when $d_{ij} = 0$, x_i is far away from x_j when $d_{ij} = 1$.

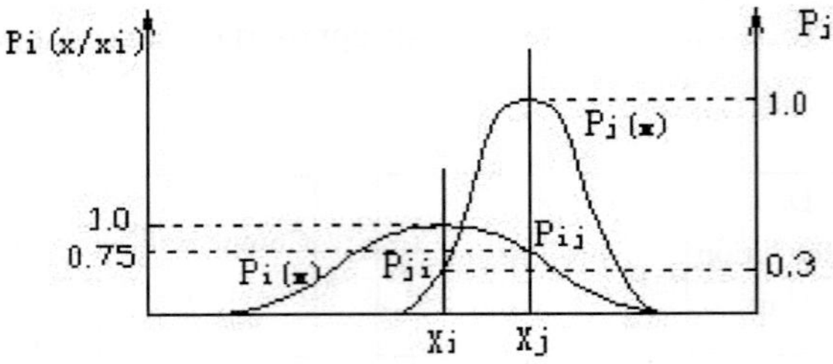

Fig. 4. Two probability distributions $P_i(x)$ and $P_j(x)$ with different variance measures

The confidence distance measure not only provides an abstract scale value, but also represents the relation of the distance versus confidence measures. Note that the higher the percentage of confidence interval, the larger the distance between x_i and x_j will be.

The confidence distance measures can be computed by the use of an error function[7]

$$erf(\theta) = \frac{2}{\sqrt{\pi}} \int_0^{\theta} e^{-u^2} du \tag{5}$$

Let $u = \frac{x - x_i}{\sqrt{2}\sigma_i}$, $du = \frac{dx}{\sqrt{2}\sigma_i}$, the formula (5) becomes

$$erf(\theta) = \frac{2}{\sqrt{\pi}\sigma_i} \int_{x_i}^{x_i + \sqrt{2}\theta\sigma_i} \exp\left\{ -\frac{1}{2}\left(\frac{x - x_i}{\sigma_i}\right)^2 \right\} dx \tag{6}$$

Let $x_j = x_i + \sqrt{2}\theta\sigma_i$, then $\theta = \frac{x_j - x_i}{\sqrt{2}\sigma_i} > 0$, The formula (6) becomes

$$erf\left(\frac{x_j - x_i}{\sqrt{2}\sigma_i}\right) = \frac{\sqrt{2}}{\sqrt{\pi}} \int_{x_i}^{x_j} \exp\left\{ -\frac{1}{2}\left(\frac{x - x_i}{\sigma_i}\right)^2 \right\} dx = 2\int_{x_i}^{x_j} p_i(x/x_i)dx \tag{7}$$

Now, from (1), (2) and (6), the confidence distance measures are

$$d_{ij} = erf\left(\frac{x_j - x_i}{\sqrt{2}\sigma_i}\right) \tag{8}$$

$$d_{ji} = erf\left(\frac{x_i - x_j}{\sqrt{2}\sigma_j}\right) \tag{9}$$

Assuming it has m sensors for measuring the same object property, the general confidence distance measures can be described as

$$D_m = \begin{bmatrix} d_{11} & d_{12} & \cdots & d_{1m} \\ d_{21} & d_{22} & \cdots & d_{2m} \\ \cdots & \cdots & \cdots & \cdots \\ d_{m1} & d_{m2} & \cdots & d_{mm} \end{bmatrix} \tag{10}$$

4.2 Creation of the Relation Matrix and Directed Graph Representation

The distance matrix can be defined as the corresponding relationships of the sensors to one another. The relation matrix R is defined as

$$R_m = \begin{bmatrix} r_{11} & r_{12} & \cdots & r_{1m} \\ r_{21} & r_{22} & \cdots & r_{2m} \\ \cdots & \cdots & \cdots & \cdots \\ r_{m1} & r_{m2} & \cdots & r_{mm} \end{bmatrix} \tag{11}$$

Where r_{ij} is the threshold value of d_{ij} and

$$r_{ij} = \begin{cases} 1, & if \quad d_{ij} \le threshold \quad value \\ 0, & if \quad d_{ij} > threshold \quad value \end{cases}$$

R_m can be conveniently represented in the form of a directed graph (or digraph for short). The complete digraph will then represent with a convenient way of visualizing the relationships among all of the used sensors, thus allowing us to look for consensus groups of sensors (i.e., with data in agreement)[8][9].

Three cases can be occurred in Fig. 5. (1) The two sensors do not support each other. The best way is to select the sensor value which has the smaller variance (i.e., the higher confidence measure). (2) Sensor 1 supports sensor 2, however, sensor 2 does not support sensor 1. The sensor 2 is chosen as the resulting of fused data and it does not attribute to sensor 1 of the status in error. (3) Sensors 1 and 2 support each other. It is probable that both sensors are near to the actual data value. It would then fuse two sensor data using the data fusion principle.

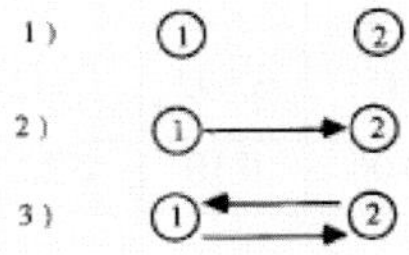

Fig. 5. Relationship between two sensor nodes in a digraph

The largest connected group in the digraph (or a clique) represents the most likely group of sensors which yield an accurate representation of the actual data. All those sensors which are weakly supported by the largest connected group sensors would then be suspected of being in error and thus would have to be subjected to either compensation or complete dismissal (if compensation is not possible).

4.3 The Strategy for Determining the Optimal Fused Sensor Data

The approach is to maximize the total probability of all sensor data. The optimal fused sensor data $\hat{\theta}$ should maximize the value of (12).

$$L_1(\theta; x_1, x_2, \cdots x_i) = \sum_{i=1}^{i} p_i(\theta \mid x_i) \tag{12}$$

Where $p_i(\theta \mid x_i) = \frac{1}{\sqrt{2\pi}\,\sigma_i} \exp\{-\frac{1}{2}\left(\frac{\theta - x_i}{\sigma_i}\right)^2\}$,

$i = 1, 2, \cdots, l$, In order to determine the optimal fused data value, taking the derivative of (12) and set it to zero. By iteration, this iteration procedure is continued until $\hat{\theta}$ converges.

5 Experiment

The experiments is established a system to verify the pressure sensor array data fusion system availability in laboratory. Experimental equipment consists of three parts: the pressure sensor array, signal processing circuit and computer data processing system. Experiments using the proposed method, with the probability distribution function as a sensor characteristic equation, and apply to distance matrix and correlation matrix of the sensor array as a sensing probe the correlation between units, and the use of the best integration of data fusion values obtained Correctness. In the sensor array devices using the same force 120N in the same line of external force sensing unit 8 under, with the LabVIEW software measurement many times on a single sensor, the measured sensor unit 8 output voltage as shown in Table 1.

Table 1. The pressure sensing unit of measurement data

Sensor number	1	2	3	4	5	6	7	8
Measurements (v)	3.136	3.152	3.092	3.942	3.247	3.184	2.665	3.278
Variance σ^2	0.08	0.06	0.10	0.23	0.15	0.09	0.25	0.12

By equation (11) and (12) to identify any credibility between the two pressure sensors spaced degrees, and with distance matrix (13) the results

$$D = \begin{bmatrix} 0.0000 & 0.0451 & 0.1236 & 0.9956 & 0.3053 & 0.1348 & 0.9041 & 0.3844 \\ 0.0521 & 0.0000 & 0.1935 & 0.9987 & 0.3019 & 0.1039 & 0.9532 & 0.3930 \\ 0.1107 & 0.1505 & 0.0000 & 0.9928 & 0.3760 & 0.2289 & 0.8231 & 0.4436 \\ 0.9072 & 0.9005 & 0.9237 & 0.0000 & 0.8527 & 0.8860 & 0.9922 & 0.8338 \\ 0.2256 & 0.1938 & 0.3110 & 0.9273 & 0.0000 & 0.1292 & 0.8671 & 0.0638 \\ 0.1271 & 0.0849 & 0.12409 & 0.9885 & 0.1663 & 0.0000 & 0.9164 & 0.2460 \\ 0.6538 & 0.6699 & 0.60969 & 0.9894 & 0.87556 & 0.7007 & 0.0000 & 0.7798 \\ 0.3181 & 0.2839 & 0.4087 & 0.9447 & 0.0713 & 0.2139 & 0.9232 & 0.0000 \end{bmatrix}$$

Select the critical value of 0.500 pairs of discrimination by the relevant distance matrix matrix (14)

$$R = \begin{bmatrix} 1 & 1 & 1 & 0 & 1 & 1 & 0 & 1 \\ 1 & 1 & 1 & 0 & 1 & 1 & 0 & 1 \\ 1 & 1 & 1 & 0 & 1 & 1 & 0 & 1 \\ 0 & 0 & 0 & 1 & 0 & 0 & 0 & 0 \\ 1 & 1 & 1 & 0 & 1 & 1 & 0 & 1 \\ 1 & 1 & 1 & 0 & 1 & 1 & 0 & 1 \\ 0 & 0 & 0 & 0 & 0 & 0 & 1 & 0 \\ 1 & 1 & 1 & 0 & 1 & 1 & 0 & 1 \end{bmatrix}$$

Discrimination related to the largest group by 6 units (serial numbers as 1,2,3,5,6,8, respectively) also is the best fusion group was 6. Finally, the total probability law to calculate the maximum voltage of the optimal integration of the measured data 3.15105V. Fusion will be using the test pressure sensor unit combines the output voltage acquisition circuit pressure sensing unit by the corresponding resistance value 16.67 kΩ.

Sensor array using this method the sensor was calibrated experiments. By measuring the output voltage sensing unit corresponds with the force device to get the pressure sensing element between the pressure and resistance, the experimental results shown in Table 2.

Table 2. Relationship between conductive rubber sensors piezoresistive experimental data

Pressure(N)	Resistance(kΩ)	Pressure(N)	Resistance(kΩ)
1	275.71	60	24.97
2	171.82	70	23.9
3	139.25	80	22.79
4	123.33	90	21.75
5	99.89	100	20.77
6	90	110	17.78
7	80.91	120	16.67
8	73.33	130	15.32
10	61.43	150	14.1
12	52.5	160	13.26
15	47.14	180	11.98
17	44.05	200	10.41
18	42.63	230	9.61
20	40	250	8.52
25	37.62	280	7.86
30	35.45	300	7.09
35	33.48	350	6.13
40	31.67	400	4.29
45	29.22	450	3.51
50	27.74	500	2.99

6 Conclusion

The pressure sensing array for multi-sensor signal processing, a fusion of the pressure sensor array signal processing methods. First intelligent array tactile sensing data for comprehensive, single-touch group of units to avoid the uncertainty of the error effects and error messages generated by the sensor fault, the effective integration of the results to ensure the accuracy and reliability, compared with a single sensor information, get more accurate, more complete, more reliable estimates and judgments. Secondly, the method is robust, can increase the pressure sensor array of spatial resolution and clarity, tactile image mapping accuracy, classification accuracy and reliability, enhanced interpretation and dynamic monitoring capabilities, reduce ambiguity, improve Tactile data utilization. Finally, the method can overcome the environment temperature, voltage disturbances and other external environmental impact, increased contact with the surface profile measurement and image reconstruction of data accuracy, so that the pressure sensor array system has good stability, fault tolerance and reliability.

Acknowledgement

This research supported by Program for Excellent Talents in Chongqing Higher Education Institutions.

References

1. Luo, Z.-z., Wang, R.-c.: Study of tactile sensor in bionical artificial hand. Chinese Journal of Sensors and Actuayors 16(3), 233–237 (2003)
2. Lee, M.H., Nicholls, H.R.: Tactile sensing for mechatronics—A state of the art survey. Mechatron 9, 1–31 (1999)
3. Pan, Z.-x., Cui, H.-l., Zhu, Z.-q.: A flexible full-body tactile sensor of low cost and minimal connections. In: IEEE International Conference on, vol. 3, pp. 2368–2373 (2003)
4. Lumelsky, V.J., Shur, M.S., Wagner, S.: Sensitive skin. IEEE Sensors Journal 1(1), 41–51 (2001)
5. Heever, D., Schreve, K., Scheffer, C.: Tactile sensing using force sensing resistors and a super-resolution algorithm. IEEE Sensors Journal 9(1), 29–35 (2009)
6. Ababou, A., Ababou, N., Chadli, S., et al.: Accuracy improvement of large area flexible piezoresistive digital tactile array sensing system. In: IEEE sensors 2008 conference, Leece, Italy, October 26-29, pp. 1048–1051 (2008)
7. Pritchard, E., Mahfouz, M., Evans III, B., et al.: Flexible capacitive sensors for high resolution pressure measurement. In: IEEE sensors 2008 conference, Leece, Italy, October 26-29, pp. 1484–1487 (2008)
8. Huang, Y., Ming, X., Xiang, B., et al.: Two types of flexible tactile sensor arrays of robot for three-dimension force based on piezoresistive effects. In: Proceedings of the 2008 IEEE international conference on robotics and biomimetics, Bangkok, Thailand, February 21-26, pp. 1032–1037 (2008)

9. Chuang, C.-H., Dong, W.-B., Lo, W.-B.: Flexible piezoelectric tactile sensor with structural electrodes array for shape recognition system. In: The 3rd international conference on sensing technology, Tainan, Taiwan, November 30 – December 3, pp. 504–507 (2008)
10. Murayama, Y., Haruta, M., Hatakeyama, Y.: Development of a new instrument for examination of stiffness in the breast using haptic sensor technology. Sensors and Actuators A 143(2), 430–438 (2008)
11. Dahiya, R.S., Metta, G., Valle, M.: Development of fingertip tactile sensing chips for humanoid robots[C]. In: Proceedings of the 2009 IEEE International Conference on Mechatronics, Malaga, Spain, April 14-17, pp. 1–6 (2009)
12. Chang, W.-Y., Fang, T.-H., Heng-Ju, et al.: A large area flexible array sensors using screen printing technology. Journal of display technology 5(6), 178–183 (2009)

A New Distributed Multi-parameter Remote Wireless Monitoring System

Bing Guo[1] and Xiaohong Zeng[2]

[1] Chongqing Technology and Business Institute, Chongqing 400050, P.R. China
[2] Chongqing College of Electronic Engineering, Chongqing 401331, P.R. China
cq_guobing@tom.com, zengxiaohong111@126.com

Abstract. A remote wireless monitoring system for multi-parameters, including temperature, humidity, illumination etc., was researched based on the embedded chip and wireless transmission technology. The Spce061A was adopted as the core processor of the system, while the nRF905 was used for wireless transmission. Since the master-slave structure was adopted, the system could be configured flexibly and expanded easily. A monitoring software was also developed on the PC platform, so it is easy for upgrade. The whole system has a simple structure, stably performance and fine portability, and could work fast and exactly in some terrible condition.

Keywords: distributed multi-parameter, wireless monitoring, embedded chip.

1 Introduction

With the continuous progress and gradually increase the productivity of the production process parameters on-site monitoring requirements gradually increase the types of parameters measured at the same time becoming diversified, specialized[1,2]. Remote monitoring and control system in industry, agriculture, defense, and many other areas of daily life is widely used, especially for the measured object and more poor working conditions (such as very high temperature, humidity, toxic environment, etc.) of the occasion, the use of wireless transmission of the measured object for remote data acquisition and control, not only to implement simple, low cost, and the system's real time and reliability are high[3,4]. In this regard, the author designed a wireless transmission based on distributed multi-parameter monitoring system, with distribution flexibility, easy maintenance, the advantages of lower cost, for the harsh environment of the site monitoring provides a new solution, and achieved good results.

2 System Structure

To meet the requirements of the different measurement parameters, the system's front-end monitoring sites set aside a number of analog signal input of the interface can be

X. Wan (Ed.): Electrical Power Systems and Computers, LNEE 99, pp. 425–432.
springerlink.com

configured according to the site corresponding module to adapt to different environmental monitoring needs[5,6]. System uses the master-slave structure, by a management station and a number of points of the front-end monitoring of distributed monitoring system. The system structure is shown in Fig.1.

Monitoring stations as the system front-end signal acquisition unit, can be flexibly arranged in the corresponding position to form three-dimensional or flat distribution[7,8]. Each site to set separate address to distinguish between different sites. Each monitoring site, including the corresponding sensors, pre-processing circuit, embedded processors and wireless transmission module. According to front-end can also set the needed actuator drive circuit to complete the corresponding control operation[9,10]. Front-end site can be the front-end site data into the sensor measurements, and simple processing, and control the wireless communication module to the measured data to the central site. Center receives part by the wireless transmission module, embedded processors, the level conversion circuit and the central processing adapter body into the PC, the main function is to control the communication system to realize the receiver front-end point data, as well as provide access to other systems Interface.

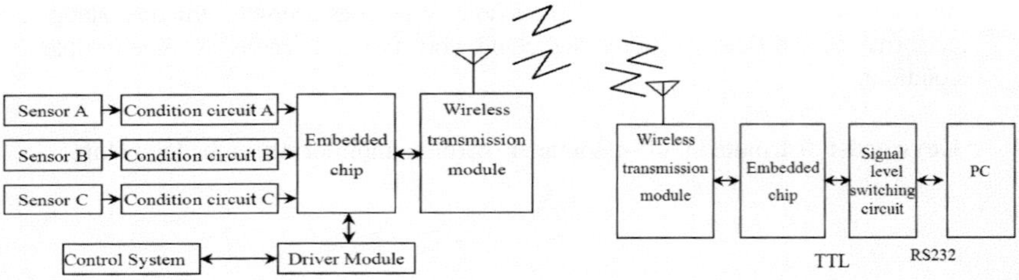

Fig. 1. Overall system block diagram

3 The Main Function Module

3.1 Temperature Measurement Circuit

According to the characteristics of AD590, AD590 the current flowing through the thermodynamic temperature is proportional to the resistance when the resistor R1 is 5kΩ, the output voltage V_O of temperature is 5mV / K[11].

However, due to the gain of AD590 biased, the resistance has a margin of error, so the circuit should be adjusted. Adjustment methods: the AD590 put in ice water mixture, adjust the potentiometer R1, the VO = 1.366V. Or at room temperature (25 ℃) under the conditions of adjustment potentiometer, so that V_O=1.366+0.125=1.491 (V). Temperature measurement circuit is shown in Fig. 2.

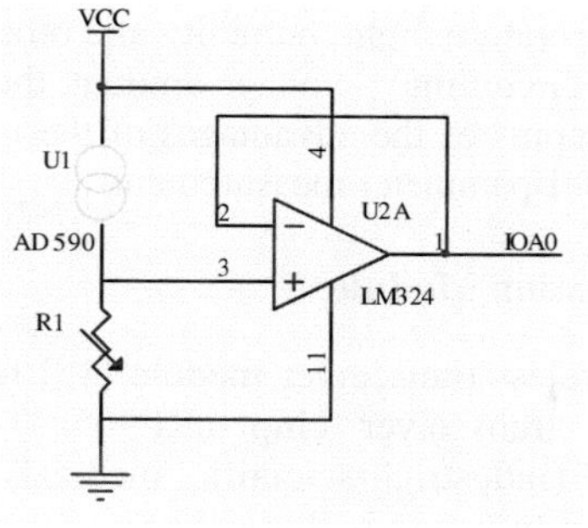

Fig. 2. Temperature acquisition circuit

Because the device causes, AD590 measurement data is not completely linear, so need to be calibrated measurement using software[12].

3.2 Illumination Circuit Design

As the partial pressure sensitive resistors style output, to reduce the measurement circuit of the sensor circuit, light detection circuit shown in Fig. 3.

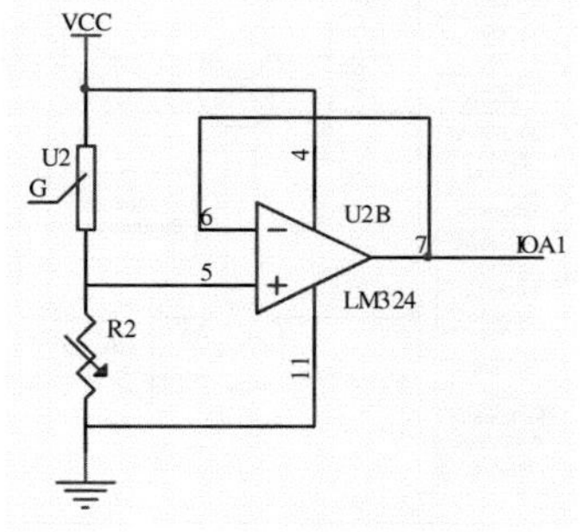

Fig. 3. Light acquisition circuit

The main characteristics of the circuit full advantage of the op amp's high input impedance measurement circuit to avoid the impact of front-end circuit, the circuit can be seen as an impedance transformation[13].

3.3 Embedded Processor

Core of the system for 16-bit microprocessor chip is SPCE061A [14]. The chip µ'nSPTM core, built-in 32K FLASH 2K SRAM and storage space, with two 16-bit can be flexibly defined in the I/O port (including 8 channels of 10-bit voltage ADC), up to 49MHz of Processing speed, to meet the needs of measurement data processing.

This system will be temperature, light, humidity and other environmental parameters with the appropriate prefix circuit into a voltage amount, the full application of the chip to support multiple ADC inputs of the advantages of the multi-channel analog signals quantified, enabling the multi-parameter measurements.

3.4 Wireless Communication Module

System uses PTR8000 wireless transceiver module [15] for remote data transfer, the core chip is monolithic transceiver chip nRF905. The module can work at 433/868/915MHz the ISM (industrial, scientific, medical) band, the working voltage of 1.9 ~ 3.6V. The chip's architecture is shown in Fig.4.

PTR8000 controlled by changing the pin MCU PWR_UP, TRx_CE, Tx_EN nRF905 level to switch the operating mode chip through SPI (Serial Peripheral Interface) chip to complete the parameter configuration and data reading and writing, according to transmission status indicator (CD / AM / DR) of the state to get the working status of data transmission.

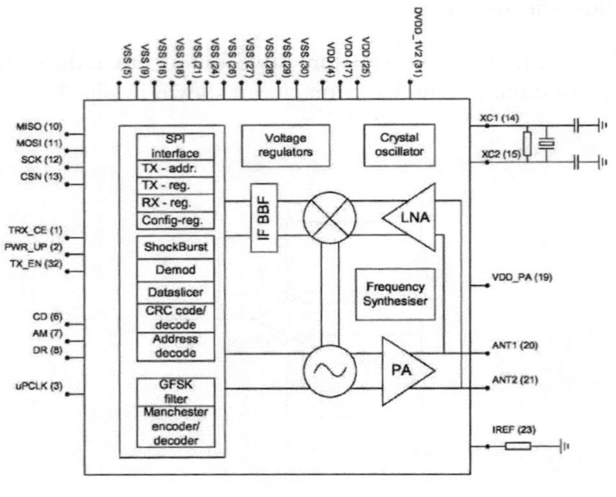

Fig. 4. Wireless communication chip nRF905 architecture

Since there is no SPI SPCE061A interface software simulation program used in a way that [15]. The modular system programs the idea of some of the features of wireless communications is divided into three levels, including hardware control, control and delivery of basic reading and writing process, simplify the main program control process of the hardware.

3.5 Data Transfer

As a result of the distributed structure, each set up an address from the machine number, in order to achieve the management of multiple front-end. Centers take a proactive approach to achieve the polling data collection front end, to avoid channel contention, improving channel transmission efficiency. Placed on each measuring site to receive state only when receiving a query command on the machine when the host response to the measured data.

In the wireless transmission process, using a Carrier Sense technology; master and slave to a handshake between the transmission confirmation. After sending the data front-end site, the host will return to wait for confirmation to verify the need to send again; if re-issued three times while still confirmed that the radio channel failure occurs if the channel failed three times, MCU driver circuit fault alarm , prompted the overhaul.

3.6 PC to Receive Data and Display

Including the central control center host PC, client and management of two parts, in the management and central control PC, using serial asynchronous communication between client. Center-side the main MCU to complete the serial conversion of data between the wireless communication. PC-side data acquisition program main function is to send and receive instructions through the serial port on the command parsing and data reception, update the display and so on.

Front-end machine with event-driven manner. Its main task is under the control of machines in the center, collecting data obtained from the local sensors and wirelessly sent to the host.

The main front-end workflow collected microcontroller shown in Fig.5, the wireless front-end machines usually in a wait state to receive instruction until the wireless receiving inquiries instruction, according to the instruction, performing the appropriate collection procedures, obtained the desired data, the Wirelessly send the data center machine.

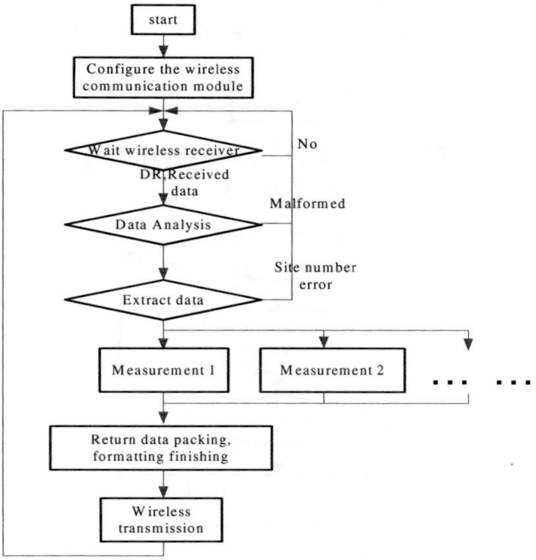

Fig. 5. The main front-end process of gathering machines

Front-end machine can be used for detecting the status of inquiries received, or interrupted manner. In order to ensure the logic of the software and reduce the complexity of the query used in this design approach, by detecting the state to check whether the DR receives a complete packet. Analysis of the main contents of the command is to check the accuracy of wireless data content and the interpretation of commands to perform the appropriate procedures.

4 Centers Program

Processing platform in the center of choice, consider the scalability from the system, using a PC, as the upper control processing center. Itself is not the PC, wireless interfaces, wireless module data from the microcontroller through the serial port to the PC.

The design goal for the realization of multiple front-end point of data collection. In order to achieve more points collected data transmission, the use of the wireless channel, only the host can take the initiative to use the wireless channel, all the front-end measured from the machine are used passively. The actual operation of the system, the host operator's control or under the procedures set out on the front of a point to send the queries. Inquiries received by the wireless signal the site, collecting complete data can send data out through the wireless module.

Central part of the software process control logic and human-computer interaction interface consists of fully functional PC, to complete. With development of software-based system design parameters can be displayed and a detailed set other advanced features. PC, the software portion of the detailed description.

Center main task of the microcontroller through the serial port for the PC, wireless data transmission expansion. The small amount of data transmission system, still use the inquiry approach to treatment. Center program of the main flow shown in Fig.6.

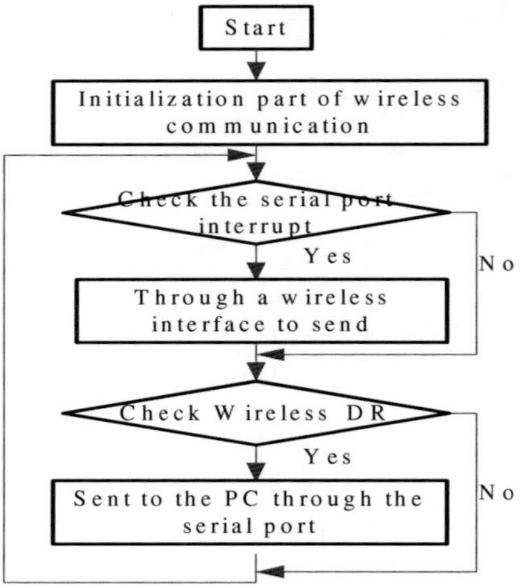

Fig. 6. Single chip receiver processing center

Center SCM process is to achieve data transmission process forward, the equivalent of the wireless channel and transmission channel between the serial port. Tested by the way check the serial transceiver, can be carried out in the 57600 baud rate reliable transmission. The design of the default serial communication baud rate is 9600bps.

5 Experiment

System testing using serial baud rate is 19200bps, 8 data bits, 1 stop bit. After testing, the system measurement site and the receiver front-end machine of the outdoor center for reliable communication distance of about 200m. The system tests the selected temperature sensor AD590; measured temperature error of less than 2%, and achieved satisfactory results, and has good repeatability, the measured data shown in Table 1.

Table 1. Temperature data and error data table

Nominal temperature (℃)	13	16	19.2	20	22.6	26.5
Measured temperature (℃)	13.25	15.73	19.53	20.23	22.24	26.22
Error (%)	1.92	1.69	1.72	1.15	1.5	1.06
Nominal temperature (℃)	28.3	30	32.3	35	38	40
Measured temperature (℃)	28.55	30.25	32.68	34.74	38.44	40.52
Error (%)	0.88	0.83	1.18	0.74	1.16	1.3

Enabled wireless transmission CRC error checking to ensure no generated. Measurement unit in front of the sensor output signal A / D conversion process accuracy 1mV, but the actual accuracy of the decision by the parameters of the sensor itself.

6 Conclusions

In this paper, the monitoring system for the realization of the harsh environment of on-site monitoring provides a new solution that can be used to forecast disasters, such as mines, rivers, toxic and hazardous situations; can also be used to the environment that require real-time multi-parameter monitoring Occasions, such as the valuable greenhouse crops, inflammable and explosive materials storage warehouse. With this system, PC, the powerful processing capabilities and large capacity of storage space, the long time data storage, query, reporting, and a series of practical features, but also combined with network technology, through the Internet from other PC, log on the appropriate page To obtain real-time monitoring data.

Acknowledgement

This research supported by Program for Excellent Talents in Chongqing Higher Education Institutions.

References

1. Mendaza, G.G., Tran, B.Q.: In-home wireless monitoring of physiological data for heart failure patients (DB/OL), October 23-26, vol. 3, pp. 1849–1850 (2002)
2. McDoniel, S.O.: Systematic review on use of a handheld indirect calorimeter to assess energy needs in adults and children. Int. J. Sport Nutr. Exerc. Metab. 17(5), 491–500 (2007)
3. Melanson, E.L., Coelho, L.B., Tran, Z.V., et al.: Validation of the BodyGem. hand-held indirect calorimeter. Abstract presented at the Nutrition Week Conference, San Antonio, Texas, January 18-22 (2003)
4. Chen, K.Y.: The Use of Portable Accelerometers in Predicting Activity Energy Expenditure. In: Presented at the Institute of Medicine, Committee on Metabolic Monitoring Technologies for Military Field Applications Workshop on Metabolic Monitoring Technologies for Military Field Applications, San Antonio, Texas, January 8-9 (2003)
5. Casa, D.J., et al.: National Athletic Trainers' Association position statement: Fluid replacement for athletes. J. Athlet. Train. 35, 212–224 (2000)
6. Carter III, R., Cheuvront, S.N., Williams, J.O., Kolka, M.A., Stephenson, L.A., Sawka, M.N., Amoroso, P.J.: Epidemiology of hospitalizations and deaths from heat illness in soldiers. Medicine and Science in Sports and Exercise, 37(8), 1338–1344 (2005)
7. Daanen, H.A.M., Van Der Struijs, N.R.: Resistance index of frostbite as a predictor of cold injury in arctic operations. Aviation Space and Environmental Medicine 76(12), 1119–1122 (2005)
8. Daanen, H.A.M.: Deterioration of manual performance in cold and windy climates. In: AGARD conference proceedings, vol. CP-540, pp.15-1–15-10 (1993)
9. O'Brien, C., Hoyt, R.W., Buller, M.J., Castellani, J.W., Young, A.J.: Telemetry pill measurement of core temperature in humans during active heating and cooling. Medicine and Science in Sports and Exercise 30(3), 468–472 (1998)
10. Lutz, R., Coker, D.T.: Heat stress - development of a personal warning device. Annals of Occupational Hygiene 20(4), 397–405 (1977)
11. Marken Lichtenbelt, W.D., Daanen, A.M., Wouters, L., Fronczek, R., Raymann, R.J.E.M., Severens, N.M.W., Van Someren, E.J.W.: Evaluation of wireless determination of skin temperature using iButtons. Physiology and Behavior 88(4-5), 489–497 (2006)
12. Epstein, Y., Moran, D.S.: Thermal comfort and the heat stress indices. Industrial Health 44(3), 388–398 (2006)
13. Benoit, H., Busso, T., Castells, J., Geyssant, A., Denis, C.: Decrease in peak heart rate with acute hypoxia in relation to sea level VO2max. Eur. J. Appl. Physiol. 90, 514–519 (2003)
14. Chakrabarti, S., Mishra, A.: A network architecture for global wireless position location services. In: ICC 1999: IEEE International Conference on Communications 1999, vol. 3, pp. 1779–1783 (1999)
15. Popa, M., Popa, A.S., Cretu, V., Micea, M.: Monitoring Serial Communications in Microcontroller Based Embedded Systems Computer Engineering and Systems. In: The 2006 International Conference on Computational Intelligence in Scheduling, November 2006, pp. 56–61 (2006)

Detection of Faint Objects between Streak-Like Stellar Images

Haifeng He[1,3,4], Qingyu Peng[1,3,4], and Zhenghong Tang[2,3]

[1] Department of Computer Science, Jinan University, 510632 Guangzhou, China
[2] Shanghai Astronomical Observatory, Chinese Academy of Sciences, 200030 Shanghai, China
[3] Sino-France Joint Laboratory for Astrometry, Dynamics and Space Science Joint Laboratory, Jinan University, 510632 Guangzhou, China
[4] Key Laboratory of Optoelectronic Information and Sensor Technologies of Guangdong Higher Educational Institutes, Jinan University, 510632 Guangzhou, China
heqihao125@163.com

Abstract. In this paper we propose a novel method based on the characteristics of the object that we are interested in to distinguish it from other stars, whose images are line-shaped trails. The point-like object and streak-like stellar images become more visible after some pre-processing, and operated by boundary tracking, then they can be preliminarily identified by their form factors. The coordinates of the object on the CCD-frame can be further calculated by a two-dimensional Gaussian fitting in the original exposure. Experiment tests show that the new method can distinguish the object and other stars rapidly and precisely. The codes developed under the environment of Windows/Visual C++6.0 can be applied to both a single exposure and series of images to identify an artificial satellite or space debris in motion.

Keywords: Faint Object, Boundary Tracking, Form Factor, Detection Method.

1 Introduction

There are some algorithms that have been proposed and used in different systems. The stacking method can only be applied if several images of the same star region are available[1]. According to Stöveken & Schildknecht[2], a star catalogue can be used to identify those pixels that belong to stars (up to a given magnitude). These pixels can then be masked, and the remaining illuminated pixels belong to the object of interest. The limiting magnitude for the star mask has to be adapted depending on the magnitude of the object of interest. And a disadvantage of this method is the comparably extensive computing effort to transform the information of the catalogue (e.g. star position in RA and Dec) into pixel coordinates. While using a median image, the image registration in preprocess is not easy for streak-like stellar images. If only two images are available, the noise introduced by image subtraction gets worse, which is especially crucial in the case of faint objects.

In this paper we propose a novel method based on the characteristics of the object that we are interested in to distinguish it from streak-like stellar images. Section 2 of

X. Wan (Ed.): Electrical Power Systems and Computers, LNEE 99, pp. 433–438.
springerlink.com

this paper describes how to use object characteristics to distinguish between object and streak-like stellar images. In section 3, some successful illustration will be given. The last section is about summary and conclusions.

1.1 A Typical CCD Image

A typical CCD image is shown in Fig. 1, a point-like object between streak-like stellar images can be seen. This type of image can be produced when the object of interest is moving fast with respect to the stars and is taken by tracking the object or by CCD drift-scan observing[3,4].

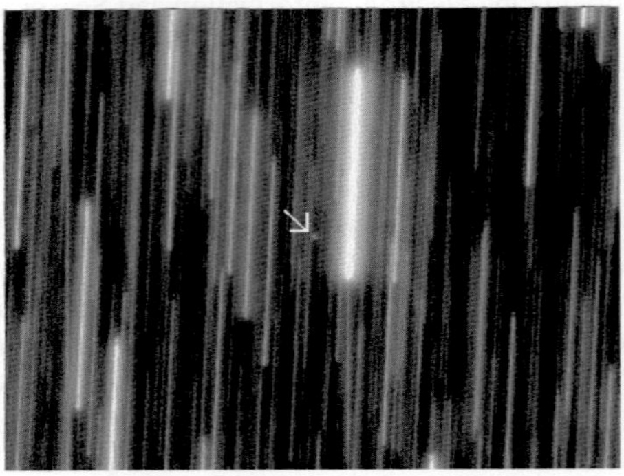

Fig. 1. A typical CCD image

2 Object Detection and Centering

An object and stars have very different characteristics (in their image), and based on these characteristics, we try to detect the object of interest from the star trails in this paper.

The image background is elevated by the blooming effect of bright streak-like stellar images, and it affects the detection of faint objects. We first use the Prewitt operator on the image to reduce its background, and strengthen the outlines of objects and stars to make them clearer and easier to identify. The pixels at the edge of objects and stars can reach the extremum after processed by the Prewitt operator, which calculates the gradient of the image intensity at each point. The image processed by the Prewitt operator is shown in Fig. 2(a). Comparing this result with the original image Fig. 1, the entire background of the image is dimmed, and the halos of streak-like stellar images get restrained. Thus, the outlines of objects and stars become sharper.

After being processed by the Prewitt operator, the center of either the object or the streak-like stellar image appears darker (called vacancy). The "vacancy" can be slightly smoothed by averaging filter, and it can significantly eliminate the noise. A major use of averaging filters is in the reduction of "irrelevant" detail in an image. By "irrelevant" we mean pixel regions that are small with respect to the size of the filter mask[5].

Therefore, we should select the filter size depending on the size of the object image. In the case of a faint object (the image shown in Fig. 1) the 3×3 filter is enough to get a good result. Comparing Fig. 2(b) with Fig. 1, it is shown that a considerable number of faint stellar images are filtered, and bright stellar images are dimmed by the smoothing operation.

We continue our work by using σ threshold to obtain the binary image. This operation can exclude those streak-like stellar images that are obviously darker than the object. In order to get the threshold σ, a Gaussian function is used to fit the image histogram, and 12 times σ is adapted. The result of thresholding is shown in Fig. 2(c).

Boundary tracking is applied after threshold, and its performance depends on two factors. The first one is the selection of a starting point, which has direct impact on tracking accuracy and probably lead to more difficult if improperly selected. The second is the selection of a tracking criteria. According to the order of reading a FITS image that is designed from left to right, bottom to top[6], we select the most bottom left pixel as a starting point, and track the boundary in a clockwise direction. Fig. 2(d) shows an image of all the object and streak-like stellar images' boundaries been tracked.

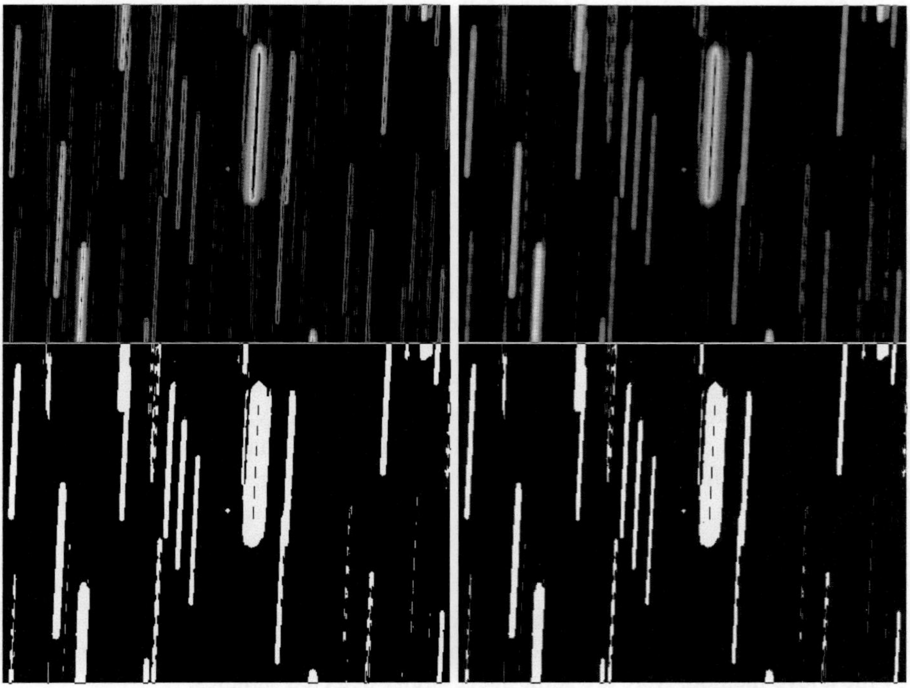

Fig. 2. (a) Image processed by the Prewitt operator. (b) Image processed by a 3×3 averaging filter. (c) Result of σ thresholding. (d) Result of Boundary tracking. (e) Result of identification by the form factor. (f) Result of the Gaussian fitting.

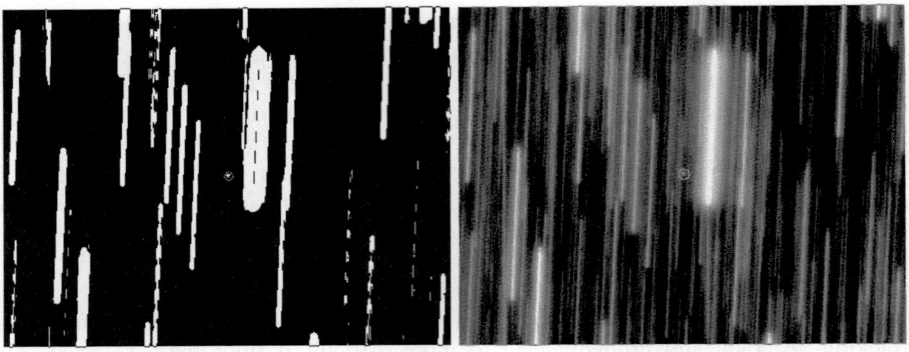

Fig. 2. (*continued*)

After the operations mentioned above, it is found that: the object of interest usually is round, and the stellar image is either streak-like (for a bright star), or segmented into some irregular adjacent regions (for a dark star), as shown in Fig. 2(d). We further use a form factor to distinguish them by the following formula

$$F = \frac{C^2}{4\pi A} \tag{1}$$

where C is the perimeter, A is the area. In realization, the perimeter is the number of pixels in boundary tracking, area is the number of pixels in that closing object. We find that the F of objects of interest are in the range of 0.7-0.8. In Fig. 2(e), the form factor of the object marked by an round F = 0.731.

An initial position of the object can be estimated by the tracked boundary after the object and streak-like stellar images had been roughly identified by their form factors. A two-dimensional Gaussian fitting can be made in a small region centered by the initial position in the original image[7]. Even there are some fragmented regions of faint streak-like stellar images falling into 0.7-0.8 fortunately, they can be removed by the failure of this two-dimensional Gaussian fitting. The final output is the measured position of the faint object of interest from a Gaussian fitting.

3 Detection Results

We have developed our own image processing software using Visual C++ 6.0 in Windows environment. The faint objects in most of the images can be quickly detected by the above method. Here are some successfully detected samples.

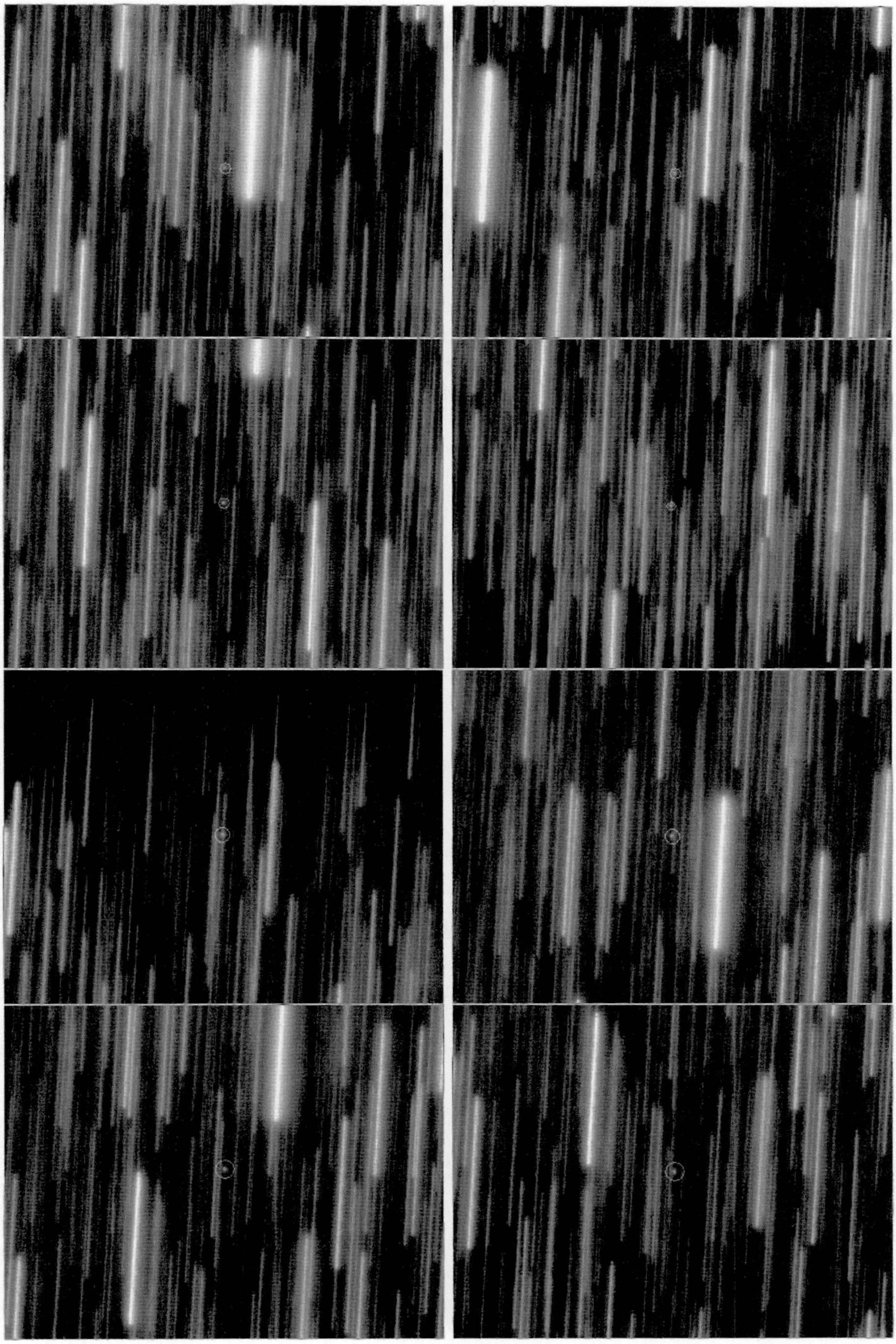

Fig. 3. Some successfully detected images

Occasionally, some images may be detected unsuccessfully when an object and a bright streak-like stellar image are overlapping. These images will be studied in further.

4 Summary and Conclusions

According to the characteristics of actual images, this paper proposes a novel method to detect an object of interest between streak-like stellar images. An original image is firstly processed by the Prewitt operator to strengthen the outlines of the object and stars, and processed by a spatial averaging filter to eliminate the noise, then it is thresholded by σ to get a binary image. Afterward, an initial position is estimated by the tracked boundary after the object and streak-like stellar images had been roughly identified by their form factors, and the measured position of the faint object is computed by a two-dimensional Gaussian fitting in the original image.

The codes are designed in Windows/Visual C++ 6.0 environment, and can be applied to a single exposure or series of images rapidly.

Acknowledgements. The authors are grateful to Zhang Qingfeng, Meng Xiaohua, Li Yan and Li Zhan at Department of Computer Science, Jinan University for their constructive suggestions. This work is financially supported by the National Natural Science Foundation of China (Grant No. 10973007) and the Fundamental Research Funds for the Central Universities of Jinan University.

References

1. Yanagisawa, T., Nakajima, A., Kimura, T., et al.: Detection of Small GEO Debris by Use of the Stacking Method. Trans. Japan Soc. Aero. S Sci. 44, 190–199 (2005)
2. Stöveken, E., Schildknecht, T.: Algorithms for the optical detection of space debris objects. In: Proceedings of the 4th European Conference on Space Debris, Darmstadt, pp. 637–640 (2005)
3. Viateau, B., Réquième, Y., Le Campion, J.F., et al.: The Bordeaux and Valinhos CCD meridian circles. A&AS 134, 173–186 (1999)
4. Stone, R.C., Monet, D.G., Monet Alice, K.B., et al.: Upgrades to the Flagstaff Astrometric Scanning Transit Telescope: A Fully Automated Telescope for Astrometry. AJ. 126, 2060–2080 (2003)
5. Rafael, C., Gonzalez, R.E.: Woods: Digital Image Processing, 2nd edn. Prentice Hall, Upper Saddle River (2002)
6. Wells, D.C., Greisen, E.W., Harten, R.H.: FITS - a Flexible Image Transport System. A&AS 44, 363–370 (1981)
7. Chiu, L.-T.G.: Astrometric techniques with a PDS microdensitometer. AJ. 82, 842–848 (1977)

Design and Realization of a Real-Time Detection Device for Insect Pests of Field Crops

Zhang Hongtao[1], Gu Bo[1], and Hu Yuxia[2]

[1] Institute of Electric power, North China University of Water Conservancy and Electric Power, Zhengzhou 450011, China
[2] College of Electric Engineering, Zhengzhou University, Zhengzhou 450001, China
zht1977@ncwu.edu.cn

Abstract. The real-time identification of insect pests on field crops was an inevitable trend of modern plant protection. The second-generation detection device for insect pests was formed with the hardware and the software system. The hardware system included the trapping, stunning and buffering unit, the even illumination unit, the scattering and transporting unit, and the image vision unit. The software system included image enhancement, image segmentation, feature selection and recognition for insect pests. The device realized the complete automation from the collecting to the identification of insect pests. The nine species of insect pests were automatically recognized, and the correct identification ratio was over 86%. The experiment showed that the system was practical and feasible.

Keywords: Plant protection, Insect pests of field crops, Detection device, Real time, Image recognition.

1 Introduction

The insect pests of field crops are of a wide species and of huge quantity. There are hundreds of common species of insect pests trapped by lamps in agriculture fields. The insect pests were manually recognized and counted after they were trapped by the black light lamp. And the forecasting method was mainly used in China for a long time. Its effect, accuracy and efficiency were closely related to the comprehensive quality of the forecasting person, and it was inevitably influenced by greater subjective factors[1]. Therefore, the real-time accurate identification of insect pests on field crops was an inevitable trend of modern plant protection. It was also an important issue that must be firstly studied and resolved for today's digital agricultural.

Scholars made a deep study of the image recognition of insect pests on field crops. The static digital camera images of 40 species (25 families, 8 orders) of insect pests fastened with needles were classified[2]. The thirty-five species of common pests (Lepidoptera) placed manually were recognized with the grayscale images[3]. The eight species of insect pests in cotton fields was captured by the car, killed and brought to laboratory for further recognition analysis[4].

X. Wan (Ed.): Electrical Power Systems and Computers, LNEE 99, pp. 439–444.
springerlink.com

The related studies at home and abroad had a positive role for the identification of insect pests on field crops. However, these studies had not yet achieved the complete automation from the trapping to the final recognition, and all required manual operation. The 2nd generation real-time detection system for insect pests on field crops was developed by authors after several years of research and exploration. The experiments showed that the system was feasible. The hardware and the software of the real-time detection system were mainly discussed in this paper.

2 Hardware Design of the Detection System

The real-time detection system consisting of the hardware and the software was developed. The hardware included four major units, and they were the trapping, stunning and buffering unit, the scattering and transporting unit, the even illumination unit and the image vision unit respectively (Fig. 1).

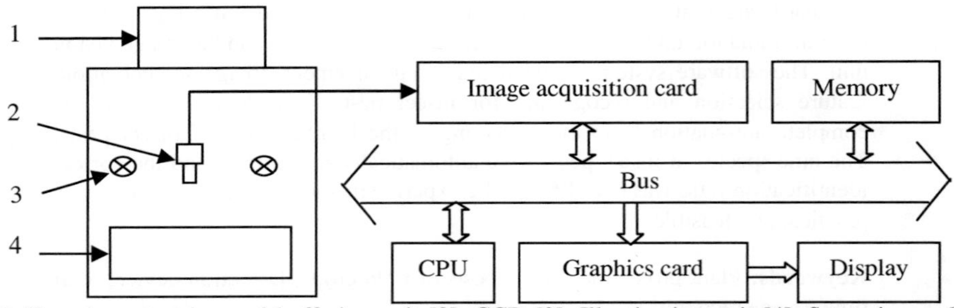

[1]. Trapping, stunning and buffering unit [2]. CCD [3]. Illumination unit [4]. Scattering and transporting unit

Fig. 1. The hardware components of the real-time detection system for insect pests on field crops.

2.1 The Trapping, Stunning and Buffering Unit

A 30W optical controlling light (Model PS-15 II) in the device was used to trap the insect pests of field crops. The area controlled by a vertical single lamp in the center of device was 4hm2. Three pieces of colorless transparent plexiglass were fixed vertically around the lamp. And the angle among the surfaces of plexiglass was 120 degrees. When they flied toward the lamp high-speedly, the insect pests would hit the plexiglass and fall into the hollow funnel under the lamp automatically. Then the stunned pests continued to drop along the connecting pipes until they fall into a specified area. The connecting pipes under the funnel were made up of three pipes that were 10cm in diameter. The angle among the three pipes connected into "S" curve was 60 degrees. When the pests passed through the "S" shaped path, the falling speed was reduced greatly, and this reduced the impact to the subsequent water flow.

2.2 The Scattering and Transporting Unit

The insect pests passed through the trapping, stunning and buffering unit and then fell into the scattering and transporting unit. Water was used as the transmission medium here. This was because that it was very easy to balance for the pests, and the gesture of the pests was either the front or the back on the water. Water had a property of natural scattering the object. The stack possibility of the insect pests was reduced greatly at time of trapping the pests in bulk, for example, from eight to ten o'clock in the evening. And the processing difficulty of the subsequent software was reduced.

The scattering and transporting unit was made up of the upper water storage tank, the lower water storage tank, 30W waterproof motor and the filter. The upper water storage tank consisted of three parts that were the overflow sink, the gathering-pests sink and the passing flow sink. The CCD camera was mounted on the top of the passing flow sink. The pump drew water from the lower water storage tank into the overflow sink of the upper water storage tank. The flow of water was as follows: when the overflow sink was full of water, water flowed into the gathering-pests sink, and over the passing flow sink, and then was filtered into the lower water storage tank by the filter. The insect pests fell slowly into the gathering-pests sink after they passed through the trapping, stunning and buffering unit. They were scattered naturally by water, and passed by the passing flow sink along the water flow, then were filtered by the filter and fell into the collecting-pests bag.

2.3 The Even Illumination Unit

The even illumination system was a key component of the image acquiring system. The pests trapped were mostly yellow, so the yellow was suitable in the color choice of the lights. It was very appropriate that the inverse color of the pests was selected as the background color of images. The dark blue was served as the color of the passing flow sink. This was because that water was colorless and transparent, and the background color was mainly decided by the color of the passing flow sink.

The self-made light box provided even illumination and was fixed over the top of the passing flow sink. The yellow plastic cloth was acted as the walls of the light box. The four 15W yellow lights were placed evenly outside of the light box. The opaque black leather was selected as the material of the upper surface of the light box. The upper surface was opened a small role in order to place the CCD lens. Therefore the image boundaries of the insect pests were relatively sharp in the illumination system. This met the requirements of the subsequent image processing and analysis.

2.4 The Image Vision Unit

An OK-C30S card was used to acquire images of the insect pests based on microcomputer PCI bus. The card had power ability in filtering saw tooth phenomenon, therefore it was quite suitable in dynamic image acquisition system. The maximum resolution of image acquisition was 768×576, the sampling period was 40ms, and the frame buffer size of the image sequence was 4M. The acquisition mode was RGB three component inputs. The complex synchronization between image acquisition and image processing was coordinated organically by calling the callback function. As the continuous images acquired were overlapping, the actual sampling period was determined by the relationship among the speed of water flow, the field of view of CCD and the time required to process a single frame by the computer. The experiment showed that 0.4 second was more appropriate to process one frame.

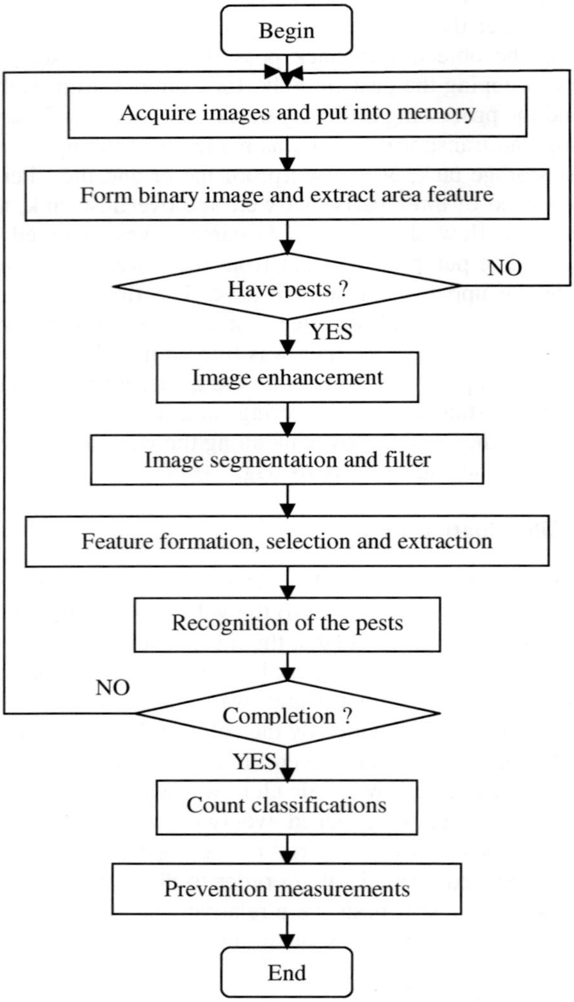

Fig. 2. The software workflow of the detection device

3 Software Design of the Detection System

On account of large amount, abundant information, high real-time performance, the burden of device for image processing was increased. The system used Windows 2003 operating platform and Visual C++ 6.0 visual development language. The application software based on MFC framework was developed with API functions provided by OK_C30S graphics acquisition card. The overall block diagram of the system was shown in Figure 2.

3.1 Image Enhancement and Segmentation

In the image acquisition process of the pests, there was a lot of noise, for example, fluctuation of water flow, body fragments of the pests, uneven illumination and others. An adaptive neighborhood average algorithm modified was used to enhance the original image. This enhancement method without any predetermined parameters can effectively smooth the noise, sharpen blurred image edges, and the calculation was simple. The image segmentation from the background was essential to form a binary image of the pests after image enhancement. The accuracy of the image segmentation was crucial to the subsequent feature extraction and recognition. It was found that the iterative threshold segmentation method had better adaptive ability after repeated experiments, and it was more suitable for image segmentation of the insect pests.

3.2 Feature Extraction and Normalization

Sixteen morphological features of the insect pests forming the original feature space were extracted from the binary image, such as area, perimeter and complexity. The difference among these features varied widely in dimension and magnitude. The classification performance of the system would be influenced greatly if these features were not preprocessed. The raw feature data would be normalized according to Equation 1 in order to eliminate the influence of dimension and magnitude[5].

$$f'_{ij} = \frac{f_{ij} - (\min_j - \Delta_{j1})}{(\max_j + \Delta_{j2}) - (\min_j - \Delta_{j1})} \qquad (i = 1,2,\cdots,m\ ,\ j = 1,2\cdots,d) \tag{1}$$

Where f_{ij} and f'_{ij} were the feature value before and after the normalization of j th feature for the i th pest separately. $\max_j$ and $\min_j$ were the maximum and the minimum of the j th feature for the N training samples separately. Δ_{j1} and Δ_{j2} were the upper limit margin and the lower limit margin of the j th feature of the pests. Taking into account the correlation degree among all the morphological features, the seven features were taken as the final classification features of the pests after the normalization data were analyzed by comparison. These features selected were area, perimeter, complexity, eccentricity, sphericity, first and second invariant moments.

3.3 Fuzzy Classification of Nine Species of Pests

The field crops were seriously damaged by the nine species of insect pests, such as Mythimna separata, Clanis bilineata, Tobacco budworm and Black cutworm. The fuzzy classifier of the pests was designed based on fuzzy decision theory[6]. The classification algorithm consisted of two parts: one was the establishment of the standard vector model library based on the feature mean and the feature standard deviation of the species of pests; the second was the identification of the samples to be recognized based on the fuzzy decision principle of minimum and maximum. The thirty images were collected for each species in the laboratory, so that there were 270 images of the samples in total. The twenty images for each species were randomly selected as the training sample set of the fuzzy classifier, and the others were taken as the testing sample set. The ninety samples were classified by the fuzzy classifier, of which 12 samples were misclassified, so the accuracy of the classifier was 86.7%.

4 Conclusions

The second-generation real-time detection device for field pests was designed. The device realized the entire automation that was from trapping, scattering, transporting, image collecting, image analyzing to image recognizing of the pests. It shortened the cycle of pest identification and improved the level of automation. The correct identification ratio of the nine species of the pests was over 86%. The further study was to increase the species of the pests, and improve efficiency of the detection.

Acknowledgement

Financial support was provided by the National Natural Science Foundation of China (No. 30871449) and the Natural Science Research Program of the Education Department of Henan Province (No. 2011B210028).

References

1. Liu, X.-y., Zhang, H.-t.: Application of Fuzzy Recognition in the Detection of Pests in Fields. Journal of Agricultural Mechanization Research 31(12), 192–194 (2006)
2. Zhao, H.-q., Shen, Z.-r., Yu, X.-w.: On Computer-aided Insect Identification Through Math-Morphology Features. Journal of China Agricultural University 7(3), 38–42 (2002)
3. Watson, A.T., O'Neill, M.A., Kitching, I.J.: Automated Identification of Live Moths (Macrolepidoptera) Using Digital Automated Identification System (DAISY). Systematics and Biodiversity 1(3), 287–300 (2003)
4. Drake, J.T., Walters, T.: RITA, the Robot, Provides Survey and Identification Support to the USDA. Cphst News 2(9), 3–4 (2006)
5. Zhang, H.-t., Hu, Y.-x., Qiu, D.-y.: Application of Simulated Annealing Algorithm in Stored-grain Pests Image Recognition. Journal of Henan Agriculture Sciences 32(7), 28–31 (2003) (in Chinese)
6. Qiu, D.-y., Zhang, H.-t., Chen, T.-j.: Application of Fuzzy Recognition Technique in Stored-grain Pests Detection. System Sciences and Comprehensive Studies in Agriculture 18(2), 122–125 (2002) (in Chinese)

A Comparison of PCA and 2DPCA in Face Recognition

Zhang Haiyang*

College of the Globe Science and Engineering of Suzhou University, 234000,
Suzhou, Anhui Province, China
seazhang188@126.com

Abstract. Principle Component Analysis (PCA) technique is of vital importance and is an efficient method in extracting features in face recognition. Generally speaking, the image always needs to be transformed into ID vector in PCA. Recently two-dimensional PCA (2DPCA) technique has been proposed. In the method of 2DPCA, PCA technique is applied directly on the original images without being transformed into 1D vector. In this paper, we will compare the two methods in face recognition based on ORL face database.

Keywords: PCA, face recognition, 2DPCA.

1 Introduction

Face recognition has obtained an increasing amount of attention in pattern recognition and computer vision over the past few years. The reason for that is face recognition technology can be applied in a wide range of fields, such as identity authentication, access control and so on [1].

Generally speaking, there are two categories of methods in face recognition [2]. One approach is based on facial feature. The most famous technique in the second approach is Principle Component Analysis (PCA).

In the all previous face recognition techniques based on, the 2D face image matrices must be previously transformed into 1D image vectors. Recently in [3], a new PCA approach called 2DPCA is developed for image feature extraction.

2 Principal Component Analyses (PCA)

Principal component analysis (PCA), which is also known as Karhunen-Loeve expansion, is a classical feature extraction and data representation technique, and this technology is widely used in the areas of pattern recognition and computer vision [4]. Turk and Pentland proposed the well-known Eigenfaces method for face recognition in 1991 in this context. Since then, PCA has been widely investigated and has become one of the most successful approaches in face recognition [5], [6].

* Foundation item: The Nature Science Research Project of Suzhou University(Anhui Province) (2009yzk03).

X. Wan (Ed.): Electrical Power Systems and Computers, LNEE 99, pp. 445–449.
springerlink.com

Principal component analysis is proposed by Turk and Pentland in 1991, which is often used for extracting features and dimension reduction. Let us have a brief view of the principle of PCA [7].

Step 1: A set of M images with size N*N can be represented by vectors of size N2 Γ1, Γ2,Γ3,…,ΓM

Step 2: The average training set is defined by

$$\Psi = \left(\frac{1}{m}\right)\sum_{i=1}^{M} \Gamma i \tag{1}$$

Step 3: Each face differs from the average by vector

$$\phi_i = \Gamma_i - \psi \tag{2}$$

Step 4: A covariance matrix is constructed as follows:

$$C = AA^T \tag{3}$$

Step 5: Finding eigenvectors of $N^2 \times N^2$ matrix is very difficult. Therefore, we use the matrix ATA of size M x M and find eigenvectors of this small matrix.

Step 6: If v is a nonzero vector and λ is a number such as Av = λv, then v is an eigenvector of A with eigenvalue λ.

Step 7: Consider the eigenvectors v_i of $A^T A$

$$A^T Av_i = u_i v_i \tag{4}$$

Step 8: Multiply both sides by A, we can obtain the result:

$$AA^T (AV_i) = u_i (Av_i) \tag{5}$$

Step 9: A face image can be projected into this face space by

$$\Omega_k = U^T (\Gamma^k - \Psi); k = 1,\ldots, M \tag{6}$$

3 Two-Dimensional PCA

In this section, we will shortly review the basic principals, essential mathematical background and algorithm of 2DPCA.

Let X denotes an n-dimensional unitary column vector. Our idea is according to the following linear transformation [8], [9] to project image A which is an $m \times n$ random matrix onto X:

$$Y = AX \tag{7}$$

From this, we can get an m-dimensional projected vector Y which is called the projected feature vector of image A. How do we determine a good projection vector X? we use the following criterion:

$$J(x) = tr(S_x) \tag{8}$$

In this place, S_x denotes the covariance matrix of the projected feature vectors of the training samples and $tr(S_x)$ denotes the trace of S_x. The covariance matrix S_x can be denoted by

$$\begin{aligned} S_x &= E(Y - EY)(Y - EY)^T = E[AX - E(AX)] \\ [AX - E(AX)]^T &= E[(A - EA)X][(A - EA)X]^T \end{aligned} \tag{9}$$

Thus,

$$tr(Sx) = X^T [E(A - EA)^T (A - EA)] X \tag{10}$$

Let us define the following matrix

$$G_t = E[(A - EA)^T (A - EA)] \tag{11}$$

The matrix G_t is called the image covariance (scatter) matrix. From its definition, it is easy to verify that G_t is an $n \times n$ nonnegative definite matrix. We can use the training image samples to evaluate G_t directly. Suppose that there are M training image samples in total, the jth training image is denoted by an $m \times n$ matrix $A_j (j = 1, 2, ..., M)$ and the average image of all training samples is denoted by $\overline{A}$.

Then, G_t can be evaluated by the following criterion:

$$G_t = \frac{1}{M} \sum_{j=1}^{M} (A_j - \bar{A})^T (A_j - \bar{A}) \tag{12}$$

Alternatively, the criterion in (8) can be expressed by

$$J(X) = X^T G_t X \tag{13}$$

In this place X is a unitary column vector. This criterion is called the generalized total scatter criterion. The unitary vector X which maximizes the criterion is called the optimal projection axis. The optimal projection axis Xopt is the unitary vector that maximizes $J(X)$. Generally speaking, it is not enough to have only one optimal projection axis. We usually need to select a set of projection axes, $X_1, ..., X_d$, subject to the orthonormal constraints and maximizing the criterion $J(X)$, that is,

$$\begin{cases} \{X_1, \ldots\ldots, X_d\} = \arg\max J(X) \\ X_i^T X_j = 0, \quad i \neq j, i, j = 1, \ldots\ldots d \end{cases} \tag{14}$$

4 Experiments and Analysis

Our experiments are based on Cambridge ORL face database and Matlab7.0 is used as programming tool. We show some images from ORL face database in Figure1.

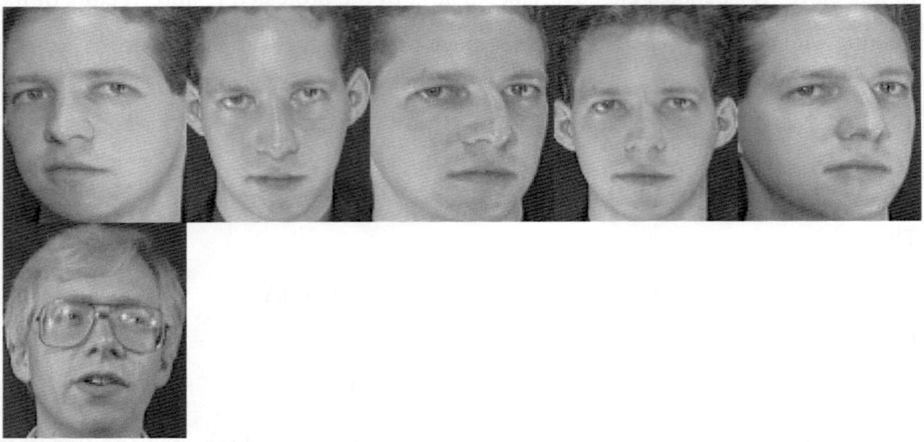

Fig. 1. Some images from ORL face database

In this experiment, we take each person's one image; two images… five images respectively as training samples and the left images are testing samples respectively.

Table 1. A comparison of PCA and 2DPCA in recognition rate (%)

Testing samples	*1*	*2*	*3*	*4*	*5*
PCA	77.5	81.25	83.33	83.13	83
2DPCA	80	81.25	86.67	91.87	92

References

1. Chellappa, R., Wilson, C.L., Sirohey, S.: Human and Machine Recognition of Faces: A Survey. Proceedings of the IEEE 83(5), 705–740 (1995)
2. Ranganath, S., Arun, K.: Face Recognition Using Transform Features and Neural Network. Pattern Recognition 30, 1615–1622 (1997)
3. Yang, J., Zhang, D., Frangi, A.F., Yang, J.-y.: Two-dimensional PCA: a new approach to appearance-based face representation and recognition. IEEE Transactions on Pattern Analysis and Machine Intelligence 26(1), 131–137 (2004)
4. Kirby, M., Sirovich, L.: Application of the KL Procedure for the Characterization of Human Faces. IEEE Trans. Pattern Analysis and Machine Intelligence 12(1), 103–108 (1990)

5. Pentland: Looking at People: Sensing for Ubiquitous and Wearable Computing. IEEE Trans. Pattern Analysis and Machine Intelligence 22(1), 107–119 (2000)
6. Grudin, M.A.: On Internal Representations in Face Recognition Systems. Pattern recognition 33(7), 1161–1177 (2000)
7. Turk, M., Pentland, A.: Eigenfaces for Recognition. J. Cognitive Neurosci. 1, 71–86 (1991)
8. Liu, K., et al.: Algebraic Feature Extraction for Image Recognition Based on an Optimal Discriminant Criterion. Pattern Recognition 26(6), 903–911 (1993)
9. Yang, J., Yang, J.Y.: From Image Vector to Matrix: A Straightforward Image Projection Technique—IMPCA vs. PCA. Pattern Recognition 35(9), 1997–1999 (2002)

Face Recognition Based on DCT and PCA

Zhang Haiyang*

College of the Globe Science and Engineering of Suzhou University, 234000,
Suzhou, Anhui Province, China
seazhang188@126.com

Abstract. In this paper we propose a new method of face recognition. DCT and PCA are combined in this method. Our experimental results show that we can get much better recognition rates based on the same face images.

Keywords: DCT, PCA, recognition rate.

1 Introduction

Face recognition is a hot topic studied by many researchers in recent years. That is because face recognition technology can be applied in a wide range of fields, such as identity authentication, access control and so on [1].Generally speaking, there are two categories of methods in face recognition [2]. One approach is based on facial feature. Firstly, the features such as eyes, nose and mouth first are located and then various feature extraction methods can be adopted to construct feature vectors of these facial features. Finally, traditional pattern recognition methods like a neural network can be used to recognize the feature vectors.

The other approach takes a holistic view of the recognition problem. It extracts the statistical characterization by the statistical method directly out of the entire training sample images instead of extracting the feature of the nose, mouth, or the eyes separately. Holistic feature extraction of face images is adopted in this approach. The most famous technique in the second approach is Principle Component Analysis (PCA).

PCA-based face recognition analysis is a method which is about the features of the whole face appearance. These features extracted are related to the whole face or even to the whole sample set. They needn't mean anything definitely. When you classify these features, you can get satisfactory results. PCA (Principal Components Analysis) is such an effective method.

The variations between the images of the same face due to illumination and viewing direction are almost always larger than image variations due to changes in face identity [3]. Two key issues should be solved in order to reduce illumination effect [4]. A face image is easily subjected to changes in viewpoint, illumination, and expression. So the first issue is what features can be used to represent a face in order to be able to deal with

* Foundation item: The Nature Science Research Project of Suzhou University(Anhui Province) (2009yzk03).

possible changes and expression. The second issue is what kind of classifier can be used to classify a new face image. In this paper DCT and PCA are combined, DCT can be considered as a preprocessing method before using PCA to extract features. Using DCT can reduce illumination effect so some extent. Our experimental results show that we can get much better recognition rates based on the same face images.

In this paper, the principles of PCA and DCT are presented. In PCA-based method within-class average faces are computed to normalize training samples in order to reduce the difference between same-class samples, while at the same time to augment the difference between different-class samples. We have done our experimental results on ORL (Olivetti Research Laboratory) face database using the method of PCA and PCA combined with DCT and compare their recognition rates.

Our paper is organized as follows: In the first chapter, we introduce the background of face recognition; in the second chapter, we introduce the principles DCT ; in the third chapter, we introduce the principles of PCA; and in the last chapter, we introduce the ORL face database and compare the recognition rates of the PCA and PCA combined with DCT.

2 Discrete Cosine Transform (DCT)

Like other transforms, the Discrete Cosine Transform (DCT) attempts to decorrelate the image data. After decorrelation each transform coefficient can be encoded independently without losing compression efficiency. In this section we will shortly introduce the theory of DCT [5]. DCT includes one-dimensional DCT and two-dimensional DCT.

In the 1D case, DCT can be defined by

$$C(u) = \partial(u)\sum_{x=0}^{N-1} f(x)\cos[\frac{\pi(2n+1)u}{2N}] \tag{1}$$

In this place, $$f(x) = \sum_{u=0}^{N-1} \partial(u)C(n)\cos[\frac{\pi(2n+1)u}{2N}] \tag{2}$$

$$\partial(u) = \begin{cases} \sqrt{\frac{1}{N}}, u = 0 \\ \sqrt{\frac{2}{N}}, u \neq 0 \end{cases} \tag{3}$$

In this place, $u = 0,1,...,N-1$

In the 2D case, DCT can be defined by

$$C(u,v) = \partial(u)\partial(v)\sum_{x=0}^{N-1}\sum_{y=0}^{N-1} f(x,y)\cos[\frac{\pi(2n+1)u}{2N}]\cos[\frac{\pi(2n+1)v}{2N}] \tag{4}$$

In this place $u, v = 0, 1, ..., N-1$

$$f(x, y) = \sum_{u=0}^{N-1}\sum_{v=0}^{N-1} \partial(u)\partial(v)C(u, v)\cos[\frac{\pi(2n+1)u}{2N}]\cos[\frac{\pi(2n+1)v}{2N}]$$

For $x, y = 0, 1, ..., N-1$.

3 Principal Component Analyses (PCA)

Principal component analysis (PCA), which is also known as Karhunen-Loeve expansion, is a classical feature extraction and data representation technique, and this technology is widely used in the areas of pattern recognition and computer vision [6]. They argued that any face image could be reconstructed approximately as a weighted sum of a small collection of images which define a facial basis (eigenimages), and a mean image of the face. Turk and Pentland proposed the well-known Eigenfaces method for face recognition in 1991 in this context. Since then, PCA has been widely investigated and has become one of the most successful approaches in face recognition [7].

Principal component analysis is proposed by Turk and Pentland in 1991, which is often used for extracting features and dimension reduction. Let us have a brief view of the principle of PCA [8].

Step 1: A set of M images with size N*N can be represented by vectors of size N^2

Γ1, Γ2,Γ3,…,ΓM

Step 2: The average training set is defined by

$$\psi = (\frac{1}{m})\sum_{i=1}^{M} \Gamma i \tag{5}$$

Step 3: Each face differs from the average by vector

$$\phi_i = \Gamma_i - \psi \tag{6}$$

Step 4: A covariance matrix is constructed as follows:

$$C = AA^T \tag{7}$$

Step 5: Finding eigenvectors of $N^2 \times N^2$ matrix is very difficult. Therefore, we use the matrix $A^T A$ of size M x M and find eigenvectors of this small matrix.

Step 6: If v is a nonzero vector and λ is a number such as Av = λv, then v is an eigenvector of A with eigenvalue λ.

Step 7: Consider the eigenvectors v_i of $A^T A$

$$A^T A v_i = u_i v_i \tag{8}$$

Step 8: Multiply both sides by A, we can obtain the result:

$$AA^T(AV_i) = u_i(Av_i) \tag{9}$$

Step 9: A face image can be projected into this face space by

$$\Omega_k = U^T(\Gamma^k - \Psi); k = 1,...,M \tag{10}$$

4 Experiments and Analysis

Our experiments are based on Cambridge ORL face database and Matlab7.0 is used as programming tool .This face database contains 40 individuals, and each individual has 10 images with variations in pose, illumination, facial expression and accessories. The size of each image is 92 ×112 pixels; with 256 grey levels per pixel .We show some images from ORL face database in Figure1.

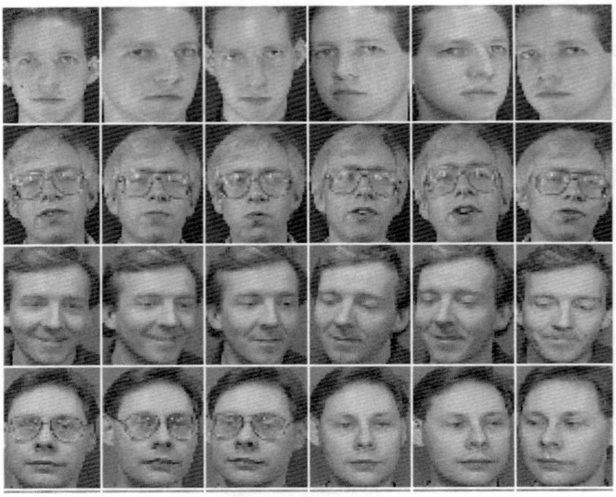

Fig. 1. Some images from ORL face database

In this experiment, we take each person's one image; two images… five images respectively as training samples and the left images are testing samples respectively. The recognition rates that we get are in Table 1.

Table 1. The comparison of PCA and PCA combined with DCT

Training samples	1	2	3	4	5
PCA	77.5%	81.25%	83.33%	83.13%	83%
PCA+DCT	90.71%	88.57%	92%	84.17%	85%

References

1. Chellappa, R., Wilson, C.L., Sirohey, S.: Human and Machine Recognition of Faces: A Survey. Proceedings of the IEEE 83(5), 705–740 (1995)
2. Ranganath, S., Arun, K.: Face Recognition Using Transform Features and Neural Network. Pattern Recognition 30, 1615–1622 (1997)
3. Adini, Y., Moses, Y., Ullman, S.: Face Recognition: The Problem of Compensating for Changes in Illumination Direction. IEE Transactions 19(7), 721–732 (1997)
4. Guodong, G., Li, S., Kapluk, C.: Face recognition by support vector machines. In: Proc. IEEE International Conference on Automatic Face and Gesture Recognition, March 2000, pp. 196–201 (2000)
5. Hafed, Z.M., Levin, M.D.: Face Recognition Using the Discrete Cosine Transform. International Journal of Computer Vision 43(3), 167–188 (2001)
6. Kirby, M., Sirovich, L.: Application of the KL Procedure for the Characterization of Human Faces. IEEE Trans. Pattern Analysis and Machine Intelligence 1(1), 103–108 (1990)
7. Pentland: Looking at People: Sensing for Ubiquitous and Wearable Computing. IEEE Trans. Pattern Analysis and Machine Intelligence 22(1), 107–119 (2000)
8. Turk, M., Pentland, A.: Eigenfaces for Recognition. J. Cognitive Neurosci. 1

Research on Monitor System of Distant Coal Mine Gas Based on Labview

Liu Hong, Huang Chaozhi, and Xiao Fayuan

School of Mechanical & Electrical Engineering,
Jiangxi University of Science & Technology, Ganzhou (341000), China
jxligonglh@163.com

Abstract. Nowadays coal mine monitor system exists some fault, such as no-perfect function, complicate technology and difficult commit, so this paper designs a monitor system of distant coal mine based on Labview. A distributed wireless sensor net is built up by Labview developments kids, Zigbee net technology and virtual instruments technology. This paper detail discusses the hardware structure and software design. This system has high stabilization, flexible control and convenient apply. Moreover, it is reliable to distantly monitor coal mine gas.

Keywords: data transmission, Wireless Sensor Network, Monitoring Systems.

0 Introduce

The gas safe accidence is one of accidences with big harmfulness and high mortality. The wire transmission mode is applied at traditional data sample in gas monitor system. However, because of complicate geological condition and distribute working surface, with working face's carrying forward, it is more difficult to place wire and build up gas monitor system, therefore some signals cannot be obtained on time on working surface and it is frequent to result to accidents taking place.

1 Monitor System Structure

As the Fig1 is shown, the system is made up of monitor station on the ground, monitor base station and mobile monitor terminal.

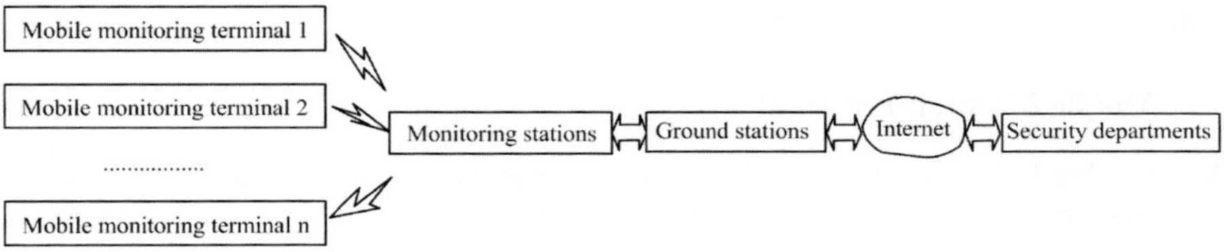

Fig. 1. System structure

X. Wan (Ed.): Electrical Power Systems and Computers, LNEE 99, pp. 457–461.
springerlink.com

(1) The monitor station on the ground puts gas monitor data together and process, which main function is to receive data, process, inquiry, transmit to internet and give an alarm if the gas's concentration exceeds its safe standard;
(2) The task of monitor base station is to receive wireless data from mobile terminal, preprocess and communicate with the monitor station on the ground.
(3) The mobile monitor terminal samples gas data and transmit them to monitor bas station, and terminal can give an alarm.

2 Hardware Design

2.1 Monitor Base Station

The monitor station on the ground controls mobile terminal through monitor base station, whose control kernel is C8051F126 microcomputer. The structure of monitor base station is shown in Fig2. The nRF9E5 is selected as wireless receive and transmit module, which is high performance single wireless chip produced by Nordic VLSI corporation. This chip is real SoC(system on chip), which includes nRF905433／868／915MHz receiver

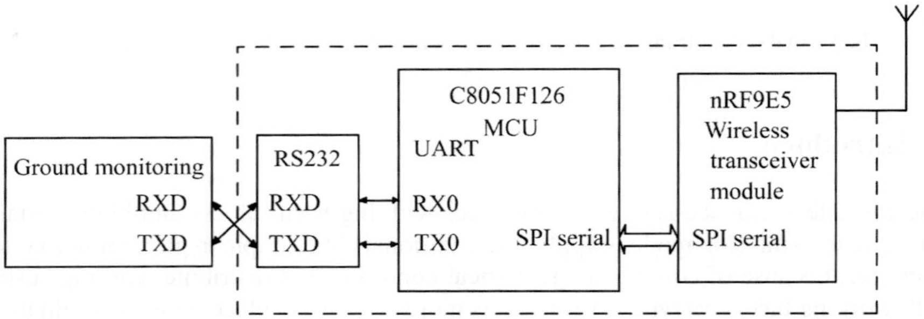

Fig. 2. Monitor base station

The monitor base station sends and receives data by SPI synchronose of C8051F126. In the nRF9E5 the pin CD, AM and DR stands respectively for carry-detect, address-matched and preparation-complete for data. These perfect function make more convenient and dependent to transmit and receive data. The monitor base station is connected to the monitor station on the ground by the URAT0 and RS232 interface. The data from mobile monitor terminal is transmitted to the monitor station on the ground, therefore data transmission is completed.

2.2 Mobile Monitor Terminal

As the Fig3 is shown, mobile monitor terminal is made up of gas sensor, microcontroller and data communication module.

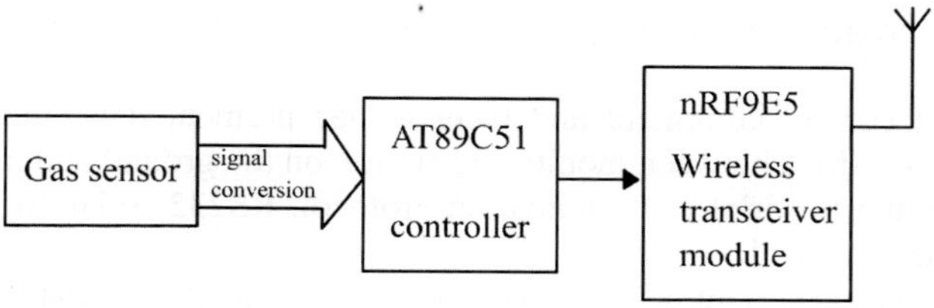

Fig. 3. Mobile monitor terminal

The gas sensor's type is KGS-20 made in China, and microcontroller is AT89C51 made in America. The signals are transferred to A/D converter in the microcontroller by gas sensor. The AT89C51 codes to correspond digital signal, converts to data according to protocol, and load to data transmitting modual.

3 Software Design

3.1 The Control Program Design by LabVIEW

LabVIEW is a perfect virtual instrument development platform, which have abundant function, such as signal sample, measurement and analysis, data process, display and store. The users spent less time and money to develop more dependent and perfect virtual instrument system.

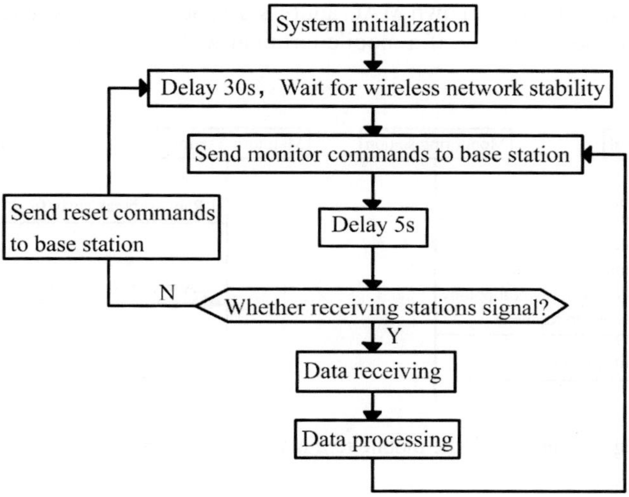

Fig. 4. Procedure diagram of the monitor station on the ground

The all sub-programs have two primary functions in control system:

1) Transmit control command and receive measurement data, that is, the serial communication is done between monitoring station on the ground and monitoring base station. According to serial communication protocol, RS232 serial communication is realized by VISA control.

2) The measurement data is processed, including comparison, analysis, restore and print. Especially, data is stored dynamically, and the file is named by time, so it is convenient for users to look over data, analysis and process.

3.2 Building the Zigbee Net

ZigBee is a kind of open wireless sensor net applied by wireless communication technology. Based on the IEEE802.15.4 protocol, this net can communicate on 3 free frequency ranges, which are general 2.400-2.484GHz in globe, 868.0-368.6MHz in Europe, and 902-928MHz in America, and transmission rate are respectively 250kbps, 20kbps and 40kbps. The communication distance range from 10 to 75 meters. Its remarkable character are low power and low cost, because Zigbee net adopts more low data transmission rate, low work frequency range and small capacity stack, and it is on sleep mode when ZigBee module doesn't work. ZigBee net is construct by coordinator, which scan a free channel other nets don't occupy, and define Cluster-Tree parameters, such as maximum mobile measure terminal, maximum quantity of route nod, and route algorithm.

Fig 5 shows how mobile terminal enter and break off net. After coordinator star, other mobile terminal can be admitted to enter net when its channel is set the same as coordinator, and the correct authentication information is supplied. After a mobile terminal enter net, it can gain own MAC address, ZigBee net address and parameter the coordinator define. Certainly, if a mobile terminal will break off net, it must apply to measure base.

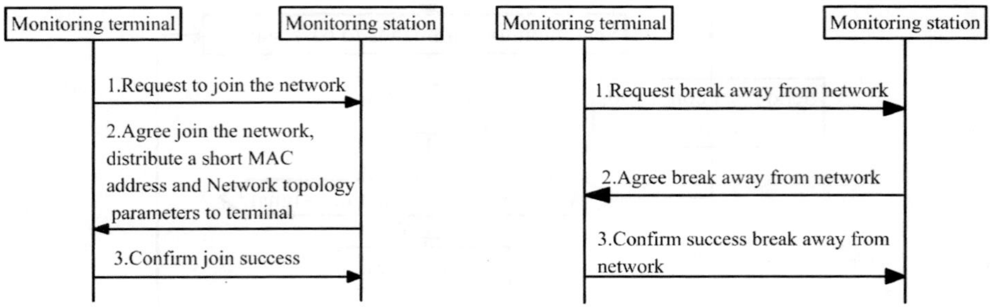

Fig. 5. Diagram of the mobile terminal's entering and breaking off net

4 Conclusion

Substituting the old cable transmission, the system adopts ZigBee technology in wireless sensor net, avoiding the effect of cable resistance and distributed capacity, and the disturb of the environment temperature and electromagnetism. Comparing cable data transmission, this system show remarkable advantage and broad application future.

References

1. Nordic VLSI ASA lnc. nRF9E5 Rf and antenna layout (2004)
2. Nordic. nRF9E5 Product Specification. Nordic corporation, 6 (2003)
3. Huang, Z.-w.: Wireless send and receive data circuit design. Press of Beihang University, Beijing (2004)
4. Yang, l.-p., Li, h.-t.: LabVIEW advanced programmer design. Tsinghua university press, Beijing (2003)
5. Wu, M.: Research on device driver development based on LabVIEW. Microcomputer Information (2), 153–155 (2006)
6. Nie, J.-n.: Multiple access communication and connection control technology. Posts&Telecom Press, Beijing (2006)

Electronics Design for an Interline CCD Camera Based on FPGA

Binhua Li[1], Jianhui Jin[1], Lin He[1], Jing Liu[1],
and Yuanyuan Shang[2]

[1] Faculty of Information Engineering and Automation, Kunming University of Science and Technology, Kunming, Yunnan 650051, China
[2] College of Electronic Information Engineering, Capital Normal University, Bejing 100048, China
lbh@bao.ac.cn

Abstract. An interline CCD camera is developed for remote outdoor monitor applications. This paper narrates the system requirements of the CCD camera first, presents the design considerations and a basic architecture of the analog circuit and digital controller in detail, such as the CCD peripheral circuit, preamplifier, the analog front end and its interface, the clock driver, the power supply, and the clock timing generator programmed with VHDL in FPGA. The simulation and testing results for some key modules and the whole system are analyzed.

Keywords: CCD Camera, Circuit Design, FPGA, Remote Monitor.

1 Introduction

CCDs widely used in the imaging systems can be divided into three types: full-frame, frame-transfer and interline according to their readout modes. Consequently, the cameras fabricated by the CCDs are called full-frame cameras, frame-transfer cameras and interline cameras. The three types of cameras have their own strengths, and are used in different imaging applications respectively. For example, the full-frame cameras with mechanical shutter are popular in the astronomical observations at night, and the frame-transfer cameras are candidates in which the mechanical shutters are not easy to use. The interline CCD has a special structure of the electronic shuttering action, thus the interline transfer cameras do not need the mechanical shutter. These cameras are suitable for high-speed imaging applications, such as solar observation, medical imaging, industrial monitoring and control, and so on.

We developed a prototype CCD camera for outdoor monitoring applications. The CCD is a Kodak KAI-04022 image sensor with 2048(H) × 2048(V) active pixels. Each pixel is a 7.4μm square one with a microlens [1]. It is an interline CCD with 2 high-speed outputs. We called this camera KAISS for simplicity.

X. Wan (Ed.): Electrical Power Systems and Computers, LNEE 99, pp. 463–469.
springerlink.com

2 System Design of the KAISS Camera

2.1 Design Requirements and Considerations

The KAISS camera will be used in the imaging experiment of outdoor monitor. Thus the system is required to have good remote imaging performance. Here the "remote" means that the camera may be far away to the computer (PC) used for imaging control and image acquisition. The remote control can be implemented by the TCP/IP internet. Through the internet, the remote PC sends commands of control and test to the camera, collects the image data from the camera, displays the image on its screen and finally saves the image to a hard disk.

The KAI-04022 is an interline CCD. Different with the full-frame and frame-transfer CCDs, after exposure it must do a charge shift so as to transfer the photoelectrons in the photodiode to its adjacent non-photosensitive vertical register (VCCD) through a interline gate. The later process of reading out the photoelectrons in the VCCD is similar to the full-transfer or frame transfer. The KAI-04022 needs 7 driving clocks: 2 for vertical transfer, 3 for horizontal transfer, 1 for electronic shutter and 1 for fast line dump [1].

Taking into account the requirements of AD bits, readout rate and power dissipation of the camera, and also the simplification of the analog circuit design and debug, we chose ADI's high-performances CCD analog front end (AFE) AD9845B. It is an improved CCD signal processor with a 3-wire serial programming interface (SPI), and features a 30 MHz single-channel architecture designed to sample and condition the outputs of interlaced and progressive scan area CCD arrays. Its signal chain consists of an input clamp, a correlated double sampler (CDS), PxGA, a digitally controlled VGA, a black level clamp, and a 12-bit A/D converter [2]. Thus the AD9845 totally needs 6 operating clocks, such as, preblanking clock, black level clamp clock, 2 CDS sampling clocks for CCD's reference and data level, input clamp clock, digital data output latch clock.

An embedded system is required to meet the requirements of the CCD control and image acquisition. The key component is an Atera's high-speed FPGA device Cyclone EP2C35 with 33216 logic elements and 4 PLLs [3]. A softcore CPU—NiosII can be downloaded to the device. It is easy to building an embedded system and to implement an internet-based transmission of the commands and image data.

2.2 System Architecture

The KAISS camera system is composed of five parts: photoelectrical imaging devices (lens and CCD), analog circuits, a digital controller and a remote terminal. All the modules and their signal flow chart are shown in Fig.1.

An optical image is formed on the CCD photosensitive area by the optical lens. After the CCD is powered up with appropriate levels of the biases and clocks, the input optical image on the CCD can be converted into a photoelectron image in the CCD pixels. With proper driving clocks the image can be readout via output amplifiers on the CCD. Then the charge image becomes into a series of analog voltage signal called analog video. The video, amplified by the preamplifier, is transmit to the AFE to do the CDS and ADC, and finally converted to a series of 12-bit data, that is, digit image. The digit image is sent to the embedded system through a buffer, and uploaded to the remote PC.

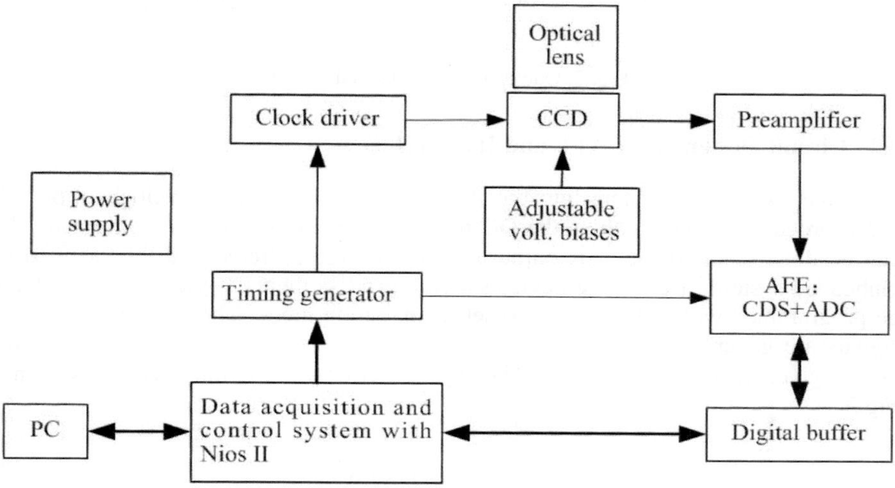

Fig. 1. System modules and signal flow chart of the KAISS camera

3 Analog Circuit Design of the KAISS Camera

The analog circuit of the camera is composed of 4 parts: CCD and preamplifier, AFE and its interface, clock driver, and power supply, which are placed in a same PCB (called Board A). The board is assembled with a commercial Altera FPGA development board (named as Board B), which is used for imaging control and data transmission.

3.1 Circuit Design of the CCD Analog Video Processing Chain

There are 2 video outputs in the KAI-04022. Consequently we should have 2 same chains of video processing. The CCD video arrives at the preamplifier first. The amplifier is made up of an emitter follower, a DC blocking capacitor and a non-inverting amplifier. In order to match the maximum CCD output signal to the AFE input range, the total gain of the preamplifier is set to be 0.78, which is composed of 2 factors: 0.6 for the emitter follower, 1.25 for the non-inverting amplifier. The preamplifier is followed by a CCD signal processor--analog front end. One channel of the analog chain is shown in Fig.2.

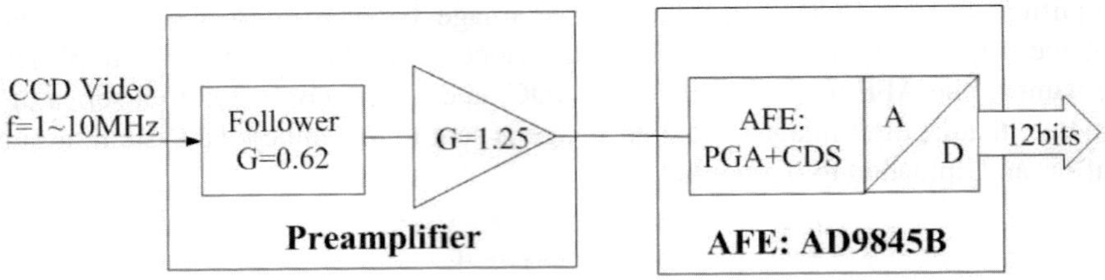

Fig. 2. One channel of the analog chain

3.2 Circuit Design of the AFE and Its Interface

After the CCD video enters into the AFE AD9845B with proper clocks input, the video processing tasks such as the DC restore, CDS, amplification, clamp and ADC can be done in the device. Its output digital data are buffered and then sent to the embedded system. Before the device starts to process the CCD video, it is needed to be programmed with a 3-wire SPI to set its operation mode. The serial programming signals are generated by NiosII. Its interface circuit is similar to the recommended circuit given in its datasheet [2]. The AFE functions and it interface are shown in Fig.3.

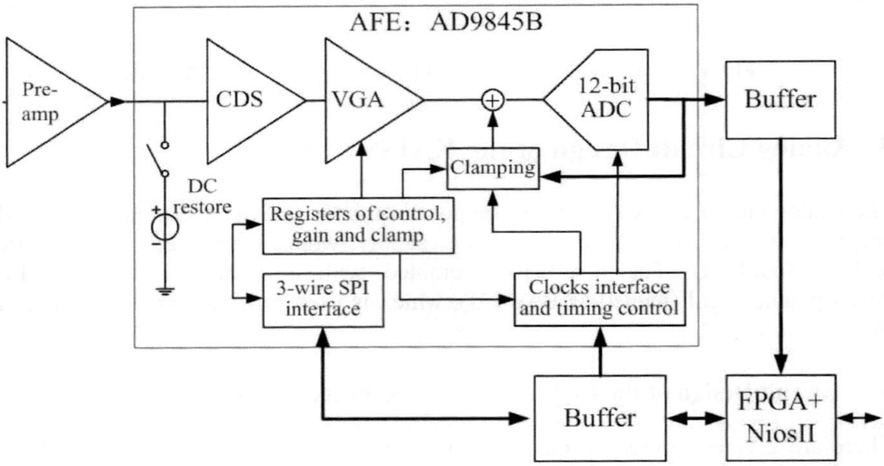

Fig. 3. AFE and its interface circuit

3.3 Circuit Design of the Clock Drivers

The clock drivers are used to convert the low volt (3.3V) TTL clock signals from FPGA to appropriate levels so as to output those vertical, horizontal, electronic shutter and fast dump clocks required by the KAI-04022. Although frequency of the vertical clocks is not high (100 kHz), but its driving loads are high capacitance loads, thus some transistors are employed for the analog switches, and some high-speed 1.5A MOSFET drivers (MAX4427) are used to drive the capacitance loads. The horizontal clocks are of higher frequency (10 MHz), but their loads are of lower capacitance, therefore some low power transistors are used for the analog switches and drivers. Block schematic diagrams of the vertical and the horizontal clock drivers are shown in Fig.4. The basic circuit structure of the clock drivers is similar to the Kodak's recommendatory drivers for the CCD [4]. But some devices are replaced, and values of some resistors and capacitors are appropriately adjusted by our PSpice simulations.

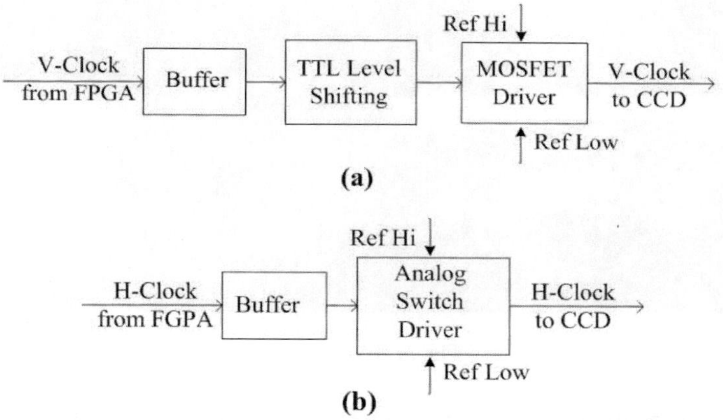

Fig. 4. Block schematic diagrams of the horizontal (a) and the vertical (b) clock drivers

3.4 Circuit Design of the Power Supply

The input power supplies of Board A are ±20V and +5V. The number of the biases and the clock driving levels required by the selected CCD is up to 19. The power supply circuit is used to generate all these voltage levels. Its main structure is similar to those presented in our published papers [5-6].

4 Design of the Timing Generator of the KAISS Camera

The most important component in the digital circuit is the FPGA. Almost all the key logic and timing modules written in VHDL can download to the FPGA. The digital logic design is to program the timing generator with VHDL according to timing diagrams of the CCD and the AFE. The design method is similar to our previous method [7-8], except for the electronic shutter, fast dump and vertical transfer for interline. The concrete implementation procedure and some test results are presented in He's paper [9].

5 Test Results and Analysis

Before the PCB design all the key circuit modules are simulated by PSpice for analog circuits or by Quartus for digital logic circuits (VHDL), and the simulation results are analyzed. The method and procedure are similar to our previous method and procedure [5-8]. Then the PCB is designed and manufactured. Board A is soldered by us. The adjusting and the test are alternant so as to make all the voltage levels and the waves meet the requirements of the CCD. A mixed-signal oscilloscope is used to record the CCD driving clocks. The results and analysis are also presented in He's paper [9]. The assembled camera is shown in Fig.5.

Fig. 5. KAISS camera

After tests of the analog, digit circuits and software debugging, the KAI-04022 CCD is plugged in its socket to do the imaging experiment. In our laboratory, thousands of images in different readout rate and exposure are acquired, of which one image is shown in Fig.6. The fundamental design requirements are met.

Fig. 6. An image taken by the KAISS camera

6 Summaries

The system requirements and architecture of the KAISS camera are presents. The design method and procedure of the CCD analog video processing chain, the clock driver, the power supply, and the clock timing generator are described in detail. After analysis of the images acquired in different rate and exposure, it is proved that the circuit design of the KAISS camera is feasible, and the fabricated camera meets our imaging demands. However, the camera is tested only in our lab environment. We will test it outdoor later.

Acknowledgments. Project funded by the Joint Fund of Astronomy of NSFC and CAS (Grant No. 10978013).

References

1. Eastman Kodak Company. KODAK KAI-04022 Imager Sensor, `http://www.kodak.com/go/imager`
2. Analog Devices, Inc. Complete 12-bit 30 MSPS CCD Signal Processor AD9845B, `http://www.analog.com`
3. Altera Corp. Cyclone II Device Handbook, `http://www.altera.com`
4. Eastman Kodak Company. KAI-4011/KAI-4021 Imager Evaluation Board User Manual, `http://www.kodak.com/go/imager`
5. Li, Y., Li, B., Wang, W., Lu, C.: Design of Analog Circuit and System for a CCD Camera with Dual-Speed Tracking Imaging on Same Frame. Astronomical Research and Technology 4, 376–382 (2007) (in Chinese)
6. Cheng, J., Li, B., Zhang, N., Li, H., Song, Q.: Design of Electronic Circuits for an Astronomical CCD Camera. Astronomical Research and Technology 6, 130–135 (2009) (in Chinese)
7. Wang, W., Li, B., Li, Y., He, C.: Design of Digital Controller for a CCD Camera with Dual-Speed Tracking Imaging on Same Frame. Astronomical Research and Technology 4, 369–375 (2007) (in Chinese)
8. Li, B., Song, Q., He, C., Jin, J., He, L.: Method to implement the CCD Timing Generator Based on FPGA. In: Proc. SPIE, vol. 7742, p. 77421Y (2010)
9. He, L., Li, B., Shang, Y., Jin, J.: VHDL Design of Digital System for a interline transfer CCD. Astronomical Research and Technology 8 (2011) (in Chinese)

Ferroresonance Evaluation at Boushehr 230/400 kV GIS Substation of Iran's Power Network

M. Majidi, H. Javadi, and M. Oskuoee

Power and Water University of Technology (PWUT)
P.O.Box 16765-1719 Tehran,Iran
Mehrdadmajidi66@gmail.com, javadi@pwut.ac.ir, moskuoee@yahoo.com

Abstract. Ferroresonance is one of the main and important reasons to distortion equipments in GIS substations. This phenomenon is not always predictable and clear. So overvoltages evaluation due to ferroresonance is necessary to analyze before designing the substation. In this paper, the Boushehr 230/400 kV GIS substation are implemented with detailed in EMTP-RV software, then possible operating scenarios simulated and possible overvoltages discussed. At the end, in worst scenarios from point of view overvoltages the effect of auto transformers hysteresis characteristic shape will be investigated.

Keywords: Ferroresonance, overvoltages, EMTP-RV, hysteresis characteristic, GIS substation.

1 Introduction

For investigating ferroresonance overvoltage in reality case should be considered the switching scenarios that are common in substations operation rules. For some reasons like maintenance or commissioning the equipments, some switching scenarios act to become out of service the failure equipment. These switching scenarios contain two steps, first shopping current at maintenance equipment and then the cutting voltage at failure equipment terminals. These two actions usually apply to power system with some time constants. Between these actions some overvoltages cause of resonance between system capacitors and transformer core nonlinear characteristics happened. In this time the only factor that limits the current of system is equipment's resistor. The simulation overvoltages contain two types of overvoltages, at the beginning of trace the transient overvoltages appear and after that the sustained overvoltages occur. Those scenarios that have the sustained peak value more that transient peak value are more critical because the overvoltages are continued for more times. However both two types of overvoltages are dangerous and should be controlled with protective equipments. According to the steady state condition, ferroresonant states can be classified into four different types as Fundamental mode, Sub harmonic mode, Quasi-periodic mode, Chaotic mode that are explained more at [1]. Boushehr GIS Substation has the unique configuration combined the GIS substation and cable system. At the 400 kV sides the one-half type of connection exists that is energized from 2 separated

X. Wan (Ed.): Electrical Power Systems and Computers, LNEE 99, pp. 471–478.
springerlink.com

overhead line with 24 Km length. Two exit lines from GIS substation are connected to Boushehr power plant generator transformers and other two lines connected to two 400/230kV auto transformers. Auto transformers with 230kV XLPE cables are connected to 12Km two circuit over head line that energize from Boushehr substation. At the 230kV side of substation is not any circuit breaker and all of the changing operational modes are handled with the circuit breakers that are in Boushehr substation. So, the overhead line capacitor and cables increase the equivalents capacitor that is challenge to nonlinear reactor of transformers. For investigating the ferroresonance study in this substation the nonlinear characteristic of autotransformer considered to analyze the various switching scenarios. In this way the parameters that influence on ferroresonance are modeled with detail.

2 System Modeling

Elements modeling in power system are main aspect of investigating the case study. The case study contains the effective elements in ferroresonance study like auto transformers, three windings transformers, GIS busbars, XLP cables, overhead lines, surge arresters, circuit breakers, Disconnectors, VTs, CTs. The system topology is shown at Figure 1.

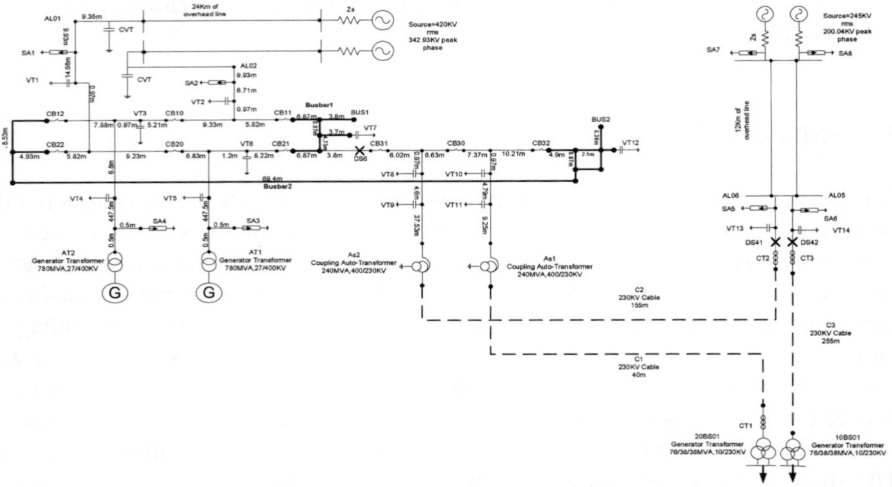

Fig. 1. Single line Diagram of Boushehr GIS substation

2.1 Hysteresis Modeling

The Hysteretic reactor is a nonlinear device designed to simulate saturation and hysteresis in the steel core of a power transformer. The theoretical background to this model is given in [2] which will be also referring to [3]-[9]. The Hysteretic reactor is modeled by a closed-form function that relates instantaneous flux to current in two steps. An intermediate flux, named unsaturated flux, is used to link these two steps. Two different equations are defined:

- A hysteresis function relating "unsaturated" flux λunsat to current. This function models the pure hysteresis effect: present state depends on previous state. Saturation is not taken into account: the more the current increases the more the intermediate flux increases.
- A saturation function relating instantaneous flux, named "saturated" flux λsat, to "unsaturated" flux λunsat.

This function is to model the saturation effect between input flux and output flux. Hysteresis and saturation functions are based on quadratic equations and represented by hyperbolic branches. Only some parts of these branches are taken to obtain the final saturation curve. The hysteresis modeling is to find a relation between the instantaneous flux and the current in the transformer. In this model saturation and hysteresis are decoupled. The saturation curve is represented by pieces of quadratic branches. It links the instantaneous flux, named saturated flux, and an intermediate flux, named unsaturated flux. In the same way pure hysteresis is modeled with a quadratic equation giving the unsaturated flux as a function of current as Eq. 1:

$$C_{hys} = \left[i - \frac{\lambda_{unsat}}{S_{hv}} - X_{hv}\right]\left[S_{hh}i - \lambda_{unsat} + Y_{hh}\right] \tag{1}$$

This equation is of the same shape as the saturation equation. The first two parts of this quadratic equation define 2 asymptotes. The vertical asymptote is defined by Eq. 2:

$$i - \frac{\lambda_{unsat}}{S_{hv}} - X_{hv} = 0 \tag{2}$$

Which is equivalent to:

$$\lambda_{unsat} = S_{hv}(i - X_{hv}) \tag{3}$$

It is a straight line with slope Shv and x-axis intercept Xhv. The horizontal asymptote is defined by Eq. 4:

$$S_{hh}i - \lambda_{unsat} + Y_{hh} = 0 \tag{4}$$

Which is equivalent to:

$$\lambda_{unsat} = S_{hh}i + Y_{hh} \tag{5}$$

It is a straight line with slope Shh and y-axis intercept Yhh. Chys is the curvature of the curve. The smaller is Chys the closer is the curve to its asymptotes. When the main shape of the hysteresis curve is defined, the convex branch will be chosen for upward trajectory and the concave branch for downward trajectory. But these two branches cannot define a closed shape. A translation of this hysteresis curve will give the width of the loop. To translate the curve it is needed to translate the asymptotes. The slopes of the asymptotes do not change, but the axis intercepts will change their sign. As a consequence the translated curve is being defined by equation (1).Figure 2 shows the initial curve (upward trajectory) and the translated curves (downward trajectory).

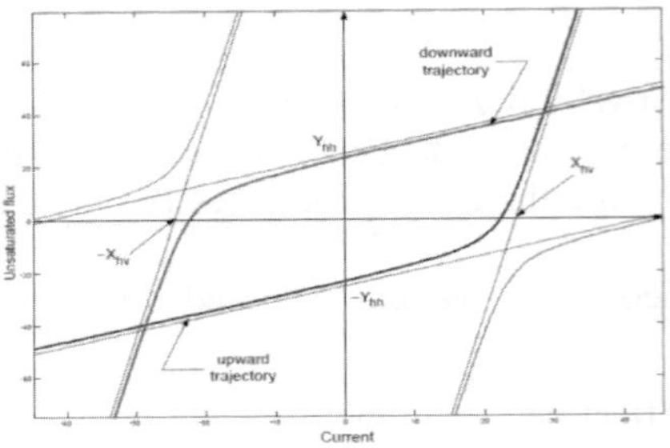

Fig. 2. Hysteresis function

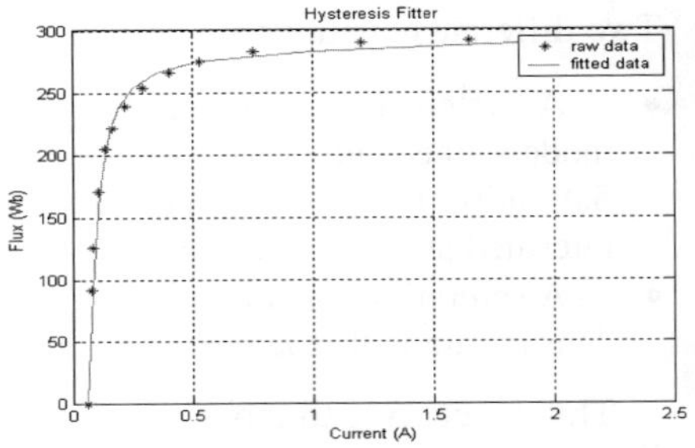

Fig. 3. Flux-Current characteristic

2.2 Autotransformer/ Three Windings Transformer Modeling

For modeling the various type of transformer the primary and secondary resistor and reactance, the nonlinear characteristic and the capacitor between HV-LV, HV-earth, and LV-earth should be considered (Fig.4). For fitting the measuring data at Flux-Current characteristic the least square method is used. The raw data and fitting data are shown on figure 3. With inputting the Fig.3 characteristic to hysteresis model the hysteresis characteristic parameters calculated on the base of equation (1-5) that are shown in table 1. Shv/ Ssv is slope of hysteresis/ saturation vertical asymptote, Shh / Ssh is slope of hysteresis/ saturation horizontal asymptote, Chyst / Csat is Curvature of the hysteresis/saturation curve.

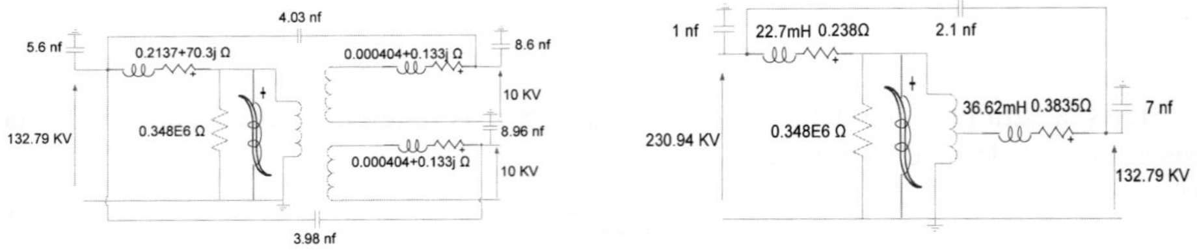

Fig. 4. Three windings transformer and Auto transformer 1-Phase model

Table 1. Hysteresis characteristic data

Parameters	Value	Parameters	Value
S_{hv}	6993.46	S_{hh}	48.29
C_{hyst}	0.24	Coer	0.06
S_{sv}	1	S_{sh}	0.0005
C_{sat}	38590.25	Y_{sh}	285.68

Ysh is Y-axis intercept of saturation horizontal asymptote and the end Coer is half width of major loop.

2.3 Cable and GIS Busbar Modeling

The PI modeled is used for modeling the XLPE cables and GIS busbar. The various types are examined like frequency depended (FD) and constant parameters (CP) model for modeling these elements. In this way, not only the results are more accuracy than the PI models but also the simulation time increase severely. It should be considered the reason of selecting this type, is low length of cables. The cable and GIS busbar lengths are shown on Figure 1.

Table 2. GIS and cable data

element	C(F/m)	L(H/m)	R(ohm/m)
GIS busbar	5.06E-11	2.43E-07	0.000231

Cable No.	L(H)	R(ohm)	C/2(F)
C1	1.07E-05	2.56E-01	2.48E-09
C2	2.93E-05	1.19E-01	1.32E-08
C3	6.85E-05	0.256	1.58E-08

2.4 Overhead Line and Switching Modeling

The 230kV overhead line is 12Km two circuit lines. For modeling this element the frequency depended model is used that responses appropriate to various range of frequency appear in system cause of the various ferroresonance modes. One of the main parameters that have the great affect on ferroresonance overvoltage is the switching angle. Previous studied believe that worst condition occur at voltage peak angle but in later studies this subject focused to analyze. In this paper, in order to verify the influence of switching angle, 50 shots of uniform distribution are used for the switch using EMTP-RV statistical approach. For the 50 operation study, the opening angle of the circuit breaker contact has normal distribution and its standard deviation is 3[ms], the open mean time is at voltage peak time. First, second and third switching mean time set at 10, 210 and 410 [ms] respectively. The circuit breaker model accommodates variations in the pole opening speed according to a normal distribution with a specified standard variation and limited ±3. For every circuit breaker the grading capacitor equal to 1400 pF is considered as modeled parallel to statistical switch.

3 Operational Scenarios Simulation

In this paper the ferroresonance regard with 230kV side operational scenarios is investigated. As it is said in introduction this scenarios contain to main step that act with circuit breaker and disconnectors. In this way, 19 different defined scenarios was simulated that here only the critical scenarios from point of view of overvoltage are listed at Table 3.

Table 3. Summaries the results of the 230&400kV ferroresonance studies

Study NO.	*CIRCUIT CONFIGURATION*	*node*	*Voltage (KV peak)*	*Current (A)*	*Ferroresonance Mode*
study 1	CB30 is switching when CB32&31 are N/O	AS2LV	221.2	220.5	Non-ferroresonant
		AS1LV	241.6	92.9	Subharmonic ferroresonance
study 2	CB30 and DS41 are switching when CB32&31 are N/O	AS2LV	221.3	220.5	Fundamental ferroresonance
		AS1LV	241.6	92.9	Subharmonic ferroresonance
study 3	AL05 is switching from Boushehr substation	10BSHV	220.4	104.2	Fundamental ferroresonance
study 4	AL06 is switching from Boushehr substation when CB31&32 are N/O	AS2LV	221.8	220.5	Fundamental ferroresonance
		AS1LV	241.7	101	Fundamental ferroresonance
study 5	AL06 is switching from Boushehr substation when CB21&10&32 are N/O	AS2LV	257.7	220.3	Fundamental ferroresonance
		AS1LV	256.2	132.5	Fundamental ferroresonance
study 6	AL06 from Boushehr substation and DS6 are switching when CB21&10&32 are N/O	AS2LV	257.7	220.3	Subharmonic ferroresonance
		AS1LV	256.1	132.5	Subharmonic ferroresonance

As it concludes, the worst condition occurs when the switching scenarios operate from Boushehr substation and some GIS busbars are in circuit configuration. In these scenarios the equivalent capacitor resonance with autotransformers nonlinear inductance contains the 12Km overhead line capacitor. For evaluating the affective parameters the scenario 5 is selected and various hysteresis parameters are changing to observing the effective gain to autotransformer terminals overvoltage. At figure 8 only 4 parameters that various range of them have the high effect to overvoltage value are reported. In contrast, Changing the various value of the Chys,Shh,Coer,Ssv have negligible effect on overvoltage.

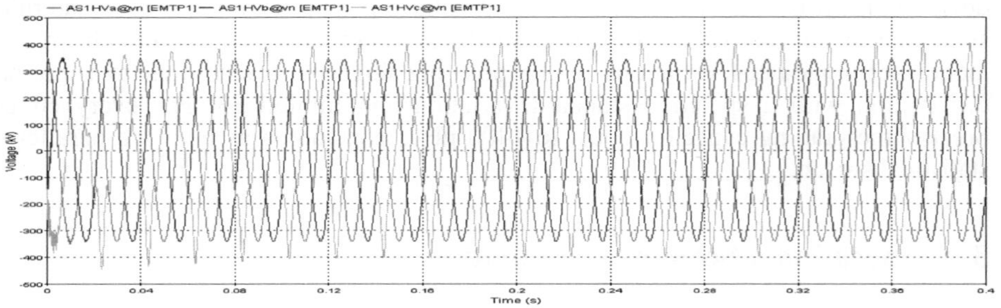

Fig. 5. Autotransformers terminals phases voltage for scenario 5

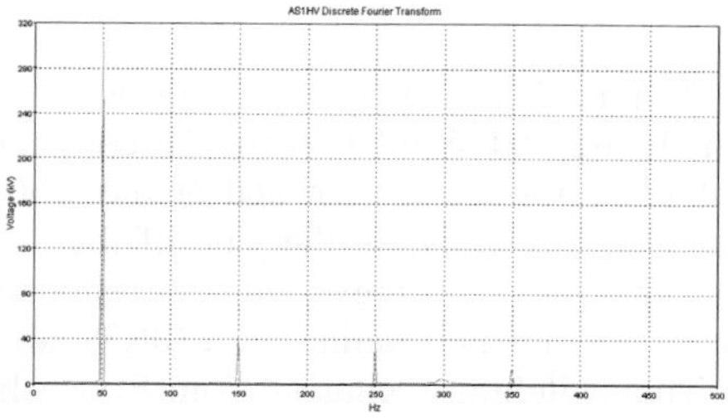

Fig. 6. Autotransformers voltage Harmonic Components for scenario 5

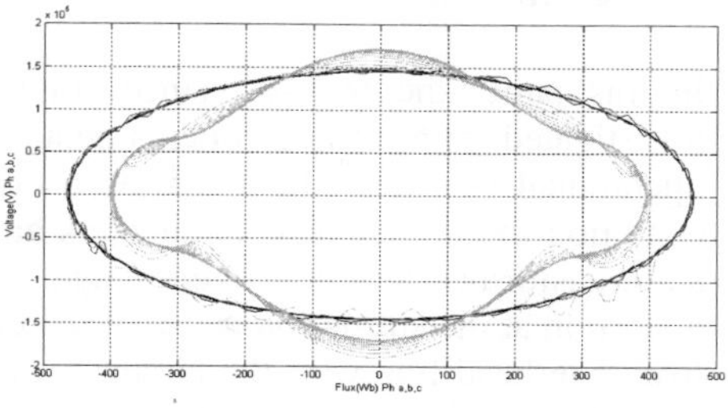

Fig. 7. Autotransformers iron Flux-Voltage Characteristic for scenario 5

With decreasing the slope of hysteresis vertical asymptote (Shv) the X-axis intercept of hysteresis characteristic at positive side of X-axis is increasing and so the area of hysteresis characteristic becomes more. This process is repeated when the Ssh is decreasing. This two parameters have the more influences on overvoltage between four reported parameters.

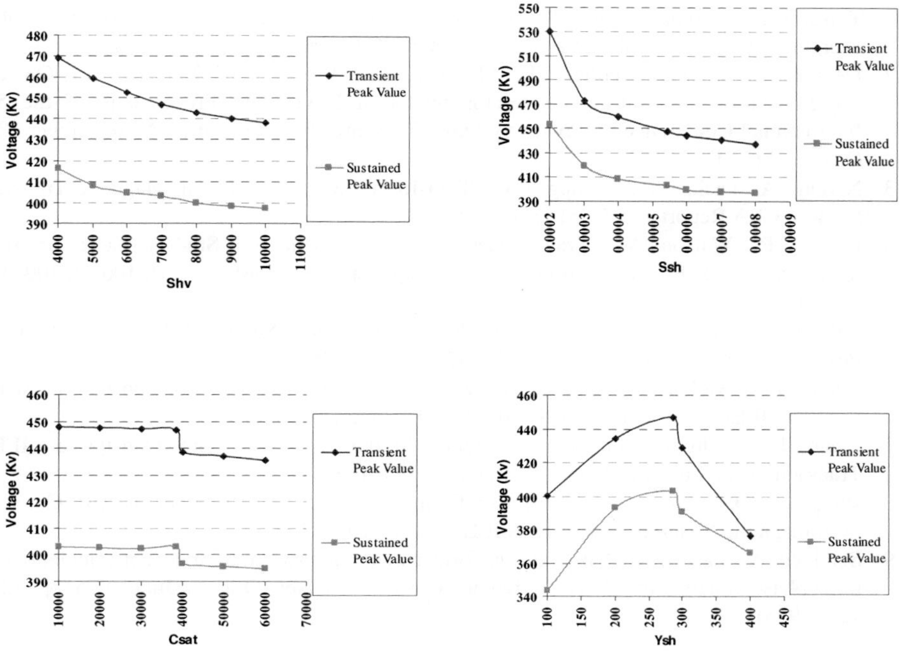

Fig. 8. Autotransformers terminal voltage for scenario 5 with changing the hysteresis parameters

4 Conclusion

In this paper the ferroresonance study at Iran nuclear energy GIS substation is investigated with all components detailed modeling and simulation the 19 common operational scenarios. In this way, 6 critical scenarios are known and at the worst condition the 8 hysteresis parameters are changed to evaluating the gain of them to autotransformer terminal overvoltage. First suggestion for preventing the element distortion at the substation cause of the critical scenarios is isolating the 230KV cable from overhead line with the circuit breaker. The switching scenarios from Boushehr substation with this circuit breaker are omitted and all of the critical operational scenarios can act without dienergize the 12 Km overhead line. Another suggestion is avoiding acting the critical switching scenarios and uses the 400 KV GIS switches to handle the common operational scenarios. With changing the hysteresis characteristic at scenario 5, changing four hysteresis parameters Shv,Ssh,Csat,Ysh have the more effect on overvoltage value and other parameters have the very low effect.

References

1. Valverde, V., Mazón, A.J., Zamora, I., Buigues, G.: Ferroresonance in Voltage Transformers: Analysis and Simulations. In: International conference on renewable energyies and power quality, Las Palmas de Gran Canaria, April 13-15 (2011)
2. Dennetière, S., Mahseredjian, J., Martinez, M., Rioual, M., Xémard, A.: On the implementation of a hysteretic reactor model in EMTP. In: Proceedings of the 5th International Conference on Power Systems Transients, New Orleans, LA, September 28 – October 2 (2003)
3. Narang, A., Dick, E.P., Cheung, R.C.: Transformer Model for ElectroMagnetic Transient Studies. CEA Report 175 T 331 G (December 1996)
4. Dick, E.P., Watson, W.: Transformer Models for Transient Studies Based on field Measurements. IEEE Transactions on Power Apparatus and Systems PAS 100(1), 409–419 (1981)
5. Talukdar, S.N., Bailey, J.R.: Hysteresis Model for System Studies. IEEE Transactions on Power Apparatus and systems PAS-95, 1429–1434 (1976)
6. Frame, J.G., Mohan, N., Liu, T.: Hysteresis Modeling in an Electromagnetic Transients Program. IEEE Transactions PAS-101(9), 3403–3412 (1982)
7. Ewart, D.N.: Digital Computer Simulation Model of a Steel-Core Transformer. IEEE Transactions on Power Delivery PWRD-1(3), 174–183 (1986)
8. Wright, A., Carneiro Jr., S.: Analysis of Circuits Containing Components with Cores of Ferromagnetic Material. Proceedings IEE 121(12), 1579–1581 (1974)
9. Semlyen, A., Castro, A.: A digital transformer model for switching transient calculations in three phase systems. In: PICA Conference Proceedings, New Orleans, June 1975, pp. 121–126 (1975)

A Research for Improved BUCK-BOOST Circuit

Jingying Shi, Huailing Wang, Xiaoxiao Peng, and Chunling Li

Electrical theory and new Technology Laboratory
Tianjin University, Tianjin, China, 300072
eesjy@163.com, hlwang0704@126.com, culiu2001@163.com,
lichunling2005@163.com

Abstract. Traditional buck-boost converter will generate switching losses when on and off, so the efficiency of the whole system is decrease. In this paper soft-switching technology is used in the buck-boost transform part, it can use soft-switching method by increasing the auxiliary switch and resonant circuit to reduce the switching loss and electromagnetic interference. This paper analyzes the principle of the main circuit buck-boost topology in detail and the conditions of soft switching operation, shows the circuit diagram for the state of the process and related waveforms. The simulation results show that the DC-DC converter achieves soft switching, reduces the switching loss and improves the efficiency of the circuit.

Keywords: soft-switching technology, switching losses, resonant circuit, DC-DC converter.

1 Introduction

DC-DC converter circuit (also known as chopper circuit) is connected to the DC power supply and the load, it becomes the uncontrollable DC input into a controllable DC output converter by controlling voltage. Traditional buck-boost Converter with simple structure, easy to implement, etc, has been widely used in various occasions[1]. But in recent years, with the switching frequency increases, the switching of power also becomes light and small, but the switching frequency and switching losses is proportional, so the switching frequency increases, switching loss also increases [2]. In order to improve conversion capacity, adjustable range and efficiency, the traditional buck-boost Converter has been improved. This paper presents a new soft-switching buck-boost converter, by adopting auxiliary switch and resonant circuit of the circuit structure to achieve the main, auxiliary switch soft switch, which not only reduces the switching loss, but also improves overall system efficiency.

2 The Buck-Boost Converter of Added Buffered Circuit

The main circuit switch of Buck-Boost converter is composed of switch, diodes, capacitors and inductors [3], as shown in figure 1.

X. Wan (Ed.): Electrical Power Systems and Computers, LNEE 99, pp. 479–485.
springerlink.com

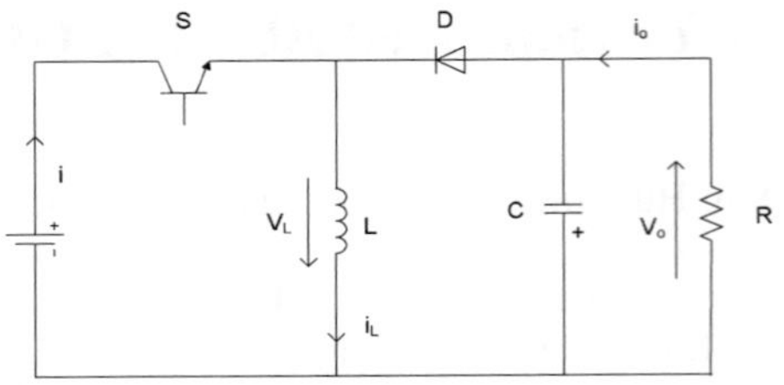

Fig. 1. The main circuit of buck-boost converter

The proposed Buck-Boost circuit adds a buffer circuit in the original basis, shown in Figure 2.

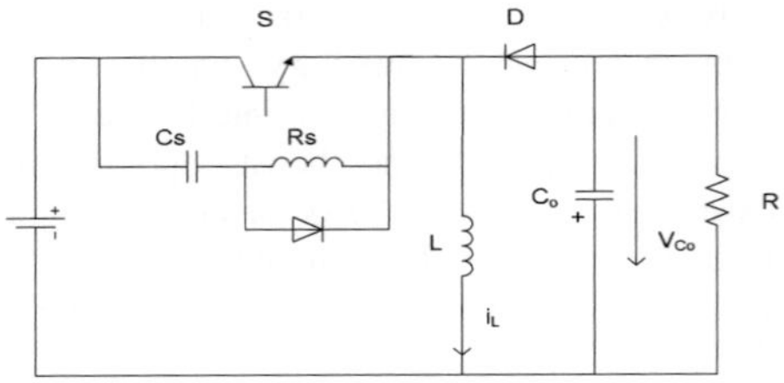

Fig. 2. The buck-boost converter of added buffer circuit

There are two ways of working of the converter. One is the current in continuous conduction mode (CCM), the other is the current discontinuous conduction mode (DCM)[4][5].Through the switch S to PWM control to regulate the output voltage. The converter is characterized by switch must be off in the inductor current reaching the maximum time. Therefore, when the switch is turned off, in order to reduce the pressure ,it should use a buffered capacitor and Buck-Boost circuit switch in parallel. But this has generated a problem, due to the increased resistance of the buffer capacitor and buffer[6][7], the Buck-Boost circuit is greatly reduced efficiency compared to previous.

3 Improved BUCK-BOOST Circuit

In order to reduce switch stress, while improving the overall efficiency of the circuit, in this chapter presents a new type of Buck-Boost circuit [8], shown in Figure 3. Improved Buck-Boost circuit is controlled mainly by the switch S_1 and S_2, the inductor Lr, Cr composition of buffered capacitor.

The working process of the improved circuit is the general with Buck-Boost circuit, by controlling the switch duty cycle to regulate the output voltage. A fixed frequency controls on and off of switches, making the switch of the improved circuit implement soft-switching by some resonant. When the switch S_1 and S_2 is suddenly opened, the circuit is the local occurrence of resonance, the inductor Lr constantly charge through the capacitor Cr, when the voltage across S_1 and S_2 is zero for the ZVS. As in the case of discontinuous inductor current weekly changes of the current is zero, so when the switch S_1, S_2, when suddenly closed, is always able to achieve ZCS. Switches S_1 and S_2 close simultaneously, through the local resonance circuit, the input voltage recharge to the buffered capacitor, although there are the voltage at this time, but are not power loss on buffered capacitor. This is a great improvement in the traditional Buck-Boost Converter.

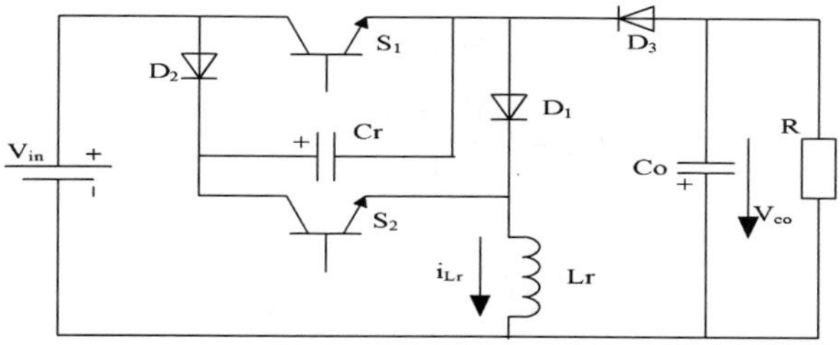

Fig. 3. Improved Buck-Boost Converter

Generally through the analysis of circuit structure can be see, improved Buck-Boost circuit achieves soft-switching with resonance part (the switch is closed when the ZCS, switching off when the ZVS), which makes switching power loss significantly reduce and the conversion efficiency of the entire circuit Greatly increase.

Circuit in a switching cycle the equivalent circuit is shown in Figure 4

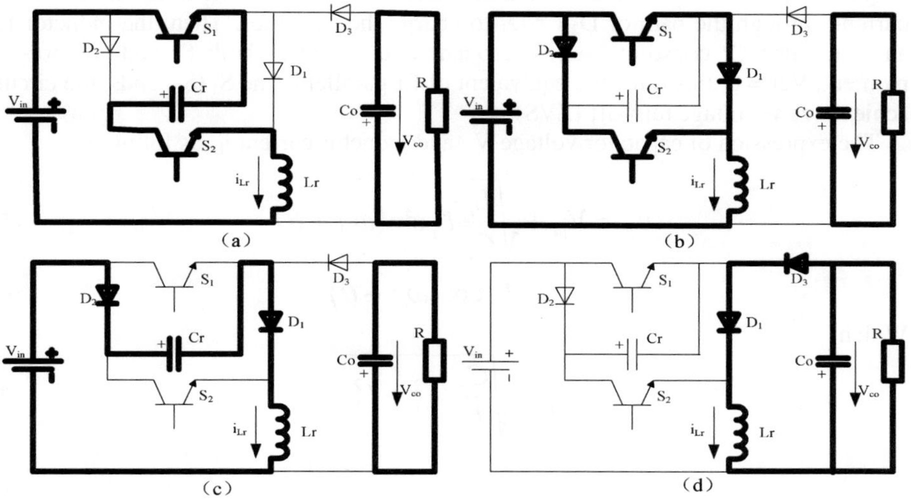

Fig. 4. Equivalent circuit of improved Buck-Boost converter in an switching cycle

In the initial conditions, the current i_{Lr} through Lr is zero, the main switch S_1 and S_2 is disconnected, the capacitor Cr is charged by the input voltage Vin and output voltage Vco, the value of both is Vcr: Vcr = Vin + Vco.

Stage 1(t0-t1): (Fig 4a) S_1 and S_2 close simultaneously, the input voltage Vin and the capacitor voltage V_{cr} superimpose on the role of the inductor Lr. Then, the capacitor Cr and inductor Lr constitute series resonant circuit, capacitor Cr discharges, while Lr constantly charges. At this time, the switch closes in the state of current to zero, so at this time is a ZCS circuit.

The expression of capacitor voltage V_{cr} and inductor current i_{Lr} as follows:

$$v_{cr} = \left(2V_{in} + V_{co}\right)\cos \omega_r t - V_{in} \tag{1}$$

$$i_{Lr} = \frac{2V_{in} + V_{co}}{X_r}\sin \omega_r t \tag{2}$$

Which

$$\omega_r = 1/\sqrt{L_r C_r}\ ,\ X_r = \sqrt{L_r / C_r} \tag{3}$$

Until Vcr reduces to zero, work 1 is completed.

Stage 2(t1-t2): (Fig 4b) Vcr is zero, diodes D_1 and D_2 conduct. Inductor current iLr not through the capacitor Cr, but by the two branches S_1-D_1 and D_2-S_2 shunt. Before the switch off, the inductor current iLr increases linearly:

$$i_{Lr2} = \frac{V_{in}}{L_r}t + I_1 \tag{4}$$

When S_1 and S_2 turn off at the same time, work 2 is the end.

Stage3(t2-t3): (Fig 4c) S_1 and S_2 both off enter another mode, the inductor Lr release current, through the path of D_2-Cr-D_1 to charge the capacitor. Then, the inductor Lr and capacitor Cr constitute series resonant circuit. When both S_1 and S_2 turn-off moment, Vcr = 0, this time, the equivalent of Cr parallel in the S_1 (S_2) ends, the circuit achieves zero-voltage turn-off (ZVS).

The expression of capacitor voltage V_{cr} and inductor current i_{Lr} as follows:

$$v_{cr} = V_{in} + \sqrt{\frac{L_r}{C_r}}I_a \sin\left(\omega_r t + \theta\right) \tag{5}$$

$$i_{Lr3} = I_a \cos\left(\omega_r t + \theta\right) \tag{6}$$

Which

$$I_a = \sqrt{\frac{C_r}{L_r}V_{in}^2 + I_2^2} \tag{7}$$

$$\theta = \sin^{-1}\left(-\frac{V_{in}}{\sqrt{V_{in}^2 + \frac{L_r}{C_r}I_2^2}}\right) \tag{8}$$

When the capacitor charge reaches Vin + Vco, the diode D_3 turns on, the work is completed.

Stage 4(t3-t4): (Fig 4d) With the diode D_3 turns on, inductor current flows into the load, the value i_{Lr} reduces linearly.

$$i_{Lr} = -\frac{V_{co}}{L_r} t + I_3 \tag{9}$$

Until the value decreases to zero, the entire switching cycle is over. With the S_1, S_2 close again at the same time, the next cycle begins.

4 Simulation Research

According to detailed analysis of the previous section of 4 working mode, in the simulation, taking the switching frequency is 40kHz, pulse width is 40%, the input voltage is 100V, the resonant inductor Lr is 100uH, resonant capacitor Cr is 50nF.

Figure 5 is the Matllab / Simulink model of Buck-Boost Converter with the addition of buffer circuit, Figure 6 is the Matllab / Simulink model of Improved Buck-Boost Circuit. Among them, in Figure 6, the buffer resistance R is 50 Ω, the buffer capacitor Cs is 0.47uF.

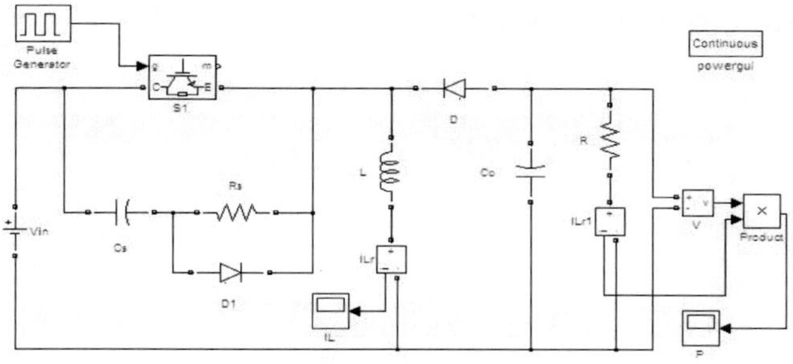

Fig. 5. The simulation module of Buck-Boost Converter with added buffer circuit

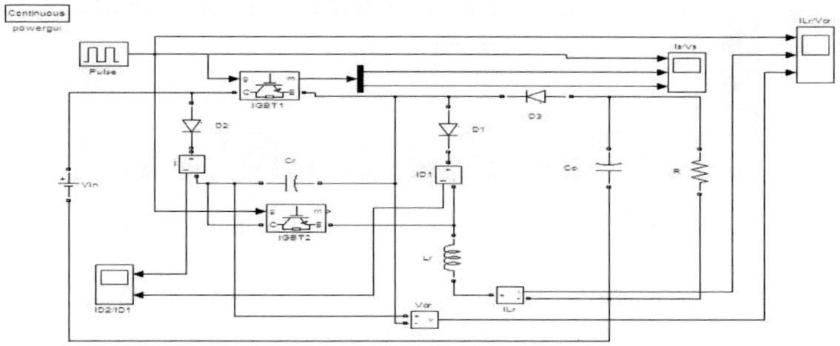

Fig. 6. The simulation module of improved Buck-Boost converter

Figure 7 and Figure 8 intercepte the waveforms of capacitor voltage V_{cr}, inductor current i_{Lr} and the current and voltage of switches S_1 and S_2 at two switching cycles. From this, we can see that inductor current works in the discontinuous mode, switches S_1 and S_2 can achieve the soft opening and soft turn-off.

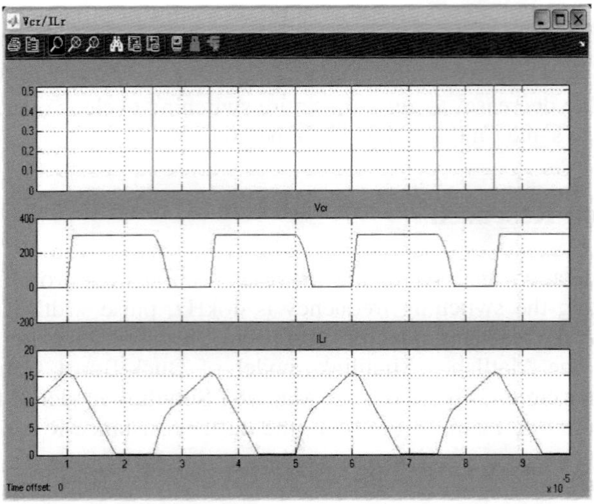

Fig. 7. The waveform analysis of capacitor voltage and inductor current

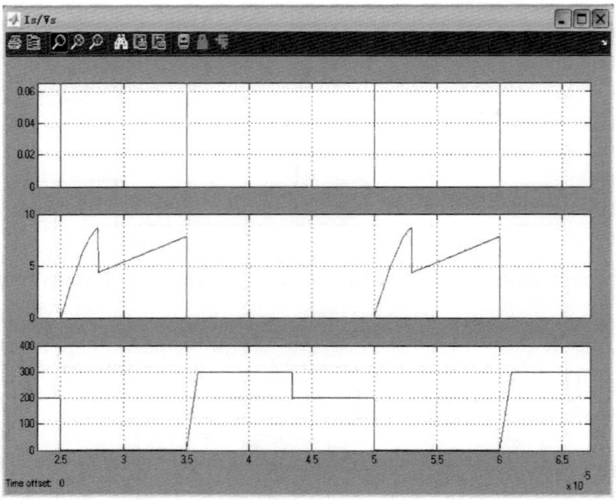

Fig. 8. The voltage and current analysis of switches S_1 and S_2

Figure 9 is the comparison of efficiency in the Buck-Boost circuit with buffered circuit and the proposed improved Buck-Boost circuit. It can be clearly seen that the conversion efficiency of the improved Buck-Boost circuit greatly improves than the former, thereby reducing the EMI and switching losses.

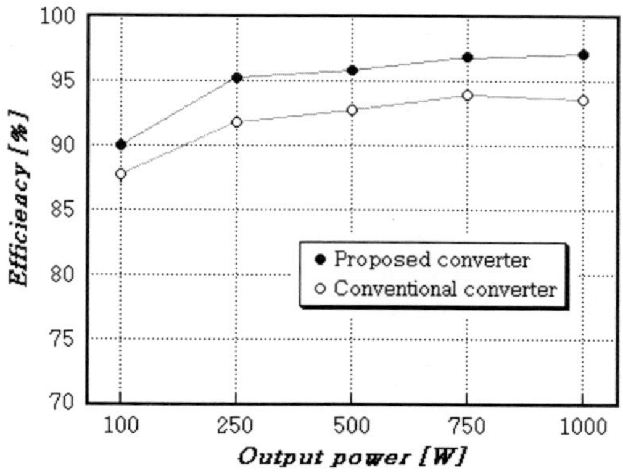

Fig. 9. The comparison of output power and output efficiency

5 Conclusion

This paper presents a new type of DC/DC converter, which is improved in the original Buck-Boost circuit. Theoretical analysis and experimental waveform analysis can show, the circuit can achieve higher conversion efficiency. The new converter applies the boost inductor and the lower loss snubber capacitors to achieve the resonance part of the circuit, reducing the loss of the resonance components and current and voltage stress. The final simulation results show that the power loss on the improved Buck-Boost converter is very low, compared to the traditional Buck-Boost circuit, the output efficiency of the system is greatly improved.

References

[1] Guo, W., Shen, Y., Yao, Z.: Integrated Broadband Voltage Controlled Oscillator design. Radio Engineering 35(5), 59–61 (2005)
[2] Luan, L.: Digital Intermediate Frequency sampling receiver Research and Design. Chinese Academy of Sciences (Space Science and Applied Research Center), Beijing (2006)
[3] Liu, Z., Deng, Y., He, X., et al.: Passive soft-switching three-level Buck / Boost circuit, Lanzhou University. Natural Science 42(1), 115–119 (2006)
[4] Wang, F.: Single-phase photovoltaic systems analysis and research. Doctoral thesis, Hefei University of Technology (2005)
[5] Papafotiou, G.A., Margaris, N.I.: Nonlinear Discrete-Time Analysis of the Fixed Frequency Switch-Mode DC-DC Converters Dynamics. IEEE Trans. on Circuit and System I 52(6), 322–326 (2005)
[6] Mazumder, S., Alfayyoummi, M., Nayfeh, A.H., Borojevic, D.: A theoretical and experimental investigation the nonlinear dynamics of DC-DC converters. In: Power Electronics Specialists' Conf., pp. 729–734 (2000)
[7] Shin, H.C., Lee, C.H.: Operation mode based high-level switching activity analysis for power estimation of digital circuits. IEICE Trans. Commum E90-B(7), 1826–1834 (2007)
[8] Zhao, G., Zhu, Z.: Buck-Boost PFC Soft Switch Circuit. Air Force Radar Academy 3(1), 60–61 (2003)

Design and Implementation of NCO in Broadband Zero-IF Digital Trunking System

Xingjing Zhou and Xiaoming Xie

College of Information Science and Technology,
Beijing University of Chemical Technology, Beijing, China
xingjingi21@163.com,
crownstar@163.com

Abstract. In this paper, we have studied and analyzed some key technologies of the NCO (numerical controlled oscillator) which based on the analysis and discussion of the NCO structure in digital trunking communication system. Currently, design of baseband digital transceiver is commonly used in some typical digital trunking system. But this way has the disadvantage of low degree of software and digital, as well as less flexibility, so it's hard to expand and upgrade the functions of the system. Broadband Zero-IF digital trunking base station can be a good solution to these problems and its structure is simpler and easier to maintain and also can reduce the system costs. In this article, the NCO is specifically designed for the broadband Zero-IF digital trunking system by using C language. And it fully adapts to the multi-carrier generation for the wideband IF with strong commonality.

Keywords: broadband Zero-IF digital trunking system, NCO, phase accumulation, LUT.

1 Introduction

Because of the current technology development of components, especially the high speed precision ADC and DAC devices and the large-scale programmable logic devices, digital trunking base station using multi-carrier access scheme based on broadband Zero-IF digital technology is more extensive application. And the NCO is an important part of the broadband Zero-IF digital trunking system. NCO also the main factors to determine the whole system performance, with its high resolution, fast frequency switching time and low phase noise.

In this paper, we introduce the structure of the broadband Zero-IF digital trunking system first. And then present the principle of digital implementation of NCO and describe how to use C language to program it. Using the MATLAB software, we design and implement the program, and apply it to engineering. The program can run on a general PC, for digital trunking stations based on broadband frequency digital technology, it can greatly reduce the cost of system.

X. Wan (Ed.): Electrical Power Systems and Computers, LNEE 99, pp. 487–494.
springerlink.com

2 The Basic Structure of Digital Trunking Base Station

The main idea of multi-carrier access solutions in broadband Zero-IF digital trunking base station is as follow. The structure showed in Figure 1.

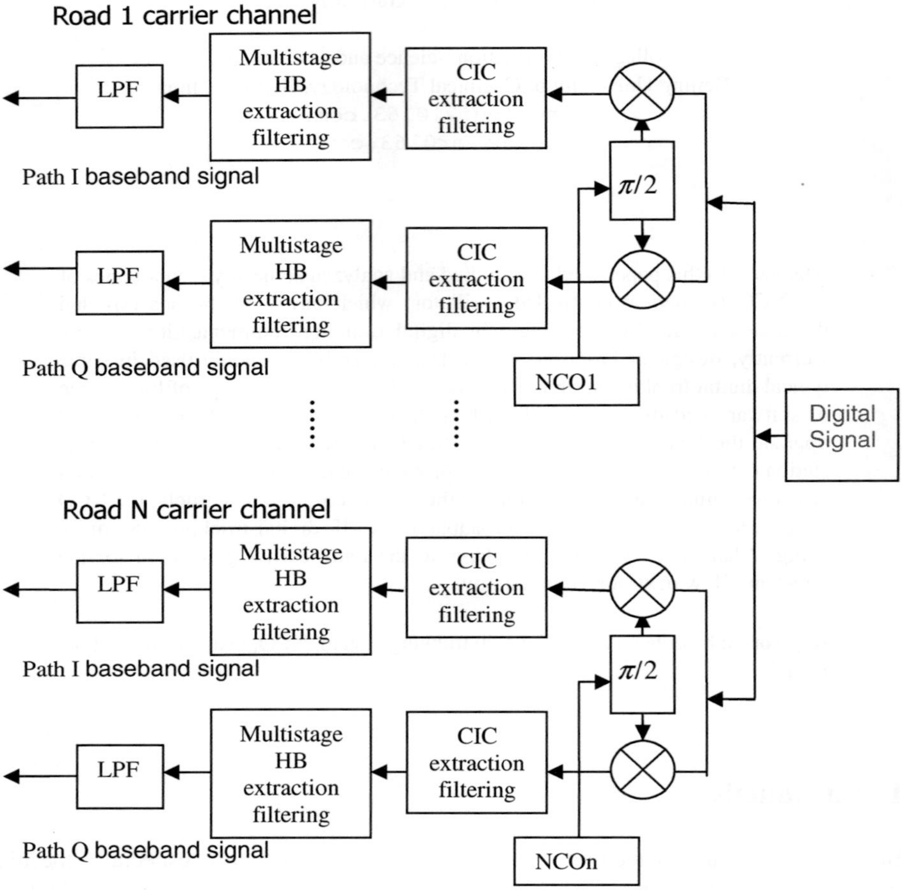

Fig. 1. Digital front-end processing algorithm of digital trunking stations at receiving channel

Firstly, move the entire frequency band of uplink signal to zero frequency position at the receiving end. Then use the broadband ADC to digitalize the whole spectrum. The subsequent signal processor and software extract the signal which in the IF processing bandwidth and at a specific carrier frequency, and complete the baseband processing work for the signal, such as synchronization, demodulation, decoding and judgments. The core structure is the front-end digital signal processing module.

Digital front-end processing algorithm at launch channel is similar as receiving channel, but plays opposite function. We can see from the figure that in the digital trunking system, NCO produces two orthogonal carrier signals which mixed by the multiplier. The operation speed and precision of NCO directly restricts the work performance of the whole system. And digital trunking system requires different carrier for different road carrier channel, which needs multiple NCOs. If each NCO is separately designed, it will waste resources. So design the NCO with general function will have remarkable significance.

3 NCO Design Principles Based on Lookup Table

The goal of NCO is to produce an ideal sine and cosine waves. In the case of high-speed signal sampling frequency in the software radio, it is impossible for NCO to real-time calculation. At this point, the easiest and the most effective way for NCO to produce sine wave, is to use Look-Up Table(LUT)[1], which pre-computed the sine value of the corresponding phase according to the sine wave phase, and used phase angle as the memory address to store the data in the sine of the phase. The structure of NCO is shown in Figure 2.

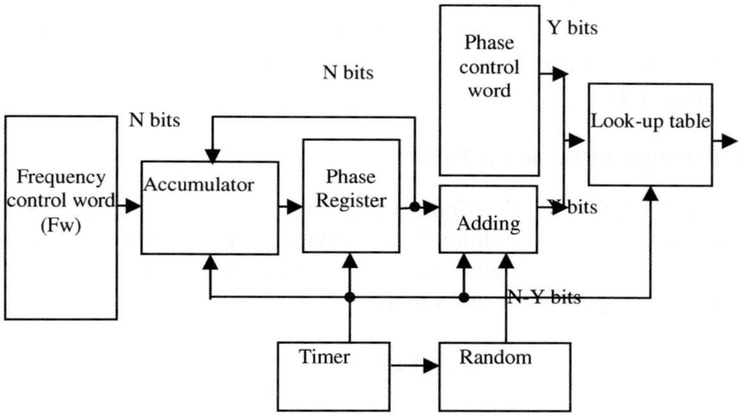

Fig. 2. NCO design principles

The Figure 2 shows that the NCO consists of three key parts: the phase accumulator, the phase adder and the sine table ROM. The role of the phase accumulator is to convert the sum of local frequency and offset frequency into phase. With each clock pulse, phase add a phase incremental on the original basis.

3.1 The Principle of Phase Accumulators

Phase accumulator accumulate the N bits Frequency Control Word (F_w), according to the reference clock frequency f_c . Output data form phase accumulator is the synthesis signal phase. Addressing the output we can get the sine or cosine of the NCO. Then the overflow output signal frequency (f_o) is the frequency of NCO. It has the following form:

$$f_o = (F_w * f_c)/2^N \tag{1}$$

When frequency control word sets 1, NCO output frequency is the lowest. So the frequency resolution of NCO is $f_c/2^N$. According to the Nyquist sampling theorem, the maximum output frequency can be achieved to $f_c/2$. However, due to the frequency stray and follow-up filter requirements, the maximum frequency is less than $f_c/2$. So the design was generated cosine frequency between $f_c/2^N$ and $0.4 f_c$.

Seeing from the above, a higher setting of phase accumulator bits would lead to a better frequency resolution. But increase the digit of accumulator will make excessive LUT data, and create resources burden. Considered both the high resolution and the resources requirements of the system, we intercept the high-Y-bit accumulator output to address the LTU. Phase interception will causes stray on output signal. The maximum stray amplitude caused by the phase interception can be expressed [2]:

$$P(dB) = 20\lg(2^{B-N}) = 6.02(B-N) \tag{2}$$

The B is the abandon digit. Spurious generated by phase interception can be suppressed by the phase jitter. In order to realize the phase jitter, we added a random number which between $(0,2^B)$ behind the accumulator and then used the intercepted value to address LTU.

3.2 The Principle of Look-Up Table

Any phase and amplitude in periodic waveform are one to one. If considering the phase as the address and the amplitude as the data, then LUT is suitable to implement this relationship. The way to generate LUT is that dividing phase π into 2N[3]. The correspondence between phase and sine is:

$$\phi = 2\pi * \frac{f_{LO}}{f_s} * n \tag{3}$$

As the actual value of the phase angle ϕ is generally not an integer, so it is very complex to take phase angle as a direct LUT address [4]. We get the transfer function to enlarge the phase:

$$F = \frac{f_{LO}}{f_s} * 2^N * n \tag{4}$$

MATLAB was used to generate 2N sine data of the corresponding phase, and then translate them into 16-bit fixed-point integer, store them in the array. The corresponding array index then used as addressing address. The size and program run-time were took into account, because we run the program on the base station system. In this paper, the corresponding relationship between sine and cosine waveforms (Figure 3) will simplify the LUT.

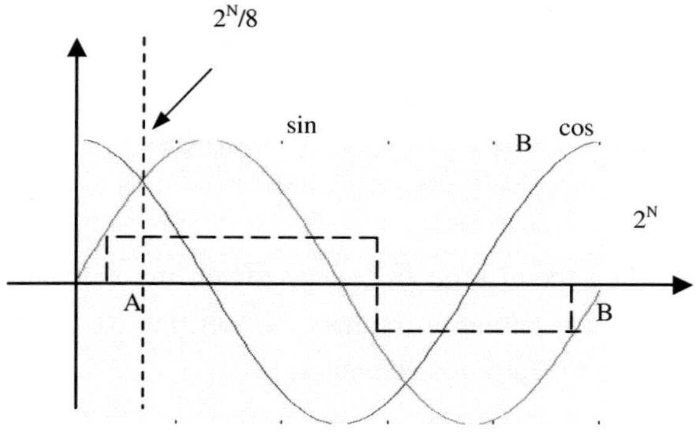

Fig. 3. Sine and Cosine Corresponding Relation Diagram

It can be seen from the Figure 3 that B point (fourth quadrant) value is the opposite number of the A point (first quadrant) value. The value of A and B have the following relationship: A+B=2N. Similarly, the cosine function also has such a corresponding relationship. Furthermore, in the first quadrant, cosine value and sine value are symmetrical between 2N / 8. So, LUT can only store the first quadrant sine value, and the other quadrants can be derived, with saving 7 / 8 space.

Since the formation of two orthogonal local oscillators is obtained by looking up the table, and the LUT values are obtained by MATLAB operators and stored in advance, so in theory, the two signals can be ensured orthogonal [5].

4 Software Realization of NCO

200MHz system clock was used to generate 75MHz local oscillator signals, and 24-bit phase accumulator was adopted, with high interception 12 bits addressing. The A / D sampling rate was 60MHz. After sampling, the input data bit was 14 bits, and the register for LUT was 16 bits. Finally, the output data bit was 15 bits.

The design used pure software implementation and without hardware-related problems. We took phase accumulation and addressing operation integrated in a function. So, we can directly addressing the cumulative output, which not only facilitate the realization of the program also reduced the time difference.

The sine or cosine value generated by the software would be multiplied with the input modulation signal, and then output to some different text files. Then the MATLAB was used to draw the relevant waveform and spectrum, and verify the related performance of the NCO software implement.

The key procedure in C language software is as follows:

```
nSum = (nSum + nFw) % MAXSUM;          // Accumulate
nSum_num = Niose(nSum);               // Add phase jitter
nSum_num = nSum_num >> sg_Bits;    // Take high 12 bits
nSum_num = (nSum_num + nPHASE) % MAXADDER;
                                 // Determine the initial phase
......
```

```
if (nLable= =2)                  // Determine the quadrant,
the third quadrant correlation processing
{
 nAddress = nSum_num - MAXADDER/ 2;
 gnSinNco [i] = -sgnREG_COS [nAddress];
 gnCosNco [i] = -sgnREG_SIN [nAddress];
}
```

If F_w=6291456, when there is no frequency offset, the software fixed produced 75MHz waveform. When receiving a frequency offset, the NCO added with the F_w, and then sent into the accumulator to accumulate.

5 Orthogonal NCO Operation Results

The following figures showed the information, which processed through the C language programming software, and then drawn through the MATLAB.

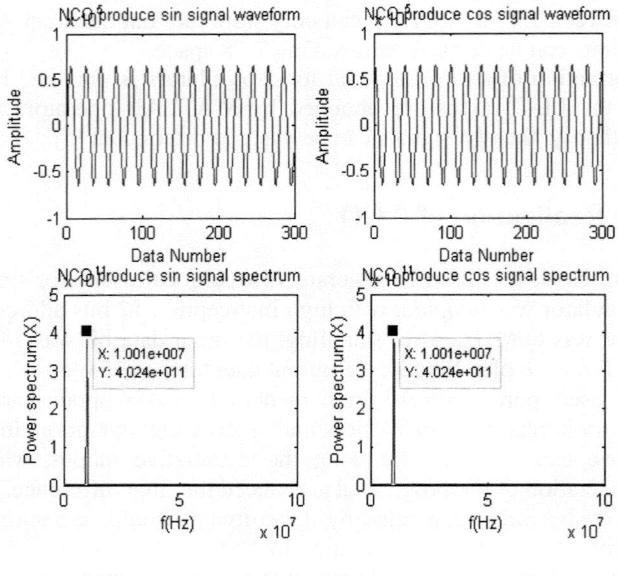

Fig. 4. The wave and spectrum diagram of sine and cosine which generated by NCO

Figure 4 shows that 10MHz NCO sine and cosine waveforms and frequency spectrum. It can be seen from the figure that the two signals have $\pi/2$ phase difference, which are totally orthogonal.

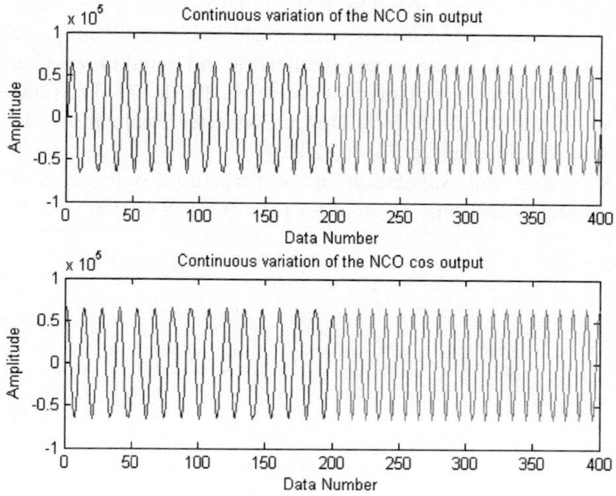

Fig. 5. Frequency change in sine and cosine wave which generated by NCO

Figure 5 shows that the NCO adjusted the frequency offset and phase offset through feedback data which returned by the frequency detector and follow-up PLL, resulting in sine and cosine waveforms with the same frequency and phase of modulated signal.

6 Conclusion

The digital design of wideband zero-IF in the digital trunking system has the advantage of high performance, flexibility and easy to modular design and the digital design of NCO introduces the feature of high precision, wide spectrum and good stability.

In this paper, we studied the design principles of NCO in detail and explained the methods of implementation of NCO. According to the work, some reference is expected in the development of digital trunking system.

References

1. Yan, J.-M., Li, J., Hu, C.-F.: Digital Down-Converter Research Based on orthogonal frequency mixing. Measuring and Controlling of Computer 17(1), 200–202 (2009)
2. Han, X.-T.: Digital Down-Converter Research Based on Software Radio. Guidance and control department (2008)
3. Shi, X.-S., Luz, H.-F.: NCO Design and Implementation in Software Radio System. The Modern Electronic Technology (15), 6–11 (2005)

4. Wu, Q.-Q.: Digital Down Converter Optimal Digital Channel Algorithm. Information and Communication Engineering in Northwestern Polytechnical University (2007)
5. Wang, F., Liu, Y.: NCO Application and Realization in Digital Down Converter. Zhongnan Forestry Science and Technology University Journals 29(3), 171–175 (2009)
6. Miao, P.-H., Yang, C., Li, Z.: Digital Down Converter Realization Method in Software Radio. Digital Communication (2), 74–76 (2010)
7. Wang, J.-J.: Modeling and Simulation of Numerically Controlled Oscillator based on Simulink. The Modern Electronic Technology (3), 102–108 (2010)

Research on Mobile Digital Health System Based on Internet of Things

Jingzhao Li, Xueqin Wu, and Hui Chen

School of Computer Science and Engineering, Anhui University of Science and Technology, Huainan 232001, China
jzhli@aust.edu.cn

Abstract. The application of the internet of things in the medical profession, the domestic and overseas research status of mobile digital medical system, the existing problems and key technology were analyzed, in view of the medical industry application characteristics, mobile digital medical system based on the internet of things was designed. This system mainly includes perception layer, network layer and application layer, the perception layer is composed of reader-writer of perception layer, mobile sensor networks, wearable sensor, and micro-actuator, network layer is consisting of all kinds of wire or wireless computer network, and the application layer is composed of service centre, medical personnel mobile terminal and patient mobile terminal. The system functions include obtaining human body multiple physiological parameters to medical service center accurately in real time micro invasive or non-invasively through miniature human wearable multi-parameter medical sensor network, developing interactive mobile medical services to realize common or serious illnesses characteristic parameter and remote diagnosis, launching cooperative medical service system with regional integration, and carrying out efficient service system to prevent and cure disease and building a healthy lifestyle system with valuation, encouragement and guarantee ,etc. It ensures the life of seamless care and overall health management.

Keywords: Internet of things, Mobile medical, Digital healthcare, Medical services, Mobile terminal.

1 Introduction

With the unceasing enhancement of social development and people's living level, people pay more attention to themselves health concern, which boosting medical care transition from "passive health" to "active prevention and monitoring ", medical industry also need the transition from " fee-oriented" to "patient-centered", this needs medical industry to take more advanced informationization means to improve management and enhance the medical technology [1].

In recent years, with the development of computer technology, sensing technology, wireless communication technology as well as radio frequency identification technology, especially the emergence of Internet of things technology, mobile digital medical systems have a certain degree of development. Internet of things technology

X. Wan (Ed.): Electrical Power Systems and Computers, LNEE 99, pp. 495–502.
springerlink.com

in the medical field have tremendous applied potential, it can help hospital to achieve the intelligent medical for people and the intelligent management for thing, support digital data acquisition, processing, storage ,transmission and sharing of the internal medical information for hospital, equipment information, drug information and personnel information and management information, realize material management visualization, medical information digitalize, medical process digitalize, treatment process scientific, service and communication humanize, can satisfy intelligent management and monitoring of the health information, medical equipment and supplies, public health and safety, etc, so as to solve the problems such as weak medical platform , the overall lower level medical service and hidden dangerous medical security, etc.

The mobile digital medical system based on the internet of things technology satisfy the health needs of people who pay attention to themselves, promote the development of medical and health information industry [2].

Mobile digital medical system can share original information system of hospital highly, and make the system more mobility and flexibility, thus achieve the purpose of improving the work efficiency of medical industry whole , mobile digital medical system can make medical process extend to each angle of the hospital , give brand-new change to medical staffs and patient doctor, can improve the quality and accuracy of medical treatment greatly, improve efficiency of the medical equipment and medicine management.

Mobile digital medical system has a huge potential market and broad application prospect, researching and developing the mobile digital medical systems has important practical significance, it can increase informationization level of the Chinese medical industry, improve medical personnel working efficiency and quality of work, improve the medical service level, improve the utilization ratio of medical information resource and improve the satisfaction of seeking treatment [3].

2 Domestic and Overseas Research Situation of Mobile Digital Medical System

Move is the key for applied internet of things technologies to solve medical informationization which need mobile computing and intelligent diagnosis. The most important object in medical industry is patient, around them is a doctor, nurse, medicines and equipment, put all the systems with patients into operation according to certain standard, which is named internet of things of digital medical strategy.

Mobile monitoring technology has many benefits, such as patients need not live at the hospital, it easy for them to get basic medical data at home after seeing a doctor, the information of patient is more comprehensive, activity is basically without restriction, and so on.

At present, most monitoring therapy instruments all need get through wired way for data transmission, a few wireless monitor are applied in clinical and daily health medical.

In May 2010, Dell put forward "mobile clinical calculation" solutions, and is helping China's health ministry enact pertinent standard, which in order to realize the sharing of standard electronic medical records, as well as the exchange of regional information [4].

Intel and Cisco have established professional mobile medical industry departments for research of mobile digital medical system. California VivoMetrics company have set up the LifeShirt subsidiary, the subsidiary is developing shirt with sensors, the sensor equipment of this shirt can monitor human respiratory function, people who with sleep apnea syndrome put on it will not have to go to the sleep clinic in hospital, sleep at home may diagnose.

Philips and Ericsson Company conduct research and development for various wearable monitoring systems. At present, a project that Philips research is MyHeart, also is developing clothes that can monitor heart, MobiHealth Ericsson developed, are testing, this system depend on PDA, transmits the collected data through wireless phone network to the doctor, for the doctor's diagnosis, consultation, treatment and monitoring, MobilHealth transmits the data getting from sensor to doctor's diagnosis room by connecting Bluetooth technology and sensor placed in humans [5].

Nortel networks (Nortel) also have launched an "end-to-end converged network" mobile digital medical system solution, providing collaborative applications of combining voice, data and video multimedia for medical diagnosis and clinical services [6].

3 Problems Existed in the Mobile Digital Medical System

In the aspects of engineering research and application of mobile digital medical systems, which has made some achievements, accumulated the massive precious experience, but set obstacles to mobile medical to a certain extent for diversity and implementation complexity of mobile applications involving technology .In addition, the realization of mobile medical have something to do with the existing informationization level of hospital, only information system of hospital have developed to a certain degree, can mobile applications implement to better. The mobile applications exists a second development of practical needs in the medical industry which is a complex and special field. Different hospitals, different users need to adapt to request that their respective work environment and mobile working, treatment or nursing of the task, according to different business flow and different groups the development of application software is more difficult than general software to a certain extent.

Besides, there still exist the following questions.

(1) The research of domestic digital mobile medical system start late, the research is seriously undercapitalized, the technical level is low, the transformation of the achievements is little, and technology application fails to get sufficient development.

(2) Researchers who engaged in mobile digital medical system development are lack, scientific research strength is weak, the knowledge innovation ability is insufficient, competitiveness is lacking.

(3) The engineering technology of domestic mobile digital medical system is not mature enough; it is necessary to further increase the strength of basic research and applied basic research.

(4) The low information level of domestic medical institutions and inadequate informatization construction fund investment, cause the further difficulty of mobile digital medical system industrialization.

(5) Countries have not a unified development plan on engineering technology of mobile digital medical systems, making basic research, applied research; market developing cultivation and construction of the mobile digital medical system are difficult to deserve the safeguard.

4 The Main Function and Key Technology of the Mobile Digital Medical System

4.1 The Main Function of the Mobile Digital Medical System

(1) The systems obtain human body multiple physiological parameters accurately in real time micro invasive or non-invasively through miniature, intelligent and digital human wearable multi-parameter medical sensor network , transmit physiological parameters data to medical service center safely, reliably and quickly, and develop different types of interactive mobile medical services anytime, anywhere.

(2) The systems have the function that realizing common or major disease characteristic parameters and remote diagnosis eventually through the internet of things technologies and human wearable multi-parameter medical sensor.

(3) The systems have cooperative medical service system with regional integration, high-quality and high-efficiency service system to prevent and cure disease, a healthy lifestyle system with valuation, encouragement and guarantee. It ensures the life of seamless care and overall health management.

(4) The digital medical system service center with perfect function and advanced technology may simultaneously monitor e.g., breathing, blood pressure, blood oxygen, pulse, temperature and physiological parameters of each service object and call the instructions of displaying comprehensive waveform signal, digital signals, and the rich image. It finish Comprehensive professional electrophysiology analysis, such as dynamic analysis, ambulatory blood pressure analysis, the dynamic breathing analysis, the dynamic the blood oxygen analysis, even sleep analysis and so on.

(5) The systems based on the medical staff and patients digital mobile terminal of embedded microcontroller processor, displaying the detected physiological parameters under the condition that daily life are not affected in the real-time, realizing the dialogue between patients and receipt of doctor's advice, and so on. The systems embed professional analysis software such as holter, ambulatory blood pressure, and sleep apnea etc, issuing detailed test reports.

(6) The systems support realizing seamless docking in the information of township, community hospital with central hospital, obtaining the expert proposal in real-time, arranging referral and accepting training, and so on.

(7) The reliable and effective information storage and testing method realize patient identification rapidly, determine the name, age, blood type, emergency contact phone number, Past Medical History, family members and other relevant details and complete Hospital registration formalities so as to obtaining precious treatment time for emergency patient.

4.2 The Critical Technology of Mobile Digital Medical System

The critical technology of mobile digital medical system is:

(1) wireless computing techniques, including mobile computing and EDA (Enterprise) technology, the mobile computing technology is using intelligent computing terminals equipment to solve different network seamless access in wireless environment, realizing mobile computing ,data transmission and sharing of resources, and providing the accurate information to any users at any time, in any place;
(2) The middleware technology includes CIS, LIS, PACS and MIS etc database services in medical service centre of hospital. In order to guarantee modularization, compatibility and expansibility of the hospital information system, use the middleware technology to shield the diversity of hardware platform and heterogeneity among operating system ,network protocol and each interface of systems ,then make application software can run on different platforms quite smoothly;
(3) The embedded software and hardware design technology, some sensory device and mobile terminal all need embedded microcontroller processor to realize them;
(4)The miniature sensor technologies need to be embedded into the monitoring personnel clothing;
(5) The wireless sensor network technology, such as Zigbee, Bluetooth, WIFI, RFID, etc.

In addition, there still include object identification, architecture, communications and networking, security and privacy, the service discovery and search, energy acquisition and storage, etc.

In medical health domain, the main application technology of the internet of things is the three aspects such as visualization networking of medical management, digitalization medical information and mobilization medical process.

5 The Architecture Design of Mobile Digital Medical System Based on the IOT

The key link of the internet of things can be summarized as comprehensive perception, reliable transport and intelligent processing. Comprehensive perception refers to the collection and acquisition of information on the people and objects anytime, anywhere by using radio frequency identification, camera, sensors, sensor network etc. Reliable transport refers to conducting interaction and sharing of reliable information through various communication networks and Internet anytime .Intelligent processing refers to analyze and deal with a vast amount of data and information to realize intelligent decision-making and control.

The mobile digital medical systems based on the internet of things include: reader-writer of Perception layer, mobile sensor networks, wearable sensor and perception layer is composed of micro actuators; Network layer consisting of all kinds of wire or wireless computer network; Application layer three-layer structure is composed of service centre, medical personnel mobile terminal and patient mobile terminal.

The component of the mobile digital medical system based on the internet of things is shown in Figure 1.

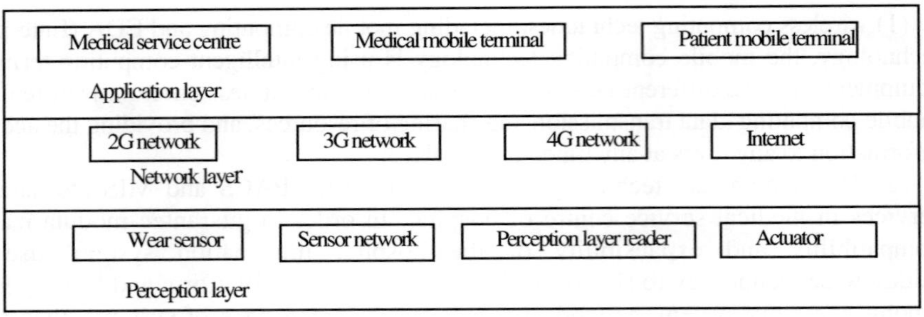

Fig. 1. The component of the mobile digital medical system based on IOT

The perception layers of the internet of things are generally including RFID tags and reader, camera, GPS, sensors, terminals, sensor network, is mainly to identify the object and gather information. The perception layers of mobile digital medical system based on the internet of things are mainly composed of medical sensors, actuators MCU, wireless communication device, data storage and output equipment parts and so on. All kinds of wear sensor including which can be composed of wear sensor communication system, seamless embedded into personal daily equipment, complete functions such as the acquisition work of medical health parameters and execution of various actuators output. The sensors which attached to the body identify the wearer's biometrics and situational state continuously, give patient the instructions that using wearable and implanted medical sensor used for distributed mobile monitoring according to the patient's current situation, ring sensors monitor the heart rate continuously and transfer the data wirelessly by using signal. Wearable sensors and actuators need miniaturization and low power consumption, the collected medical data mainly include blood pressure, sensor, blood sugar, electrocardiogram, body temperature, weight, waistline, the blood oxygen, electroencephalograph and other parameters. The information monitored by wearable monitoring are transmitted to the reader-writer of Perception layer through wireless sensor network, and are transferred to the network layer of the internet of things through reader-writer of Perception layer.

The network layer of mobile digital medical system based on the internet of things includes all kinds of wire or wireless medium-range and telecommunication networks, such as the INTERNET, 2gb GPRS, GSM, WDMA of G, CDMA2000, TD - CDMA, WIMAX and other networks, transfer the information getting from perception layers to the medical service center for intelligent processing through the fusion network of communication and internet.

The application layer of mobile digital medical system based on the interment of things was the depth fusion of the internet of things and medical professional technology; realize the intellectualized management in medical industry. The service center of mobile digital medical system based on the internet of things conduct document management and analysis showed. Personal information management mainly aimed at personal basic condition and data item need to monitor, etc.

Analysis showed module help medical workers make quick and effective judgment on data of the monitored object. This mainly provide doctors and nurses to use, can satisfy the doctor diagnosed and nurses care needs. Using it can check body features information, including the patient's temperature, pulse, respiration, blood pressure, discrepancy quantity, and weight information; Doctors can input content of orders, choose frequency using orders, input drugs specification and so on, also may disable orders or cancel orders. The doctor can look up the patient's inspection, the detailed information of inspection result, abnormal results will with red, Doctors and nurses can view patient records by using medical personnel mobile terminal, including the first of cases, duration of record, etc; collect every index such as the patient's temperature, pulse, respiration, blood pressure, discrepancy quantity, and sane information in the real time ;automatic dynamic calculate the time point that patients need to measure when combined with the provisions of the hospital, Look at the patient's checking application situation, inspection and test results, conduct assessment and health education on the patient, etc.

Patient handheld mobile terminal can be real-time check physiological monitoring situation and relative data of themselves, and have dialogue with medical staff, accept wills and Suggestions, etc, its file management realize to monitor data of the local management, facilitate in data storage when network impeded.

The component of the mobile digital medical system based on the internet of things is shown in Figure 2.

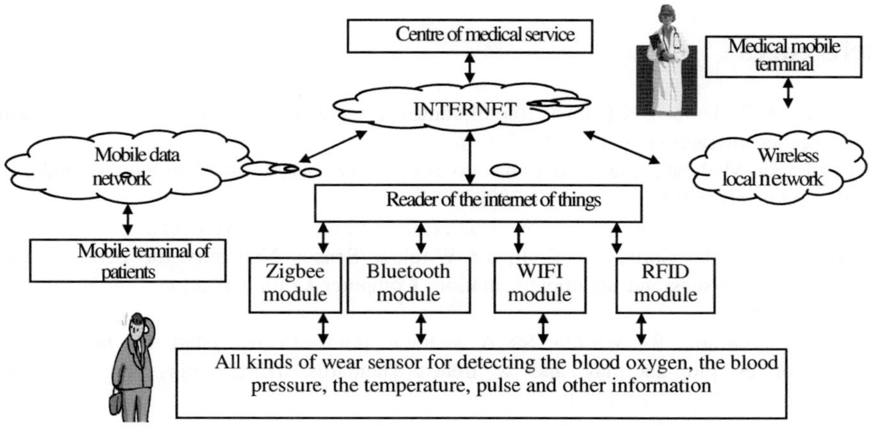

Fig. 2. Structure of mobile digital medical system based on IOT

Through the above analysis, we can conclude that the mobile digital medical system based on the internet of things have integrated advanced data technology, wireless network technology, mobile computing platforms, wireless infrastructure, mobile software and mobile services, providing a combination of remote mobile sensor integrating voice, data and video multimedia collaborative applications for medical diagnosis and clinical services .Through the wear sensor to detect blood oxygen, e.g., blood pressure, body temperature, pulse and other information. Through the sensor network: Zigbee, WIFI, Bluetooth, RFID LAN etc to transmit information.

Application solutions of Mobile digital medical system can provide wireless detection on patients' condition, mobile nursing, wireless voice, network voice, conference call and video monitoring. So to speak, patients in the hospital experienced all process, from the remote registration, treatment, can use mobile technology for optimization. It will improve the working efficiency and core competitive power of the medical institutions.

6 Conclusions

Mobile digital medical system provides a new medical model; make people accept personalization medical services anytime, anywhere. New mobile digital medical system must be able to realize the seamless links with the original information system can make full use of the existing medical information system, reduce the cost of informatization construction to the greatest degree. It will be realized detection and treatment for patients without hospitalized, through wireless network, equipped with mobile terminals, wireless radio frequency identification devices (RFID) to conduct the original system integration. With a lot of embedded and mobile information tools, there will appear intelligent hospital and intelligent doctor. Therefore mobile digital medical systems which realize the sharing of medical information are development tendency of medical information system.

References

1. Kirisci, P.T., Thoben, K.D.: The role of context for specifying wearable computers. Human Computer Interaction 5(61), 7–11 (2008)
2. Pendragon Medical presents first non -invasive "sensor wrist watch" to monitor and predict glucose levels (EB/OL) (April 19, 2008), http://www.diabetesinrontrol.com/issue179/np.shtml
3. Aleksandar, M., Chris, O., Emil, J.: Wireless Sensor Networks for Personal Health Monitoring. Issues and an Implementation. Computer Communications 29(13214), 2521–2533 (2006)
4. Yao, J., Schmitz, R., Warren, S.: A wearable point-of-care system for home use that incorporates plug-and-play and w ireless standards. IEEE Trans. Inf. Technology Biomes 9(3), 363–371 (2005)
5. Zhang, Y.T., Poon, C.C.Y., Chan, C.H., et al.: A health-shirt using e-textile materials for the continuous and cuff less monitoring of arterial blood pressure. In: 3rd IEEE/EMBS International Summer School on Medical Devices and Biosensors, Cambridge, MA (2006)
6. Texas Instruments, Incorporated. MSP430x20x1 MSP430 ×20 ×2MSP430×20×3 Mixed Signal Microcontroller(EB/OL) (August 3, 2009), http://www.ti.com

A Back Propagation Neural Network Sliding Mode Controller for Ship Course Nonlinear System*

Xiao Hai-rong[1,2], Li Yi-bin[1], Zhou Feng-yu[1], and Han Yaozhen[2]

[1] School of Control Science and Engineering, Shandong University, Jinan 250061, China
[2] Department of Information Engineering, Shandong Jiaotong University, Jinan 250357, China
Hairong.xiao@163.com, liyibin@163.com, zhoufy@163.com, hanyaozhen@yahoo.cn

Abstract. A back propagation network sliding mode variable structure control method is proposed for ship course nonlinear control. Neural network is used to simulate the functional relation between state hyperplane of the system and the exponential reaching law. A hyperbolic tangent function is applied to replace the saturation function to realize the boundary method of sliding mode control. Chattering is greatly reduced and simulation results demonstrate the presented control method has good adaptability and high robustness.

Keywords: Ship course nonlinear, sliding mode control, back propagation network, reaching law.

1 Introduction

Since 1920s and 1970s, ship autopilot based on PID algorithm and adaptive control theory has been worked out separately[1]. Unfortunately, due to ship's big inertia, nonlinearity, parameters perturbations and random external disturbances, there are severe uncertainties in ship course control. This makes it difficult for adaptive control method to control ship course perfectly[2]. Afterward, various new control algorithms[3], such as sliding mode control, H∞ control, forecast control, feedback linearization control, back-stepping control, neural network control, fuzzy control, etc, are subsequently adopted in ship course control. Although some progresses are made in these methods, many disadvantages are also exposed. For example, chatting exits in sliding mode control; high precision model is needed in H∞ control, generalized forecast control, feedback linearization and back-stepping algorithm. And generalization ability is needed further research and the problem of completeness still exists in neural network control. In fuzzy control, it is usually difficult to determine and optimize the control rules. In order to resolve these problem and satisfy the demand of ship course control, intelligent control technology is used in autopilot[4], such as combining PID and neural network, fuzzy control and neural network, fuzzy and genetic algorithm, sliding mode and fuzzy, neural network and genetic algorithm

* This work is supported by Shandong Provincial Natural Science Foundation, China under Grant # ZR2009FL013 to Xiao Hairong and "Ship Safety and Low-Carbon Intelligent Control Technology Team" Foundation of Shandong Jiaotong University.

X. Wan (Ed.): Electrical Power Systems and Computers, LNEE 99, pp. 503–510.
springerlink.com

etc. This paper is to study a back propagation(BP) neural network sliding mode variable structure controller used in ship course control.

Sliding mode control is a nonlinear control and the fundamental difference with common control is the discontinuity[5]. It forces the system state variables tracking the setting phase trajectory to desired points by using a special sliding mode control method. Because the given phase trajectory is independent with the change of controlled object and external disturbance, system has better robustness when moving on the sliding surface. Furthermore, the algorithm and application of sliding mode control is rather easy, so it offers good solution for complex industry control. In china, professor YANG Yan-sheng and SONG Li-zhong [6,7]both researched ship course sliding mode control. In fact, there exits time delay inevitably when control variable is switching. Chattering becomes more violent if control variable amplitude is bigger. So it is necessary to study new method for weakening or removing chattering.

A BP neural network sliding mode controller is designed in this paper. BP network is used to adjust switching gain of sliding mode control and saturation nonlinear section is replaced by hyperbolic tangent function. Simulation is performed on ship course nonlinear system and the results demonstrate that the chattering is greatly reduced.

2 Ship Motion Nonlinear Mathematical Model

In 1957, Professor Nomoto proposed K&T parameters to represent ship maneuverability based on maneuverability linear equation and control engineering. The ship steering linear mathematical model is established[8]:

$$\frac{\psi(s)}{\delta(s)} = \frac{K(1+T_3 s)}{s(1+T_1 s)(1+T_2 s)} \tag{1}$$

Where, T1, T2, T3 are maneuverability index of two order Nomoto mathematical model, K denotes turning ability index, then the ship model can be simplified in low frequency section:

$$\begin{aligned} T\ddot{\psi} + \dot{\psi} &= K\delta \\ T &= T_1 + T_2 - T_3 \end{aligned} \tag{2}$$

In which, ψ is course angle, δ denotes rudder angle, T, K are maneuverability index, K>0.

Ship has big inertia and its dynamic characteristic is only important in lower frequency section, so autopilot controller is often designed based on one order or two orders Nomoto model. But Nomoto model is linear and only applied in maneuvering motion when ship speed is constant. When changing course, the ship is a serious nonlinear system if big rudder angle is involved. In order to improve model precision, nonlinear term $H(\dot{\psi})$ is used to replace $\dot{\psi}$ to describe ship's nonlinear characteristic.

$$H(\dot{\psi}) = n_3\dot{\psi}^3 + n_2\dot{\psi}^2 + n_1\dot{\psi} + n_0 \tag{3}$$

In formula(3), for symmetrical ship, $n_2 = n_0 \approx 0$, so,

$$H(\dot{\psi}) = n_3\dot{\psi}^3 + n_1\dot{\psi} \tag{4}$$

For stable ship, n_1=1, and n_1=-1if ship is unstable, n_3 can be get from turning test. Take formula(4) into formula(2)

$$T\ddot{\psi} + n_3\dot{\psi}^3 + n_1\dot{\psi} = K\delta \tag{5}$$

The influence of wind, wave, and current to ship course is rather complicated. It is impossible and unnecessary to establish mathematical model exactly. Usually, equivalent rudder angle can be used to replace the wind, wave, current and other disturbance to ship course. The equivalent disturbance mainly includes direct current disturbance and periodic disturbance[9]. Direct current disturbance is equivalent to constant rudder angle acting on ship and make the ship move to a direction in constant speed. Periodic disturbance is equivalent to periodic rudder angle acting on ship and make the ship left and right periodically. Periodic disturbance is usually approximately equivalent by a series of sine wave. The equivalent rudder of wind, wave and current can be simply described as:

$$\delta_d = \sum_{i=1}^{m} \delta_i \sin \omega_i t + \delta_e \tag{6}$$

Where, δ_d is the equivalent disturbance rudder angle, δ_i , ω_i are respectively amplitude and angular frequency of periodic disturbance rudder angle. δ_e is constant disturbance rudder angle.

Considering the uncertainty of outside disturbance when ship is sailing at sea, the ship mathematical model with uncertain term is:

$$T\ddot{\psi} + \dot{\psi} + \alpha\dot{\psi}^3 = K(\delta + \delta_d) + \Delta \tag{7}$$

Δ is the uncertainty of outside disturbance and supposed bounded, $|\Delta| < J$, J is unknown positive number.

To chose state variables as $x_1 = \psi$, $x_2 = \dot{\psi}$,control variable is $u = \delta$,ship course nonlinear system mathematical model considering disturbance of wind, wave, current and uncertainty is:

$$\begin{aligned} \dot{x} &= f(x) + g(x)u + d(t) \\ &= \begin{bmatrix} x_2 \\ -\frac{\alpha}{T}x^3{}_2 - \frac{1}{T}x_2 \end{bmatrix} + \begin{bmatrix} 0 \\ \frac{K}{T} \end{bmatrix} u + \begin{bmatrix} 0 \\ \frac{K\delta_d + \Delta}{T} \end{bmatrix} \end{aligned} \tag{8}$$

3 Design of Fixed Gain Variable Structure Controller

To the ship course nonlinear system described in formula (8), choosing the switching surface:

$$s(x) = Cx = c_1 x_1 + x_2 \tag{9}$$

Where, choosing positive number c_1 to satisfy Hurwitz polynomial, then,

$$\dot{s}(x) = C\dot{x} = Cf(x) + Cg(x)u + Cd(t) \tag{10}$$

To make $\dot{s}(x)$ =0, then equivalent control variable u_{eq} on switching surface is:

$$u_{eq} = -(Cg(x))^{-1}(Cf(x) + Cd(t)) \tag{11}$$

For sliding mode control system, its control effect is composed by two parts, one is the reaching control variable u_{vss} which makes system into sliding mode hyperplane in reaching motion stage; the other is equivalent control variable u_{eq} which reacts on sliding mode hyperplane. So the sliding mode control variable is:

$$u = u_{eq} + u_{vss} = -(Cg(x))^{-1}(Cf(x) + Cd(t) \ + \eta \operatorname{sgn}(s) + ks) \tag{12}$$

In formula(12), η 、k are positive real number. Here, exponential law is used and the chattering phenomenon can be weakened by adjusting reaching coefficients. Paper[10]proofed that sliding mode exits and is accessible for system described by formula(8) if using control variable as is described in formula(12).

The discontinuous part of sliding mode control needs instantaneous switching to keep system state on sliding mode plane. But the idea sliding mode state emerges hardly because of transmission delay, actuator's limits, calculation delay and other factors. System state moves to the origin in the form of crossing sliding mode plane repeatedly. So chattering happens. For solving problem mentioned above, boundary layer conception is introduced in sliding mode control. The symbolic function in sliding mode control is replaced by saturation function to reduce the chattering. Here, boundary layer conception is used and saturation function is replaced by hyperbolic tangent activation function of neural network.

4 Design of Reaching Law for Sliding Mode Controller Based on BP Neural Network

BP network is used to design coefficient η of reaching law and hyperbolic tangent function is used as activation function. The curve shape of hyperbolic tangent function is similar to that of saturation function. In this way, continuous switching function can be realized and the chattering can be reduced[11].

The form of exponential reaching law is:

$$\dot{s}(x) = -\eta \operatorname{sgn}(s) - ks \tag{13}$$

Where, η>0, k>0. The first term of exponential reaching law produces constant speed term and the second term produces variable speed term. The state reaching speed can also be changed by adjusting coefficient η.

When s<0, from formula(13),

$$s(t)=\frac{\eta}{k}+\left(s_0-\frac{\eta}{k}\right)e^{-kt} \tag{14}$$

In which, s_0 is initial value of system state. To suppose system state reaches zero state from negative state in limited time t_0^-, then s become a monotone function about η:

$$s(\eta)=\frac{1-e^{-kt_0^-}}{k}\eta+s_0^-e^{-kt_0^-}=\alpha^-\eta+\delta^- \tag{15}$$

In formula(15), $\alpha^-=\frac{1-e^{-kt_0^-}}{k}$, $\delta^-=s_0^-e^{-kt_0^-}$, because t_0^- is a given performance index and k is a positive constant number, s can be considered as linear function about η.In the same way, when $s>0$,

$$s(\eta)=-\frac{1-e^{-kt_0^+}}{k}\eta+s_0^+e^{-kt_0^+}=\alpha^+\eta+\delta^+ \tag{16}$$

In formula(16), $\alpha^+=-\frac{1-e^{-kt_0^+}}{k}$, $\delta^+=s_0^+e^{-kt_0^+}$, then:

$$\eta=\begin{cases}\gamma^+s+\rho^+ & s>0\\ \gamma^-s+\rho^- & s<0\end{cases} \tag{17}$$

Where, $\gamma^+=\frac{1}{\alpha^+}$, $\rho^+=-\frac{\delta^+}{\alpha^+}$, $\gamma^-=\frac{1}{\alpha^-}$, $\rho^-=-\frac{\delta^-}{\alpha^-}$.

SISO neural network is used to approach the linear relation showed in formula(17). The network input is switching plane and the output is speed gain term of exponential reaching law. Here, hyperbolic tangent function is used:

$$f(s)=\frac{1-\mathrm{e}^{-\alpha x}}{1+e^{-\alpha x}} \tag{18}$$

The curve shape of hyperbolic tangent function is similar to that of saturation function. By adjusting coefficient α, the linear width of saturation function can be changed, that is, the width of boundary layer is changed. When α increases to a certain value, hyperbolic tangent activation function can be regarded as sign function approximately and it keeps the function continuity.

Here, error back propagation network algorithm is adopted which is suitable for single layer or multilayer neural network. Its principle is based on gradient descent method. SISO one layer network can realize to approach the linear function.

To suppose output of neurons $y = f(net)$, $net = WX = \sum_{j=1}^{m} \omega_j x_j$, m is input dimension, quadratic performance index is adopted to control output error.

$$E(W) = \frac{1}{2} \| D - Y \|^2 = \frac{1}{2} \sum_{i=1}^{n} (d_i - y_i)^2 \tag{19}$$

In formula(19), n is output dimension. In order to minimize $E(W)$, to calculate its derivative:

$$\frac{\partial E}{\partial \omega_j} = \frac{\partial E}{\partial net} \frac{\partial net}{\partial \omega_j} = -(d_j - y_j) f'(net) x_j \tag{20}$$

So,

$$\omega_j(k+1) = \omega_j(k) - \lambda \frac{\partial E}{\partial \omega_j} = \omega_j(k) + \lambda(d_j - y_j) f'(net) x_j \tag{21}$$

Back propagation network algorithm needs teacher sample signals. To choose a suitable value for k and determine coefficients in formula(15) and (16) according to the given reaching time t_0^+ , t_0^- and initial value s_0 of system state. Then, to calculate s according to $\eta \in (\eta^D, \eta^U)$ to construct sample data for training neural network. For approaching linear function, the network is always convergent if adopting suitable neural network structure. This method can be designed according given performance index and make the state of the whole sliding mode control system reach sliding mode hyperplane in reaching stage.

5 Simulation Research

The simulation test is based on a ship which principal dimensions is as follow[12]: ship length L is 160.9m, breadth molded d is 7.467m, block coefficient CB is 0.588. The initial speed is set as v=8.3m/s, n_3=23, n_2=0, n_1=1, n_0=0, T1=72.49, T2=8.54, T3=17.61, K=0.1141, T=63.42. The disturbance of wind, wave, current is seemed as disturbance rudder angle and uncertainty term Δ adopts uniform random distribution. Simulation results of gain-fixed sliding mode course controller are as figure1and figure2. From figure1, output course can track the given course angle rather quickly, but as is shown in figure2, the controller output does not only exit great chattering and decrease the performance seriously, but also injure actuator heavily.

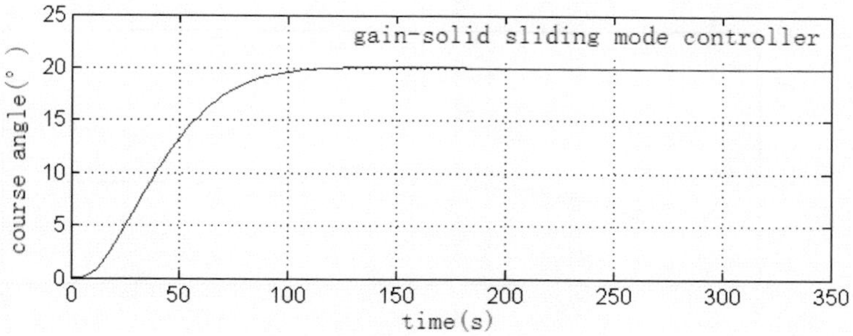

Fig. 1. Course output under solid gain

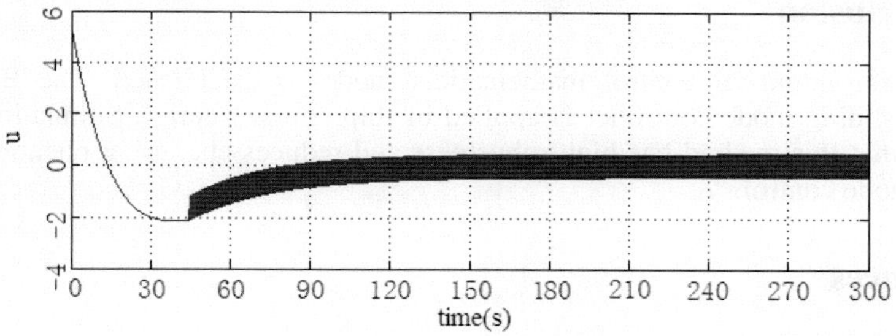

Fig. 2. Sliding mode controller output under solid gain

The ship course control simulation based on BP network sliding mode controller is shown in figure3, figure4. The course control performance does not decrease obviously, but chattering phenomenon of sliding mode control is decreased greatly.

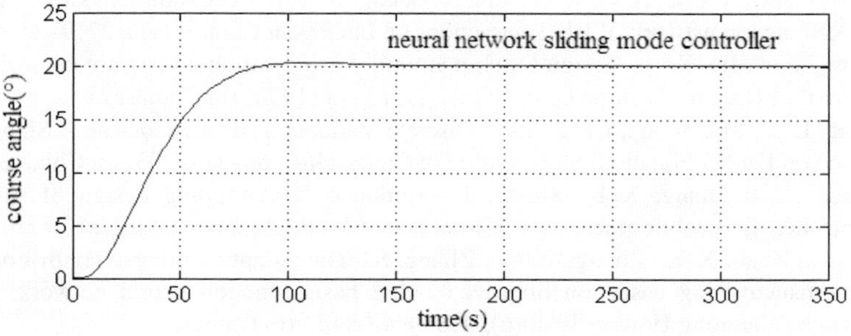

Fig. 3. Ship course angle output based on BP network sliding mode controller

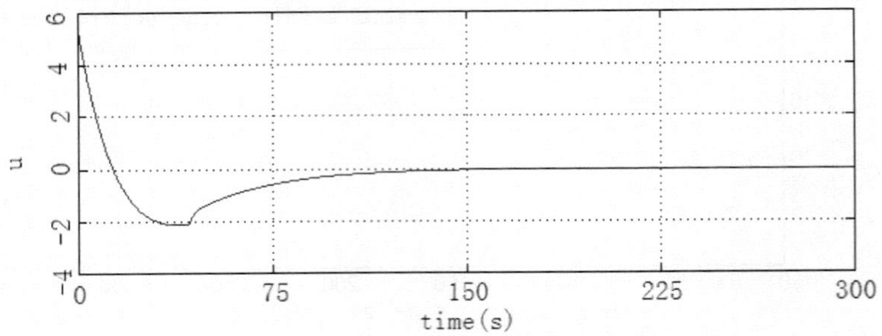

Fig. 4. Sliding mode controller output based on BP neural network

6 Conclusion

Ship course nonlinear control mathematical model is established. The BP neural network sliding mode controller is applied in ship course control. Simulation results indicate that this method has high robustness and reduces chattering phenomenon of sliding mode control.

References

1. Xu, Y., Lao, D.-z., Li, D.-h., Song, Y.-j.: Simulation study on nonlinear control of ship course. Ship Engineering 31(1), 38–44 (2009) (in Chinese)
2. He, Z.-j.: Ship's course steering controller based on adaptive neural-fuzzy inference system. Ship Engineering 30(6), 46–51 (2008) (in Chinese)
3. Luo, W.-l., Zou, Z.-j., Li, T.-s.: Robust tracking control of nonlinear ship steering. Control Theory & Applications 26(8), 893–895 (2009) (in Chinese)
4. Wu, H.-s., Huang, K., Xu, X.: Variable structure control and simulation for course-keeping of ships. Journal of naval university of engineering 16(3), 27–33 (2004) (in Chinese)
5. Ak-Ayca, G., Cansever, G.: Three link robot control with fuzzy sliding mode controller based on RBF neural network. IEEE Transactions on Intelligent Control 1(1), 2719–2724 (2006)
6. Yang, Y.-s., Jia, X.- l.: Design for ship autopilot using variable structure control algorithm. Journal of Dalian Maritime Universit 24(1), 13–18 (1998) (in Chinese)
7. Song, L.-z., Ma, W.-m., Chen, S.-c.: Discrete Variable Structure Control of Ship Autopilots Based on Bech's Equation. Shipbuilding of China 44(4), 68–72 (2003) (in Chinese)
8. Wang, X.-c., Jiang, X.-h., Zhang, J.: Nonlinear Backsetpping design of ship steering controller. Control Engineering of China 9(5), 63–68 (2002) (in Chinese)
9. Du, G., Zhan, X.-q., Zhang, W.-m., Zhang, S.: The Adaptive inverse control of nonlinear ship maneuvering based on improved radial basis function neural network. Journal of Shanghai Jiaotong University 40(6), 988–994 (2006) (in Chinese)
10. Jiang, K., Zhang, J.-g.: Design for ship autopilot using variable structure control algorithm. Journal of system simulation 14(7), 964–967 (2002) (in Chinese)
11. Wang, W., Yi, J.-q., Zhao, D.-b., Liu, X.-j.: Design of a new type of neural network sliding-mode controller. Electric machines and control 9(6), 603–606 (2005) (in Chinese)
12. Zhang, X.-k., Jia, X.-l., Liu, C.: Research on responding ship motion mathematical model. Journal of Dalian Maritime University 30(1), 18–21 (2004) (in Chinese)

Fictitious Correlation-based Tuning Integrating the Data-Based Stability Test at Each Parameter Update

Kazuhiro Yubai, Hiroki Fujii, and Junji Hirai

Mie University,
1577 Kurimamachiya, Tsu, Japan
yubai@elec.mie-u.ac.jp, hiroki_fujii@ems.elec.mie-u.ac.jp,
hirai@elec.mie-u.ac.jp
http://www.ems.elec.mie-u.ac.jp

Abstract. Recently, the model-free controller syntheses have been paid attention to as one of promising controller tuning methods. The authors proposed the Fictitious Correlation-based Tuning (FCbT) which obtains reasonable controller parameters using the input/output data set. However, the stability of the tuned closed-loop system was not guaranteed in this approach. This paper proposes the controller tuning method guaranteeing the closed-loop stability at each parameter update without any plant models. The data-based stability test is imposed at each parameter update to make the tuned closed-loop system stable at least. Moreover, Particle Swarm Optimization (PSO) is introduced to reduce the initial-value dependence in the nonlinear optimization instead of the Gauss-Newton method. The effectiveness is confirmed by experimental results.

Keywords: model-free controller synthesis, FCbT, data-based stability test, PSO, Gauss-Newton method.

1 Introduction

Model-free controller syntheses directly providing desired controller parameters using the input/output data set without any mathematical model are recently actively reseached [1,2,3,4]. The model-free controller syntheses save the time-consuming model identification and alleviate the designer's burden. Moreover, the design of the fixed structural controller such as a PID controller is easiliy addressed in comparison with the model-based controller syntheses.

The authors have proposed the model-free controller synthesis named as the Fictitious Correlation-based Tuning (FCbT) [4]. Since the FCbT is based on the correlation approach, the tuned controller parameter is robust for the measurement noise contained in the input/output data set. The distinguished feature of the FCbT is that only the one-shot intput/output data set is required for the off-line tuning even for MIMO systems and for the nonlinear optimization problem.

This paper addresses the closed-loop stability of the control system tuned by the FCbT, which has not been taken into consideration in the model-free controller syntheses. The FCbT optimizes the controller parameter so as to minimize

X. Wan (Ed.): Electrical Power Systems and Computers, LNEE 99, pp. 511–518.
springerlink.com

the 2-norm based cost function constructed by the acquired input/output data set. However, in the early works the situations are limited to the cases where only minimization of the data-based cost function leads to the appropriate solution for the given reference model, the controller structure, and the initial parameter that are determined based on partially known information on the plant. Since there is no guarantee to have enough information on the plant, we have no idea whether the control system is actually stabilized or not prior to implementaion.

This paper newly proposes the FCbT integrating the data-based stability test at each parameter update. The parameter is updated again in response to the result of the data-based stability test until the control system is expected to be stabilized, which guarantees the closed-loop stability at each parameter update. The parameter update guranteeing the closed-loop stability is called the safe update in this paper. Moreover, initial-value dependence is imroved by introducing Particle Swarm Optimization (PSO) [5] as the solver of the nonlinear optimization problem instead of the Gauss-Newton method. The effectiveness of the proposed method is confirmed by some experiments.

The following is a notation used in this paper: q represents a shift operator. For a discrete-time signal $x(t)$ at time instant t, $q^{-1}x(t) = x(t-1)$. $\boldsymbol{\rho}$ denotes the controller parameter to be tuned, especially, $\boldsymbol{\rho}_n$ denotes the controller parameter at the nth parameter update. $a^{(i)}$ denotes the ith element of the column vector $\boldsymbol{a}$. $E[\bullet]$ and $\bar{\sigma}[\bullet]$ represent a mathmatical expectation and a maximal singular value, respectively.

2 Parameter Tuning by the FCbT

This section reviews the Fictitious Correlation-based Tuning (FCbT) previously proposed by the authors [4]. Consider the controller parameter tuning problem to approximate the closed-loop system consisting of an unknown plant $\boldsymbol{P}$ and a controller described by a parameter vector $\boldsymbol{\rho}$, denoted by $\boldsymbol{C}(\boldsymbol{\rho})$, to its reference model $\boldsymbol{M}$. This problem is described as minimization of the following criterion in the frequency domain;

$$J_{\mathrm{MR}}(\boldsymbol{\rho}) = \left\|\boldsymbol{M} - (\boldsymbol{I} + \boldsymbol{P}\boldsymbol{C}(\boldsymbol{\rho}))^{-1}\boldsymbol{P}\boldsymbol{C}(\boldsymbol{\rho})\right\|_2^2, \tag{1}$$

or minimization of $\|\boldsymbol{\varepsilon}(\boldsymbol{\rho},t)\|_2^2$ in the time domain as shown in Fig. 1. Generally, minimization of $\|\boldsymbol{\varepsilon}(\boldsymbol{\rho},t)\|_2^2$ with respect to $\boldsymbol{\rho}$ requires the nonlinear optimization such as the steepest gradient method, e.g., the Gauss-Newton method. The gradient and the Hessian of $\|\boldsymbol{\varepsilon}(\boldsymbol{\rho},t)\|_2^2$ in the Gauss-Newton method must be calculated using $\boldsymbol{\varepsilon}(\boldsymbol{\rho},t)$, which means that the iterative experiments are required. In order to avoid the iterative experiments, the FCbT introduces the fictitious reference signal [6], $\tilde{\boldsymbol{r}}(\boldsymbol{\rho},t) = \boldsymbol{C}(\boldsymbol{\rho})^{-1}\boldsymbol{u}_0(t) + \boldsymbol{y}_0(t)$, where the data set $\{\boldsymbol{u}_0(t), \boldsymbol{y}_0(t)\}$ is acquired from the closed-loop system with the initial controller $\boldsymbol{C}(\boldsymbol{\rho}_0)$. The output of $\boldsymbol{P}$ is fixed to $\boldsymbol{y}_0(t)$ when $\tilde{\boldsymbol{r}}(\boldsymbol{\rho},t)$ is applied to the closed-loop system with an arbitrary controller $\boldsymbol{C}(\boldsymbol{\rho})$ in Fig. 1 instead of $\boldsymbol{r}(t)$. Since $\tilde{\boldsymbol{\varepsilon}}(\boldsymbol{\rho},t) \equiv \boldsymbol{M}\tilde{\boldsymbol{r}}(\boldsymbol{\rho},t) - \boldsymbol{y}_0(t)$ consists of the initial data set $\{\boldsymbol{u}_0(t), \boldsymbol{y}_0(t)\}$, minimization of $\|\tilde{\boldsymbol{\varepsilon}}(\boldsymbol{\rho},t)\|_2^2$ does not require the iterative experiments. In order to

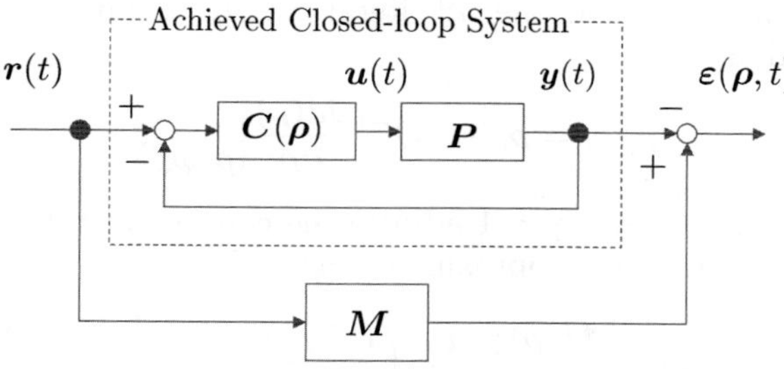

Fig. 1. Model reference control problem.

reduce the effect of the measurement noise contained in $\boldsymbol{y}_0(t)$ for the controller parameter tuning, the cost function based on the correlation approach for a $\lambda \times \lambda$ square system is newly defined in the FCbT as

$$J(\boldsymbol{\rho}) = \boldsymbol{f}(\boldsymbol{\rho})^{\mathrm{T}}\boldsymbol{f}(\boldsymbol{\rho}) = \sum_{i=1}^{\lambda}\sum_{j=1}^{\lambda} \boldsymbol{f}^{(ij)}(\boldsymbol{\rho})^{\mathrm{T}}\boldsymbol{f}^{(ij)}(\boldsymbol{\rho}), \tag{2}$$

where

$$\boldsymbol{f}^{(ij)}(\boldsymbol{\rho}) = \frac{1}{N_d}\sum_{t=1}^{N_d} \boldsymbol{\zeta}^{(j)}(\boldsymbol{\rho}, t)\tilde{\varepsilon}^{(i)}(\boldsymbol{\rho}, t), \tag{3}$$

$$\boldsymbol{\zeta}^{(j)}(\boldsymbol{\rho}, t) = [\tilde{r}^{(j)}(\boldsymbol{\rho}, t+l), \cdots, \tilde{r}^{(j)}(\boldsymbol{\rho}, t), \cdots, \tilde{r}^{(j)}(\boldsymbol{\rho}, t-l)]^{\mathrm{T}}, \tag{4}$$

N_d and l are the data length and the window length, respectively. Conventionally, the Gauss-Newton method is adopted as the solver of the minimization problem of $J(\boldsymbol{\rho})$. In this paper, Particle Swarm Optimization (PSO) is introduced to solve the above optimization problem.

2.1 Parameter Update by the Gauss-Newton Method

Without loss of generality, we describe the parameter update in the case of a 2-input 2-output square system. For sufficient large data length of N_d, (3) is rewritten in the compact form as

$$\boldsymbol{f}(\boldsymbol{\rho}) = \frac{1}{N_d}\sum_{t=1}^{N_d} \boldsymbol{L}(\boldsymbol{\rho}, t)\boldsymbol{H}(\boldsymbol{\rho}, t), \tag{5}$$

where

$$\boldsymbol{L}(\boldsymbol{\rho}, t) = \mathrm{diag}(\boldsymbol{\zeta}_{11}(\boldsymbol{\rho}, \mathrm{t}), \boldsymbol{\zeta}_{12}(\boldsymbol{\rho}, \mathrm{t}), \boldsymbol{\zeta}_{21}(\boldsymbol{\rho}, \mathrm{t}), \boldsymbol{\zeta}_{22}(\boldsymbol{\rho}, \mathrm{t})), \tag{6}$$

$$\boldsymbol{H}(\boldsymbol{\rho}, t) = [\eta_{11}(\boldsymbol{\rho}, t), \eta_{12}(\boldsymbol{\rho}, t), \eta_{21}(\boldsymbol{\rho}, t), \eta_{22}(\boldsymbol{\rho}, t)]^{\mathrm{T}}. \tag{7}$$

The controller parameter at the nth iteration, $\boldsymbol{\rho}_n$, is updated by the Gauss-Newton method as

$$\boldsymbol{\rho}_{n+1} = \boldsymbol{\rho}_n - \gamma \boldsymbol{Q}^{-1} \frac{\partial J(\boldsymbol{\rho})}{\partial \boldsymbol{\rho}}\bigg|_{\boldsymbol{\rho}=\boldsymbol{\rho}_n}, \tag{8}$$

where a positive parameter $\gamma < 1$ adjusts the convergent rate of $\boldsymbol{\rho}$. $\boldsymbol{Q}$ denotes the Hessian matrix which is approximated as

$$\boldsymbol{Q} = \frac{\partial \boldsymbol{f}(\boldsymbol{\rho})}{\partial \boldsymbol{\rho}}\bigg|_{\boldsymbol{\rho}=\boldsymbol{\rho}_n} \left(\frac{\partial \boldsymbol{f}(\boldsymbol{\rho})}{\partial \boldsymbol{\rho}}\bigg|_{\boldsymbol{\rho}=\boldsymbol{\rho}_n} \right)^{\mathrm{T}}, \tag{9}$$

where

$$\frac{\partial \boldsymbol{f}(\boldsymbol{\rho})}{\partial \boldsymbol{\rho}}\bigg|_{\boldsymbol{\rho}=\boldsymbol{\rho}_n} = \frac{1}{N_d}\sum_{t=1}^{N_d}\left(\frac{\partial \boldsymbol{L}(\boldsymbol{\rho},t)}{\partial \boldsymbol{\rho}}\bigg|_{\boldsymbol{\rho}=\boldsymbol{\rho}_n} \boldsymbol{H}(\boldsymbol{\rho},t)\right) + \frac{1}{N_d}\sum_{t=1}^{N_d}\left(\frac{\partial \boldsymbol{H}(\boldsymbol{\rho},t)}{\partial \boldsymbol{\rho}}\bigg|_{\boldsymbol{\rho}=\boldsymbol{\rho}_n} \boldsymbol{L}^{\mathrm{T}}(\boldsymbol{\rho},t)\right).$$

The derivatives of $\boldsymbol{H}(\boldsymbol{\rho},t)$ and $\boldsymbol{L}(\boldsymbol{\rho},t)$ with respect to $\boldsymbol{\rho}$ consist of the derivatives of $\tilde{\boldsymbol{\varepsilon}}(\boldsymbol{\rho},t)$ and $\tilde{\boldsymbol{r}}(\boldsymbol{\rho},t)$, respectively. As a result of some manipulation, we have the following relations:

$$\frac{\partial \tilde{\boldsymbol{\varepsilon}}(\boldsymbol{\rho},t)}{\partial \rho^{(x)}}\bigg|_{\rho_n^{(x)}} = \boldsymbol{M}\boldsymbol{C}(\boldsymbol{\rho})^{-1}\frac{\partial \boldsymbol{C}(\boldsymbol{\rho})}{\partial \rho^{(x)}}\bigg|_{\rho_n^{(x)}} \boldsymbol{C}(\boldsymbol{\rho})^{-1}\boldsymbol{u}_0(t), \tag{10}$$

$$\frac{\partial \tilde{\boldsymbol{r}}(\boldsymbol{\rho},t)}{\partial \rho^{(x)}}\bigg|_{\rho_n^{(x)}} = -\boldsymbol{C}(\boldsymbol{\rho})^{-1}\frac{\partial \boldsymbol{C}(\boldsymbol{\rho})}{\partial \rho^{(x)}}\bigg|_{\rho_n^{(x)}} \boldsymbol{C}(\boldsymbol{\rho})^{-1}\boldsymbol{u}_0(t). \tag{11}$$

Since all components of (10) and (11) are available, $J(\boldsymbol{\rho})$ is minimized using only the initial input/output data set $\{\boldsymbol{u}_0(t), \boldsymbol{y}_0(t)\}$.

2.2 Parameter Update by PSO

This subsection describes the parameter update by Particle Swarm Optimization (PSO). PSO is a robust stochastic optimization technique based on the movement and intelligence of swarms. PSO minimizes $J(\boldsymbol{\rho})$ by having a population (called a swarm) of candidate solutions (called particles), and moving these particles around in the solution space according to the simple rule. Let K be the number of particles in the swarm. The kth particle has a position $\boldsymbol{\rho}_n^k$ and a velocity $\boldsymbol{v}_n^k$ at n^{th} iteration ($n = 1, \cdots, N$). The movements of the particles are governed by their own best known position $\hat{\boldsymbol{\rho}}_n^k$ and the best known position in the swarm $\boldsymbol{\rho}_n^{\mathrm{Gb}}$ which are updated as better positions found by the particles. The parameter update law is then described as

$$\boldsymbol{\rho}_{n+1}^k = \boldsymbol{\rho}_n^k + \boldsymbol{v}_{n+1}^k, \tag{12}$$

$$\boldsymbol{v}_{n+1}^k = \mu \boldsymbol{v}_n^k + c_1 r_1 (\hat{\boldsymbol{\rho}}_n^k - \boldsymbol{\rho}_n^k) + c_2 r_2 (\boldsymbol{\rho}_n^{Gb} - \boldsymbol{\rho}_n^k). \tag{13}$$

$0 < r_1 < 1$ and $0 < r_2 < 1$ are uniform random numbers, and μ, c_1 and c_2 are non-negative weightings. PSO does not use the gradient of $J(\boldsymbol{\rho})$, which means that PSO does not require the derivative of $J(\boldsymbol{\rho})$ as is required by the Gauss-Newton method.

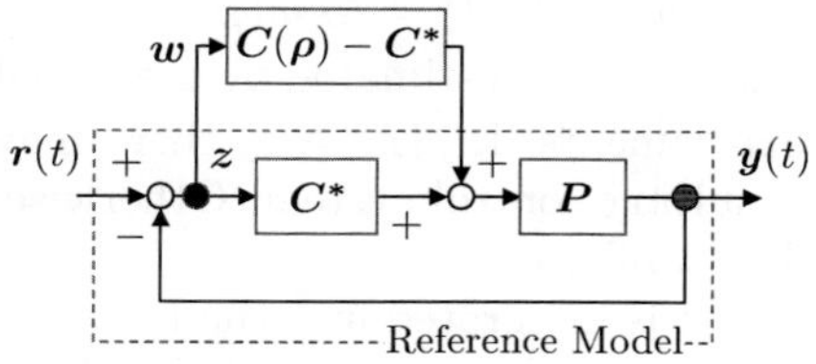

Fig. 2. Equivalent closed-loop system with the ideal controller $\boldsymbol{C}^*$.

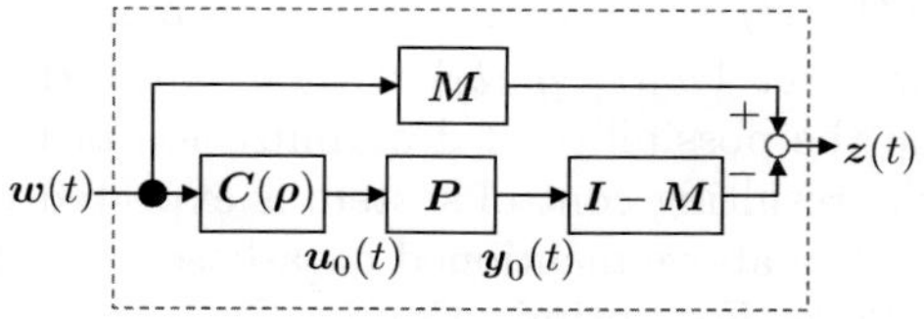

Fig. 3. Stability test.

3 Data-Based Stability Test for the FCbT

This section states the data-based test evaluating whether the updated controller parameter stabilizes the resulting closed-loop system in advance of implementation. The closed-loop system consisting of $\boldsymbol{P}$ and $\boldsymbol{C}(\boldsymbol{\rho})$ can be represented as Fig. 2 by introducing the ideal controller, $\boldsymbol{C}^*$, which is the exact solution such that $\boldsymbol{M} = (\boldsymbol{I}+\boldsymbol{P}\boldsymbol{C}^*)^{-1}\boldsymbol{P}\boldsymbol{C}^*$. If the controller mismatch $\boldsymbol{C}(\boldsymbol{\rho}) - \boldsymbol{C}^*$ is regarded as an additive perturbation for $\boldsymbol{C}^*$, the sufficient condition of the robust stability could be derived from the small-gain theorem as

$$\begin{aligned}\delta_{\text{ideal}}(\boldsymbol{\rho}) &= \left\|(\boldsymbol{I}+\boldsymbol{P}\boldsymbol{C}^*)^{-1}\boldsymbol{P}(\boldsymbol{C}^* - \boldsymbol{C}(\boldsymbol{\rho}))\right\|_\infty \\ &= \|\boldsymbol{M} - (\boldsymbol{I}-\boldsymbol{M})\boldsymbol{P}\boldsymbol{C}(\boldsymbol{\rho})\|_\infty < 1 \end{aligned} \tag{14}$$

Therefore, $\delta_{\text{ideal}}(\boldsymbol{\rho})$ is evaluated as the H_∞ norm of the transfer matrix from $\boldsymbol{w}(t)$ to $\boldsymbol{z}(t)$ shown in Fig. 3. Now, we must consider how to evaluate $\delta_{\text{ideal}}(\boldsymbol{\rho})$ using the available signals $\{\boldsymbol{u}_0(t), \boldsymbol{y}_0(t)\}$. For this purpose, the following theorem plays an important role [7].

Theorem 1. *Given time series $\boldsymbol{w}(t)$ and $\boldsymbol{z}(t)$ ($t = 0, \cdots, N$), there exists a stable, causal linear, time-invariant operator $\boldsymbol{G}(q)$ with $\|\boldsymbol{G}(q)\|_\infty \geq \gamma$ and such that $\boldsymbol{z}(t) = \boldsymbol{G}(q)\boldsymbol{w}(t)$ if and only if $\boldsymbol{Z}^{\mathrm{T}}\boldsymbol{Z} < \delta^2\boldsymbol{W}^{\mathrm{T}}\boldsymbol{W}$ or $\bar{\sigma}[\boldsymbol{Z}(\boldsymbol{W}^{\mathrm{T}}\boldsymbol{W})^{-\frac{1}{2}}] < \delta$, where $\boldsymbol{W}$ is a lower block Toeplitz matrix associated with $\boldsymbol{w}(t)$ defined as*

$$\boldsymbol{W} = \begin{bmatrix} \boldsymbol{w}(0) & 0 & 0 & \cdots & 0 \\ \boldsymbol{w}(1) & \boldsymbol{w}(0) & 0 & \cdots & 0 \\ \boldsymbol{w}(2) & \boldsymbol{w}(1) & \boldsymbol{w}(0) & \cdots & 0 \\ \vdots & \vdots & \vdots & \ddots & \vdots \\ \boldsymbol{w}(N_d-1) & \boldsymbol{w}(N_d-2) & \boldsymbol{w}(N_d-3) & \cdots & \boldsymbol{w}(0) \end{bmatrix}, \tag{15}$$

$\boldsymbol{Z}$ is also defined associated with $\boldsymbol{z}(t)$ as same as $\boldsymbol{W}$.

This paper introduces $\hat{\delta} \equiv \bar{\sigma}[\boldsymbol{Z}(\boldsymbol{W}^{\mathrm{T}}\boldsymbol{W})^{-\frac{1}{2}}]$ as a stability index. However, available information is only the input/output data set $\{\boldsymbol{u}_0(t), \boldsymbol{y}_0(t)\}$, and we do not have any information on the data set $\{\boldsymbol{w}(t), \boldsymbol{z}(t)\}$. As similar in generating

$\tilde{\boldsymbol{r}}(\boldsymbol{\rho}, t)$, we can calculate $\boldsymbol{w}(\boldsymbol{\rho}, t) = \boldsymbol{C}^{-1}(\boldsymbol{\rho})\boldsymbol{u}_0(t)$ and $\boldsymbol{z}(\boldsymbol{\rho}, t) = \boldsymbol{M}\boldsymbol{w}(\boldsymbol{\rho}, t) - (\boldsymbol{I} - \boldsymbol{M})\boldsymbol{y}_0(t)$ as shown in Fig. 3. This means that $\hat{\delta}(\boldsymbol{\rho})$ can be estimated by the initial data set $\{\boldsymbol{u}_0(t), \boldsymbol{y}_0(t)\}$ without any extra experiments. If $\hat{\delta}(\boldsymbol{\rho}) \geq 1$, there may be the possibility of destabilization of the resulting control system. Otherwise, the resulting control system is expected to be stabilized.

The above mentioned data-based stability test is integrated into the FCbT as follows: For optimization by the Gauss-Newton method, the result of the above mentioned data-based stability test is reflected in the convergent rate γ. In the case of $\hat{\delta}(\boldsymbol{\rho}_n) \geq 1$, $\gamma \leftarrow \alpha\gamma$ with $0 < \alpha < 1$, and $\boldsymbol{\rho}_n$ is updated again until $\hat{\delta}(\boldsymbol{\rho}_n) < 1$. For optimization by PSO, the choice of $\boldsymbol{\rho}_n^{\mathrm{Gb}}$ is determined according to the result of the data-based stability test. $\boldsymbol{\rho}_n^{\mathrm{Gb}}$ is chosen as the best position in terms of $J(\boldsymbol{\rho}_n)$ among the particles with $\hat{\delta}(\boldsymbol{\rho}_n) < 1$ instead of among all particles.

4 Experiment

In order to confirm the effectiveness of the FCbT integrated with the safe update, this paper addresses the speed control problem of the two-mass system encountered in many industrial processes. The two-mass system has the delay time of 20 samples (i.e., 100 ms) at the input channel simulated by the software, which makes the plant difficult to be stabilized. However, its information is not used in the tuning of $\boldsymbol{\rho}$. The sampling time, T_s, is set to 5 ms. The controller structure is set to a PI controller parameterized as $\boldsymbol{C}(\boldsymbol{\rho}) = \rho_p + \rho_i T_s/(1 - q^{-1})$, where $\boldsymbol{\rho} = [\rho_p,\ \rho_i]^{\mathrm{T}}$. Figure 4 shows the initial input/output data set $\{\boldsymbol{u}_0(t), \boldsymbol{y}_0(t)\}$ of length $N_d = 5000$ acquired from the closed-loop system consisting of $\boldsymbol{P}$ and $\boldsymbol{C}(\boldsymbol{\rho}_0)$ where $\boldsymbol{\rho}_0 = [0.035, 0.01]^{\mathrm{T}}$. Note that both the tuning of $\boldsymbol{\rho}$ and the data-based stability test can be performed using only $\{\boldsymbol{u}_0(t), \boldsymbol{y}_0(t)\}$. The reference model to be tracked is given as $\boldsymbol{M} = 40^2/(s + 40)^2$ in the continuous-time. The experiments were performed in 5 cases for comparison. The experimental conditions and the tuned parameters are listed in Table. 1. For the Gauss-Newton method (i.e., **case I**, **II**, **III**), $\gamma = 0.2$ and $\alpha = 0.5$. For PSO (i.e., **case IV**, **V**), the number of particles in the swarm $K = 20$. The initial positions and velocities of particles are randomly given. The weightings c_1 and c_2 are set as $c_1 = c_2 = 1.5$. The weighting μ is determined according to the Linearly Decreasing Weight Method [8] as $\mu = \mu_{max} - (\mu_{max} - \mu_{min})n/N$ where $\mu_{max} = 0.9$, $\mu_{min} = 0.4$, n and $N = 20$ denotes the current generation and the generation number, respectively.

Figure 5 shows the step responses for **case I**, **II**, **III** optimized by the Gauss-Newton method. In **case I**, the closed-loop system is not stabilized but the persistent vibration is observed. Although $J(\boldsymbol{\rho})$ converges to much small value, $\hat{\delta}(\boldsymbol{\rho}_{20})$ is much higher than 1. This result implies that only minimization of $J(\boldsymbol{\rho})$ may lead to the unsatisfactory parameter. On the other hand, **case II** and **III** avoid the undesirable vibration thanks to the safe update. However, the initial-value dependence is observed between **case II** and **III**. Note that the desirable

Table 1. Conditions & tuned parameters.

	case I	case II	case III	case IV	case V
opt. method	G. N.	G. N.	G. N.	PSO	PSO
safe update	w/o	w/	w/	w/o	w/
ρ_p (initial)	0.035	0.01	0.035	—	—
ρ_i (initial)	0.008	0.0005	0.01	—	—
ρ_p	0.63	0.048	0.049	0.63	0.049
ρ_i	0.13	0.0021	0.013	0.13	0.014
$J(\boldsymbol{\rho}_{20})$	59	5.9×10^8	8.8×10^6	59	6.2×10^6
$\hat{\delta}(\boldsymbol{\rho}_{20})$	9.908	0.99946	0.9999	9.908	0.9995

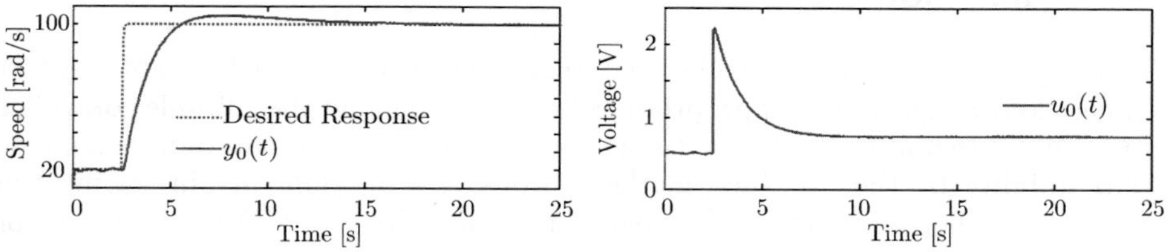

Fig. 4. Initial input/output data set $\{\boldsymbol{u}_0(t), \boldsymbol{y}_0(t)\}$.

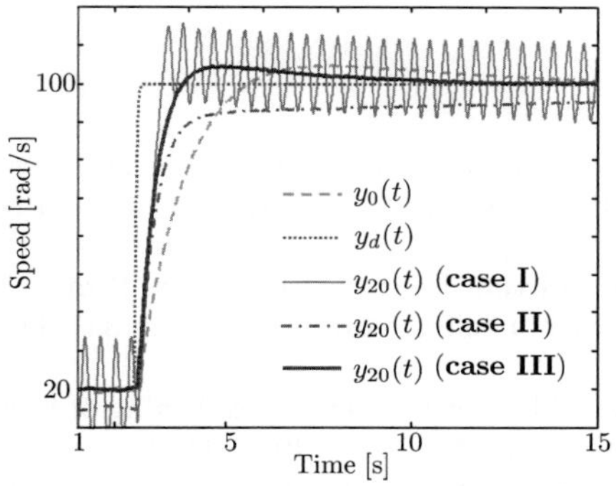

Fig. 5. Experimental results by the Gauss-Newton method (**case I**, **II**, **III**).

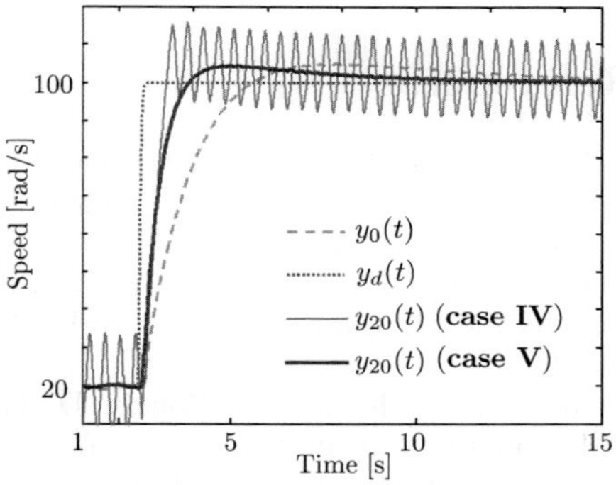

Fig. 6. Experimental results by PSO (**case IV**, **V**).

response of **case III** is obtained as a result of the time-consuming trial-and-error adjustment of $\boldsymbol{\rho}_0$.

Figure 6 shows the step responses for **case IV** and **V** optimized by PSO. From Table. 1, the tuned result of **case IV** coincides with that of **case I**. In **case V**, the least $J(\boldsymbol{\rho})$ is obtained while avoiding the undesirable vibration. Note that since the initial positions and velocities of particles are given randomly in **case V**, the time-consuming initial value setting in the Gauss-Newton method is much alleviated. Both the Gauss-Newton method and PSO can provide the stabilizing parameter with the safe update and the destabilizing parameter without the safe update, which illustrates the effectiveness of the safe update.

This paper addresses the controller parameter tuning of a SISO system with two tuning parameters (ρ_p & ρ_i). In this experimental setup, both the Gauss-Newton method and PSO provide almost same controller parameters with the safe update except for the time-consuming initial value setting in the Gauss-Newton method. However, the proposed method is not limited to SISO systems and is applicable to MIMO systems. For MIMO systems, since the number of the controller parameters drastically increases, the initial value setting in the Gauss-Newton method would not be tractable, but PSO would solve this difficulty.

5 Conclusion

This paper newly proposes the FCbT integrating the data-based stability test. If enough knowledge on the appropriate reference model and the suitable controller structure were not available, the tuned control system could maintain the closed-loop stability by the proposed method. Moreover, the difficulty in setting the initial controller parameter $\boldsymbol{\rho}_0$ is avoided by introducing PSO as the solver of the nonlinear optimization instead of the Gauss-Newton method.

As a future work, the proposed method should be applied to MIMO systems. The data-based stability test should be also improved to estimate $\hat{\delta}(\boldsymbol{\rho})$ when the initial data set contaminated by the large measurement noise is used.

References

1. Hjarmarsson, H.: Efficient Tuning of Linear Multivariable Controllers Using Iterative Feedback Tuning. International Journal of Adaptive Control and Signal Processing 13, 553–572 (1999)
2. Campi, M.C., Lecchini, A., Savaresi, S.M.: Virtual Reference Feedback Tuning: A Direct Method for the Design of Feedback Controllers. Automatica 38, 1337–1346 (2002)
3. Mišković, L., Karimi, A., Bonvin, D., Gevers, M.: Correlation-Based Tuning of Liner Decoupling Multivariable Controllers. Automatica 43, 1481–1494 (2007)
4. Wakayama, N., Yubai, K., Hirai, J.: Correlation-based Multivariable Controller Parameter Tuning by Using One-shot Experimental Data. Trans. of SICE 43, 391–399 (2007) (in Japanese)
5. Kennedy, J., Eberhart, R.: Particle Swarm Optimization. In: Proceedings of IEEE the International Conference on Neural Networks, pp. 1942–1948 (1995)
6. Safonov, M.G., Tsao, T.C.: The Unfalsified Control Concept and Learning. IEEE Trans. on Automatic Control 42, 843–847 (1997)
7. Poolla, K., Khargonekar, P., Tikku, A., Krause, T., Nagpal, K.: A Time-domain Approach to Model Validation. IEEE Trans. on Automatic Control 39, 951–959 (1994)
8. Yasuda, K., Ishigame, A.: Nonlinear Programming Algorithm: From the Practical Viewpoint. Trans. of ISCIE 50, 344–349 (2006) (in Japanese)

A Feature-Based Requirement Analysis Method

Musheng Chen and Junhua Wu

School of Software, Nanchang University, Nanchang, China
dreaminit@163.com, wujunhua110@163.com

Abstract. This paper suggests feature-based requirement analysis method. The method uses features instead of use case and user instead of actor, and introduces the generalization relationship among users, the use relationship between user and use case, the precede relationship among features , two levels view and two kinds of feature diagrams to express the requirement model.

Keywords: Feature, User, Requirement Analysis, UML.

1 Introduction

This paper suggests a feature-based requirement analysis method transfering from use-case modeling method. This method is different from UML and OML and other use case modeling approach, We use “feature” instead of “use case” and “user” instead of actor. However, their meaning is the same.

2 What Is a Feature

A feature is a description of a set of sequences of actions, including variants, which a system performs to yield an observable result of value to a user. Graphically, a feature is rendered as an ellipse, like a use case.

A feature must have a name and its name must be unique within its enclosing package in order to distinguish it from other feature. A name is a textual string. That name alone is known as a simple name; a path name is the feature name prefixed by the name of the package in which that feature lives. A feature is typically drawn showing only its name, as in Figure 1.

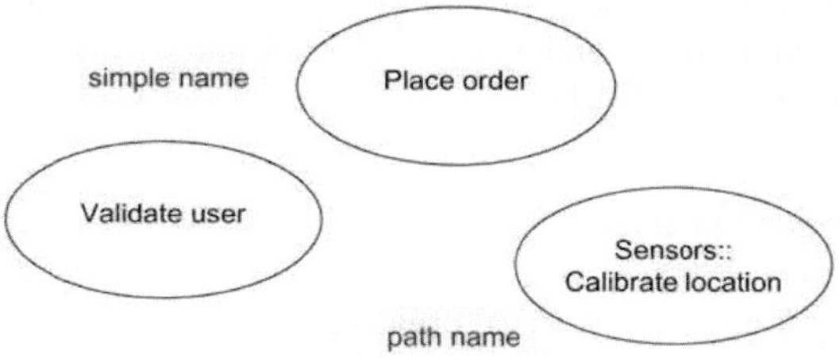

Fig. 1. Simple and Path Names

X. Wan (Ed.): Electrical Power Systems and Computers, LNEE 99, pp. 519–525.
springerlink.com

A feature name may be text consisting of any number of letters, numbers, and most punctuation marks (except for marks such as the colon, which is used to separate a class name and the name of its enclosing package) and may continue over several lines. In practice, features' names are short active verb phrases naming some behavior found in the vocabulary of the system you are modeling.

3 Features and Users

A user represents a coherent set of roles that users of features play when interacting with these features. Typically, a user represents a role that a human, a hardware device, or even another system plays with a system. For example, if you work for a bank, you might be a LoanOfficer. If you do your personal banking there, as well, you'll also play the role of Customer. An instance of a user, therefore, represents an individual interacting with the system in a specific way. Although you'll use users in your models, users are not actually part of the system. They live outside the system. As Figure 2 indicates, users are rendered as stick figures. You can use the UML's extensibility mechanisms to stereotype a user in order to provide a different icon that might offer a better visual cue for your purposes.

A user name also must be unique over the whole system. Every user also must have a name that distinguishes it from other users. A name is a textual string. Generally the users just have a simple name. The naming rules of user are the same as feature.

4 Features and Flow of Events

A feature describes what a system (or a subsystem, class, or interface) does but it does not specify how it does it. When you model, it's important that you keep clear the separation of concerns between this outside and inside view.

You can specify the behavior of a feature by describing a flow of events in text clearly enough for an outsider to understand it easily. When you write this flow of events, you should include how and when the feature starts and ends, when the feature interacts with the users and what objects are exchanged, and the basic flow and alternative flows of the behavior.

You can specify a feature's flow of events in a number of ways, including informal structured text (as in the example above), formal structured text (with pre- and post-conditions), and pseudocode. However, we just suggest specifying a feature's flow of events in informal structured text, especially in natural language. feature modeling generally happens when you capture the requirement and analyze the system. At this time, the requirement and system are ambiguous and it is very difficult to describe them formally and clearly.

5 Features and Scenarios

It is desirable to separate main versus alternative flows because a feature describes a set of sequences, not just a single sequence, and it would be impossible to express all the details of an interesting feature in just one sequence. For example, in a human resources system, you might find the feature Hire employee. This general business function might have many possible variations. You might hire a person from another company (the most common scenario); you might transfer a person from one division to another (common in international companies); or you might hire a foreign national (which involves its own special rules). Each of these variants can be expressed in a different sequence. This one feature (Hire employee) actually describes a set of sequences in which each sequence in the set represents one possible flow through all these variations. Each sequence is called a scenario. A scenario is a specific sequence of actions that illustrates behavior. Scenarios are to features as instances are to classes, meaning that a scenario is basically one instance of a feature. However, the user's instance also is a user. We do not need a different word to describe the class and instance of user.

There's an expansion factor from features to scenarios. A modestly complex system might have a few dozen features that capture its behavior, and each feature might expand out to several dozen scenarios. For each feature, you'll find primary scenarios (which define essential sequences) and secondary scenarios (which define alternative sequences).

6 The Relationship between User and Feature

In UML, the connection among things is classified into four important relationships: dependencies, generalizations, associations and realizations. However, the distinction among them is mostly orthogonal to the classification above. Both the generalization relationship and the association realization are also a kind of dependency relationship. Generalization relationship means the derived class depends on base class. Realization relationship means implementation depends on interface. Association relationship also reflects the dependency relationship. Any two things with an association relationship could be a two-way dependency or a one-way dependency. In actual fact, with regards to the relationship between these two things, the most important thing is to focus on whether a dependency relationship exists, and what direction it is. Due to the relationship between a feature and an user might be two-way dependency, it should be an association relationship. The association relationship between feature and user refers to the intercommunication between them. This means that users can send messages to features, and features can also send messages to users.

In UML, an association relationship is expressed in a straight line. Some UML modeling tools, however, use straight lines with arrows to specify the dependent direction of association relationship. If the arrow points to feature, it means that the user depends on the feature: the user sends messages to the feature, and vice versa. If there is no arrow, it means that the user depends on the feature and the feature also depends on the user The two arrows are omitted: the feature and the user send messages to each other. It does not focus too much to pay attention to the sending messages' direction between the feature and the user in the feature diagrams. When describing a feature, every message between the user and the feature is illustrated clearly. In feature diagram we only need to concern ourselves with whether there is a relationship between a user and a feature. We directly use straight line to express the relationship between a user and a feature (see Figure-2). We term the association relationship between user and feature as use relationship in order to distinguish it from the other association relationship such as the association relationship among classes. Obviously, use relationship conveys client/server relationship between user and feature.

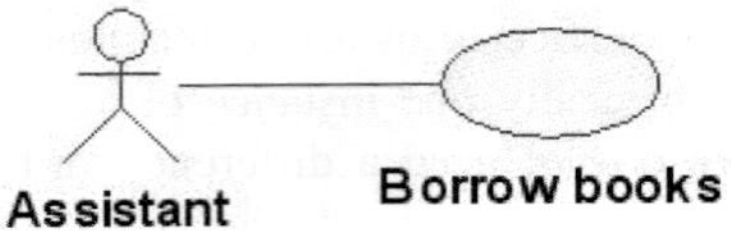

Fig. 2. The use relationship between user and feature

7 The Relationship among Users

You can define general kinds of users (such as Customer) and specialize them (such as Commercial Customer) using generalization relationships. Figure 3 shows the generalization relationship between the user Customer and Commercial Customer.

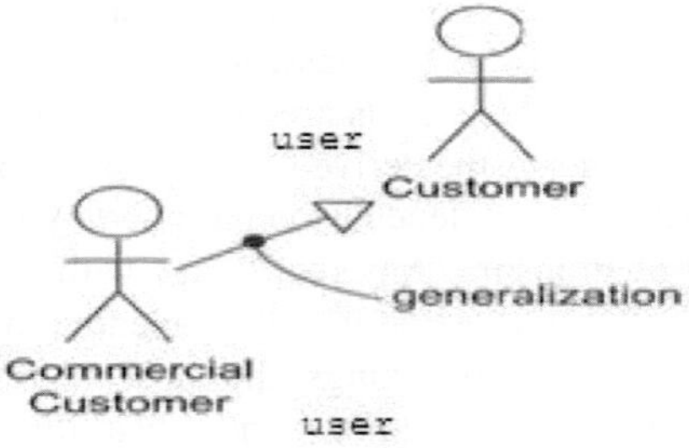

Fig. 3. Users and their generalization relationship

8 The Relationship among Features

There are three kinds of relationships among features in early UML version: generalization, extend and including. However, UML 2.0 does not support the generalization relationship among features. Some UML modeling tools do not support the generalization relationship among features at all.

In our feature based requiement analysis method, we only use the precede relationship, which is introduced in OML, among features. The precede relationship among two features means that the former feature must be executed precede the latter one. For different users, the same feature may have a different precede relationship. We use straight lines with arrows to express the precede relationship among features. The feature at the arrow tail must first execute before the feature at the arrow head (see Figure-4).

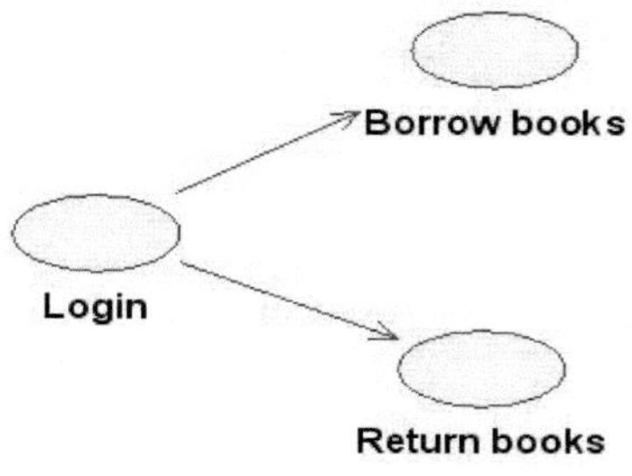

Fig. 4. The precede relationship among features

9 Feature Diagram

A feature diagram is a diagram that shows a set of features and users and their relationships. Feature diagrams commonly contain the features, the users, the generalization relationships between users, the precede relationships between features and the association relationships between users and features.

A feature diagram must have a name. A System may have many feature diagrams. Each feature diagram must have a name that distinguishes it from other feature diagram.

Like all other diagrams, feature diagrams may contain notes and constraints. Feature diagrams may also contain packages, which are used to group elements of your model into larger chunks.

The feature modeling is classified into two views. The first view is called feature and user view including several feature diagrams listing all users and features in the system but not marking the relationship among features. It only indicates the use relationship between feature and user and generalization relationship among users. This view helps to identify which users interact with a feature, capture the system requirements about user's right and define the system's boundary. The second view is called feature view. In this view, each user needs a feature diagram. We must list all features that interact with the user and precede relationships among features in the feature diagram. The user does not need to be marked in the diagram. There are different precede relationships among features for different users. Precede relationships among features convey a kind of user requirement related to the business logic order. Figure-5, 6 show two levels and two views of feature.

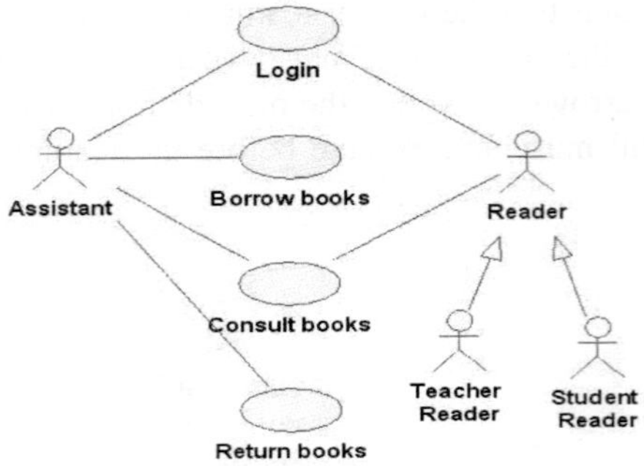

Fig. 5. The whole system's feature diagram

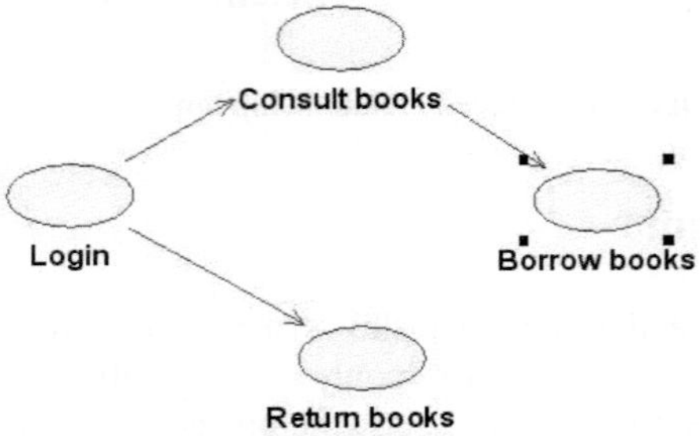

Fig. 6. The user assistant's feature diagram

10 Common Modeling Techniques

You apply feature diagrams to model the static feature view of a system. This view primarily supports the behavior of a system the outwardly visible services that the system provides in the context of its environment.

When you model the static feature view of a system, you'll typically apply feature diagrams in one of two ways, to model the context of a system and the requirements of a system.

When you model the feature view of a system, the steps are following:

Identify the users that interact with the system. Candidate users include groups that require certain behavior to perform their tasks or that are needed directly or indirectly to perform the system's functions.

Organize users by identifying general and more specialized roles.

Indentify all features for each user.

Draw feature diagrams and form the first view: feature and user view. List all users, generalization relationships among users, features, use relationships between user and feature in the feature and user view.

Draw the second view: feature view. For each user, Draw a feature diagram and list all features interactive with it and precede relationships in the feature diagram.

Describe all the event flows of every feature in detail.

We have applied the feature based requirement analysis method to our several MIS projects and have achieved good effects. The method gets rid of the complex rules of feature and ambiguity in UML and also provides enough support for capturing and describing requirements, designing, programming and testing etc.

References

1. Booch, G., Rumbaugh, J., Jacobson, I.: Unified Modeling Language User Guide, 2nd edn. Addison-Wesley Professional, Reading (2005)
2. Fowler, M., Scott, K.: UML Distilled Second Edition A Brief Guide to the Standard Object Modeling Language. Addison Wesley, Reading (1999)
3. Cockburn, A.: Writing Effective Features. Addison-Wesley Professional, Reading (2000)
4. Rosenberg, D., Scott, K.: Feature Driven Object Modeling with UML: A Practical Approach. Addison-Wesley Professional, Reading (1999)
5. Rosenberg, D., Scott, K.: Applying Feature Driven Object Modeling with UML: An Annotated e-Commerce Example. Addison-Wesley Professional, Reading (2001)
6. Wiegers, K.E.: Software Requirements, 2nd edn. Microsoft Press (2003)

Reliability Evaluation of Wind Energy Conversion Systems Incorporating Well-Being Model

Yinsha Wang[1], Wenyi Li[1], Ruigang Wang[1], and B. Bagen[2]

[1] Electric Power College, Inner Mongolia University of Technology, 010080 Huhhot, Inner Mongolia, China
[2] The System Planning Department, Manitoba Hydro, R3T0P4, Winnipeg, Canada
wangyinsha2005@126.com, Lwyyyll@vip.sina.com, ndwrg@163.com

Abstract. Along with wind energy widespread popularization and application, how to improve power supply reliability is an important issue in electric power systems including wind energy. This paper incorporates the deterministic criteria in a probabilistic framework, and builds well-being model. The model describes system reliability in terms of three different states. This paper analyzed the reliability of the three different states in small isolated power system, and researched the influence of energy storage capacity and annual peak load on small isolated wind power system reliability. This analytical approach can clearly classify system operation state; it also can improve the accuracy of reliability evaluation.

Keywords: Well-being Model, Reliability Evaluation, Wind Energy, Energy Storage, Small Isolated Power System.

1 Introduction

Owing to the random nature of the wind, reliability analysis method about wind power generation system is different from conventional units. At present, reliability analysis method of electric power system including wind energy is mainly divided into two categories of deterministic technique and probabilistic technique. Deterministic technique only can analyze the specific power failure in electric power system; it can not reflect the uncertainties effects such as random wind speed and fluctuation of load. Probabilistic technique is classified into analytic method and simulation method [1]. The main disadvantage of analytic method is that the number of system states increases exponentially as the number of system components increases. The calculating time is related tothe calculation precision closely, so simulation method also has disadvantage. Simulation method often spends a lot of time to guarantee the high accuracy [2-6]. Based on the characteristic of deterministic technique and probabilistic techniques, this paper builds well-being model, which can overcome the disadvantages of deterministic and probabilistic technique, and use the model in the small isolated power system reliability analysis. The method can improve the accurate of reliability evaluation.

2 Well-Being Model

Well-being model incorporate the accepted deterministic criteria in a probabilistic framework, which is suitable for the small isolated wind energy conversion systems. It

X. Wan (Ed.): Electrical Power Systems and Computers, LNEE 99, pp. 527–533.
springerlink.com

describes the system reliability level in terms of three different states as healthy, marginal and risk, as shown in Fig. 1. Every states adequacy indexes of generating system are calculated by some probabilistic techniques such as Monte Carlo method [7-9]. By doing this, it can accurately assess the isolated power system reliability level.

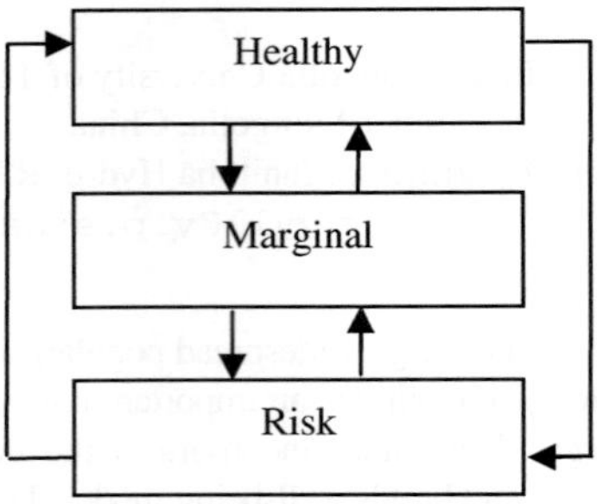

Fig. 1. System well-being model

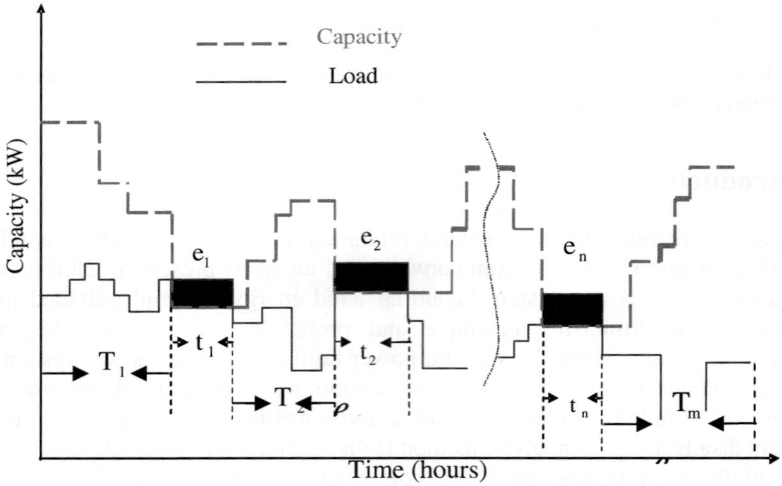

Fig. 2. Superimposition of the capacity states and the chronological load pattern

In Fig. 2, the generation model is superimposed on the load model to obtain the adequacy of generating system. T_i is the periods in which the available capacity exceeds the load and t_i is the periods in which the load exceeds the capacity. The energy curtailed in each case is e_i . The load is modeled in one hour time steps in this analysis. The basic operating strategy of energy storage is as follows: Energy storage charges when the generation power output exceeds load, and the stored energy discharges when the generation power output is shortage [10].

The conventional reliability indices of a small isolated power system can be obtained by recording the loss of load duration t_i and the energy loss e_i due to load curtailment. The well-being reliability indices can be obtained using the time duration T_i and t_i. If the energy stored in the battery (ESIB) is equal to or greater than the average load (AL) or the peak load (PL) multiplied by the accepted the number of autonomous hours (NAH), the system condition is healthy and the corresponding duration is a healthy state duration designated as $T_i(H)$ for each healthy state. On the other hand, whenever the ESIB is less than the AL or PL multiplied by the NAH, the condition is marginal and the corresponding duration is a marginal state duration and designated as $T_i(M)$ [8].

$$T_i = \begin{cases} T_i(H), & \text{if } \ \text{ESIB} \geq \text{NAH} \times \text{PL(or } \ \text{AL)} \\ T_i(M), & \text{if } \ \text{ESIB} < \text{NAH} \times \text{PL(or } \ \text{AL)} \end{cases} \tag{1}$$

The total time of the system in the healthy state, marginal state and risk state can be respectively stranded for n(H), n(M) and n(R). Well-being reliability indices are calculated by (2)–(5):

$$\text{Healthy state probability: } P(H) = \frac{1}{8760N} \sum_{i=1}^{n(H)} T_i(H) \tag{2}$$

$$\text{Marginal state probability: } P(M) = \frac{1}{8760N} \sum_{i=1}^{n(M)} T_i(M) \tag{3}$$

$$\text{Risk state probability: } LOLP = \frac{1}{8760N} \sum_{i=1}^{n(R)} t_i \tag{4}$$

$$\text{Loss of Health Expectation(h/yr): } LOHE = 8760 - \frac{1}{N} \sum_{i=1}^{n(H)} T_i(H) \tag{5}$$

3 Example System Reliability Evaluation

To evaluate and analyze the reliability of small isolated power system, example system parameters are shown in Table 1 according to the IEEE-RBTS [11]. The hourly chronological load model of the IEEE-RBTS has been used in the example system. Its peak load is 40kW.

The healthy state probability of Case 1 is zero in Table 2, because the system has no energy storage. Depending on the continuous stable of diesel generation power output and load demand, the system can run normally, but it runs in the marginal state for the most of time. Although Case 2 has energy storage devices, this system can satisfy the load demand even when the load have slight fluctuate. Case 2 is still in the marginal state and has high risk state probability. The WTG power output varies with wind speed, so little change of wind speed can cause great change of WTG power output, so Case 2 is in the low reliability level. The LOHE of Case 3 is the smallest in the three kinds of system. It can provide stable electric energy, and its system is healthy and has a high reliability level.

Table 1. Example System Data

Case	Generation and storage	No	Rating	FOR (%)	Failure Rate per year
1	WTG	2	30kW	4	4.6
	Diesel	2	20kW	5	9.2
2	WTG	2	30kW	4	4.6
	Storage	1	300kWh	-	-
3	WTG	2	30kW	4	4.6
	Diesel	2	20kW	5	9.2
	Storage	1	300kWh	-	-

Table 2. The Result of the Well-being Indices for Systems

Case	P(H)	P(M)	LOLP	LOHE(h/yr)
1	0	0.964999	0.035001	8760
2	0.115734	0.22264	0.661625	7746
3	0.988548	0.009694	0.001759	100.322

4 Analysis of Influencing Factors

4.1 Effect of Storage Capacity

Some kinds of reliability indexes in Case 2 and Case 3 are calculated in different storage capacity in order to explain the effect of storage capacity in small isolated power system. Fig. 3 shows the healthy, marginal and risk state probability for the three basic system configurations with different storage capacity levels ranging from 200 kWh to 700 kWh. The healthy state probability increases as the storage capacity increases, but the marginal and risk state probability decreases. With the energy storage capacity increasing, the marginal and risk state turn into healthy state. The reliability of Case 2 is improved.

Fig. 4 shows effect of the storage capacity on the LOHE in Case 3. With the energy storage capacity increasing reliability increases, and the incremental benefit decreases until the energy storage capacity exceeds 500kWh. Fig. 3 shows that the state indexes curves of Case 3 are smooth after the storage capacity which is greater than 300kWh. The reasonable storage capacity is 500kWh in case3 from the Fig. 3 and Fig. 4.

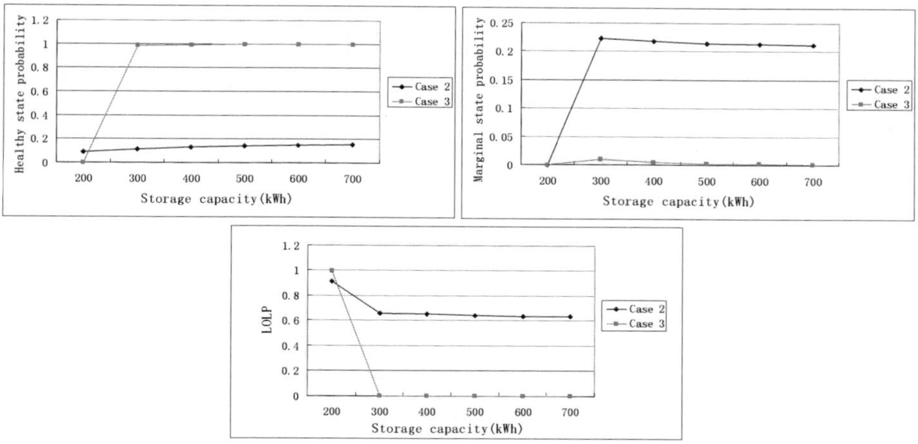

Fig. 3. Effect of the Storage Capacity on the Healthy, Marginal and Risk states

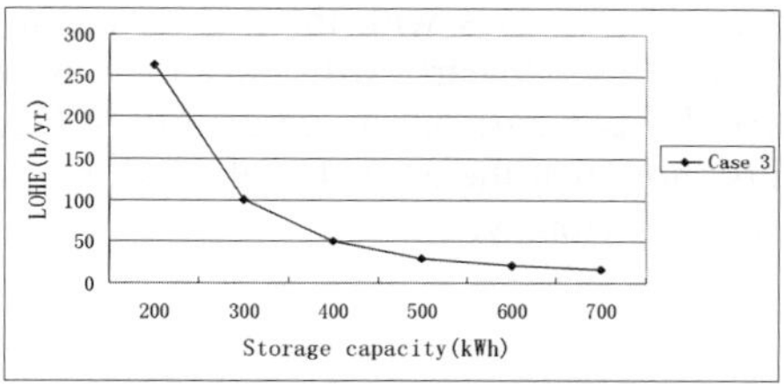

Fig. 4. Effect of the Storage Capacity on the LOHE

4.2 Effect of the Load

The system load is an important factor in reliability analysis. The annual peak load of system normally uses a single load parameter in deterministic method. However the system load often varies with the time. The annual peak load changes from 40kW to 70kW and the step value of annual peak load is 5kW as shown in Fig. 5 and Fig. 6.

Fig. 5 shows that healthy, marginal and risk state probability change with the annual peak load increasing. Healthy state probabilities about Case2 and Case 3 decrease with increasing in the annual peak load; the change curve of each state probability is smoothly when annual peak load is under 55kW. The marginal state probabilities of Case 1 and Case 2 decrease with increasing in the annual peak load. Unlike Case 1 and Case 2, the marginal state probabilities of Case 3 increases as annual peak load increases. Despite the risk state probabilities of three systems increase with increasing in annual peak load, the risk state probability of Case 3 changes slowly. Its value is the lowest in the three systems.

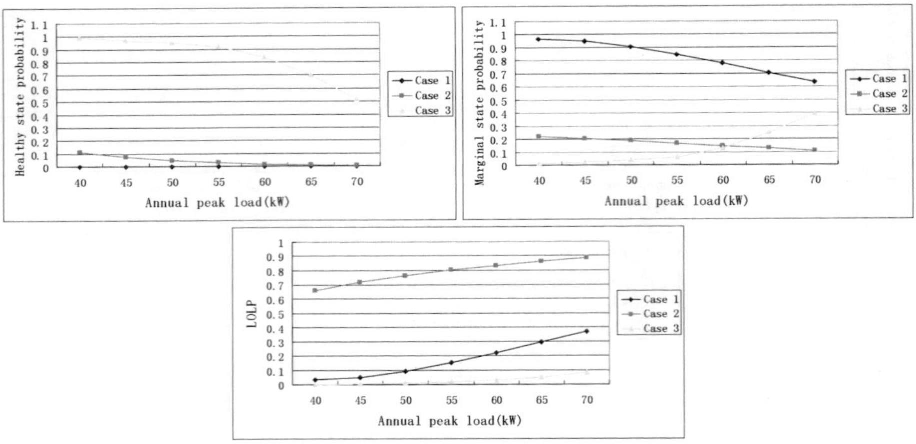

Fig. 5. Effect of the annual peak load on the Healthy, Marginal and Risk states

Fig. 6 shows the effect of the annual peak load on the LOHE in three systems. The LOHE of Case2 and Case 3 increases with the increasing in annual peak load. The LOHE of Case 3 almost increases linearly with the annual peak load when the annual peak load exceeds 55kW. By comparison, the reliability of Case 3 is better than other two systems. The results show that the annual peak load is a sensitive variable which influences system reliability seriously.

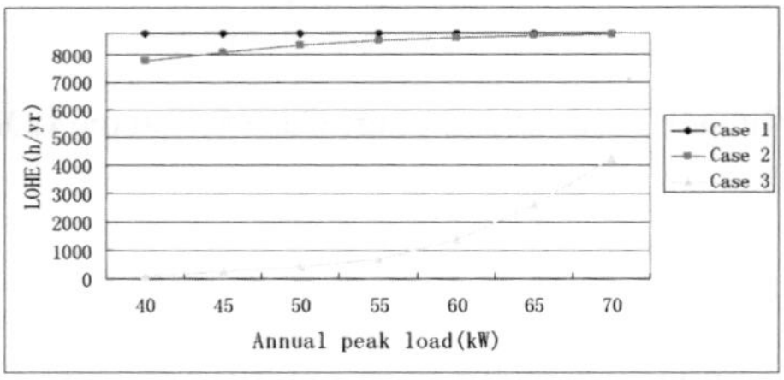

Fig. 6. Effect of the annual peak load on the LOHE

5 Conclusions

This paper developed well-being model, which can categorize the system operation state clearly. Accuracy of reliability evaluation is improved by using the well-being model in small isolated power systems. Both the energy storage capacity and the annual peak load have influence on each state probability index and LOHE, but the influence degree is different. When energy storage capacity increases, healthy state probability increases, and the reliability of system also is improved. Especially, the LOHE increases as the annual peak load increases, but the healthy state probability reduces, and the small isolated power system is in a low reliability level. The results of this paper can provide a method to choice the energy storage equipment capacity and the maximum annual peak load, which is accepted by system, in the small isolated power systems.

Acknowledgments

This work is supported by the National Natural Science Foundation of China (Project # 50967002) and the Inner Mongolia Natural Science Foundation (Project # 200711020702).

References

1. Yongji, G.: Power System Reliability Analysis. M. Tsinghua University Press (2003)
2. Billinton, R., Li, W.Y.: Reliability Assessment Electric Power Systems Using Monte-Caro Methods. M. Plenum Press (A division Of Plenum Publishing Corporation), New York (1994)
3. Li, G.-y., Gao, Y.-j., Zhou, M.: Sequential Monte Carlo Simulation Approach for Assessment of Available Transfer Capability. J. Proceedings of the CSEE 28(25), 74–79 (2008)
4. Jiang, W., Yan, Z., Feng, D.: A review on reliability assessment for wind power. J. Renewable and Sustainable Energy Reviews 13, 2485–2494 (2009)
5. Alexiadis, M.C., Dokopoulos, P.S., Sahsamanoglou, H.S., Manousaridis, I.M.: Short—Term Forecasting Of Wind Speed And Related Dlectrical Power. J. Solar Energy 63(1), 61–68 (1998)
6. Billinton, R., Chen, H.: Assessment of Risk-Based Capacity Benefit Factors Associated With Wind Energy Conversion Systems. J. IEEE Transactions on Power Systems 13(3), 1191–1196 (1998)
7. Gregor, T., Gubina, A.F.: Energy-based system well-being analysis for small systems with intermittent renewable energy sources. J. Renewable Energy 34, 2651–2661 (2009)
8. Bagen, Billinton, R.: Incorporating Well-Being Considerations in Generating Systems Using Energy Storage. J. IEEE Transactions On Energy Conversion 20(1), 225–230 (2005)
9. Zhang, X.-y., Wei, Z.-n.: Impact of Wind Engergy on the Reliability of Small Isolated Power Systems. J. Jiangsu Electrical Engineering 25(3), 58–60 (2006)
10. Bagen, Billinton, R.: Impacts Of Energy Storage On Power System Reliability Performance. In: 18th Canadian Conference on Electrical and Computer Engineering, pp. 494–497. IEEE Press, Saskatoon (2005)
11. Allan, R.N., Billinton, R., Goel, L., Sjarrief, I., So, K.S.: A reliability test system for educational purposes-basic distribution system data and results. J. IEE Transactions on Power Systems 1(2), 813–820 (1991)
12. Billinton, R., Bagen, Cui, Y.: Reliability evaluation of small stand-alone wind energy conversion systems using a time series simulation method. J. IEE Proc.-Gmrr Truasn. Dinrb. 150(1), 96–100 (2003)

Ka Band Frequency Reconfigurable Microstrip Antenna Based on MEMS Technology

Zhongliang Deng and Yidong Yao

Hongtong Building, room 315, Beijing University of Posts and Telecommunications, 100876, Beijing, P.R. China
dengzhl@bupt.edu.cn, yaoyidong@gmail.com

Abstract. A Ka band microstrip patch antenna utilizing RF MEMS switches is proposed and analyzed. The patch is dumbbell-shaped. Two planar spiral structures are arranged at both sides of the central patch. MEMS switches are used for controlling the status of connection between the planar spirals and the central patch, therefore, the structure of antenna can be switched and frequency reconfiguration is realized. The central operation frequencies are 33.2GHz and 35.1GHz, respectively.

Keywords: frequency reconfigurable, Ka band, MEMS, antenna.

1 Introduction

Due to wide frequency bandwidth and little electromagnetic interference, Ka band (26.5GHz-40GHz) is widely used in the fields of satellite television reception, airport management, armamentarium, environment detection, secure communication, etc. Some applications in these fields need to work in multi-frequency [1]. Because of wide frequency span, conventional multi-frequency communication in Ka band mostly uses a plurality of antennas operating in different frequencies for switching, therefore, the size and weight of the antenna system increases and the application in some occasions that has special requirements of size and weight, such as handheld terminals, aviation and aerospace, is limited. Reconfigurable antenna has a common aperture and can replace multi-antenna by real-time structure transformation, so reconfigurable Ka band antenna can reduce the size and lower the weight effectively. Therefore the application range of the existing antenna can be expanded. The most commonly used mechanisms of antenna structure reconfiguration are PIN diode, optical fiber switch and radio frequency micro-electromechanical system (RF MEMS) switch [2-4]. The PIN diode has fast response speed but poor isolation and high insert loss which may decrease the radiation efficiency of the antenna. The optical fiber switch has low power dissipation but high insert loss and is inconvenient to integrate with printed circuit. The RF MEMS switch has low insert loss, high isolation, small size, light weight and convenient for integrating due to the CMOS fabrication process. The only weakness is the slow response speed which is at the level of microsecond

X. Wan (Ed.): Electrical Power Systems and Computers, LNEE 99, pp. 535–541.
springerlink.com

(10-6s). This design utilizes RF MEMS switches as the reconfiguration mechanism. The designed antenna can work at 33.2GHz and 35.1GHz.

2 Numerical Simulation and Analysis of Microstrip Antenna

The designed dumbbell-shaped antenna adopts rectangle microstrip patch antenna as the basic radiation antenna. The primary radiation unit of the patch antenna can be equivalent to a magnetic stream element. The wave source is the side walls of the substrate cavity between the patch and the ground. Electromagnetic waves radiate from the open magnetic wall utilizing resonance in the cavity. According to the cavity mode theory [5], when the rectangle microstrip antenna operates at its dominant mode TM010, the length of the patch can be calculated by the following equation (1).

$$L = \frac{\lambda_g}{2} - 2\Delta l = \frac{\lambda_0}{2\sqrt{\varepsilon_e}} - 2\Delta l = \frac{c}{2f_r\sqrt{\varepsilon_e}} - 2\Delta l \tag{1}$$

Hereinto, c is the speed of light in vacuum. λ_g is the wavelength in the dielectric. λ_0 is the wavelength in free space. ε_e is the effective dielectric constant. The relation between ε_e and the relative dielectric constant ε_r is as follows:

$$\varepsilon_e = \frac{\varepsilon_r + 1}{2} + \frac{\varepsilon_r - 1}{2}\left(1 + \frac{10h}{W}\right)^{-\frac{1}{2}} \tag{2}$$

Hereinto, h is the thickness of the substrate. W is the width of the patch. Δl is the equivalent electrical length of the radiation aperture, which is corresponding to the capacitive stored energy field of the radiation reactance.

$$\Delta l = 0.412 \frac{(\varepsilon_e + 0.3)(\frac{W}{h} + 0.264)}{(\varepsilon_e - 0.258)(\frac{W}{h} + 0.8)} h \tag{3}$$

ε_e is increasing when ε_r is increasing according to equation (2). Therefore, under the condition that the operating frequency is fixed, the increasing of the effective dielectric constant ε_e will lead to decreasing of the wavelength in dielectric, which means high dielectric substrate can help reduce the size of the antenna. However, if the dielectric constant increases up to some level, the dielectric will bound the electromagnetic field and weaken the radiation, so the value of the dielectric constant shall not be too high.

The width of the patch *W* has influence on the dielectric constant. When *W* increases, the effective dielectric constant increases, which means larger *W* can help reduce the length of the patch. Also, appropriate enlargement of *W* can increase the radiation efficiency and the bandwidth [6]. However, if *W* is larger than $\frac{\lambda_0}{2}$, high order modes will appear and cause distortion of the field and lower the radiation ability of the dominant mode, so *W* shall not be larger than *L*.

The thickness of the substrate is also important to the effective dielectric constant. When *h* increases, the effective dielectric constant will decrease, so the operating frequency of the antenna will decrease. This phenomenon is particularly obvious in Ka band. The increase of *h* can cause serious loss in the substrate which is harmful for radiation in Ka band. Thus the thickness of the substrate shall not be too large. Usually, the maximum value of *h* shall not exceed h_0.

$$h_0 = 0.05 \frac{c}{f_r \sqrt{\varepsilon_r}} \tag{4}$$

3 Design and Analysis of Frequency Reconfigurable Antenna Based on MEMS Technology

The primary radiation patch of the designed antenna is the dumbbell-shaped patch. According to the equations (1), (2), (3) and (4), the size of the narrow section of the dumbbell-shape is L1=4.2mm, W1=1.2mm. The size of the rectangle at both ends is L2=6mm, W2=4.8mm. The substrate material is chosen to be Rogers RT/druid 5880(tm). The relative dielectric constant is 2.2. The size of the substrate is Wsub=Lsub=20mm, Hsub=0.5mm. The feeding method is chosen to be the coaxial cable from the bottom. The outer radius of the coaxial cable is Rout=0.28mm and the inner radius is Rin=0.08mm. The feeding point is at the center of the patch. The schematic figure of the antenna is shown in Fig. 1.

The operating frequency of the microstrip antenna is changed along with the switching of the shape and size of the patch, so the transformation of the patch shape can be used to realize frequency reconfiguration. The designed reconfigurable antenna is obtained by arranging planar spiral structures at both sides of the central patch as the additional patches, and the operating frequency of the new structure will be higher, thus size reduction can be achieved. The spiral is 3.5mm away from the central patch, which is shown by L4 in Fig 1. The width of the lines comprising the spiral is W4=0.3mm. The outermost size of the spiral is L3=3mm, W3=2mm. The inner gaps of the spiral are Gap1=Gap2=0.5mm.

Two MEMS switches are arranged between the dumbbell-shaped patch and the planar spiral patches for controlling the shape of the antenna.

The MEMS switch has the characteristics of ideal switching, high on-off ratio and good isolation of RF signals from DC to over 40GHz [7], so it can be used in the design of Ka band antenna. The structure of the capacitance switch used in the design is shown in Fig. 2.

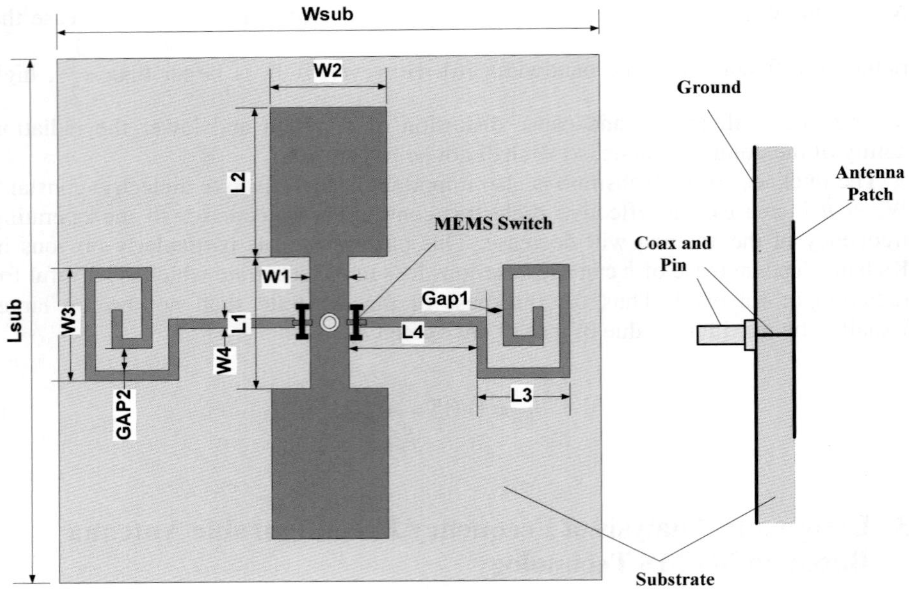

Fig. 1. The structure diagram of the designed antenna

The switches are made utilizing MEMS fabrication process such as scarification process and reaction ion etch (RIE) method, and these process are integrated with the antenna process well. The length of the MEMS bridge is 350μm. The width of the bridge is 100μm.

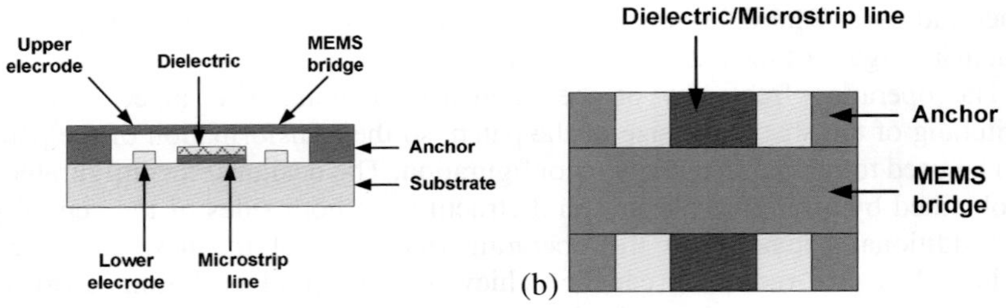

Fig. 2. Diagrams of MEMS switch: (a) sectional view, (b) top view.

The transmission coefficient (S21) and the reflection coefficient (S11) are simulated using Ansoft HFSS(tm) and shown in Fig. 3.

S21 of the ON state switch is higher than -0.11dB (more than 99.89% energy transmitted) in the frequency band from 5 to 40GHz. The isolation is higher than 17.5dB with the maximum value of 20.9dB and the insert loss is lower than 0.5dB in Ka band.

S11 of the OFF state switch is higher than -0.04dB from 26 to 40GHz (more than 99.96% energy reflected). The isolation is higher than 23.3dB with the maximum value of 32dB and the insert loss is below 0.5dB in Ka band.

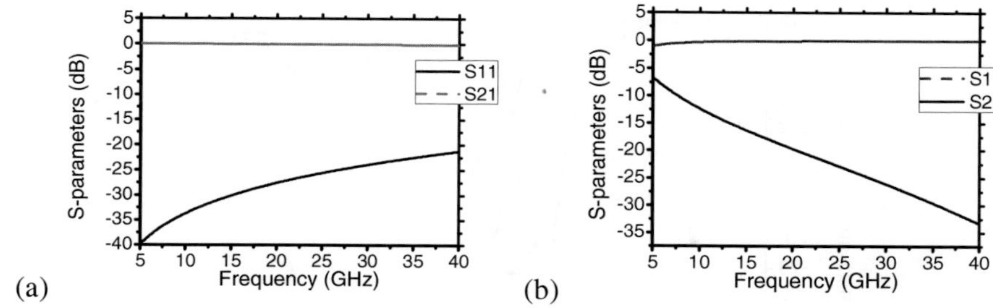

Fig. 3. S parameters: (a) switch is ON, (b) switch is OFF.

Due to the ideal performance of the designed switch in Ka band, the reconfiguration mechanism of the antenna is discussed by using an ideal model of the MEMS switch.

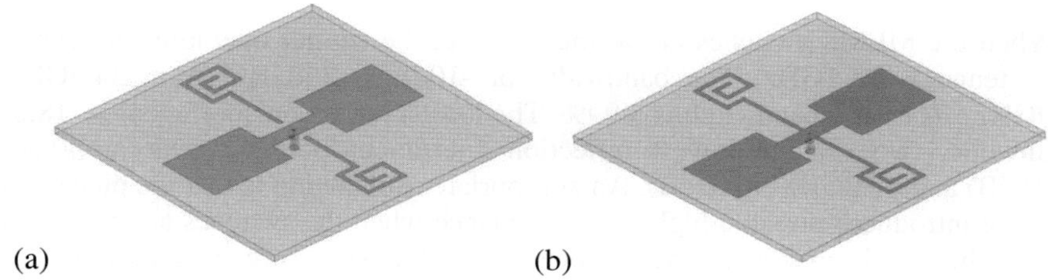

Fig. 4. Equivalent radiation structures. (a) switch is OFF, (b) switch is ON.

When DC control bias voltage is not added between the upper and lower electrodes, the switches are at the off state. The planar spiral patches are not connected with the dumbbell-shaped patch and the antenna mainly utilizes the central dumbbell-shaped patch to radiate. The equivalent radiation structure is shown in Fig. 4(a). When DC control bias voltage is added between the upper and lower electrodes, the switches are at the on state. The planar spiral patches are then connected with the dumbbell-shaped patch to form a united double wing shaped patch and the antenna utilizes the new patch to radiate, which is shown in Fig 4(b). Therefore, the on-off state of the MEMS switches changes the patch structure of the antenna, thereby realizes the frequency reconfiguration.

The reflection coefficient at the feeding port of the coax is simulated and the result curve diagrams are shown in Fig. 5.

The radiation pattern in the E-plane and H-plane are shown in Fig. 6 and Fig.7, respectively.

When the MEMS switches are at the off state, the central operating frequency of the antenna is 33.2GHz. The bandwidth of -10dB is 1.8GHz (from 32.3GHz to 34.1GHz). The ratio bandwidth is 5.42%. The lowest reflection coefficient is -26.94dB.

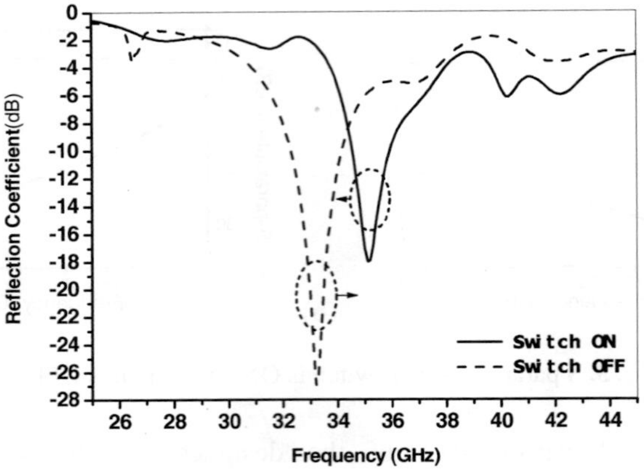

Fig. 5. The reflection coefficient curves of the antenna when switches are ON and OFF.

When the MEMS switches are at the on state, the central operating frequency of the antenna is 35.1GHz. The bandwidth of -10dB is 1.4GHz (from 34.5GHz to 35.9GHz). The ratio bandwidth is 3.99%. The lowest reflection coefficient is -18.0dB. Meanwhile, two relative lowest reflection coefficient points appear at 40.3GHz (-6.16dB) and 42.2GHz (-6.02dB). An appropriate explanation is that the planar spiral structure introduces possible high order resonance when the switches are on. At these possible high order resonance frequencies, a small amount of energy may radiate or lose, but the interference to the dominant radiation is very limited because the reflection coefficient is less than 10dB.

Analysis on the whole, under the control of MEM switches, the antenna shows a fairly good frequency switching over performance. The antenna can work at 33.2GHz and 35.1GHz, respectively.

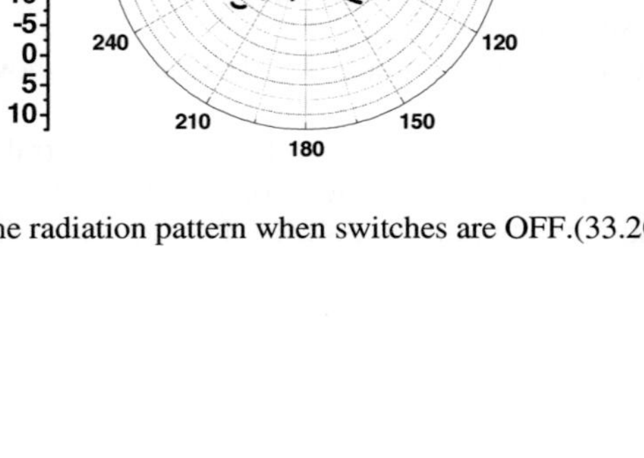

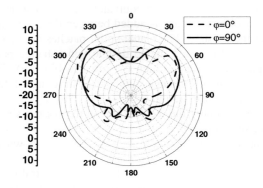

Fig. 6. The radiation pattern when switches are OFF.(33.2GHz)

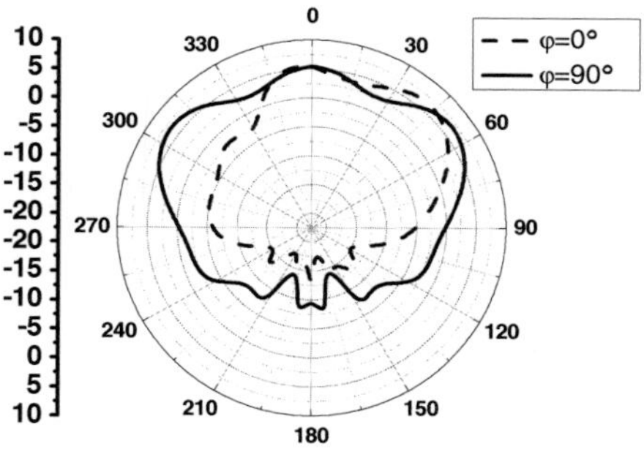

Fig. 7. The radiation pattern when switches are ON.(35.1GHz)

4 Summary

A Ka band frequency reconfigurable microstrip patch antenna has been proposed and analyzed. The reconfiguration mechanism is realized by changing the structure of the radiation patch. In this work, two planar spiral patches are arranged at the sides of the dumbbell-shaped patch. MEMS switches are utilized for controlling the connection to change the effective radiation structure of the antenna. At the two states, the designed antenna can work at 33.2GHz and 35.1GHz, respectively. The result shows that the method has reference significance to Ka band frequency reconfigurable antenna.

Acknowledgments. The work presented in this paper is supported by the Fundamental Research Funds for the Central Universities (2010PTB-03-03).

References

1. Yu, Y.C., Zuo, F.Q.: Millimeter Wave Technology and Military Applications. Weaponry Industry Press, Beijing (1991)
2. Guo, X.L., Cai, M., Liu, L., et al.: Designs for a Ku-Band Miniature MEMS Reconfigurable Antenna Based on Si Substrate. Chinese Journal of Sensors and Actuator 19(6), 2425–2427 (2006)
3. Nikolaou, S., Bairavasuvramanian, R., Lugo, C., et al.: Pattern and Frequency Reconfigurable Annular Slot Antenna Using PIN Diodes. IEEE Transaction on Antenna Propagation 54(2), 439–448 (2006)
4. Roach, T.L., Huff, G.H., Bernhard, J.T.: On the Applications for a Radiation Reconfigurable Antenna. In: 2nd NASA/ESA Conference on Adaptive Hardware and Systems, Edinburgh, pp. 7–13 (2007)
5. Lin, C.L.: Antenna Engineering Handbook, pp. 467–476. Electric Industry Press, Beijing (2002)
6. Bahl, I.J., Bhartia, P.: Microstrip Antennas, p. 46. Artech House, Boston (1980)
7. Brown, E.R.: RF-MEMS switches for reconfigurable integrated circuit. IEEE Transactions on Microwave Theory and Techniques 46(11), 1868–1880 (1998)

Optimization and Experimental Research of DRM Data Transmission Based on LT Code

Yahui Hou[1], Le Jin[2], Yi Chen[3], and Jinhui Xie[4]

[1] Enginering Research Center of Digital Audio and Video, Communication University of China
[2] Chengdu Newstar electronics Co. Ltd, Chengdu, China
[3] Chengdu Textile College, dept. EE, Chengdu, China
[4] Academy of Broadcasting Planning, SARFT, China
houyh@cuc.edu.cn, le.jin@cdnse.com, pure@cdnse.com, xiejinhui@abp.gov.cn

Abstract. This paper proposes an optimization method based on LT (Luby Transform) codes in order to resolve the lower efficiency problem caused by error code and packet-lost in Digital Radio Mondiale (DRM) data transmission. The Optimization effect is tested and analyzed. The experimental results demonstrate that this optimization method can notably improve the data transmission efficiency in DRM.

Keywords: LT code, DRM, MOT, Critical packet-lost rate.

1 Introduction

DRM (Digital Radio Mondiale, ETS ES 201 980) is the world's digital AM broadcasting standard under 30MHz, which is used in digital system for short-wave (HF), medium-wave (MF) and long-wave (LF). Because of the full use of the advanced digital encoding and transmission technology, DRM system is considered as an advanced digital broadcast system and has the ability of multimedia data transmission. Implementation of digital broadcasting in today's AM bands based on DRM technique will enable broadcasting operators to provide all kinds of multimedia services in addition to high-quality audio services. Because AM broadcast channel has time-varying feature and the available bandwidth is narrow, usually there is not only error code but also packet-lost at the terminal end as a result of burst errors in data transmission and synchronization problem of receiver's own. Packet-lost leads to that inefficient transmission of the large-capacity data object in DRM system. Under the poor channel conditions, the time required to receive a full data object successfully will be intolerable for the terminal. Therefore, using the latest data encoding technology to improve the data transmission efficiency is important for multimedia broadcasting system based on DRM technology.

LT (Luby Transform) Code is a practical digital fountain code, which is rate-independent and has linear codec complexity. Degree distribution of LT code is designed to implement the sparse coding matrix. LT code has a lower probability of

X. Wan (Ed.): Electrical Power Systems and Computers, LNEE 99, pp. 543–550.
springerlink.com

decoding failure in the massive data transmission. It is appropriate to use LT code in those data applications with large-capacity data objects [3-6]. For the sake of resolving large-capacity data transmission problem in DRM system, the paper introduces digital fountain code (LT code) technology into DRM system by inserting LT code into MOT pre-coding module. By making full use of rate-independent feature of LT codes, and making the effective use of accumulated data which has been transmitted, the terminal receiver achieves the ability of “resuming broken transfer” without ARQ (Automatic Repeat request) [7]. The experimental results demonstrate that this optimization method can notably improve the DRM data transmission efficiency.

2 Joint Pre-coding Scheme Design

The joint pre-coding scheme is specifically applied to the data transmission of DRM system. In the scheme LT code is embedded into MOT pre-coding module as outer code (Convolution code as inner code). Convolution code can effectively improve the ability of LT code symbol- recovery and the ability of system real-timing decoding. On the other hand, LT code can effectively reduce the probability of repeatedly receiving caused by packet-lost at the terminal. In addition, this scheme can maximize the retention of original codec structure in DRM system, only increase the LT codec module with no changes involved in following modulation part.

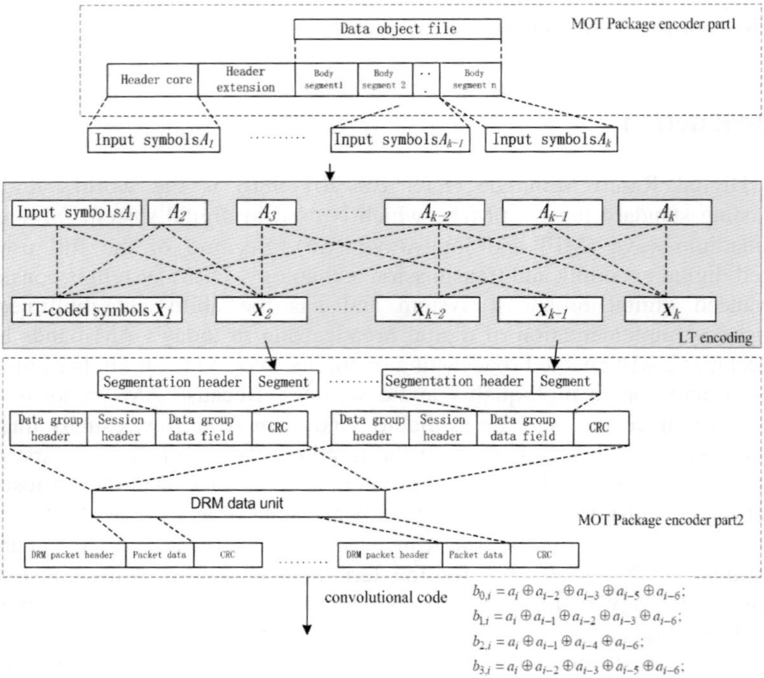

Fig. 1. Schematic diagram of joint pre-coding (LT code embedded into MOT)

The joint pre-coding process is shown as figure.1. First, the data object file is processed according to MOT Package encoder part1[2] in figure1. This step generates K input symbols Ai ($1 \leq i \leq k$). Second, input symbols Ai ($1 \leq i \leq k$) are coded to generate K LT-coded symbols Xj ($1 \leq j \leq k$), the LT-coded symbols Xj are mapped to data segment according to MOT Package encoder part2[2] in figure1, and then the following process is same as MOT original encoding process.

Considering channel characteristics and the effect of data transmission, degree distribution of the scheme adopts robust soliton.

3 Test and Data Analysis

3.1 Test of the Critical Packet-Lost Rate

Definition of critical packet-lost rate: gradually increase packet-lost rate in the test environment while keeping other parameters unchanged, when the actual time needed by the test system to transmit a data file successfully reaches 50 times of the theoretical error-free transmission time, then the packet-lost rate at the moment in test environment is called critical packet-lost rate.

Testing purposes: through extensive tests to obtain the time sample space which DRM system (without LT code) successfully transmitted data given size at different packet-lost rate, sum up the critical packet-lost rate in specific experimental conditions. The test result can be taken as comparison for the same test of DRM system (optimized by LT code).

Test conditions: DRM mode: B; the bandwidth:10kHz; MSC mapped mode: 64QAM; Data services: Slide-Show (Packet Mode); Bit-rate: 32000kbps; Data objects: 200KB JPEG image and 1MB (1024KB) image.

The figure of critical packet-lost rate test is shown as follows:

- 200KB data object (Packet size: 640 bits; Package Number:2560; Theoretical error-free transmission time: 51.2 s.)

The test shows that the critical packet-lost rate is about: 3.906E-04.

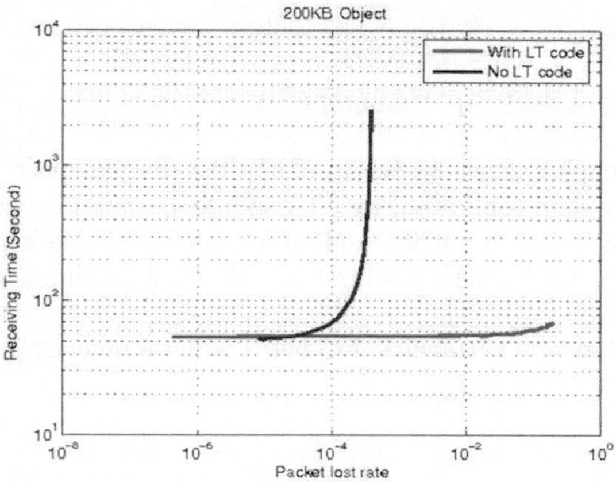

Fig. 2. The test of critical packet-lost rate (data object size: 200K Bytes)

- 1MB data object (Packet size: 640 bits; Package Number: 13108; Theoretical error-free transmission time: 262.144 s.)

The test shows that the critical packet-lost rate is about: 7.629E-05.

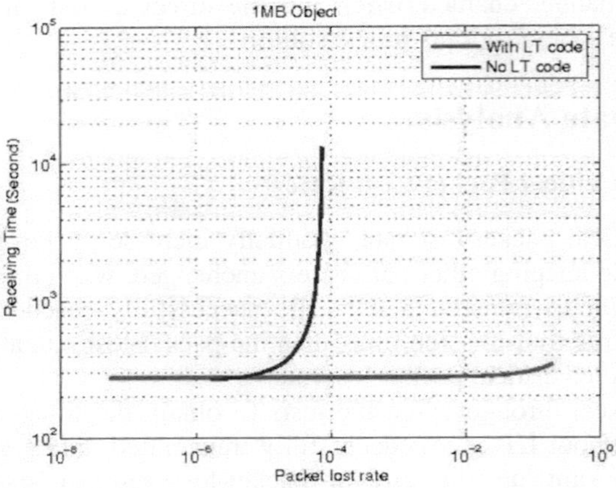

Fig. 3. The test of critical packet-lost rate (Data object size: 1M Bytes)

The test results of the critical packet-lost rate show that: In the process of packet-lost rate increasing until close to the critical packet-lost rate, the time that DRM system (without LT code) successfully transmitted data objects exponentially rise; When the packet-lost rate exceeded the critical packet-lost rate, the possibility of successfully receiving data object in DRM system (without LT code) is close to 0; Compared to DRM system (without LT code), DRM system (optimized by LT code) has a better and stable performance independent of impact of critical packet-lost rate during the process of the same data object transmission test.

3.2 The Test of Transmission Probability Distribution

Test purpose: through extensive tests to obtain the sample space of packet number needed by DRM system (optimized by LT code) to successfully transmit data object in different channel environments, and to research the relationship between the probability of successful data transmission and the distribution of packet number required to receive.

Tested Channel model: Channel 1, 2, 3 [1]; Test Data objects: 200KB JPEG image and 1024KB (1MB) JPEG image.

(1) DRM channel 1

Table 1. Environmental parameters and test condition of channel 1.

Channel no 1: AWGN		Good Typical/moderate bad	LF,MF,HF LF,var.SNR	
	Path 1	Path 2	Path 3	Path 4
Delay($\triangle k$)	0			
Path gain,rms(ρ_k)	1			
Doppler shift(Dsh)	0			
Doppler spread(Dsp)	0			
Test condition--- mode B; Bandwidth:10kHz; Service: Slide-Show; Bit-rate:32000bps; SNR: 18dB; Packet-lost rate: 11.483%				

The test figure of transmission probability distribution in channel 1 is shown as follows:

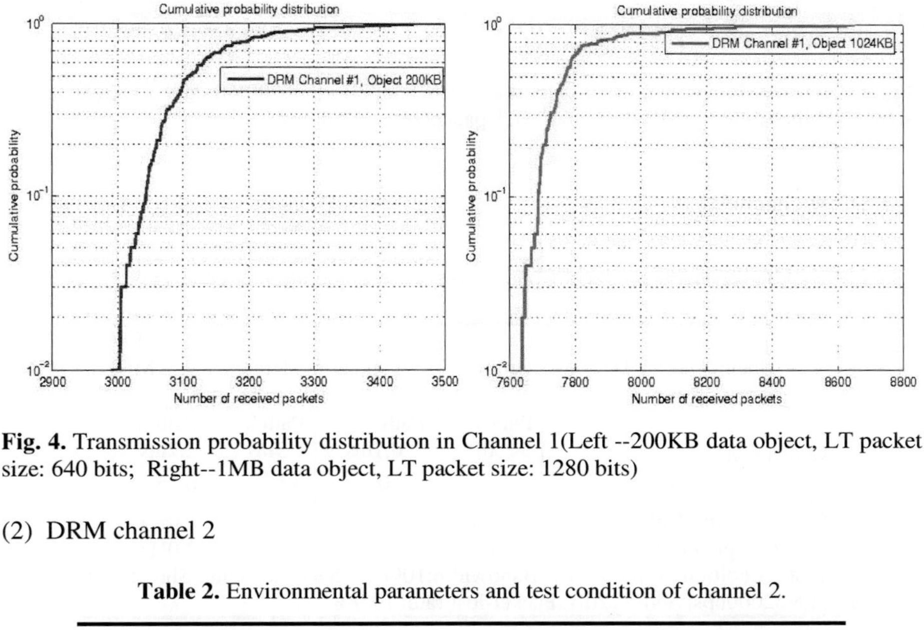

Fig. 4. Transmission probability distribution in Channel 1(Left --200KB data object, LT packet size: 640 bits; Right--1MB data object, LT packet size: 1280 bits)

(2) DRM channel 2

Table 2. Environmental parameters and test condition of channel 2.

Channel no 2: Rice with delay		Good Typical/moderate bad	MF,HF	
	Path 1	Path 2	Path 3	Path 4
Delay($\triangle k$)	0	1ms		
Path gain,rms(ρ_k)	1	0.5		
Doppler shift(Dsh)	0	0		
Doppler spread(Dsp)	0	0.1Hz		
Test condition--- mode B; Bandwidth:10 kHz; Service: Slide-Show; Bit-rate:32000bps; SNR: 20dB; Packet-lost rate: 2.286%				

The test figure of transmission probability distribution in channel 2 is shown as follows:

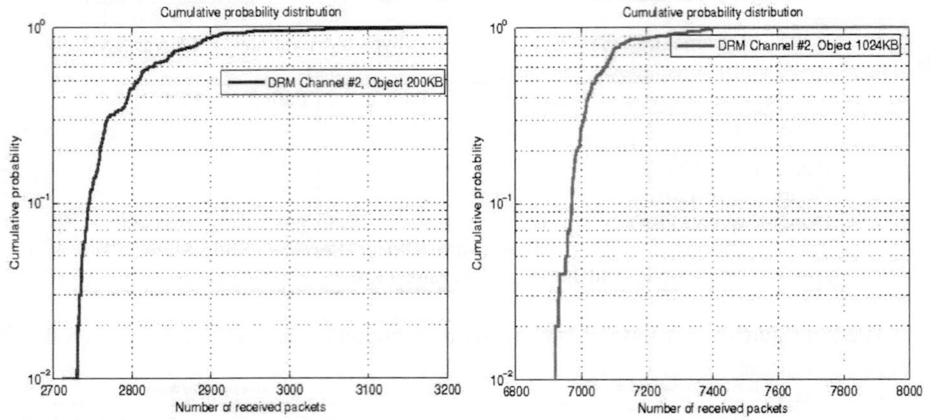

Fig. 5. Transmission probability distribution in Channel 2(Left -- 200KB data object, LT packet size: 640 bits; Right -- 1MB data object, LT packet size: 1280 bits)

(3) DRM channel 3

Table 3. Environmental parameters and test condition of channel 3.

Channel no 3: US Consortium		Good Typical/moderate bad		HF MF
	Path 1	Path 2	Path 3	Path 4
Delay($\triangle k$)	0	0.7ms	1.5ms	2.2ms
Path gain,rms(ρ_k)	1	0.7	0.5	0.25
Doppler shift(Dsh)	0.1Hz	0.2Hz	0.5Hz	1.0Hz
Doppler spread(Dsp)	0.1Hz	0.5Hz	1.0Hz	2.0Hz
Test condition--- mode B; Bandwidth:10kHz; Service: Slide-Show; Bit-rate:32000bps; SNR: 22dB; Packet-lost rate: 19.9%				

The test figure of transmission probability distribution in channel 3 is shown as follows:

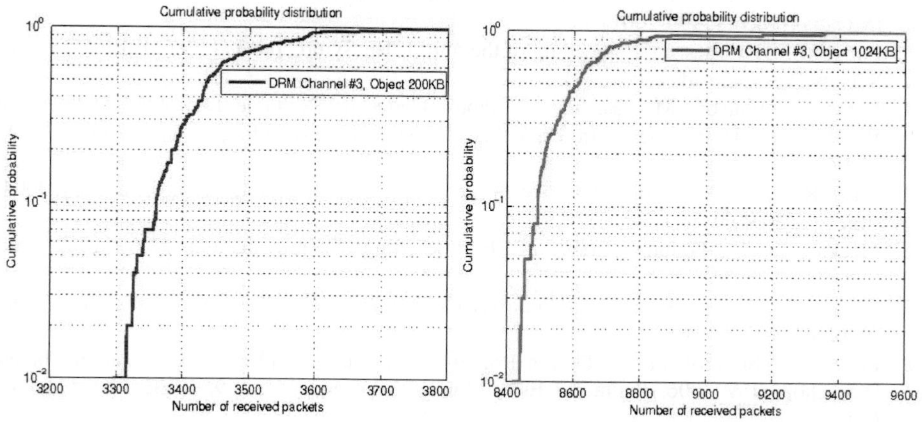

Fig. 6. Transmission probability distribution in Channel 3(Left -- 200KB data object, LT packet size: 640 bits; Right -- 1MB data object, LT packet size: 1280 bits)

Test results of the transmission probability distribution show that: in the actual transmission environment with large packet-lost rate, it is almost impossible for DRM system (without LT code) to successfully transmit large-capacity data object (>200K Bytes). Compared to DRM system (without LT code), large-capacity data object (>200K Bytes) can be transmitted successfully in DRM system (using LT coding) as long as the terminal receives enough packets.

4 Conclusion

As a result of narrow available bandwidth of Digital AM radio and the complex channel environment, a large number of error symbols will appear in the process of data transmission. Then the receiver can not make effective use of accumulated data which has been transmitted due to the high packet-lost rate, which causes the lower receiving efficiency. To improve the efficiency of large-capacity data transmission in DRM system, this paper proposes a joint coding scheme by combining LT code with the existing convolution code. The test results show that the optimization scheme can significantly improve the efficiency of DRM system for large-capacity data transmission, as well as the probability of receiver's correctly receiving data objects. Therefore the research on DRM transmission optimization based on LT code has important practical significance to the future development and application of DRM-based digital broadcast system.

References

1. ETSI ES 201 980 V2.2.1[S]. Digital Radio Mondiale (DRM), System Specification (2005)
2. EN 301 234 V1.2.1. Digital Audio Broadcasting (DAB), Multimedia Object Transfer (MOT) protocol (1999-2002)
3. Luby, M.: LT Codes. In: Proceeding of the 43rd IEEE Symp., pp. 271–280. IEEE Computer Society Press, Canada (2002)
4. Luby, M., Watson, M., Gasiba, T.: Raptor Codes for Reliable Download Delivery in Wireless Broadcast Systems. In: USA IEEE CCNC 2006 proceedings, Canada, pp. 192–197 (2006)
5. Karp, R., Luby, M., Shokrollahi, A.: Finite Length Analysis of LT Codes. In: ISIT 2004, Chicago (2004)
6. Mackay, D.J.C.: Fountain Codes. IEEE Proceedings Communications 52(6), 1062–1068 (2005)
7. Castura, J., Mao, Y.: Rateless Coding over Fading Channels. IEEE Communications Letters (S10897798) (2006)
8. Pakzad, P., Shokrollahi, A.: Design Principles for Raptor Codes. In: Information Theory Workshop ITW 2006, Punta del Este, March 13-17, pp. 165–169. IEEE, Los Alamitos (2006)
9. Puducheri, S., Kliewer, J., Fuja, T.E.: Distributed LT Codes. In: IEEE International Symposium on Information Theory, July 2006, pp. 987–991 (2006)

Coal Gas Predication Based on Improved C-C Chaotic Time Series Algorithm*

Xian-Min Ma

College of Electrical and Control Engineering
Xi'an University of Science & Technology Xi'an, China
maxm@xust.edu.cn

Abstract. An improved C-C algorithm based on the chaotic time series theory to predicate coal gas concentration sudden emission is discussed in this paper. By the improved C-C algorithm the optimization embedded dimension m and the time delay τ are selected to raise correlation integration computing speed in the gas concentration reconstructed phase space. Simulation results show that the coal gas concentration time series has a chaotic characteristic, and the best embedded dimension m and best delay τ are higher accuracy than the traditional C-C algorithm in the dimensional phase space. The results prove that the proposed C-C algorithm to determine the time delays and the embedding dimensions is effective if the time series data are polluted with noises.

Keywords: Coal Gas Predication, Chaotic Time Series, Improved C-C Algorithm, Phase Space Reconstruction.

1 Introduction

The coal gas concentration sudden emission occurred in the underground coal mine is very complex nature phenomena, which are influenced by many factors such as the outburst gas quantity size. In the actual production process, the coal gas concentration is daily measured, therefore the coal gas concentration sudden emission quantity time series can be easily obtained. It is the solid foundation for the chaotic time series theory to predict the coal gas concentration variation with these known data [1].

But these measured data are often disturbed by means of noises, which lead to the inaccuracy prediction model for gas concentration change after filter. The computing process of the embedded dimension m and the time delay τ for the reconstructed gas concentration phase space is much influenced especially when C-C algorithm is used to select the parameter embedded dimension m and the time delay τ. In this paper an improved C-C algorithm is discussed based on the chaotic time series theory to predicate coal gas concentration sudden emission with these data [2].

This paper is organized as follows. After this introductory section, in the section 2 some important impacting factors for the coal gas concentration are introduced. In section 3, the original coal gas concentration time series data are analyzed and an

* Scientific Research Program Funded by Shaanxi Provincial Education Commission (Program NO.2010JK663).

X. Wan (Ed.): Electrical Power Systems and Computers, LNEE 99, pp. 551–558.
springerlink.com

improved C-C algorithm is discussed. The chaotic time series prediction steps are introduced in section 4. A neural network model of coal gas concentration predication model is introduced in section 5.Conclusions are drawn in section 6.

2 Impact Factors for Coal Gas

The coal gas in the mining bed is produced while the plants become the spoiled organic matters during the millions years. In the biochemistry periods, the following reaction happens:

$$4C_6H_{10}O_6 \rightarrow 7CH_4 + 8CO_2 + C_9H_6O + 3H_2O \tag{1}$$

The coal is deteriorated into gas after following series responses occurred:

$$4C_{16}H_{10}O_6 \rightarrow C_{67}H_{66}O_{10} + 4CO_2 + 3CH_4 + 2H_2O \tag{2}$$

$$C_{67}H_{66}O_{10} \rightarrow C_{54}H_{42}O_6 + 2CH_4 + CO_2 + 3H_2O \tag{3}$$

$$C_{16}H_{14}O \rightarrow C_{15}H_4 + 2CH_4 + H_2O \tag{4}$$

So the main composition of the coal gas is CH_4. If there is enough oxygen, the coal gas is exploded:

$$CH_4 + 2O_2 \rightarrow CO_2 + 2H_2O + 882.6KJ/MOL \tag{5}$$

The blast concentration of the coal gas refers to 9.5% in the mixed gas and oxygen and so on.

From above reaction formulas, the some impact factors for coal gas explosion are as follows.

a) Depth of coal bed: Generally the deeper coal bed is buried, the better the coal gas is sealed. And the concentration Q of coal gas is expressed as follows:

$$Q=6.58+0.038X_1 \tag{6}$$

The parameter X1 denotes the depth of the coal bed.

b) Breathability of coal bed: Generally the permeability of coal and rock decreases, the methane gas content is the greater.

c) Lean angle of coal bed: Lean angle of the coal bed is smaller, and the methane gas density is the higher.

d) Mining technique: The mining techniques include the mining methods, speed schedule of the daily average digging up and daily production and so on [3].

e) Geologic condition: The coal bed geologic condition is one of most important factors affecting the gas concentration.

There are still some factors which are also impacting factors for the coal gas concentration, so only some important factors are listed as the model parameters.

3 Improved C-C Algorithm

The gas concentration is measured. The 500 group gas concentration data collected in half year are shown in Fig.1.

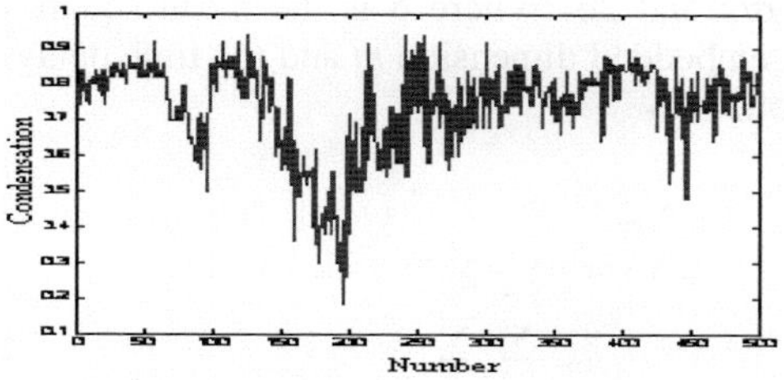

Fig. 1. Collected gas data condensation

From Fig.1, the coal gas time series data are changed in much width ranges. In fact if the appropriate embedding dimension number m and time delay τ can be selected, then reconstruction of phase space can be written as follows:

$$X(t) = \{x(t), x(t-\tau), \cdots, x[t-(m-1)\tau]\}^T \tag{7}$$

The key in the reconstruction of phase space technology is to select the embedded dimension m and the time delay τ. The C-C algorithm is often used to determine embedding delay τ and embedding window width (m-1) τ simultaneously.

The conjunction integration in embedded time series is defined as follows:

$$C(m, N, r, t) = \frac{1}{M^2} \sum_{1 \le i \le j \le M} \theta(r - \|X_i - X_j\|) \tag{8}$$

where, m is embedded dimension, N is the length of time series, r is the neighborhood radius, τ is the time delay. The Heaviside unit function is:

$$\theta(x) = \{\begin{matrix} 0, x < 0 \\ 1, x \ge 0 \end{matrix} \tag{9}$$

The conjunction dimension is expressed as:

$$D(m, \tau) = \frac{\log C(m, r, \tau)}{\log r} \tag{10}$$

The time series can be divided into the non-intersect sequence:

$$\begin{matrix} \{x(1), x(t+1), x(2t+1), \cdots\} \\ \{x(2), x(t+2), x(2t+2), \cdots\} \\ \vdots \\ \{x(t), x(2t), x(3t), \cdots\} \end{matrix} \tag{11}$$

The statistical quantity of the every sub-sequence is calculated as follows:

$$S(m, N, r, \tau) = \frac{1}{t} \sum_{l=1}^{t} \{C_l(m, N/t, r, \tau) - [C_l(1, N/t, r, \tau)]^m\} \tag{12}$$

where, Cl is the conjunction integration of the l sub-sequence.

The differential amount of the corresponding maximum and minimum radius is defined as follows:

$$\Delta S(m,t) = \max[S(m,N,r_i,t)] - \min[S(m,N,r_j,t)], i \neq j \tag{13}$$

According to statistical theory, if the range of m is between 2 and 5, then the radius r is between σ/2 and 2σ. Where σ is the mean square deviation of the time series .Therefore the embedded dimension m and the time delay τ can be calculated in the following equation.

$$\begin{cases} S_{\omega r}(t_i) = \Delta \overline{S}(t) + \left|\overline{S}(t)\right| \\ \Delta \overline{S}(t) = \frac{1}{4}\sum_{m=2}^{5} \Delta S(m,N,t) \\ \overline{S}(t) = \frac{1}{16}\sum_{m=2}^{5}\sum_{j=2}^{4} S(m,N,r_j,t) \end{cases} \tag{14}$$

where, $\overline{S}(t)$ is the mean value of the $S(m,N,r_j,t)$, the first minimum of $\Delta\overline{S}(t)$ is the first local maximum time τ and the minimum of $S_{\omega r}(t)$ is the corresponding delay time window. In practice it is difficult to compute the difference $\Delta\overline{S}(t)$ fast sometime. Therefore it is necessary to improve the traditional C-C algorithm as follows:

$$S_1(m,N,r,\tau) = C(m,N,r,\tau) - C^m(1,N,r,\tau) \tag{15}$$

$$S_2(m,N,r,\tau) = \frac{1}{t}\sum_{l}^{5}\left[C_l(m,N/t,r,\tau) - C_l^m(m,N/t,r,\tau)\right] \tag{16}$$

then,

$$\Delta S = \left|\overline{S}_1 - \overline{S}_2\right| \tag{17}$$

The △S has the same dynamic period with (14) and (15), so the noise effectiveness can be decreased and the computing speed is faster.

4 Chaotic Time Series Predication

According to the chaotic time series prediction theory, the one dimensional observation coal gas concentration data are first changed to the phase points X_T in high dimension to restore the chaos attractor, which reflects the regularity of chaotic systems. The moving states of the point X_T to X_{T+t} can be predicted by the point X_T and the points before X_T , the prediction model can be expressed as follows:

$$X_{T+t} = F(X_T) \tag{18}$$

where, X_{T+t} is the predicting state of the phase points, and the parameter t is the prediction time. The prediction function F is determined by the observation coal gas concentration data sequence. The lth predicting step is written as follows:

$$\hat{X}_{T+l} = F(\hat{X}_{T+l-1}) \tag{19}$$

In fact, it is not impossible to predict the longer time coal gas concentration. So the one step prediction is just used in this paper.

Lyapunov exponent is used to describe the orbital divergence. In phase space, the initial distance of the phase trajectory between two points is $|\delta X(t_0)|$, after the n_{th} iterations the distance between two points is defined as:

$$|\delta X(t_n)| = |\delta X(t_0)| \prod_{i=0}^{n-1} |f'[X(t_i)]| = |\delta X(t_0)| e^{\lambda t_n} \tag{20}$$

Lyapunov exponent λ can be expressed as following:

$$\lambda = \lim_{t_n \to \infty} \frac{1}{t_n} \sum_{i=0}^{n-1} \ln |f'[x(t_i)]| \tag{21}$$

If $\lambda < 0$, the system has a stable fixed point; if $\lambda = 0$, the system has bifurcation or periodic solutions; but while $\lambda > 0$, the system has chaotic characteristics.

In the practice, the maximum Lyapunov index λ_1 is gotten as:

$$\lambda_1 = \frac{1}{t_M - t_0} \sum_{i=0}^{M} \ln \frac{L'_i}{Li} \tag{22}$$

where, the $X_i(t_i)$ is a point near the point $X(t_0)$ at t_i, so distances between nearest neighbor points are $L_i' = |X(t_i) - X(t_{i-1})|$ and $L_i = |X(t_i) - X_i(t_{i-1})|$, respectively. Similarly, the sequence λ_2, λ_3, λ_4,…, and λ_n of the Lyapunov exponents can be calculated. However, only the largest Lyapunov exponent λ_1 is calculated in the practice application.

The chaotic time series prediction scheme is shown in Fig.2.

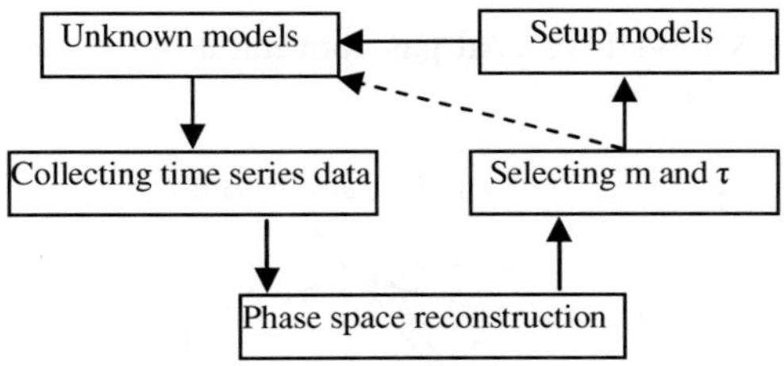

Fig. 2. Chaotic time series prediction scheme

So the prediction steps are as follows:

a) Collecting time series data: In order to predict the coal gas concentration change rule, the coal gas concentration data are collected firstly.

b) Reconstructed phase space: The collected coal gas concentration time sequence data are used to reconstruct the higher dimension phase space according to Takens principle.

c) Selecting m and τ: If the collected data are disturbed with noise, then the improved C-C algorithm can be adopted for the unknown coal gas concentration prediction mathematic model.

d) Solution for maximum Lyapunov index: Look for maximum Lyapunov index λ, and judge whether the Lyapunov index is larger than zero or not. If Lyapunov index λ>0, the theory mathematic model is established.

e) The real chaotic time series prediction coal gas concentration model is set up.

5 Neural Network Chaotic Time Series Model

In this paper, a forward neural network is used to approximate the nonlinear discrete function [4]:

$$x(t+h) = F_h[X(t)] \tag{23}$$

where *h* is the predicted step. By Takens theorem, the state evolution in reconstructive state space can be expressed:

$$X(t+1) = F[X(t)] \tag{24}$$

The state *x(t)* can be gotten by the observed state vector *X(t)*:

$$x(t) = g[X(t)] \tag{25}$$

Then the following equations are established:

$$\begin{gathered} X(t+1) = F[X(t)] \\ X(t+2) = F[X(t+1)] = F\{F[X(t)]\} \\ \vdots \\ X(t+h) = F[X(t+h-1)] = F\{\cdots F[X(t)]\} \\ x(t+h) = F_h[X(t)] = g(F\{\cdots F[X(t)]\}) \end{gathered} \tag{26}$$

The neural network model for coal gas concentration prediction based on chaotic time series is shown in Fig.3.

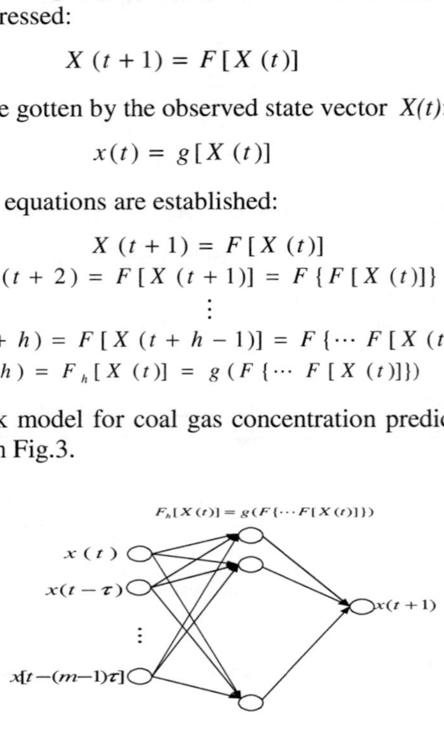

Fig. 3. The proposed neural network mode

There are five input variables. The input variable X_1 denotes the depth of the coal bed, the variable X_2 represents coal bed thickness, the variable X_3 is gas condensation, the variable X_4 is mining progress per day, the variable X_5 is the coal yield in every day, respectively. The output variable Y_1 represents gas outburst quantity.

The embedded dimension m and the time delay τ are first solved by improved C-C algorithm. The correlation function curve is shown in Fig.4.

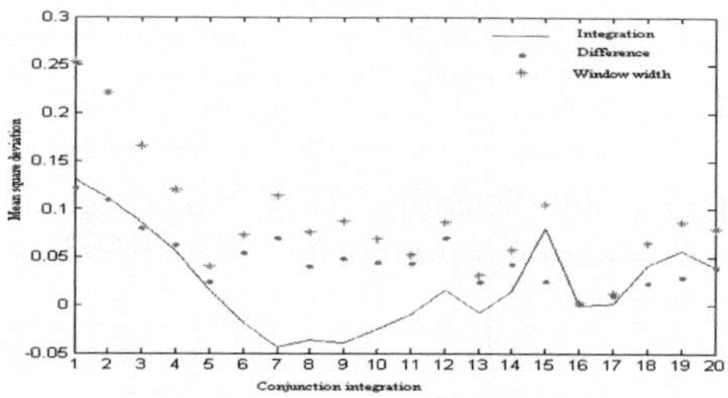

Fig. 4. The correlation function curve

In Fig. 4, the time delay τ is 5, the embedded dimension m is 5 due to the embedding window width is 16. After calculation, the gas concentration Lyapunov exponent λ is 0.2392, so that the gas concentration sequences have chaotic characteristics. The reconstruction of gas concentration in sequence can be carried out to restore the original gas concentration changes for prediction based on chaotic time series. When the training is completed, the gas concentration prediction function is expressed by:

$$X(t) = A * LW + B2 \tag{27}$$

where, $A = \dfrac{1}{1+e^{-(IW*p+B1)}}$, and p is the matrix of the gas concentration phase-space reconstruction. After the 100 new input data are reconstructed, the 80 new training data are gained. The comparison between the prediction and the real curve is shown in Fig.5.

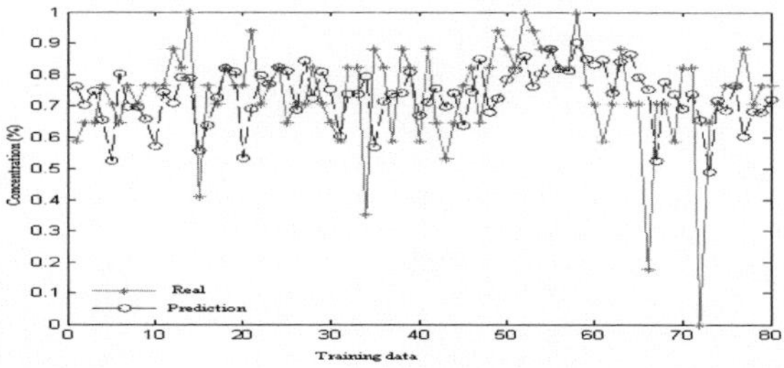

Fig. 5. The comparison between prediction and real results

In the Fig.5, the predicting coal concentration variations are better consistent with real values in short time. But the error in the longer period is larger. Therefore the model is only suitable for short time prediction.

6 Conclusion

An improved C-C algorithm is discussed to raise the speed of the optimal search for the local minimum in the coal gas concentration phase space reconstruction based on the chaotic time series theory. The simulation results show that the improved C-C algorithm has some immunity of the noise and can be used to predict the gas concentration change rule in future short time.

Acknowledgment

This project is supported by Scientific Research Program Funded by Shaanxi Provincial Education Commission (Program NO.2010JK663).

References

1. Zhao, Z., Tan, Y.: Study of premonitory time series prediction of coal and gas outbursts based on chaos theory. Rock and Soil Mechanics 30(7), 2186–2189 (2009)
2. He, H., Hu, M.: Chaos Foreca sting of Ga s Em ission Ba sed on Improved CC Method. Mining R&D 29(4), 69–71 (2009)
3. Zhang, Y., Ye, Q., Jia, Z., Jiang, W.: The analysis and forecast of gas emission in workface. China Mining Magazine 16(11), 46–49 (2007)
4. Liu, J., Liu, Z., Ma, J., Ma, J.f.: Study on Prediction System of Mine Gas Gushing Based on Neural Network and its Application. Coal Technology 27(11,31), 71–74 (November 2)

Novel Algorithm for Hand Vein Recognition Based on Retinex Method and SIFT Feature Analysis*

Hua-bin Wang, Liang Tao, and Xue-you Hu

Key Laboratory of Intelligent Computing & Signal Processing, Ministry of Education
Anhui University, Hefei, Anhui, 230039, China
huabin0584@163.com, taoliang@ahu.edu.cn, xueyouhu@hfuu.edu.cn

Abstract. Based on the Retinex method and the SIFT feature analysis, this paper presents a novel algorithm for hand vein recognition. First of all, the principle of the near-infrared hand vein image acquisition is introduced. Secondly, the Retinex method is used to normalize hand vein images, and the adaptive smoothing method is selected to estimate the illumination. Then gray cosine transform is used to enhance the discrimination of the skin and the vein in hand vein images. Thirdly, the SIFT feature analysis algorithm is used to extract the feature of hand vein. Finally, the match method of two hand vein images based on SIFT is given. A hand vein recognition system in Microsoft VC6.0 is also developed and the experimental results demonstrate the high efficiency of the proposed algorithm in runtime and correct recognition rate.

Keywords: Hand Vein Enhancement, Feature Extraction, SIFT, Biometrics, Hand Vein Recognition.

1 Introduction

Biometrics is more reliable and effective than the traditional identification technology [1]. Hand vein identification has emerged as a promising component of biometrics study [2][3]. Since hand vein is inside the skin and difficult to be copied, it is one of the safest biometric identification technologies [4]. The acquisition of the hand vein is the key step of one hand vein recognition system. A common method used for getting hand vein images is to use the low-cost near-infrared camera [5]. Since there is infrared light in the natural environment and hand vein is inside hand skin, the vein images we acquired are of less contrast and difficult to be segmented. Therefore, many algorithms were established to enhance and segment the hand vein. Ding proposed a method [6] base on threshold graph, which sets threshold on every pixel of the image and then splits together with the border. However, some noise is fetched in and the method is of higher complexity.

Because the hand vein images we acquired are of rotation, scaling and translation, the recognition rate may be affected. The feature extraction method of hand vein is mainly based on the detection of the endpoints and cross points [7], but this method

* This work is supported by the Key Project of the Natural Science Research in Anhui Provincial Higher Education Instructions under Grant No. KJ2010A011.

X. Wan (Ed.): Electrical Power Systems and Computers, LNEE 99, pp. 559–566.
springerlink.com

was only applied to fixed location identification. To extract the stable feature of hand vein, we select the SIFT (Scale-invariant feature transform) algorithm.

In this paper, we present a novel algorithm to enhance the images and extract the stable local feature of the hand vein. The proposed algorithm is based on the Retinex method and the SIFT method, and works on still images.

2 Hand Vein Image Acquisition

The principle of the near-infrared hand vein image acquisition is introduced as follows:

The near-infrared light travelling in the palm and dorsal skin is absorbed by the haemoglobin in the blood of artery and vein but can easily pass through the muscle and bones [8]. So we select an infrared camera to acquire hand vein images. The near-infrared LED sources are circularly located around the camera and peak at 850nm wavelength. Hand vein images can be acquired from the reflex light of the dorsal skin. Some hand vein images acquired by this method are shown in Fig.1.

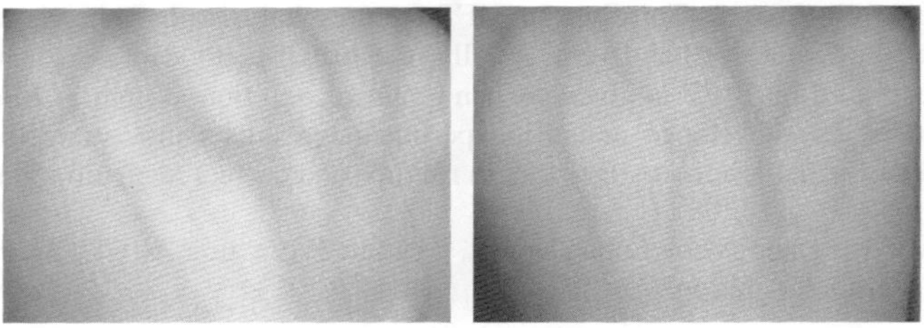

Fig. 1. Some samples of hand vein images

3 Enhancing Hand Vein Images

The illumination of hand vein images is not uniform and the vein is very thin, so it is necessary to normalize the illumination and enhance the contrast of the hand vein images.

The Retinex method proposed by Land [9] can be used to remove the influence of illumination and obtain the reflective nature of the hand vein images.

Based on the researches of Land, the Single Scale Retinex is defined by Jobson as follows [10]:

$$R(x, y) = \log I(x, y) - \log[E(x, y) * I(x, y)] \tag{1}$$

where $I(x, y)$ is the input image function, $E(x, y)$ is the estimated illumination function, usually a Gaussian function. $R(x, y)$ is the output image function which can respond the reflective nature and is of uniform grey scale value.

The key step of Retinex method is to estimate illumination, which is often estimated as a smooth version of the input image and Gaussian smooth is a common-used method. The high frequency enhanced image would be get, after the original image minus the Gaussian low-pass filtered image. But Gaussian smooth could not be applied in hand vein images, because the vein is very thin and the contrast is very low, so smoothing should especially be carried out among pixels which have homogeneous grey scale value. We select the adaptive smoothing method to estimate illumination [11]. The adaptive smoothing can be done as follows:

$$I_0(x,y)=\frac{1}{N(x,y)}\sum_{i\in D}\sum_{j\in D}I(x+i,y+j)w(x+i,y+i)$$
$$N(x,y)=\sum_{i\in D}\sum_{j\in D}w(x+i,y+i) \tag{2}$$

where $I(x,y)$ is an original input image, $N(x,y)$ is a normalizing factor, $w(x,y)$ is an adaptive smoothing filter that is a nonnegative monotonically decreasing function, which represents the amount of discontinuity at each pixel. D is the size of the filter, we select 3*3.

$w(x,y)$ can be defined as a decreasing function with the pixel gradient increasing:

$$w(x,y)=\exp\left(-\frac{|I(x+1,y)-I(x-1,y)|+|I(x,y+1)-I(x,y-1)|}{K}\right) \text{ where } K=10. \tag{3}$$

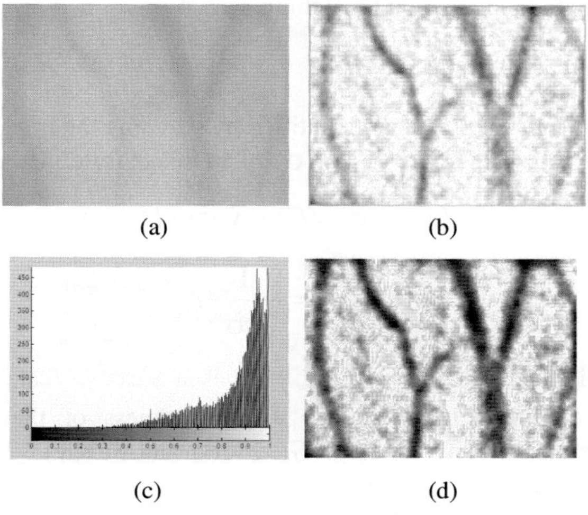

Fig. 2. (a) Source hand vein images, (b) Illumination normalized hand vein image, (c) its histogram, and (d) grey stretched hand vein image.

To get a better smoothing and enhancement, the adaptive smoothing filter should be used iteratively. With a large number of experiments, we got to know that 16 times of iteration is enough.

Once $E(x,y)*I(x,y)$ is calculated, the illumination normalized hand vein image $R(x,y)$ can be calculated easily by the definition of the Retinex method. Finally, $R(x,y)$ should be normalized to the range [0, 1]. The result is shown in Fig. 2.

We can see from Fig. 2(b) that the hand vein is clear to see and is of uniform grey scale value, but the grey scale value of hand vein image concentrates in the area from 0.8 to 1.0. To enhance the discrimination of the skin and the vein in illumination normalized hand vein image, grey cosine transform can be used as follows:

$$I_C(x,y) = 1 - \cos\left(\frac{\pi}{2} \times I_R(x,y)\right) \tag{4}$$

Where $I_R(x,y)$ is the normalized vein image $R(x,y)$, $I_C(x,y)$ is the grey stretched image (Fig.2(d)). We can see from Fig.2 (d) that there is a lot of noise after grey stretched, so the smoothing filter should be used to denoise.

4 Hand Vein Feature Extraction

In order to avoid the interference of rotation, scaling and translation of the hand vein image, we select the SIFT algorithm to extract the hand vein feature.

SIFT algorithm was proposed by Lowe [12] can be used to do stable local feature detection and representation. It consists of three major stages: key point localization, key point descriptor and key point matching.

4.1 Key Point Localization

Firstly, potential interest points are identified by scanning the image over location and scale. The Gaussian scale-space $L(x,y,\sigma)$ of an image is defined as follows:

$$L(x,y,\sigma) = G(x,y,\sigma) * I(x,y) \tag{5}$$

$$G(x,y,\sigma) = \frac{1}{2\pi\sigma^2} e^{-(x^2+y^2)/2\sigma^2} \tag{6}$$

Where $I(x,y)$ is an input image, $G(x,y,\sigma)$ is a variable-scale function, σ is the scale space factor, which determines the smoothness of the scale transformed image, $L(x,y,\sigma)$ is the convolution of $G(x,y,\sigma)$ and $I(x,y)$.

To detect the key point in the scale space efficiently, DOG (Difference of Gaussian) is defined as the difference of two nearby scales separated by a constant multiplicative factor k [13]:

$$\begin{aligned} D(x,y,\sigma) &= (G(x,y,k\sigma) - G(x,y,\sigma)) * I(x,y) \\ &= L(x,y,k\sigma) - L(x,y,\sigma) \end{aligned} \tag{7}$$

If the sample point compared to its eight neighbours in the current scale image and nine neighbours in the scale above and below is maxima or minima of $D(x,y,\sigma)$, we think it is a key point (see Fig.3).

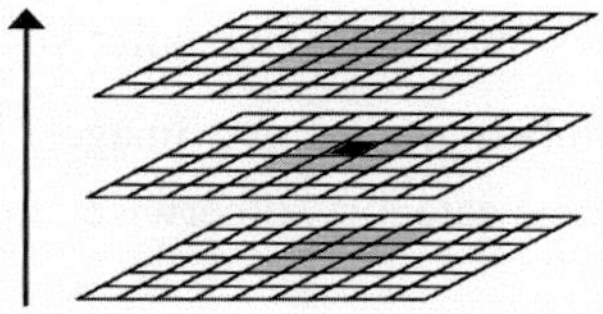

Fig. 3. Detection of feature point in DOG

The DOG is of strong edge response, so the unstable edge response points and some low contrast key points which are sensitive to noise must be deleted.

4.2 Key Point Descriptor

The key point descriptor is done by computing the gradient magnitude and orientation at each image sample point in a region. The gradient magnitude $m(x, y)$ and orientation $\theta(x, y)$ are defined as follows:

$$m(x, y) = \sqrt{(L(x+1, y) - L(x-1, y))^2 + (L(x, y+1) - L(x, y-1))^2} \tag{8}$$

$$\theta(x, y) = \alpha \tan 2\left(\frac{(L(x, y+1) - L(x, y-1))}{(L(x+1, y) - L(x-1, y))}\right) \tag{9}$$

To get the orientation of one key point, an orientation histogram should be draw with 36 orientations, and the sample points are within a region around the key point, we select 16*16 pixels. The peak orientation represents the main orientation. To achieve rotation invariance, axis should be rotated to the orientation of the key point. Then describe the key point with the 4*4 seeds around, computed the gradient orientations histogram of each seeds with eight orientations, and format the 4*4*8=128 element feature vector for each key point.

4.3 Key Point Matching

The correlation coefficient of two key points is defined as follows:

$$r = \frac{\sum_{n=0}^{N-1}[A(n) - \bar{A}][B(n) - \bar{B}]}{\sqrt{\sum_{n=0}^{N-1}[A(n) - \bar{A}]^2 \sum_{n=0}^{N-1}[B(n) - \bar{B}]^2}} \tag{10}$$

Where A and B are the feature vector of feature point with N=128, $\bar{A}$ and $\bar{B}$ respectively the mean of the elements of A and B . If $r > t_p$, we think there are the two key points matching successful. Statistics show that $t_p > 0.65$.

The match coefficient of two hand vein images is defined as follows:

$$s = Num_m / MIN(Num_A, Num_B) \tag{11}$$

Num_m is the number of the matching key point of the two image, Num_A and Num_B are the key point numbers of the two images. If $s > t_s$, we think the two hand vein images are of the same category, otherwise, on the contrary.

5 Experiments

Since there are not public hand vein databases, we established a hand vein database (40 hands, 20 hand vein images for each one, A total of 800 hand vein images with size 120×160, taken in our laboratory using an IR camera) to evaluate the performance of the proposed algorithm.

5.1 Matching Experiment

Set $t_p = 0.7$, $\sigma = 1.5$. The two hand vein images in Fig.4 (a) are from the same hand and there are 36 key points matching, the two hand vein images in Fig. 4(b) are from different hands and there are only 4 key points matching.

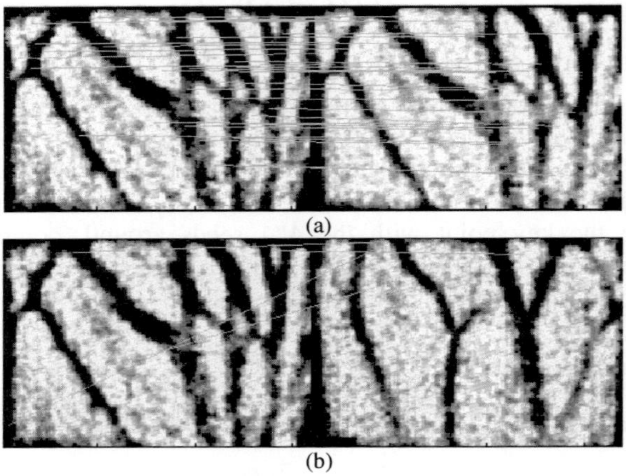

Fig. 4. Matching experiment between two hand vein images. (a) Two hand vein images from the same hand, and (b) two hand vein images from the different hands.

5.2 Recognition Rate Experiment

Taken ten images of each people, and enhanced the images according to the algorithm presented in this paper. Then got the SIFT feature vector of each feature points ($\sigma = 1.5$), and stored in the so called matching database.

In the remaining 400 hand vein images, selected one sample randomly , detected the SIFT feature points with the same algorithm as above, then computed the match coefficient s between this sample A and each sample B in the matching database.

Set $t_p = 0.7$ and $t_s = 0.3$ to 0.7.

Table 1. Experiment Results.

t_s	FAR	FRR	TOTAL
0.30	0.1172	0	0.1172
0.35	0.0551	0	0.0551
0.40	0.0123	0.0005	0.0128
0.45	0.0081	0.0014	0.0095
0.50	0.0012	0.0072	0.0084
0.55	0.0002	0.0201	0.0203
0.60	0	0.0651	0.0651
0.65	0	0.1102	0.1102
0.70	0	0.1853	0.1853

The experiment results are show in Table 1, we can see that the correct certification rate is close to 99.16%, if setting t_s=0.5.

6 Development in VC++6.0

We developed a hand vein recognition system in Microsoft Visual C++ 6.0 as shown in Fig. 5. It contains the module of enhancing and feature extracting of hand vein images, which is according to the algorithm presented in this paper. The same hand vein database was used as in the above sections to evaluate the performance of the software, which worked on a Pentium 4/3.0GHz personal computer. The average runtime for recognizing one hand vein image is about 1 second per image, so the algorithm presented in this paper fits for a real-time system.

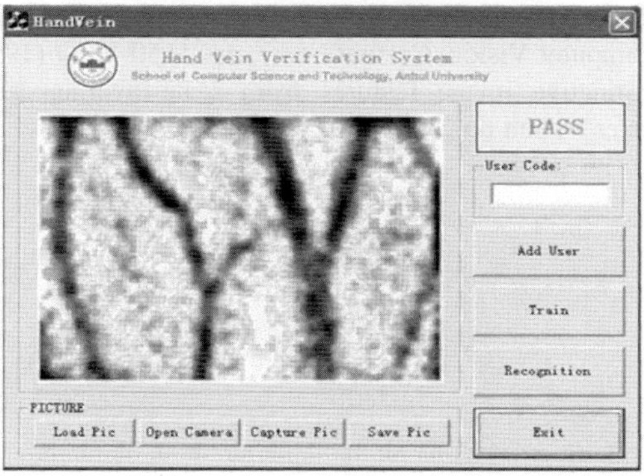

Fig. 5. The interface of our hand vein recognition software

7 Conclusions

In summary, by the Retinex method and the SIFT method, a novel algorithm was developed in this paper for hand vein recognition. The experimental results demonstrated the efficiency of the algorithm in runtime and correct certification rate.

References

1. O'Gorman, L.: Comparing passwords, tokens, and biometrics for user authentication. Proceedings of the IEEE 91(12), 2019–2040 (2003)
2. Ajay Kumar, K., Prathyusha, V.: Personal Authentication using Hand Vein Triangulation and Knuckle Shape. IEEE Transactions on Image Processing 38, 2127–2136 (2009)
3. Liu, T.G., Wang, Y.X.: Biometric Recognition System Based on Hand Vein Patten. Acta Optica Sinica 29(12), 3339–3343 (2009)
4. Lin, C.-L., Fan, K.-C.: Biometric Verification Using Thermal Images of Palm-Dorsa Vein Patterns. IEEE Transactions on Circuits and Systems for Video Technology 14(2), 199–213 (2004)
5. Tanaka, T., Kubo, N.: Biometric authentication by hand vein patterns. In: SICE 2004 Annual Conference, pp. 249–253 (2004)
6. Ding, Y.H., Zhuang, D., Wang, K.: A Study of Hand Vein Recognition Method. In: ICMA 2005, pp. 2106–2110 (2005)
7. Wang, K.J.: A study of hand vein based identity authentication method. Engineering & technology 1, 35–37 (2005)
8. Mobley, J., Vo-Dinh, T.: Biomedical Photonics Handbook. CRC Press, Boca Raton (2003)
9. Land, E.: An alternative technique for the computation of the designator in the Retinex theory of color vision. Proc. Nat. Acad. Sci. (1986)
10. Jobson, D.J., Rahman, Z., Woodell, G.A.: Properties and performance of a center/surround Retinex. IEEE Transactions on Image Processing 6(3), 451–462 (1997)
11. Saint-Marc, P., Chen, J.-S., Medioni, G.: Adaptive smoothing: a general tool for early vision. IEEE Trans. Pattern Anal. Mach. Intell. 13(6), 514–529 (1991)
12. Lowe, D.G.: Object recognition from local scale-invariant features. In: International Conference on Computer Vision, Corfu, Greece, pp. 1150–1157 (1999)
13. Lowe, D.G.: Distinctive image features from scale-invariant keypoints. International Journal of Computer Vision 60(2), 91–110 (2004)

Knowledge and Data Engineering for Analyzing the Quality of Education Using Fuzzy Logic

Sergio Valdés-Pasarón, Bogart Yail Márquez, and Luis Gaxiola

Baja California Autonomous University, Chemistry and Engineering Faculty, Calzada Universidad 14418, Tijuana, Baja California, Mexico, 22390
http://fcqi.tij.uabc.mx

Abstract. The quality of education has awakened the interest in researchers worldwide because it is an important aspect in solving the problems in education which can be studied as a complex social system; and of course, there are several ways to model a complex social system. The objective of this paper is to propose a model that deals with problems in education among knowledge societies in situations where conventional analysis is insufficient in describing the intricacies of realistic social phenomena and social actors. We use the Distributed Agency methodology that requires the use of several computational techniques and interdisciplinary theories.

Keywords: Knowledge, Data Engineering, Complex Social Systems, Data Mining, Fuzzy Logic, Education, Distributed Agencies.

1 Introduction

Social systems contain many components which depend on many relationships; this makes it difficult to construct models closer to reality. To analyze these systems with a dynamic and multidimensional perspective, we will consider data mining theory, fuzzy logic and distributed agencies. Human capital theory emerges from the contributions of Mincer (1958) [1], Schultz (1961)[2] and Becker (1964) [3], they considered education as an investment to be made by individuals which allows them to increase their human capital endowment. This investment increases productivity and, in the neoclassical framework of competitive markets in which this theory is developed, future income; thus, establishing a causal relationship between education, productivity and income, so that an increase of education produces a higher level of income and greater economic growth.

Gary Becker developed the pure model of human capital; in brief, the main hypothesis is based on the fact that as education increases so does the productivity of the individual who receives it [3].

Becker reaches two important conclusions, the first deals with theories of income distribution, rising yields, a simple model that emerges from Human capital. This can be described as:

$$G_i = f\,(QN_i, E_i) \tag{1}$$

X. Wan (Ed.): Electrical Power Systems and Computers, LNEE 99, pp. 567–573.
springerlink.com

Where G is returns, QN is innate or natural qualities, E is education or characteristics acquired through investment in human capital, and the subindex i is a person.

Becker arrives to an interesting conclusion in this first part of his study and as outlined in his article, "Human Capital." pp.62 and 63, one can assume that there is a whip positive correlation between the natural qualities and the level of educational investment.

For the purpose of satisfying all of societies expectations and needs as a whole in terms of education, there is a link between a series of qualitative and quantitative variables which together give an insight as to the differences in quality of education. By merely mentioning a few aspects we can refer to: ratio of students per teacher, student's access to technology and a county's Gross Domestic Product (GDP) allocated to education.

In recent years numerous studies have found that there is a disparity in the criteria of variables used to measure elements of education or in related factors. These variables include the cost of schooling, and the average number of paid years. For example, the number of years a person studies or the average number of years a person has to pay for education. Furthermore, these variables are imperfect measures of the educational component of human capital since they measure quantity and not quality in education, weakening the value between these comparisons [4-6].

The quality of education has begun to become a high concern among researchers of education because they believe that the expectations and needs of human beings depends on factors like the quality of curricula for which they are prepared, the countries education infrastructure, the academic environment which is developed, the faculty and the relationship between teachers and students, among others. Despite this being clearly identified, it still remains a difficult task to select the most appropriate indicators to determine which of them have a greater impact on the quality of education [4].

The motivations to incorporate these indicators to improve the quality of education imply that the factors vary from year to year. For instance, in Latin American countries the education systems vary widely in terms of the organization of resources, teacher qualifications, student-teacher ratio in classrooms, access to technology and education spending per student among other factors.

The hegemony of the positivist epistemological paradigm in the social sciences has been hindering theoretical constructions that are approximations to reality without reducing their complexity, dismissing non-scientific phenomena such as subjectivity, culture, health, social system and education.

There have recently emerged, from different disciplinary fields, a number of theories that come close to the social reality and are able to approach it in all its complexity. These have a clear epistemological emphasis.

One theory of complexity is that of fuzzy sets as a mathematical formalization of a logical model of imprecision and uncertainty [7].

2 Measurement of Quality

2.1 How to Measure the Quality of Education

Several factors have been incorporated to measure aspects involving the quality of education. Hanushek and Kim [8] proposed to measure education using skills learned from the test or tests used internationally. For example, the Program for International

Assessment Students (PISA) of the OECD or in the case of Mexico, National Assessment of Academic Achievement in Schools (LINK, from the Spanish acronym), which are intended to assess how far students near the end of their compulsory education acquired, to some degree, the knowledge and skills necessary for full participation in society.

The results of these tests show a relationship between the quality of education with the growth of gross national product (GNP) per capita. This suggests that the quality of education is a factor of great importance for the analysis of the relationship between human capital and economic growth [9-11].

However, these results have not yet reached a consensus on how to measure qualitative and quantitative factors jointly due to the heterogeneity in the capture of such data. Given the difficulty that exists in measuring the quality of education, the main contribution of this work would be a model to measure the quality of education quantitatively and qualitatively, eliminating the heterogeneity in the ways of measuring this indicator and reaching a final consensus on this controversial issue.

2.2 Fuzzy Logic

The concept of Fuzzy Logic was conceived by Lotfi Zadeh, a professor at the University of California at Berkeley, who understood the limitation of classical sets in describing the imprecision of reality. In classical sets an item either belongs or does not belong to the set. On the other hand, fuzzy sets [12] are flexible in allowing an item to have a degree of belonging, giving a smooth transition between items that do not belong and items that do.

Fuzzy logic, unlike conventional logic, can work with information that is not entirely accurate and can help classify items in a less strict fashion by not disregarding items that are near the defined boundaries of the set. This allows for a deterministic way to model and work with uncertainties, to model knowledge in the form of rules, feasible to be processed by computers.

If we compare classical and fuzzy logic we can say that classical logic provides a logical parameters for true or false, that is, using binary combinations of 0 and 1, 0 if false and 1 if true. In a different manner, fuzzy logic introduces a function that expresses the degree of membership of an attribute or variable to a linguistic variable taking the values between 0 and 1, this is called a fuzzy set and can be expressed mathematically as:

$$A = \{x / \mu A(x) \ \forall \ x \in X\} \tag{2}$$

Where *A* is the fuzzy set representing a linguistic variable and consists of ordered pairs representing the level of belonging for each point in the universe *X*. Applying classical sets to measuring the quality of education, we can analyze how much public expenditure there is on education. This approach would show either good quality or poor quality regardless of the income distribution of students or the percentage of the generations that come at a higher level, e.g., if option one then Education Quality (EQ) = 1 or if the second option EQ = 0.

Fuzzy logic allows for this classification to be made in a more natural way by not limiting the classification to two categories and allowing an item to be classified as partially belonging to multiple sets. For example, an item can be classified in the

"Quality of Education" variable with the following degree of membership for each fuzzy set: Excellent quality (EQ) = 1, good quality (GQ) = 0.8, medium quality (MQ) = 0.5, bad quality (BQ) = 0.1 and very bad quality (VBQ) = 0.

3 Methodology

The proposed methodology consists of analyzing the indicators used by the United Nations Educational Scientific and Cultural Organization (UNESCO) to measure the education of the countries that have the following input variables to consider.

a) Education expenditure as a percentage of Gross Domestic Product (EXPGDP). This variable represents the percentage of gross domestic product that countries devote to education.
b) Government Public Expenditure on Education (GPEE). This variable represents the total government spending for education.
c) Distribution of Public Spending by Level of Government (DPSLG). This variable represents the distribution of government expenditure by educational levels (primary, secondary or tertiary) of the total allocated for education.
d) Pupil-Teacher Ratios (PTR). This variable represents the number of students for each teacher at different educational levels.
e) Income rate to last grade of primary (TIUGP). Represents the rate of students who manage to conclude the primary level.
f) Percentage of students who continue to secondary school (SSPE). This variable represents the percent of students who continue their secondary studies once completed primary school.
g) Expenditure per pupil as a percentage of GDP Per Capita (EPP GDP). Represents the average expenditure per student relative to per capita gross domestic product.
h) Average per pupil expenditure (APPE). Represents the average expenditure per pupil.

3.1 Equations

The relationship between the quality of education and its determinants can be analyzed by a production function of education as:

$$Q = f\,(EF, R) + U \tag{3}$$

Where Q represents the quality of education, EF represents the economic factors, R is the resources used in schools and U is unmeasured factors that may affect the quality of education.

The system, which is divided in three blocks, makes use of fuzzy logic techniques. It is important to carefully model the inputs and outputs of the system. The former consists of the variables that are taken into account in the intended representation by the system. The output is a Quality of Education and the input variables are chosen thru selection process that involves knowing the context of the problem being addressed. To illustrate, Figure 1 shows the MATLAB module dealing with fuzzy logic, considering the following inputs and outputs given the linguistic variables: very bad, bad, medium, good, very good. The linguistic variables are made to the perception of education of a particular country.

As input, establishing the factors that influence: EXPGDP, GPEE, DPSLG, PTR, TIUGP, SSPE, EPPGDP and APPE

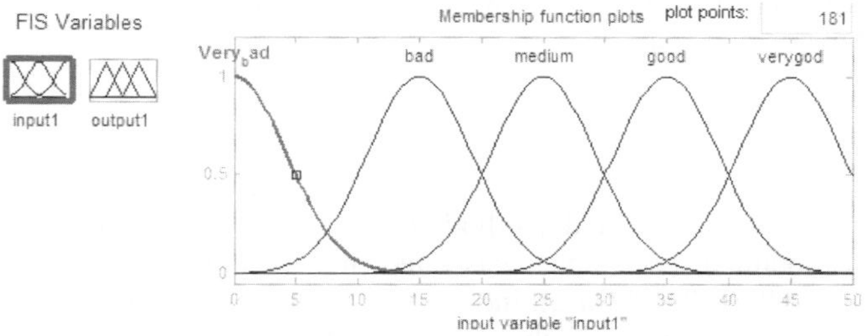

Fig. 1. Membership function

As output, gives the linguistic value that determines a country's education. The functions of each component shown in Figure 2 are:

a) Fuzzification interface: transform variables into fuzzy variables. For this interface, ranges of variation must be defined for input variables and fuzzy sets associated with their membership functions.
b) Knowledge Base: contains the linguistic rules and information relating to the membership functions of fuzzy sets.
c) Inference engine: calculates the output variables from input variables by processing the fuzzy rules in the knowledge base and, delivering the output fuzzy sets.
d) Defuzzification interface: gets a crisp output from the aggregation of fuzzy outputs.

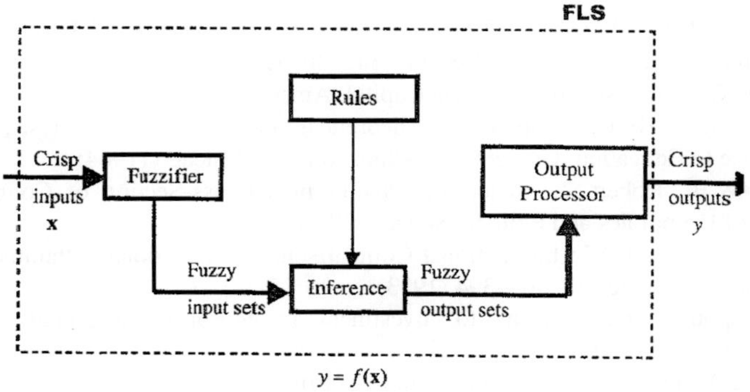

Fig. 2. Fuzzy logic system

To create the fuzzy inference system that calculates the role of education quality, we used the Matlab 2009 Fuzzy toolkit. Each education indicator was considered a linguistic variable, and each of these is associated with five fuzzy sets with membership functions of the "real" variable. Each set was labeled with the linguistic labels of "very good", "good", "medium " "bad" and "very bad" to rate the value of the indicator considered. The degree of membership of an element in a fuzzy set is determined by a membership function that can take all real values in the range [0 1]. Inputs and outputs make a total of 8 variables which correspond to the quality of education that a country has.

4 Results

With the use of new tools and methodologies in economic theory, the following results where obtained: each level was rated high or low quality depending on the limit values "good" or "bad" for all indicators, which do not always coincide with the upper and lower value ranges. In the case of PTR, EPP, and GDP, a greater disparity is shown between low and high quality, this is because of the large differences in ranges these values can have.

5 Conclusions

The methodology developed for obtaining a function of the quality of education is easy to manage, interpret and can serve for multiple sensitivity analysis of changes in the values of education indicators. The great advantage of the methodology based on fuzzy logic is that you can handle an unlimited number of indicators expressed in any unit of measurement. Like any other methodology, it is strongly dependent on the accuracy with which the indicators have been calculated or determined. This paper shows the potential of fuzzy logic to describe particular systems and public policies to determine that the education of a country is either "good "or "bad ".

References

1. Chiswick, B.R., Mincer, J.: Experience and the Distribution of Earnings. Review of Economics of the Household, 2003 1(4), 343–361 (2003)
2. Schultz, T.W.: Investment in Human Capital. American Economic Review LI, 1–17 (1961)
3. Becker, G.S.: Human Capital: A Theoretical and Empirical Analysis, with Special Reference to Education. University of Chicago Press, Chicago (1964)
4. Lee, J.-W.B., Robert, J.: Schooling Quality in a Cross-Section of Countries. London School of Economics and Political Science (2001)
5. Barro, R.J., Lee, J.-W.: International Comparisons of Educational Attainment. Journal of Monetary Economics 32, 363–394 (1993)
6. Psacharopoulos, G.: Returns to investment in education: a global update. World Development 22 (1994)
7. Zaded, L.A.: Fuzzy sets. Information and Control 8 (1965)

8. Hanushek, E.A., Kim, D.: Schooling, Labor Force Quality, and Economic Growth. National Bureau of Economic Research (1995)
9. Barro, R.: Human capital and growth in cross-country regressions. Mimeo, Harvard University (1998)
10. Mankiw, G., Romer, D., Weil, y.D.: A Contribution to the Empirics of Economic Growth. Quartely Journal of Economics, 407–437 (1992)
11. Nelson, R., Phelps, E.: Investment in Humans, Technological Diffusion, and EconomicGrowth. American Economic Review, 69–82 (1966)
12. Zadeh, L., et al. (eds.): Fuzzy sets and their applications to cognitive and decision processes. Academic Press, New York (1975)

eLearning Virtual Environments Multi-agent Model

Luis Gaxiola Vega[1], Manuel Castanon-Puga[1], Bogart Yail Márquez[1], José Magdaleno-Palencia[2], and Miguel A Cadena Alcantar[2]

[1] Baja California Autonomous University, Chemistry and Engineering Faculty, Calzada Universidad 14418, Tijuana, Baja California, Mexico, 22390
http://fcqi.tij.uabc.mx

[2] Baja California Autonomous University, Cisalud Palm Valley, Blvd. San Pedro # 1000, Tijuana, Baja California, México
http://cisaluduvp.tij.uabc.mx

Abstract. The objective of this research is to propose a multi-agent model implemented in a virtual environment to assess and demonstrate how significant the learning of users has ehanced their aquired knowledge. This paper explains how to implement the use of multi-agents. It will discuss how the curriculum can be enriched by activities involving problem-based learning, case studies simulations and virtual reality. This new model provides multiple uses for exploring knowledge and supporting learning-by-doing. It engages users in the construction of knowledge, collaboration, and articulation of knowledge.

Keywords: e-learning, Knowledge, Data Engineering, Multi-agent.

1 Introduction

1.1 E-Learning

E-learning comprises all forms of electronically supported learning and teaching. The information and communication systems, whether networked or not, serve as specific media to implement the learning process. The term will still most likely be utilized to reference out-of-classroom and in-classroom educational experiences via technology, even as advances continue in regard to devices and curriculum.

E-learning is essentially the computer and network-enabled transfer of skills and knowledge. E-learning applications and processes include Web-based learning, computer-based learning, virtual classroom opportunities, and digital collaboration. Content is delivered via the Internet, intranet/extranet, and others such as information technologies. It can be self-paced or instructor-led and includes media in the form of text, image, animation, streaming video, and audio.

1.2 Computer-Based Training

Abbreviations like CBT (Computer-Based Training), IBT (Internet-Based Training) and WBT (Web-Based Training) have been used as synonyms to e-learning. Today one can still find these terms being used along with variations of e-learning, such as, elearning, Elearning, and eLearning.

X. Wan (Ed.): Electrical Power Systems and Computers, LNEE 99, pp. 575–579.
springerlink.com

Using e-learning in virtual spaces [1], especially where the images provide most of the information needed for cases such as radiology and dermatology [2] [3].

E-learning has become common in specialties that use standardized treatment pathways, such as emergency medicine [4-6]. However, learning programs based on simulation using virtual rooms are still scarce, as they are expensive and laborious [7] [8-9].

1.3 Virtual Room

Simulation of Virtual Rooms (VR) today is a new technology applied to research of new methods, forms, techniques, and architectures that provide solutions to problems that occur in both medicine and industrial engineering, and thus create experience when actual cases are confronted[9].

Today education is heading toward virtual learning, online, and distance learning. These types of environments are incorporated into classroom practices that as we know, are greatly enhancing, with technology, the possibilities of teaching in all areas. There is no doubt that within a classroom learning environment, a process that always takes place is communication. Interaction takes place within the media.

Multi-agent systems: it consists of autonomous agents working together to solve problems, characterized in that each agent has incomplete information or capabilities for problem solving, there is no global system control, data is decentralized and computation is asynchronous. The agents dynamically decide to undertake these tasks [10].

1.4 Multi-Agent System (MAS)

MAS organizations can be understood as complex entities where a multitude of agents interact within a structured environment for a global purpose. The partner organizations are often associated with the idea of openness and heterogeneity of MAS. Heterogeneous environments which pose new challenges in the design and implementation of MAS, including the integration of global and individual perspectives and the dynamic adaptation of systems to environmental changes. As growth systems that include hundreds or thousands of agents [11].

Formal theories are needed to describe interaction and organizational structure and understand the relationship between the organizational functions of these agents. This technique is accomplished by being autonomous and intelligent agents. It is when the systems become more distributed[12]. The model using the MAS help the study of knowledge in virtual environments [9].

2 Implementation

When adopting an agent-oriented view, we sense that many problems involving multiple actors to represent the decentralized nature of the problem, such as control of multiple sites, the different perspectives and interests. Moreover, agents must interact with others to carry out their individual goals and to manage dependencies. These interactions can vary from simple semantic translations to the ability to cooperate, coordinate and negotiate on any issue. The social interaction between players is different from other paradigms. In its guidance to staff communication it occurs at a very high level (declarative). Consequently, interactions are conducted at the level of knowledge: in terms of what goals must be pursued, at what time and by whom. It should compare this with the calls or method invocations operating at the syntactic level. Secondly, agents are ideal for troubleshooting operating in an environment where they have partial control and observability. Interactions should be handled the same way. Agents can make decisions about the nature and extent of their interactions and have the capacity to initiate and respond to interactions (which may not have been foreseen at the time of design). By creating a simulation model that reproduces the behavior of a student in different scenarios, through intelligent agents, we must establish the perceptions of the environment.

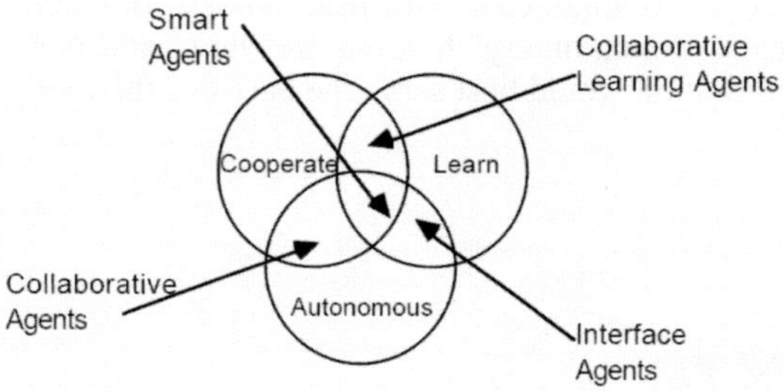

Fig. 1. Simulation model that reproduces the behavior of a student in different scenarios[13]

The virtual rooms are a consolidated solution to improve both student learning and performance of companies of any kind.

All interactions (employee, customer or partner organization) at any place and time can be combined to complete projects more efficiently.

To this end, sharing applications, results, and real-time interactions and virtual data room will be stored; the data will be placed on the multi-agent system. For our facility which will use the Netlog platform, each agent will be represented by a user, the simulation will determine the lack of knowledge and thus the agent can implement the missing knowledge and make the user learn and compare the results with simulation.

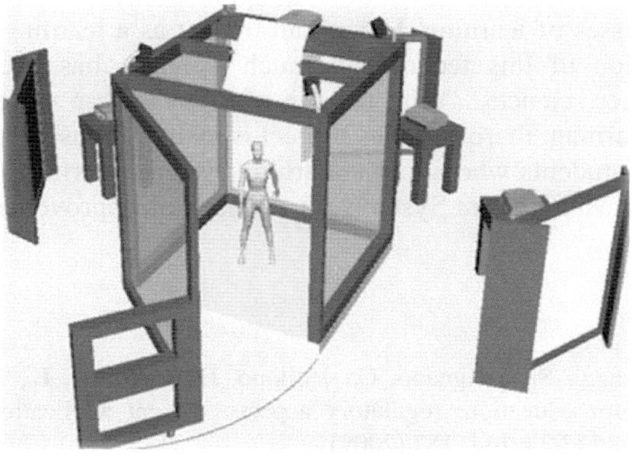

Fig. 2. Virtual Cave Simulation[14]

Using a specific case, as in the medical field, medical students use the virtual room built into a web service accessible via Netlogo [15], where students interact with each other and simulating an operation coordinated by a teacher shared within the virtual environment.

Virtual simulations in the cave where they can reproduce the behavior of students and obtain data that can compare with actual behavior.

The agents aim to assist the user by organizing the interface automatically. For this, the agent must learn the preferences and predict the user behavior. The agent captures the information while the user is performing either an operation or procedure, the agent records the user's behavior, and may proceed to give information of interest to an evaluator, performing an autonomous exploration of links that the user has at that time.

The agent is able to communicate with users and recommend the procedure that automate the process of "open outcry" learning the views and preferences of users and determine what information would best serve the needs of the user.

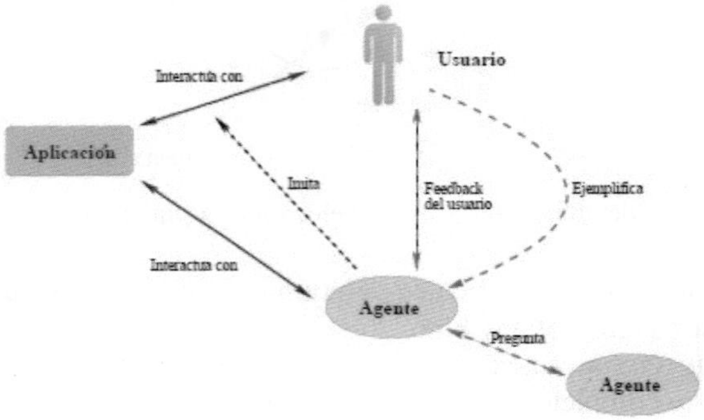

Fig. 3. Description model User-Agent

3 Conclusions

The virtual rooms have two predominant applications in both the medical and engineering processes of learning. Having an impact as a learning tool. With the use and implementation of this technology much progress has been in the areas of medicine, and other sciences. With the use of these tools a more direct application can be given to learning, therefore, this project is mainly focused on the direct support to professors and students who adopt e-learning systems learning. Also we chose the implementation of Multi-Agent Systems for the need of improvement of such tools.

References

1. Pinto, A., Selvaggi, S., Sicignano, G., Vollono, E., Iervolino, L., Amato, F., et al.: E-learning tools for education: regulatory aspects, current applications in radiology and future prospects 113(1), 144–157 (2008)

2. Scarsbrook, A.F., Graham, R.N., Perriss, R.W.: Radiology education: a glimpse into the future 61(8), 640-648 (2006)
3. Wahlgren, C.F., Edelbring, S., Fors, U., Hindbeck, H., Stahle, M.: Evaluation of an interactive case simulation system in dermatology and venereology for medical students 6, 40 (2006)
4. Freeman, K.M., Thompson, S.F., Allely, E.B., Sobel, A.L., Stansfield, S.A., Pugh, W.M.: A virtual reality patient simulation system for teaching emergency response skills to U.S. Navy medical providers 16(1), 3-8 (2001)
5. Friedl, R., Wieshammer, S., Kehrer, J., Ammon, C., Hubner, D., Lehmann, J., et al.: A case-based and multi-media computer learning program on the topic of myocardial infarct, angina pectoris and mitral valve stenosis 91(9), 564-569 (1996)
6. Weller, J., Robinson, B., Larsen, P., Caldwell, C.: Simulation-based training to improve acute care skills in medical undergraduates 117(1204), U1119 (2004)
7. Huang, G., Reynolds, R., Candler, C.: Virtual patient simulation at US and Canadian medical schools 82(5), 446-451 (2007)
8. Morin, A., Benhamou, A.C., Spector, M., Bonnin, A., Debry, C.: The French language virtual medical university 104, 213-219 (2004)
9. Zary, N., Johnson, G., Boberg, J., Fors, U.G.: Development, implementation and pilot evaluation of a Web-based Virtual Patient Case Simulation environment–Web-SP 6, 10 (2006)
10. Ovalle, D.A., Jiménez, J.A.: Ambiente Inteligente Distribuido de Aprendizaje: Integración de ITS Y CSCL por Medio de Agentes Pedagógicos. In: EIA 2006, pp. 89–104. Escuela de Ingeniería de Antioquia, Colombia (2006)
11. Galan, J.M., Lopez-Paredes, A., Olmo, R.d.: An agent-based model for domestic water management in Valladolid metropolitan area. Water Resources research 45 (2009)
12. Márquez, B.Y., et al.: Methodology for the Modeling of Complex Social System Using Neuro-Fuzzy and Distributed Agencies. Journal of Selected Areas in Software Engineering, JSSE (2011)
13. Nwana, H.S.: Software Agents: An Overview. Knowledge Engineering Review 11, 1-40 (1996)
14. USP, `http://www.lsi.usp.br/interativos/nrv/nrv.html`
15. Wilensky, U.: NetLogo Software (1999), `http://ccl.northwestern.edu/netlogo`

Study on Highway Network Operational Monitor Analysis Index Extraction

Jianjun Wang[1], Jing Zhao[2], Jingwei LI[3], and Leihong Dong[4]

[1] Key Laboratory for Special Area Highway Engineering of Ministry of Education, Chang'an University, Xi'an, China
[2] Highway College, Chang'an University, Xi'an, China
[3] School of traffic and transportation, Bejing Jiaotong University, Bejing, China
[4] Research Institute of Highway Ministry of Transport, Beijing, China
wjjun16@163.com, 396841003@qq.com, 498230213@qq.com, dlh@itsc.com.cn

Abstract. By using the relevant highway network monitoring data extracted from the national highway network management and emergency response platform, this paper summarizes the monitor and analysis index system of the highway network operation according to the demands of this industry. Then, the index extraction method under the special demands is analyzed. Finally, by making use of this method, this paper summarizes the logical relation between the index and the demand and sum up the corresponding rule between the index and the demand, provides a simple method of getting quantities and qualities indexes in the periodic analyzing report for the industry supervisors and helps to reduce human and financial resources greatly.

Keywords: Operational monitoring, Index system, Extraction method, Corresponding rule.

1 Introduction

At the end of the Eleventh Five-Year Plan, the total length of the highway network in the country has amounted to 3,984,000 km. The length of the highway has increased from 41,000 km at the end of the Tenth Five-Year Plan to 74,000 km now. Twelve main state lines (5 vertical and 7 horizontal strokes) have been completed 13 years ahead of time. Eight interprovincial main lines in the western development project area are now basically connected. It is estimated that in 2020, the total length of the road will come to 4,200,000 km and the total length of highway will break through 100,000 km. National highway network management and emergency response technique are a gradually developed and emphasized emerging technology area under the background of national highway network which takes the highway as backbone and state and interprovincial main lines as main body. At present, the basic work is to obtain highway network monitoring data of the incident and the information of traffic flow by making use of national highway network management and emergency response technique. With the

X. Wan (Ed.): Electrical Power Systems and Computers, LNEE 99, pp. 581–588.
springerlink.com

gradual establishment of highway network monitor data index system, the technical scotoma is how to obtain extract detail index in special incident, in special time and in special demand and obtain the analysis results required by actual highway network management. In the reality and the operation, a set of standardized index extraction method and model is urgently needed to be used as the junction of fundamental research and actual operation. The highway network monitoring data index extraction system proposed in this paper will provide powerful technical support for realizing this goal.

2 Current Situation of Research at Home and Abroad

Overseas: At present, developed countries like UK and Japan have conducted many researches in the aspect of highway network monitor and their technique in the relevant fields is in the leading position. British highway network center has established many acquisition and publication systems of highway network operation information. The network appraisal and index evaluation system is formed on this basis and provides real-time directions for the dispatch and emergency response for the operation of the highway network; Japanese Transport Ministry Road bureau, closely together with local road bureau, has established national traffic information service system and provided service for road users in the form of internet.

China: China has a short history of developing highway network monitoring and analysis, in addition, each level of the road administration department hasn't paid enough attention to this work, therefore, the existing technique of analyzing data cannot meet the demand of the reality. In July, 2008, highway network management and emergency response center (hereinafter referred to as highway network center) founded by the Ministry of Transport has been the important department with the function of monitoring the operation of main lines in the country. Network management platforms of each place not only monitor the highways, but also give attention to the state and provincial main lines to gradually improve the layout of the network monitoring equipment, to enrich the data collected by the network and to realize the connection and information sharing between the highway network monitoring and managing platforms of different departments and different regions. By now, the extraction of highway network monitoring data and analyzing index, especially in the choice of periodic analysis index, are basically confirmed by manual work. Therefore, extraction of automatic selection in special incident and special demand cannot be realized.

3 Research Approach

The extraction of highway network monitoring data and analyzing index and the simple computing system research work is based on the method of combining fundamental research and actual development, analyzes the advanced experience of overseas highway network operational monitor and current condition of domestic research, analyzes the foundation of the project and defined the contents of the research. The contents of the research are divided into two major parts: highway network operational monitoring and analysis index system: the data monitored from the highway network monitor system is described according to the incident properties. Every data can only describe the detained information of the incident, but it cannot describe the overall operational condition of the whole highway network in general. Index extraction method: combine information extraction and information retrieval, apply the unceasingly improved method, analyze the process of index extraction, summarize the corresponding rule of index and demand of incident and lay foundation for the development of the system.

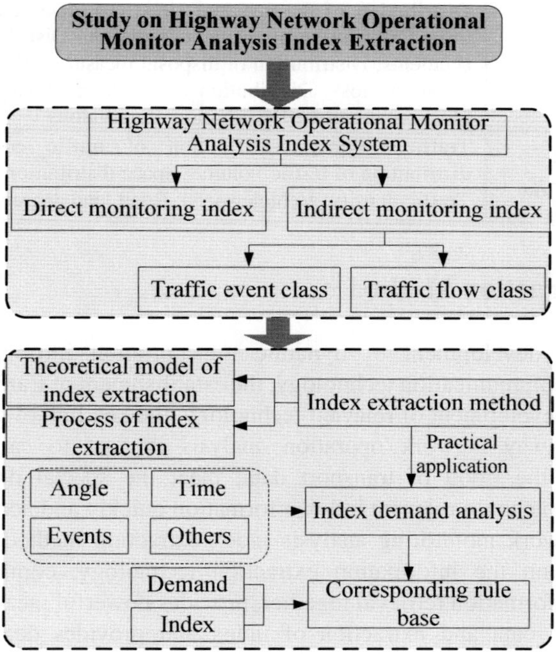

Chart 1. Technology Roadmap

4 Highway Network Monitoring Data Analysis Index System

According to different targets, highway network monitoring data analysis index system can be divided into two types: the index of incident and the index of the traffic flow. According to the data origin, it can also be divided into direct monitor index, indirect monitor index of incident, indirect monitor index of traffic flow. The highway network direct monitor index is the data directly monitored by the highway network center. It is a direct reflection of the monitored traffic accident; indirect index is the reflection of the analysis, process and manage condition of the monitor data. The index system is summarized in Table 1.

Table 1. Highway network monitoring data analysis index system

Type of the index	Index
Direct monitor index	Road compatibility of traffic; Road importance; Traffic volume; Traffic capacity; Site speed; Blocking time; Blocking reason; Disposal measures; Incidents occurring time; Site of the incident; Vehicle type in the incident; Casualties
indirect monitor index of traffic accident	Distribution of the accident sites; Accident rate; Distribution of time; Distribution of accident formation; Distribution of causes of accidents; Distribution of disposal measures; Distribution of direct economic loss; Distribution of blocking mileage; Distribution of blocking duration; Distribution of casualties
indirect monitor index of traffic flow	Traffic volume; Distribution of traffic composition; Time distribution of traffic volume; Space distribution of traffic volume; Traffic density; Vehicle speed; Road load degrees; Blocking time

5 Index Extraction Method

With the rapid development of dynamic transportation information acquisition technology and communication technology, the establishment of transportation sharing system and the development of relevant technology, the key link of giving full play of the overall highway network operation analysis is how to carry on integrated management of the flood of transport data, meet the special demand of special application and to choose relevant index information quickly and accurately[1].

Highway network monitoring analysis index extraction method proposed in this paper is based on the information extraction technology, combines information extraction and information retrieval together, provides powerful means for the analysis of transportation data and extraction of index and provides decision support for managing transportation more effectively and scientifically. The topic will construct the theory frame of highway network operational monitor analysis index extraction method on the basis of theory of information extraction[2] and information search[3], select the breakthrough point and provide model for the practical application of highway network operational monitor analysis index extraction method.

Information extraction is to extract factual information directly from the natural language document, describe the information in the structured form, provide the applications like information search, deep excavation of the document and automatic question answering and provide powerful means of getting information. The major function of information extraction system is to extract specific factual information from the document. Information retrieval is a research closely related to information extraction, but it differs from information extraction. The differences are mainly displayed in three aspects: different functions, different process technologies and different application fields. However, information extraction and information retrieval complement each other. In order to handle the flood of documents, information extraction usually uses the output of information retrieval system as input; while, information extraction technology can be used to improve the performance of information retrieval system. The combination of information extraction and information retrieval can better meet the users' demands of information processing.

On the basis of analyzing and researching the highway network monitor data index system, researching the method of index extraction is of great practical significance to the entire highway network monitoring as well as the following establishment of simple computing system. The starting point of constructing the theoretical frame of highway network monitor analysis index extraction method is to fully consider the demands of industry management and develop the advantage of information extraction in the field of transportation data processing. The extraction target of this topic is a database. The theoretical model of index extraction for this kind of database is shown in Table 2. Highway network monitor analysis index extraction method is composed of four parts: corresponding rule base of demands and index, corresponding rule base of index and parameter, extraction and result database.

Table 2. Theoretical system of highway network operational monitoring analysis index extraction method

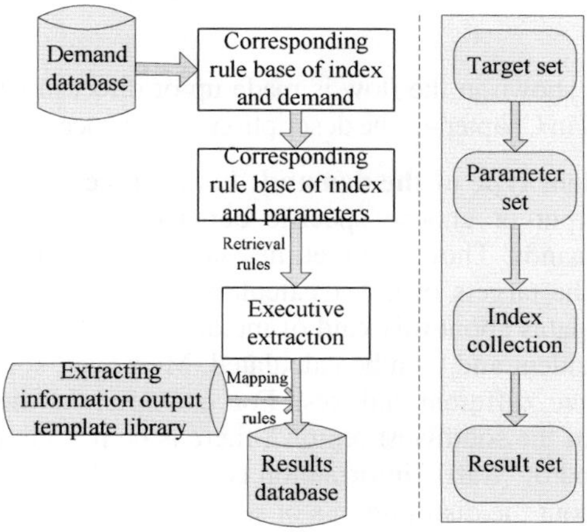

According to the theoretical method of index extraction and the inquire process of database, the process of index extraction can be summarized and extracted which is shown in Table 3.

Table 3. Process flow of index extraction

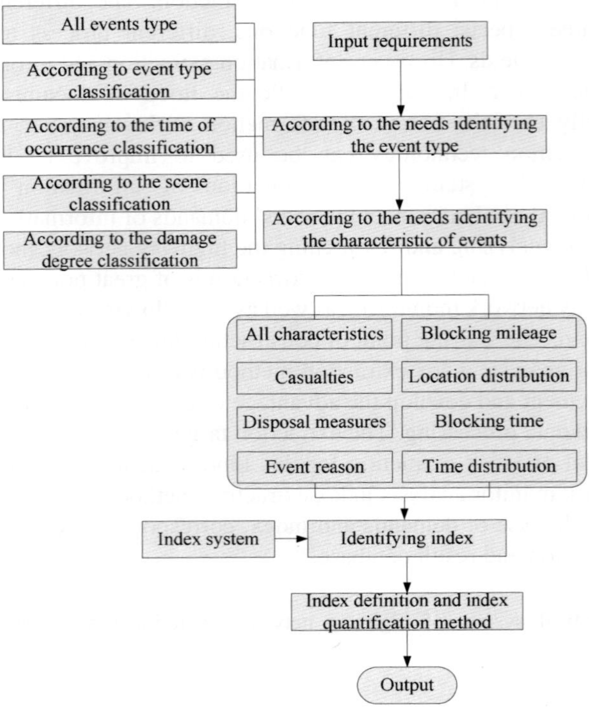

The index system shown in the flow is made up of direct monitor index and indirect monitor system listed in Chapter 4. The description of the process flow is stated as follows:

(1) **Define the incident type of the demand:** In the process of index extraction, users (system) have to input or choose specific demands and define the incident type according to the demands. Though sometimes same index can be used to describe all kinds of incidents, the targets of index calculation data are actually different. When calculate the index, datas about this kind of incident have to be selected out according to the type of the incident and then be calculated. Moreover, sometimes different type of incident may have different indexes. For example, in the demand of "traffic information census of the southwest region", there is no need to do region distribution census; in the demand of "traffic information census of May Day holiday", there is no need to do census about the distribution s of season, month and week.

There are two kinds of input modes in the system: one is choosing, users choose the options which are same with their own demands in the demands of demands; the other is automatic input, users input their own demand and language into the system. In most situations, the text which is automatically input by the users is inconsistent with the index or traffic field. In this occasion, approximate string matching has to be done in order to get the index that meets the demands.

(2) **Define the characteristics of the incident demanded:** According to the specific characteristics of the incident demanded, it can be defined which index (es) match (es)

with the demands. Under the situation that the type of the incident is defined, indexes can be searched according to the demands of the users. Due to approximate string matching, indexes may be more than expected. Therefore, all the indexes searched out will be listed to be selected.

(3) **Index set and relevant information output:** According to the two steps above, index that matches with the demand is extracted. Then, according to the definition of the corresponding index in the index system and quantities method, input all the relevant indexes and quantities formula in the special demand to help the following calculation and analysis. Define and analyze the operational condition of the monitored region (road section) correctly and take effective measures to handle the traffic accidents timely, rapidly and effectively.

6 Corresponding Rule Base of Index and Demand

According to the study above, analyze all kinds of demands of the highway network monitoring from the aspects of demand of time, demand of incident, demand of angle and summarize all kinds of demands by sorting and analyzing the datas and resources obtained. Extract the overall monitor index for special research group, extract corresponding index according to different characteristics of the demands, form the corresponding rule base of index and demand of incident, provide extraction base for index extraction and form the index extraction method based on the rules[4][5]. The details are listed in Table 4.

Table 4. Corresponding rule base of index and demand

Category of demand	Relevant index
The impact that major social incident on highway traffic	Traffic capacity in relative area; traffic volume; road load degrees; vehicle speed; traffic density
The characteristics of accident caused by bad weather	The accident formation caused by bad weather such as rain, snow, fog, high temperature; accident rate; distribution of accident site; distribution of accident time; distribution of disposal measures
The characteristics of accident caused by geological disasters	The accident formation caused by geological disasters such as earthquakes and landslides; accident rate; distribution of accident site; distribution of accident time; distribution of disposal measures
The characteristics of accident caused by accident	The accident formation caused by accident disaster such as dangerous leak, vehicle failure and wade bridge accidents; accident rate; distribution of accident site; distribution of accident time; distribution of disposal measures
The characteristics of accident in highway construction period	Accident formation; accident rate in sections; road importance
The casualties and loss and severity of accident	The number of injured people; number of deaths; million car mortality; direct economic losses
The related conditions of traffic blocking incident	The distribution of traffic blocking reasons; the distribution of disposal measures; distribution of traffic blocking mileage; the level of traffic blocking roads

7 Conclusion

This paper aims at establishing the highway network operation monitor analysis index system. By combining the technology of information extraction and information retrieval, establish highway network operation monitor analysis index extraction method. Together with the logical relation between index and demand, this system will half structurized the extraction and be applicated into the automatic extraction system of highway network operation monitor analysis index in order to save the time of selecting indexes. It will provide guaranty for fastening the establishment of highway network operation monitor analysis automatic system and getting the information of operational conditions of state highways, main lines and important stations timely.

References

1. Zhang, K., Wang, X.-j., Liu, H.: Transport Information Granular Computing Introduction Technical Architecture and Development Strategy. Journal of Highway and Transportation Research and Development (April 2007)
2. Li, Z.-y., Li, P.-y.: A Summary of Information Sampling Method. Journal of Langfang Teachers College (September 2005)
3. Li, X., Wang, H.-m.: Information Extraction Technology Exploring. Journal of Tonghua Teachers College (April 2008)
4. Jiang, D.-l.: Research on extraction of emergency event information based on rules matching. Computer Engineering and Design 31(14) (2010)
5. Shi, Q., Chen, R., Lu, M.-y.: Implementation of rule induction-based information extraction system computer. Engineering and Applications 44(21), 166–170 (2008)

Protecting Measures of Lifting Move Transversely Parking

Hong Chen, Na Bao, Meng Xue, and Juan Sun

Traffic Engineering Department
Highway College, Chang' an University
Xi'an, China
243477035@qq.com

Abstract. This paper bases on the analysis of the safety status of parking facilities especially large parking equipment, in view of the different characteristics of garage lifting equipment, puts forward parking equipment safety operation security technology composed of the active prevention technology and passive technology of prevention. But because of active prevention research has become increasingly mature, and passive prevention is often neglected, therefore this paper mainly studies the passive prevention measures, realizing parking equipment's safe, quick, continuous use function, give full play to the parking equipment social and economic benefits.

Keywords: parking equipment, active prevention, passive prevention, preventive measures.

1 Introduction

Because of the economic, technology, equipment and facilities and parking management level backward, our country in the aspect of safe operation in parking equipment is lack of adequate theory and practice experience, has not yet become effective parking equipment safety operation safeguard system. According to the development of our parking equipment and the operating characteristics, this paper proposes parking equipment safeguard technology safe operation which is composed by fault diagnosis technology, active and passive prevention technology to prevent technology. Because the research of the fault diagnosis technology and active prevention technology has a mature earlier, people in research parking equipment safety, neglect the passive prevention measures. Passive preventive measures after the incident can respond quickly and make safety measures and rescue strategy, simultaneously all aspects of coordination and management, to minimize the loss. So studying passive prevention is very necessary.

2 Safety Situation Analysis

Using parking equipment involving personal safety and vehicle intact, belongs to the greater danger hoisting machinery equipment, and has been incorporated into special

X. Wan (Ed.): Electrical Power Systems and Computers, LNEE 99, pp. 589–593.
springerlink.com

equipment safety supervisory scope. Therefore, the mechanical parallel parking equipment will inevitably have higher and more stringent security requirements than general hoisting equipment or logistics warehousing equipment higher and more stringent security requirements. To do of personnel entering and leaving no hurt, vehicle parked and take nondestructive, equipment dynamic reliability, the mechanical action being normal, motion control accurate and computer management effective and safety device perfect and emergency measures to powerful, etc. Therefore, mechanical parallel parking equipment is taken many security protection measures to ensure that the three-dimensional parking equipment is the safe and stable operation. Parking equipment have complete security systems, such as msn or car to go confirming device, the emergency brake device, preventing fall device, overload protection device, leakage protection devices, vehicle ultra-long and ultra-high detection devices, etc. From the parking equipment operation Already installed into view, overall operating is good condition, but overall operating occurred serious quality accident, if defy, fall, smashing cars, electric control cabinet explosion accidents, and due to design, installation, operation problems, causing there's a heavy security presence in some parking equipment.

3 Passive Preventive Measures of Parking Equipment

3.1 Warning Induction Device

The warnings of the parking equipment installation includes warning lamp, induced warning belt, warning marks, elevation markers, ground warning line, reflective elastic warning column, security guidance system and identify alarm function. Parking warning induction device can lead driver rapid, accurate, successfully reach their destination; in the light of the darker cases underground garage made drivers quickly and accurately read mark content and see the ground marking orientation and makes the corresponding judgement advance; safe driving in all kinds of traffic signs, facade mark, ground of line under the warning role, stop and reduce unnecessary traffic accidents or vehicle damage, improve the management ability and wheel-dreven economic benefits.

3.2 Prevent Fall Device

3.2.1 Classification of Prevent Fall Device

The prevent fall device as upper load of the floors after carrying's the protection organization, its function is to prevent the floors falling accident happens. The device generally uses hooks, adopt electromagnet drive and mechanical drive two modes [4]. The device is mainly divided into mechanical fall prevent device, traction type fall prevent device and shaft stretch type fall prevent device.

3.2.2 Prevent Fall Device in the Application of Parking Equipment

Mechanical prevent falling hook's work process schematic diagram shown as shown in figure 1. After confirmation the structure of the garage, the cars tray 9's side into the air parking frame 5 ‘s distance L is determined. Parking trays side install a4 ring, each ring 1 corresponding to a prevent falling hook. The link below side, there is a long dial the iron 8. Prevent the initial state sank hook as shown in figure 1 (a) below.

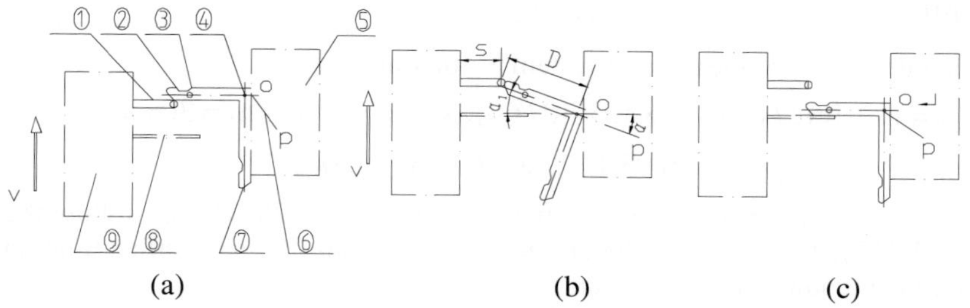

Fig. 1. Mechanical prevent falling hook of working principle diagram

Parking realization process are shown below:

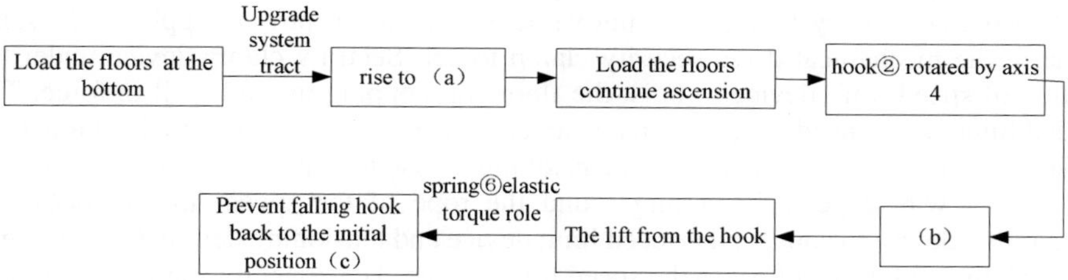

Fig. 2. Parking process

3.3 Buffering Technology

3.3.1 Classification of the Buffers

Parking system equipped with buffer device, when the floors drop or the car got into the the floors, buffer action to reducing the vibration and ensure system security. Buffer guaranteed the vehicles and garage equipment’s safety. When the floors on the limit position, it can absorb the floors energy generated. The principle is to make the movement’s energy into a harmless form or security form. Buffer will make moving load slowing and stop in a certain time travel, put in collision acceleration and collision force control in safety range.

3.3.2 The Performance Requirements of Buffers

For hydraulic buffer job performance, Gb provisions: when the elevator with rated load and operate in rated speed, the average reducing speed no greater than g, maximum reducing the speed of the duration should be no more than 1/25 SEC.

Formulas for the hydraulic buffer's buffer trip: $S = (1.15v)2/(2g)$

Type of: S ——Buffer trip m

v ——Rated speed m/s

Effect time $t = (2s)1/2/g$

The rated speed of the tractive system is 40m/min

$S = (1.15v)2/(2g) = (1.15\times 40/60)2/(2\times g) \approx 0.03\text{m} = 30mm$

$t = (2s)1/2/g = (2\times 0.03)1/2/g \approx 0.025\text{s} \le 0.04s$

In here, garage's rated speed of Hydraulic buffer is 0.25—1.0m/s, allow the total mass < 4000kg, total height is 390, buffer trip is 108mm. The slow effect time is 0.025 s, within the required 1/25 SEC range.

3.4 Overspeed Protection Device

3.4.1 Roles for the Speed Limit

In order to ensure the safe operation of hoisting system, need to add the speed limit device and security clamp, composition speeding protection device.

When ascension system got to limit value, signal cut off power supply, at the same time in the mechanical drive security clamp to act. Security clamp device under the action of speed limit device to hoist the floors or stopping in guide rail in force. The speed limit device needs to use with tensioner device and wire rope combined together. The speed device is installed on top, tensioning devise located in the abyss of the well ascension, wire rope is bypassing round the rope of the speed limit devices and tensioning devise to connect the speed limit device and tensioning devise. Garage speed through the wire rope reflect to the speed limit device. Tensioner device ensure enough friction between rope and speed device.

3.4.2 Security Clamp

Security clamp device is supporting with the speed limit. It is speeding protection device composed by stopping institutions and manipulation institutions. Both ends of the speed limit device's rope connect with drive connecting rod of the security clamp lever system. When upgrade system is in normal operation, the floor through driving connecting rod to drive speed limit device's wire rope act. This time, security clamp in not the acting state, its stopping components and rails keeping a certain gap. When the floors over the speeding allowable value, achieve speed device act then stuck wire rope. Along with the drop of the floors, drive connecting rod will be mention by wire rope to drive safely device. Security clamp mouth clamped guide, stop making the floors of load.

3.4.3 Research of Security Clamp's Requirements

In braking process of the security clamp, it can absorb all the kinetic energy of the floors of sports load and potential energy. According to the energy law:

$$E = \frac{W * V^2}{2 * g} + W * S = W\left(\frac{V^2}{2 * g} + S\right) = F * S$$

Therefore, braking force:

$$E = W * (\frac{V^2}{2 * g * S} + 1)$$

Type of: E ——The sum of the floors of kinetic energy and potential braked before (N*m);

V ——Tripping speed (m/s);

W ——Load the floors of self-respect add car weight (kg);

S ——Stopping distance (m);

g ——Gravity acceleration (9.8m/s2);

F ——Braking force (N)。

According to the design parameters, calculated the required safety pin garage for 4016N power, thus as security clamp criteria for one of selection.

4 Closing

Using vehicles warning induction device to induce the car, using the prevent fall technology to prevent floors of vehicles falling suddenly in the process of parking and taking car, using the buffering technology to control the speed of floors of vehicles, using speeding protection device protect the floors of speeding protection using the passive prevention system reduces eents; after the event, using modern communication and control technology, rapidly formulates rescue strategy, makes the corresponding, and realizes the minimum loss.

Acknowledgments. This paper is supplied by construction technology of city parking facilities (2006BAJ18B06).

References

1. Cha, M.: Parking equipment monitoring system. J. Instrumentation Journal A27, 559–560 (2007)
2. Hu, G.: PLC's application in multi-storey lifting shifting class stereo parking equipment design. J. Manufacturing automation A30, 72–75 (2006)
3. Xie, M.: Computer monitoring system research of multi-storey lifting shifting parking equipment. J. Electromechanical integration technology A9, 28–31 (2004)
4. Ding, H.: The move came of mechanical parking equipment's risk factors and protective measures. J. Modern vocational safety A86, 96–97 (2008)
5. Wu, N.: 32t ranes rope common problems and solutions. J. Metallurgy stratum A3, 26–27 (2009)

Xi'an Vehicle Exhaust Control System Design under the Trend of Low Carbon

Hong Chen, Meng Xue, Na Bao, Jibiao Zhou, Xiaowei Li, and Bin Chen

Highway Institute, Traffic Engineering Department,Chang'an University,
710064, Shanxi Xi'an, China
05110102@163.com

Abstract. The paper is based on the demand of world low carbon development. analyzes the current motor vehicle exhaust situation of Xi'an, applies the method of the combination of system engineering and environmental engineering, puts forward the overall framework of motor vehicle exhaust control and makes the system design of Xi'an, and carries on the analysis of implementing result. The application result shows that: this system design can improve the environmental quality, improve the people's living environment and realize economic sustainable development and environmental protection that is harmonious and unified, and provide decision basis for government department.

Keywords: Low Carbon, Motor Vehicle, Exhaust Control, System Design.

1 Introduction

Urban traffic is base on urban development, it is an important symbol of modern urban economic development. China is currently in a fast urban development stage, the constant development of economy, and people's standard of living rises ceaselessly, but, with the increase of motor vehicle quantity in China, the city has become the main source of air pollution. With the increasing quantity of urban motor vehicles, motor vehicle exhaust pollution problem has become an important source of air pollution in cities, in low carbon orientation, controlling vehicle exhaust pollution has become the top priority of which can improve the urban atmospheric environment quality.

On December 7, 2009, as Copenhagen world climate conference (namely the UN framework convention on climate change, the first 15 times contracting party congress) opened the curtain, sought to carbon reduction for the purpose of low carbon development mode, has become a global consensus (LIU Li-ya,2010). China announced the Chinese emission reduction target: by 2020, unit gross domestic product of carbon dioxide emission decline more than 40% ~ 45% of it in 2005, the non-fossil energy accounted for around 15% of the proportion of once energy consumption, before the treaty powers congress of the Copenhagen UN framework convention on climate change. The challenge to achieve a series of goals is very huge, various industries all need to act. In order to achieve the emission target set by China, the leading role and potential of urban transportation are very big. According to the data of international energy agency, in so far as the world, the power industry is the biggest

X. Wan (Ed.): Electrical Power Systems and Computers, LNEE 99, pp. 595–600.
springerlink.com

carbon industry undoubtedly, accounting for 40% of the total, followed by is transportation, accounted for 21% of the total (see table 1). Thus, transportation is also an important domain developing low carbon economy (SU Feng-ming,2010).

Table 1. The carbon dioxide emission of Global industries related to energy units: million tons Data sources: the international energy agency, world energy outlook

	CECD Countries		Transition economies		Developing countries		World total	
	2002 year	2030 year	2002 year	2030 year	2002 year	2030 year	2002 year	2030 year
utility industry	4793	6191	1270	1639	3354	8941	9417	16771
Petrochemical and other industries	1723	1949	400	618	1954	3000	4076	5567
transportation	3364	4856	285	531	1245	3353	4914	8739
Residents and service	1801	1950	378	538	1068	1930	3248	4417
else	745	889	111	176	605	1142	1924	2720
total	12446	19833	2444	3501	8426	18365	23579	38214

Visible, low carbon economy has become the development orientation of the whole world now, and low carbon transportation becomes city development orientation. The paper based on low carbon transportation guidance, under the background of constructing “low carbon, ecological, livable” city (ZOUDe-ci,2010). Consider the emission control technology of traffic tool, reduce carbon emission, promote energy conservation and emission reduction of urban transportation area. With rapid economic development and substantial increase in the number of urban motor vehicle under the present conditions of Xi’an using environment engineering and system engineering method, from the angle of multi-disciplinary systematically, puts forward design of motor vehicle exhaust control system of Xi’an, and carries on the environmental assessment, so as to guide and help the development of Xi’an city.

2 Analysis of Xi’an Present Motor Vehicle Exhaust Pollution

With the rapid development of social economy in Xi’an, its per capita GDP is in the sustained growth trend, until 2009 it is close to $5,000. The economy is in the midst of quick and well development period, required under the sustainable development and the scientific development concept, develop economy must be combined with environmental protection, realize environmental and economic "win-win". But, with the quickening of the process of urbanization, and the continuous increase of motor vehicle inventory, by the end of 2009, motor vehicle inventory has reached one million, vehicle increase inevitably leads to the increase in the number of vehicle exhaust emissions growth, and road traffic congestion aggravates motor vehicle exhaust emissions. Pollution trend from present line-source to area-source pollution, low level air quality of the second ring of the city center outnumber the national secondary air quality standards. From year 2004 to year 2009, the NO2/SO2 ratio in Xi’an environment air is constantly increasing, it explains that the air pollution type of Xi'an changes from "sooty and dusty" pollution type (mainly pollutant as SO2) to into "dust and exhaust mixed" pollution type (mainly pollutants as PM10 N02), motor vehicle emissions has become an important source of Xi'an air environmental pollution. Urban road traffic environment problem, not only seriously affecting the social image of Xi'an, hindering the construction of international metropolis of Xi'an, and it has become one of the key problems that cannot be ignored about the social and economic development. At present, the main pollutant of Xi'an vehicle exhausts have nitrogen oxides, carbon monoxide, hydrocarbons, total particulate, etc. Specific data analysis (Fig.1, Fig. 2 show the examples).

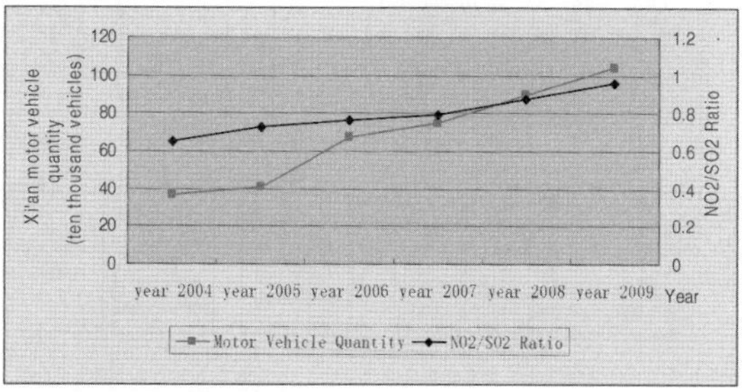

Fig. 1. Motor Vehicle Quantity and NO2/SO2 ratio in Xi'an

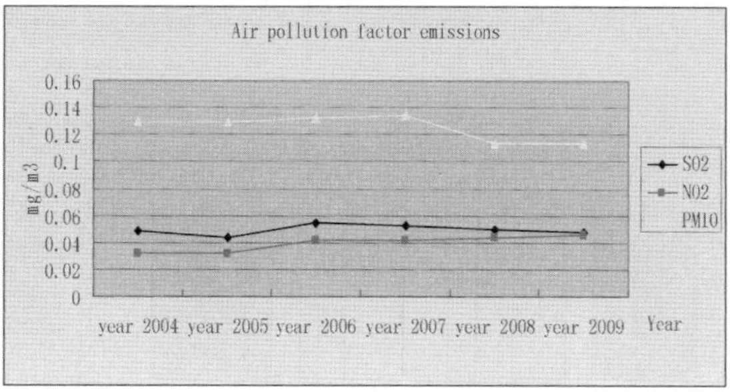

Fig. 2. Air Pollution Factor Emissions

3 Overall Framework and System Design

This paper uses system engineering and environmental engineering methods, combining the vehicle exhaust measures of environmental protection agency in Xi'an, puts forward the overall frame of management in Xi'an vehicle exhaust. (Figure 3 shows it: the labelled blue subsystems have been established by Xi'an environmental protection agency). Aiming at the city road traffic environment status of Xi'an, in order to effectively curb the traffic environment's further deterioration, and crackdown on illegal traffic action purify the city environment, improve the city people work and life condition, Xi'an city government invested a lot of people, material and financial resources. Based on the combination of dynamic monitoring and static monitoring, combination of regular monitoring and random monitoring, and combination of advanced monitoring and traditional monitoring method, with the help of environmental protection authorities and relevant government departments, Xi'an city has successively built motor vehicle exhaust pollution periodically supervision system, mobile law enforcement system, motor vehicle exhaust pollution prevention information management system in a central database, electronic map management module, new vehicle registration management system, vehicle environmental classification qualified marks management system six big systems and so on.

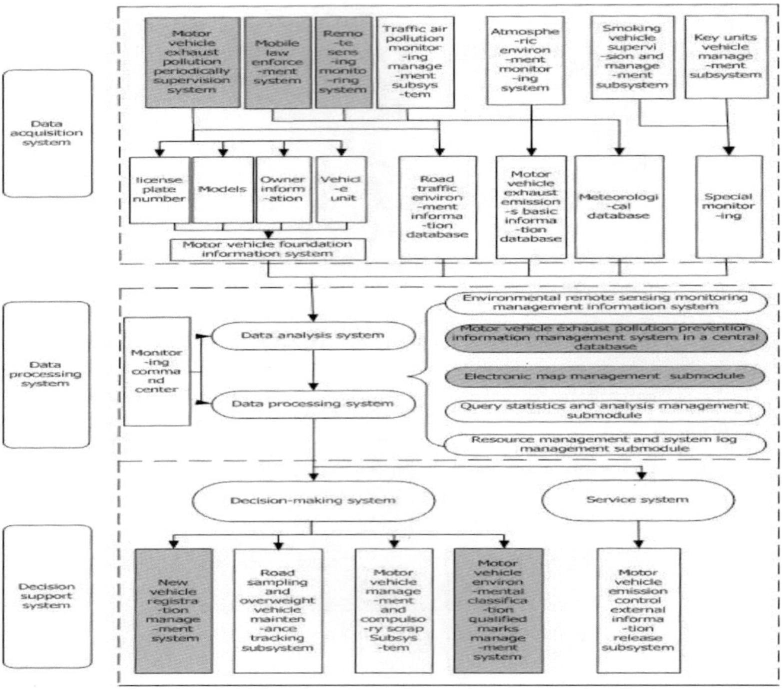

Fig. 3. Xi'an overall framework of motor vehicle exhaust control system design

3.1 Data Acquisition System

Motor vehicle exhaust pollution periodically supervision system (has been built), mobile law enforcement system (road monitoring law enforcement inspection car and law enforcement monitoring equipment) (suggested adding configuration), remote monitoring system (suggested adding configuration), traffic air pollution monitoring management subsystem and air quality monitoring substation, atmospheric environment monitoring system, "smoking" monitoring mobile law enforcement car and "smoking" vehicle fixed monitoring station, key unit vehicle management subsystem, they together make up the information collection platform of motor vehicle exhaust and air monitoring, namely, the large information collecting system of urban road traffic environment. Combining with the old data system, establish and perfect the vehicle basic information database, road traffic and environment database, basic motor vehicle exhaust emissions information database, meteorological database, etc. The monitoring data transfers through wired or wireless way to the data processing system, provide the basis for further implementing analysis and processing of pollution.

3.2 Data Processing System

This system design includes environmental remote monitoring management information system, motor vehicle exhaust pollution prevention information management system in a central database (has been built), electronic map management submodule (has been established), query statistics and analysis management submodule, resource management and system log management submodule, monitoring command center. The whole system incorporates the network technology, the communication technology and system integration technology, geographic information technology application technology, in the monitoring command center, through the system log management submodule and its correlation processing analysis equipment, judge, inspect and store the environment of collected data, then analysis, process, evaluate the the collected monitoring data. The system design uses unified interface layout and system operation, unified GIS geographic information operating system, reduce the complexity, inquire the objective through the map, and it is convenient to control.

3.3 Decision Support System

Vehicle exhaust is the main source of environmental pollution of road traffic, overweight vehicles, "smoking" vehicles are the key exhaust pollution source. This system includes new vehicle registration management system (has been built), road sampling and overweight vehicle maintenance tracking subsystem, motor vehicle management and compulsory scrap subsystem, motor vehicle environmental classification qualified marks management system(has been built), motor vehicle emission control external information release subsystem. Overweight vehicle maintenance management and tracking system is specifically targeted at overweigh vehicle that driving in the special areas; motor vehicle management and compulsory scrap subsystem aimed at the vehicles which don't comply with the national safety or environmental regulations, enforce them to scrap.

4 System Implementing Result Analysis

The negative externalities of urban transportation development is that the traffic congestion bringed to the city. Along with the increase of traffic demand, road traffic becomes deranged from unobstructed free flow state, namely exceeds the demand and can't pass bottleneck, forms the waiting rows in bottlenecks, produces traffic congestion, and gradually makes longer time to the destination. Due to the traffic congestion, cars longerly drive in low-speed condition, gasoline consumption increases nearly 50% more than normal driving, and the exhaust emissions is also far higher than normal driving. Because of congestion, the urban road resource utilization rate declines to 20%. According to the statistical data of America, it lost 84 billion hours per year caused by the road congestion, if we assume the minimum wage is 8 dollars per hour, the result was 627 billion dollars, the loss was very big, and it is about 10 billion yuan a year in our country. Thus, the economic loss of traffic congestion is very large.

Environmental benefits, mainly refer to the direct environmental benefits after the system implementation of Xi'an city roads. Through the system design; it may increase the effective number, set it the N. According to the national pollution source census statistics software, the environmental protection agency staff in Xi'an have calculated the average annual emission, it is about 0.0354 ton; the overweight vehicle's average annual emission is more than 1-3 times than normal vehicle, it may be 0.0708 ton of the average cut amount that an overweight vehicle is managed. The cost of treatment of vehicular pollutant both at home and abroad, has not uniform standard, in view of this, each city can comprehensive consideration of the per capita GNP, expense of managed and the experience of other cities, we may get a basic managed expense M. Then, this system design may bring environmental benefits

$$=N*0.0708*M \tag{1}$$

This system design can improve Xi'an city traffic environment, protect the environment, benefit the public welfare undertaking of later generations, and it will largely reduce emission of pollutants, improve the Xi'an environmental air quality, make great efforts to let the good days that above the second class of whole year environmental air grow from 304 days to 311 days.

5 Conclusion

The system design can improve the environmental quality, improve the people's living environment and realize economic sustainable development and environmental protection that is harmonious and unified, and provide decision basis for government department. The implementation can apply to other provinces and cities, and provide an experience to other managers.

Acknowledgments. The authors acknowledge that this thesis was prepared based on the National Natural Science Foundation of China (50808021), and the Science and Technology Development Foundation Program of Chang'an University (2008Q07).

References

1. Liu, L.-y.: Low carbon traffic road promotes urban sustainable development. Comprehensive transportation (January 2010)
2. Su, F.-m.: Low carbon traffic concepts and realizing ways. Comprehensive transportation (May 2010)
3. Zou, D.-c.: Low carbon, ecology, livable city – the 21st century ideal city. The urban development and planning of international convention in 2010 (May 2010)

Evaluation of Comprehensive Transportation's Development Environment Based on the SWOT

Hong Chen* and Yong-na Liu**

School of Highway, Chang'an University, Xi'an 710064, Shanxi, China
chh@gl.chd.edu.cn,
Liuyongna2005@163.com

Abstract. This paper uses SWOT method to analysis the factors that influencing development environment of comprehensive transportation. After considering all factors comprehensively, through calculating every factor's weight by AHP and setting up Fuzzy evaluation matrix, the result of comprehensive transportation's development environment can be reached. The result can determine whether it is appropriate to develop comprehensive transportation. And taking Zhejiang province as an example, this paper evaluates its comprehensive transportation's development environment.

Keywords: Comprehensive transportation, SWOT, environmental evaluation.

1 Introduction

At present, the evaluation of comprehensive transportation's development environment does not have a set of methods that is effective. It is difficult to evaluate comprehensive transportation's development environment objectively and scientifically. Therefore, it is necessary to establish a reasonable set of evaluation index system and evaluation method. According to this problem, using SWOT method, this paper analyses the comprehensive transportation's strengths, weaknesses, development environment's opportunities and challenges, to evaluate comprehensive transportation's development environment objectively and scientifically, to promote the development of the comprehensive transportation effectively. And the results can be the important basis of formulating policies, developing industry management, scientizing programming plans for transportation department.

2 Influence Factors of Comprehensive Transportation 's Development Environment

SWOT is also called situation analysis, among them, S representative strength, W represents weakness, 0 represents opportunity, T represent threat. SWOT is a kind of

* Hunan, Chang'an University doctoral supervisor.
** Shandong, Chang'an University graduate student.

X. Wan (Ed.): Electrical Power Systems and Computers, LNEE 99, pp. 601–606.
springerlink.com

strategic analysis method, which can analyse the object's strength, weakness, opportunity and threat, through the internal resources, external environment. Combining mathematical evaluation methods, it can reach the conclusion with comprehensive evaluation and analysis. The external factor refers to the factor that traffic managers cannot or hardly can change the situation through his own department or subjective effort, but this factor is an important component of comprehensive transportation's competitive. As comprehensive transportation is concerned, state and local government whether to support the comprehensive transportation's development, economic development level and transportation investment, natural conditions, location, residents of spontaneous choose travel way and so on, all those transportation management department can't change. Internal factors are those which can be changed through the traffic management department's own efforts. Improving the service level of transportation is one of the important factors, the service level of traffic indicators including transportation infrastructure, transportation coordination between the development, the quality of the traffic administrative department, etc.

Therefore, using SWOT method to analyse comprehensive transportation's development environment, it can based on internal resources and external environment. Comprehensive transportation environmental factor can be divided into two categories: the internal and external.

2.1 Internal Factors U_1

From their own perspective, internal factors analyse its strength and weakness (S, W).

(1) Transportation infrastructure l_1

Comprehensive transportation infrastructure including the infrastructure of highway, railway, aviation, water transport, pipelines and transportation hub and so on, and highway, railway, aviation and hub primarily are the main.

(2) The coordination of the transportation l_2

The coordination of the transportation can be considered from two aspects: planning research and construction management. Planning is the foundation, and construction management research is the guarantee.

(3) The quality of the traffic administrative department l_3

Traffic management department's quality is the guarantee of comprehensive transportation's development.

2.2 External Factors U_2

From the perspective of environment, external factors analyzes the opportunities and weaknesses (O, T) supplied by environment for developing comprehensive transportation.

(1) Location m_1

Location factors can be divided into the location in the domestic and international, and the domestic location is the main one.

(2) Policy m_2

It is mainly about policy that develop comprehensive transportation, and can be two levels: the national policy and local policies. The local policy affect the comprehensive transportation's environment directly.

(3) Economic m_3

It mainly refers economic strength that support comprehensive transportation's development, which can be reflected such as GDP level, traffic fixed investment, the development level of urban and rural, and so on.

(4) Natural conditions m_4

It mainly refers the natural conditions that related comprehensive transportation's development, such as land resources, geology, topography, etc.

(5) Residents travel's conditions m_5

With the economic and social development, we can see residents travel's conditions is diversity and particularity from consumption that residents spending on transportation.

2.3 Factors' Value and the Relationship with SWOT

This paper collected the indexs by the acquisition of the method, the concrete classification is as follow: (1) If the quantitative index and related material is all ready, with objective quantitative values to corresponding indexes'value are quantitative values as it's criterion, such as transportation infrastructure, economic, etc. (2) If the quantitative index data collection is not complete, ask for the expert's advice, ask them to give the evaluation results and concentrate the results. Take the results as the value, such as residents option trip mode, etc.(3) If it is qualitative indexes, design various grades comments, ask experts to choose corresponding level, and then concentrate, the comments may be as the corresponding indexes, such as location, traffic's coordination, policy, natural conditions, etc.

Multiple attribute decision making system meet the problem that the index is not male degree sexual and contradictoriness inescapable. Because of not male degree sexual, index has no unified metrics so it can't direct contrast. Therefore, before using index system to evaluate, the specific indexes must be standardized attribute values, and unified transformation to standardization within [0, 1].

When the influence factor's value is in [0,0.5], it means the factor is the weakness when develop comprehensive transportation or the factor take treat for comprehensive transportation. When the influence factor's value is in [0.5,1], it means the factor is the strength when develop comprehensive transportation or the factor take opportunity for comprehensive transportation.

3 Evaluation Method of Comprehensive Transportation 's Development Environment

Factors that influence comprehensive transportation's development environmental are many, which also can't be quantified easily. So, it is applicable to use AHP and Fuzzy comprehensive evaluation that combine qualitative and quantitative evaluating comprehensive transportation's development environmental. Considering relative importance of all factors, calculating every factor's weight by AHP and setting up Fuzzy evaluation matrix, the result of comprehensive transportation's development environment can be reached.

(1) Calculating every factor's weight by AHP

Table 1. Analysis table of total index weight

U_1	U_2
0.64	0.36

Table 2. Analysis table of internal index weight

l_1	l_2	l_3
0.62	0.33	0.05

Table 3. Analysis table of external index weight

m_1	m_2	m_3	m_4	m_5
0.32	0.22	0.25	0.09	0.12

(2) Establish fuzzy evaluation judgment matrix
Using the method of experts investigation, working out the comments set V = {strength, weakness, opportunity, threat}, this paper sets up fuzzy evaluation judgment matrix of technical evaluation index.

(3) Results of fuzzy comprehensive's evaluation
According to the method of fuzzy comprehensive's evaluation, the weight of each indicator and fuzzy evaluation judgment matrix, it can be calculated as follow:

$$P = w \cdot B \tag{1}$$

According to the principles of maximum membership degree, results of fuzzy comprehensive's evaluation about comprehensive transportation's development environment can be reached.

4 Case Study

Traffic network have been formatted basically in Zhejiang province. Economic and social development has brought more traffic demand than traffic supply, so it necessary to develop traffic network from high quantity to high quality. Taking Zhejiang province as an example, according to the actual situation of Zhejiang province, this paper uses the SWOT to analyse the internal and external environment of comprehensive transportation. And then, using experts investigation to grade internal and external factors, scope of the value is between [0,1]. When the value of internal factor is between [0,0.5], it means the factor is the weakness when develop comprehensive transportation. When the value of external factor is between [0,0.5], it means the factor takes treat for comprehensive transportation. When the value of internal factor is between [0.5,1], it means the factor is the strength when develop comprehensive transportation. When the value of external factor is between [0.5,1], it means the factor takes opportunity for comprehensive transportation. When the influence factor's value is in [0.5,1], it means the factor is the strength when develop comprehensive transportation or the factor take opportunity for comprehensive transportation. And then this paper can reach the judgment matrix of fuzzy evaluation. At last, according to the principles of maximum membership degree, the results can be reached.

First, establish an fuzzy comprehensive judgment matrix

Table 4. The fuzzy comprehensive judgment matrix

	Evaluation matrix	Strength	Opportunity	Weakness	Treat
internal factors	l_1	0.712		0.288	
	l_2	0.617		0.383	
	l_3	0.875		0.125	
external factors	m_1		0.875		0.125
	m_2		0.793		0.207
	m_3		0.674		0.326
	m_4		0.279		0.721
	m_5		0.501		0.499

Secondly, according to the fuzzy comprehensive evaluation method, the weight of the factors (table 1,2,3) and fuzzy evaluation judgment matrix, it calculates as follow:

Internal factor evaluation's results:

$$p_1 = w_1 \cdot B_1 = \{0.6888\ 0\ 0.3112\ 0\}$$

External factor evaluation's results:

$$p_2 = w_2 \cdot B_2 = \{0\ 0.70819\ 0\ 0.29181\}$$

Comprehensive factors evaluation's results :

$$p = w \bullet (p_1 + p_2)/2 = \{0.440832\ \ 0.254948\ \ 0.199168\ \ 0.105052\}$$

According to the principles of maximum membership degree, the result that fuzzy comprehensive's evaluation about comprehensive transportation development environment of external factors is strength to develop comprehensive transportation.

5 Conclusion

(1) This paper uses SWOT to analyse factors affecting comprehensive transportation development environment, and divides the factors into two parts: internal resources factor and external environment factors. Internal factors include location, transport infrastructure, transportation cohesion, etc. External factors include policy, economic and natural conditions, residents travel's conditions, etc.

(2) Considering relative importance of all factors, calculating every factor's weight by AHP and setting up Fuzzy evaluation matrix, the result of comprehensive transportation's development environment can be reached. Taking Zhejiang province as an example, the evaluation result is Strength. so, it is applicable to develop comprehensive transportation in Zhejiang province.

Acknowledgments. This paper is supplied by construction technology of city parking facilities (2006BAJ18B06).

References

1. Chen, J.-y., Wang, H.-y.: Comprehensive evaluation model of road traffic safety on modernization level. Tongji University Journals (January 2010)
2. Zhang, Y., Yu, Z.-y.: Research on transformation strategic of railway freight transportation enterprise based on SWOT. Railway transportation and economy (April 2010)
3. Pan, Y.: The development of comprehensive transportation system. Hebei traffic technology (September 2007)
4. Qiu, Y.: Comprehensive transportation system's evaluation method of hierarchy TOPSIS based on fuzzy threshold. In: 7th Word Congress on Intelligent Control and Automatic,WCICA 2008 (2008)
5. Petter, C., Simon, L.: model:current developments and future trends. ITE J., 33–38 (June1989)
6. Cheng, Y.-r.: Comprehensive transportation. People's traffic press, Beijing (2003)
7. Wu, Q.-z.: Operational research and optimal force method. Mechanical press, Beijing (2003)

Design and Realization of Heterogeneous Database Data Migration

Lixia Dong, Liming Wu, and Yaohua Deng

Faculty of Information Engineering,
Guangdong University of Technology, Guangzhou, China
dsonglixia646642@126.com

Abstract. Data migration refers to the process of moving the data from one database environment to another. This article describes the general methods and procedures of the data migration between heterogeneous databases and the implementation of data migration from Oracle to MySQL. It also analyzes the causes of heterogeneity between different databases, the principle of data migration and establishes a general model of heterogeneous data migration platform.

Keywords: data migration, Oracle, MySQL, heterogeneous databases.

1 Preface

With the development of computer application and database technology, the database is often used to store and manage large amounts of data. In the development and application of database or some integrated applications using a database, sometimes because the existing data management system of the enterprise needs to upgrade or update, or the different sites of dynamic website use different local database systems, or the historical data of old system needs to be migrated to the new or the same data needs to be applied to different sites, the data created already in a database environment usually needs to be migrated to another. So the data migration between remote or heterogeneous database systems needs to be resolved.

2 The Data Heterogeneity and Migration Methods

Data migration specifically refers to the complete migration of the database of the source server and all of the table structure, data, associated indices, triggers and stored processes of it to the purpose server, and then the new database can be fully operational. The process includes not only moving data, but also data format conversion [5]. Heterogeneous database means the database systems of source and target have heterogeneity, and this heterogeneity is reflected in three aspects (as shown in Figure 1).

X. Wan (Ed.): Electrical Power Systems and Computers, LNEE 99, pp. 607–614.
springerlink.com

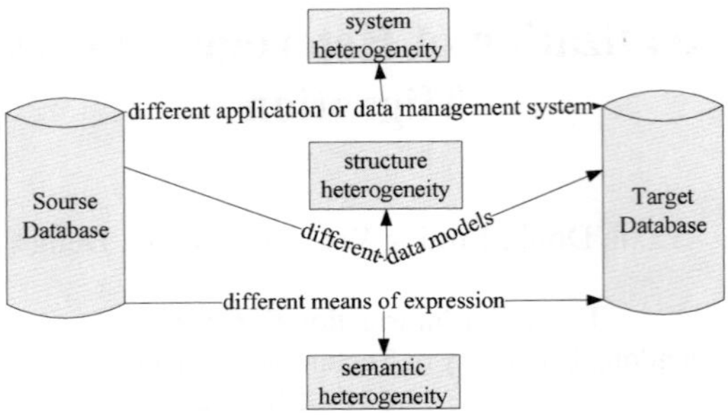

Fig. 1. Several Heterogeneities between Different Databases

It's quite difficult to achieve a smooth data migration between the heterogeneous databases successfully.

There are many ways to migrate data between heterogeneous databases: the dedicated data migration tools can be used, or combine professional tool with procedures developed, or compile dynamic SQL statements, and SHEN F X, ZHU Q M, etc. proposed the dynamic SQL migration based on JDBC and XML, and HUANG P, etc. proposed the design and implementation of cross-platform data migration based on Java and XML and developed the associated software [7]. There are a lot of data migration tools in the current. With the development of database technology, data migration tools are increasingly demanded to have a good function, simple operation and better realization of heterogeneous databases compatibility, but any one of the available data migration tools can not possess all of these properties and only can meet the general requirements. They can not handle many details of special requirements and are not suitable for large data volume of data migration; Combining professional tool with procedures developed can quickly complete the complex migration tasks containing large amounts of data, for example Data Stage and PL / SQL; Only compiling dynamic SQL statements for data migration is relatively cumbersome, especially when the data volume is very large and different syntax and functions of heterogeneous databases request to be made individual treatment; The dynamic SQL migration based on JDBC and XML has certain general effect because it is also suitable for the migration containing large amount, can well support Chinese and applies to most of relational databases [2]; the cross-platform data migration based on Java and XML also has some general effect and supports a variety of cross-platform data migration . Although the last two methods have considerable advantages in universality and quality of data migration, but they have a higher professional requirement to the migration operators. These methods have advantages and disadvantages of their own. The migration operators can select the appropriate transfer method according to their abilities and the requirements of tasks to achieve a smooth migration of the database structure, content and its system quickly, accurately, and efficiently.

3 Data Migration Model and Process

Data migration model is shown in the following figure. The data source should be fully understood and identified in identifying the data source module. All data included in the data dictionary, applications documents and text fields can be identified by refering in the JCL, application guidelines and documents and so on. The mapping of source to target data needs to be established after determining the real data source, including the conversion mapping of the metadata and the corresponding relationship of the composing fields while data integrated. The default type of the Source type in the target database requires some corresponding type matching treatment before starting the data extraction, assembly, conversion template settings, data acquisition and transmission. Log and review function should be designed to record a variety of anomalies happening in the migration and indicate the details so as to analyze and deal with these anomalies.

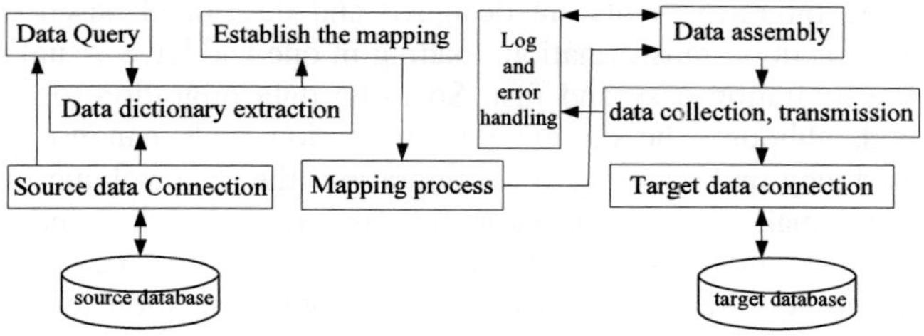

Fig. 2. Data migration model

Database Data Migration is very challenging and the valuable data resources will face the risk of loss once the inappropriate measures are adopted. We need to be well-planned and well-prepared if we are to succeed in achieving the smooth migration of database data. The process can be generally divided into three steps: preparation before data migration, migrating of data and the check after data migration [3]. Data should be made a full backup to prevent loss. And then install the relevant hardware and software facilities and finish the configuration of important parameters seriously. It is very important to analyze carefully and determine the various grammatical and semantic conflict of the two database models, for example naming conflict, if the identifier of the source model is the reserved word of the target, then renaming is needed; Format conflict, if the same data type in the two database models has different representations and semantic differences, then the transformation function between the two models needs to be defined; Structure conflict, if data definition model between the two database systems is different-such as the relational model and the hierarchical model, then the entity attributes and links need to be redefined to prevent the loss of property or contact information. As the development and application of large database Oracle are different from small database system MySQL in resource utilization, table structure design and database structure, the operation must be properly in accordance with established programs, and particularly the details need to be dealt with in the process of data migration to prevent such problems: (1)different character sets of old and new databases result in garbage problem, particularly the Chinese garbled; (2)the new database may not recognize some specific functions and technology of the old database , and even some characteristics of the old database may conflict with the new database's after migration [2]. The new database should be checked to the integrity and consistency of data after migration and the related database parameters and performance in the application procedures need to be adjusted until the application system can run successfully.

4 Data Migration between Oracle and MySQL

The database of client network center is MySQL, but we applied Oracle which is more widely used in the current when we started designing web sites. In order to enable our customers to run the website in their own application environment and save server resources, we have to migrate of the data from Oracle to MySQL. The data migration from Oracle to MySQL mainly needs six aspects: Table, including table structure and data; Triggers; Stored procedure, function and package; Job; Users and some other aspects; overcoming the differences in details happening when the related application procedure accesses with the SQL statement.

Many existing migration tools are designed and developed based on ETL. They integrate data extraction, transformation, loading in one and have visual user interface. That makes the migration easy and fast. So using data migration tool is a relatively popular method, although the current data migration tools can not guarantee data integrity, consistency and accuracy after migration, the data volume of the database involved in the small site is not large and the operator does not have a deep understanding and accurate grasp to the professional database language use in the last two methods, and so data migration tool with simple operation and relatively high speed is an appropriate choice.

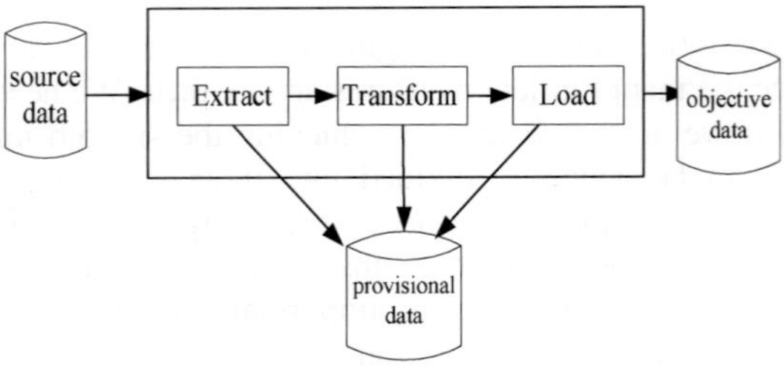

Fig. 3. ETL architecture

4.1 Implementation of Data Migration from Oracle to MySQL

MySQL GUI Tools is a MySQL database management console with visual interface provided by MySQL Official. It includes four graphical applications: MySQL Migration Toolkit, MySQL Administrator, MySQL Query Browser, MySQL Workbench, and these graphical management tools can greatly improve database management, backup, migration and query efficiency. MySQLMigrationTool of these is designed for the data conversion between MySQL and the other database. It applies to MySQL5.0 or later. It provides very powerful function and can complete data migration between Oracle, or Microsoft SQL Server, or Microsoft Access, or Sybase, or MaxDB and MySQL. It uses the JDBC connection, has a higher speed and is very suitable for the migration of the large data. Direct Migration generally includes eight steps: specify the link parameters of the source and target database server, here Oracle Database Server is selected as the source database of course, but it should be noted that we have to download a driver package ojdbc.jar14 and then select "Locate Driver on Harddisk", just after that we can restart the MySQL Migration Toolkit to continue to configure the relevant parameters; choose the objects (tables, views, sequences, etc.) to be migrated; choose the needed mapping and conversion methods for the objects(because the Chinese data is used, we need set utf8 as the default character set, otherwise garbage problem will come); edit the new objects manually to ensure proper conversion; create the conversion object in the target MySQL server and the data migrated will be saved in the object; specify the data needed to be changed in the data migration and this can be used as the basis of the data validation after migration; MySQL migration toolkit transfers the data from the source server to the target server; at last MySQL migration toolkit creates a summary report for you to review. The summary report generated in the instance is shown in the following photos. We can see 16 tables and one view has been successfully migrated from the report, but those are not all the contents of data migration.

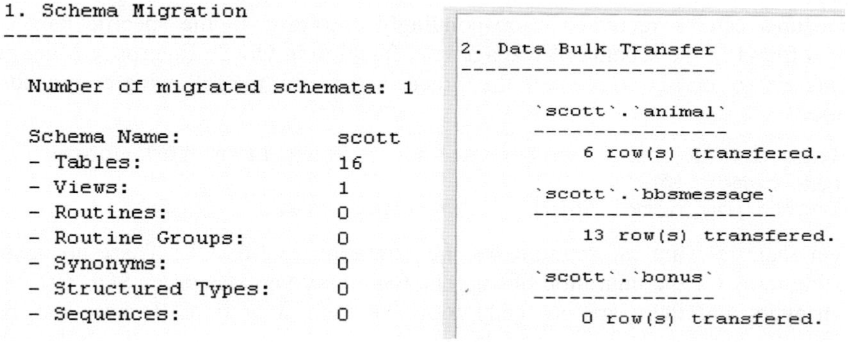

Fig. 4. Is the migration record automatically generated by MySQLMigrationTool.

Oracle to MySQL is the second tool to try. It is specially designed for the data conversion between Oracle and MySQL. The software itself is very small, but it has more powerful features. The configuration is very simple and it's easy to use. You can directly transfer the data from Oracle to MySQL or export .sql files by using it and then import it to MySQL. After exporting the Oracle's database as .sql files, in order to avoid garbage problem, we need to select gbk or utf8 as the default character set when the MySQL Server Instance is created. This is very critical. Then type "mysql -h hostname -u root -p" to enter the MySQL and after entering the password import the exported .sql file(data.sql) to MySQL using the command "mysql -h localhost -u root -p data< e:\data.sql". By default, MySQL limits the size of the imported file and the maximum is 2M, so the larger than 2M files can not be directly imported. And then you need to modify the related parameters in the file Php.ini: memory_limit=128M, upload_max_filesize=2M, post_max_size=8M to the size meeting the needs. It is generally recommended that these three values are set to be larger than the database file and the configuration value of post_max_size and memory_limit should be larger than the configuration value of upload_max_file. Only some of the objects have been migrated by using MySQL database query tool to view the new database after migration, and stored procedures, triggers, etc. are not migrated successfully.

Oracle and MySQL have some differences in structure, and although the two migration tools support different objects to migrate, they are not able to overcome these differences to achieve the complete migration of the six aspects at the same time from Oracle to MySQL. The contents not migrated successfully need to be modified manually to ensure the accuracy, completeness and consistency of the data migrated.

4.2 Solution to the Contents without Being Migrated

Use the query tools of old and new systems to check the quality of the data migrated and query the data on the same indicators, and then find only a lot of data and structure were successfully moved, purpose-built indexes, views, triggers, stored procedures and tasks were left because the two databases can not be compatible with each other. We have to recreate them one by one according to the relevant documents of source database Oracle to achieve the complete migration. Indexes, views, triggers and stored procedures can be recreated correspondingly according to the specific names and scripts found in the schema of management function in Oracle Enterprise Manager. If scripts are not chosen to recreate the indices and views, the following commands can be used:

```
select index_name, uniqueness, status from ind where
table_name='EMP'
select view_name, text  from user_views
```

to get their relevant information for the migration to MySQL; There are also not existing tools for the migration of triggers. Only later versions of MySQL 6.0 support for triggers and there are no ready tools for their migration. Because the syntax differences between the triggers of the two kinds of databases are relatively large, they are only added one by one manually according to the original logic in Oracle. The transplant package of stored procedures, functions, and packages is an object which Oracle uses to organize the logic function, and MySQL does not support it, so all the stored procedures, functions, etc. in this package should be put in the public procedures and functions of MySQL; Job is the method of Oracle achieving timing tasks while the method in MySQL6 is event.

4.3 The Heterogeneity between Oracle and MySQL

DBMS of different database systems define their own set of data types and corresponding mapping transformation needs to be made in the process of migration because of the data type differences between Oracle and MySQL(Mapping conversion table is shown in Table 1). After the map conversion some field types changed, and Data types which have changed need to be modified accordingly in the application program based on the database system. Otherwise, data type errors will occur in the process of practical application. We need to carefully deal with the default values of date field, primary key of table, the indexes of tables and so on. The bitmap indexes were converted to BTree indexes, tables and fields' notes were lost. Because Oracle has no auto-increment field, sequence attributes in Oracle need to be converted into auto_increment property in MySQL, or sometimes a separate table is created to specifically record automatic growth-oriented data.

Table 1. Data type mapping between oracle and mysql

Oracle	MySQL
number(<11)	integer
number(>11)	biginteger
varchar2(<255)	varchar
varchar2(>255)	text
clob	text
date	datetime
binary_float	float
binary_double	double precision
blob	longblob
clob	longtext
char	char

The use of the reserved words in Oracle and MySQL is different. Reserved words can be used as table names and field names in Oracle and that does not affect the application. But in MySQL reserved words can not serve as the table names and field names. It will report a syntax error if you use that. So the reserved words of sql statements in MySQL are required to use the symbol ‘`’ to quote.

Oracle and MySQL have different sensitivity to the case. Generally, case is not distinguished in Oracle but in MySQL the case sensitivity of the operating system used determines the case sensitivity of database names and table names. Case is sensitive in the operating system with Linux as the kernel. When the case is sensitive, the case of database name in MySQL should be the same as the one in Oracle, table names should be consistent with the table names in the sql string of application, the case of field names in the sql string should be consistent with the characters in double quotes if the field names use double quotes in the application. Table names and fields that are involved in and referenced by the application should be unified case.

Oracle and MySQL have different restrictions to the length of the index. The maximum length of the Oracle's index is larger than the MySQL’s. From MySQL 4.1.2, the length of the table index field of MyISAM and InnoDB can not be more than 1000 bytes. If this length is exceeded, it will report this error: ERROR 1071 (42000): Specified key was too long; max key length is 1000 bytes. If it’s encoded in UTF-8, it is equivalent to 333 characters in length (because a UTF8 character takes up 3 bytes), then the index definition or the defining length of the field needs to be modified.

Because of the syntax differences in detail between the two databases, many SQL statements of the application need to be modified. The statements which are different but achieve the same functionality should be modified one by one after the migration. And then the application reconnected with the new database can run properly.

In order to have a better compatibility and achieve a better data migration between the two databases, we must comply with the standard usage of the database, try not to use the specific usage of some database, avoid case-sensitive issue of the database and not use reserved words as table names and field names and so on.

5 Conclusion

With the development of database technology and changes of database market, multiple database systems coexist and are used in combination. The existence of heterogeneous databases is inevitable and data migration between heterogeneous databases has become a common and important issue. This article describes the general approach to solve such problems and the method of completely migrating Oracle databases to MySQL. And then it puts forward the specific operation scheme and requirements. The application program can run smoothly after reconnecting with the new database system. It has some practicality.

References

1. Shen, F.-x., Zhu, Q.-m., Liu, Z., Liu, H.: Migration method of database management system based on JDBC and XML. Computer Engineering and Design 29(20), 5376–5378, 5382 (2008)
2. Zhao, Q., Zhou, D.: Solution of data tansfer for OA information system of government. Journal of Guangxi Academy of Sciences 24(4), 354–355, 359 (2008)
3. Shi, X.-y.: The research of the data transfer. Journal of Zhejiang Business Technology Institute 6(3), 55–56 (2007)
4. Wang, J.-y.: General principle of data migration. Computer Development and Application 13(14), 31–33 (2000)
5. Luo, L.-q., Meng, Q., Li, X., Su, G.-p.: Design and implementation of patform-cross data migration based on java and XML. Computer Engineering 31(17), 74–75, 89 (2005)
6. Huang, P., Pan, Y.-r., Hu, Y.-h.: Implementation of heterogeneous DBMS migration. Computer Application Research (3), 233–238 (2006)
7. Du, B., Wang, M.-w.: Research and Method on Data Transfer over Platform. Computer and Modernization (6), 5–10 (2007)

Global Model of PMLSM Drive System Using Bond Graph Method

Teng Li, Yanjie Liu, and Lining Sun

State Key Laboratory of Robotics and System, Harbin Institute of Technology, Harbin, China
liteng_ha@126.com

Abstract. Global model of permanent magnet linear synchronous motor (PMLSM) drive system is built in this paper by bond graph method. In the global model each component of PMLSM drive system are concerned, which include motor driver, PMLSM, mechanism structure and sensor. 20-sim software is used to simulate the global model and do analysis of system's character. Simulation results illustrate that time constant of inverter in motor driver results in response lag, and sensor resolution influences the following accuracy. The process of modeling and analyzing fully shows the superiority of bond graph method and 20-sim software in multi domain system.

Keywords: PMLSM, global model, bond graph, motor driver, sensor quantitative error.

1 Introduction

Permanent magnet linear synchronous motor (PMLSM) drive system has been produced and distributed mainly through the modern industry technology, especially in semiconductor and electronic assembly industries[1], focusing on high acceleration, high speed, high precision, low noise, and simplification of driving apparatus.

PMLSM drive system is a typical mechatronic system includes sensor, driver, motor, controller, and mechanical structure, in which integrates machine, electron, magnetism and thermal energy. The performances of whole system are influenced by both electrical parameters and kinetic parameters [2], so global model should be built to obtain satisfactory performance or understand characteristics of the system, meanwhile, the multi-domain modeling method should be adopted.

Bond Graphs were used to develop the model. Bond graphs [3, 4] are a concise pictorial representation of all types of dynamic interactions. They offer a clear pictorial diagram that represents energetic interactions between elements and subsystems in physical systems. The PMLSM drive system existing in multi-domain is very suitable to modeling by bond graph technique.

This paper build bond graph model of PMLSM drive system. Unlike previous works, motor driver and sensor quantitative error are concerned. Model verification and analysis are obtained basing 20-sim software. System structure is firstly described in section 2 and sub-models of each component are built separately in section 3.

X. Wan (Ed.): Electrical Power Systems and Computers, LNEE 99, pp. 615–622.
springerlink.com

Based on these sub-models, global model of PMLSM drive system is obtained in section 4 and system's characters are also analyzed by 20-sim software in this section.

2 PMLSM Drive System Structure

A typical PMLSM drive system is as shown in Figure 1[5]. It contains PMLSM, sensor, motor driver, controller and mechanism. Slider and mechanism moving along the guide, sensor real time collects position data and transmit it to controller, controller send a control signal to motor driver according position and control algorithms, finally, motor driver gives the current signal to armature realize the motion control.

In this paper, ironless PMLSM motors are appropriate choices for its lack of detent force. Controller is not contains in global model because of there may be different algorithm in different systems and applications.

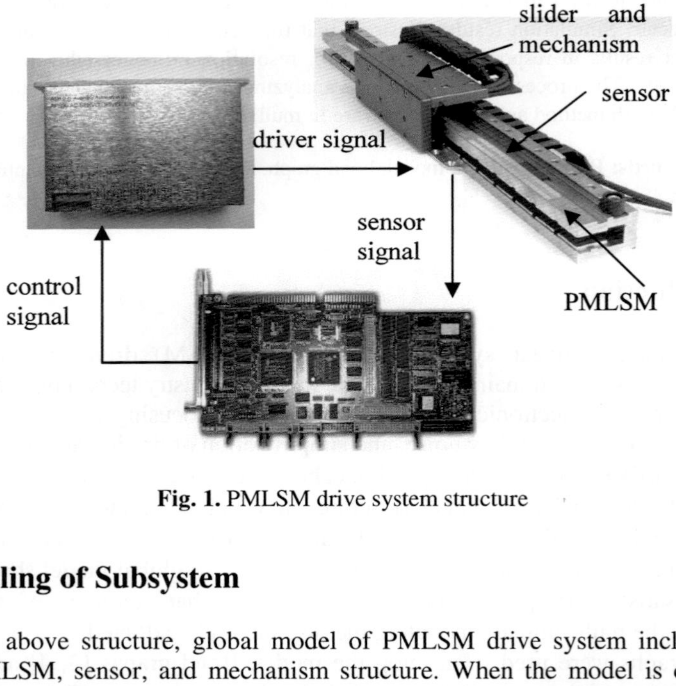

Fig. 1. PMLSM drive system structure

3 Modeling of Subsystem

According above structure, global model of PMLSM drive system includes motor driver, PMLSM, sensor, and mechanism structure. When the model is constructed, each component is modeled separately. Then these sub-models assembled into an overall system model suitable for analysis or simulation.

3.1 Bond Graphs of PMLSM

PMLSM can be thought for a normal rotating permanent magnet synchronous motor (PMSM) cut down along the radial plane and unrolled. These two kinds of motors have similar operation principle but different transmission ways. So the basic modeling of PMLSM is in accordance with that of rotating PMSM basing on electromagnetic theory.

Vector control technology in d-q coordinate that can obtain motor's constant coefficient differential equations by using coordinate transformation, is a widely used method of analyze and control motor. When field-oriented control is chosen, i_d =0, the motor force is proportional to i_q, PMLSM can be treated as DC motor, the equivalent circuit is shown in figure 2[6]. Electromagnetic induction which is the basic working principle of motor can be described by the bond graph element gyrator. The motor's bond graphs can be described directly basing the circuit as shown in figure 2.

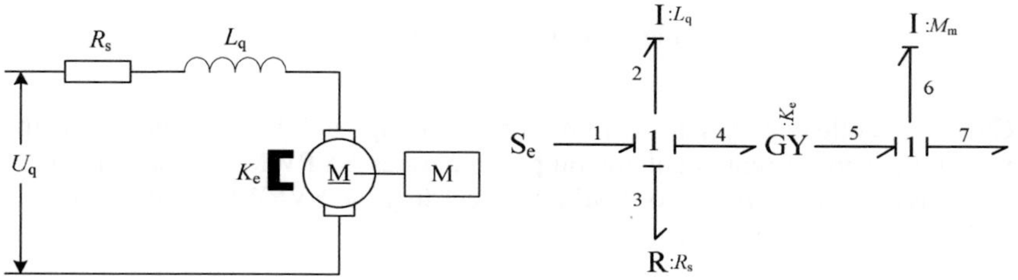

Fig. 2. Equivalent circuit and bond graphs of PMLSM

S_e denotes the input voltage in q-axis, I:L_q and R:R_s denote the inductance and resistance in circuit, GY denotes the energy conversion between magnetic energy and mechanical energy in air gap, K_e is the conversion coefficient. I:M_m denotes the slider mass. There are two 1-junctions in the bond graph, the first describes the voltage equation (1), and the second describes the force balance equation (2):

$$u_q = R_s i_q + \frac{d}{dt}\Psi_q + \frac{\pi}{\tau} v \Psi_d \quad (1)$$

Where, $\Psi_d = L_d i_d + \Psi_f$ and $\Psi_q = L_q i_q$ are flux of d-axis and q-axis. Ψ_f is the flux of permanent magnet and it is a constant. There same current flow through these elements, so the 1-junction can be used.

$$F = M_m \dot{v} + F_L \quad (2)$$

F_L is load force which has the same velocity with slider, so the 1-junction also can be used.

3.2 Bond Graphs of Motor Driver

Normally, the motor driver contains A/D converter, coordinate transformation part, current regulator ACR, PWM signal generator and inverter, as shown in figure 3.

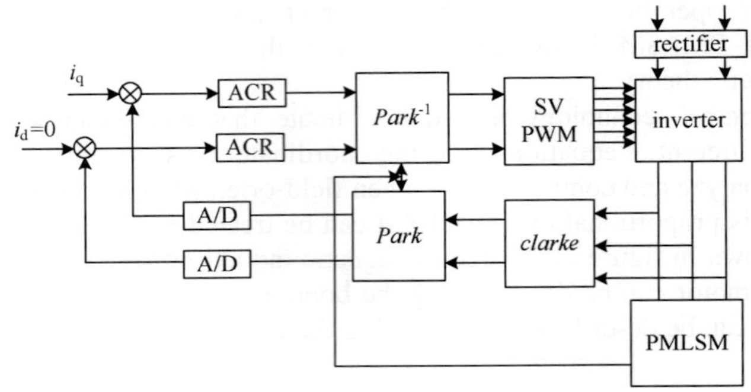

Fig. 3. Structure of motor driver

Current is collected by sensors in motor's winding, and the data subtract with input signal of i_q, then current regulator output generates SVPWM waveform, finally the inverter drive A/B/C winding of PMLSM according the SVPWM waveform.

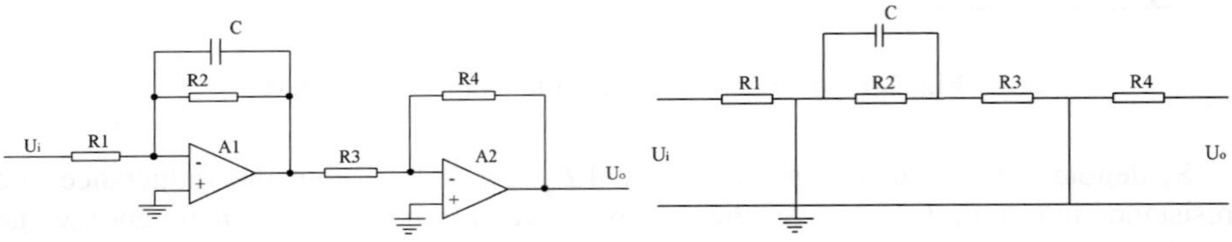

Fig. 4. Equivalent circuit of inverter

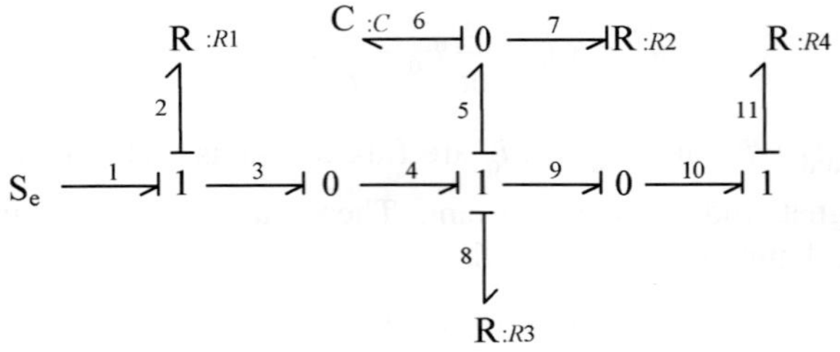

Fig. 5. Bond graphs of inverter

In motor driver, inverter can generally be treated as a first-order inertia link with time constant T_{inv} and equivalent gain K_{inv}. So it can be described by circuit in figure 4. In the circuit, A1 and A2 are amplifier, $T_{inv} = R_2C$, $K_{inv} = R_2 / R_1$, and $R_3 = R_4$. According to the circuit, its bond graphs can be obtained in figure 5.

3.3 Sub-model of Sensor

Sensor plays an important role in influencing performance of the positioning stage. A sensor's measure signal inevitably contains the quantization error which caused by the resolution, the error may degrade the control performance [7].

A general structure of sensor with resolution a is shown in Figure 6. x and y denote sensor's input and output. So the sensor can be treated as a special relay character which brings quantization error into system without taking the effect of signal processing into account. It can be modeled by block diagram.

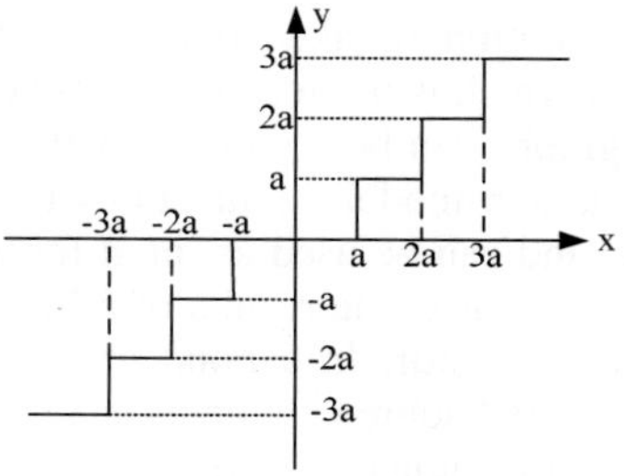

Fig. 6. Structure of general sensor

3.4 Bond Graphs of Mechanism Structure

Mechanism is simplified as a mass moving under external force and friction force, there is no deformation and rigid connection between load and slider.

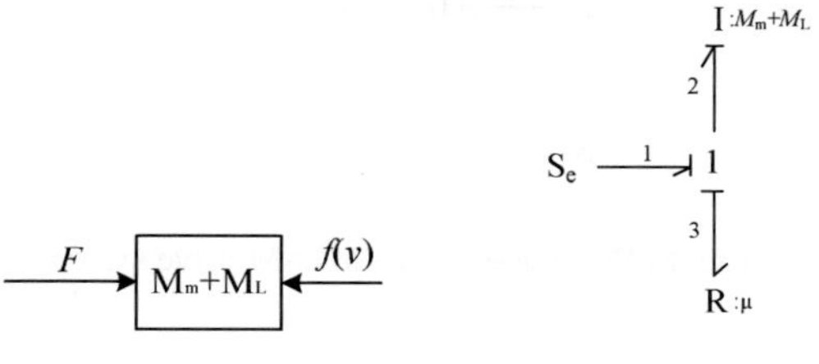

Fig. 7. Model of mechanism and its bond graphs

Model of it is shown in figure 7. $f(v) = \mu v$ is the viscous friction, it is proportional to the speed. The total moving mass is slider mass M_m plus load mass M_L, now the force balance equation (2) is become equation (3). Then the bond graphs can be built in figure 7:

$$F = (M_m + M_L)\dot{v} + f(v) \tag{3}$$

4 Global Model of System by 20-Sim Software

20-sim [8] is a modeling and simulation program for ironic diagram, bond graph, block diagram and equation models. With it you can simulate the behavior of dynamic system, such as electrical, mechanical and hydraulic systems or any combination of these. According to sub-models in section 3, the global bond graphs of PMLSM drive system is in figure 8. Basing on the bond graphs, similar model can be built in 20-sim as shown in figure 9. In the simulation model, $R_s = 0.66\Omega$, $L_q = 6.07mH$, $K_e = 25.7N/A$ are resistance, inductance and force constant of linear motor. $\mu = 0.01$, $K_{\text{inv}} = 1$, slider mass $M_{\text{m}} = 0.86kg$, load mass $M_{\text{L}} = 3kg$.

In the model, sensor resolution is taken into account. It's mainly a system that handles (powerless) signals, which is more conveniently described by block diagrams. In 20-sim software, bond graphs can be combined with block diagrams, the coupling can be performed by special sub-models. 1 junctions in 20-sim have a signal output which is equal to the flow, and can be used as input for a block diagram, such as the last 1 junction in figure 9. Similarly, the results of a block diagram can be converted into power by means of a generator. In a bond graph model, this can be done by connecting a signal to modulated source elements, such as MSe in figure 9. The first gain block that connected with 1 junction is reciprocal of resolution, and the type of its output is integer. Then the signal multiple with resolution, output signal can simulate the relay character of sensor that caused by resolution and results in quantitative error.

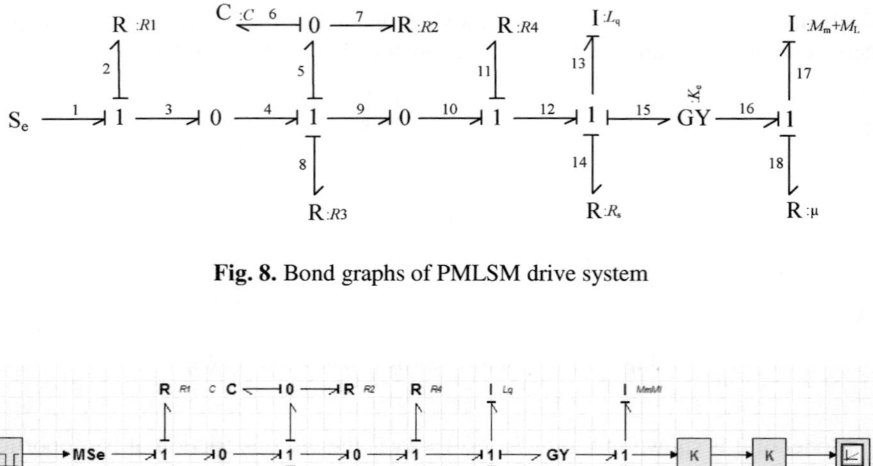

Fig. 8. Bond graphs of PMLSM drive system

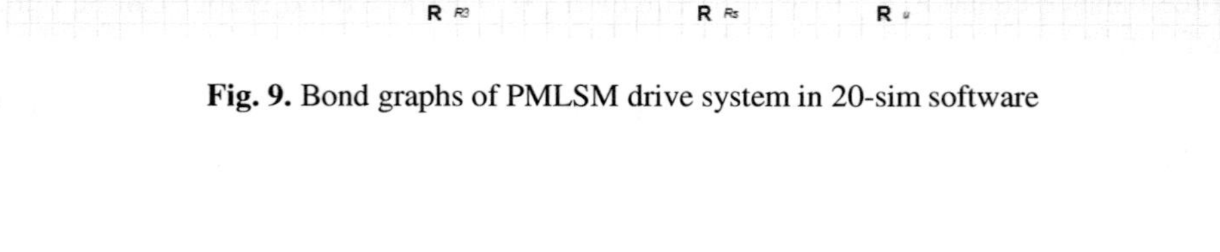

Fig. 9. Bond graphs of PMLSM drive system in 20-sim software

20-sim consists of two main functions. The first function is the Editor (as figure9) for modeling and the second is the Simulator for analyzing.

A square wave simulates the input of voltage drive. PMLSM drive system's response is plot for different inverter time constant and different sensor resolution, as shown in figure 10 and figure 11.

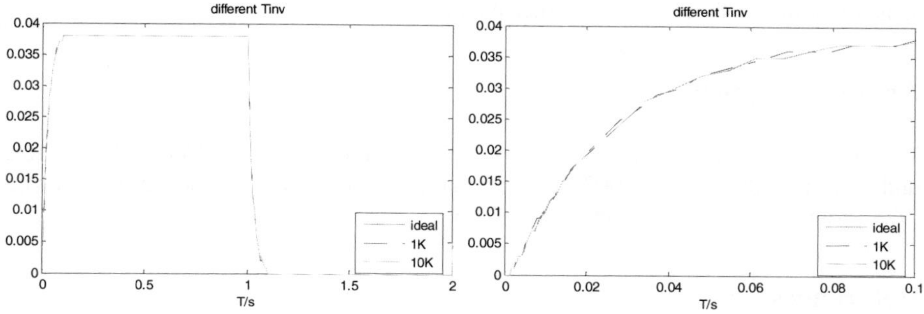

Fig. 10. PMLSM drive system's response to square wave with different inverter time constant

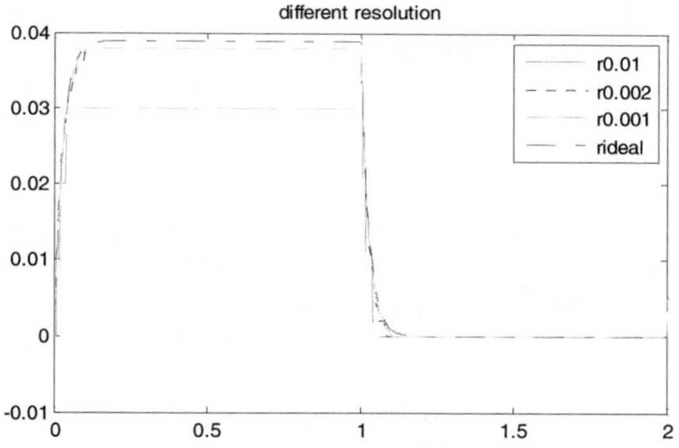

Fig. 11. PMLSM drive system's response to square wave with different sensor resolution

Figure 10 denotes the PMLSM drive system response to square wave with different inverter time constant. In the figure, switching frequency is 1 KHz, 10 KHz and infinity (in simulation it's a large enough data), the response curve lagging more with the increase of time constant. Figure 11 denotes the PMLSM drive system response to square wave with different sensor resolution. In the figure, sensor resolution is 0.01, 0.002, 0.001 and zero (that is ideal case, in simulation it's a small enough data), the positioning error becomes larger with the decrease of sensor resolution.

5 Conclusions

This paper built a global model of PMLSM drive system using bond graph method. Unlike previous work, the model not only contains mechanism structure and electromagnet transformation, but also has signal processing. Meanwhile, it takes the inverter switching frequency and sensor resolution into account. 20-sim software provides a convenient tool for modeling and analyzing. Simulation results show that time constant of inverter in motor driver results in response lag, and sensor resolution influences the following accuracy. So the switching frequency of inverter should be the faster the better, and its effect should be concerned in control algorithm. The sensor resolution should be the higher the better.

Acknowledgment

This paper is supported by the National Important Project on Technologies R&D under grant 2009ZX02021-003, the National Program on Key Basic Research Project (973 Program) under grant 2009CB724206.

References

1. Ding, H., Xiong, Z.H.: Motion Stages for Electronic Packaging Design and Control. J. IEEE Robotics and Automation Magazine 13, 51–61 (2006)
2. Zhong, J., Chen, X.L.: Coupling and Decoupling Design of Complex Electromechanical Systems. J. China Mechanical Engineering 10, 1051–1054 (1999)
3. Paynter, H.M.: Analysis and Design of Engineering Systems. MIT, Cambridge (1961)
4. Karnopp, D.C., Margolis, D.L., Rosenberg, R.C.: System Dynamics Modeling and Simulation of Mechatronic Systems. John Wiley & Sons Inc., Chichester (2000)
5. `http://www.h2wtech.com`, `http://www.deltatau.com`, `http://www.asmpacific.com`
6. Achir, A., Sueur, C., Dauphin-Tanguy, G.: Bond Graph and Flatness Based Control of a Salient Permanent Magnetic Synchronous Motor. J. Proceedings of the Institution of Mechanical Engineers. Part I: Journal of Systems and Control Engineering 219, 461–476 (2005)
7. Williamson, D.: Digital Control and Implementation: Finite Worldlength Considerations. Prentice-Hall, Englewood Cliffs (1991)
8. Kleijn, I.C.: Getting Started with 20-sim 4.1 (2009)

High-Nonlinearity Negative Dispersion Effect of Crystal Fiber with Elliptical Holes Square-Mesh Cladding

Yani Zhang

Department of Physics and Information Technology, Baoji University of Arts and Sciences, Baoji, P. R. China
Zhangyn@opt.ac.cn

Abstract. A novel photonic crystal fiber is proposed which is composed of a central defect core and a cladding with square mesh structure elliptical air holes. Its dispersion, birefringence, nonlinear and confinement loss are numerically investigated by full vector finite element method with anisotropic perfectly matched layers. And the novel PCF shows high birefringence high nonlinear negative dispersion effects which will be helpful to get admirable application in the field of polarization maintaining transmission system and dispersion compensation.

Keywords: Fiber optics and optical communications, Photonic crystal fibers, nonlinear optics fibers, fiber design and fabrication, birefringence.

1 Introduction

Photonic crystal fiber (PCF) consisting of a periodic distribution of air holes along its length and a defect region in its center, have been intensively studied in recent years due to their unique optical properties [1-5]. Now, some researches focus on the design of PCF with ultra-low and ultra-flattened dispersion for practical application [6-8]. Also, some other researches are absorbed in studying high birefringence or even single-polarization single-mode characteristics. However, few researchers discuss the property of birefringence and dispersion simultaneously. It is well known that the dispersion control of PCF can be easily achieved by adjusting its structure parameters. Also note that the birefringence (HB) of PCF can be made by changing the air-hole diameter along two orthogonal axes [9, 10], known as asymmetric core design [11-13] or by replacing the circular holes with elliptical ones in the cladding [14]. Obviously, both the dispersion and the birefringence are strongly dependent on the geometry structure of the PCF, such as the dimension, pitch and arrangement of the air-holes.

In this paper, a novel PCF is proposed which is composed of a central defect core and a cladding with square mesh structure elliptical airholes. Its dispersion, birefringence, nonlinear and confinement loss are numerically investigated by full vector finite element method (FV-FEM) with anisotropic perfectly matched layers (PMLs). Numerical results indicate that the proposed PCF shows high birefringence high nonlinear negative dispersion effect and stronger confinement ability of guided mode. It will be helpful to get admirable application in the field of polarization maintaining transmission system and dispersion compensation.

X. Wan (Ed.): Electrical Power Systems and Computers, LNEE 99, pp. 623–629.
springerlink.com

2 Design Principle and Theoretical Model

The cross section of the proposed PCF is depicted in Fig. 1, its cladding geometry structure looks like to a simple square mesh. This structure is formed by introducing another air hole between two air holes along *x*-axis for every other line to the conventional rectangular lattice PCF to improve the capability of the dispersion compensation. All cladding airholes are squeezed elliptically in the *y*-axis to induce higher birefringence. And, by introducing the added air holes, the index contrast between core and cladding gets higher than that of the conventional rectangular lattice PCF. Therefore, the higher birefringence can be achieved compared to that of the conventional rectangular lattice PCF, while maintaining the improved dispersion compensation. It can be characterized by the lattice constants Λ, air-filling fraction d/Λ, the elliptical airhole major axis d and minor axis b and ellipticity $\eta=b/d$.

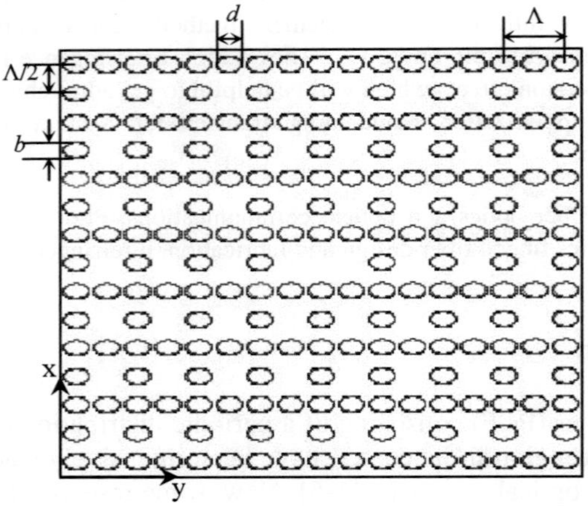

Fig. 1. Schematic cross-section of square mesh elliptical air holes PCF

In general, the waveguide dispersion $D_w(\lambda)$ of PCF, depending on the effective refractive index of the fundamental mode, can be determined according to the formula [15]

$$D_w(\lambda) = -\frac{\lambda}{c}\frac{\partial^2 |\mathrm{Re}(n_{\mathrm{eff}})|}{\partial \lambda^2}, \tag{1}$$

while, the birefringence is expressed as [16]

$$B(\lambda) = \left|\mathrm{Re}(n_{\mathrm{eff}}^{y}(\lambda)) - \mathrm{Re}(n_{\mathrm{eff}}^{x}(\lambda))\right|, \tag{2}$$

where C is the velocity of the light in vacuum, λ is the wavelength of the light, n^{y}_{eff} and n^{x}_{eff} are the effective refractive indices of two orthogonal polarization fundamental modes, and Re denotes the real part. Another, the leakage loss $L_c(\lambda)$ of PCF can be determined according to the following formulation

$$L_c(\lambda) = \frac{2\times 10^7}{\ln(10)}\frac{2\pi}{\lambda}\mathrm{Im}[n_{eff}] \tag{3}$$

where Im denotes the imaginary part. Hence, once the fundamental model refractive index n_{eff} is solved, the model birefringence $B(\lambda)$, the dispersion parameter $D_w(\lambda)$ and the leakage loss $L_c(\lambda)$ can be obtained. In order to solve the fundamental model refractive index, here, we employ the FV-FEM to accurately predict sensitive properties [17].

Meanwhile, the nonlinearity coefficient $\gamma(\lambda)$ of PCF can be defined as

$$\gamma(\lambda) = \frac{2\pi n_2}{\lambda A_{eff}}, \tag{5}$$

where n_2=3.0×10^{-20} m^2/W is the nonlinear refractive index of silica material, and A_{eff} is the effective mode area, which is calculated by [15]

$$A_{eff} = \frac{\left(\iint |E^2| dxdy\right)^2}{\iint |E|^4 dxdy}, \tag{6}$$

In what follows we investigate numerically the dependences of the values of dispersion $D_w(\lambda)$ and birefringence $B(\lambda)$ of the proposed PCF on its structural parameters in detail, with the purpose of controlling its dispersion profile in a wide wavelength range.

3 Numerical Result and Discussion

Firstly, we explore the dependences of $D_w(\lambda)$ and $B(\lambda)$ on the air-filling fraction d/Λ, The numerical results are shown in Fig. 2 by fixing Λ=2.0 μm and η=0.75. We can clearly obtain from Fig. 2(a) that the proposed PCF has negative dispersion parameter and negative dispersion slope in the wavelength range around 1.55μm, which demonstrates the excellent dispersion compensating property. Its dispersion decreases with the increase of wavelength and d/Λ. In the meantime, Fig. 2(b) shows that the dependence of the birefringence versus wavelength with different d/Λ. It is obviously that the birefringence increases with the increase of wavelength and d/Λ. According to Fig. 2, in order to gain a excellent negative dispersion parameter and a higher birefringence in a wavelength of 1.55 μm , we select the air-filling fraction d/Λ=0.5.

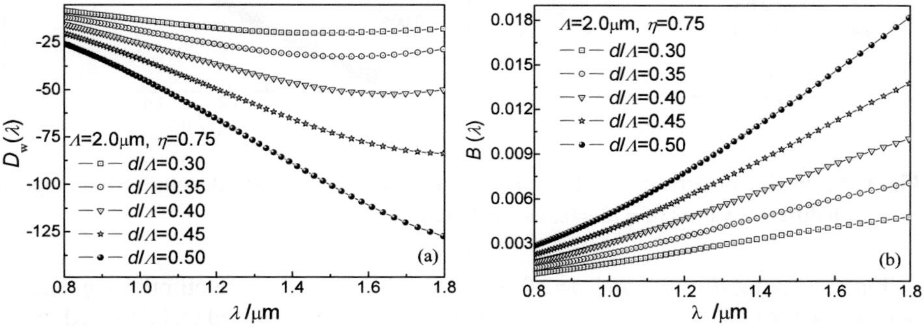

Fig. 2. Waveguide dispersion $D_w(\lambda)$ (a) and birefringence $B(\lambda)$ (b) versus wavelength for the proposed PCF with different air-filling fraction d/Λ. Λ=2.0 μm and η=0.75

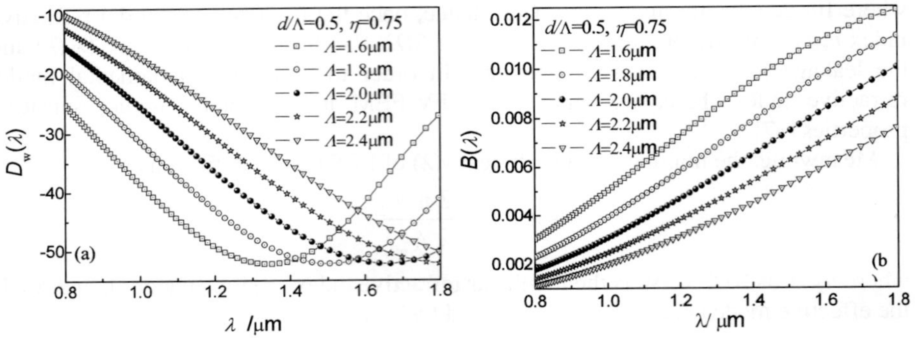

Fig. 3. Waveguide dispersion $D_w(\lambda)$ (a) and birefringence $\gamma(\lambda)$ (b) versus wavelength for the proposed PCF with different values of pitch Λ. d/Λ=0.5, η=0.75.

Next, the dependences of $D_w(\lambda)$ and $B(\lambda)$ on hole pitch Λ are simulated and shown in Fig. 3, with fixing air-filling fraction d/Λ=0.5 and hole ellipticity η=0.75, while changing Λ from 1.6 to 2.4 μm with a step of 0.2 μm. It can be seen from Fig. 3(a) that the value of $D_w(\lambda)$ decreases gradually with wavelength in a shorter wavelength range, while increases in a longer wavelength range. Meanwhile, the $D_w(\lambda)$ is enhanced by changing Λ from 1.6 to 2.4 μm in a step of 0.2 μm. Hence, the minimum dispersion wavelength is red-shift with the increase of Λ. Figure 3(b) shows that the $B(\lambda)$ increases gradually with the increase of wavelength, and the $\gamma(\lambda)$ decreases gently when Λ is increased. In order to obtain a higher $B(\lambda)$ and through an overall consideration of the dispersion compensating, here, Λ=2.0 *μm* is selected.

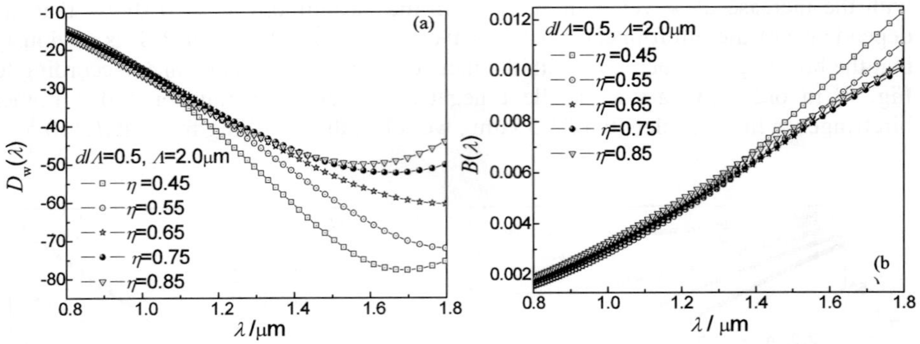

Fig. 4. Waveguide dispersion $D_w(\lambda)$ (a) and (b) birefringence $B(\lambda)$ wavelength for the proposed PCF with different values of η. d/Λ=0.5 and Λ=2.0 μm.

Finally, fixing d/Λ=0.5 and Λ=2.0 *μm* while changing air-hole ellipticity η, and the dependences of $D_w(\lambda)$ and $B(\lambda)$ on wavelength are further analyzed and depicted in Fig. 4. It is clearly seen from Fig. 4(a) that the value of $D_w(\lambda)$ increases when η is increased. While, Fig. 4(b) indicates that the value of $B(\lambda)$ will become larger when η is decreased.

Here, with the purpose of achieving higher birefringence and optimizing dispersion compensation, we fix η=0.75 for the proposed PCF.

According to the discussion above, in order to obtain a higher birefringence and a excellent dispersion compensation in a wavelength range around 1.55μm, the optimization structure parameters for the proposed PCF are found to be d/Λ=0.5, η=0.75 and Λ=2.0μm. By now, the proposed PCF has an admirable negative dispersion parameter and negative dispersion slope and higher birefringence closes to 7.9×10^{-3}. Then, the confinement leakage loss, the mode field distributions and the nonlinearity coefficient of the optimized PCF are analyzed and shown in Fig. 5. It can be seen from Fig. 5(a) that the mode field is stronger confined in the core region. The stronger mode confinement can give rise to higher birefringence and lower leakage loss. Figure 5(b) provides the leakage loss of the optimized PCF, it has been significantly reduced to the order of 10^{-2} magnitude. Figure 5(c) shows that the nonlinearity coefficient closes to 65 km^{-1} W^{-1}. Hence, the optimized PCF has admirable low-loss and the incorporation of high birefringence high nonlinear negative dispersion, which will has important applications in the field of polarization maintaining transmission system and dispersion compensation.

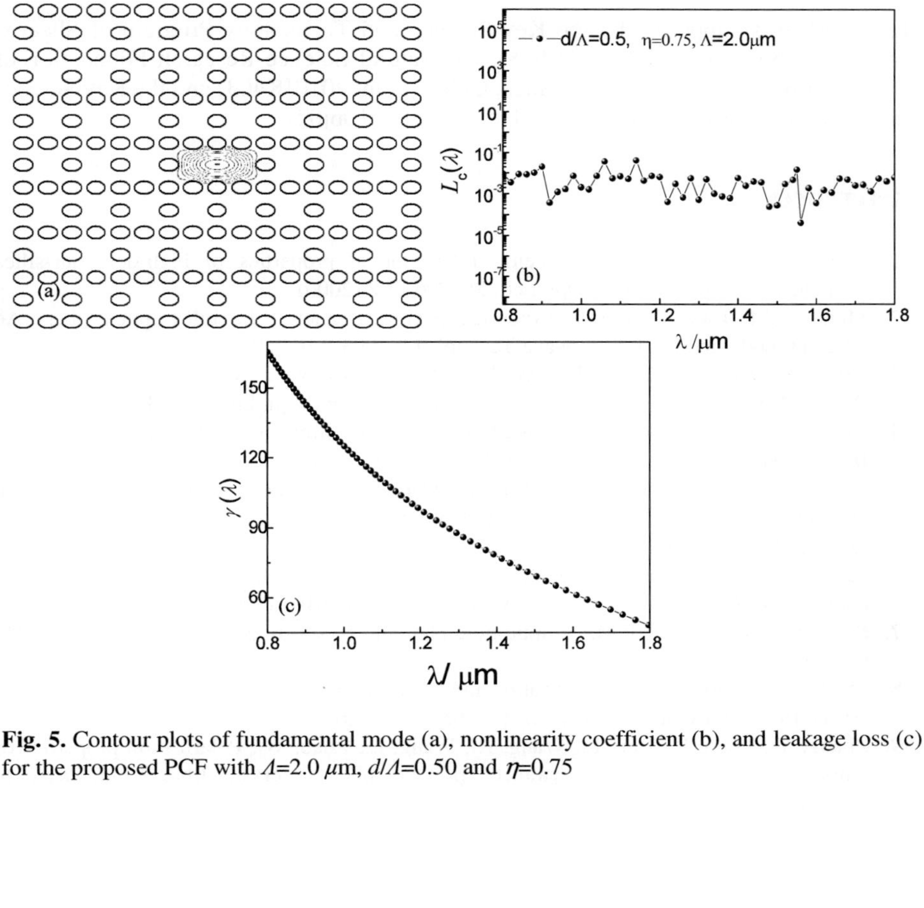

Fig. 5. Contour plots of fundamental mode (a), nonlinearity coefficient (b), and leakage loss (c) for the proposed PCF with Λ=2.0 μm, d/Λ=0.50 and η=0.75

4 Conclusion

A novel PCF is proposed which is composed of a central defect core and a cladding with square mesh structure by introducing another air hole row between two air hole rows for every other line to a conventional rectangular lattice PCF. Its dispersion, birefringence, nonlinearity and confinement loss are numerically investigated by FV-FEM with anisotropic perfectly matched layers. Numerical results indicate that the proposed PCF shows high birefringence high nonlinear negative dispersion effect, and stronger confinement ability of guided mode, in which the confinement loss is lower than 10^{-2}dB/m. The wavelength for high birefringence negative dispersion can be optimized by adjusting the parameters of proposed PCF, such as Λ and d/Λ. The nonlinearity coefficient almost close to 65 $km^{-1}.w^{-1}$ and the birefringence is higher than 10^{-3} over C band under the condition of Λ=2.0μm, d/Λ=0.5. It will be helpful to get admirable application in the field of polarization maintaining transmission system and dispersion compensation.

Acknowledgment

This work was supported by the Key Science and Technology Program of Shaanxi Province, China (grant 2010K01-078), the Natural Science Research Foundation of Education Bureau of Shaanxi Province, China (grant 2010JK403) and the Key Science and Technology Program of Baoji, China (grant 2010bj02).

References

1. Ranka, J.K., Windeler, R.S., Stenz, J.A.: Optical properties of high-delta air-silica microstructure optical fibers. Opt. Lett. 25, 796–798 (2000)
2. Monro, T.M., Richardson, D.J., Broderick, N.G.R., Bennett, P.J.: Holey optical fibers: An efficient modal model. J. Lightwave Technol. 17, 1093–1102 (1999)
3. Ortigosa-Blanch, A., Knight, J.C., Wadsworth, W.J., Arriaga, J., Mangan, B.J., Birks, T.A., Russell, P.S.J.: Highly birefringent photonic crystal fibers. Opt. Lett. 25, 1325–1327 (2000)
4. Ju, J., Jin, W., Demokan, M.S.: Properties of a Highly Birefringent Photonic Crystal Fiber. IEEE Photon. Technol. Lett. 15, 1375–1377 (2003)
5. Hansen, T.P., Broeng, J., Libori, S.E.B., Knuders, E., Bjarklev, A., Jensen, J.R., Simonsen, H.: Highly birefringent index-guiding photonic crystal fibers. IEEE Photon. Technol. Lett. 13, 588–590 (2001)
6. Kim, S., Kee, C.S., Lee, J., Jung, Y., Choi, H.G., Oh, K.: Ultrahigh birefringence of elliptic core fiber with irregular air holes. J. Appl. Phys. 101, 016101 (2007)
7. Steel, M.J., Osgood Jr, P.M.: Elliptic-hole photonic crystal fibers. Opt. Lett. 26, 229–231 (2001)
8. Steel, M.J., Osgood, R.M.: Polarization and dispersive properties of elliptical-hole photonic crystal fibers. J. Lightwave Technol. 19, 495–503 (2001)
9. Zhang, Y., Miao, R., Ren, L., Wang, H., Wang, L., Zhao, W.: Polarization properties of elliptical core non-hexagonal symmetry polymer photonic crystal fibre. Chin. Phys. 16, 1719–1725 (2007)

10. Zhang, Y.: High birefringence tunable effect of microstructured polymer optical fiber. Acta Phys. Sin. 57, 5729–5734 (2008) (in Chinese)
11. Noda, J., Okamoto, K., Sasaki, Y.: Polarization-maintaining fibers and their applications. J. Lightwave Technol. 4, 1071–1089 (1986)
12. Chen, M.Y., Yu, R.J., Zhao, A.P.: Highly birefringent rectangular lattice photonic crystal fibers. J. Opt. A 6, 997–1000 (2004)
13. Chen, M.Y., Yu, R.J.: Polarization properties of elliptical-hole rectangular lattice photonic crystal fibers. J. Opt. A 6, 512–515 (2004)
14. Wang, L., Yang, D.: Highly birefringent elliptical-hole rectangular lattice photonic crystal fibers with modified air holes near the core. Opt. Express 15, 8892–8897 (2007)
15. Saitoh, K., Koshiba, M.: Chromatic dispersion control in photonic crystal fibers: application to ultra-flattened dispersion. Opt. Express 11, 843–852 (2003)
16. Zhang, Y., Ren, L., Gong, Y., Li, X., Wang, L., Sun, C.: Design and optimization of highly nonlinear low-dispersion crystal fiber with high birefringence for four-wave mixing. Applied Optics 49, 3208–3214 (2010)
17. Poli, F., Cucinotta, A., Selleri, S., Bouk, A.H.: Tailoring of flattened dispersion in highly nonlinear photonic crystal fibers. IEEE Photon Technol. Lett. 16, 1065–1067 (2004)

Design of Monitoring Terminal System for Airport Snow Removal Vehicles

Liwen Wang[1], Guopeng Yao[2], Tao Jing[2], and Xudong Shi[1]

[1] Ground Support Equipments Research Base of Civil Aviation University of China, Tianjin, China
[2] Aeronautical Automation College of Civil Aviation University of China, Tianjin, China
wlw2885@163.com, yaoguo111@163.com,
jingtao100000@yahoo.com.cn, stone_131@sina.com.cn

Abstract. With the development of the dispatch of airport vehicles, the tracking and monitoring system of snow removal vehicles in airport becomes more and more popular to improve the efficiency of snow removal operations in the snow days. Since the small scope of airport calls for high-precision tracking and positioning of snow removal vehicles, monitoring system of snow removal vehicles based on the Global Positioning System (GPS) can play an important role. In this paper, a monitoring system of snow removal vehicle terminal was introduced, which employed pseudo-range differential methods to locate the position and adopted General Packet Radio Service technology (GPRS) for the remote wireless transmission. And it also could make alarm to alert working drivers to pay attention to driving and keeping safety when a collision was imminent between vehicles and aircrafts or among vehicles.

Keywords: Airport, Snow removal, Differential GPS, GPRS, Dispatch.

1 Introduction

With the development of Civil Aviation of China, the flight movements in large airports increase rapidly. However, it is very essential to clear the snow on the runway in the event of snowy weather. So there will be a problem that the monitoring center can not get the precise position of snow removal vehicles and its own state information timely to reduce the efficiency of snow removal system. Meanwhile, there will be a collision phenomenon among vehicles and aircrafts in the snow removal process. In addition, there will be some forbidden regions for snow removal vehicles at different times. If the vehicles invade the regions, it will affect other sectors or even cause an accident.

An ideal solution is proposed in the vehicle monitoring system, which is based on GPS. At all runways of airport, the system can collect positioning information of snow removal vehicles through the GPS positioning technology dynamically and online. Data can be transmitted via GPRS between monitoring center and vehicle terminals. So the system can complete data collection by monitoring center and arrange snow removal task reasonably.

X. Wan (Ed.): Electrical Power Systems and Computers, LNEE 99, pp. 631–636.
springerlink.com

2 System Structure

The equipment of vehicle terminal is mainly made up with ARM processor terminal, GPRS wireless transmission module, and GPS positioning module. This system employs embedded operating system $\mu c / os - \text{II}$ .The operating system of $\mu c / os - \text{II}$ begins its transplantation by the transplantation of boot-loader of the system. Based on the fundamental program code of $\mu c / os - \text{II}$, we should modify the system code corresponding to the new system hardware and this code will be called in the later program comfortably in the ARM development environment. Then we need to do the transplantation of $\mu c / os - \text{II}$ System. At the same time, we also need to pay much attention to the configuration of the system kernel to generate the file system. Finally run the $\mu c / os - \text{II}$ system and download it to FLASH when the test is OK[1].

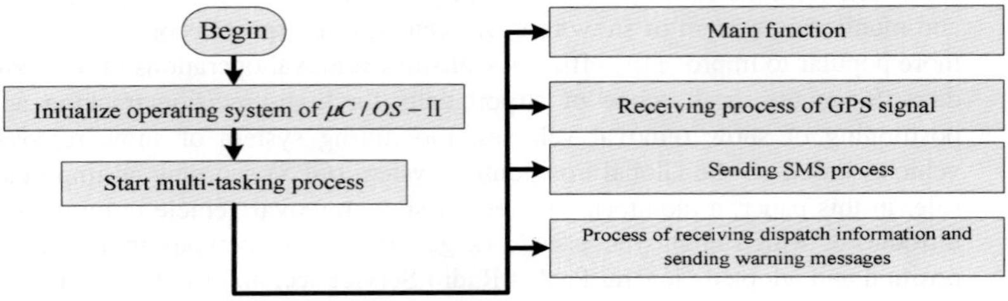

Fig. 1. This shows the main flow chart of system

Because $\mu c / os - \text{II}$ is a multi-task system, it can be divided into several parallel tasks according to the actual needs of the vehicle terminal. Likewise this system can be divided into GPS signal receiving process, SMS task process, sending dispatch information and warning information process flow, LED display process and so on[2]. It is shown as **Fig. 1.**

3 Technical Line

3.1 Differential Position with Pseudo-range

Differential GPS is a method which can improve the performance of timing and GPS positioning. It can take advantage of known coordinates obtained from the base station to get the distance from satellite to base station, and the distance will be compared with the measured one including error. Then the difference is filtered to calculate the deviation through a filter, and all satellite ranging error will be transmitted to users. Users can correct the measured pseudo-range by the ranging error. Finally, users' coordinates can be obtained by the correction of pseudo-range.

The pseudo-range between station i and satellite j at time t is:

$$\hat{\rho}_i^j = \rho_i^j + c(\delta t_i - \delta t^j) + \delta I_i^j + \delta T_i^j + d\rho_i^j \quad (1)$$

Where $c(\delta t_i - \delta t^j)$ is the equivalent distance error of the relative clock between receiver clock and satellite clock; δI_i^j is the equivalent distance error of the refraction delay of ionosphere at t time; δT_i^j is the equivalent distance error of the refraction delay of troposphere at t time; $d\rho_i^j$ is the distance error caused by satellite ephemeris error.

According to the known three-dimensional coordinates of the base station and the satellite ephemeris of GPS, we can calculate the geometric distance between the satellite and the base station.

$$\rho_i^j = \sqrt{(X^j - X_i)^2 + (Y^j - Y_i)^2 + (Z^j - Z_i)^2} \quad (2)$$

So there is the difference between the geometric distance and the pseudo-range which contains various errors and is measured by the receiver of base station.

$$\delta\rho_i^j = \hat{\rho}_i^j - \rho_i^j \quad (3)$$

In this formula, $\delta\rho_i^j$ is the correction value of pseudo-range and will be sent to the user's receiver. The user's receiver makes the correction of distance plus measured pseudo-range to obtain the corrected pseudo-range.

$$\hat{\rho}_k^{'j} = \hat{\rho}_k^j - \delta\rho_i^j \quad (4)$$

Differential positioning can improve the precision of positioning by counteracting of the errors of the two stations, so their common error is related to the distance between base station and vehicle terminal. With the distance increasing, the error is gradually decreased. Therefore, because of seeking more precise positioning, it is very important to decrease the distance between base station and vehicle terminal[3].

In this terminal, GPS receiver of base station sends pseudo-range differential information continuously, and the vehicle terminal continues to receive the differential information to correct their positioning information. Then we can obtain more precise location information. Since the scope of the monitored airport is not large, we can assume that the common system error of the two stations is the same. Before GPS module works, we should use ARM processor to initialize the module, including configuration of the module's baud rate, data output format, output parameters and interval of output time.

Establish the buffer to receive the positioning data and save the data. Then set the receiving flag and determine whether the head of string containing $GPRMC of ASCII code. If it is yes, extract the time, latitude, longitude and speed information after the $ letter[4]. It is shown as **Fig. 2.**

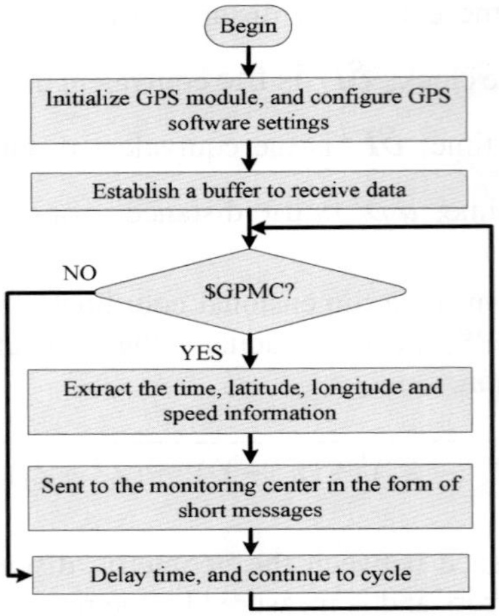

Fig. 2. This shows the receiving process of GPS signal

3.2 Wireless Transmission of Data

GPRS introduces two new network nodes based on the original GSM network of circuit-switched approach: GPRS service node (SGSN) and GPRS gateway nodes (GGSN).

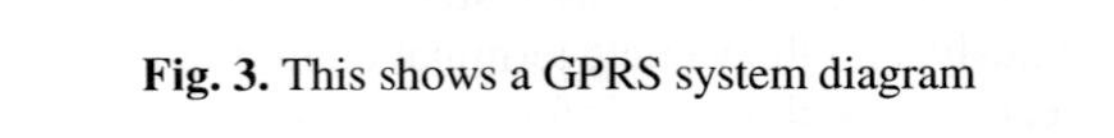

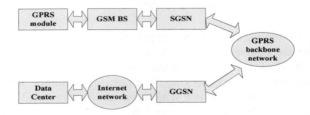

Fig. 3. This shows a GPRS system diagram

GPRS terminal can obtain data from the client system through the interface, and processed data of GPRS packet is sent to the base station of GSM. After packet data being encapsulated by the SGSN, SGSN can communicate through GGSN which is the supporting node between GPRS backbone and gateway. The packet data is processed by GGSN correspondingly and will be sent to the destination network, including the Internet[5]. System diagram is shown in **Fig. 3.**

In the design of the terminal, we should initialize the baud rate of serial communication and the system, and create the thread to read the return information of GPRS module. Configure the properties of serial port and read the content of the message to be sent in file when sending SMS. After reading it successfully, it will be sent to the destination, as shown in **Fig. 4.**

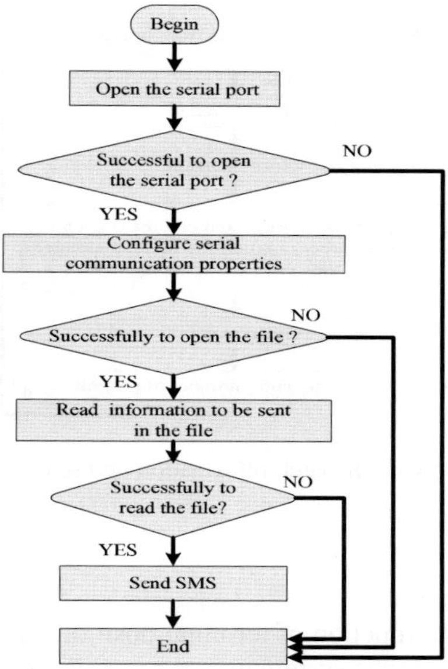

Fig. 4. This shows the sending SMS process

3.3 Sending Dispatch Information and Warning Messages

Location information and status information of snow removal vehicles can be displayed on the GIS (geographic information system) in detail. Then the monitoring center can summarize and analyze snow removal tasks for making a warning according to this information when there is an accident. Dispatch command and warning messages will be sent to the vehicle terminal via GPRS.

When the vehicle terminals receive this information, the speech data stored in FLASH will be broadcasted based on the dispatch command to control the snow removal task and warn an accident.

Speech content for broadcasting can be stored into FLASH in advance, and it can reduce amount of data for wireless transmission apparently. The monitoring center can broadcast it with a few simple commands to dispatch or alert the drivers. As shown in **Fig. 5.**

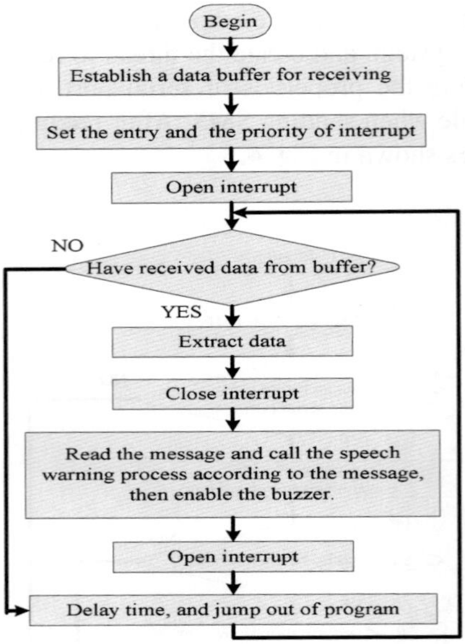

Fig. 5. Process of receiving dispatch information and sending warning messages

4 Summary

The system can achieve the function of terminal positioning, communication with the monitoring center and alarm. The system uses the positioning method of pseudo-range differential, which can improve the positioning accuracy of the vehicles. And the accuracy could reach sub-meter. At the same time, the system has characteristics of low-cost, wide system coverage, lower maintenance cost and reliable communication.

References

1. Hua, W., Zhou, Y., Chen, S., Xing, J.: The Design of Mobile Target Monitoring System Based on VHF/TDMA Technology. In: 5th International Conference on Wireless Communications, Networking and Mobile Computing, Beijing, pp. 1–4 (2009)
2. Lipin, W.: The Embedded ARM User Program Design of Vehicle Terminal. In: 2th IEEE International Conference on Computer and Automation Engineering, pp. 563–567. IEEE Press, Singapore (2010)
3. Kaplan, E., Hegarty, C.: Understanding GPS: Principles and Applications. Artech House, United States (2005)
4. Lita, I., Cioc, I.B., Visan, D.A.: A New Approach of Automobile Localization System Using GPS and GSM/GPRS Transmission. In: 29th International Spring Seminar on Electronics Technology, St. Marienthal, pp. 115–119 (2006)
5. Goetz, I.: Keeping up with GPRS. Communications Engineer 1, 46–46 (2003)

Research the Effect of Component Deterioration Based on the Cloud Particle Swarm Optimization

Wang Yonghua and Li Dong

Department of Aerocraft Engineering,
Naval Aeronautical and Astronautical University Yantai, China

Abstract. The rotors are the key components of engine which its' performance deterioration play the important role in the engine performance. The component characteristic is revised through the compressor and turbine component revised factor. The performance deterioration of engine model is established based on the revised component characteristic which presented a new cloud particle swarm optimization in order to accelerate convergence. The effects of component performance deterioration are analyzed. The results offer theoretical referenced value to rotor component even engine performance deterioration and relative influence between components.

Keywords: component characteristic, performance deterioration, cloud particle swarm optimization.

1 Introduction

The rotors are the key components of engine which include compressor and turbine. The rotors are the major composition which its' deterioration play the important role in the engine performance. The short performance deterioration and fault diagnostic model are presented for CF6 and JD9T by NASA, which provides information for assessing the effect on overall performance. A polynomial function has been specified to the variety trend of components. Much research of rotors performance focus on component performance [1-3].Professor Li has commented on the effect of the single stage compressor performance to the whole engine[4]. While the compressor and the turbine have gas dynamic contact, the compressor with small deviation may affect the other such as turbine. The present paper discusses the deterioration of jet engine performance caused by compressor deviation, turbine deviation and the two component deviation at the same time. From different view of developing the rotors performance analysis, the conclusion can supply the whole engine performance evaluation and control with a reference.

2 Engine Performance Model Based on Characteristic Revised

Deterioration of rotors performance is caused by several mechanisms, which are tip clearance increase, airfoil shape change and surface quality change. Mechanism wear and erosion can be responsible for clearance changes, erosion and corrosion can change airfoil shape and surface quality, while fouling changes mainly surface quality and secondarily shape. Turbines suffer from the same main types of changes, but in this case

X. Wan (Ed.): Electrical Power Systems and Computers, LNEE 99, pp. 637–643.
springerlink.com

shape changes have a more direct and pronounced effect on swallowing capacity. Vane trailing edge bowing and trailing edge erosion are effects that may increase swallowing capacity, while affecting efficiency at the same time. Swallowing capacity can decrease in the case deposits are formed on the airfoils surface. A common feature of all cases is the reduction in component efficiency. Swallowing capacity reduces for compressors while it can either reduce of increase in turbines. Different causes and mechanisms of performance deterioration of rotors are induced the component characteristic variation, while the component characteristic determined the precision of performance model. After revised the component characteristic such as high compressor and low turbine, the more accurate solution can be obtained at the given operating point.

Parameters employed to characterize the condition of each component are a flow factor SW and an efficiency factor SE. They determine how component flow capacity and efficiency are modified with respect to a reference condition. For a component with entrance at station i of the engine, the flow factor is defined:

$$SW_i = \left[(W_i \cdot \sqrt{T_i} / P_i)' - (W_{i0} \cdot \sqrt{T_{i0}} / P_{i0})\right] / (W_{i0} \cdot \sqrt{T_{i0}} / P_{i0}) \quad (1)$$

efficiency factor:

$$SE_i = (\eta_i' - \eta_{i0}) / \eta_{i0} \quad (2)$$

i0 is the criterion value for given condition.

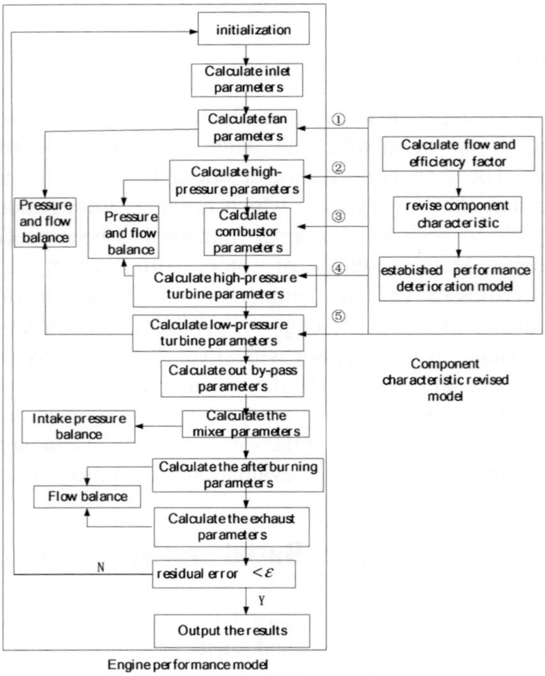

Fig. 1. Computational flow of effect of component deterioration on engine performance deterioration

The revised characteristic of rotor components is used to set up performance mode for given operating condition, which can produce the effect to performance parameter with the rotor components deterioration. The computational flow and principle is shown in Fig.1.

3 Cloud Particle Swarm Optimization Algorithm

When the characteristic parameters of components have changed, the convergence speed is slow for stability model so that the calculating time may have been increased. Taking the case into consideration, a new approach based on cloud particle swarm optimization algorithm has been produced in order to accelerate the convergence speed.

The particle swarm optimization algorithm has been advanced by Kennedy[5].. Keeping the formula of the basic particle swarm speed and position, references have advanced the improvement at the convergence aspect[6-9]. The real optimization search course can't appear through the basic particle swarm algorithm. The technique presented here based on the cloud generation implement, which can bring a new strategy creating weight factors rely on individual adaptive value.

Generally, the species group has been divided by three subgroups, each subgroup employs different strategy of weight factor. If the scale of particle swarm is m , the adaptive value f_i is the particle x_i at the *k*th iterative, so the mean adaptive value of particle swarm is $f_{avg} = \left(\sum_{i=1}^{m} f_i\right)\Big/ m$; if the adaptive value larger than f_{avg} ,then average all values , f_{avg}' is got; if the adaptive value smaller than f_{avg} ,then average all values , f_{avg}'' is got; the adaptive value of optimized is $f_{\min}$.

If f_i smaller than f_{avg}' , then the particle is excellent particle so its weight factor is defined as 0.2 which can accelerate the convergence; If f_i larger than f_{avg}' , the weight factor of the particle is defined as 0.9.

If the value of particle f_i is in the scope f_{avg}'' and f_{avg}', then the weight factor has been dynamically adjusted by cloud generation implement. The algorithm is described as follows:

$$Ex = f_{avg}'$$

$$En = \left(f_{avg}' - f_{\min}\right)/c_{11}$$

$$He = En/c_{22}$$

$$En' = normrnd(En, He)$$

$$w = 0.9 - 0.5 \times e^{\frac{(f_i - Ex)^2}{2(En')^2}}$$

Where c_{11}、c_{22} is control parameters, En can affect the cliffy of the normal cloud. the value of En is larger, the horizontal width of the cloud is more wide. In order to enhance the speed and precision, then $c_{11} = 2.9$. The scatter degree is decided by He .If the He is too small, the randomness will be lossed at a certain extent; If the He is too large, then the stable tendentiousness can be lossed, so $c_{22} = 10$. The cloud particle swarm optimization algorithm has been formed based on the adjusted strategy of weight factor, the computational flow is shown as Fig.2. Compared with the basic particle swarm optimization algorithm, the new algorithm can significantly decrease the amplitude of particle movement and accelerate the convergence of model.

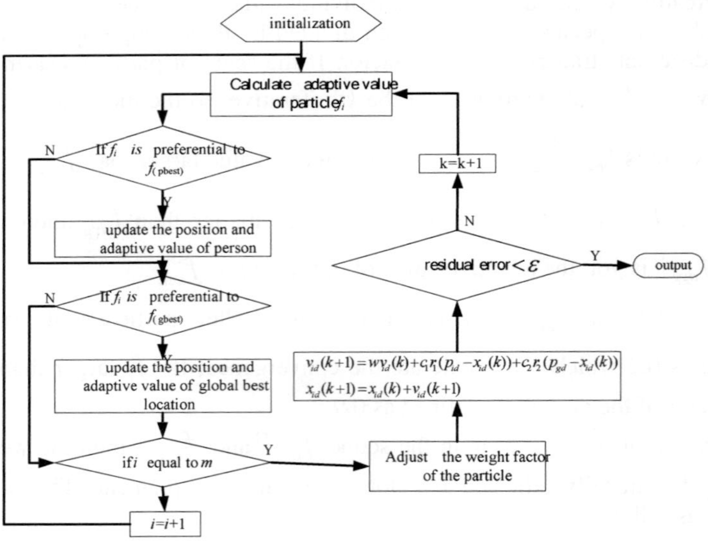

Fig. 2. Computational flow of the cloud particle swarm optimization algorithm

4 Simulation Calculation

As a case, the performance deviation caused by single component and the combined component is simulated on the performance deterioration model at the largest military condition. The results are shown as fig.3 and fig.4.

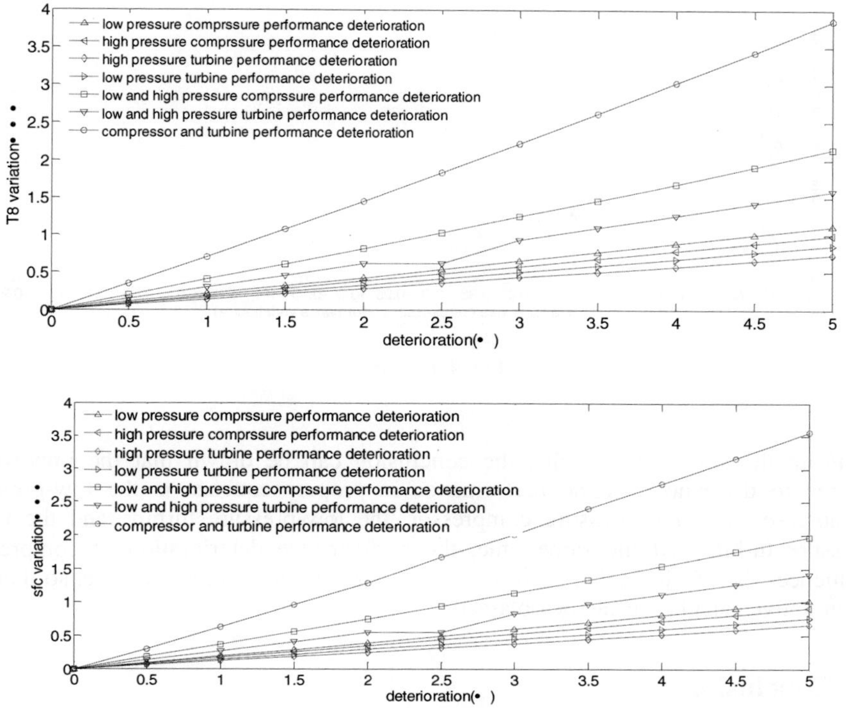

Fig. 3. The effect cure of the engine performance parameter variation

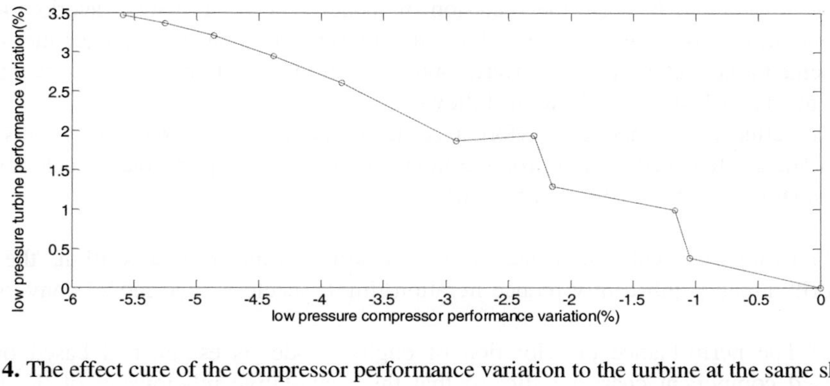

Fig. 4. The effect cure of the compressor performance variation to the turbine at the same shaft

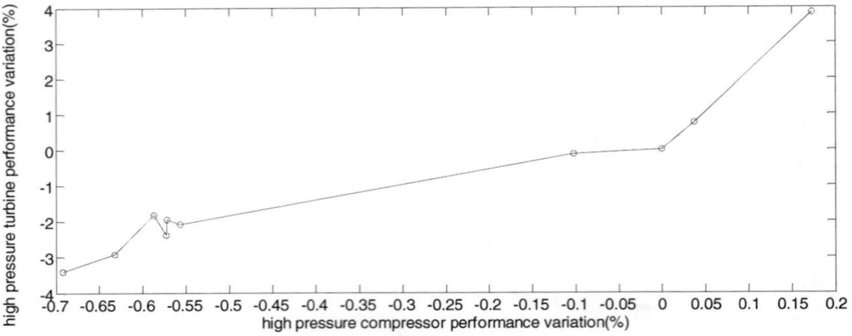

Fig. 4. (*continued*)

From the simulating results, the conclusion can be drawn that the importance degree to the whole engine performance in proper sequence is the low-pressure compressor, the high-pressure compressor, the low-pressure turbine and the high-pressure turbine. At the same time, the performance deterioration of compressor influences the whole engine performance is more critical than the turbine, so that the quantitative calculation can be obtained.

5 Conclusion

The rotors are the major composition which its' deterioration play the important role in the engine performance. The component characteristic is revised introducing the revised factor in order to increase the precision of model. The performance deterioration of engine model is established based on the revised component characteristic which presented a new cloud particle swarm optimization in order to accelerate convergence. The main conclusions are drawn as follows:

The effects of component performance deterioration are analyzed. The results offer theoretical referenced value to rotor component even engine performance deterioration and relative influence between components.

1) Compared with the basic particle swarm optimization algorithm, the new algorithm can significantly reduce iteration times, accelerate the model convergence speed.

2) The performance deterioration of engine model is established based on the revised component characteristic, so that the quantitative relationship of the rotors performance deterioration and engine performance has been obtained which can provide the basis for the engine condition monitoring.

3) The conclusion from the given turbofan engine can be extended to other models.

References

1. Kong, C.D., Ki, J.Y.: Performance Simulation of Turboprop Engine for Basic Trainer. ASME Paper No. 01-GT-391 (2001)
2. Kurzke, J.: How to Get Component Maps for Aircraft Gas Turbine Performance Calculations. ASME Paper No. 96-GT-164 (1996)
3. Kurz, R., Brun, K.: Degradation in gas turbine system. ASME Paper 20002GT2345 (2000)
4. Li, B.-w., Li, D.: Quantitative research on performance degradation of Single-stage compressor. J. Journal of Aerospace Power 7, 1588–1594 (2010)
5. Kennedy, W.M.: Spcars-'Matching Algorithms to Problems: An Esperimental Test of the Particle Swam and Some Genetic Algorithms on the Multimodal Prablom Generator. In: Proceedings of the IEEE Int'l Conference on Evolutionary Computation (1998)
6. Naka, S., Genji, T., Yura, T., Fukuyama, Y.: A hybrid particle swarm optization for distribution state estimation. J. IEEE Transaction Power System 1, 60–68 (2003)
7. Esmin, A.A.A., Lambert-Torres, G., de Souza, A.C.Z.: A hybrid particle swarm optimization applied to loss power minimization. J. IEEE Transaction Power Systems 2, 859–866 (2005)
8. Clerc, M., Kennedy, J.: The particle swarm–explosion, stability, and convergence in a multidimensional complex space. J. IEEE Trans. Evol. Comput., 58–73 (2002)
9. Shi, Y., Eberhart, R.: Parameter selection in particle swarm optimization. In: Evolutionary Programming VIZ: Proc. EP 1998, pp. 591–600. Springer, New York (1998)

Research of Engine Performance Deterioration Based on Optimal BP with GE Algorithm

Wang Yong-Hua, Li Dong, and Meng Lu

Department of Aerocraft Engineering, Naval Aeronautical and Astronautical University, 264001, China 2. The 91467th Unit of PLA, 266311, China

Abstract. This paper establishes nonlinear model of engine component performance parameters and engine measure parameters by using BP neural net, researches variation of engine performance. In the training of BP net, model precision is elevated by introducing random weight factor in the input, and optimizing key parameters using GE algorithm. This paper provides theoretical reference for performance deterioration.

Keywords: engine performance, BP net, random weight factor, optimal.

1 Introduction

As time used prolongs, main gas component performance in the engine decrease, namely compressor and turbine, which leads to whole engine experience performance deterioration [1-2]. In the process of engine performance deterioration, it can prolong engine life used and reduce maintenance cost by measures to restoring performance of deteriorated engine to some extent. Formerly, there is lack of research of engine performance deterioration, most focus on data record of performance deterioration, lacking of systematical research. Research performance variation using engine stable model, applicative means tend to modifying component characteristic parameters. But if component characteristic parameters vary too much, it maybe cause model to not converge, the results are wrong. The complexity of engine configuration, much nonlinear of gas component, jacobi matrix expresses their nonlinear relationship not very well.

BP neural net is a nonlinear mapping, nonlinear model of engine performance parameter and measure parameter is established, variation of engine performance is researched by variation of measure parameters. In the BP neural net, embedding random weight coefficient in the input of neural net is proposed, optimizing learning coefficient of BP, middle level and random weight coefficient by GE algorithm, raising train precise of BP neural net.

2 Determine Engine Component Performance Parameters

This paper uses BP neural net of optimal GE algorithm, establishes model of engine component performance parameters and measure parameters. In the primary selecting of model, component performance parameters are determined as fan flow mass deterioration ($\Delta M_{\alpha f}$), fan efficiency deterioration ($\Delta \eta_{\alpha f}$), high pressure compressor flow mass deterioration ($\Delta M_{\alpha c}$), high pressure efficiency deterioration ($\Delta \eta_{\alpha c}$), high

X. Wan (Ed.): Electrical Power Systems and Computers, LNEE 99, pp. 645–652.
springerlink.com

pressure turbine flow mass deterioration (ΔM_{tH}), high pressure turbine efficiency deterioration ($\Delta\eta_{tH}$), low pressure turbine flow mass deterioration (ΔM_{tL}), low pressure efficiency deterioration ($\Delta\eta_{tL}$). Measure parameters are determined by low pressure rotor speed ($NL0$), high pressure rotor speed ($NH0$), high pressure compressor pressure ($Pt30$), outlet temperature ($Tt8$. Performance deterioration parameters are selected which influence output parameters by further analysis.

Performance deterioration of engine mostly focus on decrease of thrust and increase of oil rate, when outlet geometrical area is constant, decrease of thrust mainly focus on decrease of flow mass in the outlet, but fuel flow mass only occupies little proportion of air, therefore decrease of thrust is mainly caused by decrease of air, therefore ΔM_{af} and ΔM_{ac} are selected. Increase of oil consumption focuses on decrease of rotor component efficiency, because turbine works in the environment of high temperature and high pressure, it makes turbine efficiency decrease, therefore, it need consider $\Delta\eta_{TH}$ and $\Delta\eta_{TL}$ [3, 4].

Qualitative analysis is work out by researching the relativity of performance parameter and output parameter [5], find out some parameters which largely influence performance parameter on output parameter. Here relative coefficient is defined:

$$r_{ij} = \left| j^{th}\ output\ parameter\ \text{var}iation / i^{th}\ performance\ parameter\ \text{var}iation \right| \tag{1}$$

Its physical means is the effect of ith performance parameter variation on j^th output parameter. Adopting disturbing method, respectively change 1.0% of performance parameter, compute corresponding percentage of output, disturbing performance parameter, variation of output parameter is showed at table 1. The results are gained by data in table 1 is divided by 1.0%.

Table 1. Variation of output parameters

	$NL0$	$NH0$	$Pt30$	$Tt8$
ΔM_{af}	-0.0313	0.0308	0.6018	0.5760
$\Delta\eta_{af}$	0.0928	-0.0616	0.4117	0.3840
ΔM_{ac}	-0.2000	-0.0616	1.0082	0.4526
$\Delta\eta_{ac}$	-0.0838	-0.0308	0.4117	0.3154
ΔM_{tH}	0.190	0.0924	1.0663	0.0960
$\Delta\eta_{tH}$	-0.0359	-0.1541	-0.6440	-0.0274
ΔM_{tL}	0.2800	-0.0616	0.7232	0.0960
$\Delta\eta_{tL}$	-0.2100	-0.1233	-0.7126	0.0411

For conveniently analysis, find performance parameter which influence some output parameter mostly, normalize relative parameters, normalized according to $r_{1j}, r_{2j}..., r_{8j}$, normalized parameters are gained:

$$\bar{r}_{ij} = |r_{ij}| / \sqrt{\sum_{k=1}^{8} r_{ij}^2}, (i = 1,2,3 \quad j = 1,2,\cdots,8) \tag{2}$$

Results are showed at table 2.

Table 2. Unitary correlative coefficient

	$NL0$	$NH0$	$Pt30$	$Tt8$
$\Delta M_{\alpha f}$	0.0672	0.1250	0.2902	0.6422
$\Delta\eta_{\alpha f}$	0.1994	0.2500	0.1986	0.4281
$\Delta M_{\alpha c}$	0.4298	0.2500	0.4862	0.5046
$\Delta\eta_{\alpha c}$	0.1801	0.1250	0.1986	0.3517
ΔM_{tH}	0.4084	0.3750	0.5142	0.1070
$\Delta\eta_{tH}$	0.0771	0.6250	0.3106	0.0306
ΔM_{tL}	0.6018	0.2500	0.3487	0.1070
$\Delta\eta_{tL}$	0.4513	0.5000	0.3436	0.0459

Table 2 is normalized relative coefficient, it is considered as strong relative when corresponding coefficients are bigger than 0.5, middle relative when corresponding coefficients are between 0.3-0.5, feeble relative when they are smaller than 0.3. Modified characteristic parameters are that this paper finally selected by this table combined with former qualitative analysis: ΔM_{af}, ΔM_{ac}, $\Delta\eta_{TH}$ and $\Delta\eta_{TL}$.

3 BP Net Optimize Parameters

BP algorithm transforms mapping of leaning input and output to nonlinear optimal problem, realizing minimalization of equal square difference of net output and expected output[6]. Considering the effect of decrease of component performance parameter on measure parameter is different, therefore introducing random weight coefficient in four selected component performance parameter of BP input. Learning coefficients influences model precision, and middle level is key parameter of model precision. Therefore, six parameters are determined in input. Enabled function $f(\cdot)$ is

Sigmoid, namely, $f(x)=\frac{1}{1+e^{-x}}$. Establish BP net input parameters{ $cM_{\alpha f}$, $cM_{\alpha c}$, $c\eta_{tH}$, $c\eta_{tL}$, mid , eta }, output parameters { $NL0$, $NH0, Pt30, Tt8$ }.Establishment of BP net is some inputs and outputs.

4 GE Algorithm Optimizing BP Input Parameter [7]

Setting population is 20, iterative times are 20. Code input six parameters, they varies 0～1. Code is adopted by eight numbers. This algorithm adopts adaptability function is

$$F = C(f - f_{\min}) + D \tag{3}$$

Therefore, $f = (\frac{1}{p*q}\sum_{i=1}^{q}\sum_{k=1}^{p}(d_i(k)-y_i(k))^2)^{\frac{1}{2}}$ is function value. $f_{\min}$ is minimal value of contemporary population.

Adopting algorithm of variable cross possibility, Therefore:

$$P_c^{'} = P_c - [P_{co} - 0.3]/G \tag{4}$$

$P_c^{'}$ is cross possibility of this generation, G is max evolutional generations.

When max adoptability of some generation $F_{\max}$ and average adoptability meets lower forum F_{avg} ,

$$\alpha F_{\max} < F_{avg} \tag{5}$$

Therefore, α is dense coefficient $(0.5 < \alpha < 1.0)$, larger mutation of this generation is done by $P_m = 5 * P_{m0}$, thus avoid local optimization.

Setting a number, namely final iterative generation G_{end} , if consecutive G_{end} generation produce equal best individuals, thus evolution stops.

5 Simulation Results and Analysis

Variation of engine performance is researched. This paper compares three conditions of former four component parameter with random weight coefficient, without random weight coefficient and expert coefficient[8]. Data is classified by two groups, one group is trained, the other group is validated. Data is dealt by normalization. Trained and validated data is showed at table 3.

Table 3. Training and validating data

Index	Component health performance deterioration				Engine measure parameter variation			
	ΔM_{af}	ΔM_{ac}	$\Delta\eta_{tH}$	$\Delta\eta_{tL}$	$NL0$	$NH0$	$Pt30$	$Tt8$
1	-3.1%	-2.1%	-3.2%	-2.5%	-1.09%	-1.11%	-3.09%	1.80%
2	-1.0%	-2.1%	-2.0%	-5.0%	-0.72%	-1.02%	-2.97%	1.06%
3	-1.0%	-1.0%	-2.0%	-2.0%	-063%	-0.31%	-1.19%	0.19%
4	-3.0%	-1.1%	-3.0%	-4.0%	-0.83%	-1.02%	-3.83%	1.51%
5	-3.0%	-2.0%	-1.3%	-5.0%	-1.10%	-0.77%	-3.09%	1.84%
6	-3.0%	-4.0%	-5.0%	-5.0%	-1.56%	-1.91%	-7.82%	1.34%
7	-5.0%	-4.0%	-2.0%	-5.0%	-1.22%	-1.20%	-5.47%	2.56%
…	…	…	…	…	…	…	…	…

BP net is trained using former group of data at table 3, net precision is validated by latter group of data. Results optimized and error are showed at table 4.

Table 4. Optimal results and error

	cm_{af}	cm_{ac}	$c\eta_{tH}$	$c\eta_{tL}$	Int(mid*30)	eta	error
1.random weight factor	0.85 10	0.61 57	0.22 75	0.63 53	27	0.34 12	0.000 260
2. without factor	0.73 33	0.47 45	0.79 61	0.09 41	5	0.6 157	0.001 192
3. expertise factor	0.05 88	0.37 65	0.01 18	0.79 22	2	0.69 41	0.001 043

They are showed at table 4, converge error with random weight factor is least and result is best. On basis of BP net, stay three parameters of ΔM_{af}, ΔM_{ac}, $\Delta\eta_{TH}$ and $\Delta\eta_{TL}$ unvaried, rest parameter (in turn for ΔM_{af}, ΔM_{ac}, $\Delta\eta_{TH}$, $\Delta\eta_{TL}$) respectively decrease from 0% to -5.0%, variation of $NL0$, $NH0$, $Pt30$, $Tt8$ is gained. They are showed at figure 1, figure 2, figure 3 and figure 4.

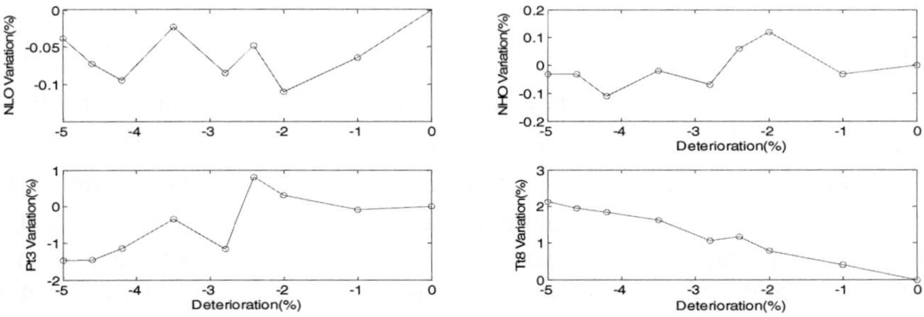

Fig. 1. Measure parameters variation with ΔM_{af}

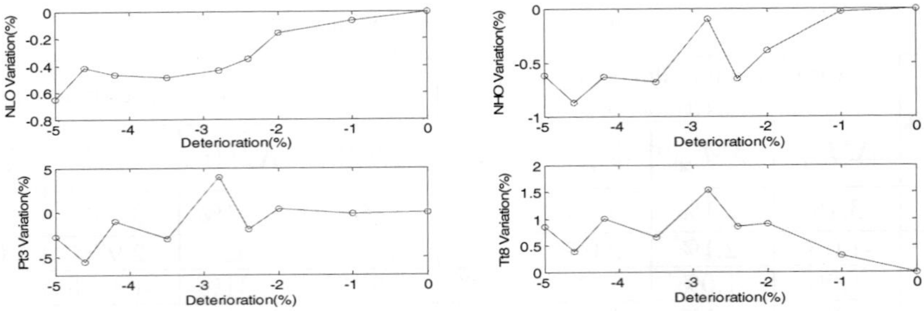

Fig. 2. Measure parameters variation with ΔM_{ac}

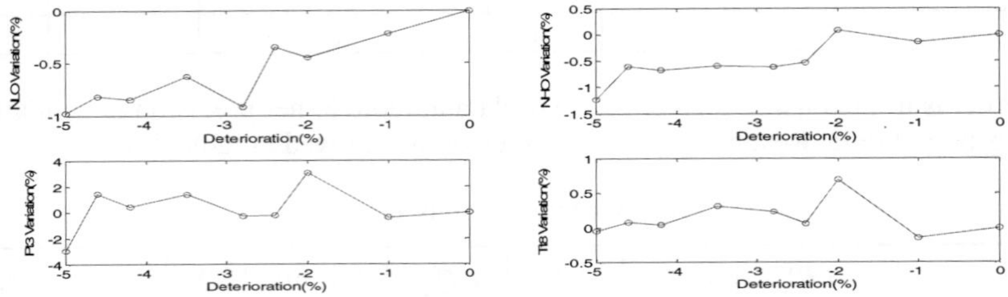

Fig. 3. Measure parameters variation with $\Delta\eta_{TH}$

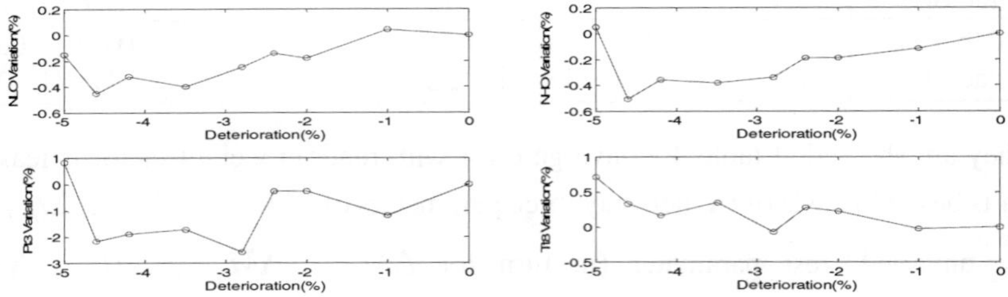

Fig. 4. Measure parameters variation with $\Delta\eta_{TL}$

They are showed at figure 1, the more increase deterioration of ΔM_{af}, the more evident of $Tt8$ increase trend. They are showed at figure 2, figure 3 and figure 4, change trend of $Tt8$ tend to increase, and magnitude is not large. Variation of $NL0$, $NH0$ tend to decrease, they are more evident at figure 2 and figure 3. Change of $Pt30$ tend to decrease, it is more prominent in figure 1 and figure3. Engine performance deterioration is showed by variation of engine measure parameters. Therefore, variation of engine performance is expressed by inspecting variation of these parameters.

Compute variation of engine measure parameter by using engine nonlinear model. When performance parameter vary larger, nonlinear model tend to not converge, correct result is not gained. when large magnitude of performance parameter deterioration appear, variation of engine measure parameter is calculated by using BP net of this paper. When engine component performance parameter deteriorates, variable results of measure parameters are showed at table 5.

Table 5. Measure parameter variation

Index	component health parameter deterioration				Variation of measure parameter			
	$\Delta M_{\alpha f}$	$\Delta M_{\alpha c}$	$\Delta \eta_{tH}$	$\Delta \eta_{tL}$	$NL0$	$NH0$	$Pt30$	$Tt8$
1	-0.50%	-0.50%	-0.50%	-0.50%	-0.13%	-0.21%	-0.30%	0.46%
2	-1.00%	-1.00%	-1.00%	-1.00%	-0.19%	-0.39%	-1.10%	0.70%
3	-1.50%	-1.50%	-1.50%	-1.50%	-0.70%	-0.56%	-1.5%	1.07%
4	-2.00%	-2.00%	-2.00%	-2.00%	-0.80%	-0.75%	-2.05%	1.24%
5	-2.50%	-2.50%	-2.50%	-2.50%	-1.14%	-0.91%	-2.17%	1.56%
6	-3.0%	-3.0%	-3.0%	-3.0%	-1.20%	-1.35%	-4.08%	1.75%
7	-5.0%	-5.0%	-5.0%	-5.0%	-2.02%	-1.95%	-6.07%	2.95%

6 Conclusion

This paper optimizes and selects BP parameter, establishes nonlinear model of engine component performance parameter and measure parameter, researches variation of performance. Main conclusions are gained:

1. In input level of BP, introduces random weight factor, optimizes and selects learning coefficient, middle level and random weight factor through GE algorithm, which enhance precision.
2. Using BP model, When engine measure parameter vary, variation of engine measure parameter is calculated. The results basically express trend of engine performance deterioration.
3. When component performance parameter appear larger deterioration, nonlinear model do not converge, the result is gained using BP net. It provided reference to research engine performance deterioration.

References

1. Diakunchak, I.S.: Performance Deterioration in Industrial Gas Turbines. ASME J. Eng. Gas Turbines Power 114 (1992)
2. Lakshminarasimha, A.N., Boyce, M.P., Meher-Homji, C.B.: Modeling and Analysis of Gas Turbine Performance Deterioration. ASME J. Eng. Gas Turbines Power 116 (1994)
3. Cheng, L., Li, Q.-T.: State Inspection and Fault Diagnosis of Aero-engine. M. Xian: Air Force Engineering Institute, pp.31-34 (1997)

4. Li, J.-G., Li, B.-W.: Solution and Application of Fault Diagnosis Equation of Aero-engine. Gas Turbine Experiment and Research 15, 8–11 (2002)
5. Yang, W.-H.: Modeling and Fault Diagnosis of Aero-engine.D.Nan Jing Aeronautic and Astronautical University (2000)
6. Wen, L.-J.: Establishment of Neural Network Model Using by Measured Data. D. Nan Jing Aeronautic and Astronautical University, pp.33-36 (2005)
7. Ma, L.: Fault Diagnosis of Aero-engine Based on Recursive Neural Net. J.
8. Sun, X.-B.: Reliability Analysis of Given Aero-engine and Spares Maintenance. D. Naval Aeronautical Engineering Institute, pp.39–40 (2010)

Research on Algorithm for Ultrasonic Array Geometric Location of Partial Discharge Based on Common Perpendicular Midpoint of Direction Finding Lines

Yanqing Li, Ling Li, Lijun Zhang, Qing Xie, and Shuyi Cheng

Department of Electrical Engineering, North China Electric Power University
Hebei Provincial Key Laboratory of Power Transmission Equipment Security Defense
liling_ncepu@163.com

Abstract. Detection of partial discharge (PD) location is critical item for quality control and insulation mechanism research in electrical equipment, and research on algorithm for ultrasonic array location of PD is an important aspect. It is easy and convenient to take the intersection of two direction finding lines as PD source in traditional algorithm for ultrasonic array location of PD. However, there are many problems that can not meet engineering needs such as bringing down the positioning accuracy, getting multiple nonunique solutions for taking two skew lines approximatively as intersecting. In this paper, a new algorithm for ultrasonic array geometric location of PD based on common perpendicular midpoint of direction finding lines is presented, which takes the common perpendicular midpoint of direction finding lines as PD source and makes the result of location unique, stable and more accurate.

Keywords: Ultrasonic phased array, partial discharge, location, direction finding.

1 Introduction

Detection of partial discharge (PD) location is critical item for quality control and insulation mechanism research in electrical equipment [1]. Currently, dual platform direction finding triangulation location method is usually used in partial discharge location. The principle of this method is to set the two sensors at different positions in space to receive ultrasonic signal of PD and form the array model, then array signal processing method is used to estimate DOA (Direction of Arrival) [2-3]. After that, the two direction finding lines are taken approximatively as intersecting, and the intersection is regarded as the location of PD source [4-5]. Dual platform direction finding triangulation location method [6] is classic in passive location, simple in principle, easy in calculation, but there are also some problems such as uncertainty of location result(unique solution can not be obtained) caused by choosing multiple nonunique parameters and the multi-level amplification of errors caused by inaccuracy of direction finding.

Accordingly, a new algorithm for ultrasonic array geometric location of PD based on common perpendicular midpoint of direction finding lines is presented in this paper,

X. Wan (Ed.): Electrical Power Systems and Computers, LNEE 99, pp. 653–660.
springerlink.com

which takes the common perpendicular midpoint of direction finding lines as PD source and makes the result of location unique, stable and more accurate. It is proven by simulations that the new method is correct and effective.

2 Location Algorithm

2.1 Direction Finding Method

For general far-field signals, there is a wave path difference when the same signal arrives at different array elements, which leads to the phase difference between the receiver array elements. And the orientation of the signal can be estimated by calculating the phase difference [7], which is the basic principle of direction finding.

In this paper, the Root-MUSIC algorithm, which finds polynomial roots of MUSIC (Multiple Signal Classification) algorithm to substitute the spectrum search, is adopted in the direction finding to estimate DOA. And it turns out to be more accurate than MUSIC algorithm [8].

Since signal the sensor receives is broadband, while the Root-MUSIC algorithm itself can only estimate DOA of narrow-band signal, hence, in order to get narrow-band signal, signal subspace transformation (SST) algorithm is adopted to focus the signal. On this basis, the Root-MUSIC algorithm is applied to estimate DOA to obtain direction finding angle (azimuth angle and elevation angle) of sensors relative to the PD source. The coordinate of PD source can be obtained by calculating direction finding angle and coordinate of the sensors (which is known).

2.2 Principle of Location Based on Dual Platform Common Perpendicular Midpoint of Direction Finding Lines

Suppose that the coordinate of the array sensor is $A_t(x_t, y_t, z_t)$, azimuth angle and elevation angle are (θ_t, φ_t), direction vector of direction finding line for sensor t is $\overline{S_t}\{m_t, n_t, p_t\}$, Where t= 1, 2.

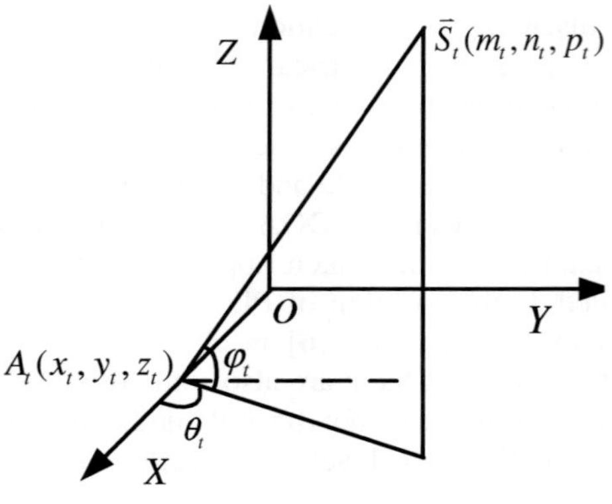

Fig. 1. Direction vector of direction finding line

According to geometric principle, the relationship between φ_t and $\overline{S_t}$ is satisfied:

$$\varphi_t = \left| \frac{\pi}{2} - (\overline{S_t} \hat{,} \overline{n_t}) \right| . \tag{1}$$

Where $\overline{n_t}$ is the direction vector of the plane XOY, which is equal to (0, 0, 1). And equation (1) can be expressed as follows:

$$\begin{aligned} \sin \varphi_t &= \left| \cos(\overline{S_t} \hat{,} \overline{n_t}) \right| \\ &= \frac{|0 \cdot m_t + 0 \cdot n_t + 1 \cdot p_t|}{\sqrt{0^2 + 0^2 + 1^2}\sqrt{m_t^2 + n_t^2 + p_t^2}} \\ &= \frac{p_t}{\sqrt{m_t^2 + n_t^2 + p_t^2}} \end{aligned} \tag{2}$$

$$\cos \varphi_t = \frac{\sqrt{m_t^2 + n_t^2}}{\sqrt{m_t^2 + n_t^2 + p_t^2}} . \tag{3}$$

Similarly, the equation below can be obtained:

$$\tan \theta_t = \frac{m_t}{n_t} . \tag{4}$$

The following function can be obtained by solving (2) to (4).

$$m_t : n_t : p_t = \cot \theta_t : 1 : \tan \varphi_t \cdot \csc \theta_t . \tag{5}$$

Hence, the direction vector of direction finding line can be determined as follows:

$$\overline{S_t} = (\cot \theta_t, 1, \tan \varphi_t \cdot \csc \theta_t) . \tag{6}$$

And then, the formula of direction finding line can be obtained:

$$\frac{x - x_t}{\cot \theta_t} = \frac{y - y_t}{1} = \frac{z - z_t}{\tan \varphi_t \cdot \csc \theta_t} . \tag{7}$$

The direction vector for sensor 1 and sensor 2 can be written by substituting the value of t into (6) as follows:

$$\begin{cases} \overline{S_1} = (\cot\theta_1, 1, \tan\varphi_1 \cdot \csc\theta_1) \\ \overline{S_2} = (\cot\theta_2, 1, \tan\varphi_2 \cdot \csc\theta_2) \end{cases} . \tag{8}$$

Suppose that the coordinate of PD source is $S(x, y, z)$, the intersection of direction finding line for sensor t and its common perpendicular is P_t', where t=1, 2.

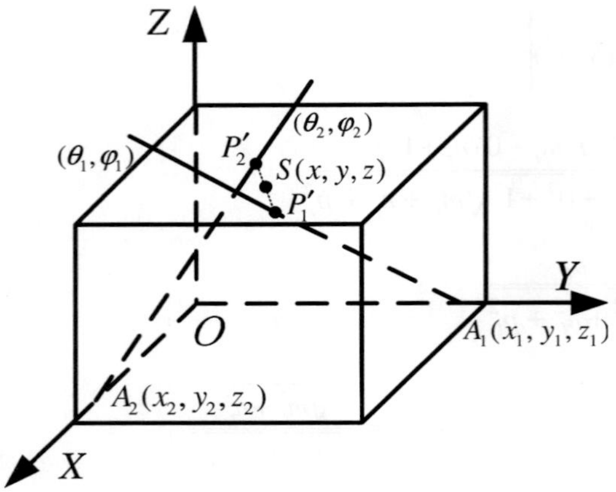

Fig. 2. Location based on common perpendicular midpoint of direction finding lines

Due to the direction finding line is perpendicular to the common perpendicular, hence, the following equations can be obtained:

$$\begin{matrix} \overrightarrow{P_1'P_2'} \bullet \overline{S_1} = 0 \\ \overrightarrow{P_1'P_2'} \bullet \overline{S_2} = 0 \end{matrix} . \tag{9}$$

The value of $P_1'(x_1', y_1', z_1')$、$P_2'(x_2', y_2', z_2')$ can be found by solving (7) to (9).

Hence, the coordinate of PD source, which is the mid point value of P_1' and P_2', can be obtained:

$$S(x, y, z) = (\frac{x_1' + x_2'}{2}, \frac{y_1' + y_2'}{2}, \frac{z_1' + z_2'}{2}) . \tag{10}$$

2.3 Algorithm Flow

Algorithm flow chart is as follows:

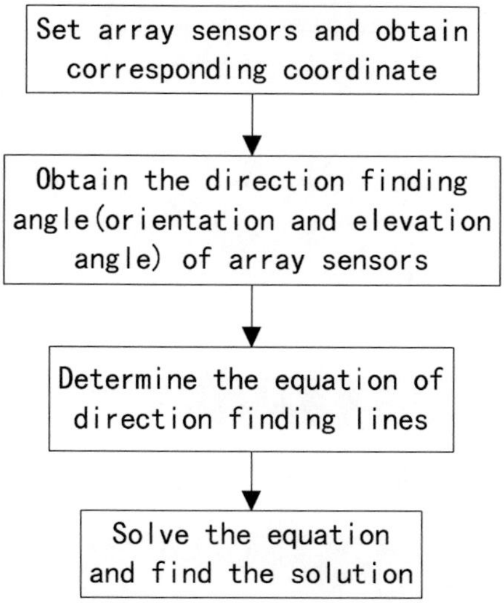

Fig. 3. Algorithm flow

3 Research on Simulation of PD Location

In the simulation, the space coordinate of PD source is setted as (30, 70, 90) cm, two space coordinates of the array sensors as (50, 0, 0) cm and (0, 50, 0) cm.

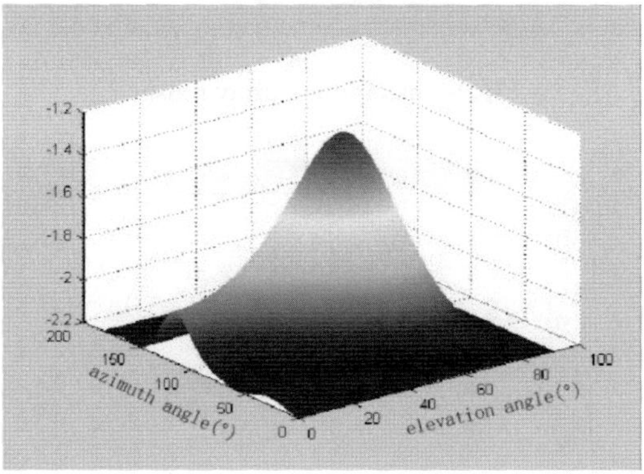

Fig. 4. Spectrum of array sensor 1

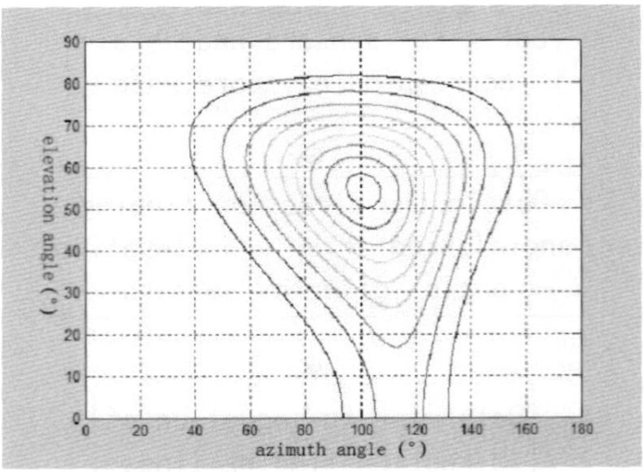

Fig. 5. Contour of array sensor 1

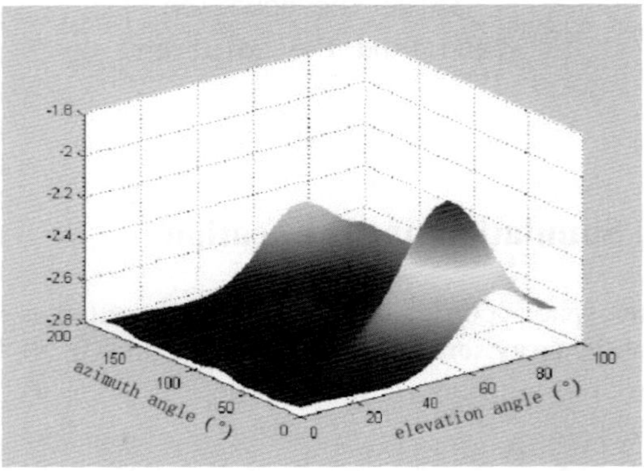

Fig. 6. Spectrum of array sensor 2

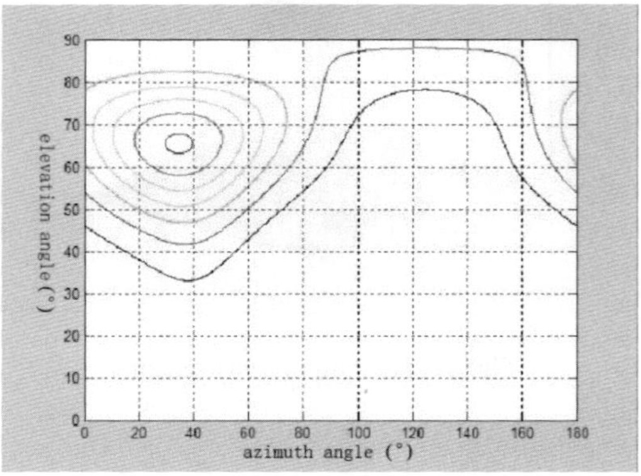

Fig. 7. Contour of array sensor 2

From Figure 6 to Figure 9, it is known that the azimuth angle and elevation angle of two array sensors are: (108.4°, 55.2°), (17.4°, 67.0°). The space coordinate of PD source obtained from the simulation is (25.7, 64.3, 94.9) cm, the error is 7.8cm. Changing the position of the simulated PD source and repeating the research on simulation, the location error is between 6cm and 9cm as table 1 shows, which proves the correctness of the method.

Table 1. Location results and errors

NO.	PD source position(cm)	location result(cm)	error (cm)
1	(25,70,90)	(22.4,74.6,94.4)	6.9
2	(80,80,50)	(84.1,85.8,54.9)	8.6
3	(60,90,60)	(55.3,94.7,57.2)	7.2
4	(25,25,100)	(20.5,20.2,104.1)	7.8
5	(30,120,40)	(25.3,116.6,44.4)	7.3

4 Conclusion

In this paper, a new algorithm for ultrasonic array geometric location of PD based on common perpendicular midpoint of direction finding lines is presented, which takes the common perpendicular midpoint of direction finding lines as PD source and makes the result of location unique, stable compared with the traditional dual platform direction finding triangulation location method. In the end, it is proven by a lot of simulation researches that the new method is correct, more effective and accurate compared with traditional location algorithm, of which error is within 9cm.

Acknowledgements

It is a project supported by the Fundamental Research Funds for the Central Universities (No. 09MG09) and Natural Science Foundation of Hebei Province (No.E2010001703).

References

1. Pedrsen, A., Crichton, G.C., Mcallister, I.W.: The Theory and Measurement of Partial Discharge Transients. J. IEEE transactions on Dielectrics and Electrical Insulation 26, 487–496 (1991)
2. Cadzow, J.A., Kim, Y.S., Shiue, D.C.: General direction of arrival estimation: a signal subspace approach. J. IEEE Trans. on AES 25, 31–46 (1989)
3. Chen, Y.M.: On spatial smoothing for tow-dimensional direction of arrival estimation of coherent signals. J. IEEE Trans. on SP 45, 1689–1696 (1997)

4. Luo, Y.F., Li, Y.M., Liu, L.C.: Simulation of PD Location Method in Oil Based on UHF and Ultrasonic Phased Array Receiving Theory. J. Transactions of China Electrotechnical Society 19 (2004)
5. Xie, Q., Li, Y.Q., Lv, F.C.: Research on Ultrasonic Array Location of two PD sources in Power Transformer. In: Asia-Pacific Power and Energy Engineering Conference, pp. 154–160. IEEE Power&Energy Society(PES), Chengdu (2010)
6. Luo, R.C., Li, W.G., Li, C.R.: A Multi-Target Method to Locate Internal Partial Discharge Sources with in Transformer Based on Array Signal Processing. J. Power System Technology 30 (2006)
7. Tolkachev, A.A., Levitan, B.A., Solovjev, G.K.: A Meawatt Power Millimeter-Wave Phased-Array Radar. IEEE AES Systerms Megazine, 25–31 (July 2000)
8. Stocia, P., Sharman, K.C.: Maximum Likelihood Methods for Direction-of-Arrival Estimation. J. IEEE Transactions on Acoustics, Speech, and Signal Processing 38, 1132–1143 (1990)

Layout Planning of Electrical Vehicle Charging Stations Based on Genetic Algorithm

Yanqing Li, Ling Li, Jing Yong, Yuhai Yao, and Zhiwei Li

Department of Electrical Engineering, North China Electric Power University
liling_ncepu@163.com

Abstract. The charging station is considered as the important infrastructure for electrical vehicle, prerequisite for the promotion of electric vehicle industry. Therefore, the layout of electric vehicle charging stations is especially important. In this paper, the conservation theory of regional traffic flow is presented, which takes electric vehicles within the district as a fixed load point of charging station. On this basis, the total amount and distribution of electric vehicles are forecasted and charging station minimal cost model of the year is proposed, and then, the genetic algorithm is applied to find the solutions of the instance in the paper. The result shows that layout planning of electrical vehicle charging stations based on genetic algorithm is simple, practical, globally optimal, highly adaptable, and economical.

Keywords: Electrical vehicle, charging stations, layout planning, genetic algorithm.

1 Introduction

Since twentieth century, electric vehicles have great advantages as a pollution-free transport in the market [1]. Meanwhile, electric vehicle charging station building is the cornerstone of the promotion of electric vehicle industry, self-contained and effective energy supply network is a necessary condition for widely promoting electric vehicles while expanding the electric power market demand. Electric vehicle charging system is an important supporting system for development of electric vehicles, but also an important part of the commercialization and industrialization of electric vehicles. The construction of the charging station should meet charging need and combine with charging mode to make appropriate planning and design [2].

At present, there are many cities and regions have started building electric vehicles charging stations, but still have not formed a complete and sophisticated layout planning system, not to mention the large scale of the construction. With the fast development of electric vehicle technology, charging stations as part of its supporting facilities is especially important, so the layout planning and construction of the charging station will enter the big scale and networking era, the research on layout planning of charging stations are also imminent. In this paper, the quantity and distribution of electric vehicles is predicted, and the minimum cost model of charging station is presented, then the genetic algorithm is applied to find the solution. Therefore, the optimal layout of the charging station, which is helpful for guiding the practical engineering, is presented.

X. Wan (Ed.): Electrical Power Systems and Computers, LNEE 99, pp. 661–668.
springerlink.com

2 Charging Station Location Model

2.1 Quantity and Distribution Forecast of Electric Vehicles

Suppose there is an area of any shape for whom the leaving vehicles and the entering ones are equal in quantity, which is called the conversation of the number of vehicles [3] (Figure 1). The large area is divided into many small area, Similarly, the number of electric vehicles in small area is a constant, so the total number of electric vehicles in small area can be viewed as an load point of charging station, and suppose that the electric vehicles is always going to the nearest charging station for charging.

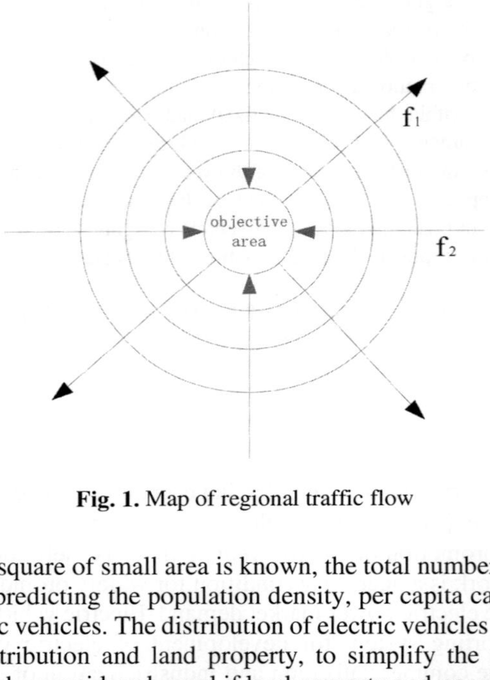

Fig. 1. Map of regional traffic flow

Suppose that the square of small area is known, the total number of electric vehicles can be obtained by predicting the population density, per capita car ownership and the proportion of electric vehicles. The distribution of electric vehicles has something to do with population distribution and land property, to simplify the model, coverage of electric vehicles can be considered equal if land property and population distribution is almost the same.

2.2 The Minimal Cost Model of Charging Stations

Suppose initial investment fixed costs of charging station is C_1, variable operating costs is C_2, and cost of wear and tear of electric vehicle charging is C_3, minimizing the sum of C_1、 C_2 and C_3 is taken as objective to establish the model:

$$\min C = C_1 + C_2 + C_3 \ . \tag{1}$$

Which

$$C_1 = \sum_{i=1}^{n} F[\frac{k(1+k)^m}{(1+k)^m - 1}] \quad (2)$$

Where: n is the number for the new charging stations, F is the investment cost of charging station i , k is investment return rate, namely the discount rate, m is the investment return period.

$$C_2 = \sum_{i=1}^{n} (1+\alpha)F \quad (3)$$

Variable costs of operating for charging station C_2 include maintenance costs, material costs, staff salaries and so on. It can be converted into the initial investment costs, of which α is conversion coefficient.

$$C_3 = t\eta cz \sum_{i=1}^{n} \sum_{j \in J_i} g_{ij} L D_{ij} \quad (4)$$

Where: t is road twist coefficient, η is smooth traffic coefficient of road, L is loss coefficient, n is the number of charging stations, c is annual charging times per vehicle, z is turnaround coefficient, J_i stands for the collection that vehicle of point j goes to charging station i for charging; g_{ij} means parameters that vehicle of point j whether goes to charging station i for charging, D_{ij} is distance between charging station i and vehicle of point j .

$$\min C = \sum_{i=1}^{n} \{[(1+\alpha) + \frac{k(1+k)^m}{(1+k)^m - 1}] F + t\eta kz \sum_{i=1}^{n} \sum_{j \in J_i} g_{ij} L D_{ij}\} \quad (5)$$

Constraints are:

$$\sum_{i=1} g_{ij} = 1 \quad (6)$$

$$D_{ij} \le R_j \quad (7)$$

$$\sum_{j \in J_i} P_j \le S_i \mathrm{e}(S_i) \cos\varphi_i \quad (8)$$

Where: $\mathrm{e}(S_i)$ is the load factor of charging station i , S_i is capacity of charging station i , $\cos\varphi_i$ is power factor of charging station i , P_j is the total load all vehicles go to station i for charging, R_j is charging radius of charging station i , $\sum_{i=1} g_{ij} = 1$ means that each vehicle goes to only one station for charging.

3 Genetic Algorithm

Genetic algorithm, which simulates the biological evolutionary process of natural selection and genetic mechanism of Darwinian's biological evolution theory, is the direct search method that is not depending on specific issue in essence, the optimal solution can be approached mainly through selection, crossover, and mutation operation [4]. Compared with traditional algorithms, Genetic algorithm has a lot of advantages such as multi-path search, implicit parallelism, stochastic operation, low demand for data, no restriction for search space, no requirement for assumptions of continuity, existence of derivation, single peak, and a variety of objective function and constraints can be take into consideration [5].

3.1 Chromosome Encoding

Binary coding is adopted to encode chromosome, encoding string for $A=\{a_j, j=1,2,\cdots,m\}$ is the m-dimensional real vector, m is the number of load point of charging stations, $a_j = z_{ij} \cdot i$ ($i=1,2,\cdots,n$), n is the number of charging stations, z_{ij} is 0-1 variables, z_{ij} stands for that point j go to charging station i for charging, the number of charging stations can be calculated as follows:

$$n=\left[\frac{ph}{24 f_1 f_2 q S e(s) \cos\varphi}\right]+1 . \tag{9}$$

Where:

p – Average charging power for each vehicle;
h – Total number of electric vehicles needed charging per day in small area;
S – Charging station capacity;
q – The charging efficiency;
f_1 – Simultaneity factor of charging station;
f_2 – Demand factor of charging machine;
$e(s)$ – Load factor of charging factor;
$\cos\varphi$ – Power factor;
[ ] – Indicating rounding.

3.2 Fitness Function

Fitness function is determined by the objective function of data model, which takes the economic efficiency as reference. After generating the population, each individual of population chromosome can be turned into corresponding data variables of feasible solution for the problem by decoding operation, and then the corresponding economic benefits of each solution can be obtained by substituting it into the objective function. Economic efficiency is the fitness of each corresponding individual, the less economical is giving priority to better economical individual, and optimal solution can be obtained by inheritance and mutation which keep its genes [6].

In this paper, the objective function itself (annual minimal operating cost) is fitness function, which can avoid complicated derivation and inverse operation. After encoding the variables and determining the initial parameters, the problem can be solved the way that has nothing to do with problem itself, it is relatively easy to realize and get the globally optimal solution [7].

3.3 The Termination Criterion of Genetic Algorithm

Termination criterion used in this algorithm is:

(1) When reaching the maximum number of iterations, the algorithm terminates.

(2) The feasible solutions have not been found when reaching pre-given number of iterations, the algorithm terminates.

3.4 The Algorithm Flow Chart

Algorithm flow chart is shown as follows:

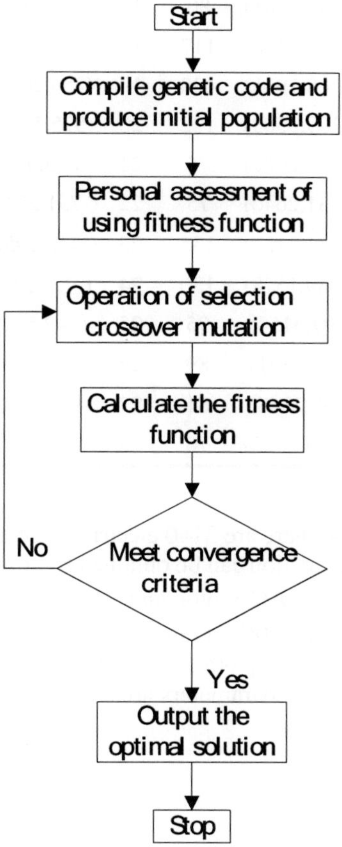

Fig. 2. Algorithm flow chart

4 Instance

There are 10.5 square kilometers in a Tianjin Development Zone, to 2020, the population density is thirty thousand/ km^2, per capita car ownership is one hundred thousandths, of which 30% is electric vehicles, the location of center of gravity and number for each small area is shown in table 1 (There are mainly commercial land, residential land and small number of green space, assuming density of automobile coverage is the same).

Table 1. Number and location of electric vehicles

No.	Coordinate (m)	number	No.	Coordinate (m)	number
1	(1733,3796)	126	15	(1235,1701)	69
2	(1938,3396)	100	16	(1474,1261)	107
3	(2083,2988)	80	17	(-132,2860)	92
4	(2301,2396)	117	18	(70,2421)	79
5	(2708,1911)	130	19	(259,2055)	62
6	(1109,3503)	122	20	(525,1523)	140
7	(1212,2989)	75	21	(835,955)	107
8	(1431,2640)	75	22	(-751,2545)	122
9	(1691,2127)	135	23	(-412,1956)	218
10	(2089,1578)	130	24	(-72,1217)	186
11	(497,3158)	95	25	(242,531)	241
12	(736,2721)	89	26	(-127,2299)	75
13	(898,2368)	81	27	(-955,1705)	117
14	(1099,2023)	70	28	(-567,1011)	100

As the above table shows, there are 3140 electric vehicles in Development Zone, the minimal volume of charging station can be obtained by solving equation (8), and taking S_i is 7200kVA. There are 15 AC motor and 21 DC motor in each station and the capacity is 7200kVA, assuming 90% of the vehicles take fast charge, 10% take normal charge, each vehicle charges every other day. The area should build 6 new charging stations by solving equation (9), parameters are shown in Table 2.

Table 2. Preferences

Name	Parameter
Initial investment (F)	Ten million
Load factor (e)	0.75
Charging station capacity(S)	7200 (kVA)
Power factor($cos\varphi$)	0.9
Capital recovery period(m)	20(years)
Discount rate(k)	0.1
conversion coefficient(α)	0.2
Road twist coefficient(t)	1.1-1.5
Turnaround coefficient(z)	1.5
Smooth traffic coefficient(η)	1.1-1.2
Loss coefficient(L)	1.3
Annual charging times per vehicle(c)	180
Simultaneity factor (f_1)	0.7
Demand factor (f_2)	0.7
Charging efficiency(q)	0.9
Charging radius(km)	1.2
Population size	50
Selection probability	0.2
Crossover probability	0.8
Mutation probability	0.01
Pre-given number of iterations	20
Maximal number of iterations	100

Equation (5) is taken as the objective function, equation (6), (7), (8) as constraints to initially search the optimal solution, and string of genetic code is randomly generated, then Matlab is adopted to program, the location of charging station and the annual minimal cost is shown as below.

Table 3. Results of genetic algorithm

Item	Results of genetic algorithm
Charging station 1	(1620 m, 3350 m)
Charging station 2	(2048 m, 2029 m)
Charging station 3	(427 m, 2653 m)
Charging station 4	(1020 m, 1441m)
Charging station 5	(-412 m, 2956 m)
Charging station 6	(-12 m, 911 m)
Annual cost	79826.7 (thousand Yuan)

As can be seen from the table, the charging stations basically locate in the gravity center of the electric vehicles, and basically meet the principles of going to the nearest charging station for charging, and the annual minimal cost, which is solved by genetic algorithm, is 79826.7 thousand Yuan in all.

5 Conclusions

In this paper, the number and distribution of electric vehicles is predicted, the conversation of the number of vehicles in area, which takes the dynamic number of electric vehicles as a constant, is presented, on this basis, annual minimal cost model is established and genetic algorithm is applied to solve it. The result of instance shows that the model and method proposed is feasible. However, the layout planning of charging station is a complicated and comprehensive work, which needs consider the actual condition based on the theory optimization to ultimately determine the capacity and locate the station sites of charging stations.

References

1. Wang, J., Jiang, J.C.: The design and realization of the information managing sys tem for the electrical vehicle charging station. J. Control & Automation 22, 13–16 (2006)
2. Xu, F., Yu, G.Q., Gu, L.F., Zhang, H.: Tentative analysis of layout of electrical vehicle charging stations. J. East China Electric Power 37, 1677–1680 (2009)
3. Zhang, C.H., Xia, A.H.: A Novel Approach for the Layout of Electric Vehicle Charging Station. In: Apperceiving Computing and Intelligence Analysis, Chengdu, pp. 16–19 (2010)
4. Li, X.B., Zhu, Q.J.: New Model Optimized by Genetic Algorithm for Distribution Substation Locating and Sizing. J. Proceedings of the CSU-EPSA 21, 21–31 (2009)
5. Wang, C.S., Liu, T., Xie, Y.H.: Substation Locating and Sizing Based on Hybrid Genetic Algorithm. J. Automation of Electric Power Systems 30, 30–34 (2006)
6. Huang, W.: Optimal allocation of water resources based Adaptive Genetic Algorithm. J. Yellow River 32, 61–64 (2010)
7. Xiao, H.Y., Yao, J.G., Qing, Z.W., Yao, P.: Substation Optimal LocatingBased on Genetic Algorithm and Multi-objective Decision-Making Method. J. Central China Electric Power 20, 13–17 (2007)

Application of ID3 Algorithm in Exercise Prescription

Quancheng Zhang[1], Kun You[1], and Gang Ma[2]

[1] Department of Physical Education, Xi'an Shiyou University, Xi'an, 710065, P.R. China
[2] School of Computer Science, Xi'an Shiyou University, Xi'an, 710065, P.R. China
zqc.123@163.com, 122284319@qq.com, jhlmg@xsyu.edu.cn

Abstract. This paper adopts the *ID3* algorithm for mining hidden classification rules from mass students' physical constitution evaluation and sports training result data. It is helpful for PE teacher on planning exercise prescriptions toward college students with different physical constitution conditions by decision support. The algorithm generates a decision tree by choosing attributes with maximum information gain ratio for classification (Fig. 1). Such process involves a classification training set *R*(Table 1) towards original data, a information gain calculation according to overall evaluation of the physical constitution, separately investigate information gain ratio between physical constitution overall evaluation and each classification attribute, eliminating classification attribute IDs which has no practical significance.

Keywords: Classification Analysis, Decision Trees, *ID3* Algorithm, Exercise Prescription.

1 Introduction

Classification analysis is one of the important methods in data mining. It can be used to discover the feature-based knowledge. Such knowledge usually reserves in common attributes of similar things and differences between different things. Classification is mainly used to predict. Therefore, one may discover model or function sets to describe typical dataset features. Those features are of important to recognize the ownership or type of unknown data [1].

Classification is to find reasonable description concepts for classes. Such descriptions are useful to construct models. Classification process involves analysis the input data, generates classification rules or classification models based on characteristics of the training data set and relevant algorithms. Then it may predicts and classify the future test data with the analysis results. In fact, classification and prediction is actually an inductive - deductive approach [2, 3].

This paper adopts the *ID3* algorithm for mining hidden classification rules from mass students' physical constitution evaluation and sports training result data. It is helpful for PE teacher on planning exercise prescriptions toward college students with different physical constitution conditions by decision support. Classification model can be obtained from a set of training data via classification algorithm. Methods for construct a classification model include machine learning, Bayesian theory and non-parametric methods as well as rough set algorithm and BP algorithms. Among them,

X. Wan (Ed.): Electrical Power Systems and Computers, LNEE 99, pp. 669–675.
springerlink.com

ID3 algorithm is the most influential classification algorithm [4]. This paper adopts the *ID3* algorithm for mining large number students' physical constitution evaluation data which was accumulated from physical education classes. Then the algorithm may discover classification rules that hidden beneath Composite Indicators of students' physical constitution. Furthermore, it may become a rational guidance for planning exercise prescriptions toward student with different physiques.

2 Decision Trees

Decision tree is a kind of directed acyclic graphics (DAG) [5]. In such tree, a node is named as root node when it has no input edges. Once when it has input edges and output edge, the node will be defined as internal node. In addition, a leaf node may be defined when it without output edges but input edges. Decision tree learning is a method for approximating discrete-valued target functions, in which the learned function is represented as a decision tree[6].The basic algorithm for decision tree is the greedy algorithm that constructs decision trees in a top-down recursive divide-and-conquer manner.

Assuming R is a raw training set, also a raw data classification. Attribute A has m different values, that is $A=\{ a_1, a_2, \ldots, a_m\}$, then set R can be divided into m classes by attribute , that is $C=\{ C_1, C_2, \ldots, C_m\}$, denoted as $C_{i,R}$, is the tuple set C_i th set in R. $|C_{i,R}|$ and $|R|$ is the number of tuples in sets $C_{i,R}$ and R, respectively [7].

Definition 1: The expectation information needed to classifying tuples in R is defined as

$$Info(R) = -\sum_{i=1}^{m} p_i \cdot \log_2(p_i) \tag{1}$$

Where $Info(R)$ is the R's Entropy, P_i is the probability of any tuple in R set, the value of which is calculated by $P_i=|C_{i,R}|/|R|$.

Definition 2: The expectation information needed to classifying tuples in R according to attribute A is defined as

$$Info_A(R) = -\sum_{i=1}^{m} \frac{|C_i|}{|R|} \cdot Info(C_i) \tag{2}$$

Definition 3: The difference between the expectation information needed to classifying tuples in R according to attribute A and the entropy of R is defined as the information gain of A.

$$Gain_A(R) = Info(R) - Info_A(R) \tag{3}$$

Definition 4: The information of classifying tuples in R according to attribute A is defined as

$$SplitInfo_A(R) = -\sum_{i=1}^{m} \frac{|C_i|}{|R|} \times \log_2\left(\frac{|C_i|}{|R|}\right) \tag{4}$$

The information gain ratio is defined as

$$GainRatio_A(R) = \frac{Gain_A(R)}{SplitInfo_A(R)} \tag{5}$$

In the process of decision tree construction, the purpose and criteria of each classification are to divide a given training data set into the best (most pure) sub-categories, ideally, all given tuples belong to the same class. But in fact it is not possible, we can only select a relatively pure attribute [8]. Usually, select the attributes of the highest information gain and the greatest information gain ratio as the property attribute selection of the current classification. Construct the first-level decision tree. Using the same way to treat circularly, we can construct second-level and multi-level decision tree until it reaches the classification objective, generate leaf nodes.

3 Process of *ID3* Algorithm

ID3 algorithm aims to calculate the information entropy. The information gain measure is used to select the test attribute at each node in the tree. Attributes with the highest information gain (or greatest entropy reduction) is chosen as the test attribute for the current node. This attribute minimizes the information needed to classify the samples in the resulting partitions. Only need to test enough attributes until all data is classified. Entropy, in general, measures the amount of disorder or uncertainty in a system in setting the classification, higher entropy (i.e., more disorder) corresponds to a sample that has a mixed collection of labels. Lower entropy corresponds to a case where we have mostly pure partitions [9-10].

ID3 algorithm Description

Algorithm: *generate_decision_tree*

Input: raw training set *R*, candidate attribute list set *attr_list*, classification criterion *attr_sele_method*

Output: decision tree

procedure

(1) create *N* as a node

(2) if $\forall r\{r \in R, r \in C\}$ then (assuming all tuples in *R* set to be class *C*, that is *R*=*C*)

(3) return $N \in C$ as a leaf node (let *N* be leaf node return, which is marked as class *C*)

(4) end if

(5) if *attr_list* $= \varnothing$ then

(6) return $N \in Most$ as a leaf node(let *N* be leaf node return, which is marked as the most class

(7) end if

(8)*split_attribute*←*attr_sele_method* (*R*, *attr_list*)(invoke *attr_sele_method* to find the best value of classification attribute)

(9) $N \in C_{split\_attribute}$ (*N* is marked as $C_{split\_attribute}$)

(10) for each $R_j \in R$ do (R_j is the jth output that *R* is classified by *split_attribute*)

(11) if $R_j = \varnothing$ then

(12) add N_j to *N* (a leaf node N_j is added to *N*)

(13) else

(14) N_j = *generate_decision_tree* (R_j, *attr_list*) (recursive function)

(15) add N_j to *N* (a leaf node N_j is added to *N*)

(16) end if

(17) end for

4 *ID3* Algorithm in Exercise Prescription

ID3 algorithm in exercise prescription requires mining valuable information from existing mass sports teaching data. For example, the student physical constitution evaluation data including maximum oxygen intake, stand-and-reach test, sitting reach test, step test, sit-up and push-up. Those data can be pre-processed before the *ID3* algorithm mining. After mining, composite physical constitution indicators of one student will be generated according to relevant evaluation rules. This would not only be a reference for PE teachers in tracing their students' physiques but also in planning exercise prescriptions.

4.1 Building Decision Trees

The core of *ID3* algorithm is attribute selection. Table 1 illustrates physical constitution evaluation result of thirty students. Physical constitution evaluation indicators are good, middle and bad.

Maximal oxygen intake has three values, stand-and-reach test and sit and reach test also has three values, step test contains four value scopes, both sit-up (female) and push-up (male) attributes have two values.

Table 1. Information table of body constitution evaluation

ID	maximal oxygen intake/ grade	step test index / grade	stand- and- reach test or sit and reach test / grade	sit-up (female) push-up (male)/ grade	comprehensive evaluation/ /grade
1	22/C	60.5/A	-1.8/C	35/A	B
2	29/B	45.3/D	-10.5/C	32/B	C
⋮	⋮	⋮	⋮	⋮	⋮

Data sets of body constitution evaluation for 30 students is defined as *R*, information entropy is

$$Info(R) = -\frac{4}{30}\log_2\frac{4}{30} - \frac{7}{30}\log_2\frac{7}{30} - \frac{19}{30}\log_2\frac{19}{30} = 1.2863$$

Attributes of maximal oxygen intake is defined as *A*, its expected information entropy is

$$Info_A(R) = \frac{13}{30}(-\frac{2}{13}\log_2\frac{2}{13} - \frac{11}{13}\log_2\frac{11}{13}) + \frac{12}{30}(-\frac{8}{12}\log_2\frac{8}{12} - \frac{3}{12}\log_2\frac{3}{12} - \frac{1}{12}\log_2\frac{1}{12})$$
$$+\frac{5}{30}(-\frac{3}{5}\log_2\frac{3}{5} - \frac{2}{5}\log_2\frac{2}{5}) = 0.8354$$

Information gain ratio that tuple in *R* is classified by attributes, is given by

$$Gain_A(R) = Info(R) - Info_A(R) = 1.2863 - 0.8354 = 0.4509$$

Attributes of step test is defined as *B*, the information gain ratio is $Gain_B(R)$=0.3702, Attributes of stand- and- reach test, sit and reach test is *C*, the information gain ratio of which is $Gain_C(R)$=0.2145, Attributes of sit-up (female) push-up (male) is named as *D*, the information gain ratio is $Gain_D(R)$=0.0167.

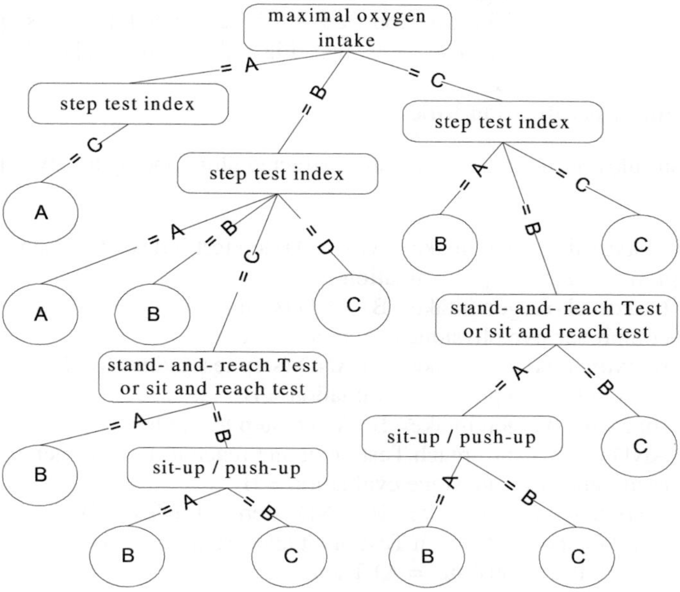

Fig. 1. One Final decision trees

Due to maximal oxygen intake has high information gain, it is considered as the first classification attribute. Obviously, value of the information gain relates to attribute that has more output. In training data set, *ID* attribute has the greatest value ($Info_{ID}(R)=0$, $Gain_{ID}=Info(R)$). Classification information of maximal oxygen intake is calculated as follows

$$SplitInfo_A(R) = -\frac{13}{30}\times\log_2\frac{13}{30} - \frac{12}{30}\times\log_2\frac{12}{30} - \frac{5}{30}\times\log_2\frac{5}{30} = 1.4884$$

Information gain ratio is

$$GainRatio_A(R) = \frac{Gain_{\text{Matri\_id}}(R)}{SplitInfo_{\text{Matri\_id}}(R)} = \frac{0.4692}{1.4884} = 0.3153$$

Similarly, we obtain

Attributes of step test is defined as *B*, the information gain ratio is

$$GainRatio_B(R) = 0.2896$$

Attributes of stand-and-reach test or sit and reach test is *C*, the information gain ratio is

$$GainRatio_C(R) = 0.2326$$

Attribution of sit-up (female) push-up (male) is called D, the information gain ratio is

$$GainRatio_D(R) = 0.0562$$

By comparison with four kinds of information gain ratio, maximal oxygen intake has the highest information gain ratio. Therefore, attribute A is selected as the first attribute for classification. The decision tree will be expanded and subsequent nodes in decision tree are obtained in a same manner. Fig.1 shows the final decision tree.

4.2 Extracting Classification Rule

Classification rules of the tree in Fig 1 are extracted along the path between root node and leaf.

Rule1: IF (maximal oxygen intake= A) AND (step test index= C) Then comprehensive physique evaluation = A

Rule2: IF (maximal oxygen intake= B) AND (step test index= A) Then comprehensive physique evaluation = A

Rule3: IF (maximal oxygen intake= B) AND (step test index= B) Then comprehensive physique evaluation = B

Rule4: IF (maximal oxygen intake= B) AND (step test index= C) AND (stand- and- reach Test or sit and reach test = A) Then comprehensive physique evaluation = B

Rule5: IF (maximal oxygen intake= B) AND (step test index= C) AND (stand- and- reach Test or sit and reach test = B) AND (sit-up / push-up = A) Then comprehensive physique evaluation = B

Rule6: IF (maximal oxygen intake= B) AND (step test index= C) AND (stand-and-reach Test or sit and reach test = B) AND (sit-up / push-up = B) Then comprehensive physique evaluation = C

Rule7: IF (maximal oxygen intake= B) AND (step test index= D) Then comprehensive physique evaluation = C

Rule8: IF (maximal oxygen intake= C) AND (step test index= A) Then comprehensive physique evaluation = B

Rule9: IF (maximal oxygen intake= C) AND (step test index= B) AND (stand-and-reach Test or sit and reach test = A) AND (sit-up / push-up = A) Then comprehensive physique evaluation = B

Rule10: IF (maximal oxygen intake= C) AND (step test index= B) AND (stand-and-reach test or sit and reach test = A) AND (sit-up / push-up = B) Then comprehensive physique evaluation = C

Rule11: IF (maximal oxygen intake= C) AND (step test index= B) AND (stand-and-reach test or sit and reach test = B) Then comprehensive physique evaluation = C

Rule12: IF (maximal oxygen intake = C) AND (step test index= C) Then comprehensive physique evaluation = C

5 Conclusions

This paper has analyzed the mass students' physical constitution evaluation data, such as maximal oxygen intake, stand-and reach test, sit-and-reach test, step test, sit-up/push-up by *ID3* algorithm. Results show that classification rules extracted from the decision tree maintain strong consistency with physical constitutions of university students. By these rules, the reasonable exercise prescription is developed to strengthening physical exercises and improving teaching activities.

References

1. Ma, G.: Development and Application Study of Hospital Management Information System Based on Data Mining Technology, Xi'an Shiyou University (2008)
2. Xu, J.-p.: Data Warehouse and Decision Support System. Science Press, Beijing (2005)
3. Han, J., Kamber, M.: Data Mining Concepts and Techniques. China Machine Press, Beijing (2007)
4. Su, X.-n., Yang, J.-l., Jiang, N.-n., Su, X.: Data Warehouse and Data Mining. Tsinghua University Press, Beijing (2006)
5. Gehrke, J., Ganti, V., Ramakrishnan, R., Loh, W.-Y.: BOAT: Optimistic Decision Tree Construction. In: Proceedings of the 1999 ACM SIGMOD International Conference on Management of Data, Philadelphia, Pennsylvania (May 1999)
6. Fayyad, U.M., Platetsky Shapiro, G., Smyth, P., Uthurusany, R.U.: Advances in Knowledge Discovery and Data Mining, pp. 83–115. AAAI, (S.1.) (1996)
7. Ma, G., Liu, T.-s., Li, J.: Application Study of *ID3* Algorithm in Mining of Doctor Classfication Regulations. International Electronic Elements 16(181), 79–81 (2008)
8. Rastogi, R., Skim, K.: Public:A Decision Tree that integrates Building and Pruning. In: Proceedings of 24th International Conference on Very Large Data Bases, New York, August 1998, pp. 404–415 (1998)
9. Zou, Y.-g., Fan, C.-h.: Improved *ID3* Algorithm Based on Attribute Importance. Computer Applications 28, 144–146 (2008)
10. Zhu, H.-d.: Research on Improvement and Simplification of ID3 Algorithm. Journal of Shanghai Jiaotong University 44, 883–886 (2010)

Application of Accessory Ingredient Semi-automatic Weighing System in Dyeing and Printing Industry

Xie Shui-ying[1], Jiang Lei[1], Yin Li[2], and Han Cheng-jiang[1]

[1] Zhejiang Industry Polytechnic College
Shaoxing, China
[2] School of Transportation & Logistics, Southwest Jiaotong University
Chengdu, China
lbyzxsy@163.com, jiangleixing@126.com, yinli209@126.com,
zjipcky@163.com

Abstract. Because the printing and dyeing chemical auxiliaries additive process exists many problems, a suit of semi-automatic weighing system has been designed. The system solves a series of problems caused by manual manipulation, verbose choose pigment, maintenance strait. It not only improves the automation degree of the corresponding links, but also effectively improves the production efficiency.

Keywords: additives, semi-automatic weighing system, dyeing and printing industry.

1 Introduction

All kinds of chemical dyes (solid) and chemical additives (liquid) are added in the dyeing process of dyeing and printing industry, which is one of the key dyeing process. If mistaken dyes or additives are added or the amount of the added material is wrong, which will significantly affect the quality of produced cloth. Then a large number of products are wasted. At the same time, it severely affects the working efficiency. Therefore, accurate and reliable weighing is very necessary in the production process of dyeing.

2 Problems

In the past years, printing and dyeing enterprises mostly used manual weighing. In the face of dozens of dyes and additives, it often faces with the following issues: (1) choosing the wrong type and dyeing the wrong color; (2) weighting is more or less used, which caused the color deviation; (3) Wasting products in the process of taking or delivering dyes and additives. At present, in order to solve these problems, most domestic enterprises has introduced automatic chemical additives auto-weighing and

X. Wan (Ed.): Electrical Power Systems and Computers, LNEE 99, pp. 677–683.
springerlink.com

automatic transmission systems from abroad. However, many of these automatic systems have brought new problems, including repair, changing planar dyes in particle dyes is not convenient, the production capacity is not meet, the cost is high, etc. A semi-automatic weighing product line has been presented, which mainly solves the problems mentioned above. The features of the system are simple equipment, low cost, simple operation, easy maintenance, and so on.

3 The Design of the System

Chemical additives are most liquid fluids, so they use canned or bottled methods. Mostly of them is viscous and a few except. The system uses gravity flow principle to rely on achieving high-altitude feeding weight. It doesn't use material feeding pump, because the life of domestic feeding pump is short or failure rate is high while imported feed pump is expensive. If the viscosity is too large, the system can pass into the high-pressure gas by way of material storage tank pressure to achieve material flowing. System diagram as shown in Fig.1.

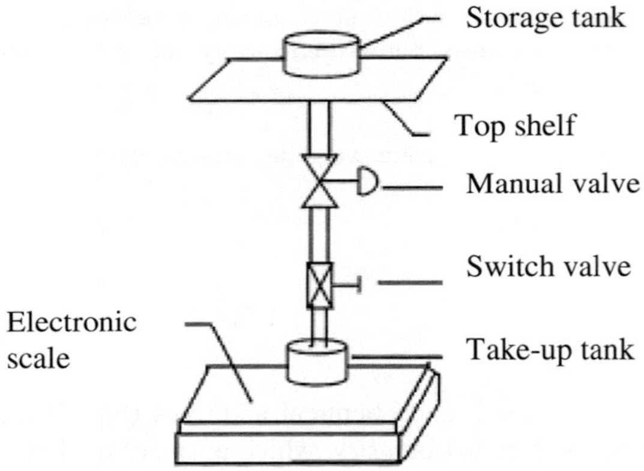

Fig. 1. System diagram

3.1 Auxiliary Automatic Feeding System

A large channel steel frame or platform has been made on the top. Top shelf must be above than the platform. A required number of large storage tanks are needed to install on the platform, and each storage tank has feeding ports and discharging ports. The individual also should have suction ports. At the bottom of storage tank, discharging ports connects the feeding system of the entrance platform. The top connecting plastic pipes of feeding ports leads to the ground stripping pump. Pumps make the corresponding auxiliary feed into the corresponding storage tank respectively. Every pump has an additive and they can not mix with each other to achieve the automatic feed. System diagram is shown in Fig.2.

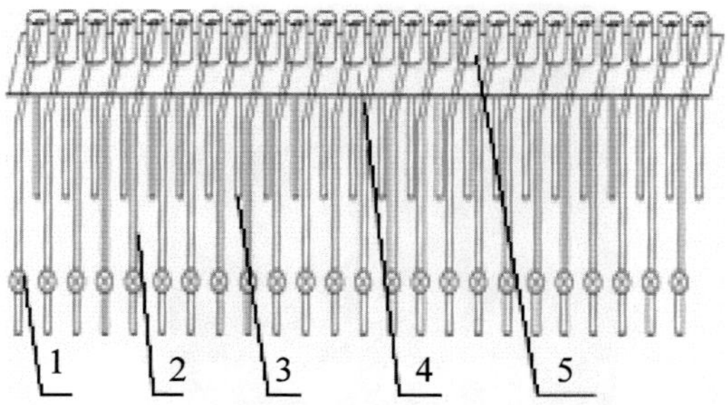

1 Stripping pump 2 Feed pipe 3 Discharge pipe 4 Platform 5 Storage tank

Fig. 2. System diagram of automatic-feeding system

Stripping system also can be adopted by manually rotating a single pump, respectively transport mode, so that we can take turns to save costs, the basic structure is shown in Fig.3.

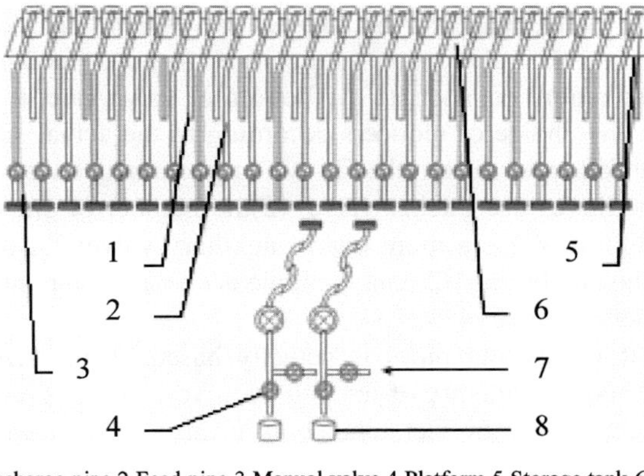

1 Discharge pipe 2 Feed pipe 3 Manual valve 4 Platform 5 Storage tank 6 Platform 7 Flushing water 8 Crude materials tube

Fig. 3. The basic structure of stripping system

Manual operation of single pump is used. Then each feeding fight pump and its subsidiary pipeline rinse need be rinsed with flushing water. The second auxiliary is carried, followed by complete stripping. Standby feed pump need leave in case the equipment damaged. The actual installation site is shown in Fig.4.

The part of the equipment automatically should be set and installed by enterprises according to size of additive from the line configuration and using time. Thickness of pipe can be adjusted if needed.

Fig. 4. The actual installation site map of system

3.2 Auxiliary Automatic Discharging System

The bottom of auxiliary storage tank connects plastic tube which leading to the platform and joining manual ball valves before the top of splice position used as opening and closing maintenance valves. Manual valves then link to automatic valves. Pneumatic valve controls valve gate, to achieve accurate feeding. A very short discharge adjustable mouth connects to the bottom of valve gate, then stretching into the top of buckets, the choice of reducers according to the actual situation. A unit of valves includes the following materials: (1) 1/2 of pneumatic valves resistant to acid, individual sticky additives use thicker valve; (2)the number of 1/2 ball valve is one, individual sticky additives use a more coarse auxiliary valve; (3) the number of 1/2 reduce is one; (4) the number of 1/2 double nipple is one; (5) the number of 1/2 Pagoda head is one, which diameter is 14.

The thickness of valves and pipes is actually based on the size of viscosity of specific auxiliaries and the quantity of feeding additives. There is a simple request for the most normal occasions. In fact, choice is always made during the designing process.

The whole weighting system is divided into three groups. Each group can add six or eight road additives, which can be arbitrary chosen. Each of the pneumatic valve is installed together closely in order to install the discharging mouth concentrated in a certain range, and then it can use the same container to weigh. Six or eight discharge mouths are placed in a sealed stainless steel box to protect themselves. An automatic opening and closing of the retractable door is only used in the lower part, when weighing material the door opened and then the door closed. The seal-door is used to prevent the remaining drops of liquid from corroding the electronic scale after feeding. The bezel is divided into six or eight small slots corresponding to the six or eight auxiliaries, and each slot does not mingle, then six or eight overflow pipes link to the back of it, passing into six or eight splice containers. The dripping of the additives on the bezel should be recycled, regularly back to charging containers for recycle use.

Practically, a role need play in savings. Material equipment, which is an open box, is under the seal-door. Electronic sale is placed at the bottom and material containers in the middle, flushing with the level of electronic cabinet of the front surface of a platform to access. It is easy for workers to take and place material container. On the pan of electronic sale, the exact placement of material containers is marked with a bright color, which is easy for operator to work. Scale feeding system structure is shown in Fig.5.

The lower part of the material is a recycled cabin, which is a drawer with a pulley. It placed eight recycling containers, each recycling containers corresponding to a pipe from the overflow baffle, recycling drops from the baffle. The drawer on a regular time is manually pulled out, and recycled residual additives. Three weighing system places in a large box on the vertical three-part. Each part made by the aforementioned structures. Therefore, an auxiliary automatic discharge system has three separated sub-system, each subsystem can weigh six or eight additives. Three groups of electronic display panels are placed on the top box, which is the electronic display instrument. It shows each weighing data through the front glass. Feeding operations light means that allowing to start charging, either manually or automatically, and only in the charging of the lamp after the button press to be effective or not allowed to feed. Actual field installation map is shown in Fig.5.

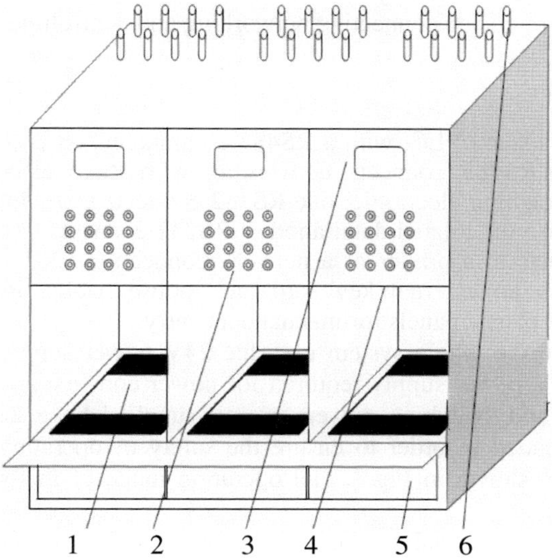

1 Electronic scale 2 Indicator light 3 Face plate of electronic scale 4 charging door 5 recycled cabin 6 feed inlet

Fig. 5. Structure of scale feeding system

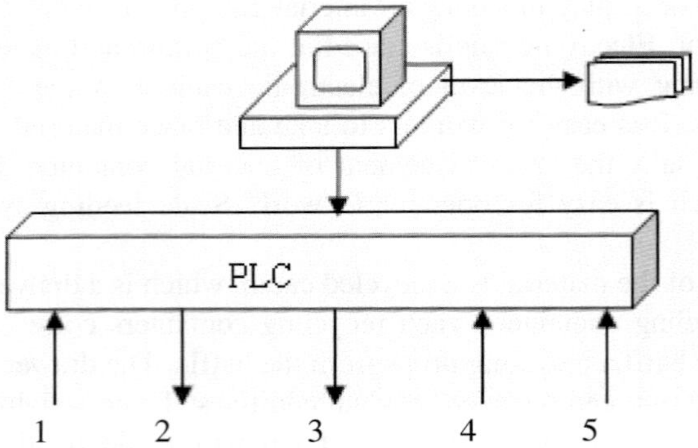

1 Scanning device 2 Air-operated valve 3 Indicator light
4 Knob 5 Electronic scale

Fig. 6. The structure of control system

Electronic sales selection standards: the scale of weight is 30kg; precision scope is 0.1g; mesa dimension is 230*300mm. Stainless steel is chosen, the return zero time is less than 1s; it includes RS485 communication mouth, telecommunication remote reset and peel function, panel display system. Three control systems have been chosen. RS485 is used to communicate connecting networking and controlling system.

3.3 Control System

Control system uses a small PLC, with a RS485 communication port and two RS232 communication ports.RS485 connects networking with three electronic sales and external high-dose weighing electronic; one RS232 connects a window scanner or the IC card reader, reading the related data; another RS232 connects to a stage crew, the host computer connects to a printer or a network connection ERP system. PC is no longer equipped with an external keyboard and mouse, which instead of simple stainless steel or ABS plastic panels for manual data entry.

For safety reasons, the whole system uses the 24V power supply by a switching power supply to 220V power supply required for power conversion. Electronic sales, computers, printers, and switching powers are all supplied by a 220V power. The system must be grounded in order to ensure the safety of operating personnel. The structure of system is shown in Fig.7. The operating mode of this system shows as follows:

The first step is that material barrels will be put on the electronic scale and the indicator will emit light if containers in right position.

The second step is inputting data with keyboard, using a scanner or access IC card reader to read material data.

The third step is pressing the reset button, to allow charging indicator and light up the corresponding indicator.

The fourth step is pressing the manually button or automatically button, which corresponding position indicator lights flashing and meaning that it is weighing and feeding.

The fifth step is feeder complete indicator light up, feeding indicator has extinguished, taking away the material barrels.

The system is called semi-automatic production lines, mainly because the material storage tank uses workers to complete feeding. The other reasons are that it needs people to input picking list before weighing the setting of barrel need to be artificial. Material barrels after the weighing should be taken by workers after weighing. The rest processes are automatic operations and the key link without artificial interference factors.

4 Conclusion

The features of this system:

a) In the future, the automatic feeding system can be extended and artificial feeding will be abolished .

b) Remaining system clean and accurate in a long time and decreasing maintenance.

c) Easy operation, no artificial interference.

d) Accuracy control , saving about 10% additives than weighing artificial.

e) Additives inventory management function can be achieved, to keep inventory information.

f) Keeping and printing auxiliary weighing unit: name, time, quantity, and other historical records.

The system has been successfully applied to the printing and dyeing enterprises and achieved remarkable results. It improves enterprises' automation degree. At the same time, the production efficiency is improved significantly.

References

1. Zhang, M.: Innovation of Digital Automatic Dispensing Technology. Dyeing & Finishing (2007)
2. Chang, L.: Ingredients for A New Device. Machinery for Cereals, Oil and Food Processing (2002)
3. Fu, J., Yan, J., Zhang, H.: Design and Research for Automatic Allocation and Compound of Dyestuff. Computer Engineering (2001)
4. Dry Automatic Batching System. Northwestern Polytechnical University (scientific research achievements), Shanxi Province (2002)
5. High-precision, Large-scale, Automatic Dyeing Ingredients Production Line. Institute of Automation, Chinese Academy of Sciences(scientific research achievements), Beijing (2007)

An Analysis of Global Exponential Stability of Genetic Regulatory Networks with Time-Varying Delay

Zhixiong Zong, Weirui Zhao, and Yunxiang Zhou

Department of Mathematics,Wuhan University of Technology 122 Luoshi road, Wuhan, Hubei, 430070, P. R. China
zzxx2005@sohu.com,
wrzhao@whut.edu.cn

Abstract. This paper presents a new sufficient condition for global exponential stability of the equilibrium point for genetic regulatory networks with time-varying delays. The conditions we obtained are weaker than the previously known ones and can be easily reduced to several special cases.

Keywords: Genetic regulatory networks, Time-varying delays, Global exponential stability.

1 Introduction

Genetic Regulatory Networks (GRN) have been intensively studied in the past decade and have been applied in the biological and biomedical sciences[[1]-[10]]. The applications crucially rely on the dynamical behavior of the GRN, thus, some useful results about the uniqueness and global asymptotic stability of the equilibrium for GRN with delays can be found in [[2],[5]-[7],[8],[9]]. As far as we know, there has not yet been a general genetic regulatory network to theoretically ensure the exponential stability and estimate the exponential convergence rate. In this paper, we consider differential equation model of genetic network, in which the variables describe the concentrations of mRNAs and proteins, as continuous values of the gene regulation systems. We focused on global exponential stability of GRN with time-varying delays. By constructing a suitable Lyapunov functional, we obtained some new criteria for global exponential stability of GRN with time-varying delays. These criteria generalized previous results.

Consider genetic regulatory networks described by the state equations [[5]-[7]]:

$$\begin{aligned} \dot{M}_i(t) &= -a_i M_i(t) + \sum_{j=1}^{n} W_{ij} f_j(p_j(t-\sigma(t))) + B_i \\ \dot{P}_i(t) &= -c_i P_i(t) + d_i M_i(t-\tau(t)), i = 1, 2, \cdots, n, \end{aligned} \tag{1}$$

where M_i and P_i are the concentrations of mRNA and protein of the ith node, respectively. The parameters a_i and c_i are the decay rates of mRNA and protein,

X. Wan (Ed.): Electrical Power Systems and Computers, LNEE 99, pp. 685–690.
springerlink.com

respectively; d_i is the translation rate, and the functions $f_j(x)$ represent the feedback regulation of the protein on the transcription in [7]. The delay $\tau(t)$ and $\sigma(t)$ is differential and bounded function with

$$0 \leq \tau(t) \leq \tau, 0 \leq \sigma(t) \leq \tau, \dot{\tau}(t) \leq \sigma < 1, \dot{\sigma}(t) \leq \sigma < 1$$

for nonnegative constants τ and σ.

Initial conditions for (1) are of the form

$$\phi = (\phi_1, \phi_2, \cdots, \phi_n, \phi_{n+1}, \phi_{n+2}, \cdots, \phi_{2n}) \in C([-\tau, 0], \mathbb{R}^{2n}).$$

For any initial value condition $\phi = (\phi_1, \phi_2, \cdots, \phi_{2n}) \in C$, systems (1) admits a unique solution, denoted $x(t, \phi)$. To simplify the notations, the dependence on the initial condition ϕ will not be indicated unless necessary.

The notation $B > 0$ means that B is symmetric and positive definite. $\|B\|_2$ represents the norm of B induced by the Euclidean vector norm. I_n denotes the $n \times n$ dimensional identical matrix. Sometimes we write $x(t)$ as x, $f(x(t))$ as $f(x)$ and the transpose of A^{-1} as A^{-T}.

We will need the following definition:

Definition 1. *[4]The equilibrium point $x^* = (m^*, p^*)$ is said to be globally exponentially stable, if there exist positive constants $k > 0$ and γ such that for any solution $x(t) = (m(t), p(t))$ of system (1) with initial function $\phi \in C([-\tau, 0], \mathbb{R}^{2n})$, there holds*

$$\|x(t) - x^*\|_2 \leq \gamma \|\phi - x^*\|_2 e^{-kt}, \textit{for all } t \geq 0.$$

The organization of this paper is as follows. In Section 2, we establish a new criterion on the globally exponential stability of the equilibrium point for GRN with delays. An example will be provided in Section 3. Section 4 presents our conclusions.

2 Main Results

Based on some facts about positive definite matrices and integral inequalities, we present the main results in this section. Firstly, we have the following lemma due to [10]

Lemma 1. *Given any real matrices X, Y, C of appropriate dimensions and a scalar $\epsilon_0 > 0$, where $C > 0$. Then the following inequality holds:*

$$X^T Y + Y^T X \leq \epsilon_0 X^T C X + \frac{1}{\epsilon_0} Y^T C^{-1} Y.$$

In particular, if X and Y are vectors, $X^T Y \leq \frac{X^T X + Y^T Y}{2}$.

Let (M^*, P^*) be an equilibrium of (1), we first shift the equilibrium point of system (1) to the origin. By the transformation $m(t) = M(t) - M^*, p(t) = P(t) - P^*$, the genetic regulatory network model (1) can be rewritten as:

$$\dot{m}(t) = -Am(t) + Wg(y(t - \sigma(t))), \dot{p}(t) = -Cp(t) + Dm(t - \tau(t)) \quad (2)$$

where $m(t) = (m_1(t), m_2(t), \cdots, m_n(t))^T$, $p(t) = (p_1(t), p_2(t), \cdots, p_n(t))^T$, $A = diag(a_1, a_2, \cdots, a_n)$, $C = diag(c_1, c_2, \cdots, c_m)$, $W = (w_{ij})_{n\times m}$, $D = diag(d_1, d_2, \cdots, d_n)$, $g(p(t)) = (g_1(p_1(t)), \cdots, g_n(p_n(t)))^T$ *with* $g_i(p_i) = f_i(p_i + P^*) - f_i(P^*)$, $i = 1, \cdots, n$. Since f_i is a monotically increasing function with saturation, it satisfies that

$$0 \le \frac{g_i(x_i)}{x_i} \le k_i \text{ and } g_i(0) = 0, i = 1, 2, \cdots, n.$$

Obviously, (M^*, P^*) is the globally exponentially stable equilibrium point of (1) if and only if the origin of (2) is globally exponentially stable. Thus in the following, we only consider the globally exponential stability of the origin of (2).

We are now ready to present our main global exponential stability result.

Theorem 1. *The origin of neural system (2) is globally exponentially stable if there exists a symmetric positive diagonal matrices* $P_1 = diag(p_1^{(1)}, p_2^{(1)}, \cdots, p_n^{(1)})$, $P_2 = diag(p_1^{(2)}, p_2^{(2)}, \cdots, p_n^{(2)})$, *a symmetric positive definite matrices* Q_1, Q_2, *and factorizations of* $W = W_1 W_2, D = D_1 D_2$ *such that:*

$$\Omega_1 = 2P_1A - P_1W_1Q_1^{-1}W_1^TP_1 - D_2^TQ_2D_2 > 0;$$
$$\Omega_2 = 2P_2CK^{-1} - P_2D_1Q_2^{-1}D_1^TP_2 - W_2^TQ_1W_2 > 0,$$

where $K = diag(k_1, k_2, \cdots, k_n)$ *and* $Q_1, Q_2, W_1, W_2, D_1, D_2$ *are constant matrices with appropriate dimensions.*

Proof. We employ the following positive-definite Lyapunov functional:

$$V(t) \equiv V(m(t), p(t), t) = \epsilon_1 V_1(m(t), p(t)) + V_2(m(t), p(t), t), \tag{3}$$

where

$$\begin{aligned} V_1(m(t), p(t)) =&\ m^T(t)m(t) + p^T(t)p(t), \\ V_2(m(t), p(t), t) =&\ 2\sum_{i=1}^{m} p_i^{(2)} \int_0^{p_i(t)} g_i(s)ds + \int_{t-\sigma(t)}^{t} g^T(p(v))R_1g(p(v))dv \\ &+ m^T(t)P_1m(t) + \int_{t-\tau(t)}^{t} m^T(s))R_2m(s))ds, \end{aligned}$$

for positive constant ϵ_1 and positive definite matrices R_1 and R_2. The positive constants ϵ_1 and positive definite matrices R_1 and R_2 will be determined later.

The derivative of V along trajectories of (2) is given by:

$$\dot{V}(t) = \epsilon_1\dot{V}_1(m(t), p(t)) + \dot{V}_2(m(t), p(t), t),$$

where $\dot{V}_1(m(t), p(t)) = 2m^T(t)[-Am(t) + Wg(p(t - \sigma(t)))] + 2p^T(t)[-Cp(t) + Dm(t - \tau(t))]$ and

$$\begin{aligned} \dot{V}_2(m(t), p(t), t) =&\ 2g^T(p(t))P_2\dot{p}(t) + 2m^T(t))P_1\dot{m}(t) + g^T(p(t))R_1g(p(t)) \\ &-(1 - \sigma'(t))g^T(p(t - \sigma(t)))R_1g(p(t - \sigma(t))) \\ &+m^T(t)R_2m(t) - (1 - \tau'(t))m^T(t - \tau(t)))R_2m(t - \tau(t))) \end{aligned}$$

Rewrite $\dot{V}_1$ as the following form.

$$\dot{V}_1(m(t),p(t)) \leq -2m^T Am - 2p^T(t)Bp(t) + 2m^T(t)A^{\frac{1}{2}}A^{-\frac{1}{2}}Wg(p(t-\sigma(t))) \\ +2p^T(t)C^{\frac{1}{2}}C^{-\frac{1}{2}}Dm(t-\tau(t)))$$

From Lemma 1, it follows that

$$\dot{V}_1(m(t),p(t)) \leq -m^T(t)Am(t) - p^T(t)Cp(t) + Mg^T(p(t-\sigma(t)))g(p(t-\sigma(t))) \\ +Mm^T(t-\tau(t))m(t-\tau(t))]$$

where $M = \max(\|W^T A^{-1} W\|_2, \|D^T C^{-1} D\|_2) \geq 0$.

Since $g(p_i(t))p_i(t) \geq k_i^{-1}(g(p_i(t)))^2$, we can get

$$-g^T(p(t))P_2Bp(t) \leq -g^T(p(t))P_2CK^{-1}g(p(t)).$$

Let the Cholesky factorization of Q_1 and Q_2 be $Q_1 = K_1^T K_1$ and $Q_2 = K_2^T K_2$. Rewriting $W = (W_1K_1^{-1})(K_1W_2)$ and $D = (D_1K_2^{-1})(K_2D_2)$, we have

$$\dot{V}_2(m(t),p(t),t) \leq -g^T(p(t))(2P_2Ck^{-1} - P_2D_1Q_2^{-1}D_1^TP_2 - R_1)g(p(t)) \\ -m^T(t)(2P_1A - P_1W_1Q_1^{-1}W_1^TP_1 - R_2)m(t) \\ -(1-\sigma)g^T(p(t-\sigma(t)))(R_1 - W_2^TQ_1W_2)g(p(t-\sigma(t))) \\ -(1-\sigma)m^T(t-\tau(t))(R_2 - D_2^TQ_2D_2)m(t-\tau(t)).$$

Since $\Omega_1 > 0$ and $\Omega_2 > 0$, there exists $\epsilon_2 > 0$ such that $\Omega_2 - 2\epsilon_2 I_n > 0$ and $\Omega_1 - 2\epsilon_2 I_n > 0$. Set $R_1 = W_2^TQ_1W_2 + \frac{\epsilon_2}{1-\sigma}I_n$, $R_2 = D_2^TQ_2D_2 + \frac{\epsilon_2}{1-\sigma}I_n$, which are symmetric positive definite matrices. It follows that

$$\dot{V}_2(m(t),p(t),t) \leq -\epsilon_2 g^T(p(t))g(p(t)) - \epsilon_2 m^T(t)m(t) \\ -\epsilon_2 g^T(p(t-\sigma(t)))g(p(t-\sigma(t))) - \epsilon_2 m^T(t-\tau(t))m(t-\tau(t)).$$

Choose $\epsilon_1 > 0$ such that $M\epsilon_1 \leq \epsilon_2$. Then $\dot{V}(t) \leq -\epsilon_1 m^T(t)Am(t) - \epsilon_1 p^T(t)Cp(t)$.

Let $a = \min\{a_1, a_2, \cdots, a_n, c_1, c_2, \cdots, c_n\}$, $k = \max\{k_1, k_2, \cdots, k_n\}$, $r = \max\{\|R_1\|_2, \|R_2\|_2\}$, and $p^* = \max\{p_1^{(1)}, p_2^{(1)}, \cdots, p_n^{(1)}, p_1^{(2)}, p_2^{(2)}, \cdots, p_n^{(2)}\}$. Choose $\epsilon > 0$ satisfying the following condition:

$$\epsilon\epsilon_1 + \epsilon p^* k - \epsilon_1 a + rk^2\epsilon\tau e^{\epsilon\tau} < 0. \tag{4}$$

We then have

$$\frac{d}{dt}(e^{\epsilon t}V(t)) \leq \epsilon e^{\epsilon t}V(t) - \epsilon_1 e^{\epsilon t}(m^T(t)Am(t) + p^T(t)Cp(t)).$$

Note that $2p_i^{(2)}\int_0^{p_i(t)} g_i(s)ds \leq 2p^*\int_0^{p_i(t)} k_i s ds \leq p^*kp_i^2(t)$, we have

$$\frac{d}{dt}(e^{\epsilon t}V(t)) \leq e^{\epsilon t}(\epsilon\epsilon_1 + \epsilon p^*k - \epsilon_1 a)(m^T(t)m(t) + p^T(t)p(t)) + \epsilon e^{\epsilon t} \\ \int_{t-\sigma(t)}^t g^T(p(v))R_1g(p(v))dv + \epsilon e^{\epsilon t}\int_{t-\tau(t)}^t m^T(v)R_2m(v)dv. \tag{5}$$

Integrating both sides of (5), we obtain

$$e^{\epsilon t}V(t)|_0^s \leq \int_0^s e^{\epsilon t}(\epsilon\epsilon_1 + \epsilon p^*k - \epsilon_1 a)(m^T(t)m(t) + p^T(t)p(t))dt + \epsilon\int_0^s [e^{\epsilon t} \\ \int_{t-\sigma(t)}^t g^T(p(v))R_1g(p(v))dv + e^{\epsilon t}\int_{t-\tau(t)}^t m^T(v)R_2m(v)dv]dt. \tag{6}$$

Estimating the second term and the third term on the right-hand side of (6) by changing the integrals and using $0 \leq \tau(t) \leq \tau, \sigma(t) \leq \tau$, we have

$$\begin{aligned} &\epsilon \int_0^s e^{\epsilon t} \int_{t-\tau(t)}^t g^T(p(v))R_1g(p(v))dvdt \leq \epsilon \int_{-\tau}^s \int_{\max\{v,0\}}^{\min\{v+\tau,s\}} e^{\epsilon t}dtg^T(p)R_1g(p)dv \\ &< rk^2\epsilon\tau e^{\epsilon\tau}(\int_{-\tau}^0 e^{\epsilon v}p^T(v)p(v)dv + \int_0^s e^{\epsilon v}p^T(v)p(v)dv) \end{aligned} \tag{7}$$

and

$$\epsilon \int_0^s e^{\epsilon t} \int_{t-\tau(t)}^t m^T(v)R_2m(v)dvdt < r\epsilon\tau e^{\epsilon\tau}(\int_{-\tau}^0 e^{\epsilon v}m^T mdv + \int_0^s e^{\epsilon v}m^T mdv) \tag{8}$$

Substituting (7) and (8) into (6) and using (4), we can obtain

$$e^{\epsilon t}V(t)|_0^s \leq rk^2\epsilon\tau e^{\epsilon\tau} \int_{-\tau}^0 e^{\epsilon v}(m^T m + p^T p)dv \equiv M_1\|\phi\|_2^2.$$

Therefore,

$$V(t) \leq (V(m(0), p(0), 0) + M_1\|\phi\|_2^2)e^{-\epsilon t}, \forall t > 0. \tag{9}$$

$$\begin{aligned} V(0) &= \epsilon_1(m^T(0)m(0) + p^T(0)p(0)) + 2\sum_{i=1}^{m} p_i^{(2)} \int_0^{p_i(0)} g(s)ds + m^T(0)P_1m(0) \\ &+ \int_{-\sigma(0)}^0 g^T(p(v))R_1g(p(v))dv + \int_{-\tau(0)}^0 m^T(v)R_2m(v)dv \\ &\leq (\epsilon_1 + p^*k + rk^2)\|\phi\|_2^2 \equiv M_2\|\phi\|_2^2. \end{aligned}$$

According to (3),(9) and the above inequality,$\forall t > 0$

$$\epsilon_1\|(m(t), p(t))^T\|_2^2 = \epsilon_1(m^T(t)m(t) + p^T(t)p(t) \leq V(t) \leq (M_1 + M_2)\|\phi\|_2^2 e^{-\epsilon t},$$

that is

$$\|(m(t), p(t))\|_2 \leq \sqrt{\frac{M_1 + M_2}{\epsilon_1}}\|\phi\|_2 e^{-\frac{\epsilon}{2}t}. \tag{10}$$

Inequality (10) implies the origin of system (2) is globally exponentially stable.

When W is nonsingular, choosing $W_1 = W, W_2 = I_n, D_1 = D, D_2 = I_n$, we have

Corollary 1. *Suppose that in systems (2), Assumptions A_1 are satisfied, W is nonsingular. The origin of neural system (2) is globally exponentially stable if there exists a positive constant γ such that:*

$$\Omega_1 = 2A - \frac{1}{\gamma}I_n - \gamma D^T D > 0; \Omega_2 = 2CK^{-1} - \frac{1}{\gamma}I_m - \gamma W^T W > 0.$$

3 Example

In this section, we present three examples to show the effectiveness and correctness of our theoretical results.

Example 1 ([5][6]). Assume that the network parameters of system (2) are given as follows:$A = B = I_5, f_i(x) = \frac{x^2}{1+x^2}, \tau(t) = \sigma(t) = 0.5 + 0.1\sin(t)$, and W as follows.

It is easy to know that the maximal value of the derivative of $f(x)$ is less than $k = 0.65$. Let $\gamma = 1$, the Ω_1 and Ω_2 in Corollary 1 becomes:

$$W = 0.5 \times \begin{bmatrix} 0 & -1 & 1 & 0 & 0 \\ -1 & 0 & 0 & 1 & 1 \\ 0 & 1 & 0 & 0 & 0 \\ 1 & -1 & 0 & 0 & 0 \\ 0 & 0 & 0 & 1 & 0 \end{bmatrix}, \Omega_1 = 0.36I, \Omega_2 = \begin{bmatrix} 1.57 & 0.25 & 0 & 0.25 & 0.25 \\ 0.25 & 1.32 & 0.25 & 0 & 0 \\ 0 & 0.25 & 1.82 & 0 & 0 \\ 0.25 & 0 & 0 & 1.57 & -0.25 \\ 0.25 & 0 & 0 & -0.25 & 1.82 \end{bmatrix}.$$

It is clear that Ω_1 and Ω_2 are positive definite. Therefore, according to Theorem 1, the origin of system (2) is globally exponentially stable. However, the origin of system (2)) is only globally asymptotically stable according to Theorem in [5] and [6]. On the other hand, our criteria are all independent of the magnitudes of delays, and so the delays under these conditions are harmless. Therefore, Theorem 1 is a generalization of Theorem in [5] and [6].

4 Conclusion

In this paper, by the technique of inequality of integral, the globally exponential stability criteria were derived. An Example implies that our results establish a new set of globally exponential stability criteria for GRN with time-varying delays.

References

1. Austin, D., Allen, M., McCollum, J., Dar, R., Wilgus, J., Sayler, G., Samatova, N., Cox, C., Simpson, M.: Gene network shaping of inherent noise spectra. Nature 439, 608–611 (2006)
2. Chen, L., Aihara, K.: Stability of genetic regulatory networks with time delay. Trans. CAS-I 49(5), 602–608 (2002)
3. de Jong, H.: Modeling and Simulation of Genetic Regulatory Systems: A Literature Review. Journal of Computational Biology 9(1), 67–103 (2002)
4. Khalil, H.K.: Nonlinear Systems. Macmilan, New York (1992)
5. Li, C., Chen, L., Aihara, K.: Stability of Genetic Networks With SUM Regulatory Logic: Luríe System and LMI Approach. IEEE Trans. CAS-I 53(11), 2451–2458 (2006)
6. Li, H., Yang, X.: Asymptotic Stability Analysis of Genetic Regulatory Networks with Time-varying Delay. In: 2010 Chinese Control and Decision Conference, pp. 566–571. IEEE Press, New York (2010)
7. Ren, F., Cao, J.: Asymptotic and robust stability of genetic regulatory networks with time-varying delays. Neurocomputing 71(4-6), 834–842 (2008)
8. Wang, Z., Gao, H., Cao, X., Liu, J.: On Delayed Genetic Regulatory Networks With Polytopic Uncertainties: Robust Stability Analysis. IEEE Trans. Manobio. 7(2), 154–163 (2008)
9. Wang, Z., Shu, H., Liu, Y., Ho, D., Liu, X.: Robust stability analysis of generalized neural networks with discrete and distributed time delays. Chaos, Solitons and Fractals 30(4), 886–896 (2006)
10. Zhou, L., Zhou, M.: Stability analyais of a class of generalized neural networks with delays. Physics Letters A 337, 203–215 (2005)

Diagnosis of Skin Diseases with Spectrum Analysis

Meng Xianjiang[1] and Wang Junjie[2]

[1] College of Communication science and engineering Shenyang Ligong University Shenyang, China
[2] College of physics Shenyang Ligong University Shenyang, China
amxj06@sohu.com

Abstract. A kind of chromatic system with self made fiber probe was designed to measure the spectrum to distinguish the skin normal or not in this paper. The testing skin's reflection spectrum was measured and the chromatic values were analyzed. Through comparing the reflection spectrum and chromatic aberration of the sick skin with the normal one, good results are achieved by the analyzing of the skin with this system, so a method to distinguish the pathological skin automatically and objectively is attained.

Keywords: Reflection spectrum;Pathological skin Diagnosis;chromatic tri-stimulating value.

1 Introduction

The spectrum analysis technology has lots of applications especially in skin diseases[1-3]. It is formal to diagnose the skin diseases through observing of the sick skin for its variation of the color, vein and state. None the less, for so many years people can only get the information with their eyes and experiences to decide the characteristics of the sick skin. In different area, the diseases are different and they can not be diagnosed with the same method.

A spectrum analysis and chromatic testing system with self-made fiber probe was devised. According the theory of the chromatics and spectrum analysis, and with the optical techniques and computers, still with the medical experiences, an automatic diagnosis system of the skin diseases was established. Through the measuring and analyzing of the general skin diseases, it is proved that the spectrum of the pathological skin over the normal one is different clearly. So the chromatic value is different, and with the different part of the body of the same diseases, the difference of the spectrum is different, so we can decided the type of the skin diseases.

2 Analysis Principle

There are many kinds of skin diseases in medical area[4-5]. The skin surface is different in color vein and smoothness according different state. Traditionally the doctors can get the first impression of the skin surface with their experiences of many

X. Wan (Ed.): Electrical Power Systems and Computers, LNEE 99, pp. 691–696.
springerlink.com

years oneself or maybe experiences from their precursors. The type and status of the skin diseases can be assured. But with this system we use the fiber sensor to get the color image of the pathological skin, comparing the difference of them with the normal, we can get the difference of them after computing.

It is proved that the skin diseases can induce the variation of the two parameters. One is the variation of the reflection spectrum of the pathological skin over the normal, and different diseases induce the variation of the spectrum area; the other is variation of the color. The chromatic aberration can be expressed with the brightness and chromatics.

2.1 The Ratio of the Relative Reflection Spectrum

The skin can be taken as a diffusing reflector. The ratio of the relative reflection spectrum is as followed.

$$\rho(\lambda) = \alpha(\lambda)\frac{\rho_T(\lambda)}{\rho_s(\lambda)} \tag{1}$$

Where $\rho_s(\lambda)$ is the reflection spectrum ratio of the standard white plate, $\rho_T(\lambda)$ is the reflection spectrum ratio of the testing skin, $f(\lambda)$ is an adjusting value, it can be decided by the reflection spectrum ratio of the standard white plate with more precious instruments.

We measure the reflection spectrum of the normal skin $\rho_n(\lambda)$ and of the pathological skin $\rho_{sb}(\lambda)$, the relative reflection spectrum ratio $\rho_r(\lambda)$ is

$$\rho_r(\lambda) = \frac{\rho_{sb}(\lambda)}{\rho_n(\lambda)} \tag{2}$$

It is obvious that the curve $\rho_r(\lambda)$ -- λ will be different with the different pathological skin.

2.2 Chromatic Coordinate and Chromatic Value

The color of the skin can be expressed by the XYZ chromatic system that is recommended by the Committee of the International Emission (CIE). It uses tri-stimulating value XYZ and corresponding chromatic coordinate x,y,z to express the color, and chromatic aberration value ΔE^*_{ab} is used to express the difference between two kinds of the color. The color of the skin is the result of the color induced by the invisible emission to the people's eyes, so the color can be expressed by XYZ, and the tri-stimulating value XYZ can be decided by the stimulated function $\varphi(\lambda)$ as followed

$$\begin{cases} X = K\int S(\lambda)\beta(\lambda)\bar{x}(\lambda)d\lambda \\ Y = K\int S(\lambda)\beta(\lambda)\bar{y}(\lambda)d\lambda \\ Z = K\int S(\lambda)\beta(\lambda)\bar{z}(\lambda)d\lambda \end{cases} \quad (3)$$

$\bar{X}(\lambda)$, $\bar{Y}(\lambda)$, $\bar{Z}(\lambda)$ are the tri-stimulating value of the standard observers, The data are form CIE1931; $\Delta\lambda$ is the wavelength interval, here $\Delta\lambda = 5nm$,K is the chromatic adjusting factor.

To the skin testing, the chromatic stimulating function $\varphi(\lambda)$ is the function of light source relative spectrum $S(\lambda)$, it is as follows

$$\varphi(\lambda) = S(\lambda)\rho(\lambda) \quad (4)$$

$\rho(\lambda)$ is the skin reflection spectrum, it is from formula (1), the light source is bromine-tungsten lamp, so the $S(\lambda)$ is definite.

The CIE chromatic coordinate value x,y,z can get from the tri-stimulating value, the formula as followed.

$$x = \frac{X}{X+Y+Z}, \quad y = \frac{Y}{X+Y+Z}, \quad z = \frac{Z}{X+Y+Z} \quad (5)$$

Additionally, CIE1976 gave a recommended (L^*, a^*, b^*) homogeneous chromatic space, it avoids the inhomogeneous chromatic space, the characteristics of the color is expressed with lightness L^* and chromatic values a^*, b^*.

$$\begin{aligned} L^* &= 116\left(\frac{Y}{Y_n}\right)^{1/3} - 16 \\ a^* &= 500\left[\left(\frac{X}{X_n}\right)^{1/3} - \left(\frac{Y}{Y_n}\right)^{1/3}\right] \\ b^* &= 200\left[\left(\frac{Y}{Y_n}\right)^{1/3} - \left(\frac{Z}{Z_n}\right)^{1/3}\right] \end{aligned} \quad (6)$$

X_n, Y_n, Z_n are the tri-stimulating value of the standard white plate, X,Y,Z are the he tri-stimulating value of the skin, in the system, the chromatic aberration can get from the followed

$$\Delta E_{ab}^* = [(\Delta L^*)^2 + (\Delta a^*)^2 + (\Delta b^*)^2]^{1/2} \quad (7)$$

ΔL^*, Δa^*, Δb^* is the chromatic aberration of the pathological skin over the normal.

As the boundary of the pathological skin and the normal is blur from the color, so the distinguishing of them can use the fuzzy mathematics method to distinguish in types, only do like this can we get the boundary of the pathological skin. In the experiment we get 50 samples of the skin, measured the spectrum, get the lightness and chromatics. So we got the mathematical expectation of the skin $\overline{L^*}$、$\overline{a^*}$、$\overline{b^*}$, mean square deviation ΔL^*, Δa^*, Δb^* and the maximum and minimum of the L^*, a^*, b^* are L^*_I, L^*_u, a^*_I, a^*_u, b^*_I, b^*_u ,They are as followed

$$
\begin{aligned}
\overline{L^*}(\overline{a^*},\overline{b^*}) &= \frac{\sum_{i=1}^{N} L^*(a^*,b^*)}{N} \\
\Delta L^*(a^*,b^*) &= \frac{\sqrt{\sum_{i=1}^{N}[L^*_i(a^*_i,b^*_i)-\overline{L^*}(\overline{a^*},\overline{b^*})]^2}}{N} \\
L^*_I(a^*_I,b^*_I) &= \overline{L^*}(\overline{a^*},\overline{b^*})-\alpha\Delta L^*(\Delta a^*,\Delta b^*) \\
L^*_u(a^*_u,b^*_u) &= \overline{L^*}(\overline{a^*},\overline{b^*})-\alpha\Delta L^*(\Delta a^*,\Delta b^*)
\end{aligned}
\tag{8}
$$

α is an adjustable parameter, it can vary from 0.5~1.5, through adjusting α, we can get the upper and lower boundary. The doctors can collect the different samples according the type of the skin diseases, from the samples, the doctors can get the α they want, and get the mathematical expectation and mean square deviation, and the boundary is the skin parameters.

Obviously it has a big chromatic aberration ΔE^*_{ab} of the pathological skin over the normal, and with the different pathological skin, the brightness and chromatic value of the skin will varied in the special boundary.

3 The Measuring System and Its Operational Principle

The framework chart for the testing system is shown in Fig.1[6]. The light from the light source irradiates into the photomultiplier. In the photomultiplier, the optical grating is driven by a step motor, so the light of different wavelength is diffracted, the output light is converted into electric signal, amplified, A/D, and put into the computer for manipulating. The final output is the intensity value according the wavelength, and we get the reflection spectrum by normalizing the intensity value. The skin spectrum was gained through the photomultiplier with the self-made optical fiber probe. Every sample at an interval of 5nm from 380nm to 780nm, so 81 samples all together.

In the system, a fiber probe is added to decrease the interference of the outside world and focus the light beam in order to increase the sensitivity of the system, and it also can improve the flexibility in measuring.

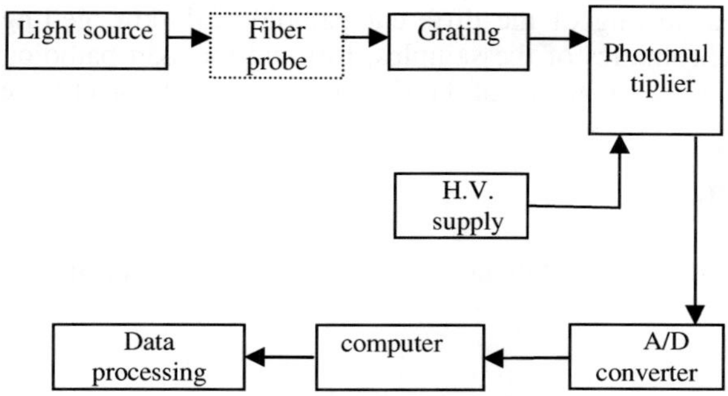

Fig. 1. The framework chart for the test system

4 Experiment Analysis

With the system, the experiment to analyze the pathological skin was accomplished. The front left arm (normal) and the right cheek were tested, the spectrum of the normal part and the pathological part is shown in Fig1. The corresponding coordinate and chromatic aberration is as followed.

Normal part:

X=25.386 Y=25.102 Z=19.569

x=0.363 y=0.358 z=0.279

pathological part:

X=9.675 Y=8.617 Z=8.284

x=0.364 y=0.324 z=0.312

chromatic aberration:

$\Delta E^* = 28.447$

The corresponding coordinate and chromatic aberration is as followed.

Normal part:

Normal part 1:

X=25.759 Y=25.628 Z=20.196

x=0.360 y=0.358 z=0.282

Normal part 1:

X=26.875 Y=26.823 Z=21.239 74.937

x=0.359 y=0.358 z=0.283

chromatic aberration:

$\Delta E^* = 1.986$

The pathological skin of another part was tested, the chromatic aberration ΔE^* =12.571

5 Conclusion

It can be seen from the curves that the spectrum of the pathological skin and the normal skin are different, and the spectrum is different to the different skin diseases. So through the medical testing of the different pathological skin, getting the standard spectrum and the chromatics of the samples, forming the skin pathological database, the doctors can make the diagnosis of the skin diseases more objective and quantizing.

Acknowlegement

This work is supported by the Science and Technology Bureau of Shenyang City of China.

References

1. Fasano, A., Catassi, C.: Current Approaches to Diagnosis and Treatment of Celiac Disease: An Evolving Spectrum. International Journal of Dermatology 42(4), 287–289 (2008)
2. Wolff, K., Stuetz, A.: Pimecrolimus for the treatment of inflammatory skin disease, vol. 5(3), pp. 643–655 (2004), http://informahealthcare.com/doi/abs/10.1517/
3. Antonsson, A., Erfurt, C., Hazard, K., et al.: Prevalence and type spectrum of human papillomaviruses in healthy skin samples collected in three continents. The new England Journal of Medicine 354, 256–263 (2006)
4. Jamison, R.E., Mendelsohn, E.: On the chromatic spectrum of acyclic decompositions of graphs. Journal of Graph Theory 56(2), 83–104 (2007)
5. Angelopoulo, E., Molana, R.: Daniilidis, Multispectral skin color modeling, vol. 2(15), pp. 635–642 (2003)
6. Meng, X., et al.: Study on the Method to Adjust Optical Fiber's Spectrum Attenuation with the Fourier Transform. In: Spectroscopy and Spectral Analysis, China, vol. 25(4), pp. 544–547 (2005)

Fixed-Frequency Quasi-Sliding Mode Controller for Single-Inductor-Dual-Output Buck Converter in Pseudo-Continuous Conduction Mode

Qing Liu, Xiaobo Wu*, and Liang Yin

Institute of VLSI design
Zhejiang University, Hangzhou, China
{liuqing,wuxb,yinlinag}@vlsi.zju.edu.cn

Abstract. Based on the quasi-sliding mode theory, a novel fixed-frequency nonlinear controller for single-inductor dual-output (SIDO) buck converter in pseudo-continuous conduction mode (PCCM) is presented in this paper. The main advantage of this nonlinear controller is that both small and large signal variations around the operating point are taken into account, which results in better transient response. To validate the feasibility of the scheme, a prototype of a SIDO buck converter with two outputs of 1.8 V and 3.3 V is developed in MATLAB SIMULINK. Simulation results show that the transient response of the SIDO buck converter with quasi-sliding mode controller is only 75 μs, contrasting to 900 μs with voltage feedback controller and 550 μs with peak current controller, when the load current of one channel changes from 100 mA to 10 mA.

Keywords: DC-DC converter, fast-response, sliding mode control, single-inductor dual-output, pseudo-continuous conduction mode.

1 Introduction

In many applications several regulated power voltages are required [1-3]. Conventionally, to generate N voltages, it is needed to use N switching converters with N inductors or transformer-based N-output DC-DC converters, which spend too many electronic components and increase the cost and volumes that is usually limited especially in portable and handheld consumer electronics [4].Thus, the single-inductor multi-output (SIMO) DC-DC converter was proposed for providing multiple output voltages by using a single off-chip inductor [5-8].

Prompted by SIMO DC-DC converter mentioned above, this paper proposed a novel nonlinear controller for PCCM SIDO converter based on sliding mode control method. The sliding mode (SM) controller was introduced for controlling variable structure system [9-12] to attain high stability and robustness against its parameter, line and load uncertainties. However, the ideal SM controller operates at infinite switching frequency

* Corresponding author.

X. Wan (Ed.): Electrical Power Systems and Computers, LNEE 99, pp. 697–704.
springerlink.com

resulting in excessive power loss and electromagnetic interference (EMI) issues [13]. Therefore, a PWM-based quasi-sliding-mode voltage controller was proposed in [14-15] to constrict the switching frequency of SM controller. Nevertheless, it has large steady state error which results from the finite switch frequency. This paper puts forward a way to reduce the steady state error and extend this theory to a converter working at pseudo-continuous conduction mode to meet the demands of SIDO buck converter.

Section 2 reviews the theoretical analysis of PCCM buck converter. In Section 3, a novel quasi-sliding controller for PCCM buck converter is presented. Furthermore, the schematic diagram of PCCM SIDO converter with quasi-sliding mode controller is introduced in Section 4. In Section 5, the simulation results under three different control methods of PCCM SIDO converter are compared. Finally, Section 6 summarized the proposed designs.

2 Analysis of PCCM Buck Converter

Being different from the current in DCM, the inductor current of the converters, in PCCM, could stay above zero as a CCM converter does to reduce inductor current ripple. When the inductor current reaches a predefined freewheeling current, the switch S3 will close (shown in Fig. 1). Meanwhile, the switch S4 will open and the inductor current stays above zero as a constant and capacitor discharges to the load. In this case the freewheeling duration can be used in SIDO converters to alleviate cross-regulation effects.

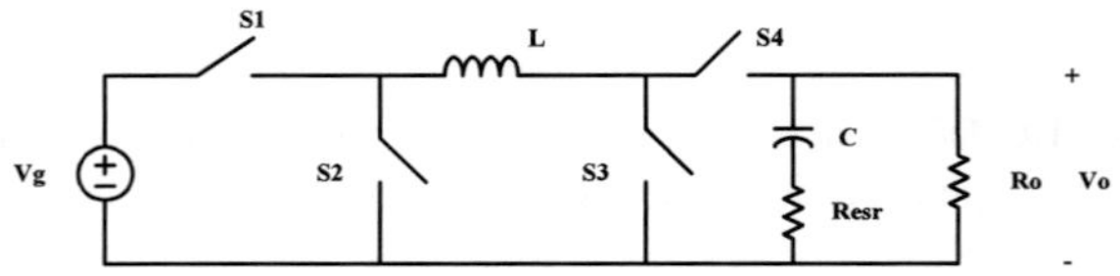

Fig. 1. Schematic diagram of PCCM buck converter

By using a well known circuit averaging technique, the equivalent circuit of PCCM buck converter can be derived and the small signal transfer function can be developed. According to the small signal model of PCCM buck converter, a voltage feedback controller can be obtained to attain adequate phase margin and good rejection of expected disturbances.

3 Quasi-Sliding Mode Controller

The basic principle of SM control is to establish a certain sliding surface S in the state space so that the trajectory of the state variable will direct toward a desired working point. In order to make the SM-controlled converters operate at a constant switching frequency for all operating conditions, a PWM based quasi-sliding-mode controller is proposed based on two key results [14-15] as follows: In the SM control, the control input u (indicates the state of the converter's power switch) can be replaced by a equivalent control signal u_{eq}, which can be derived by setting the time differentiation of sliding surface S equal to zero; At infinite switching frequency, the equivalent control signal u_{eq} is effectively equal to a duty-cycle control signal, $d=u_{eq}$.

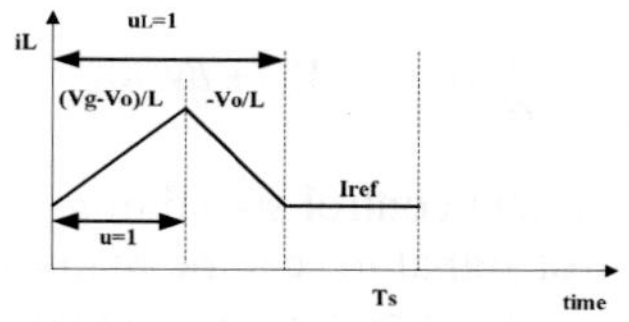

Fig. 2. Logical state of power switch S1 and S3

In PCCM buck converter, the state variables can be chosen as voltage error and its integration and differentiation as shown in Equation 1:

$$\mathrm{x} = \begin{bmatrix} x_1 \\ x_2 \\ x_3 \end{bmatrix} = \begin{bmatrix} V_{ref} - \beta V_o \\ \dfrac{d(V_{ref} - \beta V_o)}{dt} \\ \int (V_{ref} - \beta V_o) dt \end{bmatrix} . \tag{1}$$

where β denotes the feedback network ratio. The logical states of power switch S1 and S3 are defined as u and u_L respectively (shown in Fig. 2).

Substitution of the buck converter's behavioral model under PCCM into (1) produces the following state equation:

$$\begin{bmatrix} \dot{x}_1 \\ \dot{x}_2 \\ \dot{x}_3 \end{bmatrix} = \begin{bmatrix} 0 & 1 & 0 \\ 0 & -\dfrac{1}{R_o C} & 0 \\ 1 & 0 & 0 \end{bmatrix} \begin{bmatrix} x_1 \\ x_2 \\ x_3 \end{bmatrix} + \begin{bmatrix} 0 \\ -\dfrac{\beta V_g}{LC} \\ 0 \end{bmatrix} u + \begin{bmatrix} 0 \\ \dfrac{\beta V_o}{LC} u_L \\ 0 \end{bmatrix} . \tag{2}$$

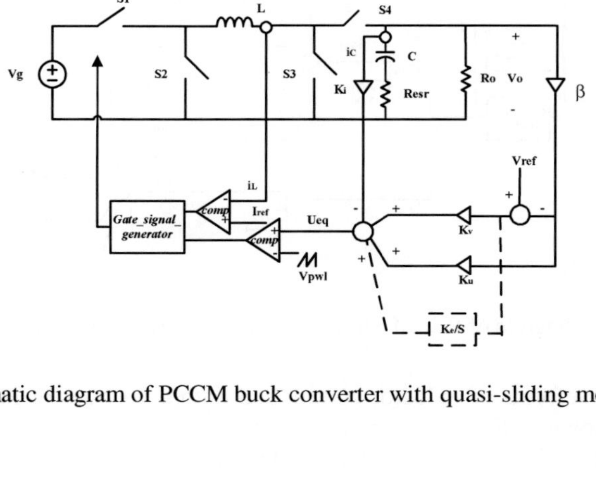

Fig. 3. Schematic diagram of PCCM buck converter with quasi-sliding mode controller

The sliding surface is defined as the linear combination of these three state variables

$$S = \alpha_1 x_1 + \alpha_2 x_2 + \alpha_3 x_3 \cdot$$

The equivalent control signal u_{eq} can be derived by setting the time differentiation of sliding surface S equal to zero

$$\beta V_g u_{eq} = -\beta L\left(\frac{\alpha_1}{\alpha_2} - \frac{1}{R_o C}\right) i_C + LC \frac{\alpha_3}{\alpha_2}(V_{ref} - \beta V_o) + \beta V_o u_{Leq} = -k_i i_C + k_v (V_{ref} - \beta V_o) + k_u \beta V_o \cdot$$

where u_{Leq} stands for the equivalent control signal of u_L.

So far, the equivalent control signal u_{eq} can be mapped onto the instantaneous duty cycle function d of the pulse-width modulation converter. However, this equalization is based on the assumption that the converter has infinite switch frequency. In the reality a fixed switch frequency is chosen which will introduce some steady error in the converter. In this paper, a linear compensator (error integration element) is added to reduce this steady state error (seen the dashed part in Fig. 3).

4 PCCM SIDO Buck Converters

Fig. 4 illustrates the basic architecture of a PCCM SIDO buck converter with quasi-sliding mode controller. It can be seen that a power switch S5 is added to attain two different output voltages, and the two quasi-sliding mode controllers are designed independently, which are synchronized and controlled by the phase-generator.

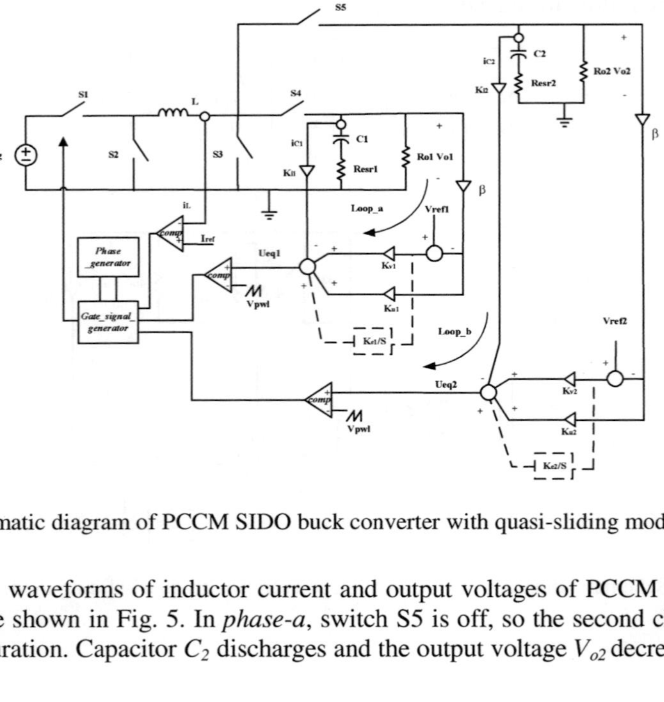

Fig. 4. Schematic diagram of PCCM SIDO buck converter with quasi-sliding mode controller

The basic waveforms of inductor current and output voltages of PCCM SIDO buck converter are shown in Fig. 5. In *phase-a*, switch S5 is off, so the second channel is in freewheel duration. Capacitor C_2 discharges and the output voltage V_{o2} decreases. At the meantime, Switch S4 is on and *loop-a* is enabled, so the first channel works as a normal PCCM buck converter, and vise versa. Because of the freewheel duration isolates these two output channels, the quasi-sliding mode controllers can be designed separately. Furthermore, this PCCM SIDO buck converter has better transient response in that the quasi-sliding mode controller preserves the large signal information of the converter.

5 Simulation Results

In this paper, three kinds of PCCM SIDO buck converters are simulated in MATLAB SIMULINK: one with voltage feedback controller, the other one with peak current controller and the third one with the novel quasi-sliding mode controller. The specifications of the PCCM SIDO buck converter are given in Table 1.

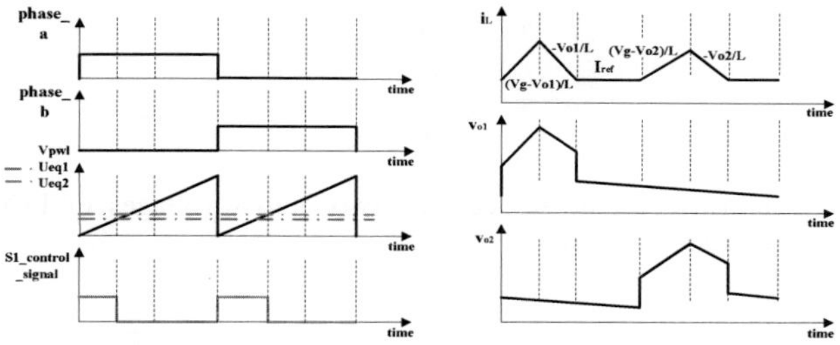

Fig. 5. Signal waveform of S1 control signal, inductor current, output voltage of two different sub-converters of PCCM SIDO buck converter

Table 1. Specifications of SIDO buck converter

Parameter Name	Value
V_g	5 V
L	2 μH
f_s	1 MHz
I_{ref}	100 mA
V_{o1}	1.8 V
V_{o2}	3.3 V
I_{o1}, I_{o2}	100 mA
C_1, C_2	10 μF
R_{esr1}, R_{esr2}	20 m Ω

Fig. 6 shows simulation waveforms of PCCM SIDO buck converter with three different controllers. The steady-state waveforms of this PCCM SIDO buck converter match the theoretically analysis. And two output voltages, 1.8 V and 3.3 V, are regulated in two different channels.

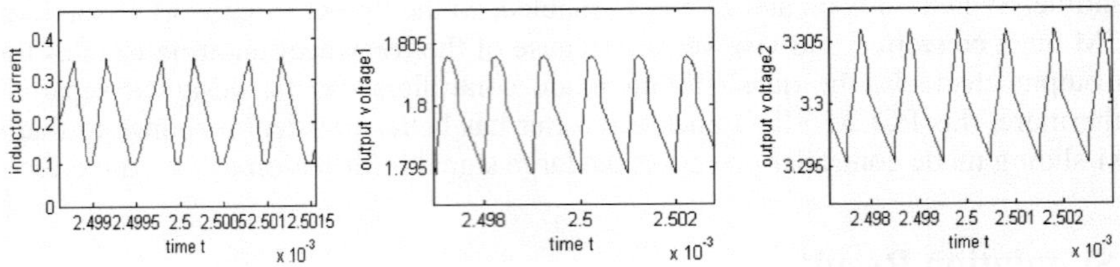

(a) Simulation waveforms of inductor current, two output voltages of PCCM SIDO buck converter with voltage feedback controller

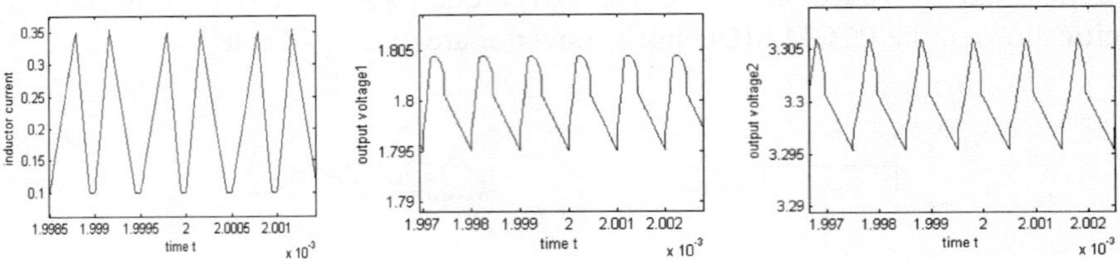

(b) Simulation waveforms of inductor current, two output voltages of PCCM SIDO buck converter with peak current controller

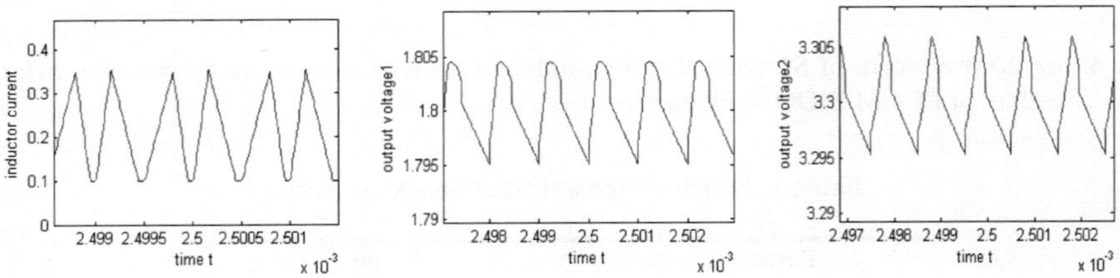

(c) Simulation waveforms of inductor current, two output voltages of PCCM SIDO buck converter with quasi-sliding mode controller

Fig. 6. Simulation waveforms of PCCM SIDO buck converters

In order to simulate the transient speed of this PCCM SIDO buck converter, a load variation is added to the first channel. And the simulation results are shown in Fig. 7. These waveforms in Fig. 7 (a) and (b) indicate that the PCCM SIDO buck converter with voltage feedback controller and peak current controller has few cross-regulation effects. However, it demands 900 μs and 550 μs respectively to reach the steady state operating point when the load changes from 100 mA to 10 mA. On the contrary, the PCCM SIDO buck converter with quasi-sliding mode controller just needs 75 μs to return to the steady-sate operating point.

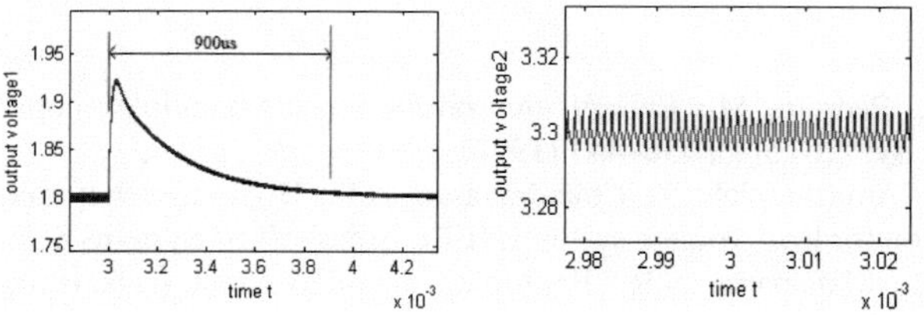

(a) Two output voltages of PCCM SIDO buck converter with voltage feedback controller

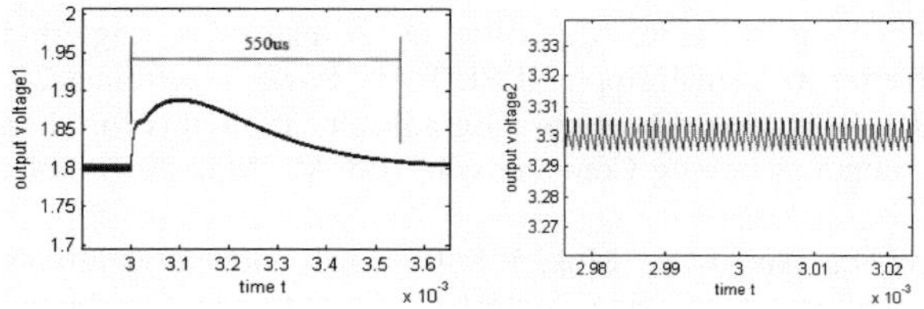

(b) Two output voltages of PCCM SIDO buck converter with peak current controller

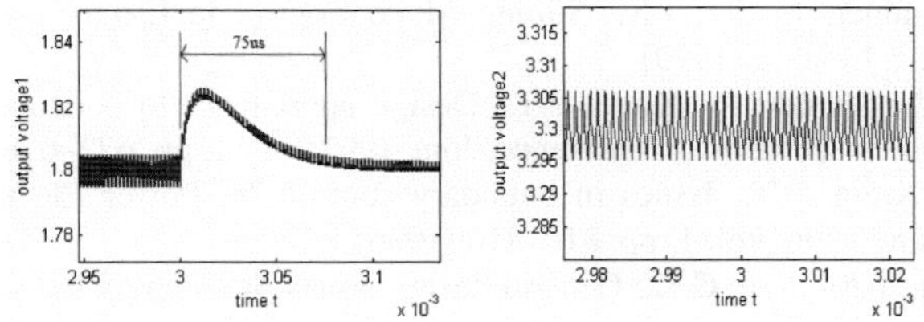

(c) Two output voltages of PCCM SIDO buck converter with quasi-sliding mode controller

Fig. 7. Transient waveforms of PCCM SIDO buck converters when the load1 changes from 100mA to 10mA

6 Conclusion

A novel fixed-frequency quasi-sliding mode controller is presented in this paper. And a new design method of the quasi-sliding mode controller for the PCCM SIDO buck converter is proposed. By using this nonlinear controller, the PCCM SIDO buck converter can achieve faster transient response than the one with voltage feedback controller or peak current controller. The simulation results indicate that the response of the converter is well in accord with the theoretical analysis.

Acknowledgments. This paper is sponsored by the National Natural Science Foundation of China under grant No. 60906012. It also gains support from the Analog Devices, Inc. (ADI). The authors would like to thank Mr. Bill Liu, the senior engineers of ADI and his colleagues, for their useful discussions and instruction.

References

1. Chang, J.M., Pedram, M.: Energy minimization using multiple supply voltages. IEEE Trans. VLSI Systems 5(4), 436–443 (1997)
2. Dancy, A.P., Amirtharajah, R., Chandrakasan, A.P.: High efficiency multiple-output DC-DC conversion for low- voltage systems. IEEE Trans. VLSI Systems 8(3), 252–263 (2000)
3. Gandhi, K.R., Mahapatra, N.R.: Exploiting data-dependent slack using dynamic multi-VDD to minimize energy consumption in datapath circuits. Design, Automation and Test in Europe 1, 1–6 (2006)
4. Lin, R.L., Pan, C.R., Liu, K.H.: Family of single-Inductor multi-output DC-DC converters. Power Electronics and Drive Systems 1, 1216–1221 (2009)
5. Xu, W., Li, Y., Gong, X., Hong, Z., Killat, D.: A dual-mode single-inductor dual-output switching converter with small ripple. IEEE Trans. Power Electronics 25, 614–623 (2010)
6. Kwan, K.T., Ki, W.H.: Freewheel duration adjustment circuits for charge-control single-inductor dual-output switching Converters. In: ISCAS, May 2010, vol. 1, pp. 2722–2725 (2010)
7. Ma, D., Ki, W.H., Tsui, C.Y., Mok, P.K.T.: Single-inductor multiple-output switching converters with time-multiplexing control in discontinuous conduction mode. IEEE J. Solid-State Circuit 38, 89–100 (2003)
8. Ma, D., Ki, W.H., Tsui, C.Y.: A pseudo-CCM/DCM SIMO switching converter with freewheel switching. IEEE J. Solid-State Circuits 38, 1007–1014 (2003)
9. Utkin, V., Guldner, J., Shi, J.X.: Sliding Mode Control in Electro-mechanical System. Taylor&Francis, London (1999)
10. Greuel, M., Muyshondt, R., Krein, P.T.: Design approaches to boundary controllers. In: Power Electronics Specialists Conference, June 1997, vol. 1, pp. 672–678 (1997)
11. Munzert, R., Krein, P.T.: Issues in boundary control. In: Power Electronics Specialists Conference, June 1996, vol. 1, pp. 810–816 (1996)
12. Tan, S.C., Lai, Y.M., Tse, C.K.: General design issues of sliding-mode controllers in DC-DC converters. IEEE Trans. Industrial Electronics 55, 1160–1174 (2008)
13. Tan, S.C.: A Unified Approach to the Design of PWM-Based Sliding –Mod Voltage Controllers for Basic DC-DC Converters in Continuous Conduction Mode. IEEE Trans. Circuits and Systems 53, 1816–1827 (2006)
14. Tan, S.C., Lai, Y.M., Tse, C.K.: A fixed-frequency pulsewidth modulation based quasi-sliding-mode controller for buck converters. IEEE Trans. Power Electronics 20, 1379–1392 (2005)
15. Tan, S.C., Lai, Y.M., Cheung, M.K.H., Tse, C.K.: On the practical design of a sliding mode voltage controlled buck converter. IEEE Trans. Power Electronics 20, 425–437 (2005)

An Optimized Low-Power and Low-Complexity Interpolation Filter for Delta-Sigma DAC

Tong Wu, Xiaobo Wu*, Menglian Zhao, and Jinchen Zhao

Institute of VLSI Design, Zhejiang University,
310027 Hangzhou, P. R. China
{wutong,wuxb,zhaoml,zhaojc}@vlsi.zju.edu.cn

Abstract. To reduce the power consumption and die area of interpolation filters, which usually determine the hardware cost of Delta-Sigma DAC systems, an improved common subexpression elimination (CSE) method is proposed in this paper. Furthermore, an improved comb filter with an optimized sharpening technique is put forward, which effectively provides sufficient sideband suppression and passband droop compensation. Using a TSMC 0.35μm Logic 1P4M process, the synthesis results show remarkable reduction in both power consumption and silicon area. Compared with several other FIR deign methods, the adder-cost is reduced significantly while the logic-depth is kept the same with the NR-SCSE, which is consistent with expectations.

Keywords: interpolation filter, FIR filter, CSE, comb filter, sharpening technique, passband droop.

1 Introduction

As is well known, Delta-Sigma digital-to-analog converters (Δ-Σ DAC) have been found extensive applications in audio/video processing systems, especially in portable battery-powered equipments which are getting extremely popular in recent years. A typical architecture of Delta-Sigma DAC is composed of an interpolation filter, a Delta-Sigma modulator, a hybrid-DAC and a reconstruction filter. Among them, the interpolation filter plays a vital role in the system, which executes the functions of signal oversampling as well as data preparation for the modulator. A common structure of interpolation filter consists of three cascaded halfband filters plus one sample/hold stage. However, this structure has long been suffered from the insufficient sideband suppression in the spectrum. An alternative choice is to employ two cascaded halfband filters plus one cascaded integrator-comb (CIC) filter, but still, the performance of sideband attenuation is suboptimal at a high expense of hardware. Besides, due to its relatively high hardware cost and logical complexity (especially the first stage halfband filter), interpolation filter significantly affects the performances of the whole system, such as power consumption, chip area, and etc.

* Corresponding author.

X. Wan (Ed.): Electrical Power Systems and Computers, LNEE 99, pp. 705–712.
springerlink.com

The adder-cost and the logic-depth are the two most important metrics that evaluate the complexity of FIR filters. The method of canonical signed digit (CSD) representation [1] allows encoding a binary number such that it contains the minimum possible number of non-zero bits. A computation reduction technique called nonrecursive signed common subexpression elimination (NR-SCSE) is put forward in [2] to simultaneously reduce the adder-cost and the logic-depth. In [3], the conventional common subexpression (CS) is extended to contain more than two non-zero bits, which is proven effective when the precision of the CSD code is extended. In [4], vertical common subexpression (VCS), as the counterpart of traditional horizontal common subexpression (HCS), is introduced to help the reduction of hardware cost.

Compared with comb filter, the sample/hold mechanism inherently lacks the capability of attenuating the stopband components of the data stream. It is well known that the comb-based structure will introduce a passband droop which may deteriorate the high-frequency performance. To fix this problem, various techniques have been invented and employed, among which the filter sharpening approach, first proposed by Kaiser and Hamming in [5], has gained favorable attentions and was introduced to the context of comb filter design in [6]. Extensive research of sharpening the comb filter was done in [7][8][9] which expands the design space significantly.

In this paper, the complete design and implementation of an interpolation filter is presented. The paper is structured as follows. In Section 2, the system specification is described. Section 3 analyzes both the previous methods and the proposed CSE method in detail. In Section 4, a novel comb filter with an optimized sharpening technique is presented. Finally, in the fifth section, some conclusions are given.

2 System Specifications

In Fig. 1, halfband filters oversample the digital input and raise the sampling rate by 2X at a time, and the rest 16X oversampling task is left to the comb filter. The passband cutoff frequency of the halfband filters is 18kHz instead of 20kHz, which can significantly reduces the number of filter taps by widening the transition band, while the loss of performance is acceptable in ordinary digital audio applications.

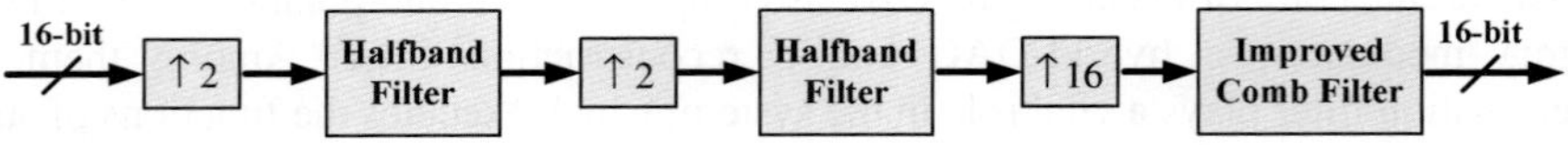

Fig. 1. The structure of the proposed interpolation filter.

Table 1. Specifications of the two halfband filters.

	Order	Passband Ripple	Stopband Attenuation
1st HB Filter	46	< 0.002 dB	> 75 dB
2nd HB Filter	14	< 0.0004 dB	> 85 dB

3 The Halfband Filters Design

All the filter coefficients are represented in CSD as shown in Table 2.

Table 2. CSD representation of partial first halfband filter coefficients.

Coefficient	Value	CSD	Relative Error (%)
C(1), C(47)	-0.00031071309037	$-2^{-12}-2^{-14}$	-1.782129333
C(3), C(45)	0.00078642716675	$2^{-10}-2^{-12}+2^{-14}-2^{-17}$	-0.076234655
C(5), C(43)	-0.0017093731612	$-2^{-9}+2^{-12}$	-0.022744373
C(7), C(41)	0.0032591548977	$2^{-8}-2^{-11}-2^{-13}-2^{-15}-2^{-17}$	-0.043061253
C(9), C(39)	-0.0056948786350	$-2^{-7}+2^{-9}+2^{-12}-2^{-14}-2^{-16}$	-0.058830310
C(11), C(37)	0.0093591951557	$2^{-7}+2^{-9}-2^{-11}+2^{-14}+2^{-16}$	-0.059379683
C(13), C(35)	-0.014747944987	$-2^{-6}+2^{-10}-2^{-13}+2^{-15}-2^{-17}$	-0.002206130
C(15), C(33)	0.022693520495	$2^{-5}-2^{-7}-2^{-10}+2^{-12}-2^{-16}$	-0.016309322
C(17), C(31)	-0.034886308995	$-2^{-5}-2^{-8}+2^{-12}+2^{-15}$	-0.013521633
C(19), C(29)	0.055679777213	$2^{-4}-2^{-7}+2^{-10}+2^{-16}$	-0.000818832
C(21), C(27)	-0.10113668206	$-2^{-3}+2^{-5}-2^{-7}+2^{-11}-2^{-14}$	-0.001412103
C(23), C(25)	0.31662384875	$2^{-2}+2^{-4}+2^{-8}+2^{-12}-2^{-15}$	-0.001255655
C(24)	0.5	2^{-1}	0

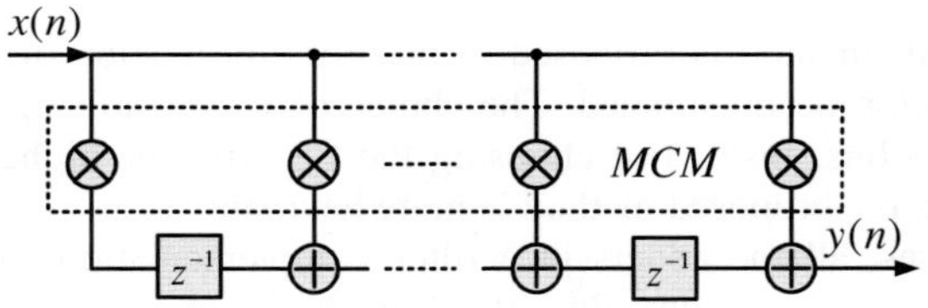

Fig. 2. Transposed-form of FIR filters.

The halfband filter could be configured in transposed-form, shown in Fig. 2, rather than direct-form, wherein one input is multiplied by all the coefficients simultaneously. This kind of configuration of multiple constant multiplication (MCM) is a transformation closely related to the widely used substitution of multiplications with constants by shifts and additions. CSE tackles the MCM problem by minimizing the number of additions/subtractions through extracting common bit patterns among the coefficients represented in CSD [3].

3.1 Direct Synthesis Using CSD Method

In the absence of application of any CSE techniques, the CSD method features its straightforward implementation, which could facilitate the whole design process. But the waste of hardware resources is extremely serious.

3.2 The NR-SCSE Algorithm

The NR-SCSE algorithm avoids high logic-depth by extracting the CS with the highest occurrence frequency according to the subexpression model matrix (SMM) associated with the given coefficient set. The NR-SCSE scans for the CSs with two nonzero bits only. For the first stage halfband filter, the SMM is shown in (1).

$$SMM = \begin{pmatrix} 7 & 4 & 6 & 3 & 5 & 6 & 1 & 3 \\ 8 & 8 & 6 & 6 & 0 & 7 & 1 & 4 \end{pmatrix}. \tag{1}$$

In the matrix above, each element indicates the occurrence number of a specific CS in the coefficient set. The CSs in the first row have two same-signed bits while the situation in the second row is otherwise. The column index represents the number of zeros between two bits. Next, the CS with the highest value in the matrix is selected and this chosen subexpression is eliminated from the coefficient set. And then, all the coefficients need to be scanned again to reconstruct the SMM and a new turn selection starts. Subexpressions belonging to different coefficients are not shared which leads to independent structures that reduce logic-depth and increase the potential operation frequency.

But the NR-SCSE algorithm is likely to undermine the possibility that one coefficient might be occupied by several short CSs completely, which may aggravate the layout complexity and the hardware cost.

3.3 The Improved CSE Method

An improved CSE method is proposed to make the utilization of CS more efficient and the optimization be maximized. The choice of CSs is still limited to those with only two nonzero bits. Instead of choosing the CS with the highest occurrence, the selection of CS in the proposed method is more heuristic.

In general cases, all the adders in a filter's implementation can be divided into three parts which are coefficient adder (CA), tap adder (TA) and subexpression adder (SA). Suppose there is a simplified FIR filter with its transfer function shown in (2),

$$y(n) = h(0)x(2) + h(1)x(1) + h(2)x(0). \tag{2}$$

Thus the number of TA equals the number of partial products minus one. Since TAs are inevitable and cannot be optimized, they are excluded in the calculation of adder-cost. Meanwhile, we can assume that the coefficients are implemented by three CSs in such a way

$$h(0) = x_1 + x_2 \gg 2, \quad h(1) = x_2 + x_3 \gg 3, \quad h(2) = x_3 + x_1 \gg 4. \tag{3}$$

So to realize the coefficients, three CAs are consumed in total. Because only CSs with two nonzero bits are used, the maximum value of CA to synthesize each coefficient (MCA) is calculated as follows

$$MCA(i) = \begin{cases} [n/2] & n \text{ is odd}; \\ n/2 - 1 & n \text{ is even}. \end{cases} \tag{4}$$

where n indicates the number of nonzero bits in one coefficient. At last, the number of SA equals the number of all the CSs used in the design.

From the discussion above, it could be concluded that the number of CA is almost determined solely by the nonzero bits in coefficients. The adder-cost incurred by CA is fixed which leaves little room for optimization. Thus the minimization of adder-cost is transformed into searching for an appropriate set of CS with lower SA.

To clarify the algorithm, the concepts of subexpression pool (SP) and bit distance (DA) are introduced. SP contains the CSs which have been selected to construct coefficients while DA is the number of zeros between two nonzero bits in one CS. The proposed algorithm is based on several principles shown as below:

Step 1. Before processing the coefficient set, the CSs 101 and 10n are selected and put into the SP in which n here represents -1. This is because these two CSs are the most commonly used ones in filter implementation.

Step 2. The algorithm processes the CSD coefficient array line by line. The CSs with shorter DA are always chosen from the SP with higher priority to construct the coefficient as completely as possible since the DA is analogous to the word length, which is also a metric evaluating the hardware consumption.

Step 3. If the current CSs in the SP cannot realize the coefficient, a new CS with the shortest possible DA which is capable of satisfying the implementation is added to the SP.

Step 4. In each line, if the number of nonzero bits is odd, then all the nonzero bits are examined one by one to check for the possibility of forming a VCS with another same-signed bit in another line.

The detailed situation of applying the instructions above to the first stage halfband filter is shown partially in Fig. 3, from which we can see that only six CSs are used and only one bit is implemented directly.

	4	5	6	7	8	9	10	11	12	13	14	15	16	17
c(1)									−1		−1			
c(3)							1		−1		1			−1
c(5)						−1			1					
c(7)					1			−1		−1		−1		−1
c(9)				−1		1			1		−1		−1	
c(11)				1		1		−1			1		1	
c(13)			−1				1			−1		1		−1
c(15)		1		−1			−1		1				−1	
c(17)		−1			−1				1			1		
c(19)	1			−1			1						1	

Fig. 3. Illustration of the proposed method on partial coefficients of the first halfband filter.

3.4 Comparison

The filter complexity of the first stage halfband filter in terms of adder-cost and logic-depth for the three methods above is shown in Table 3. The proposed method achieves the same logic-depth as the NR-SCSE and does better in adder-cost.

Table 3. Complexity comparison between three methods.

	CSD	NR-SCSE	Proposed CSE
adder-cost	124	90	84
logic-depth	5	3	3

4 Comb Filter with an Improved Sharpening Technique

Two cascaded stages of comb filter are used with additional attenuation of about 25 dB per stage. Thus the transfer function of the proposed cascaded comb filter is

$$H_o(z)=\left[\frac{1}{16}\sum_{k=0}^{15} z^{-k}\right]^2=\left[\frac{1}{16}\left(\frac{1-z^{-16}}{1-z^{-1}}\right)\right]^2=[H_1(z^{M_1})\cdot H_2(z)]^2\ , \tag{5}$$

$$H_1(z^{M_1})=\frac{1}{M_2}\left(\frac{1-z^{-M_1M_2}}{1-z^{-M_1}}\right),\quad H_2(z)=\frac{1}{M_1}\left(\frac{1-z^{-M_1}}{1-z^{-1}}\right). \tag{6}$$

$$H_s(z)=[3-2H_1^{\,2}]\cdot H_1^{\,4}\cdot H_2^{\,4}\ . \tag{7}$$

The simplest sharpening method in [5] given by ($3H^2$-$2H^3$) is applied to $H_1(z)$. We arrive at the transfer function in (7) and from Fig. 4 magnificent improvement in sideband suppression as well as passband droop compensation can be found.

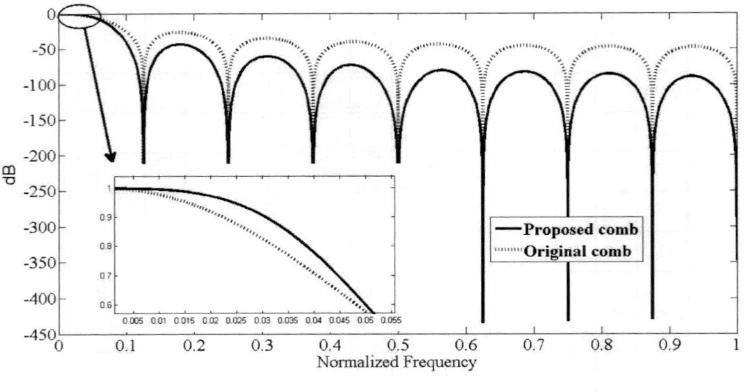

Fig. 4. Improved comb filter response.

Two possible ways of implementation correspond to the cases which are M_l=8, M_2=2 and M_l=M_2=4. Since the H_l part is responsible of the low-pass region [9], a higher value of M_l will lead to relatively poorer performance in the low frequency region. But in the mean time, due to the sequence of upsampling which is M_2 followed by M_l, in the first case, more computational operations are running in lower sampling rate compared with the second one, which may help the reduction of power consumption. Thus, a trade-off between these two performance factors must be considered seriously based on in-depth analysis and simulations of the two cases.

Table 4. Comparison between two implementations of comb filter.

	Power	**Area**
comb 2×8	1.9715 mW	889245 μm^2
comb 4×4	1.7683 mW	784560 μm^2

Using a standard TSMC 0.35μm logic 1P4M process, given the synthesis results of Design Compiler and the power estimation based on a practical testbench input, it can be shown in Table 4 that the M_l=M_2=4 case outweighs the first one in both the area and total dynamic power, which could be attributed to the two-fold poly-phase transformations. The detailed implementation diagram of the M_l=M_2=4 case is shown in Fig. 5.

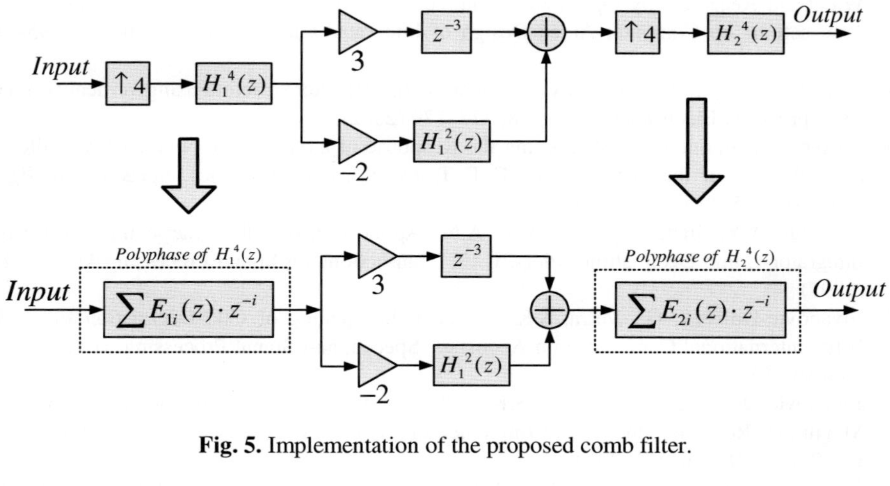

Fig. 5. Implementation of the proposed comb filter.

5 Conclusion

In this paper, the design of a 16-bit precision low-power low-complexity interpolation filter used for Delta-Sigma DAC system is discussed. An improved CSE approach as well as an optimized two-stage sharpening filter technique is proposed and the detailed implementation is provided. The function of the whole design has been verified by Altera EP2C35 FPGA with the SNR performance given in Table 5 and the synthesis result provided by Design Compiler indicates a power consumption of 1.8306mW as well as a die area of 1511280 μm^2.

Table 5. SNR performance of the proposed interpolation filter.

Frequency	495.3Hz	1098.2Hz	2174.8Hz	4414.3Hz	6481.5Hz	8677.9Hz	12941.5Hz
SNR	95.90dB	98.51dB	98.11dB	98.57dB	98.69dB	98.56dB	98.53dB

Acknowledgments. This work is sponsored by the National Natural Science Foundation of China under grant No.60906012. It also gains support from the Silergy Corp. and the authors would like to thank Dr. Isaac Chen, the CEO of Silergy Corp., and his colleagues, for their useful discussions and instructions.

References

1. Avizienis, A.: Signed digit number representation for fast parallel arithmetic. IRE Transactions on Electronic Computers EC-10, 389–400 (1961)
2. Martinez-Peiro, M., Boemo, E.I., Wanhammar, L.: Design of high-speed multiplierless filters using a nonrecursive signed common subexpression algorithm. IEEE Transactions on Circuits and Systems II 49, 196–203 (2002)
3. Vinod, A.P., Lai, E.M.-K.: On the Implementation of Efficient Channel Filters for Wideband Receivers by Optimizing Common Subexpression Elimination Methods. IEEE Transactions on Computer-Aided Design of Integrated Circuits and Systems 24, 295–304 (2005)
4. Jang, Y., Yang, S.: Low-power CSD linear phase FIR filter structure using vertical common subexpression. Electronics Letters 38, 777–779 (2002)
5. Kaiser, J., Hamming, R.: Sharpening the Response of a Symmetric Nonrecursive Filter by Multiple Use of the Same Filter. IEEE Transactions on Acoustics, Speech, and Signal Processing 25, 415–422 (1977)
6. Kwentus, A.Y., Jiang, Z., Willson Jr., A.N.: Application of Filter Sharpening to Cascaded Integrator-Comb Decimation Filters. IEEE Transactions on Signal Processing 45, 457–467 (1997)
7. Jovanovic-Dolecek, G., Mitra, S.K.: Efficient Sharpening of CIC Decimation Filter. In: IEEE International Conference on Acoustics, Speech, and Signal Processing, vol. 6, p. VI - 385–VI-388 (2003)
8. Jovanovic-Dolecek, G., Mitra, S.K.: Sharpening Comb Decimator with Improved Magnitued Response. In: IEEE Conference on Acoustics, Speech, and Signal Processing, vol. 2, pp. ii-929–ii-932 (2004)
9. Jovanovic-Dolecek, G., Mitra, S.K.: A New Two-Stage Sharpened Comb Decimator. IEEE Transactions on Circuits and Systems I 52, 1414–1420 (2005)

An Efficient Digital Front-End for 16-Bit Audio Delta-Sigma D/A Converter

Jinchen Zhao, Xiaobo Wu*, Menglian Zhao, and Tong Wu

Institute of VLSI Design, Zhejiang University,
310027 Hangzhou, P. R. China
{zhaojc,wuxb,zhaoml,wutong}@vlsi.zju.edu.cn

Abstract. In order to save area and achieve high SNDR as well, an efficient digital front-end of a 16-bit DAC for audio application, including a 4-stage interpolator and a 4th-order Δ-Σ modulator with improved DWA technique is proposed. Poly-phase structure and CSD coding method are used for interpolator to save area. An improved DWA technique named as DCSDWA is applied for the Δ-Σ modulator with a 15-level quantizer to eliminate the mismatch errors in multi-bit DAC so as to yield high SNDR. Verified by FPGA in Altera Quartus II synthesis environment, the proposed design yields 115-dB DR and 108.4-dB SNDR for a -6-dBFS sine-wave input, as well as 68-dB stopband attenuation and less than 0.008-dB passband ripple for the interpolator, which indicate the proposed work meets the design requirements well.

Keywords: DAC, interpolator, delta-sigma modulator, multi-bit, DCSDWA.

1 Introduction

By using oversampling and noise-shaping technologies, delta-sigma (Δ-Σ) modulator yields wide dynamic range (DR) as well as high signal to noise and distortion ratio (SNDR). Thus, it is especially applicable to the high resolution digital-to-analog converters (DAC) needed in portable multimedia devices, which have the increased market demands recently. Meanwhile, the multi-bit quantization for a Δ-Σ converter is widely used due to the high resolution but simple loop configurations, at the cost of additional dynamic element matching (DEM) technique to eliminate the mismatch errors of unit-elements in multi-bit DAC. The data weighted averaging (DWA) is known as one of the most effective DEM techniques. However, if a dc or a low-frequency input signal is given, the in-band signal-dependent tones will appear when a traditional DWA is employed, which degrade the SNDR seriously. The SNDR of the modulator should be high enough to relax the specifications of the post analog reconstruction filter. Therefore, it is an urgent imperative to improve the DWA technique to solve the tones issue. In addition, an interpolator is needed in Δ-Σ DAC to achieve up-sampling and suppress the out-of-band power. Finite impulse response (FIR) filter is suitable for an interpolator on account of the requirement of linear

* Corresponding author.

X. Wan (Ed.): Electrical Power Systems and Computers, LNEE 99, pp. 713–720.
springerlink.com

phase property in audio application, but which is area-cost and complex to implement. Therefore, optimization methods should be applied for the interpolator to save area.

In order to achieve area-efficient and high SNDR as well, an effective digital front-end of a 16-bit Δ-Σ DAC for audio application is proposed in this paper. The digital front-end is composed of an area optimized 4-stage FIR interpolator and a 4th-order Δ-Σ modulator with an improved DWA technique named as dual cycle shifted DWA (DCSDWA). The interpolator includes 3 stages of halfband (HB) filters and a sample-and-hold (S/H) register to carry out 64× up-sampling. Each stage of HB filters is implemented by a poly-phase structure with the coding method of Canonic Signed Digit (CSD) to realize multiplier-free to save area. The Δ-Σ modulator yields 115-dB DR via employing a 15-level quantizer and optimizing the coefficients as well as the word length within the modulation loop. DCSDWA is applied to eliminate the in-band signal-dependent tones more effectively than traditional and other modified DWA techniques. Prototype of the digital front-end is verified by FPGA in Altera Quatus II synthesis environment, and the experimental results indicate that the proposed design meets the requirements of the portable audio application well.

The paper is organized as follows. After the introduction in section 1, section 2 depicts the overall architecture of the audio DAC. The interpolator and the Δ-Σ modulator are discussed in section 3 and section 4, respectively. The DCSDWA technique is illustrated in section 5. Section 6 proposes the experimental results, and section 7 concludes the paper.

2 Architecture of the Audio DAC

The basic system diagram of the audio DAC is illustrated in Fig. 1. The proposed DAC processes a 16-bit input signal with sampling rate fs of 44.1 kHz. The whole structure can be divided into a digital front-end and an analog output stage. The digital front-end includes a 4-stage FIR interpolator, a 4th-order noise-shaping Δ-Σ modulator and a DWA block, while the analog output stage contains an internal DAC as well as a reconstruction filter.

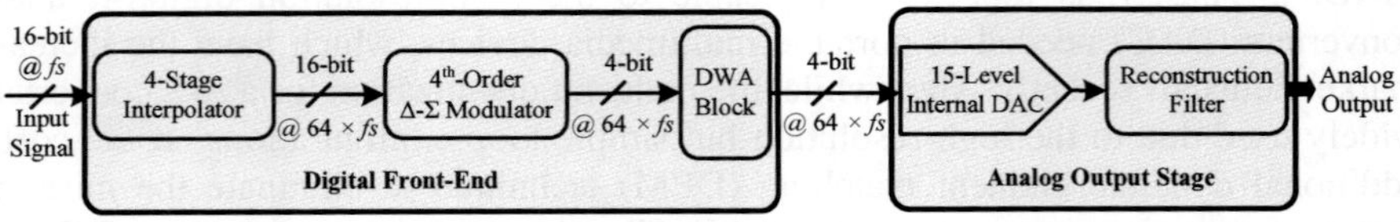

Fig. 1. Block diagram of the audio DAC.

The input signal passes through the interpolator, which raises the sampling rate from fs = 44.1 kHz to the oversampling rate (OSR) of 64×fs = 2.8224 MHz. Meanwhile, the filter suppresses the spectral replicas of the input signal in order to reduce the out-of-band power. The Δ-Σ modulator noise-shapes the 16-bit input data and reduces the word length to 4-bit. Multi-bit quantization is preferred on account of several intriguing advantages, such as simplifying the noise-shaping loop and easing the specifications of the analog reconstruction filter under the same realization of

SNDR. However, there is a drawback for multi-bit quantization. The asymmetry of unit-elements in the internal DAC due to the process deviation involves mismatch errors, which degrade the SNDR seriously. Thanks to the DWA block, the thermometer-coded output of the modulator is noise-shaped and scrambled into a random-coded form, so that the internal DAC will reproduce the digital input into the analog output without appreciable aberration. Finally, the reconstruction filter suppresses most of the out-of-band noise and hence recuperates the audible signal.

3 Interpolator

Rather than an infinite impulse response (IIR) filter, a FIR filter is suitable for the demand of linear phase property in audio application. Due to the high OSR and the narrow transition band of the input signal, the order of a single-stage FIR filter to achieve 64× up-sampling has to be exceedingly high. Therefore, a multi-stage structure is preferred to minimize area. The HB filter grants the weight of its every other tap (except the center one) to be zero, and hence which is very economical to accomplish.

The block diagram of the 4-stage FIR interpolator is depicted in Fig. 2. The interpolator comprises 3 stages of cascaded HB filters, followed by an S/H register. It increases the OSR to 64, and meanwhile realizes the stopband attenuation higher than 65-dB and the passband ripple less than 0.01-dB. Each HB filter achieves 2× up-sampling, and the S/H register raises the OSR by the other 8 times. The requirements on the first stage are the most demanding, since it should offer a flat passband with extremely small gain variation and the sharpest cutoff. Thus, the first stage is realized by a 110-order HB filter, which is much more complex than the following ones. The orders of the second and the third stages are 18 and 14, and the detailed specifications for 3 stages of HB filters are list in table 1.

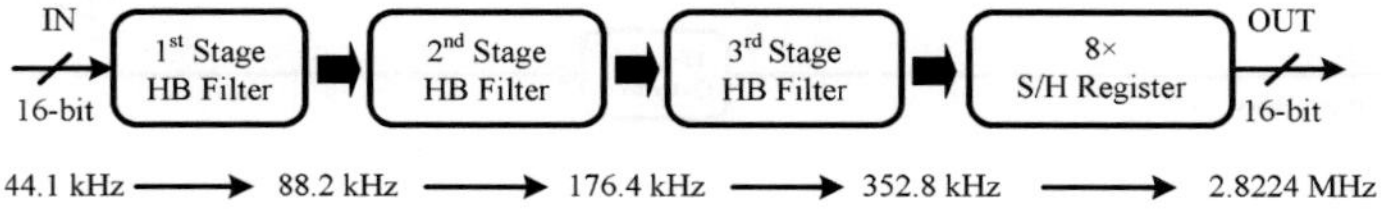

Fig. 2. Block diagram of the interpolator.

Table 1. Configurations of the HB filters.

	Passband Ripple	Stopband Attenuation	Order
1st Stage HB Filter	< 0.004 dB	> 70 dB	110
2nd Stage HB Filter	< 0.002 dB	> 72 dB	18
3rd Stage HB Filter	< 0.002 dB	> 72 dB	14

The coefficients of each HB filter can be obtained by Matlab. In order to reduce the area, every coefficient should be carried out as a sum of integer powers of two and then encoded by using Canonic Signed Digit (CSD), so that the interpolator can be optimized to be a multiplier-free realization. The CSD encodes a binary number as a specific form which includes non-zero bits as few as possible, so that it can save about 30% combinational circuit area compared with the common binary coding implementation. The poly-phase structure is employed for each stage of the HB filters. Compared with the direct-form structure, poly-phase solution can save the number of the registers by almost half.

4 Δ-Σ Modulator

The selection of the quantization level and loop order for the Δ-Σ modulator involves two basic concerns. The first one is to make the in-band quantization noise negligible in the total noise budget to fulfill the requirement of SNDR, and the second one is to make out-of-band noise low enough so that the modulator is stable and the design specifications of the reconstruction filter can be relaxed.

A 4th-order Δ-Σ modulator using a 15-level quantizer meets these two requirements well. The architecture of the modulator is illustrated in Fig. 3. A chain of 4th-order accumulators with feedforward summation is used as the basic structure. Replacing 2 zeroes out of 4 from dc to the edge of the signal band by adding a local feedback resonator guarantees the modulator to achieve high SNDR. A 15-level quantizer achieves relatively low out-of-band noise and assures the stability.

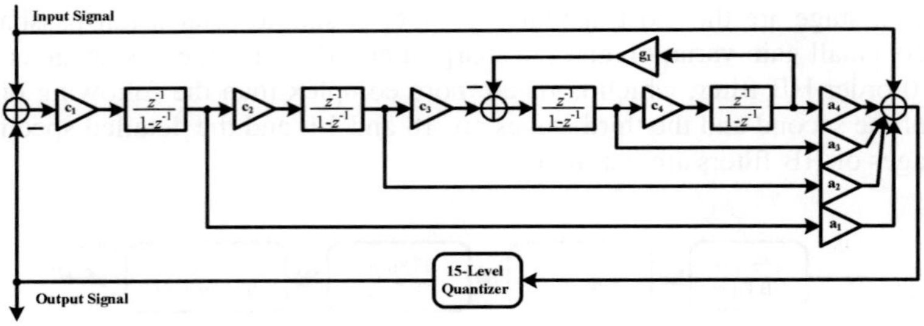

Fig. 3. Architecture of the Δ-Σ modulator.

Original internal coefficients of the proposed structure can be obtained by Matlab, then should be modified and optimized into the sums of integer powers of two for multiplier-free implementation.

For digital implementation of the Δ-Σ modulator, the word length of each accumulator should be determined to fit the in-band truncation noise caused by finite precision arithmetic lower than -120-dB. The in-band noise power P_{in} for an accumulator associated with N-bit word length is

$$P_{in} = \frac{1}{c_1^2} \cdot \frac{\left(2^{-N}\right)^2}{3} \cdot \frac{1}{OSR}, \tag{1}$$

where c_1 is the feedthrough coefficient of the first accumulator. Since the power of a full-scale sine-wave with the amplitude M is $M^2/2$, the word length N should satisfy

$$N > -\log_2\left(M \cdot c_1\sqrt{1.5\times10^{-12}OSR}\right) \approx 18.6\,. \tag{2}$$

Therefore, the word length of each accumulator is finally confirmed as 19-bit.

Feedforward structure used in the modulator configuration ensures the signal transfer function (STF) to be 1, which means the modulation loop should reproduce the input signal without any distortion. Based on the optimized internal coefficients of the modulator, the noise transfer function (NTF) H(z) can be expressed as

$$H(z) = \frac{(z-1)^2(z^2-2z+1.001)}{(z^2-1.513z+0.6189)(z^2-0.9874z+0.4203)}\,. \tag{3}$$

The NTF acting as a high-pass filter suppresses the quantization error around dc to the out-of-band.

5 DCSDWA Technique

Traditional DWA selects the unit-elements of internal DAC rotatively and achieves first-order noise-shaping to mismatch errors. However, if a dc or a low-frequency input signal is given, the selection becomes a pattern, which means that the mismatch errors of the unit-elements turn out to be a periodic sequence. Therefore, the signal-dependent tones are produced and then translated into the output of the internal DAC. In order to solve the tones problem, several modified DWA techniques are developed, e.g. pseudo DWA [1], split-set DWA (SDWA) [2], Bi-Direction DWA (Bi-DWA) [3] as well as Partitioned DWA (P-DWA) [4]. Pseudo DWA cannot meet the linearity demand for audio application. SDWA obtains higher SNDR, but the complexity of the circuit will be greatly increased when the quantization level is more than 9. Bi-DWA and P-DWA suppress the in-band tones effectively, but they achieve poor SNDR.

An improved DWA technique named as DCSDWA is applied in this paper, the principle of which is to disturb the periodic pattern of the mismatch errors of the multi-bit internal DAC due to the process deviation so that the signal-dependent tones can be suppressed without SNDR loss. As shown in Fig. 4, the DCSDWA block is composed of two main functional units, which are the traditional DWA unit and the output sequence processor. In addition, an internal k-bit counter and a pointer pt(n) are needed. DCSDWA operates as traditional DWA during the (2^k-1) clock periods. Once the $(2^k)^{th}$ output of the traditional DWA y(n) is given, the k-bit counter overflows, and the output sequence is split into two subsets based on the index of the pointer pt(n). If the value of y(n) is even, two subset-sequences are shifted counterclockwise respectively, and if the value is odd, sequences are shifted clockwise. Then, two new subset-sequences are assembled together to form a new y(n). As an example to illustrate the principle of DCSDWA, assume that y(n) is a 15-level output sequence as used in the proposed audio DAC. The $(2^k\text{ -}1)^{th}$ output y(n-1) is (00*0*111110000000) and index of pt(*n*-1) is 13 (which is indicated by '*0*' in the output sequence). When the value of y(*n*) is 8, pt(*n*) turns out to be 6 and splits y(*n*) from (111000000*0*11111) to (111000000) and (*0*11111). Because 8 is even, two subset-sequences are shifted to be (110000001) and (11111*0*), and then new y(*n*) turns out to be (11000000111111*0*). If the value of y(*n*) is 7, new y(*n*) becomes (01110000001*0*111). The split-and-shift behavior is carried out as long as the overflow happens, and noise-shapes the mismatch errors as well as suppresses the in-band tones to improve the SNDR.

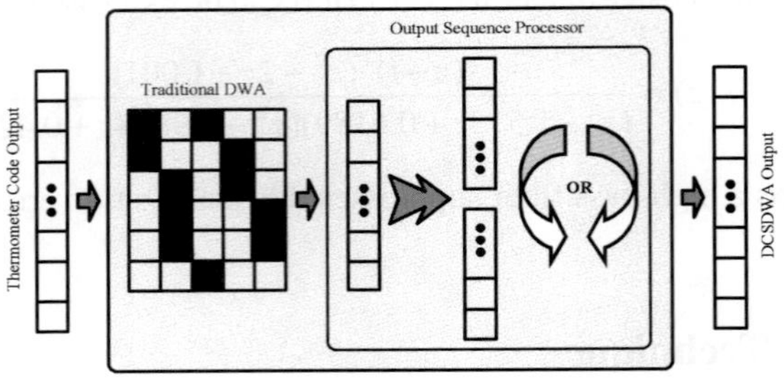

Fig. 4. Structure of the DCSDWA block.

6 Experimental Results

The proposed design is verified in Altera Quartus II FPGA synthesis environment. The output of the interpolator is shown in Fig. 5(a). It can be seen that the interpolator carries out 64× up-sampling operation effectively. The frequency response of the 3 stages of HB filters simulated by Matlab is depicted in Fig. 5(b). The simulation result shows that the passband edge frequency is 20.2 kHz, the stopband attenuation is 68-dB, and the passband ripple is less than 0.008-dB.

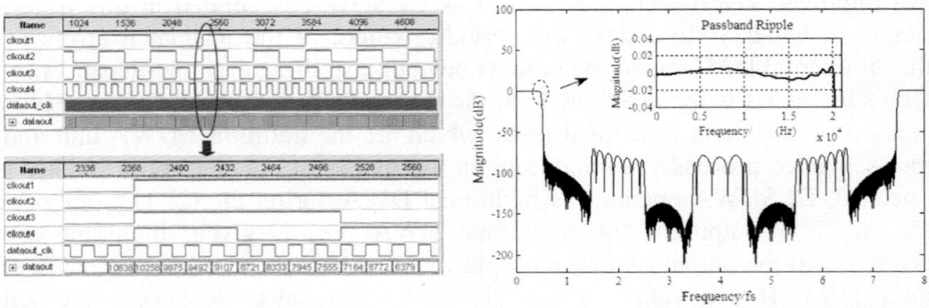

(a) Output of the interpolator (b) Frequency response of the interpolator

Fig. 5. Experimental results of interpolator: (a) FPGA output of the interpolator; (b) frequency response of the interpolator.

The areas of 3 implementations of interpolators, including direct-form structure, poly-phase structure as well as this work (poly-phase structure with CSD coding) for a standard 0.18-μm one-poly four-metal (1P4M) CMOS process are summarized and compared in table 2. All the areas are normalized, and the area of the proposed design is deemed to be 1. Table 3 demonstrates that the sequential circuit area of the poly-phase structure is only half that of direct-form structure. Compared with the common binary coding method, CSD coding can save about 30% combinational circuit area.

Table 2. Area comparison between different implementations of interpolators.

	Normalized Combinational Circuit Area	Normalized Sequential Circuit Area	Normalized Total Area
Direct-Form	1.33	1.85	1.44
Poly-Phase	1.36	1	1.29
This Work	1	1	1

The function of DCSDWA is simulated by Matlab. The mismatch errors of the unit-elements in the internal DAC are set as a Gaussian distribution with a maximum deviation value of 0.5%. Fig. 6(a) and (b) show the output spectrum of the modulator with or without DCSDWA when a -2-dBFS sine-wave input @ 5 kHz is given. According to the figures, DCSDWA technique eliminates the in-band tones effectively and yields SNDR 22-dB higher than that without any DEM technique. Comparison result between the DCSDWA and other modified DWA techniques including traditional DWA, pseudo DWA, SDWA, Bi-DWA and P-DWA is illustrated as SNDR versus input amplitude of the modulator in Fig. 6(c). A 6-bit counter is used in DCSDWA. The simulation result shows that DCSDWA yields a higher SNDR than the other DWA techniques and a good linearity.

Fig. 7(a) illustrates the FPGA output spectrum of the proposed digital front-end via SignalTap logic analyzer when a -6-dBFS sine-wave input @ 5 kHz is given, and the SNDR achieves 108.4-dB which indicates an 18-bit resolution. Fig. 7(b) shows the measured SNDR versus the amplitude of input signal. The DR of the digital front-end is 115-dB, which satisfies the requirements of portable audio application well.

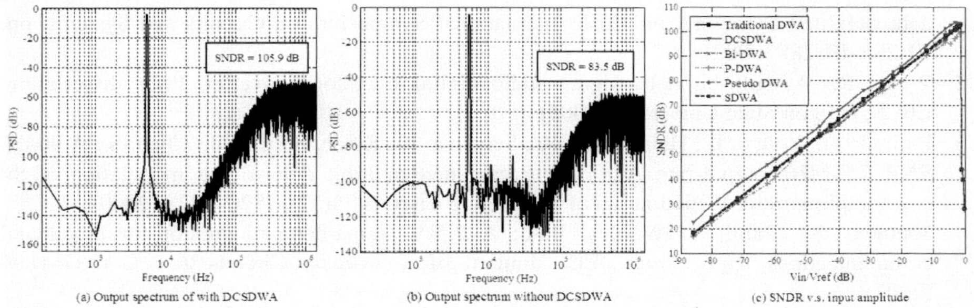

Fig. 6. Matlab simulation results: (a) output spectrum with DCSDWA; (b) output spectrum with DCSDWA; (c) SNDR versus input amplitude.

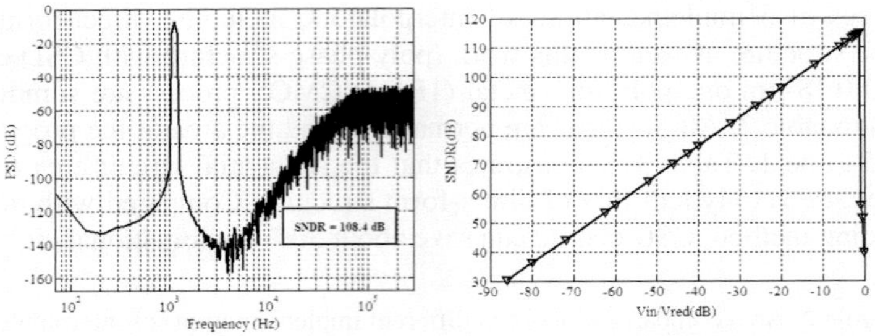

(a) Output spectrum (b) SNDR v.s. Input amplitude

Fig. 7. FPGA verification results: (a) output spectrum of the modulator; (b) SNDR versus input amplitude @5 kHz.

7 Conclusion

In order to achieve area-efficient and high SNDR, an effective digital front-end used in 16-bit audio DAC is presented. Poly-phase structure with CSD coding is used for interpolator to save area. DCSDWA technique applied to the Δ-Σ modulator with a 15-level quantizer eliminates the in-band tones effectively and yields high SNDR. The proposed design is verified by FPGA in Altera Quartus II synthesis environment and simulated by Matlab. The experimental results show that the proposed work meets the design requirements well.

Acknowledgments. This work is sponsored by the National Natural Science Foundation of China under grant No.60906012 and Fundamental Research Funds for the Central Universities. It also gains support from Silergy Corp.. The authors would like to thank Dr. Isaac Chen, CEO of Silergy Corp. and his colleagues, for their useful discussions and instruction.

References

1. Hamoui, A.A., Martin, K.: Linearity enhancement of multibit Δ-Σ modulators using pseudo data-weighted averaging. In: IEEE International Symposium on Circuits and Systems, pp. III-285–III-288 (2002)
2. Wang, R.: A multi-bit delta sigma audio digital-to-analog converter. PhD. dissertation, EECS, Oregon State University (2006)
3. Fujimori, I., Longo, L., Hairapetian, A., Seiyama, K., Kosic, S., Cao, J., Chan, S.: A 90-dB SNR 2.5-MHz output-rate ADC using cascaded multibit delta-sigma modulation at 8× oversampling ratio. IEEE Journal of Solid-State Circuits 35(12), 1820–1828 (2000)
4. Vleugels, K., Rabii, S., Wooley, B.A.: A 2.5-V sigma-delta modulator for broadband communications applications. IEEE Journal of Solid-State Circuits 36(12), 1887–1899 (2001)

Automatic Test Purpose Generation for Web Services

Sébastien Salva

LIMOS CNRS UMR 6158
PRES Clermont University, Campus des Cézeaux
Aubière, France
sebastien.salva@u-clermont1.fr

Abstract. It is now well-established that to be reliable, software have to be tested during the software life cycle, and this is particularly true with recent technologies such as Web services. Test purpose based methods are black box testing techniques which take advantage of reducing the time required for test derivation. Nevertheless, test purposes must be constructed by hand. To solve this issue, we propose, in this paper, some automatic test purpose generation methods for testing the operation existence, the critical states and the exception handling, in stateful Web services. To take into account the SOAP environment in which they are deployed, we also augment the specification with SOAP messages. We show that SOAP gives more observable reactions and helps to test specific properties.

Keywords: Stateful Web services, STS, SOAP, test purpose generation.

1 Introduction

Software testing is an important software engineering activity widely used to find defects in programs. In particular, black box testing, which is the topic of this paper, consists in testing a system implementation by means of test cases, usually constructed from a specification. This paper also focuses on Web services which represent interoperable components whose purpose is to externalize functional code in a standardized way, or the reuse of software accompanied by cost reduction.

Recently, several Web service based black box testing methods have been proposed [1,2,3,4,5]. Some of them are said exhaustive i.e. the test case selection is performed to ensure that a faulty implementation is detected by a least one test case. Nevertheless, this exhaustiveness often implies a costly test case generation which eventually may lead to a state space explosion. Moreover, the test case set is not exhaustive in practice: service oriented application specifications are often symbolic which means that these latter are composed of variables and guards. The variable domain is often infinite and impossible to test completely.

Test purpose based methods represent an interesting alternative. Test purposes are test requirements which are given by designers. They can be used to

X. Wan (Ed.): Electrical Power Systems and Computers, LNEE 99, pp. 721–728.
springerlink.com

test various properties such as the critical states, the coverage of specific actions, etc. The test selection is then guided and thereby reduced since test purposes aim to target the test of some implementation parts only. Some works dealing with test purpose based methods for Web services have been proposed recently [3,4,5]. These methods generate test cases by synchronizing test purposes with the specification to produce action sequences which respect the specification and which contain the test purpose properties. Then, test cases are experimented on the implementation under test to conclude whether test purposes are satisfied.

Although using this approach greatly reduces test costs, the main encountered issue is that test purposes are formulated manually. And, constructing them is particularly difficult when the system is large, has real-time constraints or is distributed. However, many test purposes can be generated automatically as it has been showed in some works [6] which propose test purpose generation techniques for specific untimed systems (distributed systems and protocols). But to our knowledge, none method has been proposed for service oriented applications. This is why we present, in this paper, several techniques to generate test purposes for SOAP Web services, modelled with Symbolic Transition Systems (STS [7]). Usually, Web services are deployed in specific environments, e.g., HTTP for REST Web services or SOAP [8]. We show that the latter modifies the behaviour of the tested Web services and may give new relevant information (specific messages) for testing. So, the originality of our approach is to augment the specification to take into account the SOAP environment in order to test specific properties e.g., the exception handling. From the completed specification, we propose new test purpose generation methods to test the operation existence, the critical states and the exception handling.

This paper is structured as follows: Section 2 defines the specification and test purpose modelling. We describe the advantages granted by SOAP for testing in section 3 and define the specification completion. Test purpose generation methods are given in section 4. We provide some experiment results in section 5. And finally, section 6 gives some perspectives and conclusions.

2 Web Service and Test Purpose Modelling

We formalize, in this paper, Web services with Symbolic Transition Systems (STS [7]). This extended automaton model associates a behaviour with a specification composed of transitions labelled by actions and of internal and external variables sets, which may be used to send or receive concrete values and to set guards which must be satisfied to fire transitions. Below, we only summarize the suspension STS definition where quiescence (the lack of observation) is taken into account with the δ symbol. The complete definition can be found in [7].

Definition 1. *A (suspension) Symbolic Transition System STS is a tuple* $< L, l_0, V, V_0, I, \Lambda, \rightarrow >$*, where:*

- *L is the finite set of locations, with l_0 the initial one,*
- *V is the finite set of internal variables, I is the finite set of external or interaction ones. We denote D_v the domain in which a variable v takes*

values. The internal variables are initialized with the assignment V_0*, which is assumed to take an unique value in* D_V*,*

- Λ *is the finite set of actions, partitioned by* $\Lambda = \Lambda^I U \Lambda^O$*: inputs, beginning with ?, are provided to the system, while outputs (beginning with !) are observed from it.* $a(p) \in \Lambda$ *is an action where* $p = (p_1, ..., p_k)$ *is a finite set of external variables. We denote* $type(p) = (t_1, ..., t_k)$ *the type of the variable set p.* δ *denotes the quiesence i.e. the lack of observation from a location,*
- $\rightarrow$ *is the finite transition set. A transition* $(l_i, l_j, a(p), \varphi, \varrho)$*, from the location* $l_i \in L$ *to* $l_j \in L$*, also denoted* $l_i \xrightarrow{a(p),\varphi,\varrho} l_j$ *is labelled by* $a(p) \in \Lambda$*,* $\varphi \subseteq D_V \times D_p$ *is a guard which restricts the firing of the transition. Internal variables are updated with the assignment* $\varrho : D_V \times D_p \rightarrow D_V$ *once the transition is fired.*

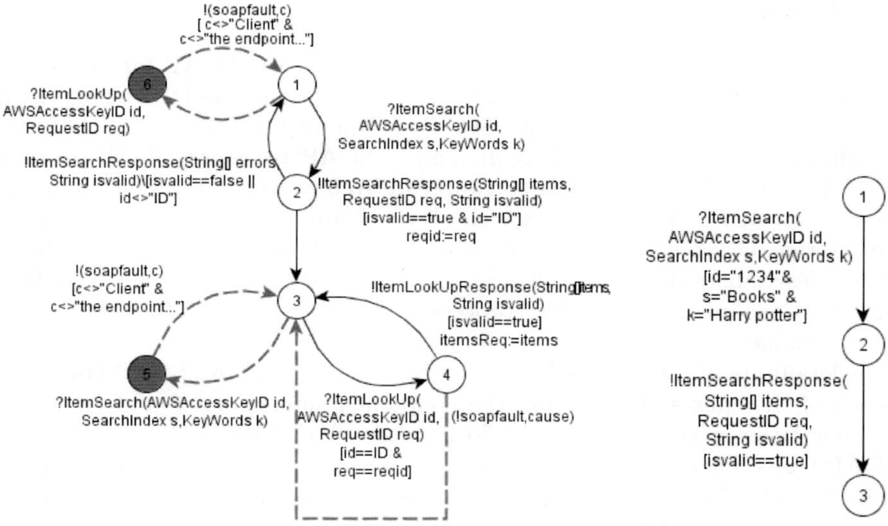

(a) A Web service specification (b) A test purpose

Fig. 1.

The STS model is not specifically dedicated to Web services. These latter may be invoked with methods called operations. This is why, for modelling, we assume that an action $a(p)$ in Λ represents either the invocation of an operation *op* which is denoted *opReq* or the return of an operation *op* with *opResp*. For an STS $\mathcal{S}$, we denote $\mathcal{OP}(\mathcal{S})$ the operation set found in Λ. A specification example, is illustrated in Figure 1(a) (black transitions). This one describes a part of the Amazon Web Service devoted for e-commerce (AWSECommerceService [9]). For sake of simplicity, we only consider two operations "ItemSearch", which aims to search for items, and "ItemLookUp", which provides more details about an item. Note that we do not include all the parameters for readability reasons.

On the other hand, test purposes describe the test intention. We assume that these ones are composed exclusively of specification properties which should be met in the implementation under test. Usually, test purposes do not represent complete specification paths. Therefore, they are often synchronized with the specification to generate executable test cases. Consequently, we also formalize a test purpose with a deterministic and acyclic STS $TP = < L_{TP}, l0_{TP}, V_{TP}, V0_{TP}, I_{TP}, \Lambda_{TP}, \rightarrow_{TP} >$ such that $\rightarrow_{TP}$ is composed of transitions modelling specification properties. So, for any transition $l_j \xrightarrow{a(p),\varphi_j,\varrho_j} l'_j \in \rightarrow_{TP}$, it exists a transition $l_i \xrightarrow{a(p),\varphi_i,\varrho_i} l'_i \in \rightarrow$ and a value set $(x_1, ..., x_n) \in D^n_{V \cup I}$ such that $\varphi_j \wedge \varphi_i(x_1, ..., x_n) \models$ true. A test purpose example is illustrated in Figure 1(b). This one aims to search for books whose description contain the keywords "Harry potter". We must obtain a valid response.

3 The Advantages Offered by the SOAP Environment for Testing

Web services are deployed in specific environments, e.g., SOAP for SOAP Web services, to structure messages in an interoperable manner and to manage operation invocations. In particular, the SOAP environment consists in a SOAP layer which serializes messages with XML and of SOAP receivers (SOAP processor + Web services) [10] which is software, in Web servers, that consumes messages. The SOAP processor is a Web service framework part which represents an intermediary between client applications and Web services and which serializes/deserializes data and calls the corresponding operations. The significant modifications involved by SOAP processors can be found in [11].

In summary, SOAP processors add new messages, called SOAP faults, which give details about faults raised in the server side. They return SOAP faults composed of the causes "Client" or "the endpoint reference not found" if services or operations or parameter types do not exit. SOAP processors also generate SOAP faults when a service instance has crashed while triggering exceptions. In this case, the fault cause is equal to the exception name. However, exceptions correctly managed in the specification and in the service code (with try...catch blocks) are distinguished from the previous ones since a correct exception handling produces SOAP faults composed of the cause "SOAPFaultException". So, SOAP faults can also be used to test whether the exception handling is correct by identifying the received causes. Consequently, taking into consideration these messages while generating test purposes sounds very interesting to check the satisfaction of specific properties e.g, the exception handling. So, we propose to augment the specification with the SOAP faults generated by SOAP processors. We denote $(soapfault, cause)$ a SOAP fault where the variable *cause* is the reason of the SOAP fault receipt.

Let $S =< L, l_0, V, V_0, I, \Lambda, \rightarrow>$ be a Web service specification. S is completed by means of the STS operation *addsoap* in S which augments the specification with SOAP faults as described previously. The result is an STS $S \uparrow$. The operation *addsoap* is defined as follow: *addsoap* in $S =_{def} S \uparrow =< L_{S\uparrow}, l_0, V, V_0, I, \Lambda_{S\uparrow}, \rightarrow_{S\uparrow}>$ where $L_{S\uparrow}$, $\Lambda_{S\uparrow}$ and $\rightarrow_{S\uparrow}$ are defined by the following inference rules:

$$R_1 : \frac{l_1 \xrightarrow{?opReq(p),\varphi,\varrho} l_2 \in \rightarrow_S, l_1 \xrightarrow{?op'Req(p),\varphi',\varrho'} l \notin \rightarrow_S,}{l \xrightarrow{?op'Req(p),\emptyset,\emptyset} l' \in \rightarrow_{S\uparrow}, l' \xrightarrow{!a(p),\varphi,\emptyset} l \in \rightarrow_{S\uparrow}, \varphi = [a(p) \neq (soapfault, \text{"}CLIENT\text{"}) \wedge \frac{l' \notin L_S}{a(p) \neq (soapfault, \text{"the endpoint reference not found"})]}}$$

$$R_2 : \frac{l \xrightarrow{?opReq(p),\varphi,\varrho} l' \in \rightarrow_S, \varphi' = \bigwedge_{l' \xrightarrow{!opResp_i(r_i),\varphi_i,\varrho_i} l'_i \in \rightarrow_S} \neg\varphi_i}{l' \xrightarrow{!(soapfault,cause),\varphi',\emptyset} l}$$

The first rule completes the initial specification on the input set by assuming that each unspecified operation request returns a SOAP fault message. The second rule completes the output set by adding, after each transition modelling an operation request, a transition labelled by a SOAP fault. Its guard corresponds to the negation of the guards of transitions modelling responses. A completed specification is illustrated in Figure 1(a) with dashed transitions.

4 Automatic Test Purpose Generation Methods

Although test purposes sound interesting to reduce test costs, these ones also raise an important drawback since they are usually formulated manually. So, we contribute to solve this issue by introducing some automatic generation techniques for Web services. We assume having a completed specification $S\uparrow$. We propose three test purpose generation approaches which aim to test the operation existence, the critical locations, and the exception handling.

Operation existence testing

This approach generates test purposes for testing whether operations in $\mathcal{OP}(S\uparrow)$, with $S\uparrow$ an STS specification, are implemented and can be invoked. With the specification completion, detailed in the previous section, it becomes possible to test the existence of any operation, even those which do not return any response, i.e. any observable reaction. Indeed, if an operation is not implemented as it is described in the specification, the SOAP processor will return a SOAP fault composed either of the cause "Client" or of the cause "the end point reference not found". So, for a specification $S\uparrow = < L_{S\uparrow}, l0_{S\uparrow}, V_{S\uparrow}, V0_{S\uparrow}, I_{S\uparrow}, \Lambda_{S\uparrow}, \rightarrow_{S\uparrow} >$, the test purpose set is given by:

$$TP = \bigwedge_{op \in \mathcal{OP}(S\uparrow)} \{tp =< L, l_0, V_S, V0_S, I_S, \Lambda, \rightarrow> \text{ where } \rightarrow = \{l_0 \xrightarrow{?opReq(p),\emptyset,\emptyset} l_1, l_1 \xrightarrow{!a(p),\varphi,\emptyset} l_2, \text{ with } \varphi = [a(p) \neq (soapfault, \text{"}Client\text{"}) \wedge a(p) \neq (soapfault, \text{"the end point reference not found"})]\}\}$$

The specification of Figure 1(a) is composed of two operations, so we obtain two test purposes. These ones will be synchronized later with the specification to test any operation invocation.

Critical location testing

The second technique aims at testing the specification critical locations. It is not obvious to set which location is critical since no general and formal definition is given in literature. So, in this paper, we suggest that the critical locations are those the most potentially encountered in the acyclic specification paths. Nevertheless, other criteria could be chosen, such as the less visited locations, or the quiescent ones. We give in [11] an algorithm which is derived from the DFS (Depth First Search) one, to detect the critical location set, denoted CS. Then, for each critical location $l \in CS$, we construct test purposes to test all the outgoing transitions of l. The test purpose set, expressed below, is composed of specification paths finished by output actions to observe the implementation reactions while testing. For a specification $S\uparrow =< L_{S\uparrow}, l0_{S\uparrow}, V_{S\uparrow}, V0_{S\uparrow}, I_{S\uparrow}, \Lambda_{S\uparrow}, \rightarrow_{S\uparrow} >$, the test purpose set is given by:

$TP = \bigwedge_{l \in CS} \{tp =< L, l_0, V_S, V0_S, I_S, \Lambda, \rightarrow>$ where $\rightarrow$ is constructed with the following inferences rules:

$$R_1 : \frac{l \xrightarrow{!a(p),\varrho,\varphi} l' \in \rightarrow_{S\uparrow}, a(p) \neq \delta}{l_0 \xrightarrow{!a(p),\varrho,\varphi} l' \in \rightarrow}$$

$$R_2 : \frac{l \xrightarrow{?a(p),\varrho,\varphi} l' \in \rightarrow_{S\uparrow}, p = l' \xrightarrow{a_1(p),\varrho_1,\varphi_1} l'_1 \ldots l'_{n-1} \xrightarrow{a_n(p),\varrho_n,\varphi_n} l'_n \in (\rightarrow_{S\uparrow})^n, a_n(p) \in \Lambda^O_{S\uparrow}/\{\delta\}}{l_0 \xrightarrow{!a(p),\varrho,\varphi} l'.p \in (\rightarrow)^{n+1}}$$

R_1 is used when an outgoing transition, from a critical location, is labelled by an output. In this case, this transition is added to the test purpose. The second rule is used when a transition is labelled by an input. The test purpose is completed with this transition followed by a specification path finished by an output. A test purpose generation algorithm is given in [11]. In the specification of figure 1(a) we have two critical locations l_2 and l_3. So, we obtain two test purposes which aim to test all the outgoing transitions of l_2 and l_3 with paths finished by output actions.

Exception handling testing

As described in Section 3, SOAP processors return SOAP faults when exception are triggered in a Web service operation at runtime. SOAP processors also enable to differentiate the exceptions resulting of unexpected Web service crashes from those which are thrown in Web service operations (with try ... catch blocks for instance). In the last case only, we obtain SOAP faults composed of the "SoapFaultException" cause.

With the specification completion described in section 3, we can construct test purposes to test whether the exception handling is correctly implemented and not managed by SOAP processors. However, to trigger exceptions, test purposes must be formulated over predefined value sets, that we denote $U(t)$. These ones are composed of unusual values well known for relieving bugs, for any simple or complex type t. For instance, $U(string)$ is composed of the values &", "$", null or "_", which usually trigger exceptions. For a specification $S\uparrow =< L_{S\uparrow}, l0_{S\uparrow}, V_{S\uparrow}, V0_{S\uparrow}, I_{S\uparrow}, \Lambda_{S\uparrow}, \rightarrow_{S\uparrow} >$, the test purpose set is given by:

$$TP = \bigwedge_{l \xrightarrow{?opReq(p),\varphi,\varrho} l' \in \rightarrow_{S\uparrow}} \{tp =< L, l_0, V_{S\uparrow}, V0_{S\uparrow}, I_{S\uparrow}, \Lambda, \rightarrow > \text{ where } \rightarrow =$$

$$\{l_0 \xrightarrow{?opReq(p),\varphi',\varrho} l_1, l_1 \xrightarrow{(!soapfault,"SOAPFaultException"),\emptyset,\emptyset} l_2 \text{ where } \varphi' = \varphi \wedge p = (p_1, ..., p_n) \text{ takes values in } U(type(p_1)) \times ... \times U(type(p_n))\}$$

The specification of Figure 1(a) contains four operation requests from locations l_1 and l_3. If we suppose that $card(U(type(p_1)) \times ... \times U(type(p_n))) = n$, we obtain at most $4n$ test purposes. It is manifest that the larger the unusual values sets, the larger the test purpose set will be. To limit it, instead of using a cartesian product, other solutions may be used such as pairwise testing [12] which constructs discrete combinations for pair of parameters only and which has been shown sufficient to cover parameter domains.

5 Experimentation

At the moment, we have implemented a preliminary tool which performs the test purpose generation from a completed STS and the synchronous products between the specification and test purposes. Then, we have manually extracted test cases and translated them into the Soapui format. Then, these ones can be executed with the Soapui tool [13] which aims to experiment Web services with unit test cases. A Soapui test case example can be found in an extended version of this paper in [11]. We applied the test purpose generation on the AWSECommerceService (09/10 version). Results are given in Figure 2. All the 22 operations handle a large number of parameters, therefore we limited the test purpose number to 10 per operation, for the exception handling method. We obtained fail verdicts only for the exception handling tests. Indeed, we obtained some SOAP faults composed of the cause *Client*, meaning that the requests are incoherent although the test cases satisfy the specification. We also received unspecified messages corresponding to errors composed of a wrong cause. For instance, instead of receiving SOAP faults, we obtained the response "Your request should have at least 1 of the following parameters: AWSAccessKeyId, SubscriptionId when we called the operation CartAdd with a quantity equal to "-1", or when we searched for a "Book" type instead of the "book" one, whereas the two parameters AWSAccessKeyId, SubscriptionId were right.

	Existence	Critical locations	Exception handling
test purposes	22	2	22
test cases	44	22	210
fail verdicts	0	0	39

Fig. 2. Test results on the Amazon AWSECommerceService Service

6 Conclusion

We have proposed, in this paper, some methods to generate automatically test purposes from a Stateful Web service specification. We believe that these latter are relevant when used in combination with existing test purpose based methods to produce test cases automatically and to prevent from writing test purposes manually.

We have also shown that taking into account the SOAP environment during the test brings new information which help to test specific properties such as the operation existence or the exception handling. An immediate line of future work is to propose other generation approaches such as the test of the location accessibility. We also intend to extend this work on service compositions to test composition properties.

References

1. García-Fanjul, J., Tuya, J., de la Riva, C.: Generating test cases specifications for compositions of web services. In: Bertolino, A., Polini, A. (eds.) Proceedings of International Workshop on Web Services Modeling and Testing (WS-MaTe 2006), Palermo, Sicily, Italy, pp. 83–94 (2006)
2. Frantzen, L., Tretmans, J., de Vries, R.: Towards model-based testing of web services. In: Bertolino, A., Polini, A. (eds.) Proceedings of International Workshop on Web Services Modeling and Testing (WS-MaTe 2006), Palermo, Sicily, Italy, pp. 67–82 (2006)
3. Lallali, M., Zaidi, F., Cavalli, A., Hwang, I.: Automatic timed test case generation for web services composition. In: The 6th IEEE European Conference on Web Services (ECOWS 2008), Dublin, pp. 53–63. IEEE Computer Society Press, Los Alamitos (2008)
4. Escobedo, J.P., Gaston, C., Le Gall, P., Cavalli, A.: Observability and controllability issues in conformance testing of web service compositions. In: Núñez, M., Baker, P., Merayo, M.G. (eds.) TESTCOM 2009. LNCS, vol. 5826, pp. 217–222. Springer, Heidelberg (2009)
5. Cao, T.D., Felix, P., Castanet, R.: Wsotf: An automatic testing tool for web services composition. In: Proceedings of the 2010 Fifth International Conference on Internet and Web Applications and Services ICIW 2010, pp. 7–12. IEEE Computer Society, Washington, DC, USA (2010)
6. Henniger, O., Lu, M., Ural, H.: Automatic generation of test purposes for testing distributed systems. In: Petrenko, A., Ulrich, A. (eds.) FATES 2003. LNCS, vol. 2931, pp. 178–191. Springer, Heidelberg (2004)
7. Frantzen, L., Tretmans, J., Willemse, T.A.C.: Test Generation Based on Symbolic Specifications. In: Grabowski, J., Nielsen, B. (eds.) FATES 2004. LNCS, vol. 3395, pp. 1–15. Springer, Heidelberg (2005)
8. WWW Consortium. Simple object access protocol v1.2 (soap) (2003)
9. Amazon. Amazon e-commerce service (2010), `http://docs.amazonwebservices.com/AWSEcommerceService/4-0/`
10. WI organization. Ws-i basic profile (2006), `http://www.ws-i.org/docs/charters/WSBasic_Profile_Charter2-1.pdf`
11. Salva, S., Rabhi, I.: Automatic test purpose generation for Web services. LIMOS Research report RR-11-04 (2011)
12. Cohen, M.B., Gibbons, P.B., Mugridge, W.B.: Constructing test suites for interaction testing. In: Proc. Intl. Conf. on Software Engineering (ICSE), pp. 38–48 (2003)
13. Eviware. Soapui (2011), `http://www.soapui.org/`

Design of Current Transformer for Power Transmission Lines Inspection Robot

Mingbo Yang, Zize Liang, En Li, Kailiang Zhang, and Guodong Yang

Institute of automation, Chinese academy of science
NO. 95, Zhongguancun east road, haidian district, Beijing, China
jlsthsdqyx@163.com,
{Zize.liang,En.li,Kailiang.zhang,Guodong.yang}@ia.ac.cn

Abstract. In order to keep power transmission lines inspection robot to perform continuous inspection work on transmission lines, a power supply system with induction charging unit for the inspection robot is introduced in this paper. Special current transformer is designed to induct power for the robot from the power transmission lines. The charging circuit for Li-ion battery pack is then presented, and a two-stage strategy method is applied in the circuit to provide precisely accessible input power for charging battery packs. Experimental results have been obtained validating the viability of this kind of power supply system.

Keywords: Field robot, power supply system, induction charging, battery management.

1 Introduction

In 1980s, Jun Sawada and his group developed a wheeled mobile robot to implement inspecting job on power transmission line (on overhead ground line). In recent years, researchers have been working on designing this kind of mobile robots to partly or fully perform the inspection tasks of power transmission line equipment since the manual inspection work are dangerous and laborious [1]. References 2~5 are some of relative researches on inspection robot posted in resent years.

Most research works in this field are concentrated on mechanical design, motion planning, and sensor information processing which could provide the inspection robot with high performance and reliability in implementing inspection task. But, there is still a problem which has serious influence on performance of inspection robot, i.e., the continuous working period of robot. Most inspection robots obtain power supply from the battery packs. Due to the limitation of the lithium battery capacity, the robot could not work for a long time and often need maintenance. Reference 2 provides an induction power system, which uses induction voltage to drive the robot directly. Because the robot does not have battery pack, it depends on power transmission line too much, and may fail to move when the current load on the power line abruptly plummets.

In this paper, an induction power supply system is introduced to solve this problem. Section2 shows the working principle of the inspection robot and the function of the induction power supply system. Section 3 introduces the three functional units of the

X. Wan (Ed.): Electrical Power Systems and Computers, LNEE 99, pp. 729–736.
springerlink.com

power supply system and gives the detail design scheme for each unit. In section 4, experiment condition and the experimental results are presented, which proves that the design in this paper has great effect on prolong continuous working period of the inspection robot.

2 Description of a System

Fig.1 shows an inspection robot powered by this induction power supply system. The robot is initially set on transmission line by human being. While walking on the transmission line, the robot inspects the transmission line with the inspection equipments, and sends the results of inspection to master PC.

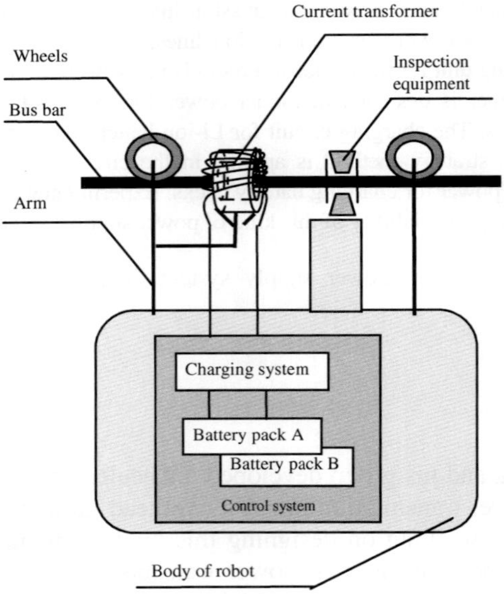

Fig. 1. Inspection robot on power Transmission line

The robot's power supply system is made up of induction unit, conditioning unit and two battery packs. As shown in Fig.2.The induction unit get power source from the power transmission line by a special current transformer, it works in magnetic field of the transmission line and output induction voltage for conditioning unit. The conditioning unit consists two parts: a switching power supply and a charging management. The switching power supply transforms the induced voltage into ideal DC output accessible to the charging management, the latter turns the DC voltage into accessible input charging for Li-battery packs and monitors the SOC of the charging battery packs.

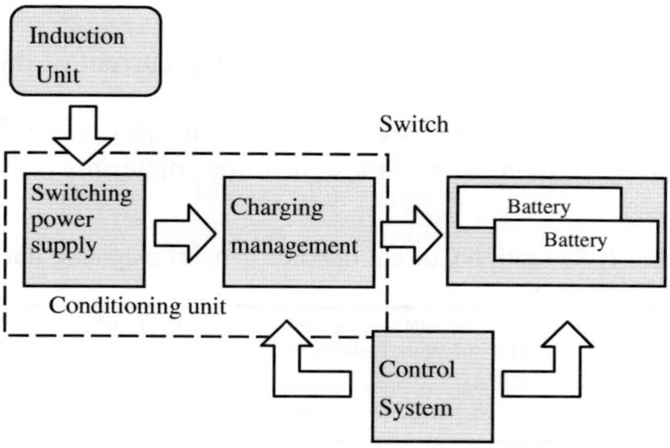

Fig. 2. Function chart of induction power Supply system

In normal working state, the robot is powered only by one of the two battery packs. The stand-by battery pack is being charged by induction charging unit. When it is full charged, charging process would be stopped and this battery pack would be cut off form the charging circuit and ready for being switched to power the robot. If the on going battery pack is in low power state, the robot could switch to the fully charged stand-by battery and power supply turns to the full state battery pack at the same time. Then this low power state battery pack is being charged in turn.

By detecting the output voltage of current transformer, robot can decide if the bus bar current is fit for charging system and turn on/off the charging switch by the control system of the robot. By doing this, circuit could be protected from high induction voltage from induction unit.

3 Hardware Design of the System

3.1 Induction Unit

Output power of the induction unit is mostly affected by induction device named current transformer. Detailed design of the transformer is introduced below.

1. Core material.

The most important parameters to choose material of the core are magnetic conductivity and saturation induction density. Material with high magnetic conductivity can get high induction density inside the core, so it could get induction power from low intensity of magnetic field. But this kind of material could become saturated easily when the material is in deep saturated state, magnetic flux in the core would not alter with exciting current, it is not possible to obtain induction voltage output. Thus, these parameters of material should be considered carefully.

According to the formula $B_R = \dfrac{\mu_0 \mu_r i}{2\pi R}$, saturation current of these materials could be calculated. Bus bar current on power transmission lines is between 5 and 1000 A, R = 4cm, saturation currents results of these different materials is as Table 1.

Table 1. Saturation Current of some general magnetic materia

	Non- brilliant alloy of iron base	Cold-rolled silicon iron	Nanocrystalline Alloy	Permalloy
Saturation current(A) (R = 5cm)	<1.2	<32	<0.96	<0.4

As shown in Table 1, Cold-rolled silicon iron has the largest saturation up to 32A; its saturation flux density and magnetic permeability are relative high and could have better performance under that current condition than any other magnetic material, so we choose silicon iron as the magnetic core of the current transformer.

2. Architecture of the core.

A magnetic core with big size could induct more power from the magnetic field than small one, and it would be helpful to broaden bus bar current window. Nevertheless, big core will increase weight of the robot and surpass the limit of motor drive capacity. So a core with enough power output and not so big size is required.

Power requirement is calculated as follow. Charging current of this system is designed up to 1.5A for Li- battery pack and voltage of the battery is 25.2V. So the output power of the charging system must be more than 37.5W (i.e. $P = U * I = 37.5W$). Besides, power consumption of the system must be considered. It is mainly distributed in the core dissipation and the switching power supply dissipation. Considering that the super limit of the core power dissipation 25%[6] and the efficiency of the switching power supply could be up to 85%[7], the real power of the system could be up to 58.8W(P=37.5/0.85/(1-0.75)). Using the semi-empirical relationship:

$$s = 1.25 * \sqrt{P} \tag{1}$$

So, $s = 1.25 * \sqrt{P} = 9.58cm^2$ s represents the cross-section of the core. The size is not so big that we can expand the area of section up to $12cm^2$. Output power could be more than 90W. Core of current transformer is divided into two semicircles and attached to one arm of the robot. When negotiates with obstacles, the core opens up, moves upward and depart from the transmission line. While working on a normal inspection, the two semicircles will move downward, and close themselves to make a magnetic circuit for induction of power. Two extra motors fixed on this arm to move up and down the current transformer, and to open or close, so the transformer could encircle the transmission line or release it whenever the robot needs. Fig.3 shows the design details and relative position of current transformer and transmission lines.

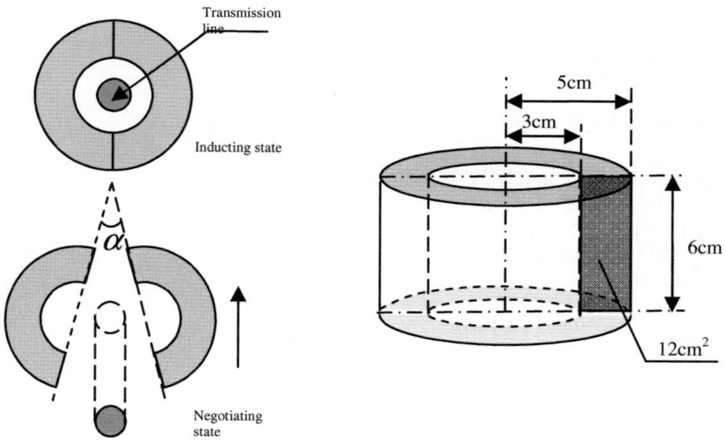

Fig. 3. Core parameters and Relative position of current transformer and transmission lines

3. Line width of current transformer.

The wider the enameled wire is, the less the copper loss is. So, at first, enameled wire must satisfy the need of secondary current. The input voltage of the switching power supply unit is in range of 85~265V and the minimum system power 58.8W, the maximum secondary current could be figured out as follows:

$$I_{MAX} = P / U_{MIN} \approx 0.7A \tag{2}$$

In terms of construction of the winding and the temperature working condition, current density J could be figured as 3A/mm2, so the minimum diameter of the enameled wire is,

$$\phi = 2 * \sqrt{\frac{I_{MAX}}{\pi * J}} \approx 0.59mm \tag{3}$$

So wire in size AWG22 is selected.

4. Induction voltage calculation

Fig.4 shows the relationship between transmission and the cross-section. The intensity of magnetic field is taper down from cylindrical surface A to B. Calculating the value of magnetic need definite integral.

$$\phi = B \bullet s$$

$$d\phi = B \bullet ds = \frac{\mu i}{2\pi R} ds = \frac{\mu i}{2\pi R} hdR$$

$$\phi = \int_{r1}^{r2} \frac{\mu i}{2\pi R} hdR = \frac{\mu i h}{2\pi} \int_{r1}^{r2} \frac{1}{R} dR$$

$$= \frac{\mu i h}{2\pi} \ln \frac{r2}{r1}$$

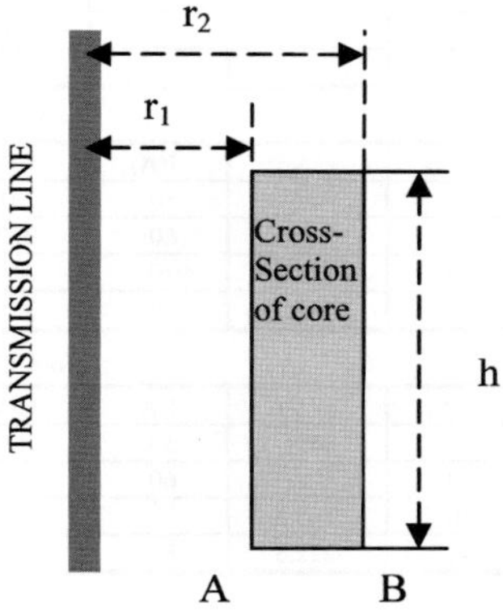

Fig. 4. Calculation of magnetic flux

Bus bar current:

$$\dot{I} = I_0 sin(2\pi f \omega)$$

So induction voltage could be:

$$\dot{E} = N\frac{d\dot{\phi}}{dt} = N\frac{\mu h}{2\pi}\ln\frac{r2}{r1}\frac{d\dot{I}}{dt}$$

$$= N\frac{\mu h}{2\pi}\ln\frac{r2}{r1}I_0 2\pi f cos(2\pi f \omega)$$

$$= N\mu h I_0 f \ln\frac{r2}{r1}cos(2\pi f \omega)$$

Then the RMS of induction voltage:

$$E = \frac{1}{\sqrt{2}}N\mu h I_0 f \ln\frac{r2}{r1}$$

4 Experiment and Results

1. Induction voltage output.

The output of current transformer is shown in Table 2. Number of turns is respectively 200, 400, 600 and 800. Current arrange is from 0 to 200A.

Table 2.

INDUCTION VOLTAGE OUTPUT OF 200 TURNS

I(A)	U(V)	I(A)	U(V)	I(A)	U(V)	I(A)	U(V)	I(A)	U(V)
10	10.8	50	65.8	90	88.8	130	94.7	170	97.4
20	25.6	60	76.0	100	91.0	140	95.6	180	97.7
30	40.6	70	82.7	110	92.3	150	96.3	190	98.1
40	54.3	80	86.4	120	93.6	160	97.0	200	98.4

INDUCTION VOLTAGE OUTPUT OF 400 TURNS

I(A)	U(V)	I(A)	U(V)	I(A)	U(V)	I(A)	U(V)	I(A)	U(V)
10	22.3	50	136.0	90	176.2	130	187.1	170	192.6
20	52.9	60	153.0	100	179.7	140	189.1	180	193.4
30	85.5	70	164.0	110	182.6	150	190.5	190	194.1
40	112.8	80	171.1	120	185.1	160	191.6	200	194.7

INDUCTION VOLTAGE OUTPUT OF 600 TURNS

I(A)	U(V)	I(A)	U(V)	I(A)	U(V)	I(A)	U(V)	I(A)	U(V)
10	33.5	50	201.2	90	263.4	130	280.4	170	288.8
20	77.7	60	227.1	100	269.1	140	283.2	180	290.0
30	123.5	70	244.8	110	273.5	150	285.6	190	291.0
40	165.5	80	255.8	120	277.3	160	287.2	200	291.9

INDUCTION VOLTAGE OUTPUT OF 800 TURNS

I(A)	U(V)	I(A)	U(V)	I(A)	U(V)	I(A)	U(V)	I(A)	U(V)
10	44.8	50	269.8	90	349.5	130	372.8	170	383.5
20	102.2	60	301.9	100	357.0	140	376.1	180	385.2
30	165.3	70	325.8	110	363.2	150	379.1	190	386.7
40	222.5	80	339.8	120	368.3	160	381.6	200	388.0

The data relationship between bus bar current and effective induction voltage is showed in Fig.5. The 4 lines with different spots reflect the different output voltage from different amount of secondary turns.

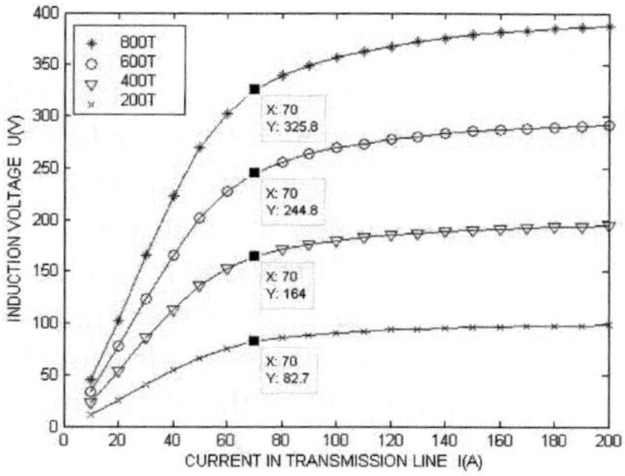

Fig. 5. Induction voltage output

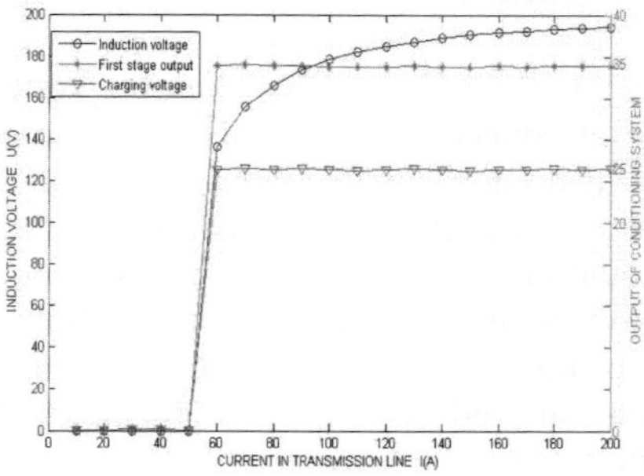

Fig. 6. Output of conditioning unit

So, as shown in Fig.5, when bus bar current is less than 50A, there is a linear relationship between induction voltage and bus bar current. When bus bar current is between 60~100A, magnetic core turns from light saturation to deep saturation. When bus bar current is higher than 100A, the magnetic core is thoroughly saturated and under this situation, current altering in bus bar could not change the value of magnetic induction intensity. So induction voltage would in form of square wave. Besides, voltage has direct ratio relationship with turns as shown in Fig.5.

2. Output voltage of conditioning unit.
The output voltage of current transformer is showed in Table 3. Turns: 400. Current arrange: 0 ~ 200A. Data in grey background is effective, as the output voltage is regulated to 25V with this bus bar current and induction voltage. From Table3, compared with no-load results, it is obvious that the induction voltage is lower. So, in load-on state, higher bus bar current is needed to get equal induction voltage with non-load situation

Table 3. Induction voltage output of 400 turns

I(A)	U(V)	I(A)	U(V)	I(A)	U(V)	I(A)	U(V)	I(A)	U(V)
10	--	50	--	90	173.7	130	186.9	170	192.3
20	--	60	136.1	100	178.6	140	188.8	180	193.3
30	--	70	155.7	110	182.1	150	190.3	190	193.6
40	--	80	166.1	120	184.9	160	191.5	200	194.1

Fig.6 shows the relationship between induction voltage, first stage transmission voltage and conditioning circuit output voltage (i.e. charging voltage in second stage transmission). When induction voltage is lower than 60V, there is no regular voltage output from first stage transmission, so does the charging voltage; this is because the input voltage of transformer is too low to start up PWM function of the switch. When induction voltage is higher than 60V, PWM starts, and regular voltages are achieved. Then voltage and current for charging Li battery packs is accessible.

5 Conclusion

In this paper, an on-line charging system for power transmission line inspection robot is introduced. Through this induction power supply, the robot can get power from the transmission line while working on it. Details about the induction power supply system are given in the paper and the experiments shows that this charging system could help the robot prolong continuous working period. The inspection robot could have better performance and reliability in implementing the task of inspecting power transmission line. In the future, research will focus on widen the accessible current in bus bar, and develop the technology of charging and supplying at the same time for the robot.

Acknowledgment

Work supported by the Major National S&T Program-"Superior NC Machine Tool and Basic Manufacturing Equipment" (2009ZX04013-011) of P. R. China.

References

1. Tang, L., Fang, L.J., Wang, H.G., Zhang, H.Z.: Research on an inspection robot control system of power transmission lines based on a distributed expert system. J. of Robot, P.R. China (3) (May 2004)
2. Peungsungwa, S., Pungsiri, B., Chamnongthai, K., Okuda, M.: Autonomous robot for a powertransmission line. In: The 2001 IEEE International Symposium on, vol. 2, pp. 121–124 (2001)
3. Sawada, J., Kusumoto, K., Maikawa, Y., Munakata, T., Ishikawa, Y.: A Mobile Robot For Inspection Of Power Transmission Lines. IEEE Transactions on Power Delivery 6(1) (January 1991)
4. Beltran, H., Segundo, S., Fuster, V., Perez, L., Mayorga, P.: Automated Inspection of Electric Transmission Lines:The power supply system. IEEE, Los Alamitos (2006), 1-4244-0136-4/06
5. Wu, G.P., Xiao, X.H., Guo, Y.L.: Development of a crawling robot for overhead high-voltage transmission line. China Mechanical Engineering 17(3), 237–240 (2006)
6. Zhu Li. Transformer. Chemical Industry Press (May 2009)
7. ON Semiconductor. Switch-mode power supply reference manual (September 1999)

Design for AC/DC Converter with High Power Factor Based on UC3854

Tingjian Zhong, Zhongqing Du, Zunnan Min, and Ye Mao

Jiang Xi Vocational &Technical College of Electricity,
Nanchang, China Nanchang, 330031 China
jxdlztj@163.com

Abstract. A 300W AC/DC converter with a universal input and fixed output voltage is analysed and then designed based on UC3854.The principle and structure of an active power factor corrector (APFC) and UC3854 is discussed in this paper. The experiment result shows that APFC can achieve a fixed output of 400V DC voltage in the universal input range of AC 85~265V. The designed EMI filter can effectively reduce the electromagnetic interference. And the experimental result verifies that the power supply can make the power factor reach over 0.99, and total harmonic distortion reach lower then 5%.

Keywords: Active power factor correction, AC/DC Converter, Total harmonic distortion, UC3854.

1 Introduction

In recent years, the high-frequency switching model power supply has been widely applied in personal computers, television sets and so on [1]. Since the weight, size and power consumption of switching power supply is less than linear power supply, the switching power supply was widely applied. With more and more use of switching power converters, the understanding to the harmonic brought by the switching power supply is growing depth. The requirements to it have been continuously improved, requiring their high efficiency, high power factor, high power density and high reliability. Moreover, the people have proposed higher and higher demands to electrical energy and electromagnetic compatibility (EMC) [2]. In order to reduce the pollution of large overshoots and undershoots from the rectifier filter circuit to the power system, and improve power factor, so that the input current harmonic can meet international standards, there is growing concern to the active power factor correction (APFC) circuit technology.

In order to solve the problem, the paper designs an electromagnetic interference (EMI) filter [3] in AC input to reduce the input electromagnetic interference, and then a Boost circuit is designed on the basis of UC3854 IC. Finally, the simulation and experimental results are presented to prove the effectiveness of the design of the converter.

X. Wan (Ed.): Electrical Power Systems and Computers, LNEE 99, pp. 737–746.
springerlink.com

2 Design of Emi Filter

Fig.1 shows the basic structure of EMI filter, which is composed by L_1, L_2, C_1, C_2, C_3, C_4. L_1 and L_2 are the common-mode inductances around the same iron core, and have the same turns and round direction. The filter capacitor is C_1 to C4. L_1 and C3, L_2 and C4, constitute a common-mode noise filter and an eliminate power line common mode noise, respectively. Since it is impossible to ensure L_1 and L_2 all the same in the inductor production process, they form a differential mode inductance between them. The differential mode inductance and C_1, C_2 constitute a differential mode noise filter to eliminate differential mode noise. The inductance of L_1 and L_2 is normally dozens of mH. Normally, C_1 and C_2 are chosen from the ceramic capacitors or polyethylene film capacitors and the capacity is generally 0.01 to 0.47 μ F. C_3 and C_4 are chosen from the ceramic capacitors and the capacity is generally 2200 pF to 0.1μF.The most voltage value is of $C_1 \sim C_4$ is 630V DC or 250V AC

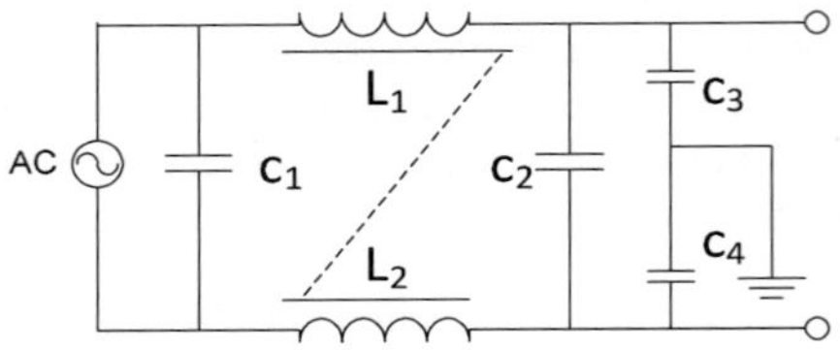

Fig. 1. The structure of EMI filter

3 Design of Pfc Double Closed-Loop Model

The design of the current control loop and voltage control loop is the key steps in the whole design. The compensation of the current error amplifier and voltage error amplifier should make the current loop stability. At the same time, the input current distortion of voltage loop is the smallest. [4] [5]

3.1 Eloped-Loop Control Structure of Current and Its Transfer Function

The current closed-loop is made up of a current error amplifier, a PWM comparator and a power amplifier. The current error amplifier shown in Fig.2(proportional plus integral controller). The transfer function is:

$$G_{CEA} = \frac{U_{CAO}}{U_{MO} - U_S} = \frac{1 + C_1 R_2 S}{R_1 C_1 S (C_2 R_2 S + 1)} \tag{1}$$

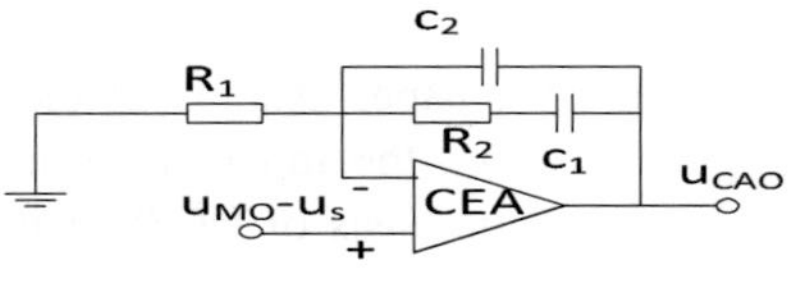

Fig. 2. Current error amplifier

The transfer function of PWM comparator is:

$$G_{PWM} = \frac{D(S)}{U_{CAO}} = K_{PWM} \tag{2}$$

The transfer function of power amplifier is:

$$G_{PS} = \frac{U_S(S)}{D(S)} = \frac{U_O D(S) R_S}{D(S) SL} = \frac{U_O R_S}{SL} \tag{3}$$

Where, Us(S) is the voltage of sampled resistor Rs;

Uo is voltage output of power circuit;
Rs is sampled resistor;
L is inductance of power circuit.

The current loop may be described by the block diagram shown in Fig.3.

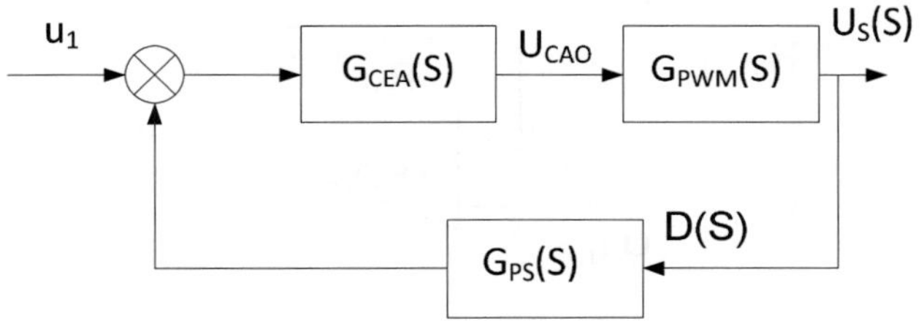

Fig. 3. Block diagram of current closed-loop control

So the current open loop transfer function is:

$$\begin{aligned} G_I &= G_{CEA}(S) G_{PWM}(S) G_{PS}(S) \\ &= \frac{K_{PWM} U_O R_S (1 + C_1 R_2 S)}{S^2 L R_1 C_1 (1 + C_2 R_2 S)} \end{aligned} \tag{4}$$

Thus the current loop is the second-order non-poor system. It can track sinusoidal input function no error.

3.2 Voltage Closed-Loop Control Structure and Transfer Function

External voltage loop, which is composed by the voltage error amplifier, a multiplier and a current loop as the core components, is described by the block diagram shown in Fig.4. H is the partial voltage network formed of R_1 and R_2 in Fig.5.

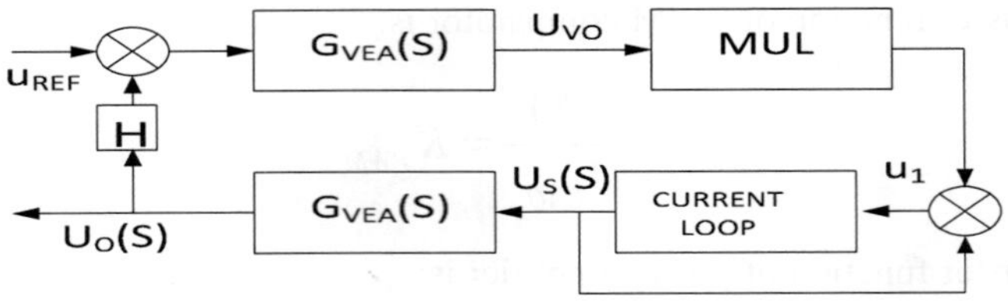

Fig. 4. Block diagram of APFC double closed-loop system

Where, the voltage error amplifier is shown in Fig.5. The transfer function is:

$$G_{VEA}(S) = \frac{U_{VO}}{U_{REF} - U_F} = \frac{R_V}{R_1} * \frac{1}{1 + SR_V C_V} \tag{5}$$

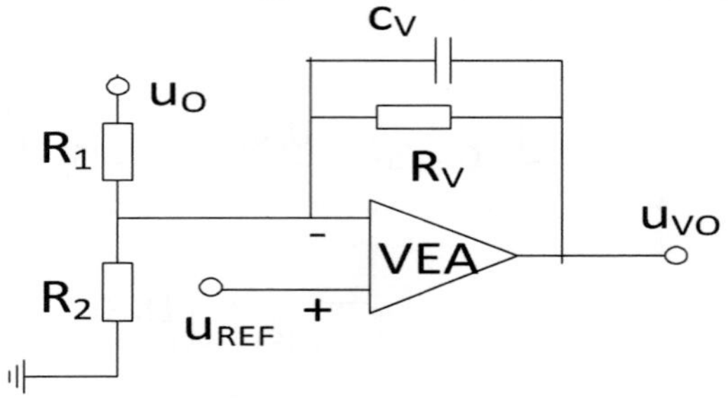

Fig. 5. Voltage error amplifier

The voltage loop posed mainly by the current loop accepts for voltage error amplifier. Therefore, the structure diagram of Figure 4 can be simplified as Fig. 6.

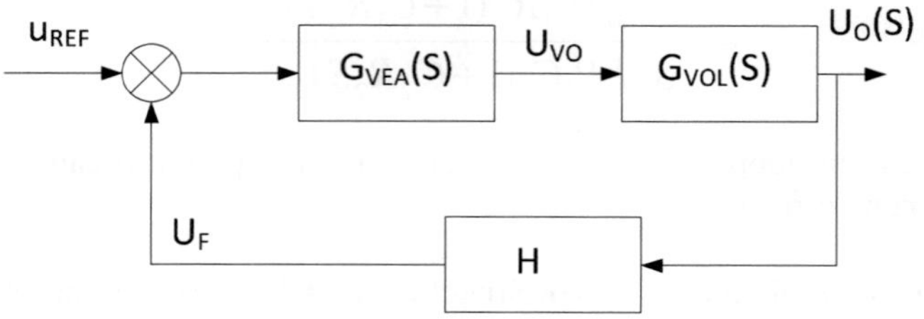

Fig. 6. Simplified structure diagram of voltage loop

The transfer function of control components including current loop can be given as:

$$G_{VOL}(S) = \frac{U_O(S)}{U_{VO}(S)} = \frac{P_{IN}}{SC_O U_O \Delta \mathrm{U}_{\mathrm{VEA}}} \tag{6}$$

Where, P_{IN} is input power;

C_o is the capacitance value of power circuit;
U_o is DC output voltage;
ΔU_{VEA} is the greatest output voltage of error amplifier.

The transfer function of voltage open-loop system is:

$$\begin{aligned} G_I &= G_{VEA}(S) G_{VOL}(S) \\ &= \frac{P_{IN} R_V}{S\ U_O R_1 C_0 \Delta \mathrm{U}_{\mathrm{VEA}} (1 + C_V R_V S)} \end{aligned} \tag{7}$$

So this system is the first-order non-poor system. It can track input signal no error, which means that the output is stable if U_{REF} be given.

3.3 Feed-Forward Voltage Filter Structure and Transfer Function

There are many high-order harmonic after rectifier and it may cause input current distortion eventually if this harmonic inflow multiplier. Therefore, the feed-forward voltage ripple should be very small. So the system uses bipolar point Filter to decrease the feed-forward voltage ripple. Figure 7 shows it (that is the K network in Fig.8). The transfer function is:

$$G_K(S) = \frac{U_2(S)}{U_1(S)} = \frac{R_3}{R_1 R_2 R_3 C_1 C_2 S^2 + R_1 C_1 (R_2 + R_3) S + R_3 C_2 (R_1 + R_2) S + R_1 + R_2 + R_3} \tag{8}$$

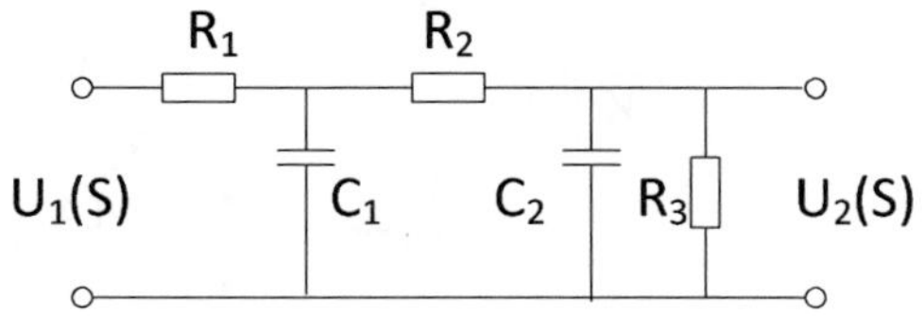

Fig. 7. Feed forward filter

4 Working Description of Boost Pfc Based On Uc3854 and Simulation

4.1 Working Principle Description

The error signal of the output voltage, which is the output of feedback voltage and the voltage reference, is amplified by the voltage error amplifier, and it is proportional to the full-wave rectifier voltage and current signals. The input feed-forward voltage is multiplied by the full-wave rectifier voltage and current signal in the multiplier, and then gets a benchmark current signal I_{MO}. The voltage U_{MO} got by resistivity multiplied I_{MO} has the same input waveform as the rectifier voltage. The inductor current I_L is multiplied the sampling resistor and get the sampling voltage U_L, which inject into the current error amplifier together with U_{MO}. Therefore, U_{MO} (the voltage difference) and U_L should also be the same, so U_{MO} forces the main circuit current to track the input voltage rectifier sine-wave. The output voltage of current error amplifier and a triangular voltage in wave inject into PWM comparator, and then produce a PWM pulse signal, which can drive the switching transistor.

4.2 PFC Converter Structure

As shown in Fig.8, the APFC system consists of EMI filter, rectifier circuit, boost circuit and control circuit based on UC3854.

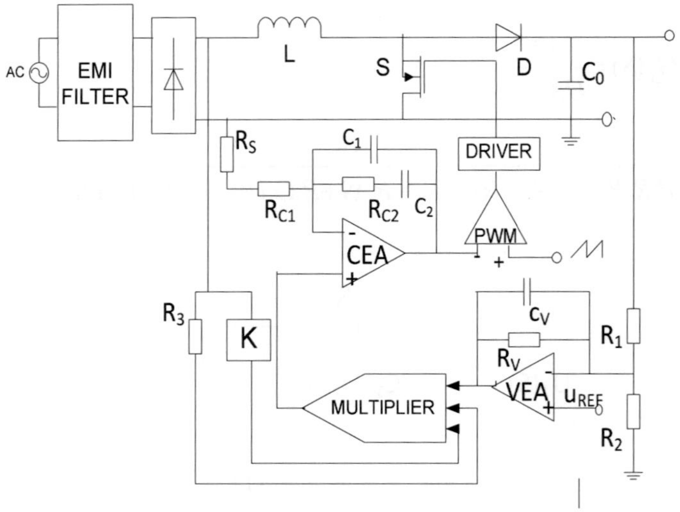

Fig. 8. Principle diagram of APFC

4.3 Simulation and Results

In order to analysis the signal mechanism further and the feasibility of realization, the simulation is carried in this paper using Pspice. The main simulation parameters are provided in Table 1.

Table 1. Simulation Circuit Parameters

Switching frequency	100KHZ
Fundamental frequency	50 HZ
Ac supply voltage	160 V
Inverter dc voltage (V_{dc})	400 V
Rectifier load resistance	5 Ω
Sampling resistance	0.25 Ω
Inverter side inductance	1 mH
Capacitor Co	1200 uF

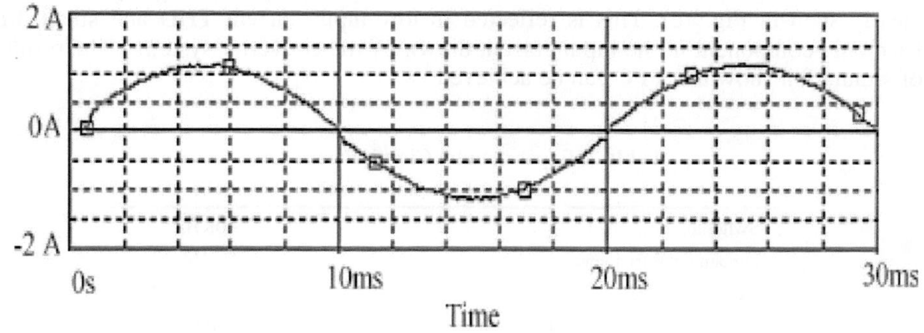

Fig. 9. (a) Input current waveform

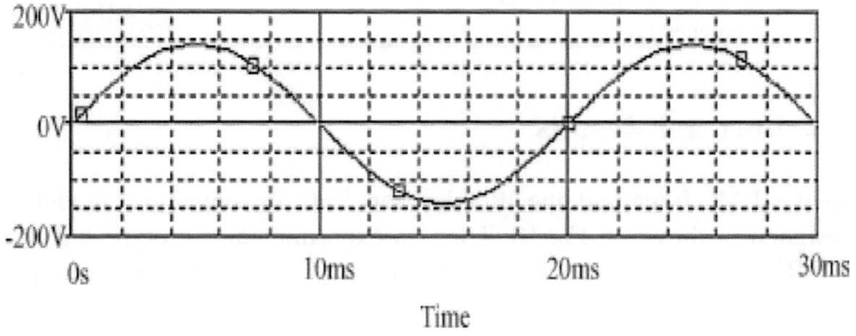

Fig. 9. (b) Input voltage waveform

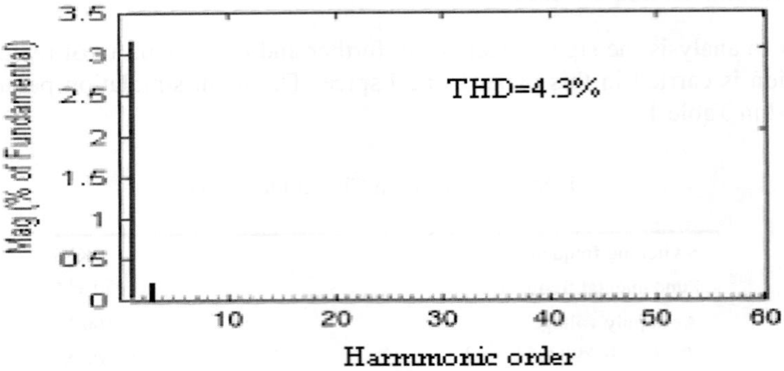

Fig. 9. (C) Harmonic spectra of the input current

Fig.9 (a) shows the results of input current waveform. Fig.9 (b) shows the results of input voltage waveform. The excellent harmonic cancellation function can be clearly seen in Fig.9(c). This is reflected in low input current THD and small 3rd harmonic component in the input current drawn by the converter systems. The results of simulation show that PFC can be achieved.

Table 2. Experimenttal Circuit Parameters

Switching frequency	100KHZ
Fundamental frequency	50 HZ
Ac supply voltage	85V~265V
Inverter dc voltage (V_{dc})	400 V
Switching transistor	IXFH26N60Q
Output power	400W
Inverter side inductance(L_1)	800 μ H
Capacitor Co	1200 μ F/450v

5 Experimenttal Results

Based on the EMI filter and the double loop PFC mathematical model, the main circuit designed is shown in Fig.10. Then an experimental prototype has been set up and tested. Regarding the above instalment, the parameters established are provided in Table 2.

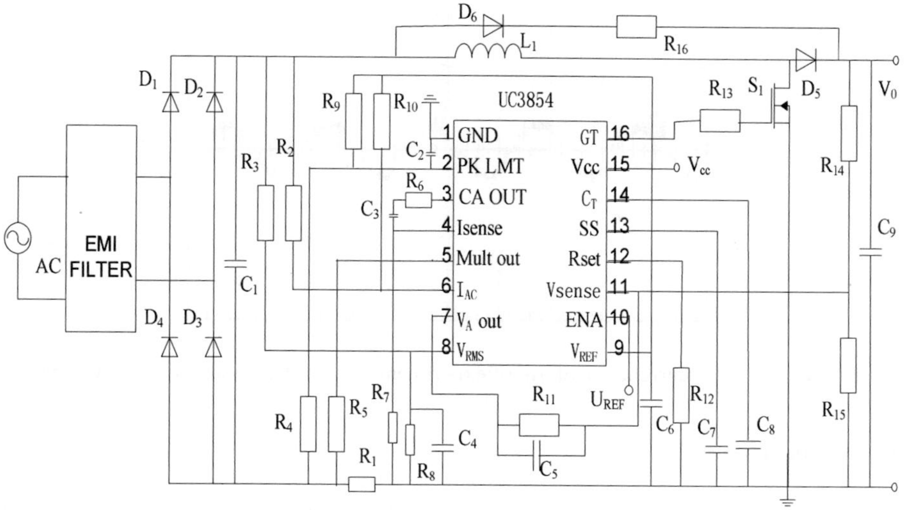

Fig. 10. Design of APFC circuit

Fig.11 shows experimental waveforms for the load condition of the controlled rectifier. The power factor is improved to unity with the designed circuit based on UC3854. It can be seen from Fig.11(b) that the supply current is almost sine waveform and follows the supply voltage shown in Fig.11.(a) in its waveform shape with almost a unity displacement power factor. From these figures, we can conclude that the effectiveness of the proposed structure for the designed circuit based on UC3854 is clear.

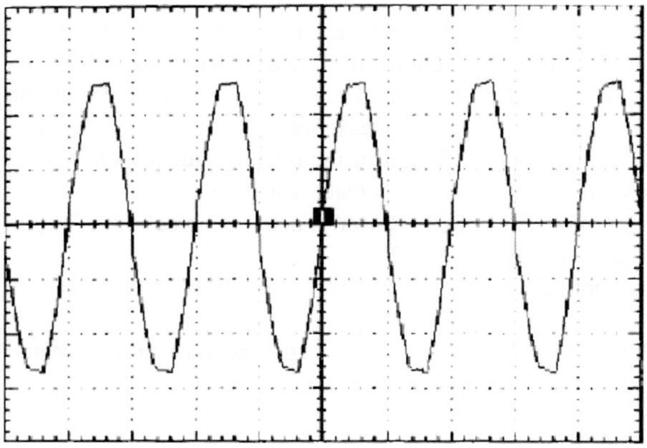

Fig. 11. (a) Input voltage waveform of experiment V [20V/div] input voltage 250v 25ms

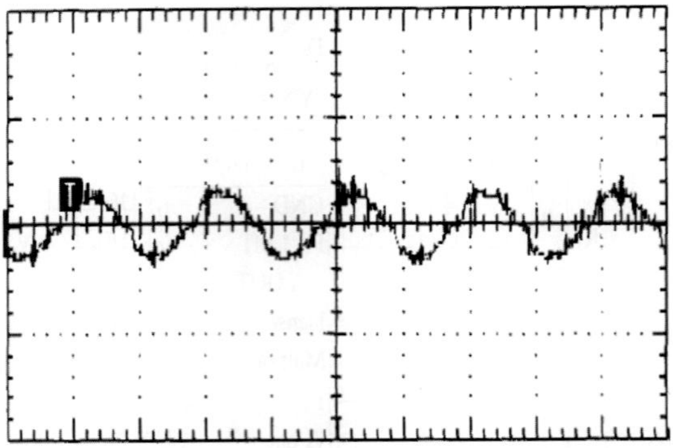

Fig. 11. (b) Input current waveform of experiment I [2.5A/div] input current 5A 25ms

6 Conclusion

In this paper, a APFC circuit based on UC3854 with high input power factor has been designed. The performance of the proposed APFC is verified through simulation studies with Pspice. An experimental APFC has been carried out on designed circuit to explore the advantages and practical implementation with the proposed control circuit. The simulation and experimental results show that APFC circuit based on UC3854 is entirely feasible. The input current THD is far below the specified standard in steady-state case. The circuit is simple and stable and it can be applied to different power systems and drivers.

References

1. Wang, z., Liu, j.: Advances of harmonic suppression and reactive power compensation technique for power electronic equipment. Power Electronics 1 (1997)
2. Lin, j., Li, j., Shen, y.: Application of active power factor correction technology in the power source of 1kW. Journal of southeast university (Natural Science Edition) 33 (2003)
3. Liu, X., Li, G.-z., Guo, D.-d.: Design of the main circuits of single-phase APFC converter. Journal of Xian university of Arts & Science 10(1) (2007)
4. Todd, P.C.: UC3854Controlled Power Factor Correction Circuit Design. Unitrode Application Note,U-134(3), 269–289
5. Andreycak, B.: Optimizing Performanc In UC3854Power Factor Correction Applications. Unitrode, Design Note, DN-39E
6. Zhou, x.-y., Zhang, x., Wu, y.: Research of active power factor corrector based on one-cycle control. Power electronic 41(1) (2007)

Research of Video Image Acquisition System Based on ARM9

Li-bo Ding[1] and Li-hua Sun[2]

[1] Dept. of Electric and Signal Engineering
School of Information Engineering
Nanchang University
Nanchang, China
dlb2046@163.com
[2] School of Software
Nanchang University
Nanchang, China
slh52@163.com

Abstract. Video image capture technology plays an important role in the security field, Qt is a mature cross-platform GUI toolkit written in C++, it provides software developers with unified elegant graphical user interface, and Qt class library can support cross-platform, also can be transplanted into different operating system platform conveniently, based on these, the paper constructs a video acquisition system based on embedded Linux and S3C2440A microprocessors, use V4L to collect camera images date, then the camera images date are compressed by MPEG-4 and transmitted to the terminal, the experimental results show that the system has good reliability and a certain practicality.

Keywords: Video4Linux, S3C2440A, RTP, Qt.

1 Introduction

Embedded video acquisition system is an integrated system, which integrated multiple technologies such as embedded technology, multimedia information, network communication, graphic display and other technology. It has a broad application perspective in industrial daily life field, such as security monitoring and control, video chat, videophone, etc. Reliable, compack and lightweight embedded video capture system has broad market demand. Since ARM is an excellent microprocessor, and can work well with Embedded Linux, this article used ARM9 and Embedded Linux as a development platform.

2 The Overall Design of the System

The system adopts C/S structure and mainly divided into two parts: monitoring terminal and client. Monitoring terminal includes video data collection, coding and

X. Wan (Ed.): Electrical Power Systems and Computers, LNEE 99, pp. 747–751.
springerlink.com

transmission in S3C2440 development board, the client is receiving, decoding and display program running on PC. Monitoring terminal capture the original image dates in real time. After coded and compressed, image dates are transmitted via Ethernet, so any PC with the client on the network can see monitoring information. The overall design is shown in Fig. 1.

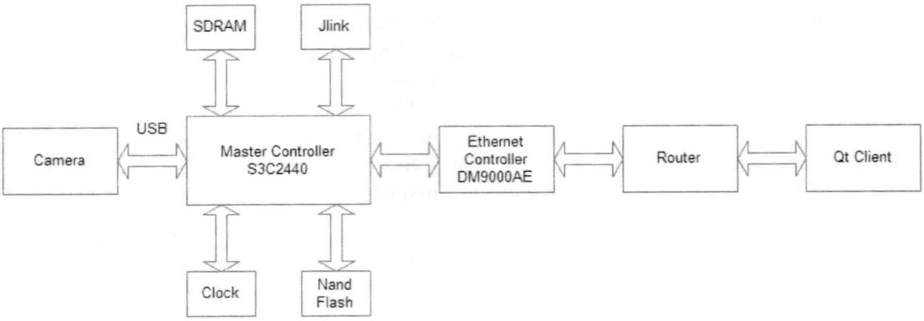

Fig. 1. The structure of the system

The outstanding feature of Samsung S3C2440[1] is its processor core,which is based on ARM920T core. ARM920T implements MMU, AMBA bus and Harvard cache architecture of the structure. This structure has a separate 16KB instruction cache and 16KB data cache.

3 Acquisition and Transmission of Image Data

Under the Linux, device driver actually is an abstract of hardware function, on the same hardware different driver can make hardware encapsulated into different function, device driver is the medium of hardware layer and the applications (or operating system), can make application or operating system use hardware. Linux OS have three kinds of major device file type: block device, character device, and network device.

Video4Linux (V4L) is the kernel driver about video device in Linux, it provides a series of interface functions for the application programming of video devices[2]. These interfaces are used to program all video device drivers in Linux, the general process of video acquisition: (1) Open the video device; (2) read device information; (3) change the current device setting (if required); (4) capture video; (5) process the video captured; (6) close video device; the most important is the fourth step, there are two ways to obtain images, directly read device and use mmp memory mapping, the work way of read() is to read date from the kernel buffer into memory, and mmap() adopts the method of memory mapping, namely map device file into memory, that is camera image buffer and image data area (which users can access) share a memory area, thus bypassing the kernel buffer, making among processes realize sharing memory by mapping the same file, the key steps is as follows:

(1) To initialize video_mbuf, to get the information of the mapped buffer. octl (vd->fd, VIDIOCGMBUF, &(vd->mbuf))

(2) To bind mmap and video_mbuf. void *mmap(void *addr, size_t len, int port, int flags, int fd, off_t offset).

(3) To initialize before image acquisition. Call function extern int v4l_grab_init(v4l_device *, int, int), here vd->frame_using[0] and vd->frame_using[1] are set to 0, means that the interception of two frames has not yet begun, to bind mmap and video_mbuf.

(4) To began collecting images. Call ioctl(vd->fd, VIDIOCMCAPTURE, &(vd->mmap), image has been obtained after call, vd.map pointer points to the first frame which has received, the position of the image stores vd.map+vd.mbuf. offsets[vd.frame_current]. When vd.frame_current=0, namely the first frame position, when vd.frame_current=1, namely the second frame position.

Then the image data collected was sent to PC machine, RTP is currently the best way to solve streaming media real-time transmission. RTP protocol is mainly used to transmit real-time audio/video data, which includes both RTP and RTCP packets[3].

In Linux platform real-time transmission programming may choose to use JRTPLIB library, it is RTP library using C++ language, the detailed operating is as follow: first, download the latest source package form JRTPLIB website, store in /tmp after download, execute the order #tar -zxvf jrtplib-2.8.tar, and then configure and compile for JRTPLIB, type order #cd jrtplib-2.8□ #./configure CC=arm-linux-g++ cross-compile=yes, modify Makefile file, change link command (ld and ar) into arm-linux-ld and arm-linux-ar, #make, finally executing the following command can be able to complete JRTLIB installation. The detailed steps to create RTP transmission are as follows.

a. before using JRTPLIB to perform real-time transmission, first, to generate an instance of RTPSession class represent the RTP session, and then to call creat() method initialize its operation, RTPSession sess; sess.Create(5000).
b. Then to set the appropriate timestamp unit, which is another important work to perform in the initialization process of RTP Session, this is realized by calling SetTimestampeUnit() method of RTPSession, this method also has only one parameter, expressed timestamp unit by the second, sess.SetTimestampUnit (1.0/8000.0).
c. When the RTP session is successfully established, the next can begin real-time transmission of streaming media data, first, it need to set correct target address to send data, RTP protocol allows the same conversation havemultiple target addresses, it can use AddDestination(), DeleteDestination() and ClearDestinations() methods of RTPSession class to complete.
d. After all target addresses are pointed, then SendPacket() method of RTPSession can be called to send streaming data to all target addresses.

4 Image Display

Qt is a signature product of Trolltech,which is a cross-platform C++ graphical user interface. It contains a class library, and used for cross-platform development and international tool[4]. The QPixmap objects can be used to implement real time video stream, which supports a variety of image formats, including JPEG format. There are tow classes for processing image int the Qt: QImage and QPixmap[5]. QImage provides a single pixel accurate image processing capabilities,QPixmap provides a double buffering mechanism. When a QPixmap object processed image is displayed on the screen, the other QPixmap object to process the next frame in the background. The process is shown in Fig. 2.

Fig. 2. Double buffering mechanism

First creat a new QPixmap object, and then load the image pixel data(use the function of loadFromData), then use the new QPixmap as a parameter to set the member variable(use the function of SetPixmap), at last use the member function of QLabel to display images.

5 Experimental Test

In experimental test, two performance index(packet drop rate and bit error rate) were selected to evaluate, and the image transmission rate slowed to 2 frames per second, the test time is about 20 to 60 minutes, through testing the packet drop rate and bit error rate are shown in Fig.3 and Fig.4.

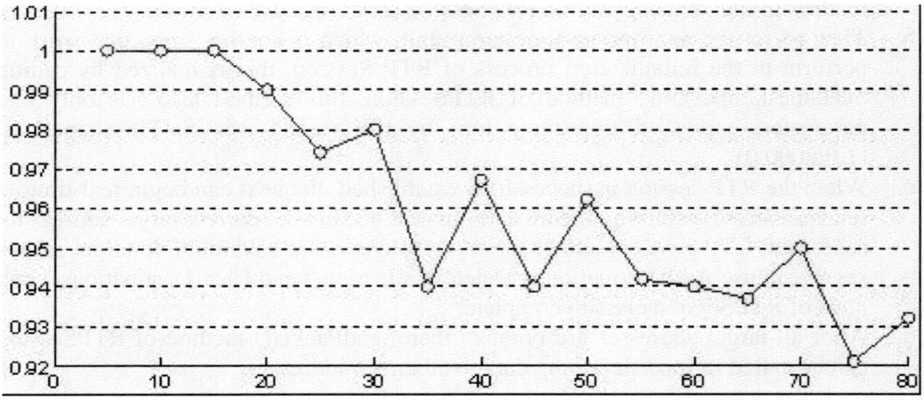

Fig. 3. Gateway packet successful receive rate

In Fig.3, the horizontal axis is time, the vertical axis is packet successful receive rate, it can be found that packet successful receive rate have declined with the increase of time.

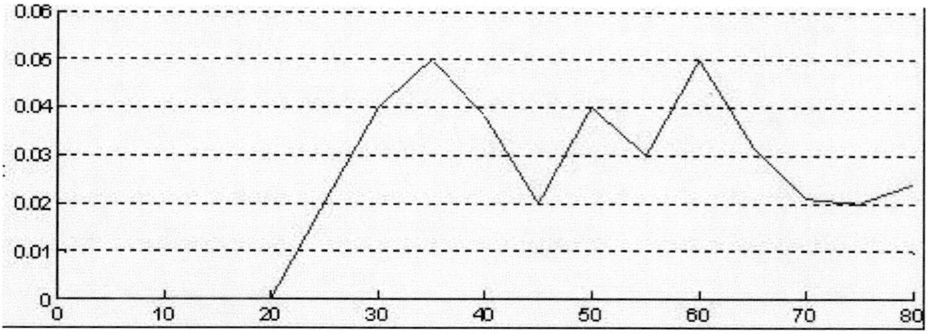

Fig. 4. The gateway bit error rate

In Fig.4, the horizontal axis is time, the vertical axis is bit error rate, it can be found that successful packet rate have declined with the increase of time. It can be found that as time tends to increase the gateway bit error rate tends to stability.

6 Conclusion

This paper introduces a video image acquisition system based on ARM9 kernel, this system can transmit the collected data to the Qt client, experimental test show the performance of the system has dropped when frame number send off frequently, but overall stability of the system can be in acceptable range. In future experiments, research will focus on the video coding algorithms in order to improve the transmission efficiency.

References

1. Samsung Corporation. S3C2440A datasheet. Samsung Corporation, 30-45 (2002)
2. Zou, J., Hu, P.: Design of embedded video monitoring system. Foreign Electronic Measurement Technology (11)(2010)
3. Zhang, H.J., Zhang, J.J., Yang, Y.G., Wu, K.J.: Research of RTP transmission control and design and application of real-time video montitoring system. Network and communication (5) (2009)
4. Zang, Y.-j., Kong, S.: The Transplantation and Application of Embedded GUI Based on Qt/Embedded and Qtopia. Journal of Tianjin Vocational Institute (1) (2010)
5. Blanchette, S.: C++ GUI Qt4 Program, pp. 150–200. Electronic Industry Press (2005)

The Application of Asterisk-Based IP-PBX System in the Enterprise

Duo Xiang[1] and Li-hua Sun[2]

[1] Dept. of Electric and Signal Engineering
School of Information Engineering
Nanchang University
Nanchang, China
xd520@163.com
[2] School of Software
Nanchang University
Nanchang, China
slh52@163.com

Abstract. Soft switch is the core technology of the next generation packet network, which realized call control function by software. IP-PBX is a typical application of soft-switching technology. Because of many advantages such as low cost, easy maintenance, etc, Asterisk-based IP-PBX system became more and more welcome to small and medium firms. The paper describes the modern soft-switching platform Asterisk, designs and implements a municipal call center.

Keywords: IP-PBX, Asterisk, Call Center.

1 Introduction

Since the emergence of telephone, PSTN (Public Switched Telephone Network) which based on circuit switching and data exchange network which based on packet provide voice and data service respectively, this kind of structure brings a lot of problems to enterprises. For the purpose of making full use of the network resources and reducing the cost of communication, people start to consider about the possibility of implementing two services in one machine. The development of IP and Soft Switch makes it possible to connect internet and telephone network, IP-PBX is proposed at this time.

2 IP-PBX System

IP-PBX is a Soft Switch that using packet switching technology, the system consists of one or more SIP phone or Internet phone and an IP-PBX server[1]. Its working principle is the client (soft phone or regular phone) first register the IP-PBX server, when need to call, it sends a connection request. Because of having all the SIP address of the users, the server can use the network gateway or VoIP service to connect to an internal call or send an external call.

X. Wan (Ed.): Electrical Power Systems and Computers, LNEE 99, pp. 753–757.
springerlink.com

Compared with the traditional PBX, IP-PBX system has many advantages:

(1) IP-PBX system built on IP protocol, the products follow a unified standard, so it's easy to connect and maintain.

(2) Convenient to extend. When you need to expand some more phones on the original PBX system, it only cost half spending as the traditional PBX.

(3) Powerful and highly integrated. Single system can do lots of things that traditional PBX system should finish them with a variety of external devices, such as automatic attendant, voice mail and so on.

(4) IP-PBX can use VoIP features to call long distance calls at local price, therefore save the cost of communication significantly.

To sum up, as the enterprise-class soft switch, IP-PBX system not only has many flexible functions of IP network, but also is compatible with the traditional PBX. So it's a hotspot currently.

3 Asterisk Platform

Because of IP-PBX system's high cost performance above, more and more businesses prepare to use the IP-PBX instead of the traditional PBX. Since entering the market in 2008, less than a year, the number of IP-PBX users was over the traditional PBX's.

Asterisk[2] is a mature platform of IP-PBX. The source code of Asterisk is open, it implement all the features of PBX in the form of software, and generally run in Linux environment. The characteristics are: (1) support for traditional analog telephony devices and digital telephone equipment; (2) support for major VoIP protocols such as SIP,H.323,MGCP,etc; (3) its own specific protocol-IAX2[3] is used to communicate between the servers of Asterisk.

4 Explore a Municipal Call Center with Asterisk

Next we use Asterisk to design a municipal call center[4].

Municipal call center is intended to provide a channel between municipal services and citizens, so that can improve quality of service. Another purpose is to achieve the network connection between Municipal Corporation and service center, afterward form a city's Public Service Network. The system we design can communicate with every department (such as transport sector, power sector, etc.) internally, make it conveniently to exchange or manage information; externally can server the general public by any way at any time.

5 Architecture of the System

We divided the system into three levels: client access layer, system control and business process layer and business management layer, as Fig.1.

Client access layer mainly provides any kind of access and interactive interface to customers. Generally speaking, it is consist of switches with the ACD, Internet access devices and other accessories. Users can connect to the server by one or more E1 or analog line, so both telephone customers and Internet customers can enjoy service.

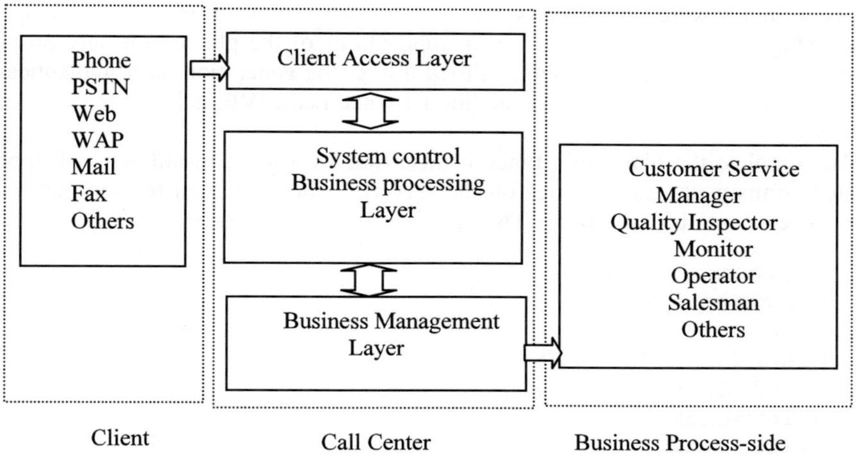

Fig. 1. The architecture of the system

System control and business process layer is the center of voice control and business process, commonly composed by CTI server, IVR/IFR, simultaneous recording system, database servers, application servers, etc. The main function is to complete call routing, call control, service logic control.

Operational management layer primarily completes the expression of business, the member include IVR / IFR and agent system.

In order to facilitate the maintenance and expansion, we design the system under B / S (Browser / Server) mode, the Internet structure shown in Fig. 2:

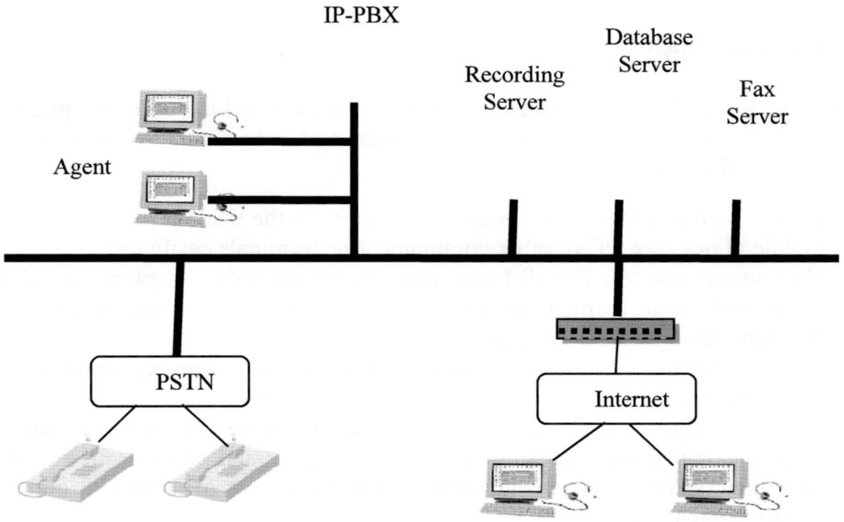

Fig. 2. Network Structure

As a middleware, Asterisk is the core of the whole system. It links up the underlying technologies and the upper applications of the phone, not only provides functions of basic call processing, call recording, log generation, user authentication and accounting, but also provides additional voice mail, IVR, call queuing and other services.

To complete the platform's functionality, first we should install Asterisk include related components and libraries on the server[5], then we need to configure some files in / etc / asterisk / directory. The following is the main code:

```
sip.conf [general]
language = en
bindport = 5060
bindaddr = 0.0.0.0
disallow = all
allow=ulaw
allow=alaw
allow=gsm
allow=g729
context=default
callerid=Unknow
...
extensions.conf [default]
exten=>*100,1,GotoIfTime(*,mon-sun,*,*?ivr1,s,1)
[ivr1]
exten=>s,1,Backgroud(custom/welcome)
exten=>1,1,Queue(7000|tT|||60)
exten=>1,n,Hangup
...
```

6 Specific Module

The municipal call center adopted a unified public service hotline to accept citizen's requests of consults, suggestions, complaints and other matters. Therefore, we realized functions as follows:

a. Access function: Public can access our system in the manner of fixed telephone, mobile phone, fax, PC or other communication terminals easily.
b. Navigation features: For different requests of the public, system can classify, guide and locate them; to different accessing methods, system is able to use different equipment for navigation.
c. Transfer function: According to user's requirements, system can transfer calls to relevant departments.
d. Processing function: Through our system, agents were able to handle problems such as complaints, suggestions, repair, consulting and so on from the citizen, after that system can generate work orders for agents to submit to correlation processing departments.

e. Management function: Including the public information management, business (complaints, inquiries, consulting) management, quality management (service time management, service error management, audit management) and general statistics (traffic statistics, business statistics, classification statistics, job log) function.

7 Summary

In this paper, we using IP-PBX system, based on the platform of Asterisk, designed and implemented a fully functional municipal call center, replacing the traditional switch, not only reduced the cost of development greatly, but also improved the quality of service. The system was praised by all the customers.

References

1. Han, Y.-y., Cai, D.-l., Wang, G., Li, H.: Study and Implementation of Embedded IPPBX. Communications Technology 43(8), 15–17 (2010)
2. About Asterisk, `http://www.asterisk.org/support/about`
3. Pan, Y.-l., Yang, G.-c., Zhou, Y.-q.: Distributed call center for enterprises based on Asterisk and OpenVPN. Journal of Computer Applications 30(3), 756–760 (2010)
4. Wang, D., Zhao, W.-d.: Research and Design of Call Center System Based on Asterisk. Computer and modernization (9), 169–172 (2009)
5. Asterisk Reference Information, `http://www.asterisk.cn`

The Measurement and Analysis for Complex Permittivity of Microwave Dielectric Based on Coaxial Measurement System*

Yi-qiang Wu[1], Chen Qian[1], and Guo-ping Du[2]

[1] Department of Electronic Information Engineering
Nanchang University
Nanchang, 330031, China
wuyiqiang@ncu.edu.cn
[2] Institute of Materials Science and Engineering
Nanchang University
Nanchang, 330031, China

Abstract. The paper based on coaxial line measurement system, analyzes the traditional algorithm of NRW, and with reference to the literature [6], sorted out an improved algorithm. By use of existing equipment in laboratory—vector network analyzer (AV3620), under 1-6GHz conditions, two kinds of low-loss and non-magnetic materials—PTFE and Air—were measured scattering parameters, improved algorithm for measured value and traditional algorithm for measured value are compared. Can be drawn from the experimental data, the method can effectively solve thickness resonance and multi-valued problems in traditional algorithms, thus the reliability and validity of the improved method have been verified effectively.

Keywords: coaxial line, complex permittivity, scattering parameter, thickness resonance.

1 Introduction

In recent years, microwave technology is developing at a rapid pace, which requires microwave circuit tend to the integration and miniaturization[5], this also promoted the development of microwave dielectric materials and device. However, for different kinds of microwave dielectric, which methods is adopt to determine dielectric parameter and how to determine and evaluate its performance more effective and precise have been one of major issue in research and development of microwave dielectric, which need to be resolved. At present the common measuring methods of dielectric parameter mainly includes: stationary wave method, transmission/reflection method, open coaxial probe method, free-space method and resonant cavity and so on. Because of wide measure band, simple operation and high precision[3], now

* The work is supported by National Natural Science Foundation of China. (Item Number: 60661001)

X. Wan (Ed.): Electrical Power Systems and Computers, LNEE 99, pp. 759–764.
springerlink.com

transmission/reflection method became the mainstream method for measuring Permittivity of the microwave dielectric material, so the research method is transmission/reflection, this paper will first introduce the basic principles and algorithms of traditional NRW transmission/reflection method.

2 The Principle of NRW

Niclon, Ross and Wire put forward the method in the 70s of 20th century, and also known as NRW transmission/reflection method. The method mainly is tested material was put into the coaxial transmission line (or waveguide), when the electromagnetic wave is transmitting in the cavity of transmission line, it will meet tested material, so part of it will were directly penetrated, the other were reflected, in this process, it accompanied by the attenuation and phase shift of energy. Using sweep function of vector network analyzer measured reflection coefficient, penetrance coefficient. Thus, according to the basic knowledge of electromagnetics, we can inverted all the electromagnetic parameters, as shown in the Fig.1.

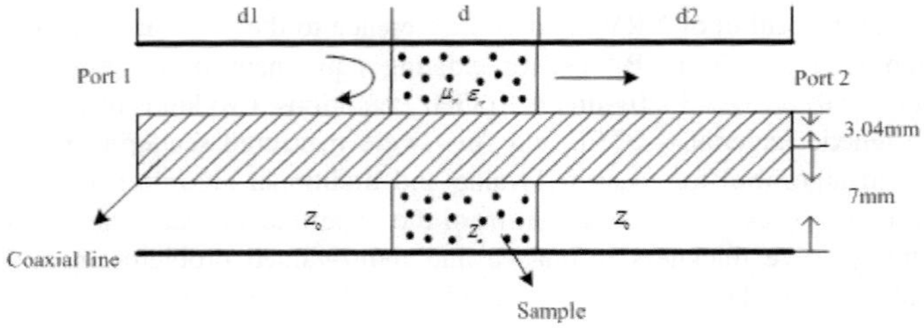

Fig. 1. The schematic of transmission /reflection method

Through the use of coaxial sampler (TEM wave), we can deduce:

$$S_{11} = S_{22} = \frac{\Gamma_c(1-T_d^2)}{1-\Gamma_c^2 T_d^2} \tag{1}$$

$$S_{21} = S_{12} = \frac{T_d(1-\Gamma_c^2)}{1-\Gamma_c^2 T_d^2} \tag{2}$$

$$\Gamma_c = \frac{Z_c - Z_o}{Z_c + Z_o} \tag{3}$$

$$Z_0 = \frac{c\mu_0}{\sqrt{1-(\frac{\lambda_0}{\lambda_c})^2}} \tag{4}$$

$$Z_c = \frac{c\mu_r\mu_0}{\sqrt{\mu_r\varepsilon_r - (\frac{\lambda_0}{\lambda_c})^2}} \tag{5}$$

In the formulas, S_{11}, S_{21} are the reflection and transmission parameters in the sample area, respectively; d is the thickness of the sample; γ is propagation constants in the sample area; Z_0 and Z_c are the wave impedance in space and the sample area, respectively; λ_0 is work wavelength in the air, λ_c is cutoff wavelength, c is the speed of the light, μ_0 is air permeability, so can be obtained from the above formulas:

$$\frac{S_{11}^2 - S_{21}^2 + 1}{2S_{11}} = \frac{1+\Gamma_c^2}{2\Gamma_c} \tag{6}$$

Let $K = \frac{S_{11}^2 - S_{21}^2 + 1}{2S_{11}}$, so we can obtain $\Gamma_c = K \pm \sqrt{K^2 - 1}$, where the value of Γ_c is less than 1. Another:

$$T_d = \frac{S_{11} + S_{21} - \Gamma_c}{1-(S_{11} + S_{21})\Gamma_c} \tag{7}$$

$$\gamma = -\frac{1}{d}\ln(T_d) \tag{8}$$

$$\varepsilon_r = \frac{[-\gamma^2(\frac{\lambda_0}{2\pi})^2 + (\frac{\lambda_0}{\lambda_c})^2]}{\mu_r} \tag{9}$$

For dielectric materials, namely μ_r =1.

In the traditional algorithm of NRW, due to need to make T_d take natural logarithm, so that:

$$\gamma = \frac{1}{d}\ln(T_d) = -\frac{1}{d}[\ln(d) + j(\theta \pm 2n\pi)], \text{ where n=0, 1, 2....} \tag{10}$$

From the above, we can see $\ln(T_d)$ have many value, therefore, corresponding ε_r also have an infinite number of value, this is the multi-value problem in the transmission/reflection method. In addition, for some low-loss materials such as PTFE, Air, etc. when thickness of the sample is greater than an integer multiple of half wavelength, in some frequency point, the absolute value of S_{11} for measured scattering parameters will be close to 0, this is the thickness resonance problem in traditional algorithm of NRW. In this paper, with reference to the literature[6], sorted out the following improved algorithm.

After improving the algorithm of NRW, the transmission/reflection method is as follow:

First, material normalized impedance is defined as:

$$\overline{Z_m} = \frac{Z_m}{Z_o} = \pm\sqrt{\frac{(1+S_{11})^2 - S_{21}^2}{(1-S_{11})^2 - S_{21}^2}} \tag{11}$$

$$\Gamma_c = \frac{Z_m - Z_0}{Z_m + Z_0} = \frac{\overline{Z_m} - 1}{\overline{Z_m} + 1} \tag{12}$$

Where $\mathrm{Re}(\overline{Z_m}) > 0$, for calculating the magnetic materials and high wastage materials, this method has high accuracy, can effectively reduce thickness resonance problem in traditional algorithm of NRW, but for the non-magnetic material, the scattering parameter of material S_{11} near the resonant frequency still have larger deviation. Therefore, combined the characteristics of non-magnetic materials ($\mu_r^{'} = 1, \mu_r^{''} = 0$), take the values of $\varepsilon_r \mu_r$ as relative permittivity of materials. Computational method is as follow:

$$\varepsilon_r \mu_r = (\varepsilon_r \mu_r)^{'} - j(\varepsilon_r \mu_r)^{''} \tag{13}$$

$$(\varepsilon_r \mu_r)^{'} = \varepsilon_r^{'} \mu_r^{'} - \varepsilon_r^{''} \mu_r^{''} \tag{14}$$

$$(\varepsilon_r \mu_r)^{''} = \varepsilon^{'} \mu_r^{''} - \varepsilon_r^{''} \mu^{'} \tag{15}$$

For multi-value problem, the paper selected the method of compensation of imaginary part in literature[1] to resolve. According to propagation law of electromagnetic wave in the medium, the change of phase shift should be continuous, according to the relationship n=int ($\frac{d}{\lambda_c}$), by the compensation of imaginary part for $\ln(T_d)$, so as to overcome the problem of multi-valued.

3 Experimental Verification

In this paper, experimental equipment is mainly vector network analyzer (AV3620), the ends of tested coaxial line are connected 1, 2 port of vector network analyzer, where the length of tested coaxial line is 100mm, impedance is 50 Ω, inner and outer conductor diameters of coaxial line are 3.04mm and 7mm, respectively. Under the frequency range of 1-6GHz, using the instrument measure the scattering parameter of material S_{11}, S_{21}, and then by using of NRW algorithm to deduce complex permittivity of material.

3.1 Verification of Complex Permittivity in PTFE

To measure scattering parameters of PTFE, the height of the sample is 5mm, and deduce complex permittivity through traditional NRW algorithm, as shown in the Fig.2, we can see from the figure, complex permittivity and reference value appeared great deviation in some frequency point, this is thickness resonance problem, and then we use improved algorithm of NRW to deduce, can obtain the following Fig.3, and we found the curve close to smooth, and agree well with the theoretical value.

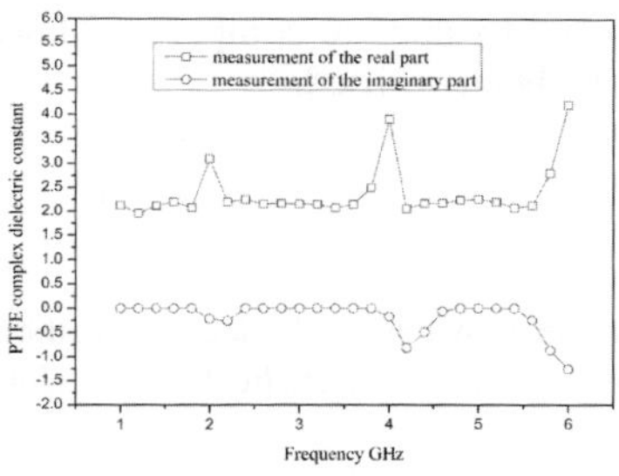

Fig. 2. Traditional NRW algorithm

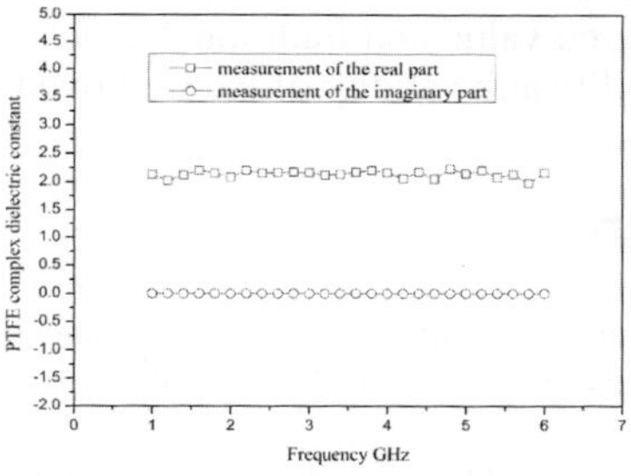

Fig. 3. Improved NRW algorithm

3.2 Verification of Complex Permittivity in Air

Under the same conditions, To measure scattering parameters of Air, the height is 100mm, the result as shown in the Fig.4, we can see from the figure, it also have thickness resonance problem, and then we use improved NRW algorithm, can obtain the following Fig.5, image is stable, no obvious deviation, it agrees well with the theoretical data.

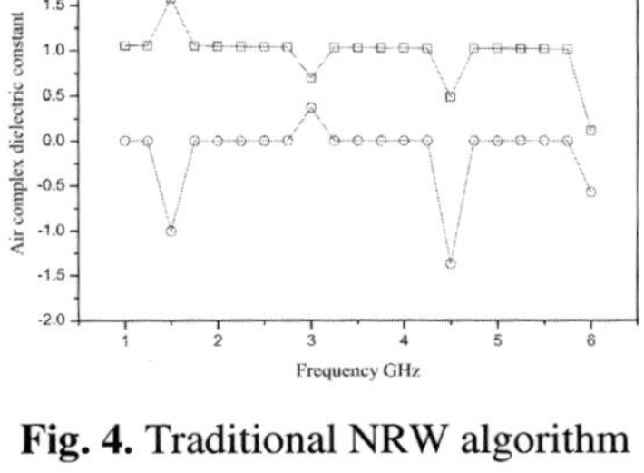

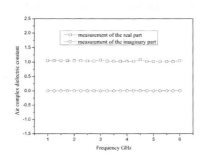

Fig. 4. Traditional NRW algorithm

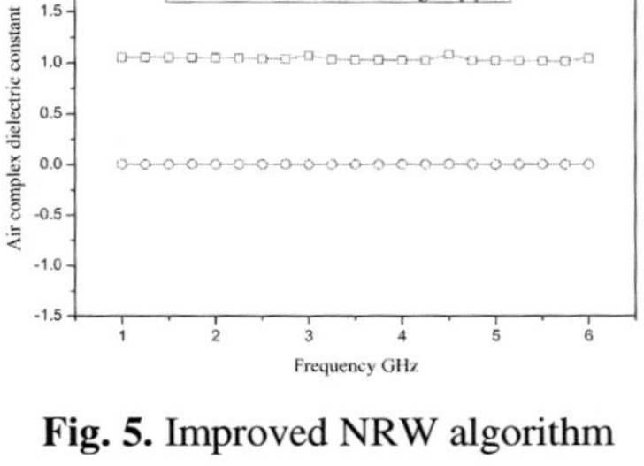

Fig. 5. Improved NRW algorithm

To sum up the above arguments, for non-magnetic and low-loss materials, it is easy to cause the problems of thickness resonance and multi-valued, through the measurement of two samples—PTFE and Air, take product value of $\varepsilon_r \mu_r$ as the value of complex permittivity, and use compensation of imaginary part to compute, experimental results agree well with the true parameters of materials. Thus, it verified improved NRW algorithm has the reliability and accuracy in the literature [6].

4 Conclusion

In this paper, on the basis of an in-depth analysis about the traditional NRW algorithm and with reference to the literature[6], the author sorted out an improved algorithm. It gives better solution for the problems of thickness resonance and multi-valued in traditional algorithm. Also, By use of existing equipment in laboratory—vector network analyzer (AV3620), two kinds of low-loss and non-magnetic materials—PTFE and Air—were measured scattering parameters, improved algorithm for measured value and traditional algorithm for measured value are compared. Thus, the reliability and validity of the improved method have been verified.

References

1. Tian, B., Yang, D., Tang, J.: Some problems of the transmission/reflection method for measuring complex permittivity of materials. Chinese Journal of Radio Science (January 2002)
2. Feng, Y., Qiu, T.: Measurement of electromagnetic parameters for microwave absorbing materials using transmission/reflection method. Chinese Journal of Radio Science (February 2006)
3. Jing, S., Jiang, Q.: Transmission/Reflection Method Based on Coaxial Line for RF Materials Characterization Measurement. Journal of Astronautics (2005)
4. Zhao, C., Jiang, Q., Jing, S.: Improved NRW Transmission/Reflection Method for the Determination of Electromagnetic Parameters of Materials with Coaxial Line. Measurement & Control Technology (November 2009)
5. Jiang, S.: The research of Electromagnetic parameters testing system. BeiJing JiaoTong University (2007)
6. Yang, G., Jiang, S., Wang, G.: Modified algorithm on electromagnetic parameters test. Foreign Electronic Measurement Technology (August 2007)

Research on the Resonance Over-Voltage of Power Distribution Network with Different Arc Suppression Coil

Jun Zhang, Fenghua Wang, Bo Xu, and Xu Cai

Key Laboratory of Control of Power Transmission and Conversion, Ministry of Education,
Department of Electrical Engineering, Shanghai Jiao Tong University,
Shanghai 200240, P.R. China
junzhang@sjtu.edu.cn

Abstract. Arc suppression coil is a main device to govern the grounding faults, but it can effect on the power network. This paper introduces 4 types of arc suppression coil and analysis their basic mechanism and control method. Based on these analyses, this paper analysis effect on power distribution system with different ASC, specifically for series resonant and ferromagnetic resonance, and give notes on the operation of the different ASC. Theoretical analysis, computer simulation and real fault recording wave verified the effectiveness of the analysis.

Keywords: arc suppression coil (ASC), power distribution system, series resonant, ferromagnetic resonance.

1 Introduction

In the medium voltage power distributed system, ungrounded, high-resistance grounded, and resonant grounded neutrals are commonly practiced in power distribution systems of some European and Asian countries and in several types of industrial systems in North America [1]. Among fault, the single-phase ground fault is more than 80% of total failure, and installation of automatic tracking neutral point arc against compensation device is an effective way to control one of the grounding faults [2]. Currently, there are many arc compensation devices in power distribution system, the traditional multi-tap arc suppression coil, but also regulating capacitor , gate-flow (also known as high-impedance transformer), bias and magnetic valve regulation and other new ASC, from the control point of view, which are divided into 2 types such as pre-tuning type and following-tuning type. And the detection of capacitive current tracking methods is different [3-4], how to properly evaluate the effect of these devices arc, but also the urgent electricity users need to be resolved.

Various types of over-voltage is the most deadly in ASC effects on power network, especially series resonant and ferromagnetic resonance, which often leads to hazards of faults. This paper will analysis effect on power distribution system with different ASC and give notes on the operation of the different ASC.

X. Wan (Ed.): Electrical Power Systems and Computers, LNEE 99, pp. 765–770.
springerlink.com

2 Model of ASC

Currently there are mainly 4 types of ASC that are the fixed multi-tap, the auto-tuning multi-tap, the auto-regulating capacitor and the phase-control ASC.

2.1 Multi-tap ASC

Fig.1(a) is a fixed multi-tap arc suppression coil that is the traditional Petersen Coil, which coil winding body by changing the number of turns to change the inductor, inductance is proportional to the square with the number of turns, so the inductance discontinuities adjustable[5]. With on-load tap-changer (OLTC), Petersen coil is changed into load multi-tap tuning ASC to instead of no excitation tap-changer. In the actual running, the working voltage of OLTC is often much lowers than the rated voltage of ASC, so the ASC voltage up to rated voltage and OLTC is not allowed operation when single phase grounding fault occurs, meanwhile inductance of ASC can not be adjusted. In this sense, load multi-tap tuning ASC can only work in pre-adjusted mode, and needs to join the damping resistor in normal running.

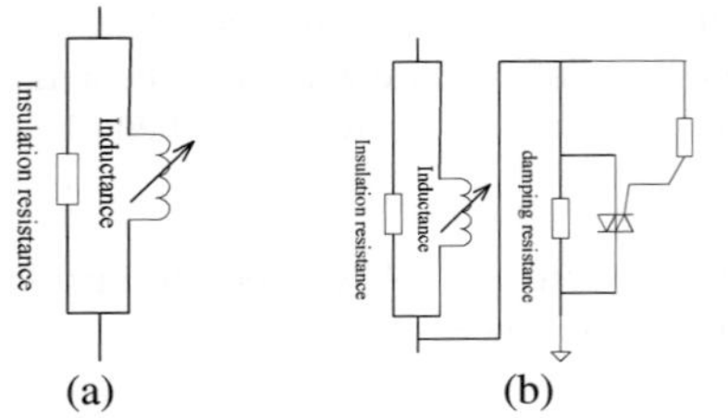

Fig. 1. Simulation Model of tuning ASC: (a) fixed; (b) Auto-tuning with damping resistor

2.2 Auto-regulating Capacitor ASC

Fig.2 (a) shows the basic structure of the auto-regulating capacitor ASC, whose L1 is high-voltage winding and L2 is low-voltage winding whose capacity is half of L1; the capacity ratio of C1, C2, C3, and C4 is 1:2:4:8 and total capacity is half of ASC. Opening-closing combination of 4 SCR switches can produce 16 different inductor currents. On this basis, increasing the capacity of L2 and compensation capacitor can satisfy the need for well regulating. To refine the current difference between the files,

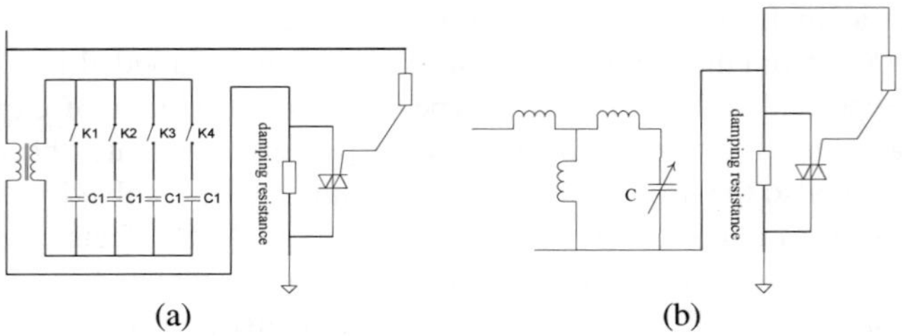

Fig. 2. Simulation Model of Regulating Capacitor ASC: (a) structure ;(b) equivalent circuit

just need to add a set of capacitor C5. Based on the different switches, the phase-control ASC is preset to work in presetting mode and following-setting mode.

2.3 Phase-Control ASC

Fig.3 (a) shows the basic structure of the phase-control ASC, whose primary winding BW connect with power network, secondary winding CW is short cut with SCR to control the current of output, and third winding FW connect with LC filter circuit to keep total harmonic current distortion controlled within the design by balancing 3rd and 5th harmonic current of primary winding. The phase-control ASC is essentially a single-phase high-impedance transformer. When the SCR is off, there is no current in primary winding, and filter circuit is capacitive at the fundamental frequency in third winding. On the other hand, equivalent circuit is inductive because of leakage reactance of secondary winding when the SCR is open. Adjusting the angle of SCR trigger control can change the secondary winding turns on and off in proportion to the ASC equivalent inductance by the amount of continuous change, so that the ASC can achieve rapid continuous adjustment of inductance. The compensation current generated by the phase-control ASC with the filter circuit changes from the capacitive status to the inductive when the SCR conduction angle changes from 0° to 90° [6]. In other words, ASC is inductive when conduction angle is 0°, so there is angle mutation between 0° and 90°.

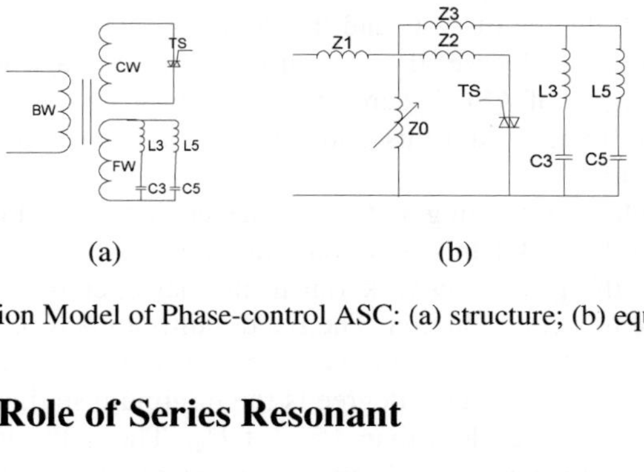

(a) (b)

Fig. 3. Simulation Model of Phase-control ASC: (a) structure; (b) equivalent circuit

3 ASC on the Role of Series Resonant

Fig.4 shows the equivalent circuit is a series LC circuit, when close to the resonant circuit current conditions large voltage on the arc suppression coil displacement

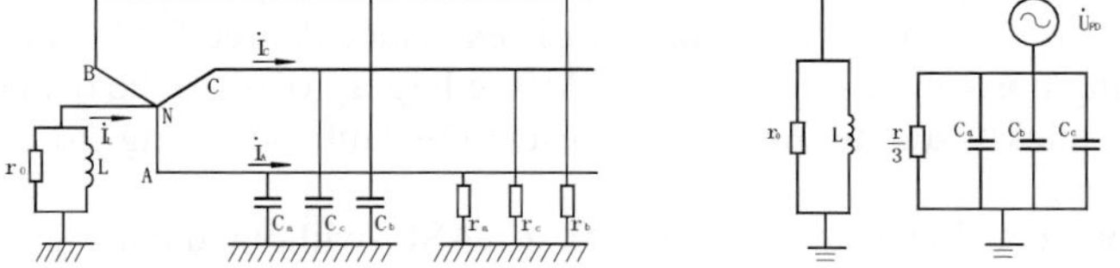

(a) (b)

Fig. 4. Power Compensation Network: (a) Equivalent Circuit, (b) Circuit of zero sequence voltage of neutral point that is great. Operation of the provisions of the neutral-to-ground voltage should not exceed 15% of phase voltage, and its expression is:

$$\dot{U}_n = -\frac{j\omega(C_a + a^2C_b + aC_c)}{j\omega(C_a + C_b + C_c) - j\frac{1}{\omega L} + \frac{1}{r_0} + \frac{3}{r}}U_\Phi = -\frac{\frac{j\omega(C_a + a^2C_b + aC_c)}{j\omega(C_a + C_b + C_c)}}{\frac{j\omega(C_a + C_b + C_c) - j\frac{1}{\omega L}}{j\omega(C_a + C_b + C_c)} - j\frac{1}{R\omega(C_a + C_b + C_c)}}U_\Phi \quad (1)$$

Where: $\frac{1}{R} = \frac{1}{r_0} + \frac{3}{r}$; $d = \frac{1}{R\omega(C_a + C_b + C_c)}$ (Damping rate)

Neutral-to-ground voltage is:

$$= -\frac{\dot{K}_c}{\upsilon - jd}U_\Phi \approx \frac{\dot{U}_{PD}}{\upsilon - jd} \quad (2)$$

Esq.(2) shows that the neutral point voltage of compensation network is related to the out-of-resonance degree of ASC and the damping rate of power network, and the smaller out-of-resonance degree, the neutral-to-ground voltage is higher. When the out-of-resonance degree of ASC is zero that is the resonant compensation, voltage of neutral point is the highest. And this voltage is the series resonance over-voltage of compensation system.

When a single-phase grounding, out-of-resonance degree for the smaller, the better the compensation effect. But this is a contradiction that there exists series resonant over-voltage when the power network run in normal operation. There are different solutions to resolve this contradiction. The fixed multi-tap ASC can only be sacrificed for the compensation effect, and set in over compensation mode in normal running, in the same time the out-of-resonance degree is the minimum setting that the neutral-to-ground voltage can be made less than 15% of U_Φ. The auto-tuning multi-tap ASC works in the best compensation. Normal to increase the damping resistance, the damping rate of the network increases, the series resonance over-voltage limit is reached less than 15% of U_Φ, after the rapid removal of damp earth resistance, the best compensation, and equivalent to adjusting the damping rate of the approach. Auto-regulating capacitor ASC and auto-tuning multi-tap ASC turn to take the same approach. Phase-control ASC is the out-of-resonance degree for by adjusting the way, while making normal out-of-resonance degree larger, to avoid the occurrence of series resonance, quickly tuning ASC after grounding fault occurring, to achieve the best compensation.

Based on the above analysis, the fixed ASC without damping resistor rely on increasing the out-of-resonance degree for the expense of compensation effect to limit series resonant over-voltage purpose, to what extent depends on the power of the natural imbalance. Auto-tuning, auto-regulating capacitor and phase-control of ASC compensation effect can be achieved without sacrificing the premise, to limit series resonant over-voltage.

4 ASC on the Role of Ferromagnetic Resonance

About Multi-tap ASC, in Fig.2(a) Zero-sequence equivalent circuit, inductance L of ASC is smaller than excitation inductance LVT of voltage transformer (VT), so the zero-sequence circuit inductance parameters are mainly determined by the ASC, and keep relatively stable neutral point potential, even if the voltage transformer magnetizing inductance change can not happen over-voltage caused ferromagnetic resonance. Therefore, the multi-tap ASC can effectively suppress the ferromagnetic resonance of electromagnetic VT.

About auto-regulating capacitor ASC, in Fig.2 (b) Zero-sequence equivalent circuit, inductance L of ASC can also play a role in inhibition of VT ferromagnetic resonance. And in the same time, the damping resistance R_{ZN} can also suppress the ferromagnetic resonance of electromagnetic VT.

About the phase controlled ASC, in Fig.2 (c) Zero-sequence equivalent circuit,

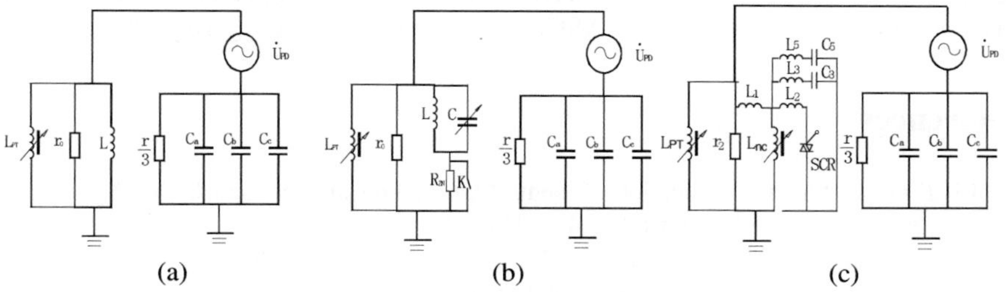

Fig. 5. Equivalent Circuit: (a) the Auto-tuning Multi-tap ASC,(b) the Auto-regulating Capacitor ASC, (c) the Phase-Control ASC

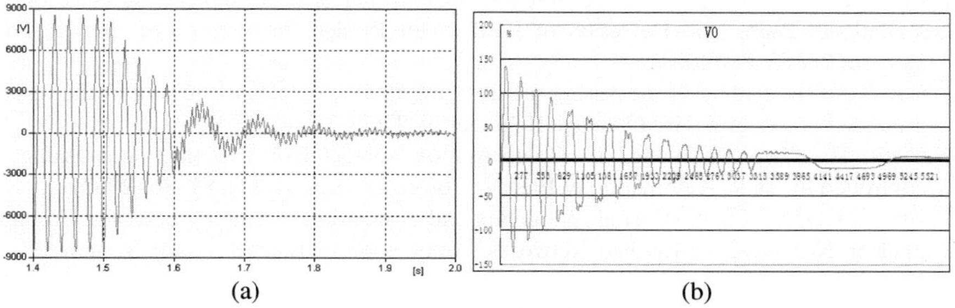

Fig. 6. Ferromagnetic Resonance: (a) Simulation ; (b) Recording Wave of Real Fault

Since the excitation reactance of ASC is also nonlinear, it can not suppress the ferromagnetic resonance of electromagnetic VT. The excitation reactance L_{nc}, C_a, C_b, C_c, C_3 and C_5 ferromagnetic resonance occurs, this phenomenon occurs after single phase grounding fault happened and SCR off will occur instantly. Fig.4 shows ferromagnetic resonance excited by the phase-control ASC in simulation of EMTP [7]. And fig.5 shows the real ferromagnetic resonance recording wave that happen in some 35kV station of P.R.China [8].

5 Summary

The auto-turning multi-tap ASC, Auto-regulating capacitor ASC and the phase controlled ASC can well avoid the series resonant when the power network works in normal operation under normal for the series resonant power over-voltage problem, except for the multi-tap Arc ASC without the damping resistance. The auto-turning multi-tap ASC and the auto-regulating capacitor ASC can suppress the ferromagnetic resonance of electromagnetic VT, but the phase controlled ASC can not suppress because of the closure of SCR. In particular, the excitation reactance of the phase controlled ASC occurs relatively ferromagnetic resonance with its own filter capacitor, the line capacitance when ground fault disappears. And it may have dangerous ferromagnetic resonance over-voltage in certain circumstances. So If the appropriate control methods, this phenomenon is likely to be avoided. So there is require that its control system must have a high anti-jamming capability in the power network with the phase controlled ASC, ensure that no false triggering phenomenon.

References

[1] IEEE Guide for the Application of Neutral Grounding in Electrical Utility Systems, Part IV -Distribution, IEEE Std. C62.92.4-1991 (1992)

[2] Li, F.: The Operation of Electric Network with Neutral Non-Effective Grounding. Hydraulic and electric power press, Beijing (1993)

[3] Cai, X.: A Study on the Capacitive Current Compensation System Following the Tracks of Operating Conditions of Electric Network. Automation of Electric Power Systems 19(8), 57–61 (1995)

[4] Cai, X., Li, S., Du, Y., et al.: An Integrated Controller of Multi-tap Arc Suppression with Variational Damp and Detection of Earth Fault Feeder. Automation of Electric Power Systems 28, 85–89 (2004)

[5] Cai, X.: New method of measuring the damping ratio and capacity current of the resonant grounded power grid. High Voltage Engineering 23(2), 38–40 (1997)

[6] Chen, H., Cai, X., Wang, J.: Compensation characteristics of arc suppressing coil controlled by TCR. Automation of Electric Power Systems 31(11), 81–86 (2007)

[7] Gu, R., Cai, X., Chen, H., et al.: Modeling and Simulating of Single-phase Arc Grounding Fault in Non-effective Earthed Networks. Automation of Electric Power Systems 33, 63–67 (2009)

[8] Chen, H.-K., Cai, X.: Experimental study on extinguishing ARC performance in different types of ASC. In: 20th International Conference and Exhibition on Electricity Distribution (CIRED 2009), Prague, Czech, June 8 - June 11 (2009)

A Distributed Speech Romote Control System Based on Web Service and Automatic Speech Recognition

Baohua Tan[1,2,3]

[1] School of Science, Hubei University of Technology
[2] The key lab of Modern Manufacture Quality Engineering in Hubei province, Hubei University of Technology
[3] School of Automobile Engineering, Wuhan University of Technology
Wuhan, P.R. China
Tan_bh@126.com

Abstract. A distributed speech remote control system based on Web Service and automatic speech recognition was introduced. The system can transit voice call via telephone into robot command by the speech recognition engine. After received and recognized the command, the system transmits the real-time command into a robot hand controller with the Web Service components, and then the robot hand moves accordingly so as to achieve the goal of controlling the robot movement. This paper introduced the system architecture and functions of each sub block, as well as the design of the programs, analyzed the system integration solution and the process of system integration. The innovation with this system lies in realize a distributed speech remote control with the sufficient application of telephone and computer network.

Keywords: AutomaticSpeech Recognition, Remote Control, Web Service Computer, Telecommunication Integration, Hidden Markov Model.

1 Introduction

Nowadays, speech control and remote control have been attached more and more importance. With the development of automatic speech recognition and the internet technology, Speech control system based on Web Service has become a burgeoning hotspot for robot control. As a new network control pattern, remote speech control system based on Web Service features of a consummate integration of telephone and computer network, inclusion of several distributed application have won favors from many technical experts and users.

Remote control based on Web Service is defined as the extension of a human sense organ to a long distance robot through internet to accomplish a task. The writers have adopted CTI technology [1] to realize a remote speech control over an industry robot hand through a variety of logic movements based on Web Service components. This not only provides a mutual interactive setting for remote control, but also inaugurates a fresh industry for internet application.

X. Wan (Ed.): Electrical Power Systems and Computers, LNEE 99, pp. 771–778.
springerlink.com

2 System Architecture

This remote control system based on speech recognition technology has extended the local speech of the robot to a far-end so that the operator is able to manipulate the robot in a direct manner either by using modern computer communication technology or the telephone network [2,3], as shown in Figure 1.

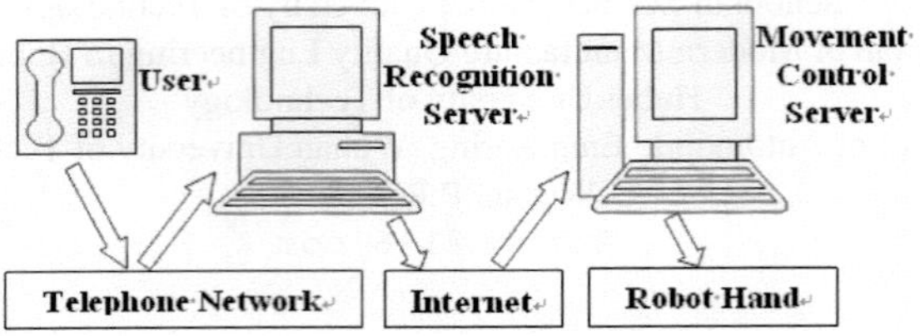

Fig. 1. System Architecture

In the system architecture, the system has integrated the telephone and computer network. The user can pass his speech command onto the recognition end through the telephone network, and then the recognition service will send the command into the control service of the robot hand through the computer network, as shown in figure 2.

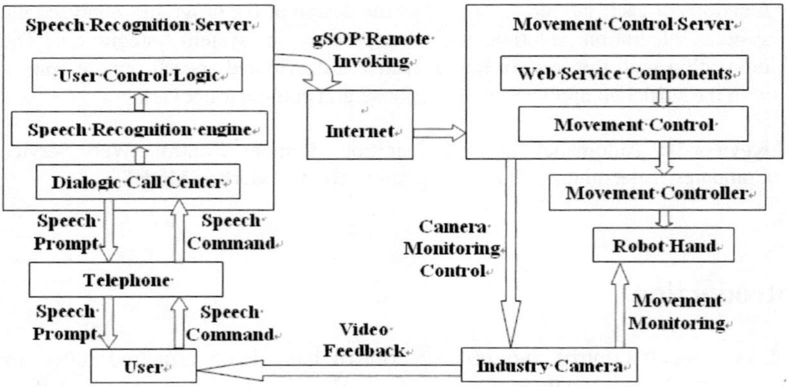

Fig. 2. System flow

The remote control system is based on Web Service technology. On the one hand, as an agent for the interaction between the operator and the robot, Web Service components' function is a bridge to receive the speech commands from the operator while collect the real-time information from the robot. On the other hand, the remote mainframe, as an individual entity, namely an independent intelligent entity which acts on itself and has a sense of environment, is able to adjust according to the present environment so as to accomplish the tasks commanded by the operator in a long distance [4,5]. In the speech control end, the computer network works as a medium for the interaction between the operator and the robot. The operator away from the sites is able to complete a task through the speech control over the robot and he can collect the parameters regarding the work conditions of the robot from the Web Service components as well.

3 Call Center Component Implement

Dialogic develops products and technologies for both the enterprise and service provider markets that enable reliable, seamless, and efficient communications and services across countless devices on a wide variety of networks.

Dialogic/12 JCT-LS PCI board is the next generation voice resource board based on Spring Ware 12-port voice resources. Developers can use the board to develop affordable, scalable, high density of multimedia resources of communications applications. The board has a lot of advanced features, and supports the latest in DSP (Digital Signal Processing) technology and industry-standard PCI bus and CT bus technology. Plate based on DSP fax enables developers to maximize the multimedia communication application system boards, such as support for Web-enabled call centers, Unified Messaging or IVR (Interactive Voice Response) applications.

The Intel Dialogic D/120JCT-LS board as shown in Figure 3.

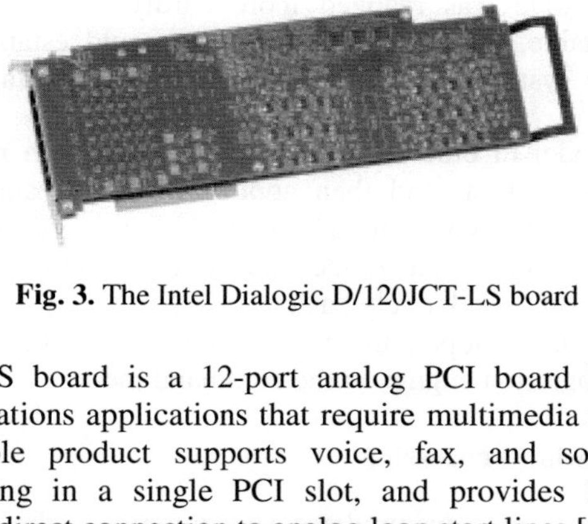

Fig. 3. The Intel Dialogic D/120JCT-LS board

The D/120JCT-LS board is a 12-port analog PCI board ideal for developing advanced communications applications that require multimedia resources. This high-performance, scalable product supports voice, fax, and software-based speech recognition processing in a single PCI slot, and provides 12 analog telephone interface circuits for direct connection to analog loop start lines [6].

We build an IVR application to construct a call center based on a Dialogic/12 JCT-LS PCI board, which can extend the existing messaging system at the same time create a unified messaging solution. The call center can response the user's call real-time. The Dialogic/12 JCT-LS PCI board takes charge in user's speech information collecting automatically.

4 Automatic Speech Recognition Design

4.1 HMM Recognition Model

The human speech is actually a dual random process, and the speech signal is a predictable time sequence, which is generated as a parameter flow by the brain according to grammar and speech rules. On the whole, human speech is an instable random process, yet when divided into several parts, it is available for linear analysis on a short-time basis.

HMM, namely Hidden Markov Model, is a statistic model for the time sequences of speech recognition which can be seen as a dual random process in math: in one aspect, It can be seen as a hidden random process simulating the changes of speech signal by using Markov chain of limited status figure, in another it is an observation sequence relevant to every single Markov chain. The latter embodies the former while the former parameters are unpredictable. Therefore HMM is an appropriate simulation of this process which has well displayed the entire instability and partial stability of speech signals.

If construct a HMM based on this speech signals, we are able to identify different transient stable signal splits and track down their transition so as to complete the model construction based on the speech velocity and acoustic changes [7,8].

4.2 Speech Recognition Engine

Nuance speech recognition engine has provided two kinds of working patterns, namely recognition mode and the command mode. Different working pattern has decisive impact on the recognition program setting [9].

The recognition mode was realized more hardly. In basis of the full speech grammar and code-table, the recognition system would establish the speech code library, and then the system can identify any external inputting voice based on the library.

The command mode, in other words, we told the speech recognition system to identify which voice at first, and then applied programs and set the recognition environment. And enter the voice; the recognition system would correctly output the result of the voice. On command mode, the system code-table was small, and the identification progressing is relatively easier.

In this system, we have adopted the command mode. As shown in figure 4 with the work flow of the recognition engine on the command mode.

4.3 Speech Recognition Processing

The user: the operator who manipulates and commands the robot away from the sites.

Speech: the voice information speaking by the user.

Speech recognition model: it translates the speech signal into test file for the robot's recognition after matching the user's command with the recognition program. In this system, we take HMM as speech recognition model.

Speech recognition application: it works to receive the speech recognition, and translates it accordingly. Then the translation will be sent to a remote control end by Web Service to manipulate the robot to conduct movements in response. In this system, we construct real speech recognition application by taking Nuance V7.0.4 [9] as development tool.

Result: the content feedback from Speech recognition Engine, which is the recognized command result to drive the Control Logic for the speech recognition application.

Command: some specific meaning which been defined beforehand for the speech recognition application, and it can correspond with the control logic and the speech recognition result unique.

Control Logic: which can drive the movement controller to apply the robot hand's moving.

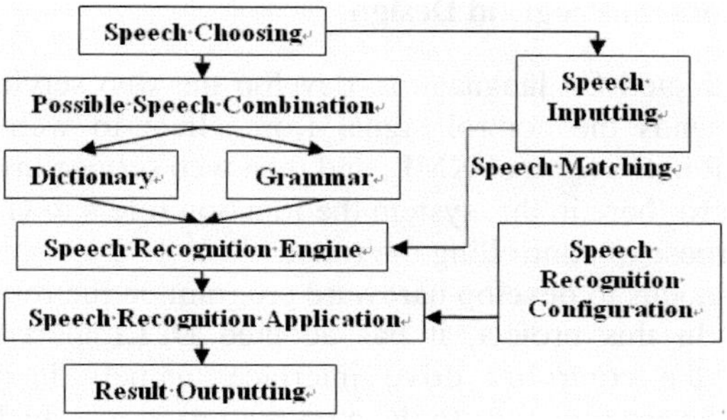

Fig. 4. Working flow of the recognition engine on the command pattern

The full speech recognition processing flow is shown in figure 5.

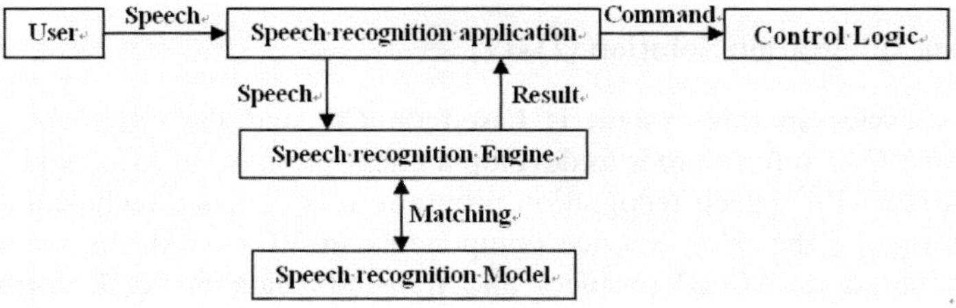

Fig. 5. Speech recognition Processing flow

5 Remote Robot Control Design

5.1 Remote Robot Control End

Remote robot control end consists of two parts.

The remote robot: This paper has utilized a four degree of freedom industry robot, which has adopted the GT moving controller that allows it to control four movement axes and realize a multiple axis movement. Meanwhile, the controller has equipped itself the C language base and window dynamic linkage library to complete sophisticated tasks.

The user then is capable of combining this language base, the data process, the interface with his control system to set up a specified application system.

The robot main control system: Start the robot control program to receive and enforce the command as well as complete the real-time video collection, coding and sending tasks; the robot receives the command, and process it with application program, if the command is correct, the robot will act accordingly and will send the movement to the operator in the form of video materials to make sure the operator is well informed of the robot's work conditions.

5.2 The Robot Control Program Design

This system has adopted C# language to develop the web service based application program [10]. It sends the control signal from client to web service application program by gSOAP in the form of XML, and then web service invokes the function in robot movement card (here in this system the function refers to GT400. DLL) so as to accomplish the purpose of controlling the robot.

It is a technical nodus to develop hardware program on the robot interface board in NET Framework. In this project, it has adopted DLLImport to import the DLL function base of the controller drive interface, namely it directly applies the DLLImport and relevant functions to develop the program, which not only solve the problem of programming in the sub-hardware but also makes it very convenient to control the robot.

6 System Integration

6.1 System Integration Solution [11-17]

The Web Service in this system is based on C#, and the command program is compiled in VC++ 6.0. In order to develop a client program in MFC style and realize the integration with speech recognition program, it becomes a technical difficulty as to how to invoke the Web Service components in VC++ 6.0. As we know, Web Service is based on SOAP protocol and transmits data in XML format, yet the traditional C/C++ is not available for SOAP protocol and XML format. Therefore we need a third party tool to realize invoke of Web Service components. The gSOAP complier is able to solve this problem.

gSOAP complier is typically used in open source project. By using gSOAP, the Client and Server programming task could be easily done in C/C++, and the programmer should not know much about xml and SOAP protocol. So the user of gSOAP could concentrate themselves on the soft programming of the Web Service client and server's, no need to badger with other specific techs. The gSOAP is a cross-platform tool for the developing of Web Service servers and clients, coded in C/C++.

By using gSOAP complier, a user Web Service function base can be defined. The user then is able to invoke relevant functions to receive and transmit data from Web Service. In this system, the user requests remote service in the XML file format and the result is backed in XML format.

6.2 The Process of gSOAP Implement

In order to integrate the gSOAP with VC++ 6.0 and invoke the Web Service, we need to start the following steps:

(1) Create a gSOAP Service in Visual Studio 2003 and a function Web Service. Mark the output functions with Web Method.

Create a client. First, click the service and store WSDL as my service wsdl file. Then with wsdl to code, create the client code. Target the wsdl file, and click so as to generate the codes.

Start a project gsoapclient, and add the ServiceSoap.nsmap, soapC.cpp, soapClient.cpp, soapH.h, soapStub.h, and stdsoap2.h, stdsoap2.cpp in the project gSOAP root directory into the project.

As shown in figure 6 with the gSOAP implement flow.

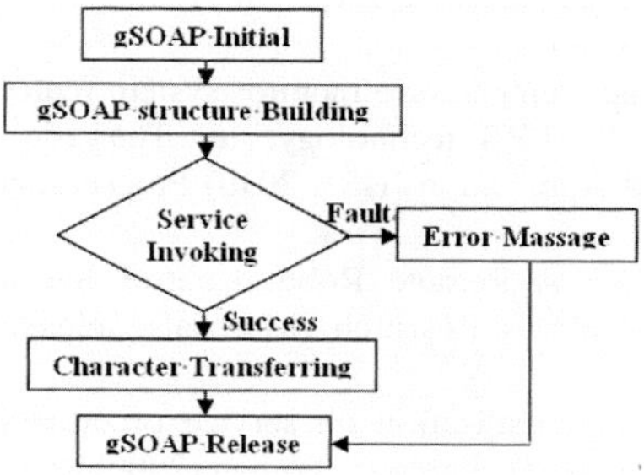

Fig. 6. gSOAP Implement flow

7 Conclusion

In this paper, Nuance speech recognition technology has been applied to realize a secondary development, and the available program for the robot's recognition has been developed. Meanwhile, it integrates with Web Service technology to realize the remote speech control over a robot.

As evidence concluded from the experiment, the robot is able to reach a rate of 98% accuracy in speech recognition. As to remote control, with Web Service technology the robots' hearing can be extended profoundly, which enables in some extent the user work in a setting far away from the sites to ensure improved speech recognition accuracy. The Web Service complexes the computer-based control system, but it gives the system the distributed deployment, better flexibility and easier extension and maintenance.

At last, we can also adopt the TTS (Text to Speech) technology in speech feedback system. As an example, jTTS (the Jie Tong Text to Speech software) is a powerful speech tool which based on real speech synthesizes technology.

Acknowledgment

Thanks a lot to the key lab of Modern Manufacture Quality Engineering in Hubei Province for providing laboratory resource and relational technology supporting.

And also we would like to thank Qingbo Zhu, Yaohe Liu, Jiangmin Xiong, Zhengxiang Yang, Tingxin Song, Jian Wu, Tianfang Zhang, Min Song, Yongming Zhou, Juntao Wang, Jintao Liu, Yuhang Lin, Jun Zhou, Changhui Hu, Zhengxing He, Kun Wang and Yang Yang for their hard work in this paper.

References

1. Jun, T.: Study on CTI technology and interactive voice response system. Harbin Institute of Technology, Harbin (2006) (in chinese)
2. Baohua, T., Xiong, J., Liu, Y.: Telephone Tax Declaration System Design Based on CT Technology. Computer Engineering and Application 41(19), 105–107 (2005) (in chinese)
3. Tianfang, Z., Liu, Y., Tan, B.: Remote Robot Control System Based on Speech Recongnition. Journal of Hubei University of Technology 22(4), 45–47 (2007) (in chinese)
4. Intel Inc. Intel Dialogic D/120JCT-LS 12-Port Analog PCI Board Datasheet, California, USA (2002)
5. Baohua, T.: Road surface temperature monitor system realization based on Rich Internet Application model and GPRS technology. In: The 2nd International Workshop on Intelligent Systems and Applications (ISA 2010) Proceedings, Wuhan,China, May 22-23, pp. 344–347 (2010)
6. Baohua, T., Qingbo, Z.: A Remote Robot Control System Based on RIA and Flex Platform. Advanced Materials Research (Materials Science and Engineering) 179–180, 320–324
7. Qingbo, Z.: Research and realization on service oriented speech romote robot control system. Hubei University of Technology, Wuhan (2009) (in chinese)
8. Rabiner, L.R.: A Tutorial on Hidden Markov Models and Selected Applications in Speech Recognition. Proceedings of the IEEE 77(2), 257–258 (1989)
9. Nuance Communications, Inc.Nuance Speech Recognition System Version7.0.4: Introduction to the Nuance System, Menlo Park, CA, USA (December 2003)
10. Microsoft.NET making next generation Internet into truth. Microsoft whitepage, 23-29 (2000) (in chinese)
11. Baohua, T., Juntao, W.: A DeviceNet Fieldbus Data Acquisition System Based on Flex Technique and RIA Model. In: 2010 International Conference on Progress in Informatics and Computing conference (PIC2010) Proceedings, Shanghai, China, December 10-12, pp. 1167–1169.
12. Xiaolu, C., Zilu, L.: Web Service Technology Architecture and Application, pp. 18–25. Electronic Industry Press, Beijing (2001) (in chinese)
13. Prosise, J.: MFC Windows Programme Design, pp. 955–958. Tsinghua University Press, Beijing (2002) (in chinese)
14. Xiaohua, L.: NET Web Service Development Guide, pp. 358–369. Electronic Industry Press, Beijing (2001) (in chinese)
15. Tapadiya, P.: A Practical Guide Using Visual C++ and ATL, pp. 117–128. China Electric Power Press, Beijing (2002) (in chinese)
16. Ying, Y., Gang, L.: Visual C++ Practice and improve - COM and COM+ Part, pp. 415–418. China Railway Press, Beijing (2001) (in chinese)
17. Maloney: Visual C++6 DCOM Development Guide, pp. 158–165. Tsinghua University Press, Beijing (2000) (in chinese)

An Accurate Equivalent Circuit Method of Open Ended Coaxial Probe for Measuring the Permittivity of Materials*

Kun-ming Liao[1], Yi-qiang Wu[1], Chen Qian[1], and Guo-ping Du[2]

[1] Institute of Information Engineering
Nanchang University
Nanchang, 330031, China
liaokunming55037@163.com
[2] Institute of Materials Science and Engineering
Nanchang University
Nanchang, 330031, China

Abstract. An accurate equivalent circuit method analysis for open-ended coaxial probe is used to determine the complex permittivity of materials from the measured input reflection coefficient and admittance. Reference materials of known dielectric properties and selected bakelite plate and water were measured in the frequency range 0.5to 6GHz. This paper offers further analysis the effect of radiation conductane at the end of coaxial probe. The comparisons between the measured results and calculated values indicate this measurement method is effective when considered the effect of radiation conductane.

Keywords: open coaxial probe, complex permittivity, radiation conductane, equivalent circuit method, reflection coefficient.

1 Introduction

Coaxial probe method is a microwave measurement method which through measuring the reflection coefficient of probe terminal by attaching probe terminal closely to the measured materials in order to capture the microwave complex permittivity of materials. This technique not only has the advantages of being Non-destructive and Non-invasive to the materials, but also the ability of measuring the bandwidth and the easiness of sample making. as a consequence, This method has been widely applied for measuring the microwave complex permittivity of dielectric materials. In this paper, An accurate equivalent circuit method analysis for open-ended coaxial lines is used to determine the complex permittivity of materials from the measured input reflection coefficient and admittance. So far, the application of coaxial probe method still has to contain the assumption that those measured samples are half the infinite medium, the measurement system was based on the AV3620 network analyzers. Reference materials of known

* The work is supported by National Natural Science Foundation of China.
(Item Number: 60661001)

X. Wan (Ed.): Electrical Power Systems and Computers, LNEE 99, pp. 779–784.
springerlink.com

dielectric properties and selected bakelite plate and water were measured in the frequency range 0.5to 6GHz. Samples have been tested with the equivalent circuit antenna model[1] and compared with the capacitor model[2], The different of both models is whether to consider the effect of radiation conductane. It shows that two models have no much difference at low frequence for low dielectric loss materials but have a little difference at high frequence for high dielectric loss materials. It suggests that the equivalent circuit antenna model be suitable to the measurement for the materials with small discrepacy in dielectric constant over the entire frequency band.

2 Modeling of the Open-Ended Coaxial Probe

The equivalent circuit antenna model for this model is presented in Fig. 1(b). The model is described in [3] and [4]. The reflection coefficient Γ at the end of the open-ended probe is obtained by considering the complex admittance of the equivalent circuit.

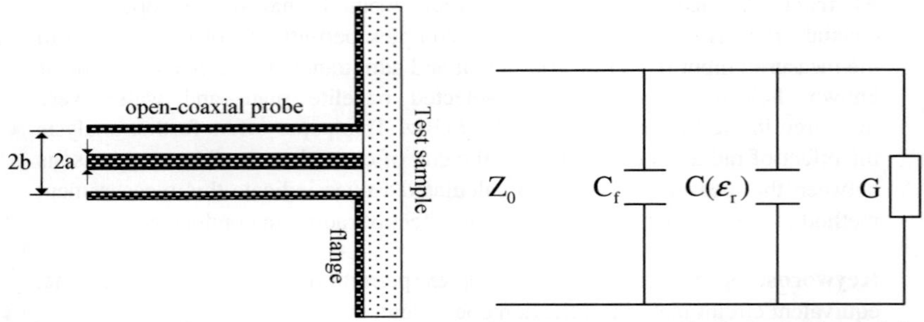

(a) Geometry (b) An accurate equivalent circuit model

Fig. 1. The open-ended coaxial probe connected with a test sample

The admittance of an open-ended coaxial line sensor terminated by a half space with a relative complex permittivity of $\varepsilon_r = \varepsilon' - j\varepsilon''$ is modeled by:

$$Y(\omega) = j\omega C_f + j\omega C(\varepsilon_r) + G \tag{1}$$

where: $C(\varepsilon_r) = \varepsilon_r C_0$ represents the capacitance of the parallel plate capacitor formed by the inner conductor and the terminating with a test sample. While C_f is the fringe field capacitance.

Generally, $C(\varepsilon_r) \gg C_f$, G is radiation conductance of the open coaxial probe, it connected in parallel to the $\varepsilon_r^{5/2}$, in the vacuum is G_0 .

$$G_0 = \frac{4\pi}{3\eta}[\frac{(b^2 - a^2)}{\lambda^2 \ln\left(b/a\right)}]^2 \tag{2}$$

where: $\eta = \sqrt{\mu_0/\varepsilon_0}$, $G = G_0\varepsilon_r^{5/2}$. So, when the geometry of coaxial line is very small, G_0 is very small, and the dielectric loss is also small, Then we can neglect G . It can simplified to the linear two-capacitor model. If we measured the reflection coefficient $|\Gamma|\exp(j\varphi)$, The dielectric constant of the tested material can be obtained. The input reflection coefficient is equal to capacitive model[5]:

$$\Gamma = |\Gamma|\exp(j\varphi) = \frac{1 - j\omega Z_0(C_f + C_0\varepsilon_r)}{1 + j\omega Z_0(C_f + C_0\varepsilon_r)} \tag{3}$$

In addition, the permittivity of the samples can be represented as:

$$\varepsilon' = \frac{-2|\Gamma|\sin\varphi}{\omega C_0 Z_0(1 + 2|\Gamma|\cos\varphi + |\Gamma|^2)} - \frac{C_f}{C_0} = A \tag{4}$$

$$\varepsilon'' = \frac{1 - |\Gamma|^2}{\omega C_0 Z_0(1 + 2|\Gamma|\cos\varphi + |\Gamma|^2)} = B \tag{5}$$

When the frequency is higher, or the dielectric loss is high, we can't neglect G . If consider the effect of G , can be obtained the dielectric constant[6]:

$$\varepsilon' = A - \frac{G_0}{\omega C_0} b \tag{6}$$

$$\varepsilon'' = B - \frac{G_0}{\omega C_0} g \tag{7}$$

$$g = (\varepsilon')^{5/2}[\alpha(1 + tg^2\sigma) - 2\beta tg\sigma] \tag{8}$$

$$b = (\varepsilon')^{5/2}[\beta(1 + tg^2\sigma) - 2\alpha tg\sigma] \tag{9}$$

$$\alpha = [\frac{1}{2}\sqrt{1 + tg^2\sigma} + 1]^{1/2} \tag{10}$$

$$\beta = [\frac{1}{2}\sqrt{1 + tg^2\sigma} - 1]^{1/2} \tag{11}$$

From the above equation, $tg\sigma$ represents the Loss tangent of the tested materials, We can get more accurate permittivity values from(6),(7)and compare with (4),(5). It indicates that considered the effect of radiation conductane can improve the accuracy.

3 Experiment Techniques

The probe employed in the experiments consisted of standard coaxial waveguide, The characteristic impedance of this probe is 50 Ohm, Between the inner and outer conductors filled with teflon. Inner radius a=0.5mm, outer conductor radius b=1.67mm, the radius of the flange , 20mm, was large enough because of the lossy media to be effectively infinite. experimental equipment is mainly vector network analyzer (AV3620), was then employed to measure the terminal reflection coefficient of the probe under the frequency range of 0.5-6GHz. The lumped-element parameters[7], C_f , C_0 and G_0 may be measured over the frequency range where the open ended coaxial probe is to be used as a sensor for permittivity measurement. Dielectric materials whose permittivities are well known, such as Teflon, air, may be measured at the frequencies in question and we can work out C_0 =0.0245 pf; C_f =0.0016 pf;The reflection coefficients of the probe connected to the samples of bakelite plate and water were measured from 0.5 to 6GH in Table 1.

Table 1. Reflection coefficient magnitude and phase at the point (0.5~6GHz)frequency

FREQUENCY (GHz)		0.5	1	1.5	2	2.5	3	3.5	4	4.5	5	5.5	6
bakelite plate	$\|\Gamma\|$	1	0.999	0.998	0.998	0.997	0.996	0.995	0.994	0.993	0.992	0.991	0.990
	$\varphi(°)$	-0.25	-0.75	-1.01	-1.26	-1.52	-1.77	-2.03	-2.29	-2.56	-2.83	-3.11	-3.46
water	$\|\Gamma\|$	0.992	0.985	0.976	0.964	0.949	0.931	0.908	0.878	0.845	0.808	0.766	0.723
	$\varphi(°)$	-3.47	-7.01	-10.65	-14.45	-18.34	-22.33	-26.48	-30.58	-34.56	-38.51	-42.35	-45.79

4 Discussion

A coaxial line open into a semi-infinite space was analyzed and employed for measurements of the permittivity of a semi-infinite dielectric samples. The permittivity of bakelite and waterl obtained experimentally is compared with the reference data. The dielectric constant reference value of bakelite is 5.5-j0.05[8], and the sample of water is 76.3323-j9.1331[9]. The reference dielectric constant of materials have neglected the impact of temperature and frequency. We have known the reflection coefficients of materials under test by experiment. When we neglect the effect of radiation conductane G, it can calculate the relative permittivity through (4)(5), and we consider the effect of radiation conductane G , it can work out dielectric constant from(6)to(7).

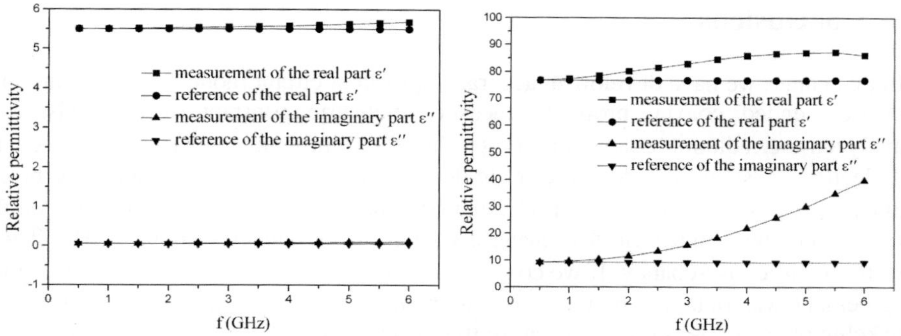

(a) bakelite plate with low dielectric loss (b) water with the high dielectric loss

Fig. 2. Permittivity of bakelite plate compared with reference values, neglected the effect of radiation conductane G .

Fig.2 shows the relative permittivity for bakelite plate, with the dielectric loss is small, from(a) compared with the theoretical reference value shows:measurement value consistent with the theoretical at low frequencies below 3GHz, but at higher frequency. For bakelite, the discrepancy compared with reference is less than 4% in the real part and imaginary part. The results of water permittivity values are shown in (b) This sample dielectric loss is more bigger. Who can see, the deviation will become more larger with the frequency increased. Particularly, with a maximum discrepancy of 20% in dielectric real part and of 400% in imaginary part at the frequency 6GHz. That is because we consider the impact of radiation conductance G .

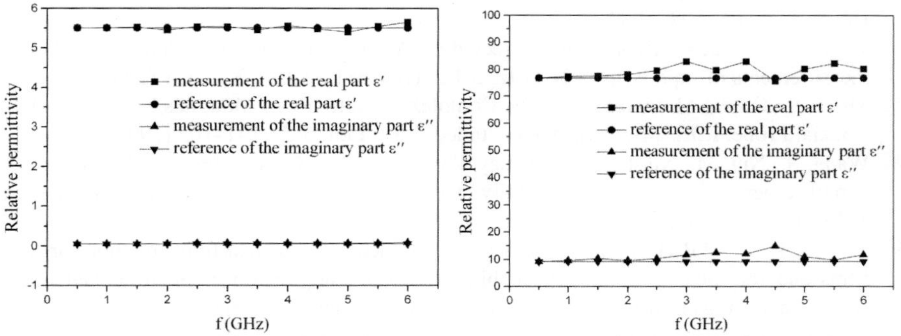

(a) bakelite plate with low dielectric loss (b) water with the high dielectric loss

Fig. 3. Dielectric constant of water compared with reference values. Considered the effect of radiation conductane G

Fig.3 shows an excellent agreement with theoretical relationship between the real part and imaginary part. That is because we consider the impact of radiation conductance G . The permittivity values are remarkably accurate with approximately a 4% maximum discrepacy in dielectric over the entire frequency band.

5 Conclusions

In this paper, we have develop an accurate equivalent circuit model for the admittance of open-ended coaxial probe to measure dielectric constant of materials, And analyzed the radiation conductane G produced the influence on Measurement of the dielectric constant. For low dielectric loss materials, with the frequencies below 3GHz, if we neglected the effect of radiation conductane G , have Little effect on the results. For high dielectric loss materials, with the frequencies above 2GHz, it will result in large discrepancy. If we consider the effect of radiation conductane G . good agreement was obtained between measurements and theoretical for the materials of bakelite plate and water over the entire frequency band.

References

1. Chen, Q.: FEM Analysis of Open Ended Coaxial line Using for Dieleetrie Constant Measurement of Mierowave biochemistry and Biological Tissue. Si Chuan University (2002)
2. Bdrube, D., Ghannouchi, F.M., Savard, P.: A comparative study of four open-ended coaxial probe models for permittivity measurements of lossy diekctric/biolocal materials at microwave frequencies. IEEE Trans.Microwave Theory Tech. (10) (1996)
3. Blanckham, D.V., Pollard, R.D.: An improved technique for permittivity measurements using a coaxial probe. IEEE Transactions on Instrumentation and Measurement (1997)
4. Misra, D., et al.: Noninvasive electrical characterization of materials at microwave frequencies using an open-ended coaxial line. Test of an improved calibration technique, IEEE Trans. MTT 38, 8–14 (1990)
5. Gajda, G., Stuchly, S.: An equivalent circuit of an open-ended coaxial line. IEEE Trans. IM. IM-32, 506–508 (1983)
6. Wei, Y., Sridhar, S.: Radiantion-Corrected open-ended coaxial line technique for dielectric measurements of liquids up to 20 GHz. IEEE Tran. MTT 39(3), 526–531 (1991)
7. Misra, S.: On the Measurement of the Complex Permittivity of Materials by an Open-Ended Coaxial Probe. IEEE Microwave and Guided Wave Lett. 5(5), 161–163 (1995)
8. Wang, S., Niu, M., Xu, D.: A frequency-Varying method for simultaneous Measurement of complex permittivity and permeability with an open-ended coaxial probe. IEEE Trans. MTT 46(12), 2145–2147 (1998)
9. Xu, Y., Bosisio, R.G.: Nondestructive measurements of the resistivity of thin conductive films and the dielectric constant of thin substrates using an open-ended coaxial line. In: IEEE Proceedings of Microwaves, Antennas and Propagation (2002)

On Image Collection and Transmission System Based on the S3C6410 and CC2430

Weiya Wang[1], Gui Hu[1], Li Gao[2], and Zhanfeng Lu[1]

[1] Academe of Information Engineering, Chang'an University, 710064 Xi'an, China
[2] Xi'an Technological University, 710032 Xi'an, China
guyue028@163.com, weiwang@chd.edu.cn

Abstract. According to the good performance in low power consumption, distribution and ad hoc pervasive of WSN, design and implement a system of image collection and transmission based on S3C6410 and CC2430, to collect and transfer image data by WSN. The test proved that the system can satisfy the needs of local area image collection and transmission both in quality and stability.

Keywords: WSN, Image Collection, CC2430, S3C6410.

1 Introduction

As the development and application of WSN, scholars both at home and abroad carried out researches on WMSN [1], Portland State University in America, Carnegie-Mellon University and Stanford University started relevant scientific efforts. In China, Institute of computer technology, Harbin Institute of Technology, Zhejiang University set out exploration in this field [2], too. Many famous companies participated in it actively, Intel brought forward prototype of sensor node Imote and Imote2 which based on ARM and Xscale, ICT center of CSIRO developed imaging sensor FleekTM [3]. In addition, companies such as TI also set out structure of WMSN of their own, some even used mobile video node which can move in small scope and adjust angle all by itself thus to realize mobile video surveillance [4]. This paper designed and realized video collection system which was based on S3C6410 and CC2430.

2 The Architecture of Image Collection and Transmission System

This system mainly consists of video collection module, transmission module, and management center, Fig. 1 shows its structure.

Image collection and processing module consists of sub-module of OV9650 imaging sensor and S3C6410 processor. Sub-module of OV9650 imaging sensor use OV9650 CIS (CMOS Image Sensor, COMS image sensor) to collect image [5], OV9650 CIS possesses the character of compact size, low costing and image that meets requirements. Sub-module of S3C6410 was comprised of S3C6410 core processor, DDR RAM, Nand Flash, reset chip and crystal oscillator, etc. While environment

X. Wan (Ed.): Electrical Power Systems and Computers, LNEE 99, pp. 785–791.
springerlink.com

information was acquired by OV9650 imaging sensor, image collection and processing module will take advantage of powerful MFC of S3C6410 to code and compress image information [6], and use the large capacity of external memory storage to save data.

ZigBee wireless transmission system mainly in charge of close range wireless transmission of data information and it was comprised of ZigBee wireless sensor network sensor node, route node and sink node. These nodes take CC2430 chip as the master control chip, CC2430 was provided with characters of low power consumption, low costing, Ad hoc network and dynamic routing. ZigBee wireless sensor network relates with image collection and processing module via serial port, thus to form wireless image collecting and transmitting nodes of the whole system, and achieve environment image information collecting and processing along with the send of ZigBee wireless network. Route node mainly takes charge of dynamic routing of network and multiple hop transmission of data. SINK mainly takes charge of the establishment and maintenance of ZigBee wireless sensor network, and the reception, summarizing and transmission of data.

If the system needs remote control and monitoring, management center can correspond with SINK via internet or satellite communication system. Management center controls ZigBee wireless sensor network and gives order. Meanwhile, we can send data information collected by ZigBee wireless sensor network to management center via internet or satellite communication, and the management center then processes and analyzes data, thus to realize environment information monitoring and management.

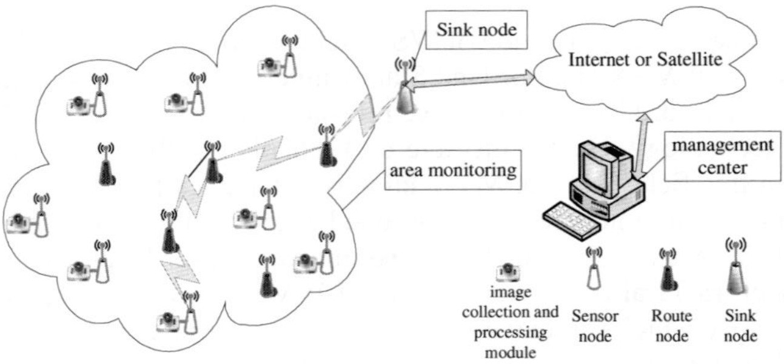

Fig. 1. General structure of image collection and transmission system

3 Design of Wireless Image Collection and Transmission Node

Wireless image collection and transmission nodes in this system mainly consist of image collection and processing module, wireless transmission module, power module, etc. Image collection and processing module was comprised of OV9650 image sensor cell and S3C 6410 disposer cell; wireless transmission module consists of CC2430 sensor cell and wireless communication cell. OV9650 image sensor cell takes charge of image collection of environment information in area coverage; S3C6410 sensor cell codes, compresses and saves image information; CC2430 sensor cell is responsible for regular work of each module in node, and does further processing to obtained information; wireless communication cell corresponds with other wireless sensor via wireless transceiver; power module is the pivotal issue of node, it provides necessary energy of regular work to node. Fig. 2 is the sketch map of the structure of wireless image collection and processing node.

3.1 Design of Image Collection and Processing Module

Due to feasibility analysis and elaborate survey, we selected and used OV9650 COMS imaging sensor to collect image information, and RSIC general purpose microprocessor S3C6410 to process image information, S3C6410 possesses the character of 16/32 bit high-performance, low power consumption, it is based on ARM1176 in SAMSUNG. S3C6410 uses 64/32 bit internal trunk framework that has mass powerful hardware accelerator, audio processor, video processing, dimensional image, show operator and scaling; an integration MFC (format codec) support code and decode of H.263/MPEG4, and decode of VC1; S3C6410 also has an optimized port that link with external memory storage; S3C6410 has a lot of peripheral asserts, such as system supervisor, camera port, TFT24 bit true color LCDC, 32 passageway DMA, 4 passageway UART, 4 passageway timer, general I/O port, IIC bus interference, USB primary device, SD primary device and high speed multimedia card port, etc [7]. Fig. 3 shows the structure of collection and processing module.

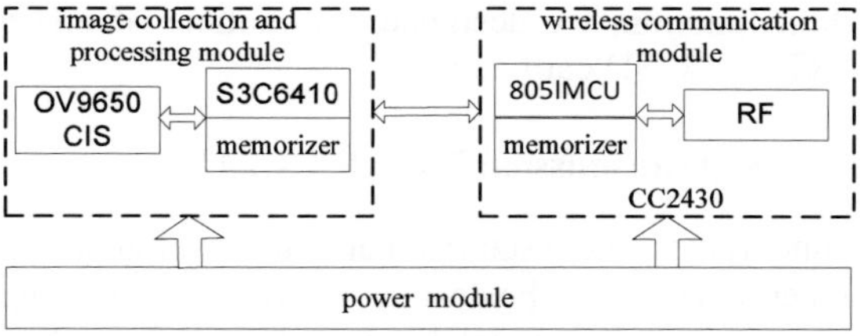

Fig. 2. Structure of wireless image collection and processing node

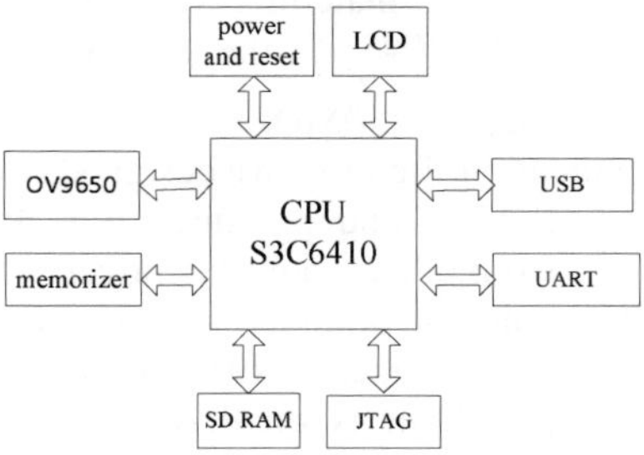

Fig. 3. Structure of collection and processing module.

As for image collection module, we used OV9650 imaging sensor chip. OV9650 is a high integrate level CMOS imaging sensor chip, there integrate sequence circuit, analog signal processing circuit, digital signal processing circuit inside of it. It supports three kind of data output format such as RGB, YUV, YCrCb. It is easy to control it via serial camera control bus [8]. Data and control signal exported by OV9650 include PCLK, HREF, VSYNC and data bus (D0-D9), they link with relevant signal pin of S3C6410 separately. PCL and HREF encounter with each other inside disposer and produce effective PCLK signal, it will lock in data at PCLK signal's rising edge and falling edge. This devise adopted YUV format, OV9650 used data cable link with S3C6410 interface circuit.

Memorizer module connects DDR and NAND FLASH memory with DRAM port and Flash/ROM/DRAM port optimized by S3C6410. S3C6410 can execute NADA starting mode. This system used NADA FLASH chip K9F2G08U0 from SAMSUNG, its 8 bit I/O port adopts the method of address, data, order multiplex and links with S3C6410. DDR uses DDR SDRAM chip K4X51163PC from SAMSUNG, its capacity is 32M×16. Owing to the DDR RAM controller inside S3C6410, address A [12:0] in processer side correspond with address A [12:0] of chip side. Column address enable signal Xm1_CASn, row address enable signal Xm1_RASn, chip selecting signal Xm1_CSn0, clock Xm1_SCLK, Xm1_SCLKn, write enable signal Xm1_WEn, Xm1_DQM0/2, Xm1_DQM1/3, Xm1_DQS0/2, Xm1_DQS1/3 of S3C6410 link with $\overline{\text{CAS}}$, $\overline{\text{RAS}}$, $\overline{\text{CS}}$, CK, $\overline{\text{CK}}$, $\overline{\text{WE}}$, LDM, UDM, LDQS, UDQS of K4X51163PC separately.

To expand system function, we designed sub-module of other function, such as power, JTAG, UART, USB, SD card, etc.

3.2 Design of Wireless Transmission Module CC2430

CC2430 module undertakes the establishment and maintenance of network, and transmission and reception of data, the design of antenna is very important. The design of CC2430 antenna optimized the typical design of CC2430 panel point provided by TI [9]. This system adopts non equilibrium single-ended, monopole antenna. To optimize the performance, we used unbalanced transformer (to use low costing electrical inductance and capacitance, coordinate with printed circuit board microwave conveyor wire (λ/2 dipole, integrated in PCB), matching RF output and input impedance 50Ω, and use computational formula $L = {}^{7125}\!/_{f}$, the result is that , the length of monopole antenna is L=7125/2450=2.9cm(f=2450MHz)) [10]. Considering the description of design requirement and application situation, we designed two kinds of antenna module to let the system be more flexible and practical. Fig. 4 shows CC2430 wireless transmission module. These two modules can switchover with each other via S3. P4 is the frequently used 2.4GHz whip antenna plug seat, the transmission distance of it is remote, and its volume is a little bit bigger. AN9520 is patch antenna whose impedance match is 50Ω and 2.4GHz, its volume is small, and transmission range is relatively small. We should choose proper antenna module that meets different requirements in different application situation.

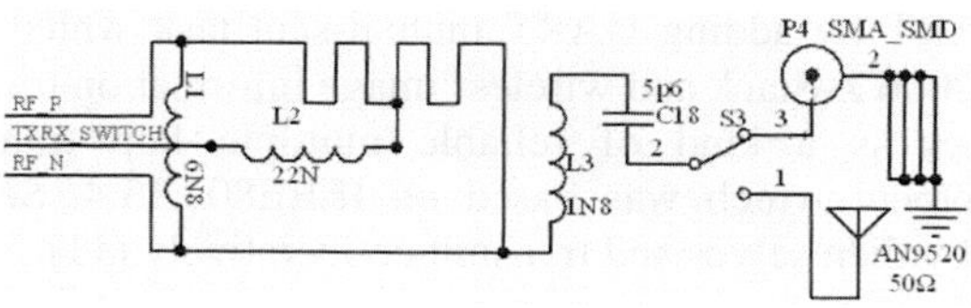

Fig. 4. CC2431 wireless transmission module

Furthermore, we should consider the problem of the integrity of system signal and compatibility of system electromagnetism, etc. it can make the system effectively collect, dispose and transmit image information in this way.

4 Image Collection and Transmission System Software Design

S3C6410 was provided with well data-handling capacity, and it has small volume, low power consumption, integrates various kinds of peripheral and supports various kinds of embedded operating system (WinCE, Linux, Android, Ubuntu, DJYOS, etc.). Comparing with various kinds of operating system, WinCE operating system was provided with advantages such as scalable core, good instantaneity, modular design, abundant API, etc. Fig. 5 shows flow chart of image collection procedure.

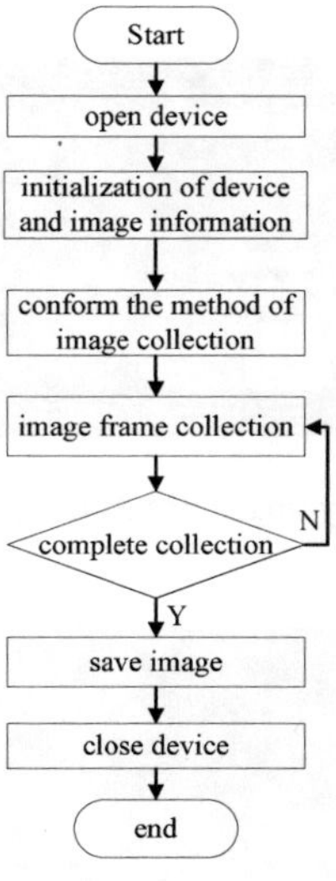

Fig. 5. Flow chart of image collection

The application development of CC2430 wireless network transport protocols is to realize effective transmission of image information in wireless sensor network by realizing the function of read UART data wireless sending and wireless receipting and sending data to UART by adding UART mutual-sent task which was based on the research of ZigBee2006 Z-Stack and wireless image information transmission from TI. ZigBee2006 Z-Stack is a kind of reliable multiple hop transmission wireless communication protocol which was based on IEEE802.15.4. Several devices can constitute network by themselves and transmit cooperatively [11].

5 Research Results and Conclusion

After experiment, the effective transmission range of whip antenna and patch antenna in open environment is about 100m and 70m, and this meets the requirement of close range transmission of CC2430.

To test and verify the collection and transmission of image, we did ordinary experiment around laboratory. Fig. 6 and Fig. 7 show the collation map of the results.

We can draw the following conclusions after experiment:

(1) The image collected is clear, this can meet the requirement of environment information collection;
(2) After compressing, we can realize network building and image transmission in local area by using CC2430;
(3) The system work is basically stable and reliable. But as the increase of panel point, network load will increase, and this will affect image data transmission to a certain degree.

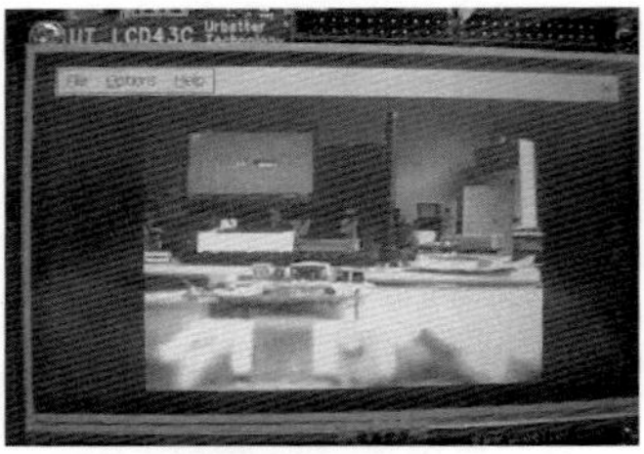

Fig. 6. LCD display image of S3C6410 collection side

Fig. 7. Image received after CC2430 wireless transmission

Supported by

The Project of Shaanxi Provincial Science and Technology Program 2010K06-11.

The Project of Shaanxi Provincial Science and Technology Program 2009K08-35.

Scientific Research Program Funded by Shaanxi Provincial Education Department 2010JK591.

References

1. Akyildiz, I.F., Su, W., Sankarasubramaniam, Y., Cayirci, E.: Wireless Sensor Networks:a Survey, Computer Networks. The International Journal of Computer and Telecommunications Networking 4, 393–4224 (2002)
2. Ma, H.-D., Tao, D.: Multimedia Sensor Network and Its Research Progresses. Journal of Software 9, 2013–2028 (2006)
3. Downes, I., Rad, L.B., Aghajan, H.: Development of a mote for wireless image sensor networks. In: Proc. of COGnitive systems with Interaetive Sensors(COGIS), Paris, France, vol. 3, pp. 1–8 (2006)
4. AKyildiz, I.F., Melodia, T., Chowdhury, K.R.: A survey on wireless multimedia sensor networks, Computer Networks. The International Journal of Computer and Telecommunications Networking 6, 32–39 (2007)
5. Liu, J.: Design of Embedded Network Video Surveillance System Based on ARM. Dalian University of Teehnology, Dalian (2009)
6. Guo, Q., Xu, X., Li, A., Mo, Q.: Design of a Wireless Video Monitoring System Based on ARM11. Computer Measurement & Control 8, 1786–1791 (2010)
7. Ye, J.: Research of Embedded System Based on ARM11 and Design of Video ProeessingTerminal. Central South University, Changsha (2009)
8. Yang, H.-s., He, D.-j.: Study and develop on the video capture system based on ARM and Linux. Microcomputer Information 11, 122–124 (2009)
9. Texas Instruments. CC2430 PRELIMINARY Data Sheet (rev.2.01) SWRS036E, pp.233-233 (2006)
10. Chen, D.: Wireless sensor network node hardware design Base on CC2431. Chang'an University, Xi'an (2009)
11. Chipcon, A.S.: SmartRF CC2420 Preliminary Datasheet rev1.2, pp. 22-25 (2004)

Analysis of Dual Three-Phase Fractional-Slot PM Brushless AC Motor with Alternate Winding Connections

Ge Qi[1], Ding Ma[2], Libing Zhou[3], and Li Shi[1]

[1] School of Electrical Engineering, Zhengzhou University, China
[2] College of Information Science and Engineering, Henan University of Technology, China
[3] College of Electrical and Electronic Engineering,
Huazhong University of Science and Technology, China
qige626@163.com

Abstract. The permanent magnet (PM) motors are applied widely in the fields of aerospace, electric ship propulsion, electric vehicles and numerical control machines because of their high performance. Conventionally, the PM brushless motors are three-phase, but it has limit in high power applications. Therefore, multiphase motors become more and more popular because of their several attractive features. This paper focuses on the dual three-phase PM brushless motor. Based on the selections of coil emf vectors and their related winding factors and mechanical arrangements for 12-slot 10-pole dual three-phase PM motor, an alternate winding connection is proposed for reducing the coupling between the two sets of three-phase windings. Furthermore, the electromagnetic performances of the dual three-phase PM brushless ac motor with different connections are analyzed and compared to those of conventional three-phase motor.

Keywords: Permanent-magnet machine, dual three-phase, fractional-slot, winding connection, electromagnetic performance.

1 Introduction

Nowadays, there has been an increasing interest in multiphase machines which can be applied in electrical vehicles, ship propulsion and wind power systems, because of their several benefits, such as achieving high performance by limited power supply levels, low rotor harmonic currents, low torque ripple and high reliability, etc. Many researches on the performances, control strategies and applications of the multiphase machines have been undertaken [1-4].

Dual three-phase machines, as one type of multiphase machines, begin more and more popular theses years. The dual three-phase machines have two sets of three-phase stator windings {A1, B1, C1} and {A2, B2, C2}, see Fig.1. The angle between the two sets of three-phase windings can be flexible for different purposes, although it is conventional 30° electrical degrees. The dual three-phase machines can be easily controlled by standard three-phase converters, so the drive system is simpler than

X. Wan (Ed.): Electrical Power Systems and Computers, LNEE 99, pp. 793–800.
springerlink.com

other multiphase drive systems (five-phase machines for example), and it is more proper for industrial applications. Works on the dual three-phase induction and synchronous machines have been presented in many aspects. The field distributions and losses of dual three-phase induction machines are analyzed and calculated in [5]. An analytical method to obtain the commutating reactance of dual three-phase synchronous machines is presented in [6] and the employment of these machines on electrical propulsion is investigated in [7]. Recently, the conception of dual three-phase stator is introduced to permanent magnet (PM) machines.

This paper focuses on the analysis of basic electromagnetic performances of dual three-phase PM motors and the reasons of high electromagnetic torque and low torque ripple are investigated. Besides, unlike other multiphase machines, there is mutual influence between the two sets of three-phase windings, and it may make the control more difficult. So the way to reduce the mutual inductances between the two sets of three-phase windings is also discussed in this paper in terms of various phase winding connections. In the paper, the electromagnetic performances of 12-slot 10-pole dual three-phase surface-mounted permanent magnet (SPM) brushless motors with different winding connections are analyzed and compared to those of conventional three-phase PM motors.

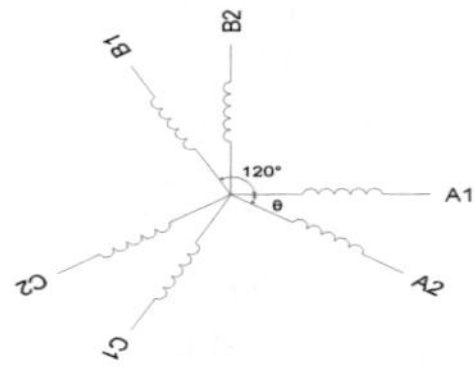

Fig. 1. Winding arrangement of dual three-phase machines, θ is the phase shift between the two three-phase.

2 Phase Winding Connections and Winding Factors

For 12-slot 10-pole motors, the distribution of the coil emf vectors is shown in Fig.2 (a). The angle between coil vectors 1 and 2 is 30° mechanical degrees due to 12 slots, and 150° electrical degrees due to 10 poles.

In a three-phase motor, each phase consists of four coil vectors. In order to obtain the highest winding factor which can result in high torque, the angles between each vector belonged to one phase should be the smallest. The emf vectors for each phase are selected and shown in Fig.2 (b). It can be seen that, the coil vectors 1, 2, 7 and 8 constitute phase A, and the current directions of 2 and 7 are opposite to those of 1 and 8. This selection can make the vector sum biggest since the vector sum of 1 & 7 and 2 & 8 are equal to their algebraic sums because of the same vector directions and the angle between 1 and 8 are the smallest. The corresponding mechanical arrangement is shown in Fig.2 (c).

In a dual three-phase motor, there are only two coils constitute one phase. The selection of emf vectors for highest winding factor is named connection 1 and shown in Fig. 3 (a1). In this connection, phase A1, B1 and C1 constitute the first three-phase winding, phase A2, B2 and C2 constitute the second three-phase winding, and the phase A2 is over A1 30° electrical degrees. The phase A1 consists of vectors 1 and 7 which have the same vector directions, and the vector sum is right their algebraic sum. The related mechanical arrangement is shown in Fig. 3 (a2). The coils belonged to one phase are distributed oppositely, and the coils of the first and second three-phase windings are interphase: A1A2'B1'B2C1C2'A1'A2B1B2'C1'C2A1, here the symbol ' means the reversed current. Thus, the influence between each two phases and between the two sets of three-phase windings is significant, that is the winding mutual inductances may be very high. Therefore, in order to reduce this influence, to reduce the mutual inductances between each two phases and especially between the first and second three-phase windings, another winding connection named connection 2 is proposed and shown in Fig.3 (b1). In the connection, the selection of coil emf vectors is similar to that of three-phase motor. There are two vectors belonged to one phase, and angle between the two vectors is 30° electrical degrees. Again, phases A1, B1 and C1 constitute the first three-phase winding, phases A2, B2 and C2 constitute the second three-phase winding, but the phase A2 is over A1 180° electrical degrees. Fig. 3 (b2) shows its mechanical arrangement. The coils in one phase and one set three-phase winding are adjacent, and the first and second three-phase winding can be separated into two nearly independent parts: A1A1'B1'B1C1C1'A2A2'B2'B2C2C2'A1. The influence between the two sets of three-phase windings can be significantly reduced.

Winding factor is an important parameter which can affect the value of back-emf and electromagnetic torque. Usually it is the higher the better for high torque. Table. 1 shows the winding factors and their related parameters of the three-phase and dual three-phase connections 1&2 motors. It can be seen that, the coil pitch factors of the three motors are the same since the same slot/pole combinations, whilst the winding distribution factor of the dual three-phase connection 1 motor is higher than others, and so is its winding factor. In the three-phase and dual three-phase connection 2 motors, the coils of each phase are distributed in different slots whereas the same slot in the dual three-phase connection 1 motor, the angle between the coils in one phase results in the smaller winding factors. The winding factors of three-phase and dual three-phase connection 2 motors are especially small for the high order harmonics, and it can restrain the harmonic components in some electromagnetic performances, such as the flux-linkage, back-emf, and torque, etc.

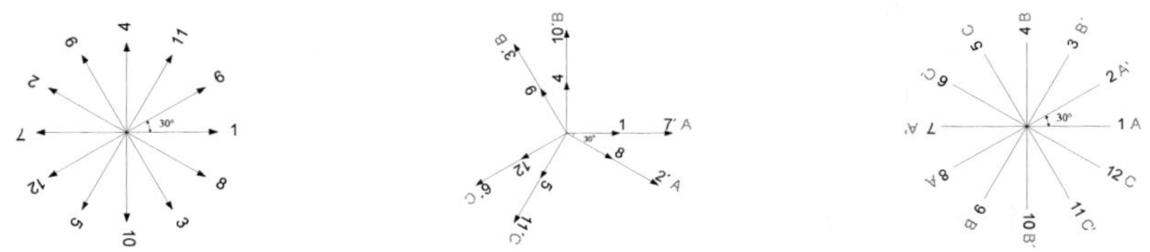

(a) Emf vectors for 12-slot 10-pole coils (electrical degree).

(b) Selection of emf vectors for each phase of three-phase winding (electrical degree)

(c) Stator coils of three-phase winding (mechanical degree)

Fig. 2. Phase winding connection of 12-slot 10-pole three-phase motor, k_w=0.933.

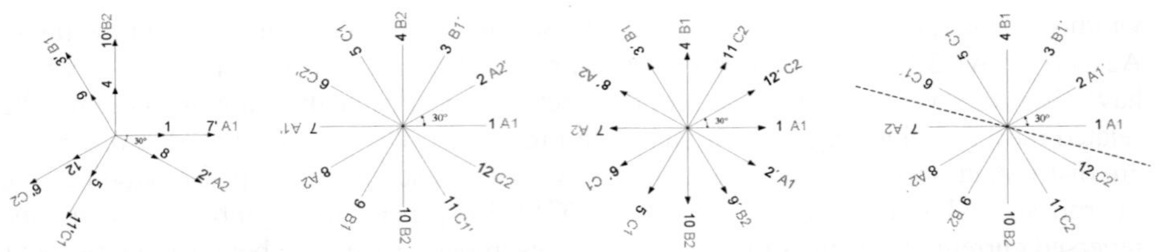

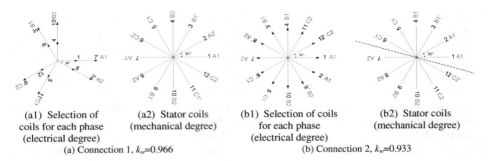

(a1) Selection of coils for each phase (electrical degree)

(a2) Stator coils (mechanical degree)

(a) Connection 1, k_w=0.966

(b1) Selection of coils for each phase (electrical degree)

(b2) Stator coils (mechanical degree)

(b) Connection 2, k_w=0.933

Fig. 3. Phase winding connections of 12-slot 10-pole dual three-phase motor.

Table 1. Winding factors of the 12-slot 10-pole three-phase and dual three-phase motors.

	Three-phase			Dual three-phase connection 1			Dual three-phase connection 2		
	m=2, σ =30°, α =30°			m=1, σ =0°, α =30°			m=2, σ =30°, α =30°		
n	k_d	k_p	k_w	k_d	k_p	k_w	k_d	k_p	k_w
1	0.966	0.966	0.933	1	0.966	0.966	0.966	0.966	0.933
3	0.707	0.707	0.4998	1	0.707	0.707	0.707	0.707	0.4998
5	0.259	0.259	0.067	1	0.259	0.259	0.259	0.259	0.067
7	-0.259	-0.259	0.067	1	-0.259	-0.259	-0.259	-0.259	0.067

3 Comparison of Electromagnetic Performances of Three-Phase and Dual Three-Phase Machines

In this paper, three-phase and dual three-phase 12-slot 10-pole surface-mounted permanent magnet (SPM) motors are analyzed and their electromagnetic performances are compared. The motor parameters are listed in the Table. 2. In order to compare the motors under similar conditions, the phase back-emf and the copper loss of the three-phase and dual three-phase motors should be the same, respectively. As the phase back-emf is proportion to the winding turns/phase, and there are four teeth belonged to one phase in three-phase motor while two teeth belonged to one phase in dual three-phase motor, the winding turns/tooth of dual three-phase motor should be twice of that of three-phase motor. The same copper loss is determined by the same current turns/tooth, thus the rated current of dual three-phase motor should be half of that of three-phase motor.

3.1 Open-Circuit Performances

The open-circuit flux-linkage and phase back-emf and its harmonic spectra of three-phase and dual three-phase SPM motors are shown in Figs.4. From the flux-linkage waveforms, the relations of each two phases in three-phase and dual three-phase motors can be seen clearly and prove the descriptions in the last section: the phase A2

Table 2. Motor parameters.

	Three-phase motor	Dual three-phase motor
Number of phases	3	6
Number of poles	10	10
Number of slots	12	12
Winding turns/phase	132	132
Winding turns/tooth	**33**	**66**
Supply voltage	36 V	36V
Rated current	**10 A**	**5 A**
Rated torque	5.5 Nm	5.5 Nm
Rated speed	400 rpm	400 rpm
Outer diameter of stator	100 mm	100 mm
Inner diameter of stator	57 mm	57 mm
Air-gap length	1 mm	1 mm
Active length	50 mm	50 mm
PM thickness	3 mm	3 mm
Magnet remanence	1.2 T	1.2 T

is over A1 30° electrical degrees in connection 1 and 180° electrical degrees in connection 2 for dual three-phase motors. For back-emf waveforms, it can be seen that, the performances of three-phase and dual three-phase motors are similar and the harmonic components in the waveform of dual three-phase connection 1 motor are quite high duo to its high winding factors. From the harmonic spectra it could be seen clearly that, the fundamental amplitude of the dual three-phase connection 1 motor is indeed higher than others own to the high winding factor. Moreover, the harmonics of the three-phase and dual three-phase connection 2 motors are much lower than that of the dual three-phase connection 1 motor though the fundamentals are correspondingly reduced, also because of their low winding factors.

3.2 Phase Winding Inductances

The phase winding self- and mutual- inductances are usually seemed as constants in many calculations and controls, however, they are actually functions with rotor positions. In order to investigate the influence of various stator phase winding connections to the values of winding inductions, especially the mutual-inductances, the phase winding inductances of dual three-phase motors are calculated and shown in Fig. 5. The phase winding self- and mutual-inductances with connections 1 & 2 in one set of three-phase winding are shown and compared in Fig. 5 (a), while the mutual-inductances between the two sets of three-phase windings are displayed in Fig. 5 (b). It can be seen that, with connection 2, the mutual-inductions, particularly those between two sets of three-phase windings decrease and close to zero. This winding connection is helpful for the de-coupling control.

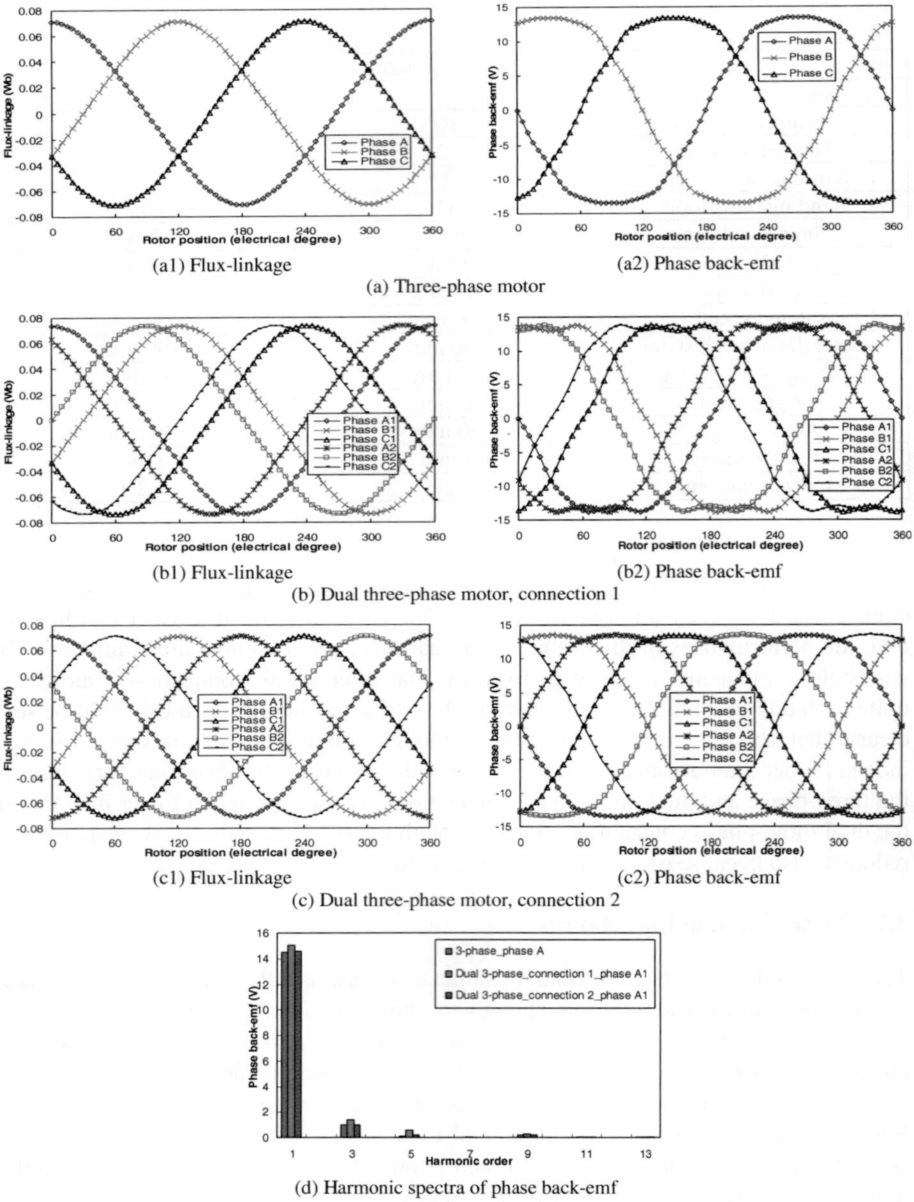

(a1) Flux-linkage (a2) Phase back-emf

(a) Three-phase motor

(b1) Flux-linkage (b2) Phase back-emf

(b) Dual three-phase motor, connection 1

(c1) Flux-linkage (c2) Phase back-emf

(c) Dual three-phase motor, connection 2

(d) Harmonic spectra of phase back-emf

Fig. 4. Open-circuit flux-linkage and phase back-emf waveforms and the comparison of harmonic spectra of phase back-emf.

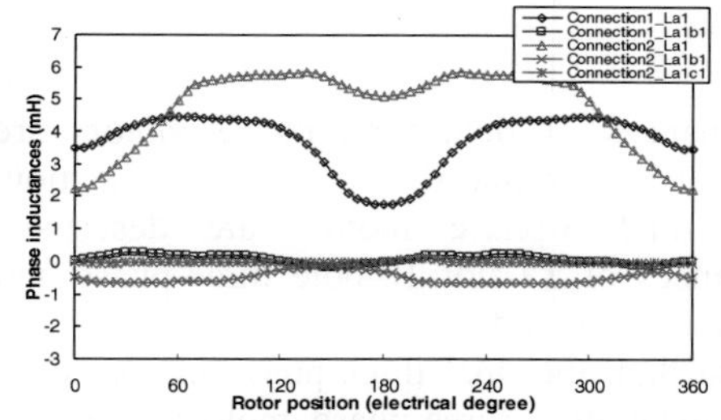

(a) Phase inductances in one three-phase winding

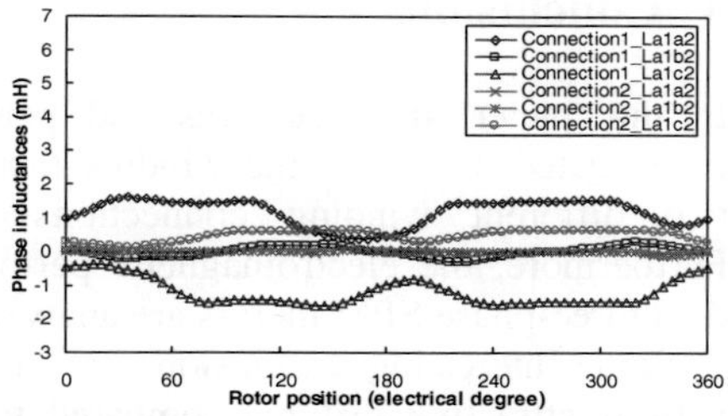

(b) Phase mutual inductances between the two sets of three-phase

Fig. 5. Phase inductances comparison of 12-slot 10-pole dual three-phase motor with connections 1 & 2.

3.3 Electromagnetic Torque

It can be seen from the open-circuit performances that the three-phase and dual three-phase motors are nearly under the same conditions, and it is the precondition for torque comparison. In the paper, the maximal electromagnetic torque and its harmonic spectra with rated current are calculated and shown in Fig. 6. The torque value of the dual three-phase connection 1 motor is about 3.55% higher than that of the three-phase and dual three-phase connection 2 motors since its high wingding factor. The increment can be estimated by the fundamental winding factors of the motors, that is, increment ≈ $(k_{w1dual1}\text{-}k_{w1}) / k_{w1}$ = (0.966-0.933)/0.933=3.54%. In addition, the torque ripple in the dual three-phase connection 1 motor is reduced. It can be explained by the harmonic spectra that the torque waveform of three-phase motor mainly consists of 6th, 12th and 18th harmonics, when phase shift angle is 30° electrical degrees in dual three-phase motor the 6th and 18th harmonics are eliminated.

In conclusion, the connection 1 winding arrangement in dual 3-phase motor can not only improve the maximal electromagnetic torque, but also reduce the torque ripple. Good for the applications that the high torque performance is required.

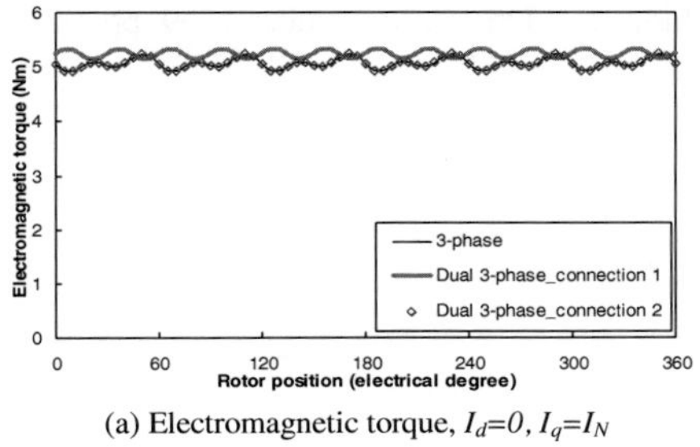

(a) Electromagnetic torque, I_d=0, I_q=I_N

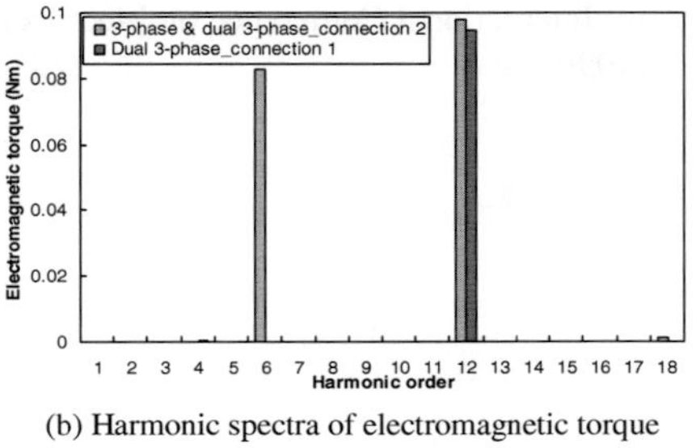

(b) Harmonic spectra of electromagnetic torque

Fig. 6. Electromagnetic torque and its harmonic spectra.

4 Conclusions

In this paper, the structures and performances of dual three-phase motors are investigated. Based on the winding factor calculation and the coil vectors selection, two different winding connections for dual 3-phase motors are described. Furthermore, the electromagnetic performances of 12-slot 10-pole three-phase and dual three-phase SPM motors are analyzed and compared.

Some interesting conclusions are derived that, the dual three-phase motors have various attractive features compared to conventional three-phase motors; the dual three-phase motor with the connection 1 winding arrangement can achieve higher electromagnetic torque and lower torque ripple, it's good for the applications that the high torque performance is required; while with the connection 2 arrangement can reduce the mutual influence between the two sets of three-phase windings and benefit for de-coupling controls.

References

1. Williamson, S., Smith, S.: Pulsating torque and losses in multiphase induction machines. IEEE Transactions on Industry Applications 39(4), 986–993 (2003)
2. Apsley, J., Williamson, S.: Analysis of multiphase induction machines with winding faults. IEEE Transactions on Industry Applications 42(2), 465–472 (2006)
3. Wen, O., Lipo, T.A.: Multiphase Modular Permanent Magnet Drive System Design and Realization. In: IEEE International Electric Machines & Drives Conference, vol. 1, pp. 787–792 (2007)
4. Levi, E.: Multiphase Electric Machines for Variable-Speed Applications. IEEE Transactions on Industrial Electronics 55(5), 1893–1909 (2008)
5. Hammache, H., Moussaoui, D., Marouani, K., Hamdouche, T.: Magnetic properties in double star induction machine. In: International Conference on Electrical Machines, September 2008, pp. 1–6 (2008)
6. Kotny, J.L., Roger, D., Romary, R.: Analytical determination of the double star synchronous machine commutating reactance. In: International Conference on Power Electronics and Variable Speed Drives, September 1996, pp. 306–310 (1996)
7. Terrien, F., Benkhoris, M.F.: Analysis of double star motor drives for electrical propulsion. In: International Conference on Electrical Machines and Drives, September 1999, pp. 90–94 (1999)

Modeling Single-Phase PV HB-ZVR Inverter Connected to Grid

Yougui Guo[1], Ping Zeng[1], Jieqiong Zhu[1], Lijuan Li[1], Wenlang Deng[1], and Frede Blaabjerg[2]

[1] College of Information Engineering,
Xiangtan University, Xiangtan 11105
guoygxtu@gmail.com
[2] Institute of Energy Technology, Aalborg University,
Aalborg DK-9220, Denmark
fbl@iet.aau.dk

Abstract. PLECS is used to model the PV H-bridge zero voltage rectifier (HB-ZVR) inverter connected to grid and good results are obtained. First, several common topologies of PV inverters are introduced. Then the unipolar PWM control strategy is described for PV HB-ZVR inverter. Third, PLECS is briefly introduced. Fourth, the modeling of PV HB-ZVR inverter is presented with PLECS. Finally, a series of simulations are carried out. The simulation results tell us PLECS is very powerful tool to real power circuits and it is very easy to simulate LCL filter. They have also verified that the unipolar PWM control strategy is feasible to control the PV HB-ZVR inverter.

Keywords: photovoltaic inverter, PLECS, unipolar modulation, LCL filter.

1 Introduction

Photovoltaic generation is developed well by many scholars in the word as a sustainable energy resource. And they have published lots of achievements with their great efforts[1-8]. It is well known the inverter plays a very important role in PV generation system. Therefore it is researched very much. It may be single-phase inverter or three-phase inverter by the requirements of users. It may also be two-level, three-level or multilevel inverter according to the topology of inverter. In addition, PLECS is very convenient to simulate the power circuits with LCL filters which are difficult to simulate without PLECS. There are 6 common PV inverter topologies[9]. We select the single-phase HB-ZVR inverter, one of them shown in Fig.1.

X. Wan (Ed.): Electrical Power Systems and Computers, LNEE 99, pp. 801–807.
springerlink.com

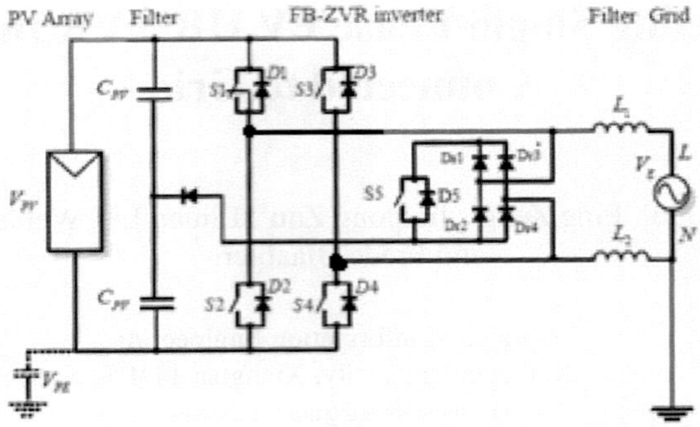

Fig. 1. Single-phase PV HB-ZVR inverter

2 Unipolar Modulation of PV HB-ZVR Inverter

Here the unipolar PWM modulation method is used to control the PV HB-ZVR inverter. It compares the same polar triangular voltage waveform with the sinusoidal reference voltage waveform to generate unipolar PWM pulses. Then they multiply the special square waveform to generate a kind of PWM pulses which are positive in half cycle of output voltage and negative in another half cycle of it to control the inverter. The special square waveform is also positive in positive half cycle of output voltage and negative in another negative half cycle of it[10]. Obviously This control strategy is used for the generation of the unipolar output voltage [9].

Here ZVR is a method to generate the zero voltage state which uses a special bidirectional switch made of one IGBT and one diode rectifying bridge shown in Fig.2. It is clamped to the midpoint of the DC-link capacitors in order to fix the potential of the PV array during the zero voltage when T_1 and T_4, T_2 and T_3 are all off. The zero voltage state is achieved by turning T_5 on when T_1, T_2, T_3 and T_4 are turned off, as is shown in Fig.2. The gate signal for T5 will be the complementary gate signal of T_1, T_2, T_3, T_4, with a small dead-time to avoid short circuit of the input capacitor. It is possible for the grid current to flow in both directions, this way the inverter can also feed reactive power to the grid, if necessary.

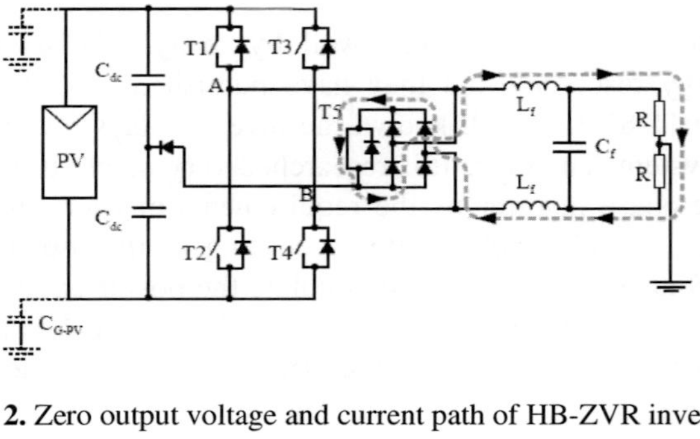

Fig. 2. Zero output voltage and current path of HB-ZVR inverter

During the positive half wave T1 and T4 are on that is used to generate a positive output voltage to the load, and T5 is controlled using the complementary signal of T_2 and T_3 shown in Fig.3. During the dead-time, between the active vector and the zero state, there is a short period while all the switches are turned off when the freewheeling current finds its path through the anti-parallel diodes to the input capacitor shown in Fig.4.

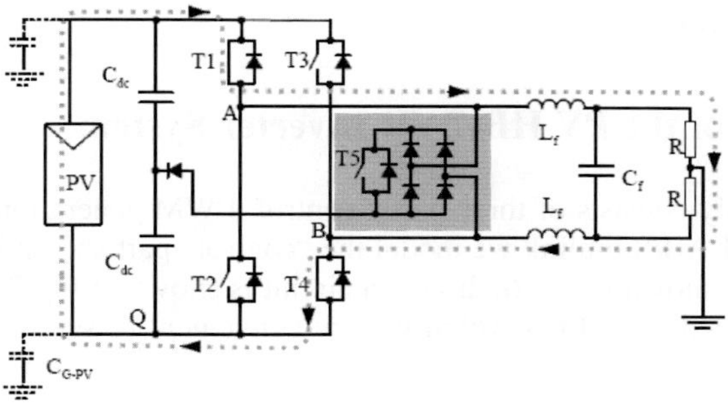

Fig. 3. Positive output voltage and current path of HB-ZVR inverter

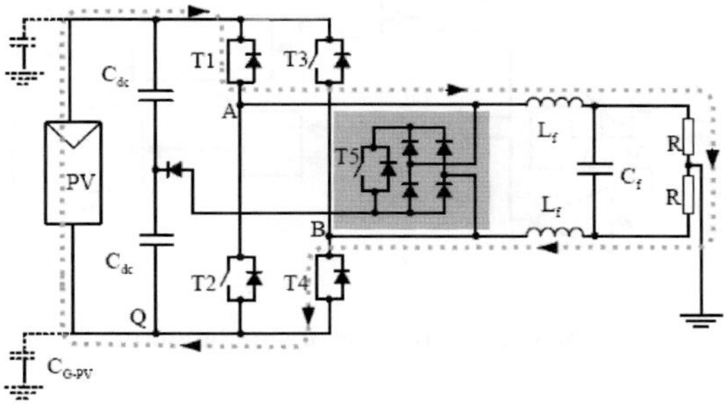

Fig. 4. Dead-time between turn-off of T_1-T4 and turn-on of T5 during positive half-wave

During the negative half wave of the load voltage, T_2 and T_3 are switched on that is used to generate the active vector. As is almost the same except for the direction of current. And the output voltage of the inverter has three levels taken into account the freewheeling part during dead-time. In this case the load current ripple is very small and the frequency is equal to the switching frequency. In all three cases this topology does not generate a varying common-mode voltage[9].

3 PLECS Tool Blockset

PLECS is a circuit simulator that makes it simple to model and simulate complex electrical systems along with their controls. Supporting a top-down approach, it lets you start with ideal component models in order to focus on system behavior. Low-level device details can be added later to account for parasitic effects.

With the intuitive, easy-to-use schematic editor, new models are set up quickly. Thanks to a proprietary handling of switching events, simulations of power electronic circuits are fast and robust. Whether you are simulating a simple power electronic converter or a complex electrical drive, PLECS is a powerful tool that will help you quickly obtain the results that you need. Here we use PLECS tool model the main circuit of PV inverter[11].

4 Modeling of the PV HB-ZVR Inverter System

The model mainly consists of three parts: control, PWM generation, main circuit. For Fig.1, use MATLAB/Simulink we model the "control" part shown in Fig.5. Similarly, the PWM part is shown in Fig.6, the main circuit is shown in Fig.7 and Fig.8. The PV inverter is modeled in PLECS which is very convenient to set up a model including the modeling of LCL filter.

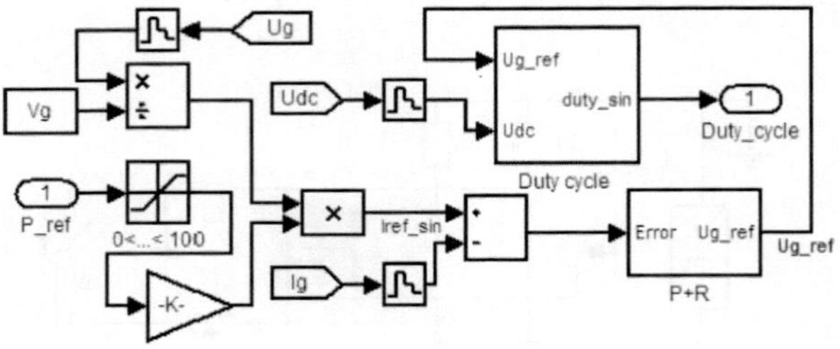

Fig. 5. The control part of simulation model

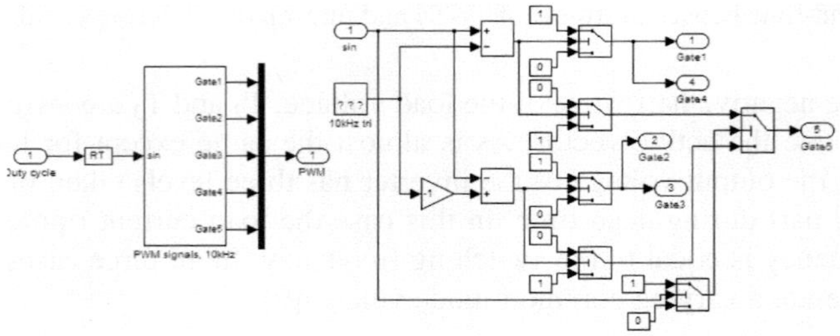

Fig. 6. The generation of PWM signals

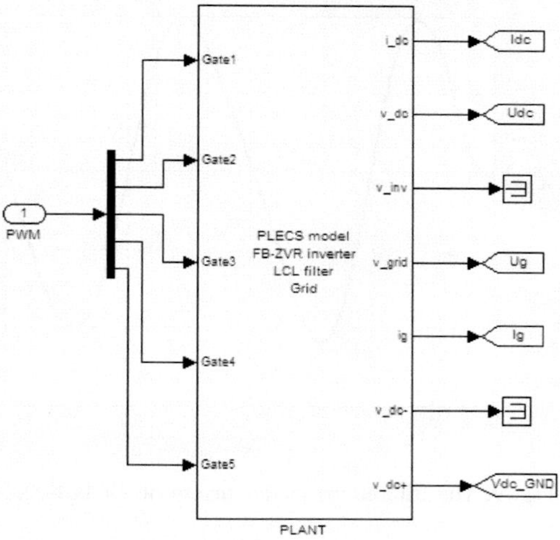

Fig. 7. The upper level diagram of main circuit

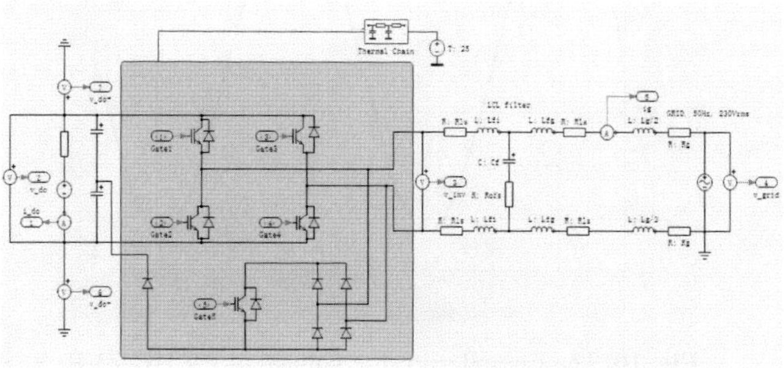

Fig. 8. The lower level diagram of main circuit

5 Simulation of PV HB-ZVR Inverter System

Here main parameters used: Nominal grid frequency, 50Hz. Grid inductance, 50uF. Single-phase DC-link voltage,400V. DC-link capacitance,900uF. Output filter inductance,1.8mH. Output filter capacitance,2uF. Grid voltage 230V(rms). Switching frequency, 10kHz. On the basis of above discussion of several parts, simulation models are integrated together consisting of the total system. On the basis of the above decision of parameters the simulation test begins.The corresponding results are obtained shown in Fig.9-11 respectively.

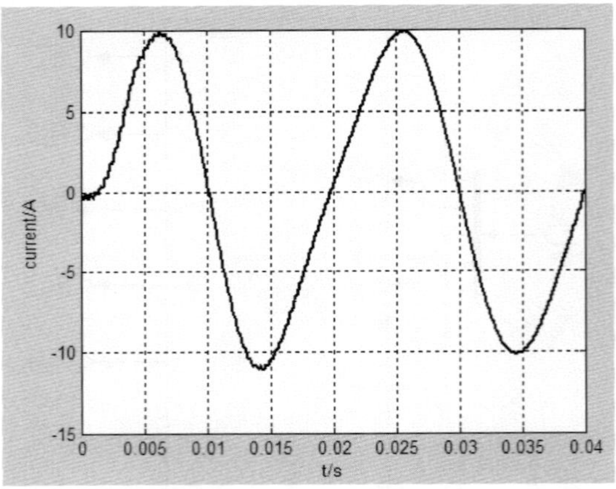

Fig. 9. The filtered grid side current of PV HB-ZVR

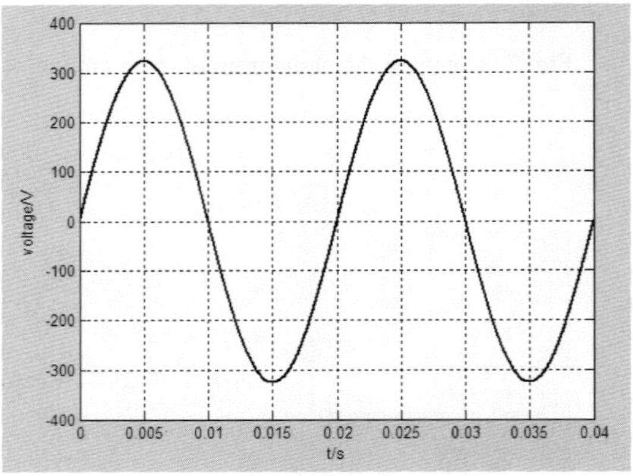

Fig. 10. The filtered grid side voltage of PV HB-ZVR

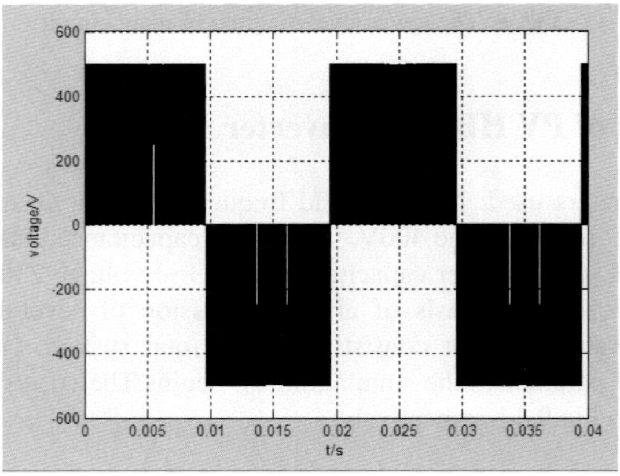

Fig. 11. The phase to neutral voltage PWM of PV HB-ZVR

From the above simulated results, unipolar modulation is suitable for control of single-phase PV HB-ZVR inverter. The performances of the LCL filter are also good. By means of PLECS simulation tool, a good filtered result is obtained which are shown in Fig.10 and Fig.11.

6 Conclusion

Unipolar modulation method is suitable for control of single-phase PV HB-ZVR inverter; and PLECS is powerful tool to model the main circuits of PV HB-ZVR inverter including the LCL filter. The simulation models in this paper almost approach to the real PV HB-ZVR inverter, which have laid a good basis for further research and the development and control of real PV HB-ZVR inverter.

Acknowledgment

I am very grateful to my supervisor, professor Frede Blaabjerg(Fellow IEEE) giving me many good opportunities and conditions while my staying in institute of energy technology, Aalborg university for one and half years.

References

1. Kjær, S., Pedersen, J., Blaabjerg, F.: A review of single-phase grid connected inverters for photovoltaic modules. IEEE Transactions on Industry Applications 41(5) (September/October 2005)
2. Myrzik, J., Calais, M.: String and module integrated inverters for single phase grid connected photovoltaic systems – a review. In: Power Tech Conference Proceedings, June 23-26, vol. 2. IEEE, Bologna (2003)
3. Calais, M., Myrzik, J., Spooner, T., Agelidis, V.: Inverters for single- phase grid connected photovoltaic systems – an overview. In: IEEE 33rd Annual Power Electronics Specialists Conference, June 23-27, vol. 4 (2002)
4. Kerekes, T., Sera, D., Teodorescu, R.: PV inverter control using a TMS320F2812 DSP. In: Proceedings of EDERS, pp. 51–57 (2006)
5. Ciobotaru, M., Kerekes, T., Teodorescu, R., Bouscayrol, A.: PV inverter simulation us- ing MATLAB. In: 32nd Annual Conference on Simulink graphical environment and PLECS blockset IEEE Industrial Electronics, IECON 2006, November 6-10, pp. 5313–5318 (2006)
6. Teodorescu, R., Blaabjerg, F., Borup, U., Liserre, M.: A new control structure for grid-connected LCL PV inverters with zero steady-state error and selective harmonic compensation. In: Record of IEEE APEC 2004, United States, vol. 1, pp. 580–586 (2004)
7. Teodorescu, R., Blaabjerg, F.: A new control structure for grid-connected LCL PV inverters with zero steady-state error and selective harmonic compensation. IEEE, Los Alamitos (2004)
8. Kerekes, T., Teodorescu, R., Borup, U.: Transformerless Photovoltaic Inverters Connected to the Grid. In: Twenty Second Annual IEEE Applied Power Electronics Conference, APEC 2007, February 25–March 1, pp. 1733–1737 (2007)
9. Teodorescu, R., Vitezslav, B., Pedro, R., Dezso, S., Tamas, K.: Photovoltaic power systems laboratory handbook, pp. 17–21 (Spring 2010)
10. Hu, Z.: Modern AC speed regulation, pp. 147–148. China machinery press (1998)
11. Allmeling, J.H., Hammer, W.P.: PLECS – Piecewise Linear Electrical Circuit Simulator for Simulink. In: PEDS 1999, Hong-Kong, July 1999, vol. 1, pp. 355–360 (1999)

Speed Loop Control of PMSM Driving Electric Vehicle

Yougui Guo[1], Ping Zeng[1], Jieqiong Zhu[1], Lijuan Li[1], Wenlang Deng[1], and Frede Blaabjerg[2]

[1] College of Information Engineering, Xiangtan University, Xiangtan 11105
guoygxtu@gmail.com
[2] Institute of Energy Technology, Aalborg University, Aalborg DK-9220, Denmark
fbl@iet.aau.dk

Abstract. Various simulation models are set up and closed speed loop control strategy of PMSM is proposed based on flux weakening control in this paper. First the model of maximum torque per ampere(MTPA) is modeled based on mathematical models and gave the corresponding simulation tests. Second the formulas are given to calculate the reference of stator current. Third the mathematical model is given to calculate q-axis current component and its detailed analyses. Fourth the modeling of PMSM is given in detail. Finally, simulation experiments are done. All the simulation experiments have verified that the models built in this paper are correct, and the closed speed loop control strategy is feasible and controlled very well. Also electromagnetic torque and three-phase stator currents are controlled well.

Keywords: PMSM, speed control, closed loop, MTPA.

1 Introduction

The electric vehicle is studied very hot[1]. And the control strategies of PMSM are its key technology[2-7]. A novel closed speed loop control structure of PMSM is put forward in this paper shown in Fig.1[7]. The system mainly consists of 6 parts: part I, part II, part III, space vector PWM, 3-phase voltage source inverter and PMSM. These 6 parts are presented in the following 6 sections.

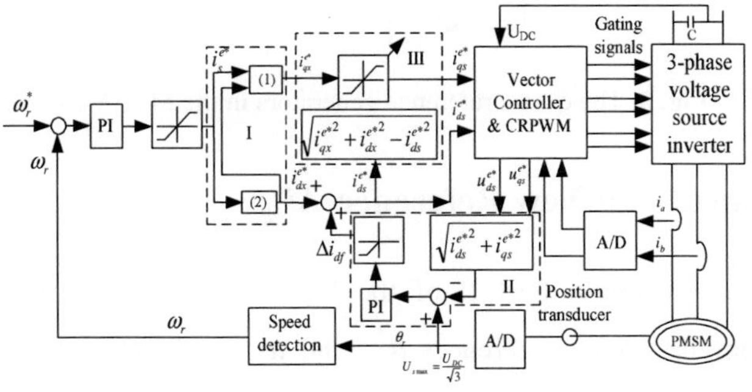

Fig. 1. Closed speed loop control structure of PMSM

X. Wan (Ed.): Electrical Power Systems and Computers, LNEE 99, pp. 809–816.
springerlink.com

2 MTPA Control of PMSM

This section is shown in Part I of the Fig.1. In order to maximize the electromagnetic torque the d-axis current reference can't be set to zero. That is to say the MTPA control algorithm must be calculated and generated the d-axis and q-axis references i^*_{sd} and i^*_{sq} from the special blocks according the following (1) and (2)[2].

$$\left|\vec{i}_s\right| = \sqrt{i^2_{sd} + i^2_{sq}} \le \left|\vec{i}_{sn}\right| \tag{1}$$

Where $\vec{i}_s$ and $\vec{i}_{sn}$ are stator current vector and nominal stator current vector respectively.

$$i_{sd} = \frac{\lambda_s}{4(L_{sq} - L_{sd})} - \frac{\sqrt{\lambda^2{}_s + 8(L_{sq} - L_{sd})^2 \vec{i}_s^{\,2}}}{4(L_{sq} - L_{sd})} \tag{2}$$

It is evident that the MTPA Control is possible to provide more torque for the same operating condition. In addition, such control algorithm has excellent dynamic performances and will make all the PMSM drive system more stable. It is not difficult to obtain the i^*_{sd} and i^*_{sq} according to (1) and (2). They are modeled in such a way shown in Fig.2 and Fig.3.

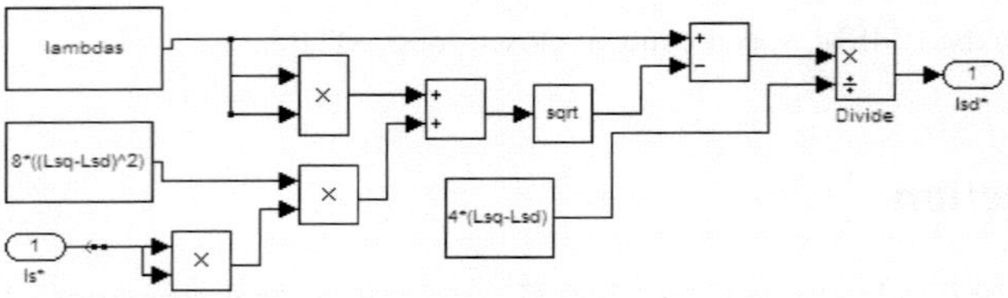

Fig. 2. The d-axis reference generators in the MTPA control

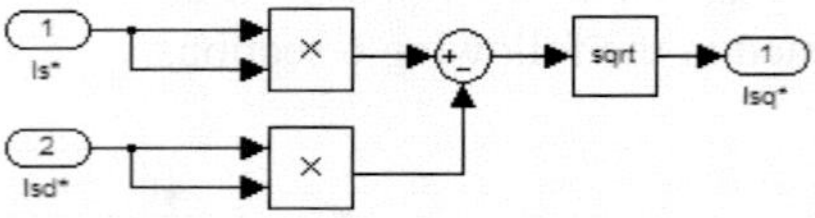

Fig. 3. The q-axis reference generators in the MTPA control

3 Obtainment of a New Reference Δi_{sdf}

This section is shown in Part II of the Fig.1. The module of stator voltage is calculated from the voltage references $u^{*\,2}_{ds}$ and $u^{*\,2}_{qs}$ in Part 'II' shown in (3). Then it is compared with the maximum voltage shown in (4) while the adder output will provide the voltage error to be integrated by the PI voltage regulator. If the error will be positive (no saturation conditions for the regulators) there isn't any type of action from the control system and the PMSM Drive System will continue to work in the constant torque region. Instead when the voltage error is negative the PI voltage regulator will provide a gradual decreased value for the d-axis current component i_{sd}, creating a new reference Δi_{sdf}. The variation of Δi_{sdf} can change from $i_{sd\max}$ to $-i_{sn}$.

$$|u_s^*| = \sqrt{u_{ds}^{*\,2} + u_{qs}^{*\,2}} \tag{3}$$

$$u_{s\max} = \frac{u_{DC}}{\sqrt{3}} \tag{4}$$

We set up the models of this part like section II.

4 Q-Axis Current Component and Its Protection

This section is shown in Part III of the Fig.1. It is used to calculate the q-axis current component according to (5) and prevent it from surpassing the maximum value of the PMSM currents, which is almost the same as (1), but the expression uses Δi_{sdf}. In this part it is possible the PMSM is working in the constant torque region which has no type of limitation for the q-axis component. The torque control will be possible to implement the field weakening control strategy shown Fig.1 adding the external voltage loop.

$$i_{qs}^{*\prime} = \sqrt{i_{qx}^{e*2} + i_{dx}^{e*2} - i_{ds}^{e*2}} \tag{5}$$

5 Modelling of PMSM

5.1 Mathematical Models of PMSM

As we all known, the DC motor has good control performances. So try to convert the PMSM into the DC motor from three-phase stationary reference frame to two-phase synchronous reference frame. The process is as follows:

(1) Three-phase abc stationary to two-phase $\alpha\beta$ stationary reference frame. For example with three-phase voltages as follows.

$$\begin{bmatrix} u_\alpha \\ u_\beta \end{bmatrix} = \begin{bmatrix} 1 & -\frac{1}{2} & -\frac{1}{2} \\ 0 & \frac{\sqrt{3}}{2} & -\frac{\sqrt{3}}{2} \end{bmatrix} \begin{bmatrix} u_a \\ u_b \\ u_c \end{bmatrix} \tag{6}$$

(2) Two-phase $\alpha\beta$ stationary to two-phase synchronous rotational reference frame. For example for the two-phase voltages as follows.

$$\begin{bmatrix} u_d \\ u_q \end{bmatrix} = \begin{bmatrix} \cos\theta & \sin\theta \\ -\sin\theta & \cos\theta \end{bmatrix} \begin{bmatrix} u_\alpha \\ u_\beta \end{bmatrix} \tag{7}$$

Note, the conversion is similar for other variables. Through a series of conversions similar like (6) or (7) we obtain the final mathematical models (8)-(11)[2].

$$u_{sd} = R_s i_{sd} + L_{sd} \frac{di_{sd}}{dt} - p\omega_m L_{sq} i_{sq} \tag{8}$$

$$u_{sq} = R_s i_{sq} + L_{sq} \frac{di_{sq}}{dt} + p\omega_m \left(L_{sd} i_{sd} + \lambda_s \right) \tag{9}$$

$$T_e - T_l = J \frac{d\omega_m}{dt} \tag{10}$$

$$T_e = \frac{3}{2} p \lambda_s i_{sq} + \frac{3}{2} p \left(L_{sd} - L_{sq} \right) i_{sd} i_{sq} \tag{11}$$

Where u_{sd} , u_{sq} are components of stator voltage u_s , T_e is electromagnetic torque of PMSM, T_l is load torque; R_s is stator electric resistance, i_{sd} , i_{sq} is components of stator current i_s , L_{sd} , L_{sq} is components of stator self-inductance L_s , p is pairs of magnetic poles, ω_m is mechanical rotational speed of rotor, λ_s is stator flux, J is rotational inertia.

5.2 Modeling of PMSM

With expression (8), (9), (10) and (11) we set up the corresponding models shown in Fig.4, Fig.5 and Fig.6.

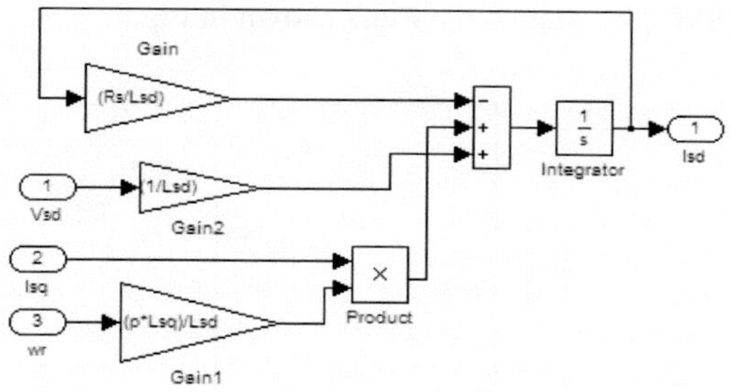

Fig. 4. Modeling of expression (8)

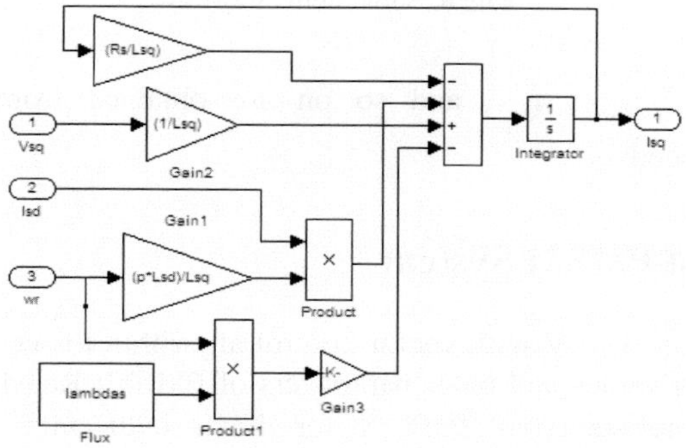

Fig. 5. Modeling of expression (9)

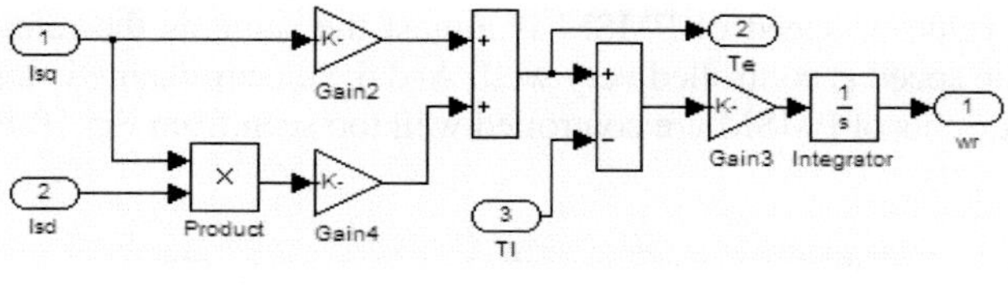

Fig. 6. Fig.5 Modeling of expression (10) and (11)

We can get three subsystems shown in Fig.7 from Fig.4-Fig6.

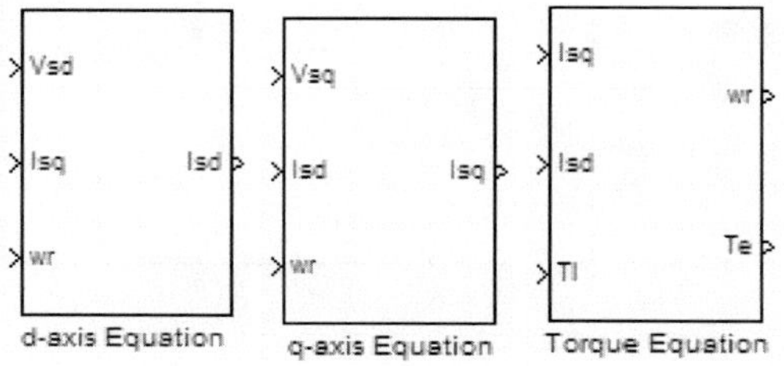

Fig. 7. Three subsystems of i_{sd} , i_{sq} and T_e

We can get the PMSM subsystem from Fig.7 shown in Fig.8.

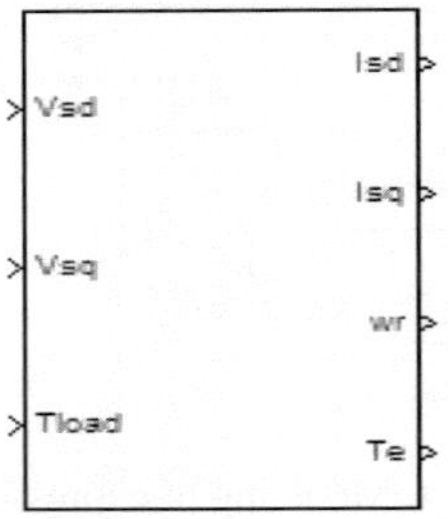

Fig. 8. Subsystem of PMSM

Here u_{sd} , u_{sq} , i_{sd} , i_{sq} , and so on are obtained from Clarke and Park transformation respectively.

6 Modelling of PMSM System

We combine section II to V with vector control algorithm to set up the model of the total system. Rated values and other parameters of PMSM: Rated current is 3A, Pole pairs is 6, Stator resistance is 2.5Ω, Stator d-axis inductance 7mH, Stator q-axis inductance 7.5mH, stator flux is 0.05Wb, rotational inertia is 0.0008kg.m^2, rated torque is 1.27N.m, rated speed is 3000rpm, output power is 400W, DC bus voltage is 200V, and so on. The simulation waveforms are shown in Fig.9-Fig.11. Seen from Fig.9 the measurement value of speed of PMSM is almost the same as the reference, which is verified that the speed is controlled very well. And the electromagnetic torque and three-phase stator currents of PMSM are controlled well too seen from Fig.10-Fig.11.

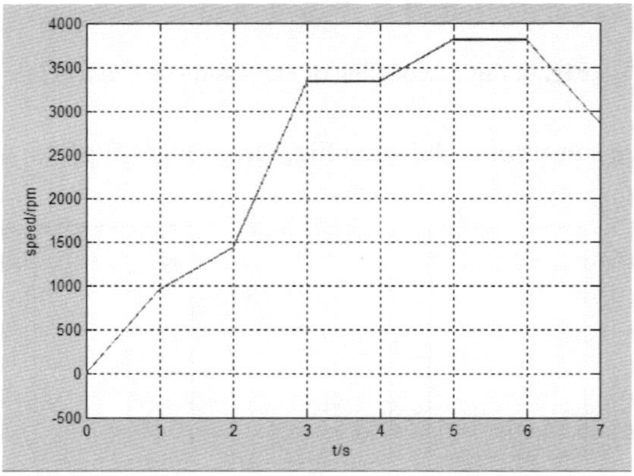

Fig. 9. The measurement and reference values of speed of PMSM

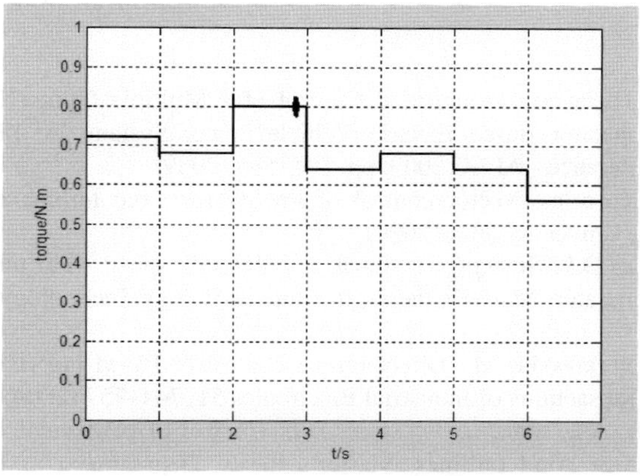

Fig. 10. The electromagnetic torque of PMSM

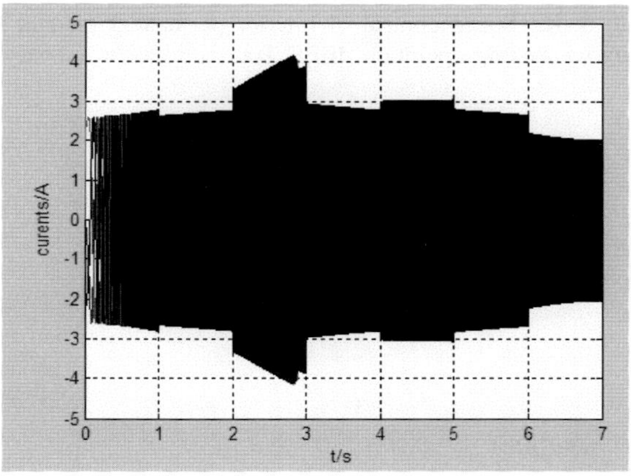

Fig. 11. The three-phase stator currents of PMSM

7 Conclusion

(1) The PMSM models built in this paper are controllable and feasible. It can be used for practical research.

(2) Speed of PMSM is controlled very well.

(3) The electromagnetic torque and three-phase stator currents of PMSM are controlled well too.

Acknowledgment

I am very grateful to my supervisor, professor Frede Blaabjerg(Fellow IEEE) giving me many good opportunities and conditions while my staying in institute of energy technology, Aalborg university for one and half years.

References

1. Erik, S., Li, Z.H., et al.: Design and Control of a Multiple Input DC/DC Converter for Battery/Ultra-capacitor Based Electric Vehicle Power System. In: IEEE Applied Power Electronics Conference - APEC 2009, pp. 591–586 (2009)
2. Mora, M.O.: Sensorless vector control of PMSG for wind turbine applications. Master Thesis, Aalborg University (June 2009)
3. Stan, A.I., Stroe, D.I., Stanciu, T., Shuai, L.: Variable speed wind turbine equipped with synchronous generator. Master's thesis, Institute of Energy Technology, Aalborg, Denmark (June 2009)
4. Buja, G., Kazmierkowski, M.: Direct torque control of PWM inverter- fed ac motors - a survey. IEEE Transactions of Industrial Electronics 51, 744–757 (2004)
5. Busca, C., Stan, A.I., Stanciu, T., Stroe, D.I.: Control of permanent magnet synchronous generator for large wind turbines. Master's thesis, Departament of Energy Technology, Aalborg University, Denmark (December 2009)
6. ABB, Direct Torque Control - the world's most advanced AC drive technology - Technical Guide No.1
7. Kim, J.-M., Sul, S.-K.: Speed control of interior permanent magnet synchronous motor-drive for the flux weakening operation. IEEE Transactions on Industry Applications 33(4), 43–48 (1997)

Achievements and Prospects of Hydropower Dispatching Management in China

Zheyi Pei[1,2], Changming Ji[2], Jun Xu[1,2], Yong Jing[3], Yanke Zhang[2], and Litong Dong[1,2]

[1] State Grid Corporation of China, West Changan Street 86, Beijing, China, 100031
[2] North China Electric Power University, Beijing, China, 102206
[3] Sichuan Power Development co., Ltd. of Sinohydro Corporation, Chengdu, Sichuan, China
Pei-zheyi@sgcc.com.cn

Abstract. China is one of the countries of most abundant water resources in the world, hydropower dispatching management have been greatly developed, as in operation management system, means and standardization etc. Which has played an important role in the safe operation of hydropower plants, the promotion of water resource use and energy savings. With the strengthening of power grid, a large number of giant hydropower station emergences and the backbone national grid with UHV as the core were constructed. To further standardize and improve the hydropower management system, a scientific, harmonious and efficient hydropower dispatching management system should be build and dispatching management innovation should be highlighted in order to fully use new energy sources such as wind power.

Keywords: achievement, prospect, hydropower, dispatching management.

1 Introduction

As we all know, China is one of the countries of most abundant water resources in the world, whose Theoretical potential hydropower is 694 million kilowatts, technically exploitable resources is 542 million kilowatts, Economic exploitation amount is 402 million kilowatts, China's water resources and technically developable capacity both rank first in the world. Since the first hydropower station in China mainland – Shilong had been constructed, water conservancy cause of China run through an extraordinary experience in almost a hundred years. Especially in recent years, as China implemented the policy to develop hydropower, water conservancy cause made development rapidly. Currently, hydropower installed capacity of China has over 2 million kilowatts, and the hydropower of China has become the first in the world. At the same time, reservoir operation and management have also been greatly developed in the management system, technical support means and standard system, and has made gratifying achievements, which has played important role in the safe operation of power plants, the promotion of water resource use and energy savings.

X. Wan (Ed.): Electrical Power Systems and Computers, LNEE 99, pp. 817–822.
springerlink.com

2 Achievement of Hydropower Dispatching Operation and Management

2.1 The Initial Formation of the Reservoir Operation Management System

Reservoir operation is the base of power plant operation, as an important link of plant produce, is also important part of power grid in dispatching and management. Hydropower dispatching is not only related to the safety and economic operation of plant, but also closely related to safety and economic operation of power grid. Therefore, hydropower dispatching is an important bridge to link power stations with power grid. With large hydropower stations such as the Three Gorges and the Ertan Hydropower Station put into operation, the safety operation of hydropower station impact on safe and stable operation of the power network is ever growing, and the importance of hydropower dispatching is increasingly outstanding. To meet the requirement of rapid development in hydropower, and make full use of comprehensive benefits of hydropower, meanwhile meet increasingly requirement of power grid to hydropower dispatching, government leaders at all levels have to pay more attention to hydropower dispatching& management system. In recent years, the dispatching center of State Grid and all provincial power grid are enhanced dispatching and management of hydropower. For power grid with a large proportion of hydropower, such as Center China, South, Northeast, Northwest and Fujian, Gansu, Hubei, Sichuan power grids have established special organizations to manage hydropower dispatching and plant operation, while most of companies without special organizations have arranged full-time persons to responsible for hydropower dispatching. Finally a relatively perfect hydropower dispatching system is formed in which national electric power dispatching and communication center plays a leading role, hydropower dispatching organizations of provincial power grid are principal part, hydropower dispatching organizations of city level are supplement, and hydropower stations are objects of dispatching. Primarily responsibility of National Electric Power Dispatching and Communication Center is professional hydropower dispatching of large reservoirs of inter-regional transmission and of state grid domain. Such as the Three Gorges power plant of which dispatching is dominated by National Electric Power Dispatching and Communication Center directly. Power dispatching of provincial power grids is responsible for large and small power stations in their regions. Power dispatching of city level power grids is responsible for small power stations in their regions. Power stations is responsible for comprehensive operation of water and power, work out utility of reservoirs and generation scheduling which are report for corresponding hydropower dispatching unit to approve or balance dispatching. Hydropower dispatching system have played positive role for using hydropower, optimizing resource distribution and supporting security of power grid.

2.2 Gratifying Achievements of Hydropower Dispatching Operation and Management Means

Hydropower dispatching has developed from one reservoir to cascade, and even compensation operation of entire grid or across grid has been developed, all of which required modern mean and processing of information collection to ensure safe and timely decision of hydropower dispatching, while to ensure safe and economic dispatching of power grid. Founding dispatching system of hydropower and power grid that utility modern advanced technologies become inevitable choice, through which to improve means of hydropower dispatching and to increasing economic efficiency and strengthen management of modern power grid.

(1) Hydrologic information automatic telemeter system with remarkable achievement

Hydrologic information automatic telemeter system is an advanced engineering system that centralized disciplines such as communication, computer, hydrology and remote sensing, which is the foundation of automatic hydropower dispatching for hydropower and power grid, and is also important technical means of supporting safety operating of hydropower stations. Construction of The forecasting system for hydropower stations was started in the early eighties, and the system has been greatly developed in 20 years of unremitting effort and practice. At present, all of large and medium hydropower plants have established forecasting system, and have become an important part of power production. Some large reservoirs such as the Three Gorges, FengMan have established automatic operation systems of water, and safety and reliable operation of these systems have became foundation that water automatic dispatching system of power grids are safety and reliable operation. Now, these systems are playing increasingly important roles for the economic operation of hydropower plants.

(2) Built a national hydropower dispatching automation system for power grid

The automatic dispatching system for hydropower is a part of power grid automatic dispatching system, and is important technical supported means of hydropower dispatching organizers to carry out dispatching. Including Southern Power Grid, China have established total of 21 automatic hydropower dispatching systems are running, these dispatching systems are National Electric Power Dispatching and Communication Center, south, northeast, central, northwest, east China, Fujian, Guangdong, Guangxi, Guizhou, Yunnan, Gansu, Hunan, Sichuan, Jiangxi, Chongqing, Zhejiang, Anhui, Xinjiang, Qinghai and Tibet though efforts for 10 years. Among them, 11 systems including the Northeast, Central, Northwest, Fujian, Gansu, Hunan, Sichuan, Jiangxi, Zhejiang, Anhui and Chongqing Power have reached practical requirements and have passed through the acceptance of experts organized by National Electric Power Dispatching and Communication Center(due to institutional changes, the a Southern Power Grid have not carried out Practical acceptance). The end of 2009, in addition to the Southern Power Grid and Xinjiang Power Grid, other hydropower automatic dispatching systems have achieved networking operation with the automatic dispatching system of National Electric Power Dispatching and Communication Center, and have formed the automatic water dispatching information network of the State Grid. Not only have these information networks achieved water information share, but also have offered a foundation for National Electric Power Dispatching and Communication Center supervising and directing operation of important reservoirs in power grid, and these timely and reliable water information have contributed leaders at all levels to make scientific decision. At same time, the automatic water dispatching information network of the State Grid has also played a important role in ensuring safe and economic dispatching of power grid, safeguarding flood safety and increasing additional power by water saving.

2.3 Formed a Relatively Complete System of Hydropower Dispatching Management Standardization

The standardization is an important way to enhance the dispatching management level and efficiency. With increasing in amount and scale of hydropower stations, especially with formed of large-scale hydropower stations to increase comprehensive utilization, task of hydropower dispatching management is increasingly heavy, and corresponding technology become increasingly complex. In order to regulate hydropower dispatching, to enhance work efficiency, experts from all aspects organized by relative state department have developed a series of standard, rules and regulations. Such as economic hydropower dispatching, ministry of water and power resources has issued "The trial incentives of hydropower increase power by saving water" and "The management ordinance of economic reservoirs operation", both of them have promoted work to develop standard of increasing power by saving water effectively in the early 80s of last century. In the 90s of last century, "standard of reservoir operation for large and medium hydropower station (GB17621-1998)" has been developed based on extensive research, as national standard, it established foundation of reservoir operation standardization. In the 21st century, the hydropower dispatching standardization is been deepening further, "Technical specifications of automatic hydrological forecasting system" and "water operation automatic specification of power grid" are developed and issued continually. At the same time, the power grid dispatching specifications such as "dispatching regulations of power grid" and "operation guideline of power grid" have also added corresponding content for hydropower dispatching management. The issued company standard of the State Grid that is "Reservoir operation specification" has became new basis for hydropower dispatching standardization in new century. the system of hydropower dispatching management standardization of continuous improvement and expansion, it is playing positive role in regulation management hydropower dispatching and improving efficiency.

2.4 Achieved Significant Social and Economic Benefits

(1) Social benefits

China is a country with more serious flood, most of reservoirs have flood control task. With developed of national economy the conflict between flood control and water uses is more prominent, it lead to reservoirs without flood control have been required to bear some flood control. Therefore, scientific and precise flood forecasting is significant for maximum effect on flood controlling and decreasing loss of water engineering. For many years in the past, scientists that take part in hydropower dispatching have ensured safety during flood season of large and medium reservoirs by strengthen accuracy of flood control, scientific analysis and optimizing reservoir operation and so on. These measures have obtained eminent social benefit by developing flood control function of reservoirs sufficiently, maximum decreasing flood controlling pressure of lower reservoirs. The summer in 1995, the flood super 0.01% with 16350m3/s maximum reservoir inflow take place in Fengman-Baishan area. Through reservoir operation, reservoir outflow of Fengman reservoir is curtailed to 4500m3/s after giving full play to integrated reservoirs operation and flood storage capacity sufficiently. Due to curtailed 72% flood peak discharge, life and property of people living downstream is defended. In July 20, 2010, the maximum flood of Three Gorges Project occurred since it is built and return period is close to 20 years, of which inflow discharge is 7000m3/s, is greater than flood in 1954 and 1998. The Maximum outflow discharge of Three Gorges Project after regulation and storage is 40000m3/s, is curtailed flood discharge 30000m3/s, and ratio of curtailment is 43% that decrease flood control pressure of Jinjiang reach and Hunan province, Hubei province extremely.

(2) Prominent economic benefit

In recent years, some measures of the organizations at all level of power dispatching have been taken to overcome adverse effects from reservoir inflow amount below normal, uneven water amount in flood season and dry season, relieved situation of power supply stress and have achieved remarkable economic and social benefit. Organizations of power dispatching implement national power policy and economic operation measures actively, meanwhile strengthen flood forecast and relative analysis, and give full play to market regulation. Reservoir operation has been carried out scientifically by elaborated coordinated dispatching between power grid and hydropower plant, and positive fulfill effective hydropower economic dispatching. Such as Fujian Power grid, in 2003 and 2004, the reservoirs such as Shuikou, Shaxikou, Ansha encountered special year in history since recording hydrology datum. Funjian Power Grid make full use of limited water amount of the two dry years by reasonable dispatching reservoirs and power grid, which increase power total 15 billion KW˙h and ratio of increasing power over 10%. It is estimated that mean total additional generation of state from water saving of hydropower dispatching over 80 billion KW˙h since 2000, that is equivalent to 2600 thousand ton standard coal, and decrease CO2 emission about 6000 thousand ton. All of these are positive contributed to sustainable development of China and relieved the conflict of power supply, saving coal resource, to the benefit of conserving energy and reducing emission and combating climate change.

Hydropower dispatching management is faces some new challenges and new problems with bringing into produce of new hydroelectric generators, Evaluation of electricity system reform, extending of power grid and hydropower scale, changing of electric power supply. First, after The Three Georges Project complete built, other giant reservoirs such as Longtan, Xiaowan, Xiluodu, Xiangjiaba have also began to be build or to be putted into operation, it is important and complicate that how to safe and effective manage these giant reservoirs, especially the multi-reservoir is made up of the giant reservoir, and to implement optimal cascade reservoir operation and cross basin reservoir operation, even union compensation dispatching of hydropower and thermal power. Second, with economy development and society progress, people have a increasingly environment requirement that lead to more reservoirs comprehensive use, while flood controlling, irrigation, water supply, stemming the tide and environmental protection have became constraint of reservoir operation, it is becoming a challenge for hydropower plants operation management and reservoir operation, how to deal with all kinds of problems to bring into full play comprehensive benefit of reservoirs. Third, it is impacting development of hydropower dispatching at some extent because the equipment and application of hydropower dispatching fall behind development of power grid, and some unit is unable to suffice for request of hydropower station management and reservoir operation due to system aging and insufficient equipment.

3 Development and Prospect

With the rapid development of hydropower and the strengthening of power grid, a large number of giant hydropower station emergences and the backbone national grid with UHV as the core constructed. The situation will provide more extensive stage to the hydropower station's operation and dispatching management. To further standardize and improve the hydropower management system, the following suggestion should be noticed:

(1) Strengthening the management of hydropower dispatching;
(2) Build a scientific, harmonious and efficient hydropower dispatching management system
(3) Improve the dispatching management of giant power plants and large scale cascade hydropower station
(4) To further improve the standard system;l
(5) Building an intelligent hydropower dispatching management support system;
(6) Give full attention to dispatching management innovation in order to fully use new energy sources such as wind power.

References

1. Pei, Z., Yao, Z., Guo, S.: Achievements and Prospects of Hydropower Dispatching in China in the Past Few Years. Hydropower Automation and Dam Monitoring 28(1) (2004)
2. Li, J., Jihuai, S., Ji, C.: Dynamic Control of Limiting Flood level of LiJiaXia Reservoir. Hydropower Automation and Dam Monitoring 31(2) (2007)
3. Li, A., Wang, L., Ji, C.: Compensation Benefit Analysis of United Operation of Inter-basin Mixed Hydropower Station. Hydropower Automation and Dam Monitoring 31(5) (2007)
4. Sun, Q., Pei, Z.: Establishment and Operation of the Automatic System of Hydrological Data Acquisition and Transmission of Electric Power Plant in China. Advances in Water Science 11(4) (2000)
5. He, G., Shu, y., Pei, z.: Algorithm and Application of Optimal Dispatching for Three Gorges Electricity Market. Automation of Electric Power Systems 27(6) (2003)

Invasive Blood Pressure Simulator Electronics Device Bed Side Monitor Testing

Marek Penhaker and Jan Kijonka

VSB – Technical University Ostrava, FEECS,
Department of Measurement and Control, 17. listopadu 15
70833, Ostrava, Czech Republic
{marek.penhaker,jan.kijonka}@vsb.cz

Abstract. The idea of this work was realization of invasive blood pressure electronics device simulator. This kind of device is ready for use for bed side monitor testing. Invasive blood sensors outputs is converting to low voltage signal which is input part to bed side monitor. This electrical input has to be checked periodically for calibrating the bed side monitor accuracy. Presented device for testing is standalone instrument with easy user interface for setting of broad scale and precise voltage continuance.

Keywords: Blood, Pressure, Testing, Simulator.

1 Introduction

The vital signs simulators are electronic devices for simulating of the ECG, invasive blood pressure (IBP), non-invasive blood pressure (NIBP), respiration, temperature, hearth sound and others signals. Some devices also allow connection to manikins and skills trainers and allowing instructors to run pre-programmed and programmable scenarios to meet their specific learning objectives. These devices are educationally effective for training in the care and treatment in hospital and pre-hospital care providers. These devices reduce the time taken to test the correct performance of a wide range of medical devices and equipment used in hospitals, operating theatres and other facilities. Some comprehensive patient simulators no longer need to use variety of different instruments for testing these functions separately.

This paper deals with a programmable IBP simulator for variety of invasive blood pressure simulations. Its parameters are tested and suitable for bed side monitors testing and calibrating. This single purpose device allowing IBP pre-programmed and programmable IBP waves simulations is a low cost but interesting solution due to its relative simple design and flexibility. This solution can be used as a part of a comprehensive device for multi vital signs simulation. However, the IBP measurement on a real patient, opposite to the ECG, NIBP, respiration and temperature measurements, is usually restricted to a hospital setting. Therefore, testing of the ECG, NIBP, respiration and temperature bed side monitor modules is not such a problem like the IBP bed side monitor module testing. [1]

X. Wan (Ed.): Electrical Power Systems and Computers, LNEE 99, pp. 823–830.
springerlink.com

Invasive blood pressure is a method of measuring blood pressure internally by using a sensitive Schwanz-Ganz catheter placed into the pulmonary artery. The proximal port of the catheter is connected to a sterile, fluid-filled system with an electronic pressure transducer, sensing the pressure changes. A low level voltage output of the IBP transducer is get to the visualization system, usually bed side monitor with an IBP module input channel. The BP invasive measurement provides a more accurate reading of the patient's blood pressure usually used where rapid variations of blood pressure are anticipated. The advantage of this system is that pressure is constantly monitored beat-by-beat and a waveform (a graph of pressure against time) can be displayed.

2 Problem Definition

Testing and calibrating of the IBP bed side monitors modules is problematic and inaccessible without any special equipment. Sometimes we don't need any of expensive comprehensive devices for all vital signs simulation and the IBP simulator satisfies to ours requirements. [2]

Common devices with IBP simulation features: input/output impedance: 300 Ω, exciter input voltage range: 2 V to 16 V, output sensitivity: switchable 5 μV/V/mmHg or 40 μV/V/mmHg, output range: -10 mmHg to 300 mmHg, accuracy ± (1 % of full range + 1 mmHg) at 80 BPM, RS232 interface for remote control via PC, several channels for generating the IBP: atmosphere (0), arterial = 120/80, central venous pressure = 15/10, left ventricle = 120/0, right ventricle = 25/0, pulmonary artery = 25/10, pulmonary artery wedge = 10/2, static = -10, -5, 0, 20, 40, 80, 100, 200, 250, 300 (manual or auto-stepping at 12-s intervals), triangle = 30.2 Hz, triangle = 300.2 Hz, Schwan-Ganz: start, insert, inflate, deflate, and remove.

The advantages of our new solution in comparison with a common devices are user programmable set of dynamic and static simulations, static pressure changeable with a step defined by the simulator resolution, RS232 / USB interface determined not only for remote control via PC, but also for editing the invasive blood pressure waves set and for uploading the device firmware, real-time changeable shape, amplitude and frequency of generated signals, comprehensive LCD visualization informing the user even about the output voltage and exciter input voltage measured in feedback circuit.

3 New Solution

The IBP intracardial measurement is shown in fig. 1. Denoted system has one output signal Vout and one input signal Vin. The Vout is a low level differential voltage signal carrying the waveform invasive blood pressure information. The Vin is an exciter input voltage generated by the bed side monitor. It is a stabilized voltage with a fixed value. However, various types of bed side monitors can vary in exciter voltage level. The IBP waveform simulation is shown also in this figure. This system has the same interface for bed side monitor connection as noted in the IBP intracardial measurement network. The electric switch point performs switching between a real IBP measurement and its simulation. It is important to acquire the Vout signal from the simulator comparable to the Vout signal from real measurements.

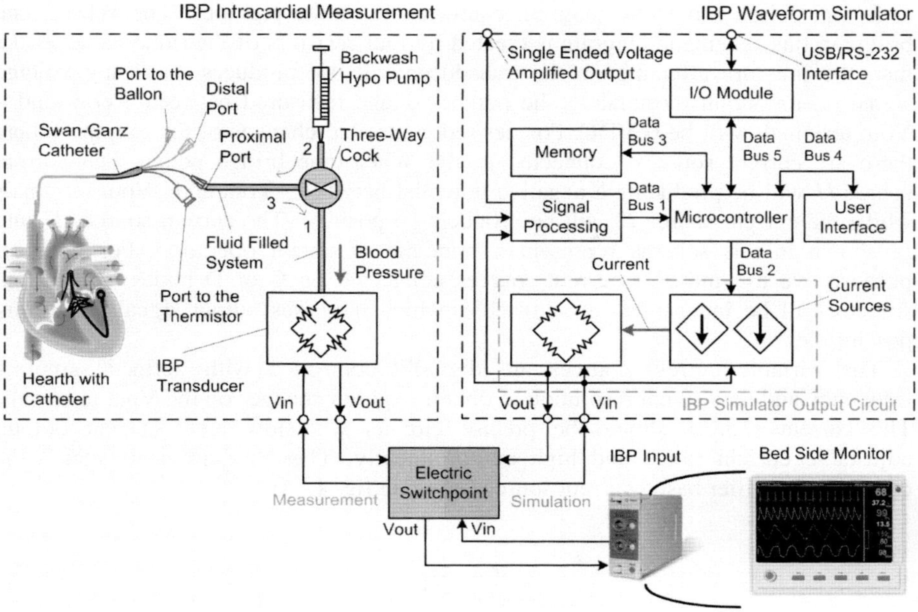

Fig. 1. Measuring network for IBP intracardial measurement and IBP waveform simulation.

The new solution is based on the principle of IBP measurement. IBP is measured by IBP transducer. It is a passive resistive element, which works as a two-port network. The output low level signal Vout [μV] is linearly dependent on the input exciter voltage Vin [V]. The IBP transducer inner circuit consists of four strain gauges interconnected to form the full Wheatstone bridge. This mounting maximizes the sensitivity of the pressure sensor and improves the non-sensitivity to ambient temperature changes. A fundamental schema is in fig. 2. A pressure affecting on the transducer produces a differential low level signal [μV] on its output terminals. [3]

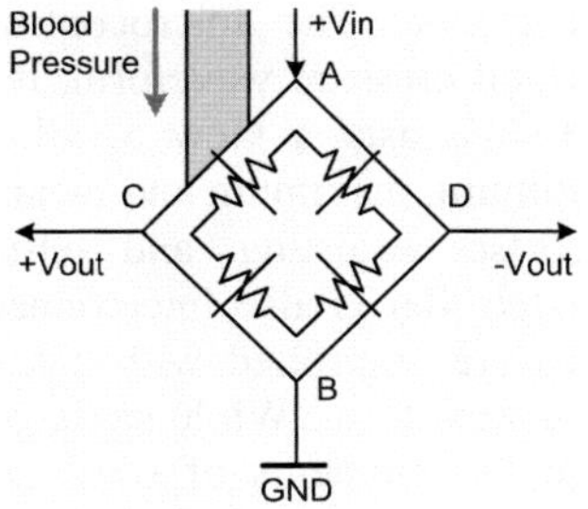

Fig. 2. IBP transducer schematic diagram.

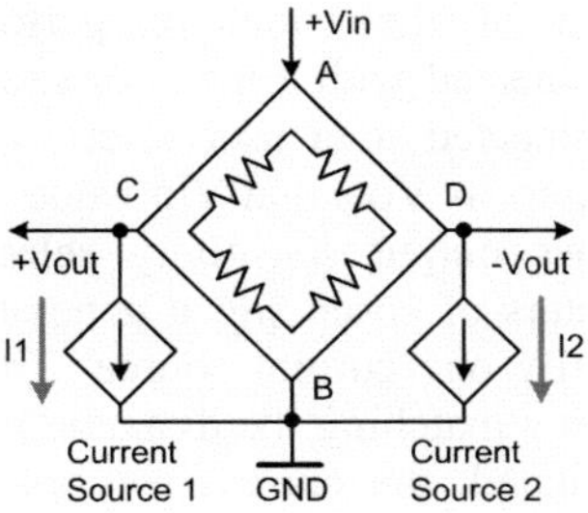

Fig. 3. IBP simulator output circuit schematic diagram.

In fig. 3 is a schematic diagram conformable to the previous. The Wheatstone bridge in this schematic diagram is created by four resistors of the same values. Like that, the pressure affecting on the transducer will not produces more any voltage signal on its output terminals. The output voltage measured between +Vout and –Vout terminals will be 0 VDC. For desired controlled changes of the output voltage there are current sources, connected to the Wheastone bridge nodes. The current source *I1* will be producing a negative potential between +Vout and –Vout terminals whilst the current source *I2* will be producing a positive. The current sources *I1* and *I2* shown in this schema represent a principle of circuit operation. For practical purpose, we use more of current sources connected on C or D nodes. All current sources will be independently adjustable, which increases output signal generating possibilities.

The variable current sources are digitally controlled with defined sampling frequency and data signal resolution to obtain required changes on the Vout terminals. The current sources should be precise circuits with low level current output, adjustable current range and high output linearly. These requirements match the operation amplifier based current source shown in fig. 4. [4]

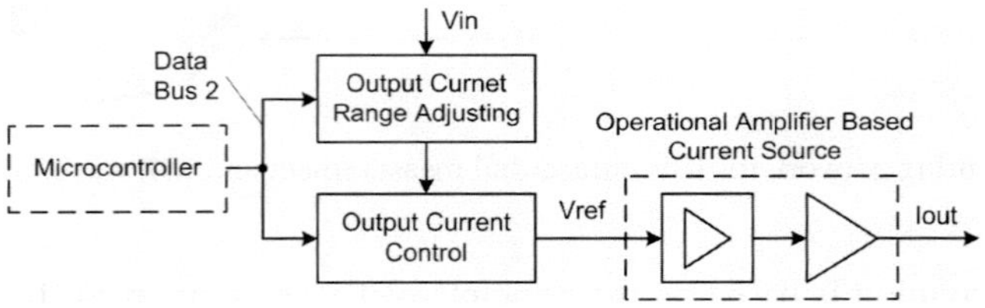

Fig. 4. Current source circuit.

The output current range adjusting and output current control is performed by digital potentiometers. Vref is an input control voltage for a current source.

3.1 Principle of Operation

The designed device block diagram is in the fig. 1 on the right side, Principal control function of the simulator performs a microcontroller. The microcontroller is interconnected with a memory stored the data of blood pressure waveform. It is also interconnected to a user interface, which gives to the user a view to all needed parameters as Vout and Vin values, IBP actual, minimum, maximum and mean value, it allows output sensitivity selecting, pressure offset adjusting, and setting the parameters of cosen signal generation. The microcontroller is also interconnected to minimally one current source. The current sources are controlled with a sampling frequency matching the dynamic range of simulated waveform. While static pressure is simulated, the current sources are controlled at the moments of static pressure changes. Some current pressures can be used for pressure offset settings. The Vin and Vout are processed in a signal processing block for its interconnection with the control unit, because the simulator has a feedback loop for monitoring the Vin and Vout. In this block, the Vout signal is amplified and converted to the single ended output suitable for measuring purposes. The microcontroller is also interconnected with an I/O module designed for simulator firmware uploading, memory data set uploading and remote control via PC.

4 Implementation of the New Solution

For testing, evaluation and bed side monitors calibration is next presented an implementation of the designed device. The block diagram is in fig. 5. You can find similarity relation to block diagram in fig. 4. For this solution, we used three simplified current source circuits controlled through the serial peripheral interface (SPI). For dynamic and static simulation, one current source is controlled with sampling frequency 200 Hz and 8-bit data resolution, or set at the moments of static pressure changes in the range from -30 to 300 mmHg. The second current source is used to offset adjusting from -25 to 25 mmHg with 8-bit data resolution. The third current source is used for adjusting the negative output voltage offset corresponding to -55 mmHg. The output current range adjusting block is simplified to fixed output sensitivity of the 5 μV/V/mmHg for chosen bed side monitor testing and calibrating. Output and input impedance are modified to match the NPC-100 transducer parameters, 300 Ω output impedance, 3710 to 3730 Ω input impedance.

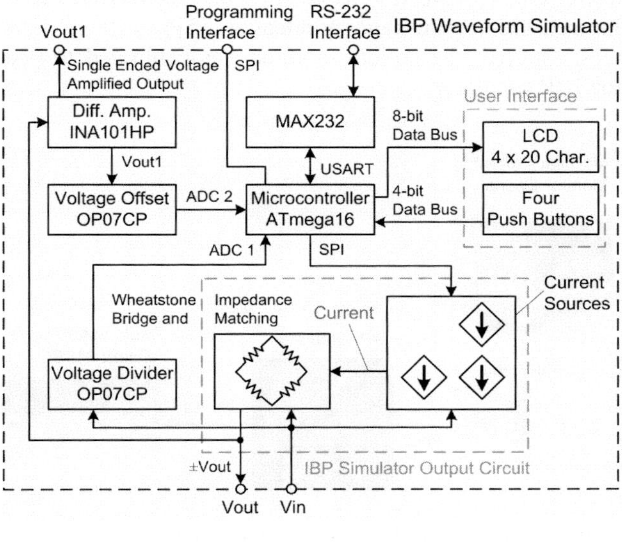

Fig. 5. IBP waveform simulator block diagram.

The user interface consists of a character LCD and four push buttons. The user menu, displayed on LCD allows to read Vout, Vin and corresponding pressure values, set the parameters for pressure waveform generation, including one pressure waveform generation, sinus function with amplitude and frequency settings generation, static pressure in the range from -30 mmHg and 300 mmHg with step of

1.3 mmHg (limited by used 8-bit resolution) generation, offset from -25 to 25 mmHg with step of 0.2 mmHg (limited by used 8-bit resolution) generation. The user interface allows additional features as LCD backlight setting and ADC range setting.

The signal processing is solved on analog platform. The Vout and Vin are adjusted to the values suitable for the internal reference of the microcontroller 10-bit analog to digital converter (ADC). The differential voltage on the Vout terminals is converted to a single ended voltage by a precise instrumentation amplifier and with the 200-multiple amplification is available on the Vout terminal for measuring and testing the simulator. The Vout1 is added to a constant voltage to ensure a positive potential on the ADC input.

The I/O module is represented by the integrated circuit MAX-232, which allows to remote control via PC, and SPI interface for uploading the microcontroller firmware and integrated flash memory data set.

5 Testing and Evaluation

The implementation mentioned above was tested to possibility of using this device for bed side IBP module testing and calibrating. Measurement networks for the static and dynamic IBP simulations testing are shown in fig. 6 and 7. The equipment used for the simulator testing was a modular patient monitor Ekona PM6000 with an IBP module, multimeter Escort 3146A, USB oscilloscope Agilent U2702A connected to a PC with Agilent measurement manager application. [5]

Before the tests started, the patient monitor was calibrated to 0 mmHg and 250 mmHg using the IBP simulator.

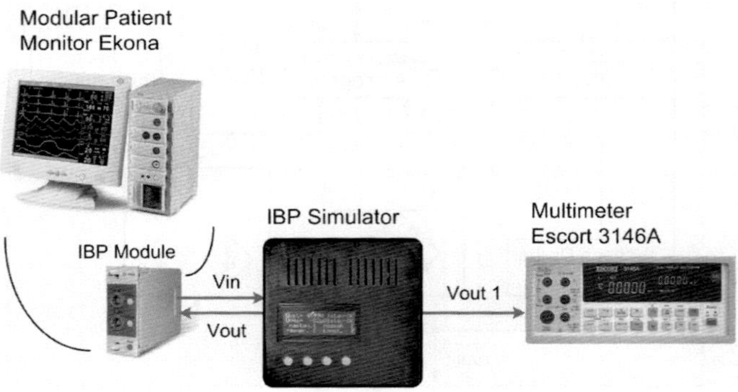

Fig. 6. Measurement network for static pressure simulation.

In the first test, the simulator was adjusted throw its user interface to generate the voltage signal corresponding to the constant blood pressure from -30 mmHg to 300 mmHg step 10 mmHg. For each adjusted value was measured the simulator output Vout1 and simultaneously it was taken the value displayed on patient monitor. The output voltage measured by the multimeter was taken as the referential value for accuracy determining. The mean deviation and standard deviation from the referential value of pressure is shown in table 1.

Table 1. Font sizes of headings. Table captions should always be positioned above the tables.

	Mean	Std.	Units
IBP simulator	0.022	0.2736	mmHg
Ekona PM6000	-0.2281	0.5548	mmHg

The dynamic testing wasn't as important for determining the simulator calibrating possibilities as for verify the accuracy of simulator minimum and maxium pressure and frequency of generated signal settings. For IBP waves simulation were important also the mean value. The output voltage measured by the oscilloscope was taken as the referential value for the accuracy determining. For IBP waveform simulation was generated the IBP waveform with defined systolic pressure = 142 mmHg, diastolic pressure = 85.18 mmHg and mean pressure = 104.16 mmHg and frequency 48.24 bpm. For function simulation was generated the sine wave with selected minimum value = 118.82 mmHg, maximum value = 218.47 mmHg and frequency = 300 bpm. The deviations from the selected frequency were quite insignificant. The significant deviation from the simulator settings were measured for the systolic pressure and maximum pressure. The maximum deviation was -2.68 mmHg. This was caused by not calibrated simulator waveform data stored in the simulator memory. The waveform simulation mode (dynamic signal generation) doesn't use the feedback measurement of the generated output signal. The best solution for solving this problem would be the feedback setting of the maximum and minimum value for generated pressure waveform just before starting the generation. Like that it would be possible to achieve the results comparable to the static pressure generating.

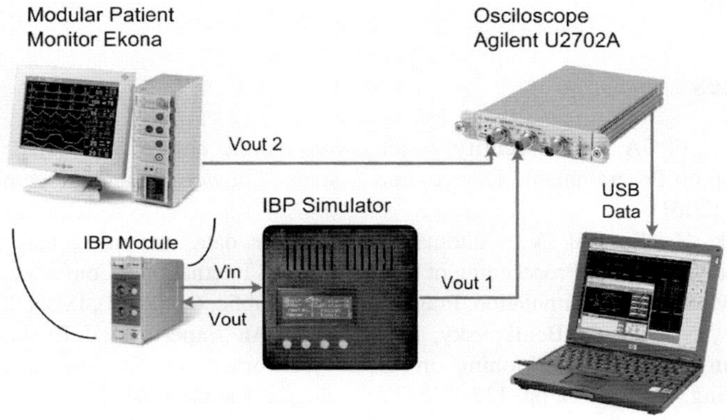

Fig. 7. Measurement network for dynamic pressure simulation.

Another test was focused on testing and calibrating the IBP module without use of the IBP simulator. Instead of that, it was used the IBP transducer connected to the BP cuff, manually inflated to a required static pressure indicated on a manometer scale, or using the BP Pump device. The main drawback is not absolutely closed pneumatic circuit and air escaping.

6 Results

The designed device has excellent results in static pressure simulation thanks to the feedback circuit. Like this, it can be easy achieved the necessary accuracy ±1 mmHg for bed side monitors calibrating. The advantages are programmable set of generated pressure waveform, programmable control unit, high flexibility, extensibility and low cost design.

7 Conclusions and Summary

The paper presents the design and realization of electronic calibrating device for invasive blood pressure bed side monitor in hospitals. The process of proposal consists of several especial sub solutions. Finally the complete device was successfully tested and provides hundred times better sensitivity for pressure wave generating then the standard requirements in medical applications.

Acknowledgments

The work and the contribution were supported by the project: Ministry of Education of the Czech Republic under Project 1M0567 “Centre of Applied Cybernetics”, student grant agency SV 4501141 “Biomedical engineering systems VII” and TACR TA01010632 “SCADA system for control and measurement of process in real time”. Also supported by project MSM6198910027 Consuming Computer Simulation and Optimization.

References

1. Kasik, V.: FPGA based security system with remote control functions. In: 5th IFAC Workshop on Programmable Devices and Systems, Gliwice, Poland, November 22-23, pp. 277–280 (2001)
2. Machacek, Z., Srovnal, V.: Automated system for data measuring and analyses from embedded systems. In: Proceeeding of the 7th WSEAS International Conference on Automatic control, Modeling and Simulation, Prague, Czech Republic, 6 p (2005), ISBN 960-8457-12-2
3. Brida, P., Machaj, J., Benikovsky, J., Duha, J.: An Experimental Evaluation of AGA Algorithm for RSS Positioning in GSM Networks. In: Electronics and Electrical Engineering, vol. 8(104), pp. 113–118. Technologija, Kaunas (2010)
4. Skapa, J., Siska, P., Vasinek, V., Vanda, J.: Identification of external quantities using redistribution of optical power - art. no. 70031R. In: OPTICAL SENSORS 2008. 7003, pp. R31–R31 (2008), ISSN: 0277-786X , ISBN: 978-0-8194-7201-4
5. Penhaker, M., Rosulek, M.: Electrodes For Biotelemetry And Home Care Applications. In: Proceedings of 3th International Conference on Systems, ICONS 2008, April 13-18, pp. 179–183. IEEE, Cancun (2008), ISBN 978-0-7695-3105-2

Audiometry for Teaching Experiment in PowerLab Systems

Marek Penhaker and Jan Kijonka

VSB – Technical University Ostrava, FEECS,
Department of Measurement and Control, 17. listopadu 15
70833, Ostrava, Czech Republic
{marek.penhaker,jan.kijonka}@vsb.cz

Abstract. The aim of this work is electronics device for audiometry measurement construction. Presented electronic device is designed as extension of ADInstruments teaching experiments PowerLab systems. Realized audiometer is suitable for standard medical testing of hearing with step by step measurement and output protocol arrangement. The original hardware design consist primary from extension electronics board containing precise tone generator and auxiliary input. Realized audiometer device can be also used as stand alone device possible supplemented by LCD outputs, I/O control panel and feedback pushbutton. Application of this device introduce accurate and reliable hearing condition testing.

Keywords: Audiometry, Control, Sound, Hearing Testing.

1 Introduction

PowerLab data acquisition units are smart peripheral devices that perform data acquisition, signal conditioning, and pre-processing. PowerLab data acquisition systems comprising hardware and software offer versatile data acquisition and analysis solutions for life science research and education applications. External signals are acquired through analog inputs, amplified, digitized and transmitted to the computer using USB connection. Software receives displays, analyzes and records the data. Some PowerLab systems provide analog and digital outputs for control or stimulation. These units are ideal for teaching and are used in field of human and animal physiology, pharmacology, neurophysiology, biology, zoology, biochemistry and biomedical engineering. Audiometry is an extending experiment for PowerLab systems. [1]

1.1 Audiometry

Audiometry is the term used to describe formal measurement of hearing. The measurement is usually performed using an "audiometer" by an "audiologist".

X. Wan (Ed.): Electrical Power Systems and Computers, LNEE 99, pp. 831–838.
springerlink.com

Pure tone audiometry (PTA) is the key hearing test used to identify hearing threshold levels of an individual, enabling determination of the degree, type and configuration of a hearing loss. PTA is a subjective, behavioral measurement. Calibration of the test environment, the equipment and the stimuli to ISO standards is needed before testing proceeds. Conventional audiometry tests frequencies lie in between 250 hertz (Hz) and 8 kHz. Whereas, high frequency audiometry tests in the region of 8 kHz - 20 kHz. The core method of pure audiometry is to present a series of tones in one ear, close to threshold (the loudness that the person can just barely detect), and keep dropping the intensity in 10 dB steps until person stops responding – pushing a button. Then the person testing the hearing goes back up in 5 dB steps until the person starts responding again. Pure tome audiometry uses both air and bone conduction. Bone conduction testing is performed in a similar way as is air conduction, but the sound is transmitted to the ear through a "bone oscillator" rather than through an earphone. For the most part, there is rarely a reason to do bone conduction if the air conduction audiogram is normal. Bone conduction testing in persons with hearing loss should be done with masking. Masking means that one puts in some "noise" in the opposite ear while testing an ear. The reason to do this is to prevent sound from the side being tested from going over to the good side.

Speech Audiometry finds out the person's ability to hear and understand speech. One of the most basic measurements is the speech reception threshold. This test determines the lowest intensity level (in dB HL) at which the patient can correctly identify 50% of common two-syllable words. The Speech Detection Threshold is the weakest intensity at which the patient demonstrates awareness that a sound is present, when that sound is speech. The Speech Discrimination Score or Word Recognition Score is the percentage of one-syllable words the patient can identify (without visual cues), when the words are heard at a loudness level that is comfortable for the patient.

Electric response audiometry (ERA) is an objective hearing testing. ERA is actually an umbrella term for a collection of techniques in which electrical potentials are recorded, usually from the scalp of the subject, evoked by a sound stimulus. The presence of the response or the response characteristics allows us to infer conclusions about the subject's hearing ability or the performance of their auditory pathways. [2], [3]

1.2 Audiometer

An audiometer is a device used for evaluating hearing loss. They usually consist of an embedded hardware unit connected to a pair of headphones and a test subject feedback button, sometimes controlled by a standard PC. Such systems can be also used with bone vibrators, to test conductive hearing mechanisms. Audiometer requirements and the test procedure are specified in IEC, ISO and AINSI standards. An alternative to hardware audiometers are software audiometers, which are available in many different configurations. Screening PC-based audiometers use a standard computer and can be run by anybody in their home to test their hearing, although their accuracy is not as high due to lack of a standard for calibration. Some of these audiometers are even available on a handheld Windows driven device. Clinical

PC-based audiometers are generally more expensive than software audiometers, but are much more accurate and efficient. They are most commonly used in hospitals, audiology centers and research communities. [4]

1.3 Audiometry Ear Phones

Earphones are one of the most significant parts of audiometry testing. Generally there are two standards of ear phones that can be used for audiometry ("ER-3" and ordinary headphones "TDK"). Different types have to be used after calibrating and frequency response identification. The ER-3 have the facility that they eliminate out surrounding noise, and they also have less drift to selective hearing by one ear. Their main disadvantage is that they has to be inserted them properly. They give wrong readings in persons who have perforations of their ear drums. [2]

1.4 Audiometry Calibration

Any audiometer should be calibrated regularly to ensure that the level evaluated by the audiometer is equal to the actual stimulus the subject is exposed to. Accurate and reliable measurements are the first stage in characterizing and quantifying hearing loss; furthermore, proper calibration ensures that the measurements are consistent, regardless of the clinic or indeed the district or nation where the measurements are carried out. Audiometers are calibrated using an Ear Simulator System (or 'Audiometric Calibration System'). For testing air-conduction hearing mechanisms, the systems include an ear simulator or acoustic coupler, conforming to either IEC 60318-1 (a.k.a. artificial ear) or IEC 60318-3 (reference coupler) respectively. These devices essentially consist of a calibrated microphone with an associated coupling volume, which is open on one side to allow application of headphones when testing. Every single part of the system that you use should be calibrated - the electrical device that produces the sound, and the headphones or speakers that deliver the sound. Practically, electrical devices (such as digital audiometers or CD players) will never drift in frequency or volume. Once their intensity is checked, formal electrical calibrations are more likely to cause trouble (i.e. noise in the calibration process) than be helpful. On the other hand, mechanical devices (such as headphones, and insert headphones in particular) nearly always break down over time. They need to be checked every day with a "sound check", and formally every 3 months. [4]

1.5 Audiometry Results

An audiogram is a standard way of representing a person's hearing loss. Most audiograms cover the limited range 100 Hz to 8000 Hz (8 kHz). Audiograms are set out with frequency in hertz (Hz) on the horizontal axis, most commonly on a logarithmic scale, and a linear dBHL scale on the vertical axis.. The hearing level (HL) is quantified relative to "normal" hearing in decibels (dB), with higher numbers of dB indicating worse hearing. 100 dB hearing loss is nearly equivalent to complete deafness for that particular frequency.

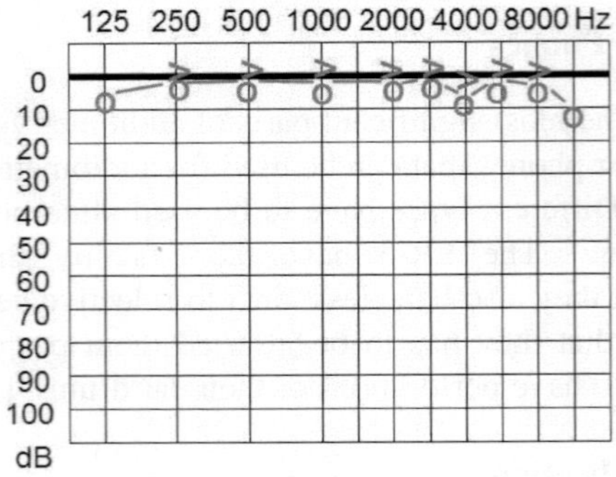

Fig. 1. An example of an audiogram in a person with normal hearing.

A score of 0 is normal. It is possible to have scores less than 0, which indicate better than average hearing. Hearing loss is often described in words as follows: normal hearing - less than 25 dB HL, mild hearing lost - between 25 and 40 dB HL, moderate hearing loss - between 41 and 65 dB HL, severe hearing lost - between 66 and 90 dB HL, profound hearing lost - 90 and more dB HL. [2]

2 Materials and Methods

The equipment, which is referred in this paper, is designed for the data acquisition system PowerLab 15T. Furthermore it can be used with any PowerLab series with a control output signal and two or three necessary analog inputs. The experiment is created for the teaching laboratory software LabTutor. Editing and writing a new LabTutor experiments is provided by the LabAuthor software.

In the fig. 6 is shown a principal scheme of PowerLab and audiometer interconnection. The audiometer is controlled from the PowerLab through an output. An audiometer output level audio signal is measured using the PowerLab input. Additional PowerLab input is available for a feedback button. [5]

For controlling the audiometer, we have to configure a control signal. PowerLab 15T has only one output, which can be configured as the control signal: the stimulator voltage output. The LabTutor software enables a stimulator pulse generation mode with an output ranges from ±200 mV to ±10 V, A user can set various stimulation parameters as pulse interval, pulse duration, pulse amplitude and baseline. An output voltage resolution is 16 bit. In fig. 2, there is a stimulator panel form in LabTutor experiment application. Possibilities for the user are limited to the amplitude and interval changing using of two slides bars.

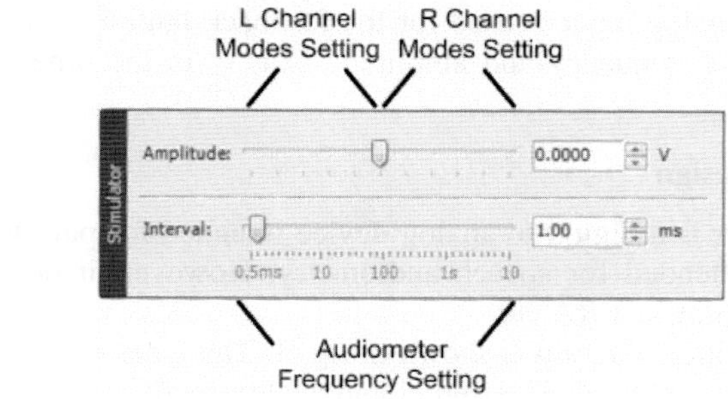

Fig. 2. Stimulator panel in LabTutor with interval and amplitude represented for audiometer control.

Controlling the audiometer is done by a control signal shape shown in the fig. 3. Ant control information consists of series of 2 pulses with defined bipolar amplitude and period. The stimulator period setting is used for audiometer frequency setting, whilst the amplitude setting is used for the audiometer mode of operation setting.

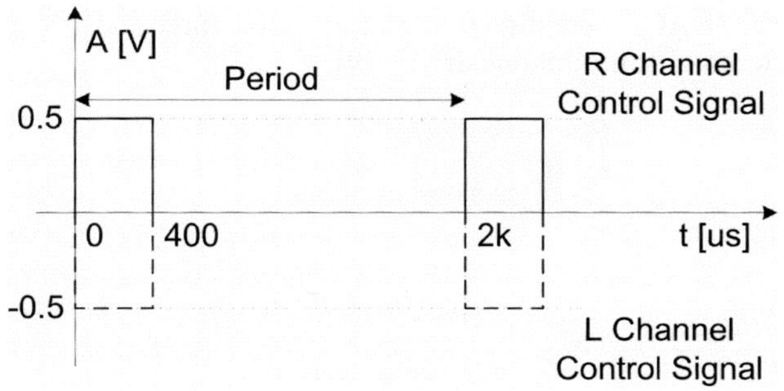

Fig. 3. Control signal shape.

Audio output signal measurement is provided by a PowerLab analog input. The PowerLab analog inputs have a selectable voltage ranges from ±20 mV to ±10 V, 16-bit resolution, 1 MΩ input impedance, 25 kHz maximum bandwidth and > 100 dB signal to noise ratio. It also allows a selectable low pass filter. [6], [7]

The analog input is used for measuring the frequency and amplitude of the audiometer output signal. If the PowerLab manages three inputs, it is possible to measure each audio channel separately, else it must be measured only one channel in one time. This can be done by switching circuit on the audiometer output.

The most important parameter for the actual stimulus determination is the amplitude. Frequency and amplitude of the input signal are evaluated by the LabTutor software. These values are continuously displayed in the graphs. The amplitude is calibrated by a frequency response for the used headphones. This frequency response must be measured according to the calibration regulations mentioned in the chapter 1.4.

Next PowerLab analog input is used for the feedback button input. An operator is able to determine the frequency and amplitude exactly in the time of the feedback button press.

2.1 Audiometer Design

Designed audiometer is a digital to analog device intended for pure tone audiometry (PTA), it can be extended for speech audiometry. However, it doesn't include an objective hearing testing as ERA etc.

The audiometer block diagram is shown in fig. 4. The inner circuit is composed of an analog and a digital section. The digital and analog sections are supplied by some separate dual power supplies. The sections are isolated from one to another using a split in the circuit board. This prevents the switching noise present on the digital supply from contaminating the analog power supply and degrading the dynamic performance of the D/A converter. [8]

The main parts of the digital section are a microcontroller, a programmable flash memory, a signal processing operational amplifier based circuit and a left / right channel switch. In the flash memory are stored the pulse code modulation (PCM) audio data of sine waves and other graphic or audio data.

The main parts of the analog section are a stereo audio sigma-delta digital to analog converter (DAC), operational amplifier based filter and I/V gain stage, mute control and a stereo hi-fi headphone driver. [9]

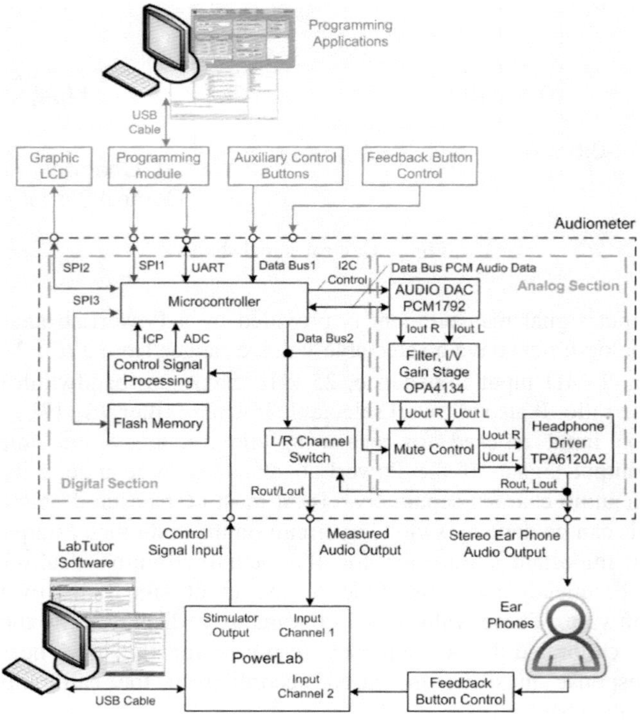

Fig. 4. Audiometry block diagram.

The microcontroller evaluates the control signal by an external interrupt and input signal measurement by an ADC. The communication between the microcontroller and DAC proceeds throw the PCM data and I2C control interfaces. The PCM data corresponding to a sine wave are transmitted in some modes of operation. For s short data packet, the data packet is loaded from the flash memory to the SRAM buffer and then cyclic transmitted to the DAC. For some longer data packets the data are continuously red from the flash memory and transmitted to the DAC.

The microcontroller serves also as a high frequency clock generator for the DAC, which includes an over sampling circuit. The DAC has a current output, which is converted to the voltage differential signal connected to the precise headphone driver. The mute control performs an output signal free of pops and clicks, which appear on power on/off DAC, PCM data flow interruption, or changing the PCM data sampling frequency. [10]

2.2 Auxiliary Functions

The blue colored blocks in the figure 5 are some extension of the designed device. For increasing the flexibility of the device, there is a possibility to use the audiometer as an independent device (without usage of PowerLab and PC) by using of some auxiliary control buttons instead of controlling the device by the PowerLab.

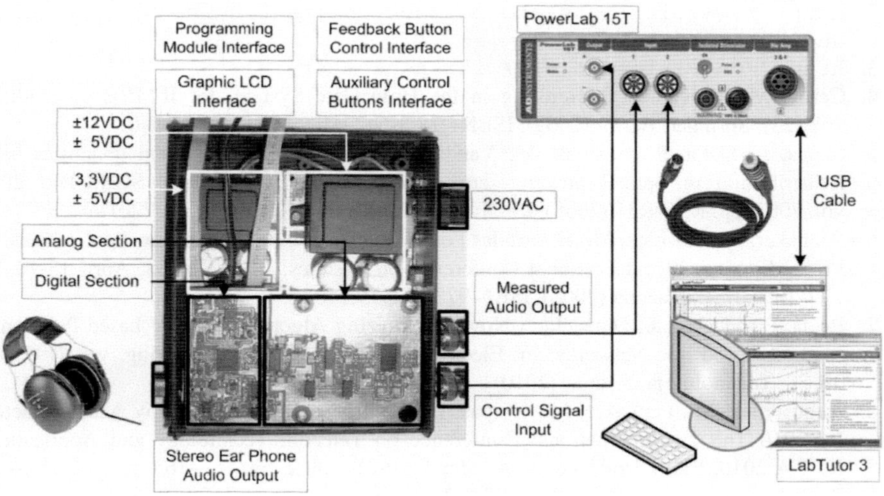

Fig. 5. Audiometry equipment setup.

However, in this case the audio output isn't measured and the audiometer must be properly calibrated for the used ear phones. For data visualization on a color LCD screen, there is an SPI data interface. There is also an input for feedback control button.

3 Conclusion

This described electro technical audiometry device for hearing performance were realized as stand alone instrument with advantage to use with PowerLab ADInstruments. There is possibility to test pure tone audiometry with excellent findings in generating tone. The device is realized 15 % better than 120 dB of dynamic range of human ear eligibility. There is also additional possibility to modify the firmware in presented hardware for better or wide capabilities which make it representative for commercial applications.

Acknowledgments. The work and the contribution were supported by the project: Ministry of Education of the Czech Republic under Project 1M0567 "Centre of Applied Cybernetics", student grant agency SV 4501141 "Biomedical engineering systems VII" and TACR TA01010632 "SCADA system for control and measurement of process in real time". Also supported by project MSM6198910027 Consuming Computer Simulation and Optimization.

References

1. http://www.adinstruments.com/products/hardware/research/DataAcquisitionSystems/
2. http://www.dizziness-and-balance.com/testing/hearing/audiogram.html
3. http://en.wikipedia.org/wiki/Audiometry
4. Cerny, M.: Movement Monitoring in the HomeCare System. In: IFMBE proceddings, vol. (25). Springer, Berlin (2009), ISBN 978-3-642-03897-6; ISSN 1680-07
5. Skapa, J., Siska, P., Vasinek, V., Vanda, J.: Identification of external quantities using redistribution of optical power - art. no. 70031R. In: OPTICAL SENSORS 2008, vol. 7003, pp. R31–R31 (2008), ISSN: 0277-786X , ISBN: 978-0-8194-7201-4
6. Penhaker, M., Rosulek, M.: Electrodes For Biotelemetry And Home Care Applications. In: Proceedings of 3th International Conference on Systems, ICONS 2008, April 13-18, pp. 179–183. IEEE, Cancun (2008), ISBN 978-0-7695-3105-2
7. Brida, P., Machaj, J., Duha, J.: A Novel Optimizing Algorithm for DV based Positioning Methods in ad hoc Networks. In: Electronics and Electrical Engineering, vol. 1(97), pp. 33–38. Technologija, Kaunas (2010)
8. Augustynek, M., Penhaker, M., Korpas, D.: Controlling Peacemakers by Accelerometers. In: 2010 The 2nd International Conference on Telecom Technology and Applications, ICTTA 2010, Bali Island, Indonesia, March 19-21, vol. 2, pp. 161–163. IEEE Conference Publishing Services (2010),ISBN 978-0-7695-3982-9, DOI: 10.1109/ICCEA.2010.288
9. Penhaker, M., Rosulek, M., Cerny, M., Martinák, L.: Design and Implementation of Textile Sensors for Biotelemetry Applications. In: IFBME Proceedings of the 14th Nordic – Baltic Conference on Biomedical Engineering and Medical Physics, June 16–20, vol. 20, pp. 54, 405–408. Springer et IFMBE, Heilderberg (2008), ISBN 978-3-540-69366-6, ISSN 1680-0737.
10. Pindor, J., Penhaker, M., Augustynek, M., Korpas, D.: Detection of ECG Significant Waves for Biventricular Pacing Treatment. In: 2010 The 2nd International Conference on Telecom Technology and Applications, ICTTA 2010, Bali Island, Indonesia, March 19-21, vol. 2, pp. 164–167. IEEE Conference Publishing Services (2010), ISBN 978-0-7695-3982-9, doi: 10.1109/ICCEA.2010.186

Phase Selection Component for Power Line Protection in Smart Grid

Wang Su-hua[1], Zhou Xin[1], Liu Zhong-jing[2], and Liu Jing[1]

[1] Shang-qiu Electric Power company of HeNan, China
shqwsh@yahoo.com.cn, zxljzxlj@sina.com,
zxjzxj@sina.com
[2] HuaZhong University of Science and Technology, China
ZEPCLZJ@126.com

Abstract. Current distribution line protection in smart grid can't be solved well in low sensitivity and unsatisfied rapidity by protection principle nowadays. The wavelet analysis has a strong feature-extracting function, which can extract the fault signatures of the fault phase. In this paper, the EMTP is used to make simulations and wavelet transformations of power line-end short-circuit failure, fault-phase supply voltage crossing zero, single-phase grounding fault of transmission line through high impedance, transitional fault, etc., showing the parameter selection of wavelet transformation used in optimization design of phase selection component for line protection.

Keywords: Line Protection, Phase selection component, Wavelet, EMTP, high-voltage line protection, HV transmission.

1 Introduction

The phase selection component is needed in HV transmission line when single-phase auto reclosing devices are widely used. The phase selection component should select the fault phase when single-phase fault occurs in order to ensure reliable removal of the fault phase, and it should determine the fault and make three-phase trip immediately when multi-phase fault occurs.

The phase selection component is also required in the distance protection; in the microcomputer distance protection, the method of "conducting measurement after fault phase selection" is generally used in order to reduce the amount of computation. To meet the requirements for correct measurement of the distance, the fault phase must be selected correctly under single-phase fault, multi-phase fault and transitional fault, and maintain correct phase selection when tripping in the opposite side first occurs. In addition, accelerating the action speed of phase selection component is conducive to improving the action speed of the whole set of protective device.

X. Wan (Ed.): Electrical Power Systems and Computers, LNEE 99, pp. 839–844.
springerlink.com

The phase selection components currently used in high-voltage line protection mainly include impedance phase selector, sequence component phase selector, fault component phase selector and so on, but each has its limitations. Greatly influenced by system operation mode and transition resistance of fault points, the impedance phase selector often fails to receive satisfactory results. For the sequence component phase selector, the extraction of positive-sequence fault current needs to be free from the impact of load current, which is not conducive to phase selection of follow-up faults; fault phase regions of negative-sequence and zero-sequence current are not partitioned reasonably, and the fault phase cannot be selected correctly under high-impedance single-phase ground fault. The fault component phase selector also has two problems: first, sensitivity on the weak source side is not enough, and the situation is most severe especially at current-carrying side of single source; second, to ensure correct phase selection when transitional fault and protective longitudinal action (first tripping on the opposite side) occurs, a number of measures should be taken to complicate the procedures of phase selection component. Comparatively speaking, the wavelet analysis method has strong feature-extracting function and can extract fault signatures of the fault phase; forming the phase selection component for microprocessor line protection with this principle can be simplify phase-selecting method and further improve the action speed and accuracy of the phase selection component.

2 Forming Phase Selection Component with the Principle of Wavelet Analysis

Collect various current and voltage signals in the appropriate sampling frequency, and make wavelet transformation of all the sampling values; if the phase where modulus maxima of wavelet transformation on several adjacent scales all exceed the setting value, the phase can be determined as fault phase. The modulus maxima of wavelet transformation only appear in the fault moment, so the phase selection component should start at the same time as the start component, and fix the phase selection results. This easy method has fast action speed and is free from the impact of load current and transition resistance of short-circuit point.

Fig1 shows A-phase current when A-phase supply voltage crosses zero, AB two-phase short circuit occurring at the end of 110k line, sampling frequency being 5k and its wavelet transformation waveform when the scale j=12,3,4,5. From the figure, it can be seen that even under the situation that the power voltage crosses zero and fault phase current does not increase in the moment of short circuit, the wavelet transformation under the scale of j=1 can still produce modulus maxima in the moment of fault.

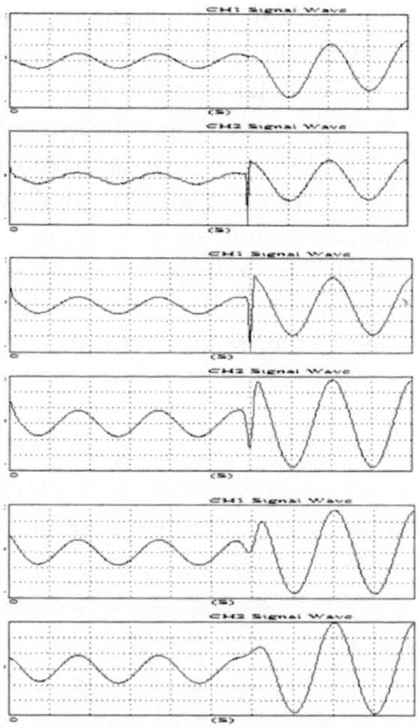

Fig. 1. i_a and its wavelet, AB two-phase short circuit u_a=0

Fig2 (a), (b), (c) respectively show A, B, C-phase current when AB two-phase short circuit occurs at the end of 110kV line and their wavelet transformation when the scale j=2.

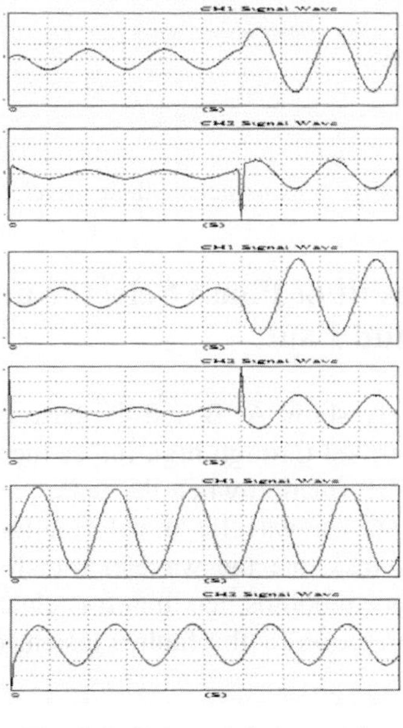

Fig. 2. ia,ib,ic and their wavelet

Fig3 (a), (b), (c) respectively show A, B, C-phase voltage under the above situation and their wavelet transformation when the scale j=2. As can be seen from the figure, in the fault moment, the wavelet transformations of fault phase current ia,ib and fault phase voltage ua,ub both produce modulus maxima, which shows obvious fault signatures; the wavelet transformations of non-fault phase current ic and non-fault phase voltage uc both fail to produce abrupt modulus maxima. Therefore, the use of wavelet transformation of current or voltage can form the phase selection component reflecting fault transient component. The use of current components together with voltage components to make phase selection can also ensure the accuracy of phase selection, and further reduce the impact of interfering signals.

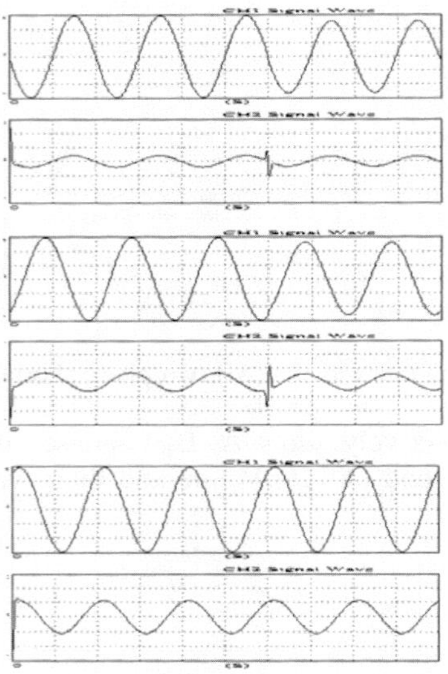

Fig. 3. ua,ub,uc and their wavelet

3 Phase Selection Result of Phase Selector under High-Impedance Single-Phase Ground Fault

Fig4 (a), (b), (c) respectively show A, B and C-phase current when there is single source, line-end A phase earthed through high impedance, sampling frequency being 5kHz and their wavelet transformation when the scale j = 2. As can be seen from the figure, in the fault moment, the wavelet transformation of fault phase current ia produces obvious modulus maxima; the wavelet transformation of non-fault phase current ib、 ic fails to produce modulus maxima. It is clear that the use of the modulus maxima of wavelet transformation of the fault transient component can make phase selection correctly for high-impedance single-phase ground fault, and the phase selection component has sufficient sensitivity.

4 Phase Selection Result of Phase Selector under Transitional Fault

Fig5 (a), (b), (c) respectively show A, B and C-phase current when there is single source, A-phase ground changed to AB two-phase ground fault at the line end, sampling frequency being 5kHz and their wavelet transformation when the scale j = 2. As can be seen from the figure, when t=0.04s, the single-phase ground fault occurs in A phase and the wavelet transformation of A-phase current ia produces modulus maxima at the same time; after one second, namely when t=0.05s, the ground fault occurs in B phase and the wavelet transformation of B-phase current ib produces modulus maxima at the same time. However, the wavelet transformation of non-fault phase current icc fails to produce abrupt modulus maxima. It is clear that the use of the modulus maxima of current wavelet transformation can make correct phase selection for transitional fault.

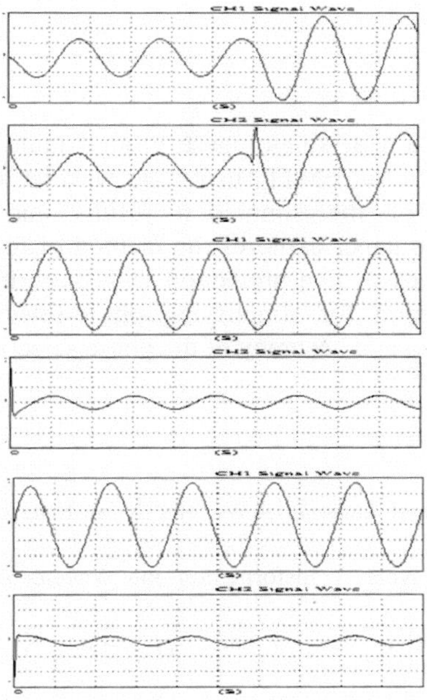

Fig. 4. ia,ib,ic and their wavelet when line-end A phase earthed through high impedance

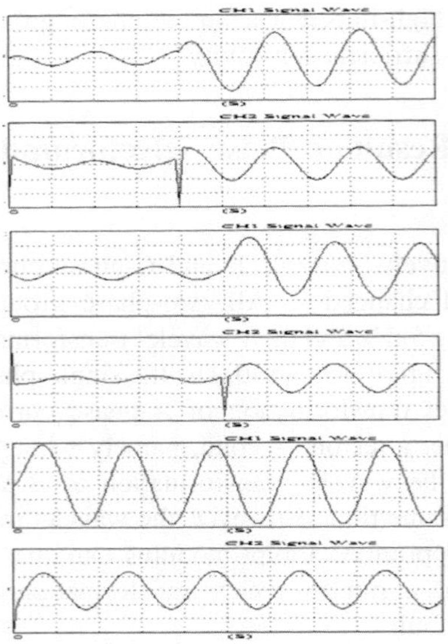

Fig. 5. ia,ib,ic and their wavelet when AB two-phase ground fault

5 Conclusion

Through focused simulation of single-phase ground fault through high impedance, transitional fault, etc. with the EMTP, using the wavelet analysis program to make wavelet transformation of the simulation results and analyzing the wavelet transformation results, the paper explores the method of forming the phase selection component for microprocessor line protection with the principle of wavelet analysis and lays certain foundation for simplifying phase-selecting method and further improving the action speed and accuracy of the phase selection component.

References

1. Qin, Q., Yang, Z.: Practical Analysis of Wavelet. Xi'an Electronic Science and Technology University Press (1994)
2. Zhao, S., Xiong, X.: Wavelet Transformation and Wavelet Analysis. Electronic Industry Press (1996)
3. Liu, G., Di, S.: Wavelet Analysis and Application. Xi'an Electronic Science and Technology University Press (1992)
4. Zhu, S.: HV Power Network Relay Protection Principles and Technology. China Electric Power Publishing House (2005)

Model-Based Vehicle Trajectory and Its Properties for Automatic Steering Systems

Xijun Zhao*, Tao Hong, and Huiyan Chen

School of Mechanical Engineering, Beijing Institute of Technology,
Beijing 100081, China
zhaoxijun318@gmail.com

Abstract. The nonlinear vehicle dynamic model which considers three degrees of freedom of vehicle motion, longitudinal, lateral and yaw was established. For the automatic steering of intelligent vehicles, however, longitudinal dynamic is not taken into consideration, thus, a reduced two DOF model is obtained which regards longitudinal velocity as time variant state of vehicle. According to such a two degrees time variant nonlinear model, a novel model-based vehicle trajectory is proposed. This model considers longitudinal velocity and front wheel steering angle as inputs and vehicle lateral dynamic states (lateral velocity and yaw rate) as outputs. This model is validated using real vehicle testing data. Based on the equations of vehicle model and trajectory, the properties of the vehicle trajectory are analyzed in detail. Finally, the proposed methods and properties are validated through comparison of experiment and simulation studies.

Keywords: vehicle trajectory, G^2-path, automatic steering.

1 Introduction

This contribution proposes a novel model-based vehicle trajectory and its properties based on the nonlinear vehicle dynamic model. Such a contribution has several benefits. It provides appropriate control algorithms for the automatic steering system of intelligent vehicles and gives a simple model for the vehicle trajectory generation or vehicle trajectory prediction. A number of research results have been reported on the vehicle trajectory properties for automatic steering systems. As pointed out in [1], the vehicle paths via continuous steering angle from time t to t+1 are the clothoid. This implies the classical pure pursuit automatic steering algorithm [2] is inapplicable in some cases. The generated desired vehicle states should satisfy vehicle dynamic constraint. Thus analysis of the vehicle trajectory properties is essential to developing an automatic steering system for intelligent vehicles.

[3] proposed a kinematical vehicle model-based vehicle trajectory and analyzed its properties. It was proved the vehicle trajectory generated by the kinematical nonholonomic model via continuous front wheel steering angle input is a G^2-path. In

* This work was supported in part by the National Natural Science Foundation of China. (Grant No. 90920304).

X. Wan (Ed.): Electrical Power Systems and Computers, LNEE 99, pp. 845–851.
springerlink.com

[4], a flatness property of vehicle dynamic model was studied and a flatness-based vehicle dynamic control law was proposed. [5] described a new path generation method based on clothidal curve for an autonomous mobile robot. However, few studies have been reported on vehicle trajectory properties for automatic steering system of intelligent vehicles. In this paper we will discuss dynamic vehicle model-based trajectory and its properties. Such a trajectory is useful for both automatic steering control system design and intelligent vehicle path generation.

2 Vehicle Modeling and Model Validation

In general, vehicle dynamic equations are expressed in planar dynamic equations, which consider three DOF vehicle motion, longitudinal, lateral and yaw respectively. This simplifies double track vehicle model using a single track one as shown in Fig. 1.

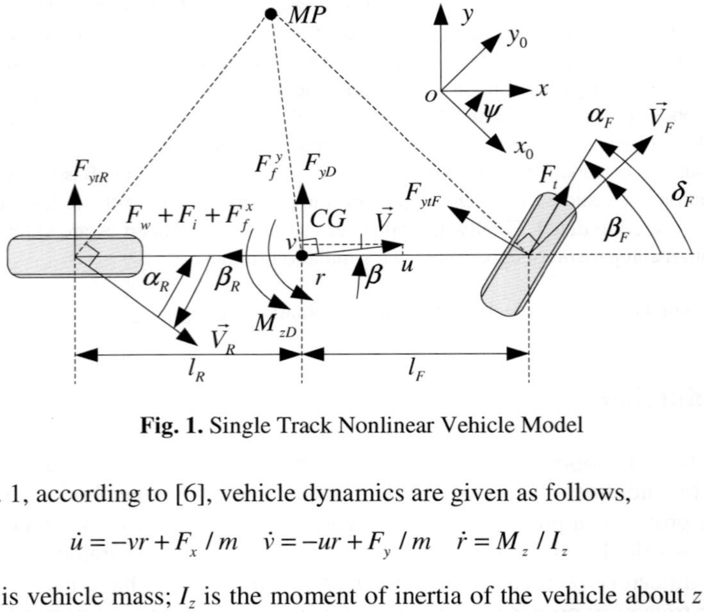

Fig. 1. Single Track Nonlinear Vehicle Model

In Fig. 1, according to [6], vehicle dynamics are given as follows,

$$\dot{u} = -vr + F_x / m \quad \dot{v} = -ur + F_y / m \quad \dot{r} = M_z / I_z \tag{1}$$

where m is vehicle mass; I_z is the moment of inertia of the vehicle about z axis; ; u is longitudinal velocity; v is lateral velocity; r is yaw rate with respect to x_0oy_0; F_x F_y and M_z are the external total force acting on x, y, and z, which is given as

$$\begin{cases} F_x = 1/\delta\left(F_t \cos\delta_F - F_f^x - F_w - F_i - F_{ytF}\sin\delta_F\right) \\ F_y = F_{ytR} + F_{ytF}\cos\delta_F + F_t \sin\delta - F_f^y + F_{yD} \\ M_z = -F_{ytR} l_R + F_{ytF}\cos\delta_F l_F + F_t \sin\delta_F l_F - F_f^y l_F + M_{zD} \end{cases} \tag{2}$$

with F_t is the longitudinal traction force from engine to wheel; F_w, F_i, F_f^x are the longitudinal wind drag, gradient resistance and rolling resistance, respectively; F_{ytF}

and F_{ytR} are lateral tire force; F_{yD} and M_{zD} are the disturbance forces; δ is rotational mass coefficient; δ_F is front wheel steering angle.

The lateral tire forces are modeled using Pacejka Magic Formula[7],

$$F_{ytk}(\alpha_k) = D\sin\left(C\arctan\left(B\alpha_k + E\left(B\alpha_k - \arctan\left(B\alpha_k\right)\right)\right)\right) \tag{3}$$

$k \in \{F, R\}$ refers to front and real axles. Thus the lateral forces are given as functions of tire side slip angles which are represented as,

$$\alpha_F = \delta_F - \arctan\left(\tan\beta + l_F r / u\right) \quad \alpha_R = -\arctan\left(\tan\beta - l_R r / u\right) \tag{4}$$

with β is the vehicle slip angle of the center of gravity CG, $\tan\beta = v/u$.

Equation (1)-(4) represent three degrees nonlinear vehicle dynamic model for planar motion. Three degrees nonlinear vehicle dynamic model regards F_t and δ_F as the inputs, and u, v and r as the outputs.

Actually, both u and δ_F could be collected from the vehicle CAN bus at a rate of 100Hz which implies that as to the vehicle trajectory analysis for automatic steering systems the degree of system may be reduced with regardless of longitudinal motion dynamics. Only lateral and yaw motion of vehicle is remained to characterize the vehicle motion while the longitudinal velocity is regarded as a system parameter changing with time. Hence, (1) and (2) become,

$$\begin{bmatrix} \dot{v} \\ \dot{r} \end{bmatrix} = \begin{bmatrix} -ur + F_y / m \\ M_z / I_z \end{bmatrix} \text{ with } \begin{cases} F_y = F_{ytR} + F_{ytF}\cos\delta_F + F_{yD} \\ M_z = -F_{ytR} l_R + F_{ytF}\cos\delta_F l_F + M_{zD} \end{cases} \tag{5}$$

Equation (5) indicates the system is reduced to be a two DOF nonlinear time variant vehicle model with the input u and δ_F and the output v and r.

In this paper we utilize model (5) to study vehicle trajectory and its property. In the following we validate the model.

Validation of the vehicle model is accomplished by comparing the test data with simulation results with same input. Experiment data including longitudinal velocity u, front wheel steering angle δ_F was collected during vehicle running from vehicle CAN bus, while lateral acceleration at CG a_y and yaw rate r measured using Inertial Measuring Unit (IMU) mounted on the vehicle CG. Mathematic model is simulated under MATLAB/Simulink. Vehicle parameters are shown in Table 1.

Table 1. Vehicle Parameters

Vehicle mass m	1385kg	Distance of CG from front axle l_F	1.02m
Moment of inertia I_z	2162kgm^2	Distance of CG from front axle l_R	1.53m

Simulation results and experiment data are shown in Fig. 2. Above figures show the input variables of model, longitudinal velocity u and front wheel steering angle δ_F and below ones depict yaw rate r and lateral acceleration a_y varying with time.

The comparison of results shows that two degree nonlinear vehicle model could characterize vehicle dynamic motion accurately, suggesting that the model is feasible for studying vehicle trajectory.

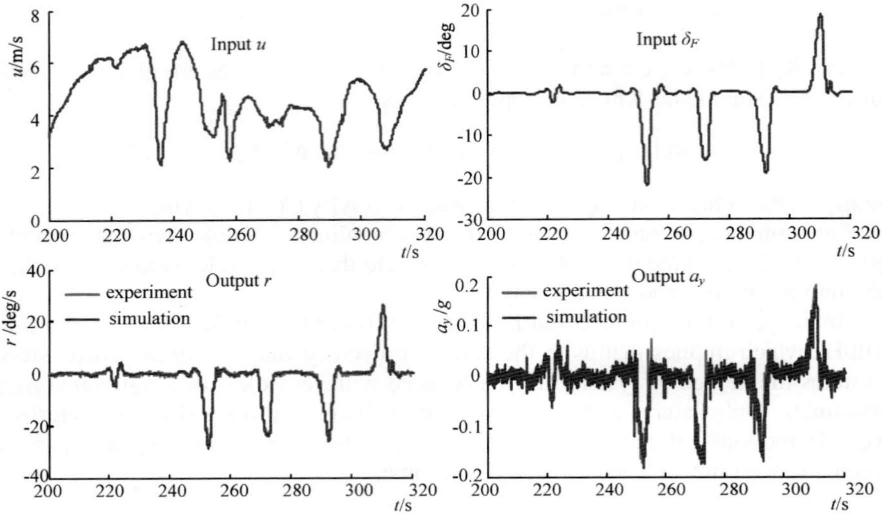

Fig. 2. Input u, δ_F and Output r, a_y

3 Vehicle Trajectory and Its Properties

Based on the measured and estimated vehicle state, the vehicle trajectory is represented with global position (x, y) of CG and orientation ψ of the z axis.

$$\dot{x} = \sqrt{u^2 + v^2}\cos(\beta + \psi) \quad \dot{y} = \sqrt{u^2 + v^2}\sin(\beta + \psi) \quad \dot{\psi} = r \tag{6}$$

Equation (6) shows that vehicle global position(x, y) and orientation ψ are varying with vehicle state u, v and r. We would use (6) to study the properties of the vehicle trajectory combining with the time variant 2 DOF nonlinear vehicle model (5).

As mentioned above, the input of the vehicle model is u and δ_F. To simplify the analysis, some assumptions are made before investigating vehicle trajectory properties. Assumptions: 1) Both u and δ_F are arbitrary smooth functions, this is to say they have the n-th derivatives. 2) Lateral tire forces are modeled as (3).

In [3], it's proved to show that a path generated by vehicle kinematical model via a continuous control input δ_F is G^2-path. Here, we generate the path by the proposed nonlinear vehicle dynamic model. Similar to the [3], let the $p(t)=[x(t), y(t)]^T$ be a point on the vehicle trajectory. Thus, the unit tangent vector is

$$\frac{\dot{p}}{\|\dot{p}\|} = \frac{\left[\dot{x}(t) \quad \dot{y}(t)\right]^T}{\sqrt{\dot{x}(t)^2 + \dot{y}(t)^2}} = \begin{bmatrix}\cos(\psi + \beta) \\ \sin(\psi + \beta)\end{bmatrix} = \left[f_x^1(u,v,r) \quad f_y^1(u,v,r)\right]^T \tag{7}$$

Because ψ and β are obtained from integral calculation, both of them are continuous. So by (7), $\dot{p}/\|\dot{p}\|$ is continuous too.

Now we investigate curvature of the vehicle trajectory, which can be expressed as

$$\kappa=\left(\dot{x}\ddot{y}-\ddot{x}\dot{y}\right)\frac{1}{\left(\dot{x}^2+\dot{y}^2\right)^{3/2}}=\left(\dot{\beta}+r\right)\frac{1}{\sqrt{u^2+v^2}}=\frac{1}{\sqrt{u^2+v^2}}\left(\frac{u\dot{v}-\dot{u}v}{u^2+v^2}+r\right) \tag{8}$$

Equation (8) implies the curvature is a function of vehicle states and their derivatives. In other words, the curvature of the vehicle trajectory is continuous if and only if $\dot{u},\dot{v}$ is continuous. According to the assumptions and (5), u is arbitrary smooth while continuity of $\dot{v}$ is equal to continuity of F_y. This is to say, if F_y is a continuous function, the vehicle trajectory generated by model (5) via assumed input is G^2-path. Furthermore, (8) means,

$$\ddot{p}/\|\ddot{p}\|=\left[f_x^2\left(u,\dot{u},v,\dot{v},r,\delta_F\right)\quad f_y^2\left(u,\dot{u},v,\dot{v},r,\delta_F\right)\right]^T \tag{9}$$

Moreover, we investigate high order property of p. the following equation could be derived if there exist high order derivatives of vehicle state u, v, r and δ_F.

$$\frac{p^{(n)}}{\left\|p^{(n)}\right\|}=\begin{bmatrix} f_x^n\left(u\cdots,u^{n-1},v\cdots,v^{n-1},r\cdots,r^{n-2},\delta_F\cdots,\delta_F^{\,n-2}\right) \\ f_y^n\left(u\cdots,u^{n-1},v\cdots,v^{n-1},r\cdots,r^{n-2},\delta_F\cdots,\delta_F^{\,n-2}\right)\end{bmatrix} \tag{10}$$

From (5), it's obvious that the existence condition for (10) is F_y has the continuity of (n-2)-order, while M_z has the continuity of (n-3)-order.

To sum up, the key properties associated with the vehicle trajectory include the following:

Remark 1. The continuity property of the vehicle trajectory mainly depends on external forces F_y, M_z, etc. If F_y and M_z are not a continuous functions, the vehicle trajectory is a G^1-path. If F_y is a continuous function, the vehicle trajectory is a G^2-path. Moreover, If F_y has (n-2)-th derivative, while M_z has (n-3)-th derivative, the vehicle trajectory is a G^n-path.

Remark 2. Under normal conditions, all the forces of (5) are varying continuously with time. Thus F_y is a continuous function of time. While under abnormal conditions (e.g., a sudden step-like external disturbance forces), F_{yD} is not varying continuously. Hence F_y is not a continuous function of time. However, with the discontinuity of F_y, we could factitious F_y to the model in order to make the model output is a G^2-path(e.g., fake step-like signal to slop-like continuous one). So we propose that under any conditions, the vehicle trajectory is always could be a G^2-path. And we design the automatic steering system and path planning generator based on the comment above.

Remark 3. Equation (10) indicates (when n=3) the change of curvature $\dot{\kappa}$ depends on front wheel steering angle rate $\dot{\delta}_F$ which is commonly a maximum constraint value in steering systems. Therefore, feed forward control law could be designed according to $\dot{\kappa}$. Meanwhile, the planed path should satisfy the restriction of $\dot{\delta}_F$. This would help us to generate a smooth desired path for intelligent vehicles.

4 Simulation Results

The proposed vehicle trajectory and its properties are validated through simulations studies which are carried out under MATLAB/Simulink. Meanwhile, the input of the vehicle (u and δ_F) and vehicle states (r, a_y, x, y and ψ) are collected using real-time data acquisition unit that is designed based on MCU (Micro Control Unit). The vehicle platform is a normal passenger car.

Fig. 3 shows the experiment vehicle trajectory and simulated vehicle trajectory for about one minute. *A* is the start point, while *B* is the end point. It depicts that model-based trajectory is close to the real vehicle trajectory. This means that the proposed methods could characterize complicated vehicle lateral dynamics accurately. The shorter the calculation time, the higher is the accuracy of the model output. The curvature of real vehicle trajectory and calculated based on (8) is illustrated in Fig. 3, respectively. The continuity of the curvature shows the vehicle path is G^2-path (*Remark 1* and *2*). Also, Fig. 3 shows $|\dot{\kappa}|$ and $|\dot{\delta}_F|$ has the similar outline varying with time, which implies that $\dot{\kappa}$ is the function of $\dot{\delta}_F$ (*Remark 3*).

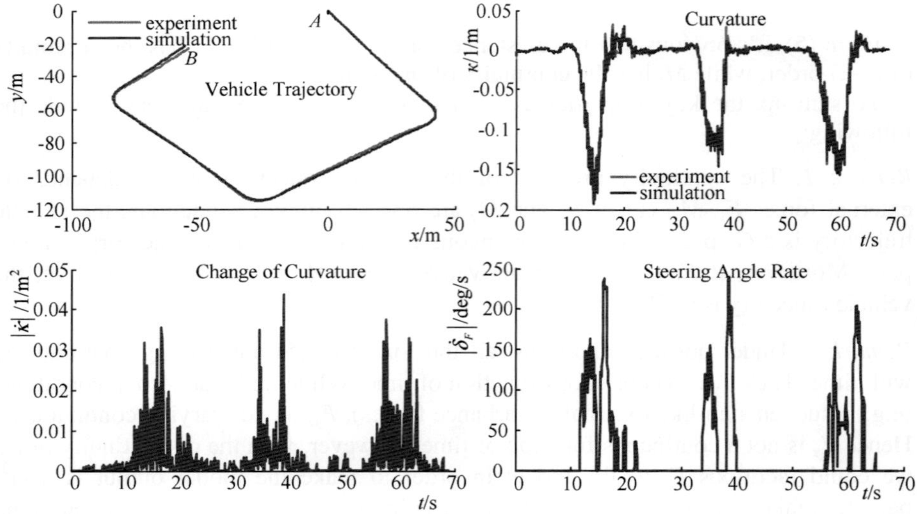

Fig. 3. Simulation Results

5 Conclusions

The paper proposed a novel model-based vehicle trajectory and its properties using the nonlinear vehicle dynamic model. As the results, three main remarks are given theoretically. Such remarks are helpful for the trajectory prediction, automatic steering control algorithm synthesis, path planning, etc. Combining the proposed method with numerical integration, the future vehicle trajectory could be predicted in the time horizon. Accordingly, new control mythologies such as model predictive control could be applicable based on model output estimation. In addition, remarks given in this paper could make the planed path more smooth, as well as satisfy vehicle dynamic constraints.

Furthermore, results of this paper should be utilized on motion planning and control subsystem of intelligent vehicle to improve trajectory tracking performances.

References

1. Kelly, A., Anthony, S.: An Approach to Rough Terrain Autonomous Mobility. In: International Conf. Mobile Planetary Robots, Santa Monica, California, pp. 1–34 (1997)
2. Coulter, R.C.: Implementation of the Pure Pursuit Path Tracking Algorithm. The Robotics Institute, Carnegie Mellon University, Pittsburgh, Pennsylvania, Rep. CMU-RI-RT-92-01 (1992)
3. Alberto, B., Massimo, B., Alessandra, F., Corrado, G., Lo, B., Aurelio, P.: The ARGO Autonomous Vehicle's Vision and Control Systems. International Journal of Intelligent Control and Systems 3, 409–441 (1999)
4. Fuchshumer, S., Schlacher, K., Rittenschober, T., Anthony, S.: Nonlinear Vehicle Dynamics Control-A Flatness based Approach. In: Decision and Control 2005 and European Control Conference 2005, Seville, Spain, pp. 6492–6497 (2005)
5. Shimizu, M., Kobayashi, K., Watanabe, K.: Clothoidal Curve based Path Generation for an Autonomous Mobile Robot. In: International Joint Conf., Busan, Korea, pp. 478–481 (2006)
6. Yu, Z.S.: Automotive Theory, pp. 103–169. China Machine Press, Beijing (2006)
7. Bakker, E., Pacejka, H.B., Lidner, L.: A New Tire Model with an Application in Vehicle Dynamics Studies. SAE Paper, 101–113 (1989)

Level Sets, the Representation Theorem and the Extension Principle for Interval Valued Fuzzy Sets

Hongmei Ju

School of Information, Beijing WUZI University, Beijing, China, 101149
jhm2000@sohu.com

Abstract. We discuss the concept of a level set of a fuzzy set and the related ideas of the representation theorem and the extension principle. We then describe the extension of these ideas to the case of interval valued fuzzy sets (IVFS). What is important to note here is that in the case of interval valued fuzzy sets, the number of distinct level sets can be greater than the number of distinct membership grades of the fuzzy set being represented. In particular, the minimum of each subset of membership grades provides a level set. Morover, the membership grades are not linearly ordered and hence taking the minimum of a subset of these can result in a value that was not one of the members of the subset.

Keywords: Fuzzy set, level set, interval-valued fuzzy set, representation theorem, extension principle.

1 Introduction

The concept of level sets associated with a fuzzy set was originally introduced by Zadeh ([1]-[3]). With the aid of level sets we are able to provide a formulation for a fuzzy set in terms of crisp subsets via the representation theorem. The importance of having such a representation is that it can allow us to extend operations defined on crisp sets to the case of fuzzy sets. This paradigm forms the basis for one version of what Zadeh calls the extension principle. Our focus here is on generalizing the idea of level sets and the associated ideas of the representation theorem and the extension principle to the case of IVFS [4].

2 Level Sets and the Representation Theorem

Assume F is a standard fuzzy subset of a space X . The level sets associated with F are defined as $F_\lambda = \{x | F(x) \geq \lambda\}$. Each F_λ is a crisp subset of X consisting of the elements of X with membership grade of at least α in F . In the case, where X is finite there exists a finite number of distinct membership grades associated with F . We denote these as k_j and index them in increasing

X. Wan (Ed.): Electrical Power Systems and Computers, LNEE 99, pp. 853–860.
springerlink.com

order $k_1 < k_2 < k_3 < \cdots < k_n$. In this case we have n distinct level sets associated with F. In particular, $F_\lambda = \{x | F(x) \geq k_i\}$, $k_{i-1} < \lambda \leq k_i$, where $i = 1$ to n and we let $k_0 = 0$ by convention. Here, we shall let $K = \{k_i \mid i = 0 \text{ to } n\}$.

The representation theorem [2], which uses the level sets, provides a method for expressing F in terms of level sets. Using this theorem we have $F = \bigcup_{\lambda \in [0,1]} \lambda F_\lambda$. Here λF_λ is a fuzzy subset such that $\lambda F_\lambda = \lambda$ for all $x \in F_\lambda$ and $\lambda F_\lambda = 0$ for $x \notin F_\lambda$. We see that $\lambda F_\lambda(x)$ is equal to $\lambda \wedge F_\lambda(x)$ as well as $\lambda * F_\lambda(x)$. We note that using the representation formulation we have for any $x \in X$ that $F(x) = \underset{\lambda \in [0,1]}{Max}\ [\lambda F_\lambda(x)]$ .In the special case, which will be of interest to us, where X is finite then the representation theorem becomes $F = \bigcup_{\lambda \in D} \lambda F_\lambda$ and $F(x) = \underset{\lambda \in D}{Max}\ [\lambda F_\lambda(x)]$.

The representation theorem plays an important role in defining the extension principle ([1]-[5]). Assume H is an operation on subsets of X resulting in values in Y, $H : 2^X \rightarrow Y$. The extension principle allows us to extend H to act on fuzzy subsets of X. In particular, if F is a fuzzy subset of X, then $H(F) = \bigcup_{\lambda \in [0,1]} \left\{ \frac{\lambda}{H(F_\lambda)} \right\}$. If F is finite then $H(F) = \bigcup_{\lambda \in D} \left\{ \frac{\lambda}{H(F_\lambda)} \right\}$ .In anticipation of presenting our approach we provide an alternative methodology for obtaining the representation theorem for standard fuzzy subsets.

Again let F be a standard fuzzy subset of the finite set X. Let 2^X be the power set of X, the set of all crisp subsets of X. For any $B \in 2^X$ we define $Val_F(B) = Min_{x \in B}[F(x)]$.

Here then $Val_F(B)$ is the minimum membership grade of the elements in B in F. We note $Val_F(B) = Min_x[F(x) \wedge B(x) \vee \overline{B}(x)]$. Let us denote $D = \bigcup_{B \in 2^X} Val_F(B)$, D is the set of all distinct values of $Val_F(B)$.

For any $\lambda \in D$ we let $P_\lambda = \{B | Val_F(B) = \lambda\}$. We see P_λ is a subset of the power set of X consisting of all the subsets of X that have $Val_F(B) = \lambda$. Using this we can obtain a form for the representation theorem. In particular, $F = \bigcup_{\lambda \in D} \lambda P_\lambda$.

Here, again λP_λ is defined as $\lambda \wedge P_\lambda(x)$. We note here that P_λ are the set of level sets of F, and $P_\lambda = F_\lambda$.

3 Level Set Representation of Interval Valued Fuzzy Subsets

Motivated by the preceding reformulation of the representation theorem in Yager [7], we extended this to the case where we have interval valued fuzzy subsets, IVFS. We now summarize the method for obtaining the level set representation of an IVFS. Let A be an IVFS on the finite set X . Let M be the set of all the distinct membership grades of elements in A . Thus $M = \{\lambda_i | i = 1 \ to \ m\}$, where $\lambda_i = [L_i, R_i]$ is the membership grade of some x in A . It is an interval value. Before proceeding we introduce some useful notation. Let G be any subset of M, it contains a collection of interval values. We define $Min[G] = Min_{\lambda \in G}[\lambda]$.

In particular, if $G = \{\lambda_1, \lambda_2, \cdots, \lambda_r\}$, then $Min[G] = [Min_{i=1 \ to \ r}[L_i]$, $Min_{i=1 \ to \ r}[R_i]]$ .Thus $Min[G]$ is also a sub-interval of the unit interval. Using this notation we define $D = \underset{G \in 2^M}{\cup} \{Min[G]\}$ .Thus D is the collection of all interval values obtained by taking the Min of the elements in some subset of M .

Note 1. $M \subseteq D$. We see that for $G = \{\lambda_i\}$ then $Min[G] = \lambda_i$.

Note 2. If all λ_i are point values, $L_i = R_i$, then $M = D$.

This follows since in this case $Min[G]$ is always some element $\lambda_k \in G$.

We shall refer to D as the set of applicable membership grades. We shall refer to an element in D as α_j and most generally each $\alpha_j \in D$ is an interval which we denote as $\alpha_j = [a_j, b_j]$. For each $\alpha_j \in D$ we define the crisp level set $A_{\alpha_j} = \{x | A(x) \geq \alpha_j\}$. We note that for any $A(x) = [L(x), R(x)]$ the condition $A(x) \geq \alpha_j$ is satisfied if $A(x) \wedge \alpha_j = \alpha_j$ that is if $[L(x) \wedge a_j, R(x) \wedge b_j] = [a_j, b_j]$. We emphasize that A_{α_j} is a crisp subset of X , $A_{\alpha_j}(x) = 1$ if $x \in A_{\alpha_j}$ and $A_{\alpha_j}(x) = 0$ if $x \notin A_{\alpha_j}$

Once having obtained A_{α_j} we can obtain the level set representation of A as $A = \underset{\alpha_j \in D}{\cup} \alpha_j A_{\alpha_j}$, where $\alpha_j A_{\alpha_j}$ is an IVFS with membership grade $\alpha_j A_{\alpha_j}(x) = [a_j * A_{\alpha_j}(x), b_j * A_{\alpha_j}(x)]$. More specifically, if $x \in A_{\alpha_j}$ then $\alpha_j A_{\alpha_j}(x) = \alpha_j$ and if $x \notin A_{\alpha_j}$ then $\alpha_j A_{\alpha_j}(x) = 0$.

We see here that for the $\alpha_j \in D$ the sets A_{α_j} can be viewed as the distinct level sets associated with A . If $\alpha_1 = [a_1, b_1]$ and $\alpha_2 = [a_2, b_2]$ are two interval membership grades then we say $\alpha_1 \leq \alpha_2$ if $Min[\alpha_1, \alpha_2] = [a_1 \wedge a_2,\ b_1 \wedge b_2]$ $= [a_1,\ b_1] = \alpha_1$. We observe here that if α_1 and $\alpha_2 \in D$ and $\alpha_1 \leq \alpha_2$ then $A_{\alpha_2} \subseteq A_{\alpha_1}$.This is a property also satisfied by the level sets of standard fuzzy subsets. One distinction between a standard fuzzy set and a IVFS is that in the case of the IVFS the elements $\alpha_j \in D$ cannot always be linearly ordered with respect to $\geq$. Thus, there can exist elements α_1 and α_2 in D such that neither $\alpha_1 \geq \alpha_2$ nor $\alpha_1 \leq \alpha_2$.

Under the above representation of the IVFS as $A = \underset{\alpha_j \in D}{\cup} \alpha_j A_{\alpha_j}$ we see that $A(x) = \underset{\alpha_j \in D}{Max}\ [\alpha_j A_{\alpha_j}(x)]$.We note that if $\alpha_j A_{\alpha_j}(x) = [u_j(x), v_j(x)]$, then $A(x) = [Max_{\alpha_{j\in D}}(u_j(x)), Max_{\alpha_{j\in D}}(v_j(x))]$.

We also observe that $A(x) = Max_{x\in A_{\alpha_j}}(\alpha_j)$ and hence $A(x) = [Max_{x\in A_{\alpha_j}}(a_j)$, $Max_{x\in A_{\alpha_j}}(b_j)]$. We note that the representation $A = \underset{\alpha_j \in D}{\cup} \alpha_j A_{\alpha_j}$ allows us to formulate an extension principle for interval valued fuzzy sets (IVFS). If $H: 2^X \rightarrow Y$ and A is an interval valued fuzzy set of X then $H(A) = \underset{\alpha_j \in D}{\cup} \{\frac{\alpha_j}{H(A_{\alpha_j})}\}$.

In the preceding we have shown that if A is an IVFS where M is the set of all distinct membership grades of A and $D = \underset{G \in 2^M}{\cup} \{Min[G]\}$ then we can represent A as $\underset{\alpha_j \in D}{\cup} \alpha_j A_{\alpha_j}$

Furthermore, under this representation if $\alpha_j A_{\alpha_j}(x) = [\mu_j(x), v_j(x)]$ then we have indicated that $A(x) = [Max_{\alpha_{j\in D}}(u_j(x)), Max_{\alpha_{j\in D}}(v_j(x))]$.In the above the A_{α_j} can be viewed as the level sets associated with A. In the following example, we illustrate the application of the above methodology to obtain a representation of an IVFS.

Example let $X = \{x_1,\ x_2,\ x_3,\ x_4\}$ and assume A is an IVFS of X where $A(x_1) = [0, 0.5] = \lambda_1$, $A(x_2) = [0.2, 0.6] = \lambda_2$, $A(x_3) = [0.4, 1] = \lambda_3$, $A(x_4) = [0.4, 0.4] = \lambda_4$.

In this case $M = \{\lambda_1,\ \lambda_2,\ \lambda_3,\ \lambda_4\}$ and

$$2^M = \{\{\lambda_1\},\{\lambda_2\},\{\lambda_3\},\{\lambda_4\},\{\lambda_1,\lambda_2\},\{\lambda_1,\lambda_3\},$$
$$\{\lambda_1,\lambda_4\},\{\lambda_2,\lambda_3\},\{\lambda_3,\lambda_4\},\{\lambda_2,\lambda_4\},\{\lambda_1,\lambda_2,\lambda_3\},\{\lambda_1,\lambda_2,\lambda_4\},$$
$$\{\lambda_1,\lambda_3,\lambda_4\},\{\lambda_2,\lambda_3,\lambda_4\},\{\lambda_1,\lambda_2,\lambda_3,\lambda_4\}\}$$

Calculating the $Min[G]$ for $G \in 2^M$, we get

$Min[\{\lambda_1\}] = \lambda_1$, $Min[\{\lambda_2\}] = \lambda_2$, $Min[\{\lambda_3\}] = \lambda_3$, $Min[\{\lambda_4\}] = \lambda_4$,

$Min[\{\lambda_1, \lambda_2\}] = [0, 0.5] = \lambda_1$, $Min[\{\lambda_1, \lambda_3\}] = [0, 0.5] = \lambda_1$,

$Min[\{\lambda_1, \lambda_4\}] = [0, 0.4]$, $Min[\{\lambda_2, \lambda_3\}] = [0.2, 0.6] = \lambda_2$,

$Min[\{\lambda_2, \lambda_4\}] = [0.2, 0.4]$, $Min[\{\lambda_3, \lambda_4\}] = [0.4, 0.4] = \lambda_4$

$Min[\{\lambda_1, \lambda_2, \lambda_3\}] = [0, 0.5] = \lambda_1$, $Min[\{\lambda_1, \lambda_2, \lambda_4\}] = [0, 0.4]$,

$Min[\{\lambda_1, \lambda_3, \lambda_4\}] = [0, 0.4]$, $Min[\{\lambda_2, \lambda_3, \lambda_4\}] = [0.2, 0.4]$,

$Min[\{\lambda_1, \lambda_2, \lambda_3, \lambda_4\}] = [0, 0.4]$.

From these we get $D = \{\lambda_1,\ \lambda_2,\ \lambda_3,\ \lambda_4,\ [0,\ 0.4],\ [0.2,\ 0.4]\}$. Thus, we have six distinct levels. Now for each $\alpha \in D$ we have $A(\alpha) = \{x | A(x) \geq \alpha\}$, that is $x \in A_\alpha$ if $A(x) \wedge \alpha = \alpha$. With $\alpha = [a,\ b]$ and $A(x) = [L(x),\ R(x)]$ this requires $[L(x) \wedge a,\ R(x) \wedge b] = [a,\ b]$.

Here, we get as our levelsets $A_{[0,0.5]} = \{x_1\}$, $A_{[0.2,0.6]} = \{x_2, x_3\}$, $A_{[0.4,1]} = \{x_3\}$, $A_{[0.4,0.4]} = \{x_3, x_4\}$, $A_{[0,0.4]} = \{x_1, x_2, x_3, x_4\}$, $A_{[0.2,0.4]} = \{x_2, x_3, x_4\}$.

Hence, $A = \cup_{\alpha \in D} \alpha A_\alpha$ where

$$D = \{[0,\ 0.5],\ [0.2,\ 0.6],\ [0.4,\ 1],\ [0.4,\ 0.4],\ [0,\ 0.4],\ [0.2,\ 0.4]\}.$$

From this we have $A(x_j) = Max_{x_j \in \alpha}[\alpha]$ and hence we see

$$A(x_1) = \max[[0, 0.5], [0, 0.4]] = [0, 0.5],$$

$$A(x_2) = \max[[0.2, 0.6], [0, 0.4], [0.2, 0.4]] = [0.2, 0.6],$$

$$A(x_3) = \max[[0.2, 0.6], [0.4, 1], [0.4, 0.4], [0, 0.4], [0.2, 0.4]] = [0.4, 1]$$

$$A(x_4) = \max[[0.4, 0.4], [0, 0.4], [0.2, 0.4]] = [0.4, 0.4].$$

Thus, we have recovered the correct membership grades for A. In the following, Fig. 1 we see the partial ordering of the elements $\alpha \in D$ .There is one computational issue that should be addressed in the preceding approach. Assume there are n distinct membership grades in A that is $B = \{\lambda_1, \ldots, \lambda_n\}$. In this case, D have $2^n - 1$ components and the calculation of $Min[G]$ for all $G \in D$ becomes computationally intense as n gets large. In the following, we describe an algorithm introduced in Walker et al. [7] which considerably reduces the amount of work in this task.

Let M be the finite set of closed subintervals of $[0,\ 1]$. Let L be the set of left end points of the elements in M and let R be the set of right end points of the elements in M . The algorithm for calculating $Min[G]$ for all $G \in 2^M$ is as follows:

1. Sort L and number the elements so that $a_1 < a_2 < \cdots < a_K$. Here K is the number of distinct left end points.

2. For each $i = 1,\ldots,k$ among all intervals in M with left end point a_i , choose one with the max right end point, to get a set $\{[a_1,\ b_1],\ [a_2,\ b_2],\cdots,[a_K,\ b_K]\}$ where each $b_i \in R$.

3. The set of intervals in D is composed of the union of the following sets

$$D_1 = \{[a_1, z] : z \in R, a_1 \le z \le b_1\},$$
$$D_2 = \{[a_2, z] : z \in R, a_2 \le z \le b_2\},$$
$$D_K = \{[a_K, z] : z \in R, a_K \le z \le b_K\}.$$

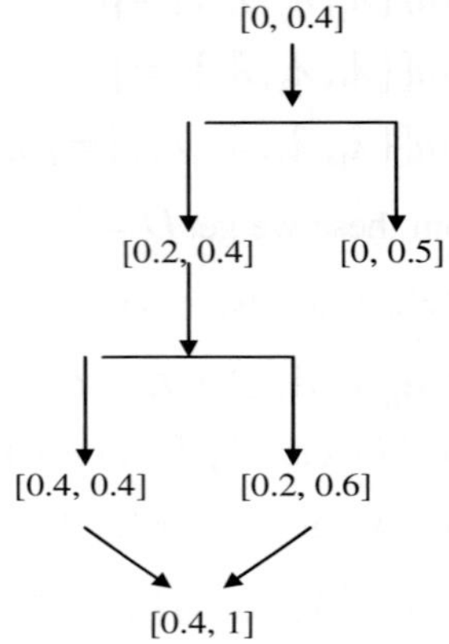

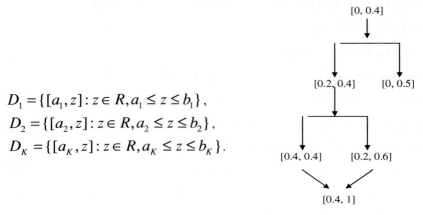

Fig. 1. Partial ordering of the elements in D

4 Extension Principle for IVFS

Let H be a set mapping $H : 2^X \to Y$ .Here H maps a subset of X into an element from Y . We can extend this to act on IVFS. Let E be an IVFS of X which we can express using the level sets as $E = \bigcup_{\alpha_j \in D} \{\frac{\alpha_j}{E_{\alpha_j}}\}$. Here D is the subset of derived membership grades which are interval-valued membership grades of the form $\alpha_j = [a_j,\ c_j]$. Using the extension principle we can extend H to act on IVFS so that $H(E) = \bigcup_{\alpha_j \in D} \{\frac{\alpha_j}{H(E_{\alpha_j})}\}$. A particularly important class of set mappings is monotonic set measures which are used to model information about uncertainty. Here, $H = 2^X \to [0,\ 1]$ and $H(\varnothing)=0$, $H(X)=1$ and if $B \subseteq A$ then $H(A) \ge H(B)$.

A notable example of these monotonic set measures is probability measures. In this case H is a probability measure and we shall denote $H(A)$ as $\Pr ob(A)$. In particular, in this case with $\Pr ob\{(x_i)\} = P_i$ we have $\Pr ob\{(x_i)\} = P_i$. Here we of course require that $\sum_i P_i = 1$.

We now illustrate the extension of a probability measure to IFS in the following example.

Example. let $X = \{x_1, x_2, x_3, x_4\}$ and assume we have a probability measure over X so that $P_1 = 0.1, P_2 = 0.2, P_3 = 0.4, P_4 = 0.3$.

Let E be an inerval valued fuzzy set where $E = \{\frac{[0.3,0.4]}{x_1}, \frac{[0,1]}{x_2}, \frac{[0.7,0.8]}{x_3}, \frac{[0.5,0.6]}{x_4}\}$. We want to calculate $\Pr ob(E)$. Since E is the same set as we used in the previous example we already have all the information needed for the representation of E in terms of level sets

$$E = \{\frac{[0,1]}{\{x_2\}}, \frac{[0.7,0.8]}{\{x_3\}}, \frac{[0,0.8]}{\{x_2,x_3\}}, \frac{[0.5,0.6]}{\{x_3,x_4\}}, \frac{[0.3,0.4]}{\{x_1,x_3,x_4\}}, \frac{[0,0.6]}{\{x_2,x_3,x_4\}}, \frac{[0,0.4]}{\{x_1,x_2,x_3,x_4\}}\}$$

Using this we have $\Pr ob(E)$ is Q where

$$Q = \{\frac{[0,0]}{0.2}, \frac{[0.7,0.8]}{0.4}, \frac{[0,0.8]}{0.6}, \frac{[0.5,0.6]}{0.7}, \frac{[0.3,0.4]}{0.8}, \frac{[0,0.6]}{0.9}, \frac{[0,0.4]}{1}\}$$

We can now consider the following question. What is the possibility that the probability of E is at least 0.7. The event a probability of greater than or equal to 0.7 is represented as the set $F = \{\frac{1}{0.7}, \frac{1}{0.8}, \frac{1}{0.9}, \frac{1}{1.0}\}$ which can be expressed as an IVFS as $F = \{\frac{[1,1]}{0.7}, \frac{[1,1]}{0.8}, \frac{[1,1]}{0.9}, \frac{[1,1]}{1.0}\}$.

The answer to our question is $Poss(\Pr ob(E)\ is\ F \mid \Pr ob(E)\ is\ Q) = Max_p[F(p) \wedge Q(p)]$.

$$Poss(\Pr ob(E)\ is\ F \mid \Pr ob(E)\ is\ Q)$$
$$= Max[[1,\ 1] \wedge [0.5,\ 0.6], [1,\ 1] \wedge [0.3,\ 0.4], [1,\ 1] \wedge [0,\ 0.6], [1,\ 1] \wedge [0,\ 0.4]]$$
$$= [0.5,\ 0.6]$$

Here then there is support of 0.5 for the proposition that the probability of E is at least 0.7 and support of 0.4 against the proposition.

A related question is to determine $\Pr ob(\{x_3\}/E)$. Using the definition of conditional probability we get $\Pr ob(\{x_3\}/E) = \frac{\Pr ob(\{x_3\} \cap E)}{\Pr ob(E)}$.

Here denoting $M = \{x_3\} \cap E$, we have $M = \left\{\frac{E(x_3)}{x_3}\right\} = \left\{\frac{[0.7,0.8]}{x_3}\right\}$. In this case since M is also the level set representation of itself we have

$$\Pr ob(M) = \left\{\frac{E(x_3)}{\Pr ob(\{x_3\})}\right\} = \frac{[0.7,0.8]}{0.4}.$$

Using this and $\Pr ob(E) = \{\frac{[0,0]}{0.2}, \frac{[0.7,0.8]}{0.4}, \frac{[0,0.8]}{0.6}, \frac{[0.5,0.6]}{0.7}, \frac{[0.3,0.4]}{0.8}, \frac{[0,0.6]}{0.9}, \frac{[0,0.4]}{1}\}$, we get $\Pr ob(\{x_3\}/E) = \Pr ob(E)$

$$= \{\frac{[0,0] \wedge [0.7,0.8]}{0.4/0.2}, \frac{[0.7,0.8] \wedge [0.7,0.8]}{0.4/0.4}, \frac{[0,0.8] \wedge [0.7,0.8]}{0.4/0.6}, \frac{[0.5,0.6] \wedge [0.7,0.8]}{0.4/0.7}, \frac{[0.3,0.4] \wedge [0.7,0.8]}{0.4/0.8}, \frac{[0,0.6] \wedge [0.7,0.8]}{0.4/0.9}, \frac{[0,0.4] \wedge [0.7,0.8]}{0.4/1}\}$$

5 Conclusion

We discussed the concept of a level set of a fuzzy set and the related ideas of the representation theorem and Zadeh's extension principle. We then described the extension of these ideas to the case of interval valued fuzzy sets.

Acknowledgments. This research work is supported by Funding Project for Academic Human Resources Development in Institutions of Higher Learning under the Jurisdiction of Beijing Municipality (PHR20101026) and Funding Project for Base Construction of Scientific Research of Beijing Municipal Commission of Education (WYJD200902).

References

1. Zadeh, L.A.: Fuzzy sets. Inf. Contr. 8, 338–353, doi:10.1016/ S0019-9958(65)90241-X
2. Zadeh, L.A.: Similarity relations and fuzzy orderings. Inf. Sci. 3, 177–200, doi: 10.1016 /S0020-0255(71)80005-1
3. Zadeh, L.A.: The concept of a linguistic variable and its application to approximate reasoning: part 1. Inf. Sci. 8, 199–249, doi: 10.1016/0020-0255(75)90036 -5
4. Zadeh, L.A.: Outline of a new Approach to the analysis of complex systems and decision processes, interval valued fuzzy sets. IEEE Trans. Syst. Man Cyber. 3(1), 28–44 (1973)
5. Yager, R.R.: A characterization of the extension principle. Fuzzy Sets Syst. 18, 205–217 (1986), doi:10.1016/0165-0114(86)90002-3
6. Walker, C.L., Walker, E.A., Yager, R.R.: Some comments on level sets of fuzzy sets. In: Proceedings of world congress on computational intelligence, Hong Kong, pp. 1227–1230 (2008)

CSFs for Service Industry SMEs Successfully Adopting E-Commerce System: A Study from China

Mingxuan Wu[1,2], Ergun Gide[1], Li Zhang[3], and Qiudan Xing[4]

[1] CQUniversity Australia Sydney
robert_wumx@hotmail.com
[2] SMEs Online International, Australia
[3] Xi'an University of Post and Telecommunications, P.R. China
[4] Shannxi Normal Univeristy, P.R. China

Abstract. This research focused mainly on understanding the common critical factors for China's small and medium enterprises (SMEs) in service industry successfully adopting e-commerce system. Based on 73 initial items discussed from previous research and literature review, focus group study and pilot test was conducted first. As a result, a total of 21 factors were explored and then catogrised into six components by strength of relationship including Web Site Effectiveness & Cost, e-Marketing, Web Site Design & Image, Managing Chang & Customer Acceptance, Knowledge, and Staff & Skills. This paper also made the recommendations with a brief guideline to be used for China's service industry SMEs successfully adopting e-commerce systems. Finally, several topics were provided for further research.

Keywords: CSFs, e-commerce success, service industry, SMEs.

1 Introduction

For China to make the move from primarily a manufacturing economy to one incorporating a strong service-based knowledge economy, it needs to significantly improve its integration of technology into all facets of business processes especially in the adoption and use of electronic commerce (e-commerce).

This research focused mainly on understanding the critical success factors (CSFs) for China's service industry SMEs successfully adopting e-commerce system. Based on 73 initial items discussed from previous research and literature review, a blend of quantitative and qualitative research methods were used, consisting of literature review, focus group studies, pilot tests, and surveys. Strategic success factors were finally identified for businesses to be successful when adopting e-commerce systems.

2 Research Background and Literature Review

Although China's economy is changing rapidly from primary and secondary industry (agriculture / manufacturing) to tertiary industry (service), China's service industry has been seen as a key factor in further economic growth [1].

X. Wan (Ed.): Electrical Power Systems and Computers, LNEE 99, pp. 861–868.
springerlink.com

The Central Government of China has stated that e-commerce is the key for promoting service industries, and will help China to compete better economically at the global level [2, 3]. One of the top research topics regarded as urgent by China's state agencies, such as [2], [4], and [5, p.94], is to understand the common critical factors for China's small and medium enterprises (SMEs) in service industry successfully adopting e-commerce system. However, china's SMEs is still lack of experienced in adopting e-commerce. Therefore, the understanding of e-commerce adopted in SMEs is becoming important, especially the understanding of CSFs.

According to previous research and literature review on existing factors for measuring e-commerce success, 73 initial items in total has been drawn and categorised [6, p.317]. Based on these, this research was then conducted in China.

3 Research Question and Research Methodologies

3.1 Hypothesis and Research Methods

The hypothesis for this research was stated as common CSFs exist for the adoption of e-commerce systems. To achieve triangulation, a blend of quantitative and qualitative research methods consisting of literature review, focus group studies, pilot tests, and surveys were used for this research.

Focus group study was adopted first for identifying a number of problems and issues that industry is facing today (current status of the e-commerce industry) and these were used to refine Critical factors identified previously. In this research, a target of nine members was adopted for each focus group.

A pilot test of the survey was then followed and conducted with open-questions. In this research, ten businesses were used to carry out the pilot test. Finally, the survey instrument used for this research comprised 50 items (see Appendix). The five-point Likert Scale (1: Strongly Disagree, 2: Disagree, 3: Neither Agree nor Disagree, 4: Agree, 5: Strongly Agree) was adopted to measure a respondent's agreement with survey statements.

The survey samples were selected randomly from China's Yellowpage online (http://www.yellowpage.com.cn). The initial survey was sent on 11 November 2006 to selected companies with attached a survey invitation letter and a survey form. A total of 1103 solicitations were emailed survey questionnaires. Data collection in China was completed at the end of December 2007.

3.2 Data Collection

Among 1103 solicitations, 20.85% (230 out of 1103) of businesses did not respond and could not be contacted, and 79.15% (873 out of 1103) of businesses could be contacted.

Among 873 businesses contacted, there are a total of 164 responses in this survey including 112 usable responses and 52 non-usable. Because of the validity reason, all

52 non-usable feedacks were removed from the following data analysis including four empty forms, eight same answers, and 40 incompleted responses which refers to those return forms with one item missed (25), two item missed (8), three item missed (4), four item missed (1), and more than four item missed (2). Therefore, a total of the response rate is 18.79% (164 out of 873). Finally, the usable response rate is 12.83% (112 out of 873).

4 Data Analysis and Results

All data was captured into an Excel spreadsheet, and then transferred into the SPSS system for conducting data analysis. These analysis procedures comprised reliability analysis, validity analysis, hypothesis testing, factor analysis and repeat reliability analysis.

4.1 Reliability and Validity Analysis

A scale measurement must be reliable and valid. Thus, this research conducted an reliability analysis first. Reliability analysis results showed that a total value of Cronbach's Alpha was 0.924. However, its value would be increased to 0.926 if the Item F14 was removed from the item statistics, and its value would be increased to 0.927 if the Item F43 was removed. Therefore, two items (F14 and F43) were eliminated. The remaining 48 items showed a total value of Cronbach's Alpha was 0.929. The results showed that this was strong evidence of very good reliability. Thus, a total of 48 items remained for further analysis.

Following reliability analysis, validity analysis was then conducted. The methods used to test for validity in this research were content validity, and construct validity. In this research, content validity was assessed subjectively but systematically to establish the appropriateness of the variables used – items not considered appropriate were rejected by focus group study and 10 pilot tests. Construct validity can be indirectly established through factor analysis discussed later in this paper.

4.2 Hypothesis Testing

Following data preparation and preliminary analysis, hypothesis testing was conducted. The most common value of 0.05 was used. This resulted in the null hypotheses of a total of 25 items (F3, F7, F9, F10, F13, F15, F19-F21, F25, F27, F28, F30, F31, F33-F40, F48-F50) within 48 items being rejected.

4.3 Results

As previously discussed, the analysis of the survey data produced the following results for the 50 research items showed:

Table 1. Summary of Data Analysis

Item	t	Mean	Std. Deviation	Std. Error Mean	Sig. (2-tailed)	Mean Difference	95% Confidence Interval		Component						Eigenvalues	Cronbach's Alpha
							Lower	Upper	1	2	3	4	5	6		
F24	0.123	4.01	0.765	0.072	0.902	0.009	-0.13	0.15	0.829						7.339	0.824
F23	0.638	4.04	0.74	0.07	0.524	0.045	-0.09	0.18	0.782							
F41	-0.446	3.97	0.636	0.06	0.657	-0.027	-0.15	0.09	0.659							
F8	-1.65	3.88	0.744	0.07	0.102	-0.116	-0.26	0.02	0.495							
F22	-0.403	3.97	0.703	0.066	0.688	-0.027	-0.16	0.1	0.471			0.401				
F26	1.892	4.13	0.699	0.066	0.061	0.125	-0.01	0.26	0.463		0.408					
F44	0.799	4.04	0.591	0.056	0.426	0.045	-0.07	0.16		0.758					1.963	0.798
F42	0.377	4.03	0.753	0.071	0.707	0.027	-0.11	0.17		0.748						
F45	1.516	4.09	0.623	0.059	0.132	0.089	-0.03	0.21		0.658						
F47	-0.649	3.96	0.728	0.069	0.518	-0.045	-0.18	0.09		0.546						
F17	-2.241	3.81	0.886	0.084	0.027	-0.188	-0.35	-0.02			0.725				1.388	0.656
F46	-0.699	3.96	0.676	0.064	0.486	-0.045	-0.17	0.08			0.654					
F18	0.943	4.06	0.701	0.066	0.348	0.063	-0.07	0.19			0.618		0.447			
F12	-0.956	3.93	0.791	0.075	0.341	-0.071	-0.22	0.08				0.819			1.361	0.646
F11	-1.491	3.9	0.697	0.066	0.139	-0.098	-0.23	0.03				0.742				
F32	-0.628	3.96	0.752	0.071	0.531	-0.045	-0.19	0.1		0.404		0.552				
F1	-0.741	3.94	0.893	0.084	0.461	-0.063	-0.23	0.1					0.81		1.047	0.672
F2	-0.371	3.97	0.765	0.072	0.712	-0.027	-0.17	0.12		0.457			0.613			
F6	-0.491	3.96	0.77	0.073	0.625	-0.036	-0.18	0.11					0.454	0.434	1.003	0.616
F5	-0.818	3.94	0.809	0.076	0.415	-0.063	-0.21	0.09						0.821		
F4	-0.47	3.96	0.805	0.076	0.639	-0.036	-0.19	0.11						0.737		
Determinant value of correlation matrix															9.85E-05	
KMO															0.819	
Bartlett's Test of Sphericity															0.000	
Average communality																0.647
Cumulative % of Variance																64.66%
The ratio for evaluating Case size requirement																5.76:1
A total of Reliability (Cronbach's Alpha)																0.888

- two items (F16 and F29) were dropped during factor analysis, and
- a total of 21 items were finally accepted as CSFs based on initial reliability analysis, validity analysis, one-sample t testing, factor analysis, and repeat reliability analysis

Exploratory factor analysis was then conducted. As a result, the 21 CSFs were categorised into six components by strength of relationship. The main outputs and the criteria used for evaluating data were analysed as follows (see Table 1).

5 Findings

Findings from the results are:

- Finding 1: Management staff knowledge and staff skills play critical roles.

This can be seen all items from F1 to F6 accepted exclude F3. The importance of Staff knowledge has been discussed internationally. These results also imply that there is an urgent need to offer related e-commerce courses into postgraduate level programs, even provide a specific program - such as “MBA program in e-commerce major”, as business management knowledge is highly relevant to e-commerce success.

- Finding 2: E-commerce systems should be flexible and work well with the existing systems.

This can be seen from F8, F11 and F12 being accepted while items F7, F9, F10, and F13 were rejected. They are not important whether businesses have the previous experience with e-commerce system or appropriate trial time in adopting system. Ease of use or learning is also not critical. However, it should be able to be easily changed in accord with the existing systems and business process changes, and able to deal with external technology changes.

- Finding 3: Security issues are highly concerned

This can be seen from items F23, F24 and F25 being accepted. A little research has been done in literature review. Further research should explore the factors important in e-commerce security.

- Finding 4: Developing an effective e-commerce system and e-marketing strategy are critical.

This can be seen from items F217, F18 and F22 being accepted and all marketing items (F42, F44 to F47) being accepted. There is almost no debate on the importance of these factors in the literature as shown by the overwhelming focus by researchers.

- Finding 5: E-commerce system is not involved in management issues.

This can be seen from all management items (from F17 to F30) being not accepted. This is different with other research. This research believes that this has been misunderstanded by China’s business managers. They considered that e-commerce should be an IT system. Therefore, further research should be conducted in this issue. This issue is not discussed much in the literature and is worthy of further research.

- Finding 6: Customer satisfaction is a key.

This can be seen from F32 being accepted while others (items F31, F33-F36) were rejected. There is almost no debate on the importance of this factor in the literature as shown by the overwhelming focus by researchers on customer satisfaction with e-commerce systems.

- Finding 7: Cost associated with keeping up to date or upgrading an e-commerce system is critical.

This can be seen from F41 being accepted only while all item F37 to F38 were rejected. This research has found that the cost associated with keeping up to date or to upgrade an e-commerce system is a highly critical factor.

- Finding 8: Most of China's businesses are still focusing on local marketing.

This can be seen from items F43, F48, F49 and F50. This implies that most of China's SMEs are still focusing on local marketing and less experience of international marketing. Little research has addressed this issue so far. There is therefore a need to conduct further research in how to use e-commere to assist China's SMEs into international marketing.

- Finding 9: Outsourcing is still a new topic.

This can be seen from items F14, F15 and F16 being rejected. Further research should be conducted to identify importance of outsourcing to China's businesses in adopting e-commerce system.

6 Conclusions, Recommendations and Further Research

As a result, 21 factors were explored and then categorized into six components by strength of relationship including Web Site Effectiveness & Cost, e-Marketing, Web Site Design & Image, Managing Chang & Customer Acceptance, Knowledge, and Staff & Skills.

In recommendations, the following brief guidelines can be used for businesses successfully adopting e-commerce systems as:

- maintain the appropriate level of e-commerce knowledge and skills for each level of staff,
- develop an effective e-commerce system satisfied with customers,
- ensure work well with existing systems,
- with the appropriate e-marketing strategy,
- keeping cost associated, and
- e-commerce security satisfied.

Further research is needed to determine whether these CSFs and results are applicable to other industries and cultures.

References

1. China Daily, Service Industry Seen as Key to Development, China Daily, March 12, p.12 (2007a)
2. EC, Strategies for Promoting E-commerce Development (2005), `http://www.ec.org.cn/2005-12/01/content_5722655.htm` (viewed April 9, 2007) (in Chinese)
3. ECW, E-commerce is Key for Promoting Service Industry(2005), `http://www.ecw.cn/2005-2/2005218113603.htm` (viewed April 9, 2007) (in Chinese)
4. GOV, Some Opinions on Accelerating The Development of The Service Industry (2007), `http://www.gov.cn/zwgk/2007-03/27/content_562870.htm` (viewed April 9, 2007) (last update March 19)
5. MOFCOM, Report for Chinese E-commerce in 2006-2007. Ministry of Commerce of the People's Republic of China, the Yearbook for Chinese E-commerce, p. 94 (2008)
6. Gide, E., Wu, M.X.: A Study To Establish E-Commerce Business Satisfaction Model To Measure E-Commerce Success In SMEs. International Journal of Electronic Customer Relationship Management 1(3), 307–325 (2007)

Appendix

- Human resource factors (six), including:
  - F.1 CEO's IT/e-commerce/e-commerce marketing knowledge.
  - F.2 Senior staff IT/e-commerce knowledge.
  - F.3 Junior staff IT/e-commerce knowledge.
  - F.4 Hiring IS/IT staff.
  - F.5 Hiring e-commerce staff.
  - F.6 Staff training regularly in the appropriate or relevant IT skills.
- Information technology factors (seven), including:
  - F.7 The previous experimental use of e-commerce system.
  - F.8 The compatibility and integration with the existing information system within business system.
  - F.9 Complexity (ease of use or learning) of e-commerce systems.
  - F.10 The ability of the existing information system to keep up to date or upgrade (internally).
  - F.11 Flexibility of e-commerce systems changes depends on business process.
  - F.12 The ability to keep up with the rate of technology change (externally).
  - F.13 Appropriate trial time in adoption of e-commerce system.
- Web site factors (nine), including:
  - F.14 Outsourcing web site development when time limited.
  - F.15 Only outsourcing the part of web site services.
  - F.16 Business control and maintenance of web site.
  - F.17 Web site design attractiveness.
  - F.18 Web site's systematic structure is clear, easily navigated, and convenient.
  - F.19 Designing web site in Multilanguage.
  - F.20 Web site's high ranking in the best known search engines.

F.21 Web site's links with other strategic web sites/pages.
F.22 The response time effectiveness/performance of an e-commerce site.

- Security factors (four), including:
  F.23 High level of security of e-commerce systems.
  F.24 Privacy of e-commerce systems.
  F.25 Trust in the interface design and information displayed in a web site.
  F.26 Reliability of web site.
- Management factors (four), including:
  F.27 Government support.
  F.28 Support from top management/decision-maker
  F.29 Support from senior management.
  F.30 Flexibility of business management changes depends on e-commerce system requirements.
- Business relationship factors (six), including:
  F.31 Competitive pressure from competitors/ industry.
  F.32 Customer pressure/acceptance/ interest.
  F.33 Supplier pressure/interest.
  F.34 Pressure/ interest from collaboration/ partnership.
  F.35 Encouragement by other agencies or government to adopt e-commerce system.
  F.36 Decision-maker's maintenance of professional links with professional associations.
- Organisational finance factors (five), including:
  F.37 Financial help from outside of business at the initial development stage.
  F.38 Return on investment (ROI) from e-commerce investment.
  F.39 Financial resources priority in e-commerce system development.
  F.40 Affordable access to e-commerce system.
  F.41 Cost associated with keeping up to date or upgrade of e-commerce system.
- Marketing factors (six), including:
  F.42 Decision-maker's effective e-commerce marketing plan.
  F.43 Firms' ability to act globally or the resources required doing business globally.
  F.44 Effective e-commerce marketing strategy.
  F.45 Adoption of different e-commerce marketing strategies based on different business requirements/ needs.
  F.46 Having a positive image with a relevant business name on the Internet.
  F.47 Having a consistent/appealing/easy to remember Internet-based brand name.
- Culture factors (three), including:
  F.48 The consistency of graphics and backgrounds with business culture used in a web site.
  F.49 E-Commerce systems' consideration of different business culture.
  F.50 E-Commerce systems' consideration of the different social culture.

Design of MRI Digital-to-Analog Conversion Card Based on USB2.0

Bei Dai, Zhaoxue Chen, and Piding Li

School of medical instrument & food engineering,
University of Shanghai for Science and Technology, Shanghai, 200093
daibei@usstvol.com

Abstract. This paper presents in detail the hardware and firmware design about frequency source digital-to-analog conversion card in magnetic resonance imaging (MRI) system. Based on USB2.0 interface, this design uses host computer to transmit digital waveform data information into controller chip CY7C68013A, which connects with CPLD by means of Slave FIFO. CPLD codes the received data and then transmits the data to digital-to-analog converter chip DAC7725UB, which is with voltage output type. Finally corresponding analog waveform signal will be obtained from outputs of DAC7725UB.

Keywords: Magnetic resonance imaging, USB2.0, CPLD, digital-to-analog conversion.

1 Introduction

Frequency source is one of the most important techniques in magnetic resonance imaging (MRI) system. Modern magnetic resonance imaging technology requires frequency source with abilities of switching frequency, phase and amplitude rapidly. Some existing analog design methods in industry have become unsatisfactory gradually for actual application requirements in MRI system. With the rapid development of large scale integrated circuit technology, the study of the design method of digital circuit is inevitable in MRI frequency source technique's development. At present, the direct digital frequency synthesis technology for the MRI frequency source has become the mainstream of current development. One important part of the MRI frequency source is about digital-to-analog conversion. In practical applications and experiments of magnetic resonance imaging, it is often necessary to have the computer connected with digital-to-analog conversion card to test some experimental data and improve performance of the whole magnetic resonance imaging system. While there are many inconveniences in the actual application of the traditional PCI bus digital-to-analog conversion card, such as inconvenient installation, expensive price, easily limited by computer slot number, address and interrupting resource constraints with poor expansibility. Especially in test environments with some strong electromagnetic interference, data distortion happens with no special electromagnetic shielding. Universal Serial Bus, i.e., USB, as a new computer peripherals bus standard, has the virtues of a hot plugging, no need of external power supply, fast speed and peripherals of large capacity. It has become a

X. Wan (Ed.): Electrical Power Systems and Computers, LNEE 99, pp. 869–875.
springerlink.com

widely used interface standards for PC peripheral expansion. While utilized to the design of MRI frequency source digital-to-analog conversion card, the USB interface will undoubtedly improve convenience of MRI applications and experiments greatly.

To meet the needs of new technological development in MRI, this paper presents a design method of MRI frequency source digital-to-analog conversion card based on USB2.0 interface.

2 The General Structure of the System

The hardware of the system is mainly divided into such two key steps as data receiving and digital-to-analog conversion together with USB interface firmware design [1]. The systematic structure diagram is presented in Figure 1.

In detail, the system is made up of USB interface chip CY7C68013A, CPLD chip EPM570T, digital-to-analog chip DAC7725UB and peripheral circuits. While processing, the corresponding waveform data sequence is first input from the host application program and received by the chip CY7C68013A based on USB2.0 interface. The received data by CY7C68013A is then encoded by CPLD and sent to D/A converter. Finally required analog signals are produced by the D/A converter. In addition, CPLD has 16 bit extension of I/O, namely 8 channel digital TTL signal input and 8 channel digital TTL signal output function. In the data processing procedure, the communication with host machine and the data exchange with CPLD are mainly controlled by chip CY7C69013A, and the CPLD is controlled by chip CY7C68013A.

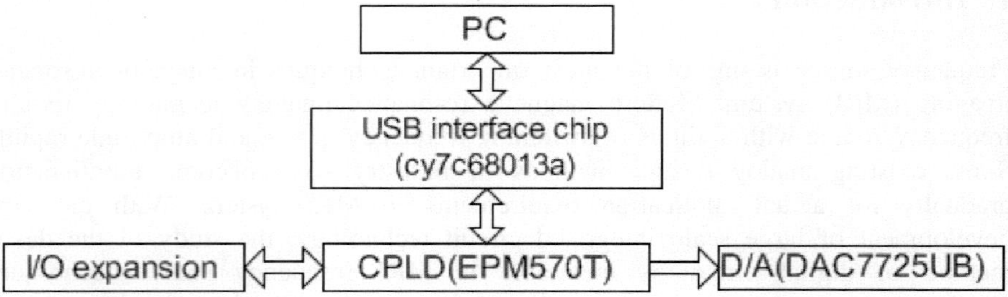

Fig. 1. System structure diagram

3 Main Parts of the System

3.1 Digital-to-Analogue Conversion Module

The digital-to-analogue conversion part of hardware design in this system uses high-performance voltage output type of D/A converter DAC7725UB. The DAC7725UB belongs to a small type of digital-to-analog converter with such features as high speed, low-power dissipation, short setting time, 12-bit parallel TTL digital input and four channel analogue signal output. Its setting time of digital conversion output is as short as 10μs and the maximum consumption is only 250 mW with a 12 bit resolution. The converter adopts double-buffered data inputs and provides function of data read back together with an interior input register based on such mode. Its working power supply takes use of external reference voltage connected means. Besides, the converter has two kinds of voltage outputs which are corresponding to unipolar or bipolar output respectively. Therefore DAC7725UB is good enough for design of our system.

There are two kinds of reference voltages in this design, by using JP3 to perform the jumper selection. When the external reference voltage is on 0V and +10V, the range of the analog output signal is from 0V to +10V; When the external reference voltage is on -10V and +10V, the range of analog output signal is from -10V to +10V and the reference voltage can be corrected by the potentiometer VR9. The design also has analog supply voltage input port and digital supply voltage input port. A0 and A1 port can be used to perform selection about four registers. The design can realize quad analog output and connecting the operational amplifier OP27 and its corresponding circuit onto each analog output channel respectively, the reference voltage can be calibrated by potentiometer VRn(n=1,2,3,4,5,6,7,8). The interface circuits between D/A and CPLD are shown in Figure 2.

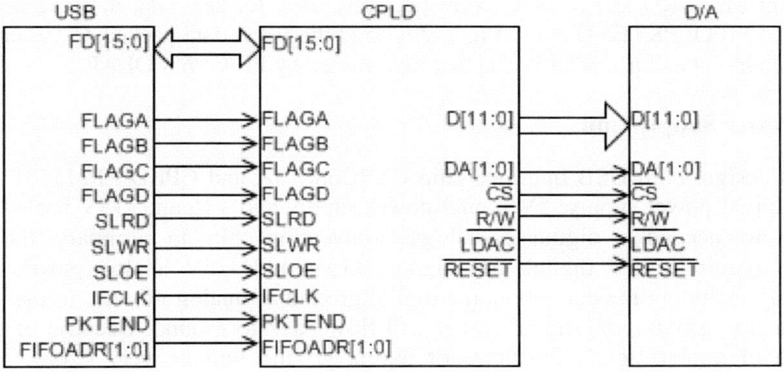

Fig. 2. Interface circuit

3.2 USB Interface Chip

In this design, the chip of CY7C68013A is selected as the USB interface chip. EZ-USB FX2LP™(CY7C68013A/14A) is a kind of highly integrated and low-power dissipation USB 2.0 microcontroller. This chip is mainly designed for USB 2.0 and is compatible with USB 1.1; without supporting low-speed rate (1.5Mbps) it supports two kinds of transmission rate: full-speed (12Mbps) and high-speed (480Mbps). CY7C68013A is realized by integrating USB 2.0 transceiver, serial interface engine, enhancement mode 8051 microcontroller, USB port, Slave FIFOs which are used in high-speed transmission and GPIF in one chip.

3.3 CPLD Logical Module

CPLD has strong logic units, simple interconnected relationship and shorter transmission delay, so it is suitable for logic systems with complex logic and multiple variable inputs. The CPLD always contains 20 ~ 256 macrocells internally, which can be considered as the integration of multiple PLD macrocells, thus can replace some common standard logic chips. The CPLD also integrates intensive internal wiring arrays which can realize the connection between the each macrocells and I/O pins. ALTERA company's MAX II EPM570T is taken as CPLD of this system. In the design, the CPLD's primary tasks include D/A module control, data coding and reading together with 8 channels digital TTL signal input and 8 channels digital TTL signal output. Moreover CPLD is also in charge of the communication to USB chip.

3.4 Hardware Connection between USB and CPLD

Figure 2 shows the connection between FX2 USB and CPLD by means of Slave FIFO connection [2].In this figure, FD[15...0] is a 16-bit bidirectional data bus; FLAGA~FLAGC are the mark pins of FIFO in FX2,which can reflect the current status of FIFO; SLCS is the CS of Slave FIFO; SLOE is used for enabling the data bus FD's output; FIFOADR[1..0] is used for choosing the connected endpoint buffer with FD(00 represents endpoint 2, 01 replacing endpoint 4, 10 represents endpoint 6, 11 represents endpoint 8); SLRD and SLWR are respectively used for selecting the read and write signal of FIFO; PKTEND is used for submitting the FIFO data packets to the input end of the endpoint and it's polarity can be programmed by FIFOPINPOLAR.5.

3.5 Power Supply Unit

In this design, both USB interface chip CY7C68013A and CPLD EPM570T require 3.3V digital power supply; 5V digital power supply and +15 and -15V analog power supply are needed in digital-to-analogue conversion chip. In summary, the whole system requires 3.3V digital, 5V digital, +15V and -15V analog power supply, certainly including the corresponding digital ground or analog ground. In the system, both analog ground and digital signal will flow back to ground. Because of the fast change of digital signal, the noise of digital ground will be very strong. But the reference work of analog signal needs a clean ground. If the analog ground and the digital ground are mixed together, the noise will affect the analog signal. So the isolation part between the digital ground and analog ground should be designed when designing the circuit board.

At first, the external provides 12V power supply. +15V and -15V analog power supply are produced by SR12D15/100. Then chip LM7805 is adopted to obtain +5V digital voltage. Finally, +3.3 digital voltage can be produced by LM1117-3.3.

4 System Software Design

The system software includes application program of upper host, USB firmware and CPLD program [3].

4.1 System Processing Flow

Figure 3 shows the system flow chart. After the system gets powered, it completes initialization, and operates through application program of upper host. At first, it performs channel selection, and then inputs 12-bit data sequence, which will be sent to CPLD through USB to decode, at the same time, the digital-to-analog converter chip will get started to perform digital-to-analog converting and output needed analog waveform.

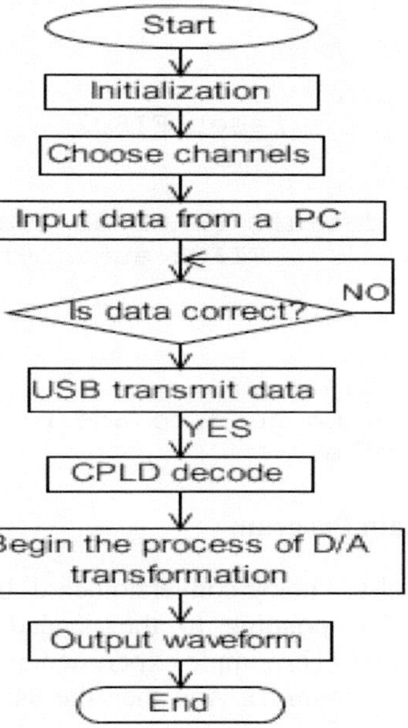

Fig. 3. System flow chart

4.2 Device Firmware Design

Here, firmware is 51 microcontroller program file in CY7C68013A chip, which is the core of USB peripherals [4]. It can be written in C language, assembly language or VHDL, etc. It operates closely with the hardware, such as connection of USB device based on USB protocol, interrupt handling and so on. It is not merely software, but the combination of software and hardware. Developer need to be very familiar with the port, interrupt, protocol as well as hardware structure. Generally, firmware program is installed in MCU. When the device is connected to the host (USB cable is plugged into the jack), upper host resets and discovers the new device, and then builds the connection. So, one of the main purpose of writing firmware program is to make the operating system capable of detecting and identifying the plugged device. The main functions of firmware program in the design are as follows: realizing of USB chip initializing and configuring; controlling the USB chip to receive and handle USB driver requirements; manipulating the chip to receive controlling-instructions from the controlling-program.

Special program is needed to add into the firmware framework in the process of realization of the device's function. Part of the key program codes of the design together with necessary comments is as below:

```
void TD_Init (void)
{
CPUCS=0x10;
CLKSPD[1:0]=10; // select the clock frequency
IFCONFIG= 0xCB; //   configure chip
FIFORESET=0x80;
SYNCDELAY;
FIFORESET=0x02;
SYNCDELAY;
FIFORESET=0x00; //   reset FIFO2
SYNCDELAY;
PORTACFG I=0x80;
SYNCDELAY; //   default FLAGD as flag pins of FIFO2
EP2FIFOCFG=0x10; //   FIFO2 auto output style
SYNCDELAY;
... ...
}
void TD_Poll (void)
{ }//    there is no need to add program because of its
//auto output style.
```

4.3 The Interface of Main Program

To use the designed digital-to-analog conversion card, input numerical information of digital sequences that corresponding to the needed waveform from application program of upper host at first (The simple experimental interface of the main program of the system is shown as Figure 4). And then the data is transmitted through USB data line based on USB interface chip and get to D/A converter DAC7725UB. And finally the needed analog waveform can be obtained under control of CPLD. Experiments results have shown that the USB transmission method in the presented design can meet requirements of the system and the output waveform can also achieve the expected result based on display of the oscilloscope.

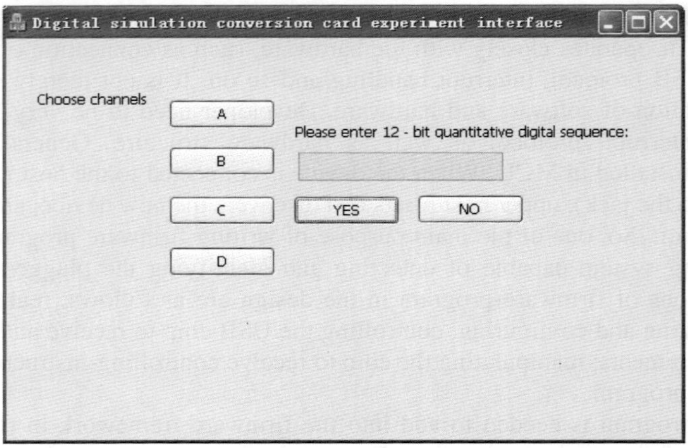

Fig. 4. The main program interface

5 Conclusion

The proposed design of digital-to-analog conversion card in this paper based on USB interface can successfully convert digital sequences to corresponding analog waveform. Adopting general USB data line to connect with computer and perform data transmission, while applied to the application and test occasions of frequency source in MRI system, the designed conversion card can make the research and development of MRI system convenient and efficient. Moreover, it can not only be put into use as MRI frequency source but also in other situations that require digital-to-analog conversion and thus this design also has a great practical significance.

Acknowledgement

This work is under the auspice of Shanghai Municipal Education Commission to Scientific Innovation Research Funds (No. 11YZ116).

Bei Dai, a postgraduate, whose research interests include design of MRI device, biomedical signal processing, is with College of Medical Instrumentation & Foodstuff, University of Shanghai for Science and Technology, Shanghai, 200093, China (phone: 86-21-18801927657; fax: 86-21-55271172 e-mail: daibei@ usstvol.com).

Zhaoxue Chen, a PH.D. and associate professor, the correspondent author of this paper, whose research interests include medical image and biomedical signal processing, is with College of Medical Instrumentation & Foodstuff, University of Shanghai for Science and Technology, Shanghai, 200093,China (phone: 86-21-18930242159; fax: 86-21-55271172 e-mail: chenzhaoxue@ 163.com).

References

1. Gu, X., Jiang, Z., Zu, D.: PCI-Based digital frequency source for MRI. Chin. J. Med. Imaging Technol. (2005)
2. Wu, Y.: Software development framework based on the USB 2.0 Slave FIFO mode. Journal of Weinan Teachers University (2007)
3. Jilin, L.: USB2.0 application system development examples. Publishing house of electronics industry, Bei Jing (2006)
4. Li, W., Wu, C.: USB system firmware design program with CY7C68013A. Electronic engineering (2007)

Overview of Low Switching Frequency Control of High Power Three-Level Converters

Xiao Fu*, Peng Dai, Qingqing Yuan, and Xiaojie Wu

School of Information and Electrical Engineering,
China University of Mining and Technology, Xuzhou 221008, Jiangsu, China
fuxiao@ieee.org

Abstract. Converters for high power drives operate at low switching frequency in order to restrain the dynamic losses of the power semiconductor devices and solve problems such as EMI, differential-mode / common-mode voltages and high cooling requirements. However, lower switching frequency will lead to current distortion and reduction in system control band; therefore specific control strategies should be adopted. This paper presents detailed discussion on PWM modulation method, system modeling, current sensing and current control with low witching frequency in the converter. Finally, prospect of low switching frequency control of high power three-level converters is viewed.

Keywords: low switching frequency, three-level converter, optimal PWM, modeling, current sensing, current control.

1 Introduction

Low voltage converters at high switching frequency (>1kHz), have a nice speed adjusting performance by the double closed-loop torque and current control. While they are not suitable for the application of high voltage and high power, such as coal hoist system, locomotive traction and so on. In part because the switching frequency of high power switching devices such as GTO or IGCT is restricted; In other part, the dynamic losses of the converters are becoming large with the high switching frequency of PWM; With the high frequency ups of carrier and high-speed switch of the power devices make the output voltage had a rapid change at a high frequency, which brought a big electromagnetic interference (EMI); And the large differential-mode and common-mode voltage would be a damage for the machine axletree; The interaction between the parasitic capacitance and parasitic inductance may cause a high frequency oscillation at the machine terminal, all of which would reduce the machine life, which also bring a high speed adjusting cost [1,2]. Medium voltage ac machines fed by high-power inverters operate at low switching frequency to restrain the switching losses of the power semiconductor devices, which can improve efficiency of the converters and reduce the command of heat dissipation. Nowadays, during the

* Project Supported by National Natural Science Foundation of China (51077124); Graduate Student Research Innovation Program of Jiangsu Province, China (CX09B_113Z); The Fundamental Research Funds for the Central Universities (2010QNB32);"Qinglan" project of Jiangsu.

X. Wan (Ed.): Electrical Power Systems and Computers, LNEE 99, pp. 877–883.
springerlink.com

application of high voltage and high power, the switching frequency of GTO usually at 200Hz, while IGBT's or GCT's are 500Hz [3].However, this kind of low switching frequency also brought about some new problems, such as high harmonic aberration rate of the output current, low bandwidth of the control system as well as a poor dynamic performance. The reported lowest switching frequency is 150Hz.

The research status and development trend of the low switching frequency control for high-power three-level converters are given in this paper: the modulation of PWM; modeling of PWM converters considering the low switching frequency ; the extraction of current fundamental at low switching frequency and the control strategy.

2 Research of PWM Modulation Method

Fig.1 shows the traditional SVPWM modulation waveforms where the carrier frequency is 50Hz.

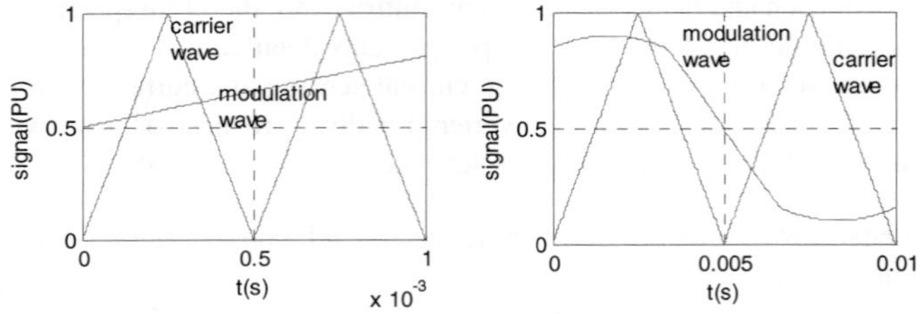

Fig. 1. Modulation and carrier waveforms at different switching frequency. (a) *fs*=2kHz, (b) *fs*=200Hz

From the Fig.1, it can be known that when switching frequency is high relative to the output frequency of the PWM converter, the modulation wave is almost linear during the carrier cycle, the modulation wave signals can be recovered precisely using the PWM converters; When the switching frequency decreased from 2kHz to 200Hz, the modulation wave changed a lot during the carrier cycle, which resulted in a lot of questions such as the common carrier modulation method could not recovery the modulation signal and that would cause a low utilization ratio of the voltage and a high harmonic aberration rate of the output current [2].

To overcome the problem that the current harmonic aberration rate increased when there are limited switching angles , three kinds of PWM modulation methods have been proposed: selective harmonic elimination PWM (SHEPWM) [5]; Synchronous harmonic optimum PWM (SHOPWM) [6];Selective harmonic mitigation PWM (SHMPWM) [7], all of these are belong to the optimal pulse width modulation techniques. However these methods need steady state results, so mostly applied at the *V/f* situation, because it's limited bandwidth and poor dynamic performance [8]. Fast control of the machine torque requires the fundamental component of the stator current, or stator flux, as a feedback signal, such a signal is inherently obtained as part of the modulation algorithm when carrier-based space-vector modulation is used. Optimal pulse width techniques do not offer a comparable feature. As a work-around, the fundamental optimal pulse pattern in actual use.

Most of the overseas manufacturers used some kind of harmonic optimal PWM strategy, using the optimal pulse sequence pre-calculated to control the system. SIEMENS company used optimal pulse sequence (OPP) strategy in its SIMOVERT MV/SINAMICS medium voltage converters, as well as the ACS6000 series converters of ABB company, the TMdrive MV TM-70 series converters of TG/GE company and the MVW01 series converters of WEG-BRAZIL company [9-11].

Compared with the overseas achievement at the domain of low switching frequency converters, there is a litter work about this in China. Tsinghua University has developed the research about the combination of optimal pulse width techniques with the carrier-based space-vector modulation [12]. Now the optimal pulsewidth techniques and the dynamic performance improving for the low switching frequency converters have become a hot issue.

3 Modeling of the PWM Converters at Low Switching Frequency

In the control system design for the traditional converters, the PWM converters are usually replaced by a delay tache. The precondition is that the frequency of PWM modulation carrier was high enough so that the frequency of input signal for the control system changed slowly compared with the PWM switching frequency. While for the high-power low switching frequency drives, low switching frequency makes the delay caused by PWM modulation much bigger, so the common modeling method would not suitable anymore[12].

The PWM converter model has been established when the carrier index is low in [14], which adopted the theory of piecewise linearization that decomposed the nonlinear model into the combination of some linear taches to get the accurate expression of model of PWM converters at low switching frequency.

The mostly used modulation methods are the selective harmonic elimination PWM (SHEPWM) and the synchronous harmonic optimum PWM (SHOPWM).

4 Current Controls at Low Switching Frequency

Torque control is essential for the high-performance speed adjusting system, which come true by controlling the current. The diagram of current control loop is showed as fig.2.

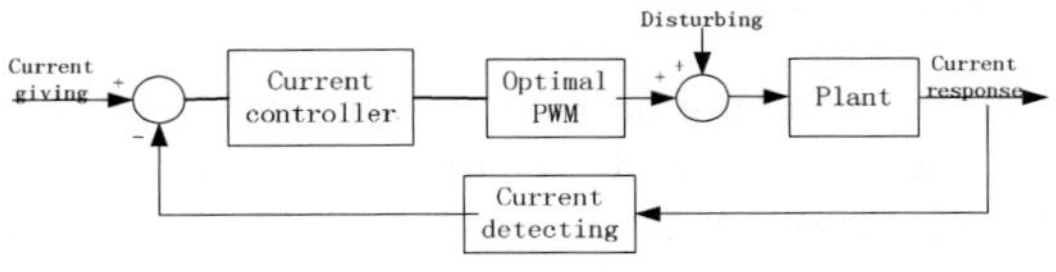

Fig. 2. Diagram of current control loop

Low switching frequency causes the large harmonic aberration rate of the output current, also with some low-order harmonics. From the Fig.2, it can be seen that there two kinds of current control measures: extracting the fundamental component of the load current to make controls or to design current controllers that can restrict the low-frequency interruption effectively.

4.1 Fundamental Current Detecting Method

A fundamental current observer was design based on the error feedback to extract the fundamental [15], this kind of method needs the parameters estimation to improve the observer accuracy.

For the power electronic devices, because of the nonlinear switching devices, there always exists harmonic components. In order to emit the harmonic component, the pre-filter (such as low-pass filter) must be used When A/D converters detecting the analogue quantity. This kind of method would increase the detecting delay.

The A/D converters based on the Σ-Δ theory could realize high accuracy digital output by the closed-loop modulation composed with the differential device, integrator and comparator, the measuring theory is given in Fig.3. It can be come true that the average of harmonic component is zero by choosing appropriate integrator cycle, so that the fundamental or the DC component would be extracted effectively [16, 17]. SIEMENS company uses this kind of technology, such as SIMOVERT MV series converters [18].

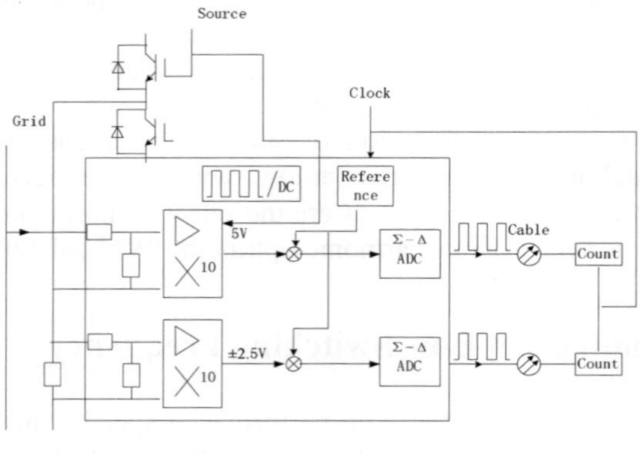

Fig. 3. Voltage and current detecting based on the Σ-Δ theory

4.2 Design of Current Controller

With the reduction of the switching frequency, there are a lot of harmonic components in current feedback signals, as well as some low-order harmonic, which would decreased the tracking performance of traditional PI regulation[19]. However the nonlinear current controller designed based on the modern control theory can work effectively even when at the low switching frequency.

A current regulation based on the combination of the state feedback and sliding mode control, which can meet the harmonic standards when the switching frequency is 500Hz[19].

Predictive control [20] was developed in 1970s. The method based on the combination of repetitive control and predictive control was given in [21], that can be applied for the PWM rectifiers whose switching frequency is 1.2kHz, with the voltage loop using common PI regulation and the current loop using predictive controller. A novel current control strategy was proposed using the model predictive control [22].

Literature [23] pointed out that the common PI regulation has a poor performance of decoupling at the low switching frequency, so it proposed a novel current regulator based on the complex vector, which can realize a nice decoupling.

The stator flux trajectory tracking combined with optimal PWM modulation method can overcome the disadvantages that PWM switch mode must be calculated off-line, which has been used in the MVW01 series converters of WEG-BRAZIL[26].

5 Summary

Medium voltage ac machines fed by high-power inverters operate at low switching frequency to restrain the switching losses of the power semiconductor devices that can improve efficiency of the converters. This paper make an overview of low switching frequency control of high power three-level converters. When the switching frequency decreased, the common SVPWM modulation method could not recovery the modulation signal and which resulted in a lot of questions such as a low utilization ratio of the voltage and a high harmonic aberration rate of the output current. Optimal PWM has been verified that it suits for the high-power low switching frequency domain. However, the traditional optimal PWM modulation method is designed for the steady state, which has a poor dynamic performance, some improvement should be taken. When the switching frequency is low, the accuracy model of PWM converters must be established to be convenient for the control system design. Low switching frequency caused a large harmonic aberration rate of the output current; some novel current observers based on the Σ-Δ theory were proposed to improve the tracking performance of the PI regulator. And some current controllers using the Predictive control, repetitive control were also designed to get a robust performance.

References

1. Bose, B.K.: Power electronics and motor drives: advances and trends. Academic Press, New York (2006)
2. Beliav, D., Weigner, A., Paes, R., et al.: Field oriented control of a synchronous drive. In: 2005 IEEE International Conference on Electric Machines and Drives (2005)
3. Wu, B.: High-power converters and AC drives. John Wiley &Sons, New Yorks (2006)

4. Bocker, J., Janning, J., Je, H.: High dynamic control of a three-level voltage-source-converter drive for a main strip mill. IEEE Transactions on Industrial Electronics 49, 1081–1091 (2002)
5. Patel, H.S., Hoft, R.G.: Generalized techniques of harmonic elimination and voltage control in thyristor inverters: part I-harmonic elimination. IEEE Trans. Ind. Appl. IA-9(3), 310–317 (1973)
6. Buja, G.S., Indri, G.B.: Optimal pulsewidth modulation for feeding AC motors. IEEE Trans. Ind. Appl. A1-13(1), 38–44 (1977)
7. Napoles, J., Portillo, R., Leon, J.I., et al.: Implementation of a closed loop SHMPWM technique for three level converters. In: 34th Annual Conference of IEEE Industrial Electronics IECON 2008, pp. 3260–3265 (2008)
8. Luiz, A.-S.A., Cardoso Filho, B.J.: Sinusoidal voltages and currents in high power converters. In: 34th Annual Conference of IEEE Industrial Electronics IECON 2008, November 10-13, pp. 3315–3320 (2008)
9. ABB Industries AG. ACS 6000SD medium-voltage drives user's manual (2004)
10. Laczynski, T., Werner, T., Mertens, A.: Active damping of LC-filters for high power drives using synchronous optimal pulse width modulation. In: Power Electronics Specialists Conference PESC 2008, June 15-19, pp. 1033–1040. IEEE, Los Alamitos (2008)
11. Laczynski, T., Werner, T., Mertens, A.: Modulation error control for medium voltage drives with LC-filters and synchronous optimal pulse width modulation. In: Industry Applications Society Annual Meeting IAS 2008, October 5-9, pp. 1–7. IEEE, Los Alamitos (2008)
12. Zhang, Y., Zhao, Z., Zhang, Y., et al.: Study on a hybrid method of SVPWM and SHEPWM applied to three-level adjustable speed drive system. In: Proceedings of the CSEE, June 2007, vol. 27, pp. 72–77 (2007)
13. Holtz, J., Oikonomou, N.: Fast dynamic control of medium voltage drives operating at very low switching frequency—an overview. IEEE Transactions on Industrial Electronics 55(3), 1005–1013 (2008)
14. Sakharuk, T.A., Standovic, A.M., Tadmor, G., et al.: Modeling of PWM inverter-supplied AC drives at low switching frequencies. IEEE Transactions on Circuits and Systems-I: Fundamental Theory and Applications 49, 621–630 (2002)
15. Holtz, J., Oikonomou, N.: Estimation of the fundamental current in low-switching-frequency high dynamic medium-voltage drives. IEEE Transactions on Industry Applications 44(5), 1597–1605 (2008)
16. Cortes, P., Rodriguez, J., Antoniewicz, P., et al.: Direct power control of an AFE using predictive control. IEEE Transactions on Power Electronics 23(5), 2516–2523 (2008)
17. Mertens, D.E.: Voltage and current sensing in power electronic converters using sigma-delta A/D conversion. IEEE Transactions on industry Applications 34(5), 1139–1146 (1998)
18. Siemens, A.: SIMOVERT MV catalog (2004)
19. Khambadkone, M., Holtz, J.: Fast current control for low harmonic distortion at low switching frequency. IEEE Transactions on Industrial Electronics 45, 745–751 (1998)
20. Camacho, E., Bordons, C.: Model predictive control. Springer, New Yorks (1999)
21. Jung, S.-L., Huang, H.-S., Tzou, Y.-Y.: A three-phase PWM AC-DC converter with low switching frequency and high power factor using DSP-based repetitive control technique. In: 29th Annual IEEE PESC 1998 Record, May 17-22, vol. 1, pp. 517–523 (1998)
22. Vargas, R., Cortes, P., Ammann, U., et al.: Predictive control of a three-phase neutral-point-clamped inverter. IEEE Transactions on Industrial Electronics 54, 2697–2704 (2007)

23. Holtz, J., Quan, J., Schmitt, G., et al.: Design of fast and robust current regulators for high power drives based on complex state variables. In: 38th IAS Annual Meeting. Conference Record of the Industry Applications Conference, October 12-16, vol. 3, pp. 1997–2004 (2003)
24. Oikonomou, N., Holtz, J.: Stator flux trajectory tracking control for high-performance drives. In: Conference Record of the 2006 IEEE Industry Applications Conference, 41st IAS Annual Meeting, October 8-12, vol. 3, pp. 1268–1275 (2006)
25. Holtz, J., Oikonomou, N.: Synchronous optimal pulsewidth modulation and stator flux trajectory control for medium voltage drives. In: Conference Record of the Industry Applications, Fourtieth IAS Annual Meeting, October 2-6, vol. 3, pp. 1748–1791 (2005)

Electrically Excited Synchronous Motor Reduced-Order Flux Observer

Peng Dai*, Fengchao Fu, Xiao Fu, Weilin Zong, and Erlei Zhou

School of Information and Electrical Engineering,
China University of Mining and Technology, Xuzhou 221008, Jiangsu, China
fuxiao@ieee.org

Abstract. Current model is sensitive to the motor parameters, there is initial value of the integral deviation and DC bias in voltage model, and the integrated model of both current model and voltage model is difficult to achieve. Taking into account these issues, this paper presents a reduced-order flux observer model of electrically excited synchronous motor based on knowledge of modern control theory. Its state equation is derived based on its voltage, current and flux equations, and reduced-order flux observer is designed according to state equation of the electrically excited synchronous motor, the system matrix is derived and the feedback matrix is determined too. Finally, simulation is done for whole system on Matlab / Simulink platform, and the simulation results are analyzed. The flux can be observed accurately in the speed range.

Keywords: Electrically excited synchronous motor, Reduced order observer, Feedback matrix, System matrix, Matlab/Simulink.

1 Introduction

In order to control the electrically excited synchronous motor, the flux must be accurately estimated. The methods of flux observer general used for electrically excited synchronous motor include current, voltage model and comprehensive model of both. When the current model is used to estimate the flux, the flux estimation is an open-loop control in the whole system, and it is sensitive to motor parameters[1]; The voltage model has shortcomings of integrator with initial value and Integral error accumulation, at low speed voltage model is not accurate enough, in addition, the motor speed is zero before the start, EMF has not been established, the output of the integrator can not be determined, therefore, the voltage model can not be used to estimate the air-gap flux when the motor startup or at low speed; a transition model was applied in [3], voltage model and current models are combined by switching, so that flux observer work in the current model at low speed and voltage model is used at high speed. However, both flux

* Project Supported by National Natural Science Foundation of China (51077124); Graduate Student Research Innovation Program of Jiangsu Province, China (CX09B_113Z); The Fundamental Research Funds for the Central Universities(2010QNB32). "Qinglan" project of Jiangsu.

X. Wan (Ed.): Electrical Power Systems and Computers, LNEE 99, pp. 885–892.
springerlink.com

amplitude and flux angle are necessary to take into account when switching, it is very difficult to achieve. in the other hand the output of voltage model passes a high pass filter, both of models built up together through a PI regulator. Although the purpose of the transition can be achieved, however, it is difficult to select the parameters of PI controller.

A reduced-order flux observer for electrically excited synchronous motor is discussed based on the knowledge of modern control theory and the equation of state deduced from the electrically its voltage, current and flux equation in d, q shaft. The design method of flux reduction observer for electrically excited synchronous motor is given by the model of reduced order observer derivation of state equations for electrically excited synchronous motor in [5] and the model of design feedback matrix in [6]. Finally, simulation is done for whole system on Matlab / Simulink platform, and the simulation results are analyzed.

2 Principles of Electrically Excited Synchronous Motor Control

Control block diagram of electrically excited synchronous motor shown in Figure 1.Traditional control method of voltage and current double closed loop is applied to the control system, while the flux and excitation current closed-loop are increased. The system uses SVPWM control. Stator voltage and current are transformed into the voltage and current under the d, q shaft through the park converter, and then send to the flux observer model. Motor speed and rotor position are measured by the encoder and H bridge control method is used in motor excitation input. The flux magnitude, flux angle and load angle output from the flux observer, In order to improve the system dynamic performance, a voltage feed-forward decoupling is added in this system.

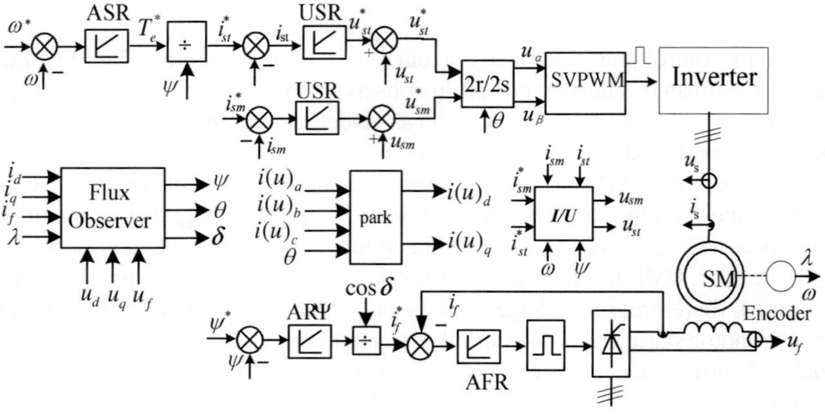

Fig. 1. Electrically excited synchronous motor control block diagram

3 Equation of State of Electrically Excited Synchronous Motor

Equation (1) and (2) are the voltage equations and current equations for electrically excited synchronous motor stator in the *d, q* shaft.

$$\begin{cases} u_{sd} = i_{sd} r_s + p\psi_{sd} - \omega_r \psi_{sq} \\ u_{sq} = i_{sq} r_s + p\psi_{sq} + \omega_r \psi_{sd} \end{cases} \tag{1}$$

$$\begin{cases} -p\psi_{sd} + (L_{sl} + \frac{L_{ad}L_{kdl}}{L_{kd}})pi_{sd} + \frac{L_{ad}L_{kdl}}{L_{kd}} pi_{fd} - \frac{r_{kd}}{L_{kd}}\psi_{sd} + \frac{L_{sd}r_{kd}}{L_{kd}} i_{sd} + \frac{L_{ad}r_{kd}}{L_{kd}} i_{fd} = 0 \\ -p\psi_{sq} + (L_{sl} + \frac{L_{aq}L_{kql}}{L_{kq}})pi_{sq} - \frac{r_{kq}}{L_{kq}}\psi_{sq} + \frac{L_{sq}r_{kd}}{L_{kq}} i_{sq} = 0 \end{cases} \tag{2}$$

Excitation flux, d-axis stator flux equation and the field voltage equation are given below

$$\begin{cases} \psi_f = L_{ad} i_{sd} + L_f i_f + L_{ad} i_{kd} \\ \psi_{sd} = L_{sd} i_{sd} + L_{ad} i_f + L_{ad} i_{kd} \\ u_f = r_f i_f + p\psi_f \end{cases} \tag{3}$$

By (3) can be obtained

$$u_f = p\psi_{sd} - L_{sl} pi_{sd} + L_{fdl} pi_f + r_f i_f \tag{4}$$

Where: $L_{sd}=L_{ad}+L_{sl}$, $L_{sq}=L_{aq}+L_{sl}$, $L_{kd}=L_{ad}+L_{kdl}$, ,$L_{kq}=L_{aq}+L_{kql}$.

Combining equation (1),(2),(4),let $\boldsymbol{x}_1=[\psi_{sd}\psi_{sq}]^{\mathrm{T}}$, $\boldsymbol{x}_2=[i_{sd}\ i_{sq}\ i_f]^{\mathrm{T}}$, $\boldsymbol{u}=[u_{sd}\ u_{sq}\ u_f]^{\mathrm{T}}$, and written as (5) form

$$\begin{bmatrix} \dot{\boldsymbol{x}}_1 \\ \dot{\boldsymbol{x}}_2 \end{bmatrix} = \begin{bmatrix} \boldsymbol{A}_{11} & \boldsymbol{A}_{12} \\ \boldsymbol{A}_{21} & \boldsymbol{A}_{22} \end{bmatrix} \begin{bmatrix} \boldsymbol{x}_1 \\ \boldsymbol{x}_2 \end{bmatrix} + \begin{bmatrix} \boldsymbol{B}_1 \\ \boldsymbol{B}_2 \end{bmatrix} \boldsymbol{u}$$
$$\boldsymbol{y} = [0 \quad \boldsymbol{I}] \begin{bmatrix} \boldsymbol{x}_1 \\ \boldsymbol{x}_2 \end{bmatrix} \tag{5}$$

$$\boldsymbol{A}_{11}=\begin{bmatrix} 0 & \omega_r \\ -\omega_r & 0 \end{bmatrix} \boldsymbol{A}_{12}=\begin{bmatrix} -r_s & 0 & 0 \\ 0 & -r_s & 0 \end{bmatrix} \boldsymbol{A}_{21}=\begin{bmatrix} \frac{L_{fl} r_{kd}}{D} & \frac{\omega_r (L_{ad} L_{kdl} + L_{fl} L_{kd})}{D} \\ -\frac{\omega_r L_{kd}}{Q} & \frac{r_{kq}}{Q} \\ \frac{L_{sl} r_{kd}}{D} & -\frac{L_{kdl} L_{ad} \omega_r}{D} \end{bmatrix}$$

$$\boldsymbol{A}_{22}=\begin{bmatrix} \frac{r_s(L_{ad}L_{kd}+L_{fl}L_{kd})+r_{kd}L_{sl}L_{fl}}{D} & 0 & \frac{L_{ad}(r_{kd}L_{fl}-r_fL_{kd})}{D} \\ 0 & \frac{r_sL_{kq}+L_{sl}r_{kq}}{Q} & 0 \\ \frac{(-r_sL_{ad}L_{kd}+r_{kd}L_{sl}L_s)}{D} & 0 & \frac{r_f(L_{ad}L_{kd}+L_sL_{sl})+r_{kd}L_{ad}L_s}{D} \end{bmatrix} \qquad \boldsymbol{B}_1=\begin{bmatrix} 1 & 0 & 0 \\ 0 & 1 & 0 \end{bmatrix}$$

$$\boldsymbol{B}_2=\begin{bmatrix} \frac{L_{ad}L_{kdl}+L_{fl}L_{kd}}{D} & 0 & -\frac{L_{ad}L_{kdl}}{D} \\ 0 & \frac{L_{kq}}{Q} & 0 \\ -\frac{L_{ad}L_{kdl}}{D} & 0 & \frac{L_{ad}L_{kdl}+L_{sl}L_{kd}}{D} \end{bmatrix} \tag{6}$$

Where : $D=(L_{sl}L_{kd}+L_{ad}L_{kdl})L_{fl}+L_{ad}L_{kdl}L_{sl}$, $Q=(L_{aq}L_{kql}+L_{kq}L_{sl})$.

4 Design of Reduced Order Flux Observer

To construct the D.G.Luenberger Observer of state for $\boldsymbol{x}_1$ according to modern control theory, then using a gain matrix $\boldsymbol{K}$ (2 × 3) of dimension, if the estimated state $\boldsymbol{x}_1$ is denoted by $\widehat{\boldsymbol{x}}_1$, then

$$\begin{aligned} \dot{\widehat{\boldsymbol{x}}}_1 &= \boldsymbol{A}_{11}\widehat{\boldsymbol{x}}_1+\boldsymbol{A}_{12}\boldsymbol{x}_2+\boldsymbol{B}_1u-\boldsymbol{K}(\dot{\boldsymbol{x}}_2-\boldsymbol{A}_{21}\widehat{\boldsymbol{x}}_1-\boldsymbol{A}_{22}\boldsymbol{x}_2-\boldsymbol{B}_2\boldsymbol{u}) \\ &= \boldsymbol{A}_{11}\widehat{\boldsymbol{x}}_1+\boldsymbol{A}_{12}\boldsymbol{x}_2+\boldsymbol{B}_1\boldsymbol{u}+\boldsymbol{K}\boldsymbol{A}_{21}(\widehat{\boldsymbol{x}}_1-\boldsymbol{x}_1) \end{aligned} \tag{7}$$

The observed and theoretical values error equation is given by

$$\dot{\tilde{\boldsymbol{x}}}_1=\dot{\widehat{\boldsymbol{x}}}_1-\dot{\boldsymbol{x}}_1=(\boldsymbol{A}_{11}+\boldsymbol{K}\boldsymbol{A}_{21})\tilde{\boldsymbol{x}} \tag{8}$$

If all the eigenvalues of matrix $\boldsymbol{A}_{11}$+$\boldsymbol{K}\boldsymbol{A}_{21}$ in the left of s plane along the entire speed range that the estimated state $\widehat{\boldsymbol{x}}_1$ approaches the actual state $\boldsymbol{x}_1$ asymptotically and the observer is stable. Therefore, as long as set the feedback matrix $\boldsymbol{K}$ reasonable that can make the system stable. From (7) can be deduced.

$$\begin{aligned} \dot{\widehat{\boldsymbol{x}}}_1 &= \boldsymbol{A}_{11}\widehat{\boldsymbol{x}}_1+\boldsymbol{A}_{12}\boldsymbol{x}_2+\boldsymbol{B}_1\boldsymbol{u}+\boldsymbol{K}(\widehat{z}-z) \\ &= \boldsymbol{A}_{11}\widehat{\boldsymbol{x}}_1+\boldsymbol{A}_{12}\boldsymbol{x}_2+\boldsymbol{B}_1\boldsymbol{u}+\boldsymbol{K}(\boldsymbol{A}_{21}\widehat{\boldsymbol{x}}_1-\dot{\boldsymbol{y}}+\boldsymbol{A}_{22}\boldsymbol{x}_2+\boldsymbol{B}_2\boldsymbol{u}) \\ &= (\boldsymbol{A}_{11}+\boldsymbol{K}\boldsymbol{A}_{21})\widehat{\boldsymbol{x}}_1+(\boldsymbol{A}_{12}+\boldsymbol{K}\boldsymbol{A}_{22})\boldsymbol{x}_2+(\boldsymbol{B}_1+\boldsymbol{K}\boldsymbol{B}_2)\boldsymbol{u}-\boldsymbol{K}\dot{\boldsymbol{y}} \end{aligned} \tag{9}$$

The observer (9) contains a derivative term of the output, i.e., $\dot{\boldsymbol{y}}$, which is unacceptable as it amplifies the noise due to quantization, measurement or otherwise. Hence, a substitution by a dummy variable is used to eliminate the derivative term as follows: $\boldsymbol{\varsigma}=\widehat{\boldsymbol{x}}_1+\boldsymbol{K}\boldsymbol{y}$, then

$$\dot{\boldsymbol{\varsigma}}=(\boldsymbol{A}_{11}+\boldsymbol{K}\boldsymbol{A}_{21})\widehat{\boldsymbol{x}}_1+(\boldsymbol{A}_{12}+\boldsymbol{K}\boldsymbol{A}_{22})\boldsymbol{x}_2+(\boldsymbol{B}_1+\boldsymbol{K}\boldsymbol{B}_2)\boldsymbol{u} \tag{10}$$

The estimated state $\hat{x}_1$ is given by

$$\hat{x} = \varsigma - Ky \tag{11}$$

The block diagram of reduction flux observer for electrically excited synchronous motor can be obtained through the equation (10) and (11), shown in Figure 2.

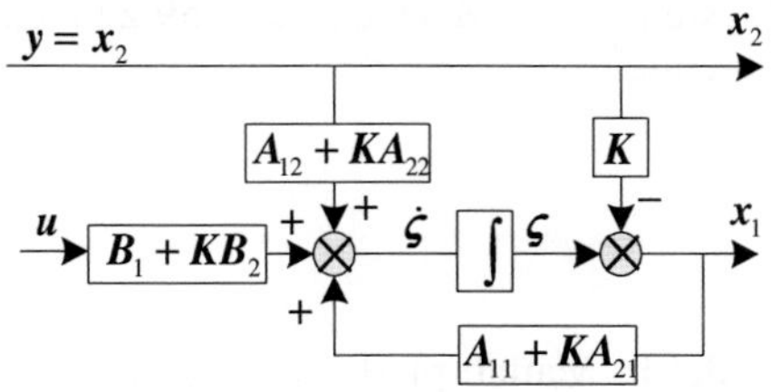

Fig. 2. Electrically excited synchronous motor reduced order observer structure

Deploy the matrix $\boldsymbol{K}$ reasonable so that all eigenvalues of $\boldsymbol{A}_{11}$+$\boldsymbol{KA}_{21}$ completely are in the negative half-plane s plane. In order to meet the requirements of fast, the eigenvalues of $\boldsymbol{A}_{11}$+$\boldsymbol{KA}_{21}$ should be as far as left without affecting the stability of the system. While let the input not affect the stability of the system, it also asked that the input matrix $\boldsymbol{B}_1$+$\boldsymbol{KB}_2$ should be stable. The motor parameters used in simulation shown in tab.1, to make a design for the feedback matrix $\boldsymbol{K}$ according to the motor parameters.

Table 1. Parameters in simulation of synchronous motor

Motor parameters	Electromagnetic parameters	
Power 8.1KW	r_s =1.62Ω	r_f =1.208Ω
Voltage 380V	x_{ad} =0.1086H	x_{aq} =0.05175H
Frequency 50Hz	x_{sl} =0.004527H	x_{fl} =0.01132H
Speed 1500r/m	r_{fd} =3.142Ω	r_{fq} =4.772Ω
Pole pairs 2	x_{fdl} =0.007334H	x_{qfl} =0.01015H

The motor parameters substituted into the matrix $\boldsymbol{A}_{11}$+$\boldsymbol{KA}_{21}$, and find its eigenvalues. In order to make the system stability not change by speed, set a number of items with ω_r and ω_r^2 to zero, for example, K_{12}=0,K_{13}=0,K_{21}=0,K_{23}=0.

$$s^2+(-1916.1K_{11}-5924.4K_{22})s+\omega_r^2(113.6059K_{11}+1)(76.85K_{22}+1)+1916.6K_{11}5924.4K_{22}=0 \tag{12}$$

If the system stability is without the interference for input in $\boldsymbol{B}_1+\boldsymbol{kB}_2$the matrix$113.6K_{11}+1>0$, $76.85K_{12}+1>0$, so

$$\begin{cases} -0.0088 < K_{11} < 0 \\ -0.013 < K_{22} < 0 \end{cases} \tag{13}$$

To meet the requirements of fast, take K_{11}=-0.00,K_{22}=-0.01, substituting into (12) test, the eigenvalues of system matrix $\boldsymbol{A}_{11}+\boldsymbol{KA}_{21}$ are -59.244 and -15.3288, respectively, meet the design requirements.

5 Simulation

Build the simulation model in Matlab platform based on the principle of electrically excited synchronous motor control system with the reduced order flux observer, and use of the motor parameters shown in Table 1, the simulation results shown in Figure 3.

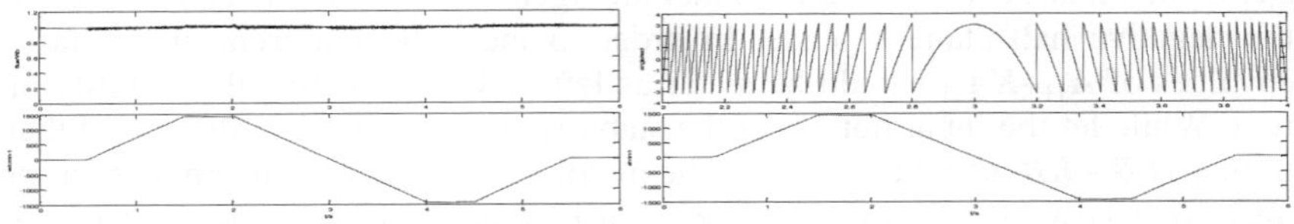

(a) Flux amplitude wave with velocity (b) Flux angle waveforms with velocity

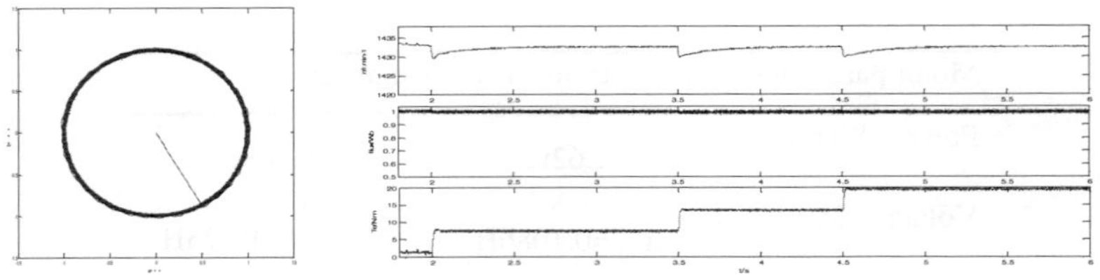

(c) Circular wave of flux (d) Speed and flux waveforms with the load change

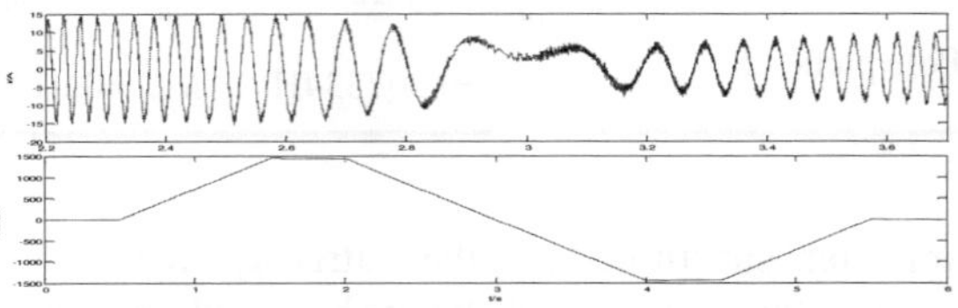

(e) Stator current waveform with the speed change

Fig. 3. Simulation output waveforms

From the output waveform can be seen in Figure (a) that the motor operate in four-quadrant, and speed can closely follow a given. When the speed changed, flux amplitude kept constant, indicating that reduced order flux observer is not affected by the speed.

Figure (b) is the waveform of flux angle during the motor in the transition from forward to reverse, if the motor is turning in positive direction, the flux angle increased by a certain slope cyclically; in opposite, if the motor is turning in reverse direction, the flux angle decrease by a certain slope cyclically. Flux angular frequency is proportional to the speed.

Figure (c) can be seen that the flux linkage circle is a rules circle whose radius is approximately equal to 1 in α, β-axis. Indicating that the flux sine is excellent and the amplitude remains constant in α, β-axis.

Figure (d) show the speed and flux amplitude waveforms when the load changing, from the figure can be seen that at the rated speed, the speed is only with a little change when motor load changed, and recover quickly, the flux amplitude almost independent to the motor load changed. So if in the rated load range, the robustness of the speed and flux for the motor load is good.

Figure (e) is the current waveform when the motor in the transition from forward to reverse, the current sine is excellent and the frequency increases with increasing speed can be seen from the figure (e).

6 Conclusion

This paper has designed a reduced-order flux observer for electrically excited synchronous motor based on the knowledge of modern control theory and its equation of state which deduced from its voltage, current and flux equation in d, q shaft. The system state observer system matrix $\boldsymbol{A}_{11}+\boldsymbol{K}\boldsymbol{A}_{21}$ has obtained too. Used the pole placement method of modern control theory, the feedback matrix $\boldsymbol{K}$ has designed according to system stability and fast requirements.

Finally, the simulation model of electrically excited synchronous motor vector control system has built in Matlab based on previous theory and the reduced order flux observer, and simulation output waveforms were ideal. Through analysis to the simulation output waveform, the reduction flux observer for electrically excited synchronous machine stabilizes the output flux quickly, and the output of the flux observer sine is excellent, amplitude fluctuation is small in a full-speed range. This is a great improvement on the traditional electrically excited synchronous motor flux observer model.

References

1. Boldea, I., Andreescu, G.D., Rossi, C.: Active Flux Based Motion-Sensorless Vector Control of DC-Excited Synchronous Machines. In: Energy Conversion Congress and Exposition, ECCE 2009, pp. 2496–2503. IEEE, Los Alamitos (2009)
2. Incze, I.I., Szabo, C., Imecs, M.: Flux Identification for Vector control of the synchronous motor drives. In: IEEE International Conference on 2008, vol. 2, pp. 105 – 110 (2008)

3. Andreescu, G.D., Pitic, C.I., Blaabjerg, F., Boldea, I.: Combined Flux Observer With Signal Injection Enhancement for Wide Speed Range Sensorless Direct Torque Control of IPMSM Drives. IEEE Transactions on Energy Conversion 23(2), 393–402 (2007)
4. Wu, X., Tan, G., Liu, M., Li, H.: Electrically Excited Synchronous Motor Three-Level DTC_SVM Control Based on Novel Flux Observer. In: Electrical and Control Engineering (ICECE), pp. 3689–3692 (2010)
5. Das, S.P., Chattopadhyay, A.K.: Observer-Based Stator-Flux-Oriented Vector Control of Cycloconverter-Fed Synchronous Motor Drive. IEEE Transactions On Industry Applications 4(33), 943–955 (1997)
6. Behera, S., Das, S.P., Doradla, S.R.: A Novel Quasi-Resonant Inverter for High Performance Induction Motor Drives. In: Eighteenth Annual IEEE APEC 2003, vol. 2(2), pp. 819–825 (2003)
7. Xu, J., Xu, Y., Xu, J.: Direct Torque Control of Permanent Magnet Synchronous Machines Using Stator Flux Full Order State Observer. In: IEEE International Symposium on Industrial Electronics, vol. 11(2), pp. 913–916 (2004)
8. Hasegawa, M., Matsui, K.: Position sensorless control for interior permanent magnet synchronous motor using adaptive flux observer with inductance identification. IET Electric Power Applications 3(3), 209–217 (2009)
9. Koonlaboon, S., Sangwongwanich, S.: Sensorless Control of Interior Permanent-Magnet Synchronous Motors Based on A Fictitious Permanent-Magnet Flux Model. In: Fourtieth IAS Annual Meeting, vol. 1(1), pp. 311–331 (2005)

Dead-Time Compensation of Three-Level Optimal PWM Based on On-Line Switching Angle Adjustment*

Xiao Fu, Peng Dai, Xiaojie Wu, and Bingjie Zhao

School of Information and Electrical Engineering, China University of Mining and Technology, Xuzhou 221008, Jiangsu, China
fuxiao@ieee.org

Abstract. The optimal PWM method is widely in high power drive systems for its capability of achieving low output distortion at low switching frequency. However, the PWM dead-time inserted into PWM pulses for avoiding "shoot through" of inverter bridge arms will introduce modulation errors between ideal optimal PWM patterns and actual outputs. This problem is extremely severe in high power drives for the large dead time, and will cause large distortion of the optimal PWM output waveforms. This paper took the selective harmonic PWM (SHEPWM) as an example. First, the basic principles of SHEPWM and dead-time effects were presented. By analyzing the polarity of three level inverter output currents and switching behaviors of the inverter, a dead-time compensation method based on on-line switching angle adjustment for the optimal PWM method is proposed. Finally, the experimental results were given for verifying the effectiveness of the proposed scheme.

Keywords: optimal PWM, dead-time compensation, on-line switching angle adjustment, three-level.

1 Introduction

With the development of power electronics and control technology, the three-level frequency converter is widely used in high power speed adjustment systems [1-3]. The PWM switching frequency is limited to low values to restrain the switching losses [4], normally below 500Hz [5, 6]. This will cause intolerable output harmonics to traditional carrier modulation based PWM methods, so the optimal PWM methods based on direct switching angle modulation which can produce satisfying output waveforms at low switching frequency are becoming focus of study [7-11].

However, the PWM dead time needed to avoid the shoot-through phenomena will distort the optimal PWM output. This becomes severely in in high power drive systems, since the dead time needed for high power electronics device can increase to tens of microseconds [5]. Current research on dead-time effect compensation is mainly

* Project Supported by National Natural Science Foundation of China (51077124); Graduate Student Research Innovation Program of Jiangsu Province, China (CX09B_113Z); The Fundamental Research Funds for the Central Universities (2010QNB32);"Qinglan" project of Jiangsu.

X. Wan (Ed.): Electrical Power Systems and Computers, LNEE 99, pp. 893–898.
springerlink.com

focused on the carrier base PWM, generally named as the error voltage correction method and the pulse edge modification method [12]. The former method detects the error vectors between the reference voltage and the actual voltage and compensates them [13, 14]. It's easy to implement, while the compensation accuracy is not satisfying because of various nonlinear factors. The latter method affine the PWM pulse edges based on current polarity detection and the switching condition. It can achieve satisfying compensation results, while the current polarity detection is difficult. It is pointed out in [15] that the dead-time will cause the optimal PWM to produce large low-order harmonics, thus severely degrade the waveform quality. There are rarely discussions in literatures on the dead-time effect compensation method for optimal PWM.

This paper is organized as follows. First, the selective harmonic elimination method, which is one of the most commonly used optimal PWM, is introduced with its basic principle. Based on the analyzing of three-level output current polarity and optimal PWM switching angles, a dead-time compensation method based on on-line switching angle adjustment for the optimal PWM method is proposed. Finally, the experimental results are given for verifying the effectiveness of the proposed scheme.

2 Basic Principles of SHEPWM

The main circuit diagram of neutral-point diode clamped three level inverter is shown in Fig.1. Fig. 2 gives the waveform that the power electronic devices of phase A switching *N* times in a fundamental period.

Based on the Dirichlet principle, phase A voltage can be presented by following Fourier series

$$v(t) = \frac{a_0}{2} + \sum_{n=1}^{\infty}\left(a_n \cos n\omega_0 t + b_n \sin n\omega_0 t\right),\ n = 0,1,2,\cdots \tag{1}$$

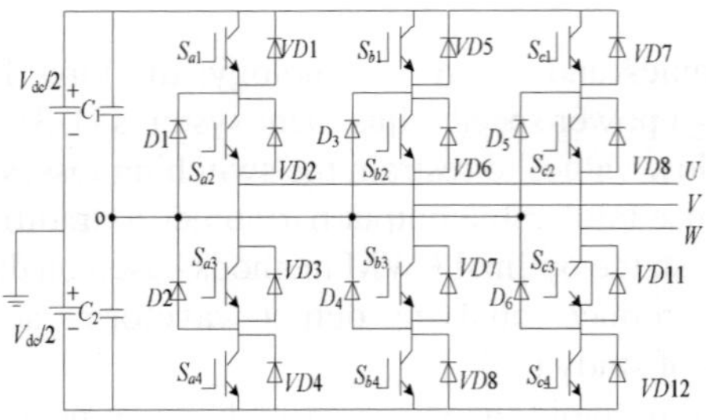

Fig. 1. The main circuit of three-level inverter

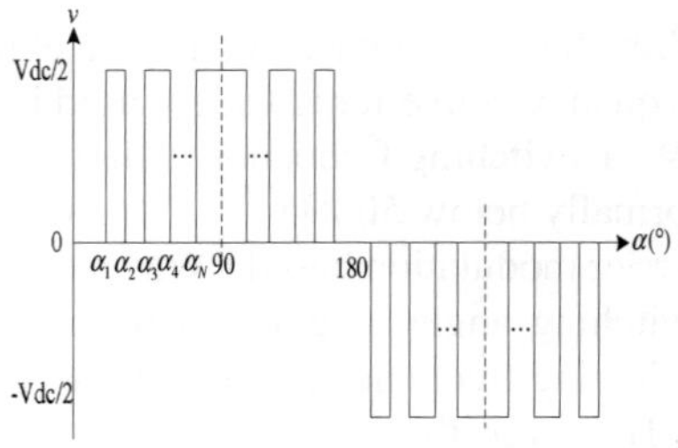

Fig. 2. The output phase voltage waveform

Where, ω is fundamental voltage angular frequency; $a_n = \frac{2}{T}\int_{-T/2}^{T/2} v(t)\cos n\omega t dt$; $b_n = \frac{2}{T}\int_{-T/2}^{T/2} v(t)\sin n\omega t dt$.

Usually the phase voltage is made to be mirror symmetry to π and even symmetry to π/2 to eliminate even order harmonics and dc elements of inverter output voltage. Equation (1) can be simplified as

$$v(t)=\sum_{n=1}^{\infty}(b_n \sin n\omega_0 t),\quad (n=1,3,5,\cdots) \tag{2}$$

Where, $b_n=\frac{4E}{n\pi}\sum_{n=1}^{\infty}(-1)^{k+1}\cos n\alpha_k$; k is the index of switching angles. Considering that the 3rd harmonics does not exist in the output line voltage in the structure shown in Fig. 1, there are

$$\begin{cases} b_1=\dfrac{4V_{dc}}{\pi}\sum_{k=1}^{N}(-1)^{k+1}\cos\alpha_k \\ b_k=\dfrac{4V_{dc}}{n\pi}\sum_{k=1}^{N}(-1)^{k+1}\cos n\alpha_k=0 \end{cases}. \tag{3}$$

Where, $4V_{dc}/\pi$ is the phase voltage amplitude.

Take N =7 as an example. Now 6 harmonics elements of phase voltage can be eliminated. The left eliminated harmonics are then the 23rd, 25th, 29th, etc. Considering that the orders of left harmonics are high and already known, an output filter can be easily designed to filter out them. Solving equation (3), then the SHEPWM switching angles will be obtained [16].

3 The Proposed Method

According to the Fig. 1 and taking the phase A voltage as an example, the switching state s is defined in Table 1.

Table 1. Definition of switching state

s	P	O	N
Description	S_{a1} and S_{a2} are on S_{a3} and S_{a4} are off	S_{a2} and S_{a3} are on S_{a1} and S_{a4} are off	S_{a3} and S_{a4} are on S_{a1} and S_{a2} are off

The single edge PWM dead-time mode is used here. That's to say, the IGBT state transition from the "off" state to "on" state will be delayed. To restrain the switching frequency, the transition of s can only occur between P and O, or O and N, while the transition between P and N is prohibited. Defining the current polarity that the current flows from the inverter to the load is positive, and take the i_a>0 as an example, then

a) When s switches from P to O (meaning that the switching angles are changing), the power electronics device Sa2 keeps its state, while Sa1 turns off and Sa3 takes on. In order to prevent shoot-through, the rising edge of pulse of Sa3 needs to be delayed a dead time Dt. After the switching, the current will flows through D1 and Sa2. It's shown that the current does not flow through Sa3, so the dead time of Sa3 will not impact the inverter output current quality. Relating to the adjusting of switching angles, we can conclude that increasing the switching angle of Sa3 Dt/ω_1 rad, the output current is not affected by the dead time effect.

b) When *s* switches from O to P, the power electronics device Sa2 keeps its state, while Sa3 turns off and Sa1 takes on. In order to prevent shoot-through, the rising edge of pulse of Sa1 needs to be delayed a dead time *Dt*. After the switching, the current will flows through Sa1 and Sa2. It's shown that the current does not flow through Sa3, so the dead time of Sa3 will not impact the inverter output current quality. Relating to the adjusting of switching angles, we can conclude that decreasing the switching angle of Sa3 Dt/ω_1 rad, the output current is not affected by the dead time effect.
c) When *s* switches from O to N, the power electronics device Sa3 keeps its state, while Sa2 turns off and Sa4 takes on. In order to prevent shoot-through, the rising edge of pulse of Sa4 needs to be delayed a dead time *Dt*. After the switching, the current will flows through VD3 and VD3. It's shown that the current does not flow through Sa4, so the dead time of Sa4 will not impact the inverter output current quality. Relating to the adjusting of switching angles, we can conclude that increasing the switching angle of Sa4 Dt/ω_1 rad, the output current is not affected by the dead time effect.
d) When *s* switches from N to O, the power electronics device Sa3 keeps its state, while Sa4 turns off and Sa2 takes on. In order to prevent shoot-through, the rising edge of pulse of Sa2 needs to be delayed a dead time *Dt*. After the switching, the current will flows through D2 and Sa2. It's shown that the current does not flow through Sa4, so the dead time of Sa4 will not impact the inverter output current quality. Relating to the adjusting of switching angles, we can conclude that decreasing the switching angle of Sa4 Dt/ω_1 rad, the output current is not affected by the dead time effect.

For conditions of i_a<0 and the switching angles adjustment of the other two phases, the same analyzing and dead-time compensation method can be used.

4 Experimental Results

After the theoretical analysis， the experimental platform is set up to verify the compensation effect for the seven-section SHEPWM. The IGBTs are SK50MLI066 three-level IPM from SEMIKRON. The current sensors are LA28-NP from LEM. The TMS320F28335 high performance digital signal processor from TI is used to implement the SHEPWM, with a 20kHz sampling frequency and 50Hz inverter output frequency. The XC3S400 FPGA from Xilinx is used for 10μs dead-time generation and fault protection. The inverter load is three-phase series connected 5Ω/50W resistor and 5mH inductor. The experimental results are obtained with Fluke 43B power quality analyzer, as shown in Fig. 3.

From the analytical above, we know that the output of ideal SHEPWM should have no low order harmonics. However, the compensated dead time will cause the 5th, the 7th, etc. harmonics to the output current. After compensation with the proposed method, it's shown that the current total harmonics distorting factor decreases from 7% to 6.3%. The experimental results verify the dead-time compensation function of the proposed method.

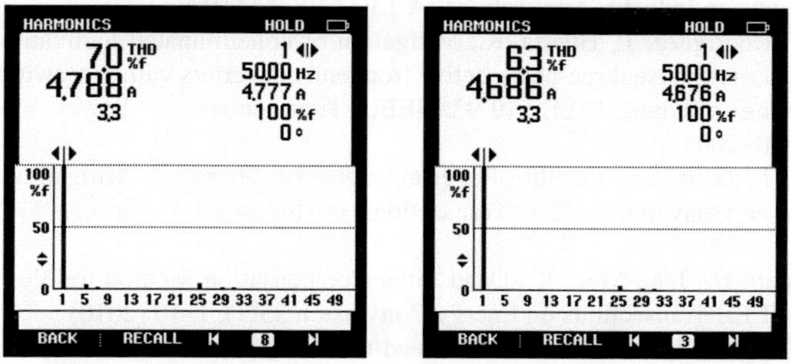

(a) Before Compensation (b) After Compensation

Fig. 3. Inverter Output Current distorting before and after Compensation

5 Conclusion

The dead-time compensation for optimal PWM with three-level inverter is researched. Taking the typical SHEPWM as an example, the switching angles are adjusted according to the PWM dead-time and the inverter output current. Experimental results are given for verification. The proposed is also suitable for other optimal PWM methods, since they are all impended based on direct switching angle modulation. Future works can be done on the accuracy of current polarity detection.

References

1. Abu-Rub, H., Holtz, J., Rodriguez, J., et al.: Medium voltage multilevel converters - state of the art, challenges and requirements in industrial applications. IEEE Transactions on Industrial Electronics 57(8), 2581–2596 (2010)
2. Rodriguez, J., Bernet, S., Steimer, P.K., et al.: A Survey on Neutral-Point-Clamped Inverters. IEEE Transactions on Industrial Electronics 57(7), 2219–2230 (2010)
3. Siemens, A.G.: Power semiconductors: for medium voltage converters - an overview. In: 13th European Conference on Power Electronics and Applications (2009)
4. Holtz, J., Oikonomou, N.: Fast dynamic control of medium voltage drives operating at very low switching frequency - an overview. IEEE Transactions on Industrial Electronics 55(3), 1005–1013 (2008)
5. Wu, B.: High-power converters and AC drives, p. 333. Wiley, Hoboken (2006)
6. Bose, B.K.: Power electronics and motor drives: advances and trends, p. 917. Elsevier/Academic Press, Amsterdam (2006)
7. Flourentzou, N., Dahidah, M., Agelidis, V.G.: On distributing multilevel SHE-PWM waveforms in HVDC systems built with conventional three-phase VSC modules. In: IEEE Power Electronics Specialists Conference (2008)

8. Watson, A.J., Wheeler, P.W., Clare, J.C.: A phase shift selective harmonic elimination method for balancing capacitor voltages in a seven level cascaded H-bridge rectifier. In: 13th European Conference on Power Electronics and Applications (2009)
9. Buja, G.S., Indri, G.B.: Optimal pulsewidth modulation for feeding AC motors. IEEE Transactions on Industry Applications IA 13(1), 38–44 (1977)
10. Pontt, J., Rodriguez, J., Huerta, R.: Mitigation of noneliminated harmonics of SHEPWM three-level multipulse three-phase active front end converters with low switching frequency for meeting standard IEEE-519-92. IEEE Transactions on Power Electronics 19(6), 1594–1600 (2004)
11. Napoles, J., Leon, J.I., Portillo, R., et al.: Selective Harmonic Mitigation Technique for High-Power Converters. IEEE Transactions on Industrial Electronics 57(7), 2315–2323 (2010)
12. Seon-Hwan, H., Jang-Mok, K.: Dead Time Compensation Method for Voltage-Fed PWM Inverter. IEEE Transactions on Energy Conversion 25(1), 1–10 (2010)
13. Munoz, A.R., Lipo, T.A.: On-line dead-time compensation technique for open-loop PWM-VSI drives. IEEE Transactions on Power Electronics 14(4), 683–689 (1999)
14. Yong-Kai, L., Yen-Shin, L.: Dead-time elimination method and current polarity detection circuit for three-phase PWM-controlled inverter. In: Energy Conversion Congress and Exposition (2009)
15. Khaligh, A., et al.: Dead-Time Distortion in Generalized Selective Harmonic Control. IEEE Transactions on Power Electronics 23(3), 1511–1517 (2008)
16. Fu, X., Wu, X., Zhao, B., et al.: Digital implement of selective harmonic elimination based on digital signal processor. In: Asia Pacific Conference on Postgraduate Research in Microelectronics & Electronics (2009)

FPGA Based Real-Time Emulation of Single Phase H-Bridge Inverter

Peng Dai*, Hongshun Zhu, Xiao Fu, and Guangzhou Wang

School of Information and Electrical Engineering,
China University of Mining and Technology, XuZhou 221008, Jiangsu Province, China

Abstract. Traditional simulation software such as MATLAB/SIMULINK has shortcomings of low speed and high demand for computer hardware. This paper presents an FPGA based real-time simulation for power electronic devices. An SPWM signal generator, an H-Bridge inverter simulator and a RL load simulator are realized on one FPGA chip, thus creating an H-Bridge inverter system. A DAC is used to show the simulation waveforms. The proposed simulation system is realized using a XC3S500E FPGA from Xilinx, Inc. Experimental results are then presented for verification.

Keywords: Real-Time Simulation, H-Bridge Inverter, Field Programmable Gate Arrays (FPGA).

1 Introduction

Real-time simulator is becoming more and more popular in AC/DC motor drive and power system [1]-[2]. IGBT based power electronic apparatus can be modeled using two types of simulation tools: system level and device level [3]. Power electronic devices in system-level based simulation software, such as MATLAB/SIMULINCK, are often modeled using three types of behavior models: ideal model, switching function model and average model [4]. System-level modeling is often fast, however, it does not take the device nonlinear characteristics into account. Thus, power losses and thermal characteristics of devices can't be estimated. SPICE and SABER simulation software use device-level models, thereby, the accuracy is improved. Although device-level models are very detail, it can be very time consuming in AC/DC motor drive simulation, for it should take the power devices nonlinear characteristics and the complexity of motor models into account.

While there are some shortcomings in traditional simulation software, real-time simulation system is developing rapidly. Due to their high clock speed and parallel hardwired architecture, FPGA are becoming more and more popular when implementing computationally algorithms in real-time simulator [5].

This paper is organized with the following sections. First, the IGBT typical turn on and turn off behaviors and the math model of RL load circuit are introduced. Detailed

* The Fundamental Research Funds for the Central Universities(2010QNB32).

X. Wan (Ed.): Electrical Power Systems and Computers, LNEE 99, pp. 899–905.
springerlink.com

FPGA based implement of a single phase H-Bridge inverter is discussed afterward. Experimental results are then provided for verification.

2 Analysis of IGBT and RL Load Circuit

2.1 Analysis of IGBT Switching Characteristics

Switching characteristics of IGBT should be considered in device-level based real-time simulation system. In this paper, switching characteristics of IGBT IGW60T120 from Infineon is modeled. Fig.1 shows the typical turn on and turn off behaviors of IGW60T120 [6]. Before the real-time simulator starts to work, the curves in Fig.1 are stored in block RAMs after sampling and quantization. When simulator is working, the values in block RAMs will be read one by one, thus modeling the switching characteristics of IGBT.

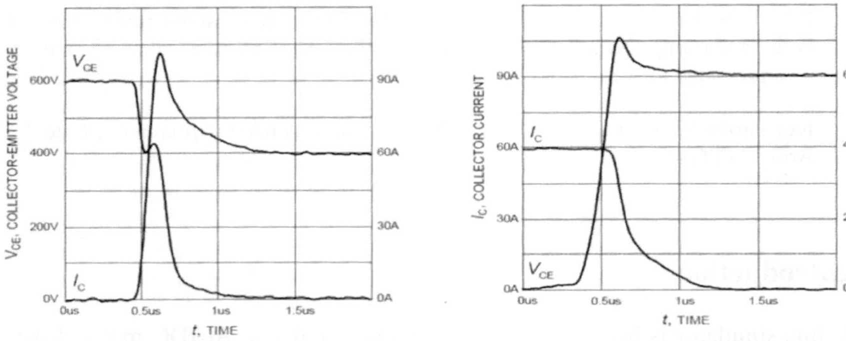

(a)Typical turn on behavior (b)Typical turn off behavior

Fig. 1. IGBT Switching Characteristics

2.2 Analysis of RL Load Circuit

The basic RL load circuit is shown in Fig.2.

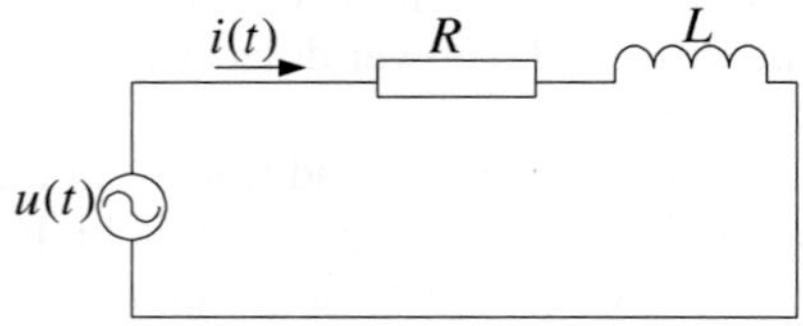

Fig. 2. RL Load Circuit

In a real-time simulation system, the values of R, L and $u(t)$ can be determined first, only the value of $i(t)$ should be calculated. According to Fig.2, we can obtain

$$L\frac{di(t)}{dt}+Ri(t)=u(t) \quad (1)$$

In order to calculate $i(t)$ by a digital processor, (1) should be converted to its corresponding difference equation. First, converter (1) to its Laplace transformation, as shown below

$$(sL+R)I(s)=U(s) \quad (2)$$

According to bilinear transformation theory, change s in equation (2) to $2(z-1)/(z+1)/T$, then obtain the corresponding discrete equation.

$$(z+\frac{RT-2L}{RT+2L})I(z)=\frac{T}{RT+2L}(z+1)U(z) \quad (3)$$

Where, T is the sample period.

From equation (3), we can get the difference equation.

$$i(k)=\frac{2L-RT}{RT+2L}i(k-1)+\frac{T}{RT+2L}[u(k)+u(k-1)] \quad (4)$$

In real-time simulation system, the current of RL load circuit can be calculated according to equation (4).

3 FPGA Based Implement

3.1 Overall Tasks

A single phase H-Bridge inverter with RL load is modeled in an FPGA chip. And an SPWM generator is also modeled to drive the four IGBTs. Then calculate the load current according the output voltage of the inverter. Finally, the voltage and current curves will be displayed by a DAC. Thus, the whole system can be divided into four main parts.

1) SPWM generator module
2) Realization of the structure of H-Bridge inverter
3) RL load current calculation module
4) DAC control logic module

The structure of the whole system is shown in Fig.3.

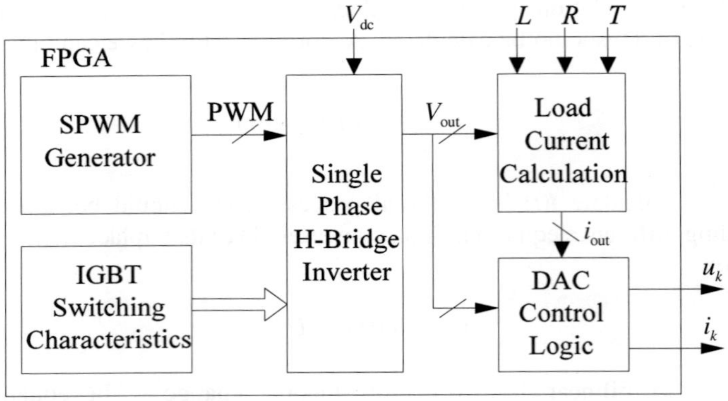

Fig. 3. FPGA Based System Structure

3.2 Implementation of SPWM Generator

An SPWM generator contains three main components: sine wave generator module; triangle wave generator module and a comparator. First, store sine wave in block RAM, and when simulating, read the values in corresponding RAM, thus generate a sine wave. Triangle wave generator can be realized by an up-down counter. Finally, compare the values of sine wave and triangle wave, thus the PWM wave can be generated.

3.3 Implementation of H-Bridge Inverter

(1) Implementation of IGBT switching characteristics

First, sample the values of IGBT turn on and turn off curves shown in Fig.1 with the interval 0.2ns. Then, get the values quantified to 12 bits, and store them in FPGA block RAMs. When simulating, the values will be read with the frequency of 50MHz, thus the IGBT switching characteristics can be modeled.

(2) Implementation of H-Bridge

The structure of a single phase H-Bridge is shown in Fig.4.

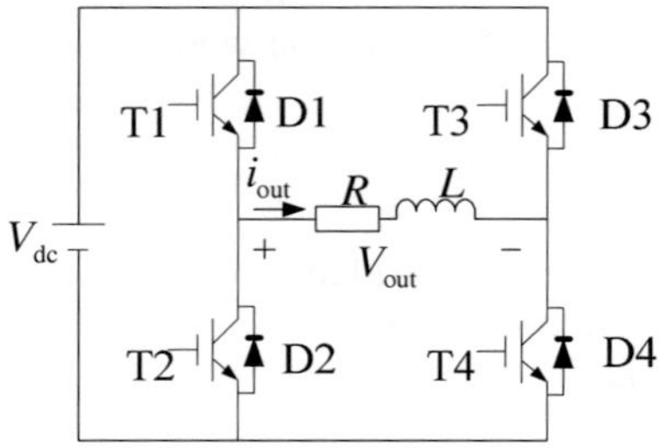

Fig. 4. The structure of H-Bridge

In Fig.4, T1 and T4 use the same triangle pulse, while T2 and T3 use another one. The two IGBTs in the same bridge have three effective states: 10, 00 and 01. Where 1 indicates the IGBT is on and 0 indicates the opposite.

The switch state and the value of load current are used to determine the inverter output voltage. For example, when the gate signals for T1, T2, T3 and T4 are 1, 0, 0, and 1 respectively. If the load current i_{out} is positive, the output voltage $V_{out}=V_d-2V_{ce}$. Similarly, the relationship between V_{out} and i_{out} is shown in table1 [7].

Table 1. Relations between V_{out} and i_{out} at different switching states

T1 T2 T3 T4	$i_{out}>0$	$i_{out}<0$
1001	$V_{out}=V_{dc}-2V_{ce}$	$V_{out}=V_{dc}+2V_d$
0000	$V_{out}=-V_{dc}-2V_d$	$V_{out}=V_{dc}+2V_d$
0110	$V_{out}=-V_{dc}-2V_d$	$V_{out}=V_{dc}-2V_{ce}$

Where, V_d is the forward voltage drop of diodes.

3.4 Implementation of Load Current Calculation Module

There are two choices when using an FPGA to calculate the load current: one is traditional HDL design method, the other is system based design method. Because equation (4) refers to float multiple, divide and add function, using HDL design method to realize these functions has difficulty in coding, debugging and other shortcomings. In this paper, a system based design tool System Generator is used to calculate the load current.

System Generator provide the interface with MATLAB/SIMULINK, thus parameters modification is very flexible. Fig.5 shows the System Generator based structure of load current calculation module.

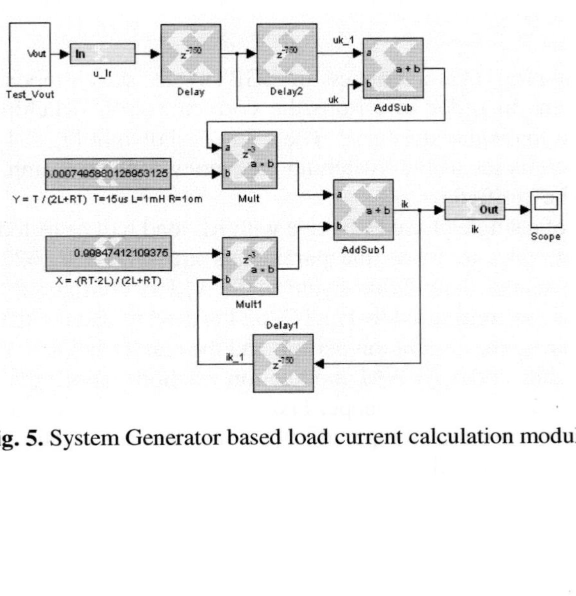

Fig. 5. System Generator based load current calculation module

The parameters in Fig.5 are set as T=15μs, L=1mH and R=1Ω. And these parameters can be changed when needed.

4 Experimental Results

In order to realize the proposed single phase H-Bridge inverter, an S3EStarter_ug320 Board form Xilinx,inc is selected. The XC3S500E FPGA on board is used to store the switching characteristics of IGBT, to realize the structure of H-Bridge inverter, and to calculate the load current. The DAC LTC2624 on board is used to display waveforms of load voltage and current. Experimental results are shown below.

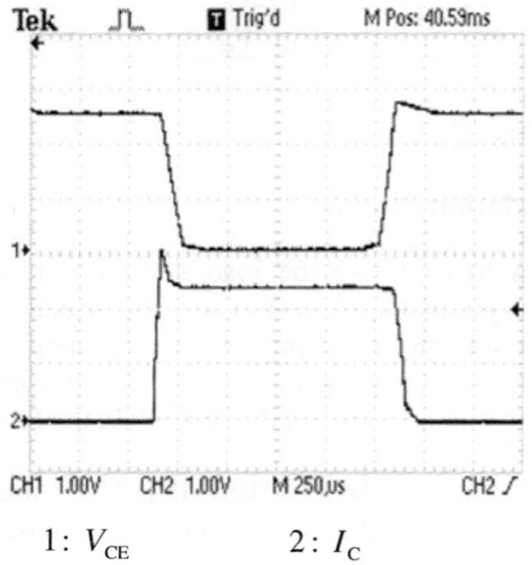

Fig. 6. Characteristics of IGBT

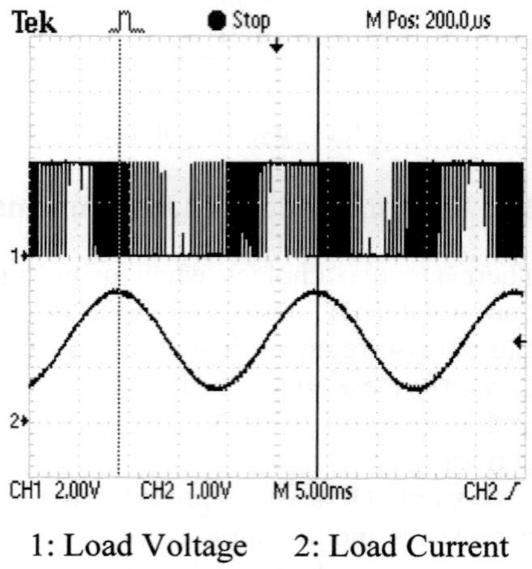

Fig. 7. Voltage and current waves of RL load

Because switching characteristics of IGBTs are very important in real-time simulation system. In order to ensure the correctness of switching characteristics, LTC2624 is used to display the curve. The result is shown in Fig.6. Fig.6 shows IGBT switching characteristics at the switching frequency 500Hz. Compared to Fig.1, the correctness can be verified.

The function of a single phase H-Bridge with RL load will be tested next. Before this real-time system starts to work, the parameters are set as: V_{dc}=220V, L=1mH and R=1Ω. In addition, the modulation cycle of SPWM is configured to 50Hz, and the modulation degree is configured to 1. LTC2624 is used to display the load voltage and current waves during the simulation period, and the results are shown in Fig.7.

Fig.7 shows that under SPWM modulation method action, the RL load current waveform of an H-Bridge is sine shape. The amplitude of the sine waveform represents the value of load current. Compared to theory, the correctness of this real-time simulation system is verified.

5 Conclusion

This paper presents detailed steps of how to set up an FPGA based real-time simulation system. And then a single phase H-Bridge inverter is modeled. Experimental results are then provided to verify the correctness of this real-time simulation system. It is believed that FPGA based real-time emulation will become a popular component in real-time simulation.

References

1. Li, H., Steurer, M., Woodruff, S., Shi, K.L., Zhang, D.: Development of a unified design, test, and research platform for wind energy systems based on hardware-in-the-loop real time simulation. IEEE Trans. Ind Electron 53(4), 1144–1151 (2006)
2. Lu, B., Wu, X., Figueroa, H., Monti, A.: A low-cost real-time hardware in-the-loop testing approach of power electronics controls. IEEE Trans. Ind. Electron 54(2), 219–931 (2007)
3. Gole, A.M., Keri, A., Nwankpa, C., Gunther, E.W., Dommel, H.W., Hassan, I., Marti, J.R., Martinez, J.A.: Guidelines for modeling power electronics in electric power engineering applications. IEEE Trans. Power Del 12(1), 505–514 (1997)
4. Jin, H.: Behavior-mode simulation of power electronic circuits. IEEE Trans. Power Electron 12(3), 443–452 (1997)
5. Chen, Y., Dinavahi, V.: FPGA-based real-time EMPT. IEEE Trans. Power Del 24(2), 892–902 (2009)
6. Infineon. IGW60T120 (2006)
7. Myaing, A., Dinavahi, V.: FPGA-Based Real-Time Emulation of Power Electronic System With Detailed Representation of Device Characteristics. IEEE Trans. On industrial electronics 58(1), 358–368 (2011)

Research on Viewing Zones of Autostereoscopic Display System Based on Lenticular Lens

Jing Yin[1,2,3], Yao-hui Hu[1,3], Guo-qiang Lv[1,3], and Juan Wu[1,2,3]

[1] Key Laboratory of Special Display Technology, Ministry of Education, Hefei Anhui, China
[2] School of Instrument Science and Opto-electronic Engineering, Hefei University of Technology, Hefei Anhui, China
[3] Academe of Opto-electronic Technology, Hefei University of Technology, Hefei Anhui, China
yinjing1004@163.com, momofarm@sina.com, guoqianglv@hotmail.com, wujuan8989@126.com

Abstract. Multi-view stereoscopic display systems based on lenticular lens work as naked display systems with the principle of lenticular splitting, forming viewing zones(VZ) in front of display. To improve the 3-D effect of the display system, the forming cause and the sensitive parameters of VZ need to be studied and analyzed, which are the optical parameters of the system. In this paper, the relations between viewing-zone and the optical parameters of lenticular focal length, sub-pixel etc. are discussed. Quality value is adopted to be a parameter of evaluating 3-D effect, in addition, the corresponding formulas and diagrams are given. Finally, using the above results, a 47-inch autostereoscopic display based on lenticular lens is produced, the above theory is verified. The experiment shows that the formula obtained is consistent with the test data.

Keywords: viewing zone, lenticular lens, central viewing zone, multiview image, sub-pixel.

1 Introduction

The naked eye stereoscopic display technology based on binocular parallax is one of the main research directions[1]. The primary principle mainly concentrated in two ways of parallax barrier and lenticular sheet. And lenticular display technology can achieve a high degree of more viewpoints and more viewers watching simultaneously, so it become a research hotspot[2].

For a two-view lenticular display, there is a 50 percent chance the viewer will receive the wrong stereo pairs, leading to perceive a stereoscopic image[3], Because of these limitations, researchers have been studying multi-view autostereoscopic display, the viewing region range of which is wide[4]. However, not all of the stereo image effects observed at regions in front of the screen are perfect, so the research on the viewing range is important. And the viewing range and display system parameters are closely related, thus, the research on the sensitivity parameters of viewing range is very helpful to enhance and improve the lenticular display system.

X. Wan (Ed.): Electrical Power Systems and Computers, LNEE 99, pp. 907–914.
springerlink.com

2 Forming of Viewing Zone

2.1 Principle of Lenticular Display System

The principle of lenticular display is lenticular beam dispersion effect and binocular parallax principle. With the multi-view image displaying on the display screen, by the effect of lenticular splitting, the viewers can observe different view image at different position, and can obtain left and right eye stereo image pairs in a certain position, finally form the 3-D effect by the integration of the brain[5]. The system works specifically with a lenticular sheet in front of the screen, which fits precisly with display screen. The LCD pixel array is located at the focus plane of a lenticular sheet. As figure 1 shows, the lenticular sheet is slanted, because the slanted lenticular sheet can alleviate the non-balance state and Moriefringe caused by LCD pixel array[6].

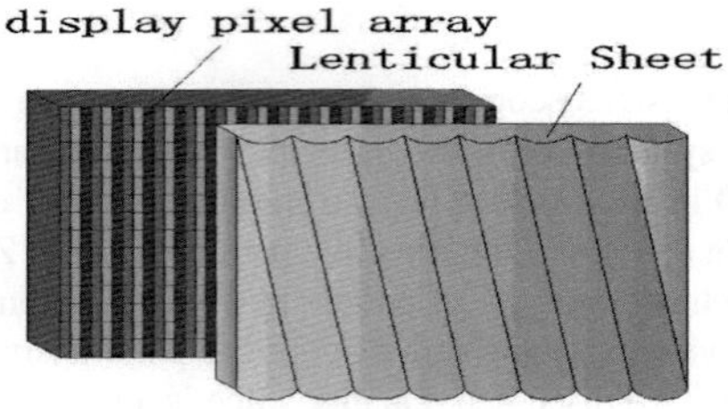

Fig. 1. Autostereoscopic display based on lenticular lens

2.2 Definition and Form of Viewing Zone

The stereo viewing zone(VZ) is a region in which the viewer can observe the 3-D image that he(she) want. For autostereoscopic display based on lenticular lens, it is reasonable to analyze the stereo viewing zone using geometrical optics. By the lenticular beam dispersion effect, in horizontal direction, the viewer can observe various images along different direction, and the rays finally form VZ. The rays from different views are separated in space, so, the VZ is composed of many sub-zones(SVZ)[7]. Each SVZ parallel with display screen has the same width in horizontal direction, side by side with adjacent VZ without overlap or gap. The horizontal width of SVZ should be less than binocular distance, in this case, the viewers can locate their left and right eyes at two different SVZ segments without failing, and will perceive depth sense by virtue of the binocular parallax, and they can move their eyes within the VZs to see other view images, so this movement provides motion parallax, hence the systems can provide both binocular and moving parallaxes to the viewer[8].

Fig.2 is the VZ horizontal section diagram of 9-view autostereoscopic display based on lenticular lens (The rays between VZs are not drawn). The width of central viewing zone(CVZ)is the region only to see a single view, that the total width of 9 sub-zone at the best viewing distance(BVD), there, the views are separated completely,

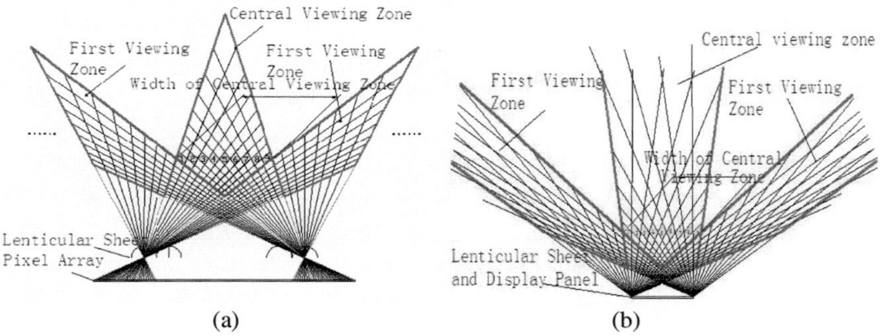

Fig. 2. VZ schematic diagram of 9-view lenticular display (a) the width of CVZ be less than the display system;(b) the CVZ be wider than the display system

the viewer's left of right eye can see only a single view image simultaneously. Fig.2(a) shows the case that the width of display is wider than the CVZ, in this case, each region is a diamond shape; fig.2(b) shows the opposite, VZ can not be a closed diamond. We can see that there are other side VZs in addition to the central one, because the light emitted by the pixels covered by adjacent lenticular unit can also pass through the lenticular unit. The boundaries of CVZ and the side VZ are drawn in thick red. These side VZs arrange symmetrically on both sides of the central one, followed by the first side VZ, the second VZ, etc. And in addition to BVD, monocular view number is proportional to the distance between the location and the best viewing position in the vertical direction. That is, the farther the viewer stand from the optimum viewing position, the more the monocular view number, and the worse 3-D effect would be obtained. Pseudoscopic effect exsits in the border between CVZ and the first side VZ and other side VZs, because the obtained view image vill suddenly change from 9 to 1 at the junction of adjacent VZ, i.e., the view order reverse, and this view order reverse causes reversal in the depth of the perceived image, hence there will be a transition zone[9].

3 Analysis of Central Viewing Zone

For 9-view display, take the display normal direction for Z direction. Take the case that display width is lager than CVZ for example[see Fig. 2(a)]. Analyzing CVZ with the geometric optics method, the light emitted from pixel units around the display panel's border is drawn. In practice, light rays emitted from other pixels intersect in the region, thus the CVZ of the 9-view stereo display system is a diamond zone.

W_p is the width of a sub-pixel, W is the width of display, W_v is the width of CVZ, D_l is the distance between lenticular centre of circle and pixel array, D_{opt} is the optimum viewing distance, D_{max} is the farthest viewing distance, D_{min} is the nearest viewing distance[as shown in Fig. 3(b)]. In the case that without considering the viewing number obtained in each sub-zone and the width of each sub-zone, the relationships between D_{max}, D_{min} and D_{opt} can be calculated. Since W_p, D_l and r is very small compared to D_{min}, D_{opt} and D_{max}, it can be seen from the similar triangular relationship:

$$D_{min} = \frac{W}{W + W_v} D_{opt} \text{ , } D_{max} = \frac{W}{W - W_v} D_{opt} \text{ .} \tag{1}$$

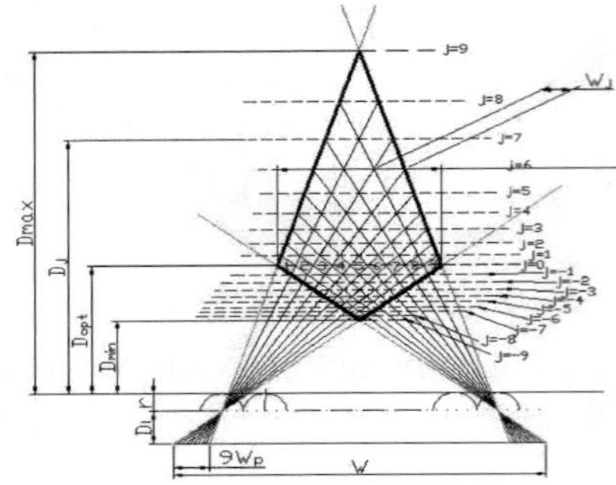

Fig. 3. CVZ analysis

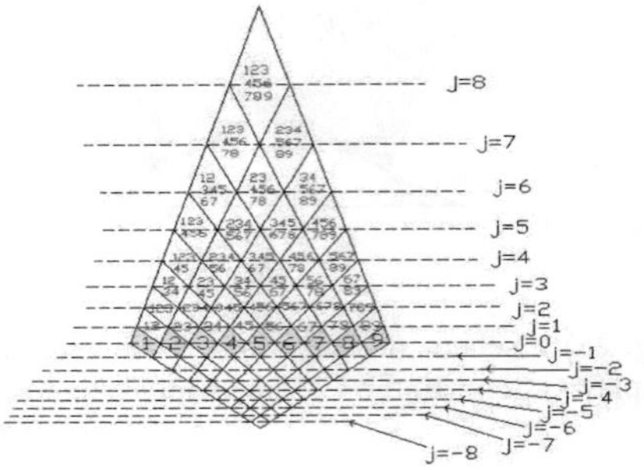

Fig. 4. View images obtained at each SVZ in CVZ

As shown in Fig. 3, where j=0,±1, ±2…±9 represents the jth column SVZs, CVZ would be analyzed now: The number in Fig.4 corresponds with the serial number of the view. As shown in Fig.4, the viewer observe only a single view at the j=0th column, and one can obtain two views at the j=±1th column(it doesn't mark view observed at j=-1,-2…-8th column), i.e. the number of mixed images in the SVZs centered at the ±jth columns in the CVZ is |j|+1. The obtained view number order in SVZs at the first column from left to right is, respectively, 1,2,3,4,5,6,7,8,9; and 12 3,234,…789 at the second column, etc., and the viewer can obtain 9 views simultaneously at the 8th column. For the case of j=-1,-2…-8, the obtained view number at the jth column is the same as the corresponding column that the j=1,2…8th column separately. The less number the viewer obtains, the better the 3-D effect would be, so the larger the numerical value of |j|, the worse the 3-D effect would be.

D_j is the distance between the jth column and the display panel in Fig. 3, i.e. the Z-distance. W_j is the width of SVZ at the jth column. Let $W_{SV}=W_v/9$, that is, W_{SV} is the width of SVZ at the zeroth column.

$$D_j = \frac{W}{W - jW_{SV}} D_{opt} \text{ , } W_j = \frac{W}{W - jW_{SV}} W_{SV} \text{ .} \tag{2}$$

4 Analysis of Factors Affected Central Viewing Zone

Without any qualification, the diamond-shaped region shown as Fig.4 is CVZ. In fact, with increasing distance from BVD, the width of SVZ will increase, but it cannot be larger than binocular distance. If the width of SVZ is bigger than one's binocular distance, the viewer can not locate one's left and right eyes at two different VZs without failing. The farther the viewers are apart from the best viewing position, the worse the image quality, therefore CVZ is restricted by these two conditions.

4.1 Width of Subzone

As we mentioned above, for each SVZ, its width can not be bigger than the viewer's binocular distance, that is $W_j \leq E$, E is the viewer's binocular distance, which is about 65mm, so the relationship can be expressed as:

$$j_{max}(W_{SV}, W) = (65 - W_{SV})W / 65W_{SV} \quad (j_{max} \in Z) \cdot \tag{3}$$

It can be seen that j_{max} is relevant to the width of SVZ and display, and j_{max} is inversely proportional to W_{SV} and proportional toW. The 3-D image is patched by 9 views, so $j_{max} \leq 9$, and the corresponding D_{jmax} can be calculated by equation(2)and(3):

$$D_{jmax}(W_{SV}, D_{opt}) = 65\, D_{opt} / W_{SV} \cdot \tag{4}$$

D_{jmax} is the Z-direction distance that the farthest position of the $j_{max}{}^{th}$ column be apart from the display panel, that is, D_{jmax} is the farthest distance of VZ. It isn't relevant to the width of display,it's proportional to D_{opt} and inversely to W_{SV}.

4.2 Image Quality Evaluation

Because the image obtained at the ±jth column is mixed by |j|+1 views, we can use the reciprocal of the view number to assess the image quality, that is, the quality factor mentioned in citation [3] $Q=1/(|j|+1)$. The quality factor Q_{jmax}corresponding with the maximum of j is relevant to W and W_{SV}:

$$Q_{jmax}(W_{SV}, W) = \frac{65W_{SV}}{65(W - W_{SV}) - WW_{SV}} \,. \tag{5}$$

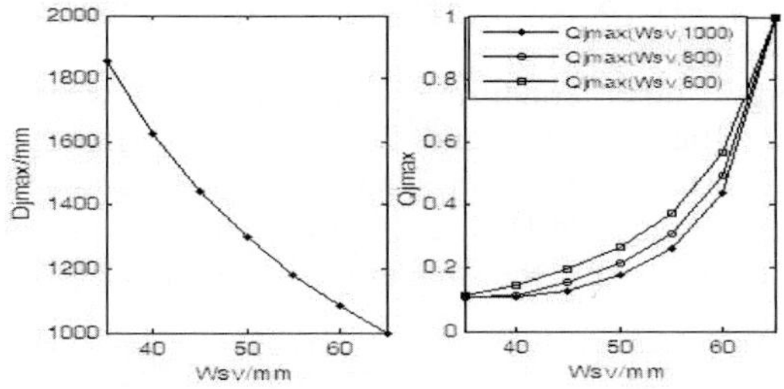

Fig. 5. Relationships among D_{jmax}, Q_{jmax} and W_{SV}

In condition that the width of the 0th column keeps changeless, the wider the display, the smaller the numerical value of Q_{jmax}; and the farthest viewing distance isn't relevant to W; and in condition that the width of display keep constant, with the numerical value of W_{SV} increasing, the farthest viewing distance will get nearer, the corresponding Q_{jmax} will get bigger(As seen in Fig. 5, D_{opt}=1000mm). Therefore, image quality and the scope of VZ are contradictory, and cann't achieve optimal. Considering these two factors, the scope of VZ and quality factor, comprehensively, the width of the 0th column can be designed according to different needs. If you want to obtain a great range of viewing zone, you can design the value of W_{SV} to be small; and if you are more focused on the image quality, the value of W_{SV} should be big.

4.3 Influence Analysis of Parameters of Display and Lenticular Lens

The choices of parameters of display and lenticular have a great influence to viewing zone, and the relationship will be analyzed next. As shown in Fig. 6, f is the image focal length of selected lenticular.

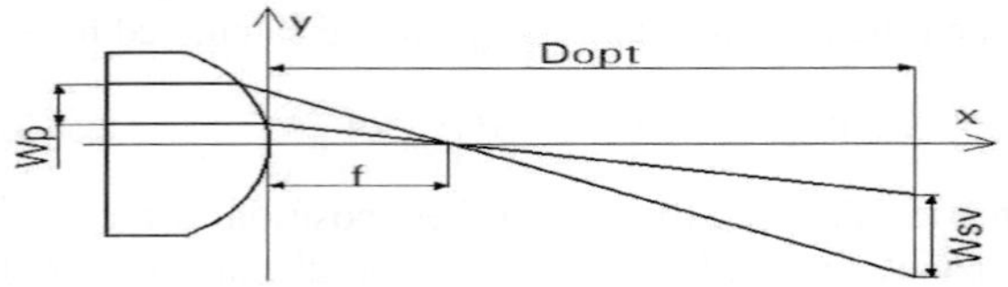

Fig. 6. Optical transmission property of a single lenticular lens

From the similar triangular relationship, it can be calculated as:

$$W_{SV} = W_p D_{opt} / f \cdot \tag{6}$$

It can be obtained by equation (1),(2), (4) and (6):

$$D_j = \frac{WfD_{opt}}{Wf - jW_p D_{opt}}, \quad D_{j\max} = \frac{65f}{W_p}. \tag{7,8}$$

$$D_{\min} = \frac{fWD_{opt}}{fW + 9W_p D_{opt}}, \quad Q_j = \frac{W_p D_j D_{opt}}{fW\left|D_j - D_{opt}\right| + W_p D_j D_{opt}}. \tag{9,10}$$

As shown in Fig.7, other conditions being constant, when f increases or W_p decreases, D_{jmax} amplifies notably, and D_{min} increases slightly. So, if the design requirement D_{opt} is constant, for a given display, W_p is determined, and the larger the value of f, the bigger the range of D_{jmax}-D_{min} in Z-direction; if focal length is constant, the larger the width of sub-pixel, the smaller the range of D_{jmax}-D_{min} in Z-direction.

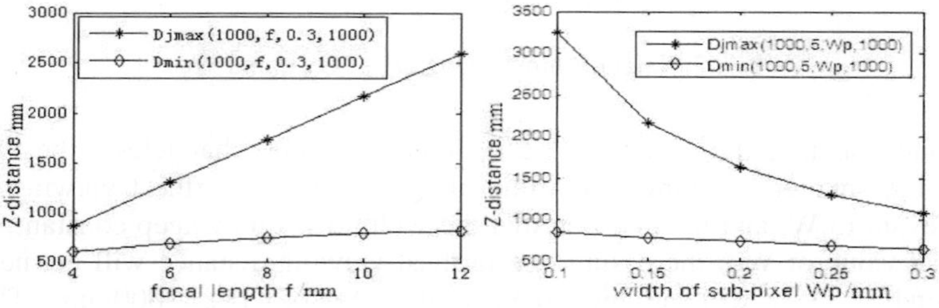

(a) (b)

Fig. 7. Relationship between (a)viewing region and f;(b) viewing region and w_p.

4.4 Experiment Result

According to the above results, we have manufactured a 47-inch auto-stereoscopic display based on lenticular lens. The parameters of this system are:W=1042.56mm; W_p=0.1805mm; f=8mm; D_{opt}=2000mm. According to the analysis above, for this display system, D_{min}=1440mm, D_{jmax}=2880mm, W_{SV}=45.125mm and j_{max}=7, the distance of each column of viewing zone and the quality value are shown as fig.8:

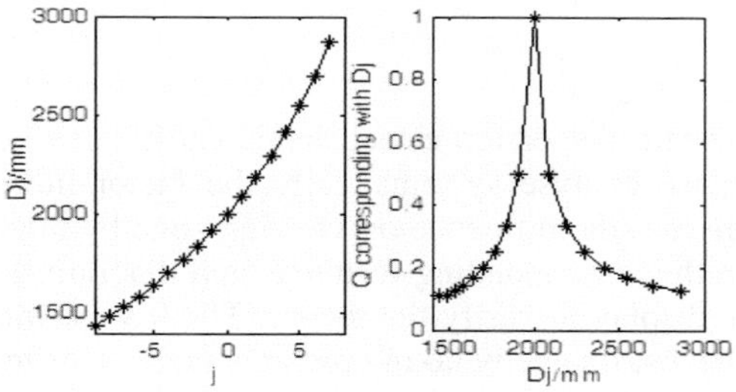

Fig. 8. Q corresponding with subzone

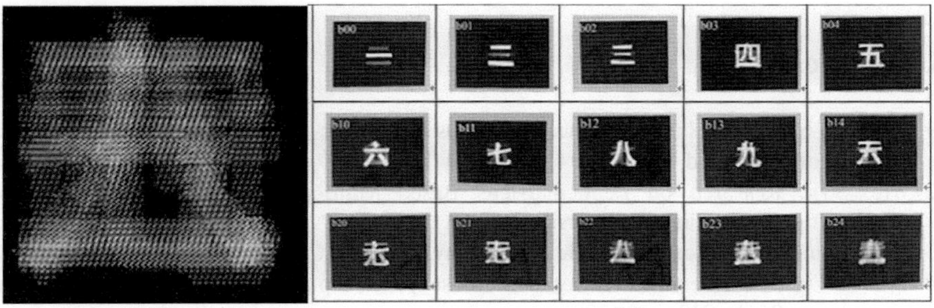

(a) (b)

Fig. 9. (a)9-view image patched by 9 figure; (b)15 images photographed in different positions

Multi-view image shown in Fig.9(a) is displayed on screen, which is patched by chinese characters(one, two, three, four, five, six, seven, eight, nine, respectively)by certain rules. The background is black, and the character is white. The images shown in Fig.9(b) are photographed in nine different position by a digital camera. The viewers may obtain its neighbour view image in the subzone due to lenticular misalignment and the number of sub-pixel covered by lenticular unit is non-integer[10]. b_{00}~b_{13}: the distance is 2000mm, and when the camera moves horizontally, nine completely separated views can be obtained, which represents 'one', 'two', 'three', 'four', 'five', 'six', 'seven', 'eight' and 'nine', respectively; b_{14}: the image is patched by the views 'five' and 'six', the shooting distance is 2090mm; b_{20}: the image is patched by the views 'five', 'six' and 'seven', the shooting distance is 1840mm; b_{21}: the image is patched by the views 'four', 'five', 'six' and 'seven', the distance is 2300mm; b_{22}: patched by 'three', 'four', 'five', 'six', 'seven' and 'eight', the distance is 2550mm; b_{23}~ b_{24}: the distances are 1580mm and 1480mm respectively, these two images look very fuzzy, it is difficult to find out of which views the image is composed.

It can be discovered that the view image completely separated, and with the distance from the best viewing position increasing, monocular image will patch with more view images. The quality value of the image taken in corresponding position is consistent with Fig.8.

5 Conclusion

Based on lenticular beam dispersion effect, lenticular display system forms viewing zone. In this paper, we analyse systematically the factor influencing the range of viewing zone, and discuss the influence on viewing zone by the optical parameters of the system, and give the corresponding formulas and diagram. Finally, we produce a 47-inch stereoscopic display to verify the theory. The 9-view image displayed on the display, we obtain the test image by taking pictures in some different positions, which were compared with the above theory, and analyze the cause of the error, finally, prove the correctness of the theory. The research on the formula about viewing zone and evaluation of 3-D effects will help to design autostereoscopic display system based on lenticular lens.

References

1. Qin, K.-h., Luo, J.-l.: Techniques for Autostereoscopic Display and Its Development. J. Journal of Image and Graphics 14(10), 1934–1941 (2009)
2. He, S.-j.: Research on Lenticular-lens Based Multi-view Auto-stereoscopic Display Technology. D. Zhejiang University (2009)
3. Dodgson, N.A.: Autostereoscopic 3D Displays. J. IEEE Computer Society, 31–36 (2005)
4. Sun, C.: A Probe into Several Stereoscopic Display Technologies. J. Computer Simulation 25(4), 213–217 (2008)
5. Huang, Y.-g., Liu, W.-w.: Geometry Modeling for Autostereoscopic Display Based on Principle of Parallax Illumination. J. Liquid Crystals and Displays 21(5), 579–583 (2006)
6. Yang, L., Song, X.-h., Hou, C.-p., Dai, J.-f.: Image Synthesizing for Autostereoscopic Display Based on Lenticular Technology. J. Journal of Tianjin University 40(9), 1105–1110 (2007)
7. Son, J.-Y., Saveljev, V.V., Kim, J.-S., et al.: Viewing zones in three-dimensional imaging systems based on lenticular,parallax-barrier,and microlens-array plates. J. Optical Society of America 43(26), 4985–4992 (2004)
8. Son, J.-Y., Saveljev, V.V., Choi, Y.-J., et al.: Parameters for designing autostereoscopic imaging systems based on lenticular, parallax barrier and IP plates. J. Society of Photo-Optical Instrumentation Engineers 42(11), 3326–3333 (2003)
9. Saveljev, V.V., Son, J.-Y., Kim, S.-H., et al.: Image Mixing in Multiview Three-Dimensional Imaging Systems. J. Journal of Display Technology 4(3), 319–323 (2008)
10. Lee, Y.-G., Ra, J.B.: Image distortion correction for lenticula misalignment in three-dimensional lenticular. J. Displays Optical Engineering 45(1), 017007-1–017007-9 (2006)

Experimental Study on Wind Turbine Characteristic Emulator System Based on the Blade Element Theory

Zhenlan Dou[1], Qiuqiong Zhang[1], Zhibin Ling[1], and Xu Cai[1,2]

[1] Key Laboratory of Control of Power Transmission and Transformation, Wind Power Research Center, School of Electronic Information and Electrical Engineering, Shanghai Jiao Tong University, China
[2] State Key Laboratory of Ocean Engineering, School of Naval Architecture, Ocean and Civil Engineering, Shanghai Jiao Tong University, China

Abstract. In the laboratory without the condition of actual wind farm or wind turbine, building a wind turbine output characteristic emulation system can provide an effective way for the research on WPGS. Firstly the wind turbine torque model was built, redistributed the bladed element by weight coefficient and calculated the wind turbine torque based on the blade element theory. Then characteristics of the wind turbine were emulated by controlling the DC motor armature winding current. So the wind turbine emulator based virtual wind was built. Finally the experiment platform of VSCF WPGS was also built. The experiments not only confirms the correctness and superiority of the detailed wind turbine model, reflects the rapid tracking performance of the emulated torque, but also simulates the wind turbine to meet the research of VSCF WPGS such as the sub-synchronous or super-synchronous running of DFIG, consequently verifies the correctness of the control strategies for variable speed constant frequency WPGS.

Keywords: wind turbine, wind turbine torque, modeling, blade element theory, weight coefficient assignment, DC motor, wind turbine emulator.

1 Introduction

Exploiting wind energy and actively developing wind power generation have great significance to solve the global energy and environmental crisis. Research and experiments on wind power generation system (WPGS) are difficult to carry out due to the limitations of severe conditions of wind farm and high cost. The wind turbine emulator system can substitute the actual wind turbine by calculating the output torque with the wind turbine torque model and rapidly producing a torque corresponding to the current wind condition. It not only has the same mechanical characteristics with the actual wind turbine, but also provides controllable equivalent wind energy. So it can be used in the design, estimation and test application of the WPGS, which greatly improves the efficiency and validity of research and development [1-4].The key of the wind turbine emulator system is the establishment of the wind turbine output torque model which can be applied to any wind speed input varying with time and space.

X. Wan (Ed.): Electrical Power Systems and Computers, LNEE 99, pp. 915–922.
springerlink.com

In order to satisfy the research requirements of the doubly-fed induction generator(DFIG) under super-synchronous and sub-synchronous status and maximum power point tracking(MPPT) control etc, Firstly the wind speed model with the average wind speed on the turbulence component was established. Secondly the model of wind turbine torque was built by the blade element theory (BET) and the weight coefficient assignment for blade element on the basis of wind speed model, considering the influence of vortex. Then the wind turbine emulator system was realized by controlling the torque of DC motor. Finally the experiment platform of WPGS with DFIG was built. The experiment of wind turbine emulator system confirms the correctness of the wind turbine torque model and the validity of the emulator system, and also verified the correctness of the control strategies for VSCF.

2 The Model of Wind Speed

Considering the wind speed sequence $v(k)$, $k = 1, 2, 3, \ldots\ldots$, which meets the Gaussian distribution whose variance is $\sigma_u^{\ 2}$ based on the average wind speed $\bar{v}$, it can be expressed as follows:

$$v(k) = \bar{v} + v_t(k) \tag{1}$$

where $\bar{v}$ is the average wind speed and a constant, $v_t(k)$ is the turbulence component and a random process, which meets the Von Karman power spectral density distribution. The turbulence intensity is defined as $T_l = \sigma_u / \bar{v}$.

On the other hand, the turbulence component of the wind speed can be regarded as a discrete-time random process[5], which can be expressed by Autoregressive (AR) model as follows:

$$v_t(k) = v_t(kT_s) = \sum_{i=1}^{n} \alpha_i v_t(k-i) + a(k) \tag{2}$$

where α_i is the autoregressive coefficient, $a(k)$ is the white noise sequence, n is the autoregressive order, k is the index value of wind speed sequence.

The turbulence component of wind speed is a random sequence with zero-mean, so its autocovariance function is equivalent to the autocorrelation function. The autocovariance function and power spectral density function is a pair of Fourier transform pairs. According to Wiener -Khintchine theory, the autocorrelation function $R_{xx}(l)$ and power spectral density function of AR model is a pair of discrete Fourier transform. So $R_{xx}(l)$ can be worked out, then all α_i and σ_a^2 can be obtained. The fluctuating wind speed model based on spectral density analysis is derived, which can realistically simulate the variation rules of actual wind speed, and provides a good input foundation for the simulation of the wind turbine torque characteristics.

3 The Model of Wind Turbine Output Torque

Modeling methods of the wind turbine output torque can be mainly divided into two types: one is under the condition that the torque coefficient is known, that is modeled based on the aerodynamic equipment sub-model, and the function can be expressed as $T = T(\omega, v, \beta; C_T, R)$; the other under the condition that the torque coefficient is unknown, that is modeled based on the blade element theory, and the function can be expressed as $T = f(\omega, v, \beta; C_l, C_d, l, \beta_0, R, r_0)$. The unknown inputs of the two wind turbine models are the same: wind speed v, wind rotor speed ω and blade pitch angle β. The known inputs of the first modeling are the torque coefficient of wind turbine and wind turbine radius R, while the known inputs of the second modeling are intrinsic parameters of the wind turbine. Many researchers have made the related research of the wind turbine modeling in [1-4, 6]. However it is difficult to model based on the torque coefficient because the wind turbine torque coefficient is unknown, and the only coefficient is the manufacturing parameters of the blade which may be obtained from the manufacturer. Although the wind turbine torque was calculated by the blade element theory in [5], but the impact of the vortex effect was ignored, and the difference between the torque on the root and the tip of the blade on the large-scale wind generator was ignored, and it also lacked of the related experiments only with simulation. The wind turbine model by the assignment of weight coefficient for blade element is based on the theoretical foundation of the aerodynamic force analysis for blade element, so the actual force and wind turbine output torque on the wind turbine can be accurately calculated.

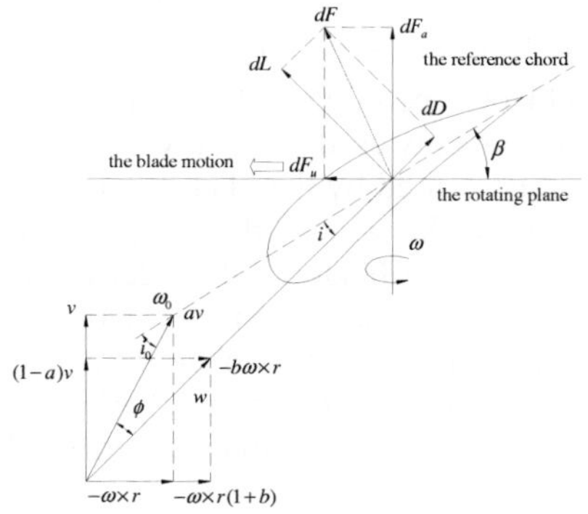

Fig. 1. Blade Element Velocities and Forces

The wind turbine model is derived by dividing the blade into several blade elements, and got the micro-element torque through the force analysis of the blade elements. Then the sum of all the micro-elements' torque is the wind turbine output torque. Considering that the blade element with the length dr has the distance r from the wind turbine shaft, the chord length is l, the pitch angle is β, its forces are shown in the Figure 1. Assuming that the wind turbine always faces the wind direction, the axial wind speed blowing through the wind turbine is v, the wind rotor speed is $u = \omega \cdot r$, and the relative speed between the airflow and the blade is w_0, so $\overrightarrow{w_0} = \vec{v} - \vec{u}$ can be deduced. In order to make the wind turbine torque model be nearer to the actual wind turbine, the vortex effect can not be negligible. According to the analysis of blade element forces, which is shown in Figure 1, the wind speed under vortex affect is $v - v_a$, the wind rotor speed is $u + u_a$, the relative wind speed w can be expressed as:

$$w = \sqrt{\left[(1-a)\cdot v\right]^2 + \left[(1+b)\cdot u\right]^2} \tag{3}$$

The attack angle can be expressed as:

$$i = \arctan\left[(1/\lambda)\cdot(1-a)/(1+b)\right] - \beta \tag{4}$$

where a is the coefficient of axial induced speed, b is the coefficient of tangential induced speed, λ is the tip speed ratio.

The aerodynamic force dF of the blade element dr in the up-oblique direction is generated by the flow of relative wind speed w. It can be decomposed as dF_a and dF_u in the direction of vertical and parallel to the rotation plane. The wind turbine torque dT is produced by dF_u. The total torque of wind turbine T is the sum of all the torque microelements dT on the blade elements, which is expressed as:

$$T = n\cdot\int \sigma dT = \frac{1}{2}\rho\int_{r_0}^{R} rw^2(r)l(r)C_l(i)\sin[\beta(r)+i]\{1-\varepsilon(i)\cot[\beta(r)+i]\}\sigma dr \tag{5}$$

where n is the number of wind turbine blade; $C_l(i)$ is the lift coefficient function relevant to the attack angle i; $\varepsilon(i) = C_d(i)/C_l(i)$; σ is used to correct the wind turbine torque with multi blades according to the Prandtl theory.

The above torque model can only used in small wind turbine because there's a small difference between the torque caused by the blade element of blade tip and root. With the increasing of the radius of the wind turbine, the wind turbine torque is mainly determined by the blade element of blade tip. In order to improve the accuracy of the wind turbine model, the limited blade elements should be re-divided by weight coefficient assignment. Assuming that the number of blade element is p, the blade can be divided into feature regions n according to the variation of the airfoil parameters, and the blade element distributed in each region is determined by the weight coefficient k_t, that is

$$p_t = k_t \times p \qquad \sum_{t=1}^{n} k_t = 1 \tag{6}$$

The weight coefficient is:

$$k_t = \frac{\left(\frac{1}{2}\rho r l_t w_t^2 C_{lt}\cdot\sin I_t(1-\varepsilon_t\cdot\cot I_t)\sigma_t\Delta r_t\right)^2}{\sum_{t=1}^{n}\left(\frac{1}{2}\rho r l_t w_t^2 C_{lt}\cdot\sin I_t(1-\varepsilon_t\cdot\cot I_t)\sigma_t\Delta r_t\right)^2} \tag{7}$$

The corrected wind turbine torque is:

$$T = \sum_{j=1}^{n}\sum_{t=1}^{k_j \times p} \frac{1}{2} \rho r_t l_t(r) w_t^2(r) C_{lt}(i) \cdot \sin\left[\beta_t(r) + i_t\right]\{1 - \varepsilon_t(i)\cot\left[\beta_t(r) + i_t\right]\}\sigma_t \Delta r_t \tag{8}$$

where *j* is the numerical number of the blade element *j* in each region.

In order to verify the correctness and precision of the wind turbine torque model, firstly the model is established by Matlab, then it is verified by wind turbine emulator experiment. The given wind speed is shown in Figure 2(a) which the average wind is 10-12-10m/s and the turbulence intensity is 0.16 and the cycle of each segment wind speed is 20s. The corresponding wind turbine toque is shown in Figure2 (b), which testifies the wind turbine torque model can reflect the real-time change of wind speed and provides theoretical foundation and support for the later experiments. Figure 2(c) shows the wind turbine torque output under in a certain range of generator speed, and the results calculated by the wind turbine torque model (thick line) is coincident with the result calculated by GH Bladed, which verifies the correctness and accuracy of the wind turbine torque model. The fluctuating curve is the calculated torque by the wind turbine torque model under the turbulence wind speed. By comparison, the calculated torque fluctuated near the standard torque value calculated by the GH Bladed, and its fluctuation is in the calculation allowable range.

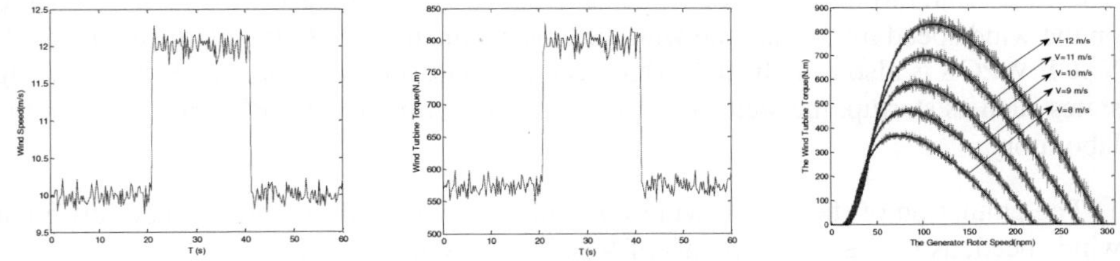

(a) The wind speed variation under certain turbulence intensity

(b) The wind turbine torque under wind speed with certain turbulence intensity

(c) The wind turbine torque–generator rotor speed under wind speed

Fig. 2. The out torque of wind turbine

4 The Emulation of Wind Turbine by DC Motor Control

As shown in Figure 2, the characteristics of wind turbine torque have great similarity to the characteristics of DC motor, so it can be simulated by a DC motor[5-8]. The control strategies of wind turbine emulation are power control and torque control according to the different control object of DC motor. The characteristics of power or torque of the wind turbine are realized by controlling the output power or the output torque of DC motor. Because the torque control strategy is easy to realize and has high accuracy, it is used in wind turbine emulator system, its control block diagram is shown in Figure 3.

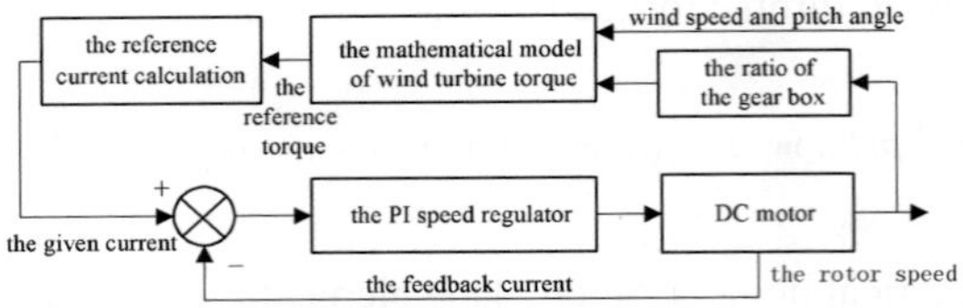

Fig. 3. Control block diagram of wind turbine emulator system

In the implementation of the wind turbine emulator system, firstly the wind rotor speed is expressed as the measured DC motor speed divided by the ratio of the gear box. Secondly the wind turbine aerodynamic torque which is considered as the reference toque of the DC motor is calculated by the wind turbine model according to the current dynamic wind speed and the blade pitch angle and the wind rotor speed. Thirdly the given armature current of the DC motor is obtained by the reference toque of the DC motor. At last the armature voltage which is used to drive the DC motor is derived by the sum of the output of the current loop regulator and the back EMF compensation.

5 Experimental Research of Wind Turbine Emulator System

An 11kW wind turbine emulator system and the virtual wind farm are established based on the wind speed model and the wind turbine torque model. So the platform of 7.5kW VSCF WPGS is also established, which can simulate various kinds of WPGS running states such as startup, the status of start, grid connection and MPPT running etc in the laboratory.

1) The emulation of the virtual wind farm and the wind turbine torque under different wind speed, as well as experiments of VSCF doubly-fed WPGS.

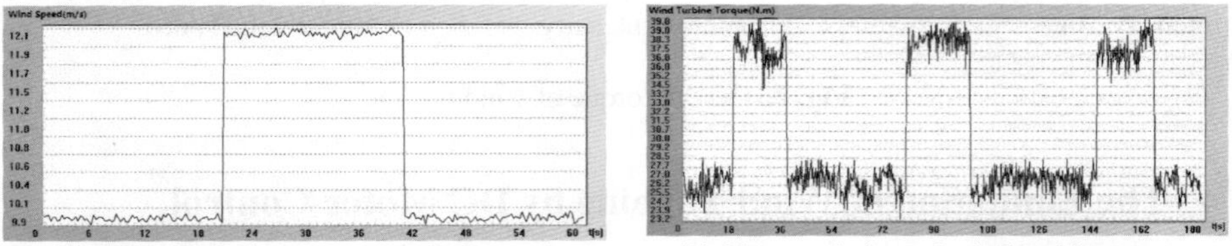

(a) The emulation of cycle wind speed **(b)** The emulation of wind turbine torque

Fig. 4. The torque emulation of wind turbine under the cycle wind speed

The wind turbine torque under periodic wind speed is shown in Figure 4, (a) is the periodic wind speed with an average value of 10-12-10m/s, turbulence intensity is 0.16, and the cycle of each segment wind is 20s; (b) is the torque of the virtual wind turbine corresponding to the wind speed and rotate speed when the WPGS is in MPPT status. Compared with the waveforms in Figure 2, the DC motor simulates accurately the variation rule of the generator high-speed shaft torque, and verifies the correctness of the wind turbine torque model and feasibility of the wind turbine emulator system.

The dynamic response of VSCF DFIG when wind stepped from 12m/s to 10m/s is shown in Figure 5. The figure shows that when the wind speed decreased, the wind turbine output torque was reduced, the energy input to the grid from the stator of DFIG was reduced, the amplitude of stator current i_{sb} of DFIG was reduced, and the frequency remained at 50Hz, and realized variable speed constant frequency. The grid current i_{gb} is reduced, but its decreasing amplitude is smaller than the decreasing amplitude of the stator current. The frequency of the generator rotor current i_{rb} is increased which is consistent with the rule that the frequency of the rotor current multiplies the frequency of the stator current frequency by the slip ratio. In the whole process, the DC bus voltage of dual PWM converter remains stably.

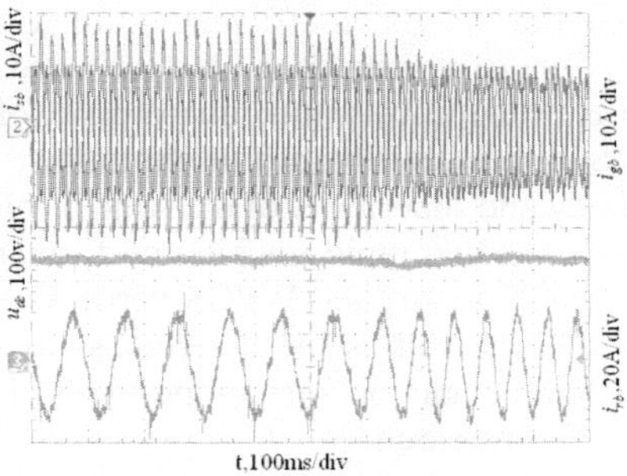

Fig. 5. The dynamic response of DFIG

2) The Operation of VSCF doubly-fed WPGS from the sub-synchronous state to the super-synchronous state.

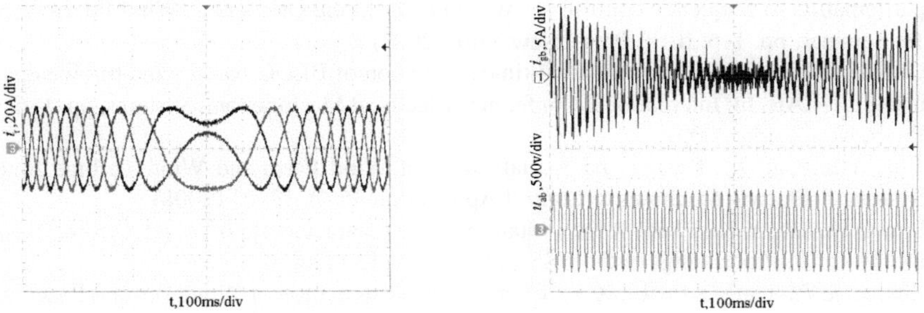

(a) The rotor current of DFIG **(b)**The grid current and the grid voltage

Fig. 6. The operation state from sub-synchronous to super-synchronous of VSCF WPGS

When the wind speed stepped from 12m/s to 15m/s, the rotor speed of DFIG increased from sub-synchronous 1050 r/min to super-synchronous 1950 r/min, the dynamic response is shown in Figure 6. The rotor current is shown in (a), the frequency of the rotor current decreased as the rotor speed of DFIG increased. Before and after the synchronous speed, the direction of the rotor current changed, which could be seen from the change of three phase current sequence. The grid current and the grid voltage are shown (b). The results show that the phase of the grid current reversed and the power flow also changed before and after the synchronous speed. In sub-synchronous, the grid absorbed power and fed into the converter of the rotor side. And in super-synchronous, the converter of the rotor side fed power to the grid.

6 Conclusion

In the paper, the detailed wind model and wind turbine torque model which can be used in different fluctuating wind speed and operating states of DFIG are built based on the blade element theory and the weight coefficient optimal assignment of the blade element considering the effect of vortex. Based on the wind turbine model, the wind turbine emulator system is realized by the torque control of the DC motor, and the experiment platform of VSCF WPGS is built. The experiment of wind turbine emulator not only verifies the accuracy and superiority of the wind turbine torque model, but also reflects the real-time simulation and rapid tracking performances of the wind turbine. The overall operation of VSCF doubly-fed WPGS verifies the operation principles of the VSCF doubly-fed WPGS as well as the correctness of the control strategies. In conclusion, the wind turbine emulator system provides a complete and effective implementation for the WPGS research.

References

1. Wang, Q.S., Hu, Y.W., Huang, W.X.: Review on The Technology of Wind Turbine Emulation. J. Electric Machines & Control Application 37(03), 1–6 (2010)
2. Monfared, M., Kojabadi, H.M., Rastegar, H.: Static and dynamic wind turbine simulator using a converter controlled dc motor. J. Renewable Energy 33(5), 906–913 (2008)
3. Courtecuisse, V., El Mokadem, M., Saudemont, C., et al.: Experiment of a wind generator participation to frequency control. In: Wind Power to the Grid - EPE Wind Energy Chapter 1st Seminar, pp. 1–6. IEEE Press, New York (2008)
4. Guo, H., Zhou, B., Li, J., et al.: Real-time simulation of BLDC-based wind turbine emulator using RT-LAB. In: International Conference Electrical Machines and Systems, pp. 1–6. IEEE Press, New York (2009)
5. Yue, Y.S., Cai, X.: Design and Actualization of Wind Farm and Wind Turbine Imitation System. J. Electric Machines & Control Application 35(4), 17–21 (2008)
6. Liu, Q.H., He, Y.K., Zhao, R.D.: Imitation of the Characteristic of Wind Turbine Based on DC Motor. J. Proceedings of the CSEE 26(7), 134–139 (2006)
7. Zhang, X.Y., Yang, Z.C., Li, X.Y.: Characteristic Simulation of Wind Turbine Based on DC Motor Closed-loop Current Control. J. Journal of Nanjing Institute of Technology (Natural Science Edition) 22(02), 17–21 (2008)
8. Ovando, R.I., Aguayo, J., Cotorogea, M.: Emulation of a Low Power Wind Turbine with a DC motor in Matlab/Simulink. In: Power Electronics Specialists Conference, pp. 859–864. IEEE Press, New York (2007)

Encoding-Based Algorithm for Minimization of Inductive Cross-Talk Based on Off-Chip Data Transmission

Souvik Singha[1] and Debarshi Saha[2]

[1] Asst. Professor, Department of Computer Science & Informatics, Bengal Institute of Technology & Management Santiniketan -731236
singha.souvik@gmail.com
[2] Pre-final year Student, Electrical Engineering Department, Bengal Institute of Technology & Management, Santiniketan -731236
deb1990.m.saha@gmail.com

Abstract. Inductive Cross-talk within IC Packaging is becoming a significant bottleneck in high speed inter chip communication. So the off-chip drivers typically source and sink 10 to 1000 drive internal loads. Thus simultaneously switching many off-chip drivers can cause large power (V_{DD}) and ground current surges. These changes in current flow induce a voltage drop on the drivers' local V_{DD} rail and a voltage rise on the drivers' local ground rail. The voltage drop is proportional to both the inductance of the V_{DD} (ground) distribution network and the rate of change of the current flow (V= L * dl/dt).

In this work, we proposed a technique to avoid the inductive cross-talk in the interconnect by encoding the data being transmitted off-chip. Bus encoding algorithms have been developed to overcome the capacitive cross-talk for on-chip buses, so the problem of on-chip capacitive cross-talk minimization for busses is very different from that of off-chip inductive cross-talk minimization. In this paper our approach also constructs cross-talk resistant CODEC algorithmically to utilize the memory-based CODEC solution.

Here we construct a set of equations which encoded the constraints that any legal vector sequence must satisfy to avoid supply bounce, signal glitching, and signal edge speed degradation. From this set of equations, we construct a set of legal vector sequences for the bus. We use this set to find the largest effective size of the bus that can be achieved by encoding, for a given physical size of the bus.

Our experimental results show that the proposed encoding based techniques result in reduced supply bounce and signal degradation due to inductive cross-talk, closely matching the theoretical predictions. As a result the overall delay of the bus actually decreases even after the use of the encoding scheme.

Keywords: Cross-talk, Off-chip bus, Bus Encoding algorithm, Delay.

1 Introduction

The limitation in package performance comes from the parasitic inductance and capacitance in the electrical interconnect [1, 3, 5] The parasitic inductance within IC

X. Wan (Ed.): Electrical Power Systems and Computers, LNEE 99, pp. 923–929.
springerlink.com

packaging causes bounce on the power supply pins in addition to glitches and rise-time degradation on the single pins [1, 2, 10].In this work we give the mathematical analysis and the coding algorithm based on the reducing cross-talk for off-chip data transmission which we help on [1, 2]. This code is commonly referred minimization of inductive cross-talk for Off-chip data transmission based on encoding.

The first step in creating the bus expansion encoder is to create a set of constraints equations. [1, 3, 7]. The constraint equations are written so that arbitrary transitions can be evaluated for noise limit violations. When the transition is evaluated using the constraint equations and violates one of the user-defined noise limits, the transition is flagged as illegal and is removed from the set of data sequences that are allowed to be driven through the package interconnects. Each of the possible off-chip transitions are evaluated against each of the constraint equations. The inductance factors that effect signal speed and integrity are as follows:-

Supply Bounce Constraints

When a pin I in segment j is a V_{DD} pin, it is required that the bounce magnitude due to the electrical parasitic in the package must not exceed the user-defined noise limit $P_{supply.}$ When the pin under evaluation is a V_{DD} pin, a constraint equation is written to determine is any transitions that occur on the bus segment will result in a violation of P_{supply} [1, 4, 6]. So the V_{DD} pin in addition to any mutual inductive or capacitive coupling that occurs due to switching signals in adjacent pins. By multiplying the coupling magnitude by the transition value v_i^j (which can be 0, 1, and -1). When the pin under evaluation is a V_{SS} pin, a constraint equation is written to determine if any transitions that occur on the bus segment will result in a violation of $P_{gnd.}$ Typically supply (V_{SS} and V_{DD}) pins are interspersed at regular intervals between signal pins. Every nth pin is a V_{SS} or V_{DD} . The supply bounce is proportional to the number of pins switching low or high. Ground bounce is expressed as:

$$V_{bnc} = L\sum_i (di / dt) \tag{1}$$

Where L is the self inductance of the V_{SS} pin and $\sum_i (di / dt)$ is evaluated over the number of signal pins switching low.

Glitch Magnitude Constraints

When a pin I in segment j is a signal pin, it is required that the coupled voltage onto that pin does not exceed any of the user-defined noise limit for signal coupling [9,11,13] if the signal pin is static $(v_i^j = 0)$ then the glitch magnitude onto the victim pin must not exceed P_0. As in the constraint equations for supply bounce, the magnitude of the coupling contribution of any neighboring pin is multiplied by the transition value V (which can be 0, 1, or -1) of the neighboring pin.

If a signal pin j is static, then a glitch may be induced in its voltage due to neighboring pins which switch.

This is governed by the expression

$$V^{j}_{glitch} = \sum_{k} \pm (M_{jk} \frac{di_{k}}{dt}) \tag{2}$$

Where i_k is the current in the Kth pin, and M_{jk} is the mutual inductance between the jth pin being consider and the Kth pin. The sign of the coupled voltage is positive or negative depending on whether the kth neighboring pin undergoes a rising or falling transition.

Switching Speed Constraints

When a signal is switching, its transition can be speed-up if the coupled voltage induced by its neighbor's mutual inductance aids the transition. We would like that a signal is not slowed down (i.e. either speed-up, or un-integrated), in this transitions due to this effect we would like that when a signal j is rising (falling), the coupled voltage on this signal (equation 2) due to its neighbor's transitions is zero or positive or negative.

In this way, the transitions of signals are not slowed down due to inductive cross-talk.

Rise time and Fall time Degradation Constraints

When a signal pin I in segment j transitions from logic 0 and logic 1, $(v_i^j = 1)$ it is required that the coupled voltage onto that pin does not hinder its rise time.

In a similar manner, when a signal pin I in segment j transitions from logic 1 to logic 0, $(v_i^j = 1)$ it is required that the coupled voltage onto the pin does not hinder its fall time.

Bus encoding algorithms have been developed to overcome the capacitive cross-talk for on-chip buses [8, 9, 10]. However the problem of on-chip capacitive cross-talk minimization for busses is very different from that of Off-chip inductive cross-talk minimization. Although our approach also constructs cross-talk resistant to [1, 8, 9], we utilize memory based CODEC solution [14].

In this work we proposed systematic encoding scheme to reduce inductive cross-talk.

2 Our Approach

Consider a bus consisting of k identical segments, each of width n. For any segment j, let j-1 represent the segment to its immediate right. Let us also denote the values if the n bits of segment j as $v_i^j (0 \le i \le n-1)$. Figure 1 shows an example of a bus configuration with k = 3 and n = 5. So the signal- to- power ratio for this bus configuration is 5/2 shown in figure1. In general, when assigning package pins for an off- chip bus, V_{DD} and V_{SS} pins are interspersed among the signal pins in a regular fashion. The over all bus arrangement consists of a repetitive pattern of segments, each with their V_{DD} and V_{SS} pins in the same relative position within the segment which shown in figure 1.

In our approach, we write equations to encode the inductive cross talk constraints for all bits of the jth bus segment. The constraints are different for the signal, V_{DD} and V_{SS} pins. Depending of the number of neighboring pins whose mutual inductive effects we want to model, the constraint equations will include pins belonging to neighboring segments as well. Since the segments are arranged in a repetitive manner, the encoding obtained for segment will be valid for all K segments within the bus. So here we used the valid sequences to construct a legal transition between bus vectors. From this digraph, we construct a memory – based CODEC which is used during the bus data transfer.

3 Coding Algorithm

If an m- bit bus can be encoded using the legal transitions in G, then there must exist a set of vertices $V_C \subseteq V$, such that each $Vs \in Vc$ has at least 2^m out going edges $e(v_s, v_d)$

(Including the self edge), such that the destination vertex $v_d \in Vc$. so the cardinality of Vc is at least 2^m. Now for any given graph G, we find m by the following given Algorithm.

Algorithm 1. Testing if G (V, E) can encode an n-bit bus

Test (M, G(V,E)) where M=n-1 (Where n is the physical bus size.)

1. Find degree $\forall v \in V$
2. For check_ degree = 1to Check_ degree = 0 do

Set Check_ degree = 0

For each $v \in V$ do

IF degree $(v) \leq 2^m$ then

2.2.1.1 Set $v \leftarrow V - v$

2.2.1.2 Set $E \leftarrow E -$ out_ degree (v)

2.2.1.3 Set $E \leftarrow E -$ in_ degree (v)

2.2.1.4 Set Check_ degree = 1

End IF

End For

3 End For

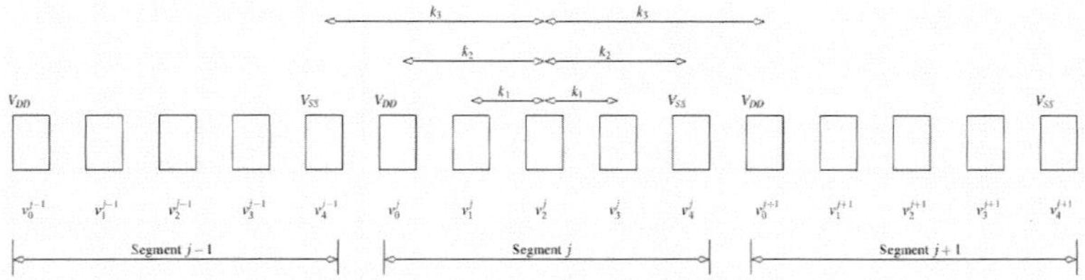

Fig. 1. Example Bus Configuration

Initially m is taken n-1 (when n is physical length of the bus). So the input of this Algorithm will be m and G. Here we first we find out- degree of each v (single vertices) ∈ V. If the out- degree of v is less than 2^m then we assigned V ⟵ V/v and delete all out- going edges rooted at V as well as all in- coming edges incident on v will continue this step until the graph is convergence. If after convergence the cardinality of v is greater than $2^{m,}$ We can construct a memory- based encoder using the legal transition of G. In this case the effective bus size can be encoded in m. If m bit bus can not be encoded using G then we decrement m, we repeat this until we find a value of m such that the m bit bus can be encoded by G.

4 Result

In this work we encode an example bus configuration to avoid inductive crosstalk. The bus configuration is shown in figure1. Here we taken from Reference1, The first step consists of writing the constraint equations for every pin in the bus. In this bus r =7 k =3 and α=7/2. For the inductive coupling with a magnitude less than 0.08.This exercise yields 7 constraints equations shown below. Note that these constraints have been simplified by removing terms with $v_i^j = 0$

$$1.\ v_0^j = v_{DD} = \frac{z}{2}(no\ of\ v_i^j\ pins\ that\ are\ 1) \le p_{bnc}$$

$$2.\ v_1^j = 1 \Rightarrow k1(v_2^j) + k2(v_3^j) + k3(v_4^j) + k4(v_5^j) \ge p_1$$

$$3.\ v_1^j = -1 \Rightarrow k1(v_2^j) + k2(v_3^j) + k3(v_4^j) + k4(v_5^j) \le p_{-1}$$

$$4.\ v_1^j = 0 \Rightarrow -p_0 \le k1(v_2^j) + k2(v_3^j) + k3(v_4^j) + k4(v_5^j) \le p_o$$

$$17.\,v_6^j = v_{ss} = \frac{z}{2}(no\,of\;v_i^j\;pins\;that\,are - 1 \leq p_{bnc}$$

5 Conclusion

In this work, we present encoding techniques which can help a designer trade off cross-talk against area overhead. Our experimental results show that the proposed techniques result in reduced delay variation due to cross- talk. As a result in reduced delay of a bus actually decreases even after the use of the encoding scheme. We also presented a technique to encode Off-chip bus data to avoid inductive cross –talk effects. The technique involves writing constraints equations which express the user-specified bounds on the amount of edge speed degradation, glitch magnitude, and supply bounce that can be tolerated. In this paper, we have introduce the concept of using data encoding to mitigate cross talk delay on buss and we presented a theoretical framework for understanding crosstalk immune coding. In this work we proposed a bus encoding technique to prevent cross talk delay for off-chip data. We incorporate all these inductive cross-talk effects in a common Mathematical framework. We construct a set of legal vector sequences with respect to of legal vector sequences with respect to inductive cross-talk, and use these to develop a CODEC for inductive cross-talk avoidance.

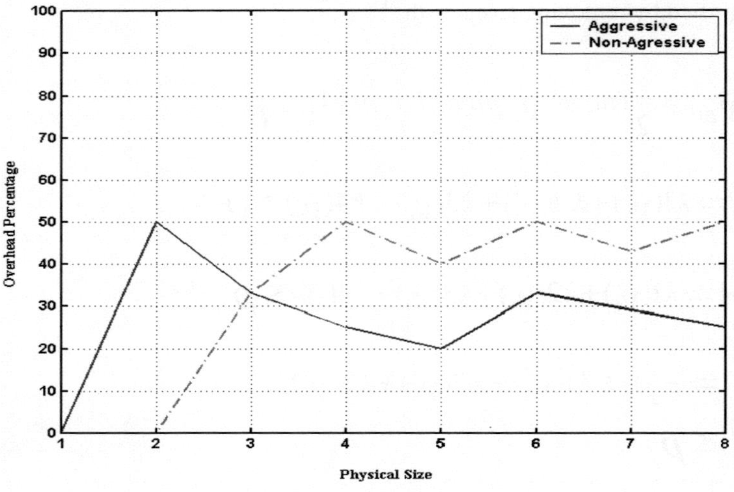

Showing the Encoding Efficiency in the above figure.

Fig. 2. Encoding Efficiency

Our experimental results show that the proposed encoding based techniques result in reduced supply bounce and signal degradation due to inductive cross-talk, closely matching the theoretical predictions. As a result the overall delay of the bus actually decreases even after the use of the encoding scheme.

We found the value of the effective bus size n as a function of physical bus size m. the results are shown in figure 2, where we plot the bus size overhead (n-m / m)as a function of n.

References

1. LaMeres, B.J., Khatri, S.P.: Encoding-based Minimization of Inductive Cross-talk for off-chip Data Transmission. In: Proceedings of the Design, Automation and Test in Europe Conference and Exhibition (DATE 2005), pp. 1530–1591. IEEE, Los Alamitos (2005)
2. Chen, C., Curran, B.: Switching codes for delta-I noise reduction. In: IEEE Transactions of the 43rd IEEE Midwest Symposium on Circuits and Systems, vol. 45, pp. 1017–1021 (1996)
3. The International Technology Roadmap for Semiconductors (2003), `http://public.ltrs.net`
4. Tummalo, R.: Fundamentals of Micro system Packaging. McGraw-Hill, New York (2001)
5. Miura, M., Hirano, N., Hiruta, Y., Sudo, T.: Electrical characterization and modeling of simultaneous switching noise for leadframe packages. In: Proceedings of 45th Electronic Components and Technology Conference, pp. 857–846 (1995)
6. Young, B.: Return path inductance in measurements of package inductance matrixes. In: IEEETransmissions on Components, Packaging, and Manufacturing Technology, vol. 20 (February 1997/August 2000)
7. Hirano, N., Miura, M., Hiruta, Y., Sudo, T.: Characterization and reduction of simultaneous switching noise for a multilayer package
8. Lopez, M., Prince, J., Cangellaris, A.: Influence of a floating plane on effective ground plane inductance in multilayer and coplanar packages. IEEE Transactions on Advanced Packaging 22, 182–188 (1999)
9. Duan, C., Tirumala, A., Khatri, S.: Analysis and avoidance of cross-talk in on-chip buses. In: IEEE Symposium on High-Performance Interconnects (HOT Interconnects), August 2001, pp. 133–138 (2001)
10. Duan, C., Khatri, S.: Exploiting crosstalk to speed up on-chip buses. In: Design Autamation and Test in Europe Conference (February 2004)
11. Victorand, B., Keutzer, K.: Bus encoding to prevent crosstalk delay. In: Proceedings, IEEE/ACM International Conference on Computer Aided Design, SanJose, CA, November 2001, pp. 57–63 (2001)
12. Powell, M., Vijaykumar, T.: Pipeline damping: a micro architectural technique to reduce inductive noise in supply voltage. In: Proceedings of 30th International Symposium on Computer Architecture, June 2003, pp. 72–83 (2003)
13. Mejia-Motta, E., Sandoval-Ibarra, F., Santana, J.: Design of cmos buffers using the settling time of the ground bounce voltage as a key parameter. In: Proceedings of 43rd IEEE Midwest Symposium on Circuits and Systems, vol. 2, pp. 718–772

Research on the Devices Fault Diagnosis Based on PSO Elman Neural Network

Jiejia Li[1,2], Hao Wu[1], and Jinxiang Pian[1]

[1] School of Information and Control Engineering, Shenyang Jianzhu University 110168 shenyang, China
[2] School of Information Science and Engineering, Northeastern University 110004 shenyang, China

Abstract. The aluminum electrolysis is a complex electrolytic process, which makes the aluminum electrolysis has many types of fault and high occurrence rate. Therefore, this paper proposes a fault diagnosis method based on PSO Elman neural network. The fault diagnosis model is established with the feedback Elman neural network, where hidden and the output layer nodes feedback have memory function. The weights of neural network are updated through the improved particle swarm optimization algorithm. The global PSO is used to find out the approximate results, and the local PSO is applied later for the specific search. These two PSO methods are switched by calculating the ratio of the distance between the two particles and the maximum distance, which results in the guarantee of the speed and accuracy, avoiding falling into local optimal values, and improving the rate of fault diagnosis.

Keywords: fault diagnosis, Elman neural network, particle swarm optimization.

1 Introduction

The aluminum electrolysis is a complex electrolytic process. Because of aluminum electrolytic current usually work under the condition of hundreds of thousands of amps, which caused strong electric field, magnetic field and strong heat field. The chemical reaction is violent, which make the aluminum electrolysis process have many types of fault and highly occurrence rate, which made electrolytic process work in an unstable condition, the production of aluminum is low and waste a lot of energy. Therefore, effective prediction of fault diagnosis and the treatment of the fault have great significance for saving energy and the improvement of the yield and quality of aluminum. This paper proposes a fault diagnosis method based on PSO Elman neural network with the mechanism and the fault of the characteristics of aluminum electrolysis process. This method is demonstrated by simulation experiments with good rate of fault diagnosis.

2 Modeling of Fault Diagnosis Based on PSO Elman Neural Network

In this paper, predictor is established by the output feedback Elman neural network. The network training adopts improved particle swarm optimization instead of gradient

X. Wan (Ed.): Electrical Power Systems and Computers, LNEE 99, pp. 931–936.
springerlink.com

descent method to update the weights. The node number in input layer is the number of characteristic quantities in a model volume. In the fault diagnosis of aluminum electrolysis, the feature of input is the tank resistance r(k),cell voltage v(k) and the cell temperature c(k) are selected according to the theoretical analysis and field data

2.1 The Structure of Elman Neural Network

The input and output feedback Elman neural network includes input layer, hidden layer, structure layer and output layer. The input layer is used for signal transmission, the output layer is used for weight linearly, and the hidden layer output data to the structural layer, structure layer and the input layer output data to hidden layer together. The hidden and the feedback nodes of the output layer have memory function. It is the recursive dynamic memory that made the network has dynamic monitoring performance. The structure of the Elman feedback network is shown in Figure 1.

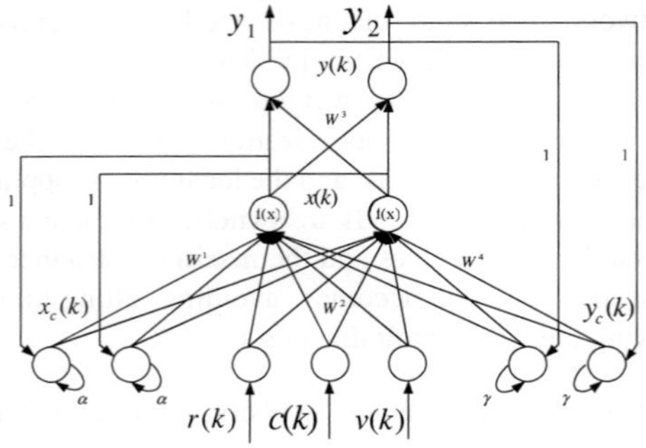

Fig. 1. Fault diagnosis model of aluminum electrolytic based of Elman neural network

In this paper, the weights of neural network are updated by the improved PSO algorithm. The PSO divided into two different types. The global PSO is used to find out the approximate results and then the local PSO is used for the specific search.

2.2 Particle Swarm Optimization

The global PSO is applied to find out the approximate the network weight, and the local PSO is used for the specific search. Switch these two PSO methods by calculating the ratio of the distance between the two particles and the maximum distance as $\frac{\|X_a - X_b\|}{d_{\max}}$.Where, $\|$ Xa - Xb $\|$ is the current distance form particle a to particle b. d_{max} is the maximum distance between any two particles in iterative method. The choice of the threshold η varies is according to the number of iterations as

$$\frac{\|X_a - X_b\|}{d_{\max}} < \eta \tag{1}$$

And then, the b becomes the neighborhood of the particles. All the particles satisfy the conditions compose a set Ni. After the improved neighborhood rules adopted, the choice of the threshold η varies is according to the number of iterations. η is expressed as follow,

$$\eta = \frac{3t + 0.6iter}{iter_{\max}} \tag{2}$$

when η> 0.9, the algorithm is adopted which use global PSO to renew the particle's velocity and location. When η< 0.9, the algorithm is adopted which use local PSO to renew the particle's velocity and location.

1) Global particle swarm algorithm

The network takes the feedback Elman neural network weight matrix as particles, which is initialized to a group of random weight particles. Through the iteration, the network follows two extreme to renew itself. The first extreme is the maximum individual weight p_{best} which is found out by the particle itself. Another is the maximum global weight g_{best} which is found out by the entire population.

$$v = w \times v + c_1 \times rand\ () \times (p_{best} - p_{present}) + c_2 \times rand\ () \times (g_{best} - p_{present}) \tag{3}$$

$$p_{present} = p_{present} + v \tag{4}$$

$$w = w_{\max} - \frac{w_{\max} - w_{\min}}{iter_{\max}} \times iter \tag{5}$$

Where, v is the speed of the weight particle, w is the inertia factor, *iter* is the number of current iteration, $iter_{\max}$ is the maximum number of iterations, $p_{present}$ is the current particle position, rand () is a random number between (0,1),c1 and c2 are learning factors.

2) Local particle swarm algorithm

The global PSO has fast convergence, but sometimes it probably fall into local optimum. Local PSO maintains multiple attractors to avoid premature. Assuming each particle is defined as a set in the neighborhood with the size of 1 as follow

Ni = {pbest i - l , pbest i - l +1 ,..., pbest i + l - 1 , pbest i + l} (6)

Neighborhood weight extreme lbest is selected from Ni which is instead of the global weight extreme gbest. The other parameters are the same as the global version of PSO. Weights particle according to the following formulas to renew their own pace and the new location,

$$v = w \times v + c_1 \times rand\ () \times (p_{best} - p_{present}) + c_2 \times rand\ () \times (l_{best} - p_{present}) \tag{7}$$

2.3 The Process of Learning Algorithm of the Improved PSO Elman Neural Networks

a. Determine the number of particles and initialize particle velocity and position.

b. Set the learning error accuracy ε , the current iteration number iter and the maximum iterations number $iter_{\max}$

c. Calculate the particle fitness and the ratio of the distance between the two particles and the maximum distance.

d. When $\varepsilon > E$ or $iter > iter_{\max}$, if $\frac{\|X_a - X_b\|}{d_{\max}} < \eta$,use the local PSO to renew the particle's velocity, location and the inertia factor w. If $\frac{\|X_a - X_b\|}{d_{\max}} > \eta$, use the local PSO to renew the particle's velocity, location and the inertia factor w.

e. If not meet the convergence criteria, go to step c, else stop the algorithm.

Suppose the network's real output is $y(k)$ and the target output is $t(k)$, the PSO fitness function can be defined as

$$E = \frac{1}{N}\sum_{j=1}^{N}\sum_{i=1}^{2}(T_{ij} - Y_{ij})^2 \tag{8}$$

3 Simulation

The model number of electrolytic process is taken as four in fault identification, that is normal, anode effect, cold tank, and hot tank. Therefore, the network output layer nodes are two. The output fault category is shown in table 1.

Table 1. The output fault category

output fault category	output	
	Y_1	Y_2
normal	0	0
anode	0	1
effect	1	0
cold tank	1	1
hot tank		

Because of parameters have different units, and the larger input value neural networks will greatly reduce the convergence rate. In order to ensure the accuracy for the network prediction, normalized the training data into the range from 0 to 1, and the normalized formula is

$$x_i = \frac{x_i - \min(x_i)}{\max(x_i) - \min(x_i)} \tag{9}$$

where xi is the actual input values of the samples, i=1,2, …, n

This paper adopts the fault data which are the aluminum plant provided to do some simulation. Take 30 minutes of the cell data in normal condition (Sampling in every 30 seconds). There are 60 groups of tank resistance, cell voltage and cell temperature data and normal input test data. System chooses three groups of various faults around 30 minutes of data for simulation.In this paper, take anode effect as an example.

At the first 8 minutes, neural network outputs are relatively stable. The cell is in a normal state. Starting from 11th minutes the neural network output Y_2 is greater than 0.5, which shows that the cell anode effect will occur. The original Elman network in the 13th minute output Y2 is greater than 0.5. That means the forecast is 2 minutes than original network. Starting from the 21th minutes the neural network output Y_2 is close to 1. We can see the cell anode effect occurred. In 33th minutes, the anode effect has been lifted, the improved network output quickly returns to 0.1 to 0.2, which respond more quickly to the original network.

When the anode effect occur, the network output are shown in figure 2.

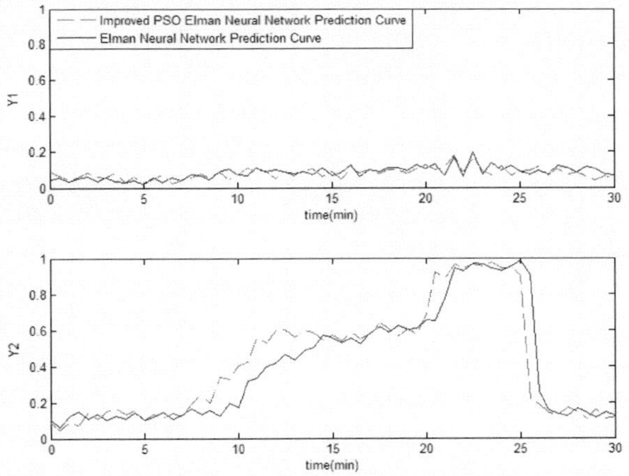

Fig. 2. The anode effect fault diagnosis results

4 Conclusion

In view of the three kinds of common faults in industrial aluminum electrolysis process, the PSO Elman neural network fault diagnosis model is proposed in this paper. The proposed method improve the fault diagnosis accuracy of the anode effect, and the simulation results show the neural network fault diagnosis method is feasibility and effectiveness.

References

1. Zhu, R.: Proficiency in Matlab 7. Tsinghua University Press (2006)
2. Dupuis, M., Bojarevics, V.: Weakly Coupled Thermo-electric and MHD Mathematical Models of an Aluminum Electrolysis Cell. Light Metals, 449–454 (2005)
3. Hinton, G.E., Salakhutdinov, R.R.: Supporting Online Material for Reducing the Dimensionality of Data with Neural Networks. Science, 313–504 (2006)

Wavelet Neural Network Process Control Technology in the Application of Aluminum Electolysis

Jiejia Li[1,2], Chengdong Wu[1], and Hao Wu[2]

[1] School of Information Science and Engineering, Northeastern University 110004 shenyang, China
[2] School of Information and Control Engineering, Shenyang Jianzhu University 110168 shenyang, China

Abstract. Aluminum electrolysis is a non-linear, time-varying and large time delay process, which interfered by the interaction of strong electric field, strong magnetic field and strong heat field. So, it is a high energy consumption process and the process control is very difficult. Therefore, the hot issue for the control system is how to save energy, improve the current efficiency, increase the yield and the quality of aluminum electrolysis. A wavelet neural network predictive control method is proposed in this paper which based on the analysis of characteristics and problems for the aluminum electrolysis process. The proposed method combines the neural network control technology and forecasting techniques. By tracking the parameter of the cell resistance which reflects the alumina concentration, the controller regulates the control strategy real timely to make the alumina concentration in an ideal range through controlling the alumina feeding quantity of the feeding device, and the system's hardware and software are also designed .The experiment results show that the method not only has a good effective control performance and an energy-saving effect, but also has an important significance of increasing the yield and quality of aluminum.

Keywords: Aluminum electrolysis, neural network, predictive control, alumina concentration.

1 Introduction

The aluminum electrolysis is a nonlinear, multivariable coupling, time-varying and large delay process. Because of aluminum electrolytic current usually work under the condition of thousands of amp. So in industry, people often choose the method which adds the alumina to the cell by timing or adopts simple adaptive mathematical model which is used to feed point method. Aluminum electrolysis is a complex electrolytic process which current efficiency is poor and it is difficult to establish accurate mathematical model. So control effect is not good enough. This paper proposes a fault diagnosis method based on wavelet neural network with the mechanism and the fault of the characteristics of aluminum electrolysis process. The experiment results show that the method not only has a good effective control performance and an energy-saving effect, but also has an important significance of increasing the yield and quality of aluminum.

X. Wan (Ed.): Electrical Power Systems and Computers, LNEE 99, pp. 937–941.
springerlink.com

2 Hardware System Structure

System hardware structure is shown in figure 1. This system adopts PC bus industrial control machine as control host, mainly include: analog input channel, digital input channel and analog output channel.

Analogue input channel collect the signal of the series current, series voltage and cell voltage in the field. The voltage isolation transmitter and A/D converter send the signal into the industry control machine to analyze. That is according to the work of the state to adopt different control strategy in order to make the cell to achieve the best working condition.

Digital input channel is used to switch the signal acquisition process and determine the operational status of the motor and manual / automatic system working state.

Analog output channel output control signal to control alumina charging device which is the key to electrolytic control techniques for saving energy, which has an important significance.

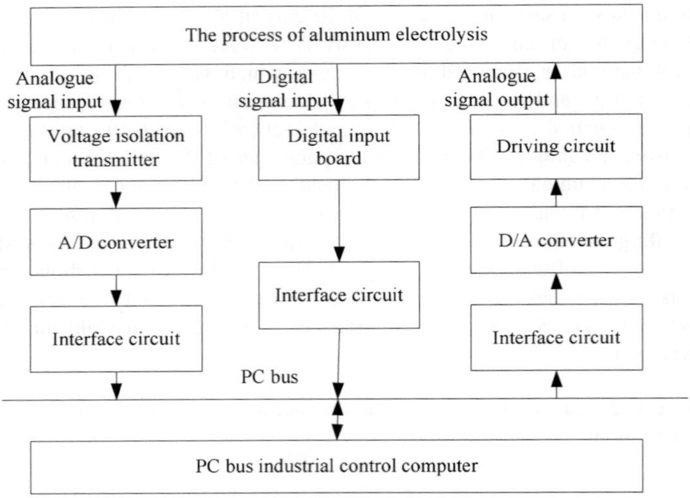

Fig. 1. Hardware system structure

3 Design of Neural Network Controller and Predictive Model

Neural network control system structure is shown in figure 2. In figure 2, NNC is the neural network controller and NNI is the recurrent wavelet neural network predictor. The given value is cell resistance x(k), the output y(k) is the actual output value cell resistance and the neural network controller output u(k) is used as the input of the predict structure of recursion wavelet neural network. The output of forecasting structure $\hat{y}(k)$ is a feedback parameter of the controller to improve the system control precision.

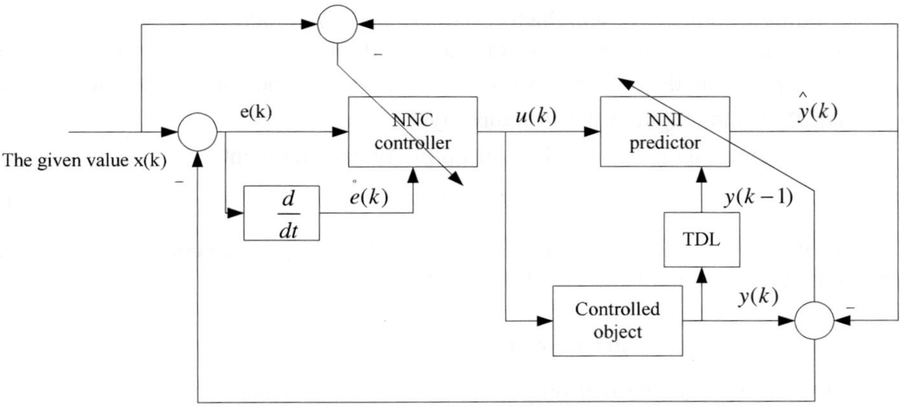

Fig. 2. neural network predictive control system structure

This system is mainly includes both the controller and the predictor. Where the controller adopt BP neural network and the predictor adopt recursion wavelet neural network which has three layers of neurons. The input of the structure layer is the output of the hidden layer. And the structure layer and the input layer output data to hidden layer together. It is the recursive dynamic memory that made the network has dynamic monitoring performance.

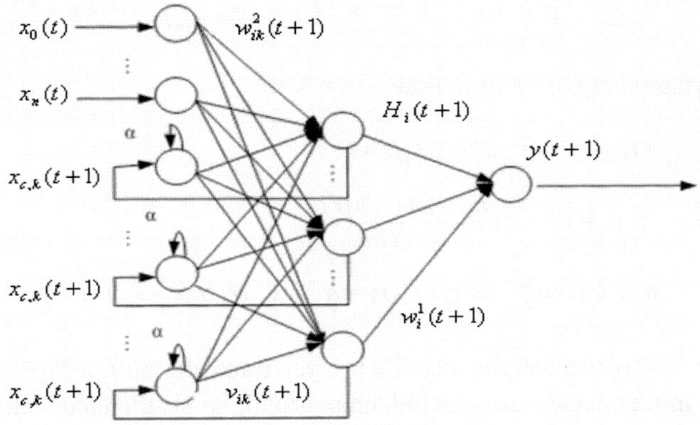

Fig. 3. the structure of recursion wavelet neural network

Figure 3 is the structure of recursion wavelet neural network. The recursive wavelet neural network has n+1 input neurons (the inputs is 3, namely n=2). Respectively $y-\hat{y}$ is the error between the outputs of controlled object and predicted, y(k-1) is the actual output value delay quantity, $\hat{y}$ is neural network controller output control value. A output neuron y(k+1) output the prediction $\hat{y}$,while the number of the hidden layer neuron is m and the number of the structure layer neuron is m. $x_0(t)=1, w_{i0}=\theta_i(t)$ .The corresponding weights represent the threshold of network hidden layer neuron i, which is not the actual training input. $y(t) \in R$ is the one-dimensional output vector under the t moment. They form the training vectors $H_i(t+1) \in R_m$ are the hidden layer output, and $v_{ik}(t+1) \in R_m$ is the output of the structure layer.

Set the feedback gain is α which is the output of each moment.

$$x_{c,k}(t+1) = aH_k(t) \tag{1}$$

This paper adopts Morlet wavelet. Set a as a wavelet expansion coefficient. Set b as a translation coefficient of wavelet. Set $t' = \frac{h_i(t+1) - b(t)}{a(t)}$. Morlet mother wavelet function is

$$\varphi(t') = \cos(1.75t')e^{-t^2/2} \tag{2}$$

Set y(t) and y_e(t) as the real output and the expected output at the time t, and the network error will be

$$e(t) = y_e(t) - y(t) \tag{3}$$

and we will get

$$E = \frac{1}{2}(y_e(t) - y(t))^2 = \frac{1}{2}(e(t))^2 \tag{4}$$

Set network works from the time 1 to time N. The total error function of each cycle is

$$E = \sum_{i=1}^{N} \frac{1}{2}(y_e(t) - y(t))^2 = \sum_{i=1}^{N} \frac{1}{2}(e(t))^2 \tag{5}$$

The dynamic equation of neural network is

$$y(t+1) = \sum_{i=1}^{m} w_i^1(t)H_i(t+1) \tag{6}$$

$$H_i(t+1) = \varphi(\frac{h_i(t+1) - b(t)}{a(t)}) \tag{7}$$

$$h_i(t+1) = \sum_{j=0}^{n} w_{ij}^2(t)x_j(t) + \alpha \sum_{i=1}^{m} v_{ik}(t)H_i(t) + \theta_i(t+1) \tag{8}$$

Where, $w_i^1(t)$ is the weight of hidden layer i and the output layer neuron. $w_{ij}^2(t)$ is the weight of input neuron j and the hidden neuron i. $v_{ik}(t)$ is the feedback weight of relevant neuron k and the hidden neuron i. $H_i(t)$ is the output of hidden neuron i. Controller and predictor in the aluminum power solutions ensure the system good dynamic and stable capacities, by changing the parameters of controller and predictor according to the load changes.

4 System Software Design

The software modular design method mainly includes: the data acquisition module, the fault processing module, the aluminum processing module, predictive control module and so on.

The data acquisition module mainly includes: analogue data acquisition and digital data acquisition, which collect the signal of the series current, series voltage and cell voltage in the field. Through the digital filter and calculation, it sends the signal into the industry control machine to analyze the signal.

The fault processing module mainly include: the anode effect, hot tank, disease cell monitoring and alarming.

The aluminum processing module mainly include: cell monitoring of aluminum, the control in the process of the aluminum electrolysis and the new control strategy of the electrolytic cell after the aluminum electrolysis.

The control module is the key to the program which include: control of cell movements, cell and system operating status, display and the control of alumina concentration cell. Alumina concentration is controlled by wavelet neural network, which control the alumina feeding device, the rate of feeding in time, adjusting the concentration of alumina in the cell. That make the concentration of the alumina in the cell achieve the best working conditions.

5 Conclusion

A wavelet neural network predictive control method is proposed in this paper which based on the analysis of characteristics and problems for the aluminum electrolysis process, which control of the alumina feeding device. That make the concentration of the alumina in the cell achieve the best working conditions.The experiment results show that the method not only has a good effective control performance and an energy-saving effect, but also has an important significance of increasing the yield and quality of aluminum.

References

1. Di., G., Ren, J.: Status of Domestic and International Aluminum Consumption. Nonferrous metallurgy energy-saving, 41–43 (2004)
2. Tian, Z., Lai, Y., Yin, G., Sun, X., Duan, H., Zhang, G.: Progress in Low Temperature Aluminum. Non-ferrous metals, 26–28 (2004)
3. Fan, L., Zhang, Y.: The Improvement of BP Neural Network and Application Based On Matlab. Journal of China West Normal University (2005)

A Fault Diagnosis Method of Switch Current Based on Genetic Algorithm to Optimize the BP Neural Network

Chengjun Tang, Yigang He, and Lifen Yuan

College of Electrical and Information Engineering, Hunan University,
Changsha 410082, China
tangchengjun123@163.com,
hyghnu@yahoo.com.cn,
6007542@qq.com

Abstract. Based on BP neural network slow convergence speed, easy to fall into the local minima and network structure is not quite stable shortcomings, the genetic algorithm (GA) optimization BP neural network is proposed on the basis of BP neural network in this paper. This paper put the GA optimized BP neural network apply to switch current circuit fault diagnosis, through comparing the result of experiment proves that the method is effective.

Keywords: Switch current circuit, wavelet transform, GA, BP neural network.

1 Introduction

Switch current (SI) technology is another kind of new simulation sampling data signal processing technology after the switch capacitance (SC) technology [1]. Being different from SC technology, which requires special double level polysilicon technology, the SI technology is a kind of sampling data network that being only consist of MOS transistors and MOS switch. It can maintain drain current ability through the charge that storied on a grid oxidation capacitance when the grid in MOS transistors opened. As it needs not strict capacitance accuracy, linear floating ground capacitance, and digital CMOS process compatible, it is easy for VLSI realization. What's more, SI circuit transfer signal using current, can work even in conditions of low voltage and low power consumption. These characteristics of SI caused extensive concern over SI technology in the academe of the world, and it has been emphasized and develops rapidly. Because of the widely use of SI technology and the possibility of fault in electronic circuit, a new project came up, that is how to diagnose the fault in SI circuit.

Although SI test technology[2] has been developed to some extent, the development of fault diagnosis technology is still in the primary stage [3]. There exist many difficulties which need to be overcome one by one.

Neural network have already been used in switch current circuit fault diagnosis. This paper aims at solving the problems existing in the BP network. How the Fault Diagnosis system (BP Neural Network that Optimized by the genetic algorithm) works: This paper adopts output current of measured circuit as the original signal, decompose

X. Wan (Ed.): Electrical Power Systems and Computers, LNEE 99, pp. 943–950.
springerlink.com

the circuit waveform of the output circuit using wavelet and obtain decomposition coefficients. Then conduct principal component analysis (PCA) and the normalized processing of fault signal, thus getting the optimal feature pattern vectors. The optimal feature pattern vectors will be sent to GA_BP network in which it will be classified and diagnosed on faults.

2 Wavelet Decomposition

As outlined in Literature [6], the wavelet decompose algorithm will not be detailed here. The use of wavelet decomposition mainly aims at the fault signal pretreatment, so as to acquire wavelet coefficients of each layer as fault feature model and offer GA_ BP corresponding fault diagnosis. Therefore, wavelet decomposition, together with subsequent PCA and the normalized processing of fault signal can get optimal characteristic pattern vectors as the input of GA_BP network through fault signal pretreatment, thus greatly reducing the number of network input terminal and finally speeding up the training of GA_BP network and improving the performance of the network.

3 Genetic Algorithm to Optimize the BP Neural Network

Genetic algorithm (GA) [7] is a heuristic global-optimization search method based on the biological evolution in nature world. Its essence is the cycle of copy-exchange-mutation operator. The problems are translated into a big population in genetic algorithm. And every member of the population has been genetic code, so each individual is actually entities with chromosome characteristics. This creates an initial population, which according to Darwin's principle of "Survival of the Fittest" can result in a new solution set population that evolved from gradual crossover, mutation. The new population can better adapt to the new environment, and it can be used as approximate optimal solution after coding. Genetic algorithm has strong search capabilities and good macro-global optimization ability.

Genetic algorithm combined with the BP network [8], can well overcome the short-comings of BP neural network, such as the slow convergence rate, tendency to fall into local minimum points and the instability of the network structure. Being operated, the weights and threshold of neural network will be firstly optimized by means of genetic algorithm. After precise solution, exact calculation would be achieved through using BP network, thus can finally achieve global search and rapid and efficient purposes.

For the three-layer BP network, w_{ih} is set to connection weights from the i -th node in input layer to the h -th node in hidden layer; w_{ho} is set to connection weights from the h -th node in hidden layer to the o -th node in input layer; θ_j is set to threshold for the j -th node in hidden layer; θ_s is set to threshold for the s -th node in input layer. The steps of genetic algorithm to optimize the BP neural network are as follows:

(1). Initialization population P , including crossover、scale、crossover probability P_c、mutation probability P_m and Initialization of any w_{ih}、 w_{ho} and θ_j、θ_s .
(2). Calculate each individual evaluation function, and sort them. Network individuals can be chosen according to following probability:

$$P_r = \frac{f_i}{\sum_{i=1}^{N} f_i} \quad (1)$$

In the formula above, f_i is the fitness value of individual i , which can be measured by the error squares E , namely

$$f_i = \frac{1}{E(i)} \quad (2)$$

$$E(i) = \sum_{k=1}^{m}\sum_{o=1}^{q}(d_o - y_o)^2 \quad (3)$$

In the formula: $i = 1, 2, ..., N$ is the number of chromosome; $o = 1, 2, ..., q$ is the number of output layer node; $k = 1, 2, ..., m$ is the number of studying samples; y is the actual output for network, and d is the expected output of the network.

(3). The crossover operation of individuals of P_i and P_{i+1} , through crossover probability P_c , generates new individuals of P_i' and P_{i+1}' , while those don't encounter crossover operations reproduce directly.

(4). The mutation of mutation probability P_m changes the individual of P_j into a new individual of P_j' .

(5). Insert the new individual into the population, and calculate the evaluation function of the new individual.
(6). How to judge whether the algorithm has ended: Another round of computation of step (3) should continue if no satisfactory individual was found.

(7). The optimized initial value which resulted from genetic algorithm is initialized weights and threshold. Meanwhile, train the network by improved BP algorithm till the error precision is decided.

4 Switch Current of GA_BP Network Diagnoses Methods

Switch current circuit entirely consist of MOS device, and in current Transfers signals, may diagnose less node. At first switch circuit is simulated and obtain the output current signal waveform in the paper, after that the corresponding feature extraction form feature vector, then GA_BP network identify and diagnosis.

This paper adopts the program of GA_BP network diagnostic methods as follows:

(1). The pretreatment fault signal and the characteristic vector form. First, OrCAD-10.5 treated with fault simulation of diagnostic circuit in different fault, under the condition of the circuit terminal output waveform. This incentive stimulates circuit, amplitude is 50uA, cycle is sinusoidal signal of 100μs, Then it will get output signal waveform, Harr wavelet is decomposed by 6 layers wavelet, It will get different levels of approximation coefficients c_j^k and detail coefficients d_j^k , Again according to the following steps, to the energy for element structure the vector of characteristic. The each frequency energy is $E = (E_0, E_1, E_2, E_3, E_4, E_5)$, which is made six floors wavelet of decomposition, $E_i (i = 1,...5)$ mean the energy of each signal layer. E_0 mean the energy of the sixth floor of about signal part, E_i mean the energy of the i floor of specific signal part, so $E_0 = \sum_{j,k} \left| c_{j,k}^6 \right|^2, E_i = \sum_{j,k} \left| d_{j,k}^6 \right|^2, i = 1,...5$,

Thus there has normalized feature vector $T = \left(\frac{E_0}{\|E\|}, \frac{E_1}{\|E\|}, ..., \frac{E_5}{\|E\|} \right)$.

Finally the characteristic vector is made PCA analysis, the redundancy of each feature vector is removed and normalized, in order to enhance the characteristics of feature vector signal, then which is feed into GA_BP networks to train and test, it could reduce the size of the neural network, decrease the time and improve the purpose of performance.

(2). On account of system architecture design of GA_BP network. Based on the result of characteristic vector decomposed and created, this paper selected network's input layers, which percepts six nodal points, according to performance requirement, the number of neurons in hidden layer determine the optional for 10 ~ 30, the number of neurons output to in the fault diagnosis is based on the number of categories and fault-free state as 4 (a binary code form output), according to classified requirements, its neuron activates function to identify logsig function, its output for [0, 1], which are easily to give the circuit diagnosis directly.

(3). GA_BP network training and testing. Input the training sample, GA_BP network is made adequate training, and input test sample to test.

5 Diagnostic Circuit Example

Fig 2 shows switch current resonance circuit fault diagnosis, for example, to explain GA_BP network can be a very good overcome BP network into the local minimum, slow convergence speed and structural instability shortcomings. The parameters set each component, including each MOS device tolerance, set to 10%. Circuit fault including component open and short-circuit fault.

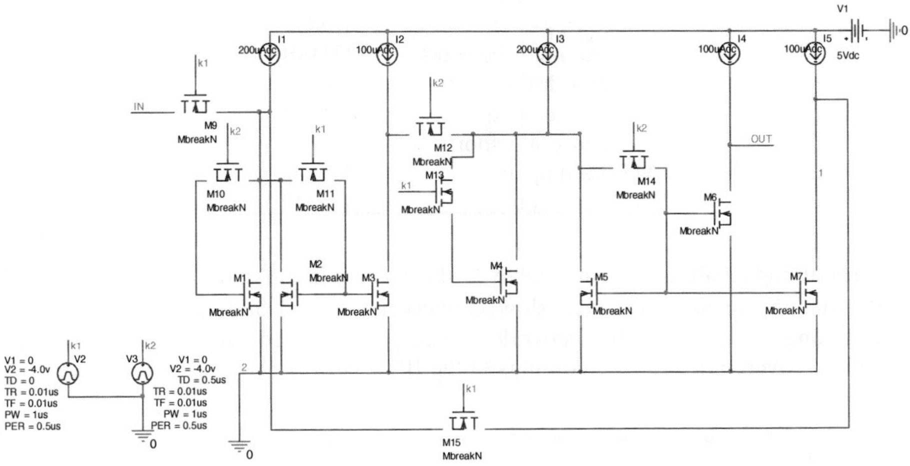

Fig. 2. switch current resonance circuit

5.1 Fault Type Determination

For simplify fault type, it don't consider the MOS switch existence fault. Consider each MOS device to the extent of the output circuit influence, so we proceeding sensitivity analysis for circuit-under-test, firstly. Through the relative sensitivity analysis and the structure characteristics of the circuit, then nine faults are determined finally, the type may be summarized in table 1. At the same time, circuit data respectively was trained and tested by BP network and GA_BP network.

5.2 The Training of GA_BP Network and BP Network Compare the Results

Two network use the same structure, input layer, hidden and output layer respectively establish six neurons 、 25 and 4, target performance index establish 0.001, learning rate establish 0.1. The simulation of MATLAB environment is for training. Extract feature vector input to the network was trained as shown in Fig3 and Fig 4 shows. After 77 steps, GA_BP network come up to the training target, after 347 steps the BP network eventually falls into the local minimum value. At the same time, through 1,000 times training, BP network in local minimum probability can achieve 46.7%, GA_BP networks can achieve 100%.

Table 1. Fault classification

fault-free	NOR
M1 source	M1SOP
M2 drain open	M2DOP
M2 source open	M2SOP
M3 drain-source short	M3DSS
M3 grid-drain short	M3GDS
M3 source open	M3SOP
M6 grid-drain short	M6GDS
M7 drain open	M7DOP
M7 source open	M7SOP

Part training results as shown in table 2, the data are intuitive reflect the BP network easily into the local minimum、slow convergence speed and structural instability shortcomings. But GA_BP network convergence speed and stable structure, completely overcome the shortcomings of the BP network.

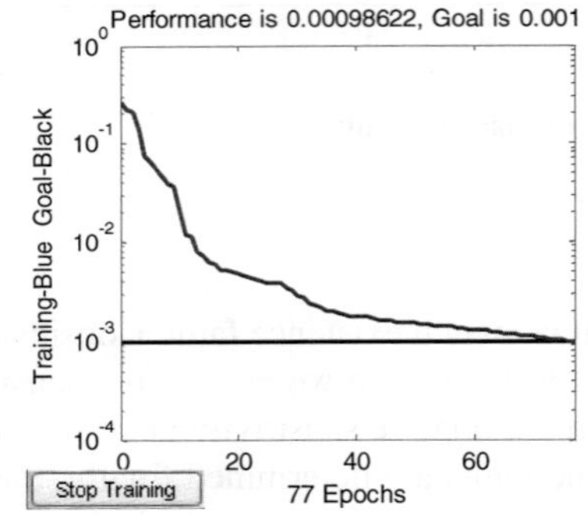

Fig. 3. GA_BP Network training times and the error

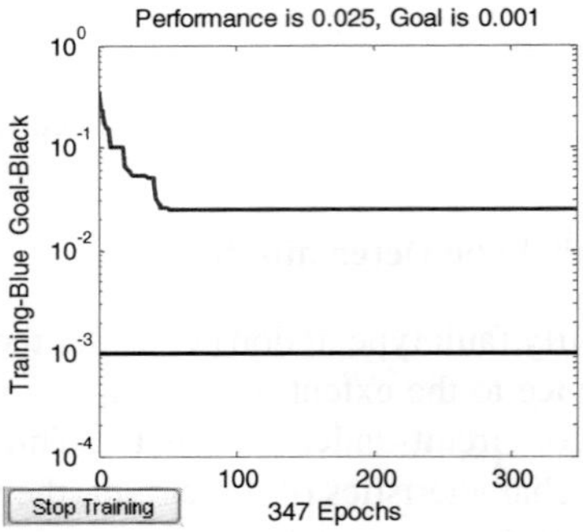

Fig. 4. BP Network training times and the error

Table 2. Part of BP network training result and GA_BP network comparison

Contrast project	BP Neural Network			GA_BP Neural Network		
Steps	107	292	41	62	73	43
Error	0.1000	0.025	0.000896	0.000958	0.000997	0.000979

Test data in put to the GA_BP neural network training, which has been trained well. Diagnosis data such as shown in table 3, get the actual output and the target have very good approximation.

Table 3. Compare with Failure actual output and target value

Fault type	Target value				actual output			
NOR	0	0	0	0	0.0027	0.0670	0.0068	0.0071
M1SOP	0	0	0	1	0.0097	0.0804	0.0328	0.9999
M2DOP	0	0	1	0	0.0034	0.0082	0.9967	0.0000
M2SOP	0	0	1	1	0.0062	0.0006	0.9842	0.9887
M3DSS	0	1	0	0	0.0030	0.9983	0.0055	0.0538
M3GDDS	0	1	0	1	0.0014	0.9978	0.0014	0.9893
M3SOP	0	1	1	0	0.0470	0.9968	0.9956	0.0422
M6GDS	0	1	1	1	0.1740	0.9961	0.9967	0.9303
M7DOP	1	0	0	0	0.9921	0.0701	0.0003	0.0045
M7SOP	1	0	0	1	0.9973	0.0038	0.0020	0.9908

6 Conclusion

In this work, we have used genetic algorithm as Optimization measures to optimize the BP neural network, this method apply to the fault diagnosis of switch current circuit. By taking advantages of the Global Search of the genetic algorithm to find the most suitable network connection weights and threshold value, the method not only can reduce the searching space of neural network, improve search efficiency, but also can reduce the network training time. The experimental results show that GA_BP network of switch current circuit faults diagnosis method can effectively classification and diagnosis. This network is very well solve the BP network entrap local minima, slow convergence speed and network structure unstable shortcomings.

Acknowledgements

This work was supported by the National Natural Science Funds of China for Distinguished Young Scholar under Grant No. 50925727, National Natural Science Foundation of China under Grant No.60876022, Hunan Provincial Science and Technology Foundation of China under Grant No.2010J4, the cooperation project in industry, education and research of Guangdong province and Ministry of education of China under Grant No. 2009B090300196.

References

[1] Toumazou, C., Hughes, J.B., Battersby, N.C.: Switch currents - digital process simulation technology (1993)

[2] Renovell, M., Azais, F., Bodin, J.-C., Bertrand, Y.: Functional and structural testing of switched-current circuits. In: IEEE European Test Workshop Proceedings, May 25–28, pp. 22–27 (1999)

[3] Long, Y., He, Y., Yuan, L.: Fault dictionary based switched current circuit fault diagnosis using entropy as a preprocessor. analog integrated circuits and signal processing 66(1), 93–102 (2010)

[4] Stopjakova, V., Malosek, P., Matej, M., Nagy, V., Margala, M.: Defect Detection in Analog and Mixed Circuits by Neural Networks Using Wavelet Analysis. IEEE Trans. Reliability 54(3), 441–448 (2005)
[5] He, Y., Liang, G.: Analog circuit fault diagnosis of the BP neural network method. Journal of Hunan University (JCR science Edition) 30(5), 35–39 (2003)
[6] He, Y., Tan, Y., Sun, Y.: Wavelet neural network approach for fault diagnosis of analougue circuits. In: IEEE Proc.-Circuits Devices Syst., vol. 8(4), pp. 379–384 (2004)
[7] Li, M., Ke, J., Li, D.: The basic theory of genetic algorithm and applied. Science Press, BeiJing (2002)
[8] Qi, X., Pan, Z.: Based on the high order fuzzy BP neural network of genetic algorithm is application in gear fault diagnosis. Mechanical Drive (4), 44–46 (2004)

Equipment Acquisition Information Service Management

Yanmei Wu[1], Kai Guo[2], Pengwen Xu[1], and Ziran Li[1]

[1] Department of Equipment Acquisition, Academy of Equipment Command & Technology, Beijing, P.R. China
[2] Management Department, Engineering Design and Research Institute, Beijing, P.R. China
wuyanmeiland@sina.com

Abstract. Information Technology Infrastructure Library (ITIL) is the most popular "best practices" framework for managing information technology services. According to the problems of IT infrastructure and application systems existed in equipment acquisition information, this paper sketches the information service management framework of equipment acquisition by using ITIL ideology, and specifies the information service processes involved in equipment acquisition.

Keywords: Information Technology Infrastructure Library (ITIL), Equipment acquisition, Information service, Business process.

1 Introduction

With the development of information technology, it is widely applied to various business phases of equipment acquisition. However, the established information systems have such problems as follows: information sharing difficult, function singularity, poor scalability, since they lack unified planning and standards, which directly impacts on the development of equipment acquisition information. The main problems are in that: (1) information service development is uneven between the military and the contractors. (2) IT infrastructure is lack of effective integration with organizational business requirements; (3) various heterogeneous systems are difficult to be integrated (4) the evaluation system of information service is not perfect. How to manage the increasing IT system to provide reliable and efficient information services is facing major challenges. Many large enterprises at home and abroad have adopted IT service management standards such as ITIL. These ITIL best practices provide a reference for our information service management.

The paper is structured as follows: Section two briefly describes the ITIL reference model. Subsequently, in Section three, the information service management of equipment acquisition based on ITIL is introduced. Following that, Section four specifies the information service processes involved in equipment acquisition. The conclusion is arrived at in Section five.

X. Wan (Ed.): Electrical Power Systems and Computers, LNEE 99, pp. 951–956.
springerlink.com

2 ITIL Overview [1, 2, 3]

Information Technology Infrastructure Library (ITIL) is a service management standard architecture developed by the Central Computer and Telecommunications Agency (CCTA) in the mid-80s aiming at the information domain. It includes six modules: IT service management implementation planning, service management, infrastructure management, business management, application management and security management. The service management module, which includes service delivery and service support, is the core component of this framework. The task of services delivery management is to plan and design service capacity, continuity, availability, service level, as well as to manage and improve the services delivery processes. It involves five processes: service level management, availability management, capacity management, service continuity management and financial management. The task of services support management is to ensure the stability and flexibility of information services. It involves service desk and five processes: incident management, problem management, change management, configuration management and release management. The relationship between these processes is shown in Fig.1.

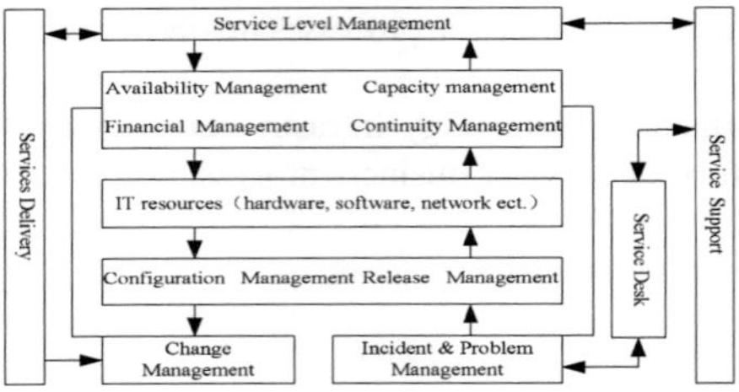

Fig. 1. Relationship between the service management processes.

Based on the idea of ITIL, we study the information service management of equipment acquisition, and construct the corresponding framework, as well as specify its service processes.

3 Information Service Management System of Equipment Acquisition

According ITIL idea, the concept of information services management in equipment acquisition can be defined that the information service management department combines the requirements of business services and network technology to manage business information systems of equipment acquisition and staff.

Information service management system can rapidly respond to requests for information services through support continued and reliable operation for information infrastructure and all kinds of information systems. According to business service levels, it can configure service resources effectively to improve the quality of information services.

Information service management system of equipment acquisition includes service management organizations, service users, management standards, information service management processes, and information services improvement. Through service level agreements, service delivery and service support as well as continuous improvement, information service system is organically integrated into a virtuous cycle. The information service management system [4] is shown in Fig.2.

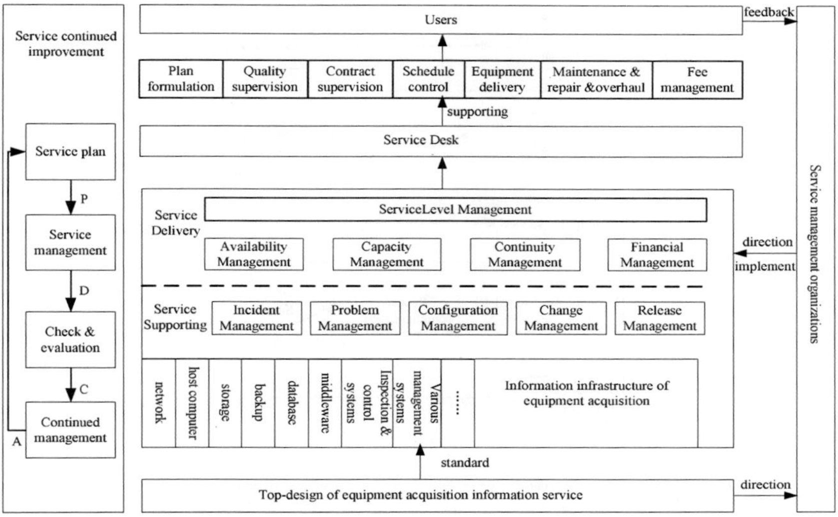

Fig. 2. The architecture of information service management for equipment acquisition.

The top-design of equipment acquisition information service is the statute and criterion of equipment acquisition information service established by the military to instruct the service management, and thus to ensure the development of equipment acquisition information with rapid and stable.

IT infrastructure includes such facilities as follows: network, storage, backup, database and middleware and so an. They are the basis conditions for the equipment acquisition information.

The Deming quality cycle of PDCA is adopted to enhance the information service of equipment acquisition. P means plan. D means do. C means check. A means act. The PDCA cycle is equivalent to closed–loop control. It introduces feedback about business results into IT service to drive IT-related decision making and thus to improve service quality.

4 The Processes of Information Service in Equipment Acquisition

4.1 Service Delivery Process

Service delivery management involves five processes: service level management, availability management, capacity management, service continuity management and financial management.

Service level management focuses on the acquisition business requirements through the coordination of equipment acquisition department, contractors, equipment used troops and information service department. To design the service level agreement, the first step is to investigate and understand business requirements of equipment acquisition by repeating communication. After that is to write a document about service level requirements (SLR). And then is to develop service level agreement (SLA). The process is shown in Fig.3.

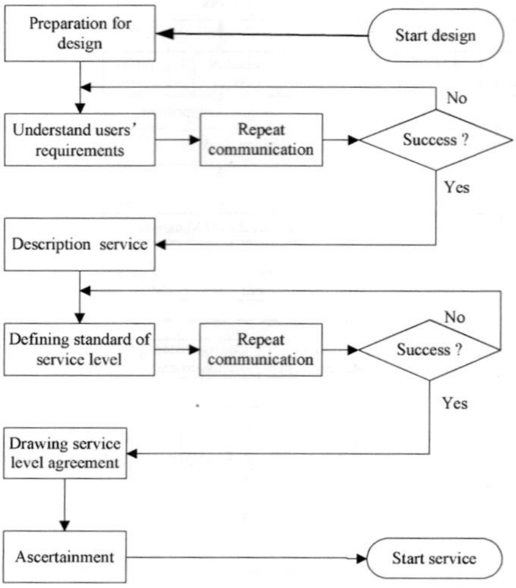

Fig. 3. Service level management process.

Service level management relies on the other areas of the service delivery process to provide the necessary support which ensures the agreed services are provided in a cost-effective, secure and efficient manner.

Financial management is to control the costs of information service at the most effective price (which does not necessarily mean cheapest). It should reflect the overall objectives of equipment acquisition information. Through the comprehensive account of operation costs and the services provided for various business departments, the detail information about the IT construction of equipment acquisition will be obtained, which is helpful to decision make and manage information assets.

Capacity management is to ensure the optimum and cost-effective IT service today and in the future by match IT resources to equipment acquisition business demands.

Continuity management is to ensure that information services can be recovered and continued even after a serious incident occurred. Therefore, the process involves the following basic step:

1) Analyzing the characteristics of equipment acquisition department, contractors and equipment used troops respectively.

2) Establishing the priorities the business activities by conducting a business impact analysis (BIA)

3) Performing the risk assessment for each business.

Availability management is responsible for ensuring the information service-availability at a justifiable cost by monitoring and improving the information service-availability. Thereby, designing the process of availability management should understand the availability of information service and components from the business perspective to ensure service goals.

4.2 Service Support Process

Service support management involves service desk and five processes: incident management, problem management, change management, configuration management and release management.

Service desk is the single point of contact between service providers and users. It is also a focal point for reporting incidents and making service requests. When the user made a service request or report incidents, it is responsible for recording these requests or incidents and try to solve them. If it cannot resolve them, it will transfer them to corresponding groups and coordinating them interaction with users. In our work, a mixed service desk is adopted (shown in Fig.4), since it combines the advantages of distributed service desk and centralized service desk. It not only facilitates the unified management but also provides the personalized service support.

Incident management is to restore normal service operation as quickly as possible and minimize the adverse effects on business operations. So designing the incident management process should determine the priority of dealing with the incidents according to the urgency and difficulty degree.

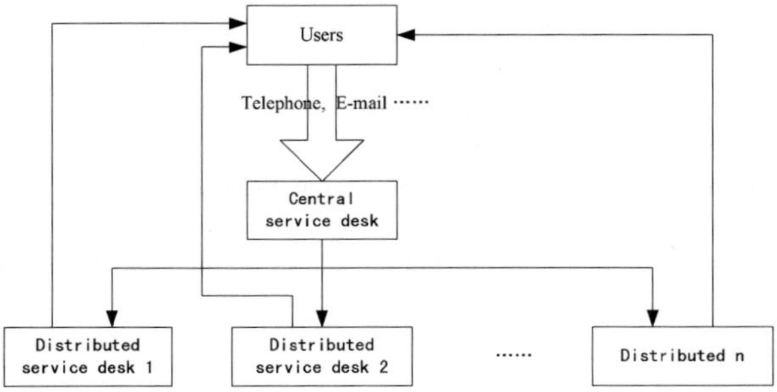

Fig. 4. Mixed service desk.

Problem management is to resolve the root causes of incidents and thus to minimize the adverse impact of incidents and problems on business that are caused by errors within the IT infrastructure, and to prevent recurrence of incidents related to these errors. Therefore, collecting and utilizing incidents are important for the problem management. Indeed, the incidents management process is not only information flow process but also problem treatment process. Through gathering, transferring and disposing of corresponding information, the management department can effectively manage business departments to be in an order state.

Change management aims to ensure that standardized methods and procedures are used for efficient handling of all changes. It has been utilized in two situations. One is in passive. The other is in active. The former is adopted when the problems occurred and have to be interrupted or the regulations demand. The latter is adopted for the sake of higher service efficiency and effect.

Release management is to protect the news or changed software and/or hardware required to implement approved changes.

Configuration management is the management and traceability of every aspect of a configuration from beginning to end. It provides a whole framework of the IT infrastructure and information service management so that the IT infrastructure can support the information service operation effectively and management staff can comprehend the changes of every configuration item.

5 Conclusion

Our investigation is an attempt to construct the information service management system of equipment acquisition based on ITIL idea. In information service framework, IT department's main task is to deliver IT services with high quality and low cost and to improve the communication between information departments and other business departments as well as to help to plan, research and implement IT systems through the coordination of operation processes. However, our research is in beginning stage. At present, we only construct an experimental environment. In the future, we will further study ITIL application in the real world of equipment acquisition information.

References

1. IT Infrastructure Library, `http://www.ogc.gov.uk/ITIL/`
2. The ITIL and ITSM Directory, `http://www.itil-itsmworld.com`
3. Lu, K.: ITIL Application Research on the Shanghai Telecom Co. information. Master thesis, Fudan University (2008)
4. Ziran, L.: Equipment Acquisition Information Service Management Based on ITIL. Master thesis, Academy of Equipment Command & Technology (2010)

Topological Variable-Density Algorithm Based Design Method for Lightweight Machine Tools

Pengzhong Li[1], Jie Liu[1], and Shousheng Liu[2]

[1] Sino-German College for Graduate Studies, Tongji University, Shanghai 200092, P.R.China
[2] Shenyang Machine Tool Co., LTD, Shenyang 110142, P. R. China
leepz@tongji.edu.cn, qinshi153@163.com, shousheng_liu@smtcl.com

Abstract. Topology optimization lends an effective tool for lightweight work of machine tools. With variable density algorithm, the mathematical model of optimization was given. After structural analysis, analyzing model for topology optimization, with practical machine tool structure, was built. The method of choices of optimization variables and constraints were related. Based on optimizing calculation, the density contour is displayed. And then the irregular structure after topology optimization is abstracted and simplified, building new improvements of machine tool structure based on topology optimization results. Re-analysis results show that, after optimization, the structure compliance is reduced; and the 1st-3rd natural frequencies are improved; the structure mass is reduced by 5.55%. Meanwhile, the new structural changes are very easy to achieve for the processing technology. With further maturity of topology optimization theory, the lightweight work of machine tools will have increasingly active role to reduce material consumption.

Keywords: Topology optimization, variable-density algorithm, lightweight design method, machine tools.

1 Introduction

To improve the economics and reduce production cost of machine tools, the lightweight design method has been gradually received more and more attention by machine tools manufacturer. Lightweight design of machine tools mainly focuses on two aspects, material and structure. Developing new functional materials to replace traditional steel and cast iron is an effective way for loss of weight. A successful sample is using new plastic instead of cast iron to manufacture slide rail. However, development and application of new materials have a long way to go, despite some great progress [1]. For structure of machine tools, especially in modern NC machine tools or Machining center, overall shape and arrangement of function units, compared with traditional machine tools, have undergone tremendous changes. Generally, machine tool design is often guided by empirical data, resulting in cumbersome structure. Based on optimizing algorithm to allocate optimized objects (design parameters such as size, shape, material and etc.) rationally to ensure the overall stiffness and other characteristics of machine tools, structure optimization design rising up recently allows structural design to get rid of blind dependence on the experience.

X. Wan (Ed.): Electrical Power Systems and Computers, LNEE 99, pp. 957–963.
springerlink.com

Structural optimization is divided into size optimization, shape optimization, and topology optimization. Relatively, compared with size optimization and shape optimization, topology optimization has more design freedom, can get greater design space. Therefore, topology optimization is the most promising optimization method. Topology optimization is an effective design tool for a variety of engineering applications, it can be used in conceptual design stage of structural, mechanical and automotive components, e.g. in [1][2]. Also, it is applicable to the synthesis of compliance mechanisms [3][4].Much research work has been done towards this design technology as mentioned in [5].

2 Topology Optimization Technology

Structure topology optimization can be traced back to the pioneer work by Kohn and Strange [6][7], Bendsoe and Kikuchi [8]. Topology optimization is based on the optimal design theory and method to solve optimization model, and finally achieve a reasonable distribution of materials to make the structure meeting the design requirements. This allows people to solve engineering problems by selecting the best possible design from the numerous design programs, thus greatly improving the efficiency of engineering design. In the last decade, the field of structure topology optimization has expanded significantly, successfully addressing many practical engineering problems [9].

2.1 Topology Optimization Theory and Its Methods

Topology optimization is a kind of structural optimization, and its research can be divided into continuum topology optimization and non-continuum topology optimization, and both areas are dependent on the finite element method. Topology optimization is under the action of certain external forces and to seek the structure layout, that is with the best power transmission path. For continuum structural topology optimization problems, the basic approach is to get design unit area, which is divided into finite elements (such as shell elements or solid elements), and delete some part of the unit area based on a certain algorithm, forming a continuum with holes in order to achieve continuous Body optimization. And it is essentially combinatorial optimization problem with 0-1 value discrete variables. Corresponding with continuum topology optimization, topology optimization of discrete structures is to build a base structure, which is formed by a finite number of beam elements within the design space. By relaxation of discrete optimization variables as a continuous variable optimization problem, the optimization algorithm based on the derivative of a continuous variable will be used. The optimization of continuous design variables model, in order to achieve topology optimization, will be used instead of the original discrete design model. At present, the main continuum topology optimization methods include homogenization method, variable density method, evolutionary structural optimization method (ESO) and the horizontal method. The non-continuum topology optimization methods are mainly based on the base structure with different optimization strategies (algorithms), such as topology optimization based on genetic algorithms and so on. Currently, the topology optimization studies have been more mature, of which variable density method has been applied to the commercial optimization software. And two famous topology optimization software are the OptiStruct module of Altair's Hyper works in the United States and the FE-design's Tosca in Germany.

2.2 Principle of Variable Density Algorithm and Its Mathematical Model

The basic principle of variable density algorithm is the introduction of a hypothetical material with variable density. After the discretization of continuum into finite element model, the density of each unit is designated as the same, and the density of each unit will be used as the design variables. When the relative density (Xe) of each unit is 1, it means that the unit contains material, so the unit (entity) should be retained or increased; in contrast, when Xe = 0, it indicates that the unit does not contain material, the unit should be removed (as holes). In topology optimization, the relative density of the material should be as much as possible closed to 0 and 1 in the design area. Taking the minimum compliance (deformation energy) of the structure as optimizing objective, and considering material mass constraint (or volume constraint) and structure balance conditions, mathematical model of variable density method for topological optimization can be expressed as following,

$$\begin{aligned} &\text{Finding} \quad X = (X_1, X_2, \cdots, X_n)^T \\ &\text{Min} \quad C = F^T D \\ &\text{St.} \quad \begin{cases} f = \dfrac{V - V_1}{V_0} \\ 0 < X_{min} \le X_e \le X_{max} \qquad e = 1, 2, \cdots, n. \\ F = K \bullet D \end{cases} \end{aligned} \tag{1}$$

Where,

C - Structural flexibility (compliance)
F - Loads vector
K - Stiffness matrix
D - Displacement vector
V - Structure volume, which is full of material
V_0 - Volume for structural design domain
V_1 - Material Volume, when the unit density is less than X_{min}
f - Percentage of surplus material
X_{min} - Lower limit of unit relative density
X_{max} - Upper limit of unit relative density

3 Case Study

3.1 Structure Analysis

Analyzed case is a five-axis Milling & Turning Center HTM40100h developed independently by Shenyang Machine Tool Co. Ltd. With high precision and efficiency advantages, it can finish multi-process high-accuracy machining by one clamping.

HTM40100h is mainly composed of the bed, upright column, tray module, spindle, tailstock, as shown in Fig.1. The Milling & Turning Center is fixed by 13 ground screws. The tray module can slide up and down along lead rails in the column, and the column can also slide right and left along the lead rail on machine bed.

The machine bed structure was designed through empirical experience and analog. Basically, the material inside the bed was hollowed out, and there are five vertical ribs and a horizontal rib inside. Nevertheless, the overall mass of the Milling & Turning center is 15,456 Kg, and only the bed mass is 7469.35 Kg.

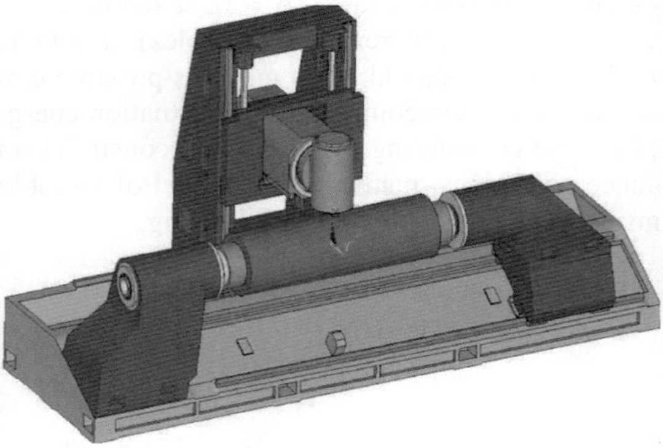

Fig. 1. Digital model of Milling & Turning Center HTM40100h

3.2 Analyzing Model Preparation

As optimizing object, the model of machine bed was appropriately simplified and imported into analysis platform. The bed is fixed by 13 ground screws; a displacement constraint condition was given. The reaction forces from processed work piece and the gravity of column and tray module are acting on the bed body through two lead rails. Therefore, a corresponding pressure can be imposed on two faces of the rails. At the same time, the pressure is also acting on the bed body through the spindle and the tailstock when the Milling & Turning Center is machining a work piece. By calculating, pressures on plane of the lead rails, spindle and tailstock are respectively 0.3MPa, 0.15MPa and 0.1MPa.

Research results have proved that performance of machine tools is mainly affected by low-order nature frequencies. Therefore, based original design, the optimized structure should maintain its low-order nature frequencies have little discrepancy with the original ones. This means that impact of optimizing work on structure dynamic performance can be reduced to minimum through change control of low-order frequencies. At the same time, one only needs to study how to improve the natural frequency of low-order modes in order to reduce the resonance possibilities effectively when the machine tool is processing with a low-speed. The 1st-3rd modes of the bed can be clearly demonstrated the dynamic nature of the bed.

3.3 Topology Optimization Process of Machine Bed

After determining loads and boundary conditions of the bed, the topology optimization also needs to define optimization parameters clearly. For example, the topology optimization variables should be defined. The green area of the bed CAE model is the interface with the other machine's module, so this part can't participate in the optimization process. The red area is the range of topology optimization variables, as shown in Fig.2.

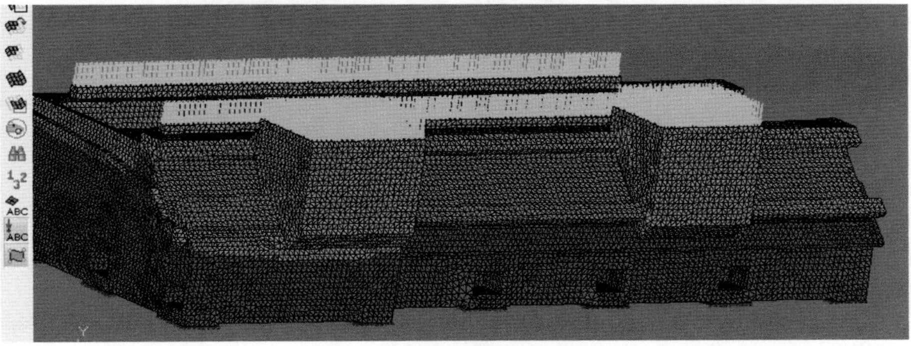

Fig. 2. Load diagram of the topology optimization

When the optimization variables were defined, the optimizing objective and constraints must be defined clearly. The stiffness increase can be taken as optimizing goal, and the constraint is that the first natural frequency should be equal or more than the original value during the optimizing process. After 13 optimization iterations, the density contour of the bed on the back is shown in Fig.3.

3.4 Result Analysis of Topology Optimization

The spatial structure after topology optimization is mostly irregular, so results of topology optimization need abstracted and simplified. New improvements of the bed structure should be built based on topology optimization results. According to density contour, long oblique panel in the front area of the bed should be removed; some holes and waist-shaped holes should be opened in the right place. Specific changes in some locations are shown in Fig.4.

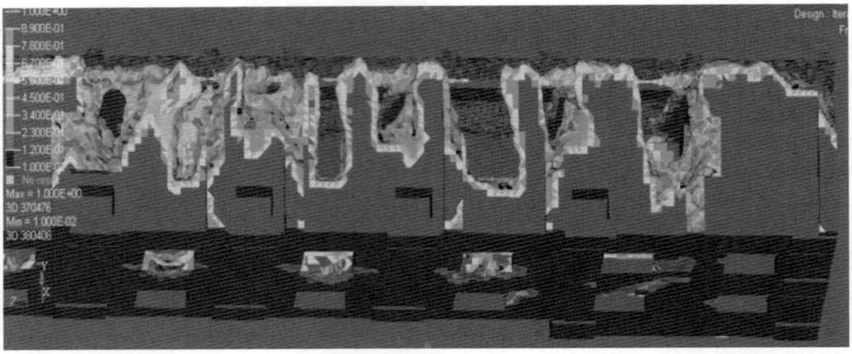

Fig. 3. Density contour of the bed on the back after optimization process

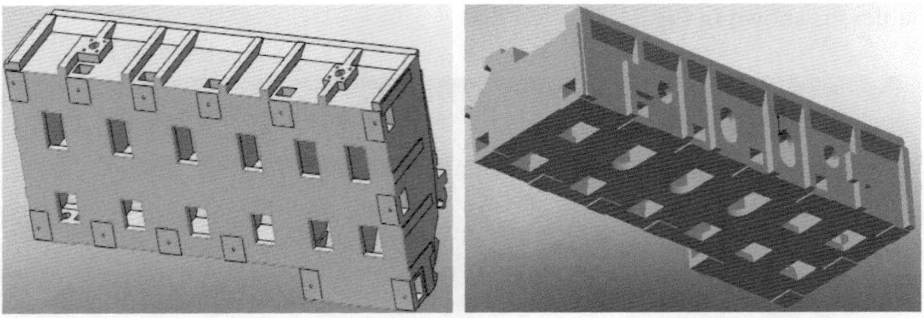

Fig. (4). a. Original structure of machine tool bed. **b.** Improved machine tool bed based on optimizing results

With re-analysis of static and dynamic characteristics before and after the optimization process, the results comparison is shown in Table 1.

Table 1. Comparison of Bed Indicators Before and After Topology Optimization Process

Indicator	Original bed	Improved bed	Change value	Change Percent
Mass	7469.35Kg	7054.84Kg	414.55Kg	5.55%
Compliance	150	137	-13	-8.67%
1st freq.(Hz)	287.80	296.24	+8.44	2.93%
2nd freq.(Hz)	334.91	342.13	+7.22	2.16%
3rd freq.(Hz)	390.18	395.54	+5.36	1.37%

From the comparison, it can be seen that the compliance of the bed is slightly reduced, which means that the stiffness increased; and the 1st-3rd natural frequencies are improved. The most impressive is that the mass y reduced by 414.55Kg, and 5.55% weight loss is a successful achievement of optimal goal. Meanwhile, the new structural changes in the bed are very easy to achieve for the processing technology, so these changes will not bring great difficulties to technology sector.

4 Discussions

Topology optimization lends an effective tool for lightweight work of machine tools. In application of variable density algorithm, however, a necessary precondition is the studied machine tool can reach the required performance index with original structures, which determines whether the optimizing results are meaningful or not. For topology optimization, choices of optimization variables and constraints are also decision factors to optimizing results, which need designers with abundant experience information. A major advantage of topology optimization is to find a best design route of lightweight associated with circumventing the computational complexity of the construction. The work presented serves as the starting point to lightweight design of structures getting rid of the blind dependence on experiences. With further maturity of topology optimization theory, the lightweight work of machine tools will have increasingly active role to reduce the material consumption.

Acknowledgements

The research was done thanks to Chinese government project of Mega-project of High-grade NC Machine Tools and Basic Manufacturing Equipment (2009zx04001-072) "HTM40100 horizontal mill-turn compound machine center"

References

1. Yildiz, A.R., Kaya, N., Ozturk, F., Alanku, O.: Optimal design of vehicle components using topology design and Optimization. Int. J. Vehicle Des. 34, 387–398 (2004)
2. Boonapan, A., Bureerat, S., Limtragool, J., Inban, S.: Multi-level design of an automotive part. In: 19th Conference of Mechanical Engineering Network of Thailand, Phuket, Thailand, pp. 172–177 (2005)
3. Bureerat, S., Boonapan, A., Kunakote, T., Limtragool, J.: Design of compliance mechanisms using topology optimization. In: 19th Conference of Mechanical Engineering Network of Thailand, Phuket, Thailand, pp. 421–427 (2005)
4. Luo, Z., Chen, L., Yang, J., Abdel-MaLek, K.: Compliance mechanism design using multi-objective topology optimization scheme of continuum structures. Struct. Multidisciplinary Optim. 30, 142–154 (2005)
5. Bendsoe, M.P., Sigmund, O.: Topology Optimization: Theory Methods and Applications. Springer, Berlin (2003)
6. Kohn, R.V., Strange, G.: Optimization design in elasticity and plasticity. Nummer. Methods Eng. 22, 183–188 (1986)
7. Kohn, R.V., Strange, G.: Optimization design and relaxation of variational problems. Comm. Pure Appl. Math. (New York), 391-25(Part I), 139-182(Part II), 353-357(Part III) (1986)
8. Bendsoe, M.P., Kikuchi, N.: Generating optimal topologies in structural design using a homogenization method. Comput. Methods Appl. Mech. Eng. 71(2), 197–224 (1988)
9. Chen, B.-C., Kikuchi, N.: Toplpogy optimization with design-dependent loads. Finite Elements in Analysis and Design 37, 57–70 (2001)

Study on Effect of Wind Power System Parameters for Hopf Bifurcation Based on Continuation Method

Li Ji and Zhou XueSong

Institute of Electrical Engineering and Automation, Tianjin University,
Tianjin, 300072, P.R. China

Abstract. In order to study the effect on voltage stability by continuous changes of the wind power system parameters. Using the active power and reactive power and network admittance of the wind power system as the bifurcation parameters, single parameter Hopf bifurcation point and two parameters Hopf bifurcation boundary are calculated based on the continuation method by using the bifurcation theory with wind power system dynamic models, and the effect of parameters on Hopf bifurcation is studied, meanwhile the control role of static var compensator (SVC) on Hopf bifurcation is analyzed. The research results show that different parameter change causes different effect on Hopf bifurcation, furthermore SVC can delay Hopf bifurcation, increase wind power system voltage stability region.

Keywords: wind power system, Voltage stability, Continuation method, Hopf bifurcation, Static var compensation.

1 Introduction

The growth of wind energy has mushroomed over the past decade. Over the next twenty years, there will be more significant growth in wind energy with the expectation of 20% wind grid penetration by 2030. However, the penetration of wind power starts to influence the power system behavior [1]. The operational experience shows that the problem generated by large-scale wind power integration basically is the voltage instability [1], which brings great challenges to large-scale wind power utilization.

It is the substantial results of researches on voltage stability of general power system based on bifurcation theory[2], combined with nonlinear dynamic characteristics of wind power system, bifurcation theory is introduced to the study of voltage stability of wind power system [3]. Moreover, most of the present bifurcation researches for wind power system are based on static bifurcation analysis [4], rarely involve dynamic bifurcation research. The researches on general power system show that the system is closed to "nasal curve tip" when Hopf bifurcation occurs, where attracting areas of the operating point have become very small. Therefore, the research on dynamic characteristics near bifurcation point of wind power system is very important.

X. Wan (Ed.): Electrical Power Systems and Computers, LNEE 99, pp. 965–971.
springerlink.com

In this paper, the equilibrium curve of wind power system is tracked based on the differential-algebra equations model by using continuation method, the one-dimensional Hopf bifurcation parameter value and the two-parameter Hopf bifurcation boundary are tracked, and the effect of parameters on voltage stability of wind power system is analyzed, and the control role of SVC on Hopf bifurcation is analyzed.

2 Continuation Method

Consider a nonlinear dynamical system described by equation (1):

$$\dot{x} = f(x,\lambda), x \in R^n, \lambda \in R^n \tag{1}$$

Where x is state vector, λ is bifurcation parameter. If the equation (1) has Hopf bifurcation at (x , λ), then the system Jacobin matrix $A \equiv \partial f / \partial x$ at (x , λ) has a pair of pure imaginary eigenvalues, notes for $j\omega$.

$$\begin{cases} f(x,\lambda) = 0 \\ Aq = j\omega q \\ \langle q, q_0 \rangle = 1 \end{cases} \tag{2}$$

The equation (2) is a plural one, namely vector q and q_0. The unknown number is (x , λ , q , ω), it is obvious that the numbers of equation are equivalent to the numbers of unknown number.

The equilibrium points of (1) satisfy $f(x,\lambda) = 0$, that is, the equation defines a one dimensional generalized curve (also named manifold) M called equilibrium curve in n+1 dimensional space $y \equiv (x,\lambda) \in R^n \times R^1$. And the continuation method can be used to track the manifold, it uses a sequence of points (y^1, y^2 , ...) satisfying the requirements of equilibrium point to approximate the curve M. In the process of tracking the manifold, whether bifurcation points exist on this manifold is judged through inspecting local bifurcation conditions. General continuation method can track and work out one-dimensional manifold, thus we can only get bifurcation values of a single parameter by the calculation.

The method of [5] to track bifurcation boundary of two-dimensional parameters of wind power system is adopted. Now assuming that we have got bifurcation point of single parameter through calculation, we can explore how to apply continuation method to figure out the mathematical equation processed by direct method. Its meaning is equivalent to using the continuation method to track manifolds satisfying the local bifurcation, rather than equilibrium manifold under the above meaning, at this time we can figure out the local bifurcation boundary of two-dimensional parameters in (1).

3 Wind Power System Models

Wind power system model shown in Fig.1 consists of a wind power bus and two generator buses, one generator bus is a slack bus, the other has a fixed voltage magnitude and swing equation dynamics as given in (3), wind power plant is simulated by dynamic Walve load.

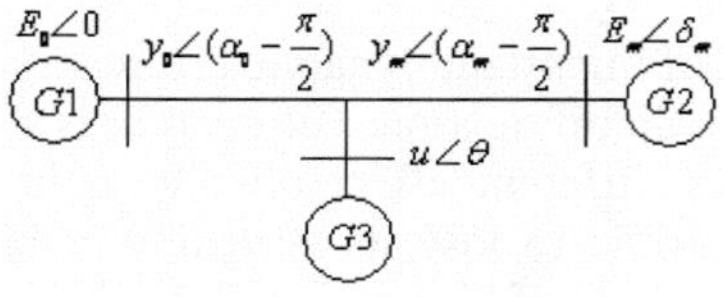

Fig. 1. Wind system model schematic.

3.1 Equivalence Generator Model

The classical model of the generator is expressed by the swing equations of the rotor:

$$\begin{aligned} &\dot{\delta}_m = \omega \\ &M\dot{\omega} = -D_m\omega + T_m + E_m u y_m \sin(\theta - \delta_m - \alpha_m) + E_m^2 y_m \sin\alpha_m \end{aligned} \tag{3}$$

Where M is the inertia constant, D_m is the damping coefficient, T_m is the input torque, E_m is the generator voltage; δ_m is the rotor angle, ω is the angular velocity , θ is the voltage phase angle , u is the node voltage, y_m and a_m are network parameters.

3.2 Wind Power Plant Mode

Nowadays induction generator is widely used in wind power plant, the induction generator at its runtime issues active power, at the same time absorbs reactive power which results in grid low voltage, increases voltage loss between wind power plant and grid contact line. In the view of the whole system, the wind power plant is special dynamic load in the power system [6]. Dynamic load models are in (4) as follows.

$$\begin{aligned} &P = P_1 + P_0 + K_{p\omega}\dot{\theta} + K_{pu}(u + T\dot{u}) \\ &Q = Q_1 + Q_0 + K_{q\omega}\dot{\theta} + K_{qu}u + K_{qu2}u^2 \end{aligned} \tag{4}$$

Where P is active power issued by wind power plant, Q is reactive power absorbed by wind power plant. The concrete parameter meaning can be referred in [6].

3.3 Wind Power System Model

A synthesis model can be described through simultaneousness of each system component model and network equations, the general form is $\dot{x} = f(x, \lambda)$, where f defines the dynamic behaviors of generators and wind power plant. State vector $x = [\delta_m, \omega, \theta, u]$, where its meaning, deduction and the concrete parameter value can be referred in [6].

4 The Effect of Wind Power System Parameters on Hopf Bifurcation

4.1 The Effect of Active Power and Reactive Power on Hopf Bifurcation

Hopf bifurcation is a kind of important dynamic bifurcation in power system, it is also the basis for the rest dynamic bifurcation. Bifurcation is related to system parameter, different parameter causes different effect on bifurcation, when system parameter changes the system is possible to lose its structural stability. The effect of several kinds of system parameters on hopf bifurcation is studied in the paper.

When the systems reactive power changes, the system voltage is in a stable or unstable equilibrium condition, the corresponding system voltage stays in stability or instability. Fig. 2 shows the voltage u of wind power system bifurcates with reactive power $Q1$ changes, the later half branch of the system equilibrium curve cannot satisfy the operational requirements, and so the analysis focuses on the first half branch of the system equilibrium curve. Fig. 2 shows from initial condition on, $Q1$ consumed by wind power plant increases, the voltage u of wind power plant drops quickly, before the system reaches the operation limit point(LP), namely $Q1$ reaches around 1.499884, hopf bifurcation H1 occurs shown in Fig.2, when system bifurcates, the corresponding voltage oscillate, the system voltage u =0.908584. Thus the large amounts of reactive power consumed by wind power system can have a great effect on the stability of system dynamic voltage.

When wind power plant woks, both active power $P1$ and reactive power $Q1$ are changing, thus it's more meaningful to make two-parameter Hopf bifurcation research of wind power system. Fig. 3 shows two-dimensional Hopf bifurcation boundary when $P1$ and $Q1$ work in common. It can be seen when $P1$ increases, the system voltage increases; when $Q1$ increases, the system voltage decreases, so with the wind plant reactive power increases, Hopf bifurcation occurs at the time of lower active power, namely to avoid the occurrence of Hopf bifurcation, the active power output must be limited when reactive power consumption is in a high level. That is the disappearance of Hopf bifurcation is at the expense of the output of wind plant active power.

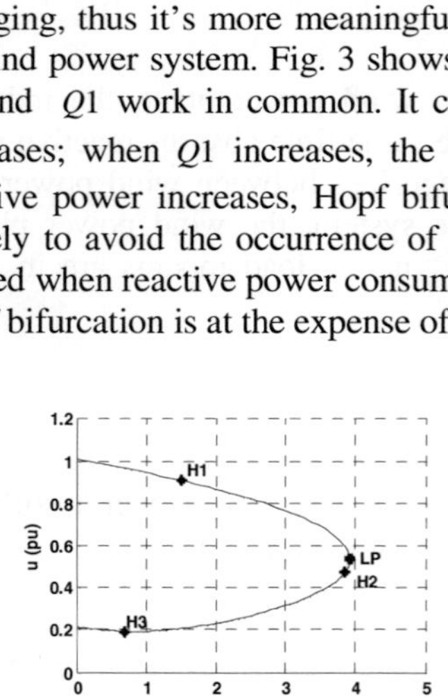

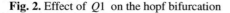

Fig. 2. Effect of $Q1$ on the hopf bifurcation

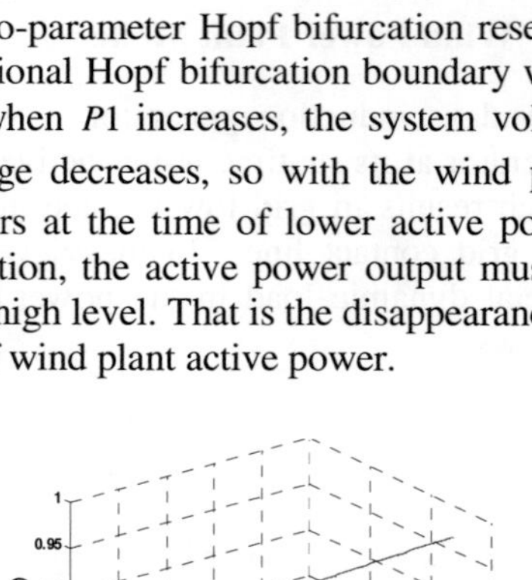

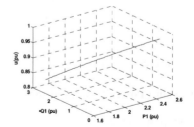

Fig. 3. Effect of $P1$ and $Q1$ on Hopf bifurcation boundary

4.2 The Effect of Admittance on Hopf Bifurcation

Fig. 4 shows that admittance parameter $y0$ has an insignificant effect on the system voltage, Fig. 5 shows a two-dimensional Hopf bifurcation boundary when $y0$ and ym work in common. Fig. 4 and Fig.5 show when $y0$ and ym increases Hopf bifurcation boundary changes little. Fig. 4 shows that Hopf bifurcation H1 occurs when $y0$ changes to 5.62661(shown in Fig.4).When $y0$ increases, electric distance of transmission lines reduces, so does the reactive power needed in lines, and the system would provide wind plant with more reactive power, resulting in the delay of Hopf bifurcation, finally the system voltage stability can be improved.

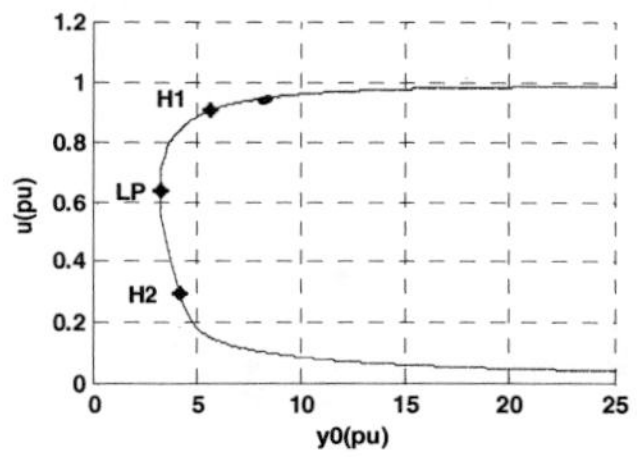

Fig. 4. Effect of y0 on the Hopf bifurcation

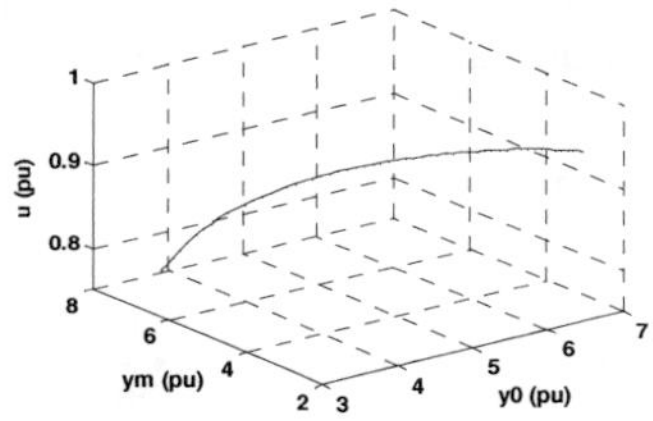

Fig. 5. Effect of $y0$ and ym on Hopf bifurcation boundary

5 The Effect of SVC on Hopf Bifurcation

Hopf bifurcation may result in voltage breakdown of power system. Thus it is a way to avoid voltage breakdown if the bifurcation can be controlled effectively. Through analyzing the effect of $P1$、$Q1$, $y0$ on voltage stability, we can conclude that Hopf bifurcation has a relation to the reactive power of wind power plant. SVC as a dynamic reactive power compensation device is an effective tool to control bifurcation.

SVC is installed on the wind power plant terminal, as shown in Fig.6, the model is shown in equation (5).

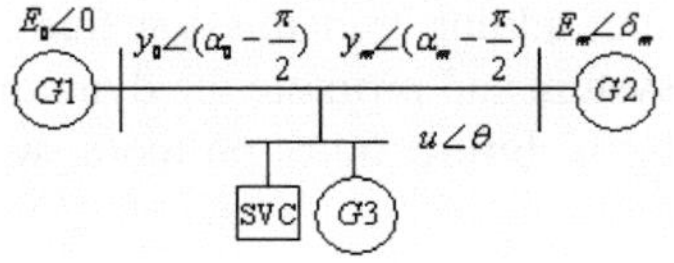

Fig. 6. Wind power system with SVC model schemes

$$\dot{b}_{SVC} = (K_r(u_{ref} - u) - b_{SVC})/T_r \tag{5}$$

Where u_{ref} is reference voltage, u is compensatory point voltage, K_r is controller magnification, T_f is Time constant, b_{SVC} is equivalent susceptance, and $Q = b_{svc}u^2$.

We take the effect of $Q1$ on Hopf bifurcation of wind power system as an instance, analyze the effect of SVC on Hopf bifurcation and voltage stability. Fig.7 shows the effect of $Q1$ on Hopf bifurcation of wind power system with SVC when K_r =1.5 and T_f =0.02. By comparing with Fig.2, it can be seen that when $Q1$ changes, the voltage of wind power system with SVC is promoted. The value of $Q1$ at the Hopf bifurcation point H1 increases ($Q1$ =11.531989) and H1 is closed to LP in the system. As is shown in Fig.8, the figure shows that SVC can effectively delay Hopf bifurcation by providing reactive power for the wind power system, and increase stability domain of wind power system.

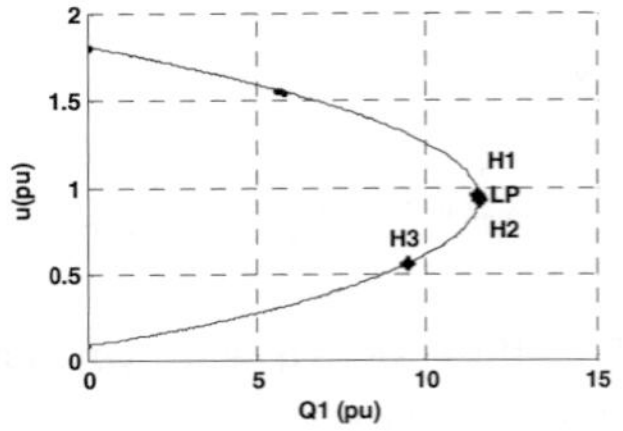

Fig. 7. Effect of SVC on Hopf bifurcation

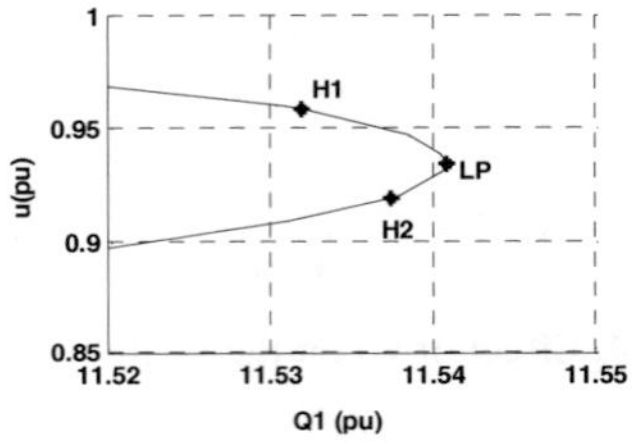

Fig. 8. Local enlargement of Fig.7

6 Conclusions

In the paper continuation method is applied to track wind power system equilibrium manifold. The effects of several parameters of wind power system and SVC on Hopf bifurcation are studied.

- When reactive power increases, wind power voltage decreases, so active power output of wind plant must be limited in order to avoid the occurrence Hopf bifurcation in wind power system.
- Admittance parameters of wind power system have insignificant effect on wind power plant voltage, but adding transmission lines admittance can delay Hopf bifurcation of wind power system and promote the dynamic voltage stability.
- SVC can delay Hopf bifurcation, promote wind power system voltage stability.

References

1. Vittal, E., O'Malley, M., Keane, A.: A steady-state voltage stability analysis of power systems with high penetrations of wind. IEEE Transaction on Power Systems 25(1), 433–442 (2010)

2. Revel, G., León, A.E., Alonso, D.M., et al.: Bifurcation analysis on a multimachine power system model. IEEE Transaction on Circuits and Systems—I: Regular Papers 57(4), 937–949 (2010)
3. Radunskaya, A., Williamson, R., Yinger, R.: A dynamic analysis of the stability of a network of induction generators. IEEE Transaction on Power Systems 23(2), 657–663 (2008)
4. Ma, Y., Wen, H., Zhou, X., et al.: Calculation and research of two-dimensional parameter bifurcation boundary with power system stability model including wind plant based on continuation method. Proceedings of the CSEE 30(19), 26–30 (2011)
5. Cao, G., Liu, L., Zhao, L., et al.: Continuation method to compute two-dimensional parameter local bifurcation boundary in power system stability differential-algebraic equation model. Proceedings of the CSEE 25(8), 13–16 (2005)
6. Abed, E.H., et al.: Dynamic bifurcations in a power system model exhibiting voltage collapse. Int. J. Bifurcation and Chaos 3(5), 1169–1176 (1993)

Computer Network: Control Infrastructure of Large Astronomical Telescopes

Lingzhe Xu[1,2] and Xinqi Xu[1,2]

[1] National Astronomical Observatories /Nanjing Institute of Astronomical Optics & Technology, Chinese Academy of Sciences, Nanjing 210042, China
[2] Key Laboratory of Astronomical Optics & Technology, Nanjing Institute of Astronomical Optics & Technology, Chinese Academy of Sciences, Nanjing 210042, China
lzhxu@niaot.ac.cn

Abstract. A high-tech Large Astronomical Telescope (LAT) integrated with cutting-edgy technologies is universally regarded as one of eye-catching events indicating comprehensive national strength in science and technology. Powerful functions and incredible performance of the telescope control system largely rest on its control infrastructure, which again is built on computer network. In the nut shell, computer network is a foundation stone for the LAT control infrastructure. This paper takes the Large Sky Area Multi-Object Fiber Spectroscopic Telescope (LAMOST) [1], a national large-scale astronomical facility, as an example to focus on its control infrastructure associated with its network configuration features. As the advent of so called 3G age and foreseeable Internet of Things (IOT) some points of analysis regarding the impact of next generation network on the design of contemporary and future astronomical telescope control system is also discussed and envisioned in this paper.

Keywords: Large Astronomical Telescope, computer network, control infrastructure, control system, LAN.

1 Introduction

Galileo built his first telescope and took it to observe the sky about 400 years ago. The telescope was merely an optic-mechanical gadget seen today, yet a revolutionary event in astronomy. Since then the persisting exploration of mysterious universe has greatly pushed the development of astronomical telescopes. The R & D of astronomical telescopes has experienced smoothly evolution and dramatic advancing at times. Astronomical telescopes have become more and more powerful and sophisticated for deep space probing. Particularly a LAT itself turns out to be a comprehensive optical-mechanical-electrical integration. The integrated telescope consists of many components at hierarchical levels to interact each other within a network system plus numerous plug-in accessorial detectors as service devices. Besides, modern LATs, from ground based to space, are capable of acquiring

X. Wan (Ed.): Electrical Power Systems and Computers, LNEE 99, pp. 973–979.
springerlink.com

enormous astronomical data at amazing high rate. These massive data require online reduction or offline analyzing, which again call for huge computing capacity. A telescope or an array of telescopes must be under complex and strict subtle control to move around for targeting celestial objects. It normally involves harmonic collaboration among multi- optical-mechanical-electrical components of telescopes. In addition, thousands of or even tens of thousands of varieties of sensors keep monitoring day and night all kinds of telescope parameters and environmental variables to ensure everything in smooth processing. What is more, remote control and observation services are definitely needed, and robot working mode is highly demanded for facility efficiency and massive scientific output. Given the highlights illustrated above, the R & D of LATs inevitably have turned to cutting-edge technologies available at the time. One of such state-of-the-art technologies is computer network, which enable to forge a powerful infrastructure in distributed form for all telescope components to interact each other consistently. From the perspective of automation a modern telescope control system is a non other than networking system with its each node, normally some hardware and software combination, to play certain functions and to communicate systematically in the topology structure. After all, computer network technology brings about revolutionary breakthrough in the design of control infrastructure for modern LATs.

This paper takes a national large scientific and engineering project LAMOST telescope as an example to focus on its network infrastructure. The telescope has been built mostly in Nanjing Institute of Astronomical Optics & Technology (NIAOT), Chinese Academy of Sciences (CAS). The Chinese ever ambitious LAMOST, which was completed and passed inspection and acceptance test by national evaluation in 2008, has become so far the world's most powerful ground based optical astronomical survey telescope with meter-class aperture. It has set the world record with the number of 4000 stars to magnitude 20.5 that can be surveyed simultaneously by the telescope. With the above facts in mind it is imaginable that the control infrastructure of the telescope with computer networking as its core structure must be powerful, flexible and robust. Throughout 10 years of R & D the control teem members in NIAOT have fully investigated and searched the network market for high cost-effective products and upgraded the hardware and software properly. The road map, from lab computer simulation to LAMOST model preliminary test in NIAOT campus and finally to the site engineering and testing and installation on Xinglong mountain of Hebei province, has witnessed how the advancing network technology has boosted and perfected the design of control infrastructure for modern LATs. The trend of networking for LAT control infrastructure currently still goes on as nowadays remote control, robot telescope, space telescope and virtual telescope and observatory all has become popular thanks largely to new generation of network comes along. Recently in the IT area the concept of IOT is frequently and loudly covered by the media. Also the problem of Internet Protocol Version 4 (IPv4) address exhaustion is much discussed. A possible solution is to make transition from IPv4 to IPv6. The implication of these big network events for LAT controls system design is presented in this paper.

2 Road Map of Network Technology in LAT Design

A computer network, often simply referred to as a network, is a collection of computers and devices interconnected by communications channels that facilitate communications among users and allows users to share resources. The concept in this regard has already been general knowledge today and widely applied in control and communication system design. Not until the 80's of last century, however, did the Single-Chip Microcomputer (SCM) at China market became commercial available then. Such as 8031/8048/8051/8098 SCMs were among very popular ones. The second author of this paper in 1992 successfully built an 8098 based development system as a front end controller in the muster-slave cascade control for first Chinese Stellar Interferometer prototype. However, this kind of control was far from networking in real sense. Most of job still had to be done painfully by connecting varieties of discrete elements and Small-Scale Integrated Circuit (SSIC). The same situation happened during the course of building 2.16 meter astronomical optical telescope, the largest one of its kind then in Far East Asia prior to the LAMOST. In the beginning of 90's of last century desktop Personal Computer (PC) became available at domestic market and hit the market instantly. In a couple of years computer control became a fashion. Most likely the small control system was only comprised of a single PS as the host machine with a drive card plug-in as the interface to external SCM slave controller, which again drives a device, say a motor. Still such a kind of control is not considered as network control. Control engineers specializing in design for LAT expected much more than that. History stepped forward, and so did the technology. In another several years to the end of last century the PC's performance amazingly boosted with the networking technology dramatically changed from embedded controller, Local Area Network (LAN) up to Internet. By the time the LAMOST project started preliminary design in 1998 control engineers fully took the advantage of new technologies of controller, computer, network, etc. in their design for LAMOST control infrastructure. A new era in the sense of control system design for LATs in China thus arrived.

3 Networklised LAMOST Control Infrastructure

Figure 1 shows the LAMOST high-level control and interfaces. Some abbreviations are explained briefly bellow [2], [3].

- The Observatory Control System (OCS) is at the top in the hierarchy responsible for supporting the on-site observer and system operator in their tasks and coordinates the activities of the other three principal systems immediately at one level below the OCS in the hierarchy, namely Telescope Control System (TCS), Instrument Control System (ICS) and Data Handling System (DHS).
- The DHS archives the observed data into the Spectroscopic Database (SDB) and implements varieties of processes such as observation planning from the input Catalogue Database (CDB), image processing, spectrum analysis and data publish.

- The ICS controls the instruments, such as the fiber positioning system, spectrographs and CCDs.
- The TCS receives observation commands with coordinates of the right ascension and declination of the sky area center to be observed and manipulates its subsystems in conjunction with ICS to complete the pointing and tracking of the target. For better tracking the TCS might also get a set of guide stars' coordinates in the target sky area as the tracking references.
- Each subsystem of the TCS is listed below.
  - ✓ Mount Control System (MCS).
  - ✓ Focal Plane Control System (FPCS).
  - ✓ Star Guide Control System (SGCS) .
  - ✓ MA Active Optics Control System (MAAOCS).
  - ✓ MB Active Optics Control System (MBAOCS)..
  - ✓ Dome Control System (DCS). .
  - ✓ Environment Monitor & Control System (EMCS)..

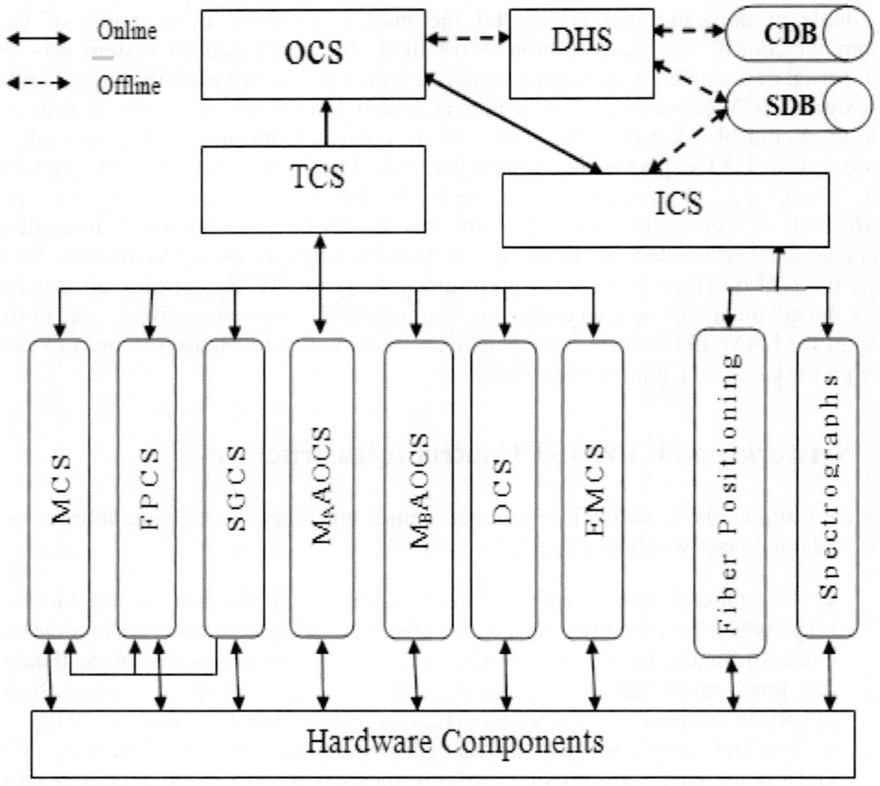

Fig. 1. LAMOST high-level control and interfaces

The TCS LAN features distributed, hierarchy and expandable. A Local Control Unit (LCU) with embedded QNX [4] microkernel and task specific codes controls each main hardware component. QNX is one of well-known real time operating systems on the current world market. The cost effectiveness of the QNX contributed to our consideration of choice of real time system under our limited budget. The LAN is built on Ethernet connection with TCP/IP standard for data traffic. In addition, there is also proprietary Fleet protocol among the QNX nodes machines, which makes them behave like a virtual computer with combined individual computer capacity, resources shareable and transparency. The idea to employee LCUs is to isolate the hardware components from the TCS main working station and lend a hand to software programmers to write high-level software module without worrying too much the specific nature of each hardware component [5], [6]. Obviously such a network layout is characterized with plain module design, economy and easy maintainability. Typically the time scale among the real time QNX nodes can reach 1ms accuracy thanks to the Fleet protocol, which is enough to meet all the real time task requirements for LAMOST. A real time distributed database has been built under QNX OS v6 environment. It is an important mechanism for a variety of system functions occurring on the TCS LAN such as recording various kinds of data for online data analysis or offline play back study. A number of advanced skills, such as the shareable memory and dynamic creation of objects and tables etc, have been implemented, which has made the database management system easy to manipulate with Windows style.

At the early R & D stage of LATs the approach of simulation is a wise and efficient way of doing, which again heavily depends on the computer network technology available at the time. Here is how it works. Generally several high performance computers hook up together making a LAN with each node of compute at hierarchical level simulating a specific function module. Often some computers at low levels have interface boards plug-in the mother board slots for driving external devices. This is similar to the master-slave application at SMC age in the 80's last century. However for high-level software package analysis and debugging more often than not it has to be implemented on network platform. In LAMOST case the concept of virtual telescope that was originally associated with pointing/tracking has been around for a couple of decades. However as the advent of networking edge the R & D of LATs have undergone a forward leap in the control infrastructure design from concept down to practice. The LAMOST simulation approach in its control system development is divided into three progressive phases from level-0 simulator up to level-2 as follows.

- Level-0 is a lowest level of the simulator giving the feel and look for the user interfaces. Virtual hardware components are represented with graphical sketches when needed to show on screens, and so is the status transition of components generated by the simulation status generator. One simple such an example is when a hand controller is shown on the screen .the user is asked to pre-select one of the three possible velocity and then keep press of the mouse on any direction key of the four possible choices, that way would make a virtual mount shown on another screen to move accordingly until the mouse is released. It is a visual simulation in its nature for user interfaces and gives responses graphically on the screens.

- Level-1 is upgrade of the level-0 simulator with major portion of the codes for the basic LAMOST control modes available. The simulator of this level can work on its own with virtual hardware components, it should also, if needed, work for the real telescope on site to receive first light manually, which can be called hardware-in-the-loop simulation. Characterization of the mount drive servo and the actively controlled optics has to be done during this phase in order to get important parameters for the simulator.
- Level-2 is upgrade of the level-1 simulator, a comprehensive working software package featuring fully automatic with all codes needed available. The simulator should work almost perfectly in reality on site, and provide simulated operation environment in lab for the users perfectly.

In a word, this chapter gives example of LAMOST to show how the computer network technology brings about dramatic advance in the LAT control infrastructure design.

4 Future Vision

3G stands for third generation and is the third generation of wireless network technology. 3G Internet refers to a type of Internet access provided by mobile phone companies that subscribers can use to access faster speeds than previous versions, such as 2.5G and 2G Internet. It is typically used from mobile devices such as cell phones. It is a set of standards, or rules, that allow mobile devices to connect wirelessly to the Internet for surfing the web and making phone calls. What this event means for our control engineers is that it is possible to wirelessly monitor the performance of control system of LASs by a cell phone [7]. The preliminary test was done in 2006 in LAMOST design when a lab simulation carried out successfully with a normal cell phone. The result is very much promising and encouraging. One thing is for sure if it is adapted to new 3G cell phone the result could be better. Another IT event is that the concept of IOT is frequently and loudly covered by the media. IOT is a network of Internet-enabled objects, together with web services that interact with these objects. Also the problem of Internet Protocol Version 4 (IPv4) address exhaustion is much discussed. A possible solution is to make transition from IPv4 to IPv6. The implication of these big IT events for LAT controls system design is anticipated. So it is possible to get an IP address for each telescope object, which is Internet-enabled, together with web services that interact with these objects. With this vision in mind we control engineers are longing for the time to arrive.

5 Summary

For the past decades we have experienced dramatic advance by leaps and bounds in the network technology. Faster speed network comes into being, and so does the network load capacity. Modern society can not be in normal operation without Web. It is also true in the points of our discussion above. .The LAMOST experience has shown fundamental transition in the way of designing for the LAT's control infrastructure as result of new era network-technology. Much more is anticipated in this regard. Full remote automatic control, wireless control, robotic control, unmanned observation scheduling, virtual telescope, intercontinental or space operation and even moon-based telescope etc. are either already realized or on the table for discussion thanks largely to the epoch-making network-technology.. Control engineers must bear this in mind to keep with the pace of technology development for best possible designing control infrastructure.

References

1. Wang, S.-g., Su, D.-q., Chu, Y.-q., Cui, X., Wang, Y.-n.: Special configuration of a very large Schmidt telescope for extensive astronomical spectroscopic observation. Appl. Opt. 35, 5155–5161 (1996)
2. Xu, X.: Control system and technical requirements - preliminary design. LAMOST Internal Technical Report (1998)
3. Wampler, S.: The software design of the Gemini 8m telescope. In: SPIE, vol. 2871, pp. 1012–1019 (1996)
4. Xu, X.: Study of QNX real time operation system and exploration on its application in large astronomical telescopes. Astronomical Instrument And Technology (1999)
5. VLT Software Telescope Control System Functional Specification. VLT-SPE-ESO-11720-0001 (1995)
6. VLT Control System Specifications. VLT-SPE-ESO-17000-1088 (1996)
7. Xu, L., Xu, X.: Remote wireless control for LAMOST telescope. In: SPIE, vol. 6274 (2006)

Radiation Analysis of Dielectric Conical Conformal Log-Spiral Antennas

Huilin Hu, Yunhua Tan, Bocheng Zhu, and Lezhu Zhou

Electronic Engineering and Computer Science School, Peking University,
100871 Beijing, China
tanggeric@pku.edu.cn

Abstract. Radiation properties of conical log-spiral antennas (CLSA) conformal on dielectric cones are analyzed by using the hybrid finite element and boundary integral equation method (FE-BI) in the paper. Firstly the radiation patterns of CLSA conformal on different dielectric cones and the ordinary CLSA are compared at the same geometry parameters and frequency. Then the influence of the spiral wrap angle and cone angle on the radiation of dielectric conical antennas are discussed. A design of geometry parameters of the dielectric CLSA with best radiation performance are finally obtained based on numerical simulations under the given dielectric parameters.

Keywords: FE-BI method, dielectric cone, conformal antenna, conical log-spiral antenna.

1 Introduction

Conical log-spiral antennas (CLSA) are commonly used in aerospace and related fields. As a class of frequency independent antennas, an important advantage of the CLSA is that the radiation direction could be regulated by adjusting the geometric parameters. As having a good aerodynamic performance, cone is often applied to the front of missiles, aircrafts etc. In these situations CLSAs usually must be conformal to this dielectric cone. Therefore investigation of radiation performances of CLSA conformal on the dielectric cone (below briefly called dielectric CLSA) possesses importance of potential applications.

From Yeh and Mei, literatures analyzing the CLSA with the moment method (MOM)[1], the finite different time domain (FDTD) method[2] and the finite element with absorbing boundary conditions (FEM+ABC) method[3] appear gradually. However, all these studies are limited to the CLSA in the air. Research on dielectric CLSA has not been seen yet and it is carried out in this paper for the first time. In order to analyze spiral antennas conformal on a dielectric cone, theoretically we can also use MOM, FDTD, FEM-ABC methods. However comparing with the above-mentioned methods, the hybrid finite element and boundary integral equation (FE-BI) method is the most suitable. The FE-BI method not only maintains the advantages of the meshing flexibility and the sparse matrix in FEM, but also strictly limits the calculating area to the surface of the cone, which minimizes the number of unknowns.

X. Wan (Ed.): Electrical Power Systems and Computers, LNEE 99, pp. 981–986.
springerlink.com

Therefore FE-BI method is applied to simulating and analyzing dielectric CLSA in this paper.

The first part of this paper gives the theoretical formula of FE-BI method and the model of CLSA. The second part shows the effect of the spiral and conical parameters on radiation characteristics of dielectric CLSA. Finally with a given dielectric parameter, we provide a typical dielectric CLSA design result. It should be pointed out that different applications for CLSA require different radiation performances. In this paper we just focus on the forward radiation demand, i.e. the maximum radiation direction of CLSA is along the cone axis and the half-power angle is smaller.

2 Formulation

2.1 FE-BI Formulations

Deriving from the differential forms of Maxwell equations,

$$\nabla\times\overline{E} = -\overline{M}^{i} - j\omega\mu_0\mu_r\overline{H} \tag{1}$$

$$\nabla\times\overline{H} = \overline{J}^{i} + j\omega\varepsilon_0\varepsilon_r\overline{E} \tag{2}$$

Then using the finite element method (FEM), we could obtain the matrix equations of the whole cone including the spiral antenna part

$$\begin{bmatrix} \boldsymbol{K}_{\mathrm{II}} & \boldsymbol{K}_{\mathrm{IS}} & \boldsymbol{0} \\ \boldsymbol{K}_{\mathrm{SI}} & \boldsymbol{K}_{\mathrm{SS}} & \boldsymbol{B} \end{bmatrix} \begin{bmatrix} \boldsymbol{e}^{\mathrm{I}} \\ \boldsymbol{e}^{\mathrm{S}} \\ \boldsymbol{b}^{\mathrm{S}} \end{bmatrix} = \begin{bmatrix} \boldsymbol{C}^{\mathrm{I}} \\ \boldsymbol{0} \end{bmatrix} \tag{3}$$

where $\{e^I\}$ are the unknown coefficients of electronic fields in the FEM area, $\{e^S\}$ and $\{b^S\}$ represents the unknown coefficients of electronic fields and surface currents of the FEM out surface, respectively. $[K_{\mathrm{II}}]$, $[K_{\mathrm{IS}}]$, $[K_{\mathrm{SI}}]$, $[K_{\mathrm{SS}}]$ and $[B]$ are all sparse matrixes, $[\boldsymbol{C}^{\mathrm{I}}]$ stands for the feed item. Then using boundary integral (BI) method for the exterior region, the fields can be expressed in terms of the surface electrical currents $\overline{J}_S = \hat{n}\times\overline{H}_S$ and the surface magnetical currents $\overline{M}_S = \overline{E}_S\times\hat{n}$. With the combined field integral equation (CFIE), we could get the matrix equation for the BI part

$$[P]\{e^{S}\}+[Q]\{b^{S}\}=0 \tag{4}$$

where $[P]$ and $[Q]$ are $N\times N$ full matrices. Combining (3) and (4), the final equation is:

$$\begin{bmatrix} \boldsymbol{K}_{\mathrm{II}} & \boldsymbol{K}_{\mathrm{IS}} & \boldsymbol{0} \\ \boldsymbol{K}_{\mathrm{SI}} & \boldsymbol{K}_{\mathrm{SS}} & \boldsymbol{B} \\ \boldsymbol{0} & \boldsymbol{P} & \boldsymbol{Q} \end{bmatrix} \begin{bmatrix} \boldsymbol{e}^{\mathrm{I}} \\ \boldsymbol{e}^{\mathrm{S}} \\ \boldsymbol{b}^{\mathrm{S}} \end{bmatrix} = \begin{bmatrix} \boldsymbol{C}^{\mathrm{I}} \\ \boldsymbol{0} \\ \boldsymbol{0} \end{bmatrix} \tag{5}$$

which can be solved together for the surface electric fields $\{e^S\}$, magnetic fields $\{b^S\}$ and interior electric fields $\{e^I\}$ including the feeding part. As a result, the pattern and the input impedance could easily be obtained. For detailed derivation, see Ref. [4].

2.2 Geometry of CLSA

Conical log-spiral antenna is also called conical equiangular spiral antenna, the radius vector on the arm can be represented as

$$r = r_0 e^{\beta\varphi}, \ \varphi = 0 \sim \varphi_{max} \tag{6}$$

where r_0 stands for the distance from the origin to the starting point of the antenna, θ_0 is the half top-angle of the cone, and α is the angle between the antenna arm and the radial line from the apex of the cone called the wrap angle, while $\beta = \frac{\sin\theta_0}{\tan\alpha}$ is the rate of wrap of arms. The geometry is shown in Fig. 1(a).

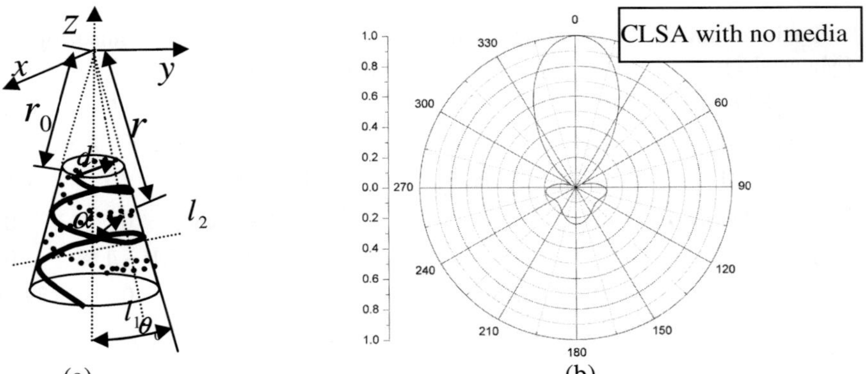

(a) (b)

Fig. 1. Conical log-spiral antenna: (a) geometry (b) pattern of CLSAwithout media

Another arm of a two-arm CLSA can be obtained by rotating the first arm by the angle 180°. The arms are governed by

$$r_1 = r_0 e^{\beta\varphi}, \ r_2 = r_0 e^{\beta(\varphi+\pi)} \tag{7}$$

when r_0, θ_0 and α are given, the total length of the antenna is determined by φ_{max}, usually $\varphi_{max} = 2n\pi$. In this paper, we take $n = 2$.

2.3 Authentication

In order to verify the methods and procedures of the paper, we firstly simulate the typical CLSA given in Ref. [5]. The given parameters are: cone angle $\theta_0 = 10^o$, wrap angle $\alpha = 73^o$, initial radius $r_0 = 78\text{mm}$. According to the formula in Ref. [5], the theoretical input impedance $Z_{in} \approx 165\Omega$, the half-power width $\Delta\varphi \approx 70^\circ$, meets the results of our paper: $Z_{in} \approx 166.78\Omega$, $\Delta\varphi \approx 60^\circ$.

3 Numerical Result

3.1 Difference between Dielectric CLSA and Ordinary CLSA

Dielectric CLSA and ordinary CLSA with the same geometry and the same working frequency are firstly discussed. Here we take the geometry as $\theta_0 = 15^o$, $\alpha = 78^o$, $r_0 = 78\text{mm}$, the same feeding point, and working frequency of 3 GHz. The only difference is the dielectric parameter. Fig. 1(b) shows the pattern of the ordinary CLSA, while Fig. 2 shows the patterns of CLSA with the different dielectric cones. The parameters are below.

Lossy isotropic media (Material 1):

$$\varepsilon_r = 25.59 - j3.89$$

$$\mu_r = 2.16 - j1.68$$

Positive uniaxial anisotropic media (Material 2):

$$\varepsilon_r = \begin{bmatrix} 8.19 - j1.3 & 0 & 0 \\ 0 & 8.19 - j1.3 & 0 \\ 0 & 0 & 15.89 - j8.19 \end{bmatrix}$$

$$\mu_r = \begin{bmatrix} 1.39 - j0.56 & 0 & 0 \\ 0 & 1.39 - j0.56 & 0 \\ 0 & 0 & 2.16 - j1.38 \end{bmatrix}$$

Nanomaterials (Material 3):

$$\varepsilon_r = 8.0 - j3.5$$

$$\mu_r = 1$$

Negative uniaxial anisotropic media (Material 4):

$$\varepsilon_r = \begin{bmatrix} 25.59 - j3.89 & 0 & 0 \\ 0 & 25.59 - j3.89 & 0 \\ 0 & 0 & 8.19 - j1.3 \end{bmatrix}$$

$$\mu_r = \begin{bmatrix} 2.16 - j1.68 & 0 & 0 \\ 0 & 2.16 - j1.68 & 0 \\ 0 & 0 & 1.39 - j0.56 \end{bmatrix}$$

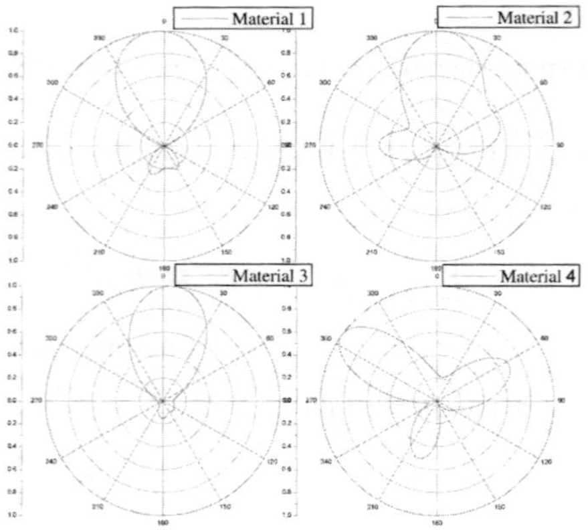

Fig. 2. Radiation patterns of CLSA with different dielectric materials

Contrasting Fig.1 (b) with Fig.2, we could see that the patterns of CLSA with anisotropic media (positive uniaxial anisotropy, negative uniaxial anisotropy) have a greater change related to those of the ordinary one. The maximum radiation direction of CLSA with isotropic media and nanomaterial slightly deviates, while the half-power angles change a bit.

3.2 Impact of Geometry of Dielectric CLSA on Its Radiation

When the CLSA is settled in the air, the cone angle and the wrap angle greatly influence the radiation pattern. Therefore the impact of these angles should also be investigated in the case of CLSA conformal on a dielectric cone. Here $\varepsilon_r = 200$ and the frequency = 3GHz are given. Fig. 3(a) shows the patterns with cone angles $\theta_0 = 5^\circ, 12^\circ, 15^\circ, 18^\circ$ respectively, while the wrap angle remain the same ($\alpha = 68^o$). Fig. 3(b) shows patterns with the wrap angles $\alpha = 70^o, 74^o, 76^o$, 78° respectively, while the cone angle keeps still ($\theta_0 = 15^\circ$).

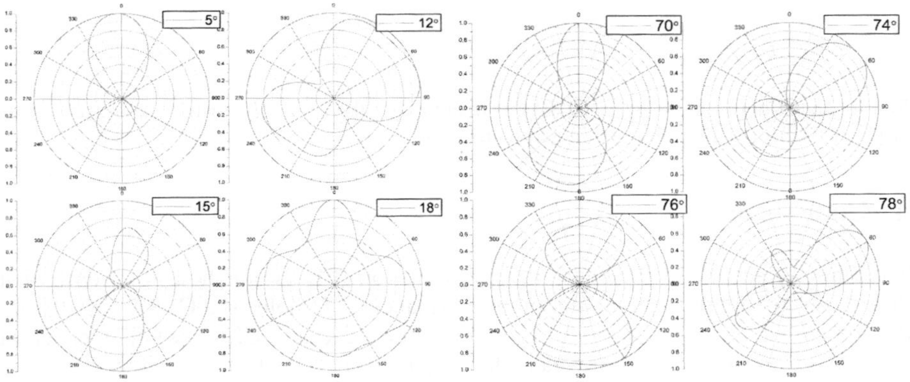

(a) (b)

Fig. 3. The dielectric CLSA pattern transformation (a)when the cone angle changes (b) when the wrap angle changes

As shown in Fig.3(a) and Fig.3(b), for dielectric CLSA, the pattern is significantly affected by the cone angle and the wrap angle. Therefore, in designing a dielectric CLSA, we could no longer directly use the design data of an ordinary CLSA in the air, especially when the dielectric constant is larger. The geometric parameters should be re-adjusted in order to get better radiation at working frequency. Next section will give a design example of a dielectric CLSA, in which one could achieve better performance through adjusting cone and spiral structural parameters.

3.3 Example of Designing Dielectric CLSA

Fig. 1(b) shows the pattern of no dielectric CLSA with the frequency of 3G, the cone angle $\theta_0 = 15^o$, the wrap angle $\alpha = 78^o$. It reveals a very good forward direction. Then maintaining the same geometric parameters of cone angle and wrap angle but plus a dielectric cone ($\varepsilon_r = 200$), the radiation pattern became Fig. 4(a). The variation of direction is obvious: the main lobe direction turned right at about 60°, and the backward radiation increases. After optimization, changing $\theta_0 = 15^o$ to 5°, $\alpha = 78^o$ to 68°, the radiation pattern is shown in Fig. 4(b). The main lobe direction is strictly forward, while prior to radiation is far larger than the backward. The directivity becomes better.

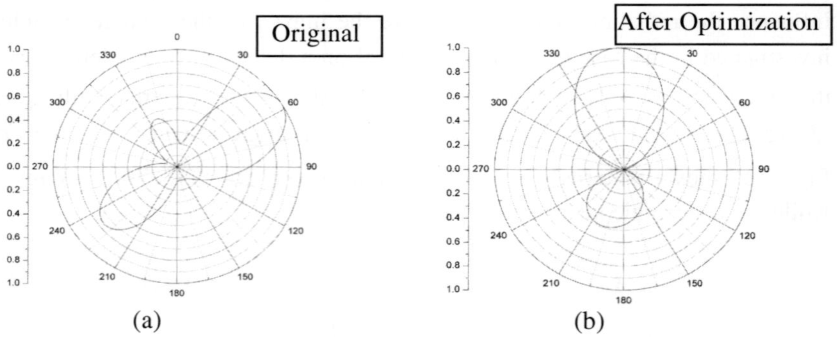

Fig. 4. The pattern on dielectric CLSA (a) maintaining the same gerometric parameters of the ordinary CLSA (b) after optimizing gerometric parameters

4 Conclusion

The finite element and boundary integral equation (FE-BI) method is introduced in this paper to analyze the radiation performances of dielectric CLSA. Firstly the differences of CLSA conformal to the dielectric cone and CLSA without dielectric cone are discussed, which reveals the impacts of dielectric constants. Then the influence of cone angle and wrap angle are investigated respectively on dielectric CLSA. Simulation results show that the structure parameters of dielectric CLSA still have a great impact on its radiation pattern. Therefore the geometric parameters of a dielectric CLSA must be re-adjusted in order to get better radiation performance at working frequency.

References

1. Yeh, Y., Mei, K.: Theory of Conical Equiangular-Spiral Antennas: Part 2—Current Distributions and Input Impedances. J. IEEE Trans. Antennas Propagat. 16(1), 14–21 (1968)
2. Penney, C.W., Luebbers, R.J.: Input impedance, Radiation Pattern and Radar Cross Section of Spiral Antennas Using FDTD. J. IEEE Trans. Antennas Propagat. 42(9), 1328–1332 (1994)
3. Xu, Y., Wang, Z.: The Design and Simulation of a Conical Spiral Antenna. J. Guidance and Fuze 25(3), 40–43 (2004)
4. Hu, H., Tan, Y., Zhu, B., Zhou, L.: A Hybrid FE-BI Method Using in Antenna Radiation Analysis. In: International Symposium on Antennas, Propagation and EM Theory, pp. 1225–1228 (2008)
5. Lin, C.L., Chen, H., Wu, W.G.: Modern Antenna Design. Posts and Telecom Press, Beijing (1990)

Influence of Fine Particle Content to Sediment Transport*

Zhang Min[1], Wu Hua-Qin[2], and Luo Li-qun[1]

[1] Yellow River Institute of Hydraulic Research, YRCC, Zhengzhou, China
[2] Henan Vocation College of Chemical Technology, Zhengzhou, China
zmiii@163.com

Abstract. The sediment transport is important in hyper concentrated flow. More studies was did about fine particle can improve sediment transport. But no results consider median grain size of fine particle was 0.006mm and coarse was 0.035mm. We aim to make it clear by flume experiment. The experiment was carried on 50m long and 0.4m width glass flume, the roughness was 0.0109~0.0114, the longitudinal slope was 0.002. The 4 groups of tests was designed to study the influence of fine particle (median diameter is 0.006mm) to the medium-coarse sediment (median diameter is 0.035mm) transport capacity. It was based on uniform flow. The 4 discharges was 30、60、90、120m^3/h. The experiment show, to add fine particle to equilibrium medium-coarse sediment, not only the fine particle was not deposited, and the medium-coarse sediment transport capacity was increased. The main reason was that the sediment concentration and viscosity as increased; the settling velocity was decreased, after the fine particle was added. And the same time the vertical velocity gradient and vertical sediment concentration gradient was decreased. For the above reason, the appropriate fine particle can improve the medium-coarse sediment transport capacity.

Keywords: fine particle content, sedimentation, medium-coarse sediment transport capacity.

1 Introduction

The fine particle was a active factor in sediment transport. It can influence the flow characteristic of coarse sediment, and then change the sediment transport capacity. There were many researches about it. ZHONG De-yu [1] proposed the fine particle content should be considered in the bed-load transport equation. Comparison of field data in the Yellow River and the calculation result of present equation shows that the equation can reproduce the phenomenon that the more the fine sediment comes from upstream, the more the coarse sediment is transported in the Lower Yellow River under the same flow conditions. Bruch and Kazanski [2] performed an experiment of sediment transport in pipe, results show when we add 0.01mm fine particle to 0.3mm

* This paper is financed by the central level, scientific research institutes for basic R & D special fund business (HKY-JBYW-2010-05) (HKY-JBYW-2008-8).

X. Wan (Ed.): Electrical Power Systems and Computers, LNEE 99, pp. 987–994.
springerlink.com

coarse sediment flow, and then the head loss remarkably decreased , the flow viscosity increased, and the turbulence intensity weaken. As a result, the coarse sediment transport capacity increased. Japanese scholar Yoshikawa and Fukuoka[3] do the experiment in flume, results indicate when add fine particle (d_{50}=0.05mm) to coarse sediment (d_{50}=0.18mm) flow, the velocity and the sediment concentration gradient all were steepened, and then total sediment carrying capacity was increased in the same hydraulic characters. FANG zong-dai[4] proposed the relationship between the fine sediment(d_{50}<0.01mm) content and coarse sediment(d_{50}>0.05mm) carrying capacity based on the field data in Yellow River, show that the higher fine sediment content, the higher the coarse sediment carrying capacity. Liu Feng[5]make the experiment in flume, result show that when add d_{50}=0.11mm fine sediment into d=0.55mm coarse sediment flow, the coarse sediment carrying capacity was increased in either low or high sediment concentration flow.

Previous research was all agreed with the fine particle can improve the sediment carrying capacity, the fine particle d_{50}=0.01~0.11mm, the coarse sediment d_{50}=0.05~0.55mm. And the main reasons was the flow viscosity increased, the velocity and the sediment concentration gradient all were steepened. However, little information was about the fine particle d_{50}=0.006mm, the coarse sediment d=0.035mm, whether the coarse sediment carrying capacity can be improved in this condition, it is uncertain. It is supposed that the fine particle can improve the coarse sediment carrying capacity, and why?

We aim to make it clear by flume experiment. For the fine particle was d_{50}<0.01mm, and the coarse sediment d_{50}=0.025 in the Lower Yellow River, if the fine particle d_{50}=0.006mm can improved the carrying capacity of coarse sediment d_{50}=0.035mm, our researches can provide technical support to XiaoLangdi reservoir regulation.

2 Methods

2.1 Flume Design

Flume is 50m long and 0.395m width. It's a self-circulation system. Out of the exit, there is an 8m long, 1m width and 1.3m depth rectangular channel acting as the mixing pool (See Fig1). For the sake of ground limited, the mixing pool was seemed to be reservoir at the same time. Through the electromagnetic flux metre, frequency converter, control value, the flow enter the flume.

It's a variable-slope flume, the sediment concentration was measured by pycnometer, the velocity by propeller-blade current metre, the discharge by electromagnetic flux metre. It arrange 5 cross-section to survey the velocity, sediment concentration, each cross-section arrange 5 survey point, they are 1.8cm to right boundary, 10.5cm to boundary, 20cm to boundary, 29cm to boundary, 37.5cm to boundary.

The discharge choose 60 m^3/h, 90 m^3/h and 120 m^3/h . To ensure it's a subcritical flow and the sediment can be lift, the longitudinal slope i should satisfied with three terms. The first is the flume slope should less than critical gradient i_k , the second is

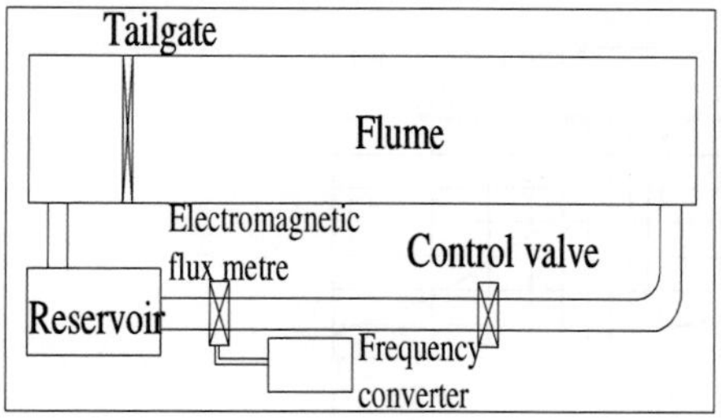

Fig. 1. Play out of flume

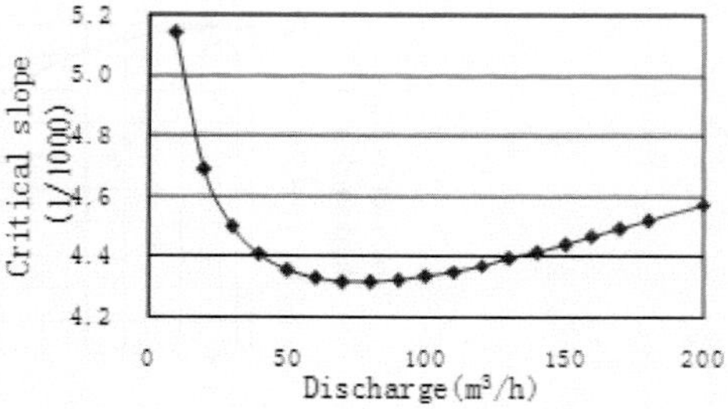

Fig. 2. The critical slope of different discharge

Froude number F_r should less than 1; the finial is is the mean velocity V should more than sediment lift velocity Vc.

The critical gradient i_k was as flow,

$$i_k = \frac{g\chi_k}{aC_k^2 B_k} \tag{1}$$

where χ_k is wetted perimeter; C_k is Chezy's coefficient, $C_k = \frac{1}{n} R_k^{1/6}$, n is synthetic roughness, n=0.011; B_k is channel width; a velocity non-uniform coefficient, a =0.0097; Figure 2 shows the critical gradient i_k in different discharge. The least i_k is 4.3‰ in the choice discharge, so the flume longitudinal slope should less than 4.3‰.

When F_r=0.8~1.2 the flow was unsteady, then F_r should less than 0.8. Figure 3 show when F_r=0.8, the flume longitudinal slope i should less than 6.5‰. Figure 4 show calculation lift velocity of three equation in different slope. When i >1.4‰ the mean velocity V> Vc, so the flume slope i should more than 1.4‰.

Overall consideration of three terms, the flume slope i was chose 2‰.

The fine particle d_{50}=0.006mm, the coarse sediment d_{50}=0.035mm, the gradation curve was shown in Figure5. The sediment were all come from the Lower Yellow River, it's prototype sediment.

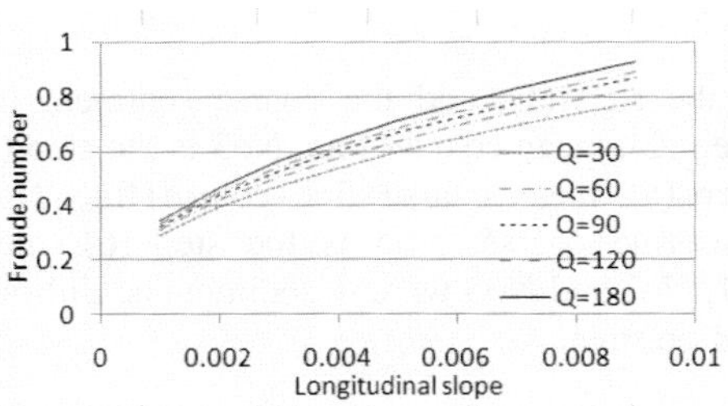

Fig. 3. Froude number of different longitudinal

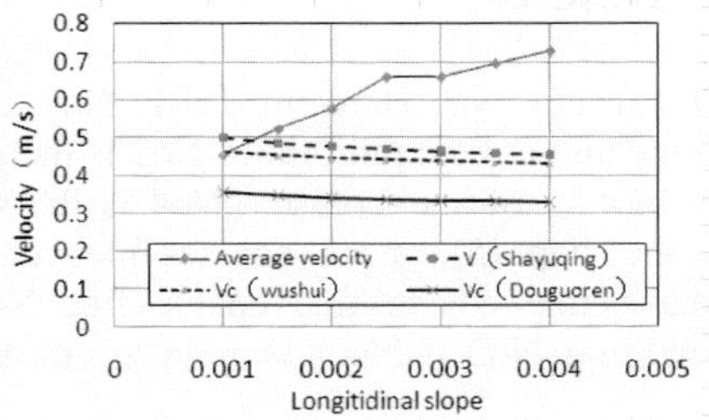

Fig. 4. Velocity of different longitudinal

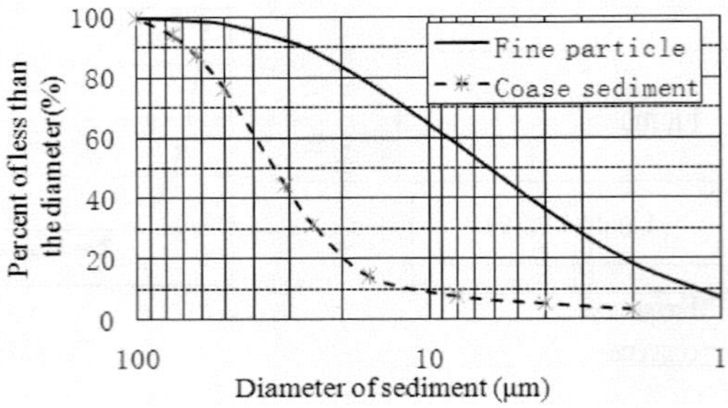

Fig. 5. The gradation curve of fine particle and coarse sediment

2.2 Experiment Content and Approach

The discharge was invariable, to reach sediment equilibrium by change the sediment concentration. So the experiment was to make the coarse sediment flow to be equilibrium at first, subsequently to add fine particle. The specific approach was as follow:

- (1)First, add coarse sediment slowly in the flume until it reaches sediment equilibrium. Then survey the velocity and sediment concentration.
- (2)Second, in the equilibrium condition (1), add appropriate fine particle to the flow. The flow was in a sub-saturation state at this time.
- (3)Third, in the condition (2), add coarse sediment slowly in the flume until it reaches sediment equilibrium.
- (4)Fourth, in the condition (3), add appropriate fine particle to the flow. The flow was in a sub-saturation state at this time.
- (5) Fifth, in the condition (4), add coarse sediment slowly in the flume until it reaches sediment equilibrium.
- (6)Sixth, in the condition (4), add fine particle until there is deposition in the base of flume , then stop the experiment.

The equilibrium criterion was the sediment concentration of inlet, on-way and outlet were same.

3 Results

The results was show in Tab1. No1 is the step (1), add the coarse sediment into flume and to be equilibrium. No2 is the step (2), mean coarse+fine. No3 is the step (3), mean coarse+fine+coarse. No4 is the step (4), mean coarse+fine+coarse+fine. No 5 is the step (5), mean coarse+fine+coarse+fine+coarse. No5 is the step (6), mean coarse+fine+coarse+fine+coarse+fine. No1, No3 and No5 were in sediment equilibrium condition. No2 and No4 were in sub-saturation state.

For the example of 60m^3/s, the first step was the coarse sediment to be equilibrium, the sediment concentration is 19.64kg/m^3(No1), then add appropriate fine particle in flow, the sediment concentration changed to 24.28 kg/m^3(No2), the sediment carrying capacity was increased more than 4.64 kg/m^3; consequently add coarse sediment in the flow until it reaches sediment equilibrium, the sediment concentration changed to 29.08 kg/m^3(No3), the sediment carrying capacity was increased 4.8 kg/m^3; and then add appropriate fine particle in flow, the sediment concentration changed to 33.79kg/m^3 (No4), the sediment carrying capacity was increased more than 4.71 kg/m^3; consequently add coarse sediment in the flow until it reaches sediment equilibrium, the sediment concentration changed to 38.96 kg/m^3(No5), the sediment carrying capacity was increased 5.17 kg/m^3. From No1 to No5, the carrying capacity was increased 19.32 kg/m^3.

So the experiment show add the fine particle in the equilibrium coarse sediment flow, can ensure not only the fine particle didn't deposited, but also can carry more coarse sediment.

Table 1. Finial equilibrium sediment concentration

Discharge (m^3/h)	**Sediment concentration (kg/m^3)**					
	No1	***No2***	***No3***	***No4***	***No5***	***No6***
60	19.64	24.28	29.08	33.79	38.96	55.83
90	25.95	30.14	35.65	41.61	44.80	53.92
120	85.43	95.23	100.37	108.48	112.79	142.86

No 1 is the step (1), No 2 is the step (2),……, No6 is the step (6).

4 Discussion

Results indicate that the d_{50}=0.006mm fine particle can improved the carrying capacity of d_{50}=0.035mm coarse sediment. We will discuss the reasons from three aspects as follows.

4.1 Average Cross-Section Velocity

The average velocity of five cross-sections was shows in Tab.2. It indicate velocity was changed little. When Q=60 m^3/h, the max velocity was 0.56~0.61m/s, the difference was 0.05. When Q=90 m^3/h, the max velocity was 0.69~0.74m/s, the difference was 0.05. When Q=120 m^3/h, the max velocity was 0.65~0.74m/s, the difference was 0.09. So we can consider the velocity was unchanged.

Table 2. The average cross-section velocity

Discharge (m^3/h)	**Average cross-section velocity (m/s)**					
	No1	***No2***	***No3***	***No4***	***No5***	***No6***
60	0.61	0.60	0.58	0.58	0.57	0.56
90	0.74	0.69	0.69	0.71	0.72	0.72
120	0.69	0.69	0.68	0.73	0.74	0.65

4.2 Influence of Viscosity

The fine particle d_{50}<0.01mm is called viscous sediment. In contrast the sediment d_{50}>0.01mm is called nonviscous sediment. In the follow discuss we distinguish the change of viscous sediment in different experiment team.

Tab.3 shows in the discharge is 60m3/h experiment, if add fine particle to the flow, then viscosity will be increased. For example from No1 to No2, No3 to No4, No5 to No6, the viscous sediment percent was increased 5, 4 and 11 percent point. In the discharge is 90m3/h experiment, the viscous sediment percent was increased 7, 2 and 7 percent point. In the discharge is 90m3/h experiment, the viscous sediment percent was increased 5, 2 and 8 percent point.

The suspend load can suspend in the water and be transport long distance, the main reason is self- gravitation and turbulence diffusion. When add the fine particle in the flow, then the flow viscosity increased, therefore the settling velocity is increased, and the downward force was decreased, then the carrying capacity increased.

4.3 Influence of Muddy Water Settling Velocity

The settling velocity of different step shows in Tab.2. It indicate when add appropriate d_{50}=0.006mm fine particle (No2) in the equilibrium flow of d_{50}=0.035mm coarse sediment, the flow was in sub-saturation state, the sediment concentration was increased, from 19.64kg/m3 to 24.28 kg/m^3. The increased sediment concentration make the muddy water settling velocity decrease from 0.00169 m/s to 0.00145m/s. For the sediment carrying capacity usually use the Zhang Rui-jin equation

$$S_* = K\left(\frac{U^3}{gR\omega}\right)^m \tag{2}$$

where S_* is the sediment carrying capacity, kg/m^3, K and m is constant, it was decided by the field or experiment data, U is velocity, m/s,$\omega$is settling velocity, m/s, R is hydraulic gradient, m. So we can know the lower the settling velocity the higher the carrying capacity. When add coarse sediment (No3) again, it is still can improve the sediment carrying capacity, then S_*=29.08 kg/m^3. And from No3 to No4, the sediment concentration increased to 33.79 kg/m^3, the muddy water settling velocity decreased from 0.00142 m/s to 0.00136m/s, then the sediment carrying capacity improved, when add coarse sediment (No5) again, S_* =38.96 kg/m^3. The experiment of Q=90 m3/hand Q=120 m3/h was the same as Q=60 m3/h.

4.4 Influence of Vertical Velocity Gradient

When Q=60m^3/h, the vertical velocity gradient of different step shows in Fig.6~Fig.11. It indicate the vertical velocity gradient in Fig.7 was less than Fig.6, Fig.9 was less than Fig8, Fig.11 was less than Fig10. The Fig.7, Fig.9 and Fig11 were the vertical velocity gradient after add fine particle. So we can conclude when add the fine particle to coarse equilibrium sediment flow, the vertical velocity gradient was decreased. The decrease of vertical velocity gradient resulted in the vertical sediment gradient decrease, and then the sediment transport was improved. The vertical velocity gradient was decreased also in experiment of Q=90 m^3/h and Q=120 m^3/h, we can't show the picture now.

Table 3. The muddy water settling velocity

Discharge (m^3/h)	Team	Settling velocity (m/s)	Sediment concentration (kg/m^3)
60	No1	0.00169	19.64
60	No2	0.00145	24.28
60	No3	0.00142	29.08
60	No4	0.00136	33.79
60	No5	0.00138	38.96
60	No6	0.00090	55.83
90	No1	0.00181	25.95
90	No2	0.00153	30.14
90	No3	0.00123	35.65
90	No4	0.00108	41.61
90	No5	0.00131	44.8
90	No6	0.00087	53.92
120	No1	0.00155	85.43
120	No2	0.00129	95.23
120	No3	0.00116	100.37
120	No4	0.00102	108.48
120	No5	0.00107	112.79
120	No6	0.00073	142.86

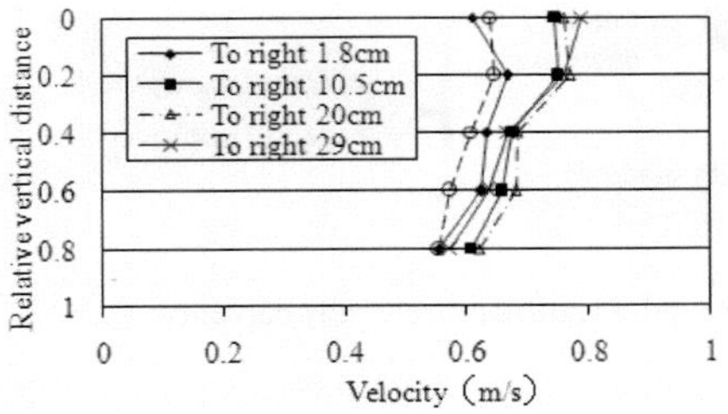

Fig. 6. Verticle

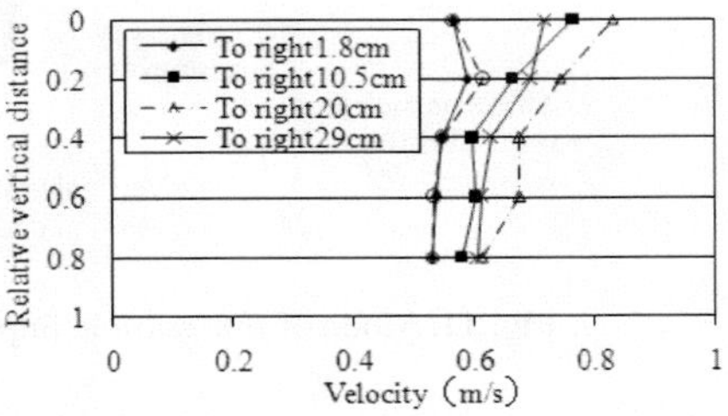

Fig. 7. Velocity of different longitudinal

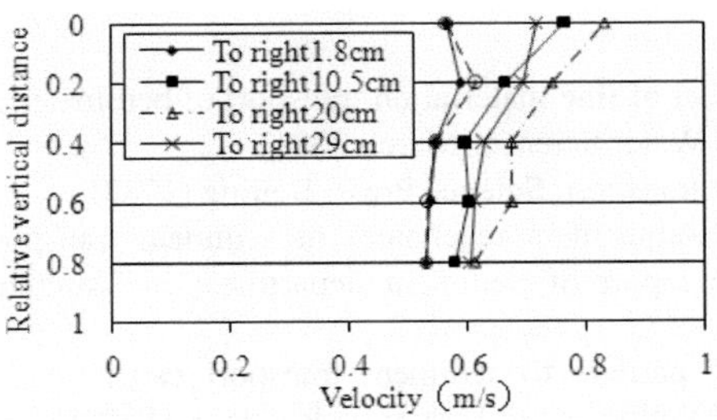

Fig. 8. Froude number of different longitudinal

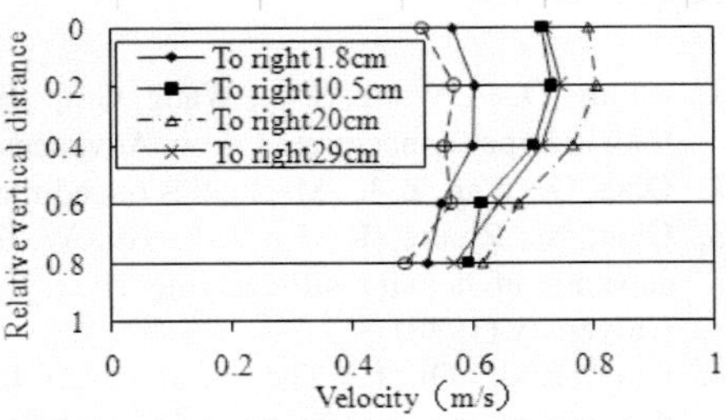

Fig. 9. Velocity of different longitudinal

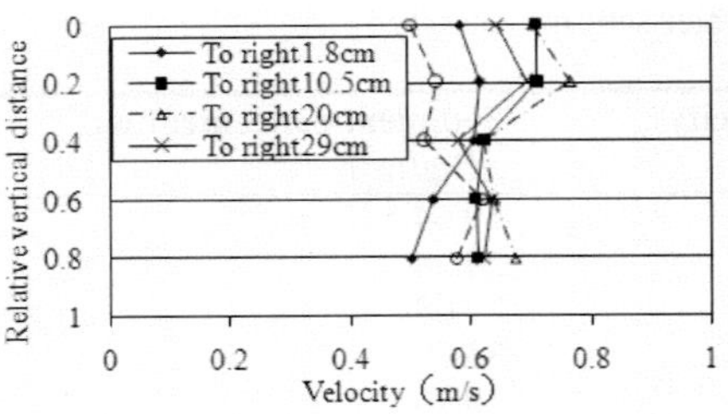

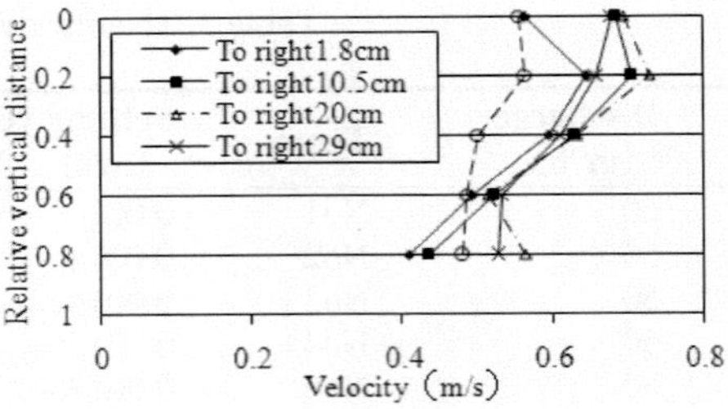

Fig. 10. Froude number of different longitudinal **Fig. 11.** Velocity of different longitudinal

Above all the analysis, we can know when add fine particle to equilibrium coarse-sediment flow, the flow structure was changed. The vertical velocity gradient was decreased, and then influences the vertical sediment concentration gradient to be decreased, and then the sediment transport capacity was improved. The same time when add fine particle, the flow viscosity was improved, then the settling velocity was decreased, and the cross-section velocity was not changed (formula 2), then the sediment transport capacity was improved. We can conclude the fine particle improves the sediment transport was according the follow mode(Fig.11):

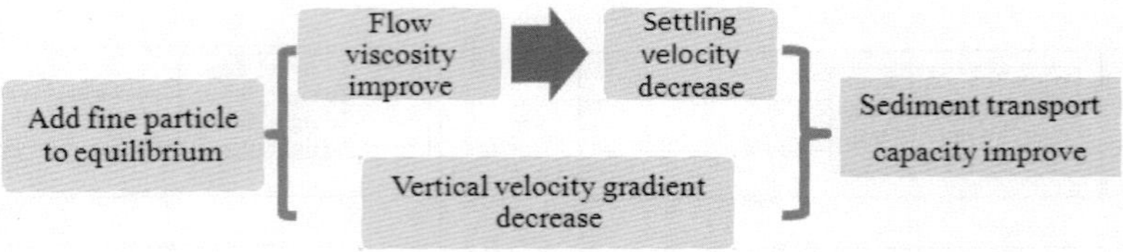

Fig. 11. Mode of fine particle improves the sediment transport capacity

Identify applicable sponsor/s here. *(sponsors)*

References

1. Zhong, D.-y., Wang, S.-q., Wang, G.-q.: Effect of fine material on transport of bed material load in hyper concentrated flow. Advancesin Water Science (March 2001)
2. Qian, Q., Wan, Z.-h.: Mechanics of sediment transport. Science Press, Beijing (1983)
3. Qian, N., Zhang, R., Li, J.-f., Hu, W.-d.: Adjustment mechnaics of sediment transport capacity in heavily silt-carrying river. The report of Sediment department in Tsinghua University (1980)
4. Fang, Z.-d., Qi, P.: The influence of fine particle to sediment transport capacity and depodition-erosion. The report of Institute of Yellow River Hydraulic Research (1978)
5. Liu, F.: Experiment on influence of fine sediment to coarse sediment transport capacity. Journey of Guangdong Water Resources and Electric Engineering (March 1999)

High Frequency Dielectric Properties of Bismuth Substituted Barium Hexaferrite

Leah M. Ridgway[1] and Ian Harrison[2]

[1] Division of Electrical Systems and Optics,
The University of Nottingham, University Park,
Nottingham, NG7 2RD, United Kingdom
[2] Department of Electrical and Electronic Engineering,
The University of Nottingham Malaysia Campus,
43500 Semenyih, Selangor Darul Ehsan, Malaysia
LeahMRidgway@gmail.com

Abstract. High frequency dielectric characterization of bismuth substituted barium hexaferrite ($BaBi_xFe_{12-x}O_{19}$ $0.0 \leq x \leq 1.5$) was performed over the frequency range 45MHz – 25GHz using an open ended coaxial probe. Results are presented for the real and imaginary parts of permittivity with respect to frequency and also for permittivity with respect to sample composition at a fixed frequency of 2.45GHz. A maximum permittivity of $\varepsilon_r = 14.69 - 1.664j$ was found thus confirming formation of a high permittivity material from a relatively low sintering temperature and duration (1100°C for three hours). The material is of interest in high frequency devices.

Keywords: Dielectric, Permittivity, Barium Hexaferrite, Material Characterization.

1 Introduction

All materials can be characterized in terms of their dielectric and magnetic properties; representing their ability to store electric and magnetic energy. Materials such as barium hexaferrite and bismuth substituted barium hexaferrite which have a permittivity and permeability greater than unity are of particular interest due to their potential applications in electronic devices. It is important to characterize the dielectric and magnetic properties of a material before it is used so that high frequency behavior can be predicted.

Barium hexaferrite, the parent compound, has been used primarily as a material for magnetic recording media [1], [2], [3]. However it has also been used in microwave devices [4] and high permittivity materials are of interest in the development of material loaded antennas [5]. The addition of bismuth to barium hexaferrite reduces the formation temperature of the compound and enhances the permittivity [6]. This makes the material a potential candidate for the production of such devices.

X. Wan (Ed.): Electrical Power Systems and Computers, LNEE 99, pp. 995–1000.
springerlink.com

This paper presents results for the dielectric properties of bismuth substituted barium hexaferrite over the frequency range 45MHz – 25GHz for different levels of bismuth addition. Comparisons are made between permittivity and sample composition at a constant frequency of 2.45GHz.

2 Material Synthesis

Bismuth substituted barium hexaferrite was produced via the conventional solid state synthesis method. The chemical formula of the material is $BaBi_xFe_{12-x}O_{19}$ where x is the bismuth doping level referred to throughout this paper. Materials were produced using x values of 0.0, 0.2, 0.5, 0.8, 1.0 and 1.5. The samples synthesized for testing were pellets measuring 20mm in diameter with thicknesses of between 3 and 5 mm. Samples were subjected to intermediate sintering and finally annealed for one or three hours at either 1100°C or 1200°C. Formation of bismuth substituted barium hexaferrite was confirmed using X-Ray Diffraction analysis. This analysis showed that after sintering at 1200°C undoped barium hexaferrite was formed, but that there were still unreacted reagents present. These reagents were not present in the doped material, confirming the action of bismuth as a sintering aid.

3 Experimental Procedure

At high frequencies the dielectric properties of materials are found using network analysis techniques which fall broadly into two categories; non-resonant and resonant methods. The measurement scheme used within this work is a non-resonant coaxial probe. This allows broadband characterization of a material when used in conjunction with a vector network analyzer.

The equipment set-up utilizes a HP 8510C VNA connected both to a PC and to the coaxial probe which is placed in contact with the sample pellet. After calibration using short, open and load conditions the samples were characterized over the range 45MHz – 25GHz. The software on the PC uses a point matching approach [7], [8] to calculate the real and imaginary parts of permittivity from the reflection of a signal at the probe-sample interface.

4 Results

4.1 Samples after Sintering at 1100ºC

Fig. 1 shows the high frequency permittivity over the range 45MHz – 25GHz of the samples from series 1 which have been sintered for one hour at 1100°C. The measured permittivity of air is also shown for reference and is within expected values.

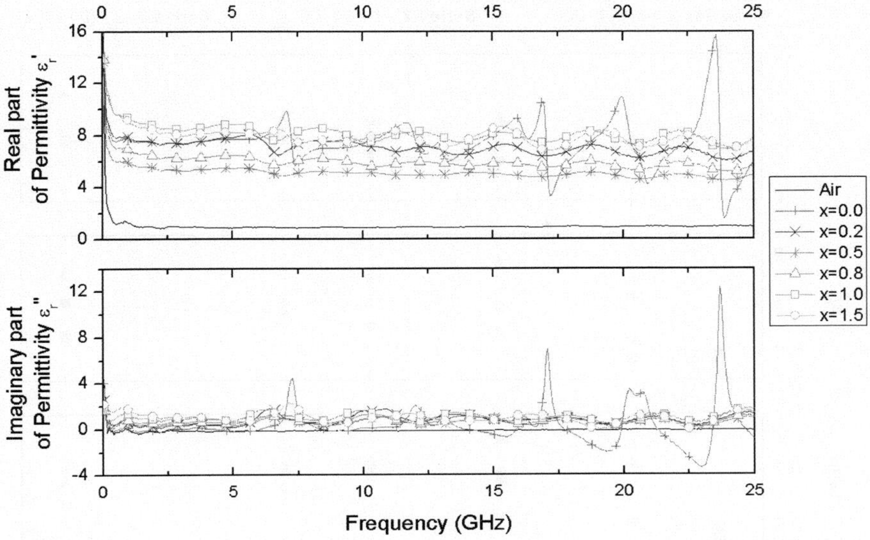

Fig. 1. Dielectric properties of bismuth substituted barium hexaferrite (formula $BaBi_xFe_{12-x}O_{19}$ where x is the value shown) for series 1 which have been sintered for one hour at 1100ºC.

Sample from series 1 display a range of values for the real part of permittivity between $\varepsilon_r' = 6$ to 9 at frequencies up to 6GHz after an exponential decay from the starting point at the lowest test frequencies. At frequencies above 6GHz the permittivity measurements show oscillations due to reflections from the back surface of the pellet where the undoped barium hexaferrite material is the most affected. Due to this, results are only considered reliable at frequencies up to 6GHz.

The reason for the variations in permittivity measurement is due to the low loss nature of the material; the sample is not appearing infinite to the probe at frequencies higher than the 6GHz limit. The highest permittivity material (x = 1.0) displays the greatest variation of the bismuth containing materials. The reasons for the large variation in the undoped and doped materials are different. The parent barium hexaferrite material has the largest reflections because it displays the lowest loss; thus the strength of the reflections are significant in the results. The high permittivity of the x = 1.0 sample means that reflections within the sample when displayed as permittivity results are proportionally larger and so are more visible in Fig. 1.

The highest permittivity sample is the x = 1.0 material with a permittivity of $\varepsilon_r = 8.54 - 0.89j$ at 2.45GHz. This was an increase relative to the pure barium hexaferrite sample, which had a permittivity of $\varepsilon_r = 7.39 - 0.01j$ at the same frequency.

High frequency characterization of samples from series 2 and 3 was also performed. The results with respect to frequency displayed similar trends to those presented for series 1 with differences in the absolute values. For this reason data is shown as bismuth content of samples plotted again the real and imaginary parts of permittivity and the loss tangent at a constant frequency of 2.45GHz. This value was selected because it allows assessment of the material for use in antennas operating at Wi-Fi and Bluetooth frequency.

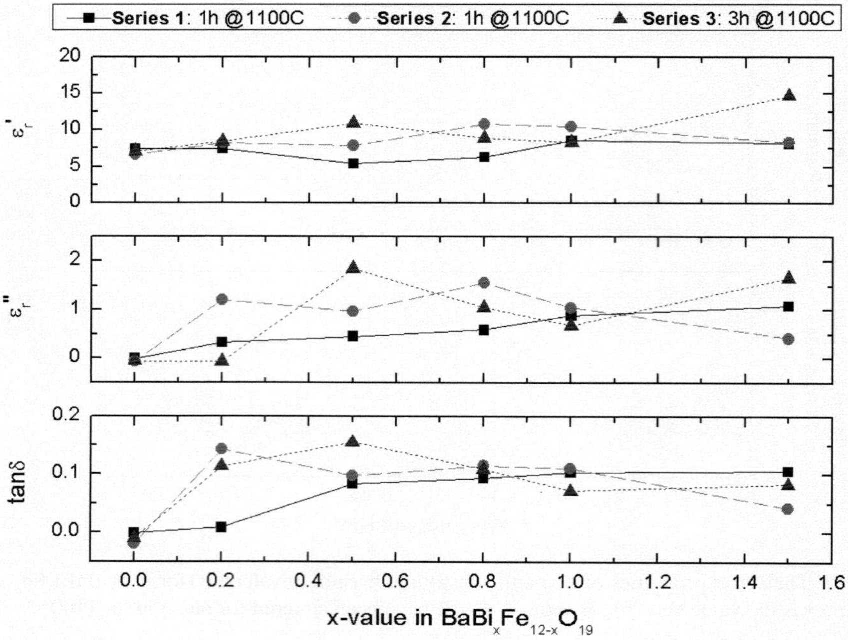

Fig. 2. Sample composition $BaBi_xFe_{12-x}O_{19}$ plotted against real and imaginary parts of permittivity (ε_r' and ε_r'' respectively) and loss tangent (*tanδ*) at a constant frequency of 2.45GHz for samples sintered at 1100°C.

All samples show similar values for the undoped barium hexaferrite material where permittivity values range from ε_r' = 6.63 to 7.39. When bismuth is added the permittivity values begin to diverge. For the lowest bismuth concentration (x = 0.2) values for the real part of permittivity are similar with clear differences in the imaginary part. Results for samples form series 1 and 2 which have been subject to the same sintering regime display similar trends. These have marked deviations from series 3 which has been subjected to a different annealing schedule.

At the x = 0.5 sample composition the highest permittivity is the series 3 material (ε_r = 10.95 – 1.852j) and the lowest is the series 1 material (ε_r = 5.39 – 0.451j) with the series 2 material falling in the centre of these values. After this point the trends seen for series 1 and 2 become more similar before converging at the final x = 1.5 value for the real part of permittivity while the losses show a small but greater degree of variation.

The trends for series 3 show an increase in permittivity with the addition of bismuth for values up to x = 0.5 before a slight decrease (the permittivity remains higher than that of the undoped material) and then a sharp increase for the final x = 1.5 composition which has the maximum permittivity of all samples with ε_r = 14.69 – 1.664j.

4.2 Samples after Sintering at 1200ºC

Samples which had been subjected to a final annealing temperature of 1200°C were also characterized using the coaxial probe over the 45MHz – 25GHz frequency range. The trends observed were similar to those seen for the samples in series 1 (shown in Fig. 1) with variation in the absolute values. For this reason results are presented at a constant frequency of 2.45GHz as shown in Fig. 3.

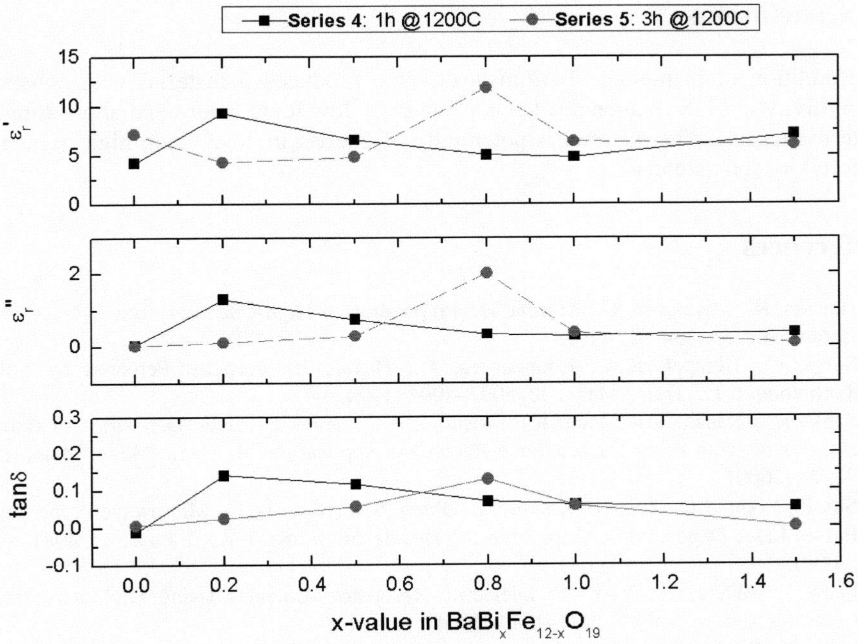

Fig. 3. Sample composition $BaBi_xFe_{12-x}O_{19}$ plotted against real and imaginary parts of permittivity (ε_r' and ε_r'' respectively) and loss tangent (*tanδ*) at a constant frequency of 2.45GHz for samples sintered at 1200°C.

The trends in the real and imaginary parts of permittivity and the loss tangent are the same for samples from each series.

Samples from series 4 show that the addition of bismuth enhances the permittivity of the material with a maximum value obtained for the x = 0.2 sample. From this point the permittivity decreases before a slight recovery for the final x = 1.5 composition.

Pellets which have been sintered for three hours at 1200°C (series 5) display a different trend. A marked increase in the permittivity of the undoped barium hexaferrite is noted ($\varepsilon_r = 7.18 - 0.039j$ at 2.45GHz). This is most likely due to the longer sintering duration at the higher 1200°C temperature allowing a greater degree of formation. When bismuth is added, the permittivity of the material initially drops to a minimum of $\varepsilon_r = 4.27 - 0.126j$ and then continues to increase to a series maximum of $\varepsilon_r = 11.81 - 1.995j$ for the x = 0.8 sample. Permittivity values then reduce with additional bismuth addition.

The reasons for the relatively low permittivity of the series 5 materials with low bismuth concentration is thought to be due to the evaporation of bismuth into the atmosphere due to the additive bismuth oxide's relatively low melting point of 817°C [9]. This is more significant when a smaller amount of bismuth is present in the mixture to begin.

5 Conclusions

The addition of bismuth to barium hexaferrite produced a material with enhanced permittivity at high frequencies from a relatively low temperature and short duration sintering regime. The material is potentially of interest in developing high frequency material loaded antennas.

References

1. Haneda, K., Miyakawa, C., Kojima, H.: Preparation of High-Coercivity $BaFe_{12}O_{19}$. J. Am. Ceram. Soc. 57, 354–357 (1974)
2. Surig, C., Hempel, K.A., Bonnenberg, D.: Hexaferrite Particles Perpared by Solgel Technique. IEEE Trans. Magn. 30, 4092–4094 (1994)
3. Sankaranarayanan, V.K., Pant, R.P., Rastogi, A.C.: Spray Pyrolytic Deposition of Barium Hexaferrite Thin Films for Magnetic Recording Applications. J. Magn. Magn. Mater. 220, 72–78 (2000)
4. Shi, P., Yoon, S.D., Zuo, X., Kozulin, I., Oliver, S.A., Vittoria, C.: Microwave Properties of Pulsed Laser Deposited Sc-Doped Barium Hexaferrite Films. J. Appl. Phys. 87, 4981–4984 (2000)
5. Peng, Z., Wang, H., Yao, X.: Dielectric Resonator Antennas Using High Permittivity Ceramics. Ceram. Int. 30, 1211–1214 (2004)
6. Ram, S., Krishnan, H., Rai, K.N., Narayan, K.A.: Magnetic and Electrical Properties of Bi_2O_3 Modified $BaFe_{12}O_{19}$ Hexagonal Ferrite. Jpn. J. Appl. Phys. 28, 604–608 (1989)
7. Grant, J.P., Clarke, R.N., Symm, G.T., Spyrou, N.M.: A Critical Study of the Open Ended Coaxial Line Sensor Technique for RF and Microwave Complex Permittivty. J. Phys. E. 22, 757–770 (1989)
8. Mosig, J.R., Besson, J.E., Gex-Fabry, M., Gardiol, F.E.: Reflection of an Open-Ended Coaxial Line and Application to Nondestructive Measurement of Materials. IEEE Trans. Instrum. Meas. 30, 46–51 (1981)
9. Kuchinskaya, E.A., Titenko, A.G., Pashchenko, V.P., Morozova, Z.P., Klochai, I.F.: Effect of Bismuth Oxide on the Phase Composition and Magnetic Properties of the $12Fe_2O_3.3BaO.2CoO$ System. Powder Metall. Met. Ceram. 16, 863–866 (1977)

Dielectric Resonator Antenna Using High Permittivity Bismuth Substituted Barium Hexaferrite

Leah M. Ridgway[1] and Ian Harrison[2]

[1] Division of Electrical Systems and Optics, The University of Nottingham, University Park, Nottingham, NG7 2RD, United Kingdom
[2] Department of Electrical and Electronic Engineering, The University of Nottingham Malaysia Campus, 43500 Semenyih, Selangor Darul Ehsan, Malaysia
LeahMRidgway@gmail.com

Abstract. A dielectric resonator antenna (DRA) was produced using a circular high permittivity bismuth substituted barium hexaferrite material. The theoretical resonant frequency was found to be 9.09GHz for the $HE_{11\delta}$ mode. S-parameters for the system located resonant frequencies at 6.6GHz, 9.55GHz and 13.6GHz. Radiation patterns for the DRA were measured at operating frequencies of 6.615GHz and 13.615GHz. Experimental results showed good agreement with simulation results published by other authors. The suitability of the material for use within a DRA was confirmed.

Keywords: Dielectric Resonator Antenna, DRA, Permittivity, Barium Hexaferrite.

1 Introduction

Materials with a large dielectric constant have previously been used in an attempt to reduce the physical size of antennas [1], [2]. A dielectric resonator antenna (DRA) is a device formed by encasing a simple antenna within or near to a dielectric material. It is designed to resonate at a specific frequency of interest where the system will be most efficient [2], [3]. The resonant frequency of the device is dependent upon the dielectric properties and the physical dimensions of the material used. Bismuth substituted barium hexaferrite has been identified as a candidate material to produce a DRA due to its high permittivity at high frequencies. This paper investigates the operation of this material loaded antenna.

2 Background Theory

A DRA composed from a circular resonant material and a coupling antenna can operate in one of two dominant resonant modes dependent upon the placement of the antenna feed. The diagram in Fig. 1 shows the two operational modes; $HE_{11\delta}$ and $TM_{01\delta}$.

X. Wan (Ed.): Electrical Power Systems and Computers, LNEE 99, pp. 1001–1005.
springerlink.com

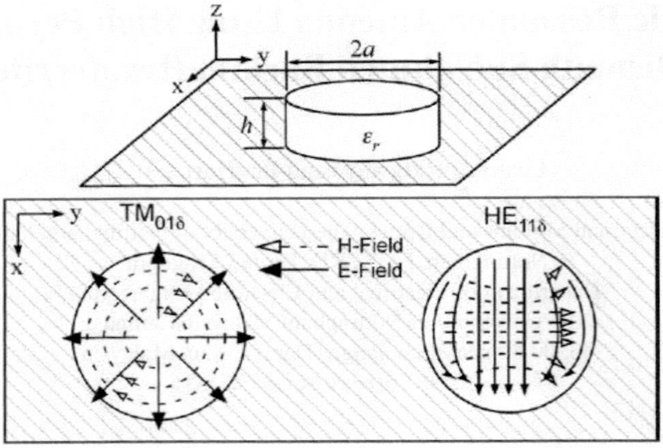

Fig. 1. Modes available in a circular dielectric resonator antenna: $TM_{01\delta}$ and $HE_{11\delta}$ [4].

To excite the $TM_{01\delta}$ mode the coaxial antenna feed is located in the centre of the material. This requires the dielectric to be machined to accommodate the feed, making this configuration more difficult to construct than the alternative $HE_{11\delta}$ system; thus the $HE_{11\delta}$ configuration was selected for study. Here the antenna feed is located outside of, but in contact with the dielectric disk as shown in the schematic in Fig. 2.

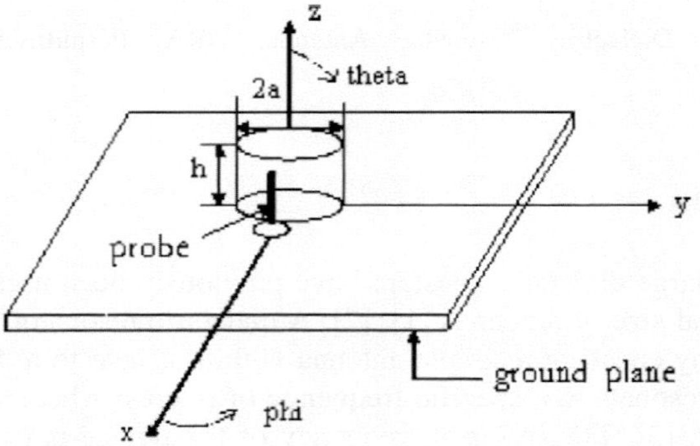

Fig. 2. Diagram illustrating the form of a DRA made using a circular resonant material and showing the *x,y* and *z* planes used in radiation pattern measurements [2].

2.1 Theoretical Calculation of Resonant Frequency

The dominant frequency of the $HE_{11\delta}$ mode can be found using equation 1 [5].

$$f_0 = \frac{3\times10^8}{2\pi\sqrt{\varepsilon_r'}}\sqrt{\left(\frac{1.841}{a}\right)^2 + \left(\frac{\pi}{2h}\right)^2} \quad (1)$$

Where f_0 is the frequency of the dominant mode, ε_r' is the real part of permittivity, a is the radius and h is the thickness of the dielectric material used in the system. Using this equation the theoretical resonant frequency of a DRA using a bismuth substituted barium hexaferrite sample (prepared by the authors) was found. The material had chemical formula $BaBi_{0.2}Fe_{11.8}O_{19}$ with a radius of 10mm and a thickness of 4.731mm. The real part of permittivity was characterized at 2.45GHz as 3.978. These values were substituted into equation 1 and the calculation produced a resonant frequency of 9.09GHz.

3 Experimental Work

3.1 Dielectric Resonator Antenna System

A device similar to that shown in Fig. 2 was constructed with a ground plane made from copper. The antenna feed to the device was formed from a coaxial SMA connector with the inner conductor extending past the grounding plate. The probe feed was in contact with the side of the high permittivity disk. The resonant material was attached to the copper plate using a thin layer of conducting epoxy (ensuring that the dielectric spacer between the inner and outer conductors of the probe feed was not shorted) this was then left to cure for 48 hours.

3.2 Frequency Response

The frequency response of the system over the range 50MHz – 20GHz was characterized by measuring the reflection coefficient (S11) of the device utilizing a HP 8510C VNA. Results from this are presented in Fig. 3 which shows three clear resonant frequencies for the system; the first at 6.6GHz, the second at 9.55GHz and a third at 13.6GHz.

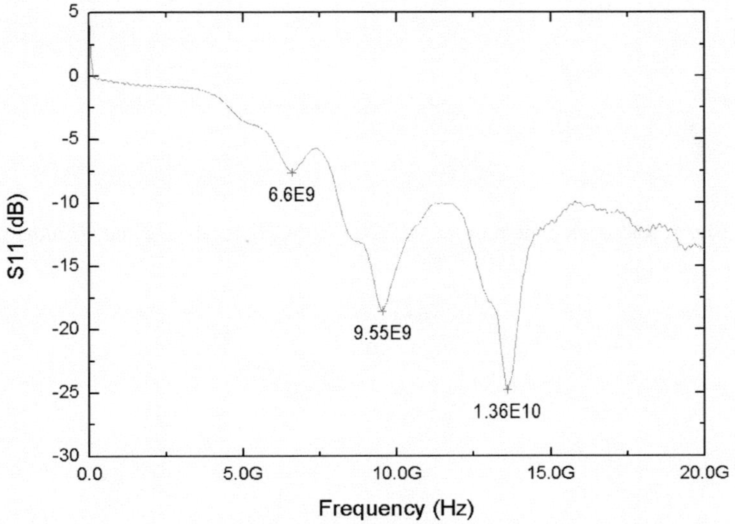

Fig. 3. *S11* (reflection coefficient) response of the DRA over the frequency range 50MHz – 20GHz.

The second resonant frequency of 9.55GHz is the closest to the theoretical value calculated from equation 1 of 9.09GHz. Differences are most likely due to deviations from the ideal conditions assumed in equation 1, such as a perfectly smooth contact surface between the resonant material and the grounding plane. There is also no mechanism in the equation to account for dielectric losses which do occur in a real material.

The lower frequency resonance at 6.6GHz was not predicted by equation 1 but is potentially of more interest for Wi-Fi applications than the higher frequency responses. A resonance around 6.6GHz was observed for several different DRA rigs incorporating different bismuth substituted barium hexaferrite materials so is a repeatable and valid result.

3.3 Radiation Patterns

The radiation patterns of the DRA were recorded at 6.615GHz and 13.615GHz using an Agilent 8757D Scalar Network Analyzer in conjunction with an Agilent E8257D PSG Analog Signal Generator. These values were selected because 6.615GHz was the closest value corresponding to the first resonant frequency found from the S11 values in the previous section and 13.615GHz is the closest value corresponding to the largest magnitude resonance.

Radiation patterns were recorded for two different orientations (as defined in Fig. 2) and are presented in Fig. 4 and Fig. 5.

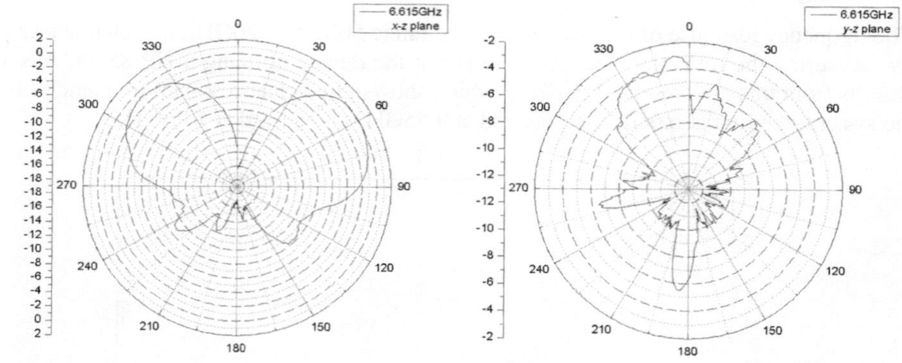

Fig. 4. Radiation pattern of the DRA at 6.615GHz for the *x-z* and *y-z planes.*

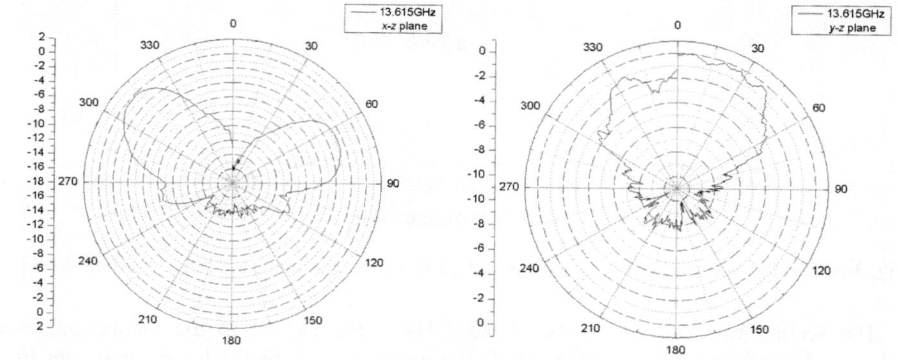

Fig. 5. Radiation pattern of the DRA at 13.615GHz for the *x-z* and *y-z planes.*

The radiation patterns for the antenna system show the same trends. In the x-z plane there are two clear main lobes centered about 60º and 315ºC with some smaller back-lobes also visible. In the y-z plane at both frequencies there is a wide lobe centered at 0º and also a series of back-lobes.

These results are in agreement with simulation work published by other authors on the subject of high permittivity dielectric resonator antennas [2]. This confirms the device can be used in applications where a DRA is suitable.

4 Conclusions

A dielectric resonator antenna was produced using a high permittivity bismuth substituted barium hexaferrite sample. The resonant frequencies of the device were located and radiation patterns for the antenna at resonance were presented. The system shows experimental results that agree with simulation work published in other literature.

Acknowledgments. The authors would like to thank Belle Ooi Pei Cheng at The University of Nottingham Malaysia Campus for her help in performing radiation pattern measurements.

References

1. Colburn, J.S., Rahmat Samii, Y.: Patch Antennas on Externally Perforated High Dielectric Constant Substrates. IEEE Trans. Antennas Propag. 47, 1785–1794 (1999)
2. Peng, Z., Wang, H., Yao, X.: Dielectric Resonator Antennas Using High Permittivity Ceramics. Ceram. Int. 30, 1211–1214 (2004)
3. Carr, J.J.: Practical Antenna Handbook. McGraw-Hill, London (2001)
4. Petosa, A.: Dielectric Resonator Antenna Handbok. Artech House, London (2007)
5. Long, S.A., McAllister, M.W., Shen, L.C.: The Resonant Cyclindrical Dielectric Cavity Antenna. IEEE Trans. Antennas Propag. 31, 406–412 (1983)

Estimation of Failure Rate with the Help of Risk Monitors

Muhammad Zubair and Zhang Zhijian

College of Nuclear Science and Technology, Harbin Engineering University,
145-1, Nantong Street, Nangang District, 150001, China
Zubairheu@gmail.com, zhangzhijian@hrbeu.edu.cn

Abstract. In order to calculate failure rate (λ), there are two methods. First is Classical Estimate Method and second Bayesian method. The objective of this paper is to use Bayes'method in a newly developed risk monitor called Risk Manager and calculate failure rate that can be used to analyzing data. The study also explains preventive replacement cycle and failure rate estimator as two main portions of Risk Manager to improve the reliability of components and equipments.

Keywords: Data analysis, Risk monitoring, Preventive replacement, Failure rate, Bath tub curve.

1 Introduction

Risk Monitoring (RM) is a plant specific real-time analysis tool used to determine the instantaneous risk based on the actual status of the systems and components. At any given time, it reflects the current plant configuration in terms of the known status of the various systems and components e.g. whether there are any components out of service for maintenance or tests. The risk monitor is used by the plant staff in support of operational decisions [1]. RM used in operational risk management and scheduling maintenance movement is based on LPSA technology development. RM provides criterions and equipment information for on-line maintenance. The objective of RM is to detect and control plant configurations and plant operational actions from a risk perspective. The utility of Risk Monitor enables us to manage risk with risk-informed methodology and make operation, more flexible and economically effective. Risk monitoring for PSA studies in recent years has become the most active part of the field.

In this paper, a newly developed risk monitor called "Risk Manager" is presented. Risk Manager has new features compared with other risk monitors such as reliability data auto update, preventive replacement cycle auto update. First we will discuss main features and functions of Risk Manager and then explain how to estimate failure rate with the help of this Risk Manager.

2 Risk Manager and Failure Rate

In order to avoid the risk in Nuclear Power Plants (NPP) by equipment failure, the preventive replacement of important and sensitive equipment will definitely achieve

X. Wan (Ed.): Electrical Power Systems and Computers, LNEE 99, pp. 1007–1014.
springerlink.com

the normalization of safety and economy. For the establishment of preventive replacement cycle and to optimize the initial value according to the safety of NPP risk level is the essential aspect of risk management. Risk Manager takes real-time calculation as well as integrated risk management into account and adopts preventive replacement method based on dynamic risk evaluation. The Preventive replacement cycle and failure rate estimator are two main parts of Risk Manager.

3 Modes and Functions

Risk Manager (nuclear power plant risk-management system) has four modes, namely: planning risk model, risk query model, risk tracking mode and professional mode.

3.1 Planning Risk Mode

This mode is related to plan makers through the risk management system input schedule. With the help of this mode we can get risk curves of a refueling cycle. With the analysis of this risk curve the value of high-risk operation can be used to modify or to give focus on monitoring and management of key equipments and systems.

3.2 Risk Query Mode

This mode deals with maintenance staff and query table. The query table is input of risk management system. This mode provides maintenance activities to be taking into account plant risk analysis and give recommendations accordingly. This mode also helps us in the production of equipment preventive replacement cycle settings and preventive replacement reminder function.

3.3 Risk Tracking Mode

Risk tracking mode is used for operating personnel, through a monitoring system. With the help of this mode we can make power plant equipment operation as well as power plant equipment information system. This mode gets the current risk value and equipment needed to focus on the prevention of incidents and make connection to receive real time status changes.

3.4 Professional Mode

This mode is special for PSA professionals. The PSA professionals establish System ET / FT Models. ET/ FT model is the basis for other models that can be achieved.

4 Six Modules for Risk Monitor

Risk Manager consists of six modules name as: system reliability analysis module, equipment reliability data update module, preventive replacement modules, fault tree analysis model, output module, and graphic modeling module as shown in Fig.1.

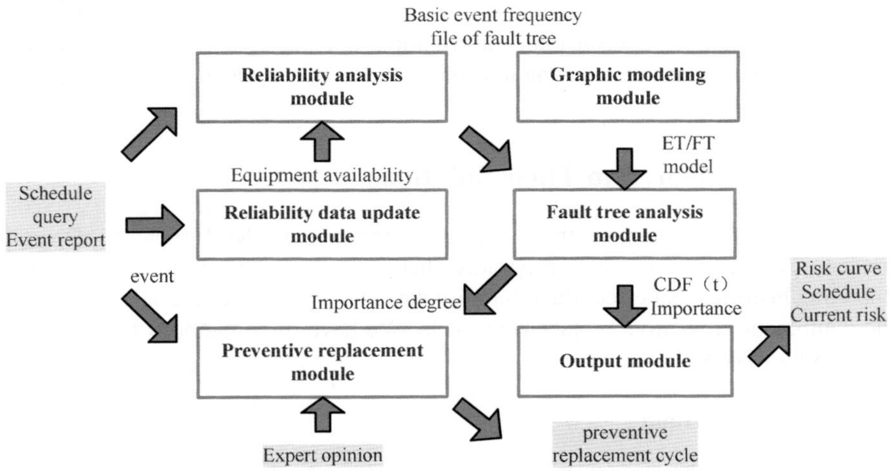

Fig. 1. Six module system block diagram

The internal events of power plant are complex and include equipment maintenance, equipment damage etc. Among them related to PSA equipment (PSA equipment refers to the ET / FT model of equipment involved). These events will affect the probability of the current risk level of the system, as well as the reliability of the equipment parameter values, so such kind of events will serve as an important basis for systematic analysis and calculation.

PSA equipment state changes may be inevitably lead to a change in system reliability model. The function of system reliability analysis module is to modify the model in order to reflect the plant's current risk level faithfully. System reliability analysis module provides input for the fault tree analysis module the risk level is determined by CDF (t) representation. By giving the risk value and the relative time, we can get time-varying risk curve.

PSA equipment state changes particularly in relation to equipment failure and maintenance that will affect the reliability data values. The function of reliability data update module is to renew original reliability data and provision of equipment unavailability for the system reliability module. So that we can reflect the impact on the system risk level that change in equipment reliability parameters.

The Preventive replacement module's function is to update the preventive replacement cycle of critical equipment according to the importance degree which is provided by fault tree analysis module.

Graphical modeling module used to establish the system ET/ FT model to provide a basis for the analysis of other modules.

System state changes are due to various types of events happened in NPP. In the planning mode, the planning table is actually formed by the undivided events and such events are a combination of the operation of power plant equipments. In query mode, the power plant's current state can be achieved by the information coming from equipments. Power plant risk monitoring system is a very important aspect because it provides connection between plant information and management system. Hence in order to start the process of monitoring system dynamics we use event driven principle.

5 Some Problems and Their Solutions

Risk Manager can manage three changes happened in NPP such as: Operation procedures and maintenance programs that impact the PSA model, equipment operating histories that effect the reliability data and experience feedback which effect the equipment preventive replacement cycle. We have three methods to reflect these changes by Risk Manager.

5.1 Modeling Method

In order to meet the requirements of real-time calculations, there are two different methods: edit ET / FT model and modify the Boolean equation obtained by the model analysis.

Boolean equations are obtained by solving sequences that lead to core damage. The accident sequence is solved by "large fault-tree" method, refer to the Fig. 2. Plant ET/FT model can be viewed as a large fault tree whose top event is core damage. The top event connects with other events with AND gate. The input of the AND gate is an initial event and a top event of a fault tree which represents safety protection system failure. If more than one initial event sequence use OR gate connected with the top event then we get a large PSA fault tree [2]. After solution of each accident sequences lead to core damage, we can obtain Boolean equation of the reactor core damage top event. In other words in order to get Boolean equation we should do two steps. First to solve the various supports Systems fault tree and front systems fault tree, and then re-solve the core damage Accident sequences.

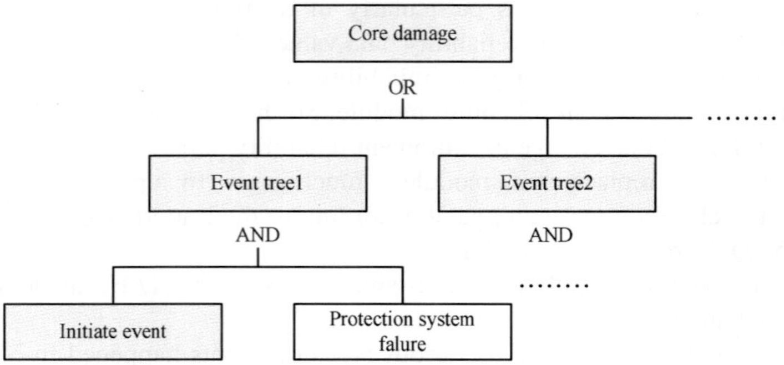

Fig. 2. Large fault tree method

After obtaining Boolean equations we establish a dynamic basic event model, we make re-quantitative calculation of the Boolean equations. Dynamic basic event model means that basic event happening frequency is assigned 0 or 1 depending on the event type. By using this model and the actual state of the plant we can calculate the real-time risk.

5.2 Data Analysis

In power plants the equipment status changes due to maintenance or failure. So in order to calculate reliability data we use historical information about the equipment states and then use Bayesian theory for updating. General reliability database store the reliability data of PSA equipment (PSA model involved in the device) including failure rate, and MTTR (mean time to repair).

For Bayesian updating Beta distribution is require, so we need to make two ultra-parameters of Gamma distribution according to general database [3]. These two ultra-parameters of Gamma distribution combined with reliability information. As a result we can update the reliability data with Bayesian theory. At last the updated equipment failure rate is used to calculate unavailability of equipment. The Unavailability of equipment become the input of large fault tree and used in recalculation so as to evaluate the influence that decrease the reliability of equipment bring in to plant risk level.

5.3 Preventive Replacement

With the formation of risk monitoring and management system in NPP we are able to obtain the impact on plant risk level and operating history of certain equipment. By using these two information we can manage important and sensitive equipment's preventive replacement cycle dynamically by including online query and update.

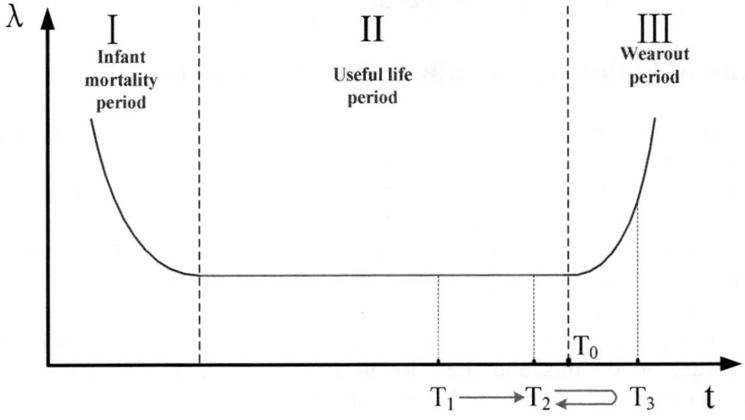

Fig. 3. Bath tub curve

This subsection proposes known as preventive replacement method which combines the experience feedback and expert advice. The different devices are given the best preventive replacement cycle and apply to Risk Manager Development.

To improve the power plant economy Preventive replacement based on experience feedback methods can get the best equipment replacement cycle time. Here we explain simple description of this method refer to "bathtub curve".

T_0 is the time that equipment enters the critical period of wear out failure point as shown in Fig.3. The ideal situation is to replace the equipment at point T_0, but we can not obtain more accurate values of T_0. Preventive replacement method based on experience feedback. In this method first we set an initial value of the replacement of time T_1. After a period of time we will get experience feedback about this equipment and operating experience will tell us the original setting is conservative or not. After several rounds of amendments finally approach to T_0.

Expert advice gathers from the equipment manufacturer, operation managers, maintenance staff, as well as the views of other users, the more extensive the better. Expert opinion is only quite good understanding of the relevant equipment or operation of operating experience with personnel, these personnel are fully aware of the equipment but they do not understand the importance of equipment when comes to risk level. So, we should give initial preventive replacement time with more or less conservative.

According to expert opinion there are different levels of equipments that can be obtained at initial stage of preventive replacement cycle. On other hand we can get p from expert's opinions which mean failure probability does not occur to preventive replacement time date equipment. We consider p as an indicator which determine whether updating preventive replacement cycle change or not. With the help of Bayesian statistical theory we can update value of p.

We get initial preventive replacement time T and P on behalf of expert opinion. Equipments will operate in accordance with the initial replacement value and obtain the operating records. We use Bayesian statistical theory to obtain P's posterior distribution and expected value. Then, we compare the new P with the old one and evaluate preventive replace time with established criteria.

6 Types of Failures and Failure Rate Estimation

A failure is defined as the loss of the ability of an item, a component or a system to perform its required function. A failure is generally a subset of a fault. It represents an irreversible state of an item, a component or a system, such that it must be replaced or repaired in order to perform its design function. An item or component failure is always defined in relation to the system in which the item or component resides [4].

In general, the failure rate of equipment depends on the environmental conditions. Therefore, these circumstances should ideally be taken into consideration in all data acquisition activities. However few data bases provide the environmental application factors needed to do this and they are generally only available for electrical and electronic components. The environmental application factor is a multiplicative constant used to modify a failure rate to incorporate the effects of other normal and abnormal environmental operating conditions.

In this study we focus on two types of failures, first is time related failure and second Demand related failures.

6.1 Time Related Failures

A time related failure is defined as a failure occurrence (e.g. per hour) of a component or system which is in the running or functioning mode of operation, continuous operation (or function).

6.2 Demand Related Failures

A failure on demand is defined as a failure occurrence of a component or system when it is demanded to start operation or to change its present state. The failure rate is a numerical value which represents the probability of specified failures of a component per time unit. The all modes failure rate of a component is an aggregate of failure rates summed over relevant failure modes.

Reliability models involve a variety of parameters such as component failure rates that need to be estimated in order to estimate the probability of specific accident sequences. If data are available and it is desired to obtain estimates that are strictly functions of the data then for the models commonly used in risk analysis, point estimators are well established. The point estimators generally used for the binomial, Poisson and lognormal models.

In case of Binomial distribution let 'f' failures in 'n' demands then probability of failure on demand can be calculated as;

$$p = f / n \tag{1}$$

In Poisson distribution case if there are 'f' failures in 't' time then failure rate 'λ' calculated as;

$$\lambda = f / T \tag{2}$$

In lognormal distribution for 'n' independent positive observations like repair time, the expected value of mean 'μ' and variance 'σ' for repair time 't' can be estimated as;

$$mean = \mu = \sum_{i=1}^{n} t_i / n = \bar{t} \tag{3}$$

$$\text{var}\,iance = \sigma = \sum (t_i - \bar{t})^2 / n - 1 = s_t^2 \tag{4}$$

The behavior of λ (failure rate) can be traced by bathtub curve and λ increases with number of failures (x) that can be observed in fig.4. In the beginning the failure rate of each device is high due to wear in failures or due to poor quality assurance during manufacturing. During middle life time failures occur with uniform rate and at late in life failure rate increase due to wear-out failures.

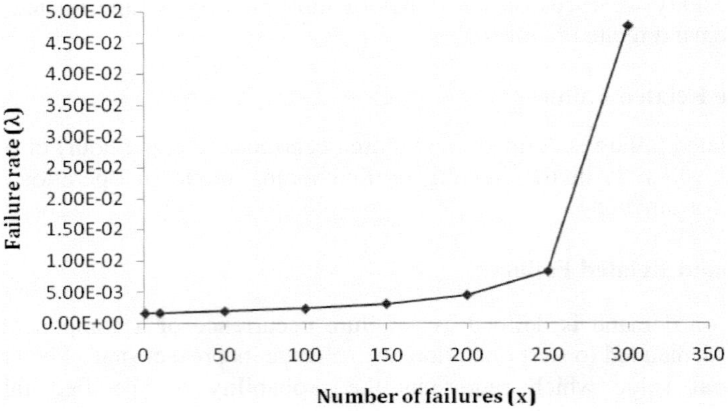

Fig. 4. Increase of failure rate (λ) with increase in number of failures (x)

7 Conclusion

It is noted that online monitoring whether coming from data acquisition system or plant computer, must first be qualified and validated. This can be done with automatic algorithms embedded in softwares and this target can only be achieved when we have high efficiency Risk monitors. The study highlights the basic requirements for the development of a Risk Monitor. In coming future on the basis of this study we can develop a Risk monitor that will be helpful in nuclear industry as well as in other growing areas of research.

References

1. CSNI Technical Opinion Papers No 7: Living PSA and its Use in the Nuclear Safety Decision-making Process, OECD. NEA No. 4411, pp. 1-21 (2005), ISBN 92-64-01047-5,
2. Johanson, G., Holmberg, J.: Safety evaluation by living probabilistic safety assessment procedures and applications for planning of operational activities and analysis of operating experience. SKI report 94, Swedish nuclear power inspectorate, Stockholm (1994)
3. Zubair, M., Zhijian, Z., Khan, S.U.D.: Calculation and updating of reliability parameters in probabilistic safety assessment. Journal of Fusion Energy 30, 12–15 (2011)
4. International Atomic Energy Agency. Manual on reliability data collection for research reactor PSAs, IAEA-TECDOC-636 (1992)

Research on FoMs of SAR ADC

Libin Hu[1,2] and Wenshi Li[1,2]

[1] School of Electronics and Information Engineering, Soochow University, Suzhou, 215006, China
[2] Key Laboratory of Modern Acoustics of MOE (Nanjing University), Nanjing, 210093, China
lwshi@suda.edu.cn

Abstract. SAR ADC is the best choice for low-power and high-resolution application in signal processing systems, since it is fit to work in midst rate per conversion (MSPS, GSPS). In order to guide trade-off design better, this paper starts with analyzing the advantages and disadvantages of known *FoMs* (figure of merits) of traditional SAR ADC. To keep the characteristic of minimum merit values and grasp the feature of bright distinctive degrees, one novel *FoM* of SAR ADC was built with five key parameters. Total four kinds of *FoMs* were compared with convinced data from 15 pieces latest literatures, including Q (based on 2-parameter), FoM_0 (based on 3-parameter), and FoM_4 (based on 2-parameter). As interesting results, the compromised design parameters region had gained. This work may provide SAR ADC designers one new practical evaluation function for efficient optimization design future.

Keywords: SAR ADC, *FoMs*, Low-Power, High-Speed, High-Resolution.

1 Introduction

ADC is a cluster of mixed signal integrated circuit paradigms and one core in portable medical instruments [1].

Back to the Lindek potentiometer configuration, first described in 1899, the early basis of most ADCs and digital voltmeters are dug out in paper [2].

Compared with other ADCs, SAR ADC, invented in 1954 [3], is an intermediate ADC model for low power consumption. It trends for high speed and high accuracy in recent studies [4].

Depending on several important parameters, *FoMs* (Figure of Merits) building can be the key evaluation paradigm for present and future optimal design of ADC.

To aim at SAR ADC, this work focuses on one novel *FoM* after further combing the known five *FoMs*. The convinced data for SAR ADC come from 15 references. Compared with five *FoMs*, our new *FoM* has more bright distinctive degree and practicability. The innovation points in this work are that it redefines SAR ADC *FoM* function with the actual factors on thermal noise and die size.

After introduction, Section 2 analyzes the thought of *FoM* formation. Section 3 illustrates five *FoMs* on SAR-ADC. Section 4 proposes our new *FoM* and compares the differences among four *FoMs*. The conclusions are thrown in Section 5.

X. Wan (Ed.): Electrical Power Systems and Computers, LNEE 99, pp. 1015–1022.
springerlink.com

2 FoM Formation Thought

SAR ADC can be used as typical representative describing the Moore's law of analog circuit, in form of that deposition between sampling rate and resolution in SAR ADC is doubled per three years, with statistics ranging from 1997 to 2008 [5].

As an evaluation function for the whole performance of SAR ADC, the thought of *FoM* making is to map multi-parameters into single-parameter with minimum merit. Equation (1) [6] defines the each conversion energy Q of SAR ADC sub-system.

$$Q = P / f_{sampling} \,. \tag{1}$$

Where P is power consumption and $f_{sampling}$ is sampling rate. Being equaled to simplified *FoM*, it evaluates the SAR ADC performance together with power and sampling rate.

According to Equation (1), the 16bit SAR ADC energy per conversion Q is shown in Fig.1 [6]. The package area also be given for detail comparison (Note: year 2007 ~ 2010 data are from authors' adding).

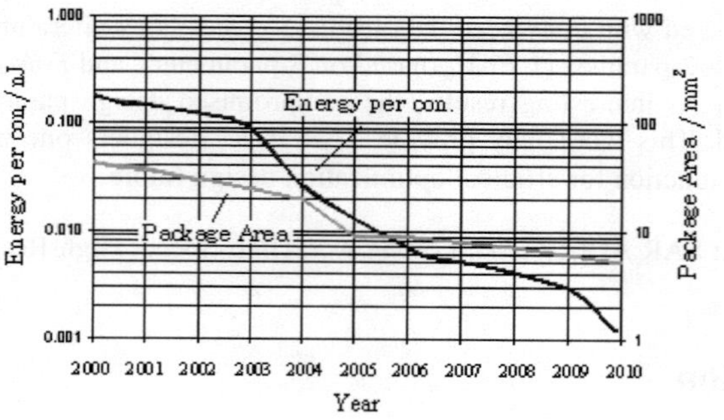

Fig. 1. The trend of energy per conversion and package area in 16bit SAR ADC

Compared with Q value and package area from 2000 to 2010, the latest converters of printed circuit board (PCB) area is reduced by 80%, and every conversion energy is reduced by 90% at least.

3 Five Major FoMs

The *FoM* function with two parameters doesn't make a good enough balance on all parameters' performance for SAR ADC, because it distinguished macroscopic superiority in respect of energy unit and linear trend. Measuring the performance of SAR ADC, it is usually based on three key indexes: resolution, sampling rate and power consumption.

3.1 Early FoM

A Merit formula with three indexes was proposed in 1993 in ISSCC:

$$FoM_0 = P/(2^{ENOB} \times f_{sampling}) \ . \tag{2}$$

Where

$$ENOB = (SNDR - 1.76)/6.02 \ . \tag{3}$$

ENOB is effective resolution; *SNDR* is signal-to-noise distortion ratio; *P* is power consumption; $f_{sampling}$ is sampling rate. In addition, the *SNDR* value can be obtained through FFT spectrum analysis, combining with formula (4) as follows:

$$SNDR = 20\log_{10}(V_{RMS} / N_{RMS}) \ . \tag{4}$$

Where V_{RMS} is the input signal RMS, and N_{RMS} is the noise and harmonic distortion of total RMS.

While applying formula (2) to evaluate products, it shows that the smaller FoM_0 value, the better performance.

The statistics of the low power, medium speed and medium resolution of ADC FoM_0 are shown in Fig.2. The shadow region S is blank one, and data concentrate in 0 ~ 10 and > 20 regions. The evolution rule shows that FoM_0 declines 20 times every ten years. For example, FoM_0 = 1pJ/con. @ 2000. The technologies are driven by reducing the power supply and choosing the appropriate structures [7].

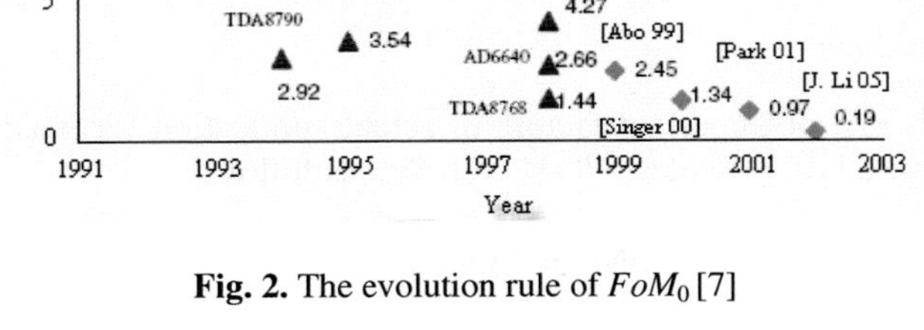

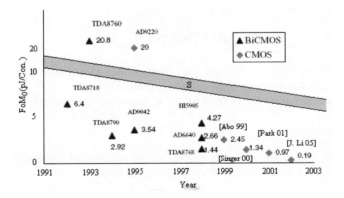

Fig. 2. The evolution rule of FoM_0 [7]

In Fig.2, the lowest FoM_0 value is 0.19pJ/con with structure of pipeline. Under conditions of 12bit resolution and 5MSPS sampling rate, this chip only consumes 4mW.

3.2 FoM with f_{ERBW}

The f_{ERBW} is signal effective bandwidth and it determines the bandwidth ADC handled. The *FoM* with f_{ERBW} is as follows [8]:

$$\mathrm{FoM}_1 = \mathrm{P}/(2^{\mathrm{ENOB}} \times \mathrm{f}_{\mathrm{sampling}}) \; . \tag{5}$$

To apply formula (5) for assessing products, it will show that the smaller FoM_1 value, the better the performance of products.

3.3 FoM with f_{in}

The f_{in} is maximum input signal frequency, determining the analog signal waveform reconstruction in no distortion in ADC. The *FoM* with f_{in} is as follows [8]:

$$\mathrm{FoM}_2 = \mathrm{P}/(2^{\mathrm{ENOB}} \times 2\mathrm{f}_{\mathrm{in}}) \; . \tag{6}$$

To evaluate SAR ADC performance, the optimal trend is single-reduction.

To apply Nyquist theory [8], it proves the significant correlation among $f_{sampling}$, f_{in} and f_{ERBW} ($f_{sampling} > 2f_{in}$, $f_{sampling} > 2\,f_{ERBW}$). Considering the signal integrity in practical application, it usually chooses the form with minimum value in ($f_{sampling}$, $2\,f_{ERBW}$, $2\,f_{in}$).

3.4 FoM with *STNR*

Dealing with ADC comprehensive performance, it is easy to overlook that ADC with resolution 8 bits or higher is largely affected by noises which come from Quantization Noise and Thermal Noise. Quantization Noise does not take much power to reduce. Hence, *FoM* processes with *STNR* is as follows [9]:

$$\mathrm{FOM}_3 = \mathrm{P}/(\mathrm{STNR}^2 \times \mathrm{f}_{\mathrm{sampling}}) \; . \tag{7}$$

The *STNR* is signal-to-thermal (or shot) noise ratio, and it is relevant to temperature *T*. *STNR* derivation can inquire literature [9].

Also the evaluation tendency for SAR ADC is towards decreasing.

3.5 FoM with *Size*

The die size is one of important factors in actual application for reason of chip cost related. Literature [10] provides a *FoM* with *Size* as follows:

$$\mathrm{FoM}_4 = \mathrm{Size}/2^{\mathrm{N}} \; . \tag{8}$$

Size is die size general with mm^2 for unit and *N* is ADC digits.

Similarly, the evaluation tendency for SAR ADC is towards decreasing.

4 Our New FoM

The latest literatures show that the development trend of SAR ADC is low power, high speed and high precision.

Combining with the above formulas, keeping single-reduction trend, considering maximum signal integrity and practical factors such as cost, we provides novel *FoM* function with five parameters as follows:

$$FoM = K_{ST} \times P/[2^{ENOB} \times \min(f_{sampling}, 2f_{ERBW}, 2f_{in})] \ . \tag{9}$$

Where *ENOB* is given by Equation (3); *P* is power consumption; $f_{sampling}$ is sampling rate; f_{ERBW} is effective input bandwidth; f_{in} is maximum input frequency. Coefficient K_{ST} is related with die size and signal to thermal noise ratio, defined as follows:

$$K_{ST} = \gamma \times Size / STNR^2 \ . \tag{10}$$

Where, *Size* is die size and *STNR* is signal-to-thermal (or shot) noise ratio. γ is transition coefficient for the welterweight mismatch by the product of chip size and conversion energy. It mainly plays a detente role, here it takes $\gamma = 1/8$. K_{ST} units use $(mm/dB)^2$ in relation of die size and thermal noise.

Compared with the above Equations, Equation (9) was found to balance the influence of die size, noise, power, resolution and sampling rate for SAR ADC. It makes up the deficiency of traditional *FoMs* functions on original *FoM* basis (e.g. ignoring the influence of thermal noise, power and die size at design cost), and improves the reliability and practical applicability.

In order to illustrate the advantages of evaluating functions, we calculated with data of the latest domestic and foreign literatures [11 ~ 25] about *Q*, FoM_0, FoM_4 and *FoM* functions' values. Based on respectively power consumption, effective resolution and sampling rate as independent variables, Fig.3 shows the relationship of *Q*, FoM_0, FoM_4 and our *FoM*.

To compare the *FoMs* functions, we use the form as follows:

$$FoM = K_{ST} \times P/(2^{ENOB} \times f_{sampling}) \ . \tag{11}$$

In order to reduce the computational complexity, let us take approximately $STNR \approx$ 50dB ($T = 300$K).

Fig.3 remind us as follows:

1. Compared with others, this *FoM* with $STNR \approx 50$dB provides a more bright degree of distinction. Then it is more favor to evaluate the combination property of SAR ADC series.

2. As far as this *FoM*, the less *FoM* value, the better performance of SAR ADC. Literature [13] provides an ultra-low power design and Literature [11] is with high speed design. However, they are just at intermediate level.

3. The region W shows the better design in Fig.3. From it we find that the literature [16] and [20] are better for considering the combination property with the optimal trade-off design, and their conceptions are worthy of high praise.

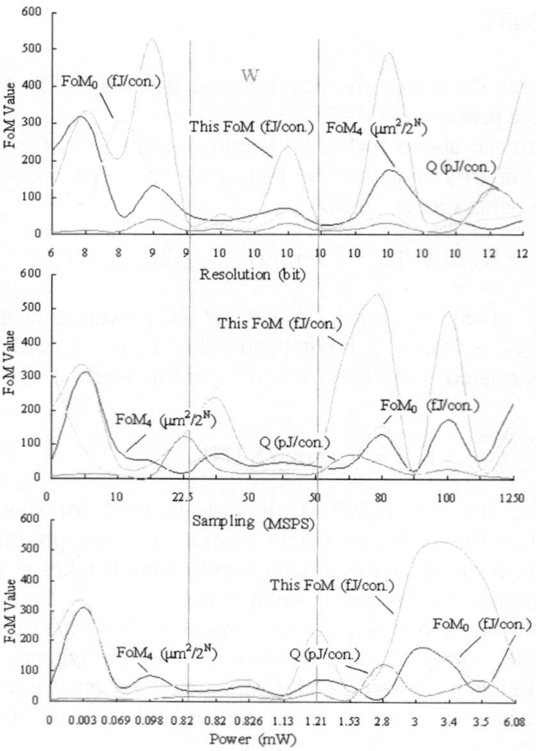

Fig. 3. The merits comparison and trade-off windows of 3-single-variable scanning

5 Summary

Advanced SAR ADC can provide 18bit resolution and GSPS sampling rate. Applying our new *FoM* of SAR ADC, it will get the reasonable trade-off design evaluation for SAR ADC. With techniques progresses such as comparator of configured threshold, input buffer, and switched capacitor DAC structure [12, 13, 17 and 22], ultra-low power SAR ADC needs our new *FoM* to help analyzing the effect of trade-off design.

Acknowledgments. This work was supported by the Natural Science Foundation of Jiangsu Province Education Hall, China in 2009 under Grant No. 09KJB510017.

References

1. Adachi, R., Landis, D., Madden, N., Silver, E., LeGros, M.: A Low Power 12-bit ADC for Nuclear Instrumentation. In: Nuclear Science Symposium and Medical Imaging Conference, vol. 1, pp. 365–367 (1992)
2. Gosling, W.: Twenty Years of ATE. In: Proceeding of International Test Conference, pp. 3–6 (1989)

3. Baker, R.J.: CMOS Circuit Design, Layout, and Simulation (Book Review, 2nd edn.). Circuits and Devices Magazine 22(3), 37 (2006)
4. Analog-to-Digital-Converters: AD7986, http://www.analog.com
5. ADC Performance Survey (1997,2008), http://www.stanford.edu
6. 21_Century SAR ADC (in Chinese), http://www.analog.com
7. Li, J., Maloberti, F.: Pipeline of Successive Approximation Converters with Optimum Power Merit Factor. In: 9th International Conference on Electronics, Circuits and Systems, vol. 1, pp. 17–20 (2002)
8. To Realize Undersampling with Reasonable Choice of High-Speed ADC (in Chinese), http://www.eaw.com.cn
9. Lee, H.-S.: Zero-Crossing-Based Ultra-Low-Power A/D Converters. Proceedings of the IEEE 98(2), 315–332 (2010)
10. Xiang, F., Srinivasan, V., Wills, J., Granacki, J., LaCoss, J., Choma, J.: CMOS 12 bits 50kS/s Micropower SAR and Dual-Slope Hybrid ADC. In: 52nd IEEE International Midwest Symposium on Circuits and Systems, pp. 180–183 (2009)
11. Jiang, T., Liu, W., Zhong, F.Y., Zhong, C., Chiang, P.Y.: Single-Channel, 1.25-GS/s, 6-bit, Loop-Unrolled Asynchronous SAR-ADC in 40nm-CMOS. In: Custom Integrated Circuits Conference (CICC), pp. 1–4 (2010)
12. Chin, S.-M., Hsieh, C.-C., Chiu, C.-F., Tsai, H.-H.: A New Rail-to-Rail Comparator with Adaptive Power Control for Low Power SAR ADCs in Biomedical Application. In: 2010 IEEE International Symposium on Circuits and Systems, pp. 1575–1578 (2010)
13. Chen, F., Chandrakasan, A.P., Stojanović, V.: A Low-Power Area-Efficient Switching Scheme for Charge-Sharing DACs in SAR ADCs. In: Custom Integrated Circuits Conference (CICC), pp. 1–4 (2010)
14. Cho, Y.-K., Jeon, Y.-D., Nam, J.-W., Kwon, J.-K.: A 9-bit 80 MS/s Successive Approximation Register Analog-to-Digital Converter with A Capacitor Reduction Technique. IEEE Transactions on Circuits and Systems II: Express Briefs 57(7), 502–506 (2010)
15. Lin, Y.-Z., Liu, C.-C., Huang, G.-Y., Shyu, Y.-T., Chang, S.-J.: A 9-bit 150-MS/s 1.53-mW Subranged SAR ADC in 90-nm CMOS. In: 2010 IEEE Symposium on VLSI Circuits (VLSIC), pp. 243–244 (2010)
16. Yoshioka, M., Ishikawa, K., Takayama, T., Tsukamoto, S.: A 10b 50MS/s 820μW SAR ADC with On-Chip Digital Calibration. In: International Solid-State Circuits Conference (ISSCC), Digest of Technical Papers, pp. 384–385 (2010)
17. Harpe, P., Zhou, C., Wang, X., Dolmans, G., de-Groot, H.: A 30fJ/Conversion-Step 8b 0-to-10MS/s Asynchronous SAR ADC in 90nm CMOS. In: International Solid-State Circuits Conference (ISSCC), Digest of Technical Papers, pp. 388–389 (2010)
18. Furuta, M., Nozawa, M., Itakura, T.: A 0.06mm2 8.9b ENOB 40MS/s Pipelined SAR ADC in 65nm CMOS. In: International Solid-State Circuits Conference (ISSCC), Digest of Technical Papers, pp. 382–383 (2010)
19. Liu, C.-C., Chang, S.-J., Huang, G.-Y., Lin, Y.-Z., Huang, C.-M., Huang, C.-H., Bu, L., Tsai, C.-C.: A 10b 100MS/s 1.13mW SAR ADC with Binary-Scaled Error Compensation. In: International Solid-State Circuits Conference (ISSCC). Digest of Technical Papers, pp. 386–387 (2010)
20. Liu, C.-C., Chang, S.-J., Huang, G.-Y., Lin, Y.-Z.: A 10-bit 50-MS/s SAR ADC with A Monotonic Capacitor Switching Procedure. IEEE Journal of Solid-State Circuits 45(4), 731–740 (2010)

21. Zhu, Y., Chan, C.-H., Chio, U.-F., Sin, S.-W., Seng-Pan, U., Martins, R.P., Maloberti, F.: A 10-bit 100-MS/s Reference-Free SAR ADC in 90 nm CMOS. IEEE Journal of Solid-State Circuits 45(6), 1111–1121 (2010)
22. Liu, C.-C., Chang, S.-J., Huang, G.-Y., Lin, Y.-Z., Huang, C.-M.: A 1V 11fJ/Conversion-Step 10bit 10MS/s Asynchronous SAR ADC in 0.18μm CMOS. In: 2010 IEEE Symposium on VLSI Circuits (VLSIC), pp. 241–242 (2010)
23. Yoshioka, M., Ishikawa, K., Takayama, T., Tsukamoto, S.: A 10-b 50-MS/s 820-μW SAR ADC with On-Chip Digital Calibration. IEEE Transactions on Biomedical Circuits and Systems 4(6), part 1, 410–416 (2010)
24. Liu, W., Huang, P., Chiu, Y.: A 12b 22.5/45MS/s 3.0mW 0.059mm2 CMOS SAR ADC Achieving over 90dB SFDR. In: International Solid-State Circuits Conference (ISSCC), Digest of Technical Papers, pp. 380–381 (2010)
25. Lee, C.C., Flynn, M.P.: A SAR-Assisted Two-Stage Pipeline ADC. IEEE Journal of Solid-State Circuits 46(4), 859–869 (2010)

Sparse Signal Representation with Dispersion Dictionary

Zhang Yanhong, Guo Jinku, and Wu Jinying

Xi'an Research Inst. of Hi-Tech
Department of Automation
Hongqing Town, Xi'an 710025, China
xinxin4457@sina.com.cn,
gjk05@mails.tsinghua.edu.cn,
wujinying1982@yahoo.com.cn

Abstract. Sparse decomposition of a signal can be obtained by decomposing signal in an overcomplete dictionary. The overcomplete dictionary function is generally all well-localized and well adapted to the signal's local structures. Although choosing an appropriate atom dictionary for analyses improves the performance of the analyses, modeling of the dispersion phenomenon will be necessary for accurate signal processing. In this paper, we introduce the dispersion dictionary which is an overcomplete dictionary composed of Gaussian-envelop functions that can simulate the pulse dispersion phenomenon. The simulation result show the proposed dictionaries successfully represent nonstationary signals whose instantaneous frequency varies nonlineary with time and model the dispersion phenomenon.

Keywords: Signal representation, Signal decomposition, Dispersion dictionary, Instantaneous frequency.

1 Introduction

In many applications, such as seismic, radar, sonar or communications, the considered signals are often nonstationary, for which the frequency content is changing in time. To analysis nonstationary signals, researchers have developed many methods, such as time-frequency analysis, wavelet analysis, Hilbert-Huang transform and so on. Signal sparse representation decomposes signal in an overcomplete dictionary and provides an interpretation of the inherent nonstationary structures of signal [1]. The atoms of the overcomplete dictionary generally have small support not only in time domain but also in frequency domain, so that the decomposition results can reveal the signal time-frequency signature. The popular decomposition algorithms are Basis pursuit [2], Matching pursuit [1] and so on [3].

Generally, the dictionary elementary function is all well-localized and well adapted to the signal's local structures. Gabor dictionary [1], [4], Chirplet dictionary [5], e-chirplet dictionary and Dopplerlet dictionary [6] are the representatives of the sparse decomposition dictionaries. It is well known that if the atoms

X. Wan (Ed.): Electrical Power Systems and Computers, LNEE 99, pp. 1023–1028.
springerlink.com

are similar to the main components of a signal, only a few of atoms are needed for a perfect signal representation. Due to the best time-frequency resolution of Gabor atoms, the significance of Gabor dictionary was recognized by many researchers and was widely used in many problems on sparse signal representation. The chirplet atom is suitable for compactly characterizing the signals whose components spectral contents vary linearly with time. The e-chirplet dictionary and dopplerlet dictionary composed of atoms whose instantaneous frequency varies nonlineary with time [9].

A truly extraordinary range of natural and artificial processes, many of which arise in a variety of relative motion scenarios, yield Dispersion effect phenomena, which are of a nonlinear time-varying nature. Although choosing an appropriate atom dictionary for analyses improves the performance of the analyses, modeling of the dispersion phenomenon will be necessary for accurate signal processing. In this paper, we introduce the dispersion dictionary which is an overcomplete dictionary composed of Gaussian-envelop functions that can simulate the pulse dispersion phenomenon.

2 Dispersion Dictionary

L. Cohen [8] models the pulse propagation in dispersive media take into account the dispersion effect and stationary phase approximation. In [7], Jin-Chul Hong and Kyung Ho Sun developed an advanced signal processing technique especially for the analysis of dispersed wave signals measured from the guided wave technology. Similarly, the physics of the dispersion is modeled by group delay and the stationary phase approximation. We select the Gabor function as the initial pulse and the dispersion dictionary elementary function is obtained.

For construct the elementary functions that can model the actual dispersive phenomena, let us begin with an initial Gabor pulse $g_{(\sigma,\xi)}(t)$,

$$g_{(\sigma,\xi)}(t) = g\left(\frac{t}{\sigma}\right) \mathrm{e}^{\mathrm{j}2\pi\xi t} \tag{1}$$

where $g\,(t)\,\mathrm{e}^{-\pi t^2}$ is the Gaussian window and ξ, σ represents the center frequency and the time width. In the frequency domain, $g_{(\sigma,\xi)}(t)$ is expressed as

$$G_{(\sigma,\xi)}(f) = G\left[\sigma(f-\xi)\right], \; G(f) = \mathrm{e}^{-2\pi^2 f^2} \tag{2}$$

where $G(f)$ denotes the Fourier transform of the Gaussian window $g(t)$. If the Gabor pulse propagates along a waveguide for the time duration of u, the Fourier transform of the pulse $g_{(u,\sigma,\xi)}(t)$ may be written as

$$G_{(u,\sigma,\xi)}(f) = G_{(\sigma,\xi)}(f) \cdot \mathrm{e}^{\mathrm{j}2\pi u(f-\xi)}. \tag{3}$$

If the pulse is dispersive, the Fourier transform of the dispersive pulse may be written as [8], [7]

$$G_{(u,\sigma,\xi,D(f))}(f) = G_{(u,\sigma,\xi)}(f) \cdot \mathrm{e}^{-\mathrm{j}2\pi D(f)} \tag{4}$$

where $D(f)$ represents the dispersion effect. The group delay $\tau(f)$ of the pulse can be expressed as

$$\tau(f) = \frac{dD(f)}{df} + u. \tag{5}$$

Approximating $D(f)/df$ as polynomials of frequency f. $\tau(f)$ can be written as

$$\tau(f) = d_n(f-\xi)^n + d_{n-1}(f-\xi)^{n-1} + \cdots + d_1(f-\xi) + u. \tag{6}$$

The order of the polynomials can be selected according special application. Here, we adopt second-order polynomials which can describe dominant dispersion phenomena sufficiently well

$$\tau(f) = \frac{dD(f)}{df} + u = d_2(f-\xi)^2 + d_1(f-\xi)^1 + u. \tag{7}$$

Hence,

$$\begin{aligned} &G_{(u,\sigma,\xi,D(f))}(f) = G_{(u,\sigma,\xi,d1,d2)}(f) \\ &= G_{(u,\sigma,\xi)}(f)\exp\left\{-\mathrm{j}2\pi\left[\tfrac{d_1}{2}(f-\xi)^2 + \tfrac{d_2}{3}(f-\xi)^3\right]\right\} \end{aligned} \tag{8}$$

The inverse Fourier transform of $G_{(u,\sigma,\xi,D(f))}(f)$ can be called as the dispersion function [7]

$$g_{(u,\sigma,\xi,d_1,d_2)}(t) = \int_{-\infty}^{\infty} G_{(u,\sigma,\xi,d_1,d_2)}(f)\mathrm{e}^{\mathrm{j}2\pi ft}d. \tag{9}$$

The dispersion dictionary is defined as a function set $D(D = \{g_\gamma(t)\}_{\gamma\in\Gamma})$ with $\gamma =_{(u,\sigma,\xi,d_1,d_2)}$, $\Gamma = \mathrm{R}^3 \times \mathrm{I}^2$.

3 Matching Pursuit Algorithm

Matching Pursuit is a practical algorithm that decomposes signal into a linear expansion of waveforms selected from an overcomplete dictionary. As opposed to other global optimization techniques, the MP is a greedy algorithm that finds suboptimal in local and its fundamental principle can be described as follows [1].

Let $\mathbf{H}$ be a Hilbert space and $\mathcal{D} = \{g_\gamma(t)\}_{\gamma\in\mathbf{\Gamma}}$ be the overcomplete dictionary in $\mathbf{H}$. The atom $g_\gamma(t)$ is defined by the index γ which is an element of the index set $\mathbf{\Gamma}$. The atoms are also normalized and, hence, $\|g_\gamma(t)\| = 1$. Let f be the signal and $f \in \mathbf{H}$. Then, f can be decomposed into

$$f = \sum_{n=0}^{m-1} \langle R^n f, g_{\gamma_n}\rangle g_{\gamma_n} + R^m f \tag{10}$$

and

$$\|f\|^2 = \sum_{n=0}^{m-1} |\langle R^n f, g_{\gamma_n}\rangle|^2 + \|R^m f\|^2 \tag{11}$$

where $R^m f$ is the residual after the mth iteration (such that $R^0 f = f$), and g_{γ_n} is the atom that best matches the residue $R^n f$. We assume that g_{γ_n} satisfies

$$|\langle R^n f, g_{\gamma_n} \rangle| \geq \alpha \sup_{\gamma \in \mathbf{\Gamma}} |\langle R^n f, g_\gamma \rangle|, \quad 0 < \alpha \leq 1. \tag{12}$$

MP algorithm searches the atom that best matches the last residual, and such process is iterated until the residual energy is below some threshold or until some halting criterion is met. While g_{γ_n} satisfying the expression (12), Mallat and Zhang have proved that the residual energy $\|R^m f\|^2$ decays exponentially in finite dimensional spaces.

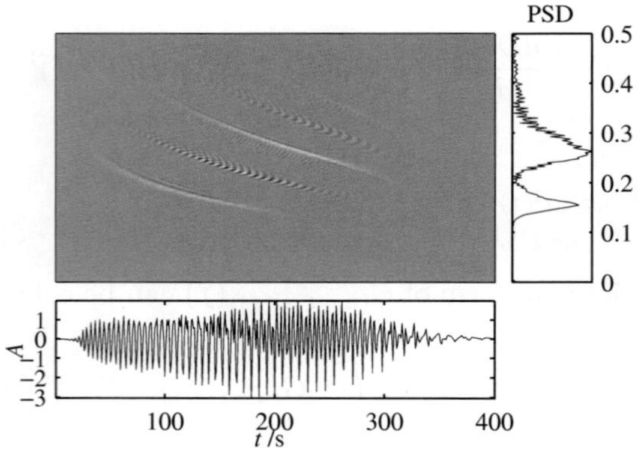

Fig. 1. The Wigner-Ville distribution of bat sonar signal.

4 Simulation Result

From the decomposition, we can derive a new time-frequency energy distribution by adding the Wigner distribution of each matched dictionary function [1].

$$Ef(t, \omega) = \sum_{n=0}^{m-1} |\langle R^n f, g_{\gamma_n} \rangle|^2 W g_{\gamma_n}(t, \omega)$$

This time-frequency distribution has well resolution by suppressing the cross-term among the elementary functions. The echolation pulse emitted by the large brown bat Eptesicus fusus is decomposed into dispersion functions with matching pursuit algorithm. Fig. 1 shows the Wigner-Ville distribution of the bat sonar signal. Fig. 2 shows the bat sonar signal time-frequency distribution which is obtained by adding the Wigner-Ville distributions of the well-matched dispersion atoms obtained by matching pursuit algorithm. We can see that the obtained energy distribution has better resolution than that of Wigner-Ville distribution.

When the frequencies of the elementary vary nonlineary with time, i.e., the instantaneous frequencies of the elementary functions are nonlinear, the final

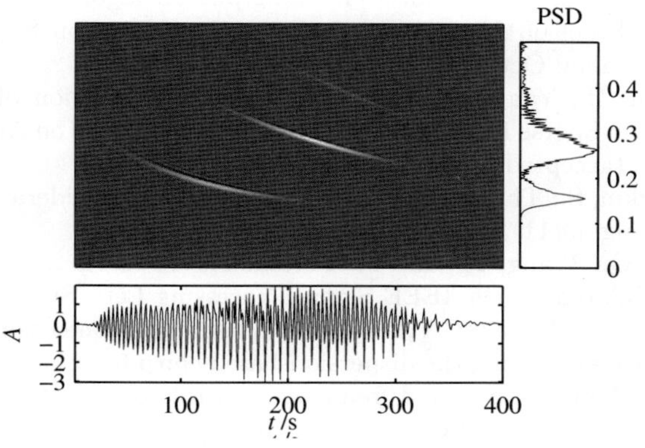

Fig. 2. The bat sonar signal time-frequency distribution which is obtained by adding the Wigner-Ville distributions of the well-matched dispersion atoms.

Wigner-Ville distribution would not be nonnegative and the Wigner-Ville distribution of the elementary function would have cross-terms. Zou [9] proposed a method for calculating a nonnegative and cross-term free time-frequency distribution. The nongetative cross-term free time-frequency distribution can be devised for a special atom dictionary in which the element function has Gaussian-envelop and frequency varies with dispersion low. Use Zou' method, we can also devise an nonnegative and cross-term free energy distribution which can adaptive to dispersion dictionary.

5 Conclusion

In this paper, we introduce the dispersion dictionary which is an overcomplete dictionary composed of Gaussian-envelop functions that can simulate the pulse dispersion phenomenon. The dispersion elementary function provides a more exact interpretation of the inherent structures of dispersion signal. It can help us to extract the sparse information from the decomposition result. The simulation result show the proposed dictionaries successfully represent nonstationary signals whose instantaneous frequency varies nonlineary with time and model the dispersion phenomenon.

References

1. Mallat, S., Zhang, Z.: Matching pursuits with time-frequency dictionaries. IEEE Trans. Signal Process 41(12), 3397–3415 (1993)
2. Chen, S.S., Donoho, D.L., Saunders, M.A.: Atomic decomposition by basis pursuit. SIAM J. Scientific Comput. 20(1), 33–61 (1999)

3. Guo, J., Liu, G., Yang, X.: A novel matching pursuit algorithm with adaptive subdictionary. In: Proc. of the 9th international Conference on Signal Process ICSP 2008, Beijing, China, October 2008, pp. 207–210 (2008)
4. Jinku, G., Zou, H., Yang, X., Liu, G.: Parameter estimation of Multicomponent Chirp Signals via Sparse Representation. IEEE Transactions on Aerospace and Electronic Systems (accepted for publication)
5. Mann, S., Haykin, S.: The chirplet transform: Physical considerations. IEEE Trans. Signal Processing 43(11), 2745–2761 (1995)
6. Zou, H., Dai, Q., Wang, R., Li, Y.: Parametric TFR via windowed exponential frequency modulated atoms. IEEE Signal Processing Lett. 8(5), 140–142 (2001)
7. Hong, J.-C., Sun, K.H., Kim, Y.Y.: Waveguide damage detection by the matching pursuit approach employing the dispersion-based chirp function. IEEE Transactions on Ultrasonics, Ferroelectrics,and frequency control 53(3) (2006)
8. Cohen, L., Loughlin, P.: Stationary phase approximation: A modification. In: Proc. SPIE, Orlando, FLA (April 2004)
9. Zou, H., Dai, Q., Zhang, X., Li, Y.: Nonnegative time-frequnency distributions for parametric time-freuqncy representations using semi-affine transformation group. Signal Processing Lett. 85, 1813–1826 (2005)

Author Index